RAPID EXCAVATION and TUNNELING CONFERENCE

2019 PROCEEDINGS

EDITED BY CHRISTOPHER D. HEBERT and SCOTT W. HOFFMAN

PUBLISHED BY THE
SOCIETY FOR MINING, METALLURGY & EXPLORATION

Society for Mining, Metallurgy & Exploration (SME)
12999 East Adam Aircraft Circle
Englewood Colorado 80112
(303) 948-4200 / (800) 763-3132
www.smenet.org

The Society for Mining, Metallurgy & Exploration (SME) is a professional society whose more than 15,000 members represent professionals serving the minerals industry in more than 100 countries. SME members include engineers, geologists, metallurgists, educators, students, and researchers. SME advances the worldwide mining and underground construction community through information exchange and professional development.

Information contained in this work has been obtained by SME from sources believed to be reliable. However, neither SME nor its authors and editors guarantee the accuracy or completeness of any information published herein, and neither SME nor its authors and editors shall be responsible for any errors, omissions, or damages arising out of use of this information. This work is published with the understanding that SME and its authors and editors are supplying information but are not attempting to render engineering or other professional services. It is sold with the understanding that the publisher is not engaged in rendering legal, accounting, or other professional services. If such services are required, the assistance of an appropriate professional should be sought. Any statement or views presented here are those of the authors and are not necessarily those of SME. Authors assumed the responsibility to obtain permission to include a work or portion of work that is copyrighted. The mention of trade names for commercial products does not imply the approval or endorsement of SME.

No part of this publication may be reproduced, stored in a retrieval system, or transmitted in any form or by any means, electronic, mechanical, photocopying, recording, or otherwise, without the prior written permission of the publisher.

ISBN 978-0-87335-470-7
eBook 978-0-87335-471-4

Copyright © 2019 Society for Mining, Metallurgy & Exploration
All Rights Reserved. Printed in the United States of America.

On the Cover: TBM "Angeli" being readied for launch by Skanska/Traylor JV on LA Metro's Regional Connector Project.

Contents

Preface xi
Executive Committee xiii
International Committee xv

Part 1: Contracting Practices and Cost 1

Are Alternative Delivery Methods and Private Financing Ways to Deliver Future Underground Projects in America? 2
    K. Bhattarai

Design-Build Construction in Underground Construction:
A Contractor's Perspective 14
    Gary A. Almeraris, Scott Hoffman

Having Difficulty Deciding on Procuring New, Refurbished or
Remanufactured TBMs? 23
    Dan Ifrim

Incentive-Based Project Delivery with Fixed Price Incentive Fee Contracts 34
    Philip Sander, Markus Spiegl, John Reilly

North American Tunneling Industry Firsts in the Midwest: The Blacksnake Creek
Stormwater Separation Improvement Tunnel 46
    Brian Glynn, Cary Hirner, Kayla Christopher, Lee Sommers, Mike Garbeth

Progressive Design-Build—Is It Coming to a Project Near You? 55
    Joe O'Carroll, Andy Thompson, Tom Kwiatkowski

A Study in the Use of Design-Build for Tunnel Projects 65
    Shawna Von Stockhausen, Erin L.D. Sibley, Derek Penrice

Tunnel Contracting Practice: One Owner's State of the Art 78
    Ted Dowey, Kate Edden, Sean McAndrew, George Schmitt, Bade Sozer

Part 2: Design and Planning 87

Crosstown Tunneling for Houston Surface Water Supply 88
    Sergio Flores, Todd Wanless, Mike McCure, Clint Wilson

Design and Construction Procurement for the Amtrak Hudson Tunnel Project 97
    Phil Rice, Richard Flanagan, Mohammed Nasim

Liner Load Estimation for Soft Ground Pressure Balance TBM Shield
Tunnel Projects 109
    Tamir Epel, Mike Mooney, Marte Gutierrez

TBM Utilization in Mixed Transitional Ground: The OCIT Experience in
Akron, Ohio 120
    Elisa Comis, Wayne Gyorgak, David Chastka

Times Square Shuttle Station Reconstruction 132
    Aram Grigoryan, David Campo, Philip Lund, Wanxing Liu

Part 3: Design/Build Projects 147

Controlling Risk of Tunneling Projects Implemented by Alternative
Delivery Method 148
    Nasri Munfah

DC Clean Rivers Project: We Built Miles of Tunnel…Now What? How to Get
the CSOs into the Tunnels in an Urban Environment 157
    Ray Hashimee, Aliuddin Mohammad, Yama Ebrahimi, Alper Ucak, Moussa Wone

Design and Construction of Montreal Express Link Tunnels and
Underground Stations 166
    Verya Nasri, Xavier De Nettancourt, Philippe Patret, Thomas Mitsch

Design and Construction of NBAQ4 Water Transfer Scheme 178
    Brendan Henry, Alessio Bianchini, Abner Bondoc, Ioannis Bournas,
    Katpakanathan Kamalachandran, Monique Quirk, Erik van der Horst,
    Kar Weng Ng, James Willey, Marcus Lübbers

Evolution of Tunnel Eyes Designs Based on Lessons Learned for the
Regional Connector Project in Los Angeles 199
    Eli Mathieu, Jose Muhr, Darren von Platen, Ivan Hee, Jake van Baarsel

LA Metro Regional Connector Transit Project: Successful Halfway-Through
Completion 206
    Mike Harrington, Tung Vu

Permitting in a Design/Build Environment—The Challenges of 120 Years
of a Design Bid Build Culture 219
    Douglas Gaffney, Frank Perrone

West Gate Tunnel Project—Evolution in Three Lane TBM Road Tunnel Design 230
    Jack Muir, Harry Asche, Ben Clarke

Part 4: Difficult Ground 241

Breaking Through Tough Ground in the Himalayas: Nepal's First TBM 242
    Missy Isaman

Challenging Geology: Blacklick Creek Sanitary Interceptor Sewer Project
in Columbus, OH 250
    Amanda Kerr, Max Ross, Ed Whitman, Ehsan Alavi

Cutting New Ground: TBM Selection for Fort Wayne Utilities Tunnel
Works Program 261
    T.J. Short, Leo Gentile, David Day

Ground Freezing to Improve Unstable Silty Ground 270
    Shawn Coughlin, Nate Long

Neelum-Jhelum Hydroelectric Project: Design and Construction Challenges
of Underground Works 277
    Peter Dickson, Francisco Tesi

Removal of Large Concrete Obstruction on 96-Inch Pipe Ram Crossing
of Highway 61, Neebing, Ontario 290
    James Carroll, Conner Beck, Travis Kraig, Spencer Shand

Unexpected Underground Obstructions: Challenges and Lessons Learned
from Angeli, the Regional Connector TBM 299
    Christophe Bragard, Bryan Hadley, Mat Antonelli

Part 5: Drill and Blast 313

Applying Lessons Learned and Taking on New Challenges at the Lake Mead
Intake No. 3 Low Lake Level Pumping Station Project 314
    Erika Moonin, Jerry Ostberg, Jordan Hoover, Stefano Alziati

Comparison of Emulsion and Anfo Usage in the Horizontal Development
Process at El Teniente 330
    Raúl Castro, Yina Herazo, Álvaro Pérez, José Medina

Connecting the TARP Des Plaines Tunnel System to the McCook Reservoir 342
    Mark White, Cary Hirner, Patrick Jensen, Carmen Scalise

Construction of the Luck Stone Inter-Quarry Tunnel, Leesburg Virginia 351
    Kyle Wooton, Matt Bauer

Drill and Blast Construction of Shafts and Starter and Tail Tunnels for
the 3RPORT CSO Project.. 357
    Stephen Miller, James Parkes, Manfred Lechner, Lance Waddell

Part 6: Environment, Health, and Safety 369

Commissioning of Tunnel Fire Life Safety Systems and Its Challenges..................... 370
    Hubert Heis, Reinhard Gertl

A Comparison of Breathing Gasses Used Under Hyperbaric Conditions................... 380
    Justin Costello

Conception and Construction of a Tailor-Made and Contractor-Built
Refuge Chamber for TBM and SEM Drives According to German Guidelines 387
    Rainer Antretter

InSAR Monitoring of Subsidence Induced by Underground Mining Operations 399
    Sara Del Conte, Giacomo Falorni

Minimizing Impacts to the Community and Commuters: Constructing
the District of Columbia's Largest Tunnel Along a Major Urban Artery 408
    William P. Levy, Moussa Wone, Justin Carl

Optimization of a Ventilation System by Using Flow Measurements
and 3D CFD Simulations 421
    Martin Schöll, Reinhard Gertl

Transit Tunnel TBM Vibration Through Glacial Till 430
    Thomas F. Bergen, Deborah A. Jue

Part 7: Future Projects... 439

Alexandria Renew Enterprises Is Now in the Tunnel Business 440
    Liliana Maldonado, Caitlin Feehan, Justin Carl, Kevin Pilong, Jennifer Jordan

DART D2 Subway Project Development 453
    Charles A. Stone, Israel Crowe, Eric C. Wang

Mountain Tunnel Improvements Project 464
    Jennifer Sketchley, David Tsztoo, Renée Fippin, Glenn Boyce

MWRA Metropolitan Boston Tunnel Redundancy Program Project Update 477
Kathleen M. Murtagh, Frederick O. Brandon

Unique Design Challenges of the Central Bayside System Improvement
Project Tunnel Connections and Shafts 490
Nick Goodenow, Mike Bruen, Steve Robinson, Michael Deutscher,
Manfred Wong

Part 8: Geotechnical Considerations 503

Anacostia River Tunnel Project Groundwater Drawdown Effects from
the Deep Well Depressurization System at the CSO-019 Drop Shaft 504
E. Gregory McNulty, Pooyan Asadollahi

Digging Deeper—Supplemental Geotechnical Site Investigation for
Parallel Thimble Shoal Tunnel 512
Amanda Wachenfeld, Frank Perrone, Scott Kibby, Jose Ballesta

How to Quantify the Reliability of a Geological and Geotechnical
Reference Model in Underground Projects 525
Guido Venturini, Gianpino W. Bianchi, Mark Diederichs

Innovative Monitoring for Urban Tunneling 538
Paul Thurlow, Glen Frank

Packer and Long-Term Aquifer Testing to Estimate Tunnel Inflows:
A Case History for the Lower Meramec Tunnel, St. Louis, MO 548
Kenneth A. Johnson, Xiaomin You, Kyle D. Williams

Underground Construction, Geology and Geotechnical Risk 564
Priscilla P. Nelson

Use and Misuse of Geotechnical Baselines to Predict Soft Ground
TBM Tool Wear 575
Ulf G. Gwildis, Michael S. Schultz

Writing the GBR So the Contractor Can Understand It 585
Barry R. Doyle

Part 9: Ground Support and Final Lining 599

7.93 m Open TBM Shotcrete System Improvement and Innovation
Jilin Project, China 600
Desiree Willis, Ya Jun Guo

Complex Tunnel Through the Abutment of the High-Risk Chimney
Hollow Dam 614
Greg Raines, Albert Ruiz, Austin Wilkes

Design Diagrams for Fiber Reinforced Concrete Tunnel Linings 623
Axel G. Nitschke

Steel Fiber Segmental Linings for Mega TBMs 636
Tom Ireland, Shu Fan Chau, Harry Asche, Jack Muir, Ben Clarke

Part 10: Grouting and Ground Modification 647

Differences in Consolidation Grouting Practices Between Near Surface and
Underground Applications 648
    Adam Bedell, Brad Crenshaw

Grouting a TBM Shotcrete Lining Under Challenging Conditions 657
    M.H. Kizilbash, Gary Peach

Grouting and Groundwater in the Greater Arncliffe Area, WestConnex
New M5 Tunnels, Sydney, Australia 668
    Jack Raymer, David Oliveira, Harry Asche, David Crouthamel

Quality Control of Secant Piles and Jet Grouting for the Ohio Canal
Interceptor Tunnel 683
    Stanley L. Worst, Matthew J. Niermann

Part 11: Hard Rock TBMs 693

Rand Park Stormwater Diversion Tunnel—Planning and Designing
to Address Stormwater Flooding for Downtown Keokuk, Iowa 694
    Mahmood Khwaja, Gregory Sanders, Michael S. Schultz,
    David R. Schechinger, Mark Bousselot

The Three Rivers Protection and Overflow Reduction Tunnel (3RPORT)—
Decision-Making During Construction 703
    Emidio Tamburri, Ludovica Pizzarotti, Paolo Perazzelli, Roberto Schuerch,
    Giuseppe Moranda

TBMs: Meeting the Challenge in Pakistan's Lower Himalayas 715
    Gary Peach

Waterway Protection Tunnel: Louisville's Innovative Solution to Four
Storage Basins 725
    Jonathan Steflik, Jacob Mathis

Part 12: Large Span Tunnels and Caverns 733

Construction Considerations for Garage Cote Vertu in Montreal 734
    Jean Habimana

First Large Diameter Hard Rock CSO Chamber in St. Louis 745
    Patricia Pride, Kevin Nelson, Clay Haynes, John Deeken

Rock Load Estimation for Shallow Rock Caverns 752
    Charles A. Stone, Changsoo Moon (Kevin)

Use of Spray Applied Waterproofing on the Downtown Bellevue Tunnel Project 764
    Mun Wei Leong, Ted DePooter, Jacob Taylor

Part 13: New and Innovative Technologies 775

Artificial Intelligence Technique for Geomechanical Forecasting 776
    M. Allende Valdés, J.P. Merello, P. Cofré

Evaluating Impact of Water Content on EPB Machine Performance Based
on Laboratory Experiments 785
    Wei Hu, Jamal Rostami

The Expanding Capabilities of Microtunneling Demonstrated in Washington DC 799
Todd Brown

High-Speed 3D Tunnel Inspection 809
Heiner Kontrus, Michael Mett

Implementation of Automation and Digitization in Tunnel Waterproofing
and Grouting Practices 817
Stefan Lemke, Andreas Heizmann, Tim Kearney

Sandwich Belt High Angle Conveyors Exclusively at Paris
Metro Expansion—2019 825
Joseph A. Dos Santos

SCMAGLEV—Fast and Innovative Mode of Transportation in
the Northeast Corridor—Tunneling Challenges 837
Vojtech Gall, Nikolaos Syrtariotis, Timothy O'Brien,
Cosema (Connie) Crawford, David Henley

Smart Office: A Data-Driven Management Tool for Mechanized
Tunneling Construction 845
Kamran Jahan Bakhsh, Jim Kabat, Roberto Bono

Tunnel Survey Control in Small Segmentally Lined Tunnels 855
Pete DeKrom, Edouard Whitman

Part 14: Pressure Face TBM Case Histories 869

Effectiveness of Risk Mitigation Strategies for Large Diameter One-Pass
Tunnels in Mixed Ground from Ohio Canal Interceptor Tunnel Construction 870
Mike Wytrzyszczewski, Christopher Caruso, David Rendini,
Geary Visca, Dan Dobbels

EPB Clogging Through Mixed Transitional Ground, Lessons Learned at
the Ohio Canal Interceptor Tunnel, Akron, OH 880
Elisa Comis, Peter Raleigh, Wayne Gyorgak

Planning, Design and Construction of the Regional Connector
Bored Tunnels—An EPB Tunneling Case History 890
Richard McLane, William Hansmire, Ron Drake,
Derek Penrice, Darren von Platen

Preparation for Tunneling, Don River & Central Waterfront Coxwell
Sanitary Bypass Tunnel Project in Toronto, ON 916
William Hodder, Ehsan Alavi, Daniel Cressman

Tunneling in Mixed Face Conditions: An Enduring Challenge for
EPB TBM Excavation 928
Jim Clark, Paul Verrall

Part 15: Pressure Face TBM Technology 941

Curved Microtunneling Alignments in the Design Toolbox 942
David Mast, Patrick Dodds, Paul Nicholas, Rob Dill, Alison Schreiber,
Frederick (Rick) Vincent

Forrestfield Airport Link—Project Challenges and TBM Solution 954
Karin Bäppler, Michael Strässer

Contents

Interpretation of EPB TBM Graphical Data 961
    Keivan Rafie, Steve Skelhorn

Mechanized Tunneling at High Pressure—More Than Just a Stronger Bulkhead 972
    Werner Burger

Proposal of Some Cuttability Indexes for Evaluating the Performance of
Mechanical Excavators Using Conical Picks 980
    Okan Su, Xiang Wang

Tuen Mun—Chek Lap Kok Link in Hong Kong—Innovative Technologies and
Methodologies for an Outstanding Project 986
    Antoine Schwob, Bruno Combe

Part 16: SEM/NATM 999

Culvert Construction Under I-89 in Vermont Using the Sequential
Excavation Method 1000
    Jon Pearson, Julian Prada, Anil Dean, Eric Eisold

Delivering Value Through the Innovative Contractor Engagement (ICE) Model
at London Underground Bank Station Capacity Upgrade Project 1010
    Enrique Fernandez, Alejandro Sanz, Juan Ares, Andy Swift, Bethan J. Haig

SEM—Single Shell Lining Application for the Brenner Base Tunnel 1028
    Thomas Marcher

SEM Cavern Construction in Downtown LA 1037
    Carlos Herranz, Christophe Bragard, Ivan Hee, Dominic Cerulli

SEM Tunnel Herrschaftsbuck: A Geological Challenge in Germany's Southwest 1051
    Roland Arnold

Shallow SEM Tunneling with Limited Clearance to Existing Structures:
Design, Construction and Observations 1061
    Hong Yang, Derek Penrice, Walter Klary, Vojtech Gall

Soft Ground Tunnelling Techniques for Mine Access Development 1074
    Ben Ablett, Joe Anderson, Jack Nolan

Steep Inclined SEM Excavation—Successful Execution Applying Drill and Blast 1088
    Richard Griesebner

Part 17: Shafts and Mining 1097

Construction of Combined Sewer Overflow 021 Diversion Facilities in
the District of Columbia 1098
    Moustafa Awad, Paul Leduc, Louise Headland, Ryan Mains,
    Ryan Payne, John Beesley

Design and Construction Considerations for Large Shafts in Hudson River
Tunnel Project: Hoboken Shaft 1114
    Arman Farajollahi, Paul Roy, Young Jin Park

Design Challenges of Deep Underground Shafts 1122
    Daniel Garcia, Ravi Jain

Merry Christmas!—Emergency Repair of the PCI-12A Interceptor Collapse
in Macomb County Michigan 1133
    Nicholas Kacynski, Fritz Klingler, Zachary Carr, Evans Bantios, Louis Urban

The Role of Mechanized Shaft Sinking in International Tunnelling Projects 1144
Andrea Fluck, Peter Schmäh

South Hartford Conveyance and Storage Tunnel Project—Drop and
Vent Shaft Construction 1153
Andrew Perham, Jim Sullivan, Scott Jacobs, Mark Careyva,
David Belknap, Clay Haynes

Structural Design of Large Diameter Shafts for the Coxwell Bypass Tunnel 1160
Gorki Filinov, Tyler Lahti

Part 18: Tunneling for Sustainability 1171

Assessment and Remediation of Gas Utility Tunnels in Chicago 1172
Chad Gailey, Brian Selph, Aswathy Sivaram, Cary Hirner

Design and Construction of Lockbourne Intermodal Subtrunk Sewer 1186
Steven Thompson, Irwan Halim, Michael Nuhfer, Jeremy Cawley

Great Hill Tunnel Inspection and Rehabilitation 1199
Brian Lakin, Joe Schrank, Daniel Ebin, Lawrence Marcik

Hecla's Mobile Mechanical Vein Miner 1210
Clayr Alexander, Mark Board, David Berberick, Wes Johnson, Marcus Eklind

LED Construction Lighting in Tunnel Projects 1222
Brian Astl, Michael Cook, Zach West, Jim Bresnen

Pumped Storage Projects in Switzerland—Challenges and Solutions 1230
Jürg Künzle, Bruno Gisi

Rebuilding TBMs: Are Used TBMs as Good as New? 1239
Doug Harding

Index 1249

Preface

The Executive Committee and SME staff extend a warm welcome to all attendees, exhibitors, authors, speakers, and guests to our biannual event. The 2019 Rapid Excavation and Tunneling Conference (RETC) returns to Chicago for the first time since 1984.

It is fitting that we are in Chicago. The Windy City has a long history of innovation in tunneling, as well as in other industries. In 1858, the first comprehensive combined sewer system in the United States was designed and constructed. In 1867, the Chicago Lake Tunnel, the first lake tap in the country, was completed. And the 1893 World's Columbian Exposition (World's Fair) included many firsts, such as the world's first Ferris wheel, the first public "moving walkway," and the first large-scale display of Nikola Tesla's newly developed alternating current electrical system.

More recently, the Chicago Tunnel and Reservoir Plan (TARP), a system of more than 100 miles of large storage and conveyance tunnels, has dramatically improved the water quality of Lake Michigan and has the capacity to store more than 17.5 billion gallons of combined sewage. As you enjoy the sights and sounds of this great city, take pride in the many contributions our industry has made in improving the quality of life for residents and visitors alike.

This year's program includes 111 papers covering such topics as contracting practices, design and planning, geotechnical considerations, hard rock tunnel boring machines (TBMs), new and innovative technologies, pressure-face TBM case histories, and tunneling for sustainability, just to name a few. The papers have been divided into 20 sessions over three days and promise to inform, challenge, and stimulate the audience.

The exhibition hall is a key component of any successful RETC and a great opportunity to reconnect with friends and colleagues, make new acquaintances, and discover what products and innovations exist that can help us improve our industry. Please visit the hall and support our exhibitors.

RETC has historically been a reunion of tunnelers from across the globe. The papers, the sessions, the exhibition hall, and yes, even the lobby bar allow us to share our experiences, our successes and failures, and our ideas and dreams, all with the goal of getting better at the work we love: building tunnels. The program chairs and RETC Executive Committee sincerely thank the authors, session chairs, exhibitors, and all attendees who make this incredible meeting possible. We are especially grateful to the authors and speakers who have so generously donated their time and expertise. Thank you!

Finally, this conference is only a reality and a success because of the leadership and tireless efforts of the amazing staff of SME in Englewood, Colorado. We are grateful for their friendship and continued support.

Christopher D. Hebert
Scott W. Hoffman

Executive Committee

Chair:

Stephen C. Redmond
Vice President, Frontier Kemper Constructors Inc

Vice Chair:

Chris Dixon
Project Executive, Tutor Perini Corp

Gregg David
Principal Engineer, McMillen Jacobs Associates

Anthony Del Vescovo
Program Manager and East Coast Tunnel Leader, Walsh Group

Michale DiPonio
Project Director, Jay Dee Contractors Inc

Chris Hebert
Vice President, Traylor Bros Inc

Scott Hoffman
Project Manager, Skanska USA Civil Northeast Inc

Mark Johnson
Global Solutions Director-Tunnels, Jacobs

Colin Lawrence
Executive Vice President, Mott MacDonald

Shemek Oginski
Project Manager, JF Shea Co Inc

Session Chairs

Phil Backers Jay Dee	**Nick Karlin** Skanska	**Jaidev "Jay" Sankar** HNTB
Rory Ball Mott MacDonald	**Matt Kendall** FKCI	**Charlie Schoch** Skanska
Jessica Buckley Jay Dee	**Steve Lottie** FKCI	**Bryce Scofield** Traylor Bros., Inc.
John Caulfield Jacobs Engineering	**David Mast** AECOM	**Mina Shinouda** Jay Dee
Nick Chen Jacobs Engineering	**Ben McQueen** FKCI	**David Smith** WSP
Liam Dalton HNTB	**AG Mekkaoui** Jay Dee	**Jonathan Steflik** Black & Veatch
Dani Delaloye Mott MacDonald	**Bianca Messina** Skanska	**Sam Swartz** McMillen Jacobs Associates
Dawn Dobson Barnard Construction	**Ami Mukherjee** WSP	**Richard Taylor** Traylor Bros., Inc.
Ken Dombroski McMillen Jacobs Associates	**Alston Noronha** Black & Veatch	**Darren VonPlaten** Traylor Bros., Inc.
Geoff Fairclough Schiavone Construction	**Tom Peyton** WSP	**Keith Ward** Seattle Public Utilities
Murray Gant Metro Vancouver	**Steve Price** Walsh Group	**David Watson** Mott MacDonald
Christian Heinz J.F. Shea	**Phil Rice** WSP	**Kim Wilson** LA Metro
Adam Hingorany Traylor Bros., Inc.	**Samer Sadek** Jacobs Engineering	**Mike Wytrzyszczewski** City of Akron
Frank Huber Metro Vancouver	**Jesse Salai** J.F. Shea	

International Committee

Argentina:	**Nestor Garavelli** Frontier-Kemper Constructor's ULC
Australia:	**Ted Nye** Mott MacDonald
	Harry Asche Aurecon
Austria:	**Norbert Fuegenschuh** Beton-und Monierbau
Brazil	**Tarcisio Barreto Celestino** Rua Pedro Américo
Canada:	**Rick P. Lovat** L2 Advisors, Inc.
Chile:	**Alexandre Gomes** GeoConsult
Czech Republic:	**Karel Rossler** Metrostav
England:	**Ross Dimmock** Normet UK Ltd.
France:	**Francois Renault** Vinci Construction Grands Projets
Germany:	**Klaus Rieker** Wayss & Freitag Ingenieurbau AG
India:	**R. Anbalagan** L&T Construction Infrastructure
Italy:	**Remo Grandori** SELI Societa Esecuzione Lavori Idraulici SpA
Japan:	**Hirokazu Onozaki** Obayashi
Mexico:	**Roberto Gonzalez Izquierdo** Moldequipo Internacional, S.A. DE C.V.
New Zealand:	**Bill Newns** Aurecon
Singapore:	**Leslie Pakianathan** Mott MacDonald (Sinagpore) Pte Ltd
Spain:	**Enrique Fernandez** Dragados S.A.
Sweden:	**Stig Eriksson** Skanska Sverige AB
Switzerland:	**Frederic Chavan** Marti Contractors Ltd.

PART

Contracting Practices and Cost

Chairs

Sam Swartz
McMillen Jacobs Associates

Mike Wytrzyszczewski
City of Akron

Are Alternative Delivery Methods and Private Financing Ways to Deliver Future Underground Projects in America?

K. Bhattarai ▪ Lane Construction

In recent years in America there has been expanding usage of full private financing in large underground transportation projects. The increasing popularity of alternative project delivery methods and private financing has provided unique opportunities to overcome chronic funding shortfalls, drive technological innovation, and deliver fast-track quality infrastructure and services to the public.

This paper represents recent examples of private financing and discusses challenges and opportunities these innovative project financing and delivery methods bring to the construction industry in terms of technological innovation and the procurement, contractual, legal, and risk management of underground infrastructure pr3ojects.

INTRODUCTION

In America, alternative delivery models are successfully implemented in roads and bridges, transit and railways, and so forth. The common alternative delivery models include: (1) design-build with or without private financing (DB or DBF); (2) design-build-operate (DBO); (3) design-build-own-operate-transfer (DBOOT); (4) design-build-operate-maintain (DBOM); (5) design-build-finance-operate-maintain (DBFOM). Payment mechanism for P3 projects include user pay (tolls), shadow payments, availability payments, construction payments, capital payments, operating cost benchmarking, and other hybrid mechanisms.

Recently ongoing, completed or in progress P3/privately funded projects with significant underground components include Purple Line Project (US $5.6 billion, DBFOM) Maryland; Elon Musk's Hawthorne Test Tunnel Los Angeles (US $ 10 million); Canada Line Rapid Transit Project (CN $2.0 billion, DBFOM) Vancouver BC; Ohio River Bridges (ORB) East End Crossing (US $763 million, DBFOM), Indiana-Kentucky; and Port of Miami Tunnel Project (US$1.1 billion, DBFOM) Florida. These projects have been partially or fully financed by the private sector capital.

WHY P-3 PROJECTS?

Due to ongoing shortfalls in public capital, partly or fully privately financed P3/alternative delivery models are actively sought after for investment and delivery of the major public infrastructure projects. The support for P3 projects is growing stronger due to the following reasons:

- Has ability to start much quicker, with the availability of funds from the private sector, than the traditional delivery methods
- Has proven track record of completing projects on-time and on-budget
- Has ability to shift construction and maintenance risks to the private sector through effective procurement and contracting agreements

- Has ability to provide creative technological-advanced concept and innovative solutions to meet the new-aged infrastructure needs rather than relying on repeated aged-old prescriptive concepts
- Has ability to achieve cost savings by lowering construction cost, reducing life-cycle maintenance cost, and other associated costs ensuring continuous quality infrastructure by proper transfer of risks and responsibilities
- Allows costs of the investment spread over the lifetime of the concession agreement and, in many cases, have a lower Net Present Value (NPV) than financed and managed by Public Sector Agencies

PROJECT FINANCING

Government

In general government can raise capital cheaper than the private sector as the private sector incurs cost of borrowing such as premium for risks of construction, capital fund raising, revenue collection, infrastructure management, and maintenance are involved in the process. The private sector can compete with the government in raising project funds in low risk projects; however, in the case of larger infrastructure projects with a significant underground segment, the cost of raising private sector capital may not to be a cheaper solution; however, the difference can be narrowed by various tax related incentives and long-term revenue support/benefits. The public agencies in the US are increasingly utilizing the concept of value for money and are going for P-3 projects with the private sector finance. In addition, the US Government provides grants to infrastructure projects via "The Transportation Infrastructure Finance and Innovation Act (TIFIA)" assistance.

In the case of MTA Purple Line Project, Federal Transit Authority (FTA) agreed to provide $875 million of federal funding to support the project (Fluor 2016). In the case of Port of Miami Tunnel Project, the concessionaire's funding sources include a $341 million TIFIA loan (USDOT 2014). Furthermore, the Federal Government provided TIFIA grants of $162 million to Ohio River Bridges (ORB) East End Crossing Project (USDOT webpage 2018).

Private Sector

Debt Financing

The private sector can secure the project debt financing by two main sources—commercial banks and bond investors. Commercial banks provide long-term loans to the project company; bond holders, such as insurance companies and pension funds, purchase long-term tradable project infrastructure bonds. Mezzanine debt financing provides funds to meet the funding shortfall. Some project financing commercial banks are City Group Inc, Bank of America, Deutsch Bank, UBS, and Macquarie Bank.

In the case of Port of Miami Tunnel P-3 Project in Florida, ten banks, BNP Paribas, Banco Bilbao Vizcaya Argentaria, RBS Citizens, N.A., Banco Santander, Bayerische Hypo, Calyon, Dexia among others provided the senior financing for the project, which totaled $341.5 million (Sigo, 2009).

Private Equity

Direct equity financing by the private sector is becoming increasingly popular, including by public employees' pension funds. Generally, majority of equity investment are provided by concession partners themselves. For example, in the case of MTA Purple Line Project, major equity shareholders are Purple Line Partners, LLC, the consortium of Meridiam, Fluor enterprises Inc., and Star America Purple Line, LLC (Fluor 2016). In the case of Port of Miami Tunnel project, the equity investment was provided by the concessionaire partners: 90% equity by Meridiam Infrastructure Miami LLC and the remaining 10% by France's Bouygues Travaux Publics SA (Cision 2014). Total private sector equity amounted to $80 million (USDOT 2014).

Some pension funds that are involved in P-3 project finance are: The California Public Employees' Retirement System (CalPERs), who bought a 10% stake in the Indiana concession, and "The California State Teachers Retirement System (CalSTRS)" (Poole 2016). Star America raises investment capital from US labor pension funds, insurance companies, asset managers, and so forth (Star America webpage 2018)

PROCUREMENT, CONTRACTUAL, AND LEGAL

P-3 project procurement generally involves two-stage process: a Request for Qualifications (RFQ) or a Request for Expression of Interest (RFEI) to screen out unqualified proponents and short list most qualified proponents, and a Request for Proposals (RFP) to bid on the project. A RFQ or RFEI states Owner's requirements for proponents' ability to finance, execute, and operate a project of similar type, size, and complexity. Also, a proponent is required to prove its ability to manage risks and deliver the project on the schedule, cost, and expected quality. The shortlisted candidates are then invited to submit a bid with a technical proposal and a financial and cost proposal.

In the RFP phase, generally, the candidates are required to meet minimum the pass/fail and best value criteria. The highest rating proposal along with the lowest NPV (Net Present Value) is considered. If the proponents do not meet the requirement, a risk trade-off analysis will be conducted. In a project with a significant underground section, geotechnical and other major risk components could adversely impact the project's NPV and Internal Rate of Return (IRR), due to higher insurance premiums added to the project cost.

Technical proposal evaluation involves rating of the following criteria and weightings, as in the case of Purple Line Project (MTA, 2014):

- Operations (about 35%)
- Project management (about 20%)
- Design and Construction (about 20%)
- Maintenance, rehabilitation, and handback (about 15%)
- Systems Integration (about 10%)

The proponents are generally evaluated based on commercial and legal; design and construction; operations and maintenance; and financial matters. The evaluation criteria include proponents' expertise on: (1) operations and maintenance; (2) design/build/construction; (3) financial; (4) community consultation; (5) public private partnership experience; (6) innovation; and (7) other general qualifications.

The requirement of these criteria is obvious—proponents (1) should be able to anticipate and manage risks regarding their ability to raise finance and absorb risks; (2) have corporate and organization expertise and experience to manage risks in design and construction phases in D/B or in EPC delivery modes, as in the case of Canada Line and MTA Purple Line Project; (3) have ability to and can generate innovative solutions to manage risks; (4) and have experienced and efficient organization structure to deliver the project. During procurement process, these criteria are heavily scrutinized regarding proponents' ability to manage risks during investigation, design, construction, and operation and maintenance. The proponents should have experience and a sound approach to communicate and manage potential conflicts with public stake holders arising out of project risks.

A public owner uses following contractual clauses/instruments to transfer/allocate risk contractually to proponents. A proponent further transfers some of these clauses to their subcontractors via a back-to-back subcontract agreement.

- Payment methods, schedule, and limitation
- Differing site condition clause
- Changes by public owner and changes by concessionaire
- Concessionaire's default and remedies, termination
- Performance bonds, escrow bid documents
- Insurance, payment and performance, and indemnity
- Dispute resolution, mediation, dispute review boards, arbitrations, or other forms of alternative dispute resolution mechanisms
- Alternative technical concepts or value engineering proposals
- Availability payment clauses
- Financing, refinancing, equity, assignment, and transfer

CASE STUDIES

Maryland Department of Transport Purple Line P-3 Project

The Purple Line Project is a 16-mile light rail that connects Bethesda in Montgomery County to New Carrolton in Prince George's County linking four Washington Metrorail lines, three MARC commuter rail lines, Amtrak Northeast Corridor, and regional and locus bus services. It also includes 21 stations and one approximately 1000-ft-long Plymouth Tunnel.

The $5.6 billion project is a public-private partnership between various levels of government and Purple Line Partners, the private sector concessionaire.

In March 2016, the concession was awarded to Purple Line Transit Partners—a consortium of Fluor, Meridium, and Star America Fund—to design, build, partially finance, operate, and maintain the system for 30 years. The construction started in August 2017 and is expected to be completed in March 2022, and the service will be commenced in February 2023. The Purple Line Transit Constructors (PLTC), a team comprising Fluor, Lane, and Traylor brothers, are responsible for design and construction of the $2.0 billion construction contract using the Design-Build delivery method. The Purple Line Transit Operators (PLTO), a team comprising Fluor, Alternate Concepts Inc. (ACI), and CAF USA inc., will perform the O&M Services (Figure 1).

The project financing includes $313 million in "Green Bonds" from private activity bonds underwritten by JP Morgan and RBC Capital Markets and an $874.6 million TIFIA loan from the USDOT. Backed by the Maryland Commonwealth Transportation Trust Fund and Fluor Corporation, the bonds were sold at the lowest interest rates ever achieved in the P3 market in the US, which created substantial savings for MTA (Cision 2016).

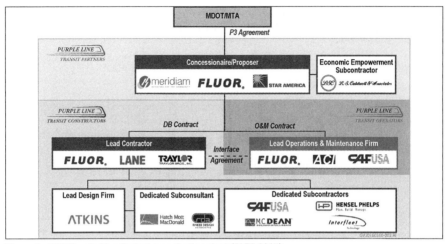

Figure 1. MTA Purple Line Project organization chart (PLTP 2017)

The Boring Company Hawthorne Test Tunnel in Los Angeles

The Boring Company's Hawthorne Test Tunnel Project includes boring of 1.1-mile long research and development transportation tunnel from Hawthorne along I-405 adjacent to the Space X headquarters and manufacturing facility in Los Angeles area. The tunnel boring was completed in November 2018. In September 2018, City of Hawthorne announced that a test spur and elevator has been proposed near the intersection of 120th street and Prairie Avenue (The Boring Company 2016 and 17).

The Boring Company's projects are privately financed with a fund of approximately US$10 million. The Boring Company has proposed a series of innovations in this project (The Boring Company 2016 and 17):

- Reduce a tunnel diameter in half; the company plans to reduce the construction cost by three to four times
- Increase TBM's speed by increasing TBM power by three times with an appropriate upgrade in cooling systems
- Build a continuous tunnel without an interruption for erecting tunnel liner segments by modifying an existing tunneling technology
- Enhance safety and efficiency by automating TBMs, eliminating the need for human operators
- Enhance clean tunnel atmospheric environment by making the locomotives electric from a diesel engine
- Increase tunneling research and development in the United States by private sector companies (A large number of R&D development and technologies for underground construction are developed and applied outside the US first.)
- Develop new technologies to recycle tunneling muck into building material such as bricks
- Build a new way of traveling using a high-speed electric skate
- Potential development of a hyperloop vacuum shell for very high-speed travel
- Make use of smaller but high-density stations to minimize traffic congestion and make easier for buying real estate for stations

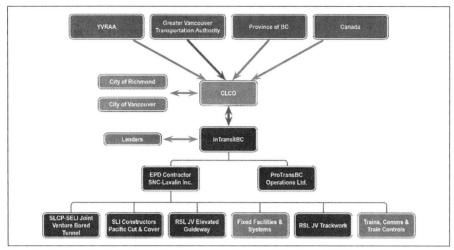

Figure 2. Canada Line Rapid Transit Project organization chart (Woodhead 2009)

Canada Line Rapid Transit Project

The 19.5-kilometre long Canada Line Rapid Transit Project is a rapid transit system that connects Downtown Vancouver with Vancouver International Airport (YVR) and Richmond City Centre.

The P-3 project was awarded to InTransitBC in July 2005 to design, build, partially finance, operate, and maintain the system for 35 years. InTransitBC is a joint venture company owned by SNC-Lavalin, the Investment Management Corporation of BC (bcIMC), and the Caisse de dépôt et placement du Québec. The concession contract required that the project must be completed by November 2009 to be ready for the 2010 Winter Olympics. The construction started in late 2005. Service commencement was completed on August 11, 2009, an outstanding 110 days ahead of the schedule, well in advance of the 2010 Winter Olympics (Woodhead, 2009).

The C$2 billion project is a public-private partnership between the government agencies and InTransitBC. SNC Lavalin Inc. (SLI) was responsible for all design, procurement, and construction (Figure 2).

The project includes three major types of structures:

- A 2.5-km long bored tunnel
- A 6.5-km long cut-and-cover tunnel
- An 8-km long elevated guideway which includes two bridges
- A 2.0-km long at-grade section near the Vancouver Airport
- Sixteen stations, 8 underground, 6 elevated, 2 at-grade

The public contributions to the project funding came from the following sources (Translink 2008, page 64):

- Government of Canada, C$450 million
- Government of British Columbia, C$435 million
- TransLink, C$334 million

- Vancouver Airport Authority, C$300 million
- City of Vancouver, C$29 million

InTransitBC, the private partner, invested C$750 million in the project (Yaffe, 2009).

Ohio River Bridges East End Crossing

The US$ 763 million Ohio River Bridges East End Crossing (ORBEEC) is a highway extension that connects Louisville, Kentucky and Utica, Indiana, crossing the Ohio River completing an outer ring expressway around the City of Louisville. The East End Crossing is part of a larger the US $2.6 billion Ohio River Bridges Project (Bhattarai et al., 2016).

The P-3 concession was awarded to Walsh, Vinci, Bilfinger (WVB) East End Partners in December 2012 to design, build, finance, operate, and maintain for 35 years. Construction started on schedule in June 2013 and opened to traffic in December 2016.

The private sector developer, WVB East End Partners, is responsible for all design, procurement, and construction. The Design-Build Contractor is the joint venture of Walsh and Vinci Construction (WVC), two of the three members of the concessionaire (Figure 3).

The project includes three main type of structures:

- Eight-miles of 4-lane limited access expressway on new alignment
- A twin-bore 1,700-foot long tunnel carrying 2-lanes in each direction
- A 2,500-foot long cable stayed bridge over the Ohio River

The project funding came from the following sources (USDOT ORBEEC Project 2018):

- TIFIA loan, $162 million
- Indiana State and Federal funding, milestone payments $392 million
- Other Indiana and Federal funding, $201.7 million
- Indiana milestone PABs (Series A), $488.9 million

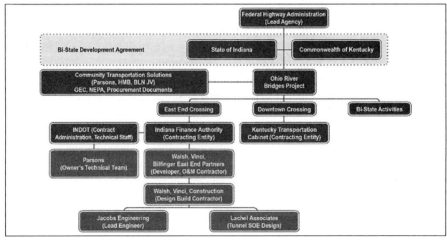

Figure 3. ORB East End Crossing organization chart (Bhattarai et al., 2016)

- Indiana long-term PABs (Series B), $18.9 million
- Kentucky and Federal funding, $94.2 million
- Developer risk capital, $78.1 million
- Relief events reserve amount, $45 million

INNOVATION IN RISK MANAGEMENT PRACTICE

All P3 projects involve risk sharing and transfer agreements. P3 project risks that are shared or transferred to a private concessionaire include financial, economic, technical, construction, legal, political, environmental, operational, and so forth. A project with a major underground component will involve geotechnical and operational risks that impact project cost and schedule.

Risk Baseline Report

Risk register has so far been used as a best management practice (BMP) vehicle to manage risks, however, not as a contract document. As Owners are now increasingly comfortable using a risk register in construction, the risks can be included in a proposed innovative contract document "Risk Baseline Report (RBR)" as a part of P-3 contract documents alongside with the GBR. It is important that all contract documents are made legally and technically consistent with the RBR, and can be used alongside the GBR when managing risks, including resolving differing site conditions claims (Bhattarai, 2017).

Risk Baseline Report as a Contract Document

Successful project executions show that the success depends upon understanding risks, formulating effective risk management strategies, and fostering win-win agreements based on sharing both risk and rewards.

Owners can use a two-stage collaborative process to prepare the RBR: Owner prepares a Risk Baseline Report—Owner (RBR-O) first and requires the Proponents to provide their input into the RBR-O.

Proponents may also suggest additional risk items that are not in the RBR-O. The Proponent's risk register shall identify major geotechnical, construction, third party risks, and so forth. The RFP should require Proponents to provide risk contingencies—cost and schedule—in the Schedule of Quantities. This submission will be referred to as the Risk Baseline Report—Proponent (RBR-P). The Owner will negotiate the final RBR with the preferred Proponent, which will be referred as Risk Baseline Report for Construction (RBR-C) (Bhattarai, 2017).

The risk register in the RBR should identify the risk owner responsible for risk management and mitigation, risk scores before and after mitigation, and contingencies—cost and schedule—allocated to an individual risk item. The risk contingency should be presented and agreed as a separate line item in the Bill of Quantities. The contingency reserve should be used to pay the risks those appear during construction.

The contingency fund should be controlled by the Owner and defined as an allowance in the Contract. The release of the contingencies, upon completion of the risk item, can be done as follows (Bhattarai 2017):

- Split contingencies for each major risk item preferably in an equal proportion (or any other proportions agreed under the contract) between the Owner and the Proponent, upon the completion without the use of contingency fund
- Split the residual contingency fund with a lower proportional share (as agreed in the contract) to the Proponent, if the contingency fund is used to mitigate the risk item

The Proponent should also sign back-to-back agreement with their designers/subcontractors to distribute the contingency fund. This will foster partnership and create a win-win situation at all levels of the design/construction organization.

Contingencies should be tracked regularly and discussed during project risk workshops. The Owner's Risk Manager and the Proponent's Risk Facilitator should jointly track the risks via the Project Risk Register (PRR) over the defined life of the risks.

The risk management steps are discussed below (Bhattarai 2017).

Pre-Procurement Stage—Public Agency

- Employ an experienced and competent engineering firm (Engineer) and a Risk Manager in the Owner's organization
- Conduct risk workshops to identify major risk items, their ownership, rankings, and mitigation methods
- Prepare a detailed PRR, perform qualitative and quantitative risk analysis, and prepare a risk management plan
- Prepare Risk Baseline Report—Owner (RBR-O) for procurement

Procurement Stage—Public Agency

- Provide risk sharing options for the Proponents, share geotechnical risks based on the GBR, and sharing other risk items using the RBR-O
- Provide an option for value engineering submission
- Require the Proponents to demonstrate their managerial and technical abilities to manage risks
- Employ a best-value hybrid selection process—both qualification and price based
- Quantify risk contingencies and establish a contingency fund
- Include a line item for risk contingencies in Schedule of Quantities and Prices Form
- Require the Proponents to demonstrate their risk management plan in their proposal
- Require the Proponents to submit Risk Baseline Report–Proponent (RBR–P) in the proposal

Procurement Stage—Public Agency Insurance Requirements

- Require multi-year term design and construction insurance (general/excess liability, professional liability, builder's risk, and so forth) starting from the financial close and ending at the construction of the infrastructure, such as until the O&M commencement date

- Require worker's compensation and commercial automobile liability insurance
- Require concessionaire controlled insurance program (CIP)

Procurement Stage—Proponents
- Prepare their own risk register and RBR-P
- Make provisions for mitigating risks or insurance
- Involve insurance companies when preparing the RBR-P, when applicable
- Propose an efficient and effective organization chart
- Provide value engineering proposals, when applicable

Procurement Stage—the Preferred Proponent and the Public Owner
- Negotiate a best-value contract, when applicable
- Negotiate a win-win Risk Baseline Report for Construction (RBR-C)

Design—Construction
- Update the PRR regularly
- Manage and mitigate risks effectively
- Foster partnering and involve all parties including the Owner in risk management process
- Involve risk insurance companies, when applicable
- Complete the project on time and within the cost

TECHNOLOGICAL INNOVATION

The private sector can bring the following, but not limited to, technological innovations in the US construction industry, especially in the unground construction:
- One-stop shop for research, innovation, manufacturing, project development, financing, construction, and operation
- 3D roadway system underneath roads and highways with heavily congested traffic
- Pollution reduction using clean electric engine vehicles; hyperloop or maglev technologies
- New technologies and designs to jack-up TBM power to increase excavation speed
- Cost effective new multi-use mix-ground TBM technologies
- Mitigation of geotechnical risks using innovative material and machine technologies
- Robotic technologies for TBM operation, support installation, and TBM face interventions, enhancing safety and productivity
- Recycling the excavated muck as building material
- Use of innovative small sized shafts/stations for vehicle/passenger entry to reduce the need for public/private land acquisition and effective use of the underground space
- Artificial Intelligence (AI) or robotic technologies to enhance fire, life, and safety
- Green sustainable environment friendly technologies

CONCLUSIONS

The private sector, via P-3 projects concession, can bring substantial benefits to the public and the construction industry by injecting much needed funds via innovative project financing, bringing industrial knowhow, fostering technological innovation, and delivering fast-track quality infrastructure. In addition, application of Risk Baseline Report as an innovative contractual tool will minimize unmitigated risk exposure to proponents as well as owners, deliver projects within the schedule and price, and create win-win situation for both the proponents and the public. Furthermore, it provides an opportunity for both the public and the private sector to work together to modernize the public infrastructure and overcome the challenges of the aging infrastructure of the United States.

Success of the privately financed P-3 projects depends upon the active support from the government and the public at large. Some potential sought-after support from the government and the public are:

- Formulate laws/regulations to facilitate expediting the environment permitting process and limit the private sector risk exposure to unknown and unmitigated environment risks
- Formulate laws/regulations to limit the private sector risk exposure to public sector grievances and minimize impact to the smooth construction execution
- Formulate laws/regulations to provide tax incentives among others to the private sector
- Assist proponents in raising project funding and reduce borrowing risks
- Assist in land purchase/acquisition and obtaining real estate easement in timely manner
- Avoid prescriptive design and construction requirements in the contract documents to foster research and innovation
- Avoid transferring unmitigated risks to the proponents and adopt Risk Baseline Report as a contractual tool share risk and rewards
- Promote technological innovation, including environmentally sustainable technology, in the in the United States construction industry

REFERENCES

Bhattarai, K. 2018. Design, Construction, and Risk Management Strategies for Shallow Tunnels in Urban Settings. *Proceedings of North American Tunneling Conference.* pp. 687–687. Washington DC, USA.

Bhattarai, K., Nicaise, S., Min, S., Thibault, K., and Trapani, R. 2016. Geotechnical Risk Management Strategies in Public Private Partnership Method for the Delivery of Tunneling Projects, a North American Perspective. *Proceedings of Tunneling Association of Canada Conference.* Ottawa, Canada.

Cision PR Newswire 2016. Meridiam Announces Financial Close on Maryland Purple Line Light Rail Transit Project. June 17 2016 Article retrieved in November 2018 from https://www.prnewswire.com/news-releases/meridiam-announces-financial-close-on-maryland-purple-line-light-rail-transit-project-300286532.html.

Fluor 2016. Fluor Announces Financial Close on Maryland Purple Line Light Rail Transit Project. Retrieved in November 2018 from Fluor website https://newsroom.fluor.com/press-release/company/fluor-announces-financial-close-maryland-purple-line-light-rail-transit-project.

Maryland Department of Transportation (MDOT) and Maryland Transit Authority (MTA). 2014–2015. Request for Proposals to Design, Build, Finance, Operate and Maintain the Purple Line Project through a Public-Private Partnership Agreement.

Nichols, C. 2011. PPP Profiles: Port of Miami Tunnel. Metropolitan Planning Council. Retrieved in November 2018 from https://www.metroplanning.org/news/6126/PPP-Profiles-Port-of-Miami-Tunnel and https://en.wikipedia.org/wiki/Port_Miami_Tunnel.

Poole, R. 2016. Pension Fund Investment in P3 Infrastructure. Reason Foundation. Retrieved in November 2018 from https://reason.org/commentary/pension-fund-investment-in-p3-infrastructure/.

Purple Line Transit Partners (PLTP) 2017. The Purple Line Project—Meet the Team: Purple Line Transit Partners. Retrieved in November 2018 from http://www.purplelinetransitpartners.com/about/team/.

Sigo, S. 2009. Miami Tunnel Reaches Closure. The Bond Buyer. Retrieved in November 2018 from https://www.bondbuyer.com/news/miami-tunnel-reaches-closure and https://en.wikipedia.org/wiki/Port_Miami_Tunnel.

Star America 2018. http://www.starinfrapartners.com/.

The Boring Company 2016–17. FAQ-The Boring Company. Retrieved in November 2018 from the Boring Company website https://www.boringcompany.com/faq/ and https://www.recode.net/2018/11/2/18053424/elon-musk-tesla-spacex-boring-company-self-driving-cars-saudi-twitter-kara-swisher-decode-podcast.

The Boring Company 2016–17. Test Tunnel-The Boring Company. Retrieved in November 2018 from the Boring Company website https://www.boringcompany.com/testtunnel.

Translink 2008. Translink 2008 Annual Report. Retrieved in November 2018 from Translink website https://www.translink.ca/-/media/Documents/about_translink/corporate_overview/corporate_reports/annual_reports/2008.pdf.

United States Department of Transportation 2014. Port of Miami Tunnel, Miami, FL. Retrieved in November 2018 from https://www.transportation.gov/policy-initiatives/build-america/port-miami-tunnel-miami-fl.

United States Department of Transportation 2018. Ohio River Bridges East End Crossing. Accessed November 2018. https://www.transportation.gov/tifia/financed-projects/ohio-river-bridges-east-end-crossing.

Woodhead, R. 2009. Building the Canada Line, An Example of a P3 Project. Innovation, a Journal of the Association of Professional Engineers and Geoscientists of BC September–October 2009.

Yaffe, B 2009. New Rapid Transit Canada Line Is Just the Ticket to Ride. Vancouver Sun 22 October 2009. Retrieved in November 2018 from https://www.pressreader.com/canada/vancouver-sun/20091022/281771330256085.

Design-Build Construction in Underground Construction: A Contractor's Perspective

Gary A. Almeraris • Skanska USA Civil
Scott Hoffman • Skanska USA Civil

The Design-Build contract delivery method, where substantial portions of the final design are developed by the Contracting entity that is also responsible for Construction, rather than by a separate Design group hired by the Owner, has been around in some form for many years. However, the modern form of this approach emerged in the late 20th century, and began to become much more popular as well as more developed in the early and mid-2000s.

This paper explores the original goals and drivers for the development of the modern Design-Build approach, reasons for the shift in underground contract delivery towards this approach, and how Design-Build has evolved in the underground contracting industry within North America, including its accommodation to some of the unique contract related challenges posed by underground project risk. It also discusses the effectiveness of this approach in achieving its original purposes since the modern version emerged in the late 20th Century, and provides a general perspective from the Contracting entity (the Design-Builder) on Design-Build, and recommended improvements going forward.

> Design-build is a method of project delivery in which one entity—the design-build team—works under a single contract with the owner, to provide design and construction services   —Design Build Institute of America

ORIGINATION OF DESIGN BUILD PROCUREMENT

In early works, a common model of major construction in history was that of the master builder. Usually an individual with special expertise and experience would be selected by the king, pharaoh, emperor, etc to oversee the design and construction. That person would bring the project from conception to construction, selecting designers, drafters, survey, craftsmen, materials, etc.

This approach reached as far back as the construction in ancient Egypt (Shaw 2003; Beard et al., 2001) and persisted through major public works throughout the world, including ancient Rome where the military would design and built major infrastructure in house, and the massive cathedrals of the middle ages.

This approach persisted in the United States through the mid-1800s, but widespread fraud and abuse of the system led to reforms (FWHA 2006), including legislation at the national level which tilted most federal and state projects towards the design-bid-build delivery system. This persisted through much of the twentieth century, with major underground projects such as the NYC subway system were designed first, and then bid out, with contractors being selected through various metrics.

Figure 1. Ancient construction mega-projects—Pyramids of Giza, Egypt

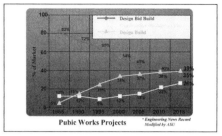

Figure 2. As presented by K Krause, City of Mesa, AZ, original citation to ENR

The situation began to change starting in the 1980s. Partly due to increasing fiscal pressures on government agencies, and due to an increased demand for reduced overall project durations (from concept to construction completion), the industry began to look at other contract delivery methods—design build was revisited. Design build offered a significant advantage in that the construction would then be able to start prior to the completion of the design, and it would move selection of the contractor to the front of the procurement process. As you can see from Figure 2, major changes began to occur in the contracting world.

RE-ENTRY OF DESIGN-BUILD PROCUREMENT

By the 1990s, the concept of design build took on new life and started to make inroads in the underground industry. In the private sector it was found that design-build would reduce the overall project duration, and mitigate scope gaps typical between the design and actual field conditions. This approach fit with a more privatized approach to government, where functions previously handled by government agencies directly from testing to design to construction management, was increasingly executed by third party design consultants. Further, advocacy groups arose to promote this means of delivery, arguing that with appropriate contract language and the support of rigorous design requirements, the risk of conflicts of interest and resultant corruption could be minimized.

As this movement began to pick up steam, government officials were aware of the shortfalls experienced in previous periods where design build was used, and thus a process of new legislation development began, to permit once again design build. These legislative efforts focused on insuring that the proper oversight, terms, and conditions were included to prevent abuse as was experienced in earlier times. Figure 3, courtesy of the Design Build Institute of America, provides a guide on the legislative status of Design-Build authorization at the state level. This trend towards design-build has only accelerated as time has progressed.

For example, a major New York area heavy/civil general contractor had a negligible number of contracts that were design build. By the early 2000s, D-B work comprised about 40% of this firm's work, and by 2018, this firm had over 90% of its revenue booked in D-B type contracts.

Another example is found in New Mexico. Based upon Engineering News Record data (as presented by Kurt Krause, City of Mesa), of public works projects, the share of D-B has increased from 5% in 1985 to almost 40% in 2015, whereas D-B-B has decreased from over 80% in 1985 to around 35% in 2015.

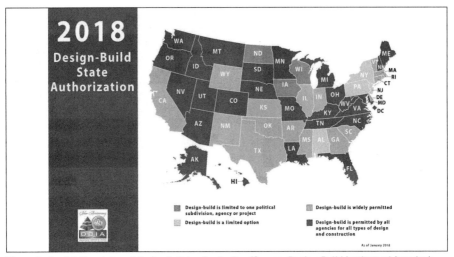

Figure 3. Legislative status of design-build authorization (Source: Design-Build Institute of America)

Why the continued trend towards Design-Build procurement and its variants (CM at risk, progressive Design-Build, etc), a trend that has been in progress for almost 40 years?

There are several drivers of this trend, a few of which are discussed below.

Reduced Project Delivery Duration

Perhaps one of the most significant drivers of this trend is that design-build projects generally have shorter durations, and the shorter duration is often significant. Data collected by a Penn State study of 351 projects showed an average schedule reduction of 12% when compared with DBB projects (see Figure 4). Other studies have been published, such as those from the FHWA, which compared schedules of DBB projects performed for the Arizona DOT. As can be seen below, savings of over 300 days was typical of DB projects, due in large part to the fact that the construction is able to start alongside the continuation and completion of design. By definition, for Design-Bid-Build projects, such activities are linear.

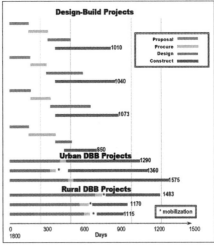

Figure 4. Arizona DBB project schedules compared (Source: FWHA, AZ DOT)

Lower Total Cost

Another key driver in the use of Design-Build is the realization of lower total costs on design-build projects. This arises from factors such as a better matching of the design with the contractor's skills and capabilities (tailoring the design to the capabilities of the builder), a much higher level of collaboration between the Designer and Builder, as both are on the same team and both stand to gain or lose from a better design, the reduction in total construction duration as noted above which results in lower program administrative and field construction oversight costs, and a better coverage of scope gaps, as such gaps are controlled by the D-B team. Cost growth from initial bid to final cost of construction has been found to average 5% lower for projects which used the D-B approach as opposed to the DBB procurement method (reference Figure 5).

COMPARISON OF PROJECT DELIVERY METHODS		
METRIC	DESIGN-BUILD VS. DESIGN-BID-BUILD	DESIGN-BUILD vs. CM@R
UNIT COST	6.1% lower	4.5% lower
CONSTRUCTION SPEED	12% faster	7% faster
DELIVERY SPEED	33.5% faster	23.5% faster
COST GROWTH	5.2% less	12.6% less
SCHEDULE GROWTH	11.4% less	2.2% less

Source: Construction Industry Institute (CII)/Penn State research comprising 351 projects ranging from 5,000 to 2.5 million square feet. The study includes varied project types and sectors.

Figure 5. Comparison of design-build with other methods of project procurement

Improved Collaboration Between Designer and Contracting Entity

One negative aspect of the DBB procurement method is the absence of the Contractor during the design development, and lack of guidance with respect to constructability considerations in the design. This is of particular concern in tunnel construction, as the means and methods can heavily influence final design, with such considerations as waterproofing, double lining, use of rock mass or temporary support in the final design, etc. While many such aspects are well known and can be captured by the larger design firms with construction experience, underground projects can be highly variable, as are the technical strengths of different firms.

Collaboration then takes place after award, and that can become complicated, especially if the schedule is tight, or major constructability issues are uncovered.

The design-build approach addresses this gap directly, by intentionally pairing the Designer with Building entity. This greatly increases early collaboration, and permits the team to more closely tailor the design to the abilities and strengths of the contractor (it is important to note that the ability to fit the design more closely to the contractor's abilities can be hampered to the degree of design proscription employed by the Owner).

Potential Improvement of Risk Allocation

With the Designer and Builder acting together in one unit, construction related risks can be mitigated, such as overall schedule, constructability, material selection and sourcing, to name a few. As the builder has a better knowledge of the market, material availability, local labor rules, and equipment capabilities, this can be shared with the designer and considered during design phase.

Reduction of Scope Gap

In a DBB procurement, both the designer and the builder have privity, or a contract, with the Owner, not with each other. As such, when the contractor is negatively impact by design errors, omissions, or other problems with the design itself, the contractor must seek compensation through the Owner, who in turn must pursue this from the Designer. In the case of a DB procurement, the Owner no longer acts as a middleman to such disputes or issues.

DESIGN BUILD AND THE UNDERGROUND CONSTRUCTION ENVIRONMENT

While Design Build in many ways appears to be well suited to the underground construction industry, there are several key aspects of the industry that need special attention in a Design-Build contract.

First, an accurate and well defined Geotechnical Baseline Report (GBR) must be included. It is well recognized within the industry that the Owner is in the best position to either know or become familiar with the geotechnical conditions at the jobsite. This is often due to permits and/or other access constraints which add time to such explorations, which Owner, often governmental agencies, are in a much better position than proposal teams to obtain permits for exploration. Further, Owners often have had earlier projects at the same locations and know well the history at these sites.

Along with this, a standard differing site conditions clause is important, as this provides important boundaries as to where the Design-Build team should price risk, and helps prevent the Design-Build proposer from carrying too much contingency for risks which are remote.

This use of unit prices mechanism has been a feature of recently successful Design-Build projects underground. Such a mechanism, which has proven useful in DBB projects, provides a way of addressing risks that have a recognized solution but may vary significantly in scale. This again helps to avoid excess contingency or risk pricing.

Another key aspect of the agreement that must be addressed in a design-build contract for underground projects is the early acquisition of rights of way, easements, and other third party access and construction constraints. As a design-build project is even more time sensitive, having these issues resolved early on will help maximize the schedule improvements that a design-build project delivery can yield. Further, the Owner is often in a much better position to handle politically difficult issues that may impact the acquisition of rights of way and easements.

Permitting responsibility, while often delegated to the design-build team, should be carefully considered. Again, when there is the possibility of third party governmental agencies raising nearly insurmountable obstacles to permitting, or where political considerations could delay or even halt the permitting process, the owner must step in early and resolve such situations so as to mitigate what could be serious risks to project viability.

Finally, the Owner should avoid being overly proscriptive in the design guidelines that are provided. This can work against the idea of a design-build arrangement, as the Owner may in fact design the project to a point where it becomes a design-design-build arrangement. At the end, this arrangement only increases costs as the design is done twice—first by the Owner making a preliminary design guideline that is actually what is expected for the design, and again by the design-build team, to verify or modify the first design. A key aspect of design-build's strength is the use of the builder's experience to inform the design—proscription by the Owner that is too heavy will dampen or even eliminate this strength.

WHEN DESIGN-BUILD IS NOT APPLICABLE

Finally, in some cases, Owners will use the D-B agreement where a DBB or hybrid DB/DBB arrangement is likely the best course of action. Such situations occur for example where a far majority of the project is covered under very specific owner requirements and designs, due to requirements for the work to have similar features to other

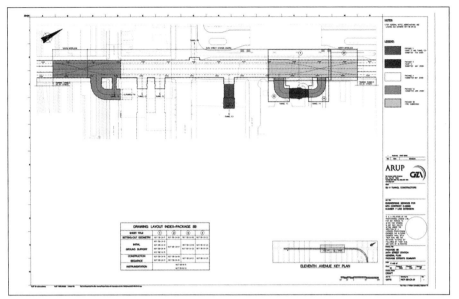

Figure 6. Temporary rock support design

parts of the system (such as new subway stations in existing systems, new wastewater facilities to be operated with the remaining system, etc). In such cases, the DB arrangement can actually lead to delays, especially if the proscriptive portions of the work are not very clearly outlined, and the DB team must try to figure out exactly what the Owner is looking for through meetings, submittal cycles, and shop drawing markups. In these cases, the Owner should rather consider more of a design-bid-build arrangement, as the design for all practical purposes is complete—all that is left may be to select rebar size/spacing or determine the section modulus of sheeting.

Figure 7. Temporary rock support—7 line 34th St cavern, NY, NY, USA

DIVISION OF DESIGN RESPONSIBILITIES IN UNDERGROUND WORK:

In a pure design-build package, all design responsibility would be borne by the D-B team. However, most owners involved in underground work desire to at least retain approval rights to the design, and in many cases wish to provide design commentary. The following is an overview of major divisions of design responsibilities to be considered in a D-B contract:

1. Temporary works/work required for construction, but not part of the permanent work. This work is typically in the hands of the D-B team, similar to what is found in a D-B-B package. This aspect of design is generally not a differentiator between D-B and D-B-B projects.

2. Determination of alignments and geometry—in a D-B project, this could be determined by either the D-B team or the owner, and is largely project specific. However, where rights of way are difficult to obtain, alignment is

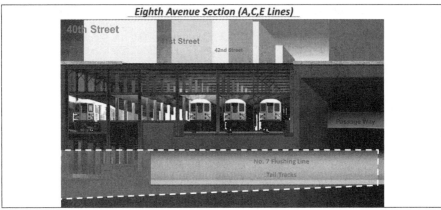

Figure 8. Connection of No 7 line into existing Times Square Station

Figure 9. Example of precast lined tunnel in DC Water sewer system (Washington, DC)

Figure 10. Example of SEM tunnel as part of Regional Connector (Los Angeles, CA)

often dictated by the Owner. An example is cited below, where the project was procured as D-B, but due to right of way constraints, difficulties in obtaining easements, and the necessary location of diversion chambers, required a rather specific alignment and facility geometry.

3. Tunnel and liner type should largely be determined by the D-B team. There are an increasing array of technologies that can be employed in the construction of tunnels and their liners, and there is also a commensurate range of capabilities amongst tunnel contractors. As such, in order to obtain the most economical solution, it is usually worthwhile to all parties involved to provide a high degree of flexibility to the D-B team in the design of these elements.

Figure 11. Diversion chambers designed to highly specific criteria DC Water sewer system (Washington, DC)

4. Proprietary or highly system specific structures and design should be retained by the Owner, rather than having the D-B team try to figure out the 'right combination'.

5. Architectural/Aesthetic features can become difficult, particularly if the Owner has specific ideas on this, but it is at first left to the D-B team. Contingencies often are advised in this case, which can be drawn upon as the Aesthetic features are defined if such features exceed what was considered initially in the proposal.
6. Temporary work that becomes permanent—such work can often realize the maximum benefit of a Design-Build arrangement by letting the D-B team fit the design to the builder's skills and abilities.

RECOMMENDATIONS GOING FORWARD

As the design-build procurement method continues to develop, the following are some recommendations to help improve on what has been achieved to date:

a. Teams generating proposals should consider carefully an evaluation that considers both technical and cost aspects of the proposals
b. Maintain a shortlist of 3 to 4 contractors—this helps draw out the best and most interested D-B entities by permitting the Owner to select only the most qualified, and encourages the selected teams to put extra effort into the proposal as they have a reasonable chance of success. In contract, having 10 or more teams will necessarily reduce the quality of the proposals as the proposing teams at the outset have a relatively low chance of success.
c. Provide sufficient time—4 to 6 months—for package development process. A reasonable procurement duration is critical in obtain quality proposals—if insufficient time is afforded the teams, the quality and competitiveness of the proposals will likely suffer.
d. Innovative Design—use emerging/most current technical approaches, rather than defaulting to 'the way we have always done it'. The temptation for governmental agencies used to DBB construction is to stay with the tried and true. While this is often advisable, recommendations to incorporate new technology must never be discouraged across the board—real improvements in efficiencies and schedule may be missed.
e. Designer of D-B team must provide adequate resources to maintain design schedule (schedule is critical).
f. Permit a reasonably aggressive design that may make major changes to the RFP documents.
g. Owners need to maintain open mind to Alternative Technical Concepts (ATCs)
h. Review process that values both best technical and cost proposals

CONCLUSIONS

The use of a Design-Build procurement approach has become prevalent in the construction industry, and the underground construction industry is no exception. There are many benefits that can be realized with such an approach, but to make full use of its benefits, several aspects of the RFP with respect to the division of design responsibility, assignment of risk, degree of proscription in the design documents, among other aspects, must be carefully considered.

A summary of benefits include

- Brainstorm opportunities between multiple designers and contractors pre-bid.
- Detailed constructability review to flush out any fatal flaws or other issues that could arise during construction.
- Re-engineering done in advance, avoid many last minute changes.
- Cost reduction review performed in advance of award.
- Alternate technical concepts are elicited up front.
- "Out of the box" concepts can be aired early, may result in a major concept change early and greatly improve cost/efficiency of achieving ultimate goals of Owner.
- Peer review of design/approaches due to efforts of multiple teams.

These benefit the Owners in particular, with no loss of time and minor cost (stipends, in-house review costs). As a result of the above, the following is realized:

- Better price certainty
- Though qualifications process, only serious teams will invest in D/B which will produce a high degree of design/construction methodology in their proposal
- Avenue for new technologies to be introduced into the wider industry.

Given the above, it is important to remember that, while the Design-Build approach is suited to many types of underground work, there are cases where it is not the best fit, particularly when there is a very high degree of proscriptive or propriety work, or where the work must be in very close alignment in terms of details, equipment, materials, etc, with a much larger system within which it operates. In such cases, a hybrid of Design-Build and Design Bid Build ought to be considered.

Finally, as with all business endeavors, trust and fairness is key. Both parties to a D-B contract are best served when risk is allocated to the party most capable of managing it, and when the division of design responsibilities and expectations are clearly laid out. Ultimately, the industry as a whole will benefit as a result.

REFERENCES

1. Shaw, Ian (2003) *The Oxford History of Ancient Egypt.*

2. Beard et al., *Design-Build: A Brief History. Design Build Planning Through Development.* McGraw-Hill, 2001.

3. FHWA—"design build effectiveness study," Dr. Keith Molenaar et al., Jan 2006.

Having Difficulty Deciding on Procuring New, Refurbished or Remanufactured TBMs?

Dan Ifrim ▪ Hatch–Mississauga

ABSTRACT

What happens to new TBMs at the end of the project? Most of these machines are recovered and stored by contractors with the hope of an upcoming project, while some are being sold or leased to other contractors or purchased back by the manufacturer.

This paper discusses the fate of a new TBM after it's intended project use and the outlook for re-employment as a used TBM on future projects.

The paper discussion focuses on the qualitative conditions to be considered in the decision to employ refurbished or remanufactured TBMs, to comply with future project opportunities and requirements. As a backgrounder, the paper will also analyze the ITA recommendations for re-manufacturing and refurbishment of TBMs.

INTRODUCTION

Tunnel projects technical specifications provides the minimal requirements for contact bid. The details and clarity of the specifications are important to ensure that all bidders are competing on equal grounds. The tunnelling and TBM specifications are often combined in one specification addressing both the TBM minimal requirements as well as the tunnelling process. Yet in some cases specially in Owner Supplied TBM projects, Design Build or P3 projects these two specifications are separated.

For the purpose of this paper will discuss the TBM technical specification and minimal requirements once for both situations since the is no difference in requirements despite the prescriptivity differences. There are various aspects of the TBM minimal requirements that are usually addressed by the TBM technical specification including the TBM type, Systems and Ancillary Equipment, Manufacturer Qualifications and last but not least important is the requirement of New, Remanufactured or Refurbished TBM.

DEFINITIONS

Before starting any discussion, we need to clarify the basis of the discussion by providing a clear understanding of what each term means.

OEM
Original Equipment Manufacturer

New TBM
A TBM manufactured specifically for the project that is intended to be used and never employed on any project.

Used TBM

A TBM that was previously employed in at least one tunnelling project that will make the object of either remanufacturing or refurbishing.

Remanufactured TBM

Encyclopedic Definition

Remanufacturing is "the rebuilding of a product to specifications of the original manufactured product using a combination of reused, repaired and new parts." It requires the repair or replacement of worn out or obsolete components and modules.

ITA Definition

The process in which the TBM systems and subassemblies are employed on a different project in the original configuration or with modifications. The process the following steps:

- Complete disassembly down to the single part level.
- Cleaning of all parts.
- Inspection and sorting of all parts.
- Reconditioning of parts or replacement by new parts.
- Reassembly.

Refurbished TBM

Encyclopedic Definition

Servicing and/or renovation of older or damaged equipment to bring it to a workable or better-looking condition.

ITA Definition

A full maintenance of systems and subassemblies that were previously employed at different projects in the original configuration or with minor modifications.

DISCUSSION

Currently there is no regulation or guidance for TBM remanufacturing or refurbishment. Each contractor addresses the topic by his own approach, either performing the refurbishment on its own or engaging a TBM manufacturer or in some cases the OEM, trying to minimize the risk of employing a used TBM with minimal cost.

To provide some confidence to owners choosing to employ remanufactured or refurbished TBMs a common ground shall be established. In 2015 ITA took the initiative to create and publish such guidelines under ITA Report No. 5 "Guidelines on rebuilds of machinery for mechanized tunnel excavation. The publication is going a review and expected to see a new revision in 2019.

Although this initial step was taken, Owners in North America are not fully aware of its content and not all the consultants are using it to define the requirements for remanufacturing or refurbishments and in most cases take the safe approach of specifying a new TBM.

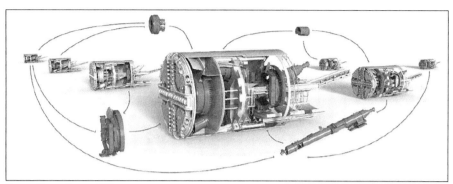

Source: Werner Burger, 2018
Figure 1. Remanufacturing, breaking down the TBM in component

Each individual project is different and selection of the type of TBM (new, remanufactured or refurbished) shall be coordinated with the project complexity, tunnel length and ground conditions.

The risk of employing a used TBM shall be assessed and quantified before any recommendation or decision is being made.

ITA Guidance Requirements

It is important to note that the ITA Guidance is intended to cover the TBMs, MTBMs, backup equipment and systems and tunnel equipment and systems such as California switches pumping stations, conveyors and jacking frames.

Remanufactured TBMs

"Remanufacturing is typically applicable for projects without special challenges where full component lifetime and a "state of the art" set of requirements for the machinery exist."

Some components such as Cutterhead, Main Drive Bearing, Gearboxes, Seals, Shield Structures and Electronic Components are mandated to be new.

Generally new components are referred to as new manufacture for the subject project or stock items that suits the subject project.

Refurbished TBMs

"Refurbishment is typically applicable for project without special challenges employing machinery that successfully completed a comparable project."

The refurbishment is viewed as a "full maintenance" process and repair or replace of defective systems or components.

Approach to TBM Reuse

In a design-built approach the Owner hires a Consultant to develop a minimum level of design, just enough to provide the basis for bidding. At this level of design, only the general type of TBM is identified and based on Owners direction a minimum criterion is set for the TBM requirements. The winning team of Contractor-Consultant continues the design to a 100% level and choses the type of TBM that satisfy the minimal

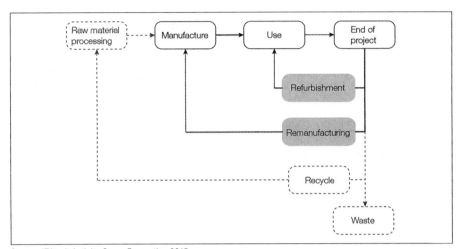

Source: ITAtech Activity Group Excavation 2015
Figure 2. TBM manufacturing process—opportunities to reduce raw material intake

requirements. There are cases when the Contractor-Consultant team proposes a completely new system or tunnelling methodology.

In the design-bid-built approach the Owner hires and rely fully for complete design and specifications. This approach is most of times very prescriptive and the required TBM is fully specified including allowance or disallowance for remanufactured or refurbished TBM.

The choice of New vs. Remanufactured vs. Refurbished TBM is an important decision and bury a lot of responsibility. Such decision should be based on project length, complexity and risk and definitely not on cost.

Consideration for Refurbished and Remanufactured are derived from the effort to reuse and recycle materials to reduce waste and ultimately to reduce manufacturing carbon footprint.

There are projects where remanufacturing or refurbishment do not apply, either due to complexity and length of the tunnel project, challenging ground conditions and risk associated with employing a refurbished or remanufactured TBM. There are also limitations arriving from the limited availability of TBMs that can be refurbished or remanufactured for the specific project conditions.

One important aspect of remanufacturing is related to the Original Manufacturer of the TBM ownership of drawings and intellectual property. In all cases of remanufacturing it is extremely important that the remanufacturer has access to the original manufacturing drawings for the structural components and especially for the Main Drive. A thorough inspection of the Main Drive sealing system and restoring manufacturing tolerances can only be achieved based on OEM drawings. Another aspect is related to the wear of the structural components including Cutterhead, Shields and associated sealing surfaces. It is expected that during the remanufacturing cycle the main structure wear is measured and a FEA is performed to understand the extent of the wear and determine the remanufacturing approach.

Source: Werner Burger, 2018
Figure 3. Unorganized TBM and used TBM components storage

Concepts and Approach

The concepts new, remanufacture or refurbishment represents the three options to explore throughout the risk assessment of the tunnelling project.

Typically, a new TBM is associated with a unique tunnel project with foreseen or unforeseen risk and challenges. In general, the TBM manufacturer quality system ISO 9001-2015 gives the consultants a minimum of comfort, however the main components brand name is a next step up in quality. A full manufacturer warranty is usually associated with new TBM and new components, however even for new a new TBM the quality may differ from supplier to supplier, for instance it is well known that Rothe Erde (Rotek) or SKF main drive bearings are the most reliable, well exceeding the threshold of L_{10} minimum life required by most consultants. Similar parallels can be drawn for gears, motors, pumps, etc.

Source: Werner Burger, 2018
Figure 4. Organized TBM used components storage

When it comes to options for remanufacturing it is important first to ensure that the proposed TBM to be remanufactured is of a recent build (max 5–8 years old depending of TBM type), fits the size of the proposed tunnel and incorporates the particular options required by the project or can be easily modified to add them. The remanufacturing process shall be viewed as an opportunity to save the environment/reduce the carbon foot print and not as an opportunity to save money. The remanufacturing process shall at minimum use new Main Drive bearing and seals, all other seals, bearings, hoses and safety components new and shall impose a strict assessment of all other TBM components during disassembly and cleaning. All components that have wear outside of the factory tolerances shall be rebuilt or replaced with new. Cutterheads are typically designed and manufactured for the specific ground conditions of each tunnel and therefore are always required to be new unless a second phase of a project (same location and ground conditions) are available. As an example of this opportunity are the Euclid Creek tunnel and Dugway tunnel in Cleveland Ohio.

Refurbishments as processes for reuse a TBM are usually referring to cleaning of all components and checking functionality only. A refurbished TBM shall be only

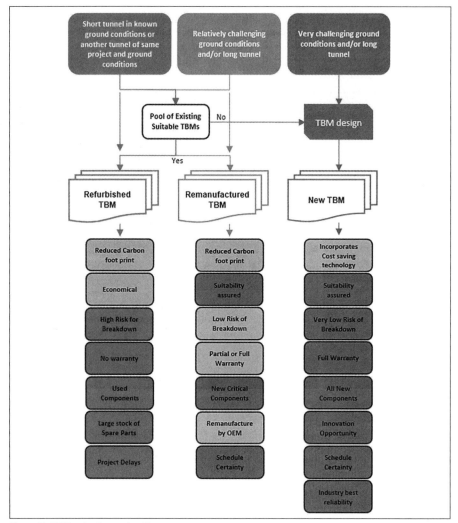

Figure 5. TBM selection and options

employed on short and low risk tunnels where access from the surface via a drop shaft is an option.

Although there is a large pool of used TBMs available not al TBMs are usable to satisfy the needs of today safety and energy standards. Also, most of these TMS are kept in the outdoor by their owners and collected a fair amount of rust.

They are however TBMs to whose main components are properly stored indoor rustproofed or conserved as per the OEM recommendations. For these reasons a thorough inspection of the proposed TBM to be reused shall be carried out by the original manufacturer to determine the feasibility for remanufacturing or refurbishment, whatever the case may be given the complexity and challenges of the tunnel project.

Deciding on Procuring New, Refurbished or Remanufactured TBMs

Table 1. Common components all TBMs new vs. remanufactured vs. refurbished

Component/System	New TBM	Remanufactured TBM	Refurbished TBM
Cutterhead	New	New	Inspect, Remanufacture
Motorplate Assy	New	Remanufacture	Reuse
Main Drive Bearing	New	Remanufacture	Inspect, Test, Reuse
Main Drive Seals	New	New	New
Main Drive Motors, Gears & Pinions	New	Remanufacture	Inspect, Test, Reuse
Hydraulic Cylinders	New	Remanufacture	Inspect, Test, Reuse
Rail Assy System	New	Remanufacture	Inspect, Test, Reuse
Oil/Grease System	New	Remanufacture	Inspect, Test, Reuse
Oil Tanks	New	Inspect, Test, Reuse	Inspect, Test, Reuse
Hydraulic Pumps	New	Remanufacture	Inspect, Test, Reuse
Electric Motors	New	Remanufacture	Inspect, Test, Reuse
Electric Cables	New	Inspect, Test, Reuse	Inspect, Test, Reuse
Grout Pumps	New	Remanufacture	Inspect, Test, Reuse
Computers, HMI & Guidance System	New	Remanufacture	Inspect, Test, Reuse
Gas Monitoring System	New	New	New
Other Pipes (Grout, Water, etc.)	New	Inspect, Clean, Reuse	Inspect, Test, Reuse
Hydraulic Pipes, Hoses and Fittings	New	New	New
Substation	New	Remanufacture	Inspect, Test, Reuse
Flow Meters and Sensors	New	Inspect, Test, Reuse/New	Inspect, Test, Reuse
Gantry Structure, Supports, Walkways, Counterweights	Opportunity for remanufacturing.	Remanufacture	Inspect, Test, Reuse
Gantry Wheels	New	Inspect, Test, Reuse/New	Inspect, Test, Reuse
Hyperbaric Chamber	Opportunity for remanufacturing	Inspect, Test, Reuse	Inspect, Test, Reuse
Refuge Chamber	Opportunity for remanufacturing	Inspect, Test, Reuse	Inspect, Test, Reuse
Hose & Cable Reels	New	Remanufacture	Inspect, Test, Reuse
Fire Safety Systems	New	Inspect, Test, Reuse	Inspect, Test, Reuse
Conveyor Belt Structure*	Opportunity for remanufacturing	Remanufacture	Inspect, Test, Reuse
Conveyor Belt Rollers*	New	New	Inspect, Test, Reuse
Conveyor Belt*	New	New	New

*Exception for Slurry TBMs

For the purpose of Refurbishment and Remanufacturing consideration, it is assumed that the used TBM is found in the pool of used TBMs, matches the proposed tunnel diameter and for the remanufacturing process specifics the used TBM is of a recent build (maximum 5–8 years old).

Table 1 is a general recommended approach to remanufacturing and refurbishing of an existing TBM. A common list of components applicable to all TBMs and particular lists of specific components for Pressurized Face TBMs, Shield TBMs and Beam TBMs are further discussed.

The table also identifies some opportunities of using remanufactured components for new TBM build.

Quality Assurance

At developing the TBM technical specification a minimum of prescriptiveness shall be considered regardless if a new or remanufactured/refurbished TBM is considered.

Table 2. Components specific to pressurized face TBMs new vs. remanufactured vs. refurbished

Component/System	New TBM	Remanufactured TBM	Refurbished TBM
Hyperbaric Chamber	Opportunity for remfg.	Inspect, Test, Reuse	Inspect, Test, Reuse
Slurry Pumps	New	Remanufacture	Inspect, Test, Reuse
Jaw Crushers	New	Inspect, Test, Reuse	Reuse
Slurry Treatment Plant	New	Inspect, Test, Reuse	Reuse
Screw Conveyor	New	Remanufacture	Reuse
Foam Pumps	New	Remanufacture	Inspect, Test, Reuse

Table 3. Components specific to shield TBMs new vs. remanufactured vs. refurbished

Component/System	New TBM	Remanufactured TBM	Refurbished TBM
Shields Skin, Bulkheads and all Other Shields Structural Components	New	Measure, Load Test, FEA	Inspect, Test, Reuse
Shields Seal Surfaces	New	Remanufacture	Inspect, Test, Reuse
Erector Bearing & Gears	New	Remanufacture	Inspect, Test, Reuse
Erector Pad	New	Remanufacture	Inspect, Test, Reuse
Segment Shuttle	New	Remanufacture	Inspect, Test, Reuse
Articulation Seals	New	New	New
Tail Seal Lines	New	Inspect, Test, Reuse	Inspect, Test, Reuse
Tail Seal Brushes	New	New	New
Tail Seal Grease Pump	New	Remanufacture	Inspect, Test, Reuse
Grout Pumps	New	Remanufacture	Inspect, Test, Reuse

Table 4. Components specific to beam TBMs new vs. remanufactured vs. refurbished

Component/System	New TBM	Remanufactured TBM	Refurbished TBM
Roof Support	New	Measure, Load Test, FEA	Inspect, Test, Reuse
Beam	New	Remanufacture	Inspect, Test, Reuse
Drill Support Ring	New	Remanufacture	Inspect, Test, Reuse
Grippers, Mini Grippers & Stabilizer Pads*	New	Remanufacture	Inspect, Test, Reuse
Dust Scrubber	New	Inspect, Test, Reuse	Inspect, Test, Reuse
Shotcrete System	New	New	New
Rock Bolting System	New	Inspect, Test, Reuse	Inspect, Test, Reuse

*Also used on Shielded Rock TBMs and Double Shields

Source: Dan Ifrim, 2013
Figure 6. Beam TBM before and after remanufacturing

Deciding on Procuring New, Refurbished or Remanufactured TBMs

Source: Werner Burger, 2018
Figure 7. Grout pump before and after remanufacturing

Source: Werner Burger, 2018
Figure 8. Grout pump before and after remanufacturing

The remanufactured TBM shall carry the same warranty as new and shall include only new main/critical components such as main drive assembly (bearing, seals, motors and gearboxes) and other critical/job stoppers components.

The quality of the refurbishing process is essential for qualifying a remanufactured TBM and as such the requirements should specify that the remanufacture process must be carried under direct supervision of the OEM and make the original manufacturer the bearer of the warranty.

In the TBM manufacturing world ISO 9001-2015 is a minimum quality assurance system requirement. Remanufacturer past experience with proven record of quality can be assessed by owners at awarding a contract that involves a remanufactured TBM. Additional conditions may be set for TBM past use, TBM age etc.

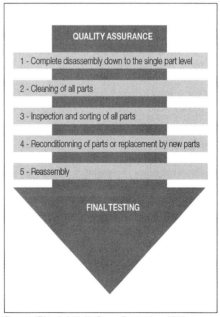

Source: ITAtech Activity Group Excavation, 2015
Figure 9. Remanufacturing process as defined by ITA guideline

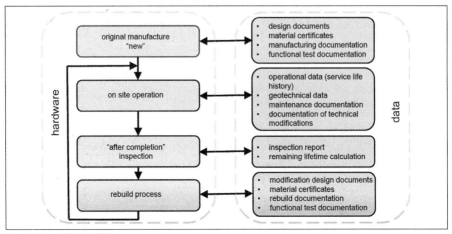

Source: Werner Burger, 2018
Figure 10. OEM manufacturer data importance

Remanufacturers Qualifications

The quality assurance requirements for the TBM rebuilder shall be identical to the TBM OEM—ISO 9001-2015 registered.

The remanufacturing process shall be carried out only by the OEM or by a facility accredited by the OEM and under direct OEM supervision. Only the OEM has access to all TBM data (design documentation, materials and equipment certificates, operational data) and can produce accurate remanufacturing drawings.

TBM Warranty

The OEM is ultimately responsible for a quality remanufacturing process. The requirements for remanufactured TBMs shall be identical to new TBMs and the OEM shall provide a full warranty of the remanufactured TBM.

In the refurbishing process the Contractor is fully responsible for completing the tunnel with the TBM selected and it is up to him to negotiate a TBM warranty with the TBM refurbishing facility as consider it fit and economical.

CONCLUSION

The global awareness of environmental impacts of manufacturing and use of resources, has gained a greater focus in the last decade. The global effort to reduce environmental impact of new manufacture has gained attention in the tunnelling industry with respect to the opportunity to reduce carbon foot print by employing new technologies, with less power consumption and reuse of TBMs and their main parts such as shields, gantry cars and other components where the nature and risk of the tunnel project provide such opportunity.

The option of Refurbishment and Remanufacturing of TBMs or TBM components shall be addressed and assessed on each tunnel project through qualitative and quantitative risk workshops.

The opportunity for reuse, recycle and remanufacturing exists in any project, from reusing the gantry cars, steel structure, rail, muck cars as major steel components to substations, pumps and motors where applicable. The applicability resides in the components fit to the new build or re-build to the technology that was applied at original manufacture and to the efficiency of that component or system.

ACKNOWLEDGMENT

The author thanks Derek Zoldy for his contribution to this paper.

REFERENCES

ITAtech Activity Group Excavation 2015, *ITA Report no 5 — Guidelines on rebuilds of machinery for mechanized tunnel excavation,* N° ISBN: 978-2-9700858-9-8/MAY2015.

Lock Home 2018, *Rebuilt TBMs — are they as good as new?* Tunnel Talk discussion.

Werner Burger 2018, *Remanufacturing,* Toronto workshop presentation.

Dan Ifrim, Derek Zoldy 2018, *TBM Procurement — Owner's Dilemma*, NAT Proceedings.

Dan Ifrim 2013, *TBM rebuild report,* AECOM Project Report.

Incentive-Based Project Delivery with Fixed Price Incentive Fee Contracts

Philip Sander ▪ RiskConsult GmbH
Markus Spiegl ▪ RiskConsult GmbH
John Reilly ▪ John Reilly International

ABSTRACT

Effective cost and risk management is essential for the success of large infrastructure projects, as demonstrated by a long history of cost overruns. In order to achieve cost transparency, risk-based probabilistic approaches are needed to determine the probability that project delivery can be accomplished within cost and schedule goals.

In addition to some owners moving to a more collaborative and incentivized project environment, a significant number of owners and agencies are also considering alternative contracting models to deliver their projects. This paradigm shift is driven by the fact that more traditional delivery methods, e.g., fixed price contracts, often fail to meet objectives due to factors which the authors have described in previous papers.

This paper describes the mechanics of fixed-price incentive fee firm target (FPIF) contracts, provides a framework to analyze such contracts and demonstrates how FPIF pricing arrangements (pain/gain mechanism, target cost, ceiling cost, etc.) can be applied with a risk-based probabilistic approach. Since, in early stages, the project's outturn cost can only be estimated using ranges, estimates of potential profit for the contractor and project price for the owner need to be made using a probability model for total project cost.

INTRODUCTION

Effective cost and risk management is essential for the success of large infrastructure projects, as demonstrated by a long history of significant cost overruns. In order to manage cost to established budgets and to achieve cost transparency, it is necessary to adequately consider cost and schedule uncertainties (risks), which means risk-based probabilistic approaches are needed. This allows us to estimate the probability that project delivery can be accomplished within cost and schedule goals and to define and manage risks that might negatively affect meeting those goals.

In addition, some owners are moving to a more collaborative and incentivized project environment, and a significant number of owners and agencies are also considering alternative contracting models to deliver their projects (Ross 2003, ICE 2018). This paradigm shift is driven by the fact that more traditional delivery methods, e.g., fixed price contracts, often fail to meet objectives due to factors that the authors have described in previous papers.

The purpose of collaborative working agreements and more integrated supply teams is to align the client, design consultants, contractors, sub-contractors, and vendors in a structure, often with incentives, to ensure that everyone works together efficiently to achieve agreed (shared) goals. Such teams are better able to create an environment where outstanding results can be achieved, with incentives leading to improved outcomes for owners and contractors.

This paper describes the mechanics of Fixed-Price-Incentive-Fee (FPIF) firm target price contracts, provides a framework to analyze such contracts, and demonstrates how FPIF pricing arrangements (pain/gain mechanism, target cost, ceiling cost, etc.) can be applied with a risk-based probabilistic approach. Since, in early stages, the project's outturn cost can only be estimated using ranges, estimates of potential profit for the contractor and project price for the owner need to be made using a probability model for total project cost.

An application similar to the example is used for the Lima Airport Extension Program.

APPROACH

FPIF and Delivery Methods

The traditional contracting approach for many project owners is to attempt to transfer as much of the risk as possible to the contractor, e.g., by a Lump-Sum-Turnkey approach (Reilly et al., 2018). This is not necessarily effective for megaprojects. Any attempt to allocate risks of complex project to different parties, no matter how well intentioned, may be little more than an illusion and can give rise to an adversarial culture that may threaten the success of the project (Ross 2003, Reilly et al., 2018).

The FPIF delivery model (using NEC- or FIDIC-type contracts) is a way to implement shared goals, related to an established, negotiated target cost. It is not a full alliancing approach, where the owner, designer, and contractor are jointly bound to meet cost and schedule targets in a pain/gain environment, but it is an option for large, complex projects with a high level of risk and uncertainty and can establish a collaborative working environment using incentives based on a pain/gain mechanism.

By convention, contracting professionals use share ratios to depict the degree of risk assumed by the owner and contractor. The share ratio for Fixed-Price (FP) contracts is depicted as 0/100. The first number is always the owner's percentage of risk, and the second number is always the contractor's percentage of risk. The 0/100 share ratio means that the contractor assumes 100 percent of risk under an FP contract. Conversely, a Cost-Plus-Fixed-Fee (CPFF) contract share ratio is depicted as 100/0. Since a CPFF contract requires a contractor's "best efforts" and they get paid the fixed fee regardless of their achievement under the contract, the government assumes 100 percent of the risk (Cuskey 2015). The Cost-Plus-Fixed-Fee delivery method allows the owner more control over his or her budget than the Cost-Plus-Percentage-Fee contract. As the total project cost increases, the fee paid to the contractor also increases. Nevertheless, these contracts are more flexible to project changes and reduce the contingency that the contractor has at the time of the bidding (Ibbs et al., 2003).

The differences between the Cost-Plus-Incentive-Fee (CPIF) and FPIF pricing arrangements occur when contract costs are substantially above or below target cost. The CPIF contract pricing arrangement must include a minimum fee and a maximum fee that define the contract range of incentive effectiveness (RIE). When costs are above or below the RIE, the Government assumes full cost risk for each additional dollar spent within the funding or cost limits established in the contract.

While there is no universal optimized FPIF model, contract parameters can be adjusted to best suit both sides. Flexibility does come with greater complexity, but

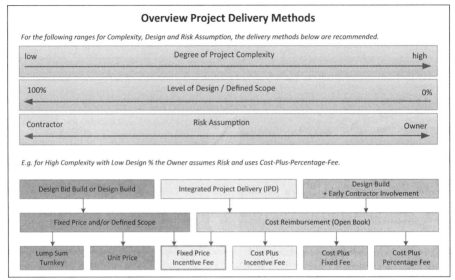

Figure 1. FPIF as hybrid delivery method

when properly executed, FPIF contracts can be highly effective in motivating contractors to control cost (Hurt et al., 2015).

PROBABILISITC METHODS

We believe the reader is familiar with basic concepts of risk, risk management, and risk mitigation and the use of probabilistic cost-risk processes versus deterministic ones (Reilly et al., 2015; Sander et al., 2015). The probabilistic approach, compared to the simpler and more common deterministic approach (unit prices times unit costs plus a contingency), offers more useful information with respect to the range of probable cost as well as cost "drivers" and better quantifies the effects of risks, opportunities, and variability. This improves understanding and leads to a better potential for profit (or loss) for contractors and added value for owners.

CEVP-RIAAT Process

To determine an accurate estimate range for both cost and schedule, significant risks must be identified and assessed. Formerly, cost estimates accounted for risk based on the estimator's experience and best judgment, without necessarily identifying and quantifying such risks—project uncertainties and risks were included in a general "contingency" that was applied to account for such uncertainties. In order to include risk and uncertainty, and to independently validate costs, the Washington State Department of Transportation (WSDOT) in the USA developed CEVP, the "Cost Estimate Validation Process," (Reilly et al., 2004) to implement better cost estimating and to include the influence of uncertainty (risk) on project delivery.

In CEVP, estimates consist of two components: the base cost component and the risk component. Base cost is defined as the planned cost of the project if everything materializes as planned and assumed. The base cost does not include contingency but does include the normal variability of prices, quantities, and like units. Once the base cost is established, a list of risks is identified and characterized, including both opportunities and threats, and listed in a Risk Register. This risk assessment replaces a general and vaguely defined contingency with explicitly defined risk events that include the associated probability of occurrence plus the impact on project cost and/or schedule for each risk event. The risk is usually developed in a CEVP Cost Risk Workshop (Sander et al., 2018).

RIAAT (Risk Administration and Analysis Tool: http://riaat.riskcon.at) is an advanced software tool that combines base costs, base variability, risks, opportunities, and schedules to indicate ranges of probable cost and schedule, plus risk management and change tracking and documentation (Sander et al., 2017).

FPIF CONTRACTING

When to Apply FPIF

As part of the application of CEVP and RIAAT to a major project in South America, the opportunity to include advanced risk management and delivery processes was evaluated. The result was the decision to apply the FPIF process using the CEVP-RIAAT process as input and to help establish an agreed target cost. The FPIF approach is based on US Department of Defense (USDOD) strategies for different types of procurement in different circumstances. The rationale for selecting this particular contract form, based on the USDOD approach, is that:

1. For projects where there are established historical data regarding outturn costs, the program is stable, and many units are to be delivered with few change requirements, a fixed-price lump sum is appropriate.
2. For projects that are uncertain, with substantial unknowns, such as new weapons systems or components that require significant research and development, a cost-plus negotiated procurement is most appropriate.
3. For projects with some unknowns, but with stable scopes, a process between a fixed-price lump sum and a cost-plus negotiated procurement—a process with characteristics of both approaches—is best. This means a firm upper-cost ceiling, with a defined target cost and a pain/gain mechanism to incentivize reduced cost for the owner, with a defined scope. This is the FPIF form of contract.

FPIF models keep a fixed-price approach but also allow for a certain degree of control over the total price by creating a more collaborative environment with the contractor.

DEFINITIONS RELATED TO FPIF

FPIF Contract: Specifies a target cost, a target profit, a ceiling price, and a profit adjustment formula. These elements are all negotiated at the outset. The profit earned by the contractor varies inversely with the project cost by application of a pain/gain mechanism. When the final project cost is negotiated, the contractor's profit is calculated, and the price paid by the owner is the final project cost plus the so-calculated contractor profit. All project transactions and costings are 100% open book and subject to audit.

Target Cost (TC): Expected total cost of the project (direct plus project-related overheads), excluding contractor profit. It should be reasonably challenging but achievable. It is based on a reasonable best-case scenario of contract performance based on an analysis of available information. It includes the contingency allocated to the risks associated with the delivery of the project, agreed by the parties.

Target Profit and Target Price (TP): The Target Profit is the profit earned by the contractor for achieving the Target Cost. The Target Price is the sum of the Target Cost plus the Target Profit.

Share Ratio (S/R): Percentage that each party shares in cost underruns and cost overruns from the negotiated Target Cost. The first number corresponds to the owner, the second to the contractor. For example, an Underrun S/R of 60/40 indicates that the contractor's profit is increased by forty cents for each dollar under the target cost. The same sharing principle applies for an Overrun S/R.

Pain/Gain Mechanism: Formula applied to calculate the final price paid by the owner, based on the agreed S/R for underruns and overruns. When the final negotiated cost of the project is lower than the target cost (i.e., there has been an underrun), application of the S/R results in a final profit greater than the Target Profit; the price paid by the owner is the final negotiated cost plus the (higher) profit so calculated. Conversely, when the final negotiated cost is higher than the Target Cost (i.e., there has been an overrun), the contractor earns a profit lower than the Target Profit, and the owner pays for the final negotiated cost plus the (lower) profit so calculated.

Ceiling Price (CP): Maximum price paid by the owner to the contractor, except for any adjustment under other contract clauses.

Point of Total Assumption (PTA): Overrun cost point at which the Pain/Gain Mechanism results in the owner paying the Ceiling Price, i.e., the negotiated cost of the project plus the profit earned by the contractor as per the Pain/Gain Mechanism equals the Ceiling Price.

CP = Cost @PTA + Profit @PTA

The S/R becomes 0/100 at the PTA because the owner no longer shares in a cost overrun. Therefore, the contractor is assuming the extra cost at the expense of his profit, dollar per dollar.

The formula to calculate the PTA is as follows:

PTA = TC + (CP − TP)/(Owner Overrun Share Ratio)

Example:

With the data: TC = 100; Target Profit = 10 (therefore TP = 110); CP = 118 and Overrun S/R = 60/40

PTA results in: PTA = 100 + (118 − 110)/0.6 = 113.3

Therefore, for this example project where the TC = 100, if the final negotiated cost is 113.3, the owner would pay the CP = 118 and the contractor would make a profit of 118 − 113.3 = 4.7, which is less than the target profit of 10. If the project cost exceeds 113.3, the owner still pays the CP = 118, therefore the contractor reduces his profit one dollar per every additional dollar of project cost.

Figure 2 shows the defined parameters applied to a probability distribution. This basic model is used to define the model for application in a project. The following example is a guide through the steps.

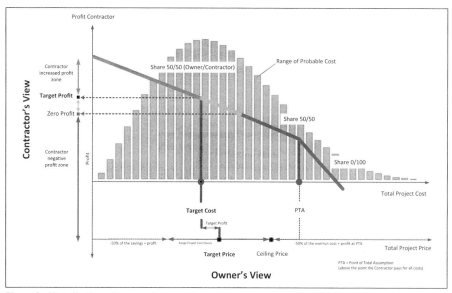

Figure 2. Visualization of terms

APPLICATION EXAMPLE

An application similar to this example is used for the Lima Airport Extension Program.

Probable Cost Range

The Probable Cost Range is the result of the CEVP-RIAAT application. As usual the range depicts the Base Cost + Risk and Escalation. It does not include the contractor's profit. Figure 3 shows a typical result using the probability distribution and the probability function.

FPIF Model Set Up

Table 1 lists all the parameters, formulas, and calculated values that are used to set up the FPIF model for our example.

Figure 4 applies the FPIF on the Probable Cost range (compare to Figure 2).

Deviation from Target Cost

Since there is a Share Ratio in the case of a cost overrun or underrun, the potential deviation from the Target Cost is essential for calculating the potential pain/gain for the owner and contractor. Figure 5 depicts the probability function that shows the potential deviation. There is a chance of about 38% that the cost will come in below the Target Cost but also a probability of 62% that the final cost will be higher than the Target Cost. For example in 42% (P80 minus Target Cost → P38) of all cases the cost overrun will not exceed $ 30.

POINT OF TOTAL ASSUMPTION—0/100 SHARE RATIO

If the final cost exceeds the PTA of $144, the contractor takes all the risk. Figure 6 visualizes a cost impact with a 9% probability that the final cost will exceed $144.

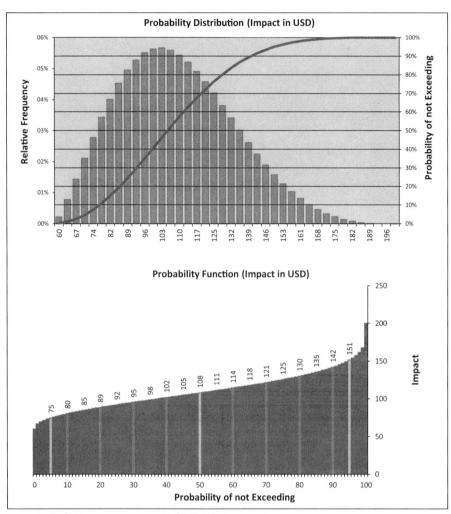

Figure 3. Probable cost as result of the CEVP-RIAAT application

Table 1. Calculation of the FPIF parameters

Parameter	Formula	Value
Probable Cost Range	Is given in a range as result from the CEVP workshops (see 0.).	
Target Cost (TC)	TC	$100
Profit	10% Profit for Contractor: Profit = TC * 0.1	$10
Target Price (TP)	TC + Profit	$110
Ceiling Price (CP)	Set to $132	$132
Owner Share Ratio (OSR)	Owner/Contractor share ratio: 50/50	50%
Point of Total Assumption (PTA)	PTA = TC + (CP − TP)/(OSR) PTA = 1.0 + (1.32 − 1.1)/0.5	$144

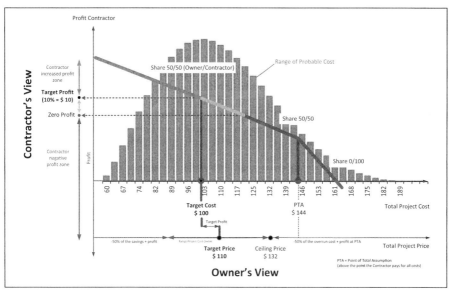

Figure 4. FPIF model applied to the probable cost range

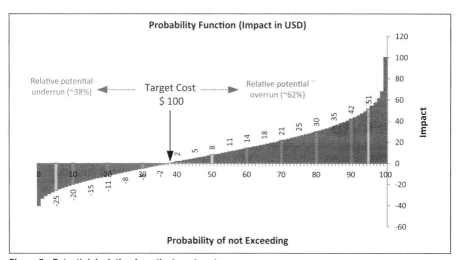

Figure 5. Potential deviation from the target cost

CONTRACTOR'S VIEW

From a contractor's view, there is a probability of 32% that he will drop into the loss zone, but also a probability of 38% that he will have increased profit above $10 (Figure 7 and Figure 8).

If the PTA is exceeded (9% probability), the contractor takes all the risk, which will rapidly increase his loss. This is depicted by the steep curve in Figure 7 and the flat tail in Figure 8. The analysis does not consider the potential increased efficiency by the contractor in order to generate higher profit.

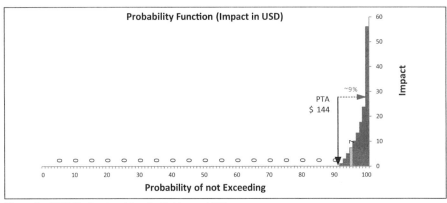

Figure 6. 100% Contractor risk potential beyond PTA

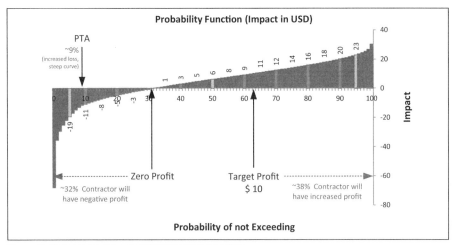

Figure 7. Contractor's view of profit/probability function

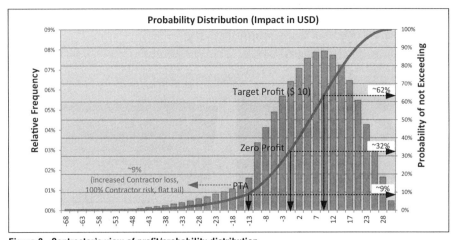

Figure 8. Contractor's view of profit/probability distribution

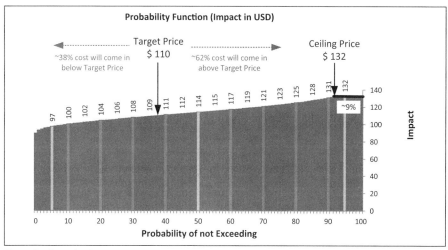

Figure 9. Owner's view of cost/probability function

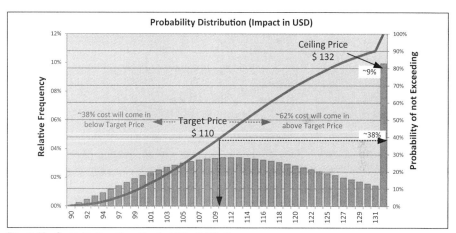

Figure 10. Owner's view of cost/probability distribution

OWNER'S VIEW

From an owner's perspective, there will be a 38% chance that his cost will be lower than the Target Price (Figure 9 and Figure 10). This chance might be higher if the contractor is incentivized to gain more profit and works with increased efficiency.

If the PTA is exceeded, the contractor takes all of the risk. This defines the Ceiling Price of $132 for the owner. There is a probability of 9% that the PTA will be exceeded and the Ceiling Price mechanism will be triggered.

CONCLUSION

The FPIF contract model is a way to implement shared cost goals and to establish a collaborative working environment, using incentives based on a pain/gain mechanism. One key to the FPIF contract is a consensual agreement on the target cost. The individual risk potential for a chosen target cost for the contractor and owner should be

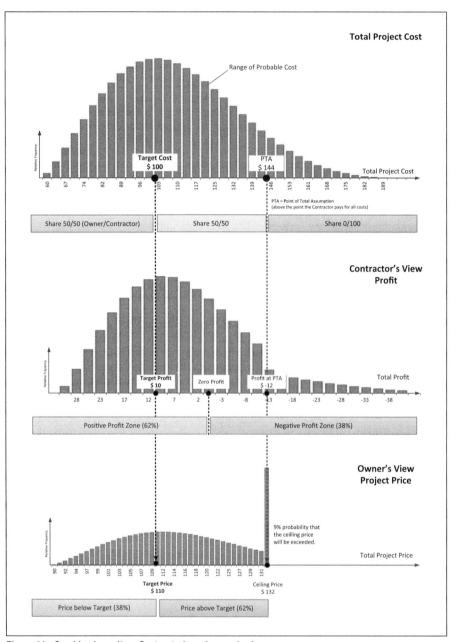

Figure 11. Combined results—Contractor's and owner's view

calculated using probabilistic methods. For the contractor, the probability is relevant in estimating the potential to increase his profit or risk to suffer loss. For the owner, the deviation from the target price with the corresponding probability is the basis for the evaluation of the contract. The probabilistic results transparently show the risk potential of both parties (Figure 11), allowing contract negotiations to be conducted from a common basis.

REFERENCES

Cuskey, Jeffrey R. (2015), "Understanding the Mechanics of FPIF Contracts," PTACS Montana State University.

Fixed Price Incentive Firm Target (FPIF) Contract Type, Acquisition Encyclopedia, https://www.dau.mil/acquipedia/Pages/ArticleDetails.aspx?aid=6794b407-22e0-4d83-aff9-80474fc70014.

Hurt, Steven, Elliot, Ryan, (2015), "Modeling Price Outcomes for Complex Government Programs," A.T. Kearney.

Ibbs, C.W., Kwak, Y.H., Ng, T., & Odabasi, A.M. (2003), "Project delivery systems and project change: Quantitative analysis." Journal of Construction Engineering and Management, 129(4), 382–387.

ICE UK (2018), Project 13, "Project 13 launch will improve how infrastructure is delivered," https://www.ice.org.uk/news-and-insight/latest-ice-news/project-13-launches.

Reilly, J.J., Essex, R, Hatem, D. (2018), "Alternative Delivery Drives Alternative Risk Allocation Methods," North American Tunnel Conference, Washington DC, June.

Reilly, J.J., McBride, M., Sangrey, D., MacDonald, D, Brown, J. (2004) "The development OF CEVP®—WSDOT's Cost-Risk Estimating Process," Proceedings, Boston Society of Civil Engineers, Fall/Winter.

Reilly, J.J., Sander, P., Moergeli, A. (2015), "Construction—Risk Based Cost Estimating," Paper and Presentation, RETC 2015, New Orleans.

RIAAT (2018), Risk Administration and Analysis Tool—http://riaat.riskcon.at.

Ross, J. (2003), "Introduction to Project Alliancing," April 2003 Update, Alliance Contracting Conference, Sydney, Australia, 30 April.

Sander, P., Reilly, J., Entacher, M. (2018), "CEVP-RIAAT Process—Application of an Integrated Cost and Schedule Analysis," North American Tunneling Conference, Washington D.C., June 2018.

Sander, P., Reilly, J., Entacher, M., Brady, J. (2017), "Risk-Based Integrated Cost and Schedule Analysis for Infrastructure Projects" Tunnel Business Magazine, August 2017, p. 43–37.

Sander, P., Moergeli, A., Reilly, J. (2015), "Quantitative Risk Analysis—Fallacy of the Single Number," Paper and Presentation, WTC 2015, Dubrovnik.

USDOD Comparison of major project acquisition types, https://www.acq.osd.mil/dpap/ccap/cc/jcchb/Files/Topical/Contract_Type_Comparison_Table/resources/contract_type_table.docx.

North American Tunneling Industry Firsts in the Midwest: The Blacksnake Creek Stormwater Separation Improvement Tunnel

Brian Glynn ▪ Black & Veatch Corporation
Cary Hirner ▪ Black & Veatch Corporation
Kayla Christopher ▪ Black & Veatch Corporation
Lee Sommers ▪ City of St. Joseph, Missouri
Mike Garbeth ▪ Super Excavators, Inc.

ABSTRACT

The Blacksnake Creek Stormwater Separation Improvement Tunnel intercepts Blacksnake Creek stream and stormwater flows up to 18 cubic meters per second (650 cfs) from the City's combined sewer system and redirects the flows to the Missouri River through a dedicated stormwater conveyance tunnel approximately 2.0 kilometers (1.2 miles) in length and 2.7 meters (9.0 feet) in finished diameter. Mixed faced conditions of soil and shale bedrock along portions of the tunnel alignment required excavation and initial support methods capable of transitioning between the varying ground conditions. An Earth Pressure Balance Machine (EPBM) was selected to provide better control over the excavation face and reduce risks associated with settlement and groundwater inflows. The particular EPBM used was designed by Lovsuns and manufactured in China by Lovsuns' parent company, their first EPBM delivered to the North American tunneling industry. The tunnel was lined with precast concrete segments solely reinforced with macro synthetic fibers (BarChip, Inc. BarChip 54 fibers) in lieu of steel fibers or traditional bar reinforcement, another first for the North American tunneling industry, to comply with the American Iron and Steel provisions of the Contract. The project also includes a baffle drop shaft, a separate shorter tunnel underneath active main line BNSF railroad tracks, and an energy dissipation structure. This paper will focus on the design decisions and lessons learned throughout construction including addressing mixed face ground conditions and groundwater during the single pass tunnel excavation and installation of the precast concrete segmental tunnel lining.

INTRODUCTION

The Blacksnake Creek Stormwater Separation Improvement Project is located in northwestern St. Joseph, Missouri, east of the Missouri River. The purpose of the Project is to redirect Blacksnake Creek stream flow and stormwater flows from the City of St. Joseph, Missouri's (City's) combined sewer system to a dedicated stormwater outfall. Currently, stream and stormwater flows are conveyed through the existing Blacksnake Creek Combined Sewer System to the City's Water Protection Facility. During storm events, the sewer system is not able to handle the excess stormwater runoff, thus leading to combined sewer overflows to the Missouri River. The Blacksnake Creek Stormwater Separation Tunnel is designed to intercept and convey stream and stormwater flows to a dedicated stormwater outfall near the Missouri River. This paper will discuss the design phase and construction phase of the project to date as well as lessons learned on the project.

DESIGN PHASE

The Blacksnake Creek Stormwater Separation Improvement Tunnel is a component of the overall Blacksnake Creek Stormwater Separation Improvement Project, which is required as part of the City's Combined Sewer Overflow (CSO) Long Term Control Plan in order to improve water quality as mandated by the Clean Water Act.

The downstream reach of Blacksnake Creek is currently directed into the City's Blacksnake Creek Combined Sewer System. Some portions of this sewer system have been in service since the 1880s. Currently, Blacksnake Creek stream and stormwater flows are conveyed through the combined sewer system to the City's Water Protection Facility (water treatment plant), and such stream and storm flows are unnecessarily treated every day of the year. During wet weather events, stormwater runoff can exceed capacity of the interceptor sewer, located between the downstream end of the Blacksnake Creek Combined Sewer System and the City's Water Protection Facility, resulting in CSOs from the Blacksnake Creek Combined Sewer System directly to the Missouri River. The CSO are a mix of stormwater and sanitary sewage and result in adverse water quality for human health and the environment.

The main goal of the project was to intercept and redirect Blacksnake Creek stream flow and stormwater flows to a new and dedicated stormwater conveyance system that flows directly to the Missouri River, and thereby reduce the frequency, volume, and impacts of CSOs to the Missouri River. The project also included repairs to existing infrastructure and other improvements (such as roadway improvements) in the project area, especially in the construction corridor. In addition, green infrastructure and community enhancements (such as park and recreation improvements) that could be cost-effectively implemented were also incorporated into the project.

The final configuration of the project is summarized as follows. Blacksnake Creek stream and stormwater flows (up to 18 cubic meters per second (650 cfs)) are diverted immediately upstream of the existing entrance to the combined sewer system to a new intake structure located at a bend in the downstream reach of Blacksnake Creek. This intake structure includes a trash rack and stop logs for operations and maintenance. The intake structure transitions the flow from Blacksnake Creek to a 2.1 meter (7 foot) wide by 1.8 meter (6 foot) tall reinforced concrete box culvert that extends to the south approximately 1.8 kilometers (1.1 miles) and parallels an abandoned railroad corridor owned by the City. The downstream end of the box culvert terminates at a Drop Shaft, which transitions the flow from the box culvert to a tunnel that extends approximately 2.0 kilometers (1.3 miles). The tunnel is lined with reinforced precast concrete segments with a finished diameter of 2.75 meters (9 feet). The tunnel terminated at junction structure which transitions the precast concrete segmental lining to steel pipe, which extends approximately 90 meters (300 feet) to the west, approximately 40 meters (130 feet) of which was installed by pipe jacking underneath active railroad tracks. The downstream end of the steel pipe terminated at an energy dissipation (outfall) structure, which transitioned the pipe flow into open channel flow. From the energy dissipation structure, the flow entered an improved channel section of Roy's Branch approximately 270 meters (900 feet) from its confluence with the Missouri River.

Bid Packages

The Project was divided into multiple bid packages in recognition of specialized area of work that contractors would have as well as to provide an opportunity to a variety of contractors to bid the work and to also encourage more local contractor participation. The bid packages are described as follows:

- Combined Sewer Crack Repair Package. The Combined Sewer Crack Repair Package included repairs to the brick and concrete lining of the existing Blacksnake Creek Combined Sewer System, which extends approximately 2.0 kilometers (6,500 feet) from end to end. Repairs to the lining were performed with crack injection with a hydrophilic grout or by application of an epoxy resin. Voids in the lining were filled and/or patched as necessary.
- Tree Clearing Package. The Tree Clearing Package included tree clearing prior to the commencement of construction for the Tunnel Package and the Conveyance Package.
- Cook Road Extension and Krug Park Drive Package. The Cook Road Extension and Krug Park Drive Package included the extension of Cook Road over Blacksnake Creek and created a new intersection at Cook Road and St. Joseph Avenue, which are both major thoroughfares in area.
- Tunnel Package. The Tunnel Package was a critical portion of the project as the topography of the area did not allow for a new dedicated stormwater conveyance system to the Missouri River via traditional open cut trench excavation. The Tunnel Package included the following components:
  - Tunnel: 2.0 kilometers (1.3 miles) of reinforced concrete segmentally lined tunnel, 2.75 meters (9 feet) inside diameter
  - Drop Shaft: 11 meter (37 foot) inside diameter, 16 meter (53 foot) deep dual baffle drop shaft
  - Box Culvert: 15 meters (48 feet) of 2.1 meter (7 foot) wide by 1.8 meter (6 foot) tall reinforced concrete box culvert
  - Open Cut Steel Pipe: 50 meters (175 feet) of 2.3 meter (7.5 feet) steel pipe
  - Jacked Steel Pipe: 38 meters (125 feet) of 2.3 meter (7.5 feet) steel jacking pipe
  - Energy Dissipation (Outlet) Structure
- Conveyance Package. The Conveyance Package connected the upstream and downstream end of the work associated with the Tunnel Package to Blacksnake Creek and the Missouri River, respectively. The Conveyance Package included the following components:
  - Box Culvert: 1.8 kilometers (5,800 feet) of 2.1 meter (7.0 foot) wide by 1.8 meter (6.0 foot) tall reinforced concrete box culvert, including four cast in place junction structures and two cast in place intake structures and associated local small diameter sewer separation or improvements.
  - Modification of an existing intake structure and rehabilitation of a screening structure and slide gate within the existing Blacksnake Combined Sewer System
  - Improvements to 270 meters (900 feet) of existing Roy's Branch channel near its confluence with the Missouri River
  - Construction of 2.3 kilometers (7,700 feet) of aggregate trail and 160 meters (530 feet) of concrete trail
  - Realignment of a portion of Northwest Parkway and Karnes Road extending nearly 0.6 kilometers (2,000 feet)
  - Demolition of an existing bridge, removal and relocation of an existing pedestrian bridge, construction of a new pedestrian bridge, basketball courts, and playground equipment
- CSO Outfall Package. The CSO Outfall Package consist mainly of construction of a large cast in place Type III stilling basin for energy dissipation immediately downstream of the existing Blacksnake Creek Combined Sewer CSO Outfall Structure as well as channel improvements, minor structural repairs

and patching of the existing wingwalls of the existing outfall structure and rehabilitation or replacement of the existing flap gate associated with the existing outfall structure. All of which will be performed immediately adjacent to the Missouri River.
- Tree Planting Package. The Tree Planting Package included tree planting and construction of green infrastructure after completion of construction in the vicinity of the Tunnel Package and the Conveyance Package as well as other areas within the City.

Tunnel Alignment and Shaft Locations

The tunnel alignment and profile, shown in Figure 1 and Figure 2, heads west from the Drop Shaft located near the Second Harvest Food Bank, following underneath the Highland Avenue right of way, and curves slightly to the south as it nears Interstate 229. The alignment crosses below both Interstate 229 and the Burlington Norther Santa Fe (BNSF) Railroad Tracks before terminating at an Energy Dissipation Structure. From the Energy Dissipation Structure, flows are conveyed by Roy's Branch, which is a tributary to the Missouri River. The tunnel alignment is approximately 2.0 kilometers (6,700 feet) in length and terminates at a junction structure located between MacArthur Drive and the BNSF Railroad Tracks.

Geotechnical Investigations

For the tunnel alignment, a total of 25 vertical borings were drilled to identify soil and rock properties and to confirm the vertical alignment of the tunnel. Twelve (12) cone penetrometer tests (CPT) were also performed to collect additional in-situ soil properties and define top of rock in areas of interest between completed borings. Ten (10) piezometers were installed in borings to collect water samples and monitor the groundwater level prior to and during construction. Soil and rock samples were collected and geologically logged at both shaft sites and along the tunnel alignment. Packer-testing was performed within rock in eight borings to estimate hydraulic conductivity. Four test pits were performed near the Drop Shaft location to investigate possible remnants of a previously existing railroad and associated structures.

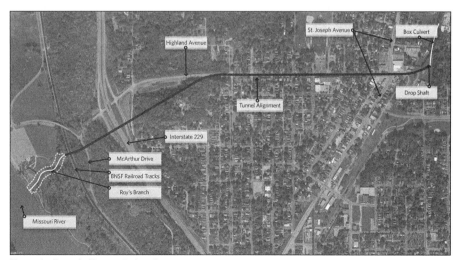

Figure 1. Tunnel alignment plan view

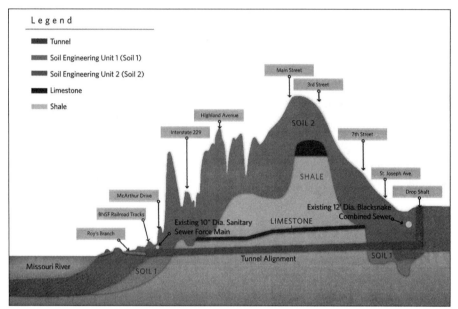

Figure 2. Alignment tunnel profile view

Geotechnical lab testing for soil encountered within the tunnel zone included: Grain Size Analysis, Moisture Content, Atterberg Limits, Unconfined Compressive Strength (UCS), and One-Dimensional Consolidation. Geotechnical lab testing for rock along the alignment included: Rock Moduli, UCS, Rock Density, Splitting Tensile Strength, Slake Durability, Cerchar Abrasivity, and Swell Testing.

Geologic Setting

Overburden at the Drop Shaft consists primarily of firm to stiff, low-plasticity clay up to depths of nearly 18 meters (60 feet). During the Geotechnical Investigation, it became evident that a buried alluvial valley was present along and perpendicular to the tunnel alignment, just west of the Drop Shaft. Material within this area consists of clayey sand and sandy clay overlying a dense sandy gravel layer. At the Launch Shaft, overburden consists of firm to stiff, low plasticity clay overlying beds of coarse-grained sand, silty clay, and sandy silt extending to depths up to 9 meters (30 feet). Overburden at both the Drop Shaft and Launch Shaft is underlain by weathered shale. Readings from piezometers near the Launch Shaft indicate that groundwater is heavily influenced by the nearby Missouri River.

Bedrock encountered along the tunnel alignment is part of the Douglas and Pedee Groups of the Upper Pennsylvanian Series. Formations encountered in the borings include the Lawrence, Stranger, Iatan Limestone, and Weston Shale. The tunnel alignment is primarily within the Weston Shale Formation with a portion of the tunnel intersecting the buried alluvial valley near the Drop Shaft. The Iatan Limestone is typically well above the tunnel profile, though it dips slightly towards the tunnel midway along the alignment.

Table 1. Baseline values of shale and limestone

	Unconfined Compressive Strength, MPa (psi)	Splitting Tensile Strength, MPa (psi)	Cerchar Abrasivity	Slake Durability, %
Weston Shale	10.2 (1,480)	1.4 (200)	1.2	70
Iatan Limestone	70.7 (10,260)	No Baseline	0.95	No Baseline

Baseline Conditions and Anticipated Rock Behavior

A Geotechnical Data Report (GDR) and a Geotechnical Baseline Report (GBR) were published after the completion of the geotechnical investigation. These reports summarized the anticipated geotechnical conditions that could be expected during construction. Both the GDR and GBR were included as Contract Documents to assist contractors during bidding. The baseline properties for the Weston Shale Formation anticipated to be encountered in the tunnel and shaft excavations, and the overlying Iatan Limestone Formation are shown in Table 1.

In addition to presenting baselines of physical properties, anticipated ground behavior and estimated ground type percentages along the tunnel alignment were discussed, with key points being summarized below:

Anticipated Ground Behavior

- Alluvial sand and gravel above the groundwater table will run
- Alluvial sand and gravel below the groundwater table will flow
- Clay and silt above the groundwater table will slowly ravel
- Clay and silt below the groundwater table will quickly ravel
- Clays encountered during tunneling have a potential for cutterhead clogging
- Shales encountered during tunneling have a potential for cutterhead clogging

Ground Type Percentages along Tunnel Alignment

- 63% encountering full face rock conditions
- 23% encountering full face soil conditions
- 14% of the encountering mixed-ground conditions

Construction Methods

To mitigate risks associated with construction, preferred construction means and methods were specified in the Contract Documents. Due to the Drop Shaft being located near commercial and residential areas, blasting was prohibited. The Launch Shaft to be designed by the Contractor was required to have a water-tight support system such due to the potential for groundwater impacts from the Missouri River.

The tunnel was required to be excavated using a fully shielded Earth Pressure Balance Tunnel Boring Machine (EPBM) with the ability to install precast concrete segmental lining in a single pass and equipped with a screw conveyor to manage pressure against the tunnel heading being excavated by controlling the rate of muck discharge.

CONSTRUCTION PHASE

The project was advertised on February 27, 2017 and the bid period was closed on April 13, 2017. Four (4) contractors submitted bids for the project, all of which had previously been prequalified. The four bids ranged from $27.0 million to $37.6 million with

Figure 3. Construction of drop shaft

Figure 4. Launch shaft

an engineer's estimate of $33.4 million. The project was awarded to the low bidder, Super Excavators, Inc. and a full Notice to Proceed was issued on August 23, 2017.

Shaft Excavation

Both the Launch Shaft and Drop Shaft were excavated using secant piles as initial support. The Launch Shaft was excavated by mechanical methods through clay and granular overburden and shale bedrock. The Drop Shaft was constructed through clay overburden and shale bedrock to a depth of approximately 15 meters (50 feet) below ground surface.

Trenchless Railroad Crossing

Jack and bore (pipejacking) methods were selected to be used to pass below the BNSF Railroad tracks to the west of the Launch Shaft. This was recommended as a best value method based on the pipe diameter, length of the crossing, ground conditions, and groundwater elevation. This method also provided a lower construction cost compared to microtunneling and was more flexible if obstructions are encountered. Due to the site's proximity to the Missouri River, it was recommended that work associated with the trenchless railroad crossing be performed during the winter months when the river elevation is historically the lowest.

Figure 5. Jack and bore pipe installation below BNSF railroad tracks

The trenchless railroad crossing consists of approximately 38 meters (125 feet) of 2.3 meters (90 inch) inside diameter jacked steel pipe. The overburden thickness above the pipe crown to the bottom of the ties of the railroad tracks is only about 3 meters (10 feet). Ground conditions at this location consist alternating firm clays and silty clays with coarse sands and sandy silts overlying shale bedrock.

Tunneling with an Earth Pressure Balance Machine

The Contractor selected an EPBM capable of operating in either open face or closed face modes for the varying ground conditions. The EPBM was designed by Lovsuns Tunneling Canada Ltd. in Toronto, Canada and manufactured and tested by its parent company Liaoning Censcience Industry Company Ltd.in Liaoyang, China. This was the first tunnel boring machine to be manufactured by Liaoning Censcience Industry Company Ltd. for use in North America. The design and manufacturing process took approximately ten (10) months to complete. Once factory testing of the EPBM was completed, the EPBM was shipped from China to Seattle, WA. From Seattle, WA, the machine was hauled by flatbed truck to St. Joseph, MO. The entire shipping process took approximately two (2) months.

As the components of the EPBM were delivered to the site, initial assembly and field testing began, and subsequently, individual components of the machine were lowered into the Launch Shaft for final assembly. As excavation began, sections of the EPBM were lowered into the shaft and connected until the entire 78 meters (256 feet) long machine was completely assembled below the surrounding ground surface. This required the aft components of the machine to be preliminarily assembled above ground surface and umbilical hydraulic and water lines, and other necessary cables ran between the aft and fore components of the EPBM. Assembling, testing, and commissioning was completed in August 2018.

Figure 6. EPBM assembly and testing at launch shaft

Table 2. EPBM manufacturer specifications

Parameter	Value
Excavation Diameter	3.3 m (11 ft)
Total Length	78 m (256 ft)
Min. Turn Radius	250 m (820 ft)
Max. Speed (Cutting Head)	8.3 rpm
Maximum Torque	993 kN·m (732 kip·ft)
Main Drive Motors	4
Max Thrust	912 Tonne (1,005 tons)
Thrust Cylinders	12

Segmental Liner and Synthetic Fiber Reinforcement

The tunnel lining consists of precast concrete segments installed by the EPBM as it is advzanced. Each tunnel lining ring consists of six trapezoidal shaped segments with rubber gaskets to create a watertight final lining with a finished diameter of 2.7 meters (9.0 feet) with a segment thickness of 190.5 millimeters (7.5 inches). The precast segments were manufactured for Super Excavators, Inc. by CSI Tunnel Systems, Inc. in their Macedonia, Ohio plant. During each advancement of the EPBM, segments are placed using a segment erector located in the trailing shield and are manually bolted together. Backfill and contact grouting of the annular space is performed through cast in grout ports in each segment after installation.

Steel fiber reinforcement of the precast concrete segmental tunnel lining was anticipated during design; however, during construction, there was minimal availability of steel fibers for manufacturing of precast concrete segments that were also compliant with the American Iron and Steel provisions of the Contract Documents. This created a need for an alternative method of reinforcement. Super Excavators, Inc.'s precast concrete segment manufacturer proposed to use synthetic fibers, specifically BarChip 54 Macro Synthetic Fibers manufactured by Barchip, Inc. for reinforcement.

Synthetic fibers were allowed to be used provided the segments still met the requirements of the contract documents, specifically, the compressive strength of concrete was to be at least 41.4 MPa (6,000 psi), the minimum flexural tensile strength at 28 days was to be at least 4.62 MPa (670 psi), and the minimum post-crack equivalent residual flexural tensile strength was to be at least 3.17 MPa (460 psi) The use of synthetic fiber reinforcement ensured that the precast concrete segments remained in compliance with the American Iron and Steel provisions of the Contract Documents while also minimizing the potential for schedule and cost impacts to the project. The Blacksnake Creek Stormwater Separation Improvement Tunnel is the first tunnel in North America to utilize synthetic fiber reinforcement as the sole method of reinforcement for the precast concrete segmental tunnel lining.

Figure 7. Initial delivery of synthetic fiber reinforced precast concrete segmental tunnel lining

CONCLUSION

At the time that this paper was finalized in December 2018, tunnel construction was underway and nearly 250 meters (825 feet) or twelve percent (12%) of the tunnel had been excavated and lined. It is currently anticipated that the Blacksnake Creek Stormwater Separation Improvement Tunnel will be completed in late 2019. Once the tunnel is completed, Blacksnake Creek will be diverted and directly connected to the Missouri River achieving the project goal of reducing the frequency, volume, and impacts of CSOs to the Missouri River, and thereby improving water quality of the Missouri River.

Progressive Design-Build—Is It Coming to a Project Near You?

Joe O'Carroll ▪ Mott MacDonald
Andy Thompson ▪ Mott MacDonald
Tom Kwiatkowski ▪ Mott MacDonald

ABSTRACT

The water/waste water industry has found progressive design-build (PDB) a good fit in delivering capital project solutions. Owners of tunneling projects are taking a keen interest in this approach for delivery of their projects. What is PDB? Is it just early contractor involvement or another method of project delivery, such as Construction Manager at Risk (CMAR) under another name? Is it a delivery method that fosters better owner-contractor relationships with potentially fewer disputes and greater certainty on the final cost? This paper examines which states allow PDB procurements, key elements of this delivery method and issues and pitfalls an owner and contractor need to understand before entering into a progressive design build contract; how pricing is managed, how risk is allocated and provides examples of tunneling projects using PDB as the preferred contracting method.

INTRODUCTION

For the tunneling industry in the United States design-bid-build delivery has been the traditional delivery method for many decades. This method gives the owner reliable price information for the project before construction starts. With proper design oversight and budgeting of the total project, costs should be predictable for the owner once the bids are received. However, design-bid-build has quite often shown itself to be prone to creating adversarial relationships between parties when issues develop, resulting in claims, delays and cost overruns. There is no contractual relationship between the contractor and the designer and no opportunity for collaboration during the design phase. There is also little incentive for design changes/improvements or value engineering during the construction phase. In an ideal scenario time and cost savings flow from a collaborative relationship between designer and builder. Fewer unforeseen problems should arise when designers and builders are on the same team from the beginning. When problems are resolved more quickly, projects have a greater chance of staying on schedule.

LOOKING FOR THE MIDDLE GROUND—DESIGN-BUILD AND CMAR PROCUREMENT

Over the last decade, the public procurement profession has seen greater use of alternative delivery methods of construction projects, including design-build, construction manager at risk (CMAR, also referred to as CM/GC), and public-private partnerships (P3). When the federal government began issuing design-build contracts—a project delivery method which combines architectural and engineering design services with construction performance under one contract, the goal was to capture the benefits of a streamlined approach to design and construction, namely: single source of accountability, faster delivery, lower costs and the reduction or elimination of claims. At the state level, quite often the need for legislative change to allow design-build was to improve emergency responsiveness to natural disasters. In 2003 only 3 states had authorized design build on public projects. By 2013 design-build was authorized in some fashion in 42 states and the District of Columbia. Today, legislature authorizes design-build for public projects in some form in all fifty states, and a majority of states permit design-build for all agencies for all types of design and construction.

There are advantages and disadvantages of design-build. From the owner's perspective the advantages, in addition to the realization of the goals mentioned above include clear allocation of risk and responsibility for design and construction and time can be saved by eliminating the procurement phase between completion of design and award of construction plus the ability to have long equipment and material lead items and site work begin before the total design is complete. From the contractor's side having control over the design allows for greater innovation which in turn should result in being more competitive.

Design-build allows owners greater opportunity to select a contractor based on qualifications, capabilities, experience and price, thus avoiding some of the pitfalls from contract awards solely based on low price associated with traditional design-bid-build. The biggest disadvantage from the owner's perspective, and one in which many owners struggle with, is using performance-based specifications for procurement and losing overall control over the design. This is somewhat less of a concern for a water or waste water conveyance tunnel project than it is for an urban transit or transportation tunnel project where there is a greater public interface and architectural, life safety, security and overall urban interface is of greater importance. Therefore, the water/waste water industry has found design-build and, as we will see further in the paper, progressive design-build a good fit in delivering projects.

Owners looking for that middle ground often turn to CMAR procurement that provides many of the benefits of earlier contractor involvement including allowing the contractor the opportunity to provide valuable constructability and engineering review but where the design control (i.e., design consultant contracts) remains with the Owner. When a project is underway, the CMAR will be selected during the design process, significantly reducing the Owner's procurement phase time. The CMAR will play a dual role as the construction manager for all of the project work—whether it is performed internally or subcontracted—and the general contractor soliciting competitive bids for subcontractor work. Once the final design is complete, and all construction prices negotiated, the owner and the contractor will agree upon a Guaranteed Maximum Price (GMP)—for the entire package of the contract. The Owner has the confidence in the price before construction begins, and although change orders are allowed, because of the structure of this procurement method, it is expected that very few changes will be made. To work properly, the CMAR and the owner need to come to an agreement on the GMP. If for some reason they cannot come to an agreement, the owner can seek other bids for the project.

So, is there a desire for some owners to find a "middle-middle ground?" A procurement approach that facilitates competitive bidding; involvement of the design-build team during the earliest stages of the owner's project development, ensuring they are part of the project team developing design solutions; maintaining some level of control on the design but at the same time devolving the responsibility for the design to the contractor; achieving the schedule benefits of long lead item procurement; and starting construction before the design is complete. Enter Progressive Design-Build.

PROGRESSIVE DESIGN-BUILD

The Water Design-Build Council (WDBC) defines progressive design-build (PDB) as a method of delivering infrastructure projects that combines the owner's direct control over project concept and detailed design with the design-builder's innovation—and creates a single point of accountability. Given the growing popularity of the PDB delivery method, the WDBC has developed a procurement guide, along with a suggested set of request for qualifications (RFQ) and request for proposal (RFP) model documents for conducting PDB procurement effectively and in accordance with best practices (WDBC PDB Procurement Guide).

Owners may choose among a variety of approaches to procuring a design-builder using the PDB method. Owners may choose to evaluate only non-price factors, basing their selection solely or primarily on qualifications; or they may evaluate a combination of price and non-price factors to determine which design-builder offers the best value. An owner may conduct a single-step process, using either an RFQ with only non-price selection factors or an RFP with both price and non-price selection factors. Alternatively, an owner may conduct a two-step process, first soliciting statements of qualifications (SOQs) and then requesting proposals from a short list of pre-qualified proposers. In the two-step process, the owner may limit the RFP to project-approach elements (non-price factors only), or may request a fee and rate proposal as well (both price and non-price factors). In any event, each owner needs to decide at the outset, on a project-by-project basis, which PDB procurement approach best fits its needs and preferences (WDBC PDB Procurement Guide).

RECENT EXAMPLES OF TUNNELING PROJECTS PRUCURED USING PDB

Brunswick Glynn Joint Water and Sewer North Mainland Sewer Transmission System

In May 2017, the Brunswick Glynn Joint Water and Sewer Commission issued an RFP for PDB procurement for improvements to the existing North Mainland Sewer Transmission System. (RFP Joint Water and Sewer Commission 2017). The RFP invited interested proposers to submit their qualifications and cost proposals to design and construct the project in two phases. Phase 1 required proposers to take the design to 30% completion and develop a GMP cost proposal for completion of Phase 2. Upon acceptance and approval of the GMP at the end of Phase 1 proposers would move into Phase 2 which included complete design, construction and post construction tasks, including performance testing, start-up commissioning and operator training and support.

The owner's stated objectives and priorities for delivery of the project included (i) maintain existing operations of the North Mainland Sewer system transmission system during construction (ii) provide a reliable and sustainable raw wastewater conveyance system (iii) achieving the agreed schedule completion date (iv) minimizing life-cycle costs and (v) implement an effective safety program using industry best practices. The owner placed high priority on the selection of equipment and the design of facilities that would provide a high level of reliability and the lowest level of maintenance costs. While capital cost was important to the owner, achieving objectives in operations, quality and schedule control were stated in the RFP as higher priorities.

By selecting PDB the owner committed to working in close collaboration with the design-builder during Phase 1 develop the design, achieve the project objectives and to obtain a mutually agreeable GMP for delivery of the project. The owner provided, in the RFP, a budget for design and construction of the project.

In their response to the RFP proposers were required to provide a conceptual description of their approach for delivering the project including a CPM schedule to meet Phase 1 Design Services and GMP development and a discussion on any major risks to achieving the schedule objectives; a narrative and drawings sufficient to describe the proposer's conceptual design and delivery approach; a discussion on how a collaborative relationship with the owner would be established and maintained throughout the entire duration of the project; a description on their use of an open book cost estimation process and a summary of plan and actions to assure delivery of a high-quality project.

The owner included a price component in the RFP requiring the proposer to provide a lump sum fixed price for all Design Services (Phase 1 and Phase 2); overhead and proposed fee expressed as a percentage of the cost of the work, for all Phase 2 services (excluding completion of the final design which is included In the Design Services fee) and a proposed fee.

Silicon Valley Clean Water Gravity Pipeline Project

In April 2017 Silicon Valley Clean Water (SVCW) invited qualified design-builders experienced in the design and construction of wastewater pipelines utilizing earth pressure balance tunnel boring machines to submit a statement of qualifications for the design and construction of a 3.3-mile-long, 15-foot outside diameter pipeline extending from Inner Blair Island south of San Carlos Airport to their wastewater treatment plant in Redwood, California (RFQ SVCW 2017). The Gravity Pipeline Project is a principal component of SVCW's overall conveyance system improvement program, which consists of major upgrades, replacements, and new facilities needed to achieve a highly reliable system.

SVCW adopted a progressive design-build approach which included Stage 1— preconstruction services to collaboratively bring the design from a 5%–10% level to a 60% to 70% level. Stage 2 included bringing the design to 100% and completing construction and start up. Multiple phases allowed the shaft construction, tunnel boring machine ordering and precast tunnel segment procurement to start before all design of the gravity pipeline is completed.

By selecting PDB procurement SVCW's objective was to receive quality submissions from capable design-builders with various team members experienced in areas such as EPB tunneling, tunnel shafts, precast concrete tunnel segments and sewer interceptors. In the RFP, SVCW stated they would give heavy consideration to the design-build team with significant experience similar to their project and a demonstrated ability to collaborate with SVCW management, engineering, operations and maintenance and consultants.

SVCW identified five key objectives for the PDB delivery process. PDB delivery should:
1. Help accelerate completion of the overall project relative to conventional design-bid-build.
2. Promote a cooperative and collaborative relationship between SVCW and the PDB team.
3. Be structured to provide the flexibility for phased design and construction.
4. Take a design-to-budget approach with the ceiling derived from the overall CIP budget for the project.
5. Provide opportunities for early and ongoing SVCW staff involvement and training that does not significantly affect the overall cost of the project.

Progressive Design-Build—Is It Coming to a Project Near You?

SVCW defined "success" as collaboratively implementing an appropriate balance of the following factors:

Provide a complete functional conveyance system that:

- Cost—meets the goals at the lowest practical capital and life-cycle costs
- Operations—easy, efficient and effective to operate
- Maintenance—minimize maintenance requirements
- Safety—safe to construct, operate and maintain
- Schedule—place into operation with best practical safe speed
- Stakeholder Impacts—meets the needs of stakeholders while reaching the program's goals.

The PDB was a 2-stage procurement process, the first step being a SOQ and short-listing; the second an RFP issued to selected responders and selection. The first step involved issuing an RFQ, receiving SOQs from Respondents, evaluating submitted SOQs conducting interviews with selected respondents, checking references and shortlisting. To reflect the importance placed on collaboration in PDB delivery SVCW's scoring criteria for the SOQ was as follows:

- Team Structure –10 points
- Qualifications of Key Personnel –25 points
- Experience Collaborating with Owners –20 points
- Experience successfully completing similar projects –35 points
- Safety experience on similar projects –10 points

SVCW performed reference check on respondents to be interviewed and scored the reference checks. The combined score of the SOQ (100 points), the background check (25 points) and the interview (100 points) was used to prepare a shortlist of interviewed Respondents that were sent an RFP.

SVCW adopted an overall risk management philosophy of reducing or mitigating risks to the extent feasible and then assigning risks to the party best able to review and comment by the responders. SVCW were receptive to considering comments and modifying the responsibility matrix/contract terms where considered appropriate before issuance of the RFP. SVCW also awarded a stipend to each shortlisted respondent that submitted an acceptable proposal but was not selected as the design-builder.

The second step involved issuing RFP to the shortlisted Respondents, conducting site visits, holding confidential meetings with each shortlisted Respondent, receiving proposals, additional reference checking, conducting interviews, selecting the winning Respondent and negotiating a Stage 1 Preconstruction Contract. Respondents were requested to furnish a firm-fixed price for Stage 1 services, mark up percentages (home office and profit) and an indicative price for stage 2 services with their proposals.

SVCW selected a design-build team based upon a best value selection criterion that considered price, anticipated life cycle costs, technical approach to project design and construction, management approach, key personnel and safety. With respect to pricing SVCW numerically scored as part of the proposal evaluation process:

- Firm-fixed price for Stage 1 services—3%
- Mark up percentages (home office and profit)—5%
- Indicative price for stage 2 services—7%.

City of Atlanta Plane Train Tunnel West Extension Phase 1

The City of Atlanta's Plane Train Tunnel West Extension Phase 1—Progressive Design Build Project (RFQ City of Atlanta, 2017) provides a crossover beyond the baggage claim terminus station to allow more efficient train movement by reducing headway and increasing capacity at Hartsfield-Jackson Atlanta International Airport. The City of Atlanta, on behalf of the Department of Aviation, issued an RFP in January 2017. The City's objectives for this project are to complete the project on schedule and with the best quality, using the design-builder's experience and skill in the best interests of the owner. The design-builder was selected through a competitive process based upon qualifications, commitments, and pricing of the project work.

The approximately 600 feet long twin 25-foot diameter tunnels will be located in rock approximately 55 feet below grade running west from the train's current terminus, beneath the Terminal West Donut area and continuing under MARTA Airport Station foundations, the Skytrain Airport Station, and ending approximately under the west limit of the Ground Transportation Center. In addition to the tunnels, the scope of progressive design build services includes an emergency stairwell, ventilation shaft and fan room, vertical circulation between baggage claim station and the boarding level, guideway, lighting, electrical and HVAC exhaust systems.

PDB for this project is structured on a phased basis with the phases referred to as components. Phase 1 of the project is comprised of four components. Each of the components are subject to a component GMP (CGMP) and the total of the four components constitute the contract amount. The owner must approve each of the CGMPs and issue a construction authorization establishing each CGMP and a notice to proceed as a condition precedent to the design-builder proceeding with any work.

The procurement included: (i) preconstruction services; (ii) a commitment to provide an overall GMP for which the design-builder will be at risk; (iii) unit rates for billing categories based on personnel working on the project; (iv) a commitment to build a complete design for those portions of the work for which the design-builder shall be the designer upon specifications and criteria provided by the owner; (v) additional requirements for the work that will be paid as a Cost of the Work; and (vi) a fixed percentage for overhead and profit to be applied to the cost of the work to be included in the GMP or component GMPs.

The draft contract terms and conditions provided with the RFP required that the design-builder accept the relationship of trust and confidence established between it and the owner and covenant with the owner to furnish its best skill and judgement and to cooperate with the owner, the owner's representative, and the construction manager in furthering the interests of the owner. It required that the design-builder, the owner, the owner's representative and the construction manager work collaboratively through completion of the project and be available thereafter should additional services be required.

All three PDB projects discussed above have been procured within three to six months prior to writing this paper. The above discussions highlight some of the nuances and commonalities of the procurement approach that are worthy of understanding for other owners and contractors to be aware of. The intent of the authors is to follow

Progressive Design-Build—Is It Coming to a Project Near You?

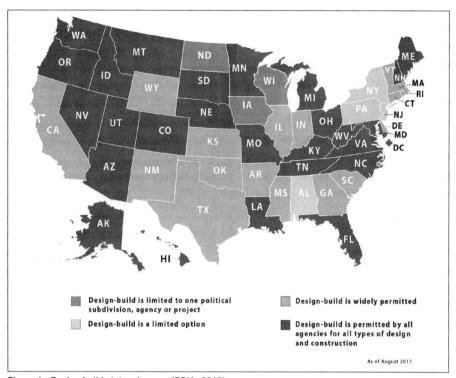

Figure 1. Design build state advocacy (DBIA, 2017)

progress of these projects in NAT 2020 regarding how successful the pre-construction design services and early procurement phase and establishments of a GMP were for each project and if the delivery method performed in accordance with the owner's expectations.

IS PDB COMING TO A PROJECT NEAR YOU?

For many in the tunneling industry the question is 'are we seeing the beginning of a limited trend?' or will PDB begin to find traction and take its place alongside other more established procurement methods. Figure 1 shows which states have passed legislation allowing design-build to be used for procurement of public works projects.

For many though there are many unique constraints. Many agencies, departments and authorities, whether state or local, have a wide variety of constraints prohibiting design-build as a procurement method. There are also cultural constraints that define an owner, experienced in designing or managing a consultant that prepare the designs, from having an appetite for change. To switch to any form of design-build procurement concept, from traditional model of designs conducted under the oversight of the owner's engineers, would require a meaningful change in mindset, training and experience of their engineering and construction management departments.

Other owners, in their beginning years of utilizing the design-build concept, have expressed a wide variety of experience. These experiences range from "good-we have concerns though" to "headaches." Large mega projects seem to lend themselves much easier to the design-build concept. The many facets of these projects permit innovation in construction sequencing by the design-builder and permit the specific methods and procedures the contractor is familiar with to be employed.

A smaller tunnel or pipeline project, which contain elements that the owner is very familiar with does not lend itself as easily to introducing design-build into the procurement toolbox. These owners generally know, based on decades of experience, what they want and how to deliver it. They don't have any interest in how projects are procured and delivered in say California, Washington, Texas or New York. Their motto "This is what we do here!"— if it's not broken don't fix it. For these owners, staff are familiar with and prefer to interact with the engineer under their control, during the design phase. Once the design is complete, the debating is over. The documents then go out for competitive bids. The construction phase is time to build it, not debate design issues. On top of this, introducing the concept of PDB becomes even more problematic including relinquishing control for the designer of record. Also, once a PDB project begins, the progressive design builder typically provides a general construction estimate on what he intends to design and build and on what schedule. Once the interaction between the owner, owners engineer, and PDB Contractor begins, the end result, specifically cost, can be considered uncertain at best.

PDB IN NEW YORK

For the current crop of megaprojects in New York the procurement strategy adopted has typically been the traditional low bid fixed price design-bid-build. On occasion a negotiated RFP strategy to optimize risk management concerns has been adopted. The use of design-build was considered during development of procurement strategy for various contracts and it is understood that some of the upcoming work may well be awarded using design-build procurement methods. Whether P DB is an option remains to be seen as there are several potential hurdles that would have to be overcome.

The author was involved in the development of a procurement strategy that could be described as a form of progressive design build for one project but ultimately discarded in favor of the design-bid-build approach with a RFP and negotiated contract with the most qualified bidder approach. The approach developed was as follows:

- Issue an RFP for prequalified contractors to bid on a 75% complete design.
- Selection based primarily on the technical approach and schedule considerations, although costs were to be provided at this stage.
- The owner would potentially look to engage the contractor through a professional services contract at this point to minimize bonding and insurance considerations as there was no actual construction occurring.
- After selection, the owner's engineer would then work alongside the selected bidder to develop the designs from the 75% to the 100% to enhance the constructability of the project.
- Upon completion of the 100% design the contractor would reprice the project. It was expected that contingency and risk would decrease leading to a lower cost, as the contractor was involved in completing the design, the opportunities for claims would be reduced.

It was at this point that the owner had some concerns with being able to demonstrate that they would be getting the best price for the 100% design as they would not be soliciting any other additional bids to be able to compare against. This issue led to the abandonment of this proposed procurement methodology.

CONCLUSIONS

Progressive design-build can probably be best described as a method of delivering infrastructure projects that combines the owner's direct control over project concept and detailed design with the design-builder's innovation—and creates a single point of accountability. Although there are some similarities progressive design build is not simply early contractor involvement and differs substantially from other alternative procurement methods such as CMAR. Owners that have adopted progressive design build on tunneling projects have done so with a common aim of implementing a delivery method that fosters better owner-contractor relationships with potentially fewer disputes and greater certainty on the final cost?

This paper looked at the procurement of three progressive design build projects—Brunswick Glynn Joint Water and Sewer North Mainland Sewer Transmission System; Silicon Valley Clean Water's Gravity Pipeline Project and the City of Atlanta's Plane Train Tunnel West Extension Phase 1.

By selecting progressive design build the owners of these projects committed to working in close collaboration with the design-builder during the initial phase of their project to develop the design, achieve the project objectives and to obtain a mutually agreeable GMP for delivery of the project. Upon acceptance and approval of the GMP the subsequent phase of the project would generally include complete design, construction and post construction tasks, including performance testing, start-up commissioning and operator training and support.

It is too early to determine whether the trend will continue and we will see more tunnel projects being delivered using a progressive design build approach however the above discussions highlight some of the nuances and commonalities of the procurement approach that are worthy of understanding for other owners and contractors to be aware of. For some public works projects legislation must still be introduced to authorize design-build as a delivery method. For some owners in a few states this may be years off. For these owners therefore, it remains business as usual irrespective of any progressive industry trends. Giving up design responsibilities that have been experienced and engrained in owner's staff is not an easy undertaking. This will take time and money. Engineers will have to be trained in the design-build process to correctly implement it and maybe more importantly the relationship of working "hand-and-hand" with the design-builder and viva-versa may be the largest obstacle.

The intent of the authors is to follow progress of these projects in NAT 2020 regarding how successful the pre-construction design services and early procurement phase and establishments of a GMP were for each project and if the delivery method performed in accordance with the owner's expectations and track other projects that adopt a similar delivery approach.

REFERENCES

Design Build State Advocacy 2017, Design Build Institute of America,.dbia.org/advocacy/state/Documents/design_build_maps.pdf.

Request for Proposal Progressive Design/Build for the 2016 SPLOST North Mainland Sewer Improvements Project Phase 1, April 12, 2017—Brunswick–Glynn County Joint Water and Sewer Commission.

Request for Qualifications (RFQ) for the Silicon Valley Clean Water Gravity Pipeline Progressive Design Build Project CIP# 6008 April 17, 2017—Silicon Valley Clean Water.

Request for Proposal FC-9277, Plane Train Tunnel West Extension Phase 1— Progressive Design Build at Hartsfield-Jackson Atlanta International Airport January 24th, 2017—City of Atlanta.

Water Design Build Council I Progressive Design-Build Procurement Guide I WDBC W-1100-2013.

A Study in the Use of Design-Build for Tunnel Projects

Shawna Von Stockhausen ▪ Mott MacDonald
Erin L.D. Sibley ▪ Mott MacDonald
Derek Penrice ▪ Mott MacDonald

ABSTRACT

Design-build has been steadily gaining traction as a preferred contracting method for large infrastructure projects in the USA over the last 25 years. This paper examines the value of the use of design-build for tunnel and large public infrastructure projects: why Owners have chosen to implement this delivery method and to what extent they are realizing its advertised benefits. The main perspectives of a design-build project will be explored: Owner, Owner's Engineer, Contractor, Contractor's Engineer, and Construction Manager. Based on these perspectives, the paper will discuss where further opportunity for improvement in the implementation of design-build exists.

INTRODUCTION

This paper examines the use of design-build as a contracting method for tunnel and large public infrastructure projects. After a brief synopsis of its history in the United States for tunnel projects, design-build as a contracting method is discussed, along with its most commonly perceived benefits. In combination with industry commentary, the paper will use experience and feedback from completed and ongoing projects to provide insight into what is working well for design-build tunnel projects, and where there is further opportunity for improvement.

In the modern era, design-build was reintroduced as a contracting method for large infrastructure projects in the United States approximately 25 years ago and has slowly been adopted for tunneling projects. The expanded use of design-build is reflected in various industry reports. Combined revenue for the Top 100 Design-Build Firms, as reported by *Engineering News-Record* (*ENR*), has steadily increased. From 2012 to 2016, the reported total revenue for design-build firms rose by 33%. Total reported revenue was $107.15 billion for 2017, up 4% from 2016. While these numbers are not specific to underground projects, they reflect a continued construction industry trend in moving from design-bid-build delivery to design-build, with this shift forecasted to continue. The Design-Build Institute of America (DBIA) in its 2018 Market Research suggests that design-build is anticipated to represent up to 44% of construction spending by 2021, which would represent 18% growth from 2018. While a proponent of the delivery method, the DBIA states that this growth has been facilitated by different factors:

- Design-build legislation, which has made it a more accessible contracting method for public Owners
- Owners receiving significant value from design-build for large and complex infrastructure projects
- Different project demands
- Changes in Owner needs
- Better education surrounding the most appropriate use of design-build

The primary objective of design-build and other alternative contracting mechanisms is to offer relief from the limitations of design-bid-build contracting—and provide greater cost and schedule certainty. Due to its sustained and increased use, design-build can no longer be considered an "alternative" delivery method.

Specific sectors of the construction industry are reporting that design-build is producing desirable results. For example, the Federal Highway Administration's 2006 Design-Build Effectiveness Study stated that managers of design-build projects estimated that design-build project delivery reduced the overall duration of their projects by 14 percent, reduced the total cost of the projects by 3 percent, and maintained the same level of quality as compared to design-bid-build project delivery. However, to date it is not clear that the tunneling industry is realizing similar benefits.

DESIGN-BUILD AS A CONTRACTING METHOD

For design-bid-build projects, the Contractor provides a price on a complete set of drawings and specifications. The award is typically thereafter made on the basis of price. In design-build projects, the contractor assumes responsibility for both construction and design, with the intent that claims against the owner based on design deficiencies are reduced. A design-build team will submit a technical and cost proposal based on a defined scope of work, design criteria and performance requirements, and drawings showing a preliminary or reference design. Contract award is typically made on the basis of a combined score involving technical and managerial approach and understanding as well as price.

The adoption of design-build results in different roles and requirements for all project parties—Owner, Engineer, Construction Manager, and Contractor—and creates additional opportunity for Engineers to work as the Owner's in preparing procurement documents or as the Contractor's engineer in preparing final design documentation on behalf of the Contractor.

ANTICIPATED BENEFITS OF DESIGN-BUILD

Commonly perceived benefits that are typically attributed to design-build versus design-bid-build are summarized below:

- Cost savings and earlier cost certainty: This results from more collaboration between designer and constructor with a focus on pricing and scheduling in the design phase, as well as the ongoing collaboration of the design-build team.
- Shortened schedule: Design-build projects can be shorter because they eliminate a bid cycle and facilitate concurrent design and construction activities.
- Innovation: Collaboration between the Contractor and Contractor's Engineer provides opportunities to try new and innovative ideas.
- Reduced litigation between Owner and Contractor: As the final design is being completed by the Contractor's Engineer, there should be fewer unanticipated issues during construction that could result in litigation involving the Owner.
- Reduced adversity: By having the final Design Engineer and Contractor on the same team, their interests are better aligned. This also offers more accountability between the final Design Engineer and Contractor.
- Improved risk management: Conflicts between Owner and Contractor are reduced because there is a single point of responsibility.

- Greater adaptability and quicker response to changing conditions: Because design and construction activities are occurring concurrently, the final Design Engineer is already mobilized and can quickly address changing conditions.
- Best value procurement: Best value (instead of lowest cost) design-build team can be selected. This can be especially valuable for complex and/or sensitive projects.

Based upon the authors' experience and understanding, in practice few of these benefits are consistently realized. Current North American experience with design-build for tunneling is erratic, with few consistent outcomes in terms of improved cost or schedule performance. Owners, particularly those experienced in design-build, have come to realize that design-build may not reduce costs, nor can they depend upon realizing significant innovation from Contractor-Designer collaboration. Expectations have been more tempered and are currently focused upon savings in schedule and risk transfer, primarily of claims related to the design. Of course, the transfer of design risk from Owner to Contractor does little to mitigate claims between Contractor and Designer.

DIFFERENT PARTIES' PERSPECTIVES

Many aspects of design-build contracts may be perceived as strengths or weaknesses of the delivery method, depending upon the role of each entity within the contractual framework. This section presents experiences and lessons learned from design-build projects from the perspective of the Owner, Owner's Engineer, Contractor's Engineer, Construction Manager, and Contractor.

Owner

From the perspective of the Owner, there are several key differences between a design-bid-build and a design-build project, including the following:

- Changed roles and responsibilities
- Opportunity for reduced litigation
- Opportunity for best-value selection

Roles and responsibilities differ for the Owner between a design-build and a design-bid-build project. In a design-bid-build project, the Design Engineer is contracted directly to the Owner and the Owner retains complete control over the final design within the bounds of limitations imposed by the Engineer's ethical obligations. The Owner provides input and direction continuously, from planning through design finalization and preparation of issued for bid documents. The Owner directs changes as they arise based upon inputs from staff: operations, maintenance, and safety personnel; external stakeholders and third parties; or public or community inputs. In design-build, the Owner should be prepared to shift from a position of design direction and control to a position of design oversight for compliance once the design-build Contractor is engaged, and entrust that the contract documents, including design criteria, scope of work, and technical performance requirements, are sufficient to allow conformance between what was conceptualized and what is actually built and that the finished construction meets the Owner's requirements and the desired quality. That is not to say that the Owner has no control over the design, but the control should be exercised more sparingly and occurs during periods of focused design submittal review.

However, to date some Owners seem to have retained control of the design, as if their project was design-bid-build—continuing to introduce significant changes to the design concept through final design and into construction. These changes become costly to implement due to the fast-tracked nature of the design-build design process and the resultant schedule impacts. This in turn negates the benefit of the delivery method and can lead to disputes.

A key lesson learned is that Owners must ensure all of their departments—engineering, construction, operations, fire and life safety—are fully engaged in the project planning, have reviewed the contract documents before the project is advertised for bid, and are in full agreement with the reference design concept. Stakeholders must also be engaged early to ensure that their feedback and requirements—including review cycle times, design standards, and potential betterments—are addressed.

Preliminary engineering typically occurs in parallel with the finalization of the project Environmental Impact Statement, which often leads to unanticipated delays, the responsibility for which is often unclear or unfairly passed onto Contractors. Issues of public concern such as acceptable construction methods—mining versus cut-and-cover construction, staging area locations, haul routes etc.—must also be driven to a conclusion early and addressed within the bid documents. On projects where there are many stakeholders with significant inputs, or where public pressure may drive significant scope and or configuration changes, design-build may not be the most appropriate delivery method.

Another key lesson learned is that Owners should have clear and workable procedures in place for design-build projects prior to embarking on the project, and all Owner staff should be trained in the application of these policies and procedures. These should provide clear definition of the Owner's and other project parties' roles during design and construction; define at what levels of design completion the Owner will receive submittals; identify to which level it will review submittals and who is responsible for reviews (Owner staff, Owner's Engineer, Construction Manager, or all of the above); and define how changes during design and construction are documented and which changes are entitled to time extension and additional compensation, all with a goal towards minimizing the potential for uncertainty and frustration in later stages of the project.

In a properly configured project, *reduced litigation* is a primary Owner-derived benefit of design-build projects. In design-bid-build contracting, claims may arise for a number of reasons such as changed or unforeseen conditions, unbuildable design, conflicts between trade packages, etc. These claims are typically borne by the Owner, unless they in turn can demonstrate error or omission by the final Design Engineer. With the integration of Contractor and final Design Engineer in a design-build environment, issues related to the constructability of the design or interdisciplinary coordination are theoretically owned by the Contractor. Risk management practices should be applied in accordance with the Underground Construction Association of the Society of Mining, Metallurgy, and Exploration (UCA of SME) Risk Management Guidelines during the preliminary engineering phase, such that the Owner can identify and transfer appropriate risk to the Contractor.

Best value selection is one aspect of design-build that offers the Owner more latitude compared to design-bid-build. With a best value model, the Owner can introduce qualifications and technical components to the evaluation and selection criteria. This benefits the Owner and the design-build team. It allows the Owner to select a qualified team that has demonstrated an understanding of the key project challenges and has the qualifications and personnel to address them, rather than selecting purely on price.

In design-build projects, the *Owner's Engineer* serves a critical function as an extension of the Owner's staff, to develop preliminary designs and related contract documents, provide engineering support to environmental documentation as applicable, and thereafter support the Owner as needed during the remainder of the design-build contract.

Roles and responsibilities of the Owner's Engineer also differ in a design-build project over a design-bid-build project. Rather than completing the final design for a design-bid-build project and performing design services during construction, in a design-build project the Owner's Engineer must develop the RFP bid documents including a scope of work (a critically important document that defines in detail the work that the design-builder is and is not responsible for), technical performance requirements that identify the Owner's minimum requirements for project durability and quality, and the bid reference design, comprising drawings and supporting reports, which should be sufficient to demonstrate concept feasibility.

For a tunnel project, a completed geotechnical data report and geotechnical baseline reports should also be part of the contract documents. The former is a factual document and the latter a contractual risk sharing mechanism, which may include provision for negotiation of baselines based upon further investigations undertaken by the design-builder. The level to which each aspect of the project is designed/developed for the RFP documents requires extensive communication and coordination with the Owner. Depending on the relative risks on the project, this could result in some areas of a project being developed to a much higher level than others and being prescriptively defined—for instance, the underpinning of structures adjacent to a tunnel or deep excavation, where appropriate mitigation of third-party impacts may need to be demonstrated to the third party's satisfaction.

The Owner's Engineer role in design-build projects requires a much more integrated and nuanced approach. An experienced team is needed, as the content and risk allocation of the RFP documents can significantly influence the number and quality of the design-build bids received, and will determine whether the Owner receives a compliant bid. For example, a typical design-build RFP document usually includes the following:

- Clear definitions of what is mandatory versus what the design-builder has the flexibility to change. For example, a project may mandate the use of a pressurized face tunnel boring machine (TBM) and minimum TBM operating requirements, but may leave the machine type selection to the design-builder based upon its interpretation of the ground conditions. The project may mandate the tunnel alignment and drive direction based upon EIS requirements but may allow the design-builder to alter the tunnel profile.
- Fully defined minimum requirements for quality and durability, and assurance that testing, verification, and submittal requirements are adequate to demonstrate these.
- Clear design criteria and definitions of minimum analysis requirements, particularly where complex or unusual structural conditions exist such as large span caverns. This can avoid protracted arguments of what is versus what is not required by the contract.
- Detailed requirements for submittal reviews and closure, which agencies will be reviewing, the submittal levels and what is required to be submitted with each, turnaround times for submittal reviews, expedited comment resolution, and requirements for independent design checks for complex work. Owner's design reviews must respect the fast-track nature of the

Contractor's design schedule. Often Owners' review periods are lengthy and can potentially impose delay to the design schedule should significant comments arrive unexpectedly to the designer. Whether comments are valid or not, risks to the Owner's overall perceived benefit of schedule certainty may be realized regardless of fault or cause. Any potential to mitigate such risks through streamlined Owner review periods will benefit the project. To expedite reviews, requiring over-the-shoulder reviews prior to submittals is recommended. These provide a useful test for aligning Owners' expectations with submittal contents and for obtaining advance comments, which ultimately help expedite submittal approval.

Design review is a major function of the Owner's Engineer. The Owner is relying upon the skills and expertise of the Owner's Engineer to assess the design-build product and verify compliance with the Owner's requirements. The ability of the Owner's Engineers to execute design review appropriately will have a significant impact on the technical and commercial success of design-build. Design reviews should address completeness and accuracy of information provided, compliance with contract requirements, and compliance with design criteria. Unless the technical criteria mandate design approaches and/or methods, these should be left to the discretion of the Contractor's Engineer, who bears ultimate responsibility for design product. Inadvertent or unconscious overreaching by the Owner's Engineer in final design decisions that are at the discretion of the Design Engineer may be a major source of claims for delay and additional work on design-build projects.

Contractor

The preparedness and qualifications of the Contractor has a significant impact on the success of a design-build project. Some of the significant differences from the perspective of the Contractor are listed below:

- Procurement
- Roles and responsibilities
- Potential for innovation

The *procurement* of a design-build project is starkly different from that of a design-bid-build project. On a design-bid-build project a Contractor will essentially provide a price based on a completed design provided by others and will be selected based on price. In design-build, Contractors are responsible for the final design of a project and must be able to provide enough understanding to address the technical requirements as well as the potential challenges associated with construction of the work.

In the best-value process, Contractors need to provide technical approach, management approach, and price. These are based on partially complete documents, requiring extensive internal coordination on risk and contingencies. Contractors can showcase their understanding of the work in a way not permitted with design-bid-build, and demonstrate how their proposed means and methods are best suited to the needs of the project. Price can be a misleading judge of which Contractor is best prepared to address the complexity of a project, and with design-build the Contractor is given a stronger voice to justify why its team is the best prepared to deliver the project for the Owner.

The *roles and responsibilities* of the Contractor also change with design-build. While on a design-bid-build project the Contractor will engage engineers for specific functions such as shaft design or tunnel initial support design, the Contractor is now in the position of managing engineers full-time. Strong and open lines of communication are required. However, care should be taken to avoid distracting key staff in excessive meetings—Owner meetings, Contractor weekly meetings, task force meetings, technical coordination meetings, third-party meetings—that can distract from the critical activity of finalizing the design.

The *potential for innovation* is high for the Contractor in design-build projects as long as the Owner's reference design is not overdeveloped or overly restrictive. This is one of the perceived benefits of design-build that Owners are trying to tap into, in the hope that the Contractor can significantly reduce project costs. Contractors are left with some freedom to select the most advantageous (for them) means of executing the project. This can lead to a Contractor pursuing a new approach to solve a problem that may not have been realized if all competitors are given the same final design, as evidenced by alternative technical concepts or value engineering recommendations. Giving Contractors freedom motivates them to find new and cheaper solutions to solve the problems posed by the project. Though much more freedom is granted to the Contractor with design-build projects, it is not without risks.

Contractor's Engineer

The role of the Contractor's Engineer under a design-build framework presents commercial and technical risks and opportunities that the consulting engineer needs to weigh carefully as this delivery method becomes more prevalent, including the following:

- Roles and responsibilities
- Potential for innovation

Roles and responsibilities are greatly changed in a design-build versus a design-bid-build project. Obviously, the Contractor's Engineer is now in the position of working directly for the Contractor rather than the Owner. This role poses significant risk for engineers, both in the pre-bid and post-award phases of a design-build pursuit. This effort is often associated with an accelerated schedule, working collaboratively with the Contractor to develop a solution that best suits the Contractor (which could be cost-driven, schedule-driven, or driven by means and methods), and a flexible and adaptable design approach to address what is urgent at any given time. All of this must be accomplished while remaining in compliance with ethical rules, regulations, and guidelines applicable to the practice of engineering and architecture.

The role of the Contractor's Engineer during the construction phase of the project is somewhat diminished, as the Contractor is self-performing QA/QC. The engineer still performs design services during construction including submittal and RFI review, but can feel somewhat detached from the construction process, until an issue with quality should arise, at which time the Engineer's input is required to verify that construction is in conformance with plans and specifications, or to recommend repair solutions in conjunction with the Contractor. As the Engineer is often not retained on a full-time basis during the construction phase and has other commitments, response time can vary.

The *potential for innovation* is a natural byproduct of strong communication within a design-build team. One such example is the application of Building Information Modeling (BIM) and common data environment. The colocation of Designer and Contractor is often cited as a benefit for design-build, ensuring timely and as-needed communication. While having appropriate key design team staff co-located with the

Contractor is certainly recommended, the use of BIM in a common data environment provides the opportunity for the Designer to engage resources nationally and internationally, resulting in significant direct cost savings for relocations, travel, and subsistence, and reducing the required size of rented project offices. Having staff in multiple locations allows design to progress across time zones, extending the duration of the working day and helping expedite production. Provided key representatives of the design team are available to the Contractor and Owner as needed, the designer should be allowed to coordinate its resources to provide Owners with the opportunity for both cost and schedule savings. This method of working can generate long-term benefits for Owner, Contractor, and Engineer if implemented in the right way.

Construction Manager

The *roles and responsibilities* of the Construction Manager (CM) on a design-build project are quite different than those on a design-bid-build project. When working as the CM, the Consulting Engineer now finds itself responsible for significantly more design coordination than it would be in a traditional design-bid-build project, since the design and construction are overlapping activities. Typically, in a design-bid-build project, the constructability review would be completed by the CM, but it is now undertaken directly by the Contractor. If engaged early, the Construction Manager may only be asked to perform a constructability review up until the receipt of design-build proposals, rather than from the 60% submittal stage and onwards in a design-bid-build project.

Additionally, the Contractor is now also responsible for QA/QC. This is no longer required of the CM, or at least much less so. The CM finds itself instead in a position of verifying Contractor compliance with its own schedule and procedures. This is still an invaluable service for Owners: submittal of daily inspection reports and provision of progress photographs can give the Owner confidence in the integrity of the design-builder.

ARE THE BENEFITS BEING REALIZED?

Each of the purported benefits of using design-build as a contract delivery method is examined in further detail to assess the extent to which that benefit is being realized and what, if any, improvements can be incorporated for future design-build projects.

Cost Savings

Design-build has been advertised as a cost-saving contract method compared to design-bid-build. This may be true for specific industry sectors such as rural highway widening projects, for example. However, changing design requirements, unanticipated ground conditions, risk transfer, or other factors can often detract from any anticipated cost savings. Instead of producing cost savings, "not all design-build projects have been successful [with reducing overall costs]" (Reilly 2011). Industry commentators have suggested that this highlights a weakness in contract writing for design-build as well, for "a design-builder will build what is required but not necessarily what is desired...with design-build, if the contract documents do not specifically require something, it will probably not be supplied" (Reilly 2011).

Clearer, more well-defined bid documents and front-end work are necessary if cost savings are to be fully realized on design-build projects. The Contractor's Engineer for its part can seek to understand better the goals and objectives of the Contractor, and the Contractor must have an appreciation for its Engineer's level of confidence with the pre-bid design. This is necessary to quantify risks and provide adequate contingency provisions in the price and schedule (Sakar et al.). Thus, quality work should be seen as a cost-saving exercise.

Shortened Schedule

Design-build is attractive for Owners looking to fast-track a project because of its purported ability to shorten the schedule to completion: because it eliminates a bidding cycle and because final design activities can be sequenced to overlap with construction.

Recent United States experience suggests that Owners can underestimate the amount of time required to get through the project bid and award phase. Regardless of the scale or complexity of a project, Owners seem to provide similar bidding duration to that of a design-bid-build project. For large, complex design-build infrastructure projects, a substantial amount of time, between 18 and 24 months, is typically required between Request for Qualifications (RFQ) and NTP. The design-builder requires sufficient time to absorb the voluminous contract documents, they need to prepare sufficient pre-bid design to enable a competitive price to be compiled and they need time to prepare the bid price, including review of required contingency based upon their perception of design and construction risk, while simultaneously preparing a technical proposal. In many if not all cases, the length of time required to ensure the best qualified team is selected for the project results in perceived schedule savings being lost versus design-bid-build.

There are several reasons for such extended proposal durations:

- Requests from Contractors to extend the bid period based upon the time actually needed to prepare a responsive bid, numbers and types of questions submitted, and numbers of addenda issued
- Clarification of risk allocation and ownership
- Time to assess "best value" or "alternative technical concepts" as part of best value selection
- Negotiations to obtain a "best and final" offer
- Re-solicitation with revised/reduced scope due to initial bids coming in over budget

Suggestions for shortening the bid and award phase durations are these:

- Improved risk management: Assess the contingency needs of the project early.
- Improved cost estimates: Clearer, more well-defined Owner-provided bid documents and front-end work are necessary if cost savings are to be fully realized on design-build projects.
- More realistic time frame into the schedule: If the duration to properly evaluate proposals is built into the project schedule from the start, "delays" can be avoided.
- Removal of the contractor qualifications phase from the critical path: This phase can be started earlier in the project development so as not to impact the project bid and award duration.
- Attention to review comments submitted by design-build teams on previous submittals: Common issues that have resulted in design-build teams

requesting a bid-extension in the past should be identified and addressed during the preliminary engineering and bid document preparation.
- Greater Owner flexibility: The Owner should have more flexibility in assessing and implementing Contractor-derived alternative concepts where schedule, cost, risk reduction, improved safety, or quality can be demonstrated.

Opportunities for Innovation

Design-build is championed as ideal for complex projects or projects with challenging subsurface conditions (i.e., higher-risk underground projects), because the teaming partnership between the Contractor and the Designer from the beginning presents an opportunity for more innovative solutions. In turn, those innovations are expected to go hand in hand with cost savings for the Owner, because the Contractor has more latitude to use its preferred means and methods and efficient design, rather than strictly conforming to a prescribed design.

The case that design-build projects lead to innovation is often tempered by the need to balance prescriptive Owner requirements, often driven by operational needs, (which can restrict innovation and change), with broad requirements that may incur additional costs and therefore be of less value to the Owner.

In reality, design-build projects can be snagged on the very aspects of contracting that they intend to avoid. Because the timeline for design is accelerated and tied into the construction, the Owner must provide specific, detailed technical performance requirements to ensure that the product they want is the product they will get. This stifles the potential for innovation.

Litigation and Reduced Adversity

With design-build projects, the Contractor is responsible for the performance of the design, its constructability, and for coordination between the various design disciplines. This risk transfer is one of the benefits from an Owner's perspective that is realized by using design-build. However, with increased use of BIM, with regular and ongoing clash detection, the risk of claims relative to interdisciplinary design coordination on design-bid-build projects will also be reduced. The potential for greater project control of design-bid-build, plus improved design risk mitigation through BIM, may prompt owners to reconsider design-build.

Risk Management and Contract Interfaces

Design-build projects demand more proactive risk management. As in design-bid-build projects, risk management must be an iterative, ongoing process and must specifically address uncertainties. For instance, right-of-way acquisition, permitting, and coordination with third-party agencies are typical risks for most underground projects. These risks are heightened for a design-build contract due to the fast-tracked schedule. Permitting agencies do not have any incentive to expedite their processes and others are not familiar with the design development process of design-build. Such risks must be managed by the Project Owner and addressed prior to design-builder Notice to Proceed.

Design-build projects can improve risk management practices. However, this depends on very well-defined contract language and roles and responsibilities of all parties. Attempts by Owners to unfairly transfer all project risks to the design-builder usually result in the need to overemphasize contingencies, thus driving up project costs and creating potential conflict between the design-builder and its Design Engineer. Failure by Owners to recognize the difference between construction risks and design risks and the differences between insurance policies that cover them also lead to protracted negotiations between Contractors and Engineers when Contractors attempt to contractually flow those risks to the designer. When the design-build team has a solid understanding of what risks were identified during the design process, and how these risks have been allocated and mitigated, the odds that risks can be more effectively managed increase. Correspondingly, in line with industry recommendations for risk management, it is recommended that the Owner's design brief is shared with the design-build contractor, and that it be reviewed periodically over the course of the design-build project.

Adaptability and Response Time

In general, adapting design requirements can be accomplished very well with design-build. Doing so quickly, because of co-location of staff in project offices, helps facilitate timely issue resolution, especially when changes are encountered during construction. Care should be taken to avoid distracting key staff in meetings addressing changes that could distract from the rest of the work at hand.

Best Value

For a design-bid-build project, the industry traditionally awards the contract to the Contractor with the lowest bid. This can present problems for the Owner and the project if the reason for the lowest bid is due to inexperience or the Contractor does not fully acknowledge or understand the risks and complexities of the project. Underground projects can be especially vulnerable to delays or other impacts when the selected Contractor is unqualified. One benefit of design-build is that the Owner can select the best-value team, rather than just the lowest-priced Contractor. While price will still be a factor in the overall selection criteria, it is no longer the only selection criterion. This gives Contractors more leeway to present new and innovative ideas that illustrate their understanding of the project and gives them more ability to display their successes on similar projects.

Collaboration and Communication

Collaboration and communication are among the successes of design-build projects and a root cause of many of the other successes. A design-build team cannot adequately meet the needs of the Owner unless everyone is actively working towards a common goal: "Design-build provides ability to optimize design and construction… [but only if] the designer and Contractor can work as a coordinated team and in a transparent manner" (Johnson 2017). This philosophy needs to be built deliberately into the management and delivery approaches. This is especially true because the design-build process is iterative. The team therefore must have a similar approach to innovation and risk and select key staff to spearhead teams in this way.

Some of the drawbacks to this intense level of iteration and communication are that key staff—at all levels of the design-build project—can be quickly distracted in meetings. This can hinder effective collaboration and reduce the available time to do the actual work. It is important that the design-build team establish a workable meeting schedule that addresses the need to come to conclusions together while not impeding the progress of the work.

For the Contractor's Engineer, having key leadership in place early—including project management and technical discipline leads—helps facilitate development of the plans, procedures, and training necessary for the efficient execution of the project. This has shown to be fundamental to the success of design-build projects. While it is recommended that appropriate key staff be co-located with the Contractor, the adoption of the basis of design report and use of common data environments can promote consistency across a dispersed design team.

Challenges can be overcome by a commitment to cooperative working relationships and when all sides make a conscious effort to understand each party's main goals and motivations. The motivations and objectives of a design-build client are not the same as those of traditional public-sector clients. This is where staffing plays a critical role. Key staff must have experience working in a design-build environment. Contractors look to their engineers to empathize, and expect the designer to be loyal to them first and foremost. Designer's must be experienced enough to understand this dynamic, yet professional enough to make sure that ethics and sound design practices are not compromised.

PROGRESSIVE DESIGN-BUILD

In addition to the use of design-build for project delivery, progressive design-build has started to have an increasing presence. While progressive design-build is closer to design-build than design-bid-build, there are still significant differences. For example, the Owner is most likely to select progressive design-build when there is a benefit to taking advantage of the knowledge and "know-how" of its Contractor and Design Engineer while being an active participant during the design. The selection of a progressive design-build team is based on qualifications and approach rather than price. In fact, the price is often not negotiated for up to a year after project award as the design is developed. This allows the Owner to be involved in all decisions and can help the Owner to keep stronger oversight on price. However, the flip side of this benefit is that it can put the Owner in the position of making decisions that it would normally not be involved with, such as equipment or means and methods selection and other decisions traditionally left up to the Contractor. The primary benefit of a progressive design-build project is the opportunity to align the Owner's needs with the Contractor's means and methods.

CONCLUSIONS

The increased use of design-build does not necessarily mean it is the "best" contracting method for tunneling projects, but it can be the "right" method for a given project and can be successful, as with design-bid-build projects, if applied in the right way.

In many instances, the benefits typically associated with design-build projects do not necessarily manifest in the ways expected. Problems with design-build arise when there is less clarity and understanding about the design-build process—such as when a design-bid-build approach is adopted to a design-build project, or when there is inappropriate transfer of risk to the design-build contractor.

Ultimately, the success of a design-build project appears to come down to the familiarity of the parties involved with design-build. If the differences between design-build and design-bid-build delivery are well understood by the project team, the roles and responsibilities of all parties are clearly defined, and risk is properly allocated to the party best able to manage it, a successful outcome is achievable. With increasing use of design-build, and with proper documentation of lessons learned, its most common pitfalls can be avoided, and more consistent outcomes will result.

REFERENCES

Design-Build Effectiveness Study, Final Report, USDOT—Federal Highway Administration, January 2006.

Drake, Ronald D. (2018) "Interlake Tunnel–A Future Design-Build Project," *North American Tunneling: 2018 Proceedings*. Society for Mining, Metallurgy, and Exploration, Inc. 839–848.

Engineering News-Record, "The 2018 Top 100 Design-Build Firms." June 21, 2018.

Engineering News-Record, "Design-Build Gets a Big Boost from New Building Systems Requirements," June 6, 2007.

FMI Corporation 2018 (2018). "Design-Build Utilization–Combined Market Study," June 2018. Design-Build Institute of America.

Greiman, Virginia. "The Big Dig: Learning from a Mega Project." Story, *Ask Magazine*, 47.

Hughes, G., Kroncke, M., Wone, Moussa (2015). "Practical Aspects of Final Design Development Using Design-Build Procurement," *Rapid Excavation and Tunneling Conference: 2015 Proceedings*, 330–337.

Johnson, Mark, Ellis, Martin. "Design-Build Project Delivery: The Importance of Successful Coordination between Designer and Contractor," *Rapid Excavation and Tunneling Conference: 2017 Proceedings*, 584–593.

Kannry, Jack S. (2016). "Unique Concerns for Design-Build Contract.".

Kramer, Steven R., (2018). "Current trends in procurement delivery of major tunnel projects," *North American Tunneling: 2018 Proceedings*. Society for Mining, Metallurgy, and Exploration, Inc., 825–832.

O'Carroll, Joe, and Goodfellow, Bob. (2015) "Guidelines for Improved Risk Management on Tunnel and Underground Construction Projects in the United States of America." *Underground Construction Association of SME.*

"Progressive Design-Build: Design-Build Procured with a Progressive Design & Price. A Design-Build Done Right Primer." DBIA, October 2017.

Reilly, John. (2011). "Alternative contracting and delivery methods," *Tunnel Talk*, September 2011.

Roach, Michael F., Lawrence, Colin A., Klug, David R., Fulcher, W. Brian (Eds.) (2017). *The History of Tunneling in the United States*. Society for Mining, Metallurgy, and Exploration, Inc.

Sakar, Subal (2014). "Risk management for design-build projects." *North American Tunneling Conference: 2014 Proceedings,* 801–807.

Smirnoff, Timothy P. (2014) "Design/Build a Panacea?—No," *North American Tunneling Conference: 2014 Proceedings,* 807–812.

Wonneberg, James, Castro, Rafael, Allen, Christopher (2014), "Tunnel Construction Management for Design-Build Delivery," *North American Tunneling Conference: 2014 Proceedings*, 782–792.

Tunnel Contracting Practice: One Owner's State of the Art

Ted Dowey ▪ New York City Department of Environmental Protection
Kate Edden ▪ New York City Department of Environmental Protection
Sean McAndrew ▪ New York City Department of Environmental Protection
George Schmitt ▪ New York City Department of Environmental Protection
Bade Sozer ▪ McMillen Jacobs Associates

ABSTRACT

The tunneling contracts issued by the New York City Department of Environmental Protection have evolved significantly over the decades. This paper presents the state of the tunneling contract as utilized for the Bypass Tunnel and includes discussion of prequalification of contractors, a choice of bid sheets and associated excavation techniques, differing site conditions, liquidated damages, incentives, mobilization, bonding a long contract, the risk process, partnering, GDR and GBR, TBM specifications, standards for contractor-designed underground elements, rolling allowances, probing and pre-excavation grout requirements, segment design, and safety. The paper does not address the administration of the ongoing contract.

NOTE

This paper presents how the contract addressed the challenges of building the Bypass Tunnel. It does not address all the New York City contracting processes and requirements, nor does it address the administration of this ongoing contract. The DEP follows a design-bid-build process, and the award of the contract goes to the lowest prequalified bidder. In addition, this paper does not discuss work in the tunnel under Wawarsing, which is part of the contract.

PROJECT OVERVIEW

The New York City Department of Environmental Protection (DEP) has a significant leak in a primary water tunnel that delivers 50 percent of the City's water from upstate reservoirs to the City. The Rondout-West Branch Tunnel (RWBT), a portion of the Delaware Aqueduct, leaks between 15 and 35 million gallons per day, primarily on the west side of the Hudson River in Roseton, NY. This portion of the 45-mile-long RWBT where most of the leaking occurs is 600 feet under the Hudson River, and it is also the low point of the tunnel. Safety issues associated with working at the low point of a tunnel, while millions of gallons of groundwater are entering the tunnel, render a traditional in-tunnel repair impossible.

After much study, the solution to the leak is to build a Bypass Tunnel to circumvent the leaking portion. The Bypass Tunnel is offset 1,750 feet to the north of the existing tunnel. The connection points at the ends of the Bypass Tunnel where it will be connected to the RWBT are located in competent rock formations. The Bypass Tunnel will include 9,200 feet of steel interliner in order to contain the internal water pressure as it passes through the same problematic geology as the original tunnel. The new tunnel is 13,543 feet long, 14 feet in diameter, and connects to the RWBT at the same elevation, about 600 below sea level.

PRIOR TO THE CONTRACT: CONDITION ASSESSMENT, TECHNICAL REVIEW COMMITTEES

The DEP had conducted two autonomous underwater vehicle (AUV) inspections of the RWBT prior to the start of the design of the Bypass Tunnel. The AUV took hundreds of thousands of pictures. Maps of cracks in the tunnel lining were developed from the photographs. The lining, although leaking substantially in the Roseton location, was found to be structurally intact overall. The AUV data plus the data from a horizontal boring along the tunnel in the leaking zone, coupled with historical documents on the RWBT tunnel construction in the 1930s and 1940s, formed a solid basis for the design of the Bypass Tunnel.

Early in the design phase of the Bypass Tunnel, the DEP convened a series of Technical Review Committees (TRCs) to evaluate various components of the Bypass Tunnel design. These TRCs consisted of consultants and contractors and the meetings were attended by the DEP design team, construction managers, and the DEP Bureau that would operate the tunnel. Consultants and contractors were assured in writing that they would not be precluded from participating in future contracts associated with the Bypass Tunnel as a result of participating in the TRCs. There were no drawings or specifications at this point in the design, only concepts.

The TRCs addressed the concept of the Bypass Tunnel and the associated drainage tunnel, the constructability of the Bypass Tunnel and the connection of the Bypass Tunnel to the RWBT, the duration of the connection, the anticipated groundwater inflows during the connection, and the structural integrity of RWBT while it is empty. These TRCs were essential to confirming that the Bypass Tunnel and the plan for building and connecting it to the RWBT were the most efficient, safest, and most likely to succeed of all the solutions available.

SAFETY

Safety is of paramount importance to the DEP. The DEP has developed extensive Environmental Health, and Safety (EHS) policies and procedures and Bureau of Engineering Design and Construction (BEDC) EHS Standards as a result of doing billions of dollars of heavy construction for decades. All bidders must follow these EHS policies and procedures and meet these standards. This includes having an Experience Modification Rating of less than 1.0, in order to receive a bid selection. There are numerous requirements during the contract, including monthly reporting containing environmental compliance, safety metrics, incident reporting, job hazard analyses, and EHS Plans. Emergency Action Plans are developed to address potential emergencies, including spill prevention plans. All EHS requirements extend to the contractor's subcontractors.

Specific to the Bypass Tunnel project, emergency rescue crews for underground rescue are required, as the local response personnel will only handle aboveground issues. An EMT is required to have an office and be on site whenever there is underground work occurring. A "No Drop Zone" is to be established at the top of shafts.

PROJECT STRUCTURE

Prior to the connection, the Bypass Tunnel will be built in its entirety, but not connected to the RWBT. The Bypass tunnel excavation will end 100 feet from the RWBT at each end, leaving a rock barrier. RWBT must then be taken out of service and pumped out prior to connection with the Bypass Tunnel. This connection work is anticipated to take 8 months. In order for the RWBT to be taken out of service for this duration, the reservoirs must be full to support the two other water delivery systems that will be used to supply the City with water. In the event that the reservoirs are not full, in order to make the planned connection in 2022, a historical analysis of the hydrology of the reservoirs indicates that in one of the next two years, the reservoirs will be full and the connection can be accomplished. In order to accommodate the timing of the connection, two distinct standby items have been included in the contract to address these scheduling issues.

The excavation of the tunnel, interliner installation, tunnel concrete lining, and shaft completion coupled with a three-year period for the connection led to a very long contract. However, DEP recognized that having the connection work as a separate contract might not attract any bidders, as the majority of the risk is in that phase of the work. Therefore, DEP insisted that a single contract be used to complete both the Bypass Tunnel and the connection.

STANDBY/DEMOBILIZATION ITEMS

A historical analysis of the weather over the past 95 years indicates that there will be an 84 percent chance the reservoirs are full in any given year including the year scheduled for the connection, 2022. If the reservoirs are not full or the Engineer states the connection cannot be performed in 2022, the contractor will be told to demobilize and stand by for one year until 2023. The same goes for the following year, which would result in a connection in 2024. A bid item is included for the contractor to price this standby item. It can be used a maximum of twice.

Hydrologic analysis states that once a connection starts, in any given year, it can be completed given the anticipated duration of the connection. The Engineer has the right, for hydrologic or other reasons, to interrupt the connection construction. There is a bid item, a demobilization item, in the contract for this purpose. This demobilization item can be used a maximum of twice.

MILESTONES, LIQUIDATED DAMAGES, AND INCENTIVES

There are two milestones in this contract. Milestone 1 is for the completion of the Bypass Tunnel prior to the connection. There are liquidated damages of $30,000 per day associated with this milestone with no cap. Milestone 2 is for the completion of the connection. This is the most critical portion of the contract as New York City is without the use of the RWBT during this time period. There is a $30,000 per day incentive payment for early completion, capped at $1,020,000.00. The liquidated damages are $30,000 per day. There are also $7,500/day liquidated damages to be assessed if substantial completion occurs later than the contractual date.

BONDING AND CONTRACT DURATION

Issues arose with bonding the entire Bypass Tunnel contract because of its long duration. During the process of pre-qualifying bidders, some contractors mentioned having difficulty bonding a 10+ year-long contract. During the bid period, several major bonding companies reached out to DEP to resolve the issue. The contract duration of more than 6 years was considered too high a risk. The suggestion by the surety companies was to divide the work into multiple contracts, but, as explained above, DEP was committed to having one contract for building the Bypass Tunnel and making the connection to the RWBT.

The contract length includes a year of mobilization intended to allow time for the contractor to procure the tunnel boring machine (TBM) and other major pieces of equipment. Milestone 1 includes the excavation of the tunnel, installation of the interliner, concrete lining of the tunnel, and completion of the shafts for a total duration of 2,346 days, or 6.4 years. The start of the connection is slated for one of the next three years following Milestone 1, primarily due to the hydrology of the upstate reservoirs. Milestone 2 includes the completion of the connection of the Bypass to the RWBT. The end result is a contract that could last 3,806 days, or 10.4 years.

The solution was to bond the contract in three phases, utilizing the milestones and the end of the excavation to define the three phases. The first bonding period begins with the Notice to Proceed and ends with the completion of all underground excavation. The second bonding period begins with the completion of the underground excavation and ends with Milestone 1. The third bonding period starts from Milestone 1 and extends through the connection to substantial completion. At the end of each bonding period, DEP will issue a partial substantial completion determination for the work completed. There was no separate bid item for the cost of the bonds.

To accommodate bonding in phases the bid documents included a schedule in which the contractors had to identify which work would be included in the three bonding phases. Lump sum items were apportioned according to the bid breakdown and placed in the three phases. Unit price bid items were assigned to phases or prorated. Allowance bid items (with one exception) are included in phase one and can be passed to subsequent phases if not utilized. In the event the contractor does not obtain bonding in the subsequent phases, an additional bonding requirement, a Guarantee Bond of $5 million, was added to the contract. The City may make a claim on the Guarantee Bond should the contractor fail to obtain the bonding for the subsequent phase.

The DEP does run a risk if the contractor does not get a new bond when required. The tunnel is a robust structure that will not deteriorate and will remain useful should the City need to procure a new contractor to finish the contract. The pre-qualification process provided well-established tunneling companies that will see this challenging job through to its end.

TBM OR DRILL AND BLAST

The length of the Bypass Tunnel (13,543 linear feet) is such that it could be excavated economically with a TBM or by drill and blast methods. The original tunnel was excavated using drill and blast, although with difficulty. The Bypass Tunnel is offset 1,750 feet to the north but must pass through the same troublesome rock formations. Because of the Environmental Impact Statement (EIS) and local commitments, it must be excavated from the west shaft to east shaft, so drill and blast from each end simultaneously was not allowed.

The DEP decided to let the market/tunneling industry decide whether to excavate the tunnel by drill and blast methods or using a TBM and a segmental lining. The solution was to allow the contractor to bid on either. Additional design was required to design the tunnel excavation, rock support, grouting, and concrete lining for both modes of excavation. Two bid sheets were created along with the additional drawings and specifications. The Geotechnical Baseline Report (GBR) was developed to address each mode of excavation. Contractors could only bid on the method for which they were pre-qualified.

CONTRACTOR PRE-QUALIFICATION

While making the connection of the Bypass Tunnel to the RWBT, the City is without its major water supply tunnel. The City and the contractor take on a great deal of risk during this time period. The City wanted experienced contractors performing the work and elected to pre-qualify contractors bidding on the work. The Agency Chief Contracting Office published the Request for Qualifications (RFQ) detailing the requirements for bidding on the Bypass Tunnel contract. The Bypass Tunnel contract allowed contractors to bid on one of two excavation methods: tunnel boring machine or drill and blast. Contractors were limited to bidding only on the excavation method(s) they qualified for. The RFQ contained numerous requirements. Some of these requirements applied to individual firms, some to joint ventures, and some pertain to any firm bidding NYC work. For a joint venture, all partners had to meet certain criteria related to environmental health and safety, quality assurance programs, financial capabilities, and contract performance and contract evaluations over the past seven years. Joint venture criteria excluded bidders with any previous law suits addressing performance, labor conflicts, or EEO or M/WBE complaints.

If qualifying as a joint venture, one of the joint venture members had to have tunneling experience including projects of a similar nature, size, and scope; deep rock tunnels of at least 16 feet in diameter driven from the base of a shaft approximately 300 feet deep, facing at least 250 gpm inflow in the tunnel; performing pre-excavation grouting at the heading or through the face of a TBM; and the use of a shielded TBM with precast segments for the TBM option. The requirements for the drill and blast option included experience in tunnel excavation with over 1,000 gpm inflows in a tunnel at least 2,500 feet long and at least 16 feet in diameter. Both excavation techniques required experience that included the placement of a cast-in-place concrete lining with a finished diameter of 12 feet or more. Pre-qualification requirements also included individual personnel qualifications for six key staffing positions on the project.

Ultimately the process was made easy, as contractors formed joint ventures first and then applied for pre-qualification for both excavation methods. Eight joint ventures were formed, and all applied for both excavation techniques. There was one exception, whereby a joint venture substituted the third member of a tri-venture and each applied for only one excavation technique. Ultimately, six of the eight joint ventures were pre-qualified. The pre-qualification process, according to the NYC Procurement Policy Board, must include five qualified firms. The DEP ultimately received three bids, two using a TBM and one utilizing drill and blast.

The RFQ additionally stated that there is a Project Labor Agreement (PLA) in place and that the PLA is binding on the contractor and subcontractors awarded the contract. The RFQ also stated that the contract is subject to government regulations, and a web link to the EIS commitments was provided.

GENERAL MOBILIZATION BID ITEM

A general mobilization bid item was not included. Therefore, the contractor must include these costs in the most relevant bid items in the contract. A mobilization item for the tunnel boring machine was provided.

TBM MOBILIZATION BID ITEM

In order to reduce TBM financing costs and to help reduce the bid price, a TBM mobilization item was included in the tunnel boring machine version of the contract. The amount designated for this item was $20 million. The $20 million was intended assist in the financing of the machine and associated equipment, but not to cover the full cost of the TBM. The Item was payable in five parts with the delivery of each of the following components: review of the TBM design ($1M), shop floor assembly ($4M), delivery of the equipment to the work site ($6M), testing of the equipment ($3M), and successfully mining of 1,000 feet of tunnel ($6M). These expenditures would, in the absence of a TBM mobilization item, be distributed in the price paid for mining the tunnel, or potentially in another contract item that would be completed early. The last amount, payable upon mining 1,000 feet of tunnel, does not function as a mobilization item because at this point in time the TBM has already been delivered and is operating. It does, however, provide a bonus for mining the first 1,000 feet of tunnel.

GEOTECHNICAL BASELINE REPORT

Substantial geotechnical work was done prior to the contract. These data are included in the Geotechnical Data Report (GDR). In addition, a significant amount of as-built historical data were available from the original tunnel construction. The City recognizes the need to include a Geotechnical Baseline Report (GBR) that will interpret the GDR and as-built conditions, but more importantly spells out what geotechnical conditions the contractor should anticipate. The GBR is a risk allocation tool and states "The baselines established in the GBR for the anticipated ground conditions are recommended to be used as the basis for developing bids. The baselines will be used for evaluating potential differing site conditions during construction." The City will accept financial responsibility for certain geologic conditions that differ from those stated in the Baseline report.

The GBR "provides baselines for rock characteristics and anticipated ground conditions for the Bypass Tunnel, drill and blast tunnel construction considerations, TBM tunnel construction considerations, geotechnical factors affecting the Bypass Tunnel final lining type and installation, and geotechnical monitoring and instrumentation requirements for the Bypass Tunnel construction," among others.

The tunnel was broken down into four ground classes. Estimates for the occurrence of each ground class to be expected were given in estimated linear feet. Estimates of the groundwater inflows are given for the various reaches of tunnel. The groundwater inflows for drill and blast (both the heading and along the tunnel) and TBM (heading only, the segments address the tunnel) are presented.

The Standard Construction Contract contains provisions for differing site conditions and dispute resolution.

TBM SPECIFICATIONS

Tunnel excavation methods are normally the contractor's means and methods. The owner/designer may have some overriding concerns regarding management of project risks and ensuring that a sufficient basis of bid exists for tunnel excavation. These concerns will manifest themselves in the TBM specifications.

The DEP's concern for groundwater heads over 600 feet coupled with the historical records of over 2,000 gpm of cumulative groundwater inflows during construction of the existing tunnel drove much of the TBM requirements to manage the risks. Some TBM requirements of note are described here. The TBM is to be a shielded machine, capable of erecting a bolted and gasketed precast segment, that excavates a full-face tunnel utilizing a rotating cutterhead equipped with disc cutters or other excavation tools. The TBM carries equipment to perform probe drilling and grouting ahead of the tunnel face. The TBM is not required to mine in closed mode (pressurized condition), but must contain a bulkhead that can be closed and withstand 20 bar (290 psi) of water pressure with a 1.5 safety factor when the machine is not mining. The machine must be able to mine with up to 800 gpm of inflow through the cutterhead. The TBM must have the pumping capacity to pump groundwater flush flows of 2,500 gpm indicated in the GBR. The machine must have tail shield brushes with grease, shield strain gauges to measure ground loads, and a programmable logic controller (PLC) system that can report out on over 30 sensors to monitor the performance of the TBM. The TBM is required to have 15 grout ports through the face to enable sufficient coverage for performing pre-excavation grouting ahead of the tunnel face. Cores of the rock must be obtained by drilling through the face of the TBM, two cores every 500 feet. A turning radius of 1,000 feet is required. The propulsion system must have a minimum of 50,000 pounds per cutter. Cutterhead motors must be such that the TBM is not torque limited at the design thrust. The cutters must be back-loaded. Several requirements are also included for the main bearing.

Safety elements that must be powered by emergency generators are defined as ventilation fans, lighting, pumps, compressors if used, and communication systems. There are performance criteria for the guidance system. The TBM was to be built to address "potentially gassy" ground.

The contract required numerous submittals for the TBM. While the DEP has had experience with main beam TBMs, it does not have experience using a single shielded TBM with a segmented liner. The TBM submittals allowed the owner to participate in discussions with the contractor concerning the TBM features, capabilities, and its safe operation. It is the DEP's practice that the TBM is the contractor's equipment, and will therefore not pre-procure the TBM.

PROBING AND PRE-EXCAVATION GROUT REQUIREMENTS

The groundwater in this tunnel is thought to be the largest risk facing the excavation of the new tunnel. The owner/designer provided prescriptive pre-excavation grouting requirements in the contract. This is based upon the historical experience with driving the existing tunnel, the results of the geotechnical program, and the designer's experience with groundwater on other projects. The grouting equipment and its performance with pipe and packer dimensions, types of cement, pumping procedures, number and angle of the pre-excavation grout holes through the TBM face, drilling equipment, minimum hole lengths, typical grouting patterns, verification holes, and grout triggers were all well developed in the contract. The drills were required to be capable of drilling 200 feet, and had to have monitoring systems with feed, rotation, and percussion pressure monitoring to interpret relative ground quality ahead of the face.

STANDARDS FOR CONTRACTOR-DESIGNED UNDERGROUND ELEMENTS

Several elements of the contract required contractor design. The bell out for assembling the TBM is one example. The DEP recognized that every contractor will have a different concept for the TBM assembly area at the base of the shaft. DEP set up the pay items in the contract to allow for different bell out configurations. The design of the ground support for the altered bell out must be approved. It should be noted that the design of underground structures is a bit of an art, and designers have different risk tolerances. The specifics of the assumptions and parameters to be included must be explicit.

Components of the segment design were left up to the contractor to ensure the segments would be compatible with contractor's segment placer and the operation of the TBM. The design delineates which components of the design the designer was responsible for and which components the contractor was required to finalize. The contract documents included minimum performance criteria for the components for which the contractor was responsible. While the delineation between the owner design of the segments and the contractor design of the handling elements was clear, it was anticipated that there may be technical discussions on the submittals from the contractor before a mutually agreed upon design was reached.

ROLLING ALLOWANCES

Best contracting practices accurately define the scope of work to be conducted and then include that work in lump sum or unit price items to be bid on in the contract. There are some cases where the scope of work cannot be determined and an allowance is used. The use of allowances is discouraged as the work is not subject to competitive bidding and project costs are increased. For this contract, there were four cases where an emergency response might be required, and each had only a generally defined scope. Each of these scenarios entailed a substantial risk that would need to be addressed immediately in order to keep the connection phase of the contract moving. Close examination of the four scenarios revealed that they could not occur at the same time, and while not entirely mutually exclusive, it is very unlikely multiple cases would occur in the contract. This allowed four possible allowance scenarios to be assigned to one allowance, substantially reducing the amount that needed to be included in the contract.

MAJOR EQUIPMENT/MATERIAL MANUFACTURERS REQUIREMENT

A requirement for bidding on the contract states that certain major equipment and material manufacturers be identified with the bid. While none of these elements were sole sourced, certain eligible manufacturers are listed in the contract. The list requiring the equipment/manufacturers includes the TBM manufacturer (if applicable), interliner manufacturer, shaft cap manufacturers, pump fabricators, and valve manufacturers, among others. The reason for requiring this disclosure is to identify up front which of the suggested critical items would be supplied, and to head off contract disputes later on.

QUALITY ASSURANCE INSPECTION AND SHOP WITNESS TESTING

The DEP has a comprehensive Quality Assurance Inspection program whereby the manufacture of critical pieces of equipment and material production are witnessed at the place of manufacture. DEP has performed the inspection of equipment and material production for decades with both in-house staff and consultant inspection. This practice was evaluated in 2010 through an internal study that reached out to a dozen major water and wastewater utilities nationwide to look at their QA processes. The rigorous practice of QA inspection was confirmed by this study. The Bypass Tunnel contract includes a list of those items subject to QA inspections. The list includes the manufacture of the interliner, shaft caps, transition pieces, valves, and pumps. The cost of the inspection is born by the DEP, independent of the Bypass Tunnel contract.

Shop witness testing is also conducted. Shop witness testing typically involves the testing of equipment at the place of manufacture witnessed by the design team. A list of equipment requiring this inspection is included in the contract. The scope of the testing is defined, and the costs for the shop witness testing are included in the contract.

PARTNERING

The concept of partnering with the contractor has recently been integrated into DEP contracts. The Bypass Tunnel contract states: "Partnering is considered important to the overall success of this project by DEP, its design engineer, and its construction manager. It is also important to this project that the Contractor be equally concerned with quality, performance, budget, and schedule and that it will endorse and adopt Partnering as an effective tool for achieving these objectives. Partnering will be effective only if all parties willingly and enthusiastically enter into this cooperative arrangement which is supported by each entity at the highest level in their organizations."

DEP has found that partnering meetings are beneficial despite the criticism that one cannot legislate cooperation. Another valuable practice ensuring continued cooperation is the morning meeting between the Construction Manager/Resident Engineer and the contractor's Project Manager to discuss each day's activities.

RISK

A risk identification process was utilized during the design. A risk register was maintained and revisited monthly. Risks were given due dates so that they would be addressed when appropriate. At the end of the design, the remaining construction risks were documented, and handed off to the construction team to continue the process once the contractor was brought on board.

ESCROW BID DOCUMENTS

The contract requires that the contractor's bid documents be placed in escrow. The contract states that the escrow bid documents (EBDs) are and shall always remain the property of the contractor subject only to joint review by the City of New York and the contractor, except as provided herein. The EBDs shall clearly itemize, organize, and identify the estimated costs of performing the work of each item contained in the Schedule Bid of Prices. Items should be separated into sub-items as required to present a complete and detailed cost estimate and allow a detailed cost review. The EBDs may be examined by both the City and the contractor at a time mutually agreeable to both the City and the Contractor to assist in the negotiation or settlement of disputes.

CONCLUSION

The use of traditional design-bid-build documents can and will evolve to address a variety of challenges faced by owners on all kinds of projects. The Bypass Tunnel Project faced a variety of challenges; the length of the contract, multiple excavation methods, bonding challenges, multiple dates for the critical connection, and uncertain ground conditions among them. In this case flexibility and a unique structure within the design-bid-build contract allowed for successful contractor procurement.

PART

Design and Planning

Chairs

David Smith
WSP

Nick Karlin
Skanska

Crosstown Tunneling for Houston Surface Water Supply

Sergio Flores • Black and Veatch
Todd Wanless • Black and Veatch
Mike McCure • Black and Veatch
Clint Wilson • Black and Veatch

ABSTRACT

As a means to comply with a mandated 60 percent reduction in ground water extraction in Houston, the West Harris County Regional Water Authority is constructing the $1.2 billion Surface Water Supply Project. The 42-mile project traverses urban Houston and includes numerous tunneled sections for the 96-inch diameter steel pipeline.

This paper is directed at the challenges involved in designing 17 soft ground tunnels that cross under major highways, city streets, bayous and extensive utilities. The paper will focus on a locally unprecedented 4.3-mile long tunnel to be constructed through a narrow pipeline corridor in clays and sands under the ground water table.

INTRODUCTION

With a current population of 6.5 million and an expected population of 9.2 million by 2030, the Houston area is projected to rank as one of the five fastest growing cities in the nation. Most of the Metro Area is located in Harris County and due to the projected growth demands, there is a need to expand public infrastructure. Most of the drinking water supplies in the Houston area come from ground water sources. Because of the large quantities of water pumped, the Houston area has experienced ground subsidence. Ground subsidence has been significant in some areas and has been a contributor to flooding. As a result of the ground subsidence experienced in the Houston area, the Harris Galveston Subsidence District (HGSD) and Fort Bend Subsidence District (FBSD) have mandated a 60 percent reduction in ground water use by 2025 within their districts. In addition, HGSD has required an 80 percent reduction in ground water use by 2035. Most of the City of Houston has converted to surface water supplies and the surrounding areas must convert to the required surface water supply in order to meet the regulatory mandates. The West Harris County Regional Water Authority and the North Fort Bend Water Authority are required to comply with both regulations. Both regional water authorities represent approximately 180 active Municipal Utility Districts and three cities. In order to meet future water demand and requirements for reduced ground water use, an additional surface water source is required. The Surface Water Supply Project will provide this additional source to regional water authorities by conveying additional potable water from the Northeast Water Purification Plant, which draws raw water from Lake Houston. The Surface Water Supply Project consists of approximately 42 miles of 96-inch, 84-inch and 66-inch pipe and two (2) large pump stations.

PROJECT OVERVIEW

The Surface Water Supply Project consists of approximately 42 miles of 96-inch, 84-inch and 66-inch pipe and two large booster pump stations approximately 170 million gallons per day (MGD) each. The project is divided into segments: Segment A is divided into 2 segments, Segment B into 3 segments, Segment C into 2 segments and KM into 6 segments. All segments have tunnel crossings; Segment B has the most tunnel crossings because it is located in the central portion of the alignment in a densely urban area.

ALIGNMENT DESCRIPTION

Segment B starts at Hopper Road, at the interface with Segment A. The alignment parallels two different petroleum pipeline corridors for the first 4.3-miles. This portion of the Segment B alignment was evaluated, and it was determined that a continuous tunnel was the most favorable alternative to install the pipeline. The 4.3-mile tunnel

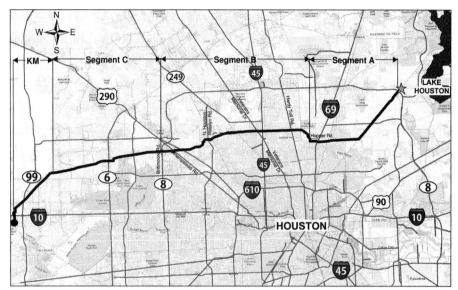

Figure 1. Surface water supply project location

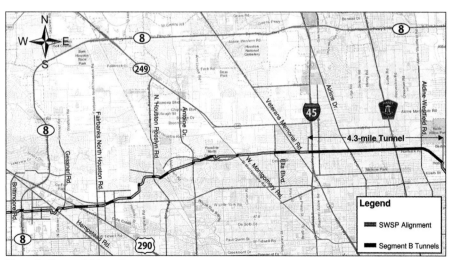

Figure 2. Segment B tunnels

Table 1. Tunnel crossings west of IH-45

Tunnel	Size (in.)	Length (ft.)	Acceptable Tunneling Method/ Machine	Tunnel Ground Support
Brittmoore Rd.	120	266	Digger Shield	Steel Ribs & Timber Lagging
Sam Houston Tollway	120	580	Dual Mode/Earth Pressure Balance	Steel Casing/Pipe Jacking
Fisher Rd.	120	350	Dual Mode/Earth Pressure Balance	Steel Casing/Pipe Jacking
Clara Rd.	120	108	Digger Shield	Liner Plate
Hempstead Highway	120	305	Dual Mode/Earth Pressure Balance	Steel Casing/Pipe Jacking
Gessner Rd.	120	265	Digger Shield	Liner Plate
Highway 290	132	1,850	Dual Mode/Earth Pressure Balance	Liner Plate under Highway, Steel Ribs and Timber Lagging
Fairbanks N. Houston Rd.	132	141	Digger Shield	Steel Ribs & Timber Lagging
W. Little York Rd.	132	170	Digger Shield	Steel Ribs & Timber Lagging
N. Houston Rosslyn	132	200	Digger Shield	Steel Ribs & Timber Lagging
White Oak Bayou	144	1,266	Dual Mode/Earth Pressure Balance	Liner Plate (BNSF), Steel Ribs & Timber Lagging
Antoine Dr.	132	143	Digger Shield	Steel Ribs & Timber Lagging
W. Gulf Bank/Vogel Creek	144	2,250	Dual Mode/Earth Pressure Balance	Steel Ribs & Timber Lagging
W. Montgomery Rd.	132	154	Digger Shield	Steel Ribs & Timber Lagging
Ella Blvd.	144	1,650	Dual Mode/Earth Pressure Balance	Steel Ribs & Timber Lagging
Veterans Memorial Dr.	132	197	Digger Shield	Steel Ribs & Timber Lagging

will start from a launch shaft at Hopper Road and will end at a reception shaft located 400 feet west of IH-45. The 4.3-mile long tunnel is the eastern section of Segment B and will be a separate construction contract. Several intermediate shafts are proposed at strategic locations to facilitate tunneling/pipe installation operations, provide corrosion protection testing stations, and house operation and maintenance structures. The average ground cover of the 4.3-mile tunnel is 33 feet.

The western section of Segment B starts west of IH-45 and continues until the interface with Segment C west of Brittmoore Road. The majority of this section of pipeline is planned to be constructed mainly by the open-cut method with 16 tunnel crossings of major highways, railroads, bayous, and flood plains. This section consists of approximately 8.6 miles of open-cut and 1.9 miles of tunnel with lengths ranging from 108 to 2,250 feet. The ground cover of the tunnel crossings ranges from 15 to 30 feet. The details of the crossings are shown in Table 1.

SUBSURFACE CONDITIONS

Houston lies in the northwestern Gulf of Mexico segment of the North American Coastal Plain. This coastal plain consists of approximately 20,000 feet of sediments which have been eroded, transported and deposited from the interior of the continent. The Beaumont and Lissie Formations are two major geologic formations in the Houston Area with most of the Segment B alignment being located within the Lissie formation.

In general, the Lissie Formation is composed of interbedded clays, silts and clayey and/or silty sands, deposited in a higher energy fluvial depositional environment than the Beaumont Clay. The sand grains of the Lissie are typically coarser than those of the Beaumont. Sand deposits are typically clayey or silty, but may also include clean, poorly graded sand, gravel, and even random cobbles and/or boulders. Interbedded with the sands are clays and sandy and clayey silts deposited during quieter periods.

Buried stream channels may occur sporadically within finer grained overbank deposits throughout the Lissie Formation.

The geotechnical investigations indicate that some clays are slickensided and this could impact the stability of the open excavations by digger shields. Exploration indicates the presence of lenses of cemented sands, this cementation is believed to be caused by the geological process of the cyclical lowering of ground water levels during ice age periods.

The ground water is generally shallow in the Houston Area. Ground water encountered at exploratory borings drilled to date ranges between 4 and 29 feet below ground surface and averages 15 feet. All tunnel crossings will be under the water table and this is typical for the Houston Area.

ALTERNATIVE ALIGNMENT STUDIES

A study to evaluate the different alternatives was conducted for the 4.3-mile section east of IH-45. The section east of IH-45 is characterized by having congested areas with petroleum pipelines and restricted construction areas where open cut construction for large diameter pipelines was found to be extremely difficult or not feasible. A total of four alternative alignments were developed for this study. Selection of these alignments was based on previous high-level studies and discussions with the Program Manager to confirm that each alternative appeared feasible for construction and that each achieved the overall Surface Water Supply Project objectives. Upon initial development of the alignment alternatives, each alternative was assessed and further advanced to clearly define alignment and grade; construction methods and sequence; staging areas; tunnel shaft locations and geometries; operations and maintenance structures; and corrosion protection systems. Furthermore, ground conditions were evaluated to confirm compatibility with the assumed construction methods. Tunneling construction methods allows for continuous working operations compared to open-cut construction where working hours are typically restricted in urban environments. Other typical open-cut construction restrictions include weather, adjacent structures and utilities. Figure 3 shows the preliminary subsurface conditions based on borings available at the time of the study. The ground conditions match the expected historical tunneling conditions in the Houston Area.

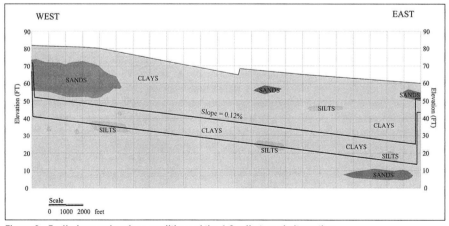

Figure 3. Preliminary subsurface conditions of the 4.3-mile tunnel alternative

All system elements and assumed construction methods for each alternative were inventoried and documented to facilitate development of construction cost estimates and schedules. Cost estimates were completed to evaluate direct costs involving materials, labor, equipment and allowances. Indirect costs include overhead and a markup for profit/contractor contingencies. Beyond construction cost and duration, other factors were incorporated into the decision-making process to facilitate selection of a preferred alternative that provides the greatest value and meets the overall project objectives. Nine criteria were considered in comparing the alternative alignments. In addition, each criterion was assigned a weighting factor (0 to 100). Based on the results of this study, the 4.3-mile tunnel represents the most favorable alternative. Since tunneling was determined to be the preferred construction method and alternative, the program performed additional borings with a spacing ranging from 100 to 500 feet.

DESIGN CONSIDERATIONS

Tunnel Design Criteria

It is imperative that the selected tunnel construction methods "match" the anticipated ground behavior to facilitate efficient tunnel production rates, limit loss of ground and associated surface settlement, and mitigate potential damages. An important process for finding the best tunneling method is considering and evaluating local experience from both the contractor and engineering side. A lack of understanding of the local experience could end in many change orders, claims and even catastrophic events. Another key component to a successful tunnel design is to apply a simple approach to constructability and tunneling equipment. In some cases, tunnel designers are inclined to specify a sophisticated machine (e.g., Earth Pressure Balance or Slurry Machine) as means to mitigate any potential risks or to give a general type of solution. It is important to note that these machines are designed for specific soil conditions and ground water pressures. Sometimes the challenge is to not specify the most sophisticated machine that would appear to mitigate all risks, but to specify the machine that best matches the soil conditions and local experience, even if it is a more simplistic machine.

Assessment of the anticipated ground conditions and associated ground behavior is important for evaluating the face stability of the tunnel and for determining appropriate construction methods and equipment. Exploratory boring data indicates that the soils anticipated to be encountered in the tunnel excavations are predominantly stiff to hard clays. Silty to clayey sands and cemented sands are also expected to be encountered at some locations.

Initial ground support will be installed directly behind the shield or tunnel boring machine to maintain ground stability until the carrier pipe can be installed and backfilled. The type of initial support system is dependent upon the soil type, tunneling method and ground water conditions. In some cases, the ground support may need to be designed as a permanent system to accommodate regulations imposed by some facility owners (e.g., railroads, highway departments, etc.). The permanent ground support system will consist of steel liner plates which will also be employed for crossings of sensitive infrastructure and/or where ground/ground water conditions are considered less favorable. Routine grouting of any voids between the excavated tunnel walls and liner plate will limit settlement realized at the surface. Three crossings will encounter thick water bearing sand layers and will require steel casing. The casing will be pushed behind the machine using pipe jacking methods. Steel ribs and wood lagging initial tunnel support will be employed elsewhere.

Upon completion of the tunnels, the welded steel pipe can be installed. Ground loads, along with installation loads and hydraulic operation loads control the design of the carrier pipe. For tunnels less than 500 feet, it is anticipated that the pipe string can be pushed into the tunnels on skids. Longer tunnels (more than 500 feet.) will require that individual pipe segments be carried into the tunnel and welded in-place to avoid damage to the pipe. A pipe carrier will be required to transport and position each pipe segment. Upon completion of the pipe installation, the pipe/tunnel annular space will be filled with cellular grout that is placed in stages to avoid pipe floatation as a result of fluid buoyancy pressures. For the longer tunnels, the annular space will be grouted with cellular grout through grout ports in the steel pipe.

Required geometry and size of the shafts will be a function of the pipeline depth, tunnel size, tunneling equipment and methods, shaft construction methods and ground conditions. Shaft sizes are anticipated to be at least 30 feet in diameter to allow a 25ft. section of pipe to be lowered and installed or pushed in the tunnel. Excavations for the shafts will be vertical (or near vertical) and will require shoring and excavation support systems to minimize the area impacted by construction, limit ground movements to acceptable magnitudes and provide for a safe working environment. Considerations will be taken for shafts in order to accommodate for the jacking forces where the contractor chooses to push the pipe. Similar considerations need to be evaluated for the three crossings that will be excavated by jacking the steel casing. Excavation support systems for shallow to moderately deep shafts may consist of trench boxes, liner plate, soldier piles/lagging and/or sheet pile systems. Deeper excavations that encounter permeable deposits may need to consider relatively water tight systems such as gasketed liner plates.

CASE STUDIES

Discussions were held with experienced local tunnel contractors so that they could provide their input on tunneling experiences. Tunneling under bayous can be challenging considering the tunnel depth, presence of ground water and potential unfavorable soil conditions. To further understand this, several papers and case studies were reviewed to document the lessons learned. Some of the reviewed papers share the same concept of the efficiency in the synergy between local contractors and design engineers. The most relevant projects found were:

Buffalo Bayou Siphon, City of Houston

This project was part of a larger sewage conveyance project. The tunnel under Buffalo Bayou was required to install the carrier pipe and the initial design involved jacking steel casing under the bayou with a closed face wheel-cutter. After nearly 150 feet of tunneling, the steel casing was not able to be pushed forward with the jacking units. The City of Houston and Black & Veatch approved the use of rib and wood lagging temporary ground support in lieu of steel casing. The reasons why the steel casing seized are not fully documented; however, it is believed that the friction between the soil and the casing was high and not mitigated properly using bentonite lubrication. The tunnel project was completed successfully by transitioning to steel ribs and lagging which the contractor was familiar with based on past experience.

72-Inch Water Main for the East Water Purification Plant, City of Houston

The 72-inch main crossed Hunting Bayou by means of tunneling with a Tunnel Boring Machine (TBM). During construction, it was determined that the ground conditions where different in some locations and the contractor decided to perform additional borings. The presence of marine clay posed a big challenge for tunnel operations.

While the contractor battled the marine clay, the tunnel support (steel ribs and wood lagging) collapsed due to the loads of the marine clay; the filter fabric behind the ribs and lagging provided some resistance to the cave-in allowing the construction personnel to exit safely. After dewatering and cleaning the material, the contractor was able to resume tunneling operations by advancing the machine without rotating the cutterhead through the difficult areas and reducing the rib spacing.

Northwest Lateral Project, Coastal Water Authority (CWA)

This project required the crossing of 2 major water bodies in soft ground: the Houston Ship Channel and Greens Bayou. This project was carried out successfully without any problems. The lessons learned are the importance of a thorough geotechnical investigation program, the importance of a design that mitigates risk, the engineer's specification for specific tunneling methods without dictating means and methods, and monitoring tunneling performance during construction. Because sufficient geotechnical information was available, the depth of the tunnel was selected to be in an impermeable clay layer. This depth was deeper than initially expected but reduced risk considerably.

Northside Sewer Relief Program (NSRP), 1989, City of Houston

The NSRP consisted of approximately 14 miles of gravity waste water mains ranging from 5.5 to 13 feet in finished diameter. The program was designed to relieve the surcharged condition of the old sewer system which exists in areas north of downtown Houston. The tunnel depths in this program are of the order of 65 feet deep. The water table is typically at a shallow depth of 6 to 16 feet below the ground surface. The tunneling techniques used in the construction of the NSRP were TBM tunneling with steel ribs and wood lagging, jacked pipe with a shield and jacked pipe with a TBM. A potential issue with the steel ribs and wood lagging was losing the fine-grained soils through the lagging openings. The problem was solved by using a geotextile filterfabric behind the ribs and lagging. The program incorporated a settlement monitoring design. One example of settlement monitoring was documented under a highway; the settlement during and approximately five months after tunneling beneath the highway ranged from 0.24 to 1.34 inches with an average of 0.67 inches. Ground water and water bearing cohesionless soils were the greatest challenges for tunneling. After a learning curve, local tunneling contractors were successful with the use of TBMs with capabilities to handle challenging ground conditions if encountered.

TUNNEL CONSTRUCTION METHODS

For relatively short and simple crossings of city streets and/or underground utilities, a digger shield or wheel cutter TBMs may be considered if ground conditions are favorable. Both shields and wheel cutters can be considered below the ground water table for heads of less than 20 feet in clays, cohesive silts and clayey/silty sands where favorable conditions are present through extended reaches. Clean sands and silts would require lowering of the ground water table below the tunnel invert to allow use of these relatively simple tunnel excavation systems.

Considering that the ground conditions are relatively favorable for most of the tunnel crossings, there is a type of TBM that may present the best option in terms of being matched to the ground conditions and providing high production rates that exceed Earth Pressure Balance (EPB) capabilities. Lovat, Inc. developed an innovative but relatively simple soft ground machine technology prior to extensive development of fully pressurized face machines for the Greater Houston Waste Water Program in the 1980s. Positive face control under the ground water table can be maintained with

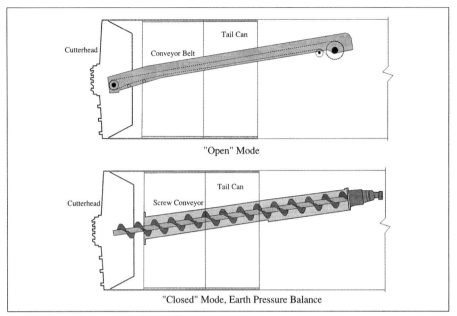

Figure 4. Dual mode tunnel boring machine

these machines by employing hydraulically operated flood doors within the cutterhead in combination with pressure relieving gates inside the cutterhead. Ground conditioning agents can also be added to treat soils and provide improved face control. As opposed to using a screw conveyor typical of EPB machines, the dual mode type machine employs a belt conveyor in open or semi-open mode to provide for more efficient tunnel production rates. As an added benefit, these machines can be converted to a fully closed pressurized mode by replacing the belt conveyor with a screw conveyor as shown in Figure 4. These machines were employed in Houston on numerous successful tunnel projects with similar ground conditions expected on this project. Allowing the use of these machines will result in more efficient tunneling and associated cost savings compared to an EPB machine.

RISK MANAGEMENT

Contractor Pre-Qualification

A contractor pre-qualification procedure will be considered to determine eligible contractors. The pre-qualifying procedure includes review and evaluation of information regarding relevant experience of prospective contractors that are interested in bidding and constructing the project. A Contractor Evaluation Team will gather information from references that have been provided by the contractor. Proven experience and actual performance on previous construction projects will be reviewed and may include many of the following:

- Previous successful projects using similar tunneling methods and ground conditions;
- Management of construction activities, subcontractors and supplies;
- Key personnel experience and availability;

- Scheduling;
- Public relations with neighbors;
- Safety record;
- Financial stability and ability to bond; and
- Installation, startup and training services for equipment.

Geotechnical Baseline Report

A Geotechnical Baseline Report (GBR) will be developed for the 4.3-mile tunnel. Another GBR will be developed for all the tunnel crossings within the project section west of IH-45. The report will focus on each of the individual tunnel crossings no matter the length.

CONCLUSION

Tunneling is a feasible and cost-effective alternative where the urban constraints for large diameter waterlines become an issue. The 4.3-mile tunnel is an example that highlights how open-cut construction methods can be impractical in congested urban environments for large diameter pipelines. Houston saw many tunneling projects in the 1980s and it will see more in the future with the region's surface water programs. Engineers will consider tunnel construction methods that best match the anticipated ground conditions and past tunnel design experiences in the Houston Area. The synergy between local tunnel contractors, design engineers and owners will contribute to the overall tunneling construction success of the projects.

REFERENCES

Ivor-Smith, D.J., S. Nandagiri, C.B. Petterson and G. Roy. 1989. Deep Soft Ground Tunneling Under Houston Ship Channel and Greens Bayou. Rapid Excavation and Tunneling Conference Proceedings, Los Angeles, CA. Littleton, CO: Society for Mining, Metallurgy & Exploration Inc., pp. 229–247.

Merritt, B.K., Crisci, A. and Klein, G.H., 1991. Houston Pipe Jacking—Large and Small. Rapid Excavation and Tunneling Conference Proceedings, Seattle, WA. Littleton, CO: Society for Mining, Metallurgy & Exploration Inc., pp. 391–407.

Merritt, B.K. and Valentine, R.M., 1988, Underground civil works in Texas. Proceedings of the symposium Tunneling '88, The Institution of Mining and Metallurgy, London, England, pp. 183–193.

Ortega, R., Henry G.J., Hosepain, H. 2003. Houston's Hidden Tunnels; Trials and Tribulations of Large Diameter Tunnels. 2003 Pipelines 2003 Proceedings, Baltimore, MD. Reston, VA: American Society of Civil Engineers, pp. 1788–1797.

Design and Construction Procurement for the Amtrak Hudson Tunnel Project

Phil Rice ▪ WSP USA, Inc.
Richard Flanagan ▪ WSP USA, Inc.
Mohammed Nasim ▪ Amtrak

ABSTRACT

Amtrak's existing North River Tunnels below the Hudson River, built in 1910 by the Pennsylvania Railroad, were flooded and damaged during Superstorm Sandy and need to be rehabilitated. Because this work cannot be done without seriously impacting regional rail service, two new tunnels are proposed along a new alignment connecting the Northeast Corridor from Secaucus, New Jersey to existing Penn Station in Manhattan. Experience gained from the cancelled similar project, Access to the Region's Core (ARC), was used to expedite the Preliminary Engineering Design for the Hudson Tunnel Project. This paper provides descriptions of the Hudson Tunnel Project key elements, geotechnical challenges, expected construction methods, and contracting approaches.

PROJECT OVERVIEW

The Gateway Trans-Hudson Partnership (GTHP), a joint venture of WSP, AECOM and STV, is providing Preliminary Engineering services to Amtrak for the Hudson Tunnel Project as well as supporting the NEPA process for getting the Record of Decision (ROD). This Preliminary Engineering design effort by GTHP includes geotechnical exploration and laboratory testing, design of structures, tunnel contract packaging and procurement planning.

The Hudson Tunnel Project includes the construction of a new Hudson River Tunnel system and rehabilitation of the existing North River Tunnel. The new tunnel system will be constructed south of the existing North River Tunnel, from North Bergen under the Palisades to Hoboken in New Jersey, below the Hudson River, and connecting directly to Penn Station New York; refer to Figure 1. The tunnel system is configured as two single-track tubes, intermediate ventilation shafts and cross passages in compliance with NFPA-130 requirements. The adjacent surface alignment west of the Palisades, will be constructed on viaducts, embankment and retained fills.

The existing North River Tunnel (Figure 2) consists of two single-track, electrified rail tunnels extending from tunnel portals in North Bergen, New Jersey to Penn Station in New York. Completed in 1910, the tunnel opened at the same time as Penn Station and the four, single-track, East River tunnels that connect Penn Station to Queens and onto Long Island and New England, respectively. Both tunnels in the existing North River Tunnel were inundated with brackish river water during Superstorm Sandy (Figure 3), resulting in the cancellation of all Amtrak and NJ TRANSIT service into New York City for five days. After the brackish water was removed from the tunnels, the residual chlorides and sulfates have had on-going detrimental impacts on the interior concrete and rail systems.

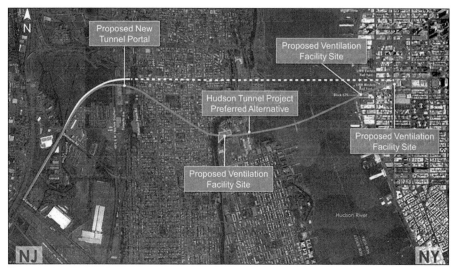

Figure 1. New Hudson River Tunnel alignment

Figure 2 and 3. Existing North River Tunnel and flooding of the tunnels during Superstorm Sandy

The poor condition of the tunnels following Superstorm Sandy as well as its continuous high level of use, has highlighted the lack of a trans-Hudson River operational flexibility necessary to conduct routine maintenance and state of good repair to ensure continued NEC operations into Penn Station at existing peak period operating levels. Any emergency or routine maintenance in the tunnels require outages that disrupt service for hundreds of thousands of passengers each day. An extended disruption, caused by natural or man-made disaster, would result in local, regional and national economic impacts. The susceptibility of the region to natural disasters and condition of this critical asset drives the urgency to strengthen resiliency of this river crossing.

The Hudson Tunnel Project is focused on tunnel resiliency and redundancy essential to reducing the risk associated with dependency on the century-old tunnel. Once the new Hudson River Tunnel is commissioned, the existing North River Tunnel will be rehabilitated and modernized. This will include renewal of the tunnel facilities, including benchwalls, catenary, communications and signals; fire and life safety; drainage facilities, as well as maintenance and emergency egress features. The existing ballasted track and drainage systems will be removed and replaced with a direct fixation track system.

The cancelled ARC Project included surface alignments and tunnels through New Jersey, under the Hudson River and into Manhattan to a new station cavern connecting to Penn Station. Most of the New Jersey alignment has been utilized by the Hudson Tunnel Project. This includes the surface works and Palisades Tunnels that have identical alignments as the ARC Project. However, the new tunnel project Hudson River and Manhattan section alignments now differ from ARC along with a direct Penn Station connection.

The ARC Project, when cancelled in 2010, had several construction packages where contracts had been already awarded or had been advertised and/or ready for award. Utilizing the knowledge and information that was collected during the ARC Project, GTHP has utilized previous information and design and adjusted and revised contract packages previously awarded and incorporating whatever changes have been required for the new tunnel alignment.

PROJECT LOCATION AND ANTICIPATED SUBSURFACE CONDITIONS

Site Development and Manmade Obstructions

Much of the New Jersey project area is set within the largely undeveloped, low lying Hackensack Meadowlands. The area on the east side of Frank R. Lautenberg Station is characterized by major transportation infrastructure, including the New Jersey Turnpike and the Northeast Corridor (NEC), and transmission towers. Industrial facilities such as the PSE&G Generating Station and warehouses and distribution centers are located north and south of the NEC. East of the Tonnelle Avenue arterial, the municipalities of North Bergen, Union City, and Hoboken are characterized by a mix of low- and high-rise residential developments, commercial corridors, and varying degrees of industrial activities.

The Hoboken and Weehawken area has increased residential development since the ARC Project was cancelled, in particular adjacent to the Hoboken Shaft area. In Hoboken, the tunnel alignment crosses two at-grade tracks of the Hudson Bergen Light Rail (HBLR) for a distance of about 400 feet. The electrified tracks were constructed in 2004. A new PSE&G substation was constructed in 2016, south of the HBLR at 19th Street adjacent to the tunnel alignment.

In Manhattan, the project alignment crosses below a heliport, Hudson River Park, 12th Avenue, W 30th Street, the elevated "Highline" urban park and new major development above the extensive rail yards west of Penn Station. The existing Manhattan Bulkhead (Figure 4) was built in the 1870s and consists of timber piles, concrete blocks, and stone rip-rap structure approximately 150 feet in width. The proposed tunnel alignment requires a tunnel excavation to mine through the lower portion of the bulkhead where timber piles and rip-rap structure will be encountered. Additionally, there are three other significant underground obstacles that will need to be addressed east of the Manhattan Bulkhead: (1) abandoned concrete filled pipe piles of the former West Side Highway, (2) a NYCDEP interceptor sewer supported by steel piles underneath 12th Avenue and (3) a NYCDEP combined sewer anticipated to be supported on timber piles underneath W 30th Street.

Anticipated Ground Conditions

The Hudson Tunnel Project is located within the Appalachian Highlands U.S. physiographic division. The project crosses two distinct physiographic provinces.

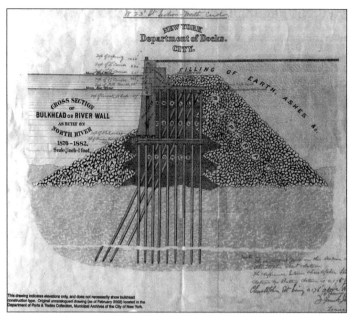
Figure 4. Typical Manhattan bulkhead construction

The New Jersey portion of the project region is located within the Piedmont Lowlands section of the Piedmont physiographic province, a broad lowland interrupted by long, northeast-trending ridges and uplands. The most prominent physiographic feature in the eastern part of the section is the Palisades, a north-south topographic ridge near the Hudson River that rises above the surrounding lowlands of the Meadowlands.

The New York portion of the project region is located within the Manhattan Prong of the New England Upland section of the New England physiographic province. Topography is largely controlled by bedrock geology. Manhattan's elongate ridges trend generally northeast.

The Hudson River portion of the project region is located between the Piedmont physiographic province to the west and the New England physiographic province to the east.

New Jersey Surface Alignment Ground Conditions

Five soil strata are identified in the New Jersey Surface Alignment section: Stratum F (Fill), Stratum O (Peat and Organic Soils), Stratum S (Sand/Silt & Sand), Stratum C (Varved Silt/Clay), and Stratum G (Glacial Till). The Stratum C can be furthered divided into two substrata, Stratum C1, an over-consolidated stiff layer, and Stratum C2, a normally consolidated or slightly over-consolidated soft layer.

From Tonnelle Avenue to the east toe of the Palisades, two soil strata are identified, Stratum F (Fill) and Stratum G (Glacial Till). From the east toe of the Palisades to the Hoboken Shaft, all five soil strata are present.

Palisades Tunnel Ground Conditions

The Palisades Tunnels are expected to be excavated and bored mainly in the igneous Palisades Diabase. There is also a limited section of Lockton Formation and the Stockton Sandstone. The diabase is generally fresh to slightly weathered. Tunneling rock mass quality classifies from poor to good with the latter predominant.

The Lockatong is encountered at the edge of the Palisades sill and typically consists of a silty argillite, laminated mudstone, very fine-grained sandstone and siltstone, and minor silty limestone. Where the Palisades Diabase sill intruded into the Lockatong Formation, it thermally metamorphosed the rock into a brittle, black, very fine-grained hornfels. Further eastward and entering the Hoboken Shaft, the Stockton Sandstone is encountered. It consists of sandstone with conglomerate and siltstone lenses. Tunneling rock quality rock mass quality is classified as very poor to poor.

The Palisades Tunnels will be mined in full-face rock with generally low groundwater inflows although there are some faulted zones. The Palisades Tunnels have the option to be supported with a one-pass precast segment lining system or an alternate two-pass support system of temporary support and permanent cast-in-place concrete lining.

Hudson River Tunnel Ground Conditions

In most of the Hoboken area there is fill underlain by a peaty and organic soil (O). It consists of fibrous peat mixed with dark gray, organic clay. The material is very soft to soft and classified as OH, Pt, CH and ML. Varying amounts of silt and sand are also present within this deposit. Methane gas is anticipated in this stratum

In the Hoboken land section of the tunnel, estuarine deposits (designated E) is the primary soil stratum through which the Hudson River Tunnels will be excavated. The deposit consists of grey, silty clay to sandy silt with estimated thickness ranging from several feet to over 200 feet. Stratum E_P is generally classified as high plasticity clays (CH); however, high plasticity silts (MH), low plasticity clays (CL), and low plasticity silts (ML) are also present. Small shells and organics have been encountered at various depths within Stratum E. Organic content is typically less than 5%.

In some locations, Stratum S locally underlies Stratum E with thickness ranging from 2 to 8 ft. and can be classified as very loose to dense silty sand (SM) or sandy silt (ML).

Glacial deposits (G) are present and will be encountered on a limited basis. The (G) soils are a heterogeneous mix of soils generally classified as CL, ML, SP, SM, SP-SM, GP, GM, GP-GM. The glacial deposits are typically reddish brown, brown or grey sands and consist of varying amounts of clay, silt, sand, and gravel. Consistency and density ranges from stiff to hard and medium dense to very dense, respectively.

Parts of the Hudson River Tunnel alignment are in bedrock of different formations. Within the tunnel excavation, the Stockton Sandstone is present for about 25% of the entire alignment. The rock mass quality is generally rated poor to very poor. As the alignment progresses, eastward towards Manhattan shore line, the rock type changes to Manhattan Schist. However, rock will not be encountered there since the alignment is shallow.

Mixed-face tunneling will be encountered at several locations in the Hoboken area as the tunnels proceed eastward towards the river.

Manhattan Tunnel Ground Conditions

The overburden soil in the Manhattan area consists of fill materials (F) overlying saturated estuarine deposits (Strata E1 and E2), silt and clay (Strata C), and glacial deposits (Strata G). The overburden soils consist primarily of very soft to medium stiff clays and silts and very loose to medium dense silty sands. The glacial deposits are stiffer and denser and have the potential to contain cobbles and boulders. The overburden soils are highly compressible and will respond to minor changes in stress, such as construction surcharge loads, changes in pore pressure (due to dewatering or lowering of the groundwater table from inflows into open excavations), or other changes from the initial in-situ stress state.

Non-TBM mined tunnel excavations in estuarine deposits will require heavy support of excavation walls due to the low strength and high compressibility characteristics of these deposits. It is generally necessary to use some form of external ground treatment to supplement the excavation support systems.

ANTICIPATED HUDSON TUNNEL PROJECT CONTRACT PACKAGES

Currently eleven contract packages are planned for the Hudson Tunnel Project, including contracts covering tunneling, other major underground work, civil work, and, and follow-on typical finish works consisting of internal concrete, track work, rail systems, ventilation systems and fan plant structures.

NJ Surface Alignment

The proposed NJ Surface Alignment contract package extends from Allied Interlocking, located east of County Road to the Tonnelle Avenue Portal, interfacing with the Palisades Tunnel contract. The double track configuration will be constructed on embankment, retained embankment, aerial viaduct structure and a bridge structure over the New York Susquehanna and Western and Conrail tracks. This contract package also includes the construction of drainage culverts, access roadways, foundations for overhead catenary structures, security fencing, access stairways and platforms for signal and communication structures.

Tonnelle Avenue Overhead Bridge

This contract was already well under construction when the ARC project was cancelled. Continued work on this contract will require further excavation of the abutment walls that were previously completed and reestablishing the bearing area for the precast concrete deck beams.

This package also includes excavation of the material that was placed within the track area between the abutment walls. A new drainage system needs to be installed to handle storm water in the overhead bridge area, and utility relocations and the provisioning of new utilities as needed to support the Tunnel Boring Machine (TBM) operations for the adjacent Palisades Tunnel contract package. The previous fabricated precast box beams need to be tested and inspected to determine whether they can be used for the new structure.

Palisades and Hudson River Tunnels

This Design/Build tunnel contract has been configured as one large contract with two TBM tunnel drives from the NJ Tonnelle Avenue portal, through Hoboken Shaft (future fan plant), under the Hudson River and into Manhattan terminating at the 12th Avenue Shaft (future fan plant). Much study went into this contract configuration to minimize potential troublesome interfacing with the adjacent Manhattan Tunnels contract. Major components of this contract are:

- Palisades portal works
- TBM power substation
- Hard rock TBM(s) excavation for Palisades Tunnels
- Construction of the Hoboken Shaft
- Hybrid (rock, soft ground) TBMs excavation for Hudson River Tunnels
- Pre-excavation ground treatment in river section with low cover
- Nine cross-passages in rock; six cross passages in soft ground utilizing ground freezing

Preparatory work within the contract for the TBM operation includes the construction of a temporary TBM power substation, stabilization of the rock face supporting Paterson Plank Road above excavation of the approach to the tunnel to establish starting line and grade for the TBM and assembly of one or two hard rock TBM's with trailing gear and supporting logistics at the Tonnelle Avenue Portal.

The Palisades Tunnel section, primarily stable rock, runs from the western slope of the Palisades to Hoboken Shaft whereby the TBM(s) will be received. This shaft will also be utilized as a future ventilation shaft. The tunnels will be supported using either single-pass pre-cast concrete segments, or, a two-pass lining system with temporary support and a follow-on permanent support of cast-in-place concrete with a waterproof membrane. The number of hard rock TBMs and the selection of the lining alternative is left to the contractor.

The Hoboken shaft excavation will utilize slurry walls through the overburden to rock, and drill & blast methods to excavate the rock to the required shaft base. During shaft construction, strict groundwater control will be required to prevent highly compressible soils reacting to pore pressure drops that would result in large settlements to nearby utilities, roadways and residences. At the shaft TBM breakout, a combination of jet grouting and rock mass rock grouting is planned.

The Hudson tunnel segment includes all work necessary to complete two TBM tunnel drives between the Hoboken Shaft and the 12th Avenue Shaft. This tunnel drive includes full-face rock, mixed-face and soft ground conditions. Because of the variable and range of ground conditions, the TBMs would be hybrids with pressurized-face capabilities, with Earth Pressure Balance technology most applicable. The contractor could propose to utilize the two hybrid TBMs to first excavate the Palisades tunnels with perhaps some TBM cutterhead modifications done at Hoboken Shaft following the Palisades rock tunnel excavations.

The Hudson TBMs will be removed at the 12th Avenue Shaft. The last several hundred feet of tunnel will be excavating though grout backfill of previously constructed SEM tunnels extending from the river edge, under 12th Avenue and into the 12th Avenue Shaft. The SEM tunnel and shaft works will be constructed by the adjacent Manhattan Tunnel contractor.

For the Hudson Tunnels, preparatory work for the TBM operation includes temporary TBM power substation (if not shared with Palisades section), underpinning of Willow Avenue Bridge and some ground treatment to protect sensitive structures located in

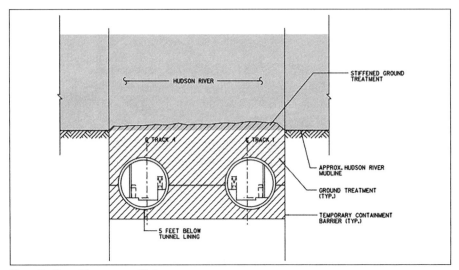

Figure 5. Ground improvement in low cover area

the mixed-face areas. Some of the latter structures are buried high voltage cables and the HBLRT light rail.

A section of the river tunnel, towards the New York side, has low ground cover in weak soils, resulting in buoyancy issues and face stability during TBM mining. Hence, prior to TBM arrival in this area, ground improvement installation (in the wet) placed within a temporary containment cofferdam is needed, see Figure 5. The current preferred ground improvement is Deep Soil Mixing (DSM).

Cross passages will be provided to house some mechanical and electrical equipment and for emergency egress between tunnels. The fifteen cross passages for this contract will be excavated and supported using two different methods depending on the existing soil or rock conditions.

Cross passages in rock are anticipated to be mined excavations by Sequential Excavation Method (SEM) with drill & blast techniques. Temporary ground support (e.g., rock dowels, shotcrete) would be used. Permanent ground support would consist of either shotcrete or cast-in-place concrete, either with a waterproof membrane,

Cross passages in the soft ground or soil areas will be pre-stabilized using ground freezing techniques. Excavation would be SEM with shotcrete/lattice girders temporary support. These cross passages would also be permanently supported with cast-in-place concrete or shotcrete with a waterproof membrane.

A separate tunnel finish contract package described herein this paper will provide the tunnel's invert concrete, high and low bench walls, including all embedded conduits required for future systems contract and a tunnel sidewall ventilation duct.

Manhattan Tunnels

This Design-Build contract entails much work centered around SEM tunnels through obstructions, two shaft excavations, major ground stabilization measures and underpinning. All these works occur in very poor ground conditions where excavations need to be strictly controlled to mitigate ground settlements. Major components of this contract are:

- Construction of the 12th Avenue Shaft using slurry walls and ground improvement
- Construction of a temporary construction access shaft west of 12th Avenue
- Ground stabilization and construction of SEM tunnels (westward) between the 12th Avenue Shaft and the Manhattan Bulkhead area including filling the tunnels with low strength grout backfill
- Ground stabilization and construction of SEM tunnels (eastward) under West 30th St running between the 12th Avenue Shaft and the presumed existing West Rail Yard Tunnels (Hudson Yard Tunnels—Segment 3) north of West 30th Street
- Installation of a permanent tunnel lining, cast-in-place concrete with waterproof membrane for the eastward West 30th St running SEM tunnels
- Building a temporary bypass for the large storm drain along West 30th Street and then finishing with a new permanent drain
- Numerous utility underpinning and reconstruction/relocations in West 30th Street and 12th Avenue areas.
- Demolition of existing structures on Block 675

The 12th Avenue Shaft excavation will utilize slurry walls in soil overburden. Overburden consists of fill underlain by weak highly compressible silts and clays. Deeper below the shaft invert are glacial deposits and rock. During shaft construction, strict groundwater control will be required to prevent the surrounding highly compressible soils reacting to pore water pressure drops that would result in large settlements to nearby utilities, roadways and buildings. Temporary construction flood gates will also be constructed in the 12th Avenue Shaft.

For this contract, there are two separate groups of SEM Tunnels connecting to the 12th Avenue Shaft, each with some slight differing construction methodologies. However, ground conditions in either group are fills underlain by weak soft clays and silts. Ground stabilization measures will be required for all tunnel excavations.

West of the 12th Ave Shaft, there is a temporary construction shaft located in Hudson River Park. This shaft facilitates construction for twin SEM tunnels running run through the Manahan Bulkhead foundation system and ending at the waterside. This temporary shaft also provides for SEM Tunnels heading towards the 12th Avenue Shaft to join with SEM tunnels from the 12th Avenue Shaft.

The Manhattan Bulkhead is founded on timber piles with cobbles and rip-rap fill. Prior to excavation through the plie/rip-rap/cobbles, permeation grouting will be applied to facilitate follow-on ground stabilization by filling rip-rap voids and locking rip-rap/cobbles in-place. More than 100 timber plies are expected to be encountered.

Heading west from the 12th Ave Shaft, SEM tunnels will be utilized to remove the steel H pile foundations and replace with underpinning of a large NYCDEP sewer below 12th Avenue. In addition, as well as removing abandoned West Side Highway concrete filled pipe piles will also be done.

These SEM tunnels, west of the 12th Ave Shaft, require ground stabilization measures, including at the bulkhead, anticipated to be primarily horizontal ground freezing.

These SEM tunnels will be backfilled with a flow fill grout and afterwards, the Hudson River contract TBMs will mine through this backfill to enter the 12th Avenue Shaft.

SEM tunnels also run eastward from the 12th Avenue Shaft, below West 30th Street to the presumed completed West Rail Yard Tunnels (Hudson Yard Tunnels) north of W 30th Street. However, the initial works in this area under West 30th St will be to build a temporary bypass for the existing large aged multicell storm drain running directly over the tunnel alignment. The storm drain is assumed to be on a pile foundation system. Post-tunneling, a new storm drain would be built.

Ground improvement will be required prior to all SEM tunneling in this latter area due to poor ground conditions. However, ground stabilization in this area is planned as jet grouting. Following excavation, the SEM will be finished with a permanent cast-in-place concrete lining with waterproofing membrane.

The 12th Avenue Shaft will be finished in a follow-on contract as a future fan plant shared with overlying commercial development.

Connection to A-Yard

A separate contract package will be prepared for the cut and cover tunnel under 10th Avenue, the connection to A-Yard and the re-profiling of the Empire Line Tunnel alignment. This package will also include the utility support and rebuilding for 10th Avenue, the underpinning of the Brookfield Building to accommodate the new fan plant and track alignment for the Hudson River Tunnels, installation of flood gates on the east side of 10th Avenue, and rock excavation at A-Yard to facilitate the new track profile.

NJ Fan Plant and Internal Tunnel Internal Concrete

Construction of the Hoboken Fan Plant will follow the completion of the Palisades and Hudson River tunnels. This contract consists of the fan plant building contrition and internal facilities except for tunnel ventilation fans and controls. It also includes the low voltage power distribution system, emergency generator fuel storage tank, lighting, plumbing and space planning for the ancillary rooms required to support the railroad systems contract.

The internal concrete finish work required for the Hudson River and Palisades Tunnels will be installed as part of this contract. This includes the high and low benches, the tunnel invert, the sidewall ventilation duct, cross passage finish and construction of the open cut section between Tonnelle Avenue and the Palisades Tunnel Portal.

Railroad Systems

Traction Power

Traction power facilities will be constructed in accordance with well-established railroad installation procedures. Each element of construction is based on previously installed facilities. Critical constructability issues that will require refinement include the scheduling of outages with Amtrak for the connection to the existing transmission network, and available staged outages for the modifications required at the existing Amtrak substations.

Overhead Catenary System

The activities in this contract include the installation of a complete overhead catenary system (OCS), including the supporting structures, for the entire alignment from the

NJ Surface works to the A-Yard. It is assumed that all new catenary structures and wire that is from the existing Northeast Corridor (NEC) will be installed by the contractor. Amtrak forces will perform the connections to the existing system at Allied Interlocking and in A-Yard.

The foundations for the catenary structures on the NJ surface alignment will be designed as part of the OCS design effort, but will be constructed under the civil package. Modification of the existing Amtrak catenary structures will need to be done as part of the force account effort. Staging plans will need to be established with Amtrak to establish nighttime and/or weekend schedules to work at the connections and adjacent to existing tracks.

Signal Work
Signal construction will follow closely after the track bed is installed. A trough for the signals and communications cables will be installed by the civil contractor along the full extent of the NJ surface contract. Platforms will also be constructed by the civil contractor along the NJ surface alignment for the new signals and communication structures for controlling the new Allied Interlocking.

Embedded conduits in the tunnel sections will be installed as part of the appropriate tunnel finish concrete contracts. Cables will be pulled into the embedded conduits or laid in the trough by the railroad systems contractor. The signal equipment installation can coincide with work on other disciplines such as Electric Traction and Communications.

Final connections of all new signal and communications equipment and cut-overs of the Allied Interlocking will be performed by the Amtrak force account.

Track Work
The railroad systems contract includes the installation of all new track except where it connects at Allied Interlocking and in A-Yard. Ballasted track construction will involve common local practice. Direct fixation track construction in the tunnel sections will be based on a booted resilient tie design.

All special trackwork for the new Allied interlocking will be procured by the railroad systems contractor and installed by the Amtrak force account.

Fan Plant MEP and Electrical Substation Installations
The Hoboken Fan Plant, 12th Avenue Fan Plant and Brookfield Building Fan Plant will all be fit-out and equipped under a common mechanical/electrical contract. This contract will also include provisioning of emergency power generators at the Hoboken Fan Plant and 12th Avenue Fan Plant sites and electrical power substations to supply operating power to the fan plants.

PROCUREMENT AND SCHEDULE CONSIDERATIONS
During the development of the ARC Project Procurement Strategy, there were several elements that had to be considered when establishing the sequence and contract package award for the various packages. Among the elements that played into the phasing of the contracts were:

- Property availability
- Construction package duration
- Agency permits
- Availability of funding
- Contract value/bonding capacity of contractor community
- Skill set required to perform the work

GTHP has created a construction sequence that is applicable to the new alignment.

While the ARC Procurement Strategy has provided a good platform for consideration and discussion during the Preliminary Design Phase, it does not preclude revisions to the packaging scheme while the design is being further developed and alternate strategies providing greater benefits are considered.

GENERAL PROJECT SCHEDULE

Assuming the required funding is secured, construction of the new Hudson Tunnels is expected to begin in perhaps 2019. The project has an overall timeline of approximately 7½ years, with an expected completion late in 2026. After the new Hudson Tunnels are operational, the existing North River Tunnels will be taken out service one at a time for complete rehabilitation, with that work expected to be complete in 2030. At that point, Amtrak will have 4 tunnels in good condition, which will make possible an increase of passenger capacity with the completion of other key projects that are part of the overall Gateway Program, including an expansion of Penn Station in New York.

CONCLUSION

There is an urgent need for a reliable and redundant backup for the aged existing North River Tunnels crossing the Hudson River into New York. These existing tunnels are critical to the need for reliable rail transportation and the economic well-being of the Northeastern U.S. The Amtrak Hudson Tunnel Project will provide relief for these existing tunnels.

Aside from work to rehabilitate the existing North River Tunnels, the Amtrak Hudson Tunnel Project construction is expected to be covered by several contract packages including design-build, design-bid-build and force account work. The major underground work contracts are currently configured as design-build type contracts.

Liner Load Estimation for Soft Ground Pressure Balance TBM Shield Tunnel Projects

Tamir Epel ▪ Colorado School of Mines
Mike Mooney ▪ Colorado School of Mines
Marte Gutierrez ▪ Colorado School of Mines

ABSTRACT

Pressure balance TBM tunneling has advanced to the point where it is routine to essentially match the face and annulus pressures around the TBM envelope. Resulting ground deformations are sensitive to the face pressure-effective stress ratio. The loading on segmental lining is impacted by this, and specifically the degree of pre-convergence, i.e., the relaxation and arching of the ground prior to lining installation. With very little information on the convergence-confinement relationship in pressure-balance TBM situations. This paper addresses the loading on precast segmental lining during pressure balance TBM tunneling based on lining load data from the Northgate Link project.

INTRODUCTION

Liner load estimation in pressure balance TBM tunneling (using both EPB and slurry machines) is highly dependent on the degree of pre-convergence prior to the installation of the segmental lining rings. The load redistribution concept is commonly taken into account in tunnel lining design by the use of the convergence-confinement method (CCM). More commonly used in conventional tunneling, the CCM is used in 2D analysis to account for the 3D effects of the excavation face, allowing for pre-convergence in front and behind the tunnel face, prior to lining installation. The pre-convergence allowed in this method results in more realistic lower liner loads as the ground stresses are redistributed and are partly taken by the ground. To implement the 3D effects in 2D numerical analysis several approaches are commonly used as: the contraction method, the stress reduction method, and the grout pressure method.

While in conventional tunneling, the geostatic stresses are not balanced at the face of excavation and at the tunnel periphery for a short time after excavation. In pressure balance TBM tunneling, the face and excavation boundary are supported by pressurized slurry or conditioned muck to reduce surface settlements. The allowance of pre-convergence prior to liner placement accomplished through the CCM is widely used for conventional tunneling. However, in most design processes involving pressurized shield tunneling projects, pre-convergence is often neglected. The rationale is that pressure at the face of the excavation reduces the pre-convergence, and that the assumption of zero pre-convergence yields more conservative (higher) lining forces.

In this paper, data from strain gages installed in segmental lining rings on the Seattle Sound Transit Northlink tunnel project (Frank et al., 2015; Epel et al., 2018a; Epel. et al., 2018b) are analyzed during the final state of geostatic loading in varying geologies along the glacially-deposited alignment. Commonly used numerical modeling methodologies are employed to predict liner loading. The experimental and model results are compared.

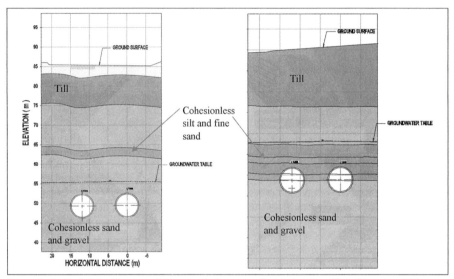

Figure 1. On the left the Northgate link geological cross section 1 at station 1335+00, on the right cross section 2 at station 1327+70

PROJECT BACKGROUND

The Northgate link extension project includes 5.6 km of twin bored tunnels and 23 cross passages. The tunnels run north from the University of Washington to the Maple Leaf portal in north Seattle. The twin bored tunnels were constructed using two EPB TBMs each with an excavation diameter of 6.64 m and supported in a single pass, gasketed segmental lining, 25 cm thick, and an extrados diameter of 6.25 m. Each ring is composed of four full size segments and key and counter key segments. The nominal ring width was 1.5 m, and the universal ring had an overall ring taper of 69.9 mm. The prefabricated segment design concrete strength was 55 MPa, with realized strengths closer to 67 MPa on average. The segments were reinforced with wire mesh with primary D14 reinforcement bars.

The geology of the area through which the tunnels were constructed consists of complex and highly variable interlayered glacial and non-glacial soil deposits. As part of the geotechnical baseline work, the geological units were grouped into engineering soil units (ESU) based on their behavioral characteristics. All tunnel excavation was conducted in the following glacially over ridden ESUs; till and till-like deposits (TLD), cohesionless sand and gravel (CSG), cohesionless silt and fine sand (CSF) and cohesive clays and silts (CCS). Figure 1 shows the variable geological conditions found at the two cross section discussed in this paper.

INSTRUMENTATION AND MONITORING

As part of the monitoring program wireless RFID strain gauges were installed in select precast concrete segments (Epel et al. 2018a, Epel. et al. 2018b) Data from two locations or cross-sections are presented in this paper. Within each selected ring, three segments were instrumented—two full-size segments plus a key or counter key. Each instrumented segment was outfitted with a set of two vibrating wire strain gauges welded to the reinforcement cage, one at the intrados and one at the extrados. The strain gauges were installed in a slightly asymmetrical configuration due to limitations in the casting process (Figure 2b). The intrados strain gauge center axis depth from the segment extrados is 58 mm as the strain gauge is installed on #3 rebar (9.5 mm) welded to the underside of the longitudinal 14 mm diameter (i.e., away from the intrados concrete face) rebar welded to the primary reinforcement (14 mm diameter), and minimum clearance is 25 mm. The extrados sister bar was installed on the exterior of the longitudinal reinforcement resulting in a depth of approximately 37 mm for the extrados stain gauge center axis. The strain gages were part of a wireless sensing system developed by Phase IV Engineering (Boulder, Colorado). The layout, implementation and recording program was developed in cooperation between the tunnel contractor JCM Northlink and designer L-7 Services (based in Golden, Colorado). Collection of the strain gauge readings required passing a flat panel RFID reading unit within 30 cm of the concrete surface in the vicinity of the embedded sensor. The monitoring schedule included an initial zero strain reading after the welding of the sister bars, and prior to casting, followed by readings every two weeks after segment installation.

The orientations of the strain gauges allowed measurement of the circumferential strain and interpretation of the stress developed in the pre-cast segments. With known geometry and strain gauge depth, the measurements collected from a set of two strain gauges allowed for the calculation of both thrust forces (hoop forces) and bending moments developed. The sign convention of the thrust forces and bending moments is positive for compressive thrust force, and positive for bending moment when the segment intrados is in tension (Figure 2d). A different arrangement of the instrumented rings was installed at the two cross section presented in this paper (Figure 3). In cross section 1 only two rings were instrumented at the second (SB) tunnel excavated. In cross section 2 one ring was installed at the first tunnel excavated and two rings at the second tunnel excavated.

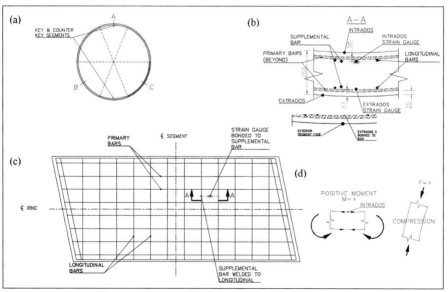

Figure 2. (a) Cross section of a typical instrumented ring, with the instrumented segments marked A, B & C; (b) Segment cross section at strain gage location (dimensions in mm); (c) Strain gage location on a typical segment plan view; (d) Sign convention for bending moments and thrust force

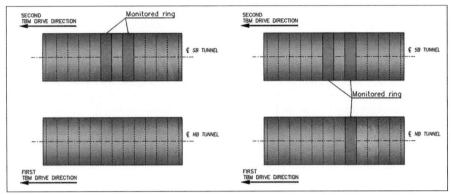

Figure 3. Instrumented rings marked in green at the two cross section, on the left cross section 1 and on the right cross section 2

NUMERICAL MODELING

An accurate prediction of lining forces requires modeling of the tunneling procedure and significant behavioral models including models for: soil constitutive behavior, grout pressure and stiffness, segmental joint rotational behavior, and 3D effects. In this study the soil constitutive model, the segmental joint rotational stiffness, and grout stiffness sensitivity are not investigated. The main purpose of this study is to examine the influence of two common 2D plain strain analysis approaches of the 3D effects of the face of excavation. The first is the 'wished into place' approach, where no pre-convergence or ground relaxation is allowed prior to liner installation. The second approach allows for pre-convergence according to the grout pressure method presented by Moeller (2006). In this study a 2D numerical model is adopted based on the work of Do et al. (2013) and is described in this section.

The 2D modeling was carried out using the finite difference program FLAC 3D. The general configuration of the 2D models is presented in Figure 7 for cross section 1. Each tunnel has an excavation diameter of 6.64 m with an annulus gap of 22.5 cm. The lining is modeled with embedded liner elements with the full thickness of 25 cm. The distance between the tunnel center lines is 12.2 m (1.85 diameters), and the depth to springline is 5.5 and 4.8 diameters. The 2D model size is 124 m in the transverse direction and extends 31 m under the tunnel SL with a total of 8016 elements. Symmetry is not used so that the influence of twin tunnel excavations can be investigated. The boundary conditions were set as fixed in the horizontal direction at the sides and fixed in both directions at the bottom (Figure 7).

The segmental lining is modeled by embedded liner elements connected by double node connections links at each node (Figure 4), to permit an interaction between the host medium (ground), and the structure element. While one side of the element is connected to the surrounding ground, the link at the other side is manipulated to connect the two separate segments. Based on Leonhardt and Reimann (1966), Janssen (1983) developed a simple quantification of the properties for these longitudinal segmental joints. The joints are described in the form of moment-rotation relationship Equations (1) and (2).

$$\text{linear:} \left\{ \phi = \frac{Mh}{EI} = 12\frac{M}{Eh^2 b} \right\} \qquad M < \frac{1}{6}F_n h \qquad \phi < \frac{2F_n}{Ehb} \qquad (1)$$

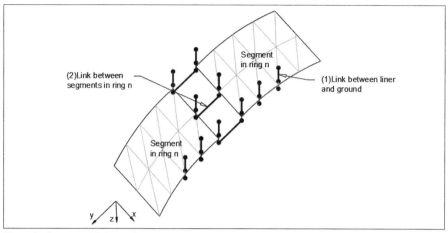

Figure 4. Segmental lining node connectivity concept (after Do et al., 2013)

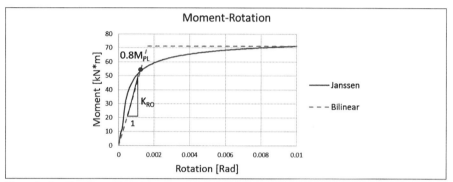

Figure 5. Bending moment—joint rotation relationship of the longitudinal joint

nonlinear: $\left\{ \phi = \dfrac{8F_n}{9bhE\left(\dfrac{2M}{F_n h} - 1\right)} \right\}$ $M < \dfrac{1}{6}F_n h$ $\phi \geq \dfrac{2F_n}{Ehb}$ (2)

This behavior of the segmental joints is significantly controlled by the normal force at the joint. With high normal force at the joint and low moments, the joint remains closed with only compression pressure on the entire cross section. However, with high bending moment or small joint thickness, a gap will form when the pressure at the outer side becomes zero, leading to significant additional rotation. The application of this method can be done by a simplified process to determine the values of the spring constants. The simplified procedure used by Do et al (2013), and Thienert and Pulsfort (2011) requires first to calculate the reference case with full hinge release at the joints. From the average normal force developed in the tunnel lining, the maximum limit bending moment is calculated for an angle rotation ϕ of 0.001 radians, which is assumed as an approximation of the maximum, permissible rotation. Do et al. (2013), and Thienert and Pulsfort (2011) also showed that the segmental joint moment-angular rotation relationship can be simplified by a bilinear model (Figure 5).

In this study, for the joint rotational stiffness, a bi-linear rotational spring is given to the joints as shown in Figure 5. While the translational connections are given a rigid

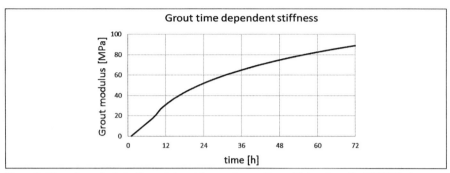

Figure 6. Time-dependent elastic modulus of the grout according a 28 day strength of 150 MPa to Kasper and Meschke (2004)

connection as Do et al. (2013) found that the axial and radial stiffness have a negligible effect on the segmental lining behavior.

The grout used in the project reached an average strength of 1.7 MPa in 28 days which correlates to an elastic modulus of 150 MPa according to a study made by Sharghi et al. (2017). In this study the grout is modeled with a time dependent function seen in Figure 6, based on the work of Kasper and Meschke (2004), where the initial stiffness of 8 hours is assumed for fresh grout properties to adjust for the fast gel time of the two-component grout.

The use of a non-linear soil constitutive model in pressure balance TBM tunneling is significant, as the ground is subjected to loading-unloading-reloading cycles in changing confinement conditions. The Plastic-Hardening model (Itasca 2009) was adopted in this study, to account for nonlinear stiffness due to the variation in confining stresses and unloading-reloading. The properties used were validated using available triaxial tests and pressuremeter tests.

Modeling the liner-ground interaction with 2D plane strain analysis requires assumptions to simulate the effects of the 3D tunneling process. This is accomplished with the well-known convergence confinement method (CCM) based on the work of Panet (1982), Vlachopoulos and Diederichs (2009), and others. For pressure balance TBM tunneling in non-linear ground behavior an incremental stress reduction procedure is required to build the ground reaction curve (GRC). This is done by the grout pressure method (Moller 2006). The transition from initial geostatic stresses to grout-pressure/machine-pressure state is expressed by Equation 3 (Moller 2006), where σ_0 is the initial normal stress and σ_g is the grout pressure.

$$\sigma = (1-\lambda)\sigma_0 + \lambda\sigma_g \tag{3}$$

The normal stress can be given as a function of the initial vertical and horizontal stresses by equation 4 (Moller 2006).

$$\sigma_0 = \sigma_{h0}\sin^2\alpha + \sigma_{v0}\cos^2\alpha \tag{4}$$

The pre-convergence at the location of lining installation can then be obtained according to the longitudinal displacement profile (LDP) of Vlachopoulos and Diederichs (2009) in the following equation.

$$u(x) = u_{max}\left[1 - \left(1 - \frac{u_f}{u_{max}}\right)e^{-\left(\frac{3x}{r_0} \cdot \frac{r_0}{2r_p}\right)}\right] \quad (5)$$

Here, x is the distance from the face and u_f is the radial displacement at the face according to equation 6.

$$u_f = u_{max}\frac{1}{3}e^{-0.15\left(\frac{r_p}{r_0}\right)} \quad (6)$$

The 3D effect of the distance from the face can be simulated by coupling the LDP and GRC and applying a fictitious internal pressure using Equation 4 in a 2D numerical model that will result in the pre-convergence estimated by Equation 5 prior to liner placement.

For the simulation of the construction process in the zero pre-convergence models (wished into place), three stages have been carried out: (1) an initial state of undisturbed ground under gravity conditions (geostatic loading); (2) the removal of the excavation material (6.25 m diameter), installation of the lining, and change of the material in the annular gap to hard grout; and (3) a repeat of stage 2 for the second bored tunnel.

For the simulation of pre-convergence, the CCM was implemented for each tunnel according to the grout pressure method (Moller 2006) before lining installation, followed by modeling the time dependent pressures and properties of the annulus grout summarized in Figure 6. In each pre-convergence model a total of 19 stages were calculated for each tunnel, with 9 pre-convergence stages prior to lining installation, followed by 9 additional calculation stages to account for the annulus grout time dependent behavior. The specific stages are: (Stage 1) the initial state of undisturbed ground; (Stages 2–10) removal of the excavation material and the annulus gap material, incremental reduction of the confinement pressure (increasing λ from 0–0.92 in Equation 3) down to the anticipated displacement (corresponding to $\lambda=0.92$) at the shield tail 1.6 diameters (10.5 m) behind the tunnel face; (Stage 11) installation of the lining and annulus grout material with fresh grout properties, and application of the grout pressure acting on both the lining and the excavation boundary; (Stage 12) removal of grout pressure; (Stages 13–19) time dependent hardening of the annulus grout to its final 28 day strength. After the completion of the first tunnel excavation sequence the same staging sequence is applied for the second tunnel.

RESULTS AND DISCUSSIONS

Cross section 1 included two instrumented rings installed in the second tunnel (SB) excavated. Strain gage measurement collection began 63 days after ring installation. Strain gage measurement continued for 500 days after SB tunnel construction at this cross section. While in cross section 1 strain monitoring was done only at the second tunnel, in cross section 2 both tunnels were monitored; one ring in the NB tunnel that was excavated first, and two rings at the SB tunnel that was excavated second. Unfortunately, data collection began 50 days after the second tunnel passed the monitored cross section. Strain gage measurement continued for 390 days beyond SB tunnel construction at this cross section. The values shown in Figures 7 and 8 are taken at 55 days after the second TBM pass having a distance greater than 20 diameters and hardened annulus grout.

Force Development in the First Tunnel

Figures 8a and 8b show results from the two analysis approaches of the first tunnel at cross section 1. For each numerical calculation two stages are displayed: the first stage is after the excavation and support of only a single tunnel, and the second stage is after the excavation and support of the second tunnel, constructed at a distance of 2D between axes. With no instrumented rings in the first tunnel excavated, comparing only between the two approaches, a significant increase of bending moments and thrust forces is seen in the model accounting for pre convergence as a result of the second tunnel excavation. However, in the 'wished into place' model a negligible change in bending moment and a slight change in thrust force is observed. While the difference in maximum absolute value of bending moment between the two methods at the final stage is only 2%, the thrust force estimated when pre-convergence is allowed is lower by about 35% from the wished into place approach. Figure 9b shows a similar thrust force behavior in cross section 2 between the two approaches. Figure 9b also shows a much better agreement between the field measurements and the approach allowing pre-convergence. Despite the good agreement between the thrust force measurements and the estimated forces using the pre-convergence approach, the bending moments measured fit the wished into place prediction (Figures 9a).

Force Development After Second Tunnel Excavation

In both cross sections liner loads were measured at the second tunnel (SB) excavated. In Figure 10b, it can be observed that the measured thrust forces at cross section 1, are in better agreement with the predicted forces using the pre-convergence approach, and are within an error of 5–30%. Where in the 'wished into place' approach the predicted values differ by 50–100%. Figure 11b showing the thrust forces at cross section 2 match the behavior observed in cross section 1 (Figure 10b) with the measured force in good agreement with the

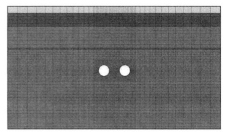

Figure 7. The 2D model used for the calculation of cross section 1

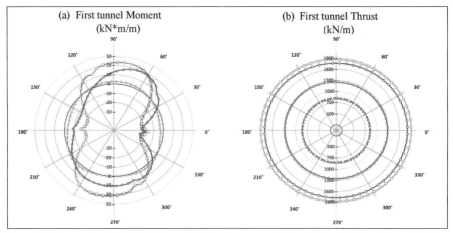

Figure 8. Cross section 1 in the first tunnel: (a) bending moment (kN-m/m); and (b) thrust force (kN/m) diagrams after first tunnel excavation, and prior to second tunnel excavation

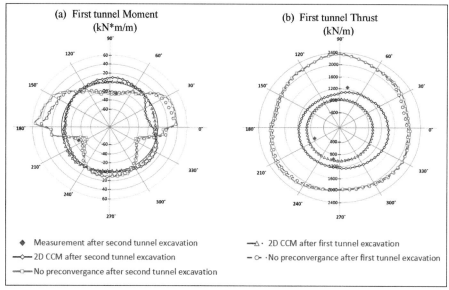

Figure 9. Cross section 2 in the first tunnel: (a) bending moment (kN-m/m) and (b) thrust force (kN/m) diagrams after first tunnel excavation, and prior to second tunnel excavation

pre-convergence model which are also lower by about 50% from the 'wished into place' approach. Similar to what was observed in the first tunnel at cross section 2, the bending moments at both cross section 1 (Figure 10a) and cross section 2 (Figure 11a) good agreement with the 'wished into place' approach.

CONCLUSIONS

This study presented a comparison of 2D plane strain numerical calculations with and without the use of convergence-confinement method (CCM) principles. The results of the analysis are validated with bending moment and thrust force from field strain gage measurements during the Northgate Link project in Seattle. From this series of analysis the following conclusions are drawn:

- Thrust forces predicted using the no pre-convergence approach (wished into place) results in forces higher than forces measured by 50–150%, while then accounting for pre-convergence (CCM) the thrust forces estimated are with 20% of the measured force in average.
- Contrary to the prediction of thrust forces, bending moments predicted using the no pre-convergence approach (wished into place) resulted in better agreement compared to the very low bending moments predicted when accounting for pre-convergence (CCM).
- The calculation of the internal lining forces without taking into account pre-convergence is commonly perceived as conservative in that it typically results in higher bending moments and thrust forces. However, the measured combination of a high bending moment with low axial force in fact finds this approach can be unconservative, as in some situations can lead to exceeding the thrust-moment (N-M) capacity envelope.

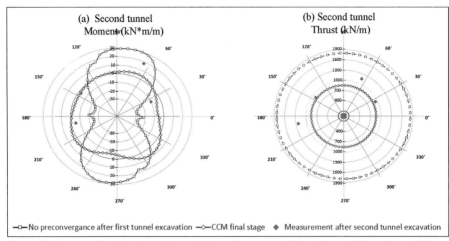

Figure 10. Cross section 1 in the first tunnel: (a) bending moment (kN-m/m) and (b) thrust force (kN/m) diagrams after first tunnel excavation, and prior to second tunnel excavation

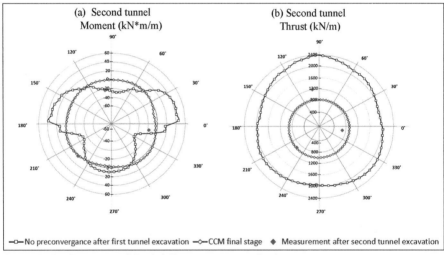

Figure 11. Bending moment (kN-m/m) digrams at cross section 2

Although some may believe that a simplified analysis that neglects the effects of relaxation is a more conservative design approach, this may not be the case. Combining the discussed approaches with different expected face pressures should be used to build the loading cases used in the design of segmental pre-cast concrete linings. This should reduce the uncertainty and potentially allow lower safety factors for design.

REFERENCES

Do, N.A., Dias, D., Oreste, P., Djeran-Maigre, I. 2013. 2D numerical investigation of segmental tunnel lining behavior. Tunneling and Underground Space Technology, 37 (2013), pp. 115–127, 10.1016/j.tust.2013.03.008.

Einstein, H.H. and Schwartz, C.W. 1979. "Simplified analysis for tunnel supports." Journal of the Geotechnical Engineering Division, 499–517.

Epel, T., Mooney, M., Gutierrez, M., Braun, K., DiPonio, M., Long, N. 2018a. Analysis Methodologies Comparison of Estimated and Measured Precast Segment Liner Loads Developed During Construction of twin EPB TBM Tunnels in Seattle. World Tunnelling Conference 2018. In submission.

Epel, T., Mooney, M., Gutierrez, M., Braun, K., DiPonio, M., Long, N. 2018b. Ground-Liner Interaction during Seattle Northgate Link Cross-Passage Construction North American Tunneling Conference 2018. In submission.

Frank, G., Hagan, B., Sandell, T., Gharahbagh, E.A., Willis, D., Diponio, M.A., Cowles, B. 2015. Preparation for Tunneling, Northgate N125 Project in Seattle, Washington, Rapid Excavation and Tunneling Conference 2015.

Itasca. 2009. FLAC Fast Lagrangian Analysis of Continua, Version 4.0. User's manual. FLAC Fast Lagrangian Analysis of Continua, Version 4.0. User's manual.

Janssen, P. 1983. Tragverhalten von tunnelausbauten mit gelenktubbings, PhD Thesis Technischen Universitat Carolo-Wilhelmina, Braunschweig, December 1983.

Kavvadas, M., Litsas, D., Vazaios, I., Fortsakis P. 2017. Development of a 3D finite element model for shield EPB tunneling Tunn. Undergr. Space Technol., 65, pp. 22–34.

Lambrughi, A., Medina Rodríguez, L., Castellanza, R. 2012. Development and validation of a 3D numerical model for TBM–EPB mechanised excavations. Comput. Geotech. 40, 97–113.

Leonhard, F., & Reimann, H. 1966. Betongelenke. Der Bauingenieur, 41, 49–56.

Moller, S. 2006. Tunnel Induced Settlements and Structural Forces in Linings, PhD Thesis, Universitat Stuttgart.

Muir Wood A. 1975. "The circular tunnel in elastic ground." Géotechnique 25(1), 115–127.

Panet M., Guenot A. 1982. Analysis of convergence behind the face of a tunnel. Proceedings, International Symposium Tunnelling '82, IMM, London, pp 197–204.

Thienert, C., Pulsfort, M., 2011. Segment design under consideration of the material used to fill the annular gap. Geomechanics and Tunneling 4, 665–679.

Vlachopoulos N., Diederichs M.S. 2009. Improved longitudinal displacement profiles for convergence confinement analysis of deep tunnels. Journal of Rock mechanics and Rock Engineering, Vol. 42, Number 2, April 2009. pp 131–146.

TBM Utilization in Mixed Transitional Ground: The OCIT Experience in Akron, Ohio

Elisa Comis ▪ McMillen Jacobs Associates
Wayne Gyorgak ▪ McMillen Jacobs Associates
David Chastka ▪ Kenny/Obayashi Joint Venture

ABSTRACT

A dual mode Rock/EPB TBM, 9.26 m diameter bore was used for the first time in North America to excavate the Ohio Canal Interceptor Tunnel (OCIT) in the downtown Akron. Tunneling was carried-out at a uniform slope of 0.15 percent through ground conditions that consisted of soft ground, mixed face soft ground over bedrock, and bedrock. The Mixed Transitional Ground (MTG) resulted in the biggest challenge for the TBM. This paper specifically addresses the influence of the MTG on the TBM Utilization Factor and present the counter-measures implemented in these special geological conditions.

INTRODUCTION

The nominal or installed operational values for a Tunnel Boring Machine (TBM) are: thrust, torque, power, and rotational speed. These are highly related to the expected geology and the tunnel diameter and will directly affect the TBM penetration rate (PR), which is the TBM progress in terms of feet/hour (m/hr) during the boring process.

However, the TBM utilization (U), which reflects the amount of time that the TBM actually excavates, has to be determined for a comprehensive TBM performance analysis. Another parameter, that is often cited as part of performance prediction, is the cutter life. This parameter is typically expressed in terms of average cutter life in hours, meter travelled on the face, cutters per meters of tunnel, or cutters per cubic meter of excavated rock.

The primary focus of performance prediction studies is the prediction of the TBM's advance rate (AR) often expressed as the amount of daily advance expressed in feet per day (m/day), feet per week (m/week). These studies have mainly tried to find out the relation-ship between the rock mass characteristics and the machine performance. Few publications can be found on utilization records for TBMs working in soft ground while typical utilization rates in rock are reported from 5%, for very difficult and complex geologies with poor site management, to around 55%, in perfect working conditions for an open type TBM in moderately strong rock with no ground support requirements. However, the most common range of TBM utilization is reported to be in the 20 to 30% range (Rostami, 2016).

Machine utilization is very sensitive to ground condition. In rock tunnels bad ground could cause face collapse and cutterhead jamming. Ground convergence and squeezing could also become an issue for shielded TBMs and water bearing formations where the in-flow of water can interrupt the operations and cause long delays for dewatering and drain-age. Often measures offered to mitigate the ground related issues, such as probe drilling or pre-excavation grouting, cause their own delays and can become source of inaccuracies in predicting machine utilization.

The impacts on machine utilization of factors such as contractor and crew experience, management approach, labor issues, site arrangement, logistical issues with supplies, repair and spare parts, electricity and power supply, transportation and site access limit muck haulage, availability of local workforce, and so on, are also very important but extremely difficult to quantify.

TBM utilization has a direct impact on TBM advance rate (AR) and ultimately project schedule and costs. The majority of case histories report the penetration rate while the advance rate or utilization, are not reported as often, or it is not clear if they are based on available time, boring days, working days, or calendar days.

The Ohio Canal Interceptor Tunnel (OCIT) Project involved the construction of a 6,200-foot-long (1,890 m) conveyance and storage tunnel with a finished inside diameter of 27 feet (8.23 m) to control combined sewer overflows for several regulators in the downtown Akron area. The OCIT Project was awarded to Kenny/Obayashi Joint Venture with a Notice-to-Proceed on November 4, 2015.

The section of the GBR related to the OCIT identified three major reaches that were defined as distinctly different ground conditions:

- Reach 1 which primarily consisted of soft ground
- Reach 2 which was a transitionary zone with soft ground overlying bedrock, defined as mixed ground conditions
- Reach 3 which was comprised of bedrock with two sections of low rock cover

The tunnel was excavated using a 30.38-foot (9.26 m) diameter dual mode type "Crossover" (XRE) Rock/EPB Tunnel Boring Machine (TBM), manufactured and supplied by The Robbins Company. The TBM was launched in October 2017 and completed the drive in August 2018. Reach 1 and 2, for a total of 810 ft (247 m), presented a challenge for the project as characterized by low penetration rate, high cutter consumption, difficulties in maintaining face pressure, and low TBM utilization. Reach 3, for a total of 5,400 ft (1,647 m), was distinguished by an exceptional high TBM utilization and steady advance rates.

IMPACT OF MIXED TRANSITIONAL GROUND ON TBM PERFORMANCE

Because of alignment restrictions and increase in demand for tunnel with larger diameters, more and more tunnel boring machines are working in mixed face conditions. As mentioned by several authors and later summarized by Oliveira and Diederichs (2016), mixed face conditions still present a challenging scenario for TBMs.

Several issues that have been reported from tunneling projects excavated along mixed ground conditions (Della Valle 2001; Thewes 2004; Zhao et al., 2007; Shirlaw 2016; Comulada et al., 2016; Silva et al., 2017) such as:

- Ground loss, settlements, and sinkholes
- Slow rates of tunneling
- Rapid tool wear, especially of the disc cutters and screw conveyor
- Damage to tools, mixing arms and other parts of the TBM
- Very frequent and lengthy interventions
- Clogging
- High temperatures inside the excavation chamber

Reach 2 of the OCIT was described in the GBR as a transitionary zone with soft ground overlying bedrock. Soft ground consisted of silty sand containing interbedded layers of silt with a variable layer of glacial till deposits following the top of the bedrock. Bedrock consisted primarily of shale with varying layers of siltstone. The top of the bedrock was identified as highly weathered at depths up to 10 feet (3 m). Groundwater level was 4 feet (1.2 m) from the crown of the tunnel at the start of the reach moving to the crown of the tunnel at the end. This geological condition can be denominated mixed transitional ground (MTG) as defined by Oliveira and Diederichs (2016).

Langmaak and Lee (2016) emphasized that a successful TBM drive in general and especially in difficult ground conditions can only be reached by combining adequate mechanical solutions together with the use of suitable chemicals as well as working with experienced TBM operators.

Especially in EPB tunneling, the correct choice and use of well adapted soil conditioners can make a considerable difference for the success of a tunneling project; both in highly permeable grounds as well as in sticky clays. Soil conditioning is determined by the following parameters defined in EFNARC (2005); Foam Expansion Ratio (FER), the ratio between the volume of foam at working pressure and the volume of the solution (with a specified the range between 5 and 30), Foam Injection Ratio (FIR), the ratio between the injected volume of foam at working pressure and the banked volume of ground (with specified range between 10% and 80%), and Concentration of surfactant (C_f), the concentration of used surfactant in the preparation of the foaming liquid (with a specified range between 0.5% and 5%).

The TBM design features adopted for the expected operating conditions were described by Comis and Chastka in 2017. Cutterhead design for proficient boring in both rock and soil conditions, adjustable main drive speed with an over-speed mode for operation in hard rock, and special screw conveyor wear protection features were addressed for these specific ground conditions. TBM specifications are listed in Table 1.

A soil conditioning assessment was also conducted prior to the construction of the OCIT. Testing varied according to soil type. For the sandy soils, testing for mobility was made using slump. For the silty soils, testing included adhesion (using a variation of the Langmaack adhesion test (NAT 2000, Langmaack)), slump, static flow, dynamic flow, and rheometry. Based on these tests the recommendation was a polymer foaming agent for Reach 1, with FER and FIR to be adjusted with actual ground conditions ant the use of an anti-clay polymer if more than expected portion of clay was found

Table 1. TBM specifications

TBM Type	Crossover Rock/EPB TBM
Bore Diameter (m)	9.26
Cutting Tools (17" Cutters/Knives)	56
Drive System	12 × 190kW VFD electric motors
Maximum Rotational Speed (rpm)	3.5
Rated Hydrostatic Pressure (bar)	3.5
Shield Length (m)	12.19
Maximum Cutterhead Torque (kNm)	23,270
Maximum Thrust Force (kN)	65,900
Screw Conveyor Capacity (m^3/h)	773 @ 16 rpm
Primary Voltage (V)	13,200

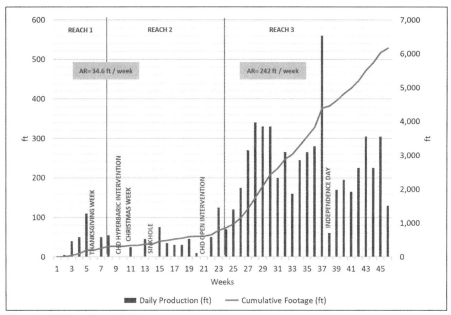

Figure 1. OCIT weekly production

in the excavated soil. For the mixed face conditions (Reach 2) it was suggested the same polymer foaming agent with particular attention to the change of rock/soil ratio. If the ratio of fine particles in the ground increased, water could be additionally injected into the chamber in order to increase fluidity and to adjust viscosity. If the ratio of fine particles in the ground decreased, foaming agent, polymer, or bentonite could be additionally injected into the chamber in order to keep necessary fluidity and viscosity. Liquid type polymers were also considered suitable for this purpose. Final set-up for Reach 2 ground conditions were considered sufficient for the bedrock zone, Reach 3.

The OCIT TBM weekly production for the whole drive is shown in Figure 1. An average weekly advance rate of 34.6 ft/week (10.5 m/week) has been recorded for Reaches 1 and 2, and 242 ft/week (73.8 m/week) for Reach 3. Reach 1 and 2 have been characterized by low penetration rate, high cutter consumption, and difficulties in maintaining face pressure.

Lack of production (high thrust with low penetration rate) prompted several cutterhead inspections which revealed an almost a full face of dry sand and showed material packed in the rear of the housings as well as the buckets. Buckets were filled of clay/sand, easily removable with a pressure washer. Most of the outer radial buckets were hard packed and needed the use of pneumatic tools to remove material. The dryer conditions at the face were thought to be related to the dewatering carried out at the mining site for the portal construction.

Material was packing into the cutter housings preventing the cutters to rotate. Three different cutter types have been tested (Figure 2): standard disk, traction cutters, and carbide tip. Standard and traction cutters both had issues with the housings clogging and then stopping the rotation of the cutter which then led to excessive wear. Carbide tip cutters provided more traction and allowed the cutter to rotate even after the housing were clogged. The original dressing of the cutterhead had carbide-tipped cutters

Figure 2. Different types of cutters being tested

Figure 3. Clogging inside the TBM cutterhead and steam in the tunnel

at the gauge position while the remaining cutters were all standard disks. During the first intervention, a decision was made to switch the transition cutters over to carbide-tipped as well. The cutterhead continued to clog developing high temperature in the plenum, baking the clay and sand material and ultimately steaming muck which was conveyed outside the tunnel (Figure 3).

During the last intervention in closed mode some grill bars were removed as well, hoping to mitigate the cutterhead clogging by increasing its opening ratio that was designed in consideration that 75% of the drive was going to be in rock. Unfortunately, the penetration rate did not improve remarkably.

Attention was brought to the soil conditioning as well. Initial foam produced a low-quality result, a liquid consistency with no stand time, when checked at the foam generators on the TBM. Foam supplier ran multiple tests to see if it was an issue with the product or the ratios being used getting good results in their lab. The medium in the generators were then evaluated and changed from the initial set-up of loosely packed larger metal cylinders to a tightly packed plastic top hat and ultimately to a tightly packed steel wool. The final result was a foam of good consistency and quality. Despite getting the desired output, the cutter-head still experienced clogging.

Focus was finally shifted to foam parameters adjustment. The initial soil conditioning evaluation suggested an FER of 15 with C_f between 3% and 5%. However, even with a FIR that has been as high as 200–300% the clogging was not mitigated. The FER

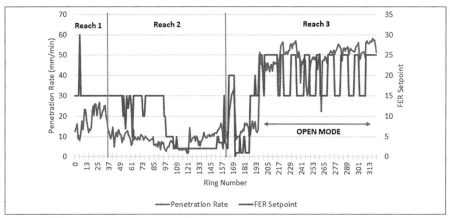

Figure 4. TBM penetration rate vs. FER setting

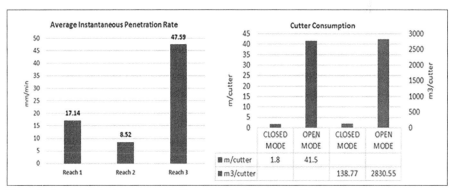

Figure 5. Average instantaneous penetration rate and cutter consumption

was reduced to 2 and the FIR set between 60% and 80%. This finally allowed the completion of Reach 2. The penetration rate increased from 5 mm/min to 15 mm/min suggesting that the cutterhead was progressively starting to clear-up, concurrently with the increase of rock ratio at the face of excavation (Figure 4).

Average Instantaneous Penetration Rate (PR) and Cutter Consumption are shown in Figure 5 for each single reach. The mixed transitional ground resulted in an average instantaneous penetration rate of 8.52 mm/min with a cutter consumption of 1.8 m/cutter. This data highlights that mixed transitional ground has to be considered a difficult ground condition by the industry.

Reach 3 had instantaneous penetration rates as high as 63.1 mm/min with an average of 47.6 mm/min. The average PR was affected by limitation in torque and thrust for the last 1,200 feet (366 m) due to failure of two main drive motors and one of the screw motors (Figure 6). Reach 1 and 2 were characterized by high torque and thrust levels as described above.

According to Comis and Chaska (2017), the main challenge in Reach 3 was expected to be the screw conveyor wear. Screw conveyor speed was in fact increased up to 14 RPM in order to increase its efficiency in rock. Special wear prevention features were adopted in the screw conveyor design. A wear monitoring plan was defined in

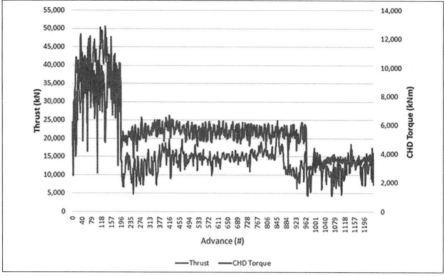

Figure 6. Torque and thrust for the whole drive

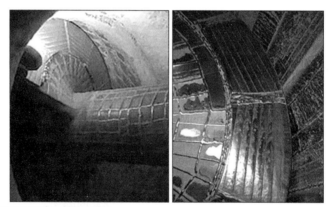

Figure 7. Screw wear visual inspection

the *Tunnel Excavation Plan* and carried-out during the mining operation with periodic visual inspection and was casing thickness measurement. The 5,400 feet (1,650 m) boring in rock did not result in measurable wear. As shown in Figure 7, the chromium carbide plates installed at the auger flights and at the casing and the hard facing in a crosshatch pattern resulted just polished by the mix of muck, soil conditioning and bentonite injected in the plenum.

TBM UTILIZATION

A mechanized tunneling project schedule and cost estimate relied on performance prediction of TBMs. This process involved a thorough understanding of the complexities in the site geology, machine specification, and site management.

Performance analysis for the OCIT TBM was carried-out from data collected from both the TBM data logger and contractor Shift Reports and categorized as shown in Table 2. The working week consisted usually of six days, five of mining and one dedicated to maintenance of the TBM and ancillary systems. The overall Operation for the length of the drive has been 44%, with an Utilization of 27% as shown in Figure 8. Major delays were observed for the cutterhead maintenance (14%) and non-machine related (20%). The Utilization was then analyzed for each tunnel Reach as identified in the GBR.

Reach 1

Reach 1 operation was 35% (Figure 9) with an Utilization of 22% and major downtimes observed in the TBM and non-machine related delays. This can be considered a common distribution, as it reflects the start-up of the job, with site final set-up for mining still to be implemented and TBM commissioning activities still on-going.

Reach 2

Reach 2 operation was 30% (Figure 10) with utilization of 23% and major downtimes due to cutterhead maintenance (29%) and TBM break-downs (17%). The non-machine delays were reduced to 16% as the site was finally configured for full mining operation. Even if namely the utilization is higher than the one recorded in Reach 1, the low penetration rate experienced by the TBM resulted in a lack of production. Furthermore, the combined effect of learning a new TBM, addressing constantly changing ground conditions in the first two reaches, and a low level of experienced workers led to poor

Table 2. Performance categories

Operation	TBM Delays	Non-TBM Delays
Boring	Cutterhead Maintenance	Probe Drilling
Ring Building	Ordinary Maintenance	Tunnel Conveyor System Delays
	TBM Downtimes	Supply Chain Delays
	Back-Up Downtimes	Surveying
		Utility Extension
		Transport
		Power Outage
		Safety

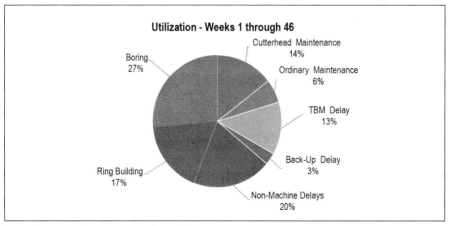

Figure 8. Total utilization for the whole drive

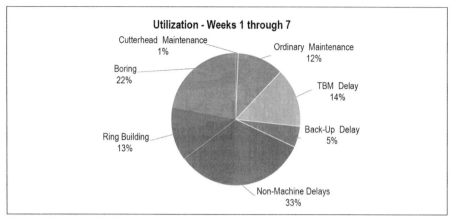

Figure 9. Reach 1 utilization

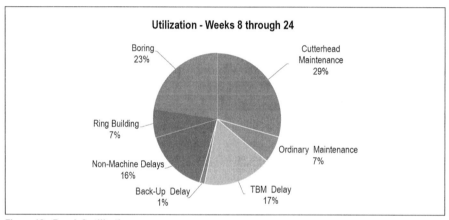

Figure 10. Reach 2 utilization

production, increased downtime, and a greater level of difficulty troubleshooting any issues that occurred.

The OCIT project requirement to hire 35% of local workforce made the project more difficult from a labor perspective. Akron had not had any TBM tunnels prior to this project thus it limited the workforce experience on the machine. The end result was a much longer learning curve due to extensive training and turnover.

The ring building time learning curve can be an example of these difficulties (Figure 11), especially in combination with the difficult ground conditions. Ring building is different on every machine as the operator and supporting staff need to understand the capabilities of the erector and the process of setting the segments. Furthermore, a relationship exists between production and ring build time. The natural assumption that during low production periods, setting of only two rings in a shift would lead to an efficient ring building with high construction quality is often misleading. The reality was in fact that in closed mode, when production was low, the ring build were around 60 to 80 minutes with gap and step issues. Almost instantaneously when the TBM switched to open mode the production increased dramatically and the ring build went to less

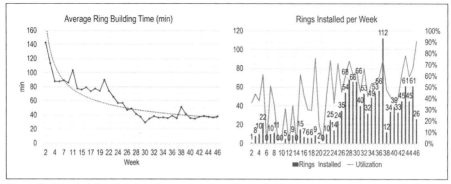

Figure 11. Ring building time learning curve and weekly production

than 40 minutes with quality issues almost negligible. This can be considered a direct result of more practice with the equipment but also the crew finding a good rhythm with a consistent cycle.

Reach 3

In Reach 3 operation was 59%, with an utilization of 31% over 20 mining weeks. This result is remarkable, especially considering that muck is removed from the mixing chamber with a screw conveyor (less efficient than a belt conveyor that is usually adopted for rock TBMs). The overall performance in this Reach was affected by the non-machine delays. The utilization analysis identified the weakness of the system as being the tunnel conveyor with 45.5% of the downtimes (Figure 12).

Explanation can be found in the attention required at the three transfer points in surface and respective hoppers (tunnel conveyor to overland, overland to stacker, and stacker to muck pile). Increases in amount of muck or changes in consistency of the material caused clogging of transfer points and large downtime to clean the system. The TBM penetration rate was also adjusted accordingly, in order to limit the muck loaded on the conveyor. Differently, probe drilling was an unavoidable contractual requirement has to be considered during future advance rate estimates.

CONCLUSION

The OCIT Project experienced the most challenging portion of the excavation in the first two Reaches of the tunnel. Although this was where the challenges were anticipated, the type and severity were under-estimated. The natural assumption of initial slower advance rates was made in consideration of inexperienced workers, typical learning curve on a TBM, and adjusting ground conditions. However, these factors combined with others such as cutterhead design, foam generation, soil conditioning, and constantly changing geological conditions resulted in an initial inefficiency of the system and a longer learning curve for the crew.

As shown in Figure 13, the anticipated advance rates were demonstrated possible by Reach 3 but were under-estimated in Reaches 1 and 2 due to the effects of a lower utilization rate and penetration rate.

As we move forward as an industry, TBM projects will only continue to grow in size and complexity. As cities grow and infrastructure expands, the demand for larger TBMs and excavation in more difficult and varying geology will increase. The need

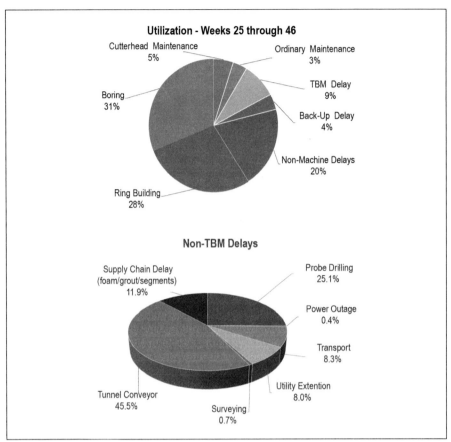

Figure 12. Reach 3 TBM utilization

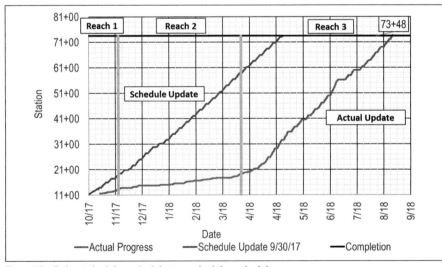

Figure 13. Estimated mining schedule vs. real mining schedule

for tunnels throughout the world is increasing at an exponential rate. This leaves the limited pool of experienced managers and engineers with the challenge of finding and training a TBM tunnels experienced workforce.

Basing anticipated production on TBM advance rate alone ignores the specific complexities of each project. Defining a real Utilization Factor, understanding the challenges and realistic capabilities of available workforce and equipment, and categorizing the actual causes of inefficiency can help to compile a database useful to make the industry advance as a whole.

REFERENCES

Comis, E. and Chastka, D. 2017. *Design and Implementation of a Large-Diameter, Dual-Mode "Cross-over" TBM for the Akron Ohio Canal Interceptor Tunnel;* Proc. RETC 2017: p. 488–497.

Comulada, M., Maidl, U., Silva, M.A.P., Aguiar, G. & Ferreira, A. 2016. *Experiences Gained in Heterogeneous Ground Conditions at the Twin-Tube EPB Shield Tunnels in São Paulo Metro Line 5;* Proc. ITA 2016 WTC, San Francisco: 1–11.

Della Valle, N. 2001. *Boring Through a Rock-Soil Interface in Singapore;* Proc. RETC 2001: p. 633–645.

EFNARC, 2005. *Specification and Guidelines for the Use of Specialist Products for Mechanized Tunneling (TBM) in Soft Ground and Hard Rock.*

Girard, D. and Chen, R. 2018. *Lessons Learned in Dry Ground Excavation Using an EPBM;* Proc. NAT 2018: p. 908–914.

Langmaack, L. and Lee, K.F. 2016. *Difficult ground conditions? Use the Right Chemicals! Chances–Limits–Requirements.* Tunnelling and Underground Space Technology 57: p. 112–121.

Oliveira, D.G.G. and Diederichs, M. 2016. *TBM Interaction with Soil-Rock Transitional Ground;* Proc. TAC 2016, Annual Conference, Ottawa: p. 1–8. .

Rostami, J. 2016. *Performance Prediction of Hard Rock Tunnel Boring Machines (TBMs) in Difficult Ground.* Tunnelling and Underground Space Technology 57: p. 173–182.

Shirlaw, N. 2016. *Pressurized TBM Tunnelling in Mixed-Face Conditions Resulting from Tropical Weathering of Igneous Rock.* Tunnelling and Underground Space Technology 57: p. 225–240.

Silva, M.A.A.P., Katayama, L.T., Leyser, F.G., Aguiar, G. and Ferreira, A.A. 2017. *Twin Tunnels Excavated in Mixed-Face Conditions.* Proc. WTC 2017, Bergen, Norway.

Thewes, M. and Burger, W. 2004. *Clogging Risks for TBM Drives in Clay.* Tunnels & Tunnelling International, June 2004: p. 28–31. .

Zhao, J., Gong, Q.M., and Eisensten, Z. 2007. *Tunnelling Through a Frequently Changing and Mixed Ground: A Case History in Singapore.* Tunnelling and Underground Space Technology 22: p. 388–400.

Times Square Shuttle Station Reconstruction

Aram Grigoryan ▪ WSP (formerly Parsons Brinckerhoff—PB)
David Campo ▪ WSP (formerly Parsons Brinckerhoff—PB)
Philip Lund ▪ WSP (formerly Parsons Brinckerhoff—PB)
Wanxing Liu ▪ WSP (formerly Parsons Brinckerhoff—PB)

ABSTRACT

The 42nd Street Times Square Shuttle Station is part of the busiest subway station complex and first subway line that was built in 1904, serving over 200,000 passengers daily. However, the station layout creates confusion, congestion, and inconvenient access to trains. The curved track alignment results in large platform gaps that preclude this key station from being ADA accessible. History of the first IRT line design and construction, and subsequent expansion and developments of the New York City (NYC) Subway directly relate to the current station configuration and operation. This project significantly modifies and structurally reconfigures the existing subway station and provides a 28-foot wide center platform for two 6-car long trains that would improve station operations, passenger circulation and safety, and would make the station ADA accessible. Station reconstruction was designed to be performed in phases to maintain station operations during the construction. All structural modifications would be performed underground without excavation from street level.

The paper discusses design of station reconstruction and suggested construction concepts by underpinning the existing station structure, removing over 120 station columns, constructing new foundations, and new structural support framing including columns and transfer beams. New foundations were developed to address varying rock quality. New underpass will be mined below operating tracks to provide an additional station entrance and connection to 42nd street passageway below the adjoining Durst building. Trains will be temporarily supported by skeletonized tracks and micro-piles.

HISTORICAL PROJECT BACKGROUND

The 42nd Street Shuttle is sometimes referred to as the Grand Central/Times Square Shuttle, since it serves these two stations. It is the shortest regular service in the system—about 2,402 feet in total length.

In 1904, the Interborough Rapid Transit Company (IRT) constructed the first subway line in New York City (NYC). It had 28 stations and ran north from City Hall on what is now the IRT Lexington Avenue Line to 42nd Street, where it turned west to run along 42nd Street. At Broadway, the line turned north to 145th Street on what is now the IRT Broadway–Seventh Avenue Line (Figure 1).

The station was originally a four-track express stop with two island platforms between the local and express tracks. This operation continued until 1918, when construction on the Lexington Avenue Line north of 42nd Street, and on the Broadway–Seventh Avenue Line south of 42nd Street was completed. The section connecting Grand Central and Times Square, via 42nd Street, was converted into a shuttle operation.

Times Square Shuttle Station Reconstruction 133

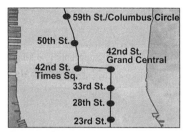 

Figure 1. Detail of 1904 first IRT subway line (left) and detail of 2018 subway map (right)

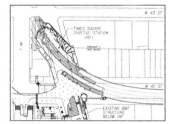 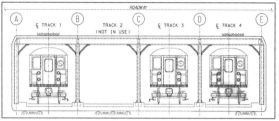

Figure 2. Existing station location plan (left) ant typical tunnel section below 42nd Street (right)

Times Square Shuttle Station (TSS) is part of the busiest subway station complex in NYC. It is composed of four interconnected stations that serve over 200,000 passengers daily (MTA, 2017). The complex serves 12 subway lines passing through this station complex (Trains 1, 2, 3, 7, A, C, E, N, Q, R, W, S) and allows free transfers between the following subway lines:

1. IRT 42nd Street Shuttle
2. BMT Broadway Line
3. IRT Broadway–Seventh Avenue Line and the IRT Flushing Line
4. IND Eighth Avenue Line one block west at 42nd Street–Port Authority Bus Terminal.

TSS is located below Times Square near the intersection of 42nd Street and Broadway, within the original IRT 4-track structure on a sharp curve that turns north from 42nd street (Figure 2).

The existing station configuration, layout and service is inconvenient for passengers. The station is served by short three-car and four-car trains on three operating Tracks 1, 3 and 4. The old southbound express track (Track 2) is not in service. At TSS, a passenger platform was built over Track 2 to service the trains on Track 3. The station layout requires passengers to board each train from a different platform, creating confusion and congestion. Additionally, the tracks are located on sharp curves, which create large platform gaps (with mechanical platform gap fillers) that preclude this key station from being ADA accessible (Figure 3).

In the past, there were several studies and attempts to modify and/or reconstruct the station to address the issues associated with the curved alignment and inconvenient station layout. In 1954, a design concept was developed for the reconstruction of TSS. The concept was based on reconstructing the station on a tangent alignment under 42nd Street (Figure 4). However, various factors, including a required station

Figure 3. Platform over Track 2 (left) and platform gap fillers at train doors, extended after train arrival (right)

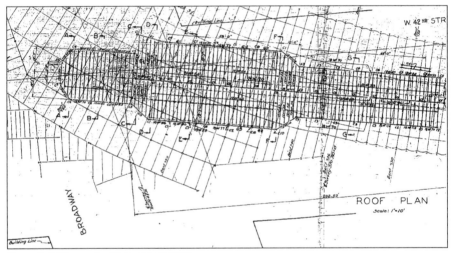

Figure 4. TSS reconstruction concept, 1954

shutdown and significant impact on street and utilities, prevented the concept from advancing into actual construction.

In the 1980s and 1990s there were several additional studies by NYCT and consultants to improve and modify TSS Station. However, none of the studies and concepts met all the requirements and constraints.

In 2014 and in 2015 NYCT engaged Parsons Brinckerhoff (now WSP) to perform two conceptual feasibility and constructability studies to address prior concerns and constraints, explore opportunities for station reconstruction and to make the station ADA accessible. These studies (CM-1050, Task Orders 1 and 3), have successfully addressed and resolved all prior concerns and confirmed feasibility and constructability of the necessary modifications without closing the station operations. The new NYCT concept was based on a modified station layout and wider center platform that would be constructed east of the current station platforms under the 42nd Street to accommodate two six-car trains with ADA accessibility. The focus of both studies was

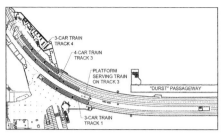 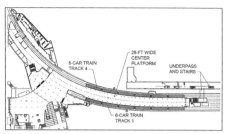

Figure 5. Existing station (left) and modified station with center platform and two 6-car trains (right)

on the feasibility and constructability of TSS structural reconstruction and modifications performed from underground without excavation from the street level.

The scope of the initial study (TO-1) was limited to (1) removing 36 select columns along the new platform edges that would align with train car doors, to eliminate obstacles at the train car doors and facilitate boarding of trains, and (2) removing an additional 20 columns in the concourse area to improve passenger circulation. The design concept and construction phasing method allow the station to remain open and minimizes impacts on operations and on passengers.

However, the remaining 5-foot spaced columns along both edges of the new platform would present safety and operational concerns. NYCT considered using Platform Screen Doors (PSDs), which separate the platform from the train. At that time, NYCT requested the challenging task (TO-3) of studying and developing conceptual design to reconfigure and reconstruct TSS to eliminate all existing columns along both platform edges, while also addressing all prior constraints and keeping the station open with minimum impact on station and train operations, and on passengers. The resulting study successfully developed a station reconstruction concept to address all identified concerns and limitations. Figure 5 shows a plan of the existing station with three trains and the plan of the modified station with two six-car trains positioned east from sharp curves eliminating platform extenders and complying with ADA accessibility requirements.

After the studies were completed, the project was funded and NYCT issued Request for Proposals (RFP) for Final Design for Structural Modifications and associated Geotechnical Engineering. NYCT in-house staff performed the Final Design for all other disciplines (Architectural, Track, Communications, MEP, etc.). WSP (PB) was awarded the Final Design Contract and worked as part of the NYCT Team. In 2018, the Design Team completed structural and geotechnical components of the Final Design for reconstruction of Times Square-42nd Street Shuttle Station. The construction is scheduled to begin in 2019.

STRUCTURAL DESIGN AND CONSTRUCTION CONSIDERATIONS

The current project will significantly modify and structurally reconfigure the existing subway station to dramatically improve station operations. The modified and reconstructed TSS Station will be shifted east approximately 315 feet from the current east end of the platform serving Track 3 into the existing IRT tunnels to accommodate two six-car long trains on Track 1 and Track 4. A new 28-foot wide center platform will be constructed to serve both trains. At the east end of the new platform new stairs and underpass would be constructed that would provide a new entry point and would connect to the existing passageway below the adjoining Durst Building that would provide a connection to the IND Sixth Avenue Line's 42nd Street–Bryant Park station.

Structural modifications include:

- Construction of a 28-foot wide center platform
- Removal of 122 existing 5-foot spaced columns along platform edges
- Removal of 11 mezzanine columns to improve passenger circulation
- Construction of 45 new station columns spaced at 15 feet and positioned at least 2 feet away from structural platform edges to accommodate tactile warning strip and for passenger safety.
- Construction of 45 new foundations to support the new station columns

All structural modifications and additional framing will be performed in the confined spaces of the existing subway tunnels and with limited headroom for framing modifications (Figure 6).

This reconstruction was designed to be performed in phases to maintain station operations during construction and minimize impact on passengers. All structural modifications would be performed underground from within the existing subway structure without affecting the integrity of the structure and avoiding excavation from the street level.

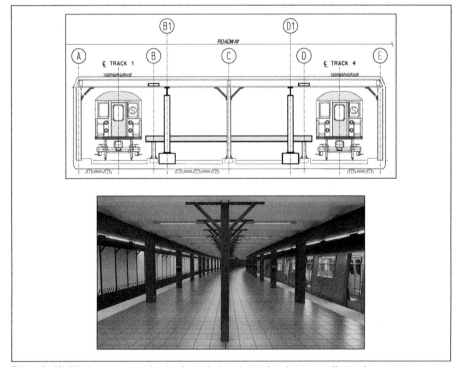

Figure 6. Modified structure section (top); rendering of completed structure (bottom)

Structural Design Criteria

Establishing structural design criteria was one of the most important steps on this project. NYCT Structural Design Guidelines DG-452 for existing NYCT subways were used for designing most of the structural modifications. Engineering evaluation of DG-452 at the start of the project confirmed that design loads, assumptions, and design methods used and implied in DG-452 were applicable to this project to ensure structural integrity and compatibility of the modifications with the existing structure and its elements. The Allowable Stress Design (ASD) method for steel and concrete construction was used for all modifications to ensure consistency and applicability of estimated loads and stresses with the existing structure. DG-452 design loads (Dead Loads and Live Loads) remain time-tested conservative loads that served well for over 100 years, and the original structures have performed satisfactorily and as intended.

All new structures and elements were designed to behave similarly to the existing structure to prevent differential deformations. The same considerations were applied in developing and designing a suggested construction concept that would maintain structural integrity during all construction phases and construction steps.

CONSTRUCTION CONSIDERATIONS AND SUGGESTED CONSTRUCTION CONCEPTS

Design and Construction Constraints

Feasibility and constructability of the proposed modifications were major parts of the design. Beginning in earlier conceptual studies and during the final design, it was apparent that some of the constraints and limitations could change and modify as project developed and subsequently moved into construction. However, recognizing that the constraints have defining impact on design, structure, structural elements, and construction methods a conservative approach was implemented. The final design was developed for the most stringent constraints and limitations that were identified in coordination with NYCT. TSS structural reconstruction had to meet the following requirements:

- The station must remain open and operational through all construction phases. Limited service reduction would be acceptable.
- All structural modifications must not affect train operations.
- All structural modifications must have minimum impact on passengers.
- All structural modifications must be performed underground from within the existing structures, without excavation from the street level.

To address all the constraints, a thorough research and study of the existing structure and its history was performed. As expected, very limited detailed information and drawings were available for the IRT subway stations and structures dating back to 1904. The Design Team focused on specific areas of the existing structure that would be affected by the modifications and new construction. Existing structures were analyzed, and it was confirmed that structural assumptions and results were consistent with project design criteria. A conservative assumption was made that all existing structural elements are fully loaded and there is no reserve capacity. The intent was to design all new structural modifications to prevent increasing the design loads and stresses on the existing structure and its elements. In most cases the loads on the existing structure and its elements remained unchanged or were reduced because of modifications.

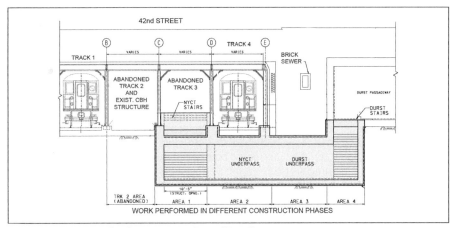

Figure 7. New stairs and underpass connection to "Durst" passageway

Constraints and requirements regarding station operations changed several times during the studies and design. These changes were driven by various other factors, such as other concurrent NYCT subway improvement projects and new constructions planned by developers that could affect the project. Design and suggested construction concepts were developed for various scenarios that would accommodate NYCT needs.

New Stairs, Underpass, and Connection to "Durst" Passageway

Construction of new stairs, a new underpass under operating tracks, and connection to existing 42nd Street passageway below the adjoining Durst building presented several structural and geotechnical engineering challenges.

Design was developed in conjunction with a suggested construction concept. After studying and evaluating the existing structures and conditions, four distinct work areas requiring individual custom design and construction methods were identified as Area 1 through 5 (Figure 7).

Construction in these four areas would be coordinated with overall station reconstruction phasing, including temporary track closures during "General Orders" (GO's). A general description of work in these areas is provided below.

Construction Area 1. Work in this area (Track 3) would be performed when Track 3 is closed, while Tracks 1 and 4 continue normal operations. A brick Circuit Breaker House (CBH) on Track 2 adjacent to the excavation area would be secured and monitored during the construction. Construction in Area 1 includes:

- Ground improvement below the CBH, Track 3 and Track 4
- Underpinning of existing columns on lines C and D
- Skeletonizing adjacent Track 4 (temporary track support in Area 2 would be performed during GO's prior to starting excavation in Area 1)
- Excavation between column lines C and D
- Construction of new stairs and part of new underpass between column lines C and D
- Re-supporting existing column lines C and D on anew underpass structure

Times Square Shuttle Station Reconstruction

Construction Area 2. Work in this area (Track 4) would be performed when Track 4 is out of service. Existing NYCT subway exterior wall and columns would be secured and underpinned prior to excavation in this area. Construction in Area 2 includes:

- Ground improvement below Track 4 and beyond the NYCT structure
- Underpinning of existing columns on line E
- Excavation between column lines D and E
- Construction of part of the new underpass between column lines D and E
- Re-supporting existing columns along line D on the new underpass structure

Construction Area 3. This area is located below the 42nd Street sidewalk and between NYCT subway and "Durst" passageway structures. Initial study considered open cut excavation from the sidewalk. However, open cut excavation would have significant impact at street level and 42nd Street lane closures that were not desirable. An old brick sewer (dating back to the original subway construction) would require support, maintenance, replacement and/or relocation, construction of new manholes, etc. Assessment of various requirements and regulations of the Department of Environmental Protection (DEP) indicated that it may not be feasible to re-construct a new sewer to meet required offset distances from the adjacent structures. Manual mining was therefore selected as the suggested construction method to avoid impacting the street and utilities (Figure 8).

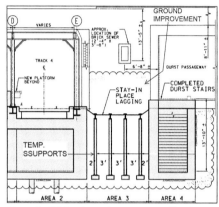

Figure 8. Mining for underpass connection to "Durst" passageway

Construction in Area 3 can start only after all work in Areas 2 and 4 are completed. This requirement is critical for maintaining structural integrity of both structures (NYCT subway and "Durst" passageway), preventing street and utilities settlements. Construction in Area 3 includes:

- Ground improvement as required
- Manual mining with closely spaced steel sets and continuous lagging to prevent ground loss.
- Construction of closing part of the underpass, between previously completed areas 2 and 4.

Construction Area 4. This area is independent of NYCT subway operations and, therefore, would allow the contractor flexibility for construction without coordination with NYCT track closures and station operations. Construction in Area 4 includes:

- Ground improvement below the "Durst" passageway
- Underpinning of the "Durst" passageway south exterior wall using drift excavation and construction of new stairs in defined increments of approximately 10 feet. The affected parts of the "Durst" passageway wall would be underpinned and re-supported on new stair walls and passageway roof.

Underpinning and Monitoring of Existing Structures

All existing structure elements affected by underpinning and structural modifications would be monitored during the construction. The main objective of monitoring would be controlling vertical deformations, deflections, and differential movements. Monitoring points would be established on the elements. Measurements would be recorded prior to start of construction to establish a "Base Line" for each location and element. Then measurements would be recorded prior to load transfer by jacking or other methods, during load transfer to temporary supports, and after final load transfer to completed structure. Taking into consideration the age of the existing subway structure, the type of structure (unreinforced concrete jack-arches) and its sensitivity to deformations, the deformation criteria was very stringent. The intent is to avoid deformations and maintain the existing structure undisturbed. The acceptable deformation criteria were established to be consistent with the existing structure deformations using Live Load deformations as guidelines. Working range of estimated acceptable deformations was established from $1/32$" to $1/16$". All new structural elements were designed with sufficient stiffness to limit deformations.

STRUCTURAL ANALYSIS

Finite Element Modeling

Three-dimensional finite element models of the TSS roof structure were developed by using a commercial analysis program SAP2000. The developed analytical model contains all significant structural components of the station roof that would be affected by new construction and modifications. Hand calculations were performed to verify the results of the finite element analysis and to conduct additional stress and deflection calculations for the modified structural members. Structural roof members were modeled using 2-node frame elements. Nodes of the frame elements were developed by using the centroidal axis of the structural elements. Columns supporting the roof structure were modeled using nodal restraints in the vertical direction. To accurately model the bearing connection between the existing roof member and the new transfer girder, compression-only truss members were used as bearing elements. The stiffness of the bearing elements was determined using a material strength in compression of 100× the Young's Modulus of steel in order to rigidly transfer all force to the transfer girder and accurately estimate the deformations in the system.

Existing roof beams were modeled between gridlines A-B and B-C. The new transfer girder was modeled along gridline B1. The nodal supports along column line A are pinned supports to model the roof beam connection with the sidewall column. The nodal supports along column line C are roller supports to model the roof beam connection with the interior station column. The nodal supports for the transfer girder along column line B1 represent the new platform columns with 15 feet spacing. The up-station column supporting the new girder was modeled as a pinned support, and the down-station column supporting new girder was modeled as a roller support. The moment is released at the J-joint for members in span A-B (at gridline B) to represent the shear connection between the two roof beam spans, which effectively acts as a hinge in the finite element model. Figure 9 illustrates the three-dimensional finite element model.

Material Properties and Section Properties

Appropriate material properties were defined for each specific structural member. Existing structural members were defined as A36 steel, while any new steel for the modified structural was defined as A992 (Grade 50) steel.

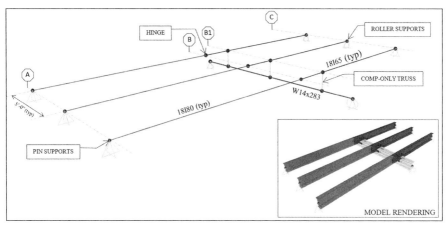

Figure 9. SAP2000 finite element model of TSS roof structure

Frame element section properties were applied based on actual section sizes of structural members.

- Span AB—18 I 80 (A36)
- Span BC—18 I 65 (A36)
- Transfer Girder—W14×283 (A992)

Section properties and other information regarding the existing structural sections were determined using "Historical Record—Dimensions and Properties Rolled Shapes" compiled by Herbert W. Ferris, 1953.

Loads and Load Combinations

The existing roof structures were analyzed using NYCT DG 452 Structural Design Guidelines. This design guide is primarily intended for projects that involve modification or rehabilitation of existing NYCT subway structures. The design guide prescribes the required loads and loading combinations to analyze the existing structure.

Per NYCT DG 452, the follow load combinations were analyzed:

- Dead Load
- Dead Load + Live Load

The dead load per NYCT DG 452 used in the finite element analysis was determinted to be 1.1 KSF, based on an equivalent cover of 11 feet. The live load per NYCT DG452 used in the finite element analysis was 0.6 KSF, which was applied on the overall structure in order to produce maximum bending on each specific member. Loads were applied to elements based on the 5-foot tributary width of each structural frame within the station.

Results

The SAP2000 finite element analysis provided design shear force and bending moments for all frame elements in the model to be used in existing member stress checks and new member design. Axial force in roof members were not considered for this analysis. Nodal support reaction forces were used to determine the design

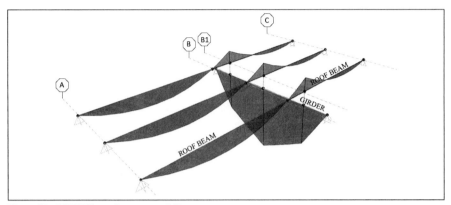

Figure 10. SAP2000 results—moment diagram

column axial forces. Frame element shear force at column line B was used to design the new shear tab connection between the two roof beam spans.

The resulting force and moment diagrams were used to verify the structural behavior of the modified structural system. In Figure 10, roof members from Span A-B exhibit a typical simply supported beam bending moment; while roof members in Span B-C have a negative bending moment at the location of the transfer girder support. From the finite element analysis, it was determined that this negative bending moment does not exceed the maximum bending moment in the beam prior to the modification of the structure.

Structural Design Methodology

The assessment and analysis of the existing structural member=s and analysis of the modified structure indicated that estimated design loads and stresses in existing structural members would essentially remain unchanged. The structure analysis of the existing structure confirmed that the original structure has been adequately designed and has performed as intended. Stress ratios from SAP2000 results were utilized to quickly draw conclusions on the adequacy of the existing roof members. Further assessment of the structural design was completed in order to check the geometry and applied loads in the SAP2000 model, in order to ensure that the forces and stresses used for design are accurate.

Additional assessment of estimated deflections in the Span A-B roof members was required in order to determine the optimal transfer beam size. A sensitivity analysis was performed to quantify the effects of using a stiffer transfer girder on the deflection of the existing roof beams. Overall depth of the roof transfer girder was limited to 18 inches to provide minimum 7-foot 6-inch clear headroom at platforms (NYCT minimum headroom requirement for public areas). Data was collected from various SAP2000 finite element analyses with transfer girder sizes ranging from W14×283 to W14×342. The design was optimized and the most economical structural section that would satisfy deflection requirements was selected. To control deformations of the existing roof beams at the new supports on line B1, the existing roof beam loads would be transferred to the new framing. The new roof framing beams would be cambered to compensate for the estimated deformations.

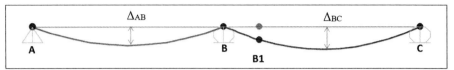

Figure 11. SAP2000 results—deformed shape of existing structure

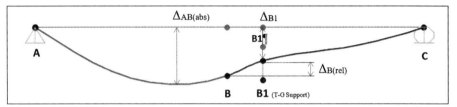

Figure 12. SAP2000 results—deformed shape of modified structure—without underpinning

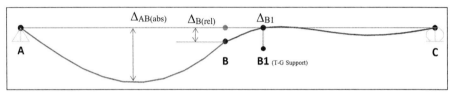

Figure 13. SAP2000 results—deformed shape of modified structure—with underpinning

Figures 11, 12, and 13 depict the displacement values critical to the assessment of the existing and modified station roof structure. The length of the specific member in question will be referred to as L for the remainder of this discussion. The absolute deformation, ΔB1, of the transfer girder was limited to L/500 for total load (Dead Load + Live Load), and L/1400 for live load deflection. These values were used to determine the required beam camber in transfer girder fabrication in order to limit overall deflection of the modified roof structure. The span deflection of roof beam A-B, ΔAB in Figure 11, is fixed from the existing structural configuration. The goal of the modification was to minimize the deflection of the roof at the removed column, or beam support (gridline B) in the below figures. The relative maximum deflection, ΔB(rel), in Figure 12 and Figure 13 represents the expected additional deflection in the TSS roof structure caused by the station modification. The relative maximum deflection, ΔB(rel), of the roof member span to the transfer girder support was limited to L/1000 for total load for the existing roof beams.

The expected deformation of the modified roof structure, after appropriate measures are taken to limit the deflection the transfer girder support at column line B1, is depicted in Figure 13. During the final design phase, a construction methodology was developed to limit the deflection, Δ_{B1}, of the roof structure at column line B1 using a combination of temporary shoring of existing roof structure and structural steel shimming at the bearing connection with the new transfer girder.

Column structural section sizes were determined using the column load from the SAP2000 analysis, with each column supporting the reaction from two roof transfer girders. Allowable axial stress from DG452 was used to appropriately size the new structural columns.

GEOTECHNICAL DESIGN AND CONSIDERATIONS

Subsurface Conditions

Subsurface conditions within the existing TSS Station were determined by six borings performed using low overhead boring equipment along the platform extension alignment. Standard Penetration Test (SPT) split-spoon sampling was performed in general accordance with American Society for Testing Materials (ASTM) Standard D1586. Bedrock was cored using a NX-size double-tube core barrel fitted with diamond bits in accordance with ASTM D2113. Based on the test boring information, the 1.5-foot thick concrete slab of the existing station box is underlain by a variable thickness of medium dense to very dense sand with silt overlying weather bedrock, that exhibits a significant dip from east to west along the platform extension. A 1 to 2-foot layer of miscellaneous fill was encountered directly below the station box. The thickness of the granular soils varied across the platform extension area, from about 1 to 4 feet at the east end of the site to about 20 feet at the west end of the proposed extension.

The bedrock is highly to slightly weathered gray Mica Schist with close to moderated fracture spacing. The retrieved rock samples have a recovery ranging from about 15 to 100 percent and RQD values ranging from zero to 86 percent indicating highly variable quality ranging from very poor to good quality rock. The noticeable rock quality variations may be attributed to an historic stream crossing under the 42nd Street Station within the limits of the proposed platform extension project as shown in Figure 14.

Foundations Design

Shallow foundations (spread footings) were considered feasible support systems for the proposed station extension. To accommodate the site conditions, caisson piles (micro-piles) socketed in bedrock, which can be installed under lower headroom and with minimal vibrations, were also considered a viable foundation alternative for this project. Both options were presented on the contract documents, allowing the contractor to select the more economical foundation based on actual field conditions and the contractor's proposed means and methods of construction.

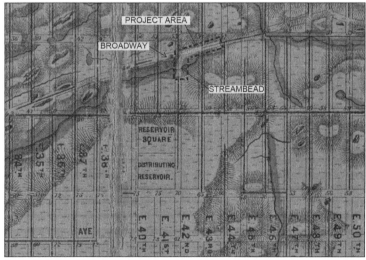

Source: Sanitary and topographical map of the City and Island of New York, 1865.
Figure 14. Historic streambed at the project area

The design dimensions and allowable bearing pressure for the proposed shallow foundation option were calculated based on a determination of the actual maximum bearing pressure of the existing spread footing column supports for the TSS box structure. The allowable bearing for the new foundations were maintained at or below the existing bearing loads for the existing spread footing foundations to ensure the consistent and acceptable settlements experienced with the original station structure.

To support the compression load of 220 kips on each caisson pile, the caisson pile was designed to be socketed into the underlying bedrock. The caisson pile consists of a 9.625-inch O.D. permanent casing with 0.592-inch wall thickness extending to 2 feet below top of rock followed by an uncased 8.5-inch diameter segment extending a minimum of 10 feet into rock. The caisson pile require filling with cement grout with a minimum compressive strength of 4,000 psi. Figure 15 shows the caisson pile detail.

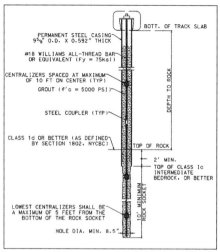

Figure 15. Caisson pile detail

The allowable structural compression capacity accounting for sacrificial corrosion thickness is estimated to be about 240 kips in accordance with NYCBC (2014), which is slightly higher than the structural load. Per Section 1810.7.3 of NYC BC (2014), the allowable bond strength of 200 psi was used for the rock socket in Class 1c rock or better. The allowable bond strength is also estimated to be about 70 to 120 psi using the unconfined compressive strength (Kulhawy et al., 2005). Based on the designer's experience with caisson piles in mica schist, an allowable bond strength of 100 psi without load test was used for this project. To support the structural load, a minimum of 10 feet rock socket will be needed.

Since the caisson piles will be installed under the third rail, corrosion from stray currents are expected to occur on these piles. To account for the corrosion, 1/16-inch sacrificial casing wall thickness was included for design of the caisson piles in accordance with FHWA NHI-05-039. The core steel reinforcement will be covered with about 3-inch grout, which can provide adequate corrosion protection as stated in the FHWA manual.

CREDITS

The authors would like to thank NYCT for dedication and valuable input during the studies and design. Special thanks to NYCT Operations Planning (OP) and David Haase (retired); Capital Planning and Budget and Planning (Tom Jablonski—project manager for study projects, and his staff); Capital Program Management (CPM), Anthony Febrizio (Program Manager) and Ashok Patel (Project Manager) and their staff—for Final Design; Maintenance of Way (MOW) and Antonio Cabrera for modified layout that met ADA requirements; NYCT CPM staff and Vikram Tadla (Construction Manager) for early involvement and input; NYCT Architects, NYCT Communication Engineers for critical input in coordinating Signals with construction phasing, and many others. The authors thank Dilip Kumar Patel for his contribution to this paper.

Last, but not the least, thanks to WSP (PB) dedicated staff that was always available to share specialized underground engineering expertise and helped during the studies and design: Lijun Shi and Victor Wu, Ray Castelli for expert geotechnical input and advice, Peter Torres for sharing his knowledge and experience with NYCT structures and operations, Giulia Gallo for structural engineering and project management help during the studies and Final Design; Artur Kasperski and Allen Berber for their contributions during the Final Design.

REFERENCES

1. MTA NYCT www.mta.info.

2. The New York Public Library, Sanitary and topographical map of the City and Island of New York; prepared for the Council of Hygiene and Public Health of the Citizens Association under the direction of Egbert L. Viele, Topographical Engineer, 1865.

3. NYCBC (2014).

4. Kulhawy et al., 2005.

5. FHWA NHI-05-039 Micro-Pile Design and Construction Manual.

6. "Historical Record—Dimensions and Properties Rolled Shapes" compiled by Herbert W. Ferris, 1953.

PART  3

Design/Build Projects

Chairs

Richard Taylor
Traylor Bros., Inc.

David Mast
AECOM

Controlling Risk of Tunneling Projects Implemented by Alternative Delivery Method

Nasri Munfah ▪ AECOM

ABSTRACT

The unpredictability inherent in underground projects and the accelerated pace of design-build or P3 tunnel projects can result in additional risks, construction delays, added costs and potentially expensive litigation if risk management strategies are not implemented early. This paper examines risks associated with alternative delivery methods of tunneling and underground projects from the owner, contractor, and engineer perspectives and identifies potential remedial measures to deal with them.

INTRODUCTION

A recent study indicates that approximately 55% to 60% of the tunneling and underground projects in North America will be delivered using Alternative Delivery Method (AD) such as Design Build (DB) or Public-Private-Partnership (P3). Tunnels and underground projects are inherently more risky than vertical projects, which typically utilize more conventional design and construction approaches and are in relatively defined conditions. However, on tunnel and underground projects, there are a host of special risk factors, including unknowns and uncertainties of the physical and behavioral characteristics of the ground, the complicated interdependence and interaction of design decisions and construction means and methods with those ground conditions, and the manner in which risks are allocated among project participants for unanticipated subsurface conditions. Tunnels delivered using alternative delivery (AD) methods such as design-build (D-B) or Private-Public-Partnership (P3), in their own respects and independent of any major subsurface component pose additional significant risk for all participants. Although the main driver of owners to procure projects using AD is financial and/or schedule improvement, often they transfer all, or substantially all, design and construction risks to the private sector consortium delivering the project including the design-build team. Often, this risk transfer includes rather onerous and aggressive contractual terms which allocate to the private sector participants substantially all risks associated with the encountering of unanticipated subsurface conditions. These aggressive risk allocation provisions impact all participants in the Alternative Delivery team and often have intensifying effect on the consulting engineer professional liability exposure.

With the increased use of D-B and P3 contracting, owners and their legal and commercial advisors may be losing sight of the importance of risk sharing especially as it relates to subsurface conditions. Furthermore, P3 projects add more complications due to the involvement of financiers, concessionaires, and sometimes the facility operators. It is recommended that risk mitigation measures be implemented with a focus on the variations between the traditional procurement of design-bid-build versus the alternative delivery methods of D-B or P3 in term of contractual arrangements and the roles and responsibilities of the various entities.

RISKS OF UNDERGROUND PROJECTS USING ALTERNATIVE DELIVERY

Tunneling and underground projects by definition carry higher risks than any traditional infrastructure projects. However, these risks have higher consequences and more impact on the owner and the alternative delivery team including the contractor, the designer and, in case of P3 projects, the financing and operation and maintenance entities. Some of the risks are transferred from the owner to the delivery team while others are flawed down by the concessionaire to the CJV and to the designer. These risks include:

- Geotechnical and geological risks
- Design development risks
- Quantities risk
- Schedule Risk
- Contractual risks

The geotechnical and design development risks will be discussed below.

Geotechnical and Geological Risk Allocation

There are several factors that influence geotechnical risk allocation in D-B and P3 tunnel projects:

- The scope of geotechnical investigation undertaken by the owner often is less comprehensive in D-B or P3 projects than in the traditional D-B-B projects because owners expect that the final design will be prepared by the D-B or P3 teams and therefore they will be responsible of the investigations, analyses and the development of design parameters.
- Owners often limit its geotechnical investigation during the preliminary design and the preparation of the bridging documents for cost saving and because they recognize that the geotechnical investigation for tunnel projects are driven by the final design of the underground facilities and by the means and methods to be used and the judgments and risk taken by the DB entity.
- The expectation that the DB or P3 team will do its own geotechnical investigations, analyses, and the development of design parameters either during the pre-award or post-award of the contract. However, the limited time and funding available during the tendering period prohibits D-B contractors from doing additional geotechnical investigations.
- Owners often transfer more of subsurface risks to D-B or P3 than in the traditional D-B-B projects. And often they include disclaimers regarding subsurface conditions and the accuracy and the reliability of data and reports furnished by the owner.

As a result, the geotechnical and subsurface risks are often shifted to D-B or P3 team rather than developing an equitable approach for risk sharing considering that the owner owns the ground whether the contract is executed by the traditional D-B-B or by alternative delivery methods.

Design Development Responsibility of Alternative Delivery Teams in Tunneling Projects

Unlike in the traditional D-B-B contract, in D-B and P3 projects the contractual requirements are that the D-B or P3 team will be responsible for the development of the final design of the permanent work elements in accordance with the project criteria, owner's requirements and standards, and the adequacy and constructability of the design. And in many cases, the design must meet "fit for purpose" standards. This last issue could be a potential significant risk for the designer as it may not be covered by its professional liability insurance.

There are several factors in the roles and responsibilities of the project owner and D-B or P3 team that may influence whether those contractual expectations relating to the responsibilities of design development are achieved:

- Whether the owner furnishes and mandates design criteria which are appropriate in the context of anticipated and actual subsurface conditions.
- Often the owner and/or the contractor does not allow sufficient time or funds for the D-B engineer to conduct its own subsurface investigations needed to support its final design or construction approaches.
- Whether the owner provided subsurface conditions and ground behavior during excavation are compatible with the D-B team's intended design approach and construction means and methods including equipment selections
- If the owner provides detailed design or detailed prescriptive specifications which are mandated that the D-B or P3 team is deprived from the development of innovative design, exercise judgment, or discretion of the design.
- Whether the owner imposes its preferences on the D-B or P3 team exceeding its rights under the contractual terms. In this situation, the owner restricts the D-B team's ability to exercise its judgment in the design and the development of innovative approaches.
- Whether the owner contractually (or otherwise) retains and exercises a dominant and controlling role over the review and/or rejection of the D-B team's proposed design submittals

It is important to acknowledge the interrelationship between the roles and responsibilities of the owner, the D-B contractor, and the D-B engineer when allocating risk sharing among the various entities. In D-B and P3 tunneling projects, given the roles and responsibilities of the D-B team with respect of the geotechnical assessment and the design development, logically it is expected that the D-B team to have greater responsibility for unanticipated subsurface conditions. However, the lack of funds and time to perform geotechnical investigations prior to bidding, the actual degree of project owner involvement in those areas, and the fact that the owner owns the ground, limit the ability of the D-B team to have a greater control and thus responsibility on the ground conditions and behavior. This in turn influences the design development that is adequate for the anticipated and actually encountered subsurface conditions. That said, fairness of risk allocation should be accomplished based on the different roles and responsibilities of the various entities.

The interrelationship between roles and responsibilities of the project participants in D-B and P3 of underground projects related to the geotechnical and subsurface conditions and design development must to be acknowledged in the decision making process regarding risk allocation or sharing.

ROLES AND RESPONSIBILITIES

The roles and responsibilities of project participants in D-B or P3 underground projects influence the risk allocation models. The risk allocation models often utilized in the traditional D-B-B projects do not precisely work with the differing roles and responsibilities of project participants in D-B and P3 projects. Unfortunately, the industry have seen the same approaches in risk allocation provided by the owner on projects delivered using alternative delivery methods as they are for the traditional D-B-B projects.

Roles and Responsibilities in the Conventional Design-Bid-Build

In the conventional DBB approach, the owner retains a design engineer; directs the scope of work and the geotechnical investigations; oversees the characterization of the ground and its behavior during excavation. He also elects to disclose (or not) the subsurface data and related reports; he defines subsurface conditions risk allocation based on the geotechnical investigations and interpretations using a Geotechnical Baseline Report (GBR). He through his designer prepares a final and complete design documents. He often also retains a construction manager to evaluate construction work for conformance with the design and the contract document requirements, and to observe and evaluate subsurface conditions encountered during construction to validate consistency and appropriateness of the encountered conditions with those contemplated in final design assumptions and approach.

In the traditional D-B-B contract, the contractor has the obligation to plan and price the work required by the contract documents; bid it based on the contract terms and conditions, to construct the work in accordance with those requirements; to plan and implement construction means and methods; and to procure the equipment appropriate for the performance of the work and to deliver the final facility in accordance with the design requirements provided by the owner.

In this situation, the risk allocation is often straight forward, but not necessarily fair and appropriate which may result in higher contingency imposed by the contractor to address potential risks whether they materialized or not.

Factors that contribute to the risk allocation/sharing in the traditional D-B-B projects include:

- The scope of geotechnical and geological investigation and their inclusion (or not) as part of the contract documents. And the presence of a provision in the contract for a potential equitable adjustment if subsurface conditions encountered during construction materially differ from the conditions indicated in the contract documents.
- The actually encountered subsurface conditions are consistent with the expected conditions in the design and in the implementation of construction approaches and methodologies.
- The adequacy and the constructability of the design provided in the contract documents consistent with the anticipated and actually encountered subsurface conditions.
- The implementation of suitable construction means and methods and the quality and conformance of construction work with the contract document requirements.

Based on the above, in D-B-B it is possible to more precisely define the roles and responsibilities regarding subsurface conditions than in D-B or P3. However, even in D-B-B of underground projects there are interdependencies in those roles and responsibilities that often result in disputes and claims. For example, the means and methods are significantly influenced by ground behavior during excavation, yet the ground behavior is the owner's responsibility while means and methods are the contractor responsibilities.

Roles and Responsibilities in Design-Build and Public-Private Partnership Projects

In D-B or P3 projects, the owner retain an engineer (or self-perform) to develop the project definition and design concept, perform preliminary geotechnical investigations and establish design parameters. The owner's engineer also prepares the preliminary design and the bridging documents for bidding. Design criteria and standards, owner's requirements, and performance specifications will also be developed and included in the contract documents.

The D-B entity develops the final design of the final work elements, performs more geotechnical investigations if desired/needed, plans construction sequencing and staging, and develops construction means and methods, and implements the work in accordance with the project design criteria, standards, and owner's requirements. In P3 contracts, the P3 team also finances and sometimes operates and maintains the facility for a number of years. This adds additional complications with respect to the relationships between the owner, the P3 concessionaire and the D-B team.

The allocation of risks in D-B or P3 underground projects is more complex and difficult than in the traditional D-B-B contract. In D-B and P3 projects there is less experience and thus less standardization of the risk allocation than in the traditional D-B-B. Also, for D-B or P3 projects, there are significant variations within the industry in the contractual terms delineating the roles and responsibilities among the entities. Furthermore, often there is inconsistency between the contract terms and the actual conduct of the various entities. This poses a greater challenge in achieving fair and effective risk allocation/sharing because the roles and responsibilities among project participants are blurred often resulting in claims and disputes.

The lack of experience and understanding of many project participants (owners and design-builders or P3 entities) of the roles and responsibilities, and resistance to change in their traditional roles often result in disputes. Such behavioral changes are critical for successful delivery of D-B or P3 tunneling projects. Furthermore, the dependency of tunneling projects on the geotechnical conditions and ground behavior complicates the risk allocation/sharing in D-B and P3 projects.

It should be acknowledged that whether the project is delivered using the traditional D-B-B or by alternative delivery method, the owner "owns" the ground, as such, ground behavior during excavation should be well defined either by the owner, or the D-B team and agreed to by the both parties via the GBR. Various approaches of reaching a final contractual GBR in D-B and P3 projects have been implemented including the use of GBR-B (bidding) and GBR-C (construction). GBR-B is usually established by the owner's engineer on the basis of the owner's geotechnical investigation program and the preliminary design prepared for the tender documents. The focus of GBR-B is the physical nature of the subsurface conditions likely to be encountered, consistent with the layouts and geometries represented in the preliminary design and the owner's anticipated construction means and methods. This will allow all bidders to bid on a common basis. The degree to which the GBR-B provides behavioral baselines will be a function of the level of specificity in the preliminary design, and the imposition of the means and methods by the owner, and the desire of the owner. GBR-C reflects the physical baselines established by the owner and its design team (as augmented by any supplemental geotechnical investigations done by the D-B team) and as clarified or modified by the D-B team, and the behavioral baselines described by the D-B team consistent with its design approach, equipment, and means and methods. GBR-C becomes part of the contract and "relied upon" document. Materially different conditions encountered from those anticipated will be legitimate changes.

RISK MANAGEMENT

Although there are generally accepted approaches for allocating or sharing risks among the various entities in the traditional D-B-B tunneling projects, such standardization rarely exist in alternative delivery projects because these projects rarely fit in the standards of D-B-B projects or with other alternative delivery methods. The variability of the site conditions, the roles and responsibilities of the various entities and the owner's desire of risk sharing require that risk allocation be implemented on a case by case basis. However, it is prudent that risk allocation be sensible and assign to the entity that is most suited to deal with the specific risks. Regardless, project specific risk allocation decision making process would be enhanced by the availability of more industry wide generic guidelines that identify relevant factors for consideration during the risk allocation. A simple example would be, regardless of the type of project delivery, unforeseen ground condition risks should be allocated to the owner, while risks related to means and methods would be allocated to the contractor. However, the situation gets murky regarding risks associated with geotechnical aspects because, although the owner owns the ground, the D-B engineer, who is often the engineer of record, should determine the ground behavior during excavation.

The challenges of risk allocation are further impacted by disconnects and deviations between the contractual definitions of roles and responsibilities, and risk allocation and the actual performance of project entities during implementation due to their lack of experience in D-B or P3 projects or their desire to retain control.

Achieving fair risk allocation for D-B and P3 underground projects needs to adequately account for the interrelationships among the geotechnical investigation and assessment performed by the owner, identification of anticipated subsurface conditions, determination of the ground behavior during excavation, and the level of design completion by the owner and how those conditions relate to the project mandated design criteria and standards.

In addition, it is critical to delineate the roles and responsibilities between the owner's engineer and the D-B team's engineer with respect to the geotechnical investigation and the development of design parameters and the anticipated ground behavior during excavation and the ability of the D-B team engineer to exercise judgment and discretion in the development of the design especially during pre-award (bidding phase) and the suitability of that design for the anticipated and/or encountered subsurface conditions.

The delineation of roles and responsibilities between project owner and the design-builder or the P3 entity for subsurface conditions and for design development has a major impact on risk allocation. Furthermore, the actual performance from contractually defined roles and responsibilities could result in more disputes and that the ultimate risk allocation determination for may not align with or conform to the contractual expectations of project participants. The potential of dispute, the dispute resolution process, and the unpredictability of the outcome will further impact negatively the relationship among the project participants, add cost, and increase contractor's contingency.

RISK SHARING

Although the main drivers of owners to procure tunneling projects using D-B or P3 are schedule and cost improvement and in case of P3 access to private funding, owners have tended to transfer most, if not all, design and construction risks to the D-B or P3 entities. This is often done using onerous contractual terms that allocate to the D-B or P3 entities substantially all the risks including unanticipated subsurface conditions risks.

Geotechnical and subsurface condition risks are by far the most important in any tunneling and underground project regardless of the method of procurement. Unknown ground conditions and unanticipated ground behavior during excavation pose serious risks. Furthermore, design solutions and construction means and methods are based on the anticipated ground conditions and its behavior during excavation. These decisions are made by the D-B or P3 contractor within short period of time and with limited geotechnical investigations relying solely on the geotechnical investigation performed by the owner's engineer and provided as part of the tender design; often are not part of the contract documents and provided as reference documents only. The geotechnical and subsurface risks are magnified when the project is being delivered in alternative delivery method especially if the owner transfers these risks to the private entity. This is further complicated in D-B and P3 projects when the D-B contractor proposes an alternative technical concept (ATC) and the owner accepts it. In this case the applicability of the owner provided geotechnical information and risk sharing tools such as the GBR (GBR-B or GBR-C) is questioned when disputes arise and when claims of differing site conditions (DSCs) are made. Owners often take a position that the D-B contractor, through its contractual obligations, should investigate and evaluate the geotechnical conditions affecting its proposed ATCs prior to its submittal. However, time limitation would prohibit the design-builder for conducting in-depth geotechnical evaluation and would rely on the owner provided geotechnical data and GBR. Therefore, it is important to clearly define where the subsurface risks are allocated under these conditions keeping in mind that contractors will add contingencies to their bids when unanticipated risks are allocated to them whether these risks are materialized or not.

Subsurface conditions are the single most cause in disputes and claims in tunneling and underground projects. Therefore, it is critical that the principles of fairness and balance in risk sharing should be adopted and implemented.

For a successful delivery of D-B or P3 tunneling projects, a fair and equitable risk sharing rather than risk shedding should be implemented. Risks should be identified during the preparation of the tender documents and assigned to the entity most suitable to deal with them. Furthermore, it is logical for the owner to assume risks related to the site and subsurface conditions provided clear definitions and delineations in the contract documents as to the limits, scope, and conditions covered.

STRATEGIES FOR SUCCESS

To achieve a successful delivery of a tunneling or underground project using D-B or P3 it is recommended that the following principles should be used when preparing the contract documents and when implementing the work.

- Fair and balanced risk-sharing, fully disclosing geotechnical information and using judicious parameters in the geotechnical baseline report (GBR) reduces risk and avoid placing large contingency budgets in the bids. The

use of GBR-B and GBR-C has proven to result in less disputes and claims and successful delivery of D-B tunneling projects.
- Early contractor involvement provides opportunities for innovative approaches, collaborative strategies, and risk sharing practice.
- Pre-qualifying the design-builder or the P3 concessionaire ensures the team's technical expertise and personnel availability aligns with the project's specific needs and provides the team financial and technical viability to deliver the project successfully.
- Implementing a comprehensive risk register through design and construction, owners and contractors work together to identify potential risks that may surface over the project's lifetime.
- Placing contingency funds by the owner to deal with unknowns reduces unallocated contingencies by contractors and allows owners to control the project contingency. However, sensible and fair allocation provisions of the contingency funds must be provided.
- Escrow bid documents, impartial dispute review board, and partnering help owners and design-builders promptly resolve disputes, claims and controversial issues

CONCLUSIONS

It is undeniable that more tunneling and underground projects will be delivered using alternative delivery methods such as DB, P3, CMAR or similar methods. To be successful, it is prudent for project participants (owners and private entities) to understand the potential risks with respect to their roles and responsibilities and to develop a fair and sensible risk sharing mechanism, rather than risk transfer, in which the risk is assigned to the entity most suitable to manage it. Similarly understanding the roles and responsibilities in the design development at the various stages of the project is critical for the success of D-B or P3 underground projects. Fair, and realistically achievable risk sharing is important to not only the primary project participants, such as the project owner and the design-builder or P3 teams, but also to other involved parties critical to the success of the project, such as designers, insurers, and financiers and above all the public who will suffer when projects are delayed and costs are escalated.

REFERENCES

1. D.J. Hatem and N.A. Munfah "Controlling Risks of Tunnel Projects" RETC 2015.

2. D.J. Hatem, "Design-Build and Public-Private Partnerships: Risk Allocation of Subsurface Conditions," Geo Strata (April 2014).

3. D.J. Hatem, "PPP and DB: Who is Responsible for Risk? A Call for Guidelines," North American Tunneling Journal (October 2014).

4. D.J. Hatem and D. Corkum, eds., Megaprojects: Challenges and Recommended Practices (ACEC 2010).

5. D.J. Hatem and P. Gary eds., Public-Private Partnerships: Opportunities and Risks for Consulting Engineers, Chapter 8 (ACEC 2013).

6. D.J. Hatem, "Public-Private Partnerships and Design-Build Subsurface Projects: Consulting Engineer Professional Liability Risk," Design and Construction Management Professional Reporter (April 2014).

7. D.J. Hatem and D. Corkum, eds, Megaprojects: Challenges and Recommended Practices (ACEC 2010).

8. D.J. Hatem "Risk Related to PPP Tunneling Projects for Design Professionals," Tunneling and Underground Construction June 2014.

9. D.J. Hatem and N. Munfah "Risk Related to P3 Tunneling Projects for the Design Professionals"—T&UC Magazine—June 2014.

10. N. Munfah et al. "Contractual Challenges of Conventional Tunneling in the US," World Tunneling Congress Proceedings, Bankok, Tailand, May 2012.

11. N. Munfah "Contracting Issues for Conventional Tunneling," ITA Conference Proceedings, Prague, Czech Republic, 2007.

12. N. Munfah "Contracting Methods for Underground Construction," NAT 2006 Proceedings, Chicago, 2006.

13. N. Munfah "Contracting Practices for Underground Construction," Underground Construction Conference Proceedings, 2003, London.

14. N. Munfah et al. "Minimizing Risks in Underground Construction Using the DBOM Approach: A Case Study." Proceedings of the International Congress of Underground Construction in Modern Infrastructure, Stockholm, Sweden, 1998.

DC Clean Rivers Project: We Built Miles of Tunnel...Now What? How to Get the CSOs into the Tunnels in an Urban Environment

Ray Hashimee ▪ EPC Consultants, Inc.
Aliuddin Mohammad ▪ EPC Consultants, Inc.
Yama Ebrahimi ▪ Belstar, Inc.
Alper Ucak ▪ McMillen Jacobs Associates
Moussa Wone ▪ DC Water

ABSTRACT

The District of Columbia Water and Sewer Authority is implementing its $2.7B Clean Rivers Program to control combined sewer overflows to the Anacostia and Potomac Rivers. The program consists of large diameter tunnels and a number of diversion facilities to be constructed under a very tight consent decree schedule, in an urban environment, in and around 100-year old live sewers. This paper focuses on the challenges and the lessons learned while constructing diversion facilities at DC Water's Main Pumping Station construction site, contracted using a combination of two delivery methods, to successfully meet the consent decree for healthier waterways in the nation's capital.

INTRODUCTION

The District of Columbia Water and Sewer Authority (DC Water) is implementing its DC Clean Rivers Project (DCCR) to control combined sewer overflows (CSOs) to the Anacostia and Potomac Rivers and Rock Creek. There are 47 active CSO outfalls; 14 empty into the Anacostia River, 10 empty into the Potomac River, and 23 empty into Rock Creek. DCCR will reduce CSOs annually by 96 percent throughout the system and by 98 percent for the Anacostia River alone.

As part of the Anacostia River sewershed CSO control objectives, a 13-mile-long tunnel system is being constructed approximately 120 feet below ground for storage and conveyance of CSOs during wet weather events and will be treated at DC Water's Blue Plains Advanced Wastewater Treatment Plant. The tunnel system consists of four tunnels: Blue Plains Tunnel (23-feet inside diameter (ID)), Anacostia River Tunnel (23-feet ID), First Street Tunnel (20-feet ID), and Northeast Boundary Tunnel (23-feet ID), as shown in Figure 1. There are 22 drop shafts located along the Anacostia River tunnel system. The drop shafts, associated diversion structures and other near-surface structures divert flow into the tunnel and are constructed in and around 100-year old combined sewers.

The Main Pumping Station Diversions contract (Division I) is an essential part of DCCR. Two diversion chambers installed near Tingey Street divert flow from the District of Columbia's antiquated sewerage system into the Blue Plains Tunnel. Furthermore, the contract had to be substantially complete to allow the Blue Plains Tunnel and the Anacostia River Tunnel to be operational.

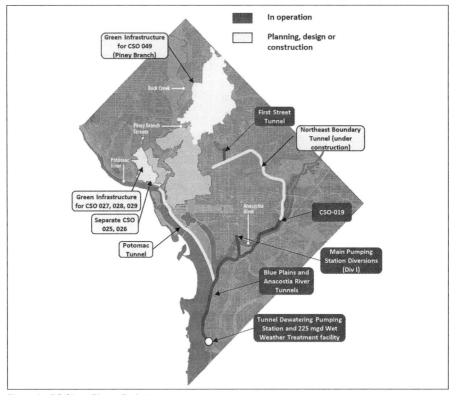

Figure 1. DC Clean Rivers Project

DIVISION I—MAIN PUMPING STATION DIVERSIONS CONTRACT

The individual components of the Main Pumping Station Diversions contract are shown in Figure 2. The project consisted of:

- Construction of two diversion chambers (CSO 009/011A and CSO 012) over two existing 14-foot high by 14-foot wide active trunk sewers to divert combined sewer overflows into the Blue Plains Tunnel via the existing 55-foot diameter by 120-foot deep Main Pumping Station drop shaft;
- Construction of ventilation control facilities within the Main Pumping Station drop shaft roof at the terminus of the Blue Plains Tunnel to regulate air flow in the tunnel system;
- Construction of the internals and the roof of the existing Main Pumping Station drop shaft;
- Extension of a 72-inch Tingey Street diversion sewer; and
- Rehabilitation of the Tiber Creek trunk sewer.

Contract Delivery and Construction Means and Methods

A hybrid project delivery method was used to procure the contract due to complexity of the project and the collaboration requirements prior to selection of the Design-Builder. Using a hybrid method where some elements of the work were designed by DC Water's designer and some elements of the work were designed by the Design-Builder

DC Clean Rivers Project: We Built Miles of Tunnel...Now What? 159

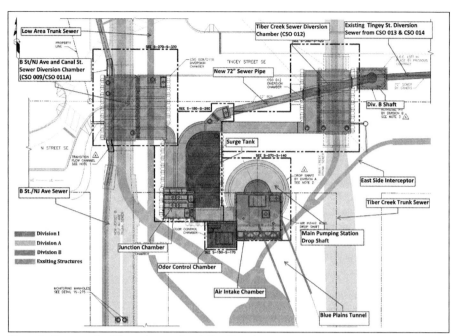

Figure 2. Components of the main pumping station diversions contract

presented a unique challenge to the project team. The team was very familiar with the traditional design-build delivery method based on lessons learned from previous projects, but the hybrid delivery method required some on-the-job learning for both parties. The permanent structures (the diversion structures, drop shaft internals and roof, and the air intake building) were designed by DC Water.

Although the new 72-inch Tingey Street diversion sewer extension is relatively short, it passes underneath the existing 100-year old Tiber Creek sewer with limited cover. The construction means and methods had to be chosen such that during mining, the structural integrity of the existing sewer was not impacted, while maintaining existing flow. Hence, to provide flexibility to the Design-Builder, all the design work associated with the extension of this short sewer tunnel (ground improvement, pipe-jacking, final lining, temporary bypass systems, protection of structures, ground improvement and instrumentation) was engineered by the Design-Builder.

The design of the CSO 009/011A and CSO 012 diversion chambers and the junction chamber utilized composite structural walls. For these structures the support of excavation (SOE) system was designed by DC Water (secant pile walls). The design of the remainder of the SOE system was the responsibility of the Design-Builder.

Rehabilitation of the Tiber Creek trunk sewer work was a change order to the contract. While constructing the Main Pumping Station drop shaft in the previous contract, large cracks were discovered in the sewer walls, and the sewer was retrofitted with temporary steel bents as shown in Figure 3a. Due to hydraulic considerations, the steel bents had to be removed during this contract, and the existing sewer needed to be retrofitted with a six-inch thick reinforced concrete (pneumatically applied) liner as shown in Figure 3b. The rehabilitation work had to be performed through an existing 36-inch manhole, while maintaining existing flow in the sewer. Given the schedule and

Figure 3. Rehabilitation of the Tiber Creek Trunk Sewer: (a) prior to rehabilitation work; (b) after rehabilitation work is complete

logistical constraints, this work was designed by DC Water. During the design phase, the Design-Builder and DC Water engaged in several collaboration sessions to identify and discuss the most suitable construction means and methods; the design was finalized accordingly.

Construction Schedule

The project team utilized its extensive design-build experience to successfully deliver the project. One of the primary advantages of design-build project delivery is that it compresses schedule by allowing elements of construction to proceed while other elements are being designed by the Engineer of Record. Conversely, the project team also kept in mind the construction adage, "schedule slippage during the final design phase is lost forever." The flowchart in Figure 4 was used to mitigate schedule slippage.

To eliminate schedule impacts associated with the design phase and, thereby eliminate multiple design review periods, the Owner/project team worked together to evaluate the different design packages and determine if they could be recategorized as construction phase submittals instead of design packages. Several packages were recategorized and this helped the project schedule enormously.

DIVERSION STRUCTURES

One of the key challenges for this project was maintaining active sewer flow and protecting the existing structures and trunk sewers passing through the project site while constructing diversion chambers in and around the B Street/New Jersey Ave. and the Canal and Tiber Creek sewers (Figure 5). These challenges presented due to the current structural conditions found at the site.

The Design-Builder was also required to ensure that its construction activities did not adversely affect the daily operation of the Main Pumping Station which handles dry weather flows. This included preventing construction debris and materials from getting into the cunette of both the Tiber Creek trunk sewer and the B Street/New Jersey Ave. trunk sewer and eventually ending up in the wet well which potentially could damage the pump screens. Operation of the pumps was vital to DC Water operations. Maintaining the structural integrity of the arches in the Tiber Creek sewer during construction and ensuring that flows were not impacted by construction activities represented a significant constructability issue. To mitigate this risk, the Design-Builder constructed a combination of flow projection shields over the cunettes at both the

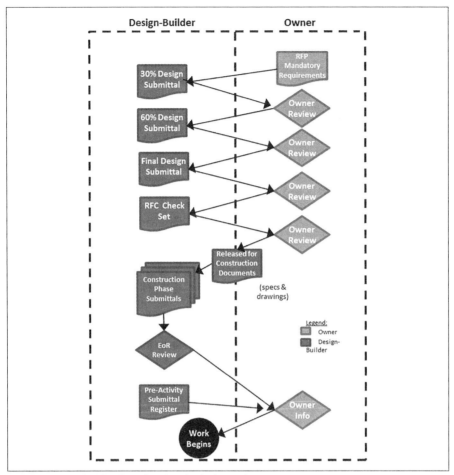

Figure 4. Design-build delivery flowchart for design packages

Tiber Creek trunk sewer and the B Street/New Jersey Ave. trunk sewer to maintain existing flows.

The flow protection shields served two primary functions: to allow maintenance of existing flows to keep them free from debris and to serve as a supplemental demolition shield during arch removal. The flow protection shields were designed to handle impact loading, thereby providing an additional protection measure during the arch demolition operation. These flow protection arch shields were comprised of liner plate arch rib segments with wood lagging nestled into the webs of the ribs. Arch demolition commenced only during the dry weather conditions. All debris, materials and equipment were removed and the arch shield anchored into the existing trunk sewer benches at the end of each work shift when the potential for forecasted rain was greater than 30 percent. These shields also provided a measure of safety to protect the working crew from falling into a cunette and being swept away.

The original contract did not foresee full replacement of Tiber Creek trunk sewer base slab within the footprint of the new diversion chamber. After an extensive preconstruction survey which also included a coring program of the existing slab, it was

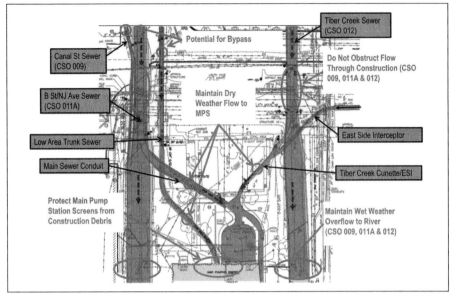

Figure 5. Maintenance of flow

determined that the existing sewer invert slab needed to be replaced. The challenge of maintaining existing flow during the demolition and replacement of the new invert concrete slab was tackled through collaboration and partnering among the Design-Builder, DC Water's representative, the Program Consultant Organization (PCO); the Engineer of Record (EOR), and the construction management team during design and constructability reviews.

The solution included installing an 11-foot diameter steel flume over the entire length of the new slab and diversion chamber. The flume was secured using steel beams traversed across the structure. The installation of the flume allowed unimpeded flow during dry and wet weather and allowed the Design-Builder to work on the diversion chamber without having to worry about rain. DC Water was also concerned about the potential for a heavy storm event during demolition of the existing arch and construction of the new diversion structures. To mitigate this risk a four-foot high protection wall made of steel beam, wood lagging and poly was installed around the perimeter of each structure.

These mitigation measures implemented by the Design-Builder resulted in a successful program to maintain existing flow and protect critical on-site sewage conveying facilities.

CONSTRUCTION CHALLENGES

Throughout the design and construction phases, the entire project team was focused on keeping this critical sewer infrastructure protected from additional damage due the construction activities and minimizing any interruption to the daily operation of the adjacent Main Pumping Station. Protection of the adjacent structures was of critical importance to the successful completion of the project. Due to the age of these sewer structures, DC Water was concerned with construction impact loading and as such required the Design-Builder to develop Construction Impact Assessment Reports for all Tier 1A, 1B, and 2 structures. Tier 1A structures were existing structures that were expected to be impacted beyond the design criteria. Tier 1B structures were existing structures that were expected to experience non-structural, minor damage at the ground. Tier 2 structures were existing structures that were not expected to be impacted beyond the design criteria.

Under a previous contract, structural concrete slabs were poured in the areas where the sewers were, so they could be protected from any equipment surcharge loading. Under this contract, the Design-Builder was responsible for:

- Performing pre-construction condition surveys,
- Assessing tier levels,
- Developing a Response Level Plan including Action and Maximum levels;
- Designing for Tier 1A protection scheme, Tier 1B repair scheme; and
- Providing a robust Instrumentation and Monitoring Plan which included surface settlement points, utility monitoring points, inclinometers, extensometers, structure monitoring, crack gauges, seismographs and tilt meters.

These mitigation measures implemented by the Design-Builder helped protect the live sewers, resulting in zero damage to these sensitive structures throughout the construction work.

The Design-Builder used an arch shield to sequentially demolish the existing arch and no flume was utilized since the existing B Street/New Jersey Ave. sewer invert slab was not required to be replaced, as shown in Figure 6. This work was performed during the dry weather condition where the potential for rain was less than 30 percent. During wet weather, all work was suspended and the flow from the sewer inundated the entire work area. After each rain event, the entire work area had to be power washed to ensure the safety of the crew prior to the recommencing work.

Figure 6. B Street/New Jersey Ave. Sewer—diversion chamber construction

As previously discussed, the Tiber Creek Sewer invert slab had to be replaced due to its deteriorating condition (Figure 7). The Design-Builder used an arch demolition shield to demolish the brick sewer arch and a 11-foot diameter flume to replace the invert slab. With the installation of the flume, the Design-Builder was able to work without any interruption due to wet weather flows.

Coordination with other projects within the Clean Rivers Program and outside stakeholders was critical to maintaining project schedule and ensuring that all the required contractual milestones were met, including removal of the tunnel boring machine from the shaft, and the final

Figure 7. Tiber Creek Sewer during flume construction

instrumentation and commissioning of the system by others. The terminus point of the Blue Plains Tunnel project was located within the construction site which meant certain areas of the site needed to be available for up to 90-days for the tunnel boring machine removal. This required a great deal of resequencing of the work and coordination with multiple internal stakeholders. Also, as part of the final commissioning to get the entire system up and running to meet the program consent decree requirements, certain structures needed to be turned over to the follow-on Contractor so that they could prepare the site for the follow-on work.

Working in an urban environment was another challenge that the project team successfully overcame. The project site was located in a busy area of the city, which has seen sustained development in the last 10 years. Key challenges included proactively dealing with the local residential, commercial and governmental agency stakeholders. This was accomplished by having a designated public relation team focused on resolving stakeholders' concerns and working with the Design-Builder to resolve issues right away. Special monitoring equipment such as seismographs, geophones, and accelerometers were installed within proximity of the site to collect noise and vibration data during the construction and this information was shared with the stakeholders.

SAFETY AND QUALITY

Working around live sewers proved to be another challenge to the project team. Several risks/challenges related to the safety of the crew had to be mitigated including: potential flows from upstream of the work area, potential accidental flows inundating the work area from the adjacent Main Pumping Station due to pump failures and human error, and malfunction of the inflatable dams that kept water from entering the sewer from the Anacostia River. Work was only performed in the sewers during the dry weather. Safety mitigation measures included the use of flow protection shields, high water alarms both upstream and downstream of the work area, and a robust lock out/tag out program which included the development of Safe Clearance Memo on a daily basis prior to anyone being able to perform any work.

The project safety performance was excellent with no lost time over 300,000 man-hours worked. The key to this success was a partnering approach which included a relentless focus on safety by the Design-Builder, the construction management and the PCO.

During the construction phase, the Design-Builder was responsible for both quality assurance and quality control (QA/QC). The Design-Builder was required to provide a sufficient number of QC representatives distinct and separate from the design and production staffs, to document inspection and testing of all work to determine conformance to the Release for Construction (RFC) documents, and use written QA/QC inspection and testing procedures for all operations. QA/QC documentation was regularly submitted to DC Water for review.

The Owner's onsite representatives provided Independent Verification and Assurance (IVA) by verifying whether the Design-Builder was complying with its approved QA/QC Plan, conducting spot checks to verify whether the Design-Builder's work complied with the RFC documents, recording all observations in the Daily Surveillance Report, and issuing Non-Conformance Notices (NCNs) for defective work and/or QC requirements. The Design-Builder's EOR or DC Water's EOR provided acceptable resolutions for NCNs, depending on which entity designed the work.

CONCLUSION

The construction of the Main Pumping Station Diversions required collaboration and close coordination among stakeholders. Site constraints, maintaining active sewer flow and protecting existing trunk sewers were the key challenges during construction. No work could be performed in the sewers when the chance of rain was greater than 30 percent, which further complicated the project given the rainy seasons of Washington, DC. Continuous communication was key for successful and timely completion of the project, which was required for the Blue Plains Tunnel and the Anacostia River Tunnel to become operational.

Design and Construction of Montreal Express Link Tunnels and Underground Stations

Verya Nasri ▪ AECOM
Xavier De Nettancourt ▪ AECON
Philippe Patret ▪ SNC Lavalin
Thomas Mitsch ▪ Dragados

ABSTRACT

Once completed, the Montreal Réseau Express Métropolitain (REM) will be the fourth largest automated transportation system in the world. For the metropolitan area, the REM also represents the largest transportation infrastructure since the Montreal metro inaugurated in 1966. The proposed solution fosters environmentally sustainable transportation. As a single, integrated transportation network, the REM will be connected to bus networks, commuter trains and the Montréal metro. The REM represents construction costs of approximately 6.3 billion Canadian dollars. This paper presents the design and construction aspects of the underground elements of this mega project.

INTRODUCTION

The Réseau Express Métropolitain (REM) is an electric and fully-automated, light-rail transit network designed to facilitate mobility across the Greater Montreal Region in Canada. This new transit network will be linking downtown Montreal, South Shore, West Island, North Shore and the airport. The project consists of 67 km of twin tracks over four branches connected to downtown Montreal. The REM system will connect with existing bus networks, commuter trains and three lines of the Montréal metro (subway). Once completed, the REM will be one of the largest automated transportation systems in the world after Singapore, Dubai and Vancouver.

The project includes 26 stations with 3 underground stations in downtown Montreal. One of the underground stations will be built using the NATM method and the two others with the cut and cover approach. The project also includes the rehabilitation and enlargement of the Mont Royal Tunnel. This 100 year old double track tunnel is about 5 km long. The REM also consists of 3.6 km new TBM tunnel connecting downtown to the Montreal International Airport through saturated soft ground and karstic rock.

The project is currently under construction by a joint venture of SNC Lavalin, AECON, Dragados, EBC, and Pomerleau and the final design is being performed by a joint venture of SNC Lavalin and AECOM. To deliver this major project, several underground works are undertaken. This paper presents the major underground developments of the REM and the solutions used for the successful achievement of the underground construction objectives which include: ensuring safety and stability of the opening during construction and for its full service lifetime, minimizing impact and disturbance to the surrounding environment, meeting Owner's technical requirements, and minimizing cost, duration and risk of underground construction.

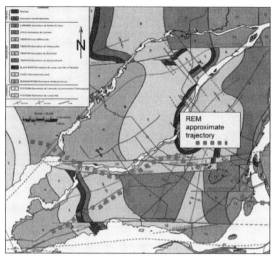

Figure 1. Montreal regional geology

GEOLOGICAL CONTEXT

Montreal geology consists of a variety of sedimentary horizons dating from the Precambrien, Cambrien and Ordovicien periods. The main associated lithologies are limestone and shale. Intrusive rocks dating from the Mesozoic/Cretaceous period are also encountered throughout Montreal, intersecting the sedimentary packages.

The strata are generally relatively sub-horizontal layers of sedimentary rock. However, events such as faulting, folding, glaciation and isostatic movement have shaped the strata differently in certain region of the island. Figure 1 shows the geological map of Montreal. The approximate path of the REM is also projected onto the map, for reference.

The various major faults that were mapped in the region can also be seen in Figure 1. Different orientation can be seen, but the main fault system has an East-West trending orientation. Faults in Montreal are characterized by fractured rock zones of centimeters to hundreds of meters. Faults are generally considered inactive in the region.

Solutions that were put forth for the successful completion of the REM underground works were selected to best match project constraints and local ground conditions. Located in different parts of the city, intersecting different strata, a total of four different types of underground works are undertaken.

DEEP UNDERGROUND STATION

To connect the deeply sitting REM track to an existing metro station located closer to the surface, an underground station accessible by an approximately 70 m deep vertical shaft is designed (Edouard Montpetit Station, EMP). Figure 2 shows the 3D model of the station with the main entrance shaft, the side platforms constructed by enlarging the existing Mont Royal Tunnel, concourse tunnel, ventilation tunnels and shafts and vertical circulation tunnels and shafts. This station configuration was optimized in order to minimize the rock excavation volume. With practically no overburden present in the area, the interchange station is almost entirely located within the Trenton formation which consists of interbedded limestone/shale packages and argillaceous

Table 1. City of Montreal vibration limits in mm/s

Building Type	Frequency (Hz)		
	<10 Hz	10–50 Hz	>50 Hz
1-Commercial	20	20–35	35
2-Residential	5	5–15	15–20
3-Historical	3	3–8	8–15

limestone. The station is also located near the intrusive Mont-Royal which consists of gabbro, monzonite and breccia, resulting in a significant amount of hard dikes and sills in the area of interest.

In this particular area, probably resulting from the contact metamorphism due to the close proximity of the Mont-Royal intrusive, the sedimentary rock package shows hard rock properties, with uniaxial compressive strength (UCS) varying between 125 and 180 MPa and Young's modulus varying between 75 and 88 GPa. Because of the nature of the work, the quality of the rock and the lower initial cost of the technique, controlled drill and blast method was selected to sink the deep vertical shaft and excavate the underground station.

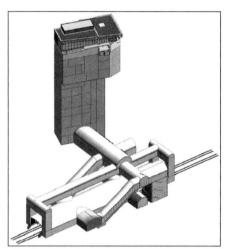

Figure 2. 3D model of the EMP deep station

The excavation takes place in a densely populated area with major infrastructure in the near vicinity. Among those infrastructures is one of the main University of Montreal pavilions. This building houses classrooms and laboratories including a recently completed state of the art and highly sensitive acoustic laboratory located within only a few meters of the excavation. Hence, several engineering control mechanisms are put in place to minimize impact and disturbance including: line drilling technique along the full entrance shaft excavation perimeter and for its entire depth, use of a maximum blast round length of 2.5 m, blasting sequences and patterns designed for low impact (specific blast hole, loading and delay patterns). Table 1 shows the City of Montreal allowable peak particle velocity for different types of building. Type 1, 2 and 3 are for commercial, residential and historical buildings respectively. The line drilling is performed using a DTH drill with the holes diameter of 140 mm placed at 250 mm center to center. The holes center is put at 200 mm from the excavation line to account for the vertical deviation of the drilling operation. To complement this effort, a comprehensive monitoring plan, counting over 150 instruments, is used.

Permanent CT bolts and shotcrete reinforced with steel fiber (Dramix 3D) are used as both initial and final liners for all shafts and tunnels. A layer of 5 cm of flashcrete is applied first for safety, the bolts are installed, spray on waterproofing membrane (BASF Masterseal 345) is added, and then another 5 cm of steel fiber reinforced shotcrete is applied. The liner is designed for a 125 year service life per the contract requirements. During development of the underground excavation, rock mapping is undertaken after each exposure of the final wall. Permanent rock bolting pattern is adjusted based on the ground condition and the need for additional rock bolting is assessed on site to

 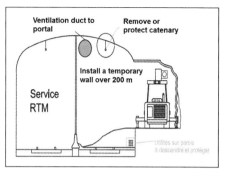

Figure 3. Enlargement of MRT to EMP Station side platforms

ensure the overall stability of the excavation. In addition to durability, the shotcrete mix is designed specifically for the cold weather application in Montreal.

The EMP Station is built within the existing double track Mont Royal Tunnel (MRT). The side platforms are built by enlarging the existing tunnel. Figure 3 shows the side platform enlargement allowing one track to remain in operation.

MONT ROYAL TUNNEL REHABILITATION

The Mont Royal Tunnel is a railway tunnel in operation since 1918, third longest in Canada, which connects the city's Central Station (Gare Centrale), located Downtown Montreal, with the north side of the Island of Montreal and Laval, passing through Mount Royal (Figure 4). The REM project will use this existing double track horseshoe tunnel (5060 m long, 9.6 m wide and 4.4 m high, and a constant 0.6% grade) and two of the project stations will be built inside this tunnel by enlarging it from a double track tunnel to side platform station at the location of these stations. To accommodate the new track system and to ensure the tunnel is to current safety standards complying with NFPA 130 fire life safety requirements, the existing tunnel conditions were assessed. Various solutions including the installation of a center wall and boring a parallel egress tunnel on one side of the existing tunnel and connected to the existing tunnel through cross passages at regular spacing were evaluated.

To accurately evaluate the current conditions and define the tunnel enlargement needs based on the new train envelops, a high resolution laser and optic scanning was performed by Dibit. Using this information, the current conditions of the existing tunnel and accurate clash analyses and interfaces requirements assessment was performed. Results from such analyses are used for the development of the optimal solution to minimize the volume of enlargement excavation.

DOWNTOWN CUT AND COVER STATION

To connect the financial district of Montreal to the REM tracks located approximately 15 m below grade, a major station is planned on McGill Avenue (Figure 5). In this area, due to previous underground works for building the MRT and the Montreal Metro Green Line, the first 15 m of ground consists of backfill. The rock located below this layer of backfill was observed to belong to the Tétrauville formation, which consists of an interbedded limestone/shaly limestone.

Given the local stratigraphic column and low water table, the soldier piles and lagging wall solution is selected as the support of excavation for this station. Drilled soldier

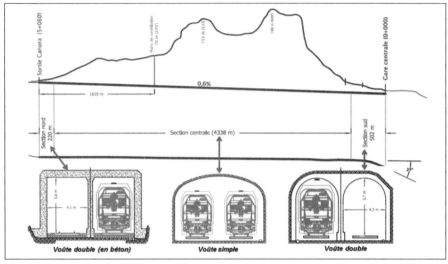

Figure 4. Existing Mont Royal Tunnel

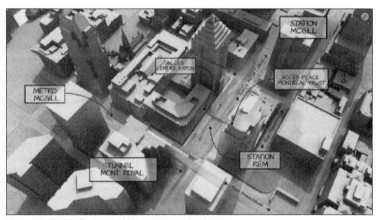

Figure 5. McGill Station in Montreal financial district

piles will be socketed into the rock and steel fiber reinforced shotcrete will be used for the lagging. The soldier piles and shotcrete lagging support of excavation walls will be considered as the permanent station walls. Once the rock is reached, controlled drill and blast will take place to cut the rock to the design level.

The station is located between two major high-rise buildings and is connected to shopping centers inside these buildings. An existing underground commercial passageway connecting these two high-rise buildings is just above the station and will be kept in place during the station construction. The station is at the intersection of McGill Avenue with two of the most important streets in Montreal and therefore the maintenance of traffic and utility relocation are among the main challenges of this station construction. In addition, the site limits of this station are at the edge of the adjacent buildings and the existing metro tunnel and therefore robust design of support of excavation in soil, controlled drill and blast in rock, and comprehensive instrumentation and monitoring program are required to prevent damage to the neighboring structures. The station new tracks will be installed at the location of the existing Mont Royal Tunnel tracks and therefore the excavation should be sequenced in a way to keep the existing tunnel liner in place as long as possible in order to minimize the existing tunnel closure.

AIRPORT TUNNEL

The REM will connect downtown Montreal to the airport, requiring the development of an entirely new underground tunnel that will run below the international airport airstrips. The overall underground stretch consists of approximately 3.6 km. Along the tunnel alignment, bedrock elevation varies significantly and a constant grade is observed at surface, resulting in an overburden thickness varying between 12 and 20 m. The overburden consists of layers of backfill, granular material and glacial till, going from grade to bedrock. The bedrock consists of interbedded limestone/shaly limestone, belonging to two different formations: the Tétrauville formation and the Montreal formation.

Within the Tétrauville formation, two different members are expected to be intersected by the underground works. The upper horizon would consist of a good quality micritic shale with UCS values varying between 75 and 185 MPa, Young's modulus varying between 25 and 65 GPa and Cherchar abrasivity index varying between 0.8 and 1.8. The lower horizon would consist of softer shaly limestone with UCS values varying between 60 and 80 MPa, Young's modulus varying between 35 and 40 GPa and Cherchar abrasivity index varying between 0.3 and 0.6. Within the Montreal formation, only the Rosemont member is expected to be intersected. This member is expected to be of good quality with UCS values varying between 55 and 145 MPa, Young's modulus varying between 45 and 65 GPa and Cherchar abrasivity index varying between 0.8 and 1.8.

To align the REM system with the Airport station to be designed by others, the REM double-track will descend from surface to approximately 40 m below grade. Given the significant constraints related to developing a tunnel underneath an international airstrip, the main portion of the underground works will be performed using a hybrid tunnel boring machine (TBM). From the surface, the REM will start its descent and enter a cut and cover underground station (Technoparc Station). A method of construction using secant piles for the sidewalls was selected as the preferred structure that will serve as both temporary and permanent support of the excavation. Wide-flange permanent roof beams will act as struts during excavation. This method is used for the approach ramp, the cut-and-cover tunnel structures and the Technoparc Station. The secant piles are drilled to rock and socketed into it. The excavation will be drained during the construction and the final structures will be tied down to the bedrock for buoyancy control during the permanent condition.

Heading out of the station, the cut and cover portion will continue and widen over a 13 m length at about 125 m away from the station to serve as the TBM launch pit. At this point, the tunnel invert is at approximately 14 m below surface, in the overburden soil. The hybrid TBM will be launched in the overburden material and will continue its descent and progress in this material for about 300 m. The ground will be improved at the break-out over the first 10 m of the drive to allow watertightness and TBM control at the launch. Once the TBM reaches the bedrock, it will continue its descent over a course of about 300 m, down to 40 m below grade. From there, it will progress at this constant elevation, totalizing approximately 2.7 km of excavation within the rock. The hybrid TBM will be able to progress within the overburden loose material in Earth Pressure Balance (EPB) mode and in open mode during its progression through competent rock.

As the TBM will advance, precast segmental lining will be installed, ensuring the stability of the opening and the safe development of the tunnel. Routine probing ahead of the face will also be performed to assess ground mechanical and hydraulic conditions prior to advancement. Depending on ground conditions, pre-excavation grouting performed ahead of the face may be required to improve mechanical and hydraulic properties of the ground before the TBM excavation.

The TBM retrieval shaft will be located within the airport property limits and will be approximately 40 m deep. The retrieval shaft will intersect approximately 20 m of overburden and will extend for an additional 20 m in rock. A combination of permanent secant pile wall in overburden soil and controlled drill and blast method in rock with permanent CT bolt and steel fiber reinforced shotcrete similar to the EMP Station shaft will be used for the excavation and support of this airport shaft. An elliptical cross section is designed to optimize the geometry of the shaft and minimize its excavation volume.

Airport Tunnel Segmental Lining Design

Based on the geotechnical challenges and evaluation of cost estimate and risk mitigation, it was concluded that shielded TBM with one-pass lining system is the preferred alternative for the REM airport tunnel. Precast segmental lining will be installed using an Erath-Pressure Balance (EPB) TBM as the initial and final lining for the main tunnel. With 6.478 m as the internal diameter of the tunnel, 300 mm was selected as the thickness of segments which is the common value used in practice for this size of tunnel. Lining thickness was verified during the design procedure. Optimized length of the ring (segment width) was selected as 1700 mm. As shown in Figure 6, a 6+1 rhomboidal system assembled ring by ring was selected for REM airport tunnel lining. It consists of a trapezoidal reverse key segment slightly larger than other five full-size rhomboidal segments (3420 vs. 3210 mm, respectively along centerline) and one small trapezoidal key segment in a ring, slightly smaller than 1/3rd of ordinary segments (882 mm length along centerline). Ring segmentation into seven afore-mentioned segments results in segment slenderness/aspect ratio (segment curved length-to-thickness) of 10.7–11.4 which is near the maximum ratio used in the world for FRC segments. This will result in less number of segment and joints, stiffer segmental ring, reduced production cost as well as less hardware for segment connection, less gasket length and less number of bolt pockets where leakage can occur. More importantly, the construction speed can increase significantly. In addition, advantages of using a rhomboidal system include staggered longitudinal joints, continuous ring building and compatibility with a dowel type connection in circumferential joints, which results in a faster ring assembly process comparing to rectangular systems.

Universal rings were selected for this project assembled from rings with circumferential joints inclined to the tunnel axis on both sides. One of the main advantages of this ring system over other systems (e.g., left/right rings) is using only one set of forms for segment production. Longitudinal and circumferential joints were designed as completely flat joints which are advantageous for load transfer between the segments and the rings compared to other types of joints. Also flat joints have been proven to have a superior sealing performance. Bolt connection was designed for longitudinal joints and dowels were chosen for connecting rings in circumferential joints as they require less work for the construction of the segment form and less manpower in the tunnel as the insertion is automatically performed by the erector when the segment is positioned. For the first time in North America, a new dowel system, SOF-FIX ANIX 60 ASY, was used as the connection device in circumferential joints. This dowel is bolted by hand in the socket of the segment to be installed and then it is pushed by the erector in the socket of previously installed segment. This will increase the precision of the dowel installation and will reduce the ring offset.

The gasket type for sealing joints between segments was designed as fiber anchored gasket. Fiber-gasket system has been successfully implemented for the first time in the world by the Designer in South Hartford Tunnel, CT which is considered as latest innovation in EPDM gasket design (Bakhshi and Nasri, 2017). This new technology offers additional pull-out resistance comparing to conventional glued gasket system and has several advantages over anchored gaskets such as reduced risk of incorrect installation and reduced risk of air entrapment in the anchorage area. Gasket Profile is DATWYLER M 802 07 "type South Hartford" providing watertightness under the maximum expected groundwater pressure of 140 psi (9.6 bar). This gasket profile guarantees watertightness for 1.5 times maximum working water pressure considering a combination of gasket differential gap of 0.2" (5 mm) and bearing surface offset of 0.4" (10 mm).

From the structural design point of view, the analysis and design of segments complies with the latest guideline (ACI 544.7R, 2016) in order to satisfy the intended objectives of the project during construction and service life of the tunnel (Bakhshi and Nasri, 2016). The design was carried out using the load and resistance factor design (LRFD) method. Accordingly, during segment production and construction, concrete segments should be able to withstand stripping (demolding), storage and handling loads, thrust force of jacks needed to drive the TBM forward during excavation, and contact/backfill grouting. After installation, segments should be able to resist loads imposed by the surrounding rocks and hydrostatic pressure. Segments were designed for final service stage and verified for production, transient and construction stages loadings.

Precast concrete tunnel segments for REM can be made of reinforced concrete (RC) or fiber-reinforced concrete (FRC). However, FRC segments are superior for crack control and are more cost effective, and therefore, are preferred solution to be used in this project. Minimum compressive strength of concrete at 28-day and at the time of stripping (demolding) is recommended to be 60, and 14 MPa, respectively. To achieve a high early-age strength segment and high performance and durable concrete lining, maximum water cement ratio was designed as 0.40 with the addition of silica fume in the level of 5 percent of cementitious materials. Minimum cement content is specified as 350 kg/m^3 and maximum aggregate size as 19 mm while chloride ion penetrability should be less than 700 coulombs following ASTM C1202. Cold-drawn wire ASTM A820 type I steel fibers, with a minimum tensile strength of 1800 MPa was specified for segment reinforcement. Other specified characteristics of fibers include double hooked ends, a minimum length of 60 mm, a maximum diameter of 0.75 mm and an aspect ratio (length/diameter) of 80. Minimum required fiber content for segments reinforced with steel fibers is recommended to be 42 kg/m^3 (71 lb/yd^3) in order to obtain a minimum first-peak flexural strength (LOP) and residual flexural strength ($f_{R,3}$) of 3 MPa, at the time of segment stripping (demolding) according to EN 14651. Specified FRC strength class at the time of stripping (demolding) is type "3d" according to fib Model Code 2010. At 28 days, minimum first-peak flexural strength (LOP) and specified residual flexural strength ($f_{R,3}$) at 28 days was considered 5 MPa, according to EN 14651. Specified FRC strength class at 28 days is, therefore, type "5d" according to fib Model Code 2010. FRC segments were designed following ACI 544.7R (2016) using these parameters and verified versus results of different performed analyses.

Note that the Designer, for the first time in the US, designed precast concrete tunnel segments with double hooked-end high-strength steel fiber (Dramix® 4D 80/60 BG). Same design will be used for the first time in Canada for REM airport tunnel

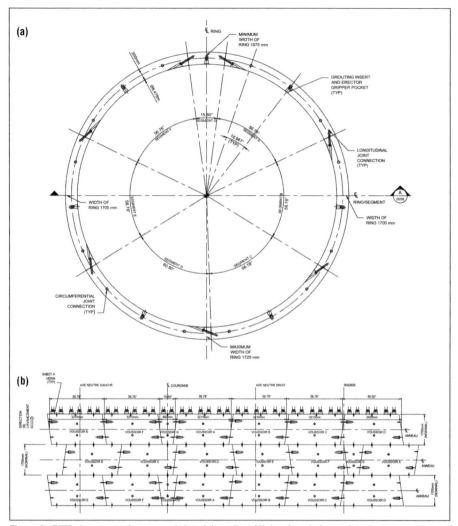

Figure 6. REM airport tunnel segmental ring: (a) section; (b) developed plan on intrados

segmental lining. This new type of steel fiber satisfies the serviceability requirements by limiting time-dependent effects of creep on crack opening and more significantly guarantees ductility requirements with conventional fiber dosage rates by providing an ultimate bending moment higher than the cracking bending moment. This is especially important for the loads applied on segments during production, storage, handling and transportation when segment is subject to pure bending loads.

BIM FOR REM TUNNELS AND UNDERGROUND STATIONS

We have implemented Building Information Modeling (BIM) system to build a full virtual tunnels and underground stations complex that incorporates existing facilities and infrastructure; utilities, ground profile and geotechnical characteristics; proposed tunnels and underground stations architecture, structure, electrical, mechanical, ventilation and FLS components. BIM system was used to support Virtual Design

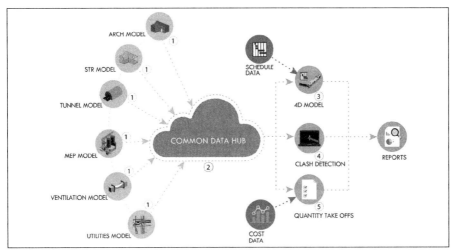

Figure 7. BIM implementation process for REM tunnels and underground stations

Construction (VDC) process by visualizing completed tunnels and underground stations and their components and systems enabling our Team to resolve identified potential conflicts and issues prior to construction. In addition, BIM system was used to assist quantity take-off estimation by providing a realistic project cost estimate and minimizing human errors and inaccuracies during cost estimation procedure. The developed BIM will also provide significant assistance to the Owner with operation and maintenance during the service life of the REM tunnels and underground stations. Figure 7 demonstrates components of virtually built REM tunnels and underground stations. This process can be described in the following steps.

1. BIM Models: Building an intelligent 3D model-based process to more efficiently plan, design, construct, and maintain the REM infrastructure system.
2. Common Data Hub: Utilizing a common data hub to coordinate and communicate more effectively and resolve issues quickly with all involved disciplines such as architectural, structural, electrical, mechanical, etc. (Figure 8).
3. 4D Model: Adding time factor, construction schedule, to 3D model to visualize planned construction sequences.
4. Clash Detection: Detecting hard and soft clashes. Hard clash detection refers to finding the conflict between two objects occupying the same space. Soft clash detection refers to finding the conflict between objects that demand certain spatial and geometric tolerances or buffer zones with other objects within their space for access, insulation, maintenance, safety, etc.
5. Quantity Take-offs: Performing object based calculated quantity take-offs based on provided units.

TUNNEL VENTILATION DESIGN

The purpose of the Tunnel Ventilation System (TVS) in transit tunnels is to perform two primary functions. First function is to maintain environmental control during normal and congested operation within the tunnel such that the electromechanical equipment in the tunnel environment can function within their operational temperature and humidity limits. The second function is to provide tenable environment along the evacuation path during a fire emergency in the tunnel.

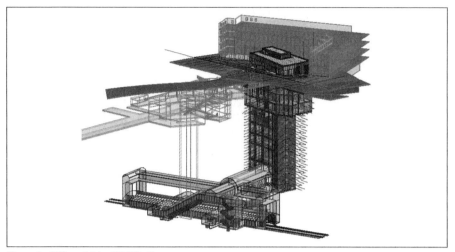

Figure 8. BIM model for EMP Station

These functions can be achieved by a variety of ventilation schemes including longitudinal, semi-transverse, and full-transverse. In the longitudinal ventilation scheme, fan plants near the ends of each ventilation zone operate in pull, push, or pull-push to generate a longitudinal air flow along the tunnel. Sometimes when the tunnel cross sectional area is large enough jet fans are used to either supplement the fan plants or completely replace them. In semi-transverse systems fan plants use an air distribution duct to either extract or supply air along the tunnel from small openings in the tunnel ceiling or tunnel walls. In this type of system air balance is achieved by air exchange with the atmosphere at the tunnel portals. In a full-transverse scheme there are two air distribution ducts along the tunnel length that provide supply and exhaust air streams to the tunnel locally through openings in the two air distribution ducts. In these systems fresh air is injected into the tunnel from openings in the supply duct either in the tunnel ceiling or in the lower portion of the tunnel wall and the warmer or polluted air exits through small openings in the tunnel ceiling. In the full-transverse system the air is treated locally within the tunnel and is not carried through the length of the tunnel. This approach eliminates the accumulation of heat or pollutants along the length of the tunnel and therefore removes the tunnel length restrictions. Therefore it is typically used in long tunnels. The disadvantage of this scheme is the increase in tunnel diameter to accommodate the supply and exhaust ducts within the excavation volume, which leads to increase in capital cost.

In the REM Project there are two tunnels, the existing Mont Royal Tunnel (MRT) in the Deux-Montagne branch and the new Airport Tunnel in the Aeroport branch. The Owner technical specification requires that a mechanical longitudinal ventilation system be provided in order to sweep the smoke and heat from a tunnel fire along the tunnel to one side of the fire and maintain tenable environment as defined by NFPA 130 on the other side of the fire, where the tunnel occupants will be evacuating the tunnel. The longitudinal ventilation system must provide sufficient air quantity to produce the Critical Velocity upstream of the fire to prevent back layering of smoke over the top of the evacuating tunnel occupants. Based on formulas provided by NFPA 130 the quantity of air to generate the Critical Velocity for the REM tunnels were calculated to be 130 m^3/s for the MRT and 60 m^3/s for the Airport Tunnel. This airflow is provided by fan plants in stations at the end of each tunnel, i.e., stations Edouard Mont-petit

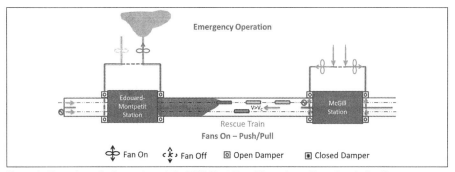

Figure 9. Tunnel ventilation system at the REM Mont Royal Tunnel, configuration during fire emergency

(EMP) and McGill within the MRT and Technoparc Station and the ventilation shaft at the southern end of the Airport Tunnel. This ventilation shaft will be housed in the TBM retrieval shaft in the airport parking area. Figure 9 shows the schematic representation of the TVS configuration for MRT under fire emergency operation.

The same TVS can be configured to provide ventilation during normal and congested operations. During normal operations the piston effect of the running trains is expected to create sufficient circulation within the tunnel to maintain the required temperature and humidity control. Therefore, fans can remain off and piston relief is achieved through bypass ducts.

During congested operations when trains are moving at reduced speed through each tunnel the heat rejection from the trains can increase the tunnel air temperature above the operating temperature of the electromechanical equipment operating within the tunnel environment such as the air conditioning condenser units on board the train. In this case the fans can operate in push-pull mode similar to the fire emergency configuration and cool the tunnel by passing fresh cool air over the trains and removing the heated air from the tunnel at the location of the fan plant operating in exhaust.

CONCLUSION

This paper presents the design and construction aspects of the underground elements of the REM mega project in Montreal. It discusses the details of one of the underground stations built using the NATM method and the two other stations constructed with the cut and cover approach. It also explains the rehabilitation and enlargement of the existing Mont Royal Tunnel. It describes as well the specifics of the airport TBM tunnel.

REFERENCES

Bakhshi, M., and Nasri, V. (2017). Design of steel fiber-reinforced concrete segmental lining for the South Hartford CSO tunnel. Rapid Excavation & Tunneling Conference (RETC) 2017. San Diego, CA, June 4–7, 2017.

Bakhshi, M., and Nasri, V. (2016). ACI guideline on design and construction of precast concrete tunnel segmental lining. ITA World Tunnel Congress (WTC) 2016, San Francisco, USA, April 22–28, 2016.

Design and Construction of NBAQ4 Water Transfer Scheme

Brendan Henry ▪ GHD Pty Ltd
Alessio Bianchini ▪ GHD Pty Ltd
Abner Bondoc ▪ GHD Pty Ltd
Ioannis Bournas ▪ GHD Pty Ltd
Katpakanathan Kamalachandran ▪ GHD Pty Ltd
Monique Quirk ▪ GHD Pty Ltd
Erik van der Horst ▪ GHD Pty Ltd
Kar Weng Ng ▪ GHD Pty Ltd
James Willey ▪ GHD Pty Ltd
Marcus Lübbers ▪ Herrenknecht AG

SCHEME

The Novaliches to Balara Aqueduct number four (NBAQ4) project is part of Manila Water's improvement and expansion initiatives and is one of the largest and most important infrastructure projects undertaken by them to date. The project involves the construction of a new intake facility at the La Mesa reservoir, a 7.3 kilometre tunnelled aqueduct and an outlet facility at the Balara Water Treatment Plant.

The existing aqueduct system (aqueducts 1, 2, and 3) is required to continuously deliver 1600 million litres per day (MLD) of raw water to Balara Treatment Plants 1 and 2. Built in 1929, 1956 and 1968 respectively, they are nearing, if not already exceeding, the service life of 50 years for concrete structures. Through the project, the construction of a fourth aqueduct, which is designed to latest seismic standards, will enable the assessment and rehabilitation of the existing aqueducts.

NBAQ4 will deliver water to approximately seven million people in Manila. The PhP 5.3 billion (USD 104 million) contract was awarded to a joint venture of CMC di Ravenna (Italy), First Balfour, Inc. (Philippines), and Chun Wo Engineering (Hong Kong) in August 2017.

TENDER DESIGN

The project was tendered at the start of 2017, based on a reference design, and pursued by three pre-qualified contractors. Manila Water's evaluation criteria was 80% technical and 20% cost. The joint venture tendered for the project with GHD as the designer. GHD provided the tender design services from Manila while niche design services for the tunnels, shafts and intake/outlet structures from Brisbane, Australia. Although not the lowest bidder, the construction joint venture, renamed NovaBala JV (NBJV), was awarded the project.

DETAILED DESIGN

The detailed design was executed through late 2017 and 2018 with a two stage review process by Manila Water's engineer, ARUP. The design of all structures, including temporary works, was to be designed to the National Structural Code of the Philippines (NSCP). However, the code mainly covers occupied building structures and as such, US Army Corps of Engineers (USACE) codes were used for the design of the Intake and Outlet Structures.

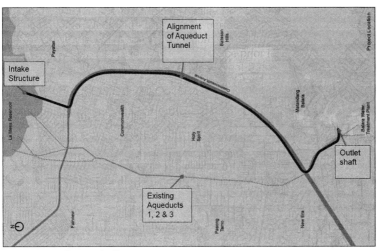

Figure 1. NBAQ4 tunnel alignment

ALIGNMENT

The tunnel commences at the upstream end at La Mesa Reservoir Intake Structure and passes beneath Commonwealth Avenue to the Balara Water Treatment Plant. It will be constructed as a bored tunnel using an Earth Pressure Balance (EPB) Tunnel Boring Machine (TBM). The tunnel length is approximately 7.3 km long with an internal diameter of 3.1 m. Ground cover varies from 8 m at the Intake Structure to a maximum depth of 52 m and an average of 34.5 m. There is an additional low point at chainage 1300 with full face of soil at that location and 26 m depth. The alignment includes three tight radius curves of 80 m radius. The alignment is depicted in Figure 1.

GEOLOGY

As evident in the above Figure, the alignment traverses Quezon City (QC) and is located west of the Marikina Valley. The tunnel drive will start at the Balara Water Treatment Plant, progressing northward to the La Mesa Reservoir.

The project site is situated within the mainland section of the Southern Sierra Madre stratigraphic group. The alignment is primarily within a pyroclastic sequence of deposits belonging to the Guadalupe Formation. It is further divided into two members, namely the Diliman Tuff which includes the pyroclastic and tuffaceous sequence and Alat Conglomerate, which is composed of the sedimentary deposits. Within the project area the rock is predominantly composed of very low to low strength, slightly jointed pyroclastic rocks, reworked volcanics, and tuffaceous units of Quaternary age, overlain by more recent deposits of clays and silts.

The following key risks were identified during the ground investigations:

- High seismicity due to proximity to the Marikina Valley Fault System;
- Weak rock strata with 'soil like' strength properties;
- Shallow ground water conditions, high ground water inflow rates in sub-vertical fractures;

- Paleosol—compacted soils characterized by the presence of organic materials and lithified relic plant structures within a sequence of geological deposits; and
- Quaternary Deposit—mixture of clay and silt materials in the face of the TBM.

The aforementioned geology is illustrated in Figure 2 through Figure 9.

Figure 2. Quaternary materials along Commonwealth Avenue

Figure 3. Mixed scoria and pumice rich ignimbrite/lithic Tuff Balara Area

Figure 4. Subvertical joints at the Balara Filtration Area

Figure 5. Tuff and tuffaceous sediments found at the Balara Filtration Area

Figure 6. Paleosol material found at Balara Filtration Plant

Figure 7. Organic material found at paleosol level

Figure 8. Fluvial deposit conglomerates

Figure 9. Volcanic flow at Balara Area

GROUND INVESTIGATION

An extensive ground investigation program was conducted by GHD, which consisted of structural mapping, borehole drilling with associated field and laboratory testing, instrumentation and geophysics. The purpose of additional ground investigation was to supplement the information from the existing boreholes, in order to confirm the elevation of soil and rock interface along the tunnel alignment. In addition, further investigation was required to identify and qualify risk and assisting in establishing a suitable tunnelling methodology.

A total of seventeen 25 m to 65 m vertical boreholes were drilled along the proposed tunnel alignment and in situ and laboratory tests were executed to derive geotechnical design parameters. Over-water boreholes were drilled at the Intake Structure and Bridge. Multi-channel analysis of surface waves (MASW) and downhole seismic surveys were conducted as well as downhole seismic surveys, onshore and offshore seismic surveys in order to obtain the seismic velocities (Vs) required for the site-specific seismic hazard assessment.

HYDRAULICS

In considering the steady state hydraulic design of the transfer system, the design flow rate was specified as 1,000 ML/d (11.6 m³/s) with a gross head operating head range of 9.78 m to 20.28 m. This head range was based on the required range of operation in La Mesa Reservoir from minimum operating level (MOL) to dam crest level (DCL). The total length of the transfer system included approximately 7.3 km of precast segmentally lined tunnel and approximately 400 m of mild steel cement lined pipe. Given the low head and long transfer distance, careful consideration was required in the geometric design and the selection of valves and fittings to minimise head losses. The friction losses in the tunnel were an important consideration and were derived based on absolute roughness heights for the precast segmentally lined section of 0.4 mm and 1 mm for new and aged surfaces.

The design arrangement included an allowance for incorporation of a mini-hydro for renewable energy generation at the outlet from the tunnel. The intention is that the flow would normally pass through the mini-hydro turbine with the bypass line including the plunger valve provided for times when the mini-hydro was offline.

With a long transfer scheme, the management of surge is important for opening and closing of the control valves, especially taking into account the operation of the mini-hydro. A range of scenarios were considered to identify the optimum arrangement for the surge structure at the outlet. In relation to the shutdown of the mini hydro due to load rejection, the surge analyses provided guidance for the design of the controls for the coupled operation of the closure of the mini hydro turbine/main insolation valve and the plunger valve on the bypass line.

The design process included CFD modelling of selected components to confirm head losses and hydraulic performance. The main focus was the section from the surge shaft through to the dissipation chamber. The layout of the outlet can be seen in Figure 10 and CFD modelling in Figure 11.

CONSTRUCTION ACCESS SHAFT

To decouple construction of the Outlet Structure with tunnel excavation works, a temporary Construction Access Shaft (CAS) was added 50 m away (Figure 12). For TBM construction logistics, a 9 m clear diameter was required with a depth to tunnel level of approximately 43 m.

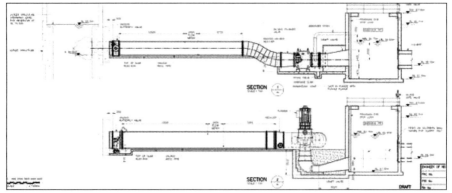

Figure 10. Revised layout of outlet

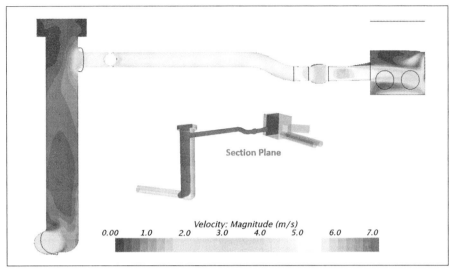

Figure 11. Velocity distribution—Scenario 3

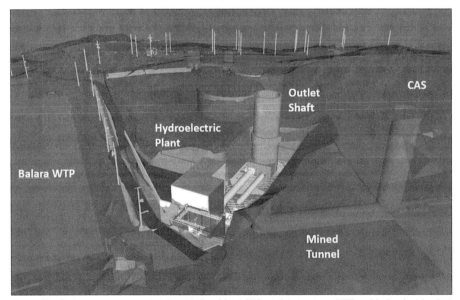

Figure 12. Outlet structure area showing relationship of CAS, outlet structure, WTP and hydroelectric plant

At the time of tender, the ground conditions interpreted from the nearest borehole allowed for the construction of a conventionally excavated shaft with rock bolt and shotcrete support. The upper 12 m portion supported by sheet piles through the soil profile. A secant pile solution was assessed but not adopted due to the depth.

NBJV engaged Bauer Foundations, to provide an alternative design. Bauer proposed a secant pile shaft with oversized piles at tight verticality tolerance (1:200, which is double the industry's normal tolerance). To achieve these tolerances, Bauer proposed to install oversized unreinforced concrete piles of 1500 mm and 50MPa around the

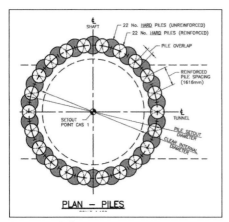 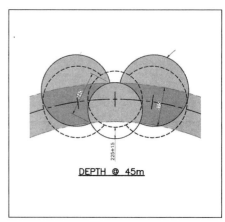

Figure 13. CAS secant pile layout and tolerance assessment

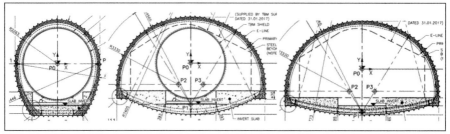

Figure 14. Cross section geometries 1 (~4.8 m span), 2 & 3 (~8 m span)

full perimeter with reinforced concrete piles of 1200 mm. The verticality tolerance of 225 mm at 45 m depth results in an available contact area of 815 mm to transfer hoop stress. The layout and tolerances are shown in Figure 13.

GHD then assessed the adequacy of the contact area for the excavation stages and for tunnel break-in stage (post base slab installation). Having confirmed the contact area to be sufficient, the detailed design of the shaft was completed utilising Rocscience program, RS2. At the tunnel break-in stage, almost 50% of the reinforced piles were removed. Although the shaft is classified temporary works, it has been designed for a seismic event commensurate with the design life of ten years.

MINED TUNNEL

At the base of the CAS, a fore-shunt is excavated for launch of the TBM (Section 1), assembly of the TBM (Section 2) and a back-shunt tunnel is required connecting CAS and the Outlet Structure to service the TBM with three muck/segment trains (Section 3). The mined tunnel construction adopts full face excavation sequencing, as denoted in Figure 14. The approximate total length of mined tunnel is 70 m.

Due to the poor ground conditions encountered in the CAS borehole, an additional borehole for the mined tunnel with specific testing at and above mined tunnel elevation was required. This hole confirmed the presence of a soil-like layer. Permeability testing indicated that strata within one tunnel diameter of the tunnel crown and invert level have permeability values between 4.7 E-09 m/s and 9.2 E-07 m/s. The water level is approximately 20 m above invert level.

The mined tunnel is expected to be excavated in Tuffaceous rock and Reworked Volcanic rock of very low strength and cohesion. The soil profile layer, which ranges in thickness between 2–7 m, is located near the tunnel crown and is expected to extend into the tunnel profile by up to 2 m.

Four different support classes covering the expected range of geological conditions were developed. The mined tunnel excavation and support were designed based on core deformation control and face stability concepts.

GSI values ranging between 35 and 65 (Class B 150 mm of shotcrete, steel sets W6×15, fiberglass face bolts) and 29 < GSI < 35 (Class C 150 mm of shotcrete, steel sets W6×20, fiberglass bolts at the face) are expected to cover all ground conditions. For better than expected geological conditions (GSI > 65 which is considered unlikely), support Class A (100 mm of shotcrete) is provided. For worse than expected geological conditions, GSI ranging between 15 and 25, support Class D (increased shotcrete thickness, larger steel sets sizes, forepole umbrella and face bolts) is provided. The implementation of Class D is also foreseen for the break-in areas at both shafts. The CAS, Outlet and Mined Tunnel layout are illustrated in Figure 15.

The tunnel was designed for drained conditions which will be achieved by predrilling drains in the face and tunnel perimeter. In case of un-manageable, excessive water ingress (more than 0.5 Lugeons), the provision of permeation grouting to manage water inflows will be required.

The floor slab was designed for temporary loads from the TBM shield and the back-up gantries as well as against uplift forces that are created during the floor excavation stage, which will experience heave.

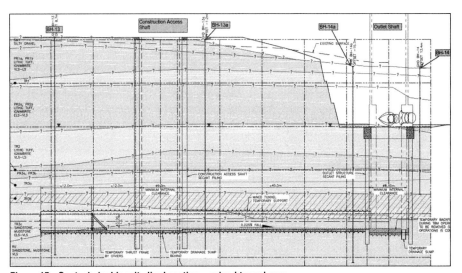

Figure 15. Geotechnical longitudinal section—mined tunnel area

Although the mined tunnel is considered a temporary structure, it has been designed for a seismic event commensurate with the design life of the structure being ten years.

EPB3100 HERRENKNECHT TBM

To cater to the ground conditions, the JV contractor's tunnelling experts alongside Herrenknecht chose an Earth Pressure Balance (EPB) shield tunnelling system as the most economical solution for the geological and hydrological project requirements. Special adaptations were specified for the TBM sealing systems to assure water tightness up to 6 bar static water pressure.

The 480 kW of main drive power will provide a maximum cutter head torque of up to 2205 kNm and rotational speed of 6.7 rpm. A total of 12 thrust cylinders will provide a combined maximum thrust force of 19,000 kN.

In order to reduce the frequency of interventions, for inspection and replacement on the 7.3 km long drive, the cutter head of the EPB3100, will be equipped with adapted excavation cutter tooling. Cutter disc wear rings with an increased diameter of 394 mm (15.5') allow for a higher wear tolerance and subsequently, longer life. The TBM is also equipped with a hydraulic wear detection system.

The 80 m turning radius is particularly challenging in small diameter machines, which tend to be quite long and which inhibits their ability to negotiate tight curves (see Figure 16).

The machine's main thrust cylinders are horizontally adjustable and compensate for the axial misalignment of both the mid-shield and tail shield. Equipment preassembly, commissioning and testing were undertaken at Herrenknecht's Guangzhou production facility prior to shipment to the Philippines (Figure 17).

Herrenknecht utilises onboard VMT hardware with installed TUnIS ring sequencing software as part of the TUnIS Navigation System enabling pre-prediction of the required rings ahead of the excavation. A reduced segment length of 800 mm is used in the tight curve sections of the alignment. During the design phase, ring sequencing for all scenarios was simulated in close co-operation between GHD and VMT specialists in order to maintain capacity for a high quality, close-tolerance finish throughout.

Rail-mounted rolling stock supply the tunnel segments and materials as well as facilitating the removal of the excavated spoil by muckskip. Despite the relatively small diameter of the machine h, a high-capacity cooling system has been installed to compensate for Manila's high atmospheric temperature and humidity levels during execution of the tunneling works.

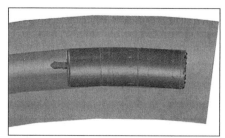

Figure 16. EPB3100 3D design drawing on R=80 m curve

Figure 17. NBAQ4 EPB3100 factory assembly

Upon completion of the tunnel drive, all tunnelling system components will be disassembled and removed back to the launch shaft. The TBM shield will left in place in the Intake Structure and incorporated into the permanent works.

ASSESSMENT OF FACE SUPPORT DURING TUNNELLING

The tunnel alignment will pass through full-face rock, mixed soils and rock, and full-face soil conditions. Where soil and mixed face conditions are encountered, the EPB will operate in closed mode. The face support pressure for such scenarios was estimated using the Anagnostou and Kovari method.

A practical approach was taken to assess whether the TBM can operate in open mode while tunnelling through the weak rock mass, which dominates majority of the alignment. Hoek and Marinos (2000), Duncan-Fama (1993) and Carranza-Torres and Fairhurst (1999) analyses were used to calculate support pressures required to support a drained face. The maximum rock pressure calculated was 98 kPa occurring at chainage 7000 m along the alignment. The cutter head under minimum thrust conditions is enough to support the face under maximum rock pressure.

Although the rock is weak and relatively permeable, the TBM will drive in open-face mode with the significant benefit of reduced cutter head wear. In general, any water ingress through the rock mass during the excavation cycle will be less than that required to condition the cut rock, along with the foam conditioning, into a workable consistency to extract through the screw conveyor.

If a water-charged and permeable soil layer is encountered, there will be significant risk of water inrush and additional material entering into the cutter head. In such locations, the TBM will be operated under either closed mode or semi-open mode conditions (compressed air in the top of the excavation chamber). If an unidentified soil layer is encountered during open mode operation, the operating procedures will allow for an immediate assessment and switch to closed or semi-open mode as appropriate.

Figure 18 Illustrates the target (approximately 2.5 bar), upper and lower bound face pressures in soil (closed mode for soil or mixed face conditions) and total rock and water pressures for the rock sections where open mode should be possible.

TBM TUNNEL LINING

The permanent tunnel lining will be precast concrete segments with EPDM gaskets. This lining will provide support of earth and hydrostatic pressures for the specified 100 year design life.

The tunnel is capable of withstanding the full range of expected internal hydrostatic loads including surge pressures from emergency shut down of the hydroelectric turbine.

The tunnel ring design has s a universal configuration comprised counter key, 4 parallelogram plus full size key, all with equal arc length at centerline (Figure 19). The segments are reinforced with steel bar. The segmental tunnel lining was designed to withstand all temporary and permanent loads including all worst case combination and effects including segment demoulding, stacking, transportation, installation, and service conditions including seismic events. Analysis using methods published by Hashash et al 2001 was adhered to. Two types of segments are required to negotiate the alignment. The majority of the alignment will utilise a 1200 mm segment with ±10 mm taper.

Design/Build Projects

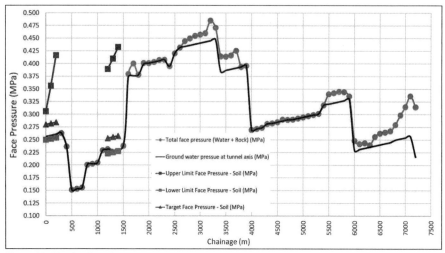

Figure 18. Total soil, rock and ground water pressure at tunnel axis against tunnel alignment

A major technical challenge encountered was the design of segments for tight radius curve of radius 80 m at three separate locations, with a total length of approximately 510 m. A specialised segment was designed for these radii and required an unconventional design approach. There are only few project examples worldwide for this limitation, and most of the cases utilised a steel lining instead of concrete.

The TBM thrust ram axis to the segment will not be concentric, meaning the segments will be forced outward at the leading edge. Additional loads on these segments are generated by the higher ram loads on the segments on the outside edges of the curve. GHD worked collaboratively with the contractor and the TBM supplier to resolve this challenge, assessing ring build sequence with VMT, which resulted in the adoption of an 800 mm segment with taper of ±30 mm.

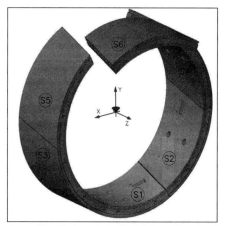

Figure 19. 3D view of standard ring

GHD developed a coupled 3D ring model in Strand 7 to analyse the effect of high eccentric thrust force on the ring through the tight curves (Figure 20). Using the 800 mm ring with the ±30 mm taper, a 10 ring model was built with flat plate segments with ring build sequence to match the 80 m radius curve. Sensitivity analysis was carried out at the recovery radius using a separate model. Each ring was coupled together at each circumferential connector location using 3 dowels. In the model, Ring 1, 2 and 2.5 remain inside the tail shield at the start of the advance. Rings 2.5 to 3 are considered to be unsupported due to the very low strength tail-shield injected grout. Ring numbers 4 to 10 are supported by varying stiffness of annulus grout to represent the increasing strength gain. Due to the uncertainty of the tail-shield brushes providing any level of restraint, they have not been modelled in the analysis. The ram forces were applied in a non-concentric manner as per information supplied by Herrenknecht. The reinforcing was detailed from the results, including bursting steel at ram pad locations. The dowel connector capacity was confirmed by the analysis. The dowels also prevent opening or ovaling of the ring.

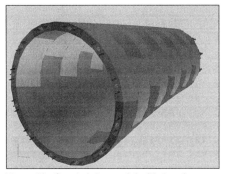

Figure 20. Coupled ring model with dowels

OUTLET STRUCTURE AND SURGE TOWER

Structural Layout

The Outlet Structure connects the tunnel invert level at RL 36.6 m to the pipe offtakes at RL 62.0 m. It has a minimum clear internal diameter of 6.0 m from tunnel invert to above the offtakes and 8.0 m (Figure 21). The tower is designed to allow for action as surge chamber. Approximately 26 m is buried 28 m freestanding above ground level.

The design development of the Outlet Structure shaft underwent the same development process as described for CAS, from bolted and shotcrete support to secant piles. The original ground level was also reduced from RL 75 m to RL 60 m to reduce shaft depth.

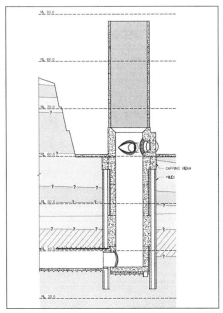 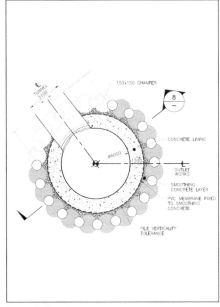

Figure 21. Cross section of outlet structure and plan at tunnel elevation

Design Approach

The Outlet Structure is a sensitive structural component during a seismic event. Combined failure of the Intake and Outlet Structure may cause an uncontrolled flood in the Balara area. According with USACE EM 1110-2-1806, a structure is defined as 'critical' if failure during or immediately following an earthquake could result in loss of life.

USACE EM 1110-2-2104 defines the performance requirements and illustrates that each event combination depends on the duration and frequency of loading. Serviceability denotes a watertight structure and resistance to potential damage from cavitation or abrasion erosion, and is adequately protected against corrosion of reinforcing steel, embedded steel and exposed steel components. USACE design guidance uses categories of Usual, Unusual, and Extreme. Each load category has a different expected performance requirement based on the frequency of loading.

The Outlet Structure was designed to resist the maximum earthquake ground motion occurring in any direction. According to EM 1110-2-2400 the orthogonal combination method was applied to take into account the directional uncertainty of the earthquake motions and the simultaneous occurrences of earthquake forces in two perpendicular horizontal directions.

The design Operating Based Earthquake (OBE) is defined as an earthquake with 144 years return period event (with PGA = 0.47 g) and design Maximum Design Earthquake (MDE) is defined as an earthquake with 1000 years return period event with PGA = 0.93 g. For critical structures such as the Outlet Structure, the MDE was set equal to the Maximum Credible Earthquake (MCE).

OBE and MCE response spectra are both defined by considering 5% damping and Vs_{30} = 357 m/s (Figure 22). GHD has assessed site specific response spectra with analyses undertaken by GHD geophysicists.

The seismic loading due to the MCE has been analysed by considering an inelastic and ductile behaviour of the structure. A ductile behaviour can be achieved avoiding fragile failure such as shear and thrust. Under a specific load condition that lead the structure into an inelastic stress condition, the structural capacity can be overcome only in bending.

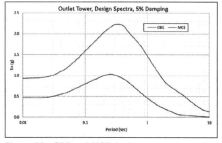

Figure 22. OBE and MCE design response spectra

The embedment has been verified against horizontal loads. The maximum horizontal load is expected during an earthquake due to the acceleration of the tower's structural mass and water inside the tower. The tower vibration may lead to the formation of a failure passive wedge down to a certain depth from ground level. This depth has been assessed by considering a 3D passive wedge in equilibrium with the maximum seismic force. The outcome was also used to assess the ground dynamic spring stiffness variation with the depth. Based on the outcome of this analysis the FEM model was set up considering the minimum ground dynamic spring stiffness for the first 12 m below ground level.

In order to allow the piles to develop tensile capacity during an earthquake, a shear connection between piles and internal lining has been designed (through roughness)

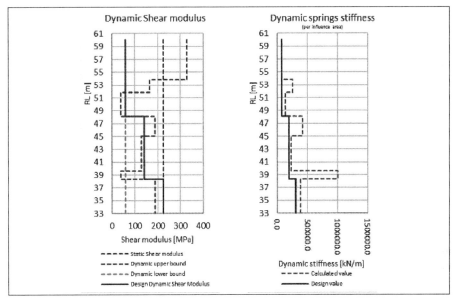

Figure 23. Dynamic lateral ground stiffness assessment

above the tunnel opening level where the shear forces at the interface due to the tower bending are expected to be minimal or nil. A deboning layer was designed between piles and internal lining where the maximum structural bending is expected (upper 12 m).

Numerical Modelling

The structural design of the Outlet Structure was performed with Strand 7. The shaft walls were modelled using Quad4 plate elements. The area around the pipe outlets was modelled using the brick elements in order to correctly capture the concrete mass, including the eccentricity of the mass to the shaft. Spring damper and point contact elements were used to model the interface between the rock mass and the shaft. The static load combinations have been analysed with a non-linear static analysis while the seismic load combinations have been analysed with a response spectra analysis. The non-linear behaviour due to the interface friction and the different axial stiffness in tension and compression was modelled in Strand 7 by using point contact elements. The dynamic axial and lateral stiffness has been assessed following the approach suggested by Y.M.A. Hashash, J.J. Hook, B. Schmidt and J I-Chiang Yao valid for underground structures. For each layer the dynamic spring stiffness has been calculated by considering the described Vs variation along the depth. For design purposes, slightly lower values of dynamic ground stiffness have been used, relying more on the structural response instead of the ground confinement (Figure 23).

The vertical static stiffness of the spring's elements simulating the interaction between base slab and foundation rock was assessed according to Bowles (1997).

When the walls of the tank accelerate back and forth, a certain fraction of the water is forced to participate in this motion, which exerts a reactive force on the tank at a certain height. The force was applied as a concentred translational mass at RL 78.30 m. The motion of the tank walls excites the water into oscillations that in turn exert an oscillating force on the tank (Figure 24).

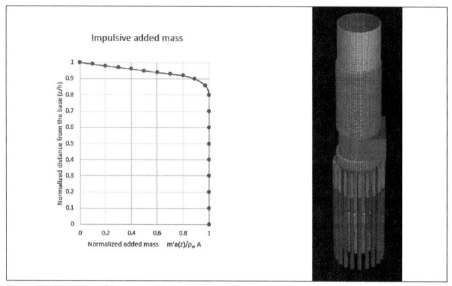

Figure 24. Impulsive added mass distribution

The pile internal forces have been assessed in accordance with the results of the two separate FEM models; geotechnical and structural. The geotechnical model was used to assess the pile's embedment and internal forces due to the interaction with the surrounding medium and water pressure, taking into account the construction stages and seismic action during construction. The same methodology to the Construction Access Shaft design was employed. The structural model aimed to assess the pile's seismic forces (mainly tension) due to the vibration of the Outlet Structure.

INTAKE STRUCTURE

Structural Layout

The Intake Structure consists of a 16 m diameter, 1.3 m high cylindrical concrete base with 6 m of added mass concrete on top and a further 2 m high octagonal pedestal supporting the 21.5 m high intake shaft of 6 m internal diameter. The base is founded at RL 51.70 m and the top of the shaft is at RL 82.50 m, which is also the dam crest level. The tunnel connects to the shaft base at an invert level of 53.00 m. There are 6 inlets at 3 levels, giving a total of 18. Each inlet consists of 1 × 1 m square clear opening. The openings are formed by cast-in stainless steel bellmouths with trash racks on the outside and a sluice gate on the inside. Each sluice gate has a rising spindle to deck level from where they can be controlled with electrical actuators that can be operated locally or remotely with manual operation also available. The layout can be seen in Figure 25.

Design Approach

The limiting factor of the design is the seismic loading. The seismic action was assessed with a site-specific analysis undertaken by GHD geophysicists.

The required performance criteria for the project has been defined according to EM 1110-2-2104. Like the Outlet Tower design, the OBE employs a 144 year return period

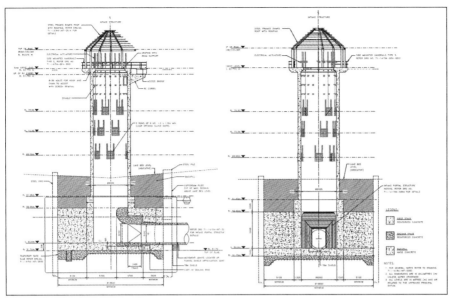

Figure 25. Cross section of intake structure

and has a PGA of 0.32 g, where the MDE employs a 1000 year return period and has a PGA of 0.69 g.

A post yield response has been considered for the MDE seismic analysis.

Numerical Modelling

The Intake Structure has been modelled and analysed using Strand7 (Figure 26). The tower base has been modelled out of 'brick' elements and the shaft and the roof of 'plate' elements. The columns supporting the roof are 'beam' elements. The base of the tower is modelled resting on springs with their spring stiffness representing the elastic properties of foundation material. The springs have also been applied to the sides of the tower base structure representing the stiffness of surrounding foundation material and the concrete filled piles of the cofferdam.

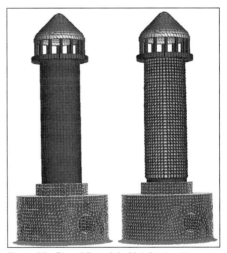

Figure 26. Strand 7 model of intake structure

Outside and inside hydrodynamic water mass (the water mass engaged in an earthquake) has been calculated in accordance with USACE EM 1110-2-2400 and added to the model. Calculated mass has been applied as 'non-structural' nodal mass.

Static loading combinations have been performed with linear static analyses while the seismic loading combinations have been performed with spectral response analyses.

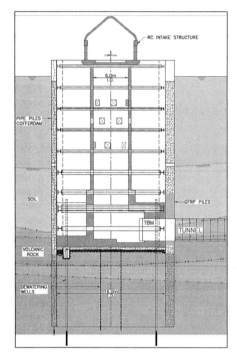 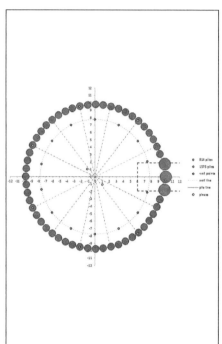

Figure 27. Cross section and plan view of the cofferdam showing steel and GFRP piles and well points.

The TBM reception seal housing and concrete structure has been designed for a face support pressure of 350kPa. The details of the TBM reception and seal will be detailed once a preferred supplier for the sealing system is appointed.

COFFERDAM

In order to build the Intake Structure 'in the dry' it was proposed that a temporary cofferdam, consisting of tubular steel piles, driven into the soil/rock below the lakebed, with circular steel walers as illustrated in Figure 27, be constructed.

The dimensions of the cofferdam are constructability driven, leaving sufficient room around the tower for a climbing form and stair access. It must allow the TBM to enter through and dock into the Intake Structure. Although temporary works, the cofferdam has been designed for a seismic event commensurate with the design life of the structure being ten years (PGA of 0.18g). It will be removed to lake bed level upon completion of the Intake Structure.

Additional boreholes were drilled for the detailed design phase which clearly indicated variable ground conditions. At the cofferdam location, a soil layer of 8 to 10 m (silt, clay, sand) depth overlies reworked volcanic rock of very low strength and variable permeability.

Structure and Dimensions

The circular shaft has a clear internal diameter of 16 m and consists of a total of 51 × 914 mm diameter steel piles spaced at 1100 mm, 3 × 1575 mm Glass Fibre Reinforced Plastics (GFRP) reinforced concrete piles from the top level of the TBM and below (to facilitate TBM entry) with 9 circular steel walers.

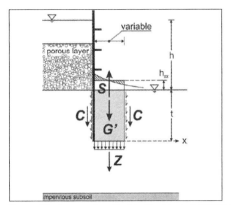 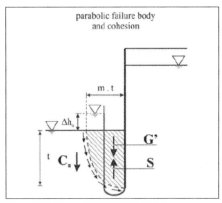

Figure 28. Proposed method by Davidenkoff (1970) (left) and Wudtke and Witt (2006) (right)

Base Stability and Pumping Requirements

Without additional measures, vertical stability is not guaranteed. Water pressures need to be lowered and managed during construction. Maximum allowable water pressures were established for each excavation stage and a de-watering system has been designed.

Depending on the permeability and strength of the soil/rock underneath and around the shaft, a hydraulic heave or an uplift problem will most likely ensue. In order to establish an appropriate factor of safety against uplift and/or hydraulic heave, a combination of hydrogeological modelling and analytical assessment was used.

Eurocode 7 describes conventional (analytical) methods to verify safety against hydraulic heave, however these checks do not take into account friction between shaft and ground, nor any cohesion of the material and for cohesive soils/weak rocks. These methods are very conservative. Various researchers have proposed methods to take cohesion and tensile strength into account for hydraulic heave checks. For example Davidenkoff (1970), Wudtke/Witt (2006) and Wudtke (2008) as shown in Figure 28.

Based on the method proposed by Wudtke and Witt (2006), a check was performed taking into account a cohesion of 25 kPa. The base stability failed both the conventional and the 'modified for cohesive soil' checks, even with the toe level at the (practical) maximum of 38.5 m RL.

To check for uplift, an analytical assessment (assuming full hydrostatic water pressure) was conducted even when taking into account friction between the soil/rock plug and the steel piles, there was insufficient safety against uplift.

In conclusion, without additional measures, vertical stability was not guaranteed and that water pressures would need to be lowered and managed during construction. Based on the uplift calculations, maximum allowable water pressures were then established for each excavation stage, based on a factor of safety of 1.5.

As part of the water pressure relief system design, FEA program Seep/W was used to model pressure relief pipes installed around the perimeter, inside the cofferdam.

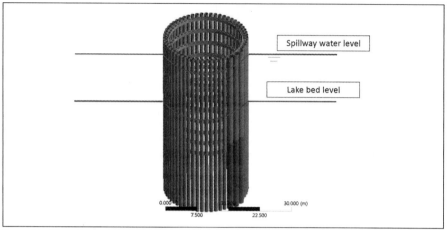

Figure 29. View of structural model of cofferdam

A radial configuration and required depth of the pipes was established to ensure water pressures at the toe level and at the centre of the shaft stay below the maximum allowable pressures (as established in the analytical assessment of vertical stability).

Based on these assessments and analyses, a monitoring regime was designed using trigger levels and associated required mitigation measures.

Structural Analysis

The structure was modelled in 3D using the FEA package ANSYS and loading conditions for each stage of water level lowering and ring beam installation were established (Figure 29). Earthquake loading was applied after the structure was fully excavated and the temporary concrete base slab was installed.

A total of nine walers were designed, with the top one (Philippines) size BH350×142, 4 nos. BH450×389, 4 nos. BH700×700 and a double stacked BH600×536 for the heaviest loaded section.

ACCESS BRIDGE

An access bridge linking to the Intake Structure provides maintenance access and is approximately 150 m in length with seven spans using ASSHTO girders and a 200 mm thick Bridge deck with guard rails. At the northern end of the bridge is a 14 m × 14 m wide platform for installation of a temporary tower crane, water pump, generator and concrete pump for construction purposes. The bridge was designed to local structural codes and the most critical loads considered were for the tower crane and the proposed construction machinery.

The bridge and relationship to the intake and cofferdam structures, is shown in Figure 30.

DOWNSTREAM DISTRIBUTION

The Downstream Distribution network is a series of pits and pipes that transfers the water from the Outlet Structure, through the mini hydroelectric plant to various parts of the Balara Water Treatment Plant. Within this network of pits and pipes there are inter

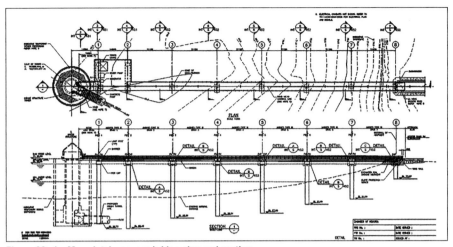

Figure 30. La Mesa intake access bridge plan and section

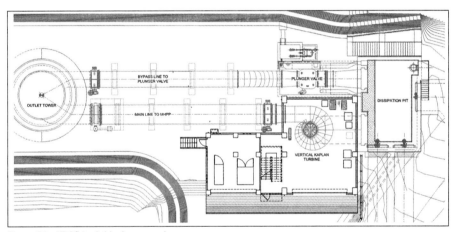

Figure 31. NBAQ4 mini hydro power house

and cross connections with some of the other existing aqueduct lines which provides Manila Water with the flexibility of shutting down some of the existing aqueducts for rehabilitation works.

MINI HYDROELECTRIC POWER PLANT

The mini hydroelectric power plant will be used as a renewable energy source which could be used throughout the Balara Water Treatment Plant. The mini hydroelectric power plant has been sized to receive flows in the range of 11.6 m^3/sec to 12.8 m^3/sec with net pressure heads ranging between 13–14.5 m. The turbine is a vertical Kaplan type model which will be housed in a dedicated large building together with electrical cabinet and components required to control the mini hydro turbine and other electrical equipment associated with the overall scheme. There are two main pipelines which come from the Outlet Structure. A DN2400 line which feeds the mini hydro turbine and a DN2000 pipeline which feeds a bypass line which has a plunger valve along the line. In the event that there is a surge or load rejection on the mini hydro turbine line, the main inlet valve on the mini hydro line shuts and the plunger valve opens to maintain flows through to the system.

The hydroelectric plant is capable of generating 1.35 megawatts of electricity as part of the scheme. Manila Water has not currently committed to installing the hydropower plant.

The overall layout of the mini hydro power plant is shown in Figure 31.

CONSTRUCTION

At this stage of writing this paper, it was estimated that construction for the mined tunnel would be complete and TBM tunnelling will be well underway at the time of presenting. Due to delays on site, the only key item underway is the Construction Access Shaft works and installation of the large diameter unreinforced piles (Figure 32).

Figure 32. Installation of 12 m casing for 1.5 m diameter secant pile shaft

Evolution of Tunnel Eyes Designs Based on Lessons Learned for the Regional Connector Project in Los Angeles

Eli Mathieu ▪ Skanska USA Civil
Jose Muhr ▪ Skanska USA Civil
Darren von Platen ▪ Traylor Bros., Inc.
Ivan Hee ▪ Skanska USA Civil
Jake van Baarsel ▪ Skanska USA Civil

ABSTRACT

Tunnel Boring Machine Angeli bored two tunnels from Little Tokyo to the Financial District of Downtown Los Angeles as part of the LA Metro's Regional Connector Project. Angeli broke into Hope Station and Flower Cut-and-Cover structure two times at each location within a period of 6.5 months. Ground conditions are similar in both locations enabling fine tuning of tunnel eye designs based on lessons learned. This paper discusses the design and construction of the eyes as well as changes to the TBM mining parameters.

INTRODUCTION

Currently, Los Angeles Metro light rail lines are split into an eastern (Gold Line) and a western (Blue, and Expo Lines) regions. Los Angeles Metro's Regional Connector Project (LA Metro) is a 1.9 mile alignment that connects these two regions, enabling a one-seat ride for travel across Los Angeles County. From Metro Gold Line, passengers will be able to travel from Azusa to Long Beach and from East Los Angeles to Santa Monica without transferring lines. LA Metro awarded this $1.0 billion project to Regional Connector Constructors (RCC)—a Skanska/Traylor joint venture with Mott MacDonald as lead design engineer. This is a turn-key project, with RCC providing all infrastructures except the trains.

Scope of Project

The Regional Connector Project (Figure 1) included a 1.1 mile twin bored tunnel, 0.5 miles of cut and cover tunnel, a mined cross over cavern to be constructed by sequential excavation methods, and cut and cover three underground stations. The alignment, from east to west consists of Central station with tunnel boring machine launch, followed by Broadway Station, Hope Station, and Flower cut and cover guideway. The stations and guideway are connected with bored tunnels, which are mined with a 21.6 feet diameter Earth Pressure Balance (EPB) Tunnel Boring Machine (TBM).

Sequence of construction are as follows: 1. Launch TBM at Central starting with the left tunnel, 2. Break-in at Hope, 3. Relaunch at Hope to Flower, 4. Retrieve TBM and transport back to Central, 5. Repeat for the right tunnel. This means the TBM mined through Broadway station which necessitate the removal of precast tunnel segments during the excavation. This change in sequence was due to the delay in utility relocation at Broadway from a previous contract.

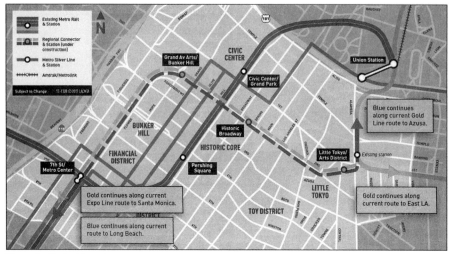

Figure 1. Project layout

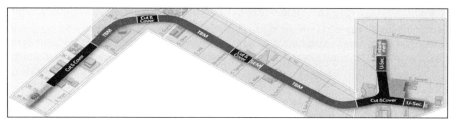

Figure 2. Alignment

Ground Conditions

Ground conditions at the two locations of break-ins, Hope Station and Flower Cut-and-cover guideways are as follows:

1. Hope Station:

    Ground cover above the tunnels crown at Hope Station break-in is about 82 feet. This consists of about 10 feet of Fill, underlain by about 50 feet of moderately to highly weather Fernando Formation, underlain by about 22 feet of slightly weathered to fresh Fernando Formation.

Figure 3. TBM Angeli

2. Flower Cut-and-Cover Guideways:

    At Flower Street, there is about 32 feet of ground above the tunnels crown. This consists of about 20 feet of Fill, underlain by about 12 feet of moderately to highly weathered Fernando Formation.

Fill typically consists of mixtures of gravel, sand, silt, and clay with construction debris, typical of an urban environment. The Fernando Formation is classified as bedrock and consists of a poorly bedded to massive clayey siltstone to silty claystone that is poorly cemented and extremely weak to very weak. Unconfined compressive strengths ranges from around 25 psi up to 300 psi, with an average of about 140 psi according to the Geotechnical Baseline Report. The moderately to highly weathered Fernando were typically wetter with pockets of perched water. Permeability of the Fernando formation is low, hence dewatering is unnecessary. Small sump pumps are sufficient control perched groundwater that flowed into the excavation.

DESIGN AND CONSTRUCTIONS

Design

The support of excavation for all the cut and cover stations and guideways are soldier pile and lagging systems. Piles are typically spaced at 7 feet centers.

Figure 4. Flower TBM retrieval

1st Design—Hope Station

Construction of the tunnel eyes occurred in multiple stages. Initial plans was to install a fiberglass reinforced shotcrete face during the excavation of the station. However, after a schedule analysis, it was determined that the critical path was to excavate to the bottom of excavation, in order to place the invert slab to enable removal of the lowest bracing level to beat the arrival of the TBM. Sequence of construction of tunnel eyes are as follows:

1. Excavate and install timber lagging to the bottom of excavation. (Figure 5)
2. Cast invert slab. (Figure 6)

Figure 5. Excavate to bottom of excavation

Figure 6. Invert slab in place and strut removed

Figure 7. Install brow steel

Figure 8. Install tunnel eye seal frame

Figure 9. Shotcrete wall with fiberglass rebar

Figure 10. Completed eye prior to pile removal

3. Remove lowest bracing level.
4. Install tunnel brow beams (to provide rigidity to the tunnel eyes). (Figure 7)
5. Install tunnel eye seal frame (Figure 8)
6. Remove timber lagging, 5 ft at a time and install a fiberglass reinforced shotcrete wall. (Figure 9)
7. Install 2 sets of bull flex seals. (Figure 10)
8. Cut soldier piles within TBM envelop

Design of the brow steel and shotcrete wall was performed in-house by RCC. Design and installation of brow steel poses some challenges due to the fact that piles were not on the same plane. Brow beams needed to be laid-out in the field and coped individually. Figures 11 and 12 shows the layout of the brow beams in section and plan.

The shotcrete wall was constructed in 5 ft lifts, top down. Thickness of the shotcrete was designed to be 8 inches, with horizontal #4 fiber glass rebars at 10 inches spacing.

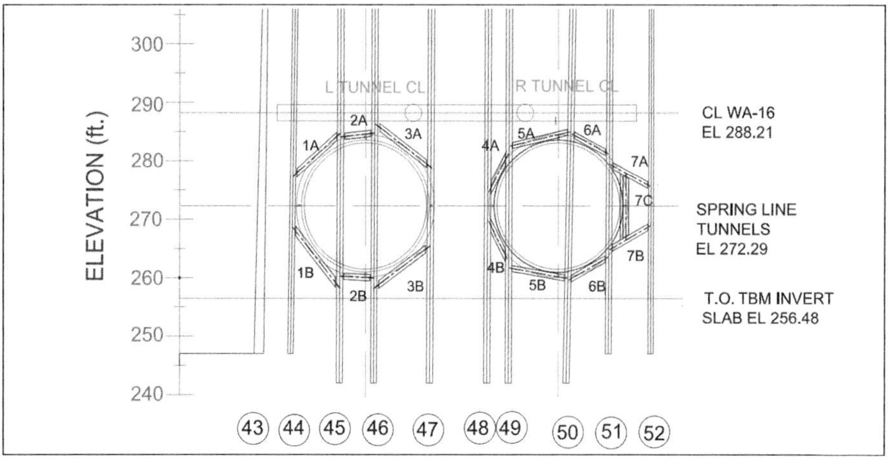

Evolution of Tunnel Eyes Designs Based on Lessons Learned

Figure 11. Brown steel layout at Hope break-in

On June 1, 2017, Angeli broke into Hope Station. The shotcrete wall did not perform as well as RCC would like as there was about 5 feet of Fernando Formation that was pushed in by the TBM. While this did not result in measurable settlement for the Hope Station break-in—where there was significant cover and the Fernando Formation was relatively massive and un-weathered—such behavior would present settlement risks during break-in to the Flower St. Cut & Cover Excavation.

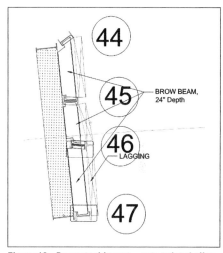

Figure 12. Brow steel layout—note twisted piles

At Flower St. Reception Shaft, there are numerous utilities that are quite close to the crown of the mined tunnels. There is a 72 inch reinforced concrete pipe storm drain only 10 feet above the crown of the tunnel, as well as multiple electrical and communication duct banks, sewer, and gas lines 20 feet above (Figures 13 and 14). Additionally, most of the cover was comprised of artificial fills, with only ½ × cut diameter (approximately 12 feet) of weathered Fernando Formation between the fills and the planned TBM cut. It was also likely that years of water flow through the bedding of the storm drain may have accelerated the weathering process of the Fernando Formation in this area. Due to the higher risks of damaging existing utilities and the challenging geology, RCC decided to provide a more robust tunnel eye design. These utilities, geological conditions and lessons learned from the first Hope break-in provided valuable information to improve on the design.

Improvements for the next three break-ins are as follows:
1. Increase thickness of shotcrete wall: 8 inches to 12 inches thick
2. Increase fiberglass reinforcement of shotcrete wall: #4 at 10 inches to #6 at 4 inches in the middle 10 feet of the eye

Figure 13. Flower St. Reception shaft after both tunnel drives completed, with shallow utilities shown above

Figure 14. Flower St. during excavation with shallow utilities including a 72 inch diameter RCP storm drain

   3. More conservative operational parameters for the TBM (described further below)

   4. At Flower, due to the low cover of Fernando Formation, spiles were installed to ensure stability of tunnel crown during break-in. Spiles consists of 1.25 inch (32 mm) hollow bars, 20 feet long, 5 degrees up from horizontal, and 8 inch on-center above brown steel. The hollow bars were then grouted and shot-creted into the sealed faced.

Figures 15 to 18 shows the four break-ins of the Regional Connector project.

TBM Mining Parameters

For both Hope and Flower break-in's, TBM face support pressure limitations were incrementally reduced over the final 50 feet of the approach to the station. The typical operational limits on the average EPB pressure of 2.0 ± 0.5 bar, were gradually stepped down to 0.3 ± 0.3 bar for the final 10 feet of advance. While this significantly reduced the pressure transmitted to the temporary support of excavation, it was clear from the experience at the first Hope break-in, that more conservative restrictions would be prudent for Flower Street.

Though face support pressure was close to zero for the first break-in, there were no direct restrictions on tooling force which could be transmitted to the face and the SOE by the cutterhead. When tunneling through concrete diaphragm walls, secant piles, or jet-grout blocks, with soft-ground machines, advance rates are often capped to protect cutting tools by reducing tool load. RCC adopted a similar approach to subsequent TBM break-ins for this project, but with the slightly altered goal of reducing load to the SOE.

Angeli operated with average advance rates of approximately 80–90 mm/min, occasionally reaching rates well over 100 mm/min. For the break-ins, advance rates were capped, and incrementally reduced as follows:

   1. 20 mm/min from 25 feet to 15 feet of break in

   2. 15 mm/min from 15 feet to 5 feet of break-in

   3. 10 mm/min within 5 feet of break-in

LESSONS LEARNED

RCC's in-house engineering team designed the components of the tunnel eye. This included the brow steel, shotcrete wall, and tunnel eye seal frame. The in-house

Figure 15. Break-in #1: Hope, June 1, 2017

Figure 17. Break-in #3: Hope, December 8, 2017

Figure 16. Break-in #2: Flower, July 19, 2017—note significantly less soil in front of TBM

Figure 18. Break-in #4: Flower, January 16, 2018

design enabled rapid re-design based on the performance of the first break-in. It was seen that minor increase in materials (in this case shotcrete thickness and fiberglass rebars) resulted in significant improvement of the behavior of the wall.

While the results of the design approach to the first TBM break-in confirmed that arching and bridging effects of the Fernando formation could be relied upon for short duration ground support, it also revealed several risks which would need to be mitigated to address the unique challenges that the Flower Street break-ins presented. By keeping the relevant design work in-house, and maintaining a hands-on approach to TBM operation, RCC was able to rapidly identify and address these risks, without realizing project delays.

°Minor weight increases to the fiberglass reinforced shotcrete eye were implemented to improve the strength of the wall, while additional operational restrictions were introduced to the TBM logic to reduce demand. An economical spile pattern was developed to ensure predictable bridging behavior where weathered Fernando formation was present, and where settlements were of immediate concern to buried utilities. This approach of incremental design improvements resulted in a shotcrete soft-eye which stood for a longer time period prior to break-in, reducing the amount of post-break-in mucking, limiting risks of further settlements below critical utilities adjacent the soldier pile and lagged excavation, and decreasing the amount of time to prepare the TBM for the next phase of the project.

LA Metro Regional Connector Transit Project: Successful Halfway-Through Completion

Mike Harrington ▪ LA Metro
Tung Vu ▪ VN Tunnel and Underground, Inc

ABSTRACT

The Regional Connector Transit Project is a 1.9-mile long underground light rail system that will connect LA Metro's Blue, Expo and Gold Lines in downtown Los Angeles. This $1.81-billion design-build project is expected to be completed in winter 2021–2022. The project consists of 21-foot diameter twin-bored tunnels, a 287-foot long crossover SEM cavern, three new underground stations (at 1st Street/Central Avenue, 2nd Street/Broadway Avenue, and 2nd/Hope Streets), and cut-and-cover tunnels along South Flower, Alameda, and 1st Streets. Final designs have been completed and the construction has reached the halfway-through completion milestone. Bored tunneling was successfully completed with little to no ground settlements. Excavation of the 36-feet high by 58-feet wide, 287-feet long SEM cavern has started, with completion scheduled by early 2019. This paper will provide overview of design elements and challenges experienced to date, as well as an update of construction progress on major components of this complex transit project.

INTRODUCTION

The Regional Connector Project (Project) is a complex subway light rail project that runs through the heart of downtown Los Angeles. The design-build contract was awarded in April 2014 to the Regional Connector Constructors (RCC), a joint venture between Skanska USA Civil West California District, Inc. and Traylor Brothers Inc., and their designer team consisting of Hatch Mott McDonald (HMM) and subconsultants. The Project broke ground on September 30, 2014 and is expected to be completed in winter 2021–2022. The main components of the Project consist of 21-foot diameter twin-bored tunnels, three crosspassages, a 287-foot long crossover SEM cavern, three new underground stations, and cut-and-cover tunnels along South Flower, Alameda, and 1st Streets. The project map is shown in Figure 1. The design and construction challenges associated with each of these main components are discussed in the following sections.

PROJECT DESIGN

Tunnels

The Regional Connector tunnels are located primarily within the Fernando formation consisting predominantly of extremely weak to very weak, massive, clayey siltstone. About 1,000 feet of tunnels on the eastern end are in alluvium and mixed face of Fernando formation and alluvium. The tightest curve of the tunnel alignment is 583 feet in radius. The tunnels were designed with a reinforced precast concrete tunnel lining (PCTL) to be used with an earth balance pressurized (EPB) tunnel boring machine. The PCTL ring is 18'-10" inside diameter, 10.5" thick, 5 feet long, and consists of 5 segments and a key. The segments were designed with 6,500 psi concrete, 80 ksi yield strength wire rebar, convex joint surfaces to enhance seismic performance, and

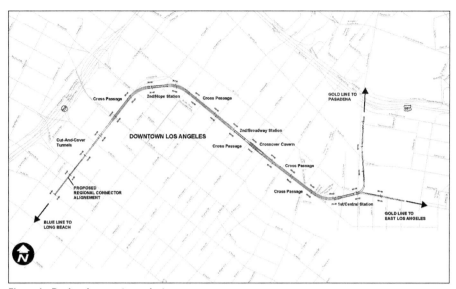

Figure 1. Regional connector project map

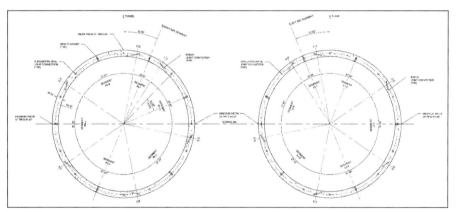

Figure 2. Tunnel typical section

a dosage of 1.7 lbs polypropylene microfibers per cubic yard of concrete for fire resistance. The rings are designed with a right ring and left ring pattern and a taper of 1.5" to allow for alignment curve negotiation. Figure 2 shows typical sections of the PCTL. The structural lining design was modelled using FLAC 3D. Four consecutive rings were modeled with 32 dowel connection. Loading considered in the lining design includes temporary ground load during excavation, long-term ground loads, seismic loads, and train loads. The seismic loads on the PCTL were simulated with the racking deformation applied at vertical boundaries of the model. Due to variation of geologic condition along the tunnel alignment and different surcharge requirements, three different types of reinforcement (Typical, Heavy, and Extra Heavy) were designed. The typical type PCTL was required in Fernando formation, the heavy type was required in alluvium or mixed-face condition, and the Extra Heavy type was required in alluvium and underneath the Japanese Village Plaza where a surcharge of 1000 psf was required.

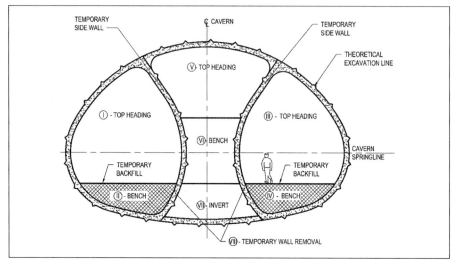

Figure 3. SEM cavern typical section

SEM Cavern

A crossover at the eastern end of the 2nd/Broadway Station is required for train operation. Since the crossover is located beneath the narrow 2nd Street where the basement of the buildings on both sides were constructed beyond the property line into the sidewalks, a cut-and-cover structure is not feasible. A sequential excavation method (SEM) cavern was designed to overcome the site constraints. The SEM cavern is 58 feet wide, 36 feet high and 287 feet long. It is located within the Fernando formation, with a depth of invert at 86 feet and the crown at 50 feet below grade. Above the cavern crown there is approximately 30 feet of Fernando formation overlaid by 20 feet of alluvium. The cavern was designed using FLAC 2D and FLAC 3D with all applicable loads required by Metro Rail Design Criteria (MRDC) and AASHTO 2012. The cavern was designed with right, left, and central drifts with top heading and bench. Figure 3 shows a typical cross section of the SEM cavern initial lining with the numbering sequence of excavation. Both TBM tunnels were completed prior to the start of SEM cavern excavation. Since the PCTL ring is 5 feet long, the SEM excavation round of 3'-4" was selected particularly to allow removal of two PCTL rings at every three SEM advances. The SEM cavern lining consists of a 12 inches thick initial fiber-reinforced shotcrete lining, a hydrocarbon resistant (HCR) membrane, and an 18 inches thick cast-in-place concrete reinforced concrete final lining with the invert slab thickness varying from 18 to 69 inches. The cavern also houses an emergency ventilation plenum located above and separated from the trainway by a 12 to 16 inches thick plenum slab.

A FLAC 3D model was first performed to simulate the SEM excavation sequence, soil properties, shotcrete lining with age-dependent strengths, and adjacent structure loading. It also serves in the prediction of ground settlements above the cavern. The cavern initial and final linings were analyzed with FLAC 2D models which were calibrated to account for the three-dimensional effects of ground relaxation by matching the ground convergence of the 2D models with that of the 3D model at locations of key performance indicators (KPIs). The linings were designed for various load combinations with different load factors specified by the MRDC and AASHTO 2012. Since load factors are not typically applied to a geo-mechanical numerical modeling, the load correction factors, which are the ratio of the load factors and a selected constant, were incorporated into the model. The selected constant was then multiplied with the lining loads at the end of the analysis to obtain the combined design loads for the lining.

The cavern final lining was designed for both the operational design earthquake (ODE) and maximum design earthquake (MDE) specified in the MRDC. The lining was originally analyzed using a simplified pseudo-static method by applying the racking displacements obtained from one-dimensional site analysis on the vertical boundaries of the FLAC 2D model. Since the cavern is a critical and complex structure, its seismic design was required to be checked with a more sophisticated method specified in the MRDC; a dynamic analysis performed with FLAC 2D and three spectra-matching time histories. The interaction between initial and final linings with the presence of a waterproofing membrane was captured by specifying interface properties to bracket the interaction range from slippage to rigid connection. The results from the dynamic analysis indicated that the pseudo-static analysis was adequate for the cavern lining, with some minor rebar modifications to the center wall and plenum slab.

1st/Central Station

The 1st/Central Station is connected to the wye structure on the east end and the twin-bored tunnels on the west end. It is a shallow underground station with the depth of invert slab being approximately 48 feet below grade and 8 feet of ground cover over the roof slab. The station houses the trainway, platform, and ancillary rooms on the south side. Due to its limited height, the station was designed with no separate concourse level between the platform and plaza, but instead uses a mid-landing for stairs and escalators. The station structure consists of typical 3 feet thick exterior walls and invert slab, and a roof slab thickness varying from 2'-6" to 3'-9". The station was designed as a typical cut-and-cover structure with applicable loads specified in the MRDC, and AASHTO with Caltrans amendments. The seismic design was done using a simplified pseudo-static method with the racking displacements obtained from one-dimensional site analysis. Figure 4 show a rendering of the station plaza.

Figure 4. 1st/Central Station

2nd/Broadway Station

The 2nd/Broadway Station is connected to the SEM cavern on the east end and the twin-bored tunnels on the west end. The trainway box is located beneath 2nd Street, and houses the trainway and platform on the first level and the concourse and ancillary rooms on the second level. The trainway box invert is located approximately 87 feet below grade with typical 4 feet thick exterior walls and a 4'-6" thick invert and roof slab.

An entrance structure is located on the south side of the station and is located within property owned by a private developer. To resolve right-of-way constraints, a subsurface easement was granted by the developer to Metro—in exchange for the new station entrance structure to also be utilized as a partial foundation for the developer's planned mix-used building of 6 to 30 stories (i.e., the overbuild). As the deal between Metro and developer went through during the bidding phase, the overbuild loads provided by the developer were included in the bid package as an addendum to be incorporated into the final design. Owing to multiple building concepts under consideration at the time, and to provide flexibility, the overbuild loads were estimated as the envelope loads from each of building concepts. As a result, the overbuild loads ended up being too large and created some constructability issues for the entrance structure. After developing and analyzing several overbuild load reduction alternatives, the entrance structure was ultimately successfully redesigned. The redesign also incorporated a new load transfer system and stem walls extending above the entrance roof, to allow for the future construction of overbuild structural elements with limited impacts to the station operations.

The revised entrance structure typically consists of a 6'-6" thick invert slab, 4 feet thick exterior and interior walls, and a 4 feet thick roof slab. It is anticipated that the entrance structure and station box will behave very differently during an earthquake event because of the overbuild structure. Therefore to allow for the anticipated seismic interaction and differential settlement between the two structures, the station box and entrance structure will be separated. The waterproofing system was also carefully designed at this interface to ensure water-tightness.

The structural design of each station structure was carried out with a 3D SAP 2000 structural model. The output from SAP model was then exported to an Excel macro for detailed design. The seismic design was performed using a simplified pseudo-static

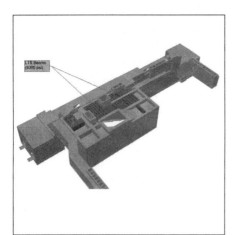

Figure 5. 2nd/Broadway load transfer system and future overbuild

LA Metro Regional Connector Transit Project

Figure 6. 2nd/Hope Station and glass railing pedestrian bridge

racking displacement method. In the 3D model, both the vertical and lateral soil springs were modelled using link elements. The lateral seismic racking displacements along the height of structures were converted into forces that were then applied at the ground ends of link elements. Figure 5 illustrates the load transfer system to be built on the roof of the entrance structure for the future overbuild.

2nd/Hope Station and Pedestrian Bridge

The 2nd/Hope Station is located adjacent to the Broad Museum and Walt Disney Concert Hall. It is the deepest station of the project with the invert slab located at depth of 105 feet below grade. The structural design of this station is similar to the 2nd/Broadway Station, however due to its great depth the vertical transportation of passengers will be accomplished using six high speed elevators between the concourse and plaza levels. The end wall of the elevator corridor will be decorated with a mosaic title artwork of 61 feet high by 17 feet wide.

A pedestrian bridge was also designed to connect the elevated station plaza level with the Broad Museum. A typical concrete structure bridge was originally envisioned and included in the bid. However, per requests from the Broad Museum and City of Los Angeles during design development, the structure type was changed to a high-end pedestrian bridge with glass railings, tree planters, and art lighting to blend in with the surrounding iconic architectural environment. Figure 6 shows a rendering of the station plaza and pedestrian bridge.

PROJECT CONSTRUCTION

Tunnels

The bored tunnels were excavated with a refurbished Herrenknecht EPB tunnel boring machine (TBM). The machine was 21'-7" in diameter, and was previously used on the Gold Line Eastside Extension Project in Los Angeles and the University Link Light Rail Project in Seattle. Figure 7 shows the TBM, named "Angeli," ready to be launched at the 1st/Central station excavation.

Geologic conditions through the tunneling alignment consisted of approximately 1000 linear feet of alluvium and mixed-face of alluvium and Fernando formation, with the rest the alignment completely within the Fernando Formation. Prior to the start of tunneling work the TBM was lowered into a launch pit located within the Mangrove site, walked through the street decking excavation underneath the Alameda/1st Street intersection, and then prepared for launching at west end of 1st/Central Station excavation.

Figure 7. TBM Angeli at launching shaft

The 1st/Central Station tunnel interface and launch points were located immediate adjacent to a three-story parking garage for the Little Tokyo malls, with the tunnels being located about 15 feet below the garage and outdoor mall building foundations. In an effort to mitigate the potential risks due to tunnel-induced ground movements, the Project installed a compensation grouting system underneath the buildings. A 60-foot tunneling demonstration zone was also established within the station footprint, to allow for necessary calibration and adjustment of the TBM operations prior to tunneling beneath the garage. A fan of compensation grout pipes up to 400 feet long were installed from the station excavation at approximate 5 feet below building foundations. Prior to the start of tunneling, grout conditioning was completed and made ready for fracturing should have building settlement occurred. A horizontal inclinometer was installed approximately 3 feet above each tunnel to capture any deep ground movement before it propagated into the building foundations and ground surface. Permeation grout was also installed beneath a large diameter storm drain along 2nd Street, where the TBM excavated in alluvium at approximately 18 feet below the pipe.

A comprehensive building protection monitoring program was additionally established using multipoint borehole extensometers (MPBXs), building monitoring points, deep surface settlement points (DSSPs), ground surface settlement points (GSSPs), water levels, tiltmeters, and crack gauges on structures along the tunnel alignment, to monitor tunneling-induced ground movements.

The tunnel excavation started from the eastern end of the alignment on February 6, 2017. To allow for installation of compensation grout tubes from within the station excavation, the upper portion of demonstration zone was excavated to provide a working platform underneath the street decking. This resulted in ground cover above the tunnels as shallow as 7 feet. To prevent tunnel blowout under TBM face pressures, the contractor installed surcharge utilizing 1-ton nylon bags filled with soil placed atop the working platform. After some minor mechanical issues and ground heaves within the demonstration zone, the TBM was able to mine beneath the garage and buildings with no measurable settlement observed. Although the compensation grouting system remained in standby-mode during tunneling operations, the system never had to be utilized.

A tunneling incident did later occur when the TBM mining the L-track tunnel struck two undocumented steel beams beneath 2nd Street. (It was subsequently determined the beams were likely abandoned in place as part of a previous adjacent construction project.) Despite the strikes though, the TBM was able to cut through and break the steel beams into pieces. Some smaller steel pieces were discharged through the end of screw conveyor. However, larger pieces became stuck inside the cutterhead muck chamber and in front of the cutterhead, which required an intervention to remove the pieces and to repair some damage to the cutter tools.

The No. 2 screw conveyor main shaft cracked shortly after the beam strikes and subsequent restart of mining. The cutterhead and screw conveyor though were temporarily repaired and the TBM safely holed-through at the 2nd/Hope Station on June 1, 2017. Some additional repair was required to the machine at the 2nd/Hope Station before it was re-launched for the L-track tunnel reach between 2nd/Hope and Flower Street. Once the TBM completed the L-track tunnel, it was transported back to the 1st/Central Station. The damaged screw conveyor was then replaced with a new one before the TBM was re-launched for the R-track tunnel.

Bored tunneling operations also had to overcome a series of abandoned steel tiebacks along Flower Street between 3rd and 4th Streets. During the preliminary engineering phase, existing tiebacks along the tunnel alignment were identified based on available record drawings. It was determined at the time that some tiebacks from the construction of the Bank of America building were located within the R-track tunnel envelope. A tieback removal pit had to be included in the bid document so that all known interfering tiebacks could be removed prior to the TBM mining through the area. Ultimately a shaft and adit were designed and constructed by the contractor, and known tiebacks were successfully removed. However, the abandoned tiebacks from the construction of another project on the other side of the street were not accurately documented or recorded. These tiebacks were struck by the TBM but, similar to the steel beam strikes along 2nd Street, the TBM was able to cut through the steel tiebacks and break most of them into pieces small enough to be discharged through the auger and conveyer. On a few occasions, pieces of tieback rod did become lodged between the auger and conveyor shaft, which required the torch-cutting of an opening in the screw cover to remove the pieces.

Finally, the TBM successfully navigated several constraints along Flower Street below the 4th Street overpass, as shown in Figure 8. The TMB mined beneath a deep brick manhole and 18 inch sewer pipe, with less than 2 feet of separation to the pipe and only 1 foot to the manhole. At the same time, the TBM traveled within a few feet above the battered piles of the 4th Street overpass. Subsequent inspections of the manhole, pipe, and overpass found no damage had occurred.

The bored tunnels were successfully completed on January 17, 2018. Even though the TBM experienced some incidents, the tunnel operations were highly successful with little to no ground and building settlements observed. The average advance rate was approximately 70 feet per day, with a Project tunneling production record of 190 feet completed on one day.

Crosspassages

There are three crosspassges along the Project alignment. Each crosspasage is approximately 17'-6" feet in outside diameter and 10 feet long, with a cast-in-place reinforced concrete final lining. The crosspassages were excavated using SEM and fiber reinforced shotcrete initial lining. Prior to removal of the PCTL segments to make openings for the crosspassages, both tunnels were supported with "hamster cages" made up of a two ring beams connected by a series of tie-beams. The cages were designed to be collapsible and were transported into the tunnels on a rail-running frame. Erection of cages into their designed positions was achieved using by hydraulic jacks mounted onto the cage frames. Figure 9 shows a hamster cage positione with a crosspassage opening to the left. All three crosspassages were successfully

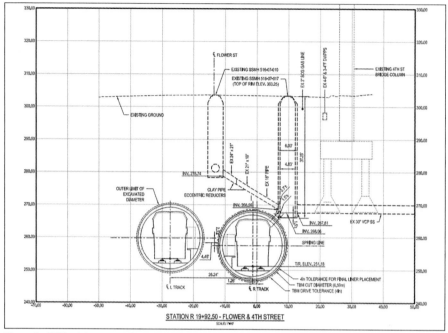

Figure 8. TBM mining constraints at Flower and 4th Streets

excavated in Fernando formation which provided excellent SEM standup time and minimal groundwater inflows.

SEM Cavern

The SEM cavern is one of the most critical components of the Project and draws a lot of attention from all affected stakeholders. Work and Action plans were carefully prepared by the contrator and approved by Metro and the City of LA Bureau of Engineering (LABOE) prior to start of SEM work. A canopy of 60 feet long grout tubes was installed from the 2nd/

Figure 9. Crosspassage excavation with hamster cage tunnel support

Broadway station excavation east headwall. Some grout tubes hit the above-mentioned abandoned steel beams under 2nd Street, and the tubes had to be terminated shorter than planned. A reinforced concrete beam was constructed at the end of canopy pipes, on face of the head wall, to provide stability for the grout pipe canopy.

The SEM excavation was performed following a left drift–right drift–center drift excavation sequence. A CAT 328D LCR excavator with a roadheader attachment was used for the left drift and right drift excavations, while an ITC was used for the center drift. Shotcrete operations were performed using a Potenza robotic sprayer. The excavation started with 3'-4" round length and top heading and bench sequence. Since the PCTL ring length is 5 feet, the construction sequence was a typical three-round cycle that includes rounds A and B to excavate and remove PCTL ring, and round C to excavate only. A typical excavation of top heading and bench of one round consisted of:

excavation; removing PCTL segments; installing 2 inches of flashcoat; installing lattice girders channels and wire mesh; installing 5 inches shotcrete to 1.5 feet from the end of current round; and installing 5 inches of shotcrete to complete the previous round. The excavation profile and shotcrete application were scaled with the Amberg system. The Fernando formation presented a very favorable ground condition with an excellent standup time. Prior to the excavation there was a concern about possible connectivity of excavation with the overlying 12 feet high by 10 feet wide storm drain with weep holes. However, no groundwater flows were observed except for some isolated damp spots on the excavation face.

Due to the significant size and critical nature of the SEM cavern, an extensive monitoring program was implemented in order to measure movements of the ground surface and adjacent buildings. This included convergence arrays inside the excavation measured with the Amberg system, automated MPBX's, utility monitoring points (UMPs), building monitoring points (BMP's), and GSSP's. The BMP's and GSSP's were monitored using total stations. Data from GSSP's were then processed to produce ground surface settlement contour maps and settlement slopes that were then used to check against project specified criteria and to determine if any adjustment to the excavation sequence was necessary. The measured ground movements were found to typically be in line with the predicted values, and no excessive ground movements were recorded.

The SEM cavern was excavated with three 8-hours shifts per day, 5 days per week. The left drift excavation was started on May 31, 2018 and completed on October 22, 2018. The right drift started on July 5, 2018 and completed on December 6, 2018. The center drift started on August 14, 2018 and is currently under excavation. Figure 10 shows the excavation of center drift in operation.

Figure 10. SEM cavern center drift excavation

1st/Central Station

The 1st/Central Station is the shallowest of the three being constructed, with an invert approximately 45-feet below finished grade. The excavated area of this station also served as the launching site for the TBM, for both the left and right tunnels. The support of excavation at this location was constructed mostly from a soldier pile and timber lagging system, with supplemental tie backs and 3-foot diameter pipe struts. However, at the TBM launch points along the west bulkhead, the SOE was constructed using an 8-inch Shotcrete facing supported by 6 rows of fiberglass soil nails which allowed the TBM to successfully penetrate the wall. Figure 11 shown the construction of the station invert slab prior to the start of TBM operations.

Figure 11. Construction of 1st/Central Station invert

 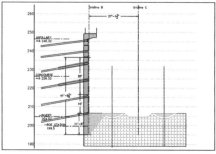

Figure 12. Underpinning of historical LA Times Building at 2nd/Broadway

2nd/Broadway Station

With an invert elevation of 199.8 located 85-ft below 2nd Street, the 2nd/Broadway Station is the second-deepest of the three stations being constructed. Excavation for the station trackway structure began in August 2016 but was then halted to install a large diameter Hobas storm sewer, and to allow the TBM to pass beneath as both the L-track and R-track tunnels. The decision to have the TBM pass beneath this area rather than walking the TBM through a completed excavation site was made to maintain the overall Project schedule.

The soil profile at this station consists of Fernando formation overlain by 15 to 30 feet of alluvium material. The TBM mined successfully through the station while the excavation was still more than 30 above. Once both bored tunnels were mined, portions of the tunnels were then backfilled to the spring-lines with spoil materials to further control convergence, and excavation was resumed. The PCTL rings were then sequentially exposed, cut or unbolted, and transported off site for demolition and disposal. Excavation to the final invert was reached in August 2018.

Immediately adjacent to the north wall of the guideway structure is an historical mid-rise building. This presented a constructability challenge as a portion of the building basement extends southward by more than 4 feet into the 2nd Street public right-of-way, and is located immediately above the trackway structure excavation. An innovative and complex underpinning system had to be erected consisting of spiling, 2 rows of precast panels, and a series of cast-in-place columns and timber lagging. The system was constructed in a top-down approach and anchored with tiebacks and 8-inch diameter pipes to control vertical movement of the system. Prior to the installation of the unpinning liquid levels, crack gauges, and monitoring points were affixed to the basement structure. However, no measurable movement or cracking of the basement structure has been observed. Figure 12 includes a photograph of the completed underpinning system a sketch of the final underpinning design.

2nd/Hope Station

With an invert depth exceeding 105-ft, the 2nd/Hope Station is the deepest of the three stations being constructed under the Project. The soil profile at this location consists mostly of Fernando formation overlain by up to 25-ft of alluvium. Mass excavation work started in March 2016 and was completed in February 2017. More than 7,100 truck loads (99,500 cubic yards) were removed and transported to a dump site located 20 miles from downtown Los Angeles. A photograph of the completed station excavation is shown in Figure 13.

Figure 13. 2nd and Hope Station excavation

A soldier pile and timber lagging system supplemented with struts and tiebacks was used for the support of excavation. Solder beams sizes ranged from W24×76 to W24×335 and were typically spaced at 7 to 8 feet on centers. Due to the upper alluvium layers and noise abatement requirements for the project, solder beams were installed in 36 inch diameter pre-drilled and cased holes, rather than being driven. Struts consisted of 36 inch diameter pipes installed at up to 5 levels. Tiebacks up to 125 feet long and angled at 15 degrees, with 8 to 14 strands each, were installed where struts could not be installed due to constructability constraints. More than 350 tiebacks were installed before the excavation was completed.

Similar to other locations along the Project corridor, a comprehensive subsurface monitoring program was established for the site. Although MPBX's were installed near the ends of the station to monitor TBM work, the station monitoring program primarily relied on a system of BMP's, GSSP's, inclinometers, tieback load cells, and strain gauges installed on the pipe struts. Settlements and wall movements were then monitored real-time using Insite GPS and web-based communication software. With few exceptions, data obtained from this monitoring system showed that settlements and movements of the ground surface and support of excavation were generally less than or consistent with predicted values.

SUMMARY

The Regional Connector is a large and complex mega-project being constructed through the urban core of Los Angeles. Major elements successfully completed at the halfway-point of construction include the TBM-bored tunnels, tunnel cross passages, cut-and-cover street decking, station mass excavations, and initial drifts for the SEM cavern. Despite the known challenges of constructing in a congested urban environment, the Contractor (RCC) has been able to keep the project on schedule. This has been achieved through the efforts of a highly experienced team, focused planning, and the development of innovative engineering and construction solutions.

Good geotechnical conditions afforded by the predominate layer of Fernando formation along the corridor has also helped to facilitate construction. The TBM tunneling was able to advance at an average rate of 70 feet per day, with 190 feet of mining achieved during one particular day—a record for LA Metro. Ground surface settlements so far have generally been less than or equal to what was predicted from the engineering modeling. The good soil conditions have also made it feasible to construct the track crossover cavern using SEM techniques rather than cut-and-cover methods.

While anticipated construction challenges were successfully overcome, some unexpected obstacles did arise. During tunnel boring the TBM struck two undocumented abandoned steel piles along 2nd Street and several incorrectly documented abandoned steel tiebacks along Flower Street. Despite these strikes, the TBM was able advance with minimal damage to the machine or impact to the construction schedule. The ability to advance past these strikes is a testament to the Herrenknecht TBM equipment and skillful operation.

The project is scheduled to be completed by winter 2021–2022. Although the project team has overcome several challenges to date, other challenges will arise. Among these challenges is the complex system integration and commissioning of the project that will integrate three operating LRT systems together. The continued focus on planning and use of innovative approaches will best position the project team to successfully complete and commission the Regional Connector Project.

Permitting in a Design/Build Environment—The Challenges of 120 Years of a Design Bid Build Culture

Douglas Gaffney ▪ Mott MacDonald
Frank Perrone ▪ Mott MacDonald

ABSTRACT

Resource agencies in the United States are not familiar with and do not like structure of design-build projects. This paper explores reasons for this and presents possible solutions for improving the permitting process in the design-build model. Benefits often attributed to design-build include cost savings, innovation in construction methods and faster project implementation. The permitting process in the United States has developed over the past 120 years, starting with Section 10 of the Rivers and Harbors Act of 1899. Over the years, new regulations have been implemented on the Federal and State level to improve the environment and protect our common heritage and natural resources. This patchwork of regulations can, however, lead to confusion, especially in the context of design-build projects. One of the guiding principles employed by permitting agencies is to review the project in its entirety, and to look at all impacts prior to granting a permit. Because of this, the permitting agencies want a "static design." In the design-bid-build world, this was relatively easy. The challenge arises with design-build projects because the contractor wants to innovate at the same time resource agencies are looking for a static design and construction methodology.

INTRODUCTION

The focus of this paper is on a subset of design-build tunnelling projects; namely, large transportation tunnels in the marine environment in the United States. Many of the permitting concepts hold true for other projects, but the marine environment presents unique challenges. The permitting process for in-water construction has developed over the past 120 years, starting with Section 10 of the Rivers and Harbors Act of 1899. Over the years, new regulations have been implemented on the Federal and State level to improve the environment and protect our common heritage and natural resources. The National Environmental Policy Act (NEPA) was enacted in 1970. NEPA's most significant impact to projects is the requirement that a lead federal agency must prepare either an environmental assessment (EAs) or an environmental impact statement (EIS). These reports state the potential environmental effects of proposed federal agency actions.

The permitting process functions reasonably well in the context of Design-Bid-Build. In its simplest form, the engineer would design the project, the engineer and owner would get permits (for the entire project), the owner would bid the project, and the winning contractor would build it. design-build on the other hand, is defined by The Design-Build Institute of America (DBIA) as a project delivery system whereby one entity performs both architectural/engineering and construction under one single contract. While this is not a new procurement model, the benefits of this delivery method to the owner include cost savings, speed of implementation, quality, innovation and flexibility. Unfortunately, speed, innovation and flexibility during the design-build phase is not well integrated with the permitting process.

The Fails Management Institute's June 2018 Design-Build Utilization Combined Market Study predicts design-build construction spending will grow 18% from 2018 to 2021 and reach over $320 billion. In addition, the design-build model is anticipated to represent up to 44% of construction spending by 2021 (FMI 2018). The FMI market study identified five factors as significant drivers of design-build utilization for highway/street and water/wastewater projects including Project Schedule, Complexity, Size, Outside Expertise and Staff experience. At times the permitting process is at odds with some of these drivers as well as some of the key benefits of this delivery model. One of the guiding principles employed by permitting agencies is to review the project in its entirety, and to look at all impacts prior to granting a permit. Because of this, the permitting agencies want a "static design." In the design-bid-build world, this was relatively easy. The challenge arises with design-build projects because the contractor wants to innovate at the same time resource agencies are looking for a static design and construction methodology.

This paper will help identify permitting challenges in the design-build framework and explore opportunities to reduce the disconnect between agency expectations for a complete, static design, and design-build innovation and flexibility.

BACKGROUND

There are five major permitting efforts to be considered when permitting a major transportation tunnel project in the United States. Broadly speaking, they are:

1. National Environmental Policy Act (NEPA) coordination
2. Nationwide Permit No. 6 for supplemental geotechnical and environmental borings
3. Section 10 and Section 404/401 permits from the United States Army Corps of Engineers (USACE) and State agencies
4. Individual National Pollutant Discharge Elimination System (NPDES) permitting for the discharge of construction process water
5. Upland construction permitting including erosion and sediment control and stormwater plans

There are often several additional permits and many third-party stakeholders involved (see Figure 1), especially when Section 408 concurrence from the USACE is required.

Each design-build project will be unique in terms of the amount and quality of pre-design work and coordination has been accomplished by the owner. Two items of great importance to permitting are the status of the Environmental Impact Statement (EIS) and whether it has a Finding of No Significant Impact (FONSI). Also, of importance is the completeness of the geo-environmental characterization of any anticipated dredging areas and the tunnel horizon.

If the project owner has completed an EIS and obtained a FONSI, it would then be likely that remaining NEPA coordination work would be limited to consultation with the National Oceanic and Atmospheric Administration (NOAA), United States Fish and Wildlife Service (USF&WS) and other resource stakeholders. An exception would be if the EIS specifically calls out the need for additional survey(s) of habitat or a biological assessment (BA) of a specific species of concern. Generally, the NEPA consultation will occur during the permitting process through the Federal sponsor, often the FHWA.

If the project is in a marine environment and supplemental borings are required for better understanding of the tunnel horizon, it is highly likely that a Nationwide Permit

Permitting in a Design/Build Environment

Environmental Permitting Summary	U.S. Navy	Local Maritime Association	State Dept of Environmental Quality or Protection	Local Port Authority	MARAD	Local Pilots	U.S. Army Corps of Engineers (USACE)	United States Coast Guard	State Marine Resources	NOAA Fisheries	NOAA Marine Mammals	NOAA Protected Resources	State Historic Preservation Office	US Fish and Wildlife Service	Environmental Protection Agency (EPA)	Federal Highways Administration (FHWA)	Local Marine Technical Advisory	State Department of Transportation (DOT)	Construction Methodology	Design	Compliance and Monitoring	Permitting
Permits and Concurrences																						
USACE Nationwide Permit #6	●	●	●				●	●	●				●					●	●	●		●
USACE Section 408 Concurrence - NWP6	●	●	●	●	●	●	●	●	●				●					●		●	●	●
Construction General Permit (incl. SWPP?)	●		●															●	●	●	●	●
NDPES Individual Permit for Construction Process Water Discharge			●																●	●	●	●
USCG Bridge Permit	●			●	●		●	●										●	●	●	●	●
Joint Permit Application or Joint Permit Process			●			●	●	●	●	●		●	●	●	●	●	●	●	●	●	●	●
USACE Section 408 Concurrence	●		●	●	●		●	●	●	●		●		●	●	●	●	●	●	●	●	●
Incidental Harassment Authority (IHA)											●								●	●	●	●
Endangered Species Section 7 Consultation										●	●	●		●					●	●	●	●
Essential Fish Habitat							●			●		●				●	●	●	●	●	●	●
Section 106 Cultural Resources													●							●		●
Dredged Material Management Plan (DMMP)							●	●	●	●		●			●		●	●		●	●	●
Coastal Zone Mgt. Certification			●	●	●	●		●	●									●		●	●	●
Revise Environmental Assessment (EA)	●	●							●	●		●	●	●		●	●	●	●	●	●	●
Mitigation/Compensation									●	●		●		●	●	●	●	●	●	●	●	●

Figure 1. Table indicating the various agencies and stakeholders required for a typical tunnel project

No. 6 will be required. And if the tunnel alignment is on the same footprint as a Federal navigation channel, then Section 408 concurrence will be required. In this case, the supplemental geotechnical investigation may be on the critical path for final design and permitting and perhaps result in impacts to the anticipated construction schedule.

The Section 10 permit comes from the USACE as they are generally the lead federal permitting agency. The Clean Water Act permits (Section 404 and 401) are generally obtained through the state having jurisdiction over the project. For this reason, many states have adopted a Joint Permit Process (JPP) or Joint Permit Application (JPA). Furthermore, the Section 10 permit cannot be issued until the USACE can assure that all applicable NEPA laws have been fully complied with, and, where applicable, Section 408 concurrence has been gained. This is how the various laws such as the Endangered Species Act, the National Historic Preservation Act (NHPA), and the Magnuson-Stevens Fishery Conservation and Management Act are enforced. Section 106 of the NHPA requires that, before approving or carrying out a federal, federally assisted, or federally licensed undertaking, federal agencies take into consideration the impact that the action may have on historic and cultural resources. Generally, the State Historic Preservation Office (SHPO) is the lead agency for consultation. Given the many touch points or interfaces between federal, state, and local laws, one can see how advanced planning from a permitting perspective could impact the success of a project.

As most states claim the bottom of tidal waterways as a public resource, dredging and other disturbances of the bottom is regulated at the state level. Royalties and licenses are often required. While a joint application to the USACE and state may seem like a way to streamline the permitting process, each entity retains its statutory authorization to regulate and permit construction thus adding layers of coordination that are needed prior to construction.

Compensatory mitigation for loss of habitat is required under the Clean Water Act. In 1990, the USACE and EPA entered into a Memorandum of Agreement (MOA) that clarified the protocol for determining the type and level of mitigation required under Section 404. This MOA has had a significant impact on the Section 404 permitting process. The 1990 mitigation MOA was developed to clarify the "appropriate and practicable measures" required to offset unavoidable impacts.

Another major permit activity involves the disposal of construction process water. In some cases, the local sewage utility will accept the waste water, but they will need to know daily quantities and the anticipated chemistry. For this reason, the characteristics of the process water to be generated on the project will need to be defined. If the Design/Builder wants to discharge treated process water into a waterbody, a NPDES permit will be required. This permit, often administered by the state, is more rigorous and has a longer lead time than other permits. In large tunneling projects the amount and characteristics of construction process water are often dependent on final equipment selection which may not be immediately clear in a design build project, adding a further complication in the permitting process.

Lastly, the permits for upland construction, outside the jurisdiction of the USACE, are provided by the state, and will include stormwater plans and erosion and sediment control plans. Often included in this effort is a Stormwater Pollution Prevention Plan (SWPPP) which implements the compliance requirements of the construction permit. Since this permit deals with stormwater, wetland delineation in staging areas may be required.

PERMITTING CHALLENGES WITH DESIGN-BID

Based on the foregoing, one can see that the design and construction methodology must be advanced to a certain point to allow the permitting agencies to evaluate the complete project. The design/builder might be tempted to use terminology such as "65 percent design" to describe the state of the project when developing a schedule or submitting a permit application. Unfortunately, there are two flaws with this from a permitting perspective, the first being potential unknowns tied to project extents and the second being unknowns tied to construction means and methods, for both permanent and temporary works.

Level of Design

The first flaw is that the level of engineering design often refers to the final configuration of the tunnel portals, land reclamation or marine protection elements or other infrastructure and does not refer to temporary structures needed for construction. The permanent footprint of the final configuration is important for the calculation of permanent impacts to natural resources, but resource agencies vary on their interpretation of permanent versus temporary impacts that require compensatory mitigation.

If a supplemental geotechnical investigation is to be conducted, it is possible that new information may lead to a change in the project's footprint, which in turn could impact the permitting process and result in delays to construction. For example, the new data may lead to the determination that more dredging is required. In that case, the footprint may increase, the dredged material disposal location may change (due to chemistry or volume limitations) or the volume of fill may increase, all affecting the final design and impacts to be mitigated. Another example often seen in marine projects involves the level of design tied to construction of temporary facilities to support permeant works construction. Often elements such as temporary docks, marine moorings and work platforms are conceptual during the procurement phase and the details of which may not be fully understood in the early phases of a design-build project.

Construction Method

The second flaw in the "percent complete" philosophy is that Agencies need to evaluate impacts to water quality, habitat and several other environmental impacts, and are therefore very interested in the manner in which a project will be constructed. This is due to the fact that construction in the water can have downstream impacts and can degrade water quality or habitat. Similarly, air quality and noise during construction can affect wildlife, and seasonality can prompt the need for construction schedules. For these reasons, construction methodology, including temporary impact footprints, dredging methodology, pile driving, foundation strengthening, treated process water discharge, moorings, navigation impacts in channels, etc., all need to be determined prior to submitting a permit application.

Also, for these reasons, the USACE tends to be very reluctant to allow the use of nationwide permits for early work items that are related to the larger project. To satisfy NEPA and the Clean Water Act, the resource agencies need to evaluate the project in its entirety to assess impacts, which is often at odds with the design-build procurement model.

Environmental impacts need to be quantified early in the design-build process to allow time for the agencies to review what the applicant has submitted and discuss internally prior to meetings to negotiate the final compensatory mitigation plan. The mitigation plan can be supported by a Habitat Condition Analysis (HCA) which describes the habitat that may be altered or lost, and any mitigating factors of the project. At the time of bid, the design-build contractor generally has little indication what the final mitigation requirements will be. This is a source of risk in the design-build model that owners generally pass onto the contractor. Once the permit applications are submitted, there is limited flexibility in the contractor's means and methods and this can stifle innovation, schedule, and increase costs. Once the permits are in place, the contract must weigh the costs (time) to the benefits (economic) of altering the footprint or construction methodology.

Compensatory Mitigation

As mentioned previously, compensatory mitigation for loss of habitat is required under the Clean Water Act and identification of appropriate and practicable measures to offset unavoidable impacts is a critical part of the permitting process.

Alternatives Analysis

Section 404 guidelines dictate that the Corps requires applicants to provide documentation that there are no practicable alternatives to the proposed project. An alternative is considered practicable after taking into consideration "cost, existing technology, and logistics in light of overall project purposes." The guidelines also provide that proposed projects may not be permitted unless "appropriate and practicable steps have been taken which will minimize potential adverse impacts of the discharge on the aquatic ecosystem."

No-Net-Loss

In 1990 the Corps and EPA clarified the "no-net-loss" of wetlands goal as "no overall net loss of values and functions." In the 2008 compensatory mitigation regulations, the agencies provide information on how to reach this goal, requiring that, when functional or condition assessments are used, compensation "must be, to the extent practicable, sufficient to replace lost aquatic resource functions. From the perspective of a design/build team, anticipating the cost of these vague requirements could be problematic.

1990 Federal Mitigation Memorandum of Agreement

Also, in 1990, the Corps and EPA entered into a Memorandum of Agreement (MOA) that clarified the protocol for determining the type and level of mitigation required under Section 404. This MOA has had a significant impact on the permitting process, and established wetland mitigation banking as an acceptable form of compensatory mitigation. Under the MOA, the agencies established a three-part process or sequencing guidelines to help guide compensatory mitigation decisions. The sequence was incorporated into regulations in 2008.

To aid the Corps in their permitting process, an Avoidance, Minimization and Mitigation Plan (AMMP) is prepared as an appendix to the Joint Permit Application. The AMMP follows the mitigation sequencing steps which are:

- Avoid: This step is in accordance with the alternatives analysis established by Section 404 guidelines, which allows permits for only the least environmentally damaging practicable alternatives. It restates, "no discharge shall be permitted if there is a practicable alternative to the proposed discharge which would have less adverse impact to the aquatic ecosystem."
- Minimize: If impacts cannot be avoided, steps must be taken to minimize the adverse impacts through project modifications and permit conditions. This

minimization step often takes place after the FEIS and during the earliest design phases of the design-build contract and may extend beyond the submittal of the permit application.
- Compensate: The final step is that the Corps, in cooperation with local resource agencies, is required to determine "appropriate and practicable compensatory mitigation for unavoidable adverse impacts which remain after all appropriate and practicable minimization has been required."

Compensatory Mitigation Mechanisms

Compensatory mitigation may be accomplished through the restoration, creation, enhancement, or preservation of wetlands or other aqueous habitats. Permittees generally fulfill this obligation through one of three available mechanisms:

- Mitigation banking: Wetland mitigation banking is the use of consolidated, off-site wetland restoration, establishment, enhancement, and/or preservation for the purpose of providing compensatory mitigation. These off-site compensatory mitigation sites generally provide compensation for multiple, smaller wetland impacts. Mitigation banks sell compensatory mitigation "credits" to permittees with regulatory requirements to offset wetland damages; the purchase of credits transfers liability for compensation from the permittee to the mitigation bank. Before selling compensation credits, a mitigation bank must acquire a compensation site, establish financial assurances for future site protection, and have an approved mitigation plan.
- In-lieu fee program mitigation: In-lieu fee programs allow permittees to satisfy compensatory mitigation obligations by donating funds to a "government or non-profit natural resources management entity." The government or non-profit organization charged with operation of the in-lieu fee program then uses these funds to complete wetland restoration, establishment, enhancement, and/or preservation, and responsibility for compensation is transferred to the in-lieu fee program operator. In order to sell credits, an in-lieu fee program must have an approved governing document, which includes a "compensation planning framework" designed to target high-priority watershed compensation sites.
- Permittee-responsible mitigation: Under permittee-responsible mitigation scenarios, a permittee maintains responsibility for achievement of compensation obligations imposed under Section 404. Permittees may opt to complete wetland compensation on their own or may contract an authorized agent or consultant to restore, establish, enhance, and/or preserve wetlands. This option is generally not preferred by design-build contractors due to the long lead time to initiate, and the long-term nature of post construction monitoring.

2008 Compensatory Mitigation Rule

In response to a Congressional mandate, on April 10, 2008 EPA and the Corps jointly issued the first federal regulation governing compensatory mitigation for impacts to wetlands and other aquatic resources (The Mitigation Rule). The Rule is focused on codifying requirements for compensatory mitigation and reiterates and codifies the sequencing steps of avoidance and minimization explained in Section 404 Guidelines and the 1990 Mitigation MOA. The Rule is intended to improve the planning, implementation, and management of compensatory mitigation, and requiring, to the extent practicable, that all mitigation decisions be made within the context of a watershed.

Individual Corps district engineers are given authority for implementation of a watershed approach, granting regional officers considerable regulatory discretion in compensatory mitigation site selection and necessitating acquisition of adequate data to assess watershed needs. Due to perceived advantages of mitigation banking and in-lieu fee programs over permittee-responsible mitigation, the Rule institutes an overall preference for use of mitigation banks and additionally prefers in-lieu fee mitigation to permittee-responsible mitigation.

DESIGN CHALLENGES IN A COMPRESSED SCHEDULE

Two additional challenges clearly emerge from the foregoing requirements. Both the design and construction methodology need to be established as soon as possible to allow the permitting agencies to evaluate the impacts of the project, and the impacts need to be quantified early in the process to allow compensatory mitigation to be implemented. Unfortunately, this works in direct contrast to a goal of design-build which is to encourage cost and schedule savings through flexibility and innovation during the project. In some cases, the increased communication between the design team and the construction team leads to flexibility and innovation which may have the unintended consequence of changing the construction method or footprint, therefore causing a disruption in the permitting process thus impacting schedule and minimizing one of the key benefits to a project Owner for selecting the Design/Build delivery method for project execution.

A solution to the challenge of integrating the construction methodology work undertaken by the contractor for "temporary' works, and the design work undertaken by the engineer for "permanent" works, has been successfully implemented by the authors. The solution is to allow the contractor and engineer to focus on the information required for the major permits during the initial months of the project. For each major permit identified on the critical path, a Marine Works Permit Report will be jointly produced by the contractor and engineer. At the proper time in the schedule, a marine works freeze is established to allow the environmental group to apply for permits. The work accomplished for the Marine Works Permit Reports will feed into the more traditional 30–60–90 percent complete design paradigm. The downside of this approach is that the marine works freeze locks the contractor into a particular construction methodology forcing early engagement between construction and design staff. Once the permit is received, then changes can be considered that may or may not trigger a permit modification.

Temporary Versus Permanent Project Elements

It is extremely important that the contractor defines how the project will be constructed during the bid phase. For example, will a dock need to be constructed to accept materials, stage operations, and receive equipment such as a Tunnel Boring Machine (TBM). If so, will it be temporary or permanent since this affects mitigation. In some cases, the type of piles chosen will be scrutinized by the permitting agencies due to hydroacoustic impacts. Will dredging be required? If so, will there be a loss of habitat or down drift sedimentation impacts? These two examples, pile driving and dredging, are used because they illustrate another point; agencies often have predetermined preferences, and sometimes offices within an agency can have conflicting opinions based on their own experience. If one agency thinks impact hammer is worse than vibratory hammer, the Design/Build team will need to be ready to describe why or why not a construction method is needed and be ready to minimize the impact.

Some states prefer mechanical dredging over hydraulic dredging, and the use of one method over the other could be considered an avoidance or minimization of impacts.

Design-Build contractors would prefer flexibility going into a permit application, but permit conditions are often written to restrict certain construction practices. Agency preferences are often identified during the pre-application meeting. This is another reason the marine works freeze needs to be in place prior to the pre-application meeting. If the agencies identify something that is a deal-breaker, the design-build team will have a chance to re-assess, re-design and re-write prior to the permit application.

Site Investigation Schedule and Early Owner Engagement

Another scheduling and permitting predicament can be found with the post-award supplemental or confirmatory geotechnical investigation. The pre-bid design is developed with the geotechnical information provided by the owner. Post-award supplemental investigations can provide exactly the kind of additional data needed to refine the most appropriate construction method. Unfortunately, marine borings require a Nationwide Permit No. 6 from the USACE. Depending on the district, and whether a Section 408 concurrence is required, this can take 4 to 6 months putting the acquisition of data on the critical path of the project. Therefore, forcing the team to weight the risks associated with proceeding exclusively with owner supplied data versus obtaining new data at the start of the project. Further, the timing of the marine permit and the unknowns tied to its approval can also impact the more mundane tasks of the site investigation such as locking in subcontracts with vessel and equipment providers and timelines for mobilization.

The more thorough the owner's pre-bid geotechnical and environmental data is the lower risk of schedule impacts. Unfortunately, this scope of work and the risks associated with it are often allocated to the Design Builder and becomes a challenge for the design/build team to manage. Therefore, increasing the importance of the Marine Works freeze. Recognizing that the laws governing the issuance of permits fit the model of design-bid-build, the contractor needs to devote significant resources up front to ensure that the project is sufficiently vetted to prevent significant changes later.

As stated above, the more work the owner can do prior to bid, the less risk there is of schedule impact. Similar to geotechnical borings, there may also be habitat assessments that are required by the Final EIS or Record of Decision (ROD). These assessments are often valid during a particular season. For example, if a benthic study is required prior to submittal of the permit application, the schedule must then accommodate this task in the critical path. In the context of design-build, the more permitting related prep work that can be accomplished, the less uncertainty there will be in the bid and delays in the permit submittal.

Innovation in Design-Build Permitting Process.

Prior to submitting the joint permit to the USACE and state agencies, a pre-application meeting is often required. This meeting will include all the relevant agencies. The purpose is to provide the agencies an opportunity to see and understand the full project and how it will be built. This in turn, allows the agencies to alert the Design-Builder to any potential permitting hurdles, issues or fatal flaws. To accomplish this, the presentation will need to describe important aspects of construction with two main purposes in mind:
1. What are the potential impacts to the environment?
2. How will the Design-Builder avoid, minimize, and monitor construction (compliance), and respond to any unforeseen impacts such as spills, breakdowns, storms, etc.

After the pre-app meeting, the Design-Builder should make final changes to their footprints, construction schedule, and means and methods prior to submitting the application. If the application is considered complete, the USACE will publish a public notice within 30 days of receipt of the application. Therefore, the information needs to be in the proper format and deemed "complete" by the USACE to go out for public notice. If construction methodology changes at this stage, the agencies may stop reviewing the permit application.

For a tunnel project in the marine environment, the marine works permitting reports described earlier should include relevant plan view and cross section drawings, construction sequence drawings, construction schedule and clarify, to within reasonable limits, the following:

- Tunnel alignment and number/location of piles in all navigable waters of the United States.
- Location and size of any temporary in water facilities including docks required to receive the TBM or other large construction components, number and type of piles, methods of pile driving, temporary and permanent impact footprints.
- Location and size of any temporary barge platform for offloading/transporting amended TBM material. Methods of protecting environmental resources during transport of TBM and other waste materials.
- Dredging (depth, footprint, and chemical analysis), including construction equipment access dredging (if needed)
- Methodology, location, and depth of in-water ground improvement, and how waste products will be stored, treated, and transported
- An estimate of potential additives to be used with the TBM or drilling equipment.
- Revetment, berm or scour protection footprints (if any) and volume of fill. This includes any in water tunnel enabling works that may be required.
- Method of land reclamation or expansion, including any beneficial use of clean excavated sand.
- Impacts to freshwater wetlands, steams, etc. during roadway construction
- Impacts to tidal habitats
- Stormwater management and nutrient credits
- Process water management and design of Water Treatment Plant (and need for VPDES permit)
- Initial assessment of mitigation requirements and compensation

It is understood by the resource agencies that some elements will be fine-tuned as the permit review progresses, especially in response to Requests for Information submitted by the agencies.

CONCLUSIONS

In summary, the following conclusions and recommendations are made to improve the permitting process on a major transportation project in the marine environment, and to reduce risk of schedule slippage and possible cost overruns due to permit delays.

1. The Design-Builder can be more understanding of the permitting process and how it developed over the years and attempt to build a team of knowledgeable construction and design staff to participate in this process. This is

in an attempt to work within the framework of laws governing how permits are processed and approved.

2. The project owner can assist by providing as much NEPA-related information and geotechnical data as possible during the bid phase.
3. Resource agencies can foster innovation by looking for ways to work within the design build framework. A recent example of this that was experienced by the authors was a decision to view project specific Dredged Material Management Plan (DMMP) as a living document opposed to a static component of the overall permit. This may not sound like much, but it in fact allows the Design-builder the flexibility to change the DMMP and obtain concurrence without having to file a permit modification.
4. Often, cost-savings attributed to design-build innovation can be environmentally friendly. A recent example of this was experienced by the authors on a Design/Build project where the contractor's ability to adapt to a smaller construction footprint reduced the amount of land reclamation, resulting in less bay bottom disturbance and a reduction of backfill materials including expensive island protection armoring stone.
5. Permitting agencies need all the relevant information for each construction element, and for the project in its entirety. The more complete the permit application is, the more likely to receive a timely review. This points to the value of the Marine Works Permitting Reports compiled early in the process, and the Marine Works freeze which allows a construction schedule to be provided, a successful pre-application meeting, rapid response to issues identified by the agencies and a timely, complete application.

REFERENCES

Design-Build Utilization Combined Market Study, Fails Management Institute for Design-Build Institute of America, Washington, D.C., June 2018.

An Introduction to Design-Build. Design-Build Institute of America, Washington, D.C., 1994.

West Gate Tunnel Project—Evolution in Three Lane TBM Road Tunnel Design

Jack Muir ▪ Aurecon
Harry Asche ▪ Aurecon
Ben Clarke ▪ John Holland

ABSTRACT

The West Gate Tunnel Project is a major upgrade for Melbourne's connectivity west of the CBD. The project includes widening the West Gate Freeway from eight to 12 lanes, a new bridge over the Maribyrnong River, connections to the Port of Melbourne and the CBD, a raised motorway and three-lane, twin tunnels (outbound 4 km, inbound 2.8 km).

The tunnel boring machine (TBM) design contains innovations in the cross section which provide major improvements for construction, operations and maintenance. The road is supported by a continuous deck along the tunnel length which allows the hydraulic, mechanical and electrical equipment to be located in space which is accessible at all times. This innovation saves considerable operational cost and provides additional flexibility for construction. The cross passages can now be provided as minimum size jacked boxes, built through a single segment opening.

Other innovations include the Egress Out and Under (EOU)—which allows the outbound tunnel to be extended an additional kilometre west of the inbound tunnel, to cross the surface freeway and simplify the ramps. The paper describes the concept and the details of the road deck, cross passages, and EOU design and construction.

INTRODUCTION

Currently, more than 200,000 vehicles a day rely on the West Gate Bridge, and a single incident can stop traffic. The West Gate Tunnel Project will reduce Melbourne's reliance on the West Gate Bridge by providing alternative links to the Port of Melbourne, CityLink and the CBD. The West Gate Tunnel Project will:

- Give people a choice—the tunnel or the bridge
- Get over 9,000 trucks off local streets in the inner west
- Allow 24 hour truck bans on six local roads
- Cut travel times to and from the west
- Better connect Melbourne's freeway network to help people and goods move around
- Create 6,000 jobs
- Make local roads and the West Gate Freeway safer and more reliable
- Provide 9 hectares of new public open space in the west
- Build and maintain better noise walls along the West Gate Freeway, making it quieter at houses, parks and sport fields.

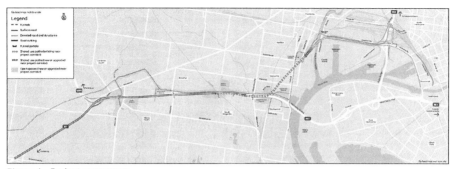

Figure 1. Project scope map

The West Gate Tunnel Project will be built over five years from 2018 until 2022. The project is a partnership between the Victorian Government and Transurban and will be built by construction contractors CPB Contractors and John Holland. The tunnel and new lanes on the West Gate Freeway, Footscray Road and Wurundjeri Way will open to traffic in 2022.

Figure 1 shows the project in a simplified representation. In the west, the existing West Gate Freeway is widened and improved. In the center and east, tunnels carry traffic to join new city and port connections east of the Maribyrnong River.

TBM TUNNELS

General Requirements

The tunnels in the West Gate Tunnel Project comprise twin, three-lane roads. The tunnels are driven from a common portal location (northern portal) with two separate Earth Pressure Balance TBMs. For most of the length of the tunneled route, the two tunnels are parallel, with cross passage connections for egress. As per Australian road tunnel convention, cross passages are spaced nominally at 120 m centers between the two tunnels. A low point sump is situated between the two tunnels at approximately the half way point of the two adjacent tunnels.

At the southern end, the portals are offset from one another by approximately 700 m. This solution was developed to allow the outbound tunnel to continue as a driven tunnel passing beneath the surface roads before breaking out at the outbound southern portal. This allowed the traffic to be maintained on the existing West Gate Freeway and have efficient traffic stages during construction. The narrow construction footprint was also desirable to avoid a high voltage power route diversion of which would have been costly and may have caused significant programme delays.

Figure 2 shows the offset tunnel extents at the Southern Portal areas. This offset leads to an alternative egress solution requirement for the last kilometer of the project, discussed in detail below.

Geology

The geology along the tunnel route is complex, characteristic of the last few million years of Melbourne's volcanic and alluvial history. The Silurian bedrock of Melbourne is deep below this tunnel. The lowest strata encountered will be the basalt flows of the Older Volcanics, while the upper strata are also basalt flows (the Younger Volcanics).

Figure 2. Offset tunnel portals at southern end

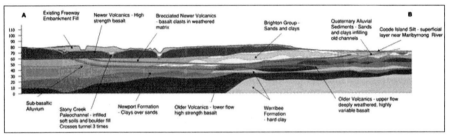

Figure 3. Geological profile along the tunnel

although with quite different characteristics. Between these two are sands and clays of various alluvial sediments. Figure 3 shows the geological profile along the tunnel.

Having the basalt cap of the Younger Volcanics over most of the tunnel length is advantageous for reducing the potential for surface settlement along the tunnel. However, the TBMs must cope with variable and mixed face conditions, ranging from a full face in basalt to a full face in soil, with a large extent of mixed face conditions. Also, the basalts are aquifers, and the control of groundwater is a significant aspect of the tunneling.

TWO LANE VERSUS THREE LANE, INVERT SOLUTIONS

As described in Asche et al (2015), the three lane roadway cross section involves challenges for a TBM. There is a significant difference between a two and a three lane tunnel when this is fitted into a TBM tunnel. Figure 4 shows both tunnels, with similar lane widths, shoulders, barriers and height allocation.

The space allocation for the three lane TBM tunnel can be seen to be sub-optimal. In particular, the space beneath the roadway can be seen to be very large. As discussed in detail in Asche et al (2015), there are a number of options that have been used in TBM driven road tunnels.

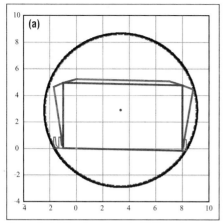 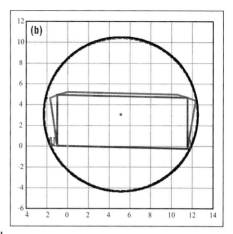

Figure 4. (a) Two-lane tunnel, (b) Three-lane tunnel

Backfill—Eastlink (Melbourne), Legacy Way (Brisbane)

- There are potential construction cost benefits if excavated material can be re-used, otherwise there is extra cost for disposal of the excavated material plus procurement of backfill material.
- It is difficult to place the backfill until the completion of the TBM drive.
- All below-pavement services (HV cables, conduits, drainage etc) may need to be excavated and laid conventionally in trenches excavated in the newly laid backfill.
- Hard spots are created under the pavement around drainage pits and soft spots in the trenches.
- By the very nature of the curved tunnel shape, the backfill methods can be complex and difficult to ensure road sub base compaction quality requirements.
- A significant issue is the "burying" a significant portion of the tunnel permanent lining. This lining will not be visible for inspection over its design life and there are leakage management and durability issues to be addressed in the design.

Box Culvert Plus Backfill—Clem7 (Brisbane), Waterview (Auckland)

- This option has the advantage that some cabling can be undertaken early.
- Some of the comments made about the complete backfill solution also apply to this solution.
- This system has the ability for the culvert installation and subsequent backfill to be completed without impacting on TBM production.
- These spaces require ventilation to avoid early equipment and services corrosion. The drainage control of any sub pavement backfill must be thoroughly designed with the ability to be maintained to ensure there is no detriment to the adjacent void space.
- Access is required from the culvert space to each cross passage for cables etc.

Road Deck—M30 (Madrid), West Gate Tunnel Project (Melbourne)

- The road deck is relatively expensive, and requires construction of corbel supports.
- The space beneath the deck allows most of the equipment to be located, rather than within the cross passages, and allows minimization of the cross passage size and breakout.
- 24-7 access can thus be provided for many maintenance tasks.
- The open under deck space provides room for emergency vehicle access.

SOLUTION FOR WEST GATE TUNNEL

Twin Tunnel Cross Section

For the West Gate Tunnel Project, where two tunnels run parallel, the tunnel cross section provides for:

- Three 3.5 m lanes,
- Shoulders: 1.0 m on the slow lane side and 0.5 m on the fast lane side,
- 4.9 m vertical clearance plus a 0.2 m soft infrastructure ("tarpaulin") allowance,
- Vehicle roll allowances on the side of each traffic barrier to a height of 4.3 m,
- 0.5 m barrier offset allowance to prevent an errant vehicle striking a sloid edge on the down-traffic side of an opening in the barrier at a cross passage,
- 2% cross fall throughout the tunnel,
- Under deck maintenance access space of 3 m height clearance,
- 20 m^2 smoke duct above the tunnel,
- 100 mm general tolerance for survey, TBM driving and segment build.

This results in segment intrados diameter of 14.1 m, and with 0.5 m segment thickness, a segment extrados diameter of 15.1 m (refer to Figure 5).

Deck

The deck consists of a precast unit with cast-insitu stitch pours providing continuity. The completed roadway solution provides a fully air/water/fire sealed separation between the upper and lower tunnel spaces. The process of construction is as follows:

1. Within the TBM backup, starter bars are cast into the segmental lining, and the segment intrados is scabbled.
2. Some distance behind the TBM, an initial correction survey is applied to the corbel. A gantry places prefabricated reinforcing cages to the corbel zone and the corbels are cast.
3. A secondary survey is applied and a mortar pad and thin elastomeric strip is placed on the corbels at the corrected design level.
4. The deck units are placed and then cast to become integral with the corbels. Sufficient reinforcing is provided to lock the roadway such that expansion joints are not required.
5. Cross passages are constructed from the road deck and the road barriers cast.
6. Finally, AC pavement is placed to meet the rideability criterion.

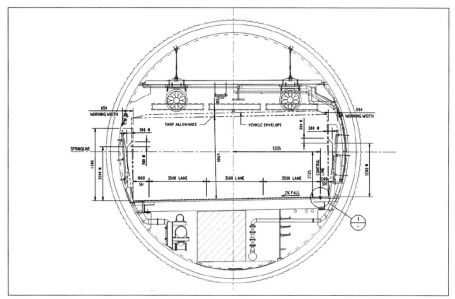

Figure 5. 3-lane tunnel

M&E Access Zone (Maintenance Tunnel)

Beneath the deck, the following services are provided:

- The drainage line, drain flush and access points
- Fire mains, connections to deluge or sprinklers and the valves for these
- Switchboards inside fireproof containers
- Substations within blockwork fire-separated enclosures
- A clear access roadway for the full length of the tunnel

At each portal, access is provided to the under-deck zone utilizing a specialized maintenance vehicle. The space is ventilated, with the ventilation built integral with each ventilation station. Note that with this arrangement, access to the majority of M&E equipment can occur during daylight hours with no tunnel shut-down, significantly reducing operations and maintenance costs (Figure 6).

Cross Passages

The cross passages are now sized for egress only. Typically the egress width required is of the order of 1200 mm. Consequently cross passage construction is now greatly simplified and an option exists to use precast boxes installed through a single segment opening. Figure 7 shows the option.

The single segment opening and jacked box solution significantly reduces the construction risks associated with poor ground scenarios or locations where groundwater inflow rates and periods need to be minimized.

The box will be jacked across with the reaction forces taken off the road deck or the segmental lining. The excavation will be carried out using a small remote controlled

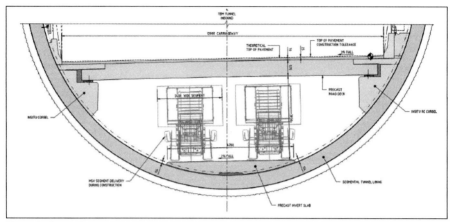

Figure 6. Under-deck zone

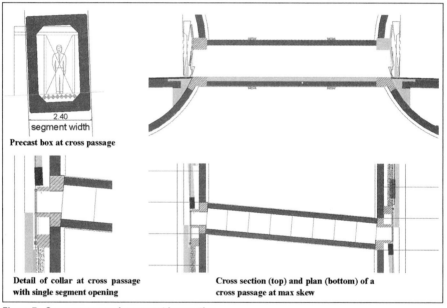

Figure 7. Cross passage using precast box section

excavator or drill-and-blast or rock-splitting techniques depending on the nature of the geology.

The box, as it is jacked between the two tunnels, is held on line and grade by steel guiderails installed as part of a pre-excavation probing and grouting regime.

A permanent structural collar is then constructed at the two openings which also provide the completed groundwater seal. This solution eliminates the often difficult and risky exercise of providing a watertight seal of sheet membrane against the back of the tunnel segments.

Such a solution approximately halves the construction costs of more conventionally sized cross passages as well as halving the construction time, primarily due to the reduction in size and removing the need to install temporary ground support, blinding, waterproof membrane and a cast in situ secondary lining.

Mechanical and Electrical Installation Under-Deck

By removing the mechanical and electrical equipment from the cross passages and placing them under the deck there are overall savings in civil works associated with the road deck solution and savings in terms of time and cost in M&E installation and commissioning works.

Egress Out and Under (EOU)

As described earlier the outbound TBM tunnel is approximately 1.2 kilometres longer than the inbound tunnel which provides various project benefits. However, this feature had a significant challenge with regards to providing a means of egress to a place of safety in a fire emergency for the extended section of the outbound tunnel. Generally, the conventional method of providing the means of egress for occupants to a place of safety on other TBM road tunnels in Australia and around the world is via cross passages. Obviously in the extended outbound tunnel cross passages were not possible which led to investigations into other methods of providing an equivalent means of egress to a place of safety.

Consideration was given to providing shafts to the surface at discrete points along the extended outbound tunnel. This had significant cost impacts as excavation volume was much larger than a conventional cross passage. Significant surface disturbance would be caused during construction and extended footprint and land requirements during operation.

In lieu of the inbound tunnel not being adjacent to this extension of the outbound tunnel a dedicated egress tunnel was considered. Cross passages would connect the mainline outbound tunnel to this dedicated egress tunnel. This also had significant cost implications as it introduced an additional mined tunnel process that would be required during construction.

A solution to the problem was developed where the available space in the invert of the tunnel below the road deck was used for occupant emergency egress. The next problem was getting tunnel occupants from the road carriageway to a Longitudinal Egress Passage (LEP) within the maintenance tunnel that was fire separated from the carriageway. A structure was designed that allowed tunnel occupants to "**E**gress **O**ut the incident tube and move **U**nder the incident tube." Hence the short form Egress Out and Under or even further abbreviated EOU was born. The LEP provided a means of

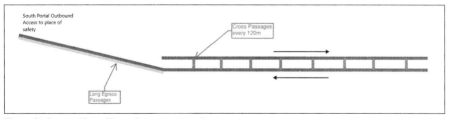

Figure 8. Long outbound tunnel egress concept

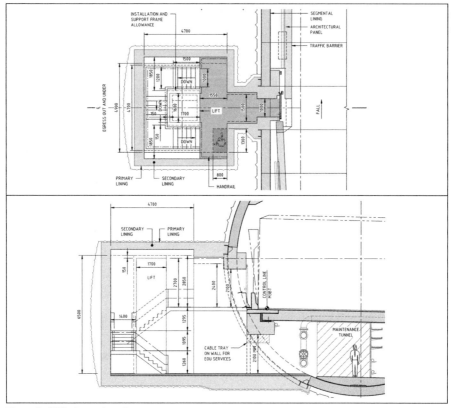

Figure 9. EOU plan and section

egress to a place of safety (the surface) through stairs and lifts at the southern outbound portal, see Figure 9. The advantage of this was that the solution used space that was already being provided in construction of the tunnel by TBM and that was used as a maintenance tunnel during normal (non-emergency) operation of the tunnel. Figure 9 shows the proposed plan and profile of the EOUs.

The EOUs have included the following features:

- Stairs that provide access from above the road deck to below the road deck
- Vertical transportation (lifts) were provided to allow emergency services personnel to provide assistance to injured or disabled occupants to get to the maintenance tunnel below road deck
- Sliding fire rated access doors were provided at road level of the EOU
- Refuge bay provided at the upper level of the egress path that allows an occupant to wait for assistance if needed
- Within the EOUs, pressurisation is achieved via the longitudinal ventilation system for the maintenance tunnel, achieving the requirement for the refuge bays within the EOUs to be separately ventilated from the mainline tunnel system (to stop smoke from entering the EOU during an incident). Hence, no dedicated ventilation elements are required within the EOUs.

As the LEP is within the maintenance tunnel that is used for M&E equipment and maintenance access during normal operation some differences were required from the standard maintenance tunnel (which did not have the provision for use as an LEP in an emergency situation). These included:

- Fire wall separating the egress envelope from M&E Equipment in the maintenance tunnel
- Automatically closing doors which would cut off the LEP from the standard maintenance tunnel to allow pressurization of the LEP and EOUs. This also removed the risk that occupants egressing the tunnel would go the wrong way to the direction of the south portal and the place of safety
- Operational systems requirements providing instructions to maintenance vehicles within the extended outbound maintenance tunnel to move to a location that would not obstruct egress for occupants via the maintenance tunnel in an emergency situation.

CONCLUSIONS AND RECOMMENDATIONS

The proposed solution has been developed to functionally optimise the available space for a three lane road tunnel in a circular cross section. The resultant tunnel internal structures have been designed to optimize their construction from economic, operational and health and safety points of view.

In terms of economy, this option is considered to be cost effective considering the whole of the activities that occur inside the tunnel during construction phase and also the costs associated with the whole of life activities during the operational/maintenance period.

Difficult Ground

Chairs

Jessica Buckley
Jay Dee

Ami Mukherjee
WSP

Breaking Through Tough Ground in the Himalayas: Nepal's First TBM

Missy Isaman ▪ The Robbins Company

ABSTRACT

Years of hard work and planning have paid off at the Bheri Babai Diversion Multipurpose Project. This 12 km tunnel is not only breaking through a historically difficult mountain range, but it has also managed to break down the notion, to the people of Nepal, that drill and blast is this only way to excavate the extreme conditions in the Himalayas. This paper will highlight the first TBM in Nepal and how it is managing to bore at an exceptional advance rate of over 700 m per month, with a high of 1202 m in one month. It will examine which design features of the Double Shield TBM are contributing to the great excavation rates, and how the crew's operational methods have maximized these results.

BHERI BABAI DIVERSION MULTIPURPOSE PROJECT

The Bheri Babai Diversion Multipurpose Project (BBDMP) is one of Nepal's 11 National Pride Projects—a prioritized plan sanctioned by the Government of Nepal to further develop the mainly rural country. This project will irrigate 60,000 hectares (almost 15,000 acres) of land in the southern region of Nepal, benefitting an estimated 30,000 households. It will divert 40 cubic meters of water (1,400 cubic feet) per second from the Bheri River to the Babai River under a head of 150 m (492 ft) using a 15 m (49 ft) tall dam, providing year-round irrigation in the surrounding Banke and Bardia districts. The water will also be used for hydroelectricity, with a generating capacity of 48 MW benefiting the country with NPR 2 billion (20 million USD) annually.

Contractor China Overseas Engineering Group Co. Ltd. Nepal Branch (COVEC Nepal Branch), represented by China Railway No. 2 Engineering Co., Ltd Chengtong Branch, is responsible for the headrace tunnel and prepared for the challenges associated with tunneling in the tough geology of the Siwalik Range, part of the Southern Himalayan Mountains, with procurement of a custom-designed Double Shield TBM. The Siwalik range consists of mainly sandstone, mudstone and conglomerate, requiring a TBM that can withstand squeezing ground, rock instability, possibly high ingress of water and fault zones. Maximum cover above the tunnel is 820 m (0.5 mi).

JOBSITE

The Bheri Babai jobsite is 56 km from Nepalgunj, which is the nearest town as well as one of the largest business hubs in western Nepal and location of the nearest airport (see Figure 1). About an hour's drive away from Nepalgunj, the jobsite is located in a river basin valley between 700 and 1000 m above sea level. The project site is a crossroads to highways that lead to much higher Himalayan towns and villages popular among trekkers and mountain climbers. The roads and bridges in the area, capable of handling heavy loads, were a very important factor when considering a TBM for the project. The area is prone to flooding during the rainy season, but overall the weather is sub-tropical and quite warm in the winter, as it is close to the Indian border.

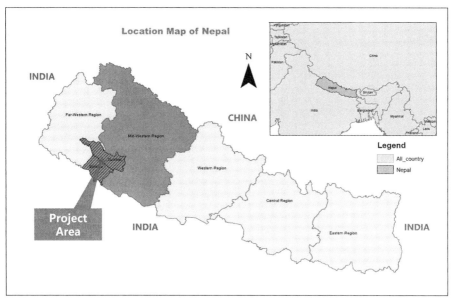

Figure 1. Map of Nepal and jobsite location

One of the most intriguing aspects of the jobsite is that it is in the middle of Nepal's largest wildlife reserve. Bardia Wildlife Reserve shelters Royal Bengal Tigers, two types of Asian Rhinos, Elephants, Asian Black Bears and many other types of vulnerable flora and fauna. Monkeys and foxes are an everyday occurrence around the jobsite, as well as colorful birds. Laborers have even spotted a tiger. The reserve is guarded by the Nepalese Army and there are many check-posts along the highway. Anyone traveling through the reserve or to the jobsite that is not a local must show valid paperwork to pass through.

TBM VS DRILL AND BLAST

Because of the notoriously difficult terrain and hard rock of the mountainous region, TBMs had not been previously used in Nepal prior to this project. With years of planning and internal lobbying for the use of a TBM, this project was finally agreed upon. The project owner, the Government of Nepal's Ministry of Irrigation (MOI), chose a TBM over the traditional method of Drill and Blast due to the faster mobilization and rate of advance offered by mechanized mining. Feasibility studies predicted an excavation time of 12 years for the tunnel, which simply wasn't an option. The TBM was also seen as an opportunity to prove the viability of the method in notoriously difficult Himalayan geology.

To put this decision into perspective, the starting portal for this tunnel was 150 m long and excavated using the Drill and Blast method. It took five months to complete and this was without any unforeseen geological difficulties. Once the TBM was up and running, under normal boring conditions in similar strata, this same length of tunneling was achieved in less than a week.

THE BEST MACHINE FOR THE JOB

In order to connect the two river valleys, the TBM will have to bore 12,210 m under a mountain range with a maximum rock cover of 800 m and gain an altitude of 152 m.

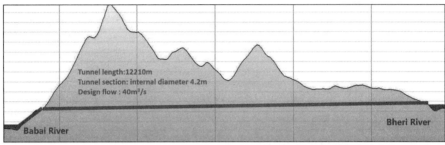

Figure 2. BBDMP tunnel alignment

The alignment is known to contain at least one large fault zone. The flow rate of water expected to be encountered is 40 m³/sec (see Figure 2).

Geological studies found the following types of rock, all part of the Himalayan Siwalik range that comprises sandstone, mudstone, and conglomerate (see Figure 3):

- Upper Siwaliks
- Siwalik Mudstone
- Middle Siwaliks 'B'
- Middle Siwalkis 'A'
- Lower Siwaliks

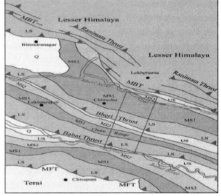

Figure 3. The geological formations along the tunnel alignment

With the amount of water and expected ground conditions, a shielded machine was necessary. Also, with the desire to complete the tunnel as quickly as possible, a 5.06 m diameter Robbins Double Shield machine was selected.

The Robbins Double Shield machine is designed to be able to bore through broken rock conditions. Because of a double thrusting system, the machine can bore forward with the auxiliary thrust cylinders while simultaneously using the rear thrust cylinders to build a segment. This process drastically reduces the time needed to bore the tunnel. The shielded machine is also beneficial to protect the workers from water and broken ground (see Figure 4).

Figure 4. BBDMP TBM in factory in Shanghai, China

In order to ensure that the machine was successful, additional features were built into the design, to prevent the machine from becoming stuck while navigating the possible squeezing ground and water ingress:

- Stepped Shield: Making the shield sections step down to smaller diameters, from the head to the tail, opens up the annular gap at the tail of the machine.

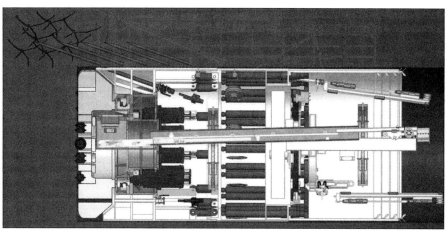

Figure 5. Forepole drilling

This helps to allow for more space around the machine for the ground to contract and lessens the chance of the shield becoming stuck.

- Probe Drilling: By probing drilling in front of the machine, the upcoming ground conditions and water content can be checked. If poor ground is found, grouting can take place to consolidate the zone ahead of the machine. This creates a solid plug to bore through. Because high water was planned for, this machine was equipped with several probe drilling locations. 14 ports in the gripper shield at seven degrees are in line with a rear probe on a ring. There are also eight ports in the forward shield at seven degrees that can he drilled by hand. In case of large amounts of water, this array of drilling and grouting gives a full 360 degrees of coverage.
- Shield Lubrication: Although this system was not used, ports were designed radially into the gripper shield that could be used to pump bentonite or other additives to the shield skin to help lubricate the surface and keep the machine moving in squeezing ground.
- Forepoling: Ports were also designed into the forward shield for the option of adding a forepoling drill in the upper forward shield area. This feature would be able to drill holes at 22 degrees, where poles can be inserted into the ground above the machine in an overlapping pattern to stabilize the ground (see Figure 5).

MACHINE PERFORMANCE

The start of boring commenced on October 15, 2017. With the exception of a few occurrences, which will be highlighted below, the current progress of this tunnel has consistently exceeded the expected excavation rates. See Table 1 for the highest excavation rates so far.

Table 1. Best TBM performance

Record Excavation Time Frame	Meters Bored	Date/Time Period
Daily Best	61.8	August 2, 2018
Weekly Best	337.8	Week #38
Monthly Best	1202	August 2018

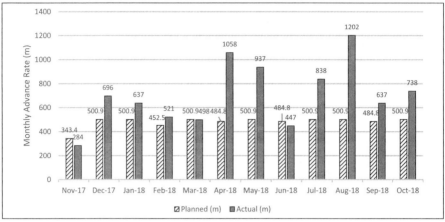

Figure 6. Average monthly excavation rates vs. planned rates

As can be seen in Figure 6, after the initial startup period, most months the excavation rates exceeded and sometimes doubled the planned rates. As of October, the tunnel had bored 8493 m compared to the planned 5756 m. This is 2736 m ahead of schedule, all despite encountering two points of high water ingress and two incidents where the machine was not able to move.

During the excavation, more or less similar types of rock were encountered as compared to those described in the Geotechnical Baseline Report (GBR); i.e., alternating beds of mudstone, siltstone, sandstone and conglomerate.

In the nearly 9 km of TBM excavated tunnel so far, 66.58% of Type 1 segments (having less reinforcement) and 33.42% of Type 2 segments (having more reinforcement) were planned to be erected. However, 92.52% of Type 1 and 7.48% of Type 2 segments have been used instead, which indicates better ground conditions than predicted.

The TBM also navigated a major fault zone shown in the GBR, known as the Bheri Thrust Zone. Clay and water ingress were expected throughout the fault, which is about 400 to 600 m wide. However, the TBM passed smoothly through this section without any problem. The rock mass in this section was found to be completely dry and belongs to rock class IV (RMR).

EXCAVATION ISSUES AND RESOLUTIONS

- On December 27, 2017 at ch 1+174.782, large amounts of water were pouring from the 8 o'clock position. As the machine advanced, the ingress of water shifted to the 11 o'clock position. An estimated rate of greater than 2000 L/min was seen. Excavation continued at a slower rate, without any interventions.
- A second occurrence of high water happened on January 6, 2018 at ch 1+337.457. Water entered at the 12 o'clock position at an approximate rate of greater than 2000 L/min. Again, the resolution was to slowly bore through the water without any drilling or grouting to stop the water (see Figure 7).
- On October 10, 2018 at ch8+588.860 the machine became stuck and could not progress. Up until this point, the axis of the tunnel was perpendicular

Figure 7. Water ingress on January 6, 2018

to the grain of the rock. This is the most favorable condition for tunneling. Around ch8+400 the ground conditions changed, and the grain became nearly parallel to the tunnel axis. The machine alignment also started to shift and reached deviation from center of around 131 mm. At this point the machine became lodged in place. A high thrust of 18,500 kN was exerted and was not able to move the machine. In order to move the machine, a bypass passage was excavated from the right side of the telescopic section up to the cutterhead. It was cleared out from around the 5 o'clock position to the 12 o'clock position. A thrust of 10,000 kN was the applied and the machine was able to start boring again. The bypass was completed and the machine was moving again in just five days.

- At ch 8+606.262 the cutterhead became jammed. Loosely cemented sandstone and high-pressure water ingress around the 11 o'clock position triggered an over-break at the left crown area and jammed up the cutterhead. To control the water ingress, 1287 kg of polyurethane was injected through a 16 m deep probe hole. This almost completely stopped the water. A torque of 440 kNm was then applied to the cutterhead and it was able to become dislodged.

REFLECTIONS ON THE PROJECT: FAST ADVANCE

On reflection, there have been multiple reasons for the extremely good performance at this tunnel site. The site staff has made it a high priority to maintain the machine daily, and they have been vigilant with their cutter changing standards. The operators have also taken the approach to boring of maintaining a continuous and stable excavation thorough the difficult areas as opposed to stopping to drill and grout. Besides the two exceptions previously discussed, this method has proven to be successful.

The use of hexagonal segments may be another contributing factor (see Figures 8–9). The hex segment design is well suited to Double Shield TBM tunneling. Only four segments are needed per ring, and these are built concurrent with boring. The hex shape of the 300 mm segments prevents cruciform intersections at radial joints, while cast-in pads on each invert segment allow it to be built directly on the invert, via the hooded tail shield. Lastly, the staggered arrangement of the hex shape allows segments to be built in two half cycles when using a long thrust jack stroke.

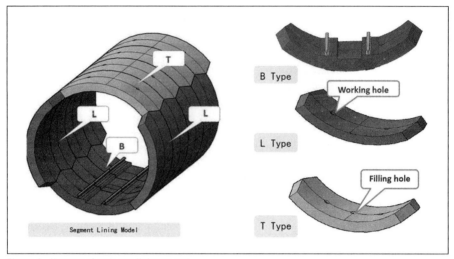

Figure 8. Hexagonal segment lining model

Figure 9. Completed lined tunnel section for BBDMP

As is the case in most long tunnels, logistics were a key to keeping advance rates high. Coordination of trains and continual supply of components is needed to keep up with the fast pace of ring builds. In addition a geologist on site conducts daily face mapping so that operational parameters and other measures can be adjusted accordingly (see Figure 10). This type of mapping is critical in mountainous tunnels where the expected geology may differ from the geological report. The geologist also analyzes the geology at the tail shield. Lastly, favorable geology has been a big factor, with less water than expected in fault zones.

SUCCESSFUL FUTURE PLANS

The success of the BBDMP, a national pride project, is paramount for the country as well as the TBM industry. It is expected to help aide the food crisis in the mid-western region of Nepal by increasing agricultural yields and invigorating socio-economic

The Rock mass from Manhole and Tail Shield

Date	Rockmass at Manhole	Photos	Geological Mapping (Through muck)	Rockmass at Tail Shield	Rockmass at Tail Shield
2018.3.23	Sandstone, slightly weathered, bedding and a set of joint can be observed. CH: 2715.1m		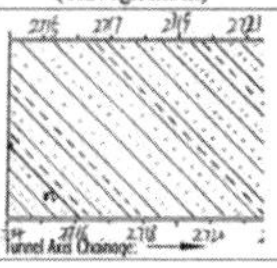	Sandstone and mudstone, slightly weathered, bedding can be observed. CH:2703m	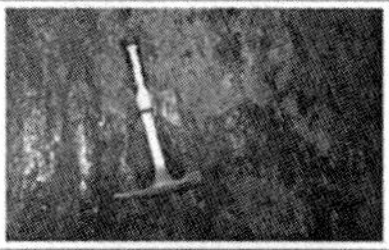
2018.3.24	Sandstone and silty mudstone, slightly weathered, bedding can be observed. CH: 2740.3m		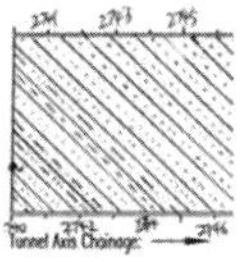	Sandstone and silty mudstone, slightly weathered, bedding can be observed. CH: 2728m	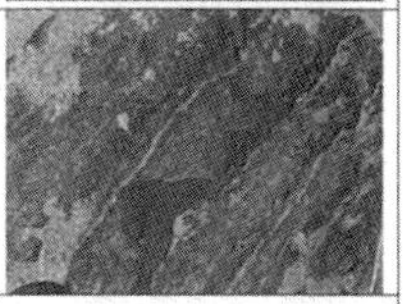
2018.3.25	Sandstone and siltstone, slightly weathered, bedding can be observed. CH: 2765.8m		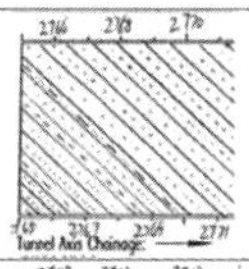	Sandstone, slightly weathered, bedding can be observed. CH: 2753m	
2018.3.26	Sandstone, slightly weathered. CH: 2797.3m		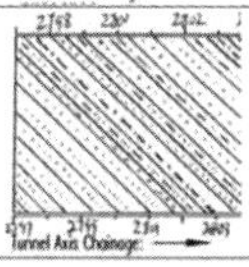	Sandstone, slightly weathered, bedding can be observed. CH: 2785m	

Figure 10. Typical face mapping reports from the onsite geologist

development in the region. Even though the project has yet to be completed, the current success of this project has proven the tunneling method to those involved and the government is planning more TBM projects. This will open up the future for areas of infrastructure such as construction in the country, water diversion and irrigation, hydropower generation, transportation and more. More than 100 km of tunneling are planned for Nepal in the next five years, of which more than 50% is considered feasible for TBM excavation. Many projects that would previously have recommended Drill & Blast only are now considering TBMs as an option.

Challenging Geology: Blacklick Creek Sanitary Interceptor Sewer Project in Columbus, OH

Amanda Kerr ▪ Michels Corporation
Max Ross ▪ Michels Corporation
Ed Whitman ▪ Michels Corporation
Ehsan Alavi ▪ Jay Dee Contractors, Inc.

ABSTRACT

The City of Columbus Blacklick Creek Sanitary Interceptor Sewer Project (BCSIS), located in northeast Franklin County, Ohio, includes 6.9 km of sewer tunnel. This project also required the excavation of two shafts at the launch point of the TBM run, 6 intermediate shafts, and two shafts at the receiving point of the TBM. A joint venture of Michels Corporation and Jay Dee Contractors, Inc. (Blacklick Constructors, LLC) has constructed the project. The tunnel was excavated using a Herrenknecht EPB TBM. This paper summarizes the preparation work that was done prior to the launch of the TBM in addition to the methodology that was utilized to successfully tunnel through the diverse geology of this project.

INTRODUCTION

Project Description

The Blacklick Creek Sanitary Interceptor Sewer (BCSIS) extends an existing 1.7 meters (66 inch) sewer with a new 3 meter (120 inch) gravity sewer from Blacklick Ridge Boulevard to Morse Road. This extension is required to support the development of the service areas of the City of New Albany, and Jefferson Water and Sewer District (JWSD). Furthermore, this extension will also enable the future connection between Rocky Fork Diversion to redirect sewer flow from the Big Walnut sewershed into the Blacklick sewershed.

The tunnel alignment initiates at the existing Manhole #12 (MH-12) outlet, just south of the Blacklick Ridge Boulevard and east of Reynoldsburg-New Albany Road. The tunnel extends in a northerly direction following the approximate alignment of Reynoldsburg-New Albany Road towards Morse Road. The total length of the alignment is 7016 meters (23,020 lf.), with 6894 meters (22,620 lf.) to be constructed using an Earth Pressure Balance (EPB) Tunnel Boring Machine (TBM) and with the initial 122 m (400 ft.) by open cut and hand tunneling, performed to avoid disrupting traffic and local residents on the Blacklick Ridge Boulevard artery. This project includes a new junction chamber between the existing sewer and the newly constructed tunnel at MH-12. The tunnel will end upstream at the intersection of Reynoldsburg-New Albany Road and Morse Road, where a drop structure will be installed to accept future flows from the New Albany, Rocky Fork, JWSD, and City of Columbus service areas. Two intermediate shafts with permanent drop structures are installed along the alignment, for future tie-ins from the Jefferson Township service area, as well as for tunnel maintenance purposes. In addition to their ultimate purpose, these shafts served as access points for TBM maintenance during the tunnel construction. Other associated work includes the installation of connection pipes, manholes and ancillary facilities necessary for the operation and maintenance of the sewer. With a bid price of $108.9 million,

Blacklick Constructors LLC (BCL) was the lowest of four bidders on November 18, 2015, with the engineer's estimate at $113.7 million.

In order to successfully tunnel through the varied geology of this project, a 3.7 meter (12 feet) diameter EPB TBM manufactured by Herrenknecht was selected. The TBM installed a six segment, fiber reinforced, precast concrete universal ring as final tunnel liner. The cutterhead was specifically designed to bore through soils, rocks, while considering the high probability of encountering major boulder clusters. The TBM's 113 meters (370 ft.) long ancillary support equipment was installed on wheeled gantries.

In addition to the TBM and necessary support equipment, BCL installed and commissioned specialty as well as peripheral tunnel support equipment at the Shaft 1 site. This equipment included a breathable compressed air plant and a medical lock for hyperbaric interventions, water treatment systems, an electrical substation and high voltage switchgear (to support the TBM), Sagami Servo automated grout plants, the TBM's closed loop water cooling system heat exchangers, a muck pit, and motion-activated wheel wash. BCL utilized muck cars in the tunnel for muck removal, and a lattice boom crane for hoisting from the main shaft. Three California switches were also installed and relocated in the tunnel as needed in order to optimize tunneling operation cycle-times.

Project Geology

A large variety of geologies were encountered along the alignment of the project. The hand tunneling and upper portion of the shaft excavations encountered fill and alluvial materials. The tunneling excavation in soil zones generally encountered glacial tills and outwash deposits. The transitions between the soils and bedrock generally consisted of highly weathered to decomposed rock; large quantities of glacially-deposited boulders were also anticipated along the transition zones. The two principal rock formations encountered during tunneling and shaft excavation were the Cuyahoga and Sunbury Shales. The contract documents categorized the ground into: Fill, Alluvial, Glaciolacustrine, Glaciofluvial, Glacial Till, Transition Material, Upper Cuyahoga Lower Cuyahoga and Sunbury Shales. For the purpose of tunnel ground classifications in the Contract Documents, these materials were organized into three main groups: soil, rock, and mixed ground.

The geotechnical baseline report (GBR) estimated that the highly variable geology along the alignment would contain 30 interfaces. However, geological investigations can prove to be extremely challenging in glacial deposits. This was compounded by the presence of ancient river systems, which carved steep valleys and ridges in the bedrock of the central Ohio region, the subsequent glaciation later filled these bedrock valleys with softer glacial deposits. The alignment of the tunnel partially skirts the contours of what is essentially a bedrock cliff, exhibited by the high variability of the geotechnical samples taken in extremely close proximity of those cliffs. It was anticipated that the tunnel alignment would encounter 39% soil, 46% rock, and 15% mixed ground. Of the soft and mixed ground reaches along the alignment, the behavior of the soils were expected to vary greatly. Approximately 14% of the alignment was to encompass Glaciolacustrine and fine grained Glaciofluvial soils and exhibit minor to moderate squeezing conditions below groundwater. Conversely, coarse grained Glaciofluvial was expected in 32% of alignment, and was anticipated to exhibit flowing behavior below groundwater. Glacial Till made up the remaining 17% of the soft ground reaches. Historically, other projects in the central Ohio region have encountered many difficulties with high groundwater flows. The geotechnical baseline documents predicted a wide range of hydraulic conductivity rates ranging from 10^{-6} (cm/sec^2) in the Glaciofluvial and glacial till geologies to 0.1 (cm/sec^2) in the Glaciolacustrine materials.

The high variability of the geology along the alignment, the potential for gassy conditions, and the extreme range of hydraulic conductivities anticipated presented a challenge, both to the design, and the construction of the BCSIS tunnel. This paper will discuss the planning undertaken to manage these challenging conditions during the construction of the BCSIS tunnel.

PROJECT PROCUREMENT AND FRONT-END PLANNING

Owner Prescribed Requirements

The BCSIS contract specified a number of design or construction requirements that had to be considered, either in the TBM design or the construction front end planning. Some of these design requirements influenced BCL's approach, and the solutions elaborated during the project planning. Some of the key design requirements that had a significant impacts on the TBM design and the tunneling means methods are listed below:

- Tunnel finished Inside Diameter (I.D.) of 10 feet,
- TBM must be a full-face, fully shielded EPBTBM, capable of excavating through all anticipated ground conditions in the GBR,
- TBM Must be equipped with a manlock, for hyperbaric tunnel face and cutterhead access,
- The shield must be articulated and able to steer through 1100' radius curves,
- The TBM guidance system must be able to record and report the machine location during excavation.

Contractor Design Criteria

In addition to the contract prescribed requirements, several additional provisions were included to ensure safe and efficient construction of the tunnel through anticipated highly variable geologic conditions:

- Engineer out or minimize construction and safety risks,
- Elaborate monitoring programs and warning systems to improve safety, and optimize tunnel construction,
- Develop a cutterhead design tailored for the extreme variability of ground types while minimizing retooling events,
- High-order survey accuracy to allow concurrent shaft and tunnel excavation.

These major design requirements, in addition to many others, were all considered in the TBM design, and the execution of tunnel construction.

TBM Design

TBM Guidance System

The tunnel's constricted finished I.D. not only precluded the use of a conveyor belt system for muck extraction, but also significantly affected the length of the TBM shields and its ancillary support equipment. Moreover, the TBM shields needed to be further

Challenging Geology: Blacklick Creek Sanitary Interceptor Sewer Project

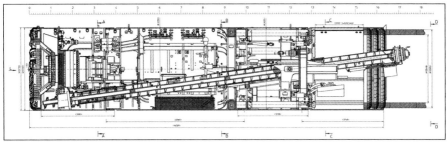

Figure 1. TBM body profile

lengthened to accommodate the installation of a hyperbaric manlock, resulting in a 16 m (53.5') overall length of the shields, consisting of four separate bodies articulated over three joints. See Figure 1 for a profile detailing the length of the TBM.

Consequently, the necessary line-of-sight for an optical based TBM guidance system (theodolite or laser) through the TBM, was severely diminished. This impediment was further compounded by the numerous tight radius curves along the alignment, which would have required an excessive amount of instrument move-ups, and likely resulted in an a tremendous amount of error propagation, rendering the TBM guidance system data unreliable at best, and further complicating the maneuverability and the accuracy of the TBM steering through an extremely narrow Right of Way (RoW).

To avoid this scenario, a gyroscope and hydraulic level based Measure While Driving TBM guidance system (MWD-II) was selected. This proved to be well adapted solution to the Tunnel and TBM configuration, as exhibited in Figure 9 later in this paper.

Cutterhead Design

Given the anticipated highly variable ground conditions along the tunnel alignment, the difficulty of obtaining unequivocal geotechnical information, the extremely complicated topology of the geology surrounding the tunnel envelope, the colossal boulder counts, and the expected high hydraulic transmissivity of the ground, increased attention and care to the conception and design of the TBM cutterhead was warranted.

With 30 distinct geological interfaces, and the wide range of geological conditions outlined above, selecting cutters specially adapted to each ground condition was less than practical. Consequently, BCL elected to use pressure compensated monoblock disc cutters with tungsten carbide inserts for the soft ground portion of the alignment.

This hybrid type of disc cutter combines the benefits of pressure compensation, typically used in hard rock excavations, with the longevity of tungsten carbide inserts, frequently found on soft ground cutters. The carbide's high abrasion resistance performs well in the softer abrasive materials, while the pressure compensation helps resist the impact loading inflicted by boulders and rocks, but also reduces the drag on the disc bearings, allowing the tool to keep spinning, even when soft ground conditions are encountered, therefore minimizing the chances of a disc getting stuck and wearing unevenly.

Furthermore, the extended length of the TBM shields, and consequently, its ponderous maneuverability, necessitated that the integrity of the gauge cutters be maintained and regularly verified, to minimize the likelihood of a hyperbaric intervention. Therefore redundant gauge cutters were installed to prevent or limit a total loss failure

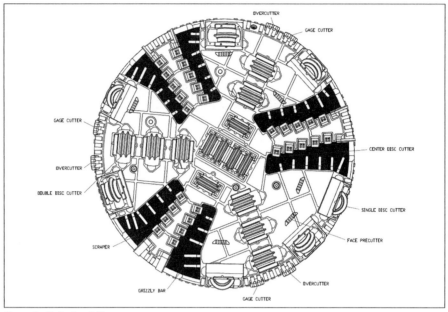

Figure 2. Cutterhead diagram

of the gauge tools, and the amount of overcut was increased to extend the wear life of those cutters. Similarly, face pre-cutters were added to supplement the disc cutters, increase their wear life, and minimize tooling changes. These face pre-cutters were designed and installed to balance both the pre-cutter and disc cutter ground engagement. Overall, the tooling selection and reinforcement measures proved effective as no emergency tool replacement interventions were performed; only two complete, preventative, scheduled disc replacements were performed and no repairs to the gauge cutters, the over-cutters, the pre-cutters, nor the cutterhead hard-facing were required. Figure 2 shows the arrangement of the cutterhead.

PROJECT TRACKING AND OPTMIZATION

TBM Data Tracking

BCL was convinced that the judicious collection, organization and analysis of TBM performance and productivity information was key to plan, predict, and if necessary, react to unexpected situations. Information from many sources was therefore collected and analyzed to facilitate adjustments to the production processes, and increase the safety and efficiency of the tunneling operation.

TBM Data Analysis in Representative Geological Reaches

With such disparate geology, continuously tracking the TBM performance was paramount. To accomplish this task, TBM Key Mechanical Performance Indicators (KMPIs) were identified and monitored. This approach offers the advantage of correlating TBM mechanical performance in previously experienced geological conditions, averaged over extended reaches, to establish a benchmark of the TBM behavior in known ground conditions. Furthermore, this method also prevents unrelated operational inefficiencies or uncertainties such as ring building or rolling stock issues, to name a few, from influencing either the data, or its analysis.

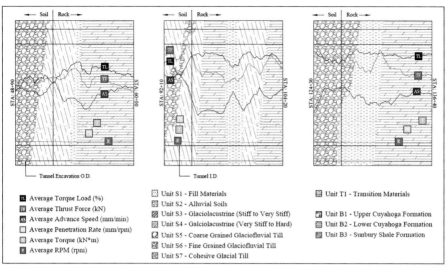

Figure 3. Example KMPI signature

When confronted with unexpected TBM performance, the KMPI signature is then compared to that of previously encountered ground conditions to assist in optimizing the ground conditioning and other operating parameters.

Figure 3 shows a comparative analysis of TBM performance in similar geologies for improved production analysis purposes, but this method can also provide indications of possible mechanical failures, or suggest deviations from the anticipated geological profile. The geological profile on the BCSIS project proved however quite accurate, despite the complexity of the soil/rock interfaces similar to that represented in the above figure. The overlay of the KMPI signatures on the geological profile in the Figure 3 clearly illustrates how the average torque load increases, while thrust and advance rate decrease through the rock interfaces.

Downtime Analysis

Interruptions greater than 5 minutes in the production cycles were collected by the heading engineers on the TBM, including explanatory details and categorization tags, which allowed for a more expeditious analysis of the information collected. The data was then organized and analyzed to highlight "improvement areas." This method has been covered in greater detail in the paper Downtime Data Collection, Analysis and Utilization on Blacklick Creek Sanitary Interceptor Sewer Tunneling Project in Columbus, OH. The collection of this information was performed concurrently to that of the automatic Data Acquisition Software on the TBM, and while slightly less precise, the "human input" proved invaluable to the analysis of the latter, as well as the identification of tunneling inefficiencies, and consequently, their improvement. Over the course of the project, over 2100 individual downtime justifications were recorded and subsequently classified by order of preventability, project system, location, and component type, resulting in 216 distinct classification possibilities.

The preventability of the downtime was the primary method to distinguish which downtimes were controllable. A Preventable event was defined as a downtime occurrence recorded for an activity that could have been mitigated without the halt of production. A Non-preventable event was a period of downtime recorded for an activity outside

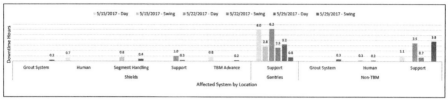

Figure 4. Three week preventable downtime comparison

the regular project controls, such as weather or gas. By focusing on these controllable events, opportunities to minimize procedural and routine downtimes were identified and communicated to field personnel (Kerr et al., 2018). For example, a recorded production stoppage caused by the routine change of a hydraulic filter of the TBM main drive would be classified as a preventable mechanical issue pertaining to the TBM advance system that took place in the heading. Changing a filter should not be cause to stop production, and with planning and communication between project personnel this can completed at a more efficient time in the tunneling cycle.

The downtime events and resulting trends were analyzed by the project staff and discussed weekly. Visual representations of the downtime were provided to management personnel to increase the awareness of preventable events, and elaborate mitigation measures. In addition, this visual representation demonstrates the effectiveness of the solutions implemented. Figure 4 shows a side-by-side comparison of the recorded weekly preventable downtimes, organized by shift, over three consecutive weeks. The decreasing trend in downtime imputed to the support system in the gantries, demonstrates efficient shift practices, ultimately resulting in a reduction of lost time events. Breaking the information down by shift identified differing practices, allowing for the more effective methods to be communicated to the lesser performing shift.

TBM Availability

The mechanized tunneling process is often compared to a mobile factory, and as such, follows many of the same objectives as manufacturing, essentially: maximizing availability and productivity. TBM availability was defined as the total hours worked without any time spent on events classified as non-preventable, shown numerically below.

$$A = \frac{E_{Production}}{E_{Production} + D_{Preventable}}$$

where

A = TBM availability (utilization time)
$E_{Production}$ = recorded production, in hours
$D_{Preventable}$ = recorded preventable events, in hours

The availability and productivity of the TBM were monitored extensively throughout the project. The average TBM utilization time for the tunnel reach between the launch shaft and shaft 2 (the first scheduled maintenance stop) was 82%. From shaft 2 to shaft 3, this utilization time dropped slightly to 81%, while from shaft 3 to shaft 4 the rate increased slightly to 84%.

While Figure 5 offers a visual representation of the TBM utilization time, various factors affecting the TBM availability were observed on a weekly basis. It is however interesting to note that, if the average availability time does not vary significantly from

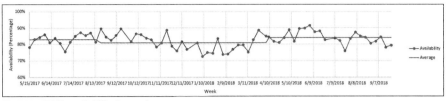

Figure 5. TBM availability by week

reach to reach, it may be slightly affected or even compounded by, reach distances (and intervals between scheduled maintenance stops), logistics efficiency in the tunnel (switch locations and train availability), as well as the possible difficulties associated with higher densities of geological interfaces to name a few. Overall, the TBM availability was fairly consistent throughout the entire tunnel excavation.

Mitigation of Downtime Due to Gas

The BCSIS tunnel was classified as potentially gassy, due to the high levels of carbon monoxide, hydrogen sulfide, and methane expected in localized areas of the alignment. The TBM was equipped with a Trolex gas detection system programmed to shut down the electrical power to the TBM, effectively de-energizing the TBM, with the exception of the Essential Services and the TBM ventilation system, should the preset concentration of combustible gas limit be reached. Unfortunately, in gassy areas, this would cause frequent TBM shut downs, which inevitably led to lengthy restart procedures and time.

The shortcoming of the Trolex system supplied with the TBM is that it did not provide the TBM operators a visual indication of gas concentrations or audible alarms, indicating that the system was detecting higher than normal gas concentrations. Instead, it simply functioned as a binary system (on/off function), de-energizing the TBM when necessary, without any warning.

A system was implemented to intercept the analog signal from selected gas detection sensors across the TBM into the Trolex gas detection system, and interpret and display the actual gas concentrations registered by the sensors in real time. This provided a remote status monitoring of the atmosphere in the heading, improving re-entry safety should an evacuation of the tunnel become necessary, but also provided the Operators real-time gas monitoring information, allowing the operators to modify the tunnel excavation procedure to maintain a safe atmosphere in the heading, below the alarm/action trigger levels. In doing so, and while improving safety, the TBM was kept running more smoothly, minimizing the possibility of gas accumulation near the screw discharge, and almost eliminated downtime due to high gas levels. Figure 6 shows the HMI screen developed as part of this improvement system.

Block Light Program

The tunnel logistics on the BCSIS tunnel relied heavily on a single track system with three California switches, for all the traffic going and coming to the heading. Each track section between two California switch was called a trunk. Considering that by definition, California switches are not stationary, their position in the tunnel is adjusted to best suit the operation. Consequently, a modular block light system was developed and implemented, using remote IO and a Programmable Logic Controllers (PLC), to allow the locomotive operators to know when a section of track or trunk was clear, regardless of the switch location in the tunnel. A set of interlocks and fail safe measures

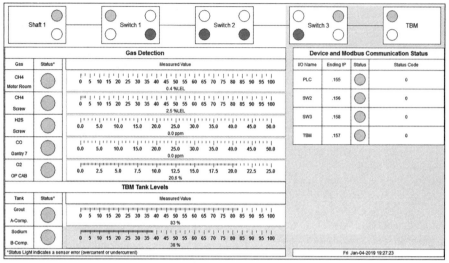

Figure 6. Additional HMI

was also programmed into the system to prevent operators on both end of the trunks from actuating the system simultaneously. Each California switch was equipped on both ends with a red and green light, toggled by a pull bottle on either end of the trunk.

In addition to improving track safety in the tunnel, the locational awareness of the trains (See Figure 6) allowed heading supervision to better estimate train arrivals into the heading, allowing for routine heading maintenance tasks to be optimally scheduled and completed without impacting production. Furthermore, the block light HMI also allowed for train studies to be performed by project supervision, tracking wait times in the shaft, heading, and on switches. This information was then used to predict and optimize the scheduling and relocation of the California switches in the tunnel. Five California switch moves were performed during the tunneling phase of the project.

Overcut Monitoring

An EAG Probe was used regularly throughout tunneling to precisely monitor the overcut of the TBM, while not impacting tunnel production. The EAG probes consisted of two cone penetration test apparatus specially outfitted to the forward shield of the TBM. The system consisted of a hydraulic jack with linear displacement and pressure transducers. The hydraulic jacks were extended simultaneously on both sides of the shield, through the overcut zone behind the TBM skin, and then into the undisturbed native ground, while the hydraulic pressure throughout the jack extension was recorded. The relationship between the increase in hydraulic pressure and the penetration depth indicates the amount of the remaining overcut on the cutterhead (*the pressure remains relatively constant through the overcut zone and increases noticeably once virgin ground is encountered* — Gharahbagh et al., 2013). Monitoring the overcut was imperative, since the shield bodies were relatively long in comparison to their diameter, and the tunnel alignment contained numerous curves.

Figure 7 shows the initial pressure signature measured only two weeks after the launch of the TBM. Notice that the pressure spiked when jack has been extended around 2.8 inches, indicating the over-cutters were intact. Weekly measurements were taken and monitored.

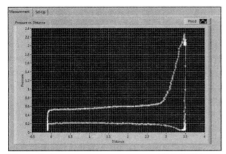 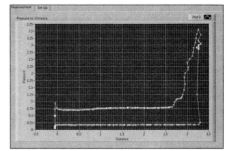

Figure 7. Overcut measurement on 5/03/2017   Figure 8. Overcut measurement on 6/22/2018

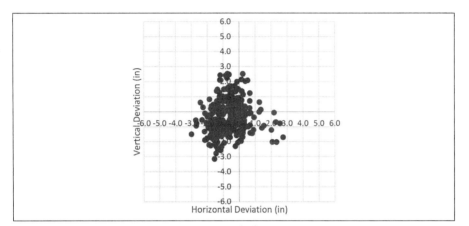

Figure 9. Tunnel as-built deviation from design tunnel alignment

The weld-on carbide over-cutters performed well, as indicated by Figure 8: pressures spikes at around 2.6 inches on reading taken over a year later shows that the over-cutters only lost 0.2 inches in approximately one year of tunneling.

Survey Controls

As discussed previously, the selection of the VMT MWD-II system proved to be well suited for this challenging application. It provided the TBM operators the real time line and grade information necessary to steer the TBM through a very narrow RoW. One of the prime advantages of this system is that, if maintained daily, it guaranteed quasi uninterrupted operation, with a relative but acceptable accuracy. Since BCL operated on two ten-hour shifts, the project survey team had a daily four hour time window to perform the necessary routine maintenance of the guidance system, which included correlating its position to the tunnel survey control, moving up the hydraulic component of the instrument, and performing the needed fine adjustments to the system.

Used in conjunction with an intensive but carefully elaborated survey program, it yielded minimal deviation from the design tunnel alignment, as presented in Figure 9. Ultimately, these survey practices resulted in a 23 mm deviation from the design tunnel alignment at the holethrough.

CONCLUSION

The significant challenges imposed by the geology of this alignment, as well as the tremendous risks associated with soft ground tunneling along a major artery, required careful planning, meticulous selection and design of essential equipment, and thorough tracking efforts throughout construction of the BCSIS tunnel.

In order to meet the prescribed design requirements, considerable efforts were engaged to engineer-out or minimize construction and safety risks both on the surface and underground, monitor and optimize the sequencing of tunnel and shaft construction activities, and minimize the TBM retooling needs to name a few. These elements all contributed significantly to the ultimate success of this project; holing through with less than an inch of deviation from the design tunnel alignment, three months ahead of schedule, with no claims or re-work.

REFERENCES

Gharahbagh, Ehsan Alavi, et al. "Periodic Inspection of Gauge Cutter Wear on EPB TBMs Using Cone Penetration Testing." *Tunneling and Underground Space Technology*, vol. 38, 2013, pp. 279–286., doi:10.1016/j.tust.2013.07.013.

Kerr, Amanda, and Ross, Max. "Downtime Data Collection, Analysis and Utilization on Blacklick Creek Sanitary Interceptor Sewer Tunneling Project in Columbus, OH." *North American Tunneling: 2018 Proceedings*, Society for Mining, Metallurgy & Exploration, 2018, pp. 99–107.

Cutting New Ground: TBM Selection for Fort Wayne Utilities Tunnel Works Program

T.J. Short ▪ Fort Wayne Utilities
Leo Gentile ▪ Black & Veatch Corporation
David Day ▪ Black & Veatch Corporation

ABSTRACT

Fort Wayne Utilities (FWU) planned to excavate its Three Rivers Protection and Overflow Reduction Tunnel (3RPORT) using an open main beam tunnel boring machine (TBM) when design began. This assumption was based on previous record-breaking TBM performance in similar Silurian and Devonian-age dolomite and limestone in Indianapolis. Based on a careful review of geotechnical and hydrogeological data and information, FWU ultimately specified a slurry-mode TBM. When finished 3RPORT will be one of a few hard rock tunnels excavated using a slurry TBM in the United States. The 19-foot diameter 25,000-foot long project is part of Fort Wayne Utilities' Tunnel Works Consent Decree program. 3RPORT is being mined in a competent dolomite that includes a reefal facies that is also a prolific aquifer. TBM selection involved a hard review of geologic and hydrogeologic conditions. The contractor, Salini Impregilo-Lane, Joint Venture procured a Herrenknecht 18.93-ft (5.77 m) Mixshield TBM that was delivered in June 2018 and will begin excavation in 2019.

PROJECT OVERVIEW

The City of Fort Wayne is located in northeastern Indiana, USA along the St. Joseph, St. Marys and Maumee Rivers, and covers 110 square miles with over 33 percent of the area being served by combined sanitary and storm sewer system. Within the combined sewer system area, there are a total of 41 outfalls that contribute combined sewer overflow (CSO) discharge into the three rivers during rain events. The goal of the City's Long-Term Control Plan is reducing the number of overflow events per year on all of Fort Wayne's three rivers by 2025. The CSO discharging into the St. Joseph River will be reduced to no more than one overflow during a rain event per typical year and no more than four overflow events per typical year into the St. Mary's and Maumee Rivers. Overall, there will be a 90-percent reduction of overflow events within a typical year. Once constructed, 3RPORT will be a 25,000-foot long 16-foot finished diameter tunnel constructed in rock about 175 to 200 feet below grade.

The key to FWU's CSO program was to convey CSO discharges to its Water Pollution Control Plant (WPCP) located east of downtown (Figure 1). The City had constructed three large impoundments, or wet weather ponds (WWP), adjacent to the WPCP in the early 1920s to collect and retain CSO flows for treatment. The three ponds provide about 350 million gallons of storage. The City also expanded capacity of its WPCP in 2016 to 90 MGD in anticipation of higher wet weather flows. 3RPORT will convey flows from CSOs to the WWP for eventual treatment by the expanded WPCP.

Once constructed, 3RPORT will consist of the following elements:

- One pump station Shaft, 68 feet excavated diameter through the overburden and 64 feet excavated diameter through rock.

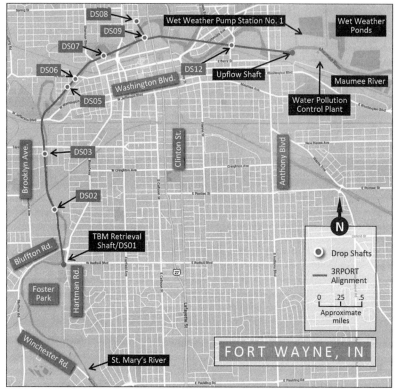

Figure 1. 3RPORT alignment, drop shafts, WPCP and wet weather ponds

- One working shaft, which will become an Upflow Shaft for a deep watering pumping station that will be constructed under a separate contract, 33 feet excavated diameter through the overburden and 29 feet excavated diameter through rock.
- Approximately 24,500 feet of 16-foot internal diameter main tunnel.
- Approximately 740 feet of 7-foot internal diameter adits for six drop shafts to the main tunnel.
- One retrieval shaft, 25 feet internal diameter through the overburden and 21 feet finished internal diameter through rock, which includes an 8-foot diameter drop shaft.
- A total of eleven drop and vent shafts ranging in finished internal diameter from 2 to 7 feet.

GROUND CONDITIONS

Recent successful tunnel projects in Indianapolis initially led the FWU's team to baseline similar ground conditions since the same geologic formations would be encountered and that an open main beam TBM would be appropriate for tunnel excavation. As the geotechnical program progressed, findings suggested that ground conditions in Allen County would be markedly different than the completed projects a hundred miles to the south.

	Formation Name		Rock Engineering Unit	Description
Devonian	Muscatatuck Group	Antrim Shale		
		Traverse Formation	1	Fine to coarse grained limestone
		Detroit River Formation	2	Dark brown to tan, dolomite. Significant vertical fracturing.
Silurian	Salina Group	Wabash Formation	3	Fine-grained, slightly dolomitic limestone and dense to fine-grained vuggy limestone. Reefal facies present in some locations
		Pleasant Mills Formation		Light-colored, thick-bedded, dolomitic limestone and mottled, dolomitic shale

Figure 2. Stratigraphy and engineering rock units to be encountered in tunnel and shafts

Stratigraphy and Engineering Rock Units

Three rock engineering units were described based on rock properties identified during the geotechnical program. The three units correspond to the geologic formations that will be encountered during tunnel and shaft excavation (Figure 2). The rock unit formations include, in descending order, the Antrim Shale, which is the upper most unit of the Devonian New Albany Shale Formation, the Traverse Formation, the Detroit River Formation, the Wabash Formation consisting of Bailey Limestone and Moccasin Springs Formation, the Pleasant Mills Formation consisting of Louisville Limestone and Waldron Shale, and the Salamonie Dolomite. The Salina Group of the Silurian Formation unconformably underlies the Muscatatuck Group. The Salina Group consists of primarily two members: the Wabash Formation and the underlying Pleasant Mills Formation.

Macro Scale Examination of Rock Conditions

A local rock quarry operator allowed the design team to inspect excavated rock conditions at the quarry. The quarry is located about 3 miles southwest of 3RPORT retrieval shaft. The north quarry high wall provided an excellent macro view of the stratigraphy. Observations in the field up close and from photographs taken from inside the quarry illustrate a reefal facies that occurs in the Wabash Formation. The rock is notably vuggy with visible coral, shells and other fossilized remains.

During winter visits, water was observed to be issuing from the quarry walls, primarily from the horizontally bedded Detroit River Formation and the contact between the Detroit River and Wabash Formations. Seepage water was also noted from sub-vertical joints in the quarry wall, which correspond to both the primary and secondary regional joint sets. However, the rate of flow from the sub-vertical features in the Wabash Formation is insignificant compared to the flow from the Detroit River Formation. A photograph of the north wall of the quarry, where the different formations, as well as water staining, can be observed, is shown on Figure 3.

Figure 3. Wabash Formation reefal facies in Hanson Ardmore quarry highwall

ANTICIPATED GROUNDWATER CONDITIONS

Characterizing groundwater conditions was a critical task of the geotechnical investigation program. Several aquifer systems are known to produce groundwater in wells located in the Fort Wayne area. The two main types of aquifers are sand and gravel aquifer systems, and bedrock aquifer systems of varying confinement. The Devonian and Silurian carbonate formations constitute the bedrock aquifer system in the Fort Wayne area. Hydrogeologists typically refer to the carbonate formations collectively as the "Silurian and Devonian Carbonate Aquifer System." From a hydrogeologic perspective, these two carbonate formations behave similarly and, therefore, are typically grouped together. However, they can be distinguished from each other upon inspection. This bedrock aquifer system underlies the unconsolidated sediments throughout the southwestern two-thirds of Allen County and is a main source for water in the area.

Packer Test Data

Hundreds of packer tests were completed and results summarized in the Geotechnical Data Report (GDR) and Geotechnical Baseline Report (GBR). Based on an analysis of these data, it was estimated that hydraulic conductivities would range from 1×10^{-5} to 3×10^{-4} centimeters per second (cm/s) in most of the bedrock tested. Similarly, most of Rock Engineering Unit 3—Wabash Formation, was expected to have hydraulic conductivities in that same range (see Figure 4). Combined with up to 240 feet of head, 3RPORT would be a wet tunnel.

A Prolific Aquifer

Until the mid-1900s the City of Fort Wayne derived its municipal water supply from wells in 19 pumping stations located in parks along the St. Joseph, St. Marys and

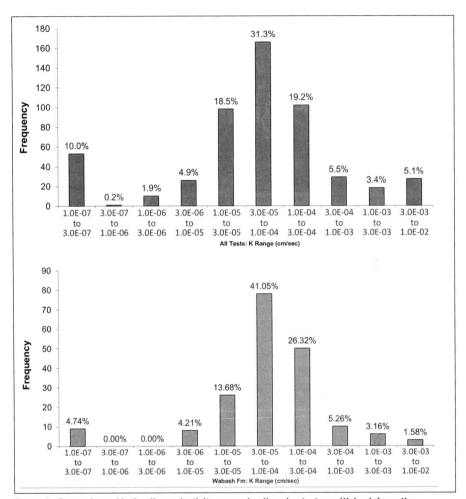

Figure 4. Comparison of hydraulic conductivity ranges in all packer tests vs. Wabash formation

Maumee Rivers. Collectively, the wells produced up to 12 million gallons per day (MGD) to the growing industrial city. Individual wells produced up to a 1 MGD from the "water lime" rock, as it was known by local drillers. Although exact locations of the wells were lost, some historical records indicated that 3RPORT working shaft is located adjacent to former Pumping Station No. 6 (see Figure 5), and the retrieval shaft is located adjacent to former Pumping Station No. 3.

Testing the Yields

Packer testing was not considered to completely define the potential for inflow into excavated rock. Experience in limestone bedrock has indicated that packer test results can be impacted if discontinuities in the bedrock are infilled. For example, if the zone of the packer test includes fractures that are infilled with clay or tightly packed cohesive silts or clays the test results can indicate reduced permeability. Pumping tests have been utilized to better define permeability of non-uniform bedrock. Water pumped out of the bedrock tends to "wash out" the infilled features where packer testing relies on water being forced into the bedrock.

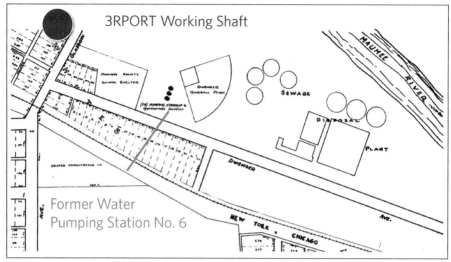

Figure 5. Historical well field adjacent to working shaft

Table 1. Pumping well testing results

Test Location/Shaft	Depth (ft)	Pumping Rate (gpm)/ Duration (days)	Estimated Hydraulic Conductivity (cm/s)	Estimated Storage Coefficient
W-1/RETRIEVAL	263	200/5 339/3	3.5×10^{-2}	3×10^{-5}
W-2/WORKING	260	780/3	2×10^{-3}	3×10^{-4}
W-3/DS07	223	111/4 140/3	2×10^{-2}	5×10^{-5}
W-4/DS-03	223	95/8	1×10^{-2}	Very small, inconclusive

Therefore, four pumping well tests were performed to better define aquifer yields at shaft locations along 3RPORT alignment. A 12-inch diameter test well was constructed and screened in the Wabash Formation. Observation wells screened in both the glacial overburden and bedrock were monitored during the pumping and recovery tests (Table 1). Each test included a step-drawdown pumping phase to determine a sustained rate at which a multi-day test could be sustained, with durations of up to eight days. Well W-2 sustained pumping rates of nearly 800 gallons per minute or 1.14 million gallons per day (gpd).

The results of these tests solidified the hydrogeological conditions in the Wabash Formation; the conclusion was that wet ground with sustained inflows would occur.

VERTICAL ALIGNMENT

The vertical alignment was selected based on a holistic view of the constructability based on geotechnical and hydrogeological data (Figure 6). The tunnel zone was proposed at a depth that 90 percent of the alignment would be in Rock Engineering Unit 3—Wabash Formation and 10 percent in the Detroit River Formation.

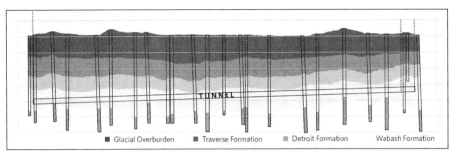

Figure 6. 3RPORT vertical alignment optimization

Table 2. 3RPORT alignment baseline percentage of rock engineering units

Rock Engineering Unit 2	10
Rock Engineering Unit 3	90

Rock Quality

One consideration for the vertical alignment was to locate the tunnel drive within a formation that maximized the life of the TBM disc cutters. Based on previous experience within limestone bedrock formations that are horizontally bedded and characterized by vertical or near vertical fractures, the wear on cutter tools can be significant due to the resulting blocks that result from the horizontal and vertical/near vertical features. During mining, the blocky ground results in impact loads to the cutters.

Also, knowing that interventions could be difficult to perform, considering the potential for inflows into the excavation heading, choosing a formation with less vertical features was deemed important. Rock Engineering Unit 3 was considered the most massive (least amount of prominent fractures) of the formations.

Hydraulic Head

An additional consideration for the vertical alignment was the hydraulic head that would be experienced at the tunnel elevation. While there was an effort to keep the tunnel as shallow as possible, the quality of the bedrock was a more significant risk than the increased hydraulic head conditions.

ANTICIPATED TUNNELING CONDITIONS

The 3RPORT will be excavated through and constructed in Rock Engineering Units 2 and 3. Table 2 lists the baseline percentages for each rock engineering unit anticipated to be encountered during tunnel excavation on a volumetric basis. Rock Engineering Units 2 and 3 will be encountered in mixed-face rock conditions, meaning there will be more than one rock engineering unit at a time at the TBM face. However, Rock Engineering Unit 3 will be encountered in a full-face condition. The starter tunnel, tail tunnel, adits, and deaeration chambers will be excavated through Rock Engineering Unit 3.

The Rock Mass Rating (RMR) for the main TBM excavated tunnel ranges from 43 to 73, with 42% of the tunnel anticipated to be Rock Mass Class II (good rock) and 58% of the tunnel anticipated to be Rock Mass Class III (fair rock). The drill and blast starter tunnel, tail tunnel, adits, and deaeration chambers are anticipated to be Rock Mass Class III (fair rock).

GROUND SUPPORT AND LINING

The tunnel initial ground support and final tunnel lining was designed for gasketed, precast concrete segmental lining system. This decision was made early in the design to mitigate post excavation groundwater inflows and to mitigate risks associated with cast-in-place lining in wet bedrock.

TBM SELECTION

FWU and its design team reviewed options for TBM in the porous bedrock under high-head conditions. Initially both non-pressurized shielded TBM and slurry TBM were considered as options for mining the tunnel. Following discussions with contractors, the design engineer and program manager, FWU decided to specify a slurry TBM as a risk mitigation measure due to anticipated high groundwater quantities (flush flows were estimated to be as high as 5,000 gpm).

The design team felt that significant effort would be required to execute pre-excavation grouting in advance of a non-pressurized face TBM. Pre-excavation grouting of the non-uniform bedrock was estimated to require approximately 2–3 days of grouting effort for every day of mining. The design team also considered the risk that if inadequately grouted (even with a tight grouting hole pattern), a significant inflow source could be missed resulting in additional delays or non-mining time. The impact of groundwater inflow on muck handling was also considered potentially significant. Experience has shown that inflows above 200 gpm often result in muck washing off the conveyor. If muck is not contained on the conveyor, additional downtime (no mining) is required to clean the invert, using hand tools, within the TBM shield.

Figure 7. Pre-cast segments for 3RPORT lining

Considerations for the use of a slurry machine included the maintenance of the slurry circuit and the potential for clogging of a slurry circuit. Clogging of the slurry line would result in obvious delays to production mining and can potentially damage the slurry circuit. As a result, the average mining advance rate is dependent on the characteristics of the muck. Optimization of the slurry circuit ensures a steady mining advance and the longest life of the slurry piping and pumps.

Based on previous projects that utilized slurry TBMs in bedrock with significant groundwater inflows, it was understood that interventions for cutter head inspection or maintenance activities may be challenging and difficult to estimate. The effort to complete the interventions was baselined by requiring that bidders include a prescribed effort in the lump sum and an additional quantity of intervention effort was included as a contingency item. This strategy was also used to baseline pre-excavation grouting and was intended to provide some certainty to the bidders and a fair and equitable method to compensate the contractor during construction for its efforts.

Interventions to inspect and change out cutters, repair cutter housing or cutter head wear, repair the rock crusher, repair worn slurry pumps and pipelines, and other tooling, is one of the risks associated with a slurry mode TBM. Groundwater inflow can present difficulties while grouting behind the segments. During stretches of dry ground or ground producing low water inflow, make up water may be needed to continue excavation operations.

The baseline conditions for the 3RPORT presented in the GBR were based upon a 16-foot minimum finished diameter tunnel excavated by TBM. The Contract Documents were prepared based on a shielded TBM, operating continuously with a closed slurry muck circuit under normal groundwater head conditions (Slurry Mode TBM). The Slurry Mode TBM was required to be designed and equipped to minimize disruptions due to interventions, operate without affecting normal groundwater head conditions, and adequately retard groundwater travel to allow complete grouting of the annular space between the segmental lining and excavated rock.

Figure 8. MamaJo Mixshield TBM named after the three rivers which confluence in Fort Wayne

Operating the TBM in closed slurry mode for the entire length of the tunnel is required in the Contract Documents. The tunnel excavated and lined using a one-pass system with gasketed segments. The baseline time (effort) for tunnel interventions was estimated to be 48 hours/intervention and 50 interventions were anticipated.

PROJECT UPDATE

The 3RPORT project was bid and awarded in 2017 and construction is ongoing towards meeting the Consent Decree deadline. The slurry TBM was delivered and is being assembled for launch in 2019.

Ground Freezing to Improve Unstable Silty Ground

Shawn Coughlin • Moretrench
Nate Long • Jay Dee Contractors

ABSTRACT

The Northgate Link light rail project included 3.4 miles of twin bore tunnels and 23 cross passages through glacially deposited soil conditions. Ground freezing was used for temporary earth support and groundwater control for construction of several cross passages: five were frozen from street level as part of the base scope, five were frozen from inside the tunnels due to unexpectedly permeable soil, and one was frozen from inside the tunnels due to unstable running silt encountered during pre-probing. This paper discusses the design and construction of the ground freezing system to improve the unstable silty soil at CP-23.

PROJECT OVERVIEW

The Northgate Link light rail extension in Seattle was completed in 2018 by JCM Northlink LLC, a joint venture of Jay Dee Contractors, Frank Coluccio Construction Co., and Michels Corp. The project extended Sound Transit's Link light rail system to Northgate Station from the previous northern terminus at the University of Washington (UW). The twin bore tunnels, each 18ft-10in finished diameter and 3.4 miles long, extend from Husky Stadium through new stations in the U District and Roosevelt neighborhoods and terminate at Maple Leaf Portal, where the rail transitions to elevated track (see Figure 1).

The tunnels are connected by 23 cross passages spaced approximately every 800ft. The cross passages provide emergency egress and house mechanical and electrical equipment. The ovoid cross passage excavations measure approximately 17ft wide by 18ft tall by 16 to 25ft long at springline and are numbered CP-21 through CP-43 from south to north, continuing the numbering from the existing light rail tunnels.

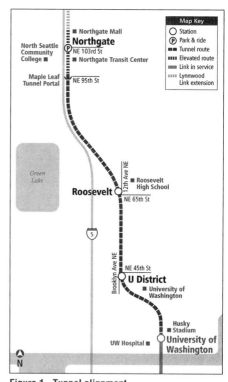

Figure 1. Tunnel alignment

The cross passages were designed to be excavated by sequential excavation methods (SEM) with three designated categories of ground support:

- Category 1, no special measures
- Category 2, dewatering prior to excavation
- Category 3, jet grouting or ground freezing prior to excavation

UNEXPECTED GROUND

Cross passage 23 is located between U District Station and Husky Stadium, beneath the Husky Union Building yard at the heart of the University of Washington. The expected soil was dense to very dense sandy silt, with SPT N-values over 50 blows per foot. Category 2 ground support was specified: JCM intended to install a vacuum wellpoint system from the tunnel interior to depressurize the soil. They had successfully used similar methods at cross passages 24 and 26.

Prior to dewatering JCM drilled probe holes through the tunnel liner to confirm ground conditions. During the probing, crews encountered loose running silt that rapidly flowed through the probe holes. Steel wellpoints installed in some of the probe holes yielded turbid gray water with high fines content in the range of 50 to 75 gpm (see Figure 2). Additional probing 100 feet to the north and south did not reveal more stable soil conditions.

JCM implemented a grouting program aimed to improve the soil consistency and reduce mobility. They injected 14.5 CY of neat cement grout in 12 grout holes at a distance up to 18 feet behind the tunnel segments. Soil inflows from the grout pipes still exceeded 50 gpm after grout injections, indicating little improvement.

The volume of soil that accumulated in the tunnel sparked concerns of surface settlement at the UW campus. Sound Transit, JCM, and their respective consulting engineers agreed that vacuum dewatering had a high potential for migration of fines, which could have caused subsidence at the surface above the cross passage. It was not feasible to dewater the cross passage as planned. Deep wells or eductors from the surface would have had limited effect in the fine-grained soil. Limited access for surface drilling equipment added another constraint.

In addition to a technical challenge, the differing site condition posed a schedule concern. CP-23 was one of the last to be excavated in the overall Sound Transit program, and JCM had expected to finish mining CP-23 by December 2016. That would have left ample time for electrical and mechanical finishings and turning over the tunnels to a follow-on contractor ahead of a May 2017 milestone date. The speedy implementation of the alternative solution was a major consideration.

Figure 2. Soil and water ingress during probing

COMPLEX CHALLENGE, UNIQUE SOLUTION

Sound Transit recognized that ground freezing was the appropriate technique to stabilize the ground for mining. The ground freezing process converts soil pore water into ice by the continuous circulation of a refrigerated coolant; the frozen pore water fuses together particles of soil to increase compressive strength and impart impermeability. The coolant (typically calcium chloride brine) is chilled by a freeze plant, pumped down a tube to the bottom of a freeze pipe and flows up the annulus, extracting heat from the surrounding soil as shown in Figure 3. When the soil temperature reaches the freezing point, ice forms around the pipes in the shape of vertical cylinders. As the cyclinders gradually enlarge with time, they interconnect to form a continuous wall. Once the frozen wall reaches design thickness, the freeze plant is typically operated at reduced capacity to maintain wall thickness rather than increase it.

Due to CP-23s location beneath the UW student union building, installing freeze pipes horizontally from within the tunnels rather than vertically from street level was an attractive option. A surface installation may have required relocation of subsurface and overhead utilities and closing lanes of traffic for several weeks or months to accommodate freeze pipe drilling. Drilling and freeze plant operation would have been very disruptive to UW students. The overall profile of third party risks and impacts was considered challenging, with potential to delay the project. Considering those constraints and risks, the ground freezing system for CP-23, including the freeze plants, was designed to be installed and operated from within the TBM tunnels.

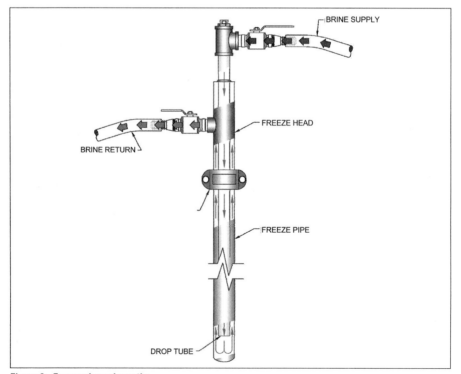

Figure 3. Freeze pipe schematic

This approach offered several advantages over surface-based installation:

- The disruption and noise of installation and operation was confined to the tunnels, rather than the campus above.
- Third party risk during installation and operation phases was eliminated.
- Fewer freeze pipes were required, their lengths were significantly reduced, and they would not have to be removed or relocated during cross passage excavation.
- Less refrigeration capacity was required to freeze the smaller volume of soil.

However, implementing this option presented unique challenges. Traditional mobile refrigeration plants for temporary construction (shafts, adits, etc.) are trailer-mounted units, 8.5ft wide by 53ft long, powered by 400HP ammonia-based compressors. Those large plants would not easily fit into a tunnel and would impede traffic during the freeze operation. The cross passages were located too far within the tunnels to seriously consider placing the large freeze plants at the surface and pumping the brine to each location. Most importantly, while ammonia is an efficient industrial refrigerant, it is dangerous to personnel and not suitable for enclosed atmospheres.

JCM contracted Moretrench to design and install a ground freezing system for CP-23 to replace the original dewatering design. Moretrench has extensive experience in challenging ground conditions and had already implemented ground freezing systems wholly installed and operated within a tunnel at five other cross passages on the Northlink project, the first such application in North America. The five systems were in various phases of operation at cross passages 34 through 38 when the issues emerged at CP-23.

The relatively small mass of frozen soil at each cross passage allowed for the use of custom skid-mounted refrigeration plants. Measuring 4ft wide by 24ft long, the smaller plants could be mounted to the side of the tunnel, allowing unimpeded access for tunneling equipment during freeze system operation. Crucially, the 60HP compressors in these plants use Freon, eliminating the health hazard posed by ammonia in a confined environment.

The freeze pipes were laid out as shown in Figure 4. The main system in the southbound tunnel included 30 freeze pipes drilled across to the northbound tunnel at angles ranging from +10° to −20° with respect to horizontal. The upper pipes extended just above the far tunnel to create a frozen canopy, but the lowest pipes were angled farther downward to avoid the excavation limits, since the cross passage centerline was beneath the tunnel springline. The lowest row of pipes was too far below the northbound tunnel to fully freeze the soil, so a single row of freeze pipes was installed from the northbound tunnel to ensure full closure of the frozen ground. Since the freeze pipes from the southbound tunnel stopped short of the northbound tunnel, a cooling ring was planned in the northbound tunnel to chill the concrete liner segments and allow the frozen soil to form a robust bond with the exterior of the tunnel.

Freezing time is dependent on freeze pipe spacing, so the freeze pipe locations with the widest anticipated spacing were inputted into a finite element thermal model to estimate heat loads and freezing time. The models determined that a freeze plant with 30 tons of refrigeration power was required for the southbound system, while a 6 ton plant was required for the northbound system. Freezing time was estimated at four weeks to reach initial closure of the frozen envelope and eight weeks to reach full structural thickness.

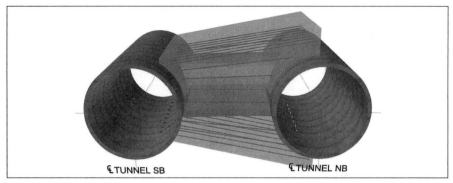

Figure 4. Freeze pipe conceptual layout

Drilling and Installation

During drilling, Moretrench's primary concern was controlling groundwater and soil to minimize the risk of ground loss, since the pipes were drilled roughly 90 feet below the water table in highly unstable soil. Each drill location was prepared by first mounting a groundwater control assembly: a steel trumpet bolted onto the tunnel lining, with a gate valve and several thick rubber gaskets sandwiched between steel flanges. The gaskets fit snugly around the outside diameter of the drill casing to prevent water infiltration, and the flange bolts were tightened as needed to maintain a tight seal. The sacrificial drill casing was left in place to prevent borehole collapse and protect the freeze pipe against damage during excavation. A roller bit and check valve were welded to the end of the casing to allow drilling by external flush methods without groundwater returning through the drill bit.

After each hole was drilled to full depth, a steel freeze pipe was capped at the end and inserted inside the casing. New sections were welded on as needed to reach the full depth of the casing. A brine leak in the surrounding soil would prevent it from freezing, so each pipe was pressure tested to 150% of working pressure to ensure a fully sealed system.

The freeze pipes were surveyed with a gyroscope to check drilling deviation. The thermal model was repeated with as-built locations to determine if any pipe-to-pipe spacing was wide enough to require a supplemental freeze pipe. One re-drill pipe was installed from the northbound tunnel to close the gap between southbound pipes.

A plastic tube was inserted to the bottom of each freeze pipe and connected to a freeze head. The freeze heads were connected to the distribution piping, pumps and freeze plant, then the system was leak tested and loaded with brine.

The freeze pipes in the northbound tunnel system were plumbed similarly but with the addition of the surface cooling loop. Rectangular steel tube was rolled to the inside diameter of the tunnel and bolted to the segmental liner, then covered with insulating blankets to protect

Figure 5. Drilling in the tunnel

against the warm ambient air in the tunnel. Brine circulated through the tubes to chill the concrete liner segments.

Lastly the instrumentation system was installed to continuously record freeze data during system operation. The instrumentation panel was connected to a remote computer terminal for monitoring and evaluation by a Moretrench technician.

Three temperature monitoring pipes were installed in the same manner as the freeze pipes and filled with brine to prevent freezing. Sensors were installed every 5 feet in the monitoring pipes to record the ground temperatures along the entire length of the frozen mass. More sensors recorded the brine parameters (supply and return temperatures, pressures, flow), the brine return temperature of the individual circuits, and the temperatures of the concrete tunnel lining.

Figure 6. Freeze plant mounted on the tunnel liner

Piezometers were installed in four locations: one in the center of the cross passage excavation at each end, and one each to the north and south of the CP-23 in the southbound tunnel. As closure was achieved, a rise in the center piezometer was anticipated due to the volumetric expansion from liquid water to ice; the unfrozen space in the center of the cross passage would squeeze down as the frozen wall continued growing inward.

The system was installed and commissioned between December 2016 and January 2017. A significant advantage of the in-tunnel system, compared with a surface installation, was a reduction in the drilling linear footage of approximately 85%, leading to a much shorter drilling and installation period.

Freeze Development

Freezing time was consistent with the thermal model: time to close the frozen wall was 19 days as indicated by evaluating the piezometer pressures. A difference in internal versus external pressure indicated that the interior of the frozen zone was isolated from the exterior. To confirm closure, Moretrench performed drain-down tests on the northbound tunnel center piezometer, draining water from the wellpoint while measuring its flow and temperature and the pressure at the interior piezometer in the southbound tunnel. When the water decreased in flow and temperature over time with no related drop in exterior pressure, there was strong evidence that closure was achieved. Armed with this test data, in conjunction with the ground temperatures from the monitoring pipes, Moretrench released the cross passage to JCM for excavation. Once the positive drain-down test was confirmed, the center piezometer on each end was left open to drain any unfrozen soil in the center of the frozen zone.

The expansion from saturated to frozen soil stressed the segmental tunnel liner, slightly egging the rings and causing some stepping between adjacent rings. The deformations had no detrimental effects on the ground freezing or excavation progress, and reduced slightly after the freeze systems were deactivated and the ground thawed and relaxed.

CROSS PASSAGE EXCAVATION

JCM drilled short probe holes before sawcutting the concrete liner segments to confirm the watertightness of the frozen structure. CP-23 was then excavated via standard SEM methods, with top heading and bench rounds, installing lattice girders every 4ft, and supporting the frozen earth with wire mesh and fiber-reinforced shotcrete. The excavation was advanced with a demolition robot with various attachments including a hydraulic breaker, roadheader, and excavation bucket. The soil in the center was unfrozen and easy to remove; excavation was more difficult around the perimeter of the excavation where the soil was colder and harder due to proximity to the freeze pipes. An additional 2" of shotcrete beyond the design thickness was applied to the frozen ground. This layer was sacrificial to ensure that the structural shotcrete lining was protected from the freezing temperatures.

The excavation was dry and stable, with no instances of groundwater seepage or raveling soil. No surface settlement or heave was recorded during freezing or cross passage excavation, and convergence deformations of the tunnel linings were minimal. After the excavation was completed, the system remained active, at a reduced capacity, during the installation of the final cast-in-place lining.

Figure 7. Excavating frozen soil

CONCLUSION

Ground freezing has long been a useful tool in the underground construction environment. The success of this unique application of ground freezing demonstrates that with the right expertise and with careful attention to detail during design and execution, the technology can be expanded to overcome new challenges. The ability to install and operate systems completely within a tunnel makes ground freezing especially applicable in urban environments where land is scarce and surface access constraints, street closures, traffic detours and utility interruptions are increasingly costly, disruptive and unpalatable to the public.

The difficult soils in the vicinity of CP-23 created difficult technical and schedule-based hurdles for the Northlink project. The successful evaluation, selection, and implementation of the ground freezing solution was the result of close cooperation between the owner, design team, tunneling contractor, and ground freezing subcontractor, demonstrating that ground freezing can be a safe and effective means for ground support in even the most challenging soil conditions.

Neelum-Jhelum Hydroelectric Project: Design and Construction Challenges of Underground Works

Peter Dickson ▪ Stantec
Francisco Tesi ▪ Stantec

ABSTRACT

Located in the Himalaya foothills of northern Pakistan, the Neelum-Jhelum project is a high-head hydro facility generating 969MW by diverting water through a long large-diameter headrace tunnel system, 26-km of which were excavated by drill-blast and 21.5 km by hard rock TBM. The total length of all hydraulic tunnels and the various access tunnels serving the underground powerhouse and surge facilities, adds up to 67.6 km (42 miles).

Design and construction of the tunnel system and powerhouse cavern had to manage many severe technical challenges, including complex geology, active faulting, high cover (up to 1900 m), anomalous rock stresses, deadly rock-burst occurrences, and massive groundwater ingress under high pressure. The project also faced many non-technical issues causing delays and inefficiencies. These challenges are described and how they were overcome by the design and construction staff, including carrying out major design modifications during the construction period, collection of *in situ* rock stress data, extensive pre-excavation grouting to control high pressure water inflows, systematic high-pressure consolidation grouting to improve rock mass modulus and reduce permeability, responsive approaches in instrumentation and rock support to reduce dangerous rock burst hazard, and methodical 2-D and 3-D numerical modeling of cavern complex during construction using actual deformation data to check and validate design or to modify design where needed. Complete re-design of many project features was required due to recognition of increased seismic hazard, major changes to the headrace tunnel alignment, introduction of TBM method of construction two years into construction. The various techniques and approaches used successfully and described in the paper are valuable guides and lessons for future large-scale underground works including the next generation of hydro-pumped storage projects.

INTRODUCTION

The Neelum-Jhelum Hydroelectric Project is in the Himalaya foothills of northeastern Pakistan where the terrain is extremely rugged with ground elevations ranging from 600 to 3200 m above sea level. The project is a high-head (420 m static head) hydro facility generating 969MW by diversion of water through a 28.6-km-long headrace tunnel system and 3.6 km of tailrace tunnel. Design capacity of the waterway system is 283 m^3/s.

The project owner is WAPDA (Water and Power Development Authority of Pakistan) and the prime contractor is a consortium of China Gezhouba Group Corporation (CGGC) and China Machinery Engineering Corporation (CMEC). The Engineer is Neelum-Jhelum Consultants, a joint venture of MWH International (now Stantec), Norplan (MultiConsult) of Norway, with local partners National Engineering Services of Pakistan (NESPAK), Associated Consulting Engineers (ACE) and National Development Consultants (NDC). Construction began in 2008 and the first unit was

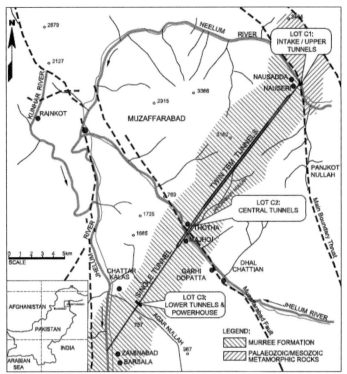

Figure 1. Project layout showing TBM twin tunnels (in bold), major faults (dashed) and simplified alignment geology

commissioned in 2017 and the entire project was completed in 2018 at a construction cost of more than $5 billion.

The project location and general arrangement of the extensive tunnel system is shown in Figure 1. The hydraulic tunnels and the various access tunnels serving the underground powerhouse and surge facilities add up to a total length of large-diameter tunnels on the job of 67.6 km. The headrace tunnels include both single bore (31%) and twin bores (69%), 26-km of which were excavated by drill-blast and 21.5 km by hard rock TBM. The tailrace tunnel consists of a single tunnel. Excavated diameter of the single bore tunnels ranged from 9.75–11.9 m and for the twin tunnels ranged from 8.0–8.55 m.

Major changes were made to the design of the headrace tunnel system early in construction. The first was introduction of a hydraulic lining for all sections of the headrace tunnel excavated by drill-blast; initial design only had shotcrete lining for most of the tunnel length. Installation of a cast-in-place concrete lining resulted in significant improvement in hydraulic performance and power generation. The concrete lining also facilitated grouting in those areas where it was needed. Negative trade-offs were serious delays in completion.

The second change was elimination of the dip in the vertical alignment under the Jhelum River. In the original design developed in the mid-1990s the headrace tunnel alignment crossed under the Jhelum River with a dip, or siphon, arrangement. This was required by the need for sufficient depth below ground surface to obtain rock stress levels that allowed a conventional pressure tunnel without steel lining. Before construction began, but after the contract had been let, the design was modified to eliminate the Jhelum dip, thus allowing dewatering by self-draining and to facilitate construction. The headrace tunnel was shifted upward to a shallower depth. From the results of in situ stress testing, it was determined that 732 m of steel lining was required in each of the twin tunnels.

The third major revision was introduction of TBM construction method in place of conventional drill-blast in the high cover sections. Tunnel construction began in 2008 using conventional drill-blast methods as indicated in the original design and construction documents. However, it soon became apparent that excavation of a 13.5-km-long section of the headrace twin tunnels would take too long if done solely by drill-blast methods (under high overburden that precluded construction of additional access adits). The construction contract was therefore amended by variation order to incorporate two 8.5-m-diameter main beam gripper hard rock TBMs to excavate approximately 10.5 km of twin headrace tunnels (Figure 1). The gripper design offered flexibility for expected conditions including possible squeezing ground given the relatively weak rock mass and overburden of up to 1870 meters, and the potential for rock bursts in the stronger rock units.

Each of these significant design modifications resulted in arduous design efforts and increased potential for delay in preparation of construction drawings. In hindsight, however, the decisions to make these design changes were correct although there were still substantial challenges and problems that had to be managed and overcome as described below.

GEOLOGIC SETTING

Geologically the project is located in the Hazara Kashmir Syntaxis, surrounded by two limbs of the Main Boundary Thrust Fault. The project area is traversed by the trace of the active Muzaffarabad Thrust, also referred to in the literature as the Himalayan Frontal Thrust Fault and Balakot-Bagh Fault (Figure 1). The 7.6 magnitude Kashmir earthquake of 8 October 2005 occurred on this fault resulting in vertical displacements of 2.5 to 6 m and more than 75,000 fatalities (Kaneda, et al., 2008).

The entire underground works of the project were constructed in molasse-type sedimentary rocks of the Murree Formation, which is of Eocene to Miocene age, and the lateral equivalent of the Siwalik Group in India. The succession comprises intercalated beds of sandstone, siltstone and mudstone that are highly deformed, tightly folded, and tectonized, with mostly steep bedding dips and a general northwesterly regional bedding strike, which was typically approximately perpendicular to the headrace tunnel alignment. Zones of weakness, shearing, and local faults were commonly observed and are invariably oriented parallel to the regional bedding strike.

Lithologies

The lithological units comprise a monotonous series of alternating beds of sandstones, siltstones, and mudstone:

Siltstones and Silty Sandstones

A suite of dark red-brown siltstones that grade to silty sandstones is the most commonly occurring rock type on the project, with about 70% of the TBM tunnels excavated in this unit; typical Uniaxial Compressive Strengths (UCS) are 50–70 MPa;

mainly distinguished by ~25% argillaceous content and lower permeability even where fractured and sheared.

Mudstones

Mudstones, which also are invariably dark red-brown, represent the weakest rocks in the Murree sequence, with UCSs in the 30 to 40 MPa range; generally highly deformed and locally sheared; low to negligible permeability; alternating with pervious sandstone units, it forms aquicludes/aquitards; concentrated groundwater discharge found at contacts with sandstone beds.

Sandstones

These are the strongest members of the Murree sequence. They are typically grey, with sharply defined contacts. About 21% of the TBM tunnels were excavated in this material. Bedding thicknesses range from a few meters to over 50 m. Of 30 samples tested during excavation, 77% fractured in the 130–170 MPa range, with the remaining 23% exhibiting higher strengths, up to 230 MPa; characterized by well-developed secondary fracture permeability and responsible for > 90% groundwater flows (mostly turbulent) into the headrace tunnel excavations.

PROJECT ISSUES AND CHALLENGES

During construction of the underground works, the project faced severe issues and challenges both technical and non-technical. Technical issues are common to most projects but, on this job, many were unique and some extreme. The non-technical issues were not only frustrating to deal with but also the cause of many delays and inefficiencies. These challenges are described briefly below.

Dearth of Geologic Information

Preliminary investigation and design had been conducted in the early to mid-1990s, before the 2005 earthquake. When the project went into construction in 2008, there was scant information on geologic conditions. Geological mapping was only available at broad regional scales, satellite imagery and aerial photos were not available, and considering the size of the project only a few exploration drill holes had been made and of these the core had been lost in the 2005 earthquake. Consequently, detailed design was severely constrained. New investigations were time-consuming and often resulted in delays to design. Attempts to conduct supplementary explorations were often met with resistance in view of pressure to meet construction schedule.

Complex Himalayan Geology

Given the geologic setting, it is obvious that bedrock units have been subject to severe tectonic processes, are highly deformed, and are folded into complex systems. Faulting and shearing are very common and active faults are now identified at two locations on the project: the Main Boundary Thrust on the right abutment of the dam site and the Muzaffarabad Thrust Fault which crosses the headrace tunnel alignment about 2 km north of where the tunnel alignment crosses under the Jhelum River (Figure 1). Taking these geologic factors into account, it is evident that the project seismic hazard is high to extreme, a situation that was not understood during preliminary design. Consequently, a new seismic hazard evaluation was performed which resulted in re-design of nearly all project structures. The design peak ground accelerations (PGAs) at the dam site and powerhouse area were concluded to be 1.16g and 1.06g respectively. Innovative seismic resistant design features were developed to manage strong shaking and ground movements (Dickson and Kovacich 2015).

In Situ Stresses

It was recognized at the start of construction that little information was available on in situ rock stresses. Some rock stress data had been collected in 1995–96 but there was uncertainty about the results since they were obtained prior to the 2005 earthquake, they provided no information on stress conditions in areas of high cover along the headrace tunnel route and were questionable in areas of low cover at the revised headrace tunnel location at the Jhelum River crossing. Concerns were raised about the length of steel lining needed in low cover areas to manage the potential for hydrojacking. In contrast in high cover areas, there was significant uncertainty about the potential for convergence or rock burst. For these reasons, investigation programs were instituted to obtain such information. However, in situ testing was time-consuming and unpopular, since, as with other quests for additional geologic investigations, there was a critical clash with the drive to maintain pace of the construction schedule.

During excavation of the twin headrace tunnels under the Jhelum River, in situ stress measurements were periodically conducted from the tunnel headings to assess hydrojacking potential where the cover would be lowest. These data were used to determine the length of steel lining required, which was eventually determined to be 732 m in each of the twin tunnels. Similar testing was also performed in the high-pressure manifold area immediately upstream of the powerhouse. Hydrojacking test results varied widely, due to the markedly anisotropic rock sequence, with large variations in stiffness between sandstones and mudstones. Most results clustered loosely around a horizontal-to-vertical stress ratio (k) of unity. High horizontal stresses might have been expected parallel to the tunnel axis since this is the major principal stress orientation indicated by the World Stress Map. Hydrojacking testing only provided data on minimum in situ stresses, with no information on magnitudes or orientations of principal stresses. Numerical modeling was used to evaluate the effects of various stress states on the ability of groundwater to move through fractured rock.

In areas of high cover along the headrace tunnel, data from over-coring tests in sandstone beds in the TBM tunnels helped identify a tectonically altered zone of high stresses (k up to 2.9) with the major principal stress oriented sub-horizontally and sub-perpendicular to the tunnel azimuth. This testing was only established after severe rock burst had already been encountered and information was needed to determine appropriate measures to manage the hazard and reduce risk (Figure 2).

It should be noted that a Geotechnical Baseline Report (GBR) was included in the variation order used for introduction of the TBMs, though it is likely that the concept was not properly understood by most project participants. The GBR listed various geological hazards and issues germane to this type of underground construction and included the potential for hazardous groundwater inflows, high in situ stresses, convergence, and rock burst. It is noted that groundwater inflows were not as critical in the TBM and high cover areas as they were in low cover areas (at Jhelum Crossing). Similarly, when tunneling in high cover areas, convergence was found to be not as important as had been estimated prior to construction.

Rock Burst

The topics of rock burst and how it was managed on this project are covered in detail in other technical papers to which the reader is referred (Ashcroft, et al., 2016; Peach, et al., 2017; Mierzejewski, et al., 2017; Dickson, et al., 2018).

No rock bursts occurred in the first 2.3 km of either of the TBM tunnel drives, after which they gradually increased in frequency and severity (Figure 2). By November

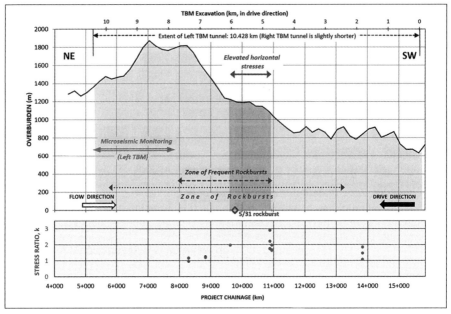

Figure 2. TBM alignment profile (vertical exaggeration × 2.75) showing overburden, zones of rock burst and elevated horizontal stresses (latter determined from over-coring). Lower chart shows H:V stress ratio (k).

2014, with 4.7 km and 4.3 km of the tunnels excavated in the left and right tunnels respectively, regular rock bursts warranted systematic recording (Figure 3). Not only had cover increased but also subtle changes in lithology and geologic structure had raised the potential for rock burst.

Rock burst events were categorized by magnitude, from "noise only" to "major rock burst." The system aimed to correlate timing and distribution of events and to facilitate selection of mitigation measures at the TBM. The categories were: (1) Noise only—a slight popping sound is heard no damage to the support or ejection of rock; (2) Noise and weak rock burst—a popping sound is heard and there may be some light damage to the support and surrounding rock; (3) Noise and medium rock burst—loud popping sounds are heard and there may be splitting, spalling or shallow slabbing to the support and surrounding rock; (4) Noise and major explosive rock burst—loud sound similar to an explosion, violent ejection of rock into the tunnel, and severe damage to the installed support. The majority of events were classified as Category 2.

The most severe rock burst occurred on 31 May 2015 on the trailing TBM. This event, equivalent to a Magnitude 2.4 earthquake on the Richter scale, severely damaged the TBM, and recovery work took seven months to complete. An unusual combination of local geological conditions, including a bedding strike parallel to tunnel direction, and unrecognized massive sandstone beds in the sidewalls masked by siltstone, contributed to its severity. The event emphasized the need for better ground investigation to identify rock burst-prone conditions ahead of each TBM to allow the TBM management to implement rock burst counter measures.

Rock bursts in sandstone beds accounted for 76.6%, with the remainder occurring in siltstone units. The more powerful rock bursts originated within sandstone beds.

Neelum-Jhelum Hydroelectric Project: Design and Construction Challenges

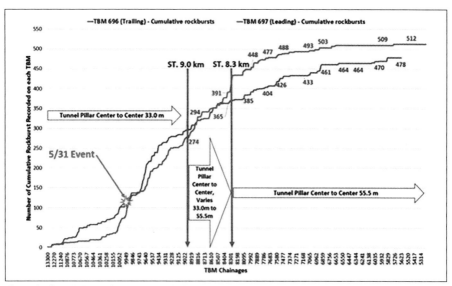

Figure 3. Cumulative rock bursts recorded between November 2014 and May 2017 for each TBM

Rock burst mitigation measures included:

- Drilling of longitudinal and radial stress relief holes using equipment on the TBMs.
- Drilling of horizontal side wall probe holes to detect possible unidentified sandstone beds that could be a source of high stress concentrations.
- Microseismic monitoring of rock burst and seismic noise.
- Installation of Shotcrete at the L1 Zone. Routine reinforcement involved installation of rock bolts and wire mesh in L1 (see Figure 4). Steel fiber-reinforced shotcrete can provide additional functionality. The TBMs used on this project allowed shotcrete application from three areas: the L1 zone as initial support, 20 m further back where the invert can be sprayed, and 60 m behind the face at the L2 zone where the permanent support installation is finalized. Where rock conditions require less support, it is preferable to apply most of the shotcrete at the L2 zone to allow quicker installation of initial rock support and faster resumption of excavation. However, 94% of rock bursts were detected at the front 10 m of the TBM. Therefore, to effectively mitigate rock burst, most of the shotcrete had to be applied as soon as possible, i.e., at the L1 zone. Once the TBMs encountered regular rock bursts, up to 62% of the shotcrete was installed at the L1 zone.
- Full Ring Steel Support. The original purpose of full ring steel supports was to support the tunnel at faults, large overbreak areas, and soft and squeezing ground. These supports are time consuming to install and can be installed at spacings ranging from 0.9 to 1.6 m. The spacing directly influenced the daily advance rate. Initially these supports had been installed in large overbreak areas adjacent to sandstone beds. As the tunnel advanced, rock bursts commenced, and Category 4 events resulted in worker casualties and caused major equipment damage and serve damage to rock bolts, mesh, and mining straps. The full ring steel supports, however, remained mostly intact even when dislodged. These elements remained rigid but certainly prevented

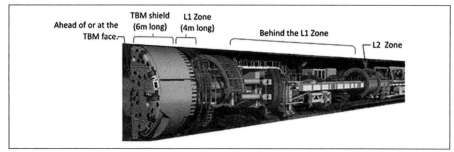

Figure 4. Illustration of different locations on the TBM

more extensive damage to rock supports and equipment and most importantly provided a degree of protection to TBM personnel.

- TBM Over-Cutters. Overcutting was achieved by extending the cutters located on the cutterhead periphery using shims to increase the effective width of the cut and by replacing the gauge cutters with larger diameter cutter wheels (from 17 to 18 inches). Both methods were employed to increase the tunnel diameter by 100 mm.
- Change of Alignment. The initial center-to-center spacing between the tunnels was 33 m. It was anticipated that increasing the spacing between the two tunnels to 55.5 m would reduce stresses in the rock pillar and thus mitigate rock burst potential. The spacing was increased gradually by 'stepping out' one of the TBMs in a tunnel widened using over-cutters. The over-cutters allowed an increased clearance from support to TBM equipment and permitted a smaller turning radius for the 190-m-long TBM. The trailing TBM experienced fewer rock bursts after the change of alignment.
- Heavy Rock Burst Support. The combination of over-cutters, full ring steel support, shotcrete in L1 followed by application of a further shotcrete layer up to a total of 350 mm allowed installation of a much heavier and stronger rock support. This became integral in dealing with the Category 1, 2 and 3 rock bursts and significantly reduced the impact of Category 4 rock bursts. Such measures provided safety for the TBM personnel and more protection to the main TBM components located in the front 100 m of the TBMs.

Groundwater Inflows and Tunnel Collapses

Severe stability and groundwater challenges were faced in construction of the twin headrace tunnels under the Jhelum River due to low cover areas (depths below ground of 175 to 225 m). Systematic procedures were developed and used to enable drill-and-blast tunnel advance through weak unstable ground that was heavily charged with groundwater at high pressures up to 20 bar. These are described in detail in another paper (Kizilbash, et al., 2017). Probe and core drilling were undertaken ahead of the faces followed by systematic pre-excavation grouting (PEG) of the tunnel face and periphery at pressures up to 50 bars wherever probe-hole drilling indicated high pressure or high-volume groundwater hazards.

Such groundwater control and tunnel stabilization were needed until final steel and reinforced concrete could be installed. About 1.1 million liters of grout were injected in each of the twin tunnels reducing initial groundwater inflows from 190,000 L/min to less than 5,000 L/min. A lower permeability result could have been achieved with microfine cement (MFC) but was not really needed, though the potential construction schedule advantages of MFC might have been beneficial if a suitable supply had been available.

Unfortunately, some serious mishaps did occur when the procedures were not followed properly, and these resulted in partial collapse of the tunnel face and periphery and uncontrolled water inflow at high pressure and flow rate. Countermeasures involved construction of temporary tunnel plugs and a lot of grouting. These were costly, very time-consuming, and seriously impacted schedule.

Groundwater inflows had to be pumped continuously and completely removed from the tunnel system. These inflows originated from an infinitely charged aquifer directly connected to the Jhelum River and was contained in highly fractured bedrock under the north bank of the river. The risks were also more critical with the dire threat of flooding the entire tunnel system under the river. Because of the depth of the tunnels, an added challenge was the need to pump drainage water vertically 230 m, with multiple stations, and for a horizontal distance of up to 5 km to evacuate it to the surface and out of the tunnel construction areas. It only needed a serious breakdown of a pump stations or the electric power feed to create a potentially catastrophic situation.

Overall, the PEG program allowed safe completion of the Jhelum River crossing but it highlighted some key lessons learned. These included the critical importance of timely installation of the right support, the risks associated with drainage water management at deep crossings under water bodies, and the potentially dire consequences of collapses that can compromise grouted tunnel perimeters.

Powerhouse Cavern Complex

Geologic conditions in the powerhouse area were poorly known at the start of construction. No exploratory drill holes had penetrated the machine hall cavern area and even surface geologic mapping was absent. Because of concerns about potentially low in situ stresses in the original cavern location, a horizontal shift into the mountain of 200 m was proposed. During construction of a pilot tunnel heading down to the cavern complex, a sub-horizontal 200-m-long core hole was drilled from the tunnel face. This penetrated through the proposed revised machine hall location and results demonstrated favorable rock units and conditions, predominantly thick-bedded sandstone with thinner ribs of siltstone and mudstone and no discernable faulting. Construction of the caverns and associated tunnel complex proceeded in a conventional manner.

Heavy rock support was installed in the caverns given the high earthquake hazard and overall concerns about long-term performance of the Murree Formation lithologies. The caverns were instrumented and monitored for movements and deformation during construction and for long-term operation. There were some unusual movements in localized areas, but these were mostly related to natural response to over-ambitious excavation shaping, particularly in the rock pillar separating the machine hall from the transformer cavern and the ribs between the bus-bar tunnels. Fortunately, these movements were controlled by installation of additional support but for a while did create an alarming scenario. An important lesson learned is that the planning and design of caverns must be closely attentive to the geology and must properly consider the anticipated performance of the geologic units as they respond to construction and operating conditions.

Design staff conducted 2-D and 3-D numerical modelling of the cavern system throughout construction. This used actual deformation data to calibrate, check, and validate design and to help modify design where needed. It was particularly important

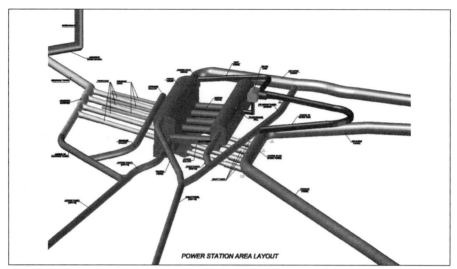

Figure 5. Power station complex showing the Machine Hall (131 m × 23 m × 52 m), Transformer cavern (125 m × 17 m × 25 m), and associated hydraulic and access tunnels. Flow during generation is from left to right in the figure, with headrace tunnel entering from upper left and tailrace exiting lower right.

in evaluation of the performance of the caverns when the hydraulic tunnels would be watered up during initial operation. The cavern complex became completely drained of groundwater during its excavation and the rock mass achieved a temporary equilibrium in this drained state. Filling of the tunnel system and during operation would result in changing the groundwater regime and re-adjustment of the effective stresses in the jointed bedrock around the excavations. This can re-initiate movements in the rock mass surrounding the caverns. Modeling was conducted to explore the potential migration of groundwater through jointed rock under various rock stress conditions. Results were used to refine support measures, grouting and drainage measures, and location of instrumentation.

Non-Technical Issues and Challenges

From the beginning, the project was beset with innumerable non-technical issues and challenges. Some of these were dealt with and abated over time, but most of them created problems throughout the construction schedule resulting in complications, inefficiencies, misunderstandings, delays, and increase in cost.

Remote Location

The project is in a remote location which had a major impact on logistics and ability to maintain a regular and dependable supply train for materials/fuel/labor, etc. needed for construction. The remoteness of the location was compounded by the fact that most roads in the region had been severely damaged by the 2005 earthquake and had to be repaired or re-built, including bridges capable of taking higher loads than previously. Mountain slopes in this region are also inherently unstable and access roads are frequently blocked by rock falls or swept away by landslides, an almost weekly situation during the monsoon season. Given the location in a developing country, there is absence of an established infrastructure or local government departments capable of managing such conditions.

Labor
There was not an adequate source of labor (semi-skilled or skilled) in the region because of the comparatively sparse rural population and general lack of prior experience in the region of work involving required skills for construction (e.g., iron-work, welding, machine operation, electricians, mechanics, heavy truck driving, etc.). Such resources had to be recruited from other regions in Pakistan, creating another snag because of language and cultural differences between local and non-local workers.

Language
Major difficulties arose because of language differences. Although English is an official language in Pakistan and was the designated project language, it was not the first or even second language of many Pakistani staff who were more familiar with Urdu, Punjabi, Sindhi, or Pashto. The Contractor was Chinese and only a very few of his staff could speak or read English. Translators were required at nearly every meeting and misunderstandings were common. Fortunately, there were fewer language problems among the expatriate design and site supervision staff despite many different countries of origin.

First Major Tunnel Project in Country
Although large dam and hydroelectric works have been built previously in Pakistan (e.g., Tarbela and Mangla), this is the first major tunnel project in Pakistan and is the largest civil works project undertaken in terms of construction cost (more than $5 billion) and it has an enormous geographic footprint. Very few engineers from the Owner or in the national professional community had any experience with design or construction of tunnels or underground works. It was therefore a steep learning curve for both the Owner and the country at large, including politicians, media, and government agencies at all levels. For these reasons, project management had also to deal with multiple social, cultural, and political issues that might not be as common on similar works in other regions of the world. The project schedule also suffered from political pressures that tended to be heedless of engineering and technical realities.

Introduction of TBMs
Two years into the construction contract, the Owner decided to request introduction of TBMs to help accelerate construction of the headrace tunnel. The idea was to change 22 km of twin drill-blast tunnels to twin TBM-excavated tunnels, each 8.5-m-dia. Several issues were immediately identified:

- Three different languages, three different currencies (dollars, rupees, yuan)
- Existing Contract Base Year was 2007; work on the change starts in 2009; work on the Contractual change (Variation Order) Starts in 2011
- No TBMs planned or ordered; no camps to accommodate additional resources; no power supply for TBMs
- No access to site for TBM main bearing and other large TBM components
- No TBM tunnel personnel mobilized, on contractor's side or with Engineer
- CGGC (Contractor) the only option available

After three years of discussions, planning, negotiations, and surmounting numerous challenges (2009 to 2012), the TBM Variation Order was executed by the Owner and the Contractor. TBM mining of the twin tunnels began in February of 2013 and after four years was completed in May of 2017, i.e., a total of six years from inception to completion. With the cooperation of all stakeholders, the means and methods of construction for the twin headrace tunnels were changed successfully to accelerate the schedule to ensure that the client could generate power as soon as possible.

Design Issues and Delays

Due to many factors, the design efforts and issuance of drawings for construction tended to become complicated, delayed, and lagged behind demands from the construction schedule. This pattern was set at the beginning of the project when the Construction Contract was let too early, not only before design was complete but also several months before a contract for the Engineer was in place. As indicated earlier in this paper, major design efforts that were not originally anticipated included:

- Upgrade of the seismic design of much of the project to account for the revised seismic hazard.
- The original design of the headrace tunnel did not include any lining other than shotcrete for most of its length. Introduction of a concrete lining for drill-blast sections was required to improve overall tunnel hydraulics.
- Re-design of the headrace tunnel was required due to the changed vertical alignment at the Jhelum Crossing.
- New design efforts in support of the introduction of TBMs.
- Finalization of design of the powerhouse complex had to consider a revised horizontal location and revised dimensions.

Financial Issues

Although beyond the scope of this paper, it is recognized that the project suffered severe financial constraints and issues, which were frequently reported and discussed in the media and of course were aggravated by political debate. Before construction started, the estimated construction cost rose from $167 million in 1989 (based on preliminary design) to $935 million in 2005. Construction costs have since increased to a current estimate of $5.1 billion. The Owner and Government of Pakistan sought funding from many different financial institutions and there were several payments crises that severely hindered the Contractor's cash flow. On this project, financial advice and consultation were not included in services to be provided by the Engineer. It is noted that such advisory services are sometimes offered as part of engineering consulting services. Given the nature of many of the non-technical issues described above, it is suggested that they be offered on future projects similar to this, especially where the Owner lacks experience in commercial and technical matters that can seriously impact cost and cash flow.

CONCLUSIONS

The Neelum Jhelum Project probably did not face any particular technical challenge that has not been encountered on another underground work elsewhere. However, the number and severity of the problems encountered were somewhat unique. In general, the project staff managed the constraints and challenges imposed by the technical issues exceedingly well, considering the remoteness of the project location and other limitations. In some cases, world-class experts were brought to the site to assist with some issues—including seismic-resistant design, rock stress testing, rock burst countermeasures, TBM performance, pre-excavation grouting, and convergence and movements in the powerhouse. By and large, however, the site staff on their own developed solutions to the issues and ways to proceed.

The non-technical issues were also not exceptional or exclusive to this project, but their occurrence point out important lessons to be learned for future large underground work. There were far-reaching repercussions due to project design being markedly immature at the outset and design work not starting until well after the construction contract was underway. The burden was made worse by introduction of major design changes in addition to all the tweaking and adjustments that need to be conducted as construction proceeds and differing site conditions are revealed. The design office and expatriate engineering staff were in a continual race and catch-up to produce design revisions and drawings for construction. Additional pressures were applied to the expatriate staff due to the inexperience of the Contractor's resources who needed more engineering support and guidance than would normally be expected, not just for the TBMs but for most activities. It is well understood how important it is in developing countries for leadership to be in the hands of national resources and for maximizing opportunities for capacity building. However, for major public works projects involving massive Capex investments and where there is potential for huge risk to public safety, it is just as important to make sure that positions for key technical and management positions are given to highly experienced individuals who have the authority or influence to help make critical decisions when needed.

REFERENCES

Ashcroft, B., G. Peach, J. Mierzejewski. 2016. Rockburst Analysis and Mitigation Measures in TBM Tunnels. Fjellsprengnings Teknikk, Bergmekanikk/Geoteknikk, 2016.

Dickson, P.A. and J.R. Kovacich. 2015. Design Solutions to Accommodate Faulting in Foundations of Critical Structures. 6th International Conference on Earthquake Geotechnical Engineering, 1–4 November 2015, Christchurch, New Zealand.

Dickson, P.A., M.H. Kizilbash, G.A. Rosario, G. Peach & B. Ashcroft. 2018. Rockburst Characteristics on Twin TBM Tunnels in the Lower Himalayas. 3rd International Conference on Rock Dynamics & Applications (RocDyn-3); "Experiments, theories and applications"; Trondheim, Norway, 25–29 June 2018.

Kaneda et al. 2008. Surface rupture of the 2005 Kashmir, Pakistan Earthquake and Its Active Tectonic Implication. Bulletin of the Seismological Society of America; April 2008; v. 98; no. 2; pp. 521–557.

Kizilbash, M.H. and P.A. Dickson, N.A. Jaffery. 2017. Neelum Jhelum Hydroelectric Project: Pre-Excavation Grouting in Twin Headrace Tunnels. Grouting 2017: 5th International Grouting Conference, Honolulu, Hawaii, July 9–12, 2017; Grouting, Deep Mixing, and Diaphragm Walls; ASCE—Geo-Institute.

Magna, R., and P.A. Dickson. 2017. Neelum Jhelum Hydroelectric Project: Construction of Tunnel System. 10th Annual Breakthroughs in Tunneling Short Course, Chicago, August 14–16, 2017.

Mierzejewski, J., G. Peach, B. Ashcroft. 2017. Short-Term Rockburst Prediction in TBM Tunnels. Proceedings of the World Tunnel Congress 2017, Surface challenges—Underground solutions. Bergen, Norway.

Peach, G., W.B. Dobbs, B. Ashcroft. 2017. Rockbursts in TBM Tunnels—Analysis and Counter Measures. RETC 2017, Rapid Excavation Tunneling Conference (RETC), San Diego; SME.

Removal of Large Concrete Obstruction on 96-Inch Pipe Ram Crossing of Highway 61, Neebing, Ontario

James Carroll ▪ MarshWagner Inc.
Conner Beck ▪ LTL Directional Drilling Services
Travis Kraig ▪ LTL Directional Drilling Services
Spencer Shand ▪ Coogar Sales & Services

ABSTRACT

The 96" pipe ram crossing of Highway 61 along Jarvis River near Neebing, Ontario was complicated when the contractor, LTL Directional Drilling, encountered a full face of concrete at the head of the 96" steal casing pipe. Unbeknown to the owner, designer, and contractor of the Jarvis River Crossing project, the original Highway 61 bridge abutment had been left in place when a new culvert and bridge were constructed. This paper will discuss the teamwork between the contractor (LTL Directional Drilling), owner (Ontario Ministry of Transportation) and blasting coordinator (Coogar Sales and Services) to advance through the 14 feet of concrete bridge abutment and successfully complete the project. Utilizing a non-detonating rock-breaking cartridge, the contractor was able to get permission from the Ministry of Transportation to blast directly under Highway 61 and do so without a burdensome impact to the traveling public above. It is noted that this stretch of highway is considered a vital link for several communities, while feeding imports crossing from the US/Canadian border (Pigeon River Border Crossing). This paper/presentation further discusses the use of Nxburst (rock-breaking cartridge) Safety Cartridges in trenchless applications when the use of high explosives is not permitted, and the protection of the public and surrounding structures is of the utmost of importance.

INTRODUCTION

The 96" pipe ram crossing of Highway 61 along the Jarvis River (the project) near Neebring, Ontario came to a complete stop when the pipe ram operations encountered an unknow obstruction 80 feet into the 161.5-foot drive. The Jarvis Creek pipe ramming operation is located in Neebring, ON near the intersection of Cloud Lake Road and Highway 61. This paper discusses the steps taken to assess the obstruction, produce an obstruction removal strategy, and finally the excavation of the obstruction and advancement of the pipe/casing to complete the job.

IDENTIFICATION OF OBSTRUCTION

The contractor on the project was utilizing TT Taurus hammer with a weight of 10,580 lb., and an approximate force output 20,000KN (4,496,178 lbf) when on July 11, 2017 the pipe ramming advance at the project was stopped by an unknown obstruction. Upon encountering the obstruction and understanding that the pipe would not be advanced further by ramming operations, the contractor proceeded with a casing cleanout via augers, remote skid-steer, and hand mucking of the excavation material in-order to reach the front of the casing pipe. When the pipe was cleared, the obstruction was observed to be a concrete wall from which approximately two feet of concrete had broken off and laid within the right side of the pipe. Figure 1 shows the obstruction as originally observed by the engineers brought to the site to assist the contractor. It is

Figure 1. Pipe obstruction investigation—initial assessment

noted that the obstruction was larger than the 96" casing pipe, and its three-inch-thick reinforced front cutting edge (98" O.D.). Shortly after understanding the impact that concrete wall will have on a pipe ram operation, the contractor set about to identify the extents of the obstruction and planning for removal.

The concrete obstruction consists of a vertical face and an angled face. The vertical face is from approximately 12:45 to below the invert at the 5:00 position. It is in this location that the two feet of concrete noted above broke free during the ramming operations. The transition line between the vertical face and the angled face was approximately 10 degrees from vertical. The angled face is sloped away from the vertical face in the direction of the pipe centerline at approximately 45 degrees.

Probe Drilling

To further identify the extent of the obstruction the contractor used a hand-held corded impact hammer to drill probe holes in the concrete wall utilizing three foot and six-foot drill rods. The holes were drilled along the vertical portion of the wall segment and into the angled portion of the wall. The holes in the vertical portion of the wall were placed approximately every 15 to 30 minutes radially, starting at 12:00 and ending at 3:45. A total of eleven (11) 3-foot holes were drilled in the vertical face at angles varying from 0 to 45 degrees. These three-foot probe holes did not penetrate past the concrete obstruction. Two six-foot holes were drilled perpendicular to the angled wall face. The holes were located above the pipe centerline on the left side of the tunnel. Again, these holes did not penetrate past the obstruction. Figure 2 shows the probe hole completed from within the pipe.

As the three- and six-foot probe holes did not penetrate the obstruction, a concrete drilling specialist was contracted in an attempt to core though the obstruction. A core (1⅝ inch in diameter) was utilized for this inspection phase. The core probe drilled was positioned at the spring line approximately 1.5 feet to the right of the centerline of the pipe and progressed parallel to the center line axis of the pipe. The core probe penetrated the obstruction at approximately 10 feet.

Vertical probe drilling from the surface was completed by another contractor in an attempt to define the limits of the obstruction. Initial indications from property maps indicated that the abutment would be 8.5 or 10 meters long. Luckily, during surface

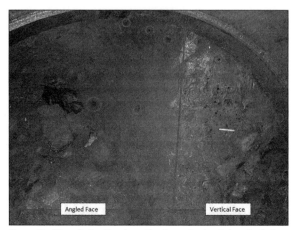

Figure 2. Pipe obstruction investigation—horizontal probe hole locations

drilling operations, discussed later in this paper, it was found that the obstruction was smaller or hit at an angle such that penetrating it with the 96" casing would be less than originally estimated. However, at the time, and without knowledge of the reduced size, additional three-foot probe holes were completed at very steep angles in an attempt to probe the extent of the obstruction. The steeply angled probe holes were drilled from the 12:00 to the 5:00 position along the perimeter of the pipe. While drilling at a steep angle, approximately half of the probe holes were able to penetrate the obstruction from within the tunnel, these were the first signs that a path forward through the obstruction may be possible.

Surface Drilling

The next step in the effort to fully identify the concrete obstruction was to locate the limits of the obstruction with a surface drill rig from the roadway above the 96" pipe ramming alignment. When a probe boring made contact with the obstruction, a core sample was collected to gather information about the vertical thickness of the obstruction. Figure 3 shows twelve (12) surface bores drilled for the investigation of the obstruction limits. Table 1 further describes the surface drilling.

Obstruction Extents

From the in-tunnel probing and surface bores, Figures 4 and 5 indicate what was estimated for the extents of the obstruction prior to excavation of the obstruction. The obstruction is believed to be trapezoidal in shape, being wider at the base than at the top, supporting the idea that the obstruction is an old highway bridge abutment. The angled wall that can be seen from within the pipe is the longer side of the trapezoid. The vertical wall is believed to be the short side of the trapezoid. The height of the trapezoid can be identified as 17 feet by the core sampling taken in SP-1. Also, from SP-1, it was found that approximal seven feet of concrete material is located above the crown of 96" pipe.

Obstruction Material

The obstruction consists of concrete with a weak cement matrix. The aggregate observed is consistent with older materials in which larger than current standard aggregate was used. Visual inspection of the material cored from surface probe

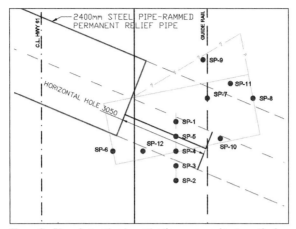

Figure 3. Pipe obstruction investigation—approximate vertical probe hole locations

Table 1. Probe hole descriptions

Surface Probe Number	Date Drilled	Drill Depth	Core Depth	Station	Offset from CL of Road
SP-1	17/07/15	11.1 m	6.1 m to 11.3 m	12+138.0	4.7 m
SP-2	17/07/17	7.5 m	NA	12+136.0	4.7 m
SP-3	17/07/17	11.6 m	NA	12+136.5	4.7 m
SP-4	17/07/18	11.1 m	NA (Auger chatter at edge of obstruction)	12+137.0	4.7 m
SP-5	17/07/18	9.9	9.9 m to 10.74 m	12+137.5	4.7 m
SP-6	17/07/18	10.5	NA	12+137.0	2.5 m
SP-7	17/07/19	6.4 m	NA	12+139	5.8 m
SP-8	17/07/20	7.6 m	NA Possibly skimmed the concrete	12+139	7.4 m
SP-9	17/07/20	8.5 m	8.5 m to 9.1 m No core	12+140	5.6 m
SP-10	17/07/20	6.4/11 m	Possibly hit bolder at 6.4 m No core	12+137.5	6.7 m
SP-11	17/07/20	6.7 m	NA	12+139.5	6.5 m
SP-12	17/07/20	8.2 m	8.2 m to 8.7 m	12+137	3.5 m

borings showed larger than normal quantities of aggregate material up to cobble size; the largest was approximately six inches in diameter. It was also noted that several different types of aggregate material were found in the core. A chisel hammer was used to remove a four-inch diameter cobble from the cement matrix by hammering at the cement surrounding the cobble. This method of removing aggregate from current standard concrete mixes is not generally possible. Due to the larger than normal aggregate, performing lab testing on the material was not seen as a viable option to determine concrete strength.

EXCAVATION OPTIONS

Numerous excavation options were investigated, such as jackhammering, hydro excavation and rock bursting but given the confined working space and need to complete

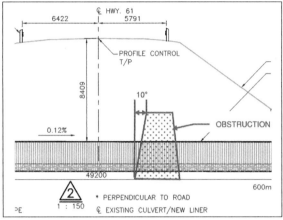

Figure 4. Cross section view along pipe heading

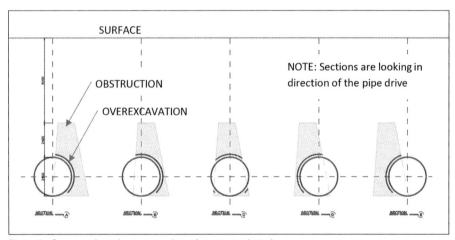

Figure 5. Cross section advancement through concrete obstacle

the work quickly, so that the pipe would not become locked in the ground, these options were not seen as viable. As the casing remains stationary the outer annulus lubrication system and surrounding material begin to settle, causing the movement resisting forces acting on the casing to increase. The contractor and the contractors engineer proceeded to investigate alternative methods. Additionally, the engineer worked with the contractor and submitted a plan to the owner outlying a plan continue bentonite lubrication of the casing pipe during the work stoppage. The applicability of Nxburst Safety Cartridges (rock-breaking cartridges) was initially introduced during the surface probe drilling with a phone call to Cooger Sales and Services in which all parties discussed how rock-breaking cartridge could be used safely within the confined spaces of the 96" casing while working under an active highway. After the initial phone call, it was decided that the deflagrating cartridges, which produce minimal vibrations, maybe the best option. Conversations were then held with the general contractor and the owner for moving forward with the rock-breaking cartridge excavation option.

The rock-breaking cartridge is classified as a 1.4S product under the United Nations hazardous substances classifications, so it does not require the same level of permitting and precautions necessary when using conventional explosives. On the Jarvis Creek project fracturing the material without displacement allowed for the safe pipe advancement while maintaining the ground support provided by the seven ft of concrete obstruction above the pipe crown. The cartridges are designed to be oxygen balanced with only harmless gasses produced. This minimizes the re-entry period and makes them more suitable for confined spaces like the Jarvis Creek project.

Rock-breaking cartridge is a self-contained cartridge that contains a non-detonating chemical compound which react quickly when ignited and produce a high volume of gases comprised mostly of water vapor, carbon dioxide, and nitrogen. This means that they are more stable than explosives, they produce a controllable pressure wave with low vibration levels instead of a destructive shock wave, and they produce minimal amounts of fly rock, dangerous gases, and dust. The functional difference between high explosives and Rock-breaking cartridge™ is the detonation following shock ignition whereas propellants are intended to burn at a steady rate determined by the design pressure of the confining structure after ignition by a flame. Rock-breaking cartridges provide an even buildup of gasses and achieves an optimal pressure to break the rock/concrete whereas high explosives release gasses in a violent event which produces a high level of fly-rock, shock waves and high overpressure. Since the gasses produced by rock-breaking cartridge achieve an optimal pressurization which results in optimal gas release at relatively low pressure the destructive effects of high explosives are negated.

When the cartridge is ignited, these gases expand rapidly to fracture the concrete obstruction. To use the cartridges in the simplest terms; a hole was drilled into the concrete, a cartridge is inserted, the hole is stemmed and then the cartridge(s) are initiated. Test holes were completed with individual cartridges to determine how the concrete would react and to help determine the appropriate powder factor, while production blast were completed with more than 25 cartridges at one time.

Rock-breaking cartridges were the method of choice to advance through the obstruction at the Jarvis Creek project. The cartridges carry less risk than conventional explosives and are more easily controlled. By working with Coogar Sales and Services, the contractor was able to bring in over 15 years of projects experience including the trench along Highway 17 in Cobden, ON and the Highway 400 Extension and many other projects. Rock-breaking cartridges also had the advantage to the Jarvis Creek project that it was already approved for use on MTO projects making conversation with MTO regarding using the rock-breaking cartridge on the project easier. Additionally, the suppler was able to provide training to the contractor's crew and the contractor's engineer on the use of the cartridges and appropriate safety procedures on the project. Additionally, the supplier also able to provide an onsite representative to assist the contractor during all blasting operations for the Jarvis Creek Project.

EXCAVATION OF OBSTRUCTION

The supplier and on-site consultant of the rock-breaking cartridges for the project has used the product to break down concrete on previous projects, their initial recommendation was to drill holes on approximately one-foot centers and place the cartridge in a drilled hole along the spring line of the tunnel. To start the process, one vertical line of holes was planned to be drilled into the angled face with the intent to break off the vertical face. See Figure 6 for the proposed hole pattern for the first stage. Note, this pattern was adjusted after the initial round by the rock-breaking cartridges consultant in the field. Later advancement patterns were determined by the supplier's onsite representatives in cooperation with the contractor and the contractor's engineer. This drill pattern purposefully created points of failure in the concrete within the excavation area of the obstruction. Perimeter holes also referred to as trim holes were drilled to limit fragmentation of the full obstruction leaving intact material bridging above the pipe for ground support. Such a method of drilling prior to igniting the cartridges is intended to break down the concrete to rubble. Once the obstruction is fragmented, the pipe was advanced through normal hammering operations after each blast round.

It is noted that work within the 96" pipe is considered a confined space work environment. Use of the rock-breaking cartridges for advancement of the pipe through the obstruction did not require increased ventilation or lighting beyond current procedures. However, a hanging system consisting of aircraft cable was installed on the inside shoulders of the pipe to decrease time required between mining operations and pipe advancement with the hammer. Ventilation remained in place when a cartridge array is ignited, and because the cartridges produce harmless gases, the standoff time after an ignition was short. A few minutes after an ignition, a technician examined the excavation site, from under the protection of the pipe, to ensure the ignition went as planned, and then hammering will commence through that section after mucking out excavated material.

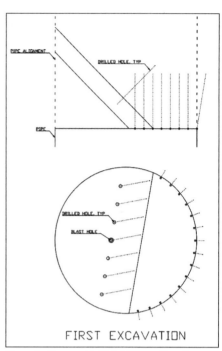

Figure 6. Proposed initial drilling pattern for use with NX Burst cartridges

Excavation of the obstruction with rock-breaking cartridges was started on July 27th, 2017 and continued for two and a half weeks. Initial charge arrays consisted of cartridges ranging from 40, 60, 80 and 120 grams. These charges were used to determine the amount of break that would occur with each charge level. Following the blasting and clearing the pipe of gasses with previously installed ventilation lines the obstruction face was inspected by the supplier's representative, contractor and contractor engineer. Figure 7 and Figure 8 shows the before and after photos of the first blasting operation. It should be noted that within Figure 8, the after photo, a larger pile of excavated material was found in front of the excavation face. However, this material simply fell off the face after the blast. Over the whole operation only a few pieces of the concrete made it to the ventilation fan placed about 25 feet from the heading at the time of most blasts.

The contractor and the contractors engineer were surprised to find an interesting aspect and the great benefit of using a deflagrant over a high explosive is the blasted material is mostly left intact after a blast. The material on initial inspection appears to be intact, however upon close review, micro cracks and fractures can be seen to be prevalent within the blasted material. When the material is left mostly intact it shows

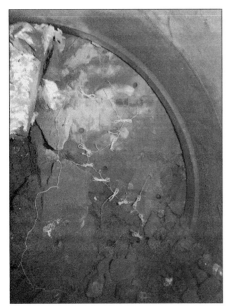

Figure 7. Before blast   Figure 8. After blast

how this method of excavation can be used safely in sensitive areas, as none of the material beyond the immediate planed excavation appears to receive any damage. When scaling the excavation face a hand-held electric jack hammer was used, and it was found that the chisel can easily be sunk at a full depth of the tool showing how the concrete was internally fractured in place.

A second day of blasting was completed on July 28th, 2017 in which 15 trim shots at 10 grams each were placed along the perimeter of the pipe and targeting an outcropping of rock which may impede advance. Once the trim shots were complete and the excavated material was mucked out of the pipe, the pipe was advanced 10.5 inches into the obstruction. It is important to note that this first advance further reinforced that the contractor would be able complete the job with the equipment on hand, as there was some great concern that the pipe may have become frozen in the ground as it had not been moved since July 11th(18 days earlier) This also shows the importance of continue lubrication, when possible, during times of work stoppage.

Over the next two weeks the contractor and the rock-breaking cartridge specialist advanced the 96" carrier pipe using the rock-breaking cartridges mining technique through the concrete obstruction utilizing a total volume of 17.69 kg NEQ of deflagrant. Table 2 summarizes the mining activities through the obstruction.

CONCLUSION

On Monday August 14th, 2018 casing advancement progressed past the known limits of the obstruction, thereby marking the successful overcoming of the bridge abutment. The utilization of rock-breaking cartridges was detrimental in the process of obstruction removal. By having the ability to control the safety cartridges degree of energy release the rock was fractured successfully without creating unwanted overcut, and the deflagrating nature allowed for the neighboring Highway 61 to maintain a safe flow of traffic while mitigating interruption. The project itself incorporated a variety of

Table 2. Description of progress of activates at Jarvis Creek

Date	Number of Charges	Total Charge (grams)	Advance (inches)	Activity and Notes
Thursday, July 27, 2017	13	980	NA	Drill and Blast
Friday, July 28, 2017	15	150	NA	Muck, Drill and Blast Trim Shots
Saturday, July 29, 2017	NA	NA	10.5"	Muck and Advance
Monday, July 31, 2017	15	2220	NA	Drill and blast
Tuesday, August 1, 2017	NA	NA	NA	Scale heading, Muck in anticipation of Advance
Wednesday, August 2, 2017	16	2260	17.75	Drill and Blast trim Shots, Scale and Muck, Advance
Thursday, August 3, 2017	28	4520	NA	Muck, drill, Blast, scale
Friday, August 4, 2017	40	400	33.5	Drill and blast trim shots, Advance, Drill
Tuesday, August 8, 2017	59	4440	NA	
Wednesday, August 9, 2017	NA	NA	18	Advance breached obstruction on right side of pipe face from 1 o'clock to 5 o'clock
Thursday, August 10, 2017	15	1980	NA	Drill and Blast, Trim shots
Friday, August 11, 2017	18	740	18	Drill Blast, Advance, Scaling, Drill and Blast
Saturday, August 12, 2017				no work
Sunday, August 13, 2017				no work
Monday, August 14, 2017	NA	NA	36", 18 ft	Advance through Final section of obstruction, and Complete a 19-foot advance

specialists carrying unique backgrounds. The atypical set of circumstances carried by the project required a multifaceted problem-solving approach. All of the engaged parties have previous experience involved on projects similar in scope, but the combination of blasting throughout a ramming operation while beneath a live highway resulted in a significant deviation from each industries norm, proving to be quite the challenge in adapting each skillset. Initial assessments of the obstruction investigation findings lead to multiple hypotheses regarding project completion feasibility. Notions of project abonnement and the remediation of completed progress were entertained considering that the potential risks of obstruction removal were substantial. The contract owner \ ultimately took a chance on the confidence of the awarded parties' ability to complete the project as originally scoped. The task of effective project completion encompassed three primary aspects with regards to overcoming the obstruction. Said aspects were defined to be; consideration for the safety of the workers and public, adhering to original contract specifications (including remaining a trenchless operation), and minimizing financial burden. Through interdisciplinary corporation and the pooling of resources the obstruction was overcome to the satisfaction of all involved parties, yet again proving the effectiveness of cooperation and teamwork.

Unexpected Underground Obstructions: Challenges and Lessons Learned from Angeli, the Regional Connector TBM

Christophe Bragard ▪ Traylor Bros Inc.
Bryan Hadley ▪ Traylor Bros Inc.
Mat Antonelli ▪ Los Angeles Metro

ABSTRACT

TBM Angeli bored two tunnels under downtown Los Angeles as part of the LA Metro's Regional Connector Project. While tunneling the first drive, the TBM encountered several unexpected obstructions: two steel piles (21×132) and 48 tie backs. While the TBM cutterhead powered through the steel and literally "chewed" the obstructions, the machine did not process (or "digest") the shredded steel as easily: the screws got stuck at multiple occasions and the shaft on screw #2 broke. Through great efforts of the tunnel team and because of fine collaboration with the owner, the project managed to minimize delays and even get ahead of schedule by the end of the second drive. This paper discusses the chain of events related to these obstructions as well as the lessons learned from this experience, both technically and contractually.

INTRODUCTION

The Regional Connector Transit Corridor project (Contract No. C0980) is a Los Angeles County Metropolitan Transportation Authority (Metro) design-build, light rail underground project that will connect three existing rail lines in Los Angeles. The 3.1-km alignment with a base contract value of $1 billion is located under the heart of downtown Los Angeles. It will provide connections between the Metro Gold Line, Metro Blue Line and Metro Expo line, providing direct routes from Long Beach to Azusa and East Los Angeles to Santa Monica. As shown in Figure 1, the project includes the construction of three underground stations with depths ranging from 40 to 110 feet, approximately a mile of twin tube tunnels, a mile of cut and cover tunnels, a 300-feet-long SEM tunneled cross-over cavern, and system-wide elements including track, traction power, train control, and communications. The running tunnels are constructed with a 6.59 meter-diameter earth pressure balance (EPB) tunnel boring machine (TBM). For broader information on this project see Hansmire and Roy (2014), and Hansmire, Roy, and Smithson (2015).

TIE-BACK REMOVAL SHAFT

In most areas, the tunnel alignment cleared existing known obstructions, but in one location, located on the south east side of the 3rd and Flower Street intersection, a conflict with existing tiebacks was unavoidable and removal of temporary tiebacks was deemed necessary. The conflicting tiebacks were located up to 57 feet below existing grade and impacted 80 feet of the right tunnel alignment.

The base contract scope required a tieback removal shaft and adit that was successfully constructed in 2016 to clear the future TBM tunnel path (Kusdogan et al., 2017). While the majority of the anticipated tiebacks were located in general agreement with the model predictions, some of the tiebacks deviated substantially from their expected

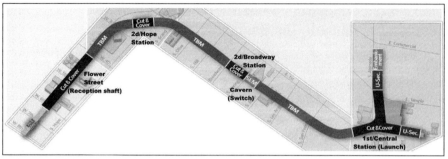

Figure 1. Project alignment

location (record drawings) which required that additional pocket excavations to be performed. A total of 27 tie-backs were removed from the right tunnel TBM alignment.

PROJECT START-UP

Angeli is a Herrenknecht completely refurbished EPB TBM that had previously been used on the LA Metro Gold Line Contract C-0800 and the Seattle Sound Transit U220 contracts. One of the specificities of the machine is that it is equipped with 3 long continuous screw augers (for a total 200 feet length) that provides better controls over the EPB pressure and the gasses entrapped in the ground.

The TBM first drive started on February 6th, 2017 with the most challenging launch Traylor Bros had ever attempted. The tunnel team faced the combination of a required demonstration zone with minimal cover and then proceeding tunneling under a sensitive building with very low cover in well graded sands. See Hansmire et Al, 2017 for details.

Once it passed the sensitive start up zone, Angeli quickly picked-up speed in the Fernando formation (soft claystone) heading towards the location of the future cavern and 2nd/Broadway Station which was to be excavated after both TBM drives passed through the station. Then, the TBM was to brush under the existing Red Line (Choi et Al, 2018), take a turn under the Broad Museum and Disney Concert Hall and breakthrough into the 2nd/Hope Station, which is the deepest Metro station in Los Angeles. However, due to unforeseen circumstances, things did not go as expected.

THE OBSTRUCTIONS EVENTS

On April 11th, 2017 at 11:30 am, about 20 feet from the 2nd/Broadway Station support of excavation (SOE), the TBM hit something hard, maximizing the cutter head torque. The tunnel team[*], expecting some isolated obstruction (most likely a tie-back) and decided to try and mine through the hard spot. Angeli proceeded forward with 2 tough advances (10 feet total), while rotating the cutter head clockwise and counterclockwise while noticing some distinct metallic sounds. Each advance took about 65 minutes, about twice as long as the 25 minute average advance experienced in the previous weeks. By 3:30 pm, the advance rate returned to a more consistent 30–35 minute advance rate, butat 4:00 pm, steel pieces and a couple of cutting knives from the cutter head started to get released at the back of screw #3 discharge, caught in the

---

[*]The tunnel team consists of the contractor personnel as well as owner's representatives. Daily coordination between parties ensured transparency and fast response to the unexpected events.

Figure 2. Steel fragments collected from the mining (left); pile section recovered in front of TBM (right)

so called "cheese grater." This grater consists of a grizzly bar grill that separates the ground into smaller fragments to optimize transport on the conveyor belt.

On April 19th, 2017, after mining 340 feet since hitting steel (6 days of mining at a regular advance rate of 30–35 min./advance, with regular collection of steel pieces at screw #3 discharge), the tunnel team parked the TBM in a safe zone within the limits 2nd/Broadway Station, which was yet to be excavated. For the next 2 days, the mining crews replaced about 40% of the cutter head tools (only 3 knives had been ripped off during mining though) and removed several fragments of steel stuck in the cutterhead. One of the steel fragments of jammed steel included an 8 feet long deformed steel (see Figure 2). This piece had to be cut in smaller sections in order to be removed through the grizzly bars and its size seemed to indicate that the TBM had intercepted a W24 or similar steel pile. The contact and impact of the steel under high load had chipped many of the carbide inserts of the cutters, cutting knives and buckets.

On May 15th, 2017 (24 days and 1,500 feet of mining since hitting steel), a strange deformed and highly polished fragment of steel was retrieved from the screw #3 discharge (see Figure 3). The shape, surface and details (corners) of the specific piece seem to indicate that it had been stuck in one of the screws for a significant amount of time.

On May 22nd, 2017 (29 days and 2000 feet of mining since hitting the steel), the drive pressure of screw #2 suddenly dropped, while the EPB pressures in screw #1 rose drastically. After quick visual inspection through screw #2 access hatches, it was found that the 8-inch diameter shaft of screw #2 had broken near the rear end of the screw (see Figure 3). The tunnel team gathered up all efforts and fast tracked the design, fabrication and installation (welding) of a bolted double flange repair on the apparently sheared shaft near the screw drive. Access to the damage area was difficult, yet the team succeeded in welding the flange through the rear access hatch in 5 working days working around the clock, and then resumed the mining operations.

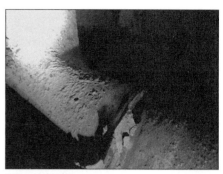

Figure 3. Polished steel piece recovered (left); screw #2 break (right)

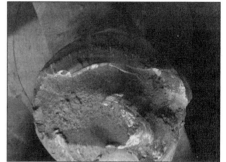

Figure 4. Break through at Hope station (left); screw #2 second break close up (right)

On June 1st, 2017 (30 days and 2140 feet of mining since hitting steel), Angeli successfully broke through into the 2nd/Hope Station which is 110 feet deep located in the Bunker Hill area of Los Angeles (see Figure 4). The screw repair was checked with dye-penetration , and looked in good condition. To confirm that no additional steel was stuck in the screw, a pod was travelled through the screws and found there were no remaining pieces. After reviewing the options and associated risks, the tunnel team decided to proceed with the 1100 feet of mining that separated the TBM from the retrieval shaft. The TBM was raised, skid across the 340 feet long station, lowered back into position and re-launched on June 14, 2017.

On June 22nd, 2017 (35 days and 2565 feet of mining since hitting steel), screw #2 broke for the second time. Using the identical design as the first repair, it took the crews 8 working days, again working around the clock, to perform the repair.

On July 5th, 2017, the TBM resumed mining during the A.M. By 9 pm, a metal piece lodged in the screws, creating high torques. The TBM operator, activating the various screws clockwise and counterclockwise, managed to avoid the screws from getting jammed. One hour later, a 1⅜" diameter tie-back rod and 8"×8"×2" steel end plate was retrieved from the screw #3 discharge grate. A total of 48 tie-backs ended up being intercepted by the TBM during the following weeks.

On July 7th, 2017, screw #2 got stalled when a tie-back plate was jammed between the screw auger and the casing. By trial and error and exposing the screw at each available hatch, the blocking debris was located and removed. The TBM was stopped for about 24 hours for repairs, before resuming its course.

Figure 5. Tie-backs and grout pieces collected at screw discharge (left); tie-back anchor wedged between screw flight and casing (right)

On July 10th, 2017, screw #2 got stuck again, this time by a tie-back rod.

On July 11th, 2017, it was screw #1 that got jammed. This time the removal process got more complex as the screw was extended forward into the excavation chamber. Again, the TBM was immobilized for 24 hours prior to resuming its course. During the retraction of the screw with additional jacks, the flight of the screw that was jammed by the tie-back section got bent and torn, meaning that screw #1 was also confirmed to be damaged.

On July 14th, 2017, the project encountered significant heave, greater than 2 inches, of the asphalt street above the TBM's location. Soon after, the asphalt surface began to crack and conditioned muck and foam found its way to the surface. It was determined that the EPB chamber material had made its way to the surface through an improperly abandoned (grouted) borehole performed under a separate contract. Thankfully, careful advance, with reduced EPB parameters, allowed the TBM to pass beyond the borehole by the end of the day, while the street clean-up was ongoing.

On July 17th, 2017, the repair on screw #2 broke again. This time the bolts at the flange connection broke which was planned by design. As a result, the repair procedures took only about 8 hours, instead of the 8 days.

On July 19th, 2017, after intercepting a steel pile[*] and 48 tie-backs, overcoming 3 screw breaks and 3 screw jams, Angeli finally broke through the Flower Street reception shaft which was a 60 feet deep soldier pile excavation hidden under the streets of downtown Los Angeles. The machine was then ready to be dismantled, transported back to the launch shaft, necessary modifications and repairs made on the machine in order to begin the 2nd drive, as soon as possible.

---

*See Epilogue for further details

DAMAGE MECHANISMS AND REPAIRS

Serrated Tools

At first, the tunnel team was definitely impressed by the somewhat unexpected power of the TBM. The machine had literally chewed through a relatively large steel pile. While the power of the machine certainly contributed to this success, this would probably not have been possible without the high performance of the cutter head tools. The disk cutters selected by Traylor Bros for the EPB machine are high performance tools that contain carbide button inserts. These inserts provide much better resistance to wear in soft ground conditions and ensure that the disk keep rotating while mining in stiff and sticky material (avoiding the flat disk syndrome). The intent for such selection was to minimize the #-number of cutter head interventions required. When the TBM encountered the pile the carbide inserts disks acted as serrated knives which, under the high thrust of the machine, imprinted the steel to gradually shred and finally slice the steel.

The downside of those inserts is that they are brittle and do not handle shocks well (e.g., hard rock or, in our case, steel). Therefore, while they managed to get the TBM to push through the obstructions, about 40% to 60% of the cutter head tools had to be replaced after each event (pile/tie-backs).

Screw Repair

When pressure built up in screw #2, the project team quickly concluded the most likely explanation for the minimal screw torque, and lack of material flow was a broken auger stem. Intuition suggested that the most likely location of the suspected break would be near the drive connection at the back of the auger stem, where the highest torque loads would be realized. Fortunately, the screw casing included a maintenance hatch immediately up-stream of the drive connection which would allow expedient access to this area. The break was found quickly, being located at the welded joint in the stem of the screw. While work was underway clearing the screw, and establishing access to the repair area, the team designed a temporary fix that could be installed in the. A number of concepts were discussed among the TBM Mechanics, Project Engineers & TBM manufacturer. The final repair concept chosen was a pair of custom flanges, which would be circumferentially welded to the auger stems, on either side of the breaks, and cross bolted to reconnect the two broken pieces. The flanges were

Figure 6. Serrated tear of the steel (left); carbide imprints in the steel pile fragment (center); carbide buttons on the cutters, some broken (right)

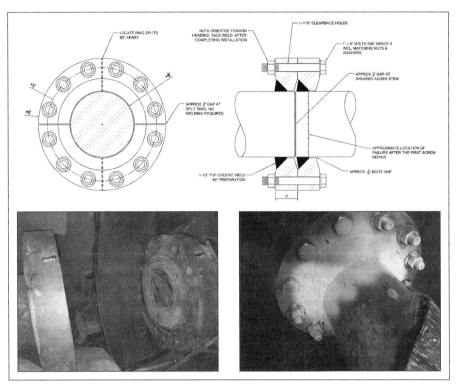

Figure 7. Screw repair design details; weld on bolted flange

designed with single bevel groove welds, so that weld access would only be from the hatch, and were each split into halves, so they could be installed without having to move the auger longitudinally.

The initial design of the repair had weaknesses which were understood beforehand: first, multiple bolts sharing a direct shear load does not represent best practice for a mechanical shear connection, and, secondly, the cramped welding conditions would not properly facilitate sufficient pre-heat and inter-pass temperature for the welding operation. These design deficiencies were accepted as necessary risks, the repair was implemented, and tunneling re-started with new drive pressure restrictions placed on the screw drives.

The repair held for approximately 140 feet of tunneling, until break-in to the Hope station excavation. At that point, the project team was faced with a decision of whether to re-launch, and complete the final 1100 feet of tunneling for the first drive—relying on the emergency repair; or to stop, and fully re-furbish or replace the auger during the station crossing. While the former option carried the risk of further extend critical path delays, in the event of a failed repair prior to the completion of the drive, the latter option required over a week of delay to perform the necessary refurbishment during the station crossing. The project team chose the calculated risk of pressing forward without a full refurbishment, knowing that the refurbishment could be made off critical path during machine extraction, and, if required, additional temporary repairs could likely be made in less time than would be required for a full refurbishment.

Roughly 400' after the re-launch, the repair failed. The break was not as clean as it was the first time, in that part of the repair and part of the original shaft were torn apart. A new approach was to the field repair was developed, in which the casing of the screw was separated to give better access to the repair area for cleaning, inspection and weld access. A 6" section of the screw shaft was fully removed to ensure the new repair would be performed on sound material. With the improved weld access, the design of the repair flange was improved to include a more robust weld design, procedure & inspection. The new repair was detailed such that the bolts would deliberately shear before the weld or stem failed.

Right after mining through the roughly 150' long stretch of the alignment where 48 steel tie-back were unexpectedly encountered, the repair failed a third time, but as planned the bolts broke in tension. This was contrary to what was originally expected, yet it would confirm the failing process of the screw. The flanges were re-bolted in a matter of hours, and mining continued until break-in to the Flower St. Cut & Cover Excavation.

Screw Break Mechanism

While evidence pointed towards a relationship between the steel ingestion and the screw break, the Owner (Metro) expressed reservation on linking both events. During subsequent meetings between the owner and contractor it was determined that more investigations were needed to evaluate the cost and schedule impacts.

RCC hired expert Snyder Engineering to analyze the facts and come to a better understanding of the failure of the screw. Snyder quickly identified from the broken fragment details that, contrary to what the tunnel team had thought all along, the failure was not a direct result of excessive torque, meaning in shear, but a bending fatigue failure due to eccentricity caused by the wedged fragment.

While the screw shaft (stem) is subject to bending stresses under normal conditions due to tolerancing and auger clearance in the casing, a steel piece wedged in the screw increases this bending stress. The closer the wedge is to the back of the screw, the higher the stress. There were clear traces of worn, gouged and deformed steel in the outer surface of the screw flight right near the break, showing where the wedge steel piece had been trapped. Under such conditions, calculations showed that it would not have taken many revolutions of the screw under the stress caused by such eccentric deflection for it to progressively develop fatigue cracks until the remaining

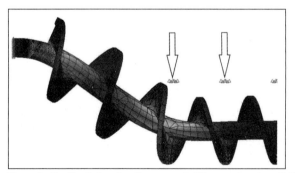 

Figure 8. Bending induced on the screw shaft by a wedge piece (left); gouged screw flight (right)

portion of the shaft could no longer transmit the torque required to rotate the screw and fail.

The mechanism and calculations provided by Snyder were further confirmed when the 3rd break occurred, in the bolts. The bolts too showed evident fatigue failure in tension first. When a sufficient number of bolt shanks worked their way out of the coupling plates the remaining bolt shanks sheared between the plates.

Screw Jams

As the TBM started intercepting the tie-backs, there were no distinct signs on the cutterhead torque nor any noise that would warn the operator that such obstruction had been hit. The first distinct signs consisted of increased torque in one of the screws. Tiebacks rods and plates were partially rusted and/or decomposed or got ripped/cut by the TBM into a real wedge that could easily get jammed in the space between the screw auger and casing. In most cases, the operator would be able slow down and/or stop the advance, eventually reverse the screw and "unjam" the screw from the tie-back fragment. On three occasions however, the screw got completely jammed.

The first occasion was on screw #2, when after fighting with one of the pieces in screw #1, screw #2 came suddenly to a complete halt. After opening some hatches, some attempts were made to unlock the screw by welding "dogs" to the flights and using small jacks to complement the thrust of the screw drive, without success. Then, by observation on position of the screw in the flights, the likely location of the wedged piece was acknowledged to be towards the front of the screw. The tunnel crews cut a hatch in the joint between screw #1 and #2, emptied out the muck, and quickly found the wedged piece. By the time the casing opening got welded back and the machine was able to restart, 24 hours had passed.

Another screw jam occurred on screw #1and this time the risks were much more significant because the screw was extended in the plenum. The screw needed to be retracted and the guillotine doors closed to isolate the screw from the pressurized plenum before any of the hatches could be opened in the casing. As the screw hydraulic jacks would not provide sufficient power to retract the screw, the tunnel team welded some brackets on the screw structure in order to add some external jacks to help with the process. With some screeching noise, the screw started moving until it was completely retracted. The guillotine door was almost completely closed, but for a 1½" remaining open gap because a tie-back rod was stuck in the gate. Pressure was released from screw #1, fortunately no material flowed out of the opened valve. A similar approach to screw #2 was conducted to extract the wedged piece out. After

Figure 9. Screw #1 jam wedge piece (left); access hatch (center); bent and cracked flight (right)

cleaning the flights and extracting the wedge, material could be observed slowly oozing out of the guillotine though. It was also noticed that the flight of the screw that was jammed by the tie-back fragment was strongly bent and had cracked under the higher load provided by the external jacks. Prior to resuming tunneling, the crack was welded back, as was the cut-off opening in the screw casing.

PROJECT IMPACTS

At each step of these events, while the operators and miners worked hard to nurse the machine through the obstacles, RCC and Metro worked together seamlessly and coordinated all project decisions. Whether or not to proceed forward after breaking through the 2nd/Hope Station or to air freight new screws for the machine's second drive, the tunnel team (owner and contractor) worked transparently together to minimize cost and delays. This allowed for fast decisions and repairs which minimized project impacts as the TBM mining was on the critical path.

While some of the pile and tie-backs encounter were immediately acknowledged to be deferring site conditions, the breaking of screw required some further analysis and explanation before all agreed that the TBM had chewed the steel pile but did not digested it well, and a piece of steel remaining stuck in screw #2 had caused it to break by bending fatigue.

The direct and indirect costs incurred resulted in a total change order of $7 million and 1 month of delay to the mining operations. Given the potential consequences these events could have caused, the cost and schedule delays remain minimal. This relatively quick settlement on the topic allowed the tunnel team to remain focused and keep the project on track and the tunnel team decided to switch the shifts from 2×10 hours shifts to 3×8 hour shifts, allowing continuous mining around the clock.

SECOND DRIVE

While excavation of the first drive was nearing completion, preparations for the full refurbishment of the TBM screw conveyor were already underway. After the steel tie-backs were encountered in the bored tunnel alignment, the project team committed to procuring brand-new replacement augers for all three of the screw conveyors in the system. Luckily, Traylor owned a spare screw #1 and the project was able to pull screw #2 and #3 from one of the Metro West Side Extension Project machines (a Skanska-Traylor-Shea JV project), modify them slightly, and install these screws before the second RCC drive.

As soon as the TBM broke through at Flower on the second drive, the tunnel team started the dismantling of the shield pieces. Concurrently, the TBM trailing gear (including the screw conveyor) was separated from the shield and reversed to the launch site, through the previously excavated left tunnel. Attacking the disassembly/reassembly from two ends also allowed for expediting the required repairs. The screws were removed from the gantries at the intermediate 2nd/Hope Station and replacement of the screws occurred quickly, despite some unexpected dimension issues that required the removal of wear bars in the screw casings.

Through relentless work in the summer of 2017, the team managed to start the second drive by the end of September, already gaining back some of the lost schedule. Equipped with the knowledge and experience of the first drive, having switched to a continuous mining 5 days a week, the team got the project back on schedule and then started to build project float in the schedule. The mining crews, working tightly

Figure 10. Archeological encounter in Athens (left); removal of expected steel piles in Hong-Kong (right)

together, broke records of 190 feet advance in a day, as well as 750 feet in a week to complete the mile-long excavation in about 3.5 months. This time, only 2 tie-backs were encountered in the vicinity of the tie-back removal shaft, yet this time not causing any damage nor delays. The 2nd drive was able to build 2 months of float back into the revised schedule.

OBSTRUCTIONS—RISKS AND MITIGATIONS

It is relatively uncommon to have TBMs encounter unexpected underground obstructions during their drives, yet a series of cases have been recorded: examples include old archeological remnants (jars) in Athens, boat anchors in Cairo, water wells in LA or even a utility tunnel in the UK.

In certain cases, the existence of the obstruction may be known in advance of the drive. In such cases, it may be decided to remove those prior to the tunnel drive (e.g., the tie-back removal shaft as per section above) or to plan for specific interventions in front of the machine (e.g., piles removal during face interventions in Hong Kong).

As Built Information and Accuracy

It is of course preferable for all parties to have accurate knowledge of the existence and location of any obstruction. This highlights how essential the gathering of information and as-builts are for the success of future projects. On the Regional Connector, the existence of piles was undocumented and therefore unknown. The presence of the tie-backs, however, was known and documented, yet the information was limited and proved to be inaccurate. From the quantity and lengths of tie-backs intercepted by the TBM, it was observed that fewer tie-back, yet longer ones had been used compared to what the (rudimentary) records showed.

This reinforces the need for detailed as-builts records on all projects, public and private (see also similar inaccuracies mentioned in Kusdogan et al.). In the present digital era, accurate records should be easily compiled in data bases using BIM and similar tools and passed on from project to project.

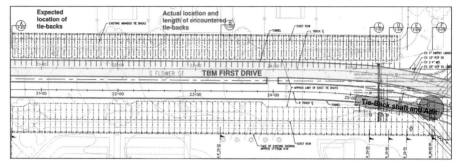

Figure 11. As-built of tie-backs (black) vs. actual tie-backs (red)

Detecting Obstructions

Ultrasonic or electromagnetic detection systems have begun being implemented on several TBM's and provide data of the quality of the ground ahead of the machine. Such system could probably have been used on this project. The complexity of the underground systems in downtown Los Angeles should not be underestimated. On the Regional Connector, the TBMs drove within 2 feet from known strand tie-backs, bridge piles, and a sewer manhole. Such detection systems may not have been accurate enough and may have created false alarms.

Contract Language Views

With the increase of tunneling in urban areas, potential unknown obstructions will constitute an increasing risk to the projects. While it would seem safe to put the onus on the contractor from an owner's perspective, it would also result in an increase in bid price. The authors view is that contractual language should prepare the Contractor for such possible obstructions in minimal quantities, while providing additional compensation if numerous obstructions are encountered.

The following example extracted from the Crossrail bid documents (London) highlights such thought: "The foundations of four (4) structures will be encountered within the tunnel face at unknown locations along the length of the running tunnels in Drives Y and Z. These foundations will consist of up to six (6) per location, unreinforced concrete piles of maximum diameter 900 mm…"

Risk Management

The probability of encountering unexpected obstructions during tunnel drives should be addressed jointly between owner and contractor in the project risk management plan.

On Flower street, 27 tie-backs were removed from the tunnel path on the south side of the street by excavating a shaft and adit prior to the TBM drive. Cost: about $1.5 Million

On the North side of the street, 48 tie-backs were intercepted by the TBM. Cost: about $1.25 Million. The risk incurred on the latter case however was much more significant. Had the tunnel team not been able to retract screw #1 when a tie-back jammed it, the project would have had much bigger operations required: possibly emptying the cutterhead chamber manually and under compressed air; or sinking a temporary shaft from the surface for example.

It is therefore important to select solutions wisely, based on the overall risks involving safety of the workers, safety of the public, cost and schedule.

Figure 12. Obstructing piles uncovered during the excavation of the SEM cavern

EPILOGUE

In March 2017, while installing canopy tubes for the start of the SEM cavern, 3 tubes had to be abandoned as they encountered a steel obstruction during the drilling/installing process. While this was in line with the expected pile obstruction location of the TBM, the piles encountered by the canopy were closer to the station than the one recorded with the TBM.

In June 2017, the excavation of the cavern started, and after 5 rounds of excavation, a steel pile was exposed, a W21×132. This pile (#1) had not been intercepted by the TBM as it stood 2 feet short of the cutter head path. The bottom of the pile was V-notched in both directions, a technique that was used when driving soldier piles, yet forbidden for many decades in downtown Los Angeles. This indicated that the obstructing piles must have dated back to the construction of the old Caltrans building (1949).

As the excavation progressed, pile #2 and #3 got exposed (and thankfully, it stopped after 3), and to the surprise of all, both piles showed torn, bent and shredded tips. Angeli, the Regional Connector TBM had not encountered and driven through one pile, but two.

REFERENCES

Choi, Crow, Baker, Drake, Jolly and Bragard, *TBM Passing under Existing Subway Tunnels in Los Angeles, California*. NAT 2018, Washington, D.C., USA.

Hansmire W.H and Roy, G. 2014. *Design-Build Tunnel Contract for Los Angeles Regional Connector Project*, North American Tunneling, 2014 Proceedings, Los Angeles, USA.

Hansmire. W.H., Roy, G. and Smithson, M.F. 2015. *Urban Challenges of the Downtown Los Angeles Regional Connector Tunnel Project*. Rapid Excavation and Tunneling Conference, SME, New Orleans, USA.

Hansmire, McLane, Frank and Bragard, *Challenges for Tunneling in Downton Los Angeles for the Regional Connector Project*. World Tunnel Congress 2017, Bergen, Norway.

Kusdogan, Brodbaek, Penrice and Bragard, *Removal of Interfering Tiebacks Using SEM in Advance of TBM Mining on the Regional Connector Project*, RETC 2017 Proceedings, San Diego, USA.

PART 5

Drill and Blast

Chairs

Charlie Schoch
Skanska

Ben McQueen
FKCI

Applying Lessons Learned and Taking on New Challenges at the Lake Mead Intake No. 3 Low Lake Level Pumping Station Project

Erika Moonin • Southern Nevada Water Authority
Jerry Ostberg • Parsons
Jordan Hoover • Barnard of Nevada, Inc.
Stefano Alziati • North American Drillers, LLC

ABSTRACT

As the drought continues in the Southwestern United States, work continues on the Lake Mead Intake No. 3 system with the construction of the Low Lake Level Pumping Station Project in Southern Nevada, USA. The project includes underground drill and blast construction of a 535-foot-deep Access Shaft, a 32-foot-wide by 34-foot high Forebay Cavern, and large-diameter drilling of 34-each, 91-inch-diameter by 500-foot-deep well shafts. This paper covers how this critical project applied lessons learned from previous adjacent underground construction, how alternative and collaborative contracting methods were implemented, and how the challenges of difficult ground were addressed for both the underground drill and blast work and the large-diameter blind bored well shafts.

INTRODUCTION

Southern Nevada and the greater Las Vegas area gets 90 percent of its water supply from Lake Mead. Severe and persistent drought conditions over the past 18 years have resulted in the dramatic lowering of Lake Mead, the largest reservoir in the United States, by more than 140 vertical feet.

The Southern Nevada Water Authority (SNWA), a nonprofit, cooperative government utility formed in 1991 to address Southern Nevada's water needs on a regional basis, responded early to these drought conditions. SNWA's proactive approach has included extensive conservation efforts that have helped reduce the per capita water use in the Las Vegas valley by more than 40 percent even as the population significantly increased, and construction of major infrastructure.

In 2005, SNWA approved the Lake Mead Intake No. 3 project, which included a deep three-mile-long tunnel beneath Lake Mead, a new intake riser, and a connector tunnel. The project also included a pumping station component that was deferred in 2008 due to economic downturn. This $817 million project was designed and constructed over a 10-year period and successfully commissioned in December 2015 to ensure a reliable and high-quality water supply for the drought-stricken communities of Southern Nevada. By all accounts, the Lake Mead Intake No. 3 project was outstanding and extraordinary in many significant areas. The team coped with a long list of significant risks associated with working deep beneath the largest man-made lake in the United States. The team faced water pressures much higher than encountered in any previous tunnel project in the world. This effort was further complicated by the mixed and uncertain geology of the area, and the need for the workers and machinery to be able to adapt and overcome these extremely difficult conditions.

However, by 2014 the reservoir's surface water level continued to fall to its lowest level since the completion of Hoover Dam in 1936. Consequently, SNWA faced severe operational challenges with its existing pumping stations because they become inoperable at lower lake levels. Discussions to pursue the new pumping station began, and in 2014 the SNWA Board of Directors approved the Low Lake Level Pumping Station (L3PS). Design of the L3PS commenced in 2015 to ensure the operability of the water delivery system despite the continued drop in lake levels.

The L3PS, located on Saddle Island at Lake Mead near Boulder City, NV, includes a 530-foot Access Shaft, a large underground Forebay cavern, a tunnel, 34 deep precision-drilled well shafts 6 feet in diameter, and a discharge aqueduct system that will tie the new pumping station into both a low-lift and high-lift transmission system to convey the raw lake water to the SNWA's two existing water treatment facilities. L3PS is located in some of the same highly variable, highly fractured and faulted geology as Intake No. 3, creating many challenges for the underground work. In addition, large submersible pumps will be installed in the deep well shafts and the pump manufacturer specified tight tolerances for their pumps to operate.

SNWA chose the Construction Manager at Risk (CMAR) delivery method to facilitate early contractor involvement in the design process and an expedited project schedule, as time is critical. Through early contractor involvement, the team could work together to address the significant challenges and risks anticipated on the $650 million project. The benefits of the CMAR agreement include a unique collaborative effort among owner, engineer, program manager, contractor, and key subcontractors that results in the best design approach; preferred underground construction methods; and a favorable project-wide culture of problem solving and teamwork among all parties involved. SNWA was the project owner and future operator of the pumping station. Parson was selected as SNWA program manager. MWHill A Joint Venture was the Design Engineer. Barnard of Nevada, Inc. was the selected CMAR which self-performed the underground drill and blast work. North American Drillers, LLC was the selected subcontractor for the well shaft drilling.

In the following sections, this paper covers the methods implemented for underground drilling and blasting and well shaft drilling. Authors explore the many technical challenges of the drill and blast excavation and the blind bore drilling method in the highly fractured/unstable ground condition and share how those challenges were overcome.

UNDERGROUND DRILL AND BLAST

To excavate the Access Shaft and underground Forebay cavern for the L3PS, the contractor used a drill and blast method with multiple benches to excavate the shaft and cavern/tunnels. Described herein is the background to the drill and blast approach and how it was implemented on at the Project.

Distances to Surrounding Structures

The nearest surrounding surface structures to the Access Shaft include the previously constructed Surge Shaft, within 430 feet; the contractors field office, within 404 feet; the temporary electrical substation, within 73 feet; and the 34 well shafts, the nearest of which is within 20 feet.

Blast Designs—Orientation, Relief, Timing

The contract stipulated the following requirements for blasting:

The primary objective for blasting rock is to construct excavations where the rock outside of the excavation will be undisturbed, nearby facilities and equipment will be unaffected, and the shape of the excavation will conform to the lines and grades indicated on the Drawings. It is the responsibility of the contractor to conduct his operations in such a manner that this objective is achieved.

Where blasting is used for excavation, the best modern practice of a controlled blasting method shall be employed. Acceptable controlled perimeter techniques include smooth wall, cushion, pre-splitting, or line drilling blasting. Controlled blasting refers to the controlled use of explosives and blasting accessories in carefully spaced and aligned drill holes to produce a uniform free surface or shear plane in the rock along the specified surface.

During excavation of the Access Shaft, the contractor drilled all blast holes vertically. Relief holes at the center of the shaft provided relief. Both millisecond timing delays and long period were used. The "Test Blast Program" and "Blast Pattern" sections provide details below.

Test Blast Program

The contract required the contractor to perform a test blast program to show how the vibrations decrease with increasing distances from the blast and increase with increasing amounts of explosives. This program was intended to provide subsequent guidance for blast design to meet the vibration controls of this project. The first blast was to be conservatively designed to produce air over pressures and ground vibrations well below the controls established for this project. Charge weight per delay and the number of holes were then to be increased until near production rounds were detonated. The contractor performed the test blast program as described below.

For approximately the first 30 vertical feet of the access shaft, the contractor performed a series of blasts with progressive hole depths of 4 feet and 6 feet. To control fly rock and minimize surface vibrations and dust, the contractor used millisecond timing delays.

As shown in Figure 1, there were 97 production holes of 1⅞ inch diameter with 30-inch burden and 30-inch spacing. There were 46 perimeter holes of 1⅞ inch diameter with 24-inch burden and 24-inch spacing. There were four relief holes of 3-inch diameter.

Figure 2 shows the test blast initiation sequence. Detonators were NONEL EZDet25/500 ms and delays were

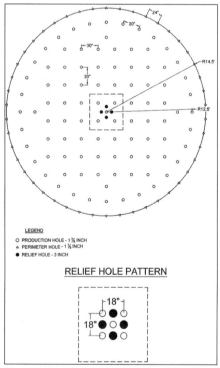

Figure 1. Test blast pattern dimensions

NONEL EZTL 42 and 17 ms. The blast holes were designed to depths of 4 feet and 6 feet, as shown in Figure 3.

Vibration and air over pressure measurements were obtained by five temporarily stationed seismographs. The test blasting operations consisted of six blasts:

1. 10/29/15, 11:07 am, 4-foot round
2. 10/30/15, 11:24 pm, 4-foot round
3. 11/3/15, 9:09 pm, 6-foot round
4. 11/5/15, 5:19 pm, 6-foot round
5. 11/9/15, 11:25 am, 6-foot round
6. 11/12/15, 11:43 am, 6-foot round

Throughout test blasting operations, zero recorded vibration velocity measurements at the contractors' field office or the existing surge shaft were beyond the prescribed limit of 2.0 inches per second (ips). The highest vibration measured at the field office occurred on Thursday, November 5, 2015 at 5:19 p.m. during blast 4, with a velocity of 0.061 ips. The highest vibration measured at the existing surge shaft occurred on Tuesday, November 3, 2015 at 9:09 p.m. during blast 3, with a velocity of 0.026 ips. The highest vibration measured during test blasting operations occurred on Monday, November 9, at 11:25 a.m. during blast 5, with a velocity of 2.74 ips measured 20 feet from the blast. Though this measurement is above the prescribed limit of 2.0 ips, blasting operations were deemed safe because no manned features existed at this distance away from the blasts.

Production blasts used for the duration of access shaft excavation were expected to produce vibrations of 0.02–0.07 ips at the field office and 0.01–0.03 ips at existing surge shaft. These expected vibrations were calculated using the testing-developed regression and a typical LP blast round defined in the access shaft detailed blasting plan: 807.3 total LBS explosives, 7.56 lbs per delay, two rounds per delay, 2.89 lbs/CYD powder factor, 10 feet round length.

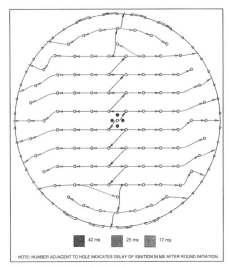

Figure 2. Test blast initiation sequence

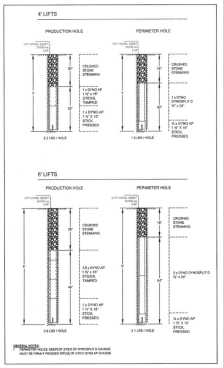

Figure 3. Cross sections for 4-foot and 6-foot hole depths

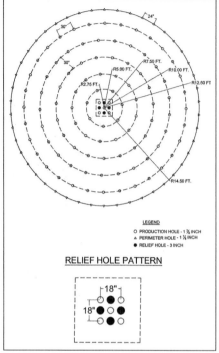 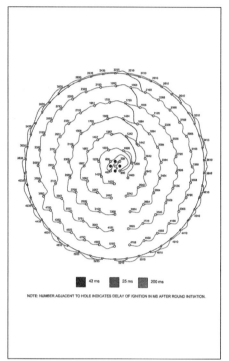

Figure 4. Access shaft blast pattern dimensions   Figure 5. Long period initiation sequence

Seismographs Monitoring

The contract required that the peak (maximum on any axis) particle velocity (PPV) should be less than 2 ips at the existing surge shaft, as monitored by a minimum of two instruments placed at the contractor field office and existing surge shaft. These seismographs are operated on a trigger of 0.03 inches per second and set up to forward vibration data automatically via cellular modem. Sensors may not be triggered if vibrations do not exceed the 0.03 inches per second trigger threshold. Transmitted vibration measurements were added to the existing regression and included to the recurring blast report provided to SNWA. If any future blast resulted in a vibration greater than 2.0 ips, the contractor would notify SNWA and modify blasting operations to reduce vibrations.

Type of Explosives

The contractor loaded the Access Shaft production holes with six tamped and one pressed stick of Dyno AP 1½ inch by 16 inches. Detonators used were NONEL EZDrifter 200/5400 ms and delays used were NONEL EZTL—42 & 17 ms.

Perimeter holes were loaded with four tamped and ½ pressed sticks of Dyno AP 1½ inch × 16 inches. Detonators used were NONEL EZDrifter 200/5400 ms and delays used were NONEL EZTL—42 & 17 ms.

Blast Pattern

Figures 4–6 show the various blast patterns implanted for Access Shaft drill and blast.

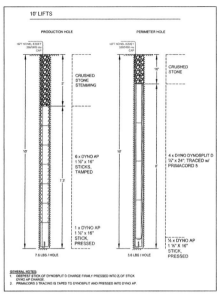

Figure 6. 10-foot blast hole cross sections

Forebay Blasting and Benching

The L3PS Forebay is 377 feet long, 33 feet wide, and 36 feet high. Because the well shafts were constructed by the blind bore method, as further described herein, the 6-foot-diameter steel casings protrude into the excavation envelope. Thirty four of these steel well casings had to be protected from blast damage during excavation of the forebay.

As shown in Figure 7, the Forebay was originally planned to be excavated with a top heading and bottom heading, however, during the Forebay cavern excavation it was determined by the contractor to proceed with a top heading, middle heading, and a bottom heading. The top heading was excavated via drilling and shooting center-cut and side slash faces. Middle and bottom headings were excavated via drilling and shooting a center-cut and side slash faces. The center-cut was advanced before the side slashes are taken.

Figures 8 through 16 show the layout and blasting sequences used for Forebay excavation. The Forebay was excavated successfully following this drill and blast plans.

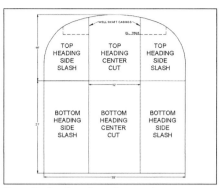

Figure 7. Forebay planned excavation sequence

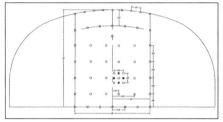

Figure 8. Top heading middle blast design

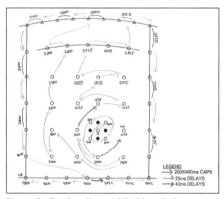

Figure 9. Top heading middle blast timing

Blasting Under Well Shaft Casings

As indicated in the Forebay excavation section above, the top heading center-cut face is advanced between the steel well shaft casings. As the top heading side slash faces reach each steel well shaft casing, the side slash alternate blasts were implemented.

Drill and Blast

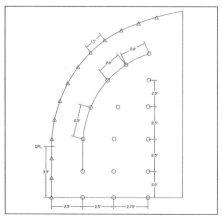

Figure 10. Top heading side slash blast design

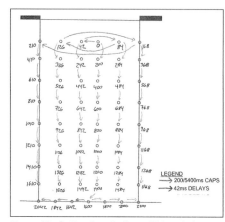

Figure 13. Bottom bench planned blast timing

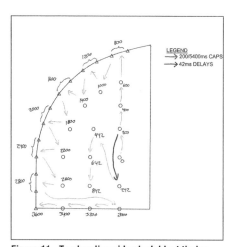

Figure 11. Top heading side slash blast timing

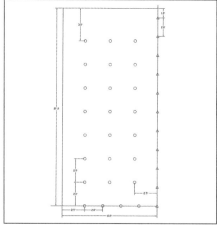

Figure 14. Bottom bench side slash planned blast design

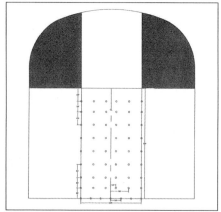

Figure 12. Bottom bench planned blast design

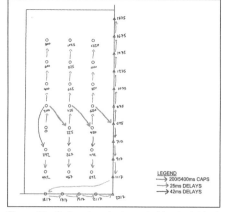

Figure 15. Bottom bench planned side slash blast timing

Lake Mead Intake No. 3 Low Lake Level Pumping

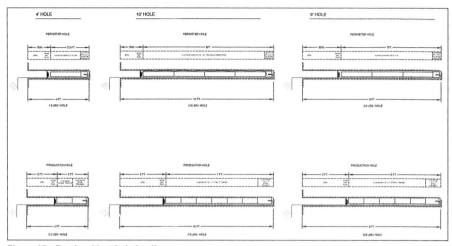

Figure 16. Forebay blast hole loading

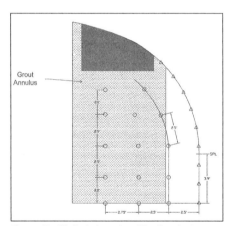

Figure 17. Well shaft casings blasting design

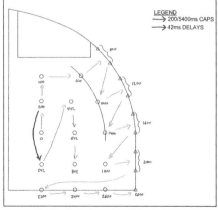

Figure 18. Well shaft casings blasting timing

Figures 17 and 18 show the blasting method employed near the well shaft casings to minimize blasting damage to the casings.

Figure 19 shows a typical scenario that well shaft casings had to be excavated around without damaging the casings.

Blasting/Mechanical Excavation Near Existing Surge Shaft Bulkhead Structure

The Surge Shaft bulkhead structure is located at the bottom of the existing Surge Shaft. It is a 16-foot diameter

Figure 19. Well shaft casings protruding into Forebay excavation

Figure 20. Surge shaft bulkhead protected by steel and timber

Figure 22. Hydraulic breaker used to mechanical excavate connector tunnel

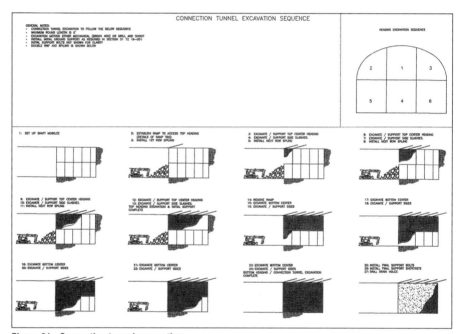

Figure 21. Connection tunnel excavation sequence

ellipsoidal steel bulkhead that will be eventually removed to connect the Forebay to the Intake No. 3 Connector Tunnel. This bulkhead is located approximately 48 feet from the face of the Forebay Connector Tunnel excavated by the contractor.

To protect the existing bulkhead from blasting damage, the contractor constructed a steel and timber bulkhead to cover it as shown in Figure 20.

Once the bulkhead was protected then the Forebay Connection Tunnel was excavated mechanically as shown in Figure 21 it was excavated with a top heading and bottom heading with side slashes.

Because of the ground conditions and the proximity of the Surge Shaft bulkhead, most of the Forebay Connection Tunnel excavation was performed by mechanical means using a hydraulic breaker.

WELL SHAFT DRILLING

The L3PS project required a completely different excavation method for the 34 deep well shafts that would hold the world's largest and deepest submersible pumps. As a means of mitigating the risk of the known difficult and unstable ground conditions, the CMAR and the SNWA jointly selected the blind bore method of drilling. Before excavation could begin on the well shafts, a large surface grouting effort commenced to prepare the ground.

Pre-Excavation Grouting

Large-diameter shaft development utilizes reverse circulation as its method of spoils removal from the development face. Reverse circulation is dependent upon the ability to maintain a flooded shaft to develop differential pressures, which result in a flow of drilling fluid that circulates the excavated material from the development face at the bottom of the shaft to the surface through the center of the drilling pipe.

Pre-excavation grouting treats the formation in which the shaft will be drilled with high mobility grout to create a sufficiently water tight cylinder or construction zone that enables the drilled shaft to remain flooded during the reaming process. In a typical pre-excavation grouting program, a series of 6-inch diameter holes is drilled in an offset pattern around the shaft's perimeter. After each hole is drilled, it is typically pressure grouted in approximately 100-foot stages starting from the bottom and moving upwards. The intent of the program is to inject grout into cracks, fractures, fissures, and other such anomalies that may be present in the formation. As the grout migrates from the bore hole out into the formation and sets, these potential sources of water loss are sealed off, making the formation more impervious. When sufficient holes around the shaft perimeter have been drilled and grouted, and it appears that grout is no longer being pushed into the formation, the pre-excavation grouting program is considered complete. In addition to sealing the formation, the grout also aids in consolidating any unstable, broken, or fractured areas of the formation that can collapse during the large-diameter reaming.

For the L3PS project, a unique pre-excavation grouting program was developed to facilitate the close proximity of the 34 well shafts. Based on previous drilling experiences in the area, the project team anticipated that certain areas of the formation would be more susceptible to taking grout than others. For this reason the pre-excavation grouting program was set up in stages with an initial primary program. Once completed and analyzed, this program could provide justification for adding secondary,

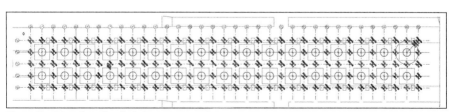

Figure 23. Plan view of well shafts pre-excavation grout pattern

and in some cases, tertiary programs. The primary program was design as shown in Figure 23.

The top 150–200 feet of the formation proved to be significantly broken and unconsolidated to the extent that the drilled bore holes kept collapsing. For this reason, a top-down grouting program was implemented. Holes were drilled to a depth of 150 to 200 feet and then grouted in a single stage from the top. Once set, the hole was then re-drilled through the grout to a total depth of 550 feet. At this point, the pre-grout holes were grouted in 100 foot stages starting from the bottom and moving upwards.

As anticipated, upon completion of the primary grout program, our analysis of the grout takes in each hole revealed different areas of the formation were still taking, while other areas appeared to have been adequately sealed. The project's well shaft drilling subcontractor, North American Drillers (NAD), analyzed the following data to create a 3D model and a heat map of the formation around the 34 shafts:

- Survey plot of every grout hole, indicating the drilled path
- Grout volume intake for each stage of each hole
- Drilling records, including penetration rates and observations from the drill cuttings

The 3D model and heat map of the formation as shown in Figures 24, 25, and 26 were then utilized to design the secondary grouting program. This program allowed NAD to focus on specific areas of the formation that still appeared to be unstable. Secondary grout holes were placed in strategic locations and drilled to varying depths to reach specific areas of the formation.

Upon completion of the secondary program, the data analysis continued to suggest that specific areas of the formation remained unstable and could potentially lead to circulation losses during large-diameter reaming. Consequently, a tertiary grout program was designed to target very specific areas of formation. Target holes were strategically positioned around the jobsite to reach areas of the formation that had not yet been addressed or continued to take grout during the secondary program. At the end of the tertiary program, there was sufficient evidence to support that additional grouting would not yield significant improvements to the formation and the pre-excavation grouting program was terminated. At the start of pre-excavation grouting, the bore

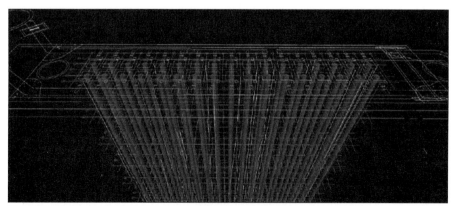

Figure 24. 3D survey of well shaft pre-excavation drilled grout holes

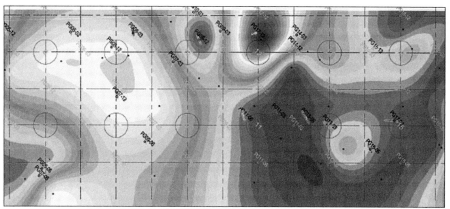

Figure 25. Heat map of well shaft pre-excavation grout volume takes

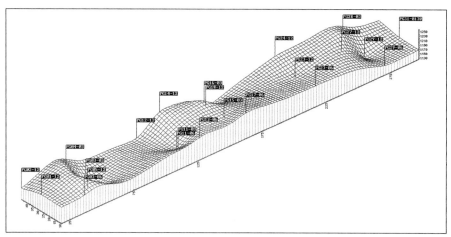

Figure 26. Grout take map

holes drilled were losing water at a rate of 300–400 gpm. Upon completion of the program that rate was dropped to an average of about 15 gpm.

The pre-excavation grouting program lasted approximately four months. Throughout the program, the project team drilled more than 200 grout holes totaling more than 145,000 feet and used 2,600 tons or 100 truckloads of cement.

Guided Pilot Hole and Reaming Drilling

For well shaft work that requires strict adherence to an alignment criteria, the pilot hole is the feature of work that provides the control of that element during the shaft reaming stage. The pilot hole provides the path in which the reamer will follow during the shaft reaming stages.

Pilot holes can be drilled in a variety of ways, but most commonly an oil and gas top head drive-type rig is used. The hole diameter can vary from 8 inches to 17.5 inches and is typically drilled to about 30 feet below the final elevation of the drilled shaft. Most importantly, the pilot hole must be drilled in such a manner as to provide a straight hole, free of any dog-legs that would represent a risk to the shaft reaming operation. Depending on the geology and the end use of the shaft, drilling the pilot hole with a stiff bottom hole assembly in a controlled manner is often an adequate method to achieve a relatively straight pilot hole. For applications where the alignment and verticality are important, different directional drilling technologies can be used to ensure a straight and vertical hole.

For the L3PS project, the verticality and straightness of the well shaft were extremely important for the efficient operation of the well pumps. The project requirements specified for the wells shafts' vertical alignment to have a lateral deviation of less than 2 inches in 500 feet. This mandated the use of directional drilling technology. NAD elected to utilize real-time optical guidance and verification system (RTOGV) developed by Coastal Drilling East, LLC for drilling the pilot hole.

Figure 27. Pilot hole drilling and real-time monitoring

The RTOGV system, as shown in Figure 27 combines the real-time monitoring of the position and location of the drilling bit with a bottom hole assembly capable of steering in real time. Simply put, a beam of light emitted from a source just above the drilling bit is captured by a high-resolution camera on surface and transmitted to the drilling operator's screen. As the drilling progresses the operator can track the path of the drilling bit and immediately correct any lateral deviation.

RTOGV technology is limited for use with 8-inch diameter drilling tools. However, for the 34 shafts of the L3PS project, a 17.5-inch diameter pilot hole was necessary for the efficient reaming of the 91-inch-diameter shaft. Therefore, upon the successful drilling of the 8-inch diameter optically drilled pilot hole, a 17.5-inch hole opener was utilized to enlarge the pilot hole.

All of the 34 each pilot holes were drilled to meet the required deviation specifications of 2 inches in 500 feet. The average deviation at the 500 foot depth was 0.28 inches and the maximum deviation registered was 1.12 inches. On two occasions the characteristics of the formation were such that steering the pilot hole back to the centerline was not possible. In both cases, the drilling tools were tripped out and a section of the hole was grouted. Once the grout reached a pre-determined strength, the tools were tripped back in the hole and pilot hole drilling was re-started and successfully steered back to the intended path.

Blind Bore Reaming

With the completion of the pilot hole, the process of reaming the shaft to its final diameter is executed using a large-diameter A-frame rig similar to the one in the below Figure 28. The drilling rig is equipped with a turntable responsible for providing the rotation and torque of the drilling tools and a hoist unit responsible for raising and lowering the drilling tools into the shaft and maintaining an optimal pressure of the drilling face.

The bottom hole assembly tools, as shown in Figure 28, are composed of a set of weights stacked on a drill mandrel. Just above the weight stack is a centralizer/stabilizer responsible for keeping the drilling tools centralized in the drilled shaft and provide additional stability to the drilling. At the bottom of the mandrel is the reamer body, which is responsible for cutting the rock and advancing the penetration. As previously mentioned, the blind bore drilling methodology relies on reverse circulation to circulate drilling cutting to the surface and to maintain a clean drilling face for the reamer to efficiently continue cutting.

The reamer bodies utilized in the L3PS project were hemi-spherical in shape and

Figure 28. Blind bore drilling bottom hole assembly tools

lined with approximately 25 13-inch-diameter tungsten carbide disc cutters, which enabled the drilling of a 91-inch diameter hole. In addition, the reamer was equipped with a 3-foot-long and 17-inch-diameter "stinger" or "snout" whose purpose was to mechanically link the reamer body to the pilot hole at all times, steering or guiding the reamer in the desired direction.

Given the large number of shafts, the close proximity of these shafts, and the time-sensitive nature of the project, the two A-frame drilling rigs used during this process were mounted on walking substructures, similar to those utilized in the oil and gas industry. Using the walking substructure, transition from one shaft to the next shaft took between 24 and 36 hours, compared to a more traditional mobilization of 5–7 days without the substructure.

Casing Installation and Grout Backfill

Upon completion of the large diameter reaming process, the drilling tools were tripped out of the hole, the drilling rig was moved to the next shaft, and a 160-ton crane was positioned next to the open hole. This was used to facilitate the lowering of the steel cans for hydrostatically lining the shaft as shown in Figure 29. Each steel can was 72 inches in internal diameter, 50 feet long, had a 1-inch thick wall and weighed approximately 38,000 lbs. The first can, equipped with a bulkhead on the bottom side, was picked up, lowered into the shaft, and landed on a strong back support frame. A second can was then picked up and positioned right above the top of the first can. The top can's alignment was verified along multiple directions and adjusted to ensure it was vertical before the start of welding. Two welders then proceeded with the welding of the two cans, making sure that they

Figure 29. Well shaft steel casing installation

worked directly opposite one another at all times. This process was repeated nine times until the bottom can reached the final depth of the shaft.

Due to the temperatures experienced in Nevada during summer time, it was soon discovered that the hanging steel can, when exposed to direct sunlight, developed temperature differences in the order of 40–50 degrees Fahrenheit in the matter of 1–2 hours. This was enough to cause elongation of the steel can to different extents on opposite sides. This caused the previously vertically-aligned steel cans to become misaligned. Despite efforts to compensate for the temperature difference effect, a notable difference was noticed between the final alignment of pipe welded during daylight hours and during night time. Consequently, to ensure the best alignment results, welding of the casing was restricted to the hours of the day when it would not be exposed to direct sunlight.

Upon reaching final depth, the casing string was landed on the strong back and the grouting of the outer annulus commenced. Grouting was done in stages from the bottom upwards using three sets of tremie tubes lowered between the drilled shaft and the outside of the steel casing. For the shafts of the L3PS project, grouting was carried out in six stages. In the first stage, a thicker grout mix was used to secure the bottom of the casing in place. Subsequent stages were approximately 100 feet in height and used a lighter grout mix.

Casing Survey and Monitoring

Due to the very tight tolerances on the verticality of the final shaft, it was also necessary to constantly monitor the vertical alignment of the shaft casing during the final grouting operation. A centralized mandrel carrying a laser device that produced a vertical beam of light, as shown in Figure 30, was positioned at the approximate elevation of operation of the well pumps. This vertical beam of light was captured at the surface and a specially developed software was used to compare the mandrels lateral position in relation to the center of the casing at the surface.

It was soon discovered that during each stage of grouting there was considerable movement of the steel casing. When the casing was left unrestrained, it was possible to observe movement in the order of 4–5 inches approximately 2 hours following the completion of each stage of grouting. This exceeded the specified alignment tolerances and for this reason a set of three air bags were lowered between the drilled shaft wall and the outside of the steel casing. These were inflated and used to push and secure the steel casing at the shaft's central axis. While the air bags provided support to restrain the steel casing's movement, it was not always possible to completely withstand the forces produced by the setting grout. However, the amount of lateral movement was significantly reduced to 1–1.5 inches.

Upon completion of the grouting of the steel casing, a final casing survey was performed on each shaft. Of the 34 well shafts, 32 were completed to the

Figure 30. Well shaft casing mandrel

specified tolerance requirements, while two were deemed to be slightly outside the specified tolerance levels. However, the magnitude of deviation in these two shafts was such that it was possible to correct the error by repositioning the well pumps at the surface.

CONCLUSION

The L3PS project team works as a cohesive team in assessing the risks, developing construction approach and working cooperatively together. The CMAR delivery method has provided the opportunity for the team to select the preferred drilling method for the well shafts, blind bore method, which has proven to be successful. In addition, the team was able to commence construction on the underground works more early in the project while proceeding with the detailed design development on the aboveground facilities. The contractor, their sub-contractors, and the SNWA, along with their consultants, have worked cooperatively to collaboratively develop the design in conjunction with the construction approach, develop the contract requirements accordingly, develop contingency plans to mitigate impacts that may arise during construction. The team had to overcome many challenges in the underground construction and has completed a major interim milestone on this critical project.

Comparison of Emulsion and Anfo Usage in the Horizontal Development Process at El Teniente

Raúl Castro • University of Chile
Yina Herazo • University of Chile
Álvaro Pérez • University of Chile
José Medina • El Teniente, Codelco

ABSTRACT

El Teniente is among the largest underground mines in the world with production of approximately 135,000 tons of copper per day (Baez, 2016). After 112 years of operation, mining is conducted in deeper and more competent rock mass conditions than when the mine was initially opened. The high production rates also require developing a larger number of tunnels in a more efficient way (rapid advance rates at a low cost). For these reasons, El Teniente has defined mine development as a key strategy area for its future, and as such it is continuously looking for technological opportunities to improve safety, efficiency and costs. For many years, ANFO has been the explosive used in this operation. In the last year, emulsion has been extensively tested at El Teniente in horizontal developments to technically quantify their benefits. Trial tests were initially conducted in the Diablo Regimiento sector followed by industrial application in the Pacífico Superior sector. Results show that emulsion has many advantages including a smaller volume of poisonous gases and, therefore, less ventilation time required, and fewer boreholes and greater efficiency in terms of advance per round when compared to ANFO. In this article the fundamentals and statistical analysis of the results derived from field tests at El Teniente are presented and compared.

INTRODUCTION

El Teniente (DET) is a mine located in the Libertador General Bernardo O'Higgins Region, 50 km from Rancagua, Chile, at a height of 2500 m. Its copper deposit, mined by the Block Caving method, has one of the highest production rates in the world.

Today the extraction and development at El Teniente occurs in what is locally termed "primary ore" under high stress conditions. The main characteristic of primary ore rock is its high hardness and brittleness. On the other hand, the high production has also meant high mine preparation rate requirements to maintain production capacity given the geotechnical conditions at the mine. This has led to the adoption of a strategic plan for mine development for its present and future (Díaz & Morales, 2008). One the strategic focus has been a review of mining practices including technological and lean management.

LITERATURE REVIEW

In terms of the technologies, one focus has been an operational review of explosives. For many years, ANFO has been the explosive used at the mine due to its low cost and familiarity of use by the operators. However, there are other types of explosives such as emulsion that have been on the market for many years. Emulsion is a liquid salt solution made of small droplets, with each droplet surrounded by a thin oil film. ANFO is a mixture of crystalline or prilled ammonium nitrate (AN) and fuel oil (FO). The emulsion properties make it viable for use in hard rock and help to obtain efficient blasting. The literature contains criteria for defining the explosive given its characteristics for a given rock mass condition for horizontal developments to achieve a given level of performance, but hard data is not available.

Explosive–Rock Relationship

Rocks could be mechanically classified as having elastic or plastic behavior. Elastic rocks are those having relatively higher compressive strength, while plastic acting rocks are those having lower compressive strength (Grant, 1970; Bhandari, 1997). The ease of generating new fractures in the medium is a function of the strength properties of the rock material. For example, with hard rock which is more elastic, a high brisance explosive is recommended (Brady & Brown, 2004).

Velocity of Detonation

To obtain good blasting performance, especially good fragmentation, the choice of a particular type of explosive must consider important aspects such as velocity of detonation (VOD) of the explosive and the P-wave velocity of rock. The VOD is affect by a number of factors like: diameter of the charged column, confinement, explosive type, density, sensitizing agent(s), among others. Regarding to the first, VOD decreases as the diameter of the charged column decreases (Cooper, 1996). The increase of confinement will increase the VOD (Persson, 1994; Esen, 2004). Is important to mention that the VOD is proportional to the detonation energy released by the rock, and if it is increasing, it produces a better stress distribution. The VOD of emulsion is greater than ANFO; it means that emulsion is suitable for hard rock whose P-wave velocity must be equal to or less than the VOD of the explosive. When VOD is less than the P-wave velocity of the rock, the P-wave could compress the explosives and result in detonation failure (Zhang, 2016).

Water Resistance

The water resistance of an explosive defines its ability to detonate after being exposed to water. Emulsion is a liquid salt solution made up of 0.005 mm droplets, and, as noted above, a thin oil film surrounds each droplet. It is this oil film that encloses the drops of salt solution and gives the emulsion its outstanding resistance to water (Johansson, 2000). A major disadvantage and limitation of ANFO is its lack of water resistance. Ammonium nitrate dissolves easily in water even with the added fuel oil. ANFO containing more than about 10% water will fail to detonate (Hustrulid, 1999).

Advance

Advance per round is affected by multiple factors including: the properties of the rock mass, geological considerations, blast design, drilling accuracy, explosive selection, the initiation system, the timing of the round, and the use of effective stemming products (Prout, 2010). Underground blasting does not have effective free face, therefore, it is necessary to generate a void to which will be released the adjacent charged boreholes; this first void is called cut. With a hard rock type and parallel cut (four-section cut), according to Persson et al. (1994) the maximum hole depth depends on the empty hole diameter, the larger the diameter, the larger the hole depth that can be achieved. If the advance achieved is less than 95% of the drilled hole depth, drifting becomes very expensive for the mine.

Overbreak

The factors influencing blast damage can be broadly categorized in three areas: rock mass features—especially discontinuities of which important considerations include orientations, aperture, frequency, filling in the joints, RQD, watery conditions and state of stress; explosive characteristics and distribution; and blast design and execution (Singh & Xavier, 2005). Regarding the latter two aspects, it was found that the smooth blasting method reduces overbreak. Persson et al. (1994), suggest an empirical relation in which the minimum linear load concentration required for contour blasting is a function of the charged hole diameter. In the smooth blast method the row of holes adjacent to the planned contour is usually drilled with an S/B ratio of 0.8 and with little delay between them (Persson et al., 1994).

Furthermore, if the charge is decoupled, this can generate losses in the shock energy delivered to the rock mass, which can restrict the damage around the borehole. In tunneling, the smooth blasting method is preferable to presplitting, since the latter is more expensive than smooth blasting, as it requires a closer spacing between contour drill charged holes; moreover, it is often more difficult to fix an extra pre-blasting operation in the underground advance cycle (Persson et al., 1994).

In summary, the literature indicates that emulsion present properties that allow efficient blasting in hard rock, given their high VOD and water resistance. However there is a lack of reported tests in the field that could be used to verify all the benefits mentioned above. For this reason full scale trials were conducted at El Teniente using emulsion to quantify and compare the performance of ANFO and emulsion. In this article the results of these experiments are shown.

EXPERIMENTAL SITE

El Teniente is a large mining complex with several productive mines or sectors located around of a pipe "Pipa Braden" where mineral is located (Figure 1). For this study, two kinds of tests were carried out in two sectors with similar characteristics. To define the baseline of blasting, trial tests were conducted in the Diablo Regimiento (DR) sector follow by industrial application at the Pacífico Superior sector (PS).

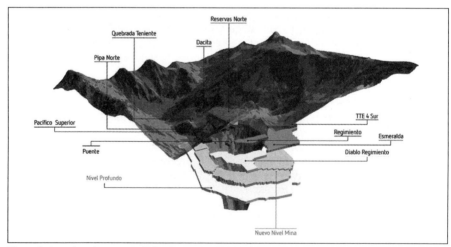

Figure 1. Location of sectors at El Teniente mine (CODELCO, 2016a)

El Teniente complex is mainly composed of two types of rock: CMET (Complejo Máfico El Teniente), which in turn is composed of Gabbros, Diabases and Basaltic Porphyry, and a Breccia Complex. The rock mechanic characteristics of the rock types for Diablo Regimiento and Pacífico Superior sectors are similar and are shown in Table 1. The CMET could be considered as rigid, fragile and hard rock.

Table 1. Geotechnical characteristics of El Teniente's rock mass (CODELCO, 2017)

Rock Type	CMET	Breccia
Percentage of area (%)	80	20
E (GPa)	57 ± 11	27 ± 4
Vp (m/s)	5646 ± 428	4287 ± 260
(MPa)	11	7
UCS (MPa)	135 ± 12	73 ± 22

The Diablo Regimiento sector is one of the 15 productive sectors in DET (CODELCO, 2014). This sector is located in the southernmost part of the deposit (Figure 1). The stress field in-situ is = 41 MPa and = 25 MPa. The geotechnical characteristics of the rock type where tests were carried out are shown in Table 1. In this case blasting tests were conducted in the production and undercut levels, having cross sections of 19.9 m² and 16.3 m² respectively.

PS (Figure 1) is located at the west of Pipa Braden between the Diablo Regimiento and the Pipa Norte sectors (CODELCO, 2017). This sector has historically been affected by water inflow with values reaching 105 ± 137 l/sec (CODELCO, 2016b). As shown in Figure 2, this sector has 80% of CMET rock type and 20% of Breccia rock. Geotechnical characteristics are shown in Table 1.

The explosives were Emulsion, a Subtek™ Charge and ANFO, and both types were initiated by a pyrotechnic detonator. The explosives used for tests were Emulsion Subtek™ Charge™ and ANFO. The properties obtained by manufactures for both explosives are presented in Table 2.

The dynamite for contour used was Softron, which has the characteristic shown in Table 3.

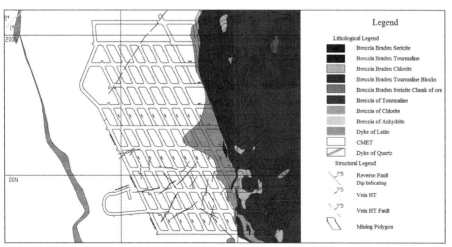

Figure 2. Rock types in the production level of Pacífico Superior sector (Codelco, 2017)

Table 2. Properties of emulsion and ANFO

Explosive	VOD (km/s)	Minimum borehole Diameter (mm)	Density (g/cm^3)	Sensitizer	Relative Weight Strength (% ANFO)
Emulsion	3.5–5.7	40	1.15–1.25	Gassing solution	87–91
ANFO	3.5–3.9	63.5	0.85–0.90	—	—

Table 3. Properties of dynamite for contour

Explosive	VOD (km/s)	Detonation Pressure (kbar)	Density (g/cm^3)	Energy (KJ/kg)	Relative Weight Strength (kg ANFO)
Dynamite for contour	3	33	1.19 ± 3%	4.480	1.13

MEASUREMENT OF PARAMETERS

During the blast tests, ventilation time, effective advance and overbreak parameters were measured to evaluate the performance of explosives. The measurement mechanism is described below:

Ventilation Time

This is quantified in terms of the ventilation time required to achieve proper air conditions for the production and undercut levels. Thirty minutes after blasting and according to Chilean standards and regulations, mine personnel enter the blasted area and measure the gas concentrations of CO, O_2 and NO_2 with a gas meter. If these concentrations are above the legal limit value, miners are not authorized to enter. After an elapsed time, mine personnel repeat the process until the poisonous gas concentration is below the legal threshold limit values. Percentage values of concentration of toxic fumes are recorded in the shift's gases control log.

Effective Advance

This is calculated as the ratio between the linear advance length in meters after blast and the drilled length. The actual linear drift advance is calculated using topographic measurement after the mucking and supporting activities are conducted, using Total Station Equipment (TSE). The information about linear advance length is stored by the operators in a common database known as the *Collaborator Platform*. To calculate the drift advance per round, two measurements of the distance were considered (Figure 3).

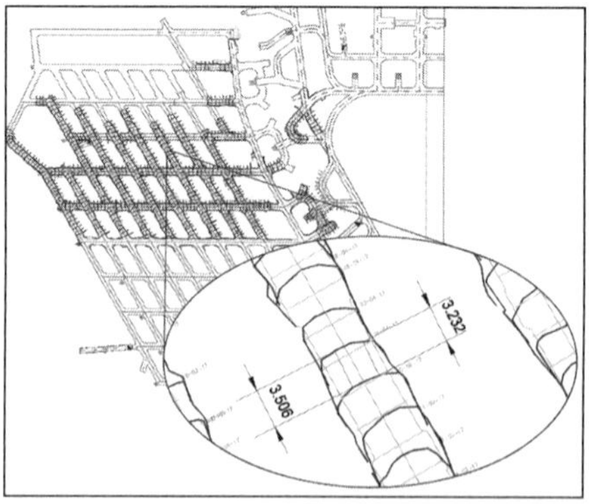

Contour Damage

This is reported as the percentage of overbreak. As in the case of the advance per round, the excavation perimeter to estimate the real cross section was calculated using the TSE measurements. Overbreak is the difference between the real cross section and designed cross section area. Figure 4 shows a cross section where the real and designed area for a drift can be observed. This value was reported in percentage terms using Equation 1:

Figure 3. Measurements of the advance on plan view

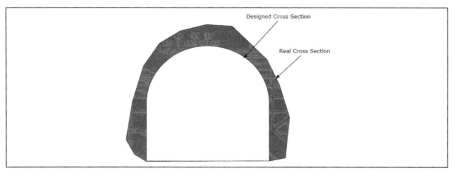

Figure 4. Example of designed and real cross section for an underground face

Table 4. Blasting trial test results using ANFO and emulsion at DR

Level	Data (un)	Explosive	Hole depth (m)	Advance length (m)	Effective advance (%)	Overbreak (%)
Undercut	5	ANFO	3.8	3.3	87.4	23.0
Production	1	ANFO	3.8	3.3	86.0	25.2
Production	5	Emulsion	3.6	3.4	93.6	7.4

$$\text{Overbreak}(\%) = \frac{\text{Real cross section} - \text{Designed cross section}}{\text{Designed cross section}} \quad \text{(EQ 1)}$$

BASELINE

The performance of ANFO and emulsion explosives was initially evaluated in two periods: from 20-07-2016 to 23-08-2016 test using ANFO and from 23-11-2016 to 18-01-2017 test using emulsion in the DR sector. The results from these tests define the baseline over which industrial tests were later conducted.

Table 4 presents a summary of the main results using ANFO and emulsion. For the first explosive, the average of effective advance reached 87% and the percentage of overbreak reached 24% on average. For the emulsion, the effective advance reached 94% and the percentage of overbreak was 7.4% on average.

During the trials, the velocity of detonation (VOD) was also measured for each explosive using a resistance wire continuous VOD measurement system using Microtrap equipment and cable to measure VOD (probe cable). In the case of ANFO, the VOD had values of 3602 m/s and 3517 m/s, in a hole diameter of 45 mm. In the case of emulsion, the VOD values reported were between 4162 m/s and 4058 m/s measured in a hole diameter of 45 mm. Therefore the VOD increased 15% when emulsion was used. Finally, the results obtained show there was an opportunity for improvement in blasting performance by using emulsion explosive. However the amount of data was not sufficient and more blasts were executed in an industrial application.

RESULTS

The blasting tests were carried out from February to June 2017. Blasting tests were conducted through the CMET rock type at production level in haulage and production drifts with cross sections of 22.9 and 17.9 m^2 respectively. The database for analysis is composed of 110 advance blasts with emulsion and 36 using ANFO.

Drilling Pattern

The blasting design at the mine depends on the section and explosive type (see Figure 5 and Figure 6 and Table 5). As can be observed, the number of boreholes increases as the section area increases. Also the required number of charged boreholes using emulsion is smaller than for ANFO (13% and 12% less in haulage drift and production drift respectively). A smaller number of boreholes play an important role in determining the development cycle, i.e., the time required between blasting procedures, as a fewer number of boreholes means savings in time and drilling costs.

Table 5. Number of boreholes for ANFO and emulsion

	ANFO			Emulsion		
	Charged Holes (d=45 mm)	Empty Holes (d=102 mm)	Total Holes	Charged Holes (d=51 mm)	Empty Holes (d=102 mm)	Total Holes
Production Drift (4.3 × 4.7 m)	52	3	55	45	3	48
Haulage Drift (5.2 × 4.82 m)	58	3	61	51	3	54

d= hole diameter

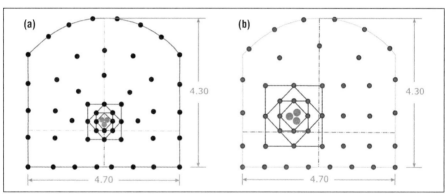

Figure 5. (a) Pattern of haulage drift charged with ANFO (b) pattern of haulage drift charged with emulsion

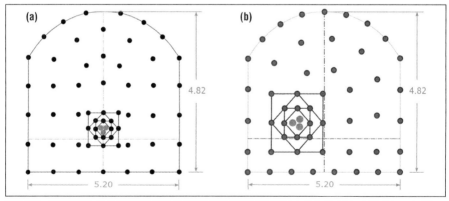

Figure 6. (a) Pattern of production drift charged with ANFO (b) pattern of production drift charged with emulsion

Gas Dilution Time

The composition of an explosive is said to be balanced when the oxygen contained in its ingredients combines with the carbon and hydrogen content to form mainly carbon dioxide and water (Bhandari, 1997). Then, the chemical composition of an explosive is one of the most important factors influencing the volume of poisonous gases after the blast. In general terms, emulsion produce less volume of poisonous gases when compared to ANFO and nitroglycerine, especially nitrous gasses (Johansson, 2000).

During the blasting tests at PS, the gas dilution time was the parameter used to indicate which one of the explosives produced the lowest poisonous gas concentration based on the assumption that the ventilation time should be smaller on the faces with lower concentrations. The average ventilation time in faces blasted with emulsion was 43 ± 12 minutes, whereas with ANFO it was 81 ± 99 minutes. Emulsion, then, required 38 minutes less for ventilation purposes, representing a 47% savings in average ventilation time. As noted, the standard deviation in ventilation time for emulsion is smaller than for ANFO indicating also a more predictable behavior. Figure 7 shows the ventilation time distribution necessary to dilute the gases post-blast using emulsion and ANFO explosives respectively. In terms of frequency, it shows that with emulsion in 93% of the cases it was possible to enter the blasted areas before 60 minutes. On the contrary in the case of ANFO, in 40% of the cases, entry was not possible until after 60 minutes had passed.

Overbreak

Overbreak in tunneling is the undesirable break of the rock due to blasting and geomechanics issues. The rock mass where tests were conducted is considered good rock quality, without important joint sets and under a relatively low stress field (σ_1=17 Mpa and σ_3=12 Mpa). From a blasting point of view, blast-induced damage is highly localized around the immediate perimeter of the blasting area (Singh & Xavier, 2005), so that the explosive charge used in perimeter holes highly influences the percentage of overbreak.

During the tests of emulsion at the mine, two different approaches were carried out over two different periods of time. In the first period, from February 7 to May 17, the explosive used in the perimeter holes was the same emulsion (Subtek™ Charge), but its density was reduced to 0.9 g/cm³ which deliver a linear charge concentration of about 1.8 kg/m. In the second period, from May 18 to June 30, the explosive used in the perimeter holes was packaged dynamite reaching a contour with a linear charge

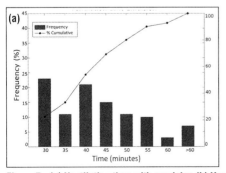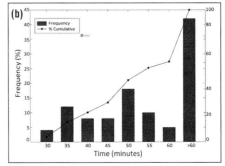

Figure 7. (a) Ventilation time with emulsion (b) Ventilation time with ANFO

Table 6. Percentage of overbreak with emulsion and ANFO

	Period 1 (February 7 to May 17)		Period 2 (May 18 to June 30)	
	Emulsion	ANFO	Emulsion	ANFO
Explosive used in contour	Emulsion of 0.9 g/cm^3	Dynamite	Dynamite	Dynamite
Data (un)	74	21	9	5
Overbreak average (%)	33 ± 18	24 ± 15	15 ± 9	16 ± 8

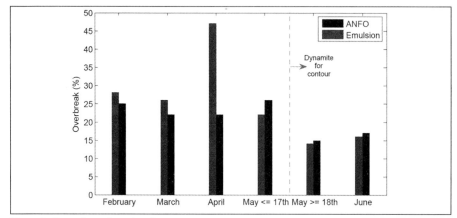

Figure 8. Percentage of overbreak

concentration of about 0.3 kg/m. For ANFO, the explosive used in the perimeter holes was packaged dynamite and the main results are shown in Table 6 and Figure 8.

The percentage of overbreak using emulsion was significantly reduced from 33% (Period 1) to 15% (Period 2). In blasted drifts using ANFO, the percentage of overbreak was reduced from 24% (Period 1) to 16% (Period 2) because of improvement accuracy in the drilling activity. The results suggest that the type of explosive used in the perimeter hole, as the literature indicates, is a key variable in the control of overbreak. Also, the use of decoupled charges in perimeter holes could help to reduce the damage surrounding the hole.

Effective Advance

Effective advance is a parameter that allows the blasting efficiency to be assessed in terms of drilling length. According to Persson et al. (1994), an effective blast should be around 95% to be considered an effective advance.

The average hole depth was measured in the case of emulsion at the front after drilling reached 3.46 ± 0.30 m. As shown in Figure 9, hole depth is distributed and shows variability. In the case of ANFO the values of hole depth is not available. All boreholes, including relief holes, were drilled at the same depth.

Figure 10 shows the effective advance between February and June 2017 for emulsion explosive. This indicates that effective advance reaches 96% ± 6% for emulsion. Therefore, emulsion is more effective than ANFO for the study case when compared to the 87% defined at baseline. This improvement corresponds to a 9% increase in blast efficiency.

From this study, the authors can identify many reasons why emulsion is more effective:

Comparison of Emulsion and Anfo Usage in the Horizontal Development Process 339

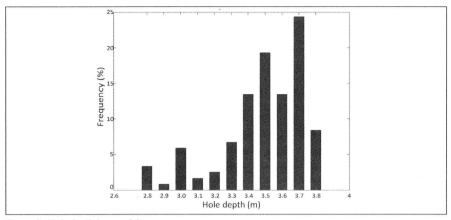

Figure 9. Hole depth for emulsion

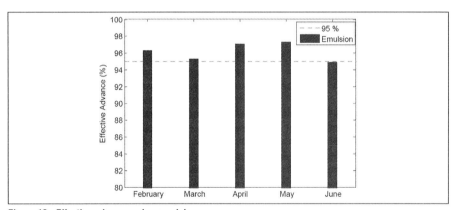

Figure 10. Effective advance using emulsion

- The brisance of emulsion is higher than ANFO. This is particularly important given that the rock at El Teniente is highly competent.
- The VOD of ANFO is lower than VOD of emulsion. This is particularly important in the case of underground developments where the diameter of drilling is small.
- Emulsion has higher water resistance than ANFO. This is particularly important at El Teniente where water is an issue and is observed at the mining fronts.

CONCLUSIONS

In this paper the performance of emulsion and ANFO was studied and compared in the horizontal development process. The focused was on the number of boreholes, poisonous gas dilution, effective advance and overbreak. For these topics, emulsion performs better than ANFO in hard competent rock.

In terms of the number of boreholes, a reduction of 13% and 12% in haulage drift and production drift respectively using emulsion was achieved. This represents a savings in the time of drilling activity and, therefore, in the mining cycle time. Poisonous gas dilution time was reduced by 38 minutes on average when the faces were blasted using emulsion; therefore, there is a decrease in the ventilation time.

The average of effective advance per round is increased in 9% when emulsion is used; this will influence the total progress of construction through an increase in advance rate and the possibility of faster access to the mineral. Furthermore, the percentage of overbreak was seen to be directly related to linear charge concentration in the perimeter hole. It has been observed that a low linear charge concentration will reduce the overbreak produced by the blast.

ACKNOWLEDGMENTS

The authors acknowledge Fidel Baez for his support in this research, as well as Rodrigo Valdés, Eliseo Ríos and Raimundo Silva from Orica and Ignacio Alvarado from DET for providing information, assistance and feedback during this study. Finally, the authors would like to acknowledge DET Codelco-Chile for permission to publish these results.

REFERENCES

Baez F. (2016). Intelligent mining—the way of the future. In: Proceeding of 7th International Conference and Exhibition on Mass Mining, Sydney, Australia, May, pp. 3–8.

Bhandari S. (1997). Engineering Rock Blasting Operations. A.A Balkema, Rotterdam, Netherlands, pp. 43, 198.

Brady B. & Brown E. (2004). Rock Mechanics for Underground Mining, Third Edition, Kluwer Academic Publishers, The Netherlands, pp. 518–532.

Cavieres P. (1999). Technology Innovation Management. Codelco, El Teniente Mine. (Report in Spanish).

CODELCO. (2017). Compilation of geologic, geothecnics and geomechanics background. El Teniente Mine. Engineering Document. (Internal document in Spanish).

CODELCO. (2016a). Plan of business and development (Internal document in Spanish).

CODELCO. (2016b). Database of water flow measurements (Internal document in Spanish).

CODELCO. (2014). Plan of business and development (Internal document in Spanish).

Cooper P. (1996). Explosives Engineering. Wiley VCH Inc, New York, USA, pp. 278, 284.

Diaz G. & Morales E. (2008). Tunneling and Construction for 140,000 tonnes per day— El Teniente Mine—Codelco Chile. In: Procedings of 5th International Conference and Exhibition on Mass Mining, Lulea, Sweden, June, pp. 83–96.

Grant M. (1970). How to Make Explosives to Do More Work. Mining Magazine 123 (2): 112–119.

Hustrulid W. (1999). Blasting Principles for Open Pit Mining General Design Concepts, Vol. 1. A.A Balkema, Rottherdam, Netherlands, pp. 165–169.

Johansson S. & Svärd J. (2000). How environmental and transport regulations will affect blasting. In: Proceedings of the 1st World Conference on Explosives & Blasting Technique, Munich, Germany, September, pp 41–45.

Orica Chile. (2016). Diagnosis of the horizontal development process P&T in Diablo Regimiento mine (Internal document in Spanish).

Persson P., Holmberg R. & Lee J. (1994). Rock Blasting and Explosive Engineering. CRC Press, Inc., Boca Raton, Florida, USA, pp. 217–218, 265–266.

Prout B. (2010). Choosing explosives and initiating systems for underground metalliferous mines. School: Drilling and Blasting 2010, Muldersdrift, South Africa, June, South African Institute of Mining and Metallurgy.

Singh S. (1995). Suggestions for Successful Cut Blasting. In: Procedings of 21st Annual Conference on Explosives and Blasting Technique, Nashville, USA, February, pp. 44–71.

Singh S. & Xavier P. (2005). Causes, impact and control of overbreak in underground excavations. Tunnelling and Underground Space Technology 20 (1), 63–71.

Zhang Z. (2016). Rock Fracture and Blasting: Theory and Applications. Elsevier. The Boulevard, Langford Lane, Kidlington, Oxford OX5 1GB, UK, pp. 190–192.

Connecting the TARP Des Plaines Tunnel System to the McCook Reservoir

Mark White ▪ Black & Veatch Corporation
Cary Hirner ▪ Black & Veatch Corporation
Patrick Jensen ▪ Metropolitan Water Reclamation District of Greater Chicago
Carmen Scalise ▪ Metropolitan Water Reclamation District of Greater Chicago

ABSTRACT

The Des Plaines Inflow Tunnel is being constructed to connect the 41km long Des Plaines Tunnel System to McCook Reservoir as part of the Metropolitan Water Reclamation District of Greater Chicago's (District's) Tunnel and Reservoir Plan (TARP). Challenging tunnel and hydraulic design and construction components to this project are being addressed using creative construction sequencing that is necessary to complete live connections to the 10 m diameter Des Plaines Tunnel and the McCook Reservoir using an elevated tunnel portal outfall. This paper will provide an update on the tunnel, high head gate shaft and energy dissipation structure construction and the challenges encountered during design and construction.

INTRODUCTION

The District is in the second and final phase of construction of the TARP to reduce flood damage and improve water quality in the greater Chicago combined sewer system area. This second phase consists of constructing reservoirs and connecting them to the existing 175 km network of deep large diameter rock tunnels. The reservoirs are designed to store combined sewage until it can be pumped to water reclamation plants (WRPs) for treatment and discharge to local waterways. The District's biosolids storage lagoons were was designated to be mined to approximately 91 meters (300 feet) below ground to create the McCook Reservoir. The Main Tunnel System connector tunnel was completed in 2017 and joins the McCook Reservoir to the TARP Mainstream Tunnel System (MTS). In addition, construction of the Des Plaines Inflow Tunnel (DPIT) between the Des Plaines Tunnel and the McCook Reservoir is in progress. The location of these connection tunnels and the McCook Reservoir are shown on Figure 1.

Computer modeling of wet weather events indicates combined sewage flows from the MTS into McCook Reservoir can exceed 850 cubic meters per second (30,000 cubic feet per second). Based on hydraulic modeling completed by the University of Illinois at Urbana-Champaign, the 10-year peak flow rate through the existing TARP Des Plaines tunnel is 317 cubic meters per second (11,200 cubic feet per second). Currently the Des Plaines tunnel is filled during wet weather events and the water is stored in the tunnel and associated shafts until there is capacity at the Stickney WRP to treat the water. The stored water in the Des Plaines tunnel is drained via a dewatering tunnel that is connected to the Mainstream Tunnel System Pumping Station that pumps the water to the Stickney WRP. The Stickney WRP is the largest waste water treatment plant in the world, with a design capacity of 5.5 billion liters per day (1.44 billion gallons per day) and the Mainstream Tunnel System Pumping Station is the largest underground sewage pumping station in the world (Hunter and Lewis 2012). Without the DPIT, two flow scenarios exist to move water from the TARP Des Plaines tunnel to

Connecting the TARP Des Plaines Tunnel System to the McCook Reservoir 343

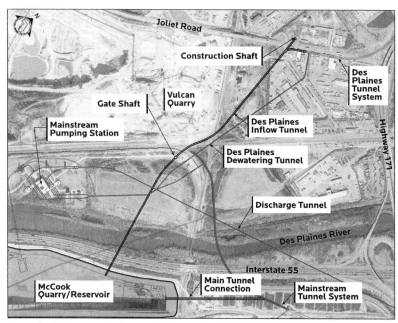

Figure 1. DPIT—overall plan

McCook Reservoir, either by pumped flow through the Mainstream Pumping Station (MSPS) or gravity flow bypassing the MSPS. Both flow scenarios would have significant limitations on allowable pressure and flow rates to prevent damage to existing infrastructure. These limitations would prevent operation at critical times during large storm events. The DPIT will provide a direct connection between the existing TARP Des Plaines tunnel and the McCook Reservoir, allowing operation at critical times during large storm events.

Stored CSOs in the McCook Reservoir are conveyed via an Inlet/Outlet Structure in the McCook Reservoir floor through a series of Distribution Tunnels to the Mainstream Tunnel System Pumping Station before being pumped to the Stickney WRP. Stage 1 of the McCook Reservoir provides 13 billion liters (3.5 billion gallons or 11,000 acre feet) of CSO storage volume. Stage 2 of the McCook Reservoir is scheduled to be completed in 2029 and, once online, will provide an additional 25 billion liters (6.5 billion gallons or 20,000 acre feet) of CSO storage volume. Upon completion of stage 2, which is currently being mined, a total of 38 billion liters (10 billion gallons) of combined sewage overflow (CSO) storage volume will be added to TARP. This additional storage volume will further reduce the risk of flooding to a significant portion of the City of Chicago and Cook County, Illinois, while also improving the quality of local waterways.

The DPIT is the focus of this paper and includes making live connections to both the existing southern terminus of the Des Plaines Tunnel and McCook Reservoir at approximately 75 meters (250 feet) below the ground surface and diverting the flow to the McCook Reservoir. Black & Veatch Corporation was retained by the District for the design of the DPIT. Walsh Construction is the General Contractor responsible for construction of all tunnel, shaft and portal facilities. At the time this paper was prepared, construction of the DPIT was nearing 75% complete.

PROJECT OVERVIEW

The DPIT includes the following project components:

- Tunnel connection between the existing southern terminus of the Des Plaines Tunnel and the McCook Reservoir, consisting of approximately 1.6 kilometers (1 mile) of 6.1 meter (20 foot) finished diameter rock tunnel.
- A tunnel adit, consisting of approximately 130 meters (420 feet) of 6.1 meter (20 foot) finished diameter rock tunnel.
- A tunnel transition from 10.1 meters (33 feet) to 6.1 meters (20 feet) to match the finished diameter of the existing TARP Des Plaines Tunnel and provide a controlled hydraulic transition.
- Construction Shaft to provide access for construction of the DPIT, consisting of an approximate 85 meter (280 feet) deep shaft with a diameter of 3.4 meters (11 feet).
- Gate Shaft for the installation and operation of a primary and backup gate, consisting of an approximate 82 meter (270 feet) deep shaft with a diameter of 13.7 meters (45 feet).
- Gate Structure and Gate Control Building to allow for the control of flow between the existing Des Plaines Tunnel System and the McCook Reservoir. The Gate Structure will consist of a high head primary and backup gate, both approximately 4.9 meters (16 feet) wide by 6.1 meters (20 feet) tall. A transition from a circular tunnel to a rectangular tunnel is required at each end of the Gate Structure. The downstream transition is steel lined to prevent damage from high flow velocities.
- Live connections between an active McCook Reservoir and the DPIT, and between the existing southern terminus of the Des Plaines Tunnel and the DPIT, including the excavation of two rock plugs and demolition of an existing concrete plug.
- Backfilling the tunnel adit and construction shaft that are not utilized following construction of the tunnel connection between the existing Des Plaines Tunnel and the McCook Reservoir.
- Tunnel Portal at the McCook Reservoir, approximately 13.7 meters (45 feet) above the floor of the McCook Reservoir. Figure 2 shows a depiction of the tunnel portal.
- Energy dissipation structure to mitigate damage to the floor of the McCook Reservoir caused by the approximate 13.7 meter (45 foot) drop of water at the DPIT outlet.
- Highwall stabilization measures at the McCook Reservoir surrounding the portal of the DPIT.

Figure 2. Tunnel portal, energy dissipation structure and highwall stabilization

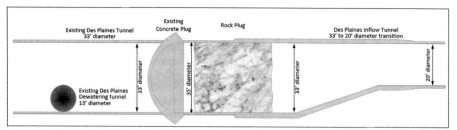

Figure 3. TARP Des Plaines Tunnel, transition and crown to crown configuration

TUNNEL AND HYDRAULIC DESIGN

Tunnel Transition

At the furthest upstream location, the DPIT will be connected to the TARP Des Plaines tunnel, requiring a transition from a 10.1-meter (33 feet) diameter to 6.1-meter (20 feet) tunnel. The major design consideration at this location governing the alignment was managing air entrapment. At high flow velocities, entrapped air has the potential to damage the concrete tunnel liner. Both invert to invert and crown to crown alignments of the transition were considered as well as a vent shaft upstream of the transition. Cost and real estate considerations eliminated the vent shaft. With the invert to invert design, the potential for air entrapment upstream of the transition would increase due to a downward sloped crown. The crown to crown design at the transition eliminates the sloped surface, as shown on Figure 3, facilitating air movement past this location.

Tunnel Bend

Approximately 30.5 meters (100 feet) downstream of the transition, the tunnel must turn 60-degrees to the south. Computational Fluid Dynamics (CFD) modeling of the bend was aimed at determining a radius that would prevent excessively high flow velocities. Several scenarios were modeled representing a miter bend and bends with radii between 4.6 and 61 meters (15 and 200 feet).

Figures 4 and 5 show the flow velocities through the bend at a horizontal plane located in the middle of the tunnel for 61 and 6.1-meter (200 and 20-foot) radius curves, respectively. Figure 5 shows significant flow separation and high velocities in excess of 15.2 meters per second (mps) (50 fps) at the bend. The 61-meter (200-foot) radius shown on Figure 4 significantly improves flow conditions with maximum velocities in the 12.2–12.8 mps (40 to 42 fps) range which represent a local increase in velocities of less than 20%. Based on these results a 61-meter (200-foot) radius was selected for the tunnel bend design and has been constructed.

Tunnel Gate Structure

A rectangular gate structure is being constructed at the approximate mid-point of the DPIT. There are two 30.5-meter (100foot) transition sections at both ends between the gate structure and the circular tunnel. CFD modeling was performed of the gate structure, transition sections and 61 meters (200 feet) upstream and downstream of the transitions. Multiple scenarios representing different gate opening widths and different head differentials across the gates were modeled. It was determined that the most critical hydraulic conditions can occur if an initial storm event occurs filling the reservoir, the tunnels are dewatered to provide storage for additional storm events, then a subsequent storm event occurs prior to dewatering the reservoir. The gate is assumed to be closed at the start of the second storm event. It is then assumed that

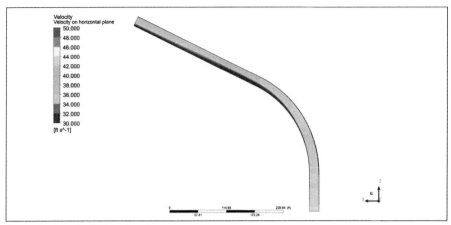

Figure 4. 61-meter (200-foot) radius curve

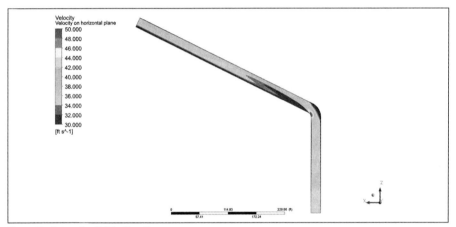

Figure 5. 6.1-meter (20 foot) radius curve

water in the tunnels rises until it reaches the same level as the water surface elevation within the reservoir, at that moment the gates are opened, and the reservoir can take on more water to relieve the tunnel system. Stormwater continues to enter the tunnel system at its maximum estimated rate of 317 cubic meters per second (11,200 cubic feet per second). Under this brief unbalanced condition flow velocity can be in excess of 30.5 mps (100 fps) through the gate and will remain high through the tunnel downstream of the gate until enough energy is dissipated resulting in the formation of a hydraulic jump. Figure 6 shows modeling results of this scenario with the gate open 30%. Under this condition, the gate and a significant portion (hundreds of feet) of the tunnel will be subject to excessive velocities and damage can be expected downstream of the gate especially areas not steelarmored. Based on the CFD modeling results, the tunnel is steel lined downstream of the gate structure to protect the tunnel from damaging high flow velocities.

Tunnel Portal

Modeling of the tunnel portal was aimed at determining flow velocities within the McCook Reservoir and evaluating the attenuation effect on velocities that different

Connecting the TARP Des Plaines Tunnel System to the McCook Reservoir   347

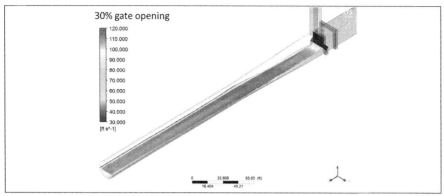

Figure 6. Water surface velocities downstream from gate 30% open in unbalanced condition

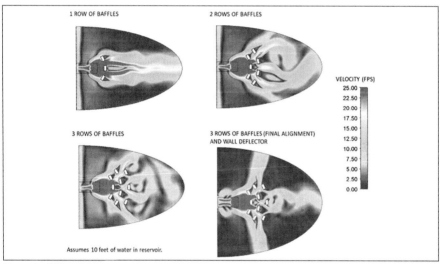

Figure 7. CFD modelling results of varying baffle block numbers and alignments

energy dissipation structures could have. The intent of the energy dissipation structures is not necessarily to eliminate high flow velocities but to prevent erosive forces along the reservoir walls. Additional consideration was given to high velocities on the floor of the reservoir and their extent. Variations in the scenarios modeled included depth of water in the reservoir and positions of the energy dissipation structures. The energy dissipation structures considered for this project included a concrete apron, baffle blocks and a backsplash wall. CFD modeling results of different baffle block numbers and alignments on the concrete apron are shown on Figure 7, with the preferential alignment as "(final alignment)." Figure 8 shows modeling results of the final alignment in elevation and isometric views.

CONSTRUCTION CHALLENGES AND SEQUENCING

Reservoir Components

Construction of components within the McCook Reservoir were completed by June 30, 2017 prior to the reservoir coming on line January 1, 2018. Mobilization to the

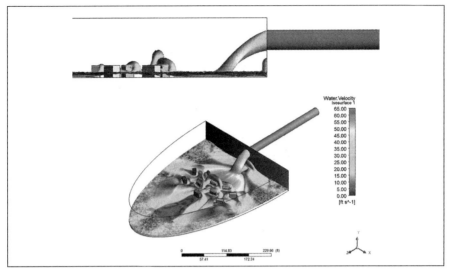

Figure 8. Elevation and isometric CFD modeling results of final energy dissipation alignment

McCook Reservoir did not begin until July 31, 2016. This required several components including highwall stabilization, the tunnel portal, and the energy dissipation apron slab with baffle blocks and backsplash wall be completed on an expedited schedule inside the reservoir all while three additional construction contracts were being completed. Construction of the McCook Main Tunnel System, the McCook Reservoir Final Preparation Inlet/Outlet Tunnel, and McCook Reservoir Weir Tunnels were progressing at the same time. Figure 9 shows a view of McCook Reservoir with the DPIT portal, highwall stabilization and energy dissipation structures that was completed prior to Stage 1 of the McCook Reservoir being placed into operation. A 10 m (32.8 ft) long rock plug remained intact approximately 15 m (49 ft) behind the tunnel portal to isolate the reservoir from the remaining DPIT construction.

Construction Shaft

The DPIT design called for installation of a construction shaft at the terminus of the adit. The construction shaft serves multiple purposes including moving equipment, materials and employees; providing an emergency egress; facilitating rapid access during the live connection to the TARP Des Plaines tunnel; and providing construction access to the DPIT from land where access could be obtained. To expedite installation of the construction shaft, it was completed by a raise bore after the adit was installed. This installation method shortened the project schedule compared to the drill and blast alternative. The raise bore reamer used to excavate the construction shaft is shown on Figure 10.

Gate Shaft and Tunnel

The gate shaft is located near the mid-point of the tunnel and, to abbreviate the construction schedule, was excavated simultaneously with construction activities in the reservoir. The gate shaft was excavated using conventional drill and blast methodology from ground surface. A shaft overburden initial support system consisted of a steel sheet pile with ring beam system and was socketed into bedrock. Pre-excavation grouting was required at the base of the gate shaft to limit groundwater infiltration from the Elwood Formation to facilitate concrete placement.

Figure 9. McCook reservoir with DPIT portal, highwall stabilization, energy dissipation structure

Figure 10. Raise bore reamer used to excavate the construction shaft

Excavation of the inflow tunnel commenced near the mid-point of the tunnel from the gate shaft. Drill and blast excavation was used and advanced from the gate shaft simultaneously to the north and south to reduce the construction schedule. The tunnel was shot in a single heading, rather than being benched. Labor for excavation occurred in three eight-hour shifts, 24 hours per day. The tunnel was excavated between March 20, 2017 and January 24, 2018. Figure 11 is a view of the tunnel and gate shaft excavation looking north.

Upon completion of excavation, the tunnel concrete liner was formed and placed and the steel liner downstream of the gate shaft was installed. Installation of the concrete liner was expedited by utilizing an Everest telescoping form system, decreasing setup and tear down duration. The steel liner was fabricated offsite at Weldall Manufacturing, Inc. in Waukesha

Figure 11. DPIT tunnel and gate shaft excavation

Wisconsin. It was manufactured in six sections, fit together in the facility for quality assurance, shipped to the site as separate sections, and reassembled in the tunnel prior to being cast in concrete. Figure 12 shows fabrication of a section of the tunnel steel liner for the transition downstream of the gate structure. Figure 13 shows the Everest telescoping form system in use.

The upstream and downstream DPIT tunnel branches are being excavated within 11.6 meters (38 feet) and 17.1 meters (56 feet) of completion, respectively. This leaves a rock plug at both ends of the tunnel so construction can proceed while the McCook Reservoir and the TARP Des Plaines tunnel are operable. A major challenge associated with the tunnel construction then becomes making live connections between the DPIT and these features. The rock plug at the McCook Reservoir will be the first to be removed because there is less risk associated with flooding from the reservoir than directly from a tunnel. After removal of the rock plug the concrete tunnel liner will be

Figure 12. Fabrication of tunnel steel liner

Figure 13. Everest form system used for tunnel concrete liner

installed. Due to a delay in the fabrication and placement of the high head gates Walsh is considering installing two concrete bulkheads to isolate the gate shaft after making the live connections, so that the gate installation can be protected. The concrete bulkheads would then be removed after the gates are placed and fully operational. This would allow the DPIT to be operational in the interim between making the live connections and installation of the gates.

The final tunnel construction activities are associated with the live connection to the existing TARP Des Plaines tunnel. This existing tunnel was designed with a concrete plug shown in Figure 3. To make the connection between the DPIT and Des Plaines tunnel the rock plug must be removed first, followed by removal of the concrete plug, and then placement of the tunnel liner. Installation of the tunnel liner requires the flow of the Des Plaines tunnel to be diverted. A diversion weir will be constructed upstream of the existing concrete plug and tunnel water will be diverted to the Des Plaines Dewatering Tunnel, shown on Figure 1. Significant coordination will be required to monitor and forecast weather conditions during this final live connection. In addition, water levels within the TARP Des Plaines tunnel will require close monitoring and maintenance by MWRDGC MSPS engineers to ensure the diversion weir is not overtopped.

CONCLUSION

The DPIT is being constructed to provide a direct connection between the existing TARP Des Plaines tunnel and the McCook Reservoir. Upon completion, the additional conveyance of the DPIT and storage capacity provided by the McCook Reservoir will further reduce the risk of flooding in the greater Chicago combined sewer system area, while also improving the quality of local waterways. At the time this paper was written, construction of the DPIT was 75% complete. Construction is scheduled to be complete in mid-2020.

REFERENCES

Hunter, P. and Lewis, S. 2012. Top 10 Biggest Wastewater Treatment Plants. Engineering News-Record. April 2. enr.construction.com/infrastructure/water_dams/2012/0328-Top10WastewaterTreatmentPlants.asp.

Construction of the Luck Stone Inter-Quarry Tunnel, Leesburg Virginia

Kyle Wooton ▪ Frontier-Kemper Constructors
Matt Bauer ▪ Frontier-Kemper Constructors

INTRODUCTION

Luck Stone is the largest family owned and operated producer of crushed stone in the United States. In 1971, Luck Stone acquired the Leesburg Plant (formerly Arlington Stone Company) in Loudon County, Virginia. The Leesburg Plant has not only played a major part in the recent development of Northern Virginia, but historically shipped rock by rail to Alexandria, Virginia since the late 1800s. In 2017, Frontier-Kemper Constructors was contracted to build the Inter-Quarry Tunnel. The Inter-Quarry Tunnel will connect the Leesburg Plant's existing East Pit to the future West Pit traveling beneath Goose Creek. Goose Creek is a state scenic river in the Potomac River basin. The Washington and Old Dominion trail crosses Goose Creek, parallel to the tunnel only 550 feet downstream from the tunnel crossing. The scenic river and trail were kept open through the duration of construction. The tunnel will accommodate both truck traffic and a conveyor belt, providing the East Pit processing plant with raw product and extending the life of mine.

METHOD OF EXCAVATION

The Inter-Quarry tunnel is a 27 feet high by 46 feet wide horseshoe shaped tunnel, 1,315 feet in length. The excavation and support of the tunnel utilizes a Sequential Excavation Method (SEM), based on the principles of the New Austrian Tunnel Method (NATM). The drilling and blasting of the tunnel is continuously monitored and the ground support system is adjusted accordingly to adequately support the changing conditions of the tunnel. The tunnel is built using "state of the art" blasting techniques, ground support systems, monitoring, and equipment by Frontier-Kemper Constructors. The IFC design and on-site SEM supervision is performed by Dr. Sauer and Partners Corp.

Luck Stone elected to utilize the SEM method primarily due to the proximity of Goose Creek to the tunnel. The tunnel passes under Goose Creek with only 50 feet of cover between the bottom of the riverbed and the top of the tunnel. Drill and blast method of excavation was selected based on the short length of the tunnel and the rock strata that is present. The tunnel passes through the same material that Luck Stone utilizes as their primary product at this facility, diabase. The diabase present in the

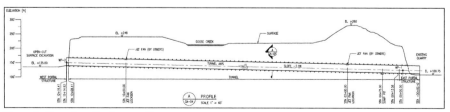

Figure 1. Tunnel profile, long section

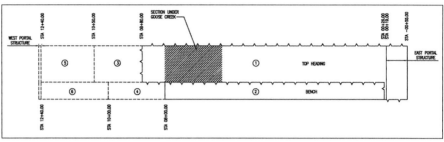

Figure 2. Sequence of excavation

Table 1. SEM support measures

	SC-1	SC-2	SC-3
A-Max Excavation Length Top Heading	10 ft.	6 ft.	4 ft.
B-Max Shotcrete Lag Top Heading	30 ft.	12 ft.	8 ft.
C-Max Excavation Length Bench	20 ft.	12 ft.	8 ft.
D-Max Shotcrete Lag Bench	40 ft.	12 ft.	8 ft.
Rockbolt Type (Roof)	#8 x 15 ft. Galvanized Threadbar		
Rockbolt Type (Ribs in Bench Area)	#8 x 10 ft. Galvanized Threadbar		
Rockbolt Spacing	eR = 8 ft.	eR = 6 ft.	eR = 4 ft.
	eL = 10 ft.	eL = 6 ft.	eL = 4 ft.
Shotcrete Thickness	2 in	4 in	8 in

Inter-Quarry Tunnel is averaged approximately 10,000 psi in the tunnel elevation in the pre-construction cores.

Frontier-Kemper elected to excavate the tunnel in two passes, a top heading and a bench. A top heading with a height of 21 feet was initially excavated and supported, and the bench was vertically drilled approximately eight feet and excavated. This heading height was selected based on the reach of the drill jumbos which were also utilized for installing rock bolts. The top heading was advanced to approximately 8 + 84 prior to excavation of the bench. The center of Goose Creek is located at approximately station 7+00 for reference. The section of tunnel from 6+00 to 8+00 encompassing the Goose Creek area was labeled the "red zone." Frontier-Kemper wanted to ensure the top heading in the "red zone" of 200 feet was fully excavated and supported prior to removing the bench.

GROUND SUPPORT

Using the SEM method of tunneling, the length of drill round and support installed in the tunnel each day was determined by the on-site SEM engineer. This daily decision was based on an intensive evaluation called the RES process. This process evaluates which type of support should be installed with tunnel advance based on the encountered field conditions. The tunnel face is mapped and the entire tunnel length monitored for movement before and after each drill round. The geologic structure, RQD, joint properties, joint pattern, and groundwater conditions in the rock strata, are some of the primary conditions evaluated daily to determine the support installed in the tunnel. The required support class (Type I, II, or III) was determined daily. A face map would be created and the Rock Structure Rating (RSR) and Tunneling Rock Quality Index (Q) would be estimated. Q is determined by the RQD and number of joint sets, joint roughness, joint alterations, and water inflow. RSR is determined by the rock type

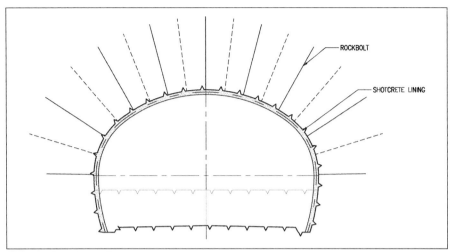

Figure 3. Tunnel ground support

and geological structure, joint spacing and direction, and water inflow. Under Goose Creek (Red zone area) two probe holes were drilled 50 feet, each at 15 degrees vertical to extend above the extent of excavation and alternated offset to the tunnel centerline. Before and beyond Goose Creek, one probe hole 30 feet long was drilled every 20 feet of tunnel excavation to allow an overlap of 10 feet. During the creek section "red zone" two probe holes were drilled, one on each side of centerline. During probe hole drilling, several things were taken into consideration including penetration rate, water color, and water flow to assist the SEM Engineer in determining rock quality as well as for the presence of ground water and geological anomalies such as voiding or fracture zones.

The primary materials used for ground support were #8 diameter galvanized threadbar, 10 feet and 15 feet in length, and fiber reinforced shotcrete. Other "toolbox items" such as welded wire fabric, self-drilling spiles, lattice girders, cementitious grout, and chemical grout were on-site in case the evaluation of the ground conditions determined they were needed for support. However, due to favorable ground conditions, these supplemental items were not used in support of the Inter-Quarry Tunnel.

UNDERGROUND EQUIPMENT

Frontier-Kemper utilized a state of the art fleet of equipment to build the Inter-Quarry Tunnel. Two Atlas Copco E2C jumbo drills were utilized for both face drilling and installing rock bolts. The E2Cs drilled the face rounds simultaneously utilizing the automated drilling program "Tunnel Manager." The entire top heading round could be reached with the drills without leaving material in the heading. The E2Cs were also utilized for drilling the vertical bench round production holes.

The explosives were delivered and loaded into the face using a custom-built machine designed to transport and load bulk emulsion from one unit. Dyno Nobel installed the "Dyno Miner," an isolated pumping system onto a Normet Himec 9905 BT. This allows the machine to pump emulsion from a mounted tank into any desired drill holes from either of the two booms. The machine has remote control capabilities which allow each of the booms to start and stop pumping emulsion independently.

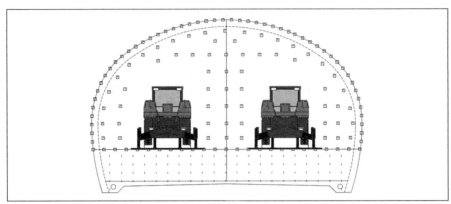

Figure 4. Drill jumbos in face

The shot rock was removed from the tunnel using three Atlas Copco ST14 Scooptrams. The ST14 Scooptrams have a 14,000 kg tramming capacity. Based on the length and width of the tunnel it was determined that utilizing three LHDs was more productive than using a loader and trucks to haul the material out of the tunnel. The LHDs were able to pass each other within the tunnel and at times, all three machines were digging in the face simultaneously. This enabled Frontier-Kemper to remove the shot rock from the tunnel at a rate of 186 cubic yards per hour.

Figure 5. Spraymec applying shotcrete to tunnel walls

Scaling of the tunnel roof and walls was performed using a tunnel excavator. A Caterpillar 328D LCR was custom modified by Zepplin Caterpillar in Stuttgart, Germany to operate in underground conditions.

The fiber-reinforced shotcrete was delivered to the project site by ready-mix trucks. The trucks transferred the material into Normet Ultimec LF 600 transmixers for transport into the tunnel. The transmixers have a capability of over 7 cubic yards of carrying capacity. The transmixers transported the shotcrete to the tunnel face and discharged into a Normet Spraymec 8100 VC. The Spraymec is a remote controlled mobile concrete sprayer with an onboard air compressor. The shotcrete in the Inter-Quarry tunnel was typically applied without having to move the Spraymec after the initial set-up.

BLASTING

Blasting was conducted in three sections; the portal section, a top heading and a bench.

The portal section was divided into a left and right heading. The left heading pattern included a burn cut and had a total heading size of 30 feet wide by 29 feet high. The burn incorporated four reamer holes with a 4 inch diameter and were drilled 6 inches deeper than the production holes. Production holes were drilled on a 36 inch by 36 inch spacing. The production holes in both the left and right headings were also

Construction of the Luck Stone Inter-Quarry Tunnel, Leesburg Virginia

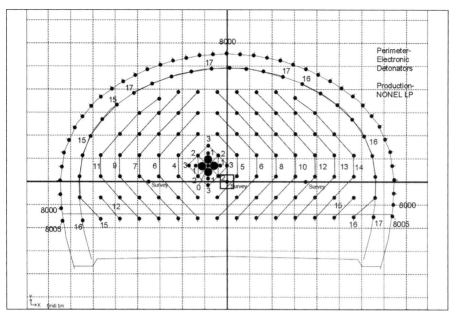

Figure 6. Blasthole diagram

drilled with varying drill depths due to the 50 degree angle of the tunnel centerline in relation to the face of the highwall. The production holes on the left side of the face were also drilled deeper than the right side to gradually square up the face, which was accomplished once the tunnel was excavated 40 feet. The left heading was shot using Dyno Digishot electronic detonators in combination with NONEL LPs and was advanced ahead of the right heading by 16 feet.

The right heading was shot with Dyno EZDET's and slashed to the left free face. Perimeter holes were drilled at 12 inch spacing and 24 inch burden from the rib relievers with a total heading size of 19 feet wide by 29 feet high. During the majority of the portal section, the perimeter holes were left unloaded in order to allow the relievers to break to a set perimeter line.

The top heading pattern was drilled with 2 inch diameter holes and was approximately 48 feet wide by 21 feet high with a production hole spacing of 40 inches by 40 inches. The pattern including a burn cut comprised of four, 4 inch diameter reamer holes (6 inches deeper than the production holes), rib reliever holes, and perimeter holes. The perimeter holes were on 1.5 feet spacing and a 2 feet burden from the rib relievers. The top heading was shot using Dyno NONEL LP, a single 90 grain booster, and Titan 7000 bulk emulsion in the production holes, while the perimeter was loaded with Digishot, 90 grain booster, 400 grain detonation cord and 1 pound of packaged emulsion to secure the booster. Typical drill depth was approximately 13 feet with a pull of 13.5–14 feet.

The bench was 7 feet high by 47 feet wide, and shot in 200 feet sections to achieve finished tunnel excavation height of 28 feet height. The pattern had 2.5 inch drill holes and a 5 feet spacing with 4 feet of burden. It was shot with 25/500 EZDET's, 42 millisecond EZTL's and 350 grain boosters. The bench was shot towards the free face at the east end of the tunnel.

MINING CONDITIONS

The mining conditions in the tunnel were more favorable than anticipated. The diabase reacted well to the blasting techniques employed by Frontier-Kemper Constructors. Only minimal scaling was required and little overbreak was encountered. Two water bearing fault zones were encountered to the east of Goose Creek with minor to moderate water inflow (50 gpm on average). The rock strata beneath Goose Creek in the designated "red zone" was more favorable than expected with no significant water inflows. The majority of the tunnel was supported using the SC II support type. This consisted of #8 galvanized threadbar 10 feet and 15 feet in length fully encapsulated with resin and 4 inch thick fiber reinforced shotcrete. Water inflow was controlled by lowering the left side corner holes in the blast pattern to create a ditch line down the left side of the tunnel. Since the tunnel had a positive grade of 2%, water control was not a construction issue. Water inflows encountered would then be directed to the subfloor drainage system using MIRADRAIN drainage mats secured to the rock and encapsulated with a shotcrete layer.

CONCLUSION

The Luck Stone Inter-Quarry Tunnel project shows that large diameter tunneling utilizing the drill and blast method can be effective in a busy suburban area and in challenging mining conditions such as under Goose Creek. The SEM process was not only a successful tool used to mine under Goose Creek, but also was used to communicate strategy regarding blasting and material deliveries to minimize negative impact to the nearby residents, pedestrians on the Washington and Old Dominion trail, and production in the operating mine. The SEM process created an environment that enabled the mine operator, the resident engineer, and the contractor to all evaluate the daily risks and impacts of the project through thorough communication including detailed daily technical reports and frequent "face to face" meetings.

ACKNOWLEDGMENT

The authors would like to thank the following individuals for their specialized technical assistance during the project:

Bryan Smith, Luck Stone

Elmar Feigl, Dr. Sauer and Partners Corp.

Nikolaos Syrtariotis, Gall Zeidler Consultants

John MacGregor, Dyno Nobel

Drill and Blast Construction of Shafts and Starter and Tail Tunnels for the 3RPORT CSO Project

Stephen Miller ▪ Schnabel Engineering
James Parkes ▪ Schnabel Engineering
Manfred Lechner ▪ Lane Construction
Lance Waddell ▪ Lane Construction

ABSTRACT

The 3RPORT Project consists of approximately 25,000 feet of TBM-mined tunnel with minimum inside diameter of 16 feet and at a depth of 230 feet, constructed primarily through limestone. The project includes three shafts with finished inside diameters ranging from 26 to 64 feet in diameter. The TBM will be launched from a starter tunnel extending from a working shaft, with trailing gear extending back through a tail tunnel that connects to the pump station shaft. Construction of these shafts requires up to 170 feet of drill and blast shaft excavation. The starter and tail tunnels will have a combined length of approximately 250 feet and are also constructed using drill and blast methods, with several enlarged sections to accommodate the TBM launch and support equipment.

This paper describes the drill and blast construction for these shaft and tunnel elements. This includes the design of the initial rock support consisting of rock dowels and shotcrete for both the shafts and starter and tail tunnels and development of alternative testing procedures for these elements. Procedures for the drill and blast excavation rounds are also presented, including the staging of the blast rounds between the adjacent shafts, details of the blasting including round lengths and perimeter controls, methods used for preventing fly-rock out of the shafts, and staging of the work in the starter and tail tunnels. Issues encountered in the field and resolutions for these operations are also presented.

INTRODUCTION

The Three Rivers Protection and Overflow Reduction Tunnel (3RPORT) is located in Fort Wayne, Indiana and is part of the City's Long Term Control Plan for reduction of combined sewer overflows (CSO's) discharged into the St. Joseph, St, Mary's, and Maumee Rivers during rain events. Once completed the tunnel and drop shafts will provide an estimated 90% reduction in the number of overflows per year for the City of Fort Wayne and is a significant part of the effort to clean up the neighborhoods and rivers, as well as prevent local flooding. The estimated volume of overflow that can be stored in the tunnel is approximately 37 million gallons, which will then be treated by the Water Pollution Control Plant. The project includes a Pump Station Shaft that will have a minimum finished inside diameter of 64 feet, a Working Shaft that will have a minimum finished inside diameter of 29 feet, and retrieval shaft with a minimum finished inside diameter of 26 feet. Each of the shafts will be installed through soft ground with slurry walls and then supported in rock with rock bolts and shotcrete. The tunnel is required to have a minimum inside finished diameter of 16 feet and will be at least 24,500 feet in length and approximately 200 feet beneath the ground. The tunnel will be excavated using a Tunnel Boring Machine and lined with a segmental precast liner to be installed as tunneling progresses.

In 2017, the Herrenknecht Company was selected to build a new Slurry Mode TBM for the excavation of the tunnel. The TBM has a 19 foot cutter head as shown in Figure 1 and has trailing gear that will stretch over 400 feet long behind the cutterhead. The tunnel will be finished using precast concrete segments with a final inside diameter of 16 feet. The TBM will be launched from a starter tunnel approximately 100 feet long with trailing gear extending though a tail tunnel approximately 100 feet long. The Tail Tunnel connects the Pump Station Shaft to the Working Shaft. The Starter Tunnel extends approximately 100 feet from the Working Shaft and will include an expanded section to enable the installation of the TBM launch frame.

Figure 1. The TBM cutterhead assembly

The focus on this paper will be on the design and construction of the shafts and starter and tail tunnels in hard rock using drill and blast methods, which are currently under construction at the time this paper was written. The Working Shaft is near completion, with excavation currently under way on the Starter Tunnel and Pump Station Shaft, with the Tail Tunnel excavation to start once the Starter Tunnel excavation has completed.

PROJECT LOCATION AND LAYOUT

The project is located in Allen County, Indiana in the city of Fort Wayne, and follows along the Maumee River and the St. Mary's River. The tunnel will run from the Water Pollution Control Plant, where the Pump Station Shaft and Working Shaft are located, to Foster Park, where the Retrieval Shaft is located in the south west portion of the city as seen in Figure 2.

The tunnel features eleven drop shafts and ventilation shafts ranging from two feet to seven feet in finished inside diameter connected throughout locations of the city to aid in the diversion of storm water to the tunnel system. The tunnel will stretch almost five miles in length once complete.

PROJECT GEOLOGY AND GEOTECHNICAL CONDITIONS

The bedrock geology of Allen County consists mainly of sedimentary and carbonate rocks, overlain with glacial and alluvial deposits. The subsurface rock has been subjected to groundwater solution activity, which has created open joints and cavities along the tunnel alignment.

For this project, a Geotechnical Data Report and a Geotechnical Baseline Report were both provided as part of the contract documents based on a comprehensive geotechnical investigation. During the geotechnical investigation there were deep vertical borings to depths of 250 feet, and shallow borings down to 50 feet at specific areas that were critical to the project. Also included as part of the geotechnical investigation were water pressure tests in rock borings, aquifer pumping tests, and soil and rock laboratory tests. There were piezometers installed for groundwater monitoring along the tunnel alignment. The GBR categorized the subsurface rock into three Rock Engineering units:

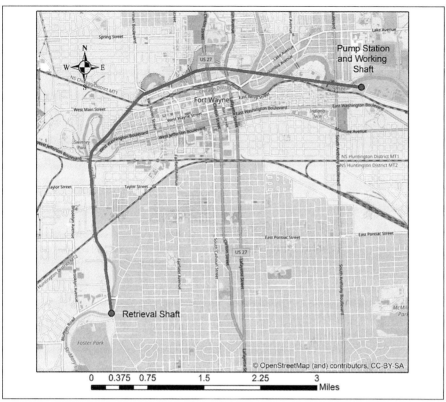

Figure 2. Project location

Rock Engineering Unit 1: Ranges from thin beds (2 inches) to thick beds (3 feet), with near horizontal bedding. The rock consists of light to medium gray, medium-grained, fossiliferous shaly limestone with chert nodules and layers. This rock has an average unit weight of 167 pcf, an average Unconfined Compressive Strength of 17,000 psi, an average Point Load Index of 750 psi, and an average Brazilian Tensile Strength of 1,100 psi.

Rock Engineering Unit 2: Ranges from thin beds (4 inches) to thick beds (3 feet), with near horizontal bedding, and is typically blocky in nature. The upper zone of Rock Engineering Unit 2 consists of light to medium gray, fine- to medium-grained, massive, conglomeritic dolomite. There are argillaceous layers of varying thickness present in the upper zone. The lower zone consists of dark brown to tan dolomitic limestone or dolomite, banded with white, crystalline calcite and shale beddings. Also present within the lower zone are some coarse sandy layers. Vugs are present in both the upper and lower zones of Rock Engineering Unit 2. This rock has an average unit weight of 153 pcf, an average Unconfined Compressive Strength of 10,700 psi, an average Point Load Index of 530 psi, and an average Brazilian Tensile Strength of 700 psi.

Rock Engineering Unit 3: Thickly bedded to massive and free of chert nodules and layers. It consists of medium to dark gray, dense, vuggy, massive argillaceous dolomite with greenish-gray shale partings. There are near vertical features in this layer. Also present are pressure dissolution features such as stylolites, and evidence of bioturbation. Reef facies are prevalent in this formation and compose approximately 75 percent of this unit. This rock has an average unit weight of 155 pcf, an average Unconfined Compressive Strength of 9,900 psi, an average Point Load Index of 510 psi, and an average Brazilian Tensile Strength of 759 psi.

The construction of the shafts will occur in all three Rock Engineering Units, while the TBM tunnel is anticipated to be excavated in Rock Engineering Units 2 and 3. The Starter Tunnel and Tail tunnel will be excavated in Rock Engineering Unit 3 only.

Horizontal features in the rock as well as sub-vertical features may contribute to groundwater inflows that will need to be managed in the shafts and tunnels during construction. The hydrostatic conditions in the TBM portion of the tunnel will vary as the construction progresses.

INITIAL ROCK SUPPORT DESIGN FOR SHAFTS AND TUNNELS

The initial rock support for the three shafts, the starter tunnel, and the tail tunnel were designed based on the results of the geologic site investigation, while factoring in constructability to provide common design elements between the different shafts and tunnels.

The rock loading for the shafts was determined through Terzaghi's method, as modified by Deere et al. (1970) which added in the consideration of RQD. Also considered was guidance from the U.S. Army Corps of Engineers EM 1110—2-2901 (1978), on application of this method for shaft design. The initial and final loading for each of the three Rock Engineering Units and each of the shafts was calculated. The initial loading was determined to be zero in all cases. The final loading varied between the shafts and each of the Rock Engineering Units by as much as 20%.

While the initial rock support for the shafts could have been designed using different support classes for the different quality of rock in each of the shafts based on the different loading and conditions, it was decided to produce a single design for each shaft based on the highest load calculated in each shaft. This provides additional safety to the design and increases long term stability. Additionally, it simplifies the construction methodology and allows for an increase in construction efficiencies while reducing the room for errors associated with rock support installation.

The rock load calculations for the Starter and Tail tunnel utilized the Terzaghi method, similar to the calculation for the shafts. An additional method for determining rock loading from Rock Mass Rating was also used, based on Bieniawski's methods and Unals formula (1989). The Norwegian Q-System analysis was also performed on the data as an additional check on the empirical design requirements for the tunnels (Barton et al., 1980). The rock loading was only determined for Rock Unit 3, as this was the anticipated Rock Engineering Unit where the Starter and Tail Tunnels were to be excavated.

Using a similar approach to the shaft support design, calculations for the Starter Tunnel, the Starter Tunnel expansion for the TBM frame, the Tail Tunnel, the Tail Tunnel with a lowered invert, and the Tail Tunnel with the slurry pump equipment chamber were analyzed for design overlap. A systematic design that was applicable to all of the Starter and Tail Tunnels sections was developed. As with the design for the shafts, this provides additional safety to the design and increases long term stability, while simplifying construction.

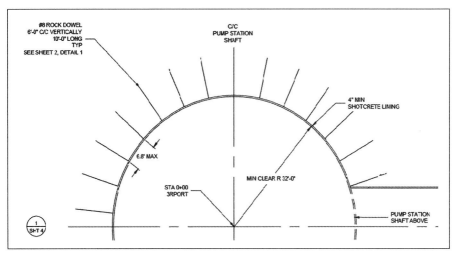

Figure 3. Pump station shaft typical dowel pattern

The resulting initial rock support designs for the shafts and tunnels featured #8 threadbar dowels embedded 10 feet into rock at a downward angle of 15 degrees to provide a slight downward force and to intersect the horizontal bedding. The dowels are designed to withstand the entire final rock loading that was calculated. To supplement the dowels, 4 inches of unreinforced shotcrete will be added to the shafts, as well as the tunnels. The shotcrete was designed to withstand any larger blocks or wedges that could develop between the dowels. In the tunnels, welded wire mesh will be utilized to provide additional support prior to the installation of the shotcrete. While not necessary from a loading standpoint, it provides additional safety from overhead rock fall until the shotcrete can be installed.

Pump Station, Working, and Retrieval Shafts Initial Rock Support Design

As previously mentioned, the initial rock support for the shafts is designed based on the information for the three Rock Engineering units presented in the GBR. The rock dowels in the shafts are spaced 6 feet vertically in each of the shafts. The Pump Station Shaft dowel pattern, as seen in Figure 3, and Retrieval Shaft dowel pattern have radial dowel spacings of 6.8 feet, and the Working Shaft dowel pattern has a radial spacing of 6.7 feet. The Working shaft has a slightly smaller radial spacing due to the number of dowels that are required to provide complete support around the circumference of the shaft.

The intent of the design was to use a similar spacing for all three shafts for simplicity for the construction crews.

The dowels in the shafts are designed to withstand the largest block that could potentially develop based on the information in the GBR. The dowels are also designed to generate a rock arch once the rock loading has developed that is capable of withstanding the rock loads in their entirety.

Starter and Tail Tunnel Initial Rock Support Design

In a similar manner to the shaft initial rock support design, the designs for the Starter and Tail Tunnels utilize a systematic design that is consistent between the two tunnels and all the enlarged sections. The dowels and shotcrete used in the tunnel sections are the same as that used in the shaft designs. This simplifies the procurement, storage, and installation of the support elements. The typical Starter Tunnel Cross Section as seen in Figure 4 shows the placement of the dowels in the crown and the walls of the tunnel.

The Starter and Tail tunnels all have the same crown shape and dimensions and therefore utilize the same rock dowel spacing in the crown, with 7 dowels spaced 5 feet radially, and 5 feet horizontally from the next row of dowels. The dowel patterns in the walls of the tunnels are all 4 feet vertical spacing, and 5 feet horizontal spacing. The tunnel heights dictate the number of vertical rows of dowels, with the lowest row of dowels placed no farther than 4 feet from the invert.

The Starter tunnel is 95 feet long, 27 feet wide by 23 feet high horseshoe shaped. There is an enlarged section with a height of 31 feet to provide room for the TBM launch frame. The tunnel will be excavated with a top heading and bench. Although analyses indicate that the Starter Tunnel could be excavated in a single heading, the height is too tall based on consideration for the equipment to be used for this excavation.

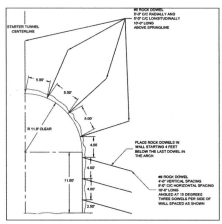

Figure 4. Typical starter tunnel cross section

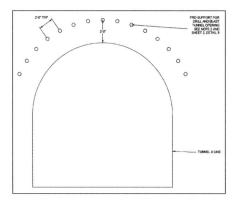

Figure 5. Typical spile installation above tunnels

The Tail tunnel is 97 feet long and is horse shoe shaped and will initially be 27 feet wide and 26 feet high. A section of the tail tunnel 18 feet long next to the Pump Station Shaft will be excavated down to the same elevation as the Pump Station Shaft, for a total tunnel height of 41 feet. There is an additional expanded section in the tail tunnel that will be excavated to provide room for the slurry pump necessary for the TBM. The tail tunnel will be excavated with a top heading and bench, and an additional excavation of the section with the invert down to the level of the Pump Station Shaft invert.

The initial crown support for the tunnels from the shafts was provided by 20 foot long fully cement grouted spiles above the excavation of the tunnel crown as shown in Figure 5. The spiles will provide additional support to the intersection during prior to the excavation of the tunnels. This minimizes the overbreak at the tunnel brow and allows for the initial rock support installation to occur with less risk of failure of the crown of the tunnel. The spiles are to be installed in four locations in the shafts. Spiles will be installed in Pump Station Shaft and the Working Shaft above the crown of the Tail Tunnel, in the Working Shaft above the Starter Tunnel, and in the Retrieval Shaft above the TBM exit location.

EXCAVATION METHODS

The Pump Station Shaft and the Working Shaft were excavated concurrently, due to their proximity to each other of less than 100 feet from shaft wall to shaft wall. This allowed for resources to be shared for the two shafts through the use of a staggered excavation cycle between the shafts. The construction cycle for the shafts followed a typical drill, blast, muck, and initial rock support installation construction cycle. The Pump Station Shaft sequence would take on average 5 days to complete a cycle, while the Working Shaft would take an average of 3 days to complete a cycle. This resulted in the completion of the Working Shaft ahead of the Pump Station Shaft. The Starter Tunnel excavation from the Working Shaft began prior to the excavation of the Tail Tunnel. The Starter Tunnel excavation proceeded first in order to expedite the installation and launch of the TBM, whereas the Tail Tunnel will be used for trailing gear and TBM utilities.

The overall construction sequence for the different excavations was as follows:

1. Concurrent excavation of Working Shaft and Pump Station Shaft
2. Excavation of Starter Tunnel once the Working Shaft is complete
3. Excavation of Tail Tunnel once the Pump Station Shaft is complete
4. Excavation of Tail Tunnel expansion with the remaining Pump Station Shaft
5. Excavation of Retrieval Shaft once Pump Station, Working Shaft, Starter Tunnel, and Tail Tunnel are complete

Common items that were installed in all the shafts and tunnels included weep drains in the shotcrete, drainage board to aid in the installation of shotcrete, drainage tubing to divert any flowing water to a sump, ventilation lines, and piping for the pumps.

Shaft Excavation by Drill and Blast

The Working Shaft and the Pump Station Shaft (Figure 6) both utilized a typical production hole that is 2½ inches in diameter and drilled to a depth of 11 feet. The hole is loaded with 7 feet of explosives and stemmed with 4 feet of crushed angular gravel. A prepackaged explosive emulsion, with 2.02 lbs per foot for a total of 14.1 lbs per hole was used.

The typical perimeter hole is 2½ inches in diameter and drilled to a depth of 11 feet. The hole is loaded with 8.5 feet of explosives and stemmed with 2.5 feet of crushed angular gravel. The powder used was a rigid paper cartridge with couplers, initiated with a non-electric detonator, with a total of 4.39 lbs per hole.

Also shared between the blast patterns for the shafts was the burn design (Figure 7). The burn consists of four, 4 inch relief holes spaced 17 inches from each other in a square pattern, with a production hole in the center and four production holes 24 inches from the center production hole. The burn pattern is surrounded by the first production ring with an 8 foot diameter that has nine production holes.

Figure 6. Pump station shaft construction

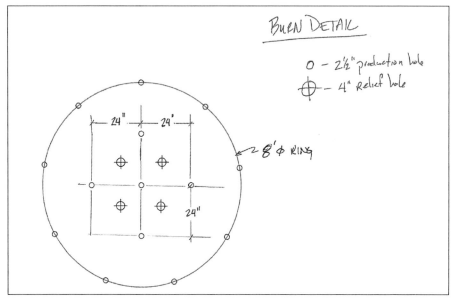

Figure 7. Typical shaft burn detail

The typical Pump Station Shaft blast layout consists the burn, eight rings of production holes, and a perimeter ring. The number of holes per ring is shown in Table 1. The production holes are arranged in rings around the burn, with the spacing between the first ring and the second ring at 3 feet, and the second and the third ring at 3 feet. The remaining rings are spaced at 4 feet, with the exception of the perimeter holes, which are spaced at 2.5 feet from the last production ring.

Table 1. Pump station shaft blast holes

Ring No. @ Radius	# of Holes
Burn	5
Ring 1 @ 4'	9
Ring 2 @ 7'	11
Ring 3 @ 10'	16
Ring 4 @ 14'	22
Ring 5 @ 18'	28
Ring 6 @ 22'	35
Ring 7 @ 26'	41
Ring 8 @ 30'	48
Perimeter	102
Total	317

The typical Pump Station blast round uses 3479 total pounds of explosive to blast 1352 cubic yards of material in one blast, which results in a typical powder factor of 2.57 lbs per cubic yard. The timing and pattern between each hole is also shown.

To reduce the Pump Station excavation cycle time, the entire Pump Station round was shot as one blast, which presented some challenges due to the number of holes in the pattern. The pattern was split into two sides to allow for proper timing and optimal detonation of the holes. Each of the holes utilized a 5400 ms delay cap at the bottom. To achieve proper detonation, the entire round had to be sequenced to initiate all the bottom caps prior to the detonation of the first hole, which would occur 5400 ms after the round detonation was triggered. Shooting the round as a single shot reduced the cycle time by an estimated 24 hours.

The typical Working Shaft blast round consists of the same burn used in the Pump Station Shaft, surrounded by three rings of production rounds, and the perimeter ring.

The number of holes per ring is shown in Table 2. The production holes are arranged in rings around the burn, with the spacing between the rings at 4', with the, with the exception of the perimeter hole ring, which is spaced at 3' from the last production ring.

The typical Working Shaft blast round uses 855 total pounds of explosive to blast 288 cubic yards of material in one blast, which results in a typical powder factor of 2.97 lbs per cubic yard. The size of the Working shaft did not present any challenges to the configuration of the blast round pattern as compared to the Pump Station Shaft

Table 2. Working shaft blast holes

Ring No. @ Radius	# of Holes
Burn	5
Ring 1 @ 4'	9
Ring 2 @ 8'	13
Ring 3 @ 12'	19
Perimeter	47
Total	90

The Retrieval Shaft blast round design will use a similar pattern to the Working Shaft that is optimized for the smaller 26 foot diameter size of the shaft. The blast round design for the Retrieval Shaft will be finalized after making any adjustments based on the knowledge gained from the construction of the Pump Station and Working Shafts.

A removable shaft cover was used to prevent flyrock during blasting for both shafts as seen in Figure 8. The cover for the Pump Station Shaft consisted of two pieces; a larger frame piece to cover most of the opening as seen in Figure 9, and smaller center piece.

Figure 8. Removable shaft cover

Figure 9. Removable shaft cover

The center piece is the same size as the Working Shaft, which allows that piece to be used for both shafts for flyrock prevention. The use of this cover system eliminated the need for blasting mats to be lowered down into the shafts with each blast, which was more time consuming.

Tunnel Excavation by Drill and Blast

Tunnel excavation began with the Starter Tunnel excavation. The Starter Tunnel top heading (Figure 10) was excavated first using 10 foot long rounds. The Starter Tunnel blast round was designed with 3' by 3' pattern and burn as shown in Figure 11. The burn pattern for the tunnels is similar to the burn for the shafts, with the use of four relief holes arranged in a square pattern. The Starter Tunnel was excavated with a top heading consisting of the upper 17 feet. Once the entire top heading is excavated, the remaining bench will be shot in either a single round, or in two rounds, which will

Figure 10. Starter tunnel heading excavation

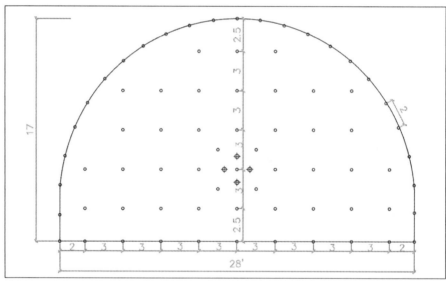

Figure 11. Starter tunnel typical heading blast pattern

depend on the condition of the invert. An assessment will be made once the Starter Tunnel Heading is complete on the decision to shoot the bench as one or two rounds.

The typical Starter Tunnel Heading blast round used 586 total pounds of explosive to blast 174 cubic yards of material in one blast, which results in a typical powder factor of 3.37 lbs per cubic yard. The rock encountered is massive, with no visible joints or discontinuities. Overbreak in the Starter Tunnel has been minimal.

The Tail Tunnel will utilize similar blast rounds to that of the Starter Tunnel, except that it will be a full face round instead of a top heading and bench. The round length will typically be 12 feet long. The typical Tail Tunnel Heading blast round will use 668 total pounds of explosive to blast 299 cubic yards of material in one blast, which results in a typical powder factor of 3.15 lbs per cubic yard. The Tail Tunnel will be excavated from the Pump Station Shaft, with the exception that one round for the Tail Tunnel will be excavated from the Working Shaft. This will allow for the initial support for the tunnel turn under to be installed and also provide a slight buffer for the Tail Tunnel break through.

The overall quality of the rock has been consistently good for blasting. The trim powder in the perimeter holes has produced constant half-casts in both the shafts and tunnels, with minimal over break. The results have been fairly smooth and consistent shaft and tunnel walls, which will reduce the volume of cast in place concrete needed for the final liner.

ADJUSTMENTS DURING CONSTRUCTION

During the drill and blast construction of the shafts and tunnels, there were adjustments made to methods of construction to improve efficiencies on the project from what was originally planned.

One of the initial issues that was realized earlier on in the project was the lack of local expertise in underground construction for the greater Fort Wayne area. To overcome this issue, specialized crew members with expertise in shaft sinking and underground excavation were brought in on the project. These crew members were placed on different shifts to work with and train the less experienced workforce. This allowed for the local work force to become trained specifically for this project.

Early on in the construction process, the 28 day strength of the shotcrete used in the shafts was coming in below the minimum 6,000 psi strength required in the specifications. The issue was resolved by switching the supply of the cement from a different source, with no changes made to the mix design or any of the processes involved in the application of the shotcrete. The shotcrete 28 day strengths are now above 6,000 psi based on the results of in-situ core tests. Design analyses were re-run to confirm that the lower strength shotcrete already in place was still adequate for the anticipated rock block sizes and loads.

The original plan for the Starter Tunnel was to excavate the tunnel using a full face round. It was decided to switch to a top heading and bench approach based on the size of equipment that was available that would easily fit into the Working Shaft. A smaller jumbo was used that could be raised and lowered into the shaft with lower risk of the jumbo swinging into the shaft walls, damaging the jumbo, the shaft rock support, or other utilities. Because the Tail Tunnel will be excavated from the larger Pump Station Shaft, a larger jumbo can be used in the excavation of that tunnel. This allows for the Tail Tunnel to be excavated in a single round.

An adjustment to the rounds for the Starter Tunnel top heading was made after the first few rounds of excavation. The rounds were not pulling as cleanly as desired, so the blast pattern spacing was reduced to 29 inches by 29 inches. This resulted in a slight increase in the powder factor, but the rounds pulled cleanly.

The lower portion of the Working Shaft was modified (Figure 12) based on changes to the final design of the interior of the finished Working Shaft structure, which increased the diameter of half the shaft by four feet. Due to the robustness of the original initial rock support for the shafts, the original initial rock support for the Working Shaft was sufficient for the modification, with the main difference being the addition of extra dowels in each horizontal row due to the change in the shaft diameter.

PROJECT STATUS

As of the time of writing of this paper, the excavation of the Pump Station Shaft, Working Shaft, and the Starter Tunnel have neared conclusion, with preparations for the excavation of the Tail Tunnel, installation of the TBM and support equipment in preparation for the launch of the TBM excavated tunnel in 2019. Once the construction has completed at the Pump Station Shaft, Working Shaft, Starter Tunnel, and Tail Tunnel, work will shift over to the construction of the Retrieval Shaft using similar means and methods, while excavation of the tunnel with the TBM is underway.

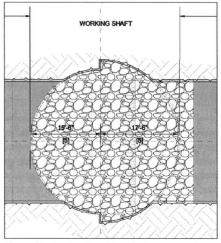

Figure 12. Modified working shaft cross section

CONCLUSION

The drill and blast construction of the Pump Station Shaft and Working Shaft, as well as the Starter Tunnel and Tail Tunnel, are expected to be completed in early 2019, with the Retrieval Shaft to be completed in late 2019. The TBM tunneling portion of the 3RPORT project is expected to be completed in 2021, with initial connections to many neighborhoods by 2023. The entire project will be fully completed in 2025.

Many factors were considered in the design of the initial rock support for the shafts and drill and blast excavated tunnels, which allowed for the consistency in the design of support elements that simplified construction between the shafts and tunnels. This provided additional benefits to the construction schedule as well as an increased level of safety and stability of the excavations.

REFERENCES

Deere, D.U. et al. (1970), "Design of Tunnel Support Systems," Highway Research Record No. 339, pp. 26–33.

US Army Corps of Engineering, EM 1110-2-2901, "Tunnels and Shafts in Rock," 15 Sep 1978. Section 3.

Bieniawski, Z.T. (1989). Engineering Rock Mass Classifications. John Wiley & Sons, New York, NY.

Barton, N., Løset, F., Lien, R. and Lunde, J. 1980. Application of the Q-system in design decisions. In Subsurface Space, (ed. M. Bergman) 2, 553–561. New York: Pergamon.

PART  6

Environment, Health, and Safety

Chairs

Kim Wilson
LA Metro

Liam Dalton
HNTB

Commissioning of Tunnel Fire Life Safety Systems and Its Challenges

Hubert Heis ▪ ILF Consulting Engineers Austria GmbH
Reinhard Gertl ▪ ILF Consulting Engineers Austria GmbH

ABSTRACT

The rehabilitation of the electrical and mechanical systems of a 2.8 km long twin-tube motorway tunnel in Austria was performed in two phases, by closing one tube for works and operating the other in bidirectional traffic mode. The design services included the rehabilitation of more than 20 systems, encompassing the upgrade of existing systems and the addition of new systems, such as a high-pressure water mist system for active fire protection of the civil structure. The test procedure as part of final commissioning focused on the water mist and the ventilation system. The final assessment of all systems, functions and interfaces was performed during the live fire tests, conducted for two scenarios: emergency operation with and without activation of the water mist system.

INTRODUCTION

The tunnel consists of two tunnel tubes with two lanes per tube, which were opened to traffic in 1993. The operator and owner of the tunnel is the Austrian publicly owned corporation ASFINAG. The north tube of the tunnel has a length of 2,842 m (9,324 ft) and the south tube is 2,808 m (9,213 ft) long. Most of the tunnel was mined according to the principles of the New Austrian Tunneling Method (NATM). The east portal sections of 93 m (305 ft) of the north tube and 82 m (269 ft) of the south tube were built as cut-and-cover tunnels.

The west portal sections of 730 m (2,395 ft) and 731 m (2,398 ft) were built using the cut-and-cover method, as a rectangular cross section with ribbed ceiling. The height of clearance is 4.7 m (15 ft 4 in) above the road. Jet fans and other equipment are mounted above the clearance.

The average longitudinal slope of the tunnel tubes is 1.22%, with a minimum of 0.59% and a maximum of 1.84%. The tunnels rise continuously from the east portal to the west portal. The difference in altitude from the east portal to the west portal is roughly 39 m (128 ft).

The tunnel has 11 cross passages. Nine are for passengers and two are suitable for vehicles, such as fire-fighting vehicles. Except for the cross passages in the cut-and-cover section all cross passages are constructed as air locks. Each tube has two emergency lay-bys next to the cross passages, which are suitable for vehicles.

The Annual Average Daily Traffic (AADT) for the year 2033 is predicted to be 28,200 vehicles. The share of heavy goods vehicles is 6,090 trucks, which results in about 21.6% of the total vehicles being heavy goods vehicles.

The construction works were divided into two main construction phases. Preparations for the first construction phase required work such as road markings and traffic signs.

Commissioning of Tunnel Fire Life Safety Systems and Its Challenges

During the first construction phase, the south tube was closed. The entire traffic was routed in the north tube in bidirectional mode. The second construction phase was divided into several sub-phases. In the first phase, the north tube was closed and the entire traffic was routed in the south tube in bidirectional mode.

In the next stage of construction, the traffic was routed in the right-hand lane in both tubes. The left-hand lane was used for finishing work.

In the last phase of construction, the traffic was routed in the left-hand lane in both tubes. The whole construction work took about two years.

The following systems were modernized and renewed:

- Extension of the medium-voltage system
- Renewal of energy distributors
- Renewal of the Uninterruptible Power Supply (UPS) system
- Extension of the lightning protection system
- Renewal of the entrance lighting
- Renewal of the road lighting
- Renewal of the tunnel ventilation system
- Renewal of the air quality and air velocity measurement system
- Renewal of the traffic control system
- Renewal of the traffic detection system
- Renovation of the video system
- Renewal of the emergency call system
- Renewal of the public address system
- Renewal of the fire alarm system
- Renewal of the Supervisory Control and Data Acquisition (SCADA) system
- Renewal of the low-voltage and low-voltage cabling
- Renewal of the electrical, emergency and fire doors
- Renovation of the building ventilation and air conditioning
- Renewal of the doors in cross passages
- Installation of an acoustic tunnel monitoring system
- Installation of a high-pressure water mist system

COMMISSIONING OF THE BIDIRECTIONAL MODE (TEMPORARY)

All systems needed for the proper use of the tunnel were commissioned according to the client's design handbooks. These handbooks define all systems and components which have to be tested as well as the required tests. For example not all components have to be tested by a Factory Acceptance Test (FAT) or a Site Acceptance Test (SAT).

The systems of the south tube were put in operation during the first construction phase. In this phase the entire traffic was routed in the north tube in bidirectional mode. The commissioning of the north tube took place in the second construction phase. It was a challenge to keep one lane per direction open to traffic and to fulfill the requirements of the emergency concept at the same time. To decrease the risk of an accident during the bidirectional mode, the speed limit was reduced from 100 km/h (about

60 mph) to 60 km/h (about 35 mph). The speed limit after construction is 100 km/h (about 60 mph). If an alarm is activated the installed variable message signs reduce the speed limit, or the tunnel will be closed based on the alarm level.

In order to fulfill actual *guidelines (1)* and hence reduce the distance to the nearest emergency exit six new cross passages were built. Five cross passages already existed. In total there are 11 cross passages. The new cross passages were built using the NATM. In the first construction phase the cross passages were mined next to the existing tunnel tube, but the work was not completed. The connection between both tubes was built in the second construction phase. The entire equipment for the cross passage was installed. The existing cross passages were also equipped with the required components.

At the existing cross passages the emergency doors and the ventilation system were installed. The commissioning of the systems installed in the cross passages was carried out in the second construction phase. Especially the ventilation system inside the cross passage would have caused problems, because smoke-free operation would have to be ensured. To ensure smoke-free air in the cross passage, air from the smoke-free tube has to be pushed into the cross section, and at this construction phase this is not possible. A smoke-free operation is mandatory for running a system properly. To fulfill the requirements both tubes had to be closed. Therefore the commissioning of the ventilation system of the cross passages took place during the night when the tunnel was closed.

FIRE LIFE SAFETY SYSTEMS/SCADA SYSTEM

More than 20 systems are installed in the tunnel. The number of installed systems is listed in the chapter "Introduction." Only a few systems are used for fire life safety. The systems which can detect fire or an accident are presented in the following chapters. Systems which are needed to run the ventilation system are also presented below.

Video System

Both tunnel tubes and the cross passages are equipped with Closed Circuit Television (CCTV). Two types of cameras are installed (fixed cameras and Pan-Tilt-Zoom (PTZ) cameras). Only fixed cameras in the tunnel are connected to the detection system where smoke, accidents or other irregularities can be automatically detected. If an incident is detected by this system, the SCADA system will receive an alarm.

Fire Alarm System

According to the relevant *guideline (2)*, a linear heat detector cable is required in the tunnel. The mounted system in the tunnel consists of a fiber optic cable. An alarm will be sent to the SCADA system in case of an increasing temperature or a detected high temperature. The fiber optic cable is installed at the tunnel ceiling.

The high-pressure water mist system is only installed in the western cut-and-cover section. In this section an additional linear heat detector cable is mounted to reduce the distance between the measurement points. The linear heat detector cable can only detect high temperatures and temperature increases. To detect fire at low temperatures (cold smoke) smoke detectors are mounted at the western cut-and-cover section. The smoke detectors are equipped with internal heaters to avoid false alarms caused by fog.

Acoustic Tunnel Monitoring System

As an additional detection system an acoustic tunnel monitoring system is installed. For this purpose microphones are placed near the CCTV cameras in the tunnel. The microphones record all noises inside the tunnel. The control unit with its special software registers abnormal sounds and sends an alarm to the SCADA system. A tire blow-out or an accident causes a characteristic noise which can be detected. In most situations the acoustic tunnel monitoring system has a shorter detection time than the CCTV or the fire alarm system.

SCADA System

All alarms, notifications, commands and so on are collected in the SCADA system. The operation building is located next to the tunnel. Most of the control cabinets and the main SCADA system components are situated in this building. In the operation building there is also a small control room where minor tasks can be performed. This control room is staffed only for maintenance or in case of an emergency. During regular operation, alarms, notifications and commands are shown in the centralized control room and the operator has to take further action or stop the alarm if it is a false alarm. The notifications of all highway tunnels in Austria are collected in nine permanently manned control centers all over Austria.

If the operator marks the alarm as a fire alarm, the automatic event schedule will be initiated and all systems will operate according to this schedule. The commands are transmitted by the SCADA system.

HIGH-PRESSURE WATER MIST SYSTEM

The high-pressure water mist system is only installed in the western cut-and-cover section. Above this tunnel section there are buildings and according to the guidelines fire protection is required. For this tunnel the authorities chose a high-pressure water mist system. According to the guidelines the protection time has to be at least 90 minutes. The water reservoir was planned to guarantee enough water for 90 minutes.

From the western portal the high-pressure water mist system is installed over a length of 627 m (2,057 ft) into the tunnel and is divided into 25 m long (82 ft) sections. In case of fire three sections will be activated and water will be released through the nozzles. Each tube has 25 sections.

The systems described in the previous chapters can detect fire and can locate the place of fire. The SCADA system sends an alarm to the operator and the operator has to confirm the exact location. Otherwise the wrong section could be activated. The time between detection of fire and activation of the high-pressure water mist system is called intervention time, and is 60 seconds. During the intervention time the operator can deactivate the alarm in case of false alarm or can move the place of fire so that the correct sections will be activated.

The high-pressure water mist system was tested using event simulations where the water mist system was activated. Several tests took place according to a testing schedule. Issues such as whether the back-up pump is ready in time or whether all sections work properly were addressed. These tests were part of the commissioning.

TUNNEL VENTILATION

The tunnel is ventilated longitudinally during normal operation and in case of fire. The tunnel ventilation system is dimensioned for unidirectional traffic. During construction

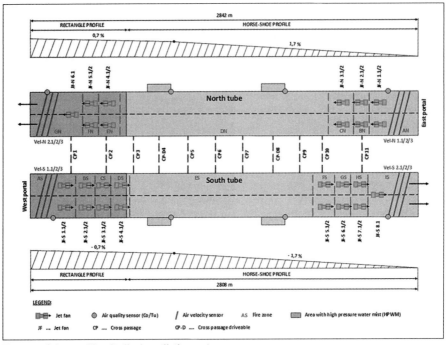

Figure 1. Scheme of longitudinal ventilation system

works the tunnel is temporarily operated with bidirectional traffic in combination with a reduced speed limit. The design in case of fire defines the required number of jet fans. The tunnel ventilation system has pairwise installed jet fans for longitudinal ventilation of both tubes. The falling tube (south tube) has 15 jet fans and the rising tube (north tube) has 11 jet fans. Every pair of jet fans is located in a separate fire zone. A jet fan, which is located in an effected fire zone, must not be activated during the emergency. The design fire size is 30 MW, which is specified in the design *guideline (3)*. The jet fans are temperature resistant up to a temperature of 400°C for a time period of 120 minutes. Air velocity sensors are installed to measure the longitudinal air velocity in case of an emergency. At each measurement location three air velocity sensors are installed to enable a plausibility check of the measured values. If the tunnel is operated with unidirectional traffic the smoke will be transported to the exit portal at an air velocity of 1.5 to 2.0 m/s upstream of the fire. In case of an emergency with bidirectional traffic the smoke will be transported in direction of the initial flow at an air velocity of 1.0 to 1.5 m/s upstream of the fire. Carbon monoxide (CO) and turbidity sensors are used to control the air quality during normal operation. The values are sent to the SCADA system. The SCADA system will force the ventilation system to increase or decrease the air velocity.

Nine cross passages are pressurized by axial flow fans during an emergency. The fan which is adjacent to the non-affected tube draws fresh air into the cross passage. The air flows through the cross passage and further through the pressure relief damper into the affected tube.

The remaining two cross passages of the tunnel are located in the cut-and-cover section. They cannot be individually pressurized, because the tubes are separated by a

Commissioning of Tunnel Fire Life Safety Systems and Its Challenges

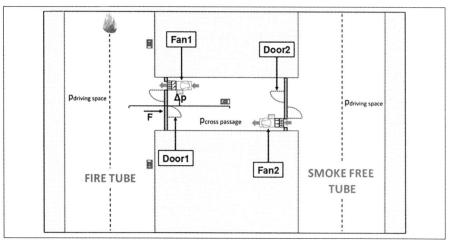

Figure 2. Scheme of ventilation system of a cross passage (4)

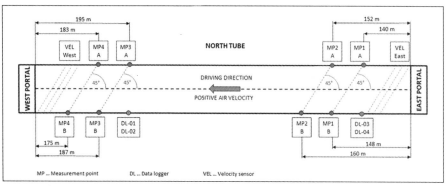

Figure 3. Schematic figure of the arrangement of the air velocity reference sensors

single dividing wall. The overpressure at those cross passages is reached by proper control of the jet fans in the tubes.

Since the air velocity sensors are used to measure the air flow in case of an emergency, their functionality and hence their calibration and integrated testing is of paramount importance. An air velocity measurement location consists of three ultrasonic sensor pairs. The measuring section between an ultrasonic sensor pair is located at a height of approximately 5 m (16 ft 5") above road surface and at an angle of 45° to the tunnel axis. Additional ultrasonic sensors were used to perform reference air velocity measurements. They were orientated crosswise within the tunnel's cross section in order to measure the mean air velocity of the cross section. The values of the reference air velocity measurements were compared to the values of the installed air velocity sensors. Afterwards, a calibration factor was calculated and input into the logic controller. The measurement setup was based on the suggestions of the client's design handbook for commissioning of the logic controller of the ventilation system (5). The air velocity reference measurements were performed for positive and negative air velocities.

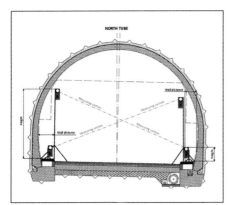 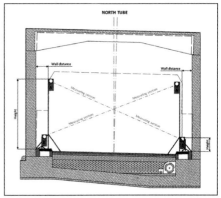

Figure 4. Setup of the air velocity reference sensors in the horseshoe (left) and rectangular cross section (right)

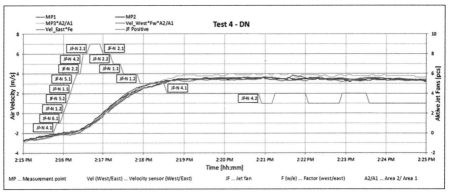

Figure 5. Test 4 in fire zone DN, development of air velocity and number of jet fans used (6)

It was found that the installed air velocity sensors in the horseshoe cross section had a calibration factor close to 1 (±2%), but those in the rectangular cross section had deviations of ±6%. The reasons are the beams, which cause an additional asymmetric flow resistance. For low flow velocities close to 0 the deviations were relatively high, but for higher air velocities of 1 to 2 m/s, which are in the range of the velocity setpoint during an emergency, the deviations were within the acceptable range.

After the calibration of the installed air sensors, the PI-controller was tested. The tests for the parameters of the logic PI-controller of the ventilation system were also performed according to the client's design handbook for commissioning of the logic controller of the ventilation system (5). The tests include testing of the leading and fault behavior of the ventilation system. The following diagram shows Test 4—which is an emergency in the DN fire zone. Once the fire is detected by the linear heat detector cable the jet fans are subsequently activated. As soon as the gradient of the flow velocity decreases, the jet fans are deactivated in order to avoid overshooting of the velocity setpoint. The following diagram shows the development of the air velocity once the fire is detected and the activation/deactivation of the jet fans. The velocity setpoint for this test was 3.5 m/s. The natural flow was about −3 m/s (opposite direction). After approximately four minutes the velocity setpoint was reached.

Commissioning of Tunnel Fire Life Safety Systems and Its Challenges 377

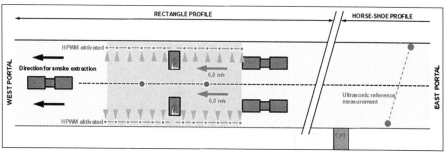

Figure 6. Scheme of the arrangement of the live fire test with the HPWM system (7)

Figure 7. Mobile jet fan in front of the tunnel portal (left) and pool fire (right)

FIRE TEST

The fire tests are required to demonstrate the integrated functionality of all tunnel systems. The responsible representatives of the Ministry of Transportation and Technology observed the live fire test carefully. After successful completion of the tests, the representatives approved the reopening of the tunnel to traffic. The fire test was chosen to be located within the section with high-pressure water mist (HPWM). The fire test was performed twice, with and without activation of the HPWM system.

During normal operation the piston effect of the moving vehicles produces relatively high flow velocities. In case of fire the cars downstream of the fire are able to leave the tunnel and the cars upstream of the fire get stuck in the traffic jam. Therefore, when the fire starts a relatively high air velocity is present in the tunnel. The air velocity decreases once the cars downstream of the fire have left the tunnel and other cars in the tunnel have stopped. To simulate the piston effect of the moving cars, some of the installed jet fans were operated manually and an additional powerful mobile jet fan in front of the portal was used.

The following diagram shows the air velocity and the jet fans used. It can be seen that the values of the installed air velocity sensors match very well with the reference air velocity measurements. The diagram also shows the number of jet fans used. Activation of the jet fan in driving direction has a positive algebraic sign; against driving direction, a negative algebraic sign. The main alarm was triggered 63 seconds after ignition of the fire. After a further 71 seconds the velocity setpoint was reached. The results of the live fire test with a relatively high base flow (simulation of piston effect of moving vehicles) showed that the distance between the fire location and the

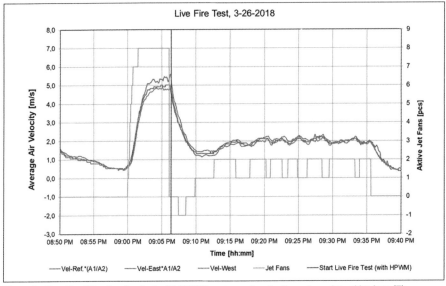

Figure 8. Results of the live fire test with HPWM system; air velocity and number of jet fans (7)

detection point is rather low (<1 m). This is because the linear heat detection cable does not detect smoke but radiated heat.

CONCLUSION

Commissioning is always a big challenge, especially when you have to operate several different fire life safety systems which are dependent on each other. In the current project the authorities gave special attention to the interaction of the HPWM system and the tunnel ventilation system. Therefore, the test procedure was prepared precisely and all possible influences have been emulated as realistically as possible.

The calibration of the installed air velocity sensors showed that the location of the installed sensors must be chosen carefully, giving consideration to the entrance length, for a more or less fully developed flow, and any obstacles such as beams, overhead signs, etc. In the case of this project a better solution could have been to move the location of the installed sensors, from the rectangular to the horseshoe cross section, because in the horseshoe cross section there are no beams like at the ribbed ceiling.

In order to get realistic results from the fire test, it is crucial to select the basic flow accordingly; the selection of the fire location must also be well thought out.

The results of the live fire test with a relatively high base flow (simulation of piston effect of moving vehicles) showed that the distance between the fire location and the detection point is rather short (<1 m), because the linear heat detection cable does not detect smoke but heat from radiation.

REFERENCES

1. Austrian Guideline for Tunnel Construction, RVS 09.01.24, 2014.

2. Austrian Guideline for Tunnel Safety Design, RVS 09.02.22, 2014.

3. Austrian Guideline for Tunnel Ventilation Design, RVS 09.02.31, 2014.

4. Design Handbook of the client for tunnel ventilation systems, PLaPB 800.542.1000 TLü V3.00 Technical Guideline, 2016.

5. Design Handbook of the client for commissioning of the logic controller of the ventilation system, PLaPB 800.542.1604 TLü V2.00 PI Controller, 2017.

6. Report of Flow Measurements of North Tube, ILF Consulting Engineers, 2018.

7. Report of Live Fire Tests of North Tube, ILF Consulting Engineers, 2018

A Comparison of Breathing Gasses Used Under Hyperbaric Conditions

Justin Costello ▪ Ballard Marine Construction

ABSTRACT

The following paper provides a comparison of the various breathing media commonly used in commercial diving and in hyperbaric support of pressure-faced tunnel boring machines. The effects of various inert gasses, the role of hyperbaric oxygen, and saturation at both high and low pressures are explained. The decision-making behind selecting a proper breathing medium for a given pressure and work scope is explored. Equivalent air depths, equivalent narcotic depths, oxygen toxicity, and air saturation are discussed, as well as a comparison of efficiency and productivity based on gas blends, associated costs, and decompression profiles.

The inspection, maintenance, or repair of a TBM's cutterhead in hyperbaric conditions is a source of ire for most contractors. When hyperbaric interventions take place, mining has ceased. Often, site workers who would otherwise be hoisting or transporting segments, building rings, or being productive in general are idle yet on the clock. From a perspective of analyzing breathing gasses and decompression profile selections over the course of several thousand hyperbaric interventions, I've been able to identify inefficiencies and shortcomings in processes, as well as ways to reduce downtime and optimize the hyperbaric interventions needed.

The air that we breathe normally is composed *mainly* of two elements—oxygen and nitrogen. Each of these pose distinct problems when pressure is elevated and thus, the partial pressure of each. Oxygen, at elevated partial pressures, poses physiological risk to the body in two forms: pulmonary oxygen toxicity and central nervous system oxygen toxicity.

Pulmonary oxygen toxicity manifests in the form of painful breathing and substernal irritation. It occurs after prolonged exposure to higher-than-normal levels of oxygen. Although most often seen in hospital patients with nasal cannulas, it can occur in prolonged hyperbaric conditions as well. For this reason, an "oxygen clock" is kept, tracking a worker's exposure. (Weinke, 2008)

Central nervous system oxygen toxicity occurs when the partial pressure of oxygen is elevated and causes symptoms including disturbances to hearing or vision, anxiety, confusion, nausea, dizziness, twitching facial muscles, fatigue, or convulsions and seizure. (Weinke, 2008)

Nitrogen makes up approximately 79% of air. It is metabolically inert and is the main concern when dealing with decompression from hyperbaric environments. When the partial pressure of nitrogen is elevated, it has a narcotic effect. This narcosis can become a limiting factor when using air or designing another nitrogen-bearing breathing gas for use in increased pressure. Nitrogen narcosis usually becomes noticeable at around 3.0 bar above atmospheric pressure when breathing air. The effects

Table 1. Partial pressures of oxygen at various pressures (depths) with common nitrox blends.

Depth (fsw)	(msw)	atm abs	21%	28%	29%	30%	31%	32%	33%	34%	35%	36%	37%	38%	39%	40%
0	0	1.00	0.21	0.28	0.29	0.30	0.31	0.32	0.33	0.34	0.35	0.36	0.37	0.38	0.39	0.40
35	11	2.05	0.43	0.57	0.59	0.62	0.64	0.66	0.68	0.70	0.72	0.74	0.76	0.78	0.80	0.82
40	12	2.21	0.46	0.62	0.64	0.66	0.69	0.71	0.73	0.75	0.77	0.80	0.82	0.84	0.86	0.88
50	15	2.52	0.53	0.71	0.73	0.76	0.78	0.81	0.83	0.86	0.88	0.91	0.93	0.96	0.98	1.01
60	18	2.82	0.59	0.79	0.82	0.85	0.87	0.90	0.93	0.96	0.99	1.02	1.04	1.07	1.10	1.13
70	22	3.12	0.66	0.87	0.90	0.94	0.97	1.00	1.03	1.06	1.09	1.12	1.15	1.19	1.22	1.25
80	25	3.42	0.72	0.96	0.99	1.03	1.06	1.09	1.13	1.16	1.20	1.23	1.27	1.30	1.33	1.37
90	28	3.73	0.78	1.04	1.08	1.12	1.16	1.19	1.23	1.27	1.31	1.34	1.38	1.42	1.45	1.49
100	31	4.03	0.85	1.13	1.17	1.21	1.25	1.29	1.33	1.37	1.41	1.45	1.49	1.53	1.57	1.61
110	34	4.33	0.91	1.21	1.26	1.30	1.34	1.39	1.43	1.47	1.52	1.56	1.60	1.65	1.69	1.73
120	37	4.64	0.97	1.30	1.35	1.39	1.44	1.48	1.53	1.58	1.62	1.67	1.72	1.76	1.81	1.86
130	40	4.94	1.04	1.38	1.43	1.48	1.53	1.58	1.63	1.68	1.73	1.78	1.83	1.88	1.93	1.98
140	43	5.24	1.10	1.47	1.52	1.57	1.62	1.68	1.73	1.78	1.83	1.89	1.94	1.99		
150	46	5.55	1.17	1.55	1.61	1.67	1.72	1.78	1.83	1.89	1.94	2.00				
160	49	5.85	1.23	1.64	1.70	1.76	1.81	1.87	1.93	1.99						
170	52	6.15	1.29	1.72	1.78	1.85	1.91	1.97								

PO$_2$ (atm) based on depth and percentage of oxygen. The body of the chart has PO$_2$ values for various mixes at a range of depths. Standard 32% and 36% mixes are in light grey. PO$_2$ levels higher than 1.6 ata are in dark grey and considered exceptional exposures and should be avoided.

increase with pressure and eventually become debilitating based on each individual's tolerance. This has been colloquially known as "martini's law," as the level of impairment is roughly that of a martini on an empty stomach with each bar of pressure over atmospheric.

There are a few notable blends of breathing gasses which are common in both commercial diving and hyperbaric work on tunnel boring machines, including nitrox, trimix, and heliox. Typically, if a large work scope is called for at moderate pressure, nitrox can be used to increase work time over that with air. As pressure increases above 4.8 bar, air and nitrox are no longer safe to use. Trimix is then the preferred breathing gas. At the operational limit of trimix, heliox becomes the only option, although its use can overlap the pressure ranges of the others as well. Each blend is explained in further detail below.

Nitrox is a non-air blend of nitrogen and oxygen. The gas typically contains an elevated fraction of oxygen compared to that in air. With decreased nitrogen, the body absorbs less than usual in hyperbaric conditions and allows for greater work time with less decompression. The increased oxygen, however, imposes a reduced maximum operational pressure due to the risk of central nervous system oxygen toxicity. See Table 1 for an outline of the partial pressures of oxygen at various pressures (depths) with various common blends of nitrox. One of the considerations when designing a nitrox breathing medium is equivalent air depth. This calculation (see Equation 1) adjusts the actual pressure to that which the body "thinks" it's being exposed to. The reduced theoretical pressure is used when determining decompression requirements. Standard air decompression profiles can be applied once this calculation is complete, simplifying the process (Pechuga, 2018).

Trimix is a blend of helium, nitrogen, and oxygen. The inert gas percentage is split between the nitrogen and helium portions of the blend. Trimix is most often used when pressure is between 4.5 and 6.0 bar over atmospheric. The benefits of its use are:

Equivalent Air Depth of Nitrox Calculation

The equivalent air depth can be computed from the following formula:

$$EAD = \frac{(1 - O_2\%)(D + 33)}{0.79} - 33$$

where:

EAD = equivalent depth on air (fsw)
D = diving depth on mixture (fsw)
$O_2\%$ = oxygen concentration in breathing medium (percentage decimal)

For example, while breathing a mixture containing 40 percent oxygen ($O_2\%$ = 0.40) at 70 fsw (D = 70), the equivalent air depth would be:

$$\begin{aligned} EAD &= \frac{(1 - 0.40)(70 + 33)}{0.79} - 33 \\ &= \frac{(0.60)(103)}{0.79} - 33 \\ &= \frac{61.8}{0.79} - 33 \\ &= 78.22 - 33 \\ &= 45.2 \text{ fsw} \end{aligned}$$

Equation 1. The calculation for determining the equivalent air depth of nitrox

- Reduced thermal conductivity due to a reduced partial pressure of helium;
- A greater operating range than nitrox or air; and
- Reduced nitrogen narcosis due to a lower partial pressure of nitrogen.

The process of developing the correct trimix blend begins with the planned working pressure. The fraction of oxygen in the mix is determined by identifying the maximum operational partial pressure of oxygen (see Figure 1). With this not-to-exceed partial pressure and the maximum operational pressure intended, the percentage is established. This creates the functional pressure range of the gas while still supporting life—typically 0.195 to 1.6 atmospheres of partial pressure. The balance of the blend is then determined by establishing an acceptable Equivalent Narcotic Depth (END). This END is calculated for the maximum planned pressure and establishes the fraction of nitrogen in the blend. The workers' tolerance of narcosis may allow for a slightly higher fraction of nitrogen if commercial divers are used. Helium completes the remaining fraction.

A major drawback to using trimix is its bespoken nature. Gas suppliers are not likely to offer a buyback for unused portions. The odds that a given blend of trimix can be used at a future location along an alignment are also somewhat low. This creates waste. The decompression profiles for trimix do not necessarily improve productivity over air—only giving access to greater pressures but requiring lengthy decompression with limited work time.

A blend of helium and oxygen only is known as heliox. The operational range of pressures for this gas is large. Due to the small molecular size of helium, its viscosity is much lower than other inert gasses. This improves the ease of breathing at great

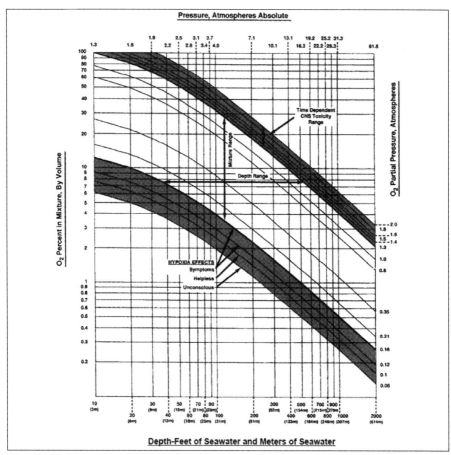

Figure 1. Minimum and maximum partial pressures of oxygen at various pressures (depths)

pressures. Narcosis is also avoided through the use of helium as a diluent as well. While allowing for work to take place at over 30 bar above atmospheric pressure, helium does have some drawbacks. The thermal conductivity is quite high. Unless inhaled heliox is heated, a worker's lungs will warm the gas as quickly as it's inhaled and then lose body heat when the helium is exhaled. This can cool the core body temperature quickly, causing hypothermia. The voice change from helium is also a challenge and communication devices with electronic descramblers are needed. Helium is a finite resource and procuring enough of it to support a tunnel project can be a challenge, depending on the location and duration of the project. (Engineer's Edge, 2008)

The decompression process from hyperbaric exposures can be divided into two categories, bounce and saturation. In bounce mode, a worker is exposed to hyperbaric conditions for a limited amount of time. The pressure and inert gas or gasses being breathed are taken into consideration when determining the maximum time of exposure. As time at pressure increases, so does the decompression obligation of the worker. Upon reaching the maximum allotted time, the worker must be decompressed through a slow reduction of pressure and usually a switch to pure oxygen. This process removes excess inert gasses from the body. This method of exposure and decompression is simple to achieve and requires little in the way of equipment.

The ratio of work time to decompression time is great at low pressures, up to around 1.5 bar. As pressure increases beyond this point, the amount of work time allowed goes down and decompression requirements increase. By switching from air to nitrox, it is possible to regain some of the productivity found at lower pressures but due to central nervous system oxygen toxicity, the gas has an operational limit.

In saturation mode, the tissues are allowed to come to equilibrium with ambient inert gas concentrations. Once this equilibrium is reached, no additional inert gasses can enter the tissues, thus no additional decompression obligation is incurred. This allows for nearly unlimited time at nearly any pressure up to 30 bar. The limitations of work shift length are the same as when workers are in free air. There is no physiologically-driven maximum duration for this type of exposure, although a 28-day limit is typically enforced for the mental wellbeing of the workers. Upon completion of the work, decompression occurs slowly, usually over the course of days.

Saturation is extremely equipment intensive. The habitat in which workers live must be a well-maintained, large, hyperbaric chamber. Transport to and from the cutterhead must be done in a hyperbaric shuttle with onboard life support. Connections to the excavation chamber must be available, which do not permit the pressure surrounding the workers to drop. This usually limits the size of TBM which can accept a saturation shuttle to only those of significant diameter. If non-air breathing media is used, it must be supplied from high-pressure cylinders, through pressurized hoses and delivered to helmets.

The saturation habitat contains the living quarters for the workers. They eat, sleep, and groom in a pressurized environment. All supplies must be sent in to them through pressurized ports. As workers breathe, oxygen levels are depleted, and carbon dioxide is produced. Metabolic oxygen must be replenished, and chemicals are used to absorb and remove the carbon dioxide. Their atmosphere is monitored around the clock by technicians who add oxygen, remove humidity and carbon dioxide, and often heat the gas. Restroom facilities are located in the habitat, as well with potable water connected to the systems. Due to the increased ability of compressed gas to hold humidity, and the need for elevated temperatures when helium is in used, the decontamination is essentially a constant task within the chamber. The prophylactic use of antibiotic ear drops is prescribed by project physicians to avoid infections in the workers. (Command, 1993)

Although saturation demands more sophisticated equipment and highly-trained workers, the benefits are easy to measure. When work on the TBM's cutterhead takes place in bounce mode, the crew is large, and teams must rotate often. In some scenarios, 24-hour coverage with nonstop work may require 11 teams of two or three workers plus support personnel. Depending on the pressure, a team's exposure may be limited to as little as 45 minutes. This causes information to be lost through multiple exchanges from one team to the next, with much downtime created in waiting for the decompression of each team. When a hyperbaric team is brought in through a services provider, the transportation and lodging cost of a large team can be high as well. By utilizing saturation, the team remains small and highly skilled. Work on the TBM is nearly uninterrupted around the clock and, when large amounts of work is needed, days of TBM downtime can be saved easily offsetting the added cost of saturation. Saturation is also widely regarded as being safer for the workers, as only one decompression occurs instead of daily when utilizing the bounce method.

Heliox is traditionally the most common gas used in saturation, as saturation in general is typically only employed when the pressure mandates its use. The cost of this gas can be high, and its availability is not always certain. Storage areas for large volumes of gas is needed on-site near the saturation habitat. With high pressure, the atmosphere holds moisture well. Reducing humidity in the saturation habitat is a constant battle, as the expensive gas must be maintained rather than simply flushed out and replaced with arid gas.

Air saturation is possible and generally considered safe when storage pressure is no greater than 1.5 bar above atmospheric pressure. Alternatively, storage in a non-air nitrogen-oxygen blend with an oxygen partial pressure in the range of 0.3–0.5 atmospheres can be safely conducted to pressures up to 3.7 bar. Excursions on air are possible in this case as long as they are planned carefully to avoid pulmonary oxygen toxicity. (NOAA)

While the composition of the atmosphere in the saturation habitat would not be air per se, it is still a nitrogen-oxygen blend with a reduced level of oxygen. The term nitrox is usually reserved for blends with higher-than-air levels of oxygen, up to 40%, so this atmosphere is technically a hypoxic nitrox. With this atmosphere, high-priced compressed gas cylinders are not needed to achieve such a blend, as natural air is used to initially pressurize the system, then the workers "breathe down" the O2 content to lower levels. Because the human body utilizes oxygen based on its partial pressure and not percentage, this is not only considered safe but mandatory to maintain health and avoid oxygen toxicity. By monitoring the workers' "O2 clock," appropriate excursions in the cutterhead can be planned.

Instead of dosing the atmosphere with metabolic oxygen from high-pressure cylinders, ventilation of the habitat with ambient air from a compressor increases the partial pressure of oxygen. Although backup oxygen supplies are still needed on hand, its use is limited to emergencies thus none is typically consumed. Likewise, this flushing of the habitat with natural air removes carbon dioxide and humidity created by the saturation workers. This limits the quantity of carbon dioxide scrubbing chemicals used on the project.

"Storage pressure" is the pressure at which the habitat is set. This is the pressure at which all workers live during their non-working hours. This may be different than the pressure at which hyperbaric interventions take place in the excavation chamber.

An "excursion" is a deviation in pressure from the storage pressure. By planning an excursion in the excavation chamber, the effects of various gasses can be reduced. For example, the storage pressure and atmospheric composition can be set so that nitrogen narcosis is the limiting factor, but oxygen toxicity is not a risk. During work on the cutterhead, breathing natural air is advantageous, as cumbersome helmets, supply hoses, and high-pressure cylinders are not needed in the tunnel. The only caveat is the workers' O2 clocks must be monitored carefully so that pulmonary oxygen toxicity is avoided.

While this scenario deviates from the unlimited work time of a helium-based breathing medium, work time is still quite lengthy and likely near the limit of productivity for a work team. By maintaining an equivalent air depth storage of 1.7 bar in a 12% oxygen/78% nitrogen environment (technically hypoxic nitrox), a six-hour work shift can be attained per team at pressures up to 3.18 bar and only slightly decreases as working pressure approaches 4 bar, although storage depth must be increased slightly (See Figure 2).

Although size constraints may limit its use on many projects, saturation offers vastly greater work times than traditional bounce methods. The crews in saturation are small,

No-stop Descending Excursion Times Incorporating NOAA 1991 Oxygen Limits

Habitat Storage Depth (fsw) / Depth	80	85	90	95	100	105	110	115	120	125	130	135	140	145	150	155	160	165	170	175	180	185	190	195	200	205	210	215
PO_2	.72	.75	.78	.81	.84	.88	.91	.94	.97	1.01	1.03	1.07	1.10	1.13	1.16	1.19	1.22	1.26	1.29	1.32	1.36	1.38	1.42	1.45	1.48	1.51	1.54	1.57
30	350	267	156	113	91	78	68	60	55	50	45	40	36	32	28	24	22	18	15	13	12	11	10	9	8	8	7	7
35	*	*	283	229	143	108	89	77	68	61	54	46	41	37	34	31	28	25	22	20	16	14	13	11	10	9	9	8
40	*	*	*	301	240	202	147	112	92	80	70	59	50	44	39	35	32	30	28	25	23	21	17	15	13	12	11	10
45	*	*	*	*	323	253	210	181	137	108	91	69	56	48	42	38	34	31	29	27	25	23	22	21	18	16	14	12
50	*	*	*	*	*	350	267	219	187	164	140	86	64	53	45	40	36	33	30	28	26	24	22	21	20	19	18	16
55	*	*	*	*	*	*	314	245	203	174	153	137	86	63	52	45	40	36	32	30	27	25	24	22	21	20	19	18
60	*	*	*	*	*	*	330	284	224	187	161	142	127	85	63	52	45	39	35	32	29	27	25	23	22	21	19	18
65	*	*	*	*	*	*	330	315	236	191	162	145	128	111	85	63	51	44	39	35	32	29	27	25	23	22	20	19
70	*	*	*	*	*	*	330	330	279	213	174	148	129	114	103	84	62	51	44	39	35	31	29	26	25	23	21	20
75	*	*	*	*	*	*	330	330	300	270	270	240	228	191	165	145	95	66	53	45	40	35	32	29	27	25	23	22
80	*	*	*	*	*	*	330	330	300	270	270	240	240	225	210	210	195	122	70	55	47	41	36	32	29	27	25	23
85			*	*	*	*	330	330	300	270	270	240	240	225	210	210	195	180	180	95	66	54	46	40	36	32	29	27
90			*	*	*	330	330	300	270	270	240	240	225	210	210	195	180	180	165	150	97	68	55	47	41	37	33	
95				*	*	330	330	300	270	270	240	240	225	210	210	195	180	180	165	150	150	113	80	62	52	45		
100				*	330	330	300	270	270	240	240	225	210	210	195	180	180	165	150	150	135	135	120	83	83	45		
105					330	330	300	270	270	240	240	225	210	210	195	180	180	165	150	150	135	135	120	83	83	45		
110						330	300	270	270	240	240	225	210	210	195	180	180	165	150	150	135	135	120	83	83	45		
115							300	270	270	240	240	225	210	210	195	180	180	165	150	150	135	135	120	83	83	45		
120								270	270	240	240	225	210	210	195	180	180	165	150	150	135	135	120	83	83	45		

Note: Values are in minutes
* No time limit (up to six hours) for that depth - time combination

Figure 2. Excursion times on air with various storage pressures

reducing overall cost, and highly-trained. Decompression following saturation occurs off the TBM, allowing it to resume mining faster. The use of hypoxic nitrox saturation removes the need for high pressure cylinders of breathing gasses, hoses and helmets, voice modifying electronics, simplifies oxygen dosing and carbon dioxide and humidity removal. It also has an extensive operational range. All of these elements contribute to an improved means and method resulting in reduced downtime and optimized hyperbaric intervention process. In comparison with traditional bounce methods during the project planning phase of upcoming tunnels, saturation has recently been shown to eliminate over one month of TBM downtime due to hyperbaric interventions during the course of an 18-month project. In addition to this time savings, the cost of specialty hyperbaric services providers is reduced. By adopting saturation procedures at lower pressures, especially within the nitrogen-based breathing gas range, projects can save significant amounts of time and money.

REFERENCES

Command, N.S. (1993). *US Navy Dive Manual.* Washington, D.C.

Engineer's Edge. (2008). Retrieved from Engineer's Edge: https://www.engineersedge.com/heat_transfer/thermal-conductivity-gases.htm.

NOAA. (n.d.). *NOAA Diving Manual.*

Pechuga, H. (2018). *Rates of Inert Gas Uptake Relative to Partial Pressure.* (Unknown, Ed.) Panama City, FL, USA: Best Publishing.

Weinke, B.R. (2008). *Hyperbaric Physics with Bubble Mechanics and Decompression Theory in Depth.* (K.L. James T. Joiner, Ed.) Flagstaff, Arizona, United States of America: Best Publishing Company.

Conception and Construction of a Tailor-Made and Contractor-Built Refuge Chamber for TBM and SEM Drives According to German Guidelines

Rainer Antretter ▪ BeMo Tunnelling GmbH

ABSTRACT

Very different and partially missing guidelines, different points of view by manufacturers and users, as well as a critical self-assessment including a risk analysis led to the decision to develop and build a refuge chamber by ourselves as a tunnelling contractor. The goal was to use the refuge chamber on a 9,3 m diameter TBM in Karlsruhe, Germany as well as in conventional SEM tunnels where long standalone periods were required. The chamber was tested and approved by simulation of design conditions under expert monitoring. Knowledge gained from this application was helpful for development of a national guideline and will contribute to further improvements of future rescue concepts.

INTRODUCTION

During preparation of a slurry shield TBM for the Tunnel Kaiser Strasse Karlsruhe, Germany (Combined Solution Karlsruhe, Heavy/Light Rail Tunnel and Light Rail Boulevard[1]) the risk assessment for emergency procedures resulted in the requirement of an eight-hour standalone rescue chamber. Emergency rescue of persons in case of fire smoke in the tunnel through rescue by railbound equipment was estimated to be possible within two hours after an extinguished fire within this 2,1 km long tunnel. The only significant fire load in the tunnel was a supply locomotive for the TBM. According to published results of trials of tunnel fires and their behavior in terms of heat production, duration and fading taken as a base for estimation, time consumption for man rescue could be thoroughly determined with a decent degree of safety. Thus, the requirement of an eight-hour shelter on the TBM was intended to be taken into consideration for the conception of the chamber.

On the other hand, national guidelines in place in 2014 did explicitly call for a safe location every 1,000 m of tunnel (category B according to DAUB—German Committee for Underground Construction) which was defined to be a 24 hours standalone rescue

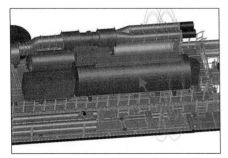 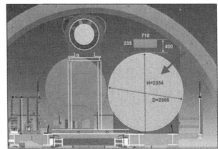

Figure 1. Design phase of TBM—Space available for the rescue chamber on backup gantry

chamber, with tremendous requirements on temperature limitation under 30°C within the chamber with an outside temperature of 60°C over eight hours.

It quickly became obvious that these requirements would not trigger a design which could fit into the available space on the backup gantry of the TBM. Beside the temperature requirements the regulation also called for a breathing air supply of 40 lt/min throughout the entire standalone period of 24 hours.

Recalculation of the quantities of both bottled air and battery capacity for cooling to achieve the required values showed they weren't possible given the space available on the TBM (see Figure 1), the dilemma was born.

Due to a tunnel length of more than 2,000 m two 24 hours rescue chambers were required according to the guideline. As a first step it was decided to have one at station 1,000 m and the second chamber on the last backup gantry of the TBM (70 m behind the cutting wheel), the latter replacing the second chamber at 2,000 m and no further chamber for the extension of the tunnel to 2,100 m.

Both chambers were proposed to be designed for a standalone period of 12 hours and the heat load of 60°C over a reasonable period of 1.5 hours which meant one third more standalone time capacity than the risk assessment revealed but also 50% less than the guideline demanded. With this conception the authorities have been approached and after a thorough proof of feasibility of man rescue both fire brigade and health & safety authorities gave their consent to the 12 hours stand alone concept. However, the trade-off stipulation therefore was the combination of nearly no fire load in the tunnel, installation of a fire suppression system on all locos and rescue via a dedicated railbound train system for operation under fire fume polluted atmosphere for quick approach of the fire brigade to the TBM via the portal.

DESIGN PARAMETERS

Fire Load

Investigation of fire behavior of the locos used for supply and personal transport for the TBM operation led to interesting trial results by the publicly-funded European project "EUREKA EU 499 Firetun." Test fires had been conducted by EUREKA in the abandoned Hammerfest tunnel in Norway and both the timely process of temperature development and the process of temperature development over distance had been recorded for different fire loads (truck, coach car and a SUV car), see Figure 2.

The findings, (published by Haack A, 1998 Fire Protection in Traffic Tunnels[2]) are interestingly quite different compared to the standard fire curves which were to be taken into consideration during engineering of tunnel construction. By comparing a 35 tons/220 HP tunnel loco with an SUV car regarding the fire load and the development of such a fire, parameters could be determined for the impact of a loco fire somewhere in the tunnel for the design of the rescue chamber. With a reassuring degree of safety, a loco fire would not exceed a temperature of 60°C at a distance of 50 to 70 m and last not longer than 90 min (1.5 hrs), see Figure 3.

Breathing Air Volumes and Gas Fractions

The guideline's second demand to fulfill respectively to question was the breathing air quantity. It was not clear where the demand for 40 l/min per person came from. Doubts were in place and made literature research necessary.

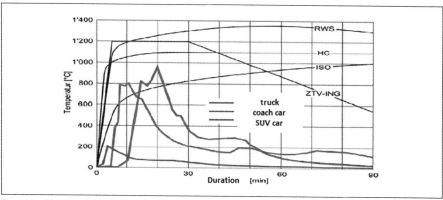

Figure 2. Fire curves of different fire loads (EUREKA EU 499 Firetun)

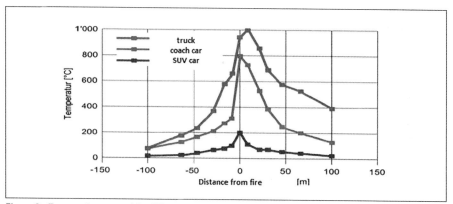

Figure 3. Temperature spread for different fire loads (EUREKA EU 499 Firetun)

Several published investigations on human respiratory and gas consumption mainly for design of safe locations in mining could be found to this topic. All those reports were mentioning of respiratory rates between 7.5 and 12.5 lt/min air when sitting and not under mental stress. This amount of air results with safety factor considered in 0.5 lt/min pure O_2 consumption per person and was one of the design parameters for the redundant gas supply.

Not so much the decrease of the O_2 level within a chamber but the increase of the CO_2 level would be the trigger for increasing the breathing frequency and must be considered as the driver for the gas supply and the design with breathing air. Therefore, the breathing fraction of CO_2 the 3.2% generation respectively 0.4 lt/min per person must be disposed of and this was the second important design parameter. Figure 4 shows the final gas control panel.

It also could be learned that for flushing the chamber with compressed air a rate of 85 lt/min per person (at 12.5 lt/min breathing rate) would be required to keep the CO_2 level under the TLV (time-weighted threshold limit value) of 5000 ppm. This also explained the guidelines requirement of 40 lt/min because for emergency situations the CO_2 level can be accepted to be higher than a TLV for normal working atmosphere, but it is doubtful whether it would be bearable for 12 hours, not to mention for 24 hours.

All the findings from the literature compared to the values in the guidelines (some international guidelines could be found on the Internet), did not give a picture that was entirely conclusive based on the laws of physics, but rather an indecisive selection of different views, which were laid down in these guidelines.

Temperature and Climatical Conditions

Temperature respectively the climate within the rescue chamber are by no means negligible. This topic showed the most significant deviations amongst the guidelines. Further, proposed and offered solutions of air conditioning system by manufacturers of rescue chambers led to the conclusion that there is a lack of practical experience (or proof by testing) with proper function at an ambient temperature of 60°C, required by the guidelines. Investigation on air conditioning systems showed that 60°C ambient temperature is a level at which standard chillers are at their absolute limit of capacity or would even fail with proper testing.

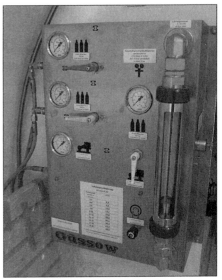

Figure 4. Final layout of gas control panel

If there is not appropriate air conditioning inside the rescue chamber when temperature and humidity increase, then human respiratory rate increases significantly as a reaction to it and all previously calculated values of respiration no longer correspond with the reality. Thus, the designed standalone period of the rescue chamber could be reduced significantly.

The correlation of basic effective temperature (BET) and humidity in the chamber is shown on the graph in Figure 5. The more humid the air in the chamber is the lower the measured temperature must be. The guideline required a BET of 30°C.

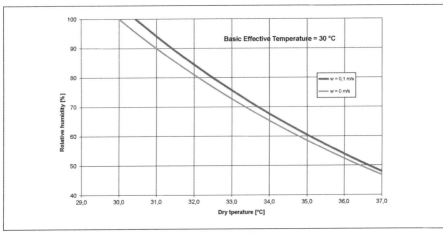

Figure 5. Basic effective temperature—influence of humidity in the chamber

Figure 6. Heavy duty air conditioning compressor during test run in workshop

Figure 7. Installation of insulation panels

It was therefore considered extremely important to design a bespoke air conditioning system (see Figure 6) which demonstrably works properly at such a high ambient temperature and provides the required cooling capacity to keep the level of 30°C effective temperature inside the chamber (not dry temperature and not wet bulb globe temperature).

Cooling by spraying water on the outside hull of the rescue chamber—within the guideline listed as an option—is very doubtful if not to say unbelievable that it would have the desired effect or that it could be controlled in any reliable way. The same is valid trying to only insulate the chamber without installing air conditioning.

In combination with the above, to achieve the best insulation of the interior against ambient temperature the steel hull must be thoroughly insulated, and the quality of insulation must be captured within the thermal engineering calculation. Installation of insulation panels see Figure 7.

CHAMBER CONSTRUCTION

A rescue chamber on the backup of a TBM for a medium sized tunnel section is always in the way from the standpoint of TBM-technique. It was therefore decided to stay with the design of a tubular chamber body in order to fit into the dedicated space which must be agreed with the TBM manufacturer. The tubular design derived from the first chambers, developed in the 1990s and made of retired compressed air locks when such were necessary by legislation as a shelter in drill and blast tunnels to protect from blast fumes. On the newly developed chamber the only protruding parts on the circumference of the body were the air condition compressor and the battery rack on top of the tube body. This envelope fit well into the available space on the TBM-backup.

A tight front door with a window capable to resist the slight overpressure of about 100 Pa inside the chamber and an emergency exit via a man hole on the opposite side, as well as a technique compartment of the chamber formed the chambers steel construction elements and resulted in dimensions of about 8.30 m in length, 2.40 m in width and 3.0 m in height and a weight of 11.5 metric tons.

Cooling and Energy Supply

Energy supply systems for standalone purposes require the principal decision on how cooling should be provided. There is the option of cooling with a calculated bulk of chilled water ($T_0 < 5°C$) which acts as the energy carrier for cooling (i.e., 2,000 litres in an insulated tank incorporated in the container which is constantly chilled by a small chiller). In the emergency case the bulk of water would provide the chilling energy which would be introduced in the chamber by flow through a heat exchanger with fan. Advantage with this option is that no big storage of electrical energy is required, the downside is that if the chiller doesn't work (i.e., due to power interruption after unplugging it for a move of the chamber in the tunnel) no cooling energy is available in emergency case.

Figure 8. Battery rack on top of rescue chamber

The other option is to install sufficient battery capacity on or in the chamber which provides the energy for the AC-compressor. Advantage is that if the batteries aren't loaded constantly by mains power, it would take many weeks until they are empty. Therefore, for emergency purposes the chamber is all-time ready to operate. The downside is that a big battery pack is relatively expensive and would only be suitable for an ambient temperature up to 60°C.

The risk assessment for the TBM drive tunnel Kaiser Strasse resulted in the opinion that the safer and more reliable

Figure 9. Compressed air cylinders

option of energy storage for cooling would be the battery system. Maintaining the temperature of chilled water at all-time was considered to be uncertain.

Energy calculation of all consumers showed a total power requirement of 800 amperehours at 48 Volts for the standalone period of 12 hours—which was then chosen as the design base, see Figure 8.

Breathing Gas Management

As with the electrical energy, breathing gases were also required by the guideline to be available redundantly for the entire duration of 12 hours.

Compressed Air—First Gas Supply System

Additional space on the TBM's backup was dedicated to supply compressed air, such huge space requirements could not be within the chamber. The calculation of compressed breathing air in cylinders resulted in four cylinder-bundles of 12 no. cylinders with 50 lt content each at a pressure of 300 bar, see Figure 9. The withdrawal of air from the cylinder-bundles was solved via a heated high-pressure relief valve inside the chamber. The expansion cold can unfortunately not be used for cooling purposes.

Gas Dosing (O_2)/(CO_2) Scrubbing—Second Gas Supply System

The second supply system works via refilling of the used O_2 by respiration on one hand and scrubbing of produced CO_2 of the occupants on the other hand.

Dosing of both, compressed air and O_2 were designed via flowmeters in the chamber and a table for the operator with the calculated flow in dependence of the number of occupants inside, see Figure 10. Control of the correct flow to be adjusted was intended manually via the gases metering system by the operator which constantly measures and displays the gas concentration, see Figure 11.

Adsorption of CO_2 can be solved via a chemical reaction of soda lime and CO_2 in the air. Soda lime is a mixture of sodium hydrate and calcium hydroxide and adsorbs CO_2 by changing its white colour into purple as an indicator of consumption, see Figure 12.

1 kg of soda lime can absorb 100 litres of CO_2. Space for 115 kg of soda lime is required to be able to absorb the CO_2 produced by 20 persons over a time period of 12 hours.

A chemical reaction takes place within a circular steel barrel with fine mesh on the bottom and a blower installed underneath the mesh. Soda lime is filled in the barrel and CO_2 which is due to density collected on the floor of the chamber is blown through the mesh and soda lime, see Figure 13.

Figure 10. O_2 dosing flowmeter

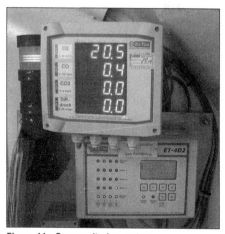

Figure 11. Gas monitoring system

Figure 12. Soda lime adsorber during testing

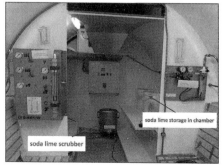

Figure 13. Scrubber on back wall of chamber

 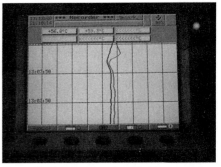

Figure 14. Thermal test under expert supervision   Figure 15. Data logging of ambient temperature

Significant exothermic heat development with scrubbing was detected during the test run on the TBM and with 20 people inside the chamber. Recalculation showed 1,248 kJ/hour which was then introduced into the thermal calculation of the chamber.

TESTING AND COMMISSIONING ON SITE

Since it was not easy to test the chambers proper function with 20 persons over 12 hours on the back of a TBM on site it was decided to conduct tests under expert supervision under simulated conditions as close as the reality would be. The independent third-party expert (by Walasiak Engineering/Consulting) was also checking the calculations and the one providing the certificate for the chamber, in accordance with the guideline.

Thermal Test on the TBM

Temperature behaviour during the heat phase of the operation (1,5 hours @ 60°C according to the guideline) was considered as the most desired parameter to test.

A tent of duct fabric on a scaffold erected around the chamber served as encapsulation to heat the ambient space of the chamber to 60°C and to conduct the 1.5 hours heat test, see Figure 14. Temperature sensors on the outside and inside the chamber, connected to data loggers, continuously monitored the temperature gradient, see Figure 15.

The thermal test duration was conducted on day one (Oct.21, 2014) over 2 hours and 15 minutes with more than one hour under ambient temperatures between 54 and 64°C. A 1,500 W oil radiator together with a three persons test team (including the expert) inside the chamber simulated full occupancy. Temperature increase inside was about 5.2°C at the end of the test. Two sensors showed a temperature of 26.2°C, see Figure 16.

The goal on day two (Oct.21, 2014) was to find out if the gradient inside the pre-heated chamber could be kept at 30°C over an hour if the outside temperature is kept at 60°C. The test started after the outside temperature of 60°C was reached and stable, see Figure 17. It could be verified that the thermal calculations were correct, the expectations of temperature behaviour were met.

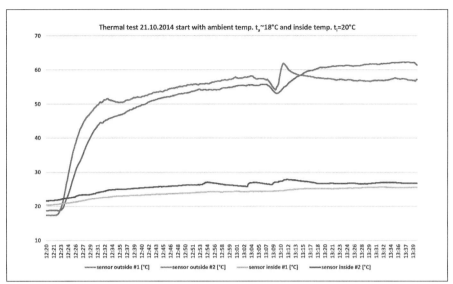

Figure 16. Test 1, graph from ambient temperature inside

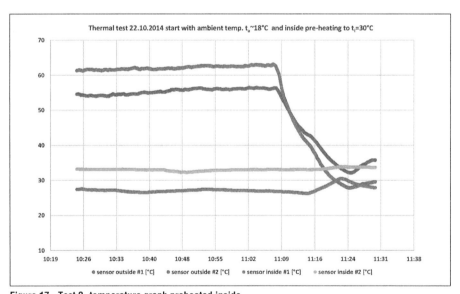

Figure 17. Test 2, temperature graph preheated inside

Breathing Gas Test in the Shop with Full Capacity of People

The effectiveness of scrubbing of CO_2 and the redosing of O_2 was the other important issue. A test was done in the shop with 20 people inside the chamber over one hour, see Figure 20.

The results showed expected values and an unexpected phenomenon. The temperature increased during the 1 hour test period slowly from 14°C to 19.6°C, humidity

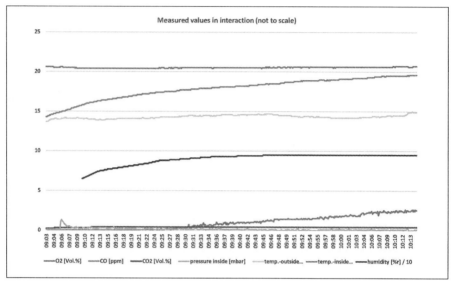

Figure 18. Recorded gas and temperature curves over the test period (CO = orange colored)

increased rather quickly from below 60% to 95% which was expected. The operator inside the chamber adjusted redosing of O_2 to the value on the table close to the flowmeter and turned the CO_2 scrubber on. The 15kg of soda lime in the scrubber were calculated to be sufficient for 154 min (@10lt/min breathing air consumption) operation so it wasn't necessary to replace consumed soda lime.

Thus, the gas management systems were adjusted for nearly automated operation and corrections weren't necessary throughout the test hour. Everybody was satisfied, and the test was considered as been successful.

An unexpected phenomenon occurred due to the increase in the measured value of CO. It had been expected that this value would be almost zero and barely measurable. The reading showed a linear increase with no saturation towards the end of the test duration, but fortunately it was only at a maximum of 2.7ppm, well below the TWA of 30ppm. The surprising result made it necessary to study literature with the goal to find out where the CO came from, see Figure 18.

Not being physicians, it took a while of investigation to come to the result that the CO would be a respiration product of the smokers amongst the test persons. A typical action occurred among the smokers before they took the test in the chamber, they were still smoking one or more cigarettes, and the after-effects of CO exhaled within about 20 minutes caused the increase in the reading. The article by Salame et al.[2] was very helpful to discover where the CO could come from and that the measured value would never be a problem for the occupants.

Commissioning on Site

All required materials for 12 hours of operation were stored in the chamber, first aid medical equipment as well as drinking water, sanitary items and 20 no. self-rescuers were placed in the chamber. After installation the chamber on the TBM-backup the miners and the employees of the client were trained. Personnel of the local fire brigade

Figure 19. Refuge chamber in tunnel

Figure 20. Visible humidity during breathing gases test

Figure 21. New battery compartment

Figure 22. Refuge chamber installed at 1,000 m in tunnel

got a proper induction before the chamber disappeared in the tunnel behind the TBM drive, see Figure 19 and 20.

Refuge Chamber as a Safe Location at 1,000 m of Tunnel

Lessons learned from construction of this first chamber were implemented into construction of the next chamber, which had to be installed according to the directive at station 1,000 m in the tunnel.

Batteries and air condition were relocated from the roof of the chamber into an extended part of the chamber to achieve better protection and to enable future use of the chamber also in drill and blast tunnels, see Figure 21 and 22.

CONCLUSION

Intensive examination of the real risks in case of a fire in the tunnel Kaiserstaße, Karlsruhe and the comparison with the requirements of the directive for certain characteristics of the refuge chamber led to a tailor-made compromise solution, which would have done its job properly in an emergency. The basis for this was a carefully considered design and realistic practical tests.

The knowledge gained was very helpful for a tunnel construction company in order to make the most accurate assessment of the necessary technology for complex projects in the future and to provide employees with the best possible protection.

REFERENCES

1. Göhringer H., Schimmelpfennig M., Nenninger F., Combined Solution Karlsruhe, Heavy/Light Rail Tunnel and Light Rail Boulevard, Germany Inner-City Tunnel Advance at Little Overburden in Difficult Geology, RETC2015 Proceedings.

2. Haack A., 1998 Fire Protection in Traffic Tunnels: Resu of EUREKA Research Project EU 499 FIRETUN.

3. Pascale Salameh, Pharm. D., MPH, Ph.D., Jdeidet El Meten, Chalet Suisse Street, Ramza Azzam bldg, 5th floor, Beirut, Lebanon: Saliva Cotinine and Exhaled Carbon Monoxide in Real Life Waterpipe Smokers: A Post Hoc Analysis. Tobacco Use Insights 2009:2 1–10.

InSAR Monitoring of Subsidence Induced by Underground Mining Operations

Sara Del Conte ▪ TRE ALTAMIRA Inc.
Giacomo Falorni ▪ TRE ALTAMIRA Inc.

ABSTRACT

Subterranean activities such as underground mining and tunnel excavations can produce extensive subsidence of the ground surface, which in severe cases can have serious consequences. In mining, excessive subsidence can lead to blocks in production and safety issues while in urban tunneling contexts it can lead to stoppages, damage to above-ground structures and extensive project delays. The timely mapping of the extent and magnitude of surface movement is usually one of the main challenges faced by geotechnical managers in these cases.

The instrumentation used for monitoring surface deformation has generally been based on conventional survey techniques (total stations, levelling, GPS, extensometers), which provide spatially sparse measurements. The advent of satellite SAR Interferometry (InSAR) significantly changed this scenario by providing wide coverage and high density of accurate information without the need to install ground instrumentation. It is also possible to assess historical deformation by processing satellite data archives going back to the early 1990s. Furthermore, recent advances in processing algorithms have significantly reduced computational time and the advent of newer satellites with increased spatial resolution and acquisition frequency have increased information density. Near-real time InSAR monitoring is now widely applied in different applications to highlight incipient movements in areas not visible to in-situ instrumentation.

Some case studies of InSAR applied to underground mining and urban tunneling will be shown, highlighting the advantages of combining different InSAR techniques to monitor both slow and fast movements.

INTRODUCTION

Ground subsidence often accompanies underground excavation activities and monitoring surface subsidence is essential to increase safety, avoid activity stoppages and provide timely detection of incipient damage to buildings and infrastructure. It has become an essential tool for mitigating the socio-economic risks related to activities that produce ground surface deformation.

The instrumentation used for monitoring surface deformation in and around excavation operations is generally based on conventional survey techniques (total stations, levelling, GPS receivers, ground-based radars) but none of these usually offer the high density, bird's-eye view of the movement areas provided by satellite InSAR (Synthetic Aperture Radar Interferometry).

After a brief introduction of the technique, selected case studies are presented in this paper to illustrate the use of InSAR to monitor both slow and fast surface deformation induce by underground mining and support mitigation strategies in urban contexts.

SAR INTERFEROMETERY

SAR satellites acquire images of the Earth's surface by emitting electromagnetic waves and analyzing the reflected signal. InSAR consists of the phase comparison of SAR images, acquired at different times with similar looking angles from space or airborne platforms (Gabriel et al., 1989; Massonnet and Feigl, 1998; Rosen et al., 2000; Bamler and Hartl, 1998). As SAR satellites are continuously circumnavigating the globe, a number of SAR images are collected for the same area over time.

The phase difference calculated between two SAR images acquired over the same area at different times is proportional to the surface deformation that occurs during that time interval (Figure 1) but also contains topographic and atmospheric contributions. Differential InSAR (DInSAR) refers to the interferometric analysis of a pair of SAR images to identify and quantify movement by removing the topographic contribution using a Digital Elevation Model.

Figure 1 shows the relationship between ground displacement measured along the satellite Line of Sight (LOS) and signal phase shift. This is the basic principle of InSAR for measuring ground movement.

In the mid-90s, after extensive application of the DInSAR technology, it became evident that the atmospheric contribution to the signal phase was significant, particularly in tropical and temperate areas. This led to the advent of Advanced DInSAR (A-DInSAR) techniques in the late 1990s, which are based on the statistical processing of multiple images to remove atmospheric noise and reach a higher accuracy of deformation measurements. Permanent Scatterer Interferometry, the first A-DInSAR technique, identifies and monitors point-wise permanent scatterers (PS), pixels that display stable amplitude and coherent phase throughout every image of the dataset (Ferretti et al. 2000, 2001). Permanent scatterers are objects such as rocky outcrops, boulders, manmade structures (buildings, street lights, transmission towers, etc.), and any structure that consistently reflects a signal back to the satellite. In 2011, Ferretti et al. presented a new technique known as SqueeSAR™ that also extracts information from distributed scatterers (DS), areas with homogeneous ground surface characteristics that can be grouped to extract ground surface information from non-urban areas with limited infrastructure and light vegetation.

The existence of data archives going back to the 1990s initially led to the extensive use of InSAR data to perform historical ground deformation analyses. Recent advances in processing algorithms have significantly reduced computational time and the advent of newer satellites with increased spatial resolution and acquisition frequency have

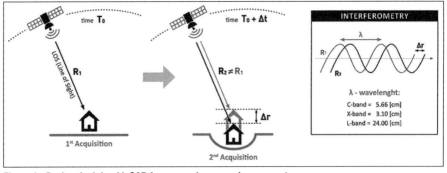

Figure 1. Basic principle of InSAR for measuring ground movement

increased information density. Near-real time InSAR monitoring is now widely applied in different applications: mining, civil engineering, natural hazard and oil&gas.

MINING APPLICATIONS

The complementary use of space-based InSAR with traditional systems has proved to be strategic for operational monitoring and risk assessment in mining operations. Successful applications to underground and open pit mines are presented by Carnec & Delacourt (2000), Raucoules et al. (2003), Colesanti et al. (2005), Jung et al. (2007), Herrera et al. (2007), Herrera et al. (2010), Espinosa et al. (2014) Iannacone et al. (2014), Paradella et al. (2015), Sanchez et al. (2016), Carla' et al. (2018).

Two case studies of InSAR applied to longwall mining are presented, illustrating the use of SqueeSAR historical data to characterize the extent of long-term subsidence (Metropolitan Mine, Australia) and the use of DInSAR to detect short-term fast deformation (Bytom City, Poland).

Metropolitan Mine (Australia)

The Metropolitan mine is an underground coal mine located in the Southern Coalfields of New South Wales, about 40 km South of Sydney and in operation since 1888. A 3 m coal seam has been mined from the Bulli Seam (DeBono & Tarrant 2011) using longwall mining. The Bulli Seam is the top seam in the Illawarra Coal Measures (Hutton 2009), and is overlain by the Arrabeen Group (300 m of sandstones, claystones and shales), a Middle Triassic quartz sandstone and the Hawkesbury Sandstone. The depth of the cover varies from 400 to 520 m depending on the local surface topography (DeBono & Tarrant 2011).

Mining started in the area of study with Longwall 1 in 1995 and finished in April 2010 with the extraction of Longwall 18 (Figure 2). Mining progressed from the southeast towards the northwest while the extraction direction of the individual longwalls proceeded from southwest to northeast (DeBono & Tarrant 2011; Morgan et al., 2013).

The processed radar imagery consists of two archives of ENVISAT radar imagery that were processed using SqueeSAR (Iannacone et al., 2014). The data sets comprise 44 images acquired from an ascending orbit and 43 images acquired from a descending orbit and cover the period June 2006 to September 2010, corresponding to the period in which panels from Longwall 13 to Longwall 18 were mined.

Longwall mining can induce large, rapid displacements in the weeks after panels are mined out, which can lead to a loss of measurement points above the longwalls. However, the focus here was to determine the extent of the subsidence area and to correlate the timing of deformation with mining operations. The analysis of the timing of ground movement within the subsidence bowl showed measurement points located close to the mining front accelerating followed by deceleration once the panel has been mined (Figure 3). InSAR highlighted a kilometer scale subsidence bowl with displacements that extend well beyond the surface area above the longwalls, and reached a cumulative value of over 150 mm between 2006 and 2010. The extent of the subsidence bowl is usually defined by means of an angle of draw which extends upwards and outwards from the working face and varies from 8° to 45° off the vertical. The SqueeSAR results indicate a much higher angle of draw of around 64° with a wider than expected subsidence bowl. Furthermore, a significant number of measurement points located over older mining areas denoted a multiyear linear deformation trend, indicating that residual subsidence can last for many years after the mining has terminated.

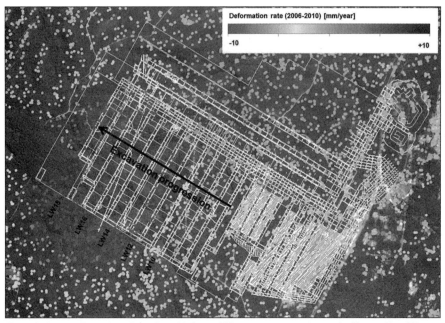

Figure 2. Longwall panel configuration with SqueeSAR measurement points at the Metropolitan Mine. The points are color-coded according to their annual displacement rate between 2006 and 2010.

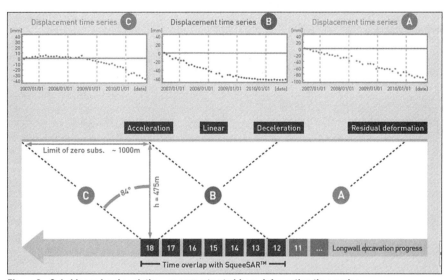

Figure 3. Subsidence bowl evolution, as reconstructed from deformation time series

Bytom City (Poland)

Bytom City is located in the north-western part of Upper Silesian Coal Basin in southern Poland. Mining activities in this area are carried out by longwall mining. The excavated coal layer is 2.5 m thick, 250 to 400 m long, and about 680 m deep. Historical subsidence of up to 27 m over 33 years has been has been measured, affecting an area of nearly 300 km^2.

InSAR Monitoring of Subsidence Induced by Underground Mining Operations 403

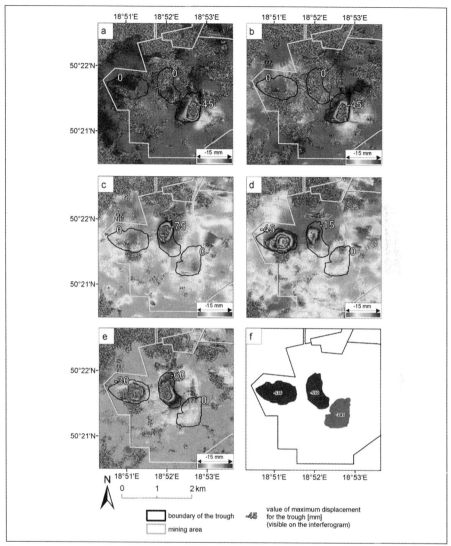

Figure 4. Interferograms over longwall coal-mining area in Bytom municipality. Acquisition dates are reported in Table 1.

To monitor the rapid subsidence over the mine, SAR images acquired by the Terrasar-X satellite every 11 days were processed using a DInSAR approach (Colombo et al., 2018). Figure 4 shows some examples of 11-day interferograms (maps of difference in phase between two images), where coloured bands, referred to as fringes, highlight areas with

Table 1. Acquisition dates for interferograms in Figure 4

Interferogram	1st Image	2nd Image
a	5 Jul 2011	16 Jul 11
b	16 Jul 2011	27 Jul 2011
c	23 Oct 2011	3 Nov 2011
d	17 Dec 2011	28 Dec 2011
e	14 Mar 2012	25 Mar 2012

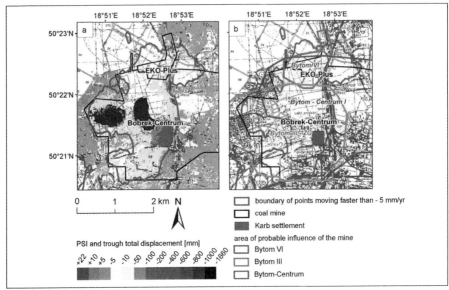

Figure 5. SqueeSAR cumulative displacement overlain on the DInSAR results

phase variations where deformation can be measured. Figure 4f shows estimates of the cumulative deformation on the surface during the observations.

While interferograms outline areas with centimetre-scale subsidence bubbles, SqueeSAR was applied to detect more subtle millimetre-scale displacements, thereby providing a more complete map of the area affected by mining-induced subsidence (Figure 5).

URBAN TUNNELING APPLICATIONS

InSAR monitoring of ground deformation has been applied to all phases (design, construction, and operation) of tunneling projects in both urban and nonurban areas (Bock et al., 2012; Iannacone et al., 2014; Hoppe et al., 2015). In urban contexts, where there is a higher need to monitor ground deformation associated with tunneling activities, InSAR significantly improves the quality of any monitoring program by refining and extending *in situ* observations. This is recognized by the inclusion of InSAR in the "ITAtech Guidelines for Remote Measurements Monitoring Systems" (2015), which provide recommendations and examples for monitoring projects in support of tunnel designers, contractors and owners to understand the benefits and limitations of remote measurement systems.

Rail Tunnel (South Italy)

The case study presented here regards a single-track 60 m^2 cross-section rail tunnel being excavated in a city in southern Italy. The excavation caused surface displacements above the alignment, affecting a number of buildings. InSAR was used to investigate the correlation between excavation progress and induced surface settlements (Barla et al., 2016). This information was of particular interest along a stretch of tunnel where jet grouting from the ground surface was performed in order to deal with challenging geologic and hydrogeologic conditions. These involved a highly heterogeneous calcarenite formation, ranging from well cemented to poorly cemented rock, that reaches from the surface down to the tunnel crown, and is underlain by a very fine sandy soil with silt, which in turn overlies the substratum (silty claystone with sandstone intervals). The water table is above the tunnel crown. Based on geological and hydrogeological studies, a low structural substratum locally acts as a drainage axis for the water flow.

Following completion of the jet grouting consolidation work, at the beginning of June 2014 tunneling advanced into this stretch. After 1 m of penetration into this area, 300 m^3 of water and silty sand collapsed into the tunnel, leading to the development of a subsidence trough at the surface and causing significant damage to the surrounding buildings.

The conventional topographic measurements (including a robotic total station) used along the tunnel axis were augmented by two InSAR data sets covering a time span of about 5 years before June 2014. The data were processed with the SqueeSAR algorithm to produce vertical and E-W horizontal displacement movements. An analysis of the vertical deformation progression along the tunnel alignment (Figure 6) and of the subsidence bowl development allowed to estimate the total volume of displaced material and to correlate this to amount to the volume of material that collapsed into the tunnel. The findings indicated the presence of previously unknown voids in the upper calcarenite formation that triggered further *in-situ* investigations to identify the possible mitigation strategies for tunnel completion.

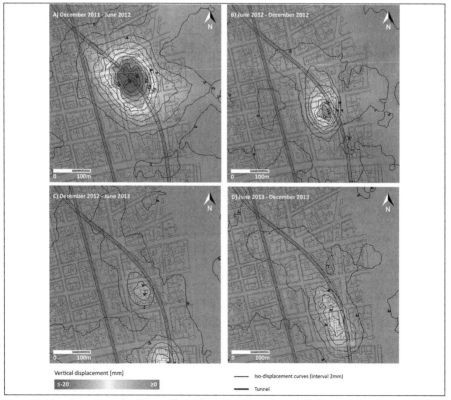

Figure 6. Time-lapse analysis of the vertical deformation along the tunnel alignment from January 2012 to December 2013. Each map represents cumulative vertical displacement in subsequent six-month periods.

CONCLUSION

The two cases presented here highlight the use of InSAR surface deformation monitoring to mitigate the socio-economic impact of ground subsidence induced by underground excavations.

In the Metropolitan longwall mining example the InSAR analysis of historical satellite archives revealed a significantly higher angle of draw and larger extent of the subsidence bowl than previously thought. It also highlighted that residual subsidence over mined out areas continues for many years. In the Bytom City mining example, the combined use of traditional and advanced InSAR techniques allowed both rapid and slow deformation to be precisely monitored for a complete characterization of ground deformation associated with the longwall mining activities.

These longwall mining examples highlight the capability of InSAR surface monitoring to provide precise, spatially dense data high accurate information without the need for ground instrumentation and the ability to monitor both slow and rapid movement (from millimetres to metres) by applying different InSAR techniques.

The use of InSAR in the railway tunnel excavation example in southern Italy highlighted the advantage of using highly dense, precise spatial coverage in measuring far-field deformation in an urban environment, the support provided in accurately defining possible mitigation strategies.

REFERENCES

Bamler, R & Hartl, P 1998. Synthetic Aperture Radar Interferometry. *Inverse Problems*, 14, R1–R54.

Barla, G, Tamburini, A., Del Conte, S., Giannico, C. 2016. InSAR monitoring of tunnel induced ground movements. *Geomechanics and Tunnelling*, 9, 15–22.

Bock, Y., Wdowinski, S., Ferretti, A., Novali, F., Fumagalli, A. 2012. Recent subsidence of the Venice Lagoon from continuous GPS and interferometric synthetic aperture radar, *Geochem. Geophys. Geosyst.*, 13, Q03023, doi:10.1029/2011GC003976.

Carlà, T., Farina P., Intrieri, E., Ketizmen, H., Casagli, N. (2018). Integration of ground-based radar and satellite InSAR data for the analysis of an unexpected slope failure in an open-pit mine, *Engineering Geology*, Vol. 235, 39–52.

Carnec, C. & Delacourt, C. 2000. Three years of mining subsidence monitored by SAR interferometry, near Gardanne, France. *Journal of Applied Geophysics*, 43, 43–54.

Colesanti, C., Mouelic, SL., Bennani, M., Raucoules, D., Carnec, C., Ferretti, A. 2005. Detection of mining related ground instabilities using the Permanent Scatterers technique: a case study in the east of France. *International Journal of Remote Sensing*, 26 (1), 201–207.

Colombo, D. & MacDonald, B. 2018. Using advanced InSAR techniques as a remote tool for mine site monitoring. *Slope Stability Proceeding 2015*. The Southern African Institute of Mining and Metallurgy.

Ferretti, A., Prati, C., Rocca, F. 2000. Nonlinear subsidence rate estimation using permanent scatterers in differential SAR interferometry. *IEEE Transactions on Geoscience and Remote Sensing*, 38(5), 2202–2212.

Ferretti, A., Prati, C., Rocca, F. 2001. Permanent Scatterers in SAR Interferometry. *IEEE Transactions on Geoscience and Remote Sensing*, 39(1), 8–20.

Ferretti, A., Fumagalli, A., Novali, F., Prati, C., Rocca, F., Rucci, A. 2011. A new algorithm for processing interferometric data-stacks: SqueeSAR. *IEEE Transactions on Geoscience and Remote Sensing*, 49(9), 3460–3470.

Gabriel, A.K, Goldstein, R.M., Zebker, H.A. 1989. Mapping small elevation changes over large areas: differential radar interferometry. *Journal of Geophysical Research*, 94, 9183–9191.

Herrera, G., Tomas, R., Vicente, F., Lopez-Sanches, J.M., Mallorquí, J.J., Mulas, J. 2010. Mapping ground movements in open pit mining areas using differential SAR interferometry. *International Journal of Rock Mechanics & Mining Sciences*, 47 (2010), 1114–1125.

Hoppe, E.J., Kweon, YJ., Bruckno B.S., Acton S.T., Bolton, L., Becker, A., Vaccari, A. 2015. Historical Analysis of Tunnel Approach Displacements with Satellite Remote Sensing. *Transportation Research Record*, 15–23.

Iannacone, J.P., Corsini, A., Berti, M., Morgan, J., & Falorni, G. 2014. Characterization of Longwall Mining Induced Subsidence by Means of Automated Analysis of InSAR Time-Series. In: *Engineering Geology for Society and Territory*—Volume 5, edited by Giorgio Lollino, Andrea Manconi, Fausto Guzzetti, Martin Culshaw, Peter Bobrowsky, and Fabio Luino, 973–77. Springer International Publishing.

Jung, H.C., Kim, S.W., Jung, H.S., Min, K.D., Won, J.S. 2007. Satellite observation of coal mining subsidence by persistent scatterer analysis. *Engineering Geology*, 92, 1–13.

Massonnet, D. & Feigl, K.L. 1998. Radar Interferometry and Its Application to Changes in the Earth's surface. *Reviews of Geophysics*, 36(4), 441–500.

Paradella, W.R., Ferretti, A, Mura, J.C., Colombo, D., Gama, F., Tamburini, A., Santos, A.R., Novali, F., Galo, M., Camargo, P.O., Silva, A.Q., Silva, G.G., Silva, A., Gomes, L.L. 2015. Mapping Surface Deformation in Open Pit Iron Mines of Carajas Province (Amazon Region) Using an Integrated SAR Analysis. *Engineering Geology*, vol. 193, 61–78.

Raucoles, D., Maisons, C., Carnec, C., Mouelic, S.L. King, C., Hosford, S. 2003. Monitoring of slow ground deformation by ERS radar interferometry on the Vauvert salt mine (France): comparisons with ground-based measurement. *Remote Sensing of Environment,* 88, 468–478.

Rosen, P., Hensley, S., Joughin, I., Li, F., Madsen, S.N., Rodriguez, E., Goldstein, R. 2000. Synthetic Aperture Radar Interferometry. *Proceedings of the IEEE*, 88(3), 333–382.

Sánchez, C., Conde, A., Salvà, B., Colombo, D. 2016. Use of SAR radar satellite data to measure ground deformation in underground and open pit mine sites, El Teniente case study, Chile. *The First Asia Pacific Slope Stability in Mining (APSSIM) Conference 2016*, Brisbane, Australia 6–8 September 2016.

Minimizing Impacts to the Community and Commuters: Constructing the District of Columbia's Largest Tunnel Along a Major Urban Artery

William P. Levy ▪ AECOM (formerly DC Water)
Moussa Wone ▪ District of Columbia Water and Sewer Authority
Justin Carl ▪ Brown and Caldwell (formerly DC Water Program Consultant Organization)

ABSTRACT

Construction is underway on the District of Columbia Water and Sewer Authority's largest component of its project portfolio in one of the most heavily populated areas of the Nation's Capital. The Northeast Boundary Tunnel is aligned along a major arterial roadway and targeted at relieving areas plagued by chronic flooding due to under-capacity sewers. This paper discusses the processes used to design community and traffic mitigation measures in the design-build contract and the actual implementation of these measures as part of the construction efforts. Focus will be placed on project milestones, mitigation of impacts, and safety.

BACKGROUND

Chronic Flooding in the Northeast Boundary Sewershed

The Northeast Boundary sewershed is served primarily by a combined sewer system and has a long history of chronic flooding problems during moderate to severe storm events. The sewer system that serves the Northeast Boundary drainage area was constructed by the Federal Government in the late 1800s when developed areas of the District terminated at what is now Florida Avenue. Since construction of the system, population within the sewershed area has grown at an exponential rate. In meeting the growing population needs, the District experienced development that transformed previously low-density rural areas into new communities. Impervious areas have increased drastically since the sewer system was constructed. Consequently, the existing sewer system does not have the capacity to convey storms with return frequencies beyond the 2-year storm, without flooding. As a result of the capacity limitations, many of the low-lying areas have become chronic flooding areas due to the surcharging of sewers.

First Street Tunnel Project

In 2012, four major storm events swept through the District and hit the Bloomingdale neighborhood particularly hard. The neighborhood experienced severe flooding. The Mayor assembled a Task Force to evaluate a range of potential actions to address combined sewer flooding. This included accelerating construction of the First Street Tunnel (FST) project, a 2,700-foot long, 20.5-foot inside diameter precast concrete tunnel and associated infrastructure that provides storage for approximately 9 million gallons of combined sewage. The First Street Tunnel ranges in depth from 80 to 160 feet below ground (to the tunnel crown). DC Water procured the First Street Tunnel project through the design-build methodology, with a intense collaboration period focused on mitigation. Notice to Proceed on the project was issued in October 2013 and it was substantially complete in October 2016 at a construction cost of $157 million.

Figure 1. Example of tight FST construction site in residential area

The project was located within the highly urbanized neighborhood of Bloomingdale—with some structures being less than 10 feet from residents' front door steps (see Figure 1). FST would provide medium term flood relief until connected to the Northeast Boundary Tunnel (see Figure 2).

The FST project was successfully implemented by the owner making mitigation of construction impacts a primary focus of the work. The robust mitigation program was critical to success of the First Street Tunnel and was the subject of a previous RETC paper by the authors (Levy et al., 2016). This paper is meant to serve as a sequel to the previous paper, as the Northeast Boundary Tunnel uses and builds on many of the successful mitigation elements from the FST project. Once again it was critical that DC Water collaborate closely with the impacted community and stakeholders during the Northeast Boundary Tunnel's multi-year construction period.

Northeast Boundary Tunnel

The Northeast Boundary Tunnel (NEBT) is one of the largest capital projects in the District since the Metro subway construction in the late 70s and 80s. The project was awarded in July 2017 to the Salini Impregilo/Healy Joint Venture (SIH) design-build team at just under $580 million. The project Notice to Proceed was given in September 2017. It must be substantially complete by March 2025 in accordance with a consent decree entered into by the District, the Department of Justice, the Environmental Protection Agency and DC Water. The NEBT project consists of the following primary components (see Figure 2):

- Design and construct approximately 26,700 feet of 23-foot minimum inside diameter soft-ground tunnel, with bolted/gasketed precast concrete liner from a mining shaft at CSO 019 to a drop shaft at the intersection of R Street NW and 6th Street NW near Rhode Island Avenue.

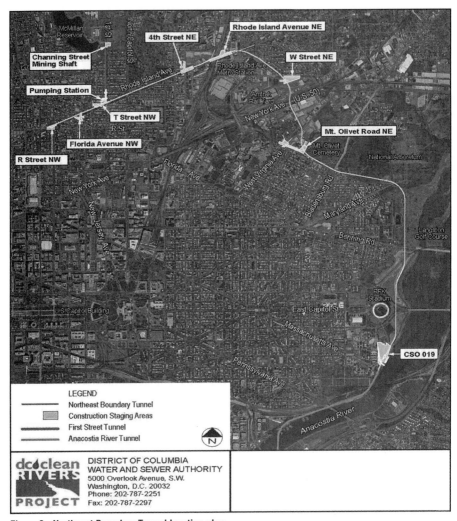

Figure 2. Northeast Boundary Tunnel location plan

- Construct six diversion chambers to relieve the existing sewers during large storm events and divert flow to the tunnel for storage. The diverted flows will be directed to the tunnel via six drop shafts and five adits to the tunnel at approximately 60 to 140 feet below grade.
- Design and construct an underground connection between the First Street Tunnel and the Northeast Boundary Tunnel. Decommission the temporary First Street Tunnel Pumping Station.
- Construct an above grade ventilation control facility at 1241 W Street NE.

Construction of the shafts, diversion chambers and other supporting facilities is being accomplished at 11 construction sites along the tunnel alignment (see Figure 2). At the time of this paper's publication, SIH is mobilized at five of the 11 construction sites, two of which are on Rhode Island Avenue.

COMMUNITY AND COMMUTER IMPACT MITIGATION BY DESIGN

While the DC Water NEBT design team learned many valuable construction impact mitigation lessons from the FST project, it was recognized that the two projects differed in significant ways. NEBT has a much larger geographic footprint than FST, which encompassed about six city blocks. While FST impacted primarily Bloomingdale residents, NEBT would impact not only residents, but commuters, bicyclists, pedestrians, bus transit and local businesses as well. During the required Environmental Assessment process for NEBT and associated public outreach efforts, there was much less community concern and opposition than there was for FST. Part of this was the result of favorable word of mouth in the District about the positive way DC Water conducted itself during the FST project as far as impact mitigation and communication with stakeholders was concerned. The NEBT project had to scale up some of the mitigation toolbox measures incorporated in the FST design (see Levy et al., 2016) and also add some measures specific to the unique NEBT impacts. For example, due to the large project footprint DC Water's Program Consultants Organization provided three public outreach staff dedicated to the NEBT project, and the contract documents required that SIH also provide three public outreach staff. This allows a more "boots-on-the-ground" approach to outreach at 11 construction sites with very different constituents and concerns. FST, by contrast, utilized the same approach but had only one or two dedicated public outreach staff throughout the project.

Pre-Construction NEBT Utility Relocation Contract

As the NEBT design team thought of lessons learned on FST, one of the problems encountered early in the project that resulted in unnecessary impacts to residents was the required relocation of utilities that conflicted with the proposed construction. There were several problems encountered such as:

- Local electric, gas and telephone utilities did not feel obligated to cooperate with DC Water's fast track construction schedule. Delays to schedule were costly and, in some cases caused a "domino effect" on activities located on already tight construction sites. The FST design-builder had assumed simultaneous construction of several CSO components on most of these sites to meet the fast track schedule that was impacted by these delays.
- Several utilities extended the lengths and scopes of the relocations actually needed, instead using the project as an excuse to upgrade old pipelines and services long in need of replacement.
- Several utilities did no outreach prior to digging up a resident's yard or shutting down street parking and blamed DC Water's FST project to the impacted parties. This ran counter to the DC Water outreach team's intent for the project, especially at project outset.

With this in mind, for NEBT an early utility relocation contract was negotiated with the local utilities including DC Water and a Notice to Proceed given in May 2016 (see Evans et al., 2017). The $18 million contract was substantially complete in November 2017 in advance of NEBT construction. In general, this contract was deemed very successful by the owner in that the contractor found many instances of mismarked, abandoned, newly discovered and mis-located utilities, typical of older urban areas. While these discoveries caused delays and redesigns to the utility relocation project, it would have cost the much more complex NEBT project millions of dollars to deal with and would have crippled an already ambitious construction schedule. However, the utility relocation contract also provided a look ahead to the NEBT project team from a public outreach and maintenance of traffic perspective which is discussed more below.

Construction Staging Area (Traffic) Study

As mentioned above, one major difference between FST and NEBT is that five of the construction sites are located along Rhode Island Avenue, a principal arterial connecting the District with the suburbs of Maryland. It has a 100-foot cross section consisting of six travel lanes plus one on each side for parking (see Figure 3). Over 31,000 vehicles traverse it daily and over 75 businesses could be directly impacted by the work. The NEBT design team spent about two years in meetings with the District Department of Transportation (DDOT) to hammer out a detailed traffic study that covered both the early utility relocation contract and the NEBT project. It was the largest traffic study in the Clean Rivers Project.

This study presented the anticipated impacts to vehicular, pedestrian, bicycle and transit operations as a result of the construction of both projects and proposed mitigation measures to minimize impacts. Given the complexity of the surface work, the NEBT design team prepared the study to outline the construction staging area limits and their associated durations required to construct the NEBT surface components for each construction site. Significantly, by breaking each construction site into several staging areas, it also envisioned how the construction would be phased (see Figure 3 and associated Table 1). This level of detail was possible as the team was able to draw on the previous Clean River Project experiences on similar CSO components and their actual schedules. The traffic study submittal and the DDOT letter of April 13, 2016 to DC Water, stating its formal acceptance of the study, marked the end of a collaborative effort between DC Water and the District to resolve concerns and reach a consensus on the construction staging areas, their impacts and proposed mitigation measures. This consensus from this collaboration effort allowed DC Water to establish contractual requirements for the proposed construction work and ensure timely delivery of permits from the District DOT.

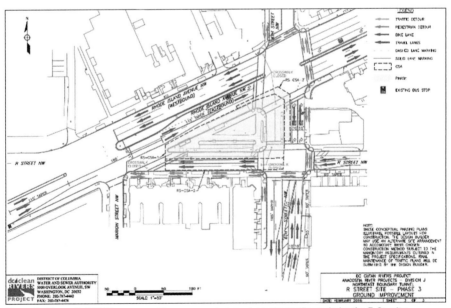

Figure 3. Example MOT for a ground improvement phase from the traffic study for the R St. site

Table 1. Summary of mandatory traffic, pedestrian and bicycle requirements for the R St. site from the traffic study

Site	Construction Staging Area Phase	Est Duration (months)	Traffic	Pedestrian	Bicycle
R Street	RS-CSA1 • Phase 2	30	**Rhode Island Avenue NW Eastbound** • Maintain two (2) 11-ft travel lanes open at all times **Rhode Island Avenue NW Westbound** • Maintain all lanes open at all times • Provide protected/permissive operations for the westbound left turn movement at 7th Street NW **R Street NW** • Maintain 11-ft travel lane open at all times **6th Street NW Northbound** • Maintain 10-ft travel lane and 10-ft turning lane open at all times **6th Street NW Southbound** • Maintain 10-ft travel lane open at all times Adjust signal timings from 3rd Street NW and Rhode Island Avenue NW to 7th Street NW and Rhode Island Avenue NW and along Florida Avenue NW	**Rhode Island Avenue NW** • Maintain the westbound sidewalk open at all times • Maintain the eastern crosswalk at the intersection of 6th Street NW open at all times • Detour pedestrians as outlined in the Conceptual TMP's **R Street NW** • Maintain southern sidewalk open at all times • Maintain the eastern crosswalk at the intersection of 6th Street NW open at all times • Detour pedestrians as outlined in the Conceptual TMP's **6th Street NW** • Maintain northbound and southbound sidewalks open at all times south of R Street NW • Maintain northbound sidewalk open at all times north of R Street NW • Maintain southern crosswalk at the intersection of R Street NW open at all times • Maintain the northern crosswalk at the intersection of Rhode Island Avenue NW open at all times • Detour pedestrians as outlined in the Conceptual TMP's	**R Street NW** • Maintain 5-ft bicycle lane at all times
	RS-CSA2 • Phase 1	3	**Rhode Island Avenue NW Eastbound** • Maintain two (2) 11-ft travel lanes open at all times • Bag signal at the intersection of Rhode Island Avenue NW and R Street NW **Rhode Island Avenue NW Westbound** • Maintain all lanes open at all times • Provide protected/permissive operations for the westbound left turn movement at 7th Street NW **R Street NW** • Detour traffic as outlined in the Conceptual TMP's **6th Street NW Northbound** • Maintain 10-ft travel lane and 10-ft right turning lane open at all times **6th Street NW Southbound** • Maintain 10-ft travel lane open at all times Adjust signal timings from 3rd Street NW and Rhode Island Avenue NW to 7th Street NW and Rhode Island Avenue NW and along Florida Avenue NW Restrict parking on 7th Street NW south of Rhode Island Avenue to provide for adequate storage for the detoured left turn movement	**Rhode Island Avenue NW** • Maintain the westbound sidewalk open at all times • Maintain the eastern crosswalk at the intersection of 6th Street NW open at all times • Detour pedestrians as outlined in the Conceptual TMP's **R Street NW** • Maintain southern sidewalk open at all times • Maintain the eastern crosswalk at the intersection of 6th Street NW open at all times • Detour pedestrians as outlined in the Conceptual TMP's **6th Street NW** • Maintain northbound and southbound sidewalks open at all times south of R Street NW • Maintain northbound sidewalk open at all times north of R Street NW • Maintain southern crosswalk at the intersection of R Street NW open at all times • Maintain the northern crosswalk at the intersection of Rhode Island Avenue NW open at all times • Detour pedestrians as outlined in the Conceptual TMP's	**R Street NW** • Detour westbound bicycle traffic as outlined in the Conceptual TMP's **6th Street NW Northbound** • Create shareable bicycle/travel lane (center lane). Stripe lane for bicycle traffic
	RS-CSA3 • Phase 3	3	**Rhode Island Avenue NW Eastbound** • Maintain one (1) 11-ft travel lanes open at all times • Maintain left/through bay at intersection of 6th Street NW **Rhode Island Avenue NW Westbound** • Maintain all lanes open at all times • Provide protected/permissive operations for the westbound left turn movement at 7th Street NW **R Street NW** • Maintain 11-ft travel lane open at all times open at all times **6th Street NW Northbound** • Maintain 10-ft travel lane open at all times **6th Street NW Southbound** • Maintain 10-ft travel lane open at all times Adjust signal timings from 3rd Street NW and Rhode Island Avenue NW to 7th Street NW and Rhode Island Avenue NW and along Florida Avenue NW No work or staging to be performed between the hours of 7:00AM–9:30AM and 3:30PM–7:30PM, Monday through Friday	**Rhode Island Avenue NW** • Maintain the westbound sidewalk open at all times • Maintain the eastern crosswalk at the intersection of 6th Street NW open at all times • Detour pedestrians as outlined in the Conceptual TMP's **R Street NW** • Maintain southern sidewalk open at all times • Maintain the eastern crosswalk at the intersection of 6th Street NW open at all times • Detour pedestrians as outlined in the Conceptual TMP's **6th Street NW** • Maintain northbound and southbound sidewalks open at all times south of R Street NW • Maintain northbound sidewalk open at all times north of R Street NW • Maintain southern crosswalk at the intersection of R Street NW open at all times • Maintain the northern crosswalk at the intersection of Rhode Island Avenue NW open at all times • Detour pedestrians as outlined in the Conceptual TMP's	**R Street NW** • Maintain 5-ft bicycle lane at all times **6th Street NW Northbound** • Create shareable bicycle/travel lane (center lane). Stripe lane for bicycle traffic

Two important outcomes came of the study. The first was to include milestones in the contract to minimize construction durations and lane closures in the roadways, including DDOT requirements for construction sites that could not be constructed simultaneously. A second important outcome was to provide funding and guidance in the contract to meet DC Water's commitment to devise a Commuter Outreach Plan as part of the public outreach efforts for the project. The design-build contractor was still responsible for developing and submitting site and means and method-specific maintenance of traffic plans (MOTs) to DDOT. The hope was that if the NEBT contractor followed the MOT described in the traffic study, which in essence was "pre-approved" by DDOT, their permit path would be short. If their MOT submission needed to differ significantly from the traffic study to better accommodate their means and methods, the permit path and review by DDOT would be longer.

Business Mitigation Plan

During the preparation of the Request for Proposal for the Northeast Boundary Tunnel project, the team realized that unlike the FST project, there would be significant impacts from the surface construction on businesses, especially in the vicinity of the construction sites along Rhode Island Avenue. A placeholder was put in the contract contingency allowances to provide funding for some sort of Business Mitigation Plan. As there has never been such a plan developed by DC Water on previous works, this would have to be created for NEBT once the contractor's means and methods were known.

COMMUNITY AND COMMUTER MITIGATION DURING CONSTRUCTION

Responses to Contract Mitigation Measures During Design-Build Collaboration

DC Water's Clean Rivers Project has discussed it's unique design-build procurement methods for its CSO Long Term Control Plan using machine-mined tunnels as the system backbone, in previous papers (Wone et al., 2015, 2016) and DC Water received ENR's 2016 Project of the Year award for the Blue Plains Tunnel primarily for how successful these methods have been for the agency. The design-build procurement for NEBT included four shortlisted firms and about eight months of intensive collaboration with the design-build proposers. Besides discussions on the many technical aspects of constructing the NEBT, the collaboration meetings also focused on getting the teams to describe any innovations that might reduce the construction impacts to commuters, businesses, transit riders, bicyclists and residents along the alignment (see Figure 4).

 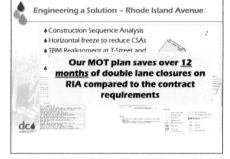

Figure 4. Example figures from shortlisted team presentations during collaboration on impact mitigation

For example, it provided an excellent opportunity to discuss and examine the duration limitations proposed by the NEBT design team in the traffic study and reach consensus on if they were achievable or needed to be increased prior to finalizing the contract term. Some teams came up with innovations like using ground freezing from the drop shafts as temporary support of adit excavations to minimize lane closures. If the consensus on a given site was that the contract duration milestone was inadequate, it was changed prior to finalizing the contract.

Another innovation involved moving the tunnel alignment closer to the drop shaft at key locations to minimize the length and duration of adit constructions and lane closures. DC Water incorporated these changes into the final contract documents. Approved innovations that SIH proposed during collaboration were incorporated into their final contract as well.

Minimizing Lane Closures During Baseline Schedule Preparation

As part of the baseline schedule preparation by the SIH design-build team, construction phasing diagrams were tied to a spreadsheet showing that the contract staging area duration milestones were met (see Figure 5). The way the traffic study was set up allowed the contractor to not count certain types of roadway mobilizations toward the allowable construction staging area (CSA) occupation durations specified in the contract. If they barricaded the site with temporary fencing, and picked up and plated the roadway work before rush hour, it was not counted towards duration. Similarly,

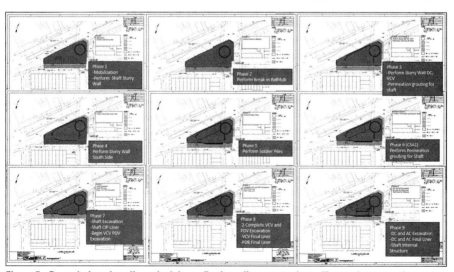

Figure 5. Example from baseline schedule confirming adherence to the traffic mandatory requirements for each CSA by construction phase

if the occupation did not take up a full lane or was not blocking traffic, it was also exempt. SIH has creatively phased the work, sometimes using these exemptions so that float was created in their schedule ("additional CSA occupation" in Figure 5). This approach benefits users of the public right-of-way by having less lane closures for shorter durations than originally agreed upon with DDOT.

Implementation of the Commuter Outreach Plan

Another new element not in FST, but added to NEBT is the Commuter Outreach Plan. The Commuter Outreach Plan is intended to provide commuters in the area of the Northeast Boundary Tunnel Project with advanced notice of potential impacts to their commute as well as information on alternate routes to avoid construction impacts. The plan is intended to keep motorists, bicyclists, pedestrians, and users of public transit (mainly buses) along the Rhode Island Avenue corridor informed. Commuters are one of the most difficult audiences to reach since they are geographically dispersed throughout the District and its outlying suburbs. Given this challenge, a detailed outreach plan utilizing television, web, mobile, print, and radio was outlined in the contract documents. SIH is responsible for implementing the plan to proactively educate commuters on the impacts the project will have on their commute during construction.

Currently, the plan includes an initial six-month media buy from a local DC broadcast media group that is able to provide the following:

- Access to two local FM radio stations that rank consistently in the top five commercial radio stations in the DC metropolitan area during AM and PM drive times.
- Their own digital agency that is a premier Google partner creating customized online and social media strategies. The agency will attach NEBT messaging to its associated websites and social media sites automatically.
- Assistance in getting a presence on Waze, a navigation software that works on smartphones and tablets with GPS support. The application provides turn-by-turn navigation information and user-submitted travel times and route details while downloading location-dependent information over a mobile telephone network.

SIH is also required to take baseline travel time readings at construction sites both before and after lane closures to confirm that impacts are not greater than those modeled as part of the Traffic Study. SIH will take a snapshot of the traffic patterns now and at specific intervals after the paid portion of the information campaign is initially deployed over a six-month period. They will also be looking

Figure 6. "Businesses seek financial help to weather losses during Purple Line construction," from Washington Post

at the number of hits to the media website to help evaluate the plan's effectiveness. Commuter surveys are also being considered.

Implementation of the Plan to Mitigate Construction Impacts to Businesses

While DC Water is under no legal obligation to mitigate impacts from construction on businesses along the NEBT alignment, there were several compelling reasons to consider pursuing some level of mitigation:

- Risk of businesses closing due to loss of revenue directly attributable to the project, which would be viewed unfavorably by the Mayor, DC government, and the local officials.
- Lawsuits against DC Water by businesses that have lost revenue they attributed directly to NEBT construction activities.
- Risk of negative media coverage (see Figure 6); many of these businesses provide essential services to the surrounding residents, provide local jobs, and enhance property values. Additionally, the owners may live in the surrounding neighborhoods and/or be active in the community and civic organizations.
- Risk of loss of public and political support for the project.

The Clean Rivers team performed research on local and nationwide projects that involved mitigation of construction impacts to businesses and found the following:

- Providing compensation to businesses due to impacts from long term construction projects is not common locally but has been employed in other parts of the country.
- The majority of projects offering business mitigation are transportation projects usually with long term durations and with many work zones that impact a large or important commercial corridor (much larger than NEBT).
- Most public agencies do some combination of business-directed public outreach along with some road closure/work zone best practices for businesses near roadway work zones.
- If some compensation is considered by the agency, it is usually not handled by the agency directly. The agency partners with a business organization or finance entity which provides the assistance or grants to the impacted businesses.
- Criteria for compensation varies with each project.

With this model in mind, DC Water felt that partnering with local organizations that could assist with essential activities to keep the existing businesses viable throughout the four to five years of construction was what was needed. As it turned out, there were three of the 17 DC government-funded Main Street organizations whose purview boundaries covered all five construction sites on Rhode Island Avenue. The staff recommendation was to partner with these organizations to implement a program to mitigate the impacts from NEBT construction on the businesses they are already created to serve.

DC Main Streets is a program that promotes the revitalization of traditional business districts. Created in 2002 through the National Trust for Historic Preservation, Main Streets serves as the citywide coordinating program that provides services and funding for the 17 Main Streets found in the District of Columbia. Main Street's mission is to support the traditional retail corridors in the District. DC Department of Small and Local Business Development is the primary funding mechanism for the 17 Main Street organizations established in the District.

DC Main Streets are built on the Main Street Approach® that was developed by the National Main Street Center in 1980 to assist commercial revitalization efforts nationwide. The Main Street Approach® consists of four points which work together to build a sustainable and complete community revitalization effort in the following areas:

- **Organization** of commercial revitalization efforts (e.g., developing and sustaining financial and volunteer resources)
- **Promotion** of neighborhood commercial districts (e.g., branding campaigns and special events) and individual businesses operating therein;
- **Designs** affecting the physical environment of the commercial district (e.g., clean teams and streetscape improvements) and the appearance of business storefronts and interiors;
- **Economic vitality** including business retention, recruitment, and expansion

DC Water used the Main Street Approach® to guide the development of its Business Mitigation program and identified the following goals:

- Keep as many small businesses in business during construction as possible.
- Balance the competing project needs to reduce traffic on Rhode Island Avenue during construction and keep the customers that patronize the local businesses.
- Provide standard best practices during roadway closures (signage, alternative parking, maintain as much access as possible, advertising assistance, phase construction if possible, etc.).
- Provide programs, marketing and education through the Main Street organizations that maintain the viable businesses near the project Construction Sites.
- The program cannot keep already failing businesses afloat, nor provide direct assistance with mortgage or rent payments, compensate for loss of business during construction, relocate businesses during construction, or provide reductions in water or sewer bills.

In November 2018, DC Water's General Manager signed the first Memorandums of Agreement (MOAs) with the first two Main Street organizations that cover the first four construction sites on Rhode Island Avenue. These agreements are the first of their kind for the agency and the District. The MOAs provide funding specifically to increase the Main Street resources and capabilities to assist these impacted businesses and are renewable on an annual basis dependent on how successful the program is at meeting program objectives. As of this writing, the Main Street organizations are preparing work plans for DC Water review, specific to the needs of the impacted businesses under their jurisdiction based on the four-point approach cited above.

CONCLUSIONS

DC Water has made a continuing commitment to funding and requiring construction impact mitigation to the commuters and local stakeholders on their tunnel projects in the densely urban District of Columbia. To be successful, this commitment needs to be shared by DC Water's designers, construction managers and design-build contractors

Table 2. Business mitigation impact responsibility matrix

Service	Service Definition	Lead Responsible Party	
		DC Water	Main Street
Communication	Mailings, email, fliers, phone with stakeholders	X	
Public Meetings	Open meetings with stakeholders	X	
Signage	Postings to inform public	X	
Website/Hotline	Updates and information available on the internet/Informational telephone operator	X	
Program Liason	Position designated to interact with business stakeholders during construction	X	X
Parking	Efforts to ensure adequate parking by business staff and customers	X	X
Economic Vitality	Assist with social media/web presence, branding, marketing strategy		X
Promotions	Assist with loyalty cards, paid advertising, events		X
Design	Storefront improvements, signage, lighting, visibility		X

throughout the project, and these teams meet regularly on public outreach and mitigation issues. The Northeast Boundary Tunnel project has furthered a robust outreach effort on the First St. Tunnel and improved on this approach as follows:

- Requiring complementary public outreach staff from the design-builder to mirror the owner's public outreach staff on projects that cover large geographic areas is important to an effective "boots on the ground" collaborative effort with the diverse stakeholders.
- The collaboration period with the owner and short-listed design-build teams provided extremely useful "ground truthing" of many design elements and criteria including milestone durations for occupation of the many subdivided construction sites into staging areas. It also set the stage for more detailed construction phasing and associated MOT submissions that demonstrated the work met the contract milestones and duration requirements.
- The collaboration period also allowed the teams to propose innovative ways to further mitigate impacts from construction beyond the contract requirements.
- Good faith efforts to mitigate difficult impacts from NEBT construction have been made for two key stakeholders. Included in the contract documents were requirements for a Commuter Outreach Plan and a Mitigation of Construction Impacts to Businesses Plan. DC Water feels that these are necessary measures given the size and duration of the NEBT project. As these programs are just beginning, the authors will report further on the results at the conference.

ACKNOWLEDGMENTS

The authors wish to recognize the citizens of the northeast District, including the Tunnel Forum members, ANC Commissioners, Ward 5 Councilmember and project's direct abutters, since without their understanding and cooperation this important project could not be built. The authors also wish to recognize the contributions of DC Water's Office of Marketing and Communications, Greeley and Hansen, McMillan Jacobs Associates, EPC, the Salini Impregilio/Healy JV and Brierley Associates.

REFERENCES

Shaver, Katherine, "Businesses seek financial help to weather losses during Purple Line construction," Feb. 24, 2018, Washington Post.

Evans, G., Ray, C., Dilego T., Carl J., Bealby S., Mohammad, A. (2017). Reduce Urban Tunnel Utility Relocation Risk through Early Relocation by Specialty Contractor. Rapid Excavation and Tunneling Conference: San Diego, CA.

Levy, W.P., Lindberg, T. Carl, J. and Ray, Carlton (2016). "They Want to Dig a 100-foot hole in Front of my House for Two Years!" Community Mitigation and Outreach for DC Water's First Street Tunnel. World Tunnel Congress: San Francisco, CA.

Wone, M., Corkum, D., Ray, C., Edgerton, W. (2016). Procurement, program management, risk, and financing of underground projects: Changing the paradigm. World Tunnel Congress: San Francisco, CA.

Wone, M., Levy, W., Allen, C., Ray, C. (2015). The planning and the procurement of the Northeast Boundary Tunnel (NEBT): DC water—District of Columbia. Rapid Excavation and Tunneling Conference: New Orleans, LA.

Optimization of a Ventilation System by Using Flow Measurements and 3D CFD Simulations

Martin Schöll • Affiliation ILF Consulting Engineers Austria GmbH
Reinhard Gertl • Affiliation ILF Consulting Engineers Austria GmbH

ABSTRACT

This paper investigated the influence of different geometries for straight and rotating flow on the total pressure loss of a fan's diffuser. Investigations were conducted using Three-Dimensional Computational Fluid Dynamics (3D CFD). The goal was to change the diffuser's geometry in such a way that the total pressure loss is substantially reduced. The simulations showed that diffuser geometries with smooth widening in the diffuser's shape in combination with a straight flow can decrease the pressure loss significantly.

INTRODUCTION

During normal operation the ventilation system shall be capable of supplying fresh air to the drivers in the tunnel and keeping the concentration of toxic gases below critical values. In case of fire the ventilation system shall be designed to enable a safe evacuation for all persons in the tunnel and to assist the firefighters.

Tunnel Geometry

The investigated road tunnel has two tunnel tubes with a length of approximately 6 km and a maximum slope of 1.5%. The tunnel tubes are connected by cross passages at regular distances.

Tunnel Ventilation System

The road tunnel is equipped with a longitudinal ventilation system. The ventilation system is designed for normal and emergency operation and has three smoke extraction points. Jet fans help to regulate the longitudinal air flow in case of normal and emergency operation. The smoke extraction points are used only in case of fire. Each smoke extraction point has three identical fans. In case of an emergency operation, only two fans at one of the smoke extraction points operate in parallel, the third fan is not operating and is installed for redundancy reasons.

Each ventilation building consists of an inflow section (ventilation shaft), followed by three fans and respective parts, a downstream diffuser, splitter sound attenuator and bird screens at the exit to the outside air. A schematic sketch of the tunnel ventilation system is provided in Figure 1.

Flow measurements were performed in each of the three ventilation sections of the tunnel. During the flow measurements a maximum of two fans were operating at the same time, just as they would be operated in case of an emergency.

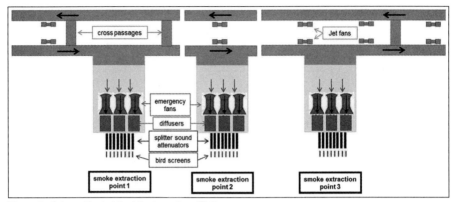

Figure 1. Schematic sketch of the tunnel ventilation

METHODOLOGY

This chapter includes a short description of the on-site flow measurements, the 3D model used for the simulation and the general approach for increasing the ventilation performance by changing the original diffuser geometry.

Flow Rate of Fans

Air flow measurements were performed in order to measure the fans' flow rates and their pressure rise. The measurements were performed pairwise for two of three fans of each smoke extraction point. All three combinations of the fans for each smoke extraction point were measured (in total nine combinations for the three extraction points).

The flow rate was determined by measuring the dynamic pressure p_{dyn} which is the difference of the total pressure p_{tot} and the static pressure p_{stat} (Equation 1).

$$p_{tot} + p_{stat} + p_{dyn} \tag{1}$$

whereas:
p_{tot} = total pressure [Pa]
p_{stat} = static pressure [Pa]
p_{dyn} = dynamic pressure [Pa]

$$p_{dyn} = \frac{\rho_i}{2} \cdot u_i^2 \tag{2}$$

whereas:
ρ_i = local density [kg/m^3]
u_i = local velocity [m/s]

The pressure rise of a fan is determined by measuring the difference of the static pressure at the diffuser outlet and the fan inlet and a further calculation process. Each measurement cross section has a total of four pressure probes, which are evenly distributed along the perimeter of the fan. For the analysis of the fans' pressure rise the mean value of the four probes per measurement location were used. Figure 2 shows the differential pressure measurements at the fans.

The ambient air temperature, humidity and pressure were measured continuously. The results of the measurements were converted to a density of 1.18 kg/m^3, which was

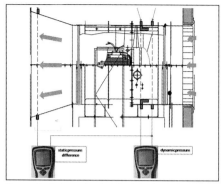 

Figure 2. Schematic sketch of the fan indirect velocity measurement (left) and installation of the measurement equipment at one fan (right)

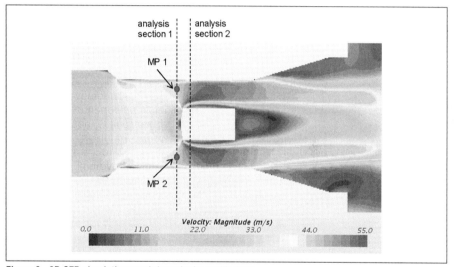

Figure 3. 3D CFD simulation result for velocity profile factor

the reference pressure of the fan curves. For the calculation of the density, the ideal gas air was used neglecting humidity. Measurements were performed over a time period of 15 minutes in order to reach steady-state conditions.

By measuring the dynamic pressure, the velocity can be calculated according to rearranged Equation 2 with respect to velocity. Since the calculated velocities for the different measurements probes at the fans are only point probes, the average cross section velocity u_{avg} has to be determined with the aid of a velocity profile factor obtained by 3D CFD simulations.

The velocity profile factor is the ratio between the mean velocity in the annular cross section close to the impeller (analysis section 2) and the velocity at the measurement points (MP1 and MP2, analysis section 1). Figure 3 shows the velocity plot along a longitudinal section through the fan, the location of the measurement devices and the used section for data analysis. The velocity profile in front of the impeller is assumed to be equal for straight flow and rotating flow through the fan.

The following equation shows the calculation of the velocity profile factor:

$$f_{vel} = \frac{u_{avg}}{u_{MP}} = \frac{45 \text{ m/s}}{43.65 \text{ m/s}} = 1.03 \tag{3}$$

f_{vel} = velocity profile factor, —
u_{avg} = average velocity in annular cross section, m/s
u_{MP} = velocity at measurement point, m/s

A velocity profile factor of 1.03 was used for the analysis of the measurement data. By means of the velocity, the velocity factor and the known cross section of the fan it was possible to calculate the air flow rate of the fans. For every smoke extraction point, all three combinations of the three fans were measured. The final air flow rate can be calculated according to Equation 4.

$$V_{dot} = A * u_{avg} * f_{vel} \tag{4}$$

A = cross section, m^2

Optimization of Installed Geometry with 3D CFD Simulations

Either a higher engine speed or adapting the geometry of the ventilation system's respective parts can increase the overall capacity of the fans. 3D CFD simulations were conducted of the installed geometry and the flow situation was analyzed. In the next step were improvements of the flow topology by changing the installed diffuser's geometry investigated. The simulation results of the installed geometry were used as a reference value for the optimizations. For symmetrically reasons only one of the two operating fans was simulated. For all 3D CFD simulations the commercial software program Star-CCM+ was used.

The installed geometry was modeled realistically according to the drawings which were provided by the client. The diffuser was assumed to be an economical part to be changed with significant improvement potential at relatively low investment costs compared to other measures like increasing the electrical power of the motor or structural works etc. The diffuser's geometry was changed according to theoretical considerations and empirical experience from literature [1]. The whole diffuser shape was extended, in order to achieve a smoother transition from the active part of the fan to the exit of the diffuser. Sharp angle enlargements at the end of the diffuser still exist, but overall the diffuser's shape was smoothed. For a good diffusor performance it is crucial that the diffusion process in the first part of the diffusor (within a length of appr. two times the hydraulich diameter) from kinetic energy into static pressure occurs and large scale mixing effects and stalls need to be avoided.

Monitor cross sections were placed along the computational model and values of total pressure were recorded during the simulation. The mass flow average equation (Equation 5) was used to obtain the cross section's average of total pressure. Monitoring and evaluating several cross sections of total pressure along the model geometry shows the overall curve of total pressure.

$$\text{Mass Flow Average} \equiv \frac{\oint \rho \phi |\mathbf{v} \cdot d\mathbf{a}|}{\oint \rho |\mathbf{v} \cdot d\mathbf{a}|} = \frac{\sum_f \rho_f \phi_f |\mathbf{v}_f \cdot \mathbf{a}_f|}{\sum_f \rho_f |\mathbf{v}_f \cdot \mathbf{a}_f|} \tag{5}$$

ρ = density, kg/m^3

Optimization of a Ventilation System

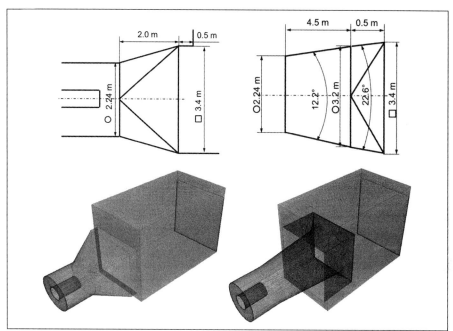

Figure 4. Installed geometry with hub, diffuser and fan room (left); improved geometry with extended diffuser with reduced diffuser angle (right)

v = velocity, m/s
ϕ = specific variable, *
a = area, m^2

According to the conservative equations of mass, momentum and energy the total pressure in a pipe must be constant along a flow path in an ideal environment, assuming no friction or other losses. In a real environment wall friction decelerates the flow and triggers turbulence and total pressure losses occurs along the flow path. Energy is partly converted into friction, turbulence, inner friction and dissipation. Finally energy is converted into thermal energy. Total pressure losses cannot be avoided, but should be minimized to obtain the fan's maximum capacity. Thus, evaluation of total pressure along the flow path is the key variable for judging the ventilation's performance.

Simulations were conducted for a straight and a rotating flow, both for the installed and for an improved diffuser geometry. The general model setup is shown in Table 1. For all simulations an implicit unsteady approach was used. The simulation time was set to 30 s with a time step of 0.01 s. The base mesh size was 0.3 m, at the diffuser the mesh size was 0.15 m. For the evaluation of the simulation results the values with a sampling frequency of 0.1 s were averaged between 20 s and 30 s.

RESULTS

The installed geometry was simulated first, and the results represent the basis for the comparison of the installed and the improved diffuser. The evaluation of the total pressure along the flow path is shown in the following figures in this chapter. All simulations were performed with the general setup shown in Table 1. Figure 5 and 6 show velocity simulation results of the installed diffuser geometry.

Table 1. 3D CFD model setup

Numerical Method: Implicit Unsteady	
Turbulence model	k-ε RANS realizable two layer model
Time step	0.01 s
Base cell size	0.3 m
Simulation time	30 s
Density	1.18 kg/m^3
Ambient pressure	1,013.0 hPa
Flow rate	150 m^3/s

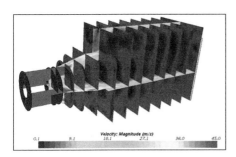
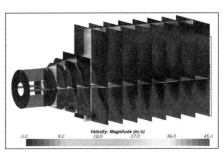
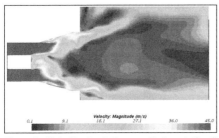
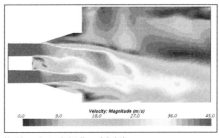

Figure 5. Velocity simulation results for rotating flow (left) and straight flow (right)

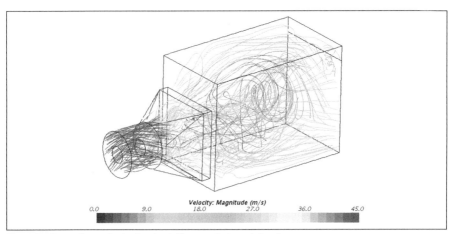

Figure 6. Streamlines of velocity simulation results for rotating flow

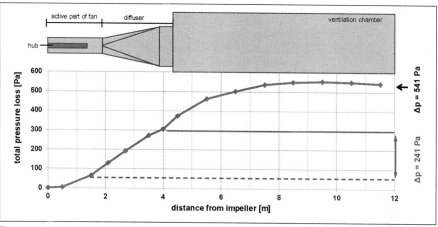

Figure 7. Total pressure loss of the installed geometry with straight flow

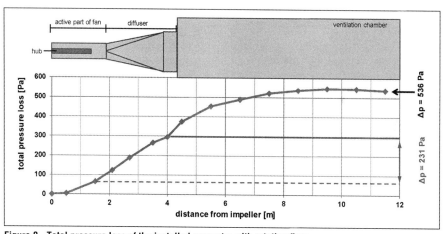

Figure 8. Total pressure loss of the installed geometry with rotating flow

Installed Diffuser Geometry with Straight Flow

Figure 7 shows the simulation results of the installed diffuser with a straight, non-rotating flow. Total pressure decreases constantly from the inlet of the fan, the diffuser and further downstream to approximately 8 m distance from the impeller. The total pressure loss in the ventilation chamber is greatly influenced by the diffuser and amounts to 140 Pa. The geometry of the installed diffuser causes a total pressure loss of 241 Pa in the diffuser and 541 Pa in total.

Installed Diffuser Geometry with Rotating Flow

Figure 8 shows the simulation results of the installed diffuser with rotating flow. Total pressure decreases constantly over the active part of the fan and the downstream end of the diffuser, which is quite similar to the straight flow. In the ventilation chamber the total pressure loss is approximately 140 Pa. The geometry of the installed diffuser with rotating flow results in a total pressure loss of 231 Pa in the diffuser and 536 Pa in total. In summary there is no clear difference between a straight and a rotating flow for the installed geometry.

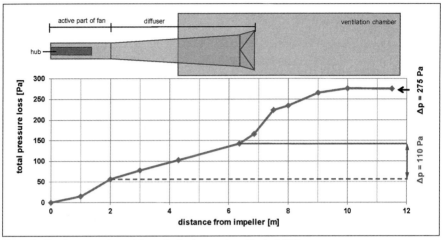

Figure 9. Total pressure loss of the optimized geometry with straight flow

Optimized Diffuser Geometry with Straight Flow

Figure 9 shows the simulation results of the optimized diffuser with a straight, non-rotating flow. The total pressure curve of the improved geometry is clearly different from the installed geometry. Total pressure decreases constantly along the fan, but now the curve of the pressure loss is flatter than that of the installed geometry. In the ventilation chamber the total pressure loss increases quickly, since the sharp transition of the diffuser exit to the ventilation chamber is now shifted downstream. The installed geometry of the optimized diffuser results in a total pressure loss of 110 Pa in the diffuser and 275 Pa in total. This means that the overall pressure loss is now 51% compared to the existing geometry with straight flow.

Optimized Diffuser Geometry with Rotating Flow

Figure 10 shows the simulation results of the optimized diffuser geometry with rotating flow. The total pressure's curve to the end of the diffuser is nearly the same as for the optimized geometry with straight flow. Beginning from the downstream end of the diffuser and following the ventilation chamber, total pressure loss increases quickly by approximately 150 Pa between 6.5 m to 8.0 m distance from the impeller. Overall, the geometry of the optimized diffuser with rotating flow results in a total pressure loss of 159 Pa in the diffuser and 437 Pa in total. This is now an overall pressure loss of 18% compared to the existing geometry with rotating flow.

Summary of Simulation Results

The pressure losses of the different diffuser geometry for straight and rotating flow are summarized in Table 2. The pressure difference between the diffusers' inlet and its outlet are indicated in the column "Diffuser." The overall pressure loss over the whole geometry is shown in the column "Total."

For the installed geometry the simulation results show that the straight flow and the rotating flow cause approximately the same total pressure loss in the diffuser and in total.

The optimized diffuser's total pressure loss for a straight flow is 51% of the installed diffuser's total pressure loss. The total pressure loss of the optimized diffuser with rotating flow is 82% of the installed diffuser's total pressure loss.

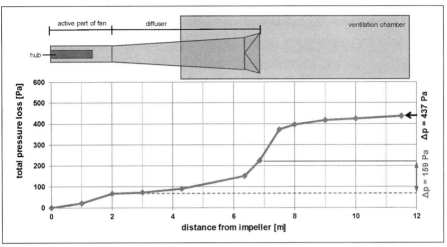

Figure 10. Total pressure loss of the optimized geometry with rotating flow

Table 2. Summary of the diffuser's simulation results for straight and rotating flow

Geometry/Flow	Diffuser [Pa]	Total [Pa]
Installed geometry Straight flow	241	541
Installed geometry Rotating flow	231	536
Optimized geometry Straight flow	110	275
Optimized geometry Rotating flow	159	437

DISCUSSION AND CONCLUSION

Total pressure curves of straight and rotating flow for different diffuser geometries were shown in the previous chapter. The results show almost no difference between a straight and rotating flow for the installed geometry. The difference of 10 Pa lies in the range of the model's numerical inaccuracy. The total pressure loss along the fan, the diffuser and the exit to the outside air is quite high.

The optimized geometry case shows significantly different results for the straight and rotating flow. The straight flow shows the lowest total pressure loss. Hence the type of flow, straight or rotating, has a significant impact on the performance of the diffuser. If only the geometry is changed, the total pressure loss can be reduced by approximately 18%, when comparing the rotating flow of the installed and the optimized diffuser. If guiding vanes are installed in such a way that straight flow enters the diffuser, the total pressure loss is reduced by 51%. From a physical point of view it is also advisable to guide the flow straight through the diffuser, since rotation leads to enhanced turbulence and inner friction.

REFERENCE

Idelchik I.E., 1996. Handbook of Hydraulic Resistance 3rd Edition. CRC Press, ISBN 81-7992-118-2.

Transit Tunnel TBM Vibration Through Glacial Till

Thomas F. Bergen ▪ Wilson Ihrig
Deborah A. Jue ▪ Wilson Ihrig

INTRODUCTION

Tunnel excavation for the Central Puget Sound Regional Transit Authority (Sound Transit) University Link Extension (U-Link) was completed in 2012 and tunneling for the Northgate Link Extension was completed in 2016 in Seattle, Washington. In both cases, the twin bore tunnels were advanced under the University of Washington (UW) main campus, and there was concern that the vibration could be detected at nearby buildings and potentially interfere with sensitive research activities. This paper will describe a field vibration measurement program and compare the predicted and measured TBM vibration at the ground surface above the tunnel alignment. The over-consolidated glacial till that underlies much of the region is a particularly efficient transmission medium for low frequency vibration. The predictions were based on available data collected in advance of the TBMs reaching the campus. Measurements were conducted outside and within the buildings. The U-Link and Northgate Link tunnel construction used a total of four 21-foot diameter Earth Pressure Balance TBMs. The TBMs were refurbished and relaunched multiple times to complete a total of about 13 miles of tunnel. The U-Link tunnel construction used small diesel mining locomotives on rails to supply the TBMs which were found to produce significant vibration at the surface. Rubber tired supply vehicles were subsequently used during construction of the Northgate Link tunnels to minimize vibration through the UW central campus.

SUMMARY OF FIELD MEASUREMENTS

Field measurements of ground surface vibration produced by TBM activity were made between February and July 2015 at seven separate locations. The field measurement scenarios at test locations north of the UW campus in Seattle are summarized in Table 1. For both tunnels, boring proceeded from north to south.

The measurements were configured to capture vibration associated with tunnel mining and concrete ring construction as well as the TBM supply vehicles. The measurement equipment included seismic accelerometers or piezoelectric accelerometers and vibration signals were recorded continuously on portable digital data recorders. Measurements were typically made during the third (night) shift when ambient vibration was lowest. All recorded data were analyzed with one-second contiguous samples of one-third octave band spectra and integrated to determine overall vibration velocity levels. Samples of the resulting time series were carefully selected for calculating one-third octave band vibration velocity levels for the TBM cutterhead, supply vehicles, and the ambient vibration at the measurement location. The selected samples excluded vibration from nearby surface sources such as cars, bicycles and pedestrians.

VIBRATION PREDICTION PROCEDURE

Predictions of ground vibration produced by TBM activity were derived from measurement data taken from the NB and SB TBMs at locations on NE 72nd, NE 69th, and NE 52nd Streets in Seattle north of the UW campus. Data from the NE 58th Street and NE

Table 1. Summary of tunnel construction vibration measurements north of UW (Feb–July 2015)

Test Location	Tunnel (depth)	Soil Description	Measurement Positions Offset from Tunnel centerline
NE 72nd St	NB (110 ft)	Sand	0, 33, 100 & 185 ft
NE 72nd St	SB (110 ft)	Sand	20, 75, 145 & 225 ft
NE 69th St	NB (90 ft)	Sand	0, 55, 110 & 220 ft
NE 69th St	SB (90 ft)	Sand	20 & 100 ft
NE 58th St	NB (85 ft)	Mixed (clay/silt/sand)	20, 60, 100 & 180 ft
NE 56th St	NB (115 ft)	Mixed (clay/silt/sand)	0 ft
NE 52nd St	NB (115 ft)	Sand	0 ft

56th Street locations were excluded since the vibration levels were measurably lower in the soils with greater clay content. A purely empirical approach was employed to develop the prediction model.

The basic steps used to estimate vibration in UW buildings were as follows:

1. Determine source reference vibration level from TBM measurements taken on the ground surface near the tunnel centerline
2. Determine vibration attenuation with respect to offset distance from tunnel centerline by curve fitting data obtained in field measurements out to 225 feet from centerline
3. Calculate attenuation to UW buildings based on distance from near tunnel centerline and extrapolated curve fit
4. Estimate TBM one-third octave band vibration levels at each UW building by applying the attenuation factor to the reference vibration level
5. Plot predicted TBM vibration levels and compare with criteria

Additional details about the major elements of the procedure are provided below. Included are a summary of the measured source vibration levels for the TBM, an empirically-based propagation model to determine how these levels were expected to attenuate with distance, and predicted spectra of TBM vibration levels at select UW buildings on campus.

TBM Reference Vibration Levels

TBM vibration predictions focused on the cutterhead operation which was the dominant source associated with tunnel construction on Northgate Link Extension. A major conclusion drawn from the TBM field measurement results was that different vibration levels were generated depending on whether the cutterhead was boring through clay/silt or sandy soils. In the over-consolidated sandy soils, relatively consistent vibration levels were generated at the surface during mining for each ring, as well as from one ring to the next. In clay/silt, vibration levels varied substantially during cutter head operation, but were generally lower than the levels produced in sandy soils. At their highest, the clay/silt vibration levels approached but did not exceed the sandy soil vibration levels for brief periods.

Underneath the UW campus, the advancing TBMs were expected to alternate boring through clay/silt and sandy sections based on geotechnical survey data. One-third octave band reference vibration levels on the ground surface in the vicinity of the tunnel centerlines (NB and SB) were derived from measurement data taken in sandy soils where the highest vibration levels were measured (NE 72nd, NE 69th and NE 52nd). The reference vibration levels were produced from samples taken at locations close

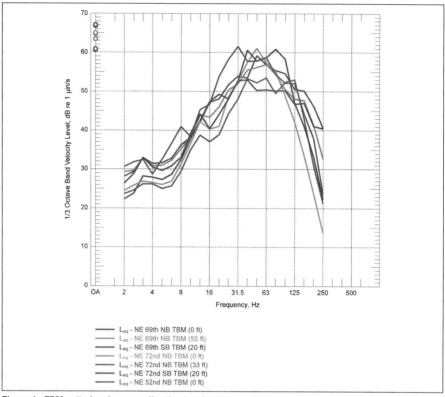

Figure 1. TBM cutterhead source vibration levels on ground surface near tunnel centerline

to the tunnel centerlines (0 to 55feet) to provide a range of values near the tunnels. These are plotted together in Figure 1. The vibration levels measured in the clay/silt at NE 56th and NE 58th Streets were measurably lower and excluded from the group of spectra to provide conservative reference levels for the entire campus. The vibration spectrum measured at NE 52nd Street were slightly different from that measured at NE 69th and 72nd Streets (peak in 31.5 Hz band) but was included since the soils were similar. Also, the sandy soils at NE 52nd Street were expected to be representative of the geology near Wilcox and More Halls. The average and range of the one-third octave band levels were used to develop the reference levels. Variations due to tunnel depth were assumed to be minimal and are captured in the range of source reference levels. The variations in vibration levels appeared to be influenced much more by the soil conditions than by the tunnel depth.

Empirical Vibration Propagation Model

The attenuation of vibration levels with increasing distance from the tunnel centerline were estimated from the measured data where vibration was measured up to an offset distance of 225 feet. For each one-third octave frequency band, measured vibration levels were plotted against the horizontal offset distance from the tunnel centerline for the TBM measurements made at NE 69th and NE 72nd Streets. In all frequency bands, these data points were observed to fall approximately on a linear regression trend line for each band from 12.5 to 100 Hz. The slope of each line was taken to be the attenuation factor for each frequency in terms of dB/offset distance. As examples, the data

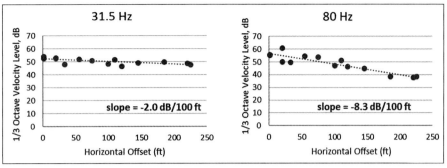

Figure 2. Examples of surface vertical vibration velocity level vs. offset distance from tunnel centerline for TBM cutterhead measured at NE 69th and NE 72nd Streets: 31.5 Hz and 80 Hz one-third octave bands

points and trend lines for the 31.5 and 80 Hz frequency bands for the NE 69th and NE 72nd Street data are shown in Figure 2. Below 12.5 Hz, the vibration levels were dominated by the ambient vibration at the measurement locations, and, as such, it was not possible to compute an attenuation factor for the cutterhead at these lower frequencies. The attenuation factors for the one-third octave bands are presented in Table 2.

Vibration data associated with the TBM cutterhead was largely influenced by the ambient vibration at the measurement locations at frequencies below 12.5 Hz, and due to the influence of the ambient, particularly at the larger distances from the tunnel, it was difficult to determine the true rate of attenuation at these frequencies. Thus, for the purposes of predicting vibration levels at UW, the one-third octave band ambient vibration levels in these bands were conservatively assumed to be the upper bound of TBM vibration. Using this assumption, the average rate of attenuation would appear to be zero or positive due to the interference of the ambient conditions. The measured data was used to develop the rate of TBM vibration attenuation with distance from the 12.5 Hz to 100 Hz band. At the lower frequencies the rate was fixed down to 2 Hz at −0.2 dB per 100 feet so that the predicted values were presumed to be conservative.

Table 2. Estimated TBM vibration attenuation with distance

One-third Octave Band Center Frequency (Hz)	Distance Attenuation Factor (dB/100 feet)
12.5*	−0.2
16	−0.5
20	−0.8
25	−1.0
31.5	−2.0
40	−2.3
50	−4.1
63	−6.0
80	−8.3
100	−9.9

*Constant attenuation factor applied down to 2 Hz (−0.2 dB/100 feet)

UW Building Vibration Predictions

The predicted TBM vibration level for each UW MIA building was computed based on the reference vibration level, distance from the tunnel, and attenuation with distance, as follows:

$$L_{TBM-UW} = L_{TBM-REF} + A \times D/100$$

Where

L_{TBM-UW} = predicted one-third octave band vibration level at UW building
$L_{TBM-REF}$ = range of TBM reference vibration levels at tunnel centerline (Figure 1)

A = attenuation factor (in dB/100 feet) for each one-third octave band (Table 2)
D = horizontal offset distance (in feet) between tunnel centerline and UW building

The results of these calculations included an average and range of predicted surface vertical vibration levels at selected UW buildings. The predicted levels represent the highest expected vibration in each building as the TBMs pass in closest proximity. In lieu of measured data collected at greater distances from the tunnels, the empirical attenuation model included here was expected to provide a conservative approximation of maximum TBM vibration.

UW VIBRATION PREDICTION RESULTS

Using the method described above, the predicted TBM vibration level spectrum for the Wilcox Hall building is plotted in Figure 3 as an example. Similar plots were generated for the 23 sensitive buildings identified by UW. The predicted average and range of TBM vibration levels were plotted with the international standard Vibration Criteria curve E (VC-E) which is the most stringent criteria for vibration sensitive research activity. Any one-third octave band with a value greater than 42 VdB is considered as an exceedance of the criteria. Even though the predicted levels in the 8 Hz and lower bands are generally low, these predictions are based largely on ambient vibration levels at the field measurement locations and not on identifiable TBM vibration. Vibration in the 12.5 Hz and higher bands are based on clearly discernable, measured TBM vibration levels.

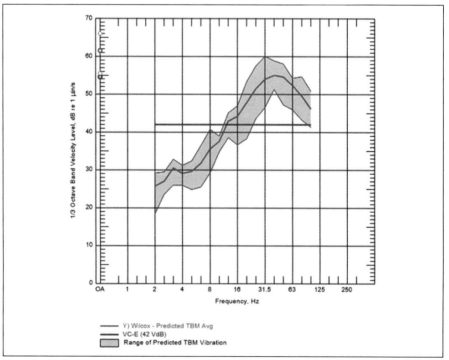

Figure 3. Predicted range of vibration levels at Wilcox Hall (75 feet offset) during TBM passage

Figure 4. Predicted maximum overall outdoor ground surface vibration level contours for TBM cutterhead

The prior measurements north of campus indicated that the TBMs may produce overall vibration levels at the surface directly above the tunnel in the range of 61 to 67 dB re 1 µ-in/sec (VdB) as shown in Figure 1. The overall vibration levels at each building computed from the predicted spectra were plotted with respect to the offset distance from the tunnel to provide predicted vibration contours as shown in Figure 4. The overall levels roughly follow logarithmic attenuation rate with distance from the source.

Depending on the rate of tunneling progress, the highest cutterhead vibration levels at the buildings closest to the tunnels were predicted to persist for up to 4 or 5 workdays, and the cutterhead vibration was estimated to be discernable from other sources for up to 36 workdays. These estimates were based on the long term TBM monitoring results and an average tunnel construction progress of 70 feet per day. The expected duration of exposure to UW buildings was determined by estimating the length of tunnel over which the cutterhead would have been within 1,000 feet of any point on the building perimeter.

Comparison of Predictions and Direct Measurements

As the tunnel construction proceeded though the UW campus, the TBM progress was tracked daily and numerous vibration measurements were made at the sensitive buildings closest to the tunnels. Measurements were made for both the northbound and southbound tunnels. Where possible, the directly measured vibration levels were compared with the predictions based on the data collected north of campus.

One area of campus of the greatest concern was the Engineering complex where the tunnel alignment was directly below some of the sensitive buildings. Monitoring revealed that the TBM was indistinguishable from the ambient vibration through this part of campus where the soil was found to be mostly clay. Presumably, the TBMs cut through the clay more smoothly than over-consolidated sand and gravel and therefore produced less vibration.

At the south end of campus near the end of the tunnels, the soils transitioned back to over-consolidated glacial till, and TBM operations were measured and found to be clearly detectible at the ground surface and inside the buildings. The last group of buildings with exposure to Northgate Link tunnel construction vibration were More, Roberts and Wilcox Halls. The vibration measurement locations conducted at those buildings as the TBM bored the NB tunnel are shown in Figure 5. In Figure 6, the TBM vibration measured outside Wilcox Hall is compared with the predicted average level from Figure 3. The prediction provided a conservative envelope of the vibration in the 20 to 100 Hz frequency range where the levels are highest and otherwise agree with the measured time-averaged vibration levels at lower frequencies. In general, predictions were observed to be conservative for locations within 200 feet of the tunnel centerline, and the model over predicted the vibration at greater distances and was deemed invalid.

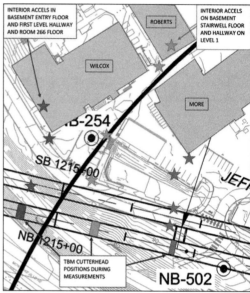

Figure 5. Ground vibration measurements at UW More, Roberts, and Wilcox Halls during boring of NB tunnel (stars indicate vibration sensor locations on surface and inside buildings)

TBM SUPPLY VEHICLE VIBRATION

During construction of the U-Link tunnels, small diesel mining locomotives were used to transport personnel, tunnel ring segments and other supplies to the TBMs as they advanced to the south of the UW campus. Vibration levels measured in the neighborhoods south of campus were observed to be higher than for the cutterheads, and the cause was identified as vibration from the locomotives traveling over the jointed mismatched rail segments laid in the bottom of the tunnels. This equipment was a concern for UW since, unlike the TBM which passed each building once, the supply trains would pass many times per day for a period of several months. This scenario was deemed excessive for UW and vibration sensitive research, so the Northgate Link contractor opted for rubber tired TBM supply vehicles. An image of the rubber tired vehicle fully loaded with concrete tunnel ring segments is shown in Figure 7. Note the wheels have reverse camber which is adjustable to match the curvature of the tunnel.

The rubber tired vehicles (RTVs) produced vibration with a distinct signature as determined by measurements at NE 72nd and NE 58th Streets, and later at the UW campus. The unloaded RTVs produced a low frequency vibration peak in the 3.15 and 4 Hz one-third octave bands, and the loaded vehicles produced peaks at 2.5 and 6.3 Hz. The low frequency peaks were attributed to bending modes of the flatbed and vehicle suspension. RTV vibration was not detectable at higher frequencies. This is to be expected, as vibration from rubber-tired vehicles do not often generate substantial higher frequency components. The measurements showed on average about 4 dB attenuation for the four peak frequencies over a 185 foot distance from the tunnel.

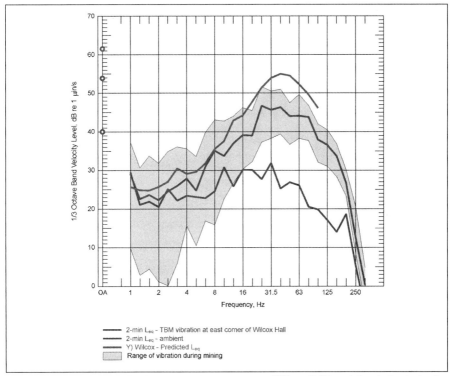

Figure 6. TBM cutter head vibration measured at east corner of Wilcox Hall foundation during boring for Ring 3581 compared with prediction

The low frequency vibration peaks for the loaded and unloaded RTVs measured near the tunnel centerline were enveloped for comparison with VC-E levels. These maximum expected RTV vibration levels are plotted in Figure 8 and apply to areas outside buildings with minimum offset from the tunnels (Mechanical Engineering & Annex, Wilcox, More and Roberts Halls). RTV vibration levels inside these buildings were presumably lower due to coupling loss from the soils to the building interior and were not found to be a disturbance to UW researchers during tunnel construction.

Figure 7. Rubber tired tunnel supply vehicle with tunnel ring segments

CONCLUSIONS

Tunnel construction vibration results were substantially different from the previous characterization that was based on the U-Link measurements south of the UW campus. Previously, the cutterhead vibration levels were relatively low and the dominant vibration source was the supply trains operating on jointed rail sections. For Northgate Link, significantly higher vibration levels were associated with the cutterhead,

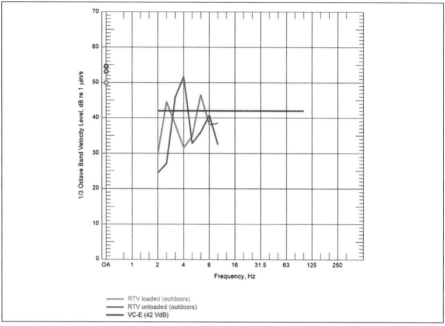

Figure 8. Maximum RTV outdoor vibration levels at UW buildings close to the tunnel

especially in homogeneous sandy soils. Observations and conclusions derived from this effort are summarized below.

- Tunnel construction vibration associated with cutterhead operation at many of the UW buildings was predicted to be exceed the VC-E criteria when the TBM passed in closest proximity.
- The predictions suggested that the TBM would be measurable out to 1,000 feet offset from the tunnel, perhaps more. In reality, this was not the case.
- Since all field measurements were made within 225 feet of the tunnel centerlines, there was significant uncertainty associated with the prediction of vibration levels at greater distances.
- The vibration produced by the TBM through the UW campus was generally lower than that measured north of campus.
- The 2015 prediction model was overly conservative for buildings more than 200 feet from the tunnel alignment
- TBM vibration was substantially lower and mostly undetectable in the center of campus while cutting through predominantly cohesive clay soil type. Vibration increased in the vicinity of Wilcox, More, and Roberts Halls where the soil type transitioned back to sand and gravel.
- In general, ground vibration in the vicinity of tunnel construction activity can vary substantially depending on geometric factors, but it varies primarily due to the local geology.
- Vibration and groundborne noise issues in residential areas associated with the supply trains can be essentially eliminated with the use of rubber tired tunnel supply vehicles.

PART

Future Projects

Chairs

John Caulfield
Jacobs Engineering

Matt Kendall
FKCI

Alexandria Renew Enterprises Is Now in the Tunnel Business

Liliana Maldonado ▪ Alexandria Renew Enterprises
Caitlin Feehan ▪ Alexandria Renew Enterprises
Justin Carl ▪ Brown and Caldwell
Kevin Pilong ▪ Brown and Caldwell
Jennifer Jordan ▪ JCK Underground, Inc.

ABSTRACT

Alexandria Renew Enterprises, the local wastewater agency for Alexandria, VA is implementing a large-scale deep storage and conveyance tunnel program to capture and treat combined sewer overflows from four outfalls in the City of Alexandria by 2025. The preliminary planning and engineering studies include approximately two miles of 12-foot diameter soft ground reinforced concrete tunnel, a ½-mile, 6-foot diameter diversion sewer, three shafts, five diversion chambers, a tunneling dewatering pumping station, and odor control facilities. This paper will discuss the technical approach, construction considerations, contract delivery plan and schedule, and challenges of constructing in a historically-significant and densely-occupied urban environment.

INTRODUCTION

Project Overview

Alexandria Renew Enterprises (AlexRenew), with support from the City of Alexandria (City), is implementing RiverRenew, a major infrastructure program designed to substantially reduce discharges of sewage mixed with rainwater from Alexandria, Virginia's combined sewer system to the Potomac River, Hooffs Run, and Hunting Creek. RiverRenew is essential to comply with the Commonwealth of Virginia's 2017 legislation which requires that Alexandria's four existing combined sewer outfalls be brought into compliance with the 2017 law by July 1, 2025.

RiverRenew features a series of storage and conveyance tunnels coupled with upgrades to AlexRenew's Water Resource Recovery Facility (WRRF) to treat the additional flows delivered during rain events.

History and Background

AlexRenew is a political subdivision of the Commonwealth of Virginia that was created in 1952 under the Virginia Water and Wastes Authority Act. AlexRenew owns and operates a WRRF that provides sanitary and combined sewage treatment services to the City of Alexandria and sanitary sewer treatment services to Fairfax County.

Like many older cities in the United States, Alexandria is served by two (2) types of sewer systems: a separate sewer system and a combined sewer system. Alexandria's combined sewer system was built in the late 1800s through the early 1900s.

During dry weather and most rainfall conditions, the combined sewer system conveys sewage to the AlexRenew's WRRF for treatment and discharge. During intense wet weather events, the capacity of the combined sewer system may be exceeded which results in combined sewer overflow (CSO) discharges to Alexandria's waterways via four (4) permitted outfalls. Figure 1 illustrates the location of the combined sewer system in Alexandria as well as the outfall locations, and proposed tunnel system.

Virginia's 2017 CSO Law and Response

In April 2017, the General Assembly of Virginia enacted a law, referred to as the "2017 CSO Law" that requires all outfalls discharging to the Chesapeake Bay, but not under consent decree to be brought into compliance with Virginia law, the Clean Water Act, and the Presumption Approach described in the EPA CSO Control Policy by July 1, 2025. In response, AlexRenew and the City submitted the Long Term Control Plan Update (LTCPU) to the Virginia Department of Environmental Quality (VDEQ) to meet the 2025 deadline. The LTCPU was approved by VDEQ in June 2018 and sets forth a publicly vetted alternative consisting of the following major elements:

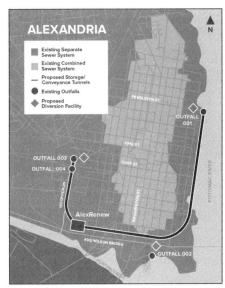

Figure 1. Alexandria's existing sewer systems and proposed tunnel system

- Expansion of the AlexRenew WRRF's primary treatment system from 108 to 116 million gallons per day (MGD).
- Construction of a tunnel system to provide additional storage and conveyance of wet weather flows including:
  - A storage and conveyance tunnel serving Outfalls 001/002;
  - A conveyance tunnel serving Outfalls 003/004; and
  - A Tunnel Dewatering Pumping Station (TDPS) to dewater the tunnel after a rain event.
- Reconfiguration of existing primary settling tanks for a dual-use 40 MGD wet weather treatment facility to provide primary treatment and disinfection of wet weather flows.

Figure 2 shows the schematic of the unified tunnel system. The unified tunnel system provides several benefits that minimizes impacts to the community, centralizes most of the operation and maintenance activities at the WRRF, and minimizes the costs with respect to other options.

Outfall Transfer Agreement

The City and AlexRenew recognized that many of the proposed RiverRenew facilities will be located on AlexRenew property and would require close operational oversight between the RiverRenew facilities and the WRRF to achieve the performance objectives. As a result, AlexRenew and the City executed an Outfall Transfer Agreement that transferred ownership of the four existing combined sewer system regulators and outfalls from the City to AlexRenew. Transferring ownership of the assets to AlexRenew provides the following implementation and operational advantages for meeting the mandated July 2025 deadline:

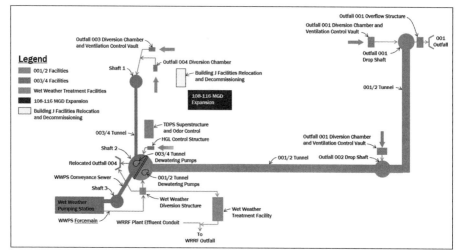

Figure 2. Schematic of the tunnel system from the long term control plan update (approved June 2018)

- Implementation Advantages:
  - Efficiencies of single entity owning and administering the program which results in simplified permitting.
  - Ability to leverage planned AlexRenew WRRF projects to assist in meeting deadline
  - Tunneling operations can be conducted from the WRRF to limit construction impacts to the community
- Operational Advantages:
  - Integration of operations and maintenance under a single entity (AlexRenew)
  - AlexRenew has expertise in treatment technology and innovation

TECHNICALLY PREFERRED ALTERNATIVE

A Parallel Approach to Regulatory Requirements

In October 2018, AlexRenew submitted to VDEQ a draft Preliminary Engineering Report (PER) advancing the recommended plan in the LTCPU. The PER, required by Virginia Code, analyzed alternatives for the design and construction of the recommended plan and presented a technically preferred alternative.

At the time of the submission of this manuscript, the National Park Service (NPS) and AlexRenew are preparing an Environmental Assessment (EA) to consider potential environmental effects of the RiverRenew program. RiverRenew is subject to an EA since it requires permits and approvals from the NPS to potentially construct within their property. Due to the accelerated timeline required to comply with the schedule mandated by the 2017 CSO Law, the EA is being developed simultaneously with the PER and is estimated to be finalized in Spring 2019.

It should be noted that the Virginia Administrative Code requires Preliminary Engineering Reports to identify a "selected alternative." In order to maintain the integrity of the EA process, the PER did not identify a selected alternative, but recommended a technically preferred alternative based on engineering judgement, cost, and schedule. The EA will further study the environmental and community impacts for all alternatives and will conclude with a decision document recommending a preferred alternative. Therefore, until the EA is complete, any recommendation will be solely technical in nature and referred to as the "technically preferred alternative."

Selecting the Technically Preferred Alternative

While all identified alternatives will achieve the required degree of performance, the relative merits of each were evaluated with respect to the following criteria:

- Property Acquisition and Permits: Locate new facilities within public land and rights-of-way, where feasible
- Community and Environmental Impacts: Avoid or minimize potential impacts on the community and cultural and environmental resources.
- Constructability: Facilitate constructability of the facilities, including the provision of sufficient space to stage the construction of the tunnels and surface features
- Cost: Minimize overall program cost
- Operations and Maintenance: Complexity in maintaining and operating the permanent facilities

These alternatives are summarized in Table 1 and generally illustrated in Figure 3.

Generally, the 001/2 tunnel alternatives include two main east west tunnel branches, along Church or Green Streets, and three north south branches along the Potomac River, Union Street, and Lee Street. Four proposed surface facility alternatives were studied at Outfall 001, while two proposed surface facility alternatives were analyzed at Outfall 002.

The 003/4 alignment alternatives primarily consist of an evaluation of three approaches to connect Outfalls 003 and 004 to the WRRF that includes deep tunnel construction, trenchless diversion sewer construction, and traditional open-cut construction. Each of these are alternatives are illustrated in Figure 3.

The main factors to consider when developing and evaluating the tunnel alignments included:

Table 1. Alternatives analyzed for 001/2 system

Tunnel Alignment	Tunnel Dia. (ft)	Tunnel Length (ft)	Outfall 001 Diversion Facility Location	Outfall 002 Diversion Facility Location
Potomac-Church	12	10,960	Robinson Terminal North	South Royal
		11,620	Oronoco East	South Royal
		12,070	Oronoco North	South Royal
Union-Church	12	10,690	Oronoco East	South Royal
Lee-Church	12	10,720	Oronoco West	South Royal
Potomac-Green	12	10,690	Robinson Terminal North	Green Street
		11,320	Oronoco East	Green Street
		11,770	Oronoco North	Green Street
Union- Green	12	10,320	Oronoco East	Green Street
Lee-Green	12	10,090	Oronoco West	Green Street

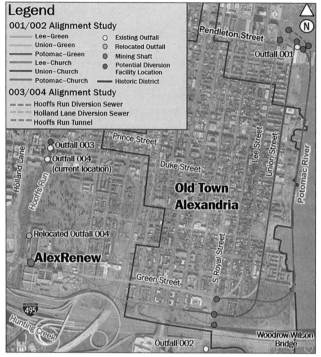

Figure 3. Overview of alternatives analyzed

Table 2. Alternatives analyzed for 003/4 system

Alternative	Tunnel Dia. (ft)	Tunnel Length (ft)	Depth (ft)	Shaft Needed	Means and Methods of Construction
Deep Tunnel	12	2,648	120–130	Yes	Tunnel boring machine
Holland Lane Diversion Sewer	6	3,386	20–40	No	Trenchless
Hooffs Run Diversion Sewer	6	2,458	10–20	No	Open-cut

Settlement Risk

The alignment selection process considered structures that may be affected by ground movements and surface settlement associated with soft-ground tunneling and open excavations. The amount of ground subsidence is a function not only of the existing geotechnical conditions, but also the horizontal and vertical distance from the construction to the affected structures, the specified tunneling approach, the workmanship of the builder, and the ability to monitor the work and make adjustments during construction. In addition to the amount of settlement, the other major factor is the ability of the structures to withstand the stresses induced by this settlement. For instance, unreinforced masonry structures on slab foundations will show distress at lower levels of angular distortion than will steel buildings on pile-supported foundations.

Based on empirical methods and recent tunneling projects in similar geology using the latest Tunnel Boring Machine (TBM) technology and risk management measures, surface settlements induced by tunneling are expected to be largest over the tunnel crown, and approach zero at a lateral distance of about 100 feet from the tunnel centerline. Therefore, for planning purposes, a 200-ft buffer zone (100-ft from centerline of tunnel) was considered reasonable for identifying utilities and structures that may be impacted by construction. In general, the number of structures within the buffer area increases as the alignment moves westward of the Potomac alignment and northward of the Church Street alignment.

Location of Historic Structures

In the vicinity of the proposed alignments, a significant proportion of structures are historic brick/masonry buildings which typically have a lower capacity to resist exhibiting distress than steel or wooden frames structures. Minimizing the potential risks associated with historic structures is another major factor when considering tunnel alignments. Figure 3 also shows the location of the Old Town Historic District where many brick/masonry buildings exist.

TBM Turning Radius Limitations

For a 12-ft inside diameter tunnel, a minimum turning radius for standard TBM is about 800-ft. Specialty TBMs may be designed to navigate tighter turning radii, down to about 450-ft to 500-ft. Tighter radius tunnel alignments are typically more expensive and take longer to construct.

Tunnel System Project

The outcome of the proposed facility evaluations was the selection of the technically preferred alternative. The preferred 001/2 Outfall System and 003/4 Outfall System alignment are illustrated in Figure 4 and summarized in Table 3 and are known collectively as the Tunnel System project.

The major components of the Tunnel System are illustrated in Figure 4 and provided in Table 4.

Other components associated with the Tunnel System project include:

- Five (5) diversion chambers
- 40 MGD tunnel dewatering pumping station with an associated above-grade structure to house equipment

Table 3. Summary of technically preferred alternative

RiverRenew Facilities	Technically Preferred Alternative
001/2 Alignment	Potomac-Church Alignment with Outfall 001 drop shaft located at Robinson Terminal North
003/4 Alignment	Hooffs Run Diversion Sewer Alignment

Table 4. Major components of the technical preferred alternative

Component	Minimum Diameter (feet)	Approximate Length (feet)	Approximate Depth (feet)
Potomac-Church Tunnel	12	11,500	115–135
001 Drop Shaft	30–40	—	115
002 Drop Shaft	30–40	—	125
Mining Shaft	50	—	135
Hooffs Run Diversion Sewer	6	2,400	10–40

- 130 MGD wet weather pumping station
- 90,000 cfm (cubic feet per minute) odor control facility
- Two, 3,000 cfm odor control systems
- Electrical and instrumentation equipment

Potomac-Church Tunnel

Of the alternatives evaluated for the portion of the tunnel system collecting flows from Outfalls 001 and 002, the alignment located under Alexandria's Church Street and the Potomac River emerged as the most technically favorable alternative. The alignment, called the Potomac-Church alignment, was coupled with the most technically favorable alternatives for drop shaft locations, with Outfall 001 located on a private property and Outfall 002 within National Park Service land. A number of technical factors contributed to this decision including:

- The tunnel alignment significantly reduces the coordination needed with third party property owners, particularly those within the Alexandria's historic Old Town neighborhood.
- Compared to some alignment alternatives that would like require tighter turning radii for the tunnel boring machine, the alignment geometry along the Potomac River and Church Street allows the potential use of a larger radii, reducing the degree of difficulty for tunnel excavation methods.
- The current Outfall 001 Drop Shaft location on the private property provides the advantage of being the furthest of the alternatives from residences in the area therefore lowering the degree of impacts during construction to adjacent neighbors. Additionally, the location is close to the proposed outfall extension, and the site's use (during both construction and permanently) aligns with the intended redevelopment of the waterfront.
- The current Outfall 002 Drop Shaft location provides the advantage of being the furthest of the alternatives from residences in the area therefore lowering the degree of impacts during construction to adjacent neighbors. Additionally the location is in close proximity to the existing regulator and minimizes impact to Royal St.

Hooffs Run Diversion Sewer

Outfalls 003 and 004 will be remediated by the construction of a six-foot diameter conveyance pipeline coupled with a 130 million gallons per day (MGD) wet weather pumping station and 40 MGD wet weather treatment facility. The RiverRenew team is currently analyzing two alternatives for the pipeline: trenchless or open cut. Both options are shallower than the proposed deep tunnel serving Outfalls 001 and 002, with a depth ranging from 10 to 40 feet.

The open cut alternative, known as the Hooffs Run Diversion Sewer, was considered the most technically favorable because of the following factors:

- A significant reduction in the potential for impacts to traffic and businesses in the vicinity of the construction area.
- The deep tunnel option would likely result in work for Outfalls 003 and 004 being performed before or after work for Outfalls 001 and 002. A non-tunnel alternative can be scheduled such that it does not drive the program's critical

Alexandria Renew Enterprises Is Now in the Tunnel Business

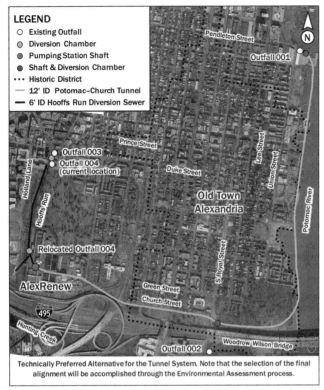

Technically Preferred Alternative for the Tunnel System. Note that the selection of the final alignment will be accomplished through the Environmental Assessment process.

Figure 4. Overview of the technically preferred alternative for the tunnel system

path i.e., it can be performed concurrent to construction of the 001/002 facilities.
- Due to its location, there is minimal need for property acquisition
- The shallower options both result in more simplified TDPS and drop shaft designs at the plant which are more favorable for design, construction and operation.
- Following installation of the new facilities, there is an opportunity for surface restoration within Hooffs Run to improve public accessibility and provide an amenity to the Hooffs Run vicinity wetlands and adjacent park.

Tunnel Dewatering Pumping Station and Drop Shaft

The Pumping Station Shaft at the WRRF will be the mining shaft for the Potomac-Church Tunnel. Following the completion of mining, the Pumping Station Shaft will contain the 003/4 drop shaft and Tunneling Dewatering Pumping Station (TDPS). The 003/4 drop shaft will include an approach channel, vortex generator, a vortex drop pipe, deaeration chamber, baffle wall and orifice to minimize air entrainment and the potential for geysers in the tunnel. Future design will consider ways to optimize the TDPS sizing and design for operation and maintenance while reserving space on the WRRF to allow for future plant expansion. Submersible type pumps are currently under evaluation over other types of pumps to minimize the cost and footprint of the TDPS.

Screening and Debris Removal

Due to typical nature of CSO characteristics, grits, rags, and large debris can be expected in the flow stream. Based on storm event and flow velocity, some of the grit and debris will eventually reach the TDPS. Most pumps can pass 2- to 4-inch diameter solids and some grit if they are fully in suspension. However, the quantity of debris in a CSO can overwhelm the pumps and can accumulate in the pumping station or clog the pumps if no removal mechanism is provided. To prevent excessive accumulation of debris, automatic bar screens with a raking mechanism and clam shell bucket type debris removal equipment will be provided at the inlet to the pump stations.

The screening equipment will include a bar rack and cable system grabber-type raking mechanism. The bar screen in this application may get fully flooded and/or reverse flow through the screen during tunnel filling events. The bar screen clear opening should be small enough to protect the pumps and large enough to prevent rapid blinding of screen during wet weather events.

Procurement Method and Project Packaging

Four major procurement methods were analyzed for the tunnel system project including design-bid-build, construction management at risk, progressive design-build, and fixed-price design-build. Ultimately, a two-step fixed-price design-build process was recommended for the packaging of the Tunnel System project.

GEOLOGICAL SETTING

Local Geology

Selection of both the horizontal and vertical alignment and construction methods is influenced by the geological and geotechnical conditions, as well as site constraints. The type of ground encountered along the alignment affects the selection of the tunnel type and its method of construction. The proposed tunnel system is located within the Atlantic Coastal Plain Physiographic Province of Virginia. The Atlantic Coastal Plain (Coastal Plain) consists of a seaward thickening wedge of unconsolidated to semi-consolidated sedimentary deposits from the Cretaceous Geologic Period to the Holocene Geologic Epoch. These deposits represent marginal-marine to marine sediments consisting of interbedded sands and clays. The Coastal Plain is bordered to the east by the Atlantic Ocean and to the west by the Piedmont Physiographic Province (Piedmont).

According to USGS geologic maps, the project sites are mapped in the Shirley Formation deposited during the Holocene Geologic Epoch of the Quaternary Geologic Period. Underlying sediments are composed of the Potomac Group clays deposited during the Cretaceous Geologic Period, which are the oldest sedimentary deposits in the Washington, D.C. area. These soils are known to be highly over-consolidated as a result of the weight of a substantial thickness of overlying soils that have since been eroded away. Underlying these sedimentary deposits is the older Piedmont residual soil and crystalline bedrock

Subsurface Exploration Program

To gain a better understanding of the depth of the Potomac formation, a subsurface exploration program is being conducted in three phases as part of the planning and design. The phases include:

- Phase A: Preliminary Planning Subsurface Exploration
  - Using the multiple alternatives of tunnel system routes, the purpose of Phase A was to verify the local geology and ground conditions to support development of the alternative alignment considerations and to facilitate the identification and planning of a technically preferred tunnel alignment.
  - Started in summer 2018 and ongoing as of the submittal of this manuscript, initial borings was spaced 1,000 feet along the tunnel alignment
- Phase B: Detail Design Subsurface Exploration
  - The purpose of Phase B was to narrow focus of the geotechnical investigation on the area of the technically preferred tunnel system alignment. Phase B is anticipated to be complete by early summer 2019 and is estimated to consist of up to 30 borings.
  - It is proposed that the borings will be located so that final boring spacing is roughly 500 feet along the tunnel alignment.
- Phase C: Contractor Conducted Subsurface Exploration
  - The tunnel system will be procured as a design-build contract. Additional borings may be performed by the design-build team to support the final design of the tunnel, shafts, and near-surface structures.

Vertical Alignment

The tunnel vertical alignment will be selected based on the combination of the following construction and operational objectives:

- Depth of the Potomac formation, which is favorable for tunneling.
- Depth of support of excavation for shafts
- Tunnel slope that provides reliable transport of solids to the TDPS (terminal shaft)
- Range of operating head conditions for the TDPS pumps

HYDRAULIC DESIGN

Diversion Facility Design

The design of the diversion facilities and performance evaluations of the proposed tunnel system are based on 15-minute rainfall data collected from Ronald Reagan Washington National Airport(DCA). The 15-minute DCA data recently became available for the entire calendar years of the 2000 to 2016 study period. While hourly rainfall data was appropriate for evaluating performance in the LTCPU, 15-minute data is being used and processed through the hydraulic and hydrologic model for design purposes. Rainfall data collected over 15-minute periods provides conservatism in the evaluation of peak flows (typically greater than hourly data) and more detailed resolution of each storm event, especially those with durations less than one hour (hr). Storage/conveyance tunnels typically dampen the peak flows associated with the 15-minute rainfall.

The hydraulic and hydrologic model was used to determine the peak diversion rate that must be conveyed from the existing combined sewer system into the proposed tunnel system to meet the performance requirements set by the 2017 CSO Law. The peak diversion rates were established using 15-minute rainfall data collected from Ronald Reagan Washington National Airport. A safety factor of 1.2 was applied to each peak diversion rate established by the model to develop design diversion rates used for the tunnel system. The safety factor was used to account for uncertainties associated with the hydraulic and hydrologic model.

Tunnel Hydraulic Transients

Tunnels with high filling rates relative to tunnel diameter have the potential to have significant transient conditions. Transients can create high pressures on the tunnel lining (both positive and negative), spill wastewater to the ground surface, and create undesired combined sewer overflows from the connecting trunk sewers. Transients can also lead to ventilation issues such as pipeline choking and geysers. As part of the planning and design, AlexRenew is utilizing the Transient Analysis Program (TAP) developed by Applied Science Incorporated (ASI) to develop a model to perform the computational hydraulic modeling of the tunnel system. The analysis will evaluate the expected worst-case tunnel filling conditions to determine if unacceptable transient or ventilation issues are likely to occur. If so, mitigative measures will be developed and incorporated into the design.

The major objectives of the transient analysis are to assess the following:

- Hydraulic/transient conditions in the tunnel system;
- Emergency type conditions such as pump shut offs;
- Drop shaft and tunnel ventilation rates;
- Potential for air trapping in the tunnel during filling; and
- Tunnel overflow layout and configuration, and recommend surge control measures

Diversion Chambers

The diversion chambers consist of a weir and orifice to control diversions to the tunnel. The weir provides the diversion capability and the orifice restricts the diversion of flow into the tunnel at the design diversion rate. When the tunnel is full or when flows in the diversion chamber exceed the design diversion rate, flows are conveyed through the diversion chamber through an outfall to discharge to an adjacent receiving water body. The diversion chamber also includes metering and sampling equipment, stop log grooves for isolation and manholes to provide access for flow monitoring required for regulatory compliance. The major components of a typical diversion chamber are illustrated in Figure 5 and 6.

Approach Channel and Vortex Generator Drop Shaft

Combined sewer flows that are diverted into the diversion chambers are conveyed through an adjustable orifice plate. Once flows are conveyed through the diversion chamber orifice, the flows are "dropped" into the tunnel via an approach channel and vortex generator drop shaft following the Milwaukee approach. The approach channel and vortex generator drop shaft reduce the potential for air entrainment and facilitate energy dissipation.

SCHEDULE

RiverRenew Schedule

Planning and design will continue through early 2020 with construction starting by 2021 to meet the July 1, 2025 compliance deadline. The construction schedule for

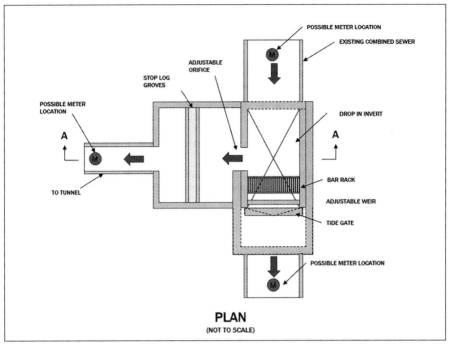

Figure 5. Typical diversion chamber—plan

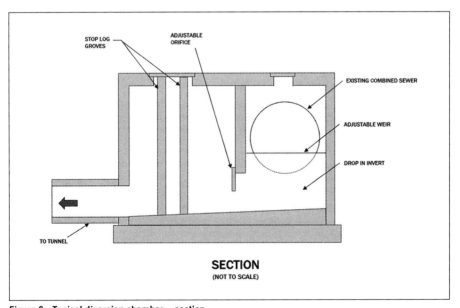

Figure 6. Typical diversion chamber—section

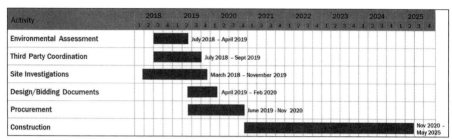

Figure 7. Projects scheduled in coordination with RiverRenew

Table 5. Summary of reductions in number of combined sewer overflow events and volume of combined sewer discharges following the completion of RiverRenew

Outfall	Overflow Events/Year		Volume/Year (million gallons)	
	Now	After RiverRenew's Implementation	Now	After RiverRenew's Implementation
001	34	2	63	8
002	78	2	38	5
003	60	2	31	1
004	w	< 1	8	2
Total			140	16

implementing the technically preferred RiverRenew facilities was developed based on the following criteria.

- Meet the July 1, 2025 compliance deadline
- Minimize community impacts
- Comply with permit requirements
- Concurrent construction of both the 001/002 and 003/004 tunnel systems

RiverRenew High-Level Critical Path Schedule

The RiverRenew planning, design and construction activities will be performed in close coordination with other projects at AlexRenew. AlexRenew will be in various stages of design and construction of the projects as illustrated in Figure 7.

CONCLUSIONS

Once completed, the new tunnel system will connect to the existing outfalls and significantly reduce the frequency and volume of combined sewer discharges. Overflows occur approximately 60 times per year, or anytime there is a rain event greater than a quarter-inch. These discharges result in approximately 140 million gallons of combined sewage pollution from the 540-acre combined sewer area each year. The tunnel system and wastewater plant upgrades will reduce the frequency of discharges to less than four per year and reduce the volume discharged to 16 million gallons. In addition, the new system will provide approximately 97 percent capture of combined sewage.

The significant reduction in the frequency and volume of these discharges will achieve cleaner, healthier waterways by reducing the amount of bacteria, trash, and other pollutants that currently impact Hooffs Run, Hunting Creek and the Potomac River.

DART D2 Subway Project Development

Charles A. Stone • HNTB
Israel Crowe • HDR
Eric C. Wang • HNTB

ABSTRACT

The DART D2 Subway Project Development effort in Dallas, Texas is currently proceeding with preliminary engineering to support a Design-Build procurement and implementation. A new locally preferred alternative is the 2.34-mile long Victory-Commerce-Swiss Streets alignment. The alignment includes an at grade station near the Perot Museum, three underground subway stations at Metro Center Station, Commerce Street, and the east end of downtown, and ties into the existing light rail system at Victory Station and Good Latimer Expressway. This complex project will encounter geotechnical, constructability, interface and community challenges during its two-year preliminary design period.

INTRODUCTION

The Dallas Area Rapid Transit (DART) Central Business District (CBD) Second Light Rail Alignment (D2 Subway) project development effort in Dallas, Texas is currently proceeding with preliminary engineering and Environmental Impact Statement (EIS) to support a Design-Build (DB) procurement and implementation. The project sponsor for the D2 subway project is DART. The design development is being carried out in several phases. The initial phase of concept engineering up to 30% design is currently being performed by HDR as the prime engineering consultant with HNTB as the tunnel sub-consultant. Final Design is currently anticipated to be completed under a design build contract.

LOCALLY PREFERRED ALIGNMENT (LPA)

The Locally Preferred Alignment (LPA) alternative travels at-grade through the Victory development with a proposed at grade station adjacent to the Perot Museum (Museum Way Station). The LPA then proceeds under Woodall Rodgers Freeway viaduct before transitioning from surface-running to below-grade in a tunnel via a train portal immediately south of Woodall Rodgers.

The underground segment of this DART D2 extension begins east of the Woodall Rodgers Freeway viaduct as a U-section, the West Portal, and transitions underground below North Griffin Street adjacent to the Dallas World Aquarium. The underground section proceeds about 930 feet until it reaches the Metro Center Station, with connections to the West Transfer Center and West End light rail station. From this station the alignment continues underground along North Griffin to Belo Garden where it turns onto Commerce Street and continues along under the heart of the Dallas CBD, about 2,020 feet to the Commerce Station. This station is located three blocks south of the existing Akard light rail station. The alignment then continues along Commerce until it reaches Main Street Garden Park. There it turns northeast and runs 1,655 feet to the CBD East Station, including a section under a parking structure on Elm Street. This station is located one block south of the East Transfer Center. From the CBD East

Station, the alignment continues northeast 825 feet to the East Portal structure where it returns to grade.

The LPA then turns northeast parallel to Swiss Avenue and begins transitioning from subway to at-grade via a train portal under and immediately east of IH-345. The LPA continues parallel to Swiss Avenue at-grade before tying back in to the existing light rail system at Good Latimer. The LPA would result in the removal of the existing Deep Ellum light rail station.

The Commerce/Victory/Swiss alignment is presented in schematic form in Figure 1.

PROJECT HISTORY

In May 2007, DART launched a study for the Dallas CBD Second Light Rail Alignment (D2 Subway). Between 2007 and 2010 a host of alignment options were investigated by the DART team to find the best second alignment through downtown. This three-year effort resulted in the Alternative Analysis/Draft Environmental Impact Statement (AA/DEIS) in May 2010. The AA/DEIS presented four alignment options. At this point the project slowed down due to the recession in the country and the outlook that funding would not be available for the project. By the beginning of 2013 the next phase began to select an LPA. This phase incorporated public comments from the AA/DEIS as well as other local changes. Between 2010 and 2013 the City of Dallas began to move forward with a streetcar system with plans for future extensions into downtown. The Downtown Dallas 360 Plan was published and active discussions regarding a downtown high-speed rail station commenced. With this additional information the DART D2 team evaluated the four alignments and any possible refinements. The process resulted in a LPA being selected in late 2015.

In September 2015, the Dallas City Council and the DART Board selected an LPA with two design options based on the work that was done between 2007 and 2015. Upon selection of the LPA the project moved into Project Development where the EIS and Preliminary Engineering plans are prepared.

The September 2015 LPA begins south of Victory Station. It then proceeds within DART Right of Way (ROW) in the center of Museum Way and through the parking lot adjacent to the Perot Museum. The alignment crosses under Woodall Rodgers then begins transitioning to below grade. It does so until it crosses south of Commerce Street. At that point, it begins angling in a southeasterly direction diagonally across City blocks until it resurfaces near the Yang/Field Street intersection and southwest to Lamar Street. It remains underground under Lamar Street then returns to at-grade before the intersection of Young Street and Field Street. It remains at-grade along Young Street to Ervay Street. From this point the alignment follows one of three alignments; generally following Jackson, Wood or Young Streets before running down Good Latimer and joining the existing light rail alignment. See Figure 2 for a map of the September 2015 LPA.

During Project Development additional analysis of the corridor was conducted and it was determined that shifting the alignment from Lamar Street to Griffin Street would result in less impact. The Lamar Street corridor has a narrower ROW in addition to existing Oncor transformer vaults that would require relocation. There are utility services that lie directly below the underground pedestrian crossing connecting to the Bank of America tower that would conflict with the tunnel. Anticipated geological conditions also improve marginally with the move from Lamar Street to Griffin Street. The Austin Chalk layer tends to thicken moving east through downtown away from the

DART D2 Subway Project Development

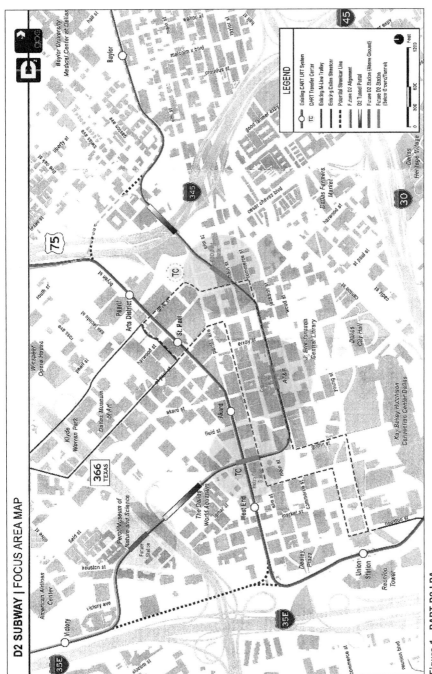

Figure 1. DART D2 LPA

river. Additionally, the shift would avoid the impact to a future museum that is planned at McKinney Avenue and Old Griffin Street. Beyond enhancing capacity and operational flexibility, the second downtown light rail alignment would provide the Victory area with another light rail station. This additional station would introduce light rail service to the museums and residents in the area. Additionally, it would provide another transportation option for events at American Airlines Center.

Between September 2015 and October 2016, the DART D2 team advanced project development for the LPA and two design options. During this period there was significant stakeholder and community input about the LPA. It was acknowledged that parts of downtown and the surrounding areas have changed or are changing significantly since the completion of the AA/DEIS in 2010.

In October 2016, direction from the Dallas City Council and DART Board was to pursue the second downtown light rail alignment as a tunnel from Woodall Rodgers to IH 345. This direction led to the LPA Refinement Phase which focused on establishing an LPA that is underground through downtown. This phase took place from December 2016 to June 2017. During this phase, the DART D2 team worked with stakeholders and the community to define the optimal location for the underground alignment and connection points to the existing light rail system. Many stakeholder meetings and workshops were held to identify issues and opportunities for the corridor. Ultimately, the LPA alignment refinement phase resulted in three alignments to be advanced. They were all identical in the Victory area, proceeded down Griffin then turned east

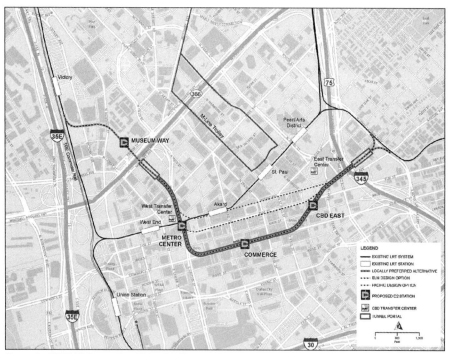

Figure 2. DART D2 September 2015 LPA (Source: DART)

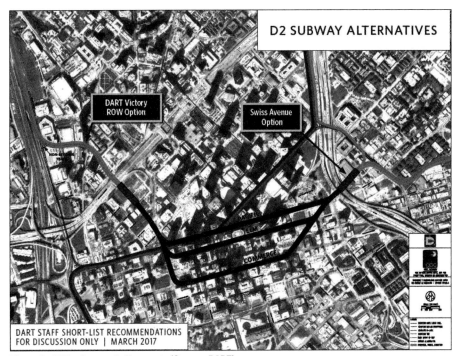

Figure 3. LPA refinement alignments (Source: DART)

along Pacific, Elm Street, or Commerce Street with a connection at Swiss Avenue and Good Latimer. See Figure 3 for a map of the three alternatives from the LPA refinement phase.

In September 2017, the Dallas City Council and the DART Board selected the Victory/Commerce/Swiss alignment as the LPA. See Figure 1 for a map of the LPA. The Project Development phase recommenced in May 2018. The DART D2 team is advancing the EIS and Preliminary Engineering with completion anticipated by March 2020.

The DART D2 Subway will add capacity and operational flexibility to the DART light rail system. It will allow two of the light rail lines to move to the new alignment and two light rail lines to remain operating on the existing alignment, allowing more trains to move through downtown and increasing capacity throughout the entire system. The D2 Subway alignment will provide new stations in downtown giving riders more access opportunities in the downtown area.

PROJECT SCHEDULE

The current project development phase began in May 2018 and will conclude in March 2020. By February 2020, the Preliminary Engineering plans, reports and the EIS will be complete. It is anticipated that DART would receive the Record of Decision (ROD) in March 2020. Upon receipt of the ROD, DART will request from the Federal Transit Administration (FTA) to enter engineering allowing final design to be completed. The Design-Builder is expected to begin final design and construction activities in 2020. Final Design, Construction and Testing may be complete in 2024 with a revenue service date of December 31, 2024. The project schedule is depicted in Figure 4.

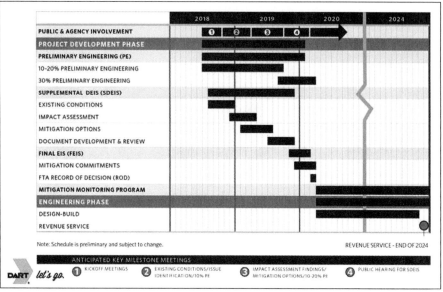

Figure 4. Proposed DART D2 subway schedule (Source: DART)

ANTICIPATED CONTRACTING STRATEGY

It is planned that DART will advance the project beyond Project Development with a Design-Build strategy. Project Development will end with completion of Preliminary Engineering, publication of the Final EIS and receipt of the ROD. The Design-Builder will advance and complete Final Design, construct the project and test the corridor prior to revenue service.

GEOTECHNICAL CONSIDERATIONS

Site Conditions

The project area is a densely populated urban area with significant high occupancy commercial structures immediately adjacent to and overlying the underground alignment. There is a seven-foot diameter storm sewer tunnel and a 24-inch sewer utility line near the alignment under Commerce Street.

Geologic Setting

The Dallas area is located within the Blackland Prairie Physiographic Province, an area of low rolling terrain topography with bedrock consisting of chalks and marls with bedding dipping to the southeast. The ground surface elevation in Dallas ranges from about 400 to 700 feet above mean sea level. The city is drained by the Trinity River and its tributaries. The prominent topographic feature in the area is the Austin chalk cuesta. This escarpment was formed through differential erosion of the beds at the contact between the Austin Chalk and the underlying Eagle Ford shale. The land surface along the underground section of the alignment is generally flat to gently rolling. The highest elevation of the ground surface on the alignment is approximately elevation 475 feet near Swiss Avenue and Interstate 345.

Several regional tectonic events have created the gently sloping monoclinal structure of the Gulf Coastal plain. Structures resulting from these events include the Ouachita

Fold Belt, the East Texas Embayment, the Sabine Uplift, the Llano Uplift, and the Northeast Texas Fault System, which includes the Balcones Fault Zone. (Leech 1995), (Moore and Teetes 2009). The city of Dallas is located near the northwestern margin of the East Texas Embayment of the Blackland Prairie Physiographic Province of Central Texas, the eastern margin of the Ouachita Fold Belt, and is north to northwest of the currently mapped Balcones Fault Zone. Because of these events, faulting in the Dallas area typically includes normal faults as well as some reverse faults with displacements generally less than 15 feet.

The DART D2 project is in North-Central Texas in Dallas County. North-Central Texas is underlain by a thick sequence of sedimentary rocks that dip gently southeastward towards the Gulf of Mexico. These rocks form broad northeast-trending outcrops. The outcropping units decrease in age from Paleozoic inland to Quaternary at the coast. The outcropping units in Dallas County belong to the Upper-Cretaceous Gulf Series except for thin deposits of alluvial material along streams and on terraces. The Upper-Cretaceous Gulf Series includes the Taylor, Austin, Eagle Ford, and Woodbine Groups.

Portions of overlying sediments were eroded by the ancestral Trinity River exposing the underlying Austin Chalk and Eagle Ford Shale. (Fugro 2004) Due to the southeasterly regional dip and erosional truncation, the Eagle Ford subcrops throughout much of the greater Dallas vicinity to the west of the Austin Chalk subcrop, and the Taylor Marl subcrops to the east of the Austin Chalk subcrop. There are no Triassic, Jurassic, or Tertiary rocks affecting excavation along the DART D2 Project alignment. Triassic and Jurassic rocks lie well below the ground surface and tend to outcrop to the west of Dallas.

East Texas faults share numerous indicators of low seismic risk. Normal displacements ensure that stresses are neutralized by tensile fracture at low stresses because the tensile strength of materials is generally much lower than their compressive strength. Most of these faults are related to the slow gravitational creep of salt and its sedimentary overburden rather than to movement of lithospheric plates. Future movement on the Lulia-Mexia-Talco zone is extremely unlikely because undeformed Pleistocene terraces cross them. Similarly, only a few central-basin faults extend upward to the Lower Tertiary stratigraphic units exposed at the surface. (Jackson 1982) The DART D2 project site can be generally characterized as a relatively low seismically active region.

Austin Chalk

The Austin Chalk is primarily light to medium gray chalk (microgranular calcite) with interbedded calcareous claystone. The calcium carbonate content of the chalk averages about 85 percent. The Austin Chalk is subdivided by geologists based on fossil zones and characteristics of surface exposures. The primary lithologic characteristics that distinguish subdivisions in the chalk are variations in bed thickness, concentrations of fossil material, and thin marly zones containing bentonite reported in the literature as developed from volcanic ash falls. The subdivisions recognized at the surface are not identified by these lithologic variations in the subsurface. Except for the local variations mentioned previously, the physical characteristics of Austin Chalk are quite uniform. The average thickness of the Austin Chalk is reported as 600 feet in Dallas County. (Roberts 1953) However, since the Austin Chalk outcrops in Dallas County, only the lower division of the Austin Chalk will be encountered in station cavern and running tunnel excavations.

The unweathered Austin Chalk is light to medium gray, thin-bedded, moderately hard and fine grained, with clay-rich layers. In outcrops, bedding thickness ranges from about two-inches to two-feet. Bedding is much less pronounced in cores. Clay-rich chalk layers occur between the thicker clay-poor chalk beds and contain varying amounts of bentonite.

The Austin Chalk is a relatively competent soft rock. Engineering properties of Austin Chalk have been shown to be consistent in a wide variety of projects completed with excavation into chalk from Austin through the outcrop belt to the north of Dallas. Unconfined compressive strength is typically about 2,200 psi, varying from about 500 to 4,500 psi. Rock strength generally increases with decreasing water content and increasing carbonate purity. The failure mode of the rock at low confinement is brittle. Rock Quality Designation (RQD) values typically range from 85 to 100 percent. The compositional variations are not expected to affect the overall performance of Tunnel Boring Machines (TBM) or other common excavating equipment operated in the chalk.

Eagle Ford Shale

The Eagle Ford Shale underlies the Austin Group. The shale has an average thickness of 475 feet within Dallas County. Due to the southeasterly regional dip and erosional truncation, the Eagle Ford subcrops throughout much of the greater Dallas vicinity to the west of the Austin Chalk subcrop. The top of the Eagle Ford shale generally lies approximately 35 to 80 feet below the surface along the alignment. It will therefore be encountered in excavations and affect mining excavation productivity as well as means and methods at some locations on the alignment.

This shale is divided into two members in the Dallas-Fort Worth area. Only the upper Arcadia Park member is relevant to construction of the station caverns and running tunnels. The Arcadia Park is a dark gray to black, calcareous to non-calcareous, laminated marine shale, that may contain pyrite along bedding. The upper part of the section contains bentonite seams, while flaggy limestone beds are more common toward the base. The formation is a montmorillonitic shale with high shrink-swell potential. Surles described the Arcadia Park in the Dallas area as consisting of gray to dark gray, fissile, calcareous mudstone or clay shale with thin laminae of siltstone, sandstone, and fragmental limestone. (Surles 1987)

The Eagle Ford Shale is soft rock with a typical unconfined compressive strength of 300 psi and a range of 15 to 700 psi. The rock is very susceptible to alteration if left exposed to air and/or water. Core samples may appear quite uniform but tend to split along horizontal partings when allowed to dry. The shale contains a relatively high percentage of bentonite and exhibits low shear strength and marked slaking and swelling tendencies. Occasional pyrite nodules, very hard septarian nodules, very thin limestone, and fine sandstone interbeds may also be present. The RQD of the Eagle Ford shale core can vary from 0 to 100 percent.

The Eagle Ford Shale may exhibit significant slaking and swelling tendencies. It is extremely susceptible to degradation upon cyclical drying and wetting sequences. This characteristic will affect the construction of station caverns, running tunnels, and other underground structures where it is encountered in the invert.

Geotechnical Design Considerations

It is necessary to perform geotechnical exploration and testing to determine the geology, hydrology, appropriate design parameters, design loadings, and appropriate analytical models for designing underground structures. Several design challenges will have to be considered.

Ground Response

The ground response and behavior due to the location, thickness, geotechnical parameters, and strength of bentonite seams, shale partings, argillaceous zones, and arenaceous zones in proximity to underground excavations must be determined by geotechnical investigation. In particular, clay seams and shale partings must be identified and their effects upon mining determined. Joint planes, shear zones, and possible faults, when combined with the bentonite seams and shale partings, may present potentially unstable conditions in the crown and upper arch, particularly where these features are near the station caverns and running tunnel excavations. Tendencies for swelling, slaking, and long-term deterioration of the rock mass due to exposure to air and water must also be accounted for in the design.

Argillaceous Zones

Some areas of underground excavation will encounter significant zones of argillaceous or marly limestone that may be subject to slaking and deterioration upon exposure to the atmosphere and moisture. These materials are likely to produce trafficability problems in the invert of the underground excavations, and the deterioration of these materials in the crown and sidewalls of the excavations may induce slabbing and separation of material along bedding planes.

Combustible and Toxic Gasses

The Austin Chalk and Eagle Ford Shale produce natural gas and oil. Investigation performed during the DART construction project identified and confirmed a deep source of methane existing more than 700 meters below the tunnel horizon of the NC-1B tunnel alignment. (Rogstad 1995) (Doyle 2001) This source of methane may also be capable of reaching the mining horizon of the DART D2 project via open faults and other geotechnical considerations. The presence of hazardous gases including both benzene and methane discharging from the bedrock was reported during construction of the existing DART light rail tunnel project. (Kwiatkowski 2015)

Swelling Pressure

Two considerations that affect the swelling pressure of clays are the quantity of and the physical, mineralogical makeup of the clay deposit and also the degree to which mining activities will change the in situ moisture content of the clay. In particular, higher percentages of montmorillonite clay generally suggest increasing swelling potential. These parameters are reflected in Plasticity Index and the Activity Index tests. The potential for mining activities to result in swelling pressures must be considered in the design.

Vertical Swelling Pressure

The potential impact of vertical swelling uplift pressure due to underlying soils and rock under horizontal slabs must be taken into consideration in the design of U-wall, cut-and-cover structures, and large span cavern inverts. The design must incorporate a strategy to address the potential for swelling clays below the invert such as over-excavation and replacement of the sublayers including clay for a limited depth.

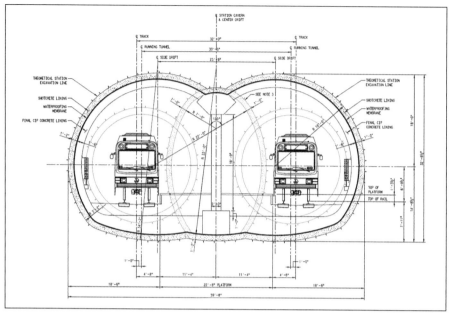

Figure 5. Proposed station section—center pillar—center platform option

SUBWAY STATION CAVERN DESIGN

Considerable excavation experience with respect to tunneling in the Austin Chalk near Dallas is available. Tunnel boring machines have been used successfully in both the Austin Chalk and Taylor Marl in the Dallas and San Antonio areas. (Nelson and Lundin 1990) There has also been one large subway station cavern excavated in the Dallas area, City Place Station, on the DART NC-1B contract. (Sauer, et al., 1996) Experience on several large underground construction projects in the Dallas area have shown that the Austin Formation is an excellent medium for tunneling. (Leech, et al., 1998) Quarry excavations are also common in the Austin Chalk for cement production.

The binocular shaped cross section for the Commerce Station has been developed considering the requirement for a center platform configuration and the respective minimum platform width, vertical circulation elements, architectural finishes, and very low rock cover due to the proposed LPA track elevation. Space-proofing of station facilities will be conducted after development of the architectural, ventilation/fire life safety systems and rail systems are advanced to a sufficient level of completion to allow coordinated interdisciplinary review. The minimum internal width to accommodate the track and 20'-8" wide center platform is 52'-6" as shown in Figure 5.

Top down or bottom up cut-and-cover construction is required for CBD East Station due to the extreme shallow cover conditions of the alignment dictated by top-of-rail elevations based on limitations of acceptable track grades and available distance from approved portal location. Minimum internal width of the station is 57'-0" to accommodate a 30'-8" wide center platform. This construction method and support of excavation, although temporarily disruptive to streets, traffic, utilities, businesses and public, are considered conventional and widely implemented throughout the industry.

The Metro Center Station could be constructed as a mined cavern or cut and cover station depending on design development as the project progresses.

CONCLUSION

The DART D2 subway project promises to be a challenging, complex project that will encounter many geotechnical, constructability, interface and community challenges during its two-year preliminary design period. This challenging project is expected to be procured as a Design Build contract in 2020, with an expected revenue service date of December 2024.

REFERENCES

Doyle, Barry, 2001, Hazardous Gasses Underground: Applications to Tunnel Engineering, Marcel Dekker, Inc., New York.

Fugro, May 2004, Phase 1 Geotechnical Baseline Report, IH-635 (LBJ Freeway) Corridor, Section 4-West, Contract 18-2XXP0004, Texas Department of Transportation.

Jackson, M.P.A., 1982, Fault Tectonics of the East Texas Basin, Bureau of Economic Geology, University of Texas at Austin, Geological Circular 82-4.

Kwiatkowski, April 2015, Geotechnical Baseline Report, Mill Creek/Peaks Branch/State Thomas Drainage Relief Tunnel Project, City of Dallas, Texas, Trinity Watershed Management.

Leech, William D., C. Richard Lineman, and Daniel S. Hubenak, 1998, Addison Airport Tunnel Project, Tunnelling Association of Canada Annual Publication.

Moore, J.F., and G.R. Teetes, June 2006, IH-635 Managed Lanes Project, Geotechnical Interpretive Report for the Reference Schematic—Segment A, Texas DOT, Lachel, Felice, and Associates.

Nelson, P.P., and T.K. Lundin, June 1990, Geotechnical Characterization and Construction Methods for SSC Tunnel Excavation, Unique Underground Structures Symposium, Denver, CO.

Roberts, Carl N., 1953, Geology of the Dallas Quadrangle.

Rogstad, Dave, 1995, The Dallas Area Rapid Transit Subway Tunnel Project, Rapid Excavation and Tunneling Conference, Proceedings.

Sauer, G., Ugarte, E., and Gall, V., April 1996, Instrumentation and its Implications—DART Section NC-1B, City Place Station, Dallas, TX, North American Tunneling Conference, Washington, DC.

Surles, Milton A., 1987, Stratigraphy of the Eagle Ford Group (Upper Cretaceous) and Its Source-Rock Potential in the East Texas Basin, Baylor Geological Studies, Fall 1987 Bulletin No 45.

Mountain Tunnel Improvements Project

Jennifer Sketchley ▪ McMillen Jacobs Associates
David Tsztoo ▪ San Francisco Public Utilities Commission
Renée Fippin ▪ McMillen Jacobs Associates
Glenn Boyce ▪ McMillen Jacobs Associates

ABSTRACT

The almost century-old, 18.9-mile Mountain Tunnel is part of the San Francisco Public Utilities Commission water system. Rather than construct a new tunnel, an intensive rehabilitation of the existing tunnel will be performed. As part of this work, five tunnel outages (starting in winter 2021–2022) are planned. The major challenge will be performing all the work during 60- and 100-day outages. This paper discusses the components of the Mountain Tunnel Improvements Project, including a new downstream Flow Control Facility, a new Priest Adit, an extension of the South Fork Siphon; the rehabilitation components of the existing tunnel consisting of control lining repairs, contact grouting, groundwater cutoff grouting, installation of steel lining, invert paving, rock bolting, and debris removal; and the intensive roadway improvements required along each access road.

INTRODUCTION

San Francisco Public Utilities Commission (SFPUC) provides drinking water to 2.7 million residential, commercial, and industrial customers in the San Francisco Bay Area. The SFPUC's water system includes 160 miles of transmission pipelines and tunnels from the SFPUC's largest reservoir (the Hetch Hetchy Reservoir in Yosemite National Park constructed in the early 1920s, known collectively as the Hetch Hetchy Regional Water System)—before crossing the San Joaquin Valley and feeding into the numerous downstream pipelines, tunnels, and reservoirs. The tunnels within the Hetch Heathy system are mostly nonredundant and serve as a singular route for 85% of the

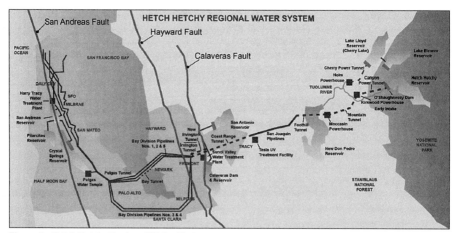

Figure 1. Hetch Hetchy Regional Water System

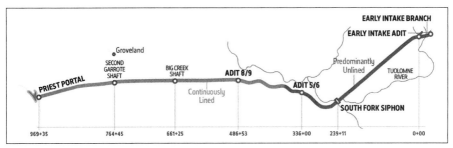

Figure 2. Existing mountain tunnel system

SFPUC's water supply. See Figure 1 for an overview of the Hetch Hetchy Regional Water System.

The Mountain Tunnel is one of the longest tunnels in the Hetch Hetchy system at 18.9 miles, with an internal diameter ranging between 10 feet within the lined sections and 15 feet within the unlined sections. The eastern 7+ miles of tunnel are predominantly unlined, and the downstream western 11 miles are continuously lined with unreinforced concrete. Water flows from east to west under gravity conditions between Kirkwood Powerhouse to Priest Reservoir (Figure 2). Priest Reservoir serves as a regulating reservoir, and the tunnel outlet is below the water line. This outlet is also a primary access point into the tunnel's lined section and currently requires draining the reservoir for entry. Because the tunnel provides water for the town of Moccasin and for local fisheries, this draining of the reservoir reduces the time available during tunnel outages.

MOUNTAIN TUNNEL IMPROVEMENTS PROJECT

The Mountain Tunnel has been in service for 93 years and has exhibited defects within the tunnel concrete lining since the 1930s. These defects range from concrete washout to large holes within the lining that expose the rock behind the lining. Further, the tunnel has localized water infiltration issues that contribute to increased water turbidity.

The purpose of the Mountain Tunnel Improvements Project (MTIP) is to improve reliability of daily water delivery to customers, ensure water quality, and provide continued capacity to meet future demands. The improvements in both the lined and unlined portions of the Mountain Tunnel will extend the service life for another 100 years, improve the operational hydraulic performance, and improve access to the tunnel for future maintenance.

The MTIP consists of a series of improvements to the existing 18.9-mile-long water Mountain Tunnel. These improvements will be complicated by the project's remote location, limited access points, and the time-critical nature of water-supply shutdowns. Below is a discussion of the geology within the MTIP area and the project's various components—including the new downstream Flow Control Facility shaft and tunnels, the new Priest Portal and Adit, the new South Fork access shaft and siphon extension, the rehabilitation of the existing tunnel lining, and the major roadway improvements required along each access road (Figure 3). The critical in-tunnel activities require completion during a series of five winter shutdowns ranging between 60 and 100 days each.

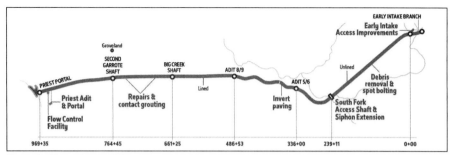

Figure 3. Mountain tunnel improvements project

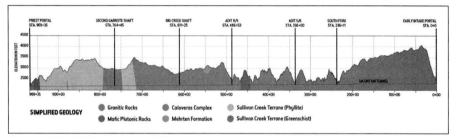

Figure 4. Simplified geological profile

Geology

From east to west, the downstream tunnel alignment generally traverses metasedimentary rocks of the Paleozoic Calaveras Complex as well as metasedimentary and metavolcanic rocks of the Mesozoic Foothills terrane. Small intrusive bodies of Mesozoic granitic rocks also are mapped along the alignment, in both the Paleozoic rocks and the Mesozoic rocks. See Figure 4 for a simplified geological profile.

Near the middle of the alignment, the previously mapped Sonora Fault juxtaposes the Paleozoic Calaveras Complex on the east against the Jurassic Sullivan Creek terrane on the west. The Sullivan Creek terrane includes a greenschist belt (to the west) composed of metavolcanic rocks and a phyllite belt (to the east) composed of metasedimentary rocks. The largest granitic body along the downstream tunnel alignment occurs directly west of the Sonora Fault in the phyllite belt of the Sullivan Creek terrane. This granitic body generally coincides with the Second Garrote basin, a relatively large, bowl-shaped depression within the Mesozoic metamorphic rock.

The following geologic characteristics exist along the tunnel alignment:

- Granite was encountered from the Early Intake to South Fork.
- Granodiorite was encountered from South Fork to near Adit 5/6.
- Siliceous schist and quartzite were encountered from Adit 5/6 to approximately Station 461+00 (which is between Adit 7/8 and Adit 8/9).
- Slates, quartzite, and mica schist, with some hard diorite intrusives (near Second Garrote Shaft), were encountered from approximately Station 461+00 to Priest Portal.
- The slate, quartzite, greenshist, and mica schist formations as shown on the geologic profile have high-angle/near-vertical bedding or foliation features.

Flow Control Facility

A major component of the MTIP is the introduction of the new downstream Flow Control Facility (FCF). The addition of the FCF at the tunnel outlet was proposed to reduce damage to the tunnel lining by eliminating the stress on the concrete lining caused by the cycling between the open channel condition and pressurized flow condition inside the Mountain Tunnel. The FCF will house two 72-inch flow control valves and will be designed to maintain the tunnel in a full condition by continuously throttling the flow at the downstream end of the tunnel. This facility will allow tunnel inflows to match tunnel outflows and for the tunnel to remain in a full pressurized condition, which reduces flow velocities, turbulence, and pressure variations.

To minimize the work to be done during a time-critical water outage, the shaft and bypass tunnels are offset from the existing in-service tunnel. The necessary tie-in of the FCF will be required to occur during the first tunnel outage, which is limited to 60 days in length. The reason for this requirement is so that the isolation valves in the FCF can be installed and operational to isolate Priest Reservoir from the tunnel for subsequent outages. Isolating Priest Reservoir allows the water levels in the reservoir to remain high and provide supplemental water for both the town of Moccasin and local fisheries for the longer 100-day outages. The construction of Priest Adit will facilitate tunnel entry in the 100-day outages (Figure 5).

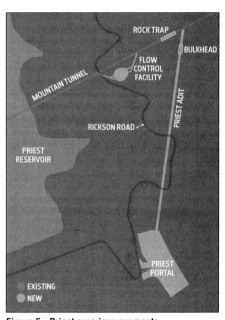

Figure 5. Priest area improvements

The FCF consists of an approximately 160-foot-deep, 55-foot-diameter circular shaft from the surface to a depth of 110 feet, at which point the shaft bells out to a larger 55-foot by 69-foot ellipse to accommodate the valves and mechanical equipment. The shaft floor will house the isolation and flow control valves and associated mechanical, electrical, and instrumentation at an invert elevation, which provides maximum back pressure to minimize adverse cavitation at the valves (Figure 6).

The FCF bypass will connect at 45-degree angles to the existing Mountain Tunnel (Figure 7). The portion of the existing Mountain Tunnel between these two connections will be backfilled with concrete to divert the water into the upstream bypass tunnel and into the FCF pipelines. Within the bypass tunnels, new steel lining will bifurcate into two main flow lines, each with a flow control valve and two isolation valves, to provide redundancy when one valve is taken out-of-service. The bypass tunnels will be oversized to accommodate the placement of 107.2-inch steel lining through the shaft prior to facility tie-in during a tunnel outage. The final bypass tunnel lining will be either be 72- and 107.2-inch polyurethane-lined steel pipe backfilled with cement grout. The tie-in angles were optimized to minimize critical outage work associated with welding.

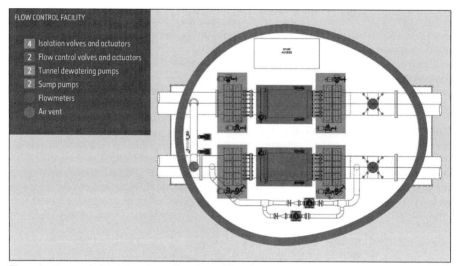

Figure 6. Flow control facility valve and equipment layout

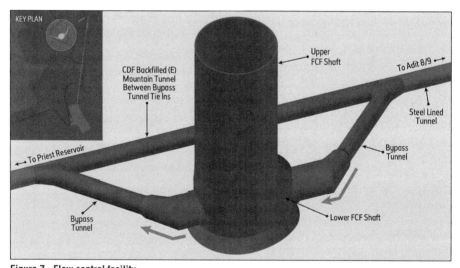

Figure 7. Flow control facility

Construction of the underground elements of the FCF is anticipated to be primarily by drill and blast methods with rock dowel and shotcrete initial support. The FCF shaft will incorporate a double-shelled PVC waterproofing to reduce the maintenance requirements and avoid the need to pump groundwater from the bottom of the shaft to the ground surface. A plastic sheet membrane system will be applied to the shaft initial support to prevent water migration into the shaft, and then a cast-in-place concrete lining will be installed to encase the membrane and provide structural support for permanent loads. The shaft base slab is designed in conjunction with the final lining to resist groundwater uplift pressures, and to support necessary facility components including the sump and pipe supports.

The addition of the FCF will create greater internal pressures inside the tunnel than current operations. For example, the downstream end of the tunnel would be exposed to a pressure as high as 80 psi (static), whereas under current operation the maximum pressure at this location is closer to 30 psi. As a result, the new bypass tunnels and a short segment of the existing tunnel upstream of the FCF will be steel lined to prevent excessive outflows caused by hydrojacking of the surrounding rock.

New Priest Adit and Portal

As noted above, a primary goal of the MTIP is to provide equipment and personnel access to the Mountain Tunnel without draining Priest Reservoir. To accomplish this, the addition of a New Priest Adit was selected to facilitate ingress and egress of equipment and personnel for the planned tunnel rehabilitation, long-term inspections, and maintenance activities. The tie-in location of New Priest Adit will be upstream of the new steel lining associated with the FCF as previously discussed. Additionally, there will be a new rock trap between the tie-in location and the steel lining to collect debris before it reaches the mechanical equipment in the new FCF. Entry into the tunnel from the New Priest Adit will be via a new 11-foot by 13-foot steel bulkhead door, which will allow for large pipe entry into the tunnel.

The New Priest Adit has a horizontal length of approximately 1,075 feet and maintains a uniform slope of approximately 8% towards Mountain Tunnel. The adit is a straight-legged horseshoe section providing a clear finished width of 14 feet at the invert and a clear finished height at the adit centerline of 11.5 feet (Figure 8). Prior to constructing the New Priest Adit, a New Priest Portal construction pad will be excavated and graded about 1,100 feet to the south of the existing tunnel. This new, approximately 1-acre portal area will serve for both temporary access and staging during construction and long-term access and staging for inspection and maintenance activities (Figure 9).

Most of the adit construction will be conducted prior to the first tunnel outage up to a set distance away from the existing tunnel. It will be tied-in during the first 60-day tunnel outage. As with the FCF, most of the excavation is anticipated to be accomplished by drill and blast with rock dowel and shotcrete initial support.

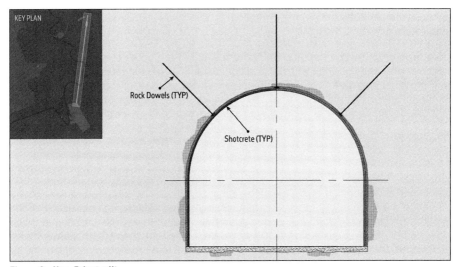

Figure 8. New Priest adit

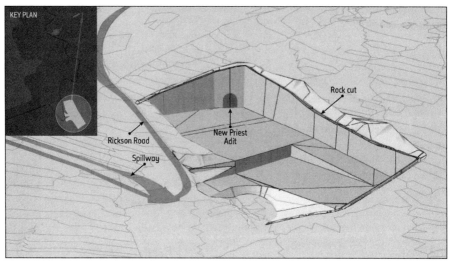

Figure 9. New Priest portal

South Fork Siphon Access Shaft and Extension

Between approximately Mountain Tunnel Sta. 233+50 and Sta. 233+75, a large volume of water flows into the existing tunnel. This location produces approximately 2,000 to 3,000 gallons per minute (gpm) when the Mountain Tunnel is drained or partially drained. The inflows are due to the fractured ground above and to the side of the tunnel and the proximity of the river as a sustained water source to the tunnel. Ground cover above the tunnel is estimated to range from 20 feet to near zero within the fractured rock zone producing the water inflows.

The amount of inflow is dependent on the flow elevation of the South Fork of the Tuolumne River (e.g., there is greater water pressure in the winter during storm events than in the summer). The winter and spring river flows contain sediment. The infiltration of water at this location during these periods creates higher turbidity within the tunnel water. In 2009 a temporary manifold solution was installed to capture this water and return it back to the river.

This area of infiltration needs a permanent solution to address water quality associated with turbidity. Further, with the addition of the FCF, the tunnel operating pressures will increase, and water would be subject to continuous excessive exfiltration (i.e., loss of tunnel water into the river) under this new scenario because of the fractured rock and low cover. This further necessitates a permanent solution for this stretch of tunnel.

The South Fork site is complicated by steep and narrow access, upslope rock hazards, the adjacent river and constricted (30-inch-diameter) tunnel access.

For these reasons, the recommended project improvement at South Fork is to construct a short extension of the existing siphon to bypass the fractured rock zone and abandon the section of tunnel currently subject to high inflows. Construction of this siphon extension will primarily occur during non-outages, off the critical path, with final tie-ins constructed during a 100-day outage. Construction of the siphon extension allows for improved access to the existing tunnel and the siphon. Additionally, the construction of the new siphon extension will improve the hydraulics by optimizing the

Mountain Tunnel Improvements Project

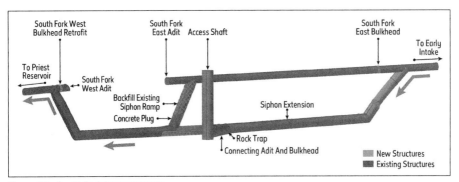

Figure 10. South Fork Improvements

eastern inclined shaft angle and locally increasing the tunnel cross-sectional area, which is currently constricted.

These tunnel improvements are founded in granitic rocks and will be constructed via drill and blast methods with rock dowel and shotcrete initial support. To construct these improvements, a new access shaft will be constructed to facilitate the mining of a connecting adit to mine the siphon extension. Improvements will also be made to the South Fork East tunnel access. The improved access into both the tunnel and siphon will allow for easier long-term maintenance. Refer to Figure 10 for a schematic of the South Fork improvements.

Figure 11. South Fork access shaft

The siphon access shaft will be constructed from an existing concrete pad area adjacent to the South Fork of the Tuolumne River and will provide access to the underground works during construction as well as permanently. The 17-foot-diameter shaft will be approximately 105 feet deep to reach the existing siphon elevation. The construction staging area at the South Fork site is minimal and the shaft is located at the base of the narrow South Fork access road. See Figure 11 for a photo with projection of the access shaft.

From the access shaft, an approximately 60-foot connecting adit will be excavated towards the siphon extension tunnel alignment. The connecting adit will create a wye with the new siphon extension tunnel. This wye will also be the location of a rock trap. The connecting adit will have a 15.5-foot-wide, horseshoe-shaped profile with a permanent 8-foot by 8-foot steel bulkhead door entry.

The siphon extension tunnel will extend from the current east end of the siphon eastwards, where it will connect into the Mountain Tunnel via an inclined shaft. The extension will have a 14.5-foot inside diameter (ID) to match the existing siphon profile. Probe holes will be advanced from the extension tunnel heading to identify fracture zone(s) and determine water infiltration quantities. Where the siphon tunnel section encounters any fracture zones, the tunnel will be over excavated to accommodate additional ground support and a steel final lining such that river water is cut off from the tunnel water. If water is encountered during construction, pre-excavation grouting will be initiated to help cut off infiltration.

The inclined shaft between the siphon extension and the existing Mountain Tunnel will connect into the existing tunnel at 45 degrees from the vertical. Where the inclined shaft meets the tunnel, the existing tunnel will be enlarged on one side to provide walking access around the inclined shaft. Additional in-tunnel work includes a new equipment-sized bulkhead access within the existing Mountain Tunnel South Fork East access, backfilling the existing east siphon inclined shaft, and the installation of new air vents above the siphon inclined shafts.

In the permanent condition, the access shaft will have a watertight concrete lid with removable panels to prevent water ingress from river flooding events while allowing for future access. A steel stairwell will be constructed for shaft bottom access.

REHABILITATION OF THE EXISTING MOUNTAIN TUNNEL

The existing Mountain Tunnel will undergo an intensive rehabilitation program to extend its service life and allow for easier long-term access. All in-tunnel work will need to occur during time-critical winter outages. The following items will be completed:

- Tunnel lining repairs of damaged concrete tunnel lining (i.e., structural patching of small and large defects)
- Contact grouting of 11 miles of concrete-lined tunnel
- Localized cutoff grouting in areas of turbid water infiltration
- Localized steel pipe or reinforced concrete lining at locations of hydrojacking risk
- A new bulkhead at Early Intake Adit to improve vehicular access at the upstream end of tunnel
- Invert paving to improve in-tunnel vehicular access between Adit 5/6 and the lined tunnel
- Removal of existing rock debris and localized spot dowels in the unlined tunnel

The tunnel lining repairs will occur within the 11 miles of continuously lined tunnel. Repair sites have been identified with LiDAR imaging (see Figure 12). All defects identified for repair will be repaired with weld-wire fabric reinforcement and shotcrete with a smooth finish (see Figure 13). The New Priest Adit will be the primary access points for the repairs on the western end of the tunnel, and Adit 8/9 will be the primary access point for the repairs on the eastern end of the lined section.

Contact grouting will be performed systematically throughout the lined tunnel after the lining has been repaired. This will ensure the tunnel lining and rock are in full contact, which is important both for durability and because of the modified operations that will be created by the addition of the FCF.

Groundwater cutoff grouting will be performed at locations within the lined tunnel section where groundwater inflow rates exceed the acceptable criteria after completion of the contact grouting program and water has been measured to be turbid. Although other locations may require cutoff grouting, the anticipated areas of focus are the stretches of tunnel immediately upstream and downstream of the Second Garrote Shaft and elsewhere in the Second Garrote Basin. The cutoff grouting program will

Figure 12. LiDAR imaging of defects within lined tunnel

Figure 13. Installation of tunnel lining repair

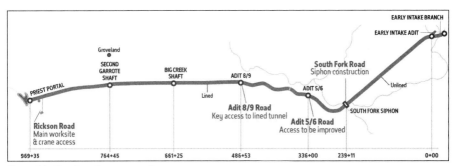

Figure 14. Roadway improvements

focus on injecting fine and ultrafine cement grouts into the rock formation to cut off infiltration into the Mountain Tunnel.

To complete the grouting operations within the scheduled outages, a continuous grout line is proposed from each of the existing shafts: Second Garrote (approximately 750 feet deep) and Big Creek (approximately 575 feet deep). This will necessitate a grout plant at the ground surface at each shaft location, the pumping of grout down each shaft, and pumping along the tunnel to reach each section of the tunnel.

A cast-in-place concrete invert will be placed between Adit 5/6 and the end of the continuously lined tunnel to aid in vehicular access to the lined portion of the tunnel.

Early Intake Adit is at the upstream end of the Mountain Tunnel and is the only access at this end of the tunnel. It currently does not provide adequate personnel and equipment access with its existing bulkheads. As such, all the existing structures inside the adit will be removed and a new concrete bulkhead with an 8 foot by 8 foot opening and steel access door will be installed.

ROADWAY IMPROVEMENTS

The access roadways to the Mountain Tunnel adits were originally constructed in the 1910s through the mountainous site terrain. These roads supported small sled equipment and railways for the original construction. The roadways are narrow, single-lane, and primarily gravel-surfaced, and have deteriorated to different levels over the years of supporting larger vehicular access. To support construction and provide safe long-term access, these access roadways need improvements. Roadways requiring improvements are Rickson Road at the Priest Reservoir Site, the Adit 5/6 access road, the Adit 8/9 access road, the South Fork East Adit access road, and Second Garrote Shaft Road.

The roadway improvements generally consist of increasing roadway width, adding turnouts to facilitate single lane access where topographically feasible, installing shoulder reinforcement, slope stabilization, rockfall mitigation, and drainage improvements.

The design strategy is to incorporate retaining wall systems that are flexible, durable, permeable, and constructible, as well as repeatable from site to site. Rock-filled gabion baskets will be used as the primary downslope road widening element to reinforce soft shoulders, stabilize slopes, and increase width through downslope retention. Taller gabion walls will be anchored into the hillside with grouted anchors.

Slope stabilization and rockfall mitigation measures will reduce exposure for construction traffic and operations staff to this hazard. Measures will include scaling loose rock, stabilizing key blocks with rock bolts, slope cuts reinforced with grouted anchors, and various draped slope mesh and post-mounted cable net systems.

Rickson Road

Rickson Road is an approximately 3.2-mile paved road that circles the Priest Reservoir. It will provide access for in-tunnel improvements, the FCF, and New Priest Portal and Adit. Rickson Road improvements will be completed on the east side of the reservoir. This site is moderately hilly without extreme topographic relief. For this reason, most of the widening will be accomplished by constructing upslope cuts into the hillside to add an additional 2.5 feet of road width and accommodate large crane access to the FCF and new Priest Portal. The improved road will have a minimum of 14 feet of drivable width and will be much wider at the curves.

Adit 8/9 Road

The Adit 8/9 access road is an approximately 2-mile-long unpaved mountain road, leading to the Adit 8/9 Portal from Ferretti Road. Widening of the roadway will occur mainly along the last ¼ mile along a spur road off of the main roadway. The spur road will be widened to 12 feet via downslope gabion walls. Four turnouts along the main stretch of road will be added, and a large rock cut for a new laydown area is planned at the hairpin turn onto the spur road.

Adit 5/6 Road

The Adit 5/6 access road is a 1.9-mile unpaved mountain road leading to the Adit 5/6 portal (Figure 15). The primary improvements include widening the road to 12 feet minimum as well as the addition of surface runoff management features since many slipouts can be attributed to poor drainage. Similar to Adit 8/9 access road, widening will be accomplished by downslope gabion walls. One portion of roadway is particularly challenging because of a near vertical downslope and unstable boulders on the upslope.

Figure 15. Adit 5/6 existing roadway

At this location, the design includes a micropile supported concrete deck to obtain the desired road width. Other improvements include the addition of seven turnouts to facilitate single-lane access and rock fall mitigation measures at critical locations. The turnouts include cut slopes as well as grouted anchor walls. Drainage management features consist of an in-sloped road surface with concrete waterbars placed at approximately 100- to 200-foot spacings and at other identified critical locations on the road.

South Fork Road

The South Fork access road is the most extreme of the roadways as it travels down a narrow steep canyon along the South Fork of the Tuolumne River. It is an 0.8-mile partially paved road from Old Big Oak Flat Road to the South Fork East Adit portal. The gravel portion of the roadway experiences repeated and severe erosion and rutting because of inadequate drainage and the steep road gradient.

The improvements to this portion of the road consist of replacing the existing roadbed with properly sized base and surface aggregate and the addition of concrete waterbars to promote appropriate drainage, as well as minor improvements to direct drainage into already existing inlets and culvert. Although this roadway will not be consistently widened, the existing downslope walls will be reinforced with grouted anchors and shotcrete and localized turnouts will be added in a few locations. The South Fork access road contains numerous substantial rockfall hazards due to the extreme topography. The improvements to provide safer access include scaling, pattern and spot rock bolts, draped mesh, elevated cable net, anchored high strength steel mesh at the Siphon Shaft site, and rockfall attenuator systems (Figure 16).

Figure 16. South Fork locations for scaling, spot bolting, and rockfall protection

CONSTRUCTION SCHEDULE AND OUTAGES

The anticipated construction schedule is shown in Figure 17 at the end of this paper. The schedule indicates NTP in March 2020 and construction continuing through 2026 with five tunnel shutdowns shown in red. The red text indicates in-tunnel work that must occur during an outage such as tie-ins or tunnel repairs.

The critical path items include the improvements to Rickson Road at the Priest Reservoir site, the FCF, the new Priest Portal, and new Priest Adit. It is paramount that all the pre-outage work associated with making the two tie-ins for the FCF and one tie-in for New Priest Adit be completed prior to the first 60-day outage. The primary components of this pre-outage work include the Rickson Road improvements to facilitate crane access, the FCF shaft and bypass tunnels to a predetermined offset, and the New Priest Portal and Adit to a predetermined offset. Critical outage work will require 24/7 work.

CONCLUSION

The MTIP consists of an intensive rehabilitation project with new construction components including the FCF, Priest Portal, Priest Adit, South Fork Siphon Extension, and

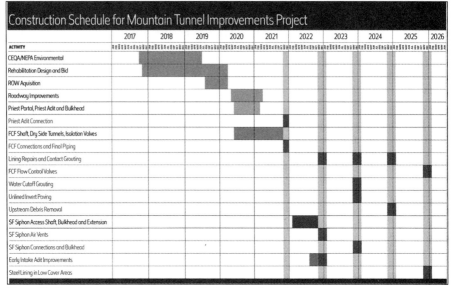

Figure 17. Mountain tunnel improvements project construction schedule

thorough access roadway improvements. Careful consideration to the construction sequencing needs to be given to ensure that all critical path items and outage work are completed on time to ensure project completion. Through this project, Mountain Tunnel will be improved to provide drinking water for the SFPUC's customers for the next century.

REFERENCES

Fippin, R., J. Sketchley, T. Redhorse, and D. Tsztoo. 2018. In-depth inspection of a century-old San Francisco water tunnel. In *Proceedings of the North American Tunneling Conference*. Englewood, CO: Society for Mining, Metallurgy & Exploration, Inc. 236–245.

Tsztoo, D., A. Yu, and T. Redhorse. 2018. Rehabilitation of tunnels: An owner's perspective. In *Proceedings of the North American Tunneling Conference*. Englewood, CO: Society for Mining, Metallurgy & Exploration, Inc. 99–704.

MWRA Metropolitan Boston Tunnel Redundancy Program Project Update

Kathleen M. Murtagh ▪ Massachusetts Water Resources Authority
Frederick O. Brandon ▪ Massachusetts Water Resources Authority

ABSTRACT

The Massachusetts Water Resources Authority (MWRA) provides wholesale water and wastewater services to over 3.1 million customers in 61 communities in eastern and central Massachusetts with most service communities located in the Boston area. The Quabbin Reservoir and Wachusett Reservoirs, which are the main water supply sources, are located 65 and 35 miles west of Boston, respectively. A redundant water transmission system exists for approximately 25 miles from the Wachusett Reservoir to the beginning of existing Metropolitan Tunnel System. The Metropolitan Tunnel System carries approximately 60% of the total system daily demand the remaining 10 miles. With no redundancy to the Metropolitan Tunnel System, partial system shutdowns for planned maintenance of the aged infrastructure or unplanned emergencies cannot take place. The planned Metropolitan Tunnel Redundancy Program is proposed to consist of approximately 14 miles of 10-ft internal diameter deep rock tunnel at an estimated cost of approximately $1.5 billion. This paper discusses the need for a redundant tunnel system, alternatives considered, the current plan, and project outlook.

THE MWRA

In 1984, legislation was enacted to create the Massachusetts Water Resources Authority (MWRA). The MWRA is a public authority that provides wholesale water and sewer services to 3.1 million people and more than 5,500 large industrial users in 61 communities in eastern and central Massachusetts. The primary mission of the MWRA was to clean up Boston Harbor and modernize the area's water and sewer systems. Other key elements have included a major capital program to repair and upgrade the systems, increase staff to improve operations and maintenance, promote water conservation, and plan for the future to meet growing demand.

Boston's water system has been governed by various water authorities over the years starting with the Cochituate Water Board (1845–1875), followed by the Boston Water Board (1875–1895), Metropolitan Water Supply Commission (1926–1946), Metropolitan District Commission (1946–1985), and finally the MWRA (1985–Present).

The MWRA's water system currently has more than 200 separate facilities, including the John J. Carroll Water Treatment Plant, with a capacity of 405 million gallons per day (mgd), 11 pump stations, and 14 below- or above- ground storage tanks. The water transmission system includes 105 miles of active tunnels and aqueducts (mostly 10 to 14 feet in diameter) and 39 miles of standby aqueducts.

HISTORY OF BOSTON'S WATER SYSTEM

Although the MWRA's current water system of interconnected reservoirs, tunnels, aqueducts and pipelines provides an abundant supply of clean drinking water to millions of customers, the system was not always sufficient to meet the growing needs of the City of Boston and the surrounding communities.

When the Boston area cities and towns first faced the problems of providing clean water sources in the 1600s, their methods were primitive, relying on local wells, rain barrels, and a spring on the Boston Common. By 1795 wooden pipes delivered water from a centralized water supply at Jamaica Pond to Boston proper. By the late 1840s, however, Jamaica Pond was too small and too polluted to provide water to Boston's 50,000 residents. So, the pattern of continually moving westward in search of larger fresh water sources began.

Cochituate System: 1848–1951

In 1845, the Cochituate Water Board began construction of a new water supply and transmission system. A tributary of the Sudbury River was impounded, which created Lake Cochituate. Lake Cochituate, with its 17 square miles of watershed, 2 billion gallons of storage and yield of 10 mgd, became the cornerstone of the Boston water system.

The Cochituate Aqueduct, extending 14.5 miles, was completed to transport water to the Brookline Reservoir and then to smaller distribution reservoirs in all parts of the city. This aqueduct was in service for approximately 100 years, until 1951, when water quality had declined and alternate methods of transporting water to the hub of the distribution system had been constructed.

The Sudbury Aqueduct and Chestnut Hill Reservoir: 1878

Boston grew rapidly after the Irish Potato Famine of 1843–45 and by 1870, its population exceeded 200,000 and consumed 17 mgd of water. Planners had not anticipated this rapid growth; they thought that the Cochituate system would be adequate for many years. In order to provide the growing city with the water it needed, the process of diverting water from a western pure upland source was repeated.

In 1878, the mainstream of the Sudbury River was diverted via the Sudbury Aqueduct to the Chestnut Hill Reservoir. Between 1875 and 1898, seven major reservoirs were constructed in the Upper Sudbury River Watershed. The Sudbury and Cochituate Aqueducts were designed to operate by gravity to fill the Chestnut Hill and Brookline Reservoirs.

Wachusett Reservoir: 1897

The Boston metropolitan area continued to grow rapidly through the 1890s. Indoor plumbing became commonplace. Planners had not foreseen this development and the current water supply had become inadequate. Under the leadership of Frederick Stearns, Chief Engineer of the Boston Water Board, it was decided that a new water source that could be gravity-operated and not require filtration was required.

In 1897, the Nashua River above the Town of Clinton in Central Massachusetts was impounded by the Wachusett Dam. Six and 1/2 square miles were flooded and water conveyed by the Wachusett/Weston Aqueduct to the Weston Reservoir and then by pipeline to the Chestnut Hill and Spot Pond Reservoirs. Work was completed in 1905 and the reservoir first filled in May 1908. The Wachusett system was built to service the 29 municipalities within the 10 mile radius of the Massachusetts State House. At the time, the Wachusett Reservoir was the largest public water supply reservoir in the world with a capacity of 65 billion gallons.

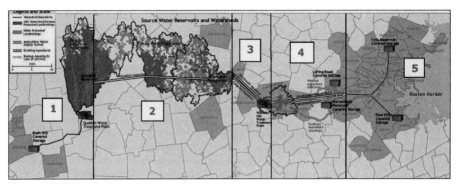

Figure 1. MWRA water transmission system

Quabbin Reservoir, Ware River Intake and Hultman Aqueduct: 1926–1946

Eventually the Wachusett System became inadequate for the increasingly industrialized city and a westward focus for a new water source resumed. The Quabbin Reservoir was Boston's fourth westward reach for a pure upland source of water that could be delivered by gravity and not require filtration. Construction of the Quabbin Reservoir required impoundment of the Swift River and the taking of the towns of Dana, Enfield, Greenwich, and Prescott.

In 1926, construction began on the Wachusett-Coldbrook Tunnel, which is now the eastern section of the Quabbin Tunnel. During the 1930s, the Wachusett-Coldbrook Tunnel was extended to the Swift River. It is a two-way tunnel: water flows west from the Ware River to the Quabbin during the high-water months and then east from the Quabbin Reservoir to Wachusett at other times of the year.

Construction on the Quabbin Reservoir began in 1936. Filling commenced in 1939 and was completed in 1946 when water first flowed over the spillway. At the time, the 412-billion-gallon reservoir was the largest man-made reservoir in the world which was devoted solely to water supply. The existing reservoirs, located at sufficiently high elevations, could now supply an abundant supply of water to the metropolitan Boston area by gravity through pressurized aqueducts or tunnels.

In the 1940s, planners believed that the Quabbin Reservoir will be sufficient to supply the metropolitan area into the foreseeable future, and at the time, was the last major investment in the water system with no plans in place for upgrades to carry the system into the next century. Many of the previous expansions used gravity for supplying water instead of costly pumping. Fortunately, these crucial foundations laid by the early water engineers provide the backbone of the system we run today.

Origins of the Pressure Aqueduct System (Hultman Aqueduct): 1937–1941

In 1937, a plan was developed for a high service pressure aqueduct system to deliver water to the Metropolitan area. A portion of the plan included two parallel aqueducts to carry water from the Wachusett Aqueduct to the new Norumbega Reservoir and the terminus of the Weston Aqueduct in the Town of Weston. Work began on schedule in 1939 and by the outbreak of World War II in 1941, one of the two proposed parallel Pressure Aqueducts had been built. This portion of the Pressure Aqueduct is the Hultman Aqueduct.

Pressure Aqueducts and Tunnels: 1950–1978

After WWII, additional segments of the pressurized transmission system came online with the construction of the Chicopee Valley Aqueduct, Metropolitan Tunnel System and the Cosgrove Tunnel. As these sections of the pressure transmission system have come online, the need for pumping from open reservoirs was reduced since more of the service area could be supplied by this pressurized transmission system. Older facilities which originally provided a level of redundancy to the new pressure tunnels were eventually retired from use. More reliance was placed on the newer pressurized system to the point where it is now relied upon to deliver 85% of the Metropolitan area demand.

REDUNDANCY IN THE TRANSMISSION SYSTEM

Transmission System Overview

The current Water Transmission System can be divided into five major segments as shown in Figure 1. Redundancy projects for segments 1 through 4 have been completed. The fifth segment, the Metropolitan Tunnels, represents the next challenge for the MWRA in improving the reliability of this great water system.

Metropolitan Tunnel System (Segment 5)

The Metropolitan Tunnel System includes the City Tunnel (1950), the City Tunnel Extension (1963), and the Dorchester Tunnel (1976). These three tunnels interconnect at Shaft 7 at Chestnut Hill. Together, these tunnels carry approximately 60% of the total system daily demand with no redundancy.

Condition of Metropolitan Tunnel System

Each tunnel comprising the Metropolitan Tunnel System consists of concrete-lined deep rock tunnel sections linked to the surface through steel and concrete vertical shafts. At the top of each shaft, cast iron or steel pipe and valves connect to the MWRA surface pipe network. These pipes and valves are accessed through subterranean vaults and chambers. The tunnels and shafts, themselves, require little or no maintenance and represent a low risk of failure. However, many of the valves and piping are in poor condition.

Valve reliability for the Metropolitan Tunnels is a concern. As an example, the City Tunnel (1950) appurtenances are 68 years old and can't be adequately maintained or replaced until a back-up exists. Failure of some valves can cut off a majority of the system's capacity to supply water and, due to the physical condition, age, and environment in which they were installed, have not been exercised for fear of failing in a closed position. These valves should be, but cannot be, replaced because shut down of the City Tunnel would be required.

Access to some of the valve structures and chambers is hampered by high ground water or damp conditions (see Figure 2). Original protective pipe coatings are gone and pipes and valves are coated in thick layers of rust. Loss of metal thickness and structural strength is a concern. Bolts and fasteners have corroded and are planned to be replaced where feasible. Some chambers must be pumped down to allow access, which impedes any emergency response and aggravates further corrosion concerns.

At many of the top-of-shaft structures are piping and valves of varying diameters (ranging from less than an inch to several inches in diameter) provide air and vacuum relief, along with drains, flushing connections, valve by-passes, and control piping for hydraulic valve actuators (see Figure 3). Some of these pipes and valves are in a similar deteriorated condition as the main pipes and valves themselves. Failure of one of these smaller diameter connections could require a tunnel shut down to allow for a safe repair in some of these confined spaces. The amount of water that can flow out of a modest opening under high pressure can be significantly more than one might think; potentially over 100 million gallons per day (mgd).

Some of these concerns can be mitigated somewhat through replacement of corroded bolts, wrapping or coating corroded pipeline segments, replacement of air valves, and installation of cathodic protection systems. A program is being developed to implement some of these

Figure 2. Valve chamber filled with ground water

measures to reduce the risk of certain failures that would require complete tunnel shut down. However, all the potential failure points cannot be addressed without tunnel isolation and complete replacement or maintenance of failed or failing components at some point in the future.

Water Main Break of May 1, 2010

MWRA experienced a major break on a ten-foot diameter pipe connection at Shaft 5 of the City Tunnel on May 1, 2010. The break occurred at a coupling on the surface pipe interconnection between the recently constructed MetroWest Water Supply Tunnel and the City Tunnel. The MWRA had a redundant pipe (Hultman Aqueduct) at this location, but at the time of the break, the Hultman Aqueduct was being rehabilitated and was out of service.

The incident resulted in a release of approximately 250 mgd over a period of eight hours until the break was isolated (see Figure 4). During this time, an emergency water source was activated to maintain water supply prior to shutting down the affected pipe. While the pipe was being repaired over the following two days, the Boston metropolitan area was supplied through alternate lower capacity mains with augmentation from an emergency raw water reservoir with chlorination. The water service area was

Figure 3. Hydraulic valve actuators

Figure 4. 250 MGD flow at Shaft 5 break

issued a boil water order during these two days. This boil water order affected approximately 2 million people in 30 serviced communities.

After the water main break, the MWRA performed an economic impact analysis of a failure and forced shut down of the Metropolitan Tunnel System. The analysis estimated that the economic loss to businesses and residents within the Metropolitan area would be approximately $208 million and $102 million per day, respectively for a total estimated economic impact of approximately $310 million per day.

HISTORY OF REDUNDANCY PLANNING FOR METROPOLITAN AREA

1937 Plan

A redundant tunnel system was proposed as early as 1937. The plan included a proposed pressure aqueduct and tunnel system with a tunnel loop beginning in Weston near the Charles River and running east into Boston, turning north to Everett, looping west to Belmont and connecting back to Weston (see Figure 5).

While much of the 1937 plan for pressure aqueducts and tunnels was implemented from 1937 to present day, the proposed tunnel loop was never completed.

Redundancy Planning 1990 through 2016

In 1990, a plan was proposed to construct a tunnel from Marlborough to Weston (the MetroWest Water Supply Tunnel) to provide redundancy for the Hultman Aqueduct and a future northern tunnel loop from Weston to Stoneham and Malden (see Figure 6). The MetroWest Water Supply Tunnel was approved for construction and was completed in 2003. However, the proposed northern tunnel loop was not constructed.

In 2011, the MWRA completed a new evaluation of alternatives for redundancy within the Metropolitan Boston area. This evaluation included surface pipe alternatives in addition to tunnel alternatives with an objective of incorporating redundancy planning into the existing pipeline asset management program (ie. allocating funds already budgeted for rehabilitation of existing pipelines toward replacing the existing pipelines with larger pipelines). The result of that evaluation was a plan of constructing primarily large diameter surface pipes to provide redundancy (see Figure 7). However, as the planning for this program progressed, it became apparent that the construction of

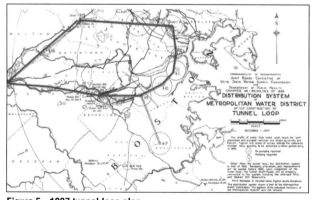

Figure 5. 1937 tunnel loop plan

Figure 6. 1990 northern tunnel loop plan

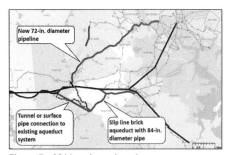

Figure 7. 2011 surface pipe plan

large diameter pipelines through dense urban areas would cause unacceptable community disruption and had serious implementation challenges.

Finally, in 2016 MWRA revisited the all-tunnel approach to providing redundancy to the Metropolitan area. Over 30 alternatives were screened based on level of redundancy, constructability, cost and operation and maintenance. Based on this evaluation, an all-tunnel alternative was recommended for redundancy.

PROPOSED PLAN

Given the difficulties associated with the construction and significant community impacts associated with large diameter surface pipe together with operational reliability concerns, MWRA staff are pursuing a preferred all-tunnel redundancy alternative. The preliminary alignment, which will be subject to more detailed review and alternatives analysis during the public review period, is shown in Figure 8.

This alternative consists of two deep rock tunnels beginning at the same location in Weston near the Massachusetts Turnpike/Route 128 interchange. The Northern Tunnel generally follows the route of MWRA's existing Weston Aqueduct Supply Main (WASM) 3 transmission main to a point about midway along the pipeline near the Waltham/Belmont border, which will allow flow in WASM 3 in both directions. The length of the Northern Tunnel would be approximately 4.5 miles and the tunnel would have a finished inside diameter of approximately 10 feet. It would include one connection shaft to provide a redundant supply to MWRA's Lexington Street Pump Station and to allow isolation of the WASM 3 line in segments for repair and maintenance. The Northern Tunnel has an estimated midpoint of construction cost of $472 million.

The Southern Tunnel would run east to southeast to tie into the surface connections at Shaft 7C of the Dorchester Tunnel and about midway down the southern surface mains allowing flow in both directions. The length of the Southern Tunnel would be approximately 9.5 miles and would have a finished inside diameter of 10 feet. The estimated midpoint of construction cost of the Southern Tunnel is approximately $1.003 billion.

The proposed plan limits community disruptions and construction impacts to the locations of the tunnel construction and connection shaft sites. The all-tunnel alternative meets the strategic objective of being able to make a seamless transition to a backup supply, allowing maintenance to be scheduled for the Metropolitan Tunnels, without use of a boil order, without impacting the ability to provide for local fire protection, and without noticeable changes in customers' water quality, flow or pressure. It has the ability to meet high demand conditions which extends the potential time frame for future maintenance and rehabilitation activities.

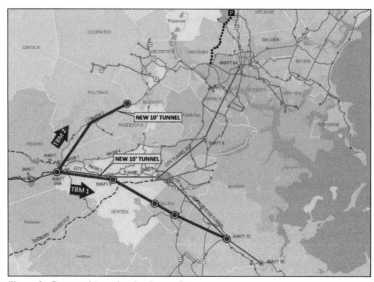

Figure 8. Proposed tunnel redundancy plan

To the north, the all-tunnel alternative will provide redundancy for the critical WASM 3 pipeline. To the south, it will eliminate the need for the Chestnut Hill Emergency Pump Station during Metropolitan Tunnel shut downs, thereby reducing operational risks associated with extended use of the Emergency Pump Station at higher system pressures. The estimated total midpoint of construction cost for both the recommended north and south tunnels is $1.475 billion with an estimated time to completion of 17 years. This estimate includes 30% contingency and 4% annual construction cost escalation.

GEOLOGIC CONDITIONS

The new redundancy tunnels will be hard rock pressure tunnels, similar to the seven existing tunnels which currently make up the main MWRA water distribution system. These existing tunnels ranging in size from 10 to 14 feet in finished diameter are primarily concrete lined with reliance on the overlying rock for confinement. Existing tunnel depths range from approximately 50 feet to approximately 660 feet below ground surface. Tunnels within the existing Metropolitan Tunnel System (City Tunnel, City Tunnel Extension and Dorchester Tunnel) are approximately 150 to 350 feet below grade. The seven existing tunnels, constructed from the 1930s to the early 2000s, were mined using methods that progressed from drill and blast, drill jumbos, to modern tunnel boring machines (TBMs). It is anticipated that the new redundancy tunnels will be mined using TBMs with only short segments (e.g., tail and/or starter tunnels) constructed using alternate methods.

The alignments, both horizontal and vertical, of the new redundancy tunnels are not finalized at this time: however, it is possible that both tunnels could cross the Northern Boundary Fault and extend into the Boston-Avalon Terrace and the Boston Basin. Within the Boston-Avalon Terrace, the primary rock type is anticipated to be the Dedham Granite and within the Boston Basin, the primary rock types are anticipated to be the Cambridge Argillite and the Roxbury Conglomerate. The existing MetroWest Tunnel, City Tunnel, City Tunnel Extension, and Dorchester Tunnel were mined in these same bedrock formations (see Figure 9).

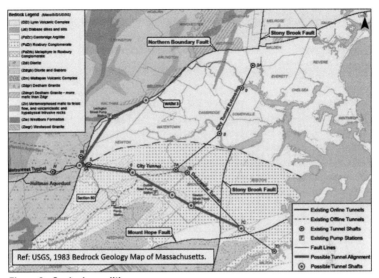

Figure 9. Geologic conditions

The current proposed alignment of the North Tunnel generally follows the Northern Boundary Fault which could pose challenges for tunneling. The existing MetroWest Tunnel intersects this fault near its eastern limit. Extensive geotechnical investigations were conducted to determine the location of this fault as well as provide information on the anticipated behavior of the bedrock during mining. This level of investigation is largely credited for the preparedness that occurred during construction when the fault was actually encountered. It is worth noting that the only location along the MetroWest Tunnel where a steel liner was installed is along approximately 2,000 feet of tunnel where it intersects this fault.

The current alignment of the South Tunnel places this tunnel primarily within the Roxbury Conglomerate formation; crossing the Stoney Brook Fault but not extending sufficiently south to cross the Mount Hope Fault. Both the existing City Tunnel and Dorchester Tunnel are located primarily within the Roxbury Conglomerate formation with only the southern limit of the Dorchester Tunnel extending past the Mount Hope Fault and/or into the Cambridge Argillite. The geology of these tunnels is well documented (Tierney et al., 1968 and Richardson, 1977).

The quality of rock encountered along the City Tunnel is noted by Tierney as being "excellent for tunneling." Just 16 feet of the 4.78-mile tunnel required structural steel support with the remaining approximately 4.7 miles of tunnel mined with no need for temporary supports. The bedrock through which the City Tunnel extends was considered "unusually good" as compared to that encountered along other tunnels previously mined through similar geology in the greater Boston area.

The 6.4-mile-long Dorchester Tunnel was mined through bedrock consisting of Roxbury Conglomerate and Cambridge Argillite. This tunnel was excavated using drill and blast for approximately 90 percent of the tunnel length. A "Mole" (i.e., rotary boring machine) was used to mine the remaining 10 percent of the tunnel length. Tunnel excavation using the Mole was considered experimental at the time since the contractor had not used similar equipment in the past and it had not previously been used in

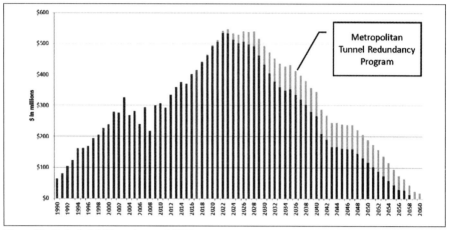

Figure 10. Projected debt service as of 2017

the Boston area prior to this project. Steel supports were required over a fraction of the total tunnel length; primarily where the tunnel crosses the Stoney Brook Fault.

PREVIOUS LARGE PROJECTS

The MWRA has planned, designed, and constructed a number of large projects, including mega-projects, in the past. The largest and most notable project in recent years is the Boston Harbor Project (BHP) which spanned from the mid-1980s to the early 2000s. This nearly 20-year program focused on the clean-up of a much polluted Boston Harbor and involved numerous significant program elements including construction of a new Deer Island Wastewater Treatment Plant, a 9.5 mile by 24-foot diameter outfall tunnel, and a 5-mile by 11.5-foot diameter inter-island tunnel. The overall BHP cost was $3.8 billion.

Overlapping the BHP was the Integrated Water Supply Improvement Program which occurred between 1995 and 2005 and cost approximately $1.7 billion. This 10-year program included construction of the 17.6 mile by 12 to 14-foot diameter MetroWest Water Supply Tunnel, seven covered storage tanks, and the new state-of-the-art John J. Carrol Water Treatment Plant.

Following the completion of the BHP and Integrated Water Program, the MWRA moved on to other significant wastewater projects including the planning, design and construction of the Braintree-Weymouth Relief Facilities Project. This $200 million project, executed between 2002 and 2010 included an intermediate pump station, 2.7-mile by 12-foot diameter deep rock tunnel and shafts.

Alongside the Braintree-Weymouth Project is the South Boston CSO Storage Tunnel and Related Facilities. This $260 million project, executed between 2006 and 2011, included a 2.1-mile by 17-foot diameter soft ground CSO storage tunnel, shafts, pump station, sewer and storm drains, and ventilation building.

The MWRA currently executes approximately $100 million in capital programs each year to add redundancy to, improve, and maintain its current water and wastewater assets.

APPROVAL OF TUNNEL REDUNDANCY PLAN BY MWRA BOARD OF DIRECTORS

After the May 2010 water main break and during the mid-2010s, it became apparent that executing another large water tunnel program would be needed in the near future. On October 6, 2016, the MWRA Board of Directors held a special meeting where MWRA staff provided a briefing on the status of the existing MWRA water transmission system and the lack of redundancy for the Metropolitan Tunnel System. The preferred alternative of constructing two tunnels, one to the north and one to the south, was recommended.

At the conclusion of the special meeting, staff were directed to brief member communities and state and local officials on the Metropolitan Tunnel Redundancy initiative in order to build consensus and support for the preferred project approach. On December 8, 2016, a Long-Term Water Redundancy Forum hosted by the MWRA Advisory Board for the customer communities was held at Boston College. MWRA staff presented the history of the MWRA waterworks system, the need for Metropolitan Tunnel redundancy and the challenges, both implementation and financial, of building redundancy.

On January 19, 2017, the MWRA Advisory Board met and voted to support moving forward with a deep-rock, two-tunnel project. They voted also to recommend: a program management division approach to manage the program similar to the model used for the BHP; concurrent construction of both tunnels rather than a phased approach; and allocation of any revenue from non-typical or one-time water users (e.g., emergency drought connections) towards the cost of the program.

On February 15, 2017, the MWRA Board of Directors approved the preferred alternative of construction of northern and southern deep rock tunnels from the Hultman Aqueduct and MetroWest Water Supply Tunnel to the Weston Aqueduct Supply Main 3 (WASM3) and to the Southern Spine water mains for the purpose of providing redundancy for the Metropolitan Tunnel System (City Tunnel, City Tunnel Extension and Dorchester Tunnel), and directed staff to proceed with preliminary design, geotechnical investigations and Massachusetts Environmental Policy Act (MEPA) review of the project.

In June 2018, the MWRA Board of Directors approved the fiscal year 2019 Capital Improvement Program, which includes $1.4 billion (2019 dollars) for the tunnel redundancy program.

PROJECT GOALS

The Metropolitan Tunnel Redundancy Program was conceived to address several outstanding challenges, most notably the fact that the existing Metropolitan Tunnel System cannot be maintained or repaired nor can an emergency be readily addressed because shut down of the system is not currently possible without imposing a boil-water order.

The first and foremost goal of the program is an operational goal; to protect public health, provide sanitation, and provide fire protection. The MWRA exists to provide these services. In support of this overall goal, the Tunnel Redundancy Program is intended to:

- Provide full redundancy for the Metropolitan Tunnel System:
    - Provide normal water service and fire protection when the existing tunnel system is out of service

- Provide the ability to perform maintenance on existing tunnels year-round
- Provide uninterrupted service in the event of an emergency shut down
- Meet high day demand flow with no seasonal restrictions
- Avoid activation of emergency reservoirs
- Meet customer expectations for excellent water quality
- Preserve sustainable and predictable rates at the water utility level
- Minimize cost of borrowing
- Be constructible
- Result in no future boil water orders!

The selected tunnel alternative is expected to meet all of these goals.

PROJECT COSTS AND FINANCING

The cost of the Metropolitan Tunnel Redundancy Program is being allocated in the MWRA's Capital Improvement Program (CIP) with the goals of: preserving sustainable and predictable rates, ensuring adequate capital is available when necessary; and minimizing the cost of borrowing. Since 1985, MWRA has spent approximately $8.4 billion to upgrade the wastewater and water systems. The majority of these improvements were funded through the issuance of tax-exempt bonds. The MWRA is projected to reach the peak of its debt service payments in fiscal 2022 (refer to Figure 10), which provides an opportunity to mitigate water rate impacts of financing the proposed tunnel program.

MWRA's uses a multi-year rate management strategy to provide sustainable and predictable assessments to our communities. The impact on the CIP and the debt service on the Current Expense Budget (CEB) were evaluated for a variety of options for the Metropolitan Tunnel Redundancy Program. The options evaluated ranged from "do nothing" to the most expensive tunnel option.

MWRA communities are either combined water and sewer users, only sewer users or only water users. The projected average annual increase on the combined water and sewer assessment of the preferred alternative is 1.3%. The projected average annual increase on the water-only assessment of the preferred alternative is 4.0%.

The rate impacts of the preferred option on both the combined and water-only assessments are within the MWRA's long-term rates management strategy. The preferred option is both consistent with the Authority's core mission of providing reliable, cost-effective and high quality water, and its goal of providing sustainable and predictable assessments.

PROJECT OUTLOOK

The Metropolitan Tunnel Redundancy Program is currently at the very early stages of planning and design. The organizational framework to manage the program within the MWRA is in place in the form of the Tunnel Redundancy Department. Procurement of initial consultant contracts for Program Support Services and Preliminary Engineering are underway.

It is expected that the next several years will include a number of program wide activities including risk management planning, quality management planning, health and safety planning, design criteria and standardization, document management and project controls, work breakdown planning, procurement planning, construction package planning, field investigation procedures, rock core storage, critical path scheduling, and budget planning and management.

The Preliminary Design Phase of the program will involve significant efforts on geotechnical investigations, preliminary route and shaft site alternative evaluations, preliminary design, an assessment of environmental permits needed and preparation of the Massachusetts Environmental Policy Act (MEPA) review for the project. This phase of the project will initiate actual design. Preliminary design is anticipated to be complete by 2023. It is envisioned that final design(s) will follow on the heels of preliminary design with the first tunnel construction package issued in or around 2027.

REFERENCES

Tierney, F.L., Billings, M.P. and Cassidy, M.M. (1968). *Geology of the City Tunnel, Greater Boston, Massachusetts,* Jour. Boston Soc. Civil Eng., V. 55, pp. 60–96.

Richardson, S.M. 1977. *Geology of the Dorchester Tunnel, Greater Boston, Massachusetts,* Jour. Boston Soc. Civil Eng., V.63. pp. 247–269.

Unique Design Challenges of the Central Bayside System Improvement Project Tunnel Connections and Shafts

Nick Goodenow ▪ Stantec Consulting
Mike Bruen ▪ Stantec Consulting
Steve Robinson ▪ Stantec Consulting
Michael Deutscher ▪ Jacobs Engineering
Manfred Wong ▪ San Francisco Public Utilities Commission

ABSTRACT

The Central Bayside System Improvement Project is a critical element of San Francisco Public Utilities Commission's Sewer System Improvement Program, to upgrade its aging, seismically vulnerable combined sewer system. The project is comprised of small and large diameter conveyance tunnels, two deep large diameter shafts, two deep smaller diameter shafts, and several complex and challenging soil ground tunnel connections within a network of existing infrastructure. The key design challenges include tunnel connections mined through, under, and around existing pile foundations and utilities located within San Francisco's Young Bay Mud and alluvial deposits. Several unique and innovative design solutions are described to address timber and steel pile obstructions, protection of sensitive utilities in a highly vulnerable seismic area.

INTRODUCTION

The Central Bayside System Improvement Project (CBSIP), located on eastern side of the San Francisco Peninsula, is a major component of the San Francisco Public Utilities Commission's (SFPUC's) 20-year, city-wide $6.9 billion Sewer System Improvement Program (SSIP). The project's primary objective is to provide seismic resiliency, system reliability, and operational flexibility not currently provided by the existing Channel Force Main (CHFM), a vital link in the Bayside wastewater system, shown in Figure 1, has ruptured on multiple occasions since 1981.

The existing Channel Pump Station (CHS) currently conveys flow to the Southeast Water Pollution Control Plant (SEP) for treatment via an existing 66-inch OD CHFM. During dry weather, the CHFM conveys approximately two-thirds of the Bayside's total flows of approximately 35 million gallons per day (MGD). With the Bayside collection and treatment system in continuous operation, there is no alternate means to bring flows from the northeast watersheds (North Shore and Channel) to the Southeast Treatment Plant.

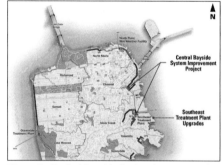

Figure 1. Central Bayside System Improvement Project (CBSIP) components

The five main components of the CBSIP are as follows:

- The Northern Connection, between the existing CHS and a new drop structure at the proposed Channel Tunnel (CHTL) Retrieval Shaft, includes two short tunnels (a hand-mined section and a bored section), an open-cut section, a new gate shaft, and the Berry Street Drop Shaft.
- The Southern Connection, which includes two new force mains to convey dry weather (DW) and wet weather (WW) flows from the proposed Central Bayside Pump Station (CBS) to the new Southeast Treatment Plant Headworks.
- The Intertie Connection, which consists of a connection between the existing Islais Creek Transport/Storage Facilities (ICT) and the CHTL via a drop shaft and hand-mined adit connection. The connection will provide flexibility in operating the two facilities.
- The gravity-flow Channel tunnel is a proposed 1.7 mile long 24 foot internal diameter tunnel located 110 to 150 feet below ground surface, designed to convey dry weather and wet weather flows from the existing CHS via the North Connections to the CHTL Receiving Shaft at the Mission Bay Parcel P7 site, through the CHTL, to the CBS, which will be constructed in the CHTL Launch Shaft following completion of tunneling.
- The 120 MGD capacity CBS will be used to lift both DW and WW flows from the Channel tunnel in twin pressurized pipelines to the new SEP headworks.

The major components of the CBSIP are shown in Figure 2.

With San Francisco's urban environment, potentially unfavorable geologic conditions, known and unknown obstructions, and a complex and a dense network of existing infrastructure to work around and/or avoid impacting during construction, all of which present constructability challenges that will need to be addressed in design and in the approach to construction.

This paper focuses primarily on the Northern Connection, Southern Connection and ICT Intertie Connection. For discussion of the CHTL, including the CHTL Launch and Retrieval Shafts, see Deutscher et al. (2018).

Geology and Geotechnical Information

The CBSIP is located within the Coast Ranges geomorphic province, with northwesterly trending ridges and valleys. Rugged hills consisting of Jurassic- to Cretaceous-aged rock are juxtaposed with low, flat-lying areas underlain by Quaternary sedimentary deposits. Bedrock consists of consolidated rocks of the Franciscan Complex.

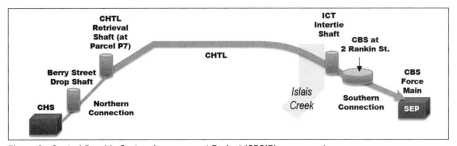

Figure 2. Central Bayside System Improvement Project (CBSIP) components

In the San Francisco Peninsula area, Quaternary sediments lie on the eroded bedrock surface and consist of an alternating sequence of terrestrial and estuarine deposits that reflect major sea-level fluctuations. Low stands of sea level are recorded by the deposition of terrestrial sediments, whereas estuarine sediments such as the Old Bay Clay (OBC) and the Young Bay Mud (YBM) were deposited during high sea-level stands (Sloan, 1992).

Geology Along the Northern Connection

The soil stratigraphy for Northern Connection between the CHS and the Berry Street Drop Shaft were found to be comprised of:

- 0 to 37 feet depth: Fill (predominantly loose to medium dense silty sand with construction debris).
- 37 to 69 feet depth: Bayside Sand, very dense poorly graded sand with some silty sand varying to Young Bay Mud, a very soft to soft highly plastic clay.
- 69 feet depth to termination of explorations at 121.5 feet depth: Old Bay Clay, very stiff clay, low to high plasticity.

Locally, depth of the Old Bay Clay can vary from approximately 65 to 75 feet below ground surface. In some locations the Young Bay Mud transitions directly into Old Bay Clay, while in other locations the Old Bay Clay is interlayered with the Bayside Sand as shown in Figure 3. The near surface gravity pipeline for dry weather flow is anticipated to be constructed in the Fill and Young Bay Mud while the deep tunnel connection between the Berry Street to CHTL Retrieval Shafts is anticipated to be constructed in the Old Bay Clay.

Geology Along the Southern Connection

The generalized geologic profile for the Southern Connection is presented in Figure 4. The force mains comprising the Southern Connection are expected to be constructed within the Young Bay Mud (very soft to soft highly plastic clay). Launching and receiving pits will require excavation of the overlying fill, but they are not expected to extend

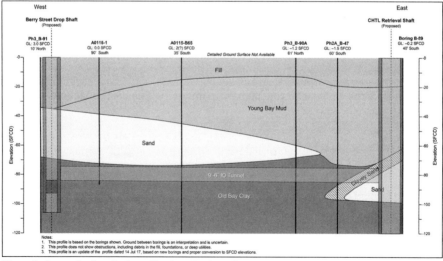

Figure 3. Generalized geologic profile along Northern Connection

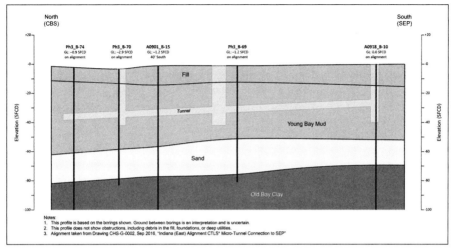

Figure 4. Generalized geologic profile along Southern Connection

through the Young Bay Mud into to the underlying Bayside Sand. Fill at this location is approximately 10 feet thick and consists of silty sand with gravel and cobbles.

Geology at the ICT Intertie Connection

At the ICT connection to the CHTL, the anticipated soil stratigraphy at the ICT connection to the CHTL consists of:

- 0 to 7 feet depth: Fill.
- 7 to 58 feet depth: Young Bay Mud (very soft to stiff highly plastic clay, consistency increasing with depth).
- 58 to 143 feet depth: alternating layers of Bayside Sand (medium-dense medium grained sand with silt and gravel) and Old Bay Clay (stiff highly plastic clay).
- 143 to 153 feet depth: Granular Undifferentiated Sediments (very dense silty sand and gravel).
- 153 to 187 feet depth: Cohesive Undifferentiated Sediments (very stiff sandy clay or sandy silt).
- 187 to 195 feet depth: Residual soil (very stiff to hard sandy clay and gravel).
- Below 195 feet depth: Bedrock (serpentinite/mélange, moderately to completely weathered, part of the Franciscan Complex).

The ICT shaft excavation will encounter Fill, Young Bay Mud, Bayside Sand and Old Bay Clay layers. The open-cut connection to the ICT will be made in Fill and Young Bay Mud, while the adit connection from the base of the ICT shaft to the CHTL is anticipated to be in Old Bay Clay with possible zones of Bayside Sand.

Northern Connection Constructability Challenges

The new Northern Connection structures will require the design and construction of near surface DW and WW pipeline, connections from the CHS to the Berry Street Drop Shaft, and relocation of the 60-inch gravity sewer. Deeper excavations will include the Berry Street Drop Shaft, and the 9.5-internal diameter tunnel connection from the

Berry Street Drop Shaft to the CHTL Retrieval Shaft. The Northern Connection design and construction is very complex, due to the extensive array of existing infrastructure and ground conditions, and all components are directly linked to one another, as well as to the wider CBSIP as a whole. To more efficiently describe the constructability considerations, the following sections break down the Northern Connection into finite components, working upstream to downstream and detailing the implications and constraints associated with component interaction.

Dry Weather Connection

The Dry Weather (DW) flow is designed to be intercepted at the CHS influent channel and diverted into a 48-inch OD gravity fiberglass reinforced polymer mortar pipe (FRPMP) which will cross under the Division Street Outfall (DVO) structure. An access hatch at the ground surface will allow access into the gate structure for inspection or maintenance. The access structure invert at the junction with the CHS influent channel is planned at elevation -35.5 feet. DW flow will be conveyed by gravity in the 48-inch ID FRP to the new Berry Street Drop Shaft, shown in Figure 5, and descend through the shaft using a vortex drop structure, shown in Figure 5. The flow will then be combined with WW flows and conveyed under the southwest corner of Mission Creek and Interstate 280 (I-280) through a 9.5-foot connector tunnel (the 9.5-foot Tunnel) to the CHTL Retrieval Shaft where it will drop once more using a vortex drop structure in the CHTL Retrieval Shaft through the CHTL.

The open-cut excavations for launching and retrieval of the trenchless portion between the CHS and the Berry Street Drop Shaft will likely be excavated first to facilitate ground improvement and pilot tube probing beneath the Division Street Sewer to determine a clear path and to confirm that the existing piles supporting the sewer will not obstruct the tunnel excavation.

The proposed DW Connection (48-inch ID pipe) is designed to "thread the needle" through a small corridor under the existing 50-ft wide by 35-ft deep Division Street Sewer's existing, which is supported by five rows of concrete-filled steel piles which are on a 6-foot spacing crossing between piles in the same row. The DW connection

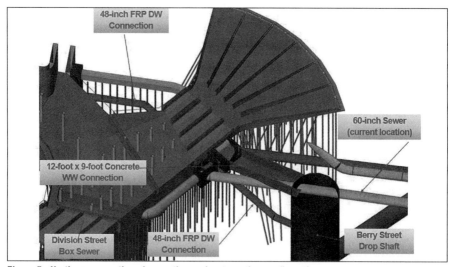

Figure 5. Northern connection: dry weather and wet weather configuration

alignment. The DW connection alignment crosses all five rows in a perpendicular orientation. Pilot tube probing is planned to be performed by the Contractor to determine if a path between the piles is wide and straight enough to allow placement of the pipe under the Division Street Sewer. Based on the pile locations and dimensions indicated in the as-built record drawings, a maximum corridor of 58 inches in width is available for installation of the DW connection, assuming the piles are installed in the location shown in the record drawings and assuming they achieve a true vertical orientation.

In order to clear the constraints described, the DW connection will comprise a nominal 48-inch internal diameter (ID) FRP, which has an outside diameter of approximately 51 inches (OD). This allows for a clearance of approximately 3.5 inches on either side of the outside of the single pass FRP without a steel casing. The actual clearance is reduced further when accounting for the overcut from the excavation equipment, which is typically one inch radially for this size of pipe. The actual clearance is reduced to 2.5 inches during installation of the new pipeline.

With the tight tolerances for the 48-inch ID tunnel, the likely installation would involve a series of pilot tubes being pushed within the proposed excavated tunnel outside diameter (48-inch ID). The pilot tubes would act as highly accurate four to five-inch probes which would detect obstructions such as piles or confirm a clear path for tunneling. The use of an open shield pipe jacking method could be used in conjunction with ground improvement under the Division Street Sewer requiring hand excavation or potentially a closed face machine to balance the face pressures acting on the tunnel. If potential obstructions are encountered at the tunnel alignment depth of 45 ft during the pilot tube probing, alternative installations have been considered such as such as moving the crossing over one pile bay or bifurcating the flow to reduce the diameter of the pipe and excavated diameter, in which case alternative trenchless methods would be used.

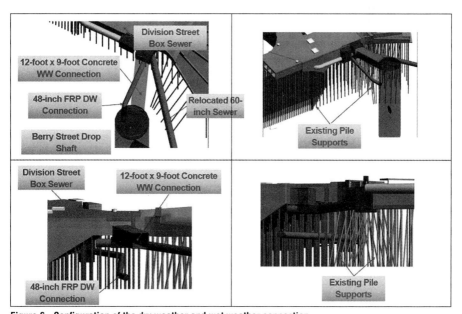

Figure 6. Configuration of the dry weather and wet weather connection

Ground improvement methods such as ground freezing or jet grouting are considered in advance of tunneling beneath the Division Street Sewer to stabilize the ground, thereby enabling tunneling with using hand mining techniques such as hand spades, shovels, and small muck boxes at the heading under the cover of an open-face shield. Alternatively, a pressurized-face Microtunnel Boring Machine (MTBM) could be used if there is confidence in the pilot tube probing operation.

Ground freezing would provide a temporary solution to pile removal if encountered during open shield pipe jacking operations. If a pile is removed, the permanent lateral and axial load resisting components must be replaced. Inclined jet grouting would be used to replace the structural components needed for the pile that was cut.

Potentially liquefiable soils as well as the consolidation of the Young Bay Mud requires several pile supported structures for the Northern Connection. The sequencing of piles to support the new control gate structure and DW pipeline is difficult due to the existing pipeline connection at this location as well as existing timber piles to support the pipeline. New pile supports for the new open cut portion of the 48-inch ID DW sewer will be installed to mitigate future settlement of the new and existing infrastructure.

Wet Weather Connection

The WW Connection will convey flow through a pile-supported reinforced concrete box culvert section constructed between the existing Division Street Sewer and the Berry Street Drop Shaft. The WW Connection, shown in Figure 6, will be constructed by open-cut means and will require breaking into and modifying the existing Division Street Sewer.

Relocation of an existing 60-inch-diameter RCP sewer will be incorporated into the WW diversion structure. The relocation of this sewer is necessary to facilitate construction of the Berry Street Drop Shaft.

Wet Weather (WW) flow (a combination of DW flow and stormwater) is planned to be diverted through a 10-foot-tall-by-12-foot-wide junction structure constructed adjacent to the Division Street Sewer. A 12-foot-wide-by-6-foot-tall opening will be cut into the side of the Division Street Sewer at an invert elevation of -11.0 feet. WW flow will be conveyed through a 12-foot-wide-by-9-foot-tall pile supported box culvert to the Berry Street Drop Shaft, where it will descend through the drop structure and be combined with flow from the DW Connection. Figure 7 depict the DW and WW connections as well as a 60-inch gravity sewer requiring relocation. Numerous pile supports are needed for the existing structures because of the differential movements and increased stresses imposed on the structures from the consolidation of the Young Bay Mud and potential liquefaction mitigation measures.

The WW Connection will be constructed directly above the DW Connection for approximately half of its alignment, with the WW Connection directly above the WW Connection for approximately 35 feet. The WW Connection box will be constructed on pile caps, each supported by a pair of piles. The longitudinal pile cap spacing is anticipated at approximately 8 feet on center. The pile configuration will be modified at the location where the DW Connection exits the WW Connection trench to provide sufficient clearance between the piles and the DW Connection. In order to keep the existing 66-inch CHFM in service during construction and avoid damage, the piles will not be driven within 5 feet of the CHFM. Ground improvement measures are being considered to provide underpinning support to the CHFM, to increase the bearing capacity and resistance to consolidation of the Young Bay Mud and reduce

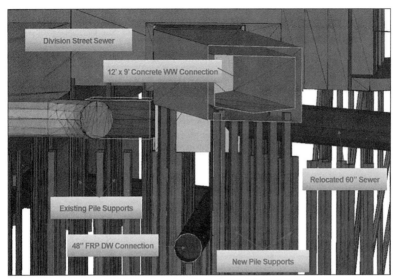

Figure 7. Wet weather connection detail

the liquefaction potential in the high sand content areas near Mission Creek for the relocated 60-inch gravity sewer.

Existing 60-Inch Sewer Relocation

Relocation of the existing gravity 60-inch sewer is required because of conflicts with the Berry Street Drop Shaft and the new WW Connection to the Division Street Sewer. The relocation of the 60-inch sewer will require pile supports and is located at the margin of Mission Creek, a hydraulically connected body of water with the San Francisco Bay.

The existing Channel Force Main (CHFM) will likely require underpinning or some other support where the trench for the relocated 60-inch sewer crosses under the CHFM. The trench will require backfill with controlled density fill to prevent settlement under the CHFM after completion. It will also necessitate underpinning of an existing thrust block located immediately adjacent to the existing Channel Transport/Storage Box where the closure details will be installed.

Berry Street Shaft

The Berry Street Drop Shaft, shown in Figure 8, has been developed as 24-foot ID circular shaft constructed with slurry walls for initial support of excavation. The diameter is driven by a combination of functions, namely:

1. Retrieval of the tunnel boring machine (TBM) for the 9.5-foot Tunnel;
2. Permanent maintenance access from ground surface;
3. Permanent DW drop assembly; and
4. Permanent WW drop assembly.

The drop structure has been designed to incorporate a vortex assembly for DW flows to minimize odor release. At the point of connection to the Berry Street Drop Shaft, a

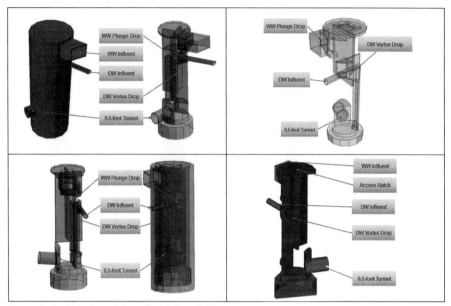

Figure 8. Configuration of Berry Street drop shaft

vertical 6-foot-radius bend will be constructed within the final liner, to transition WW flow into the drop shaft. The drop structure also includes a WW plunge drop arrangement and an access shaft sized to accommodate a 6-foot-by-10-foot clear opening for operation and maintenance (O&M) access.

9.5-Foot Diameter Connector Tunnel

The 9.5-foot (ID) tunnel is planned be installed using an Earth Pressure Balance Tunnel Boring Machine (EPB TBM) including the installation of either a precast segmental lining or a FRP pipe (i.e., pipe jacking). The selection of an EPB TBM is primarily because of three locations along the tunnel alignment where historic abandoned timber piles are expected to occur, however verification of the pile depth has not been confirmed. The EPB TBM's ability to excavate through timber piles or perform an intervention at 85-ft depth under Mission Creek is considered to be an advantage over a slurry MTBMs.

The horizontal alignment of the 9.5-foot Tunnel was selected to cross between existing structural bents for the elevated I-280 freeway and the locations of the shafts, which were selected with consideration for the required internal geometry of the Berry Street Drop Shaft and the CHTL Retrieval Shaft at Parcel P7. The locations of these two shafts are driven by a number of factors, namely constructability and real estate considerations.

The vertical alignment of the 9.5-foot tunnel was determined based on the geologic conditions in the area (i.e., avoiding the Young Bay Mud), the need to avoid obstacles (existing timber piles, etc), and the ability to connect to the CHTL Retrieval Shaft at Parcel P7 at depth. The elevation of the CHTL invert is much lower than the connection of the 9.5-foot diameter tunnel connecting to the CHTL Retrieval Shaft.

Historic timber piles exist in this area, especially near the edge of Mission Creek. These are not expected to extend into the tunnel zone, but as a contingency the tunneling method must accommodate potential excavation or removal of the timber piles.

Several locations exist along the planned 9.5-foot tunnel alignment where historic piles have been driven as a means of support for utilities, as a slope retaining structure, or to support wharves. Regardless of their intended use, a number of abandoned piles are anticipated to a depth of approximately 70 feet below ground surface near the edge of Mission Creek. The piles are not anticipated to extend to the tunnel zone, but the tunnel methods selected must accommodate potential excavation or removal of the timber piles. It will be important to specify a TBM to provide face access to facilitate removal of any timber piles encountered.

Southern Connection

The Southern Connection, comprising the twin 48-inch OD Central Bayside Force Main (CBFM) pipes, includes a number of complex interfaces with existing and proposed structures. These interfaces have largely driven the design of the alignments. The Southern Connection alignment connects the CBS to the SEP—Headworks on Evans Avenue and includes an 11-ft diameter pipeline within the 2 Rankin Street site, twin pipelines under the ICT transport box that connect to the channel force mains at the SEP. In addition, the intertie connection includes an upper connection to the ICT transport box, the intertie drop shaft and a lower connection adit to connect to the CHTL.

The upstream connection point of the CBFM pipes is the connection to the CBS. The location of this connection is set by the exit point of the force main pipes leaving the CBS at approximately 35 feet below the ground surface. A tunneled connection is currently included in the design to connect the CBFM into the CBS because of unbalanced loading on the initial support of the CBS structure. Currently, an 11-foot diameter casing is envisioned to will allow workers to access to the face and fit the two 48-inch OD force main pipes.

Figure 9 shows the dual force main pipes spacing within the 11-foot-diameter casing in plan while Figure 10 shows the arrangement in cross section. The 11-foot casing will only be used for force main installation on the 2 Rankin Street site. Dual 60-inch casings will be used for installation under the ICT and along the remaining alignment to Evans Avenue.

11-Ft Tunnel

An attempt was made to minimize conflicts between the tunnel alignment and existing pile assemblies at the 2 Rankin Street site. However, it is estimated that four abandoned pile support systems from a building that was demolished will be encountered and will require removal or cutting out the portion of the timber pile lengths encountered on the 2 Rankin Street site. An open-shield, jacked, large diameter steel casing, currently planned as 11-foot ID, will be used to allow worker access to the face of the tunnel while the casing is being advanced to readily cut and remove the timber piles at the face.

The purpose of the large diameter steel casing shown in Figure 10 is to provide adequate access to the tunnel face to remove approximately 48 timber piles from within the tunnel alignment between Rankin Street and the CBS. Prior to excavation and installation of the 11-foot-diameter steel casing, the intent is to have ground improvement

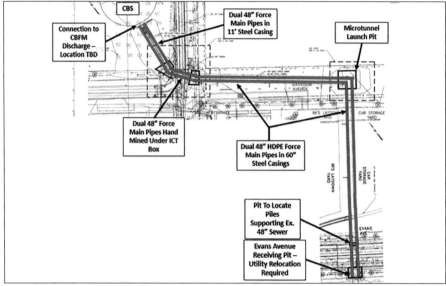

Figure 9. Southern connection overview

(e.g., jet grouting) will likely be required in advance of tunneling to maintain heading stability in the Young Bay Mud in front of the tunnel. The casing will be launched from a launching pit immediately west of the ICT in Rankin Street. Muck removal and pipe insertion will be done from this pit. Penetration of the CBS wall will only include the two force main pipes, the casing is planned to terminate external to the CBS shaft.

ICT Box Crossing

At the location of the existing ICT box structure, the tunneling methodology for installation of the twin 48-in diameter

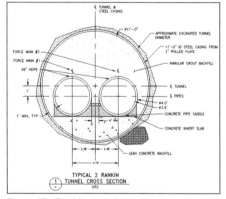

Figure 10. Dual force main spacing

CBFM pipes switches from a single mined tunnel, performed within an 11-ft diameter casing, to twin mined tunnels, each performed within a 60-in diameter casing. A pit is planned on either side of the ICT Box Structure, which is supported on 14-inch timber piles, to enable probing for (pile) obstructions in advance of tunneling. Record drawings indicate sufficient clear space exists between the piles to accommodate the twin tunnels.

Crossing Evans Ave.

After clearing the existing ICT box structure, the tunneling methodology for the twin 48-in diameter CBFM pipes switches to microtunneling. The microtunnel under Evans Avenue requires a small pit in Evans Avenue for access to the planned 48-inch ID sewer and to locate the existing piles supporting the sewer; alternatively, pilot tube probing from within the planned reception pit can be performed to confirm that obstructions such as piles are not located within the planned microtunnel alignment.

Pile location information is needed to align the microtunnel so that it passes between existing support piles.

The microtunnel drive lengths are approximately 270 lineal feet along Davidson Avenue and 305 lineal feet from Davidson Avenue, across Evans Avenue to the SEP—Headworks retrieval pit. After microtunneling is completed, contact grouting is expected to occur immediately after completion to mitigate the over-excavated area around the pipe casing and over-excavated material at the MTBM face.

ICT Intertie Connection Adit and Drop Shaft

The ICT Intertie Connection is designed to provide maintenance access to the CHTL and to hydraulically balance the Bayside's northeast and the east central watersheds to allow a system wide approach to permitted Combined Sewer Discharge (CSD) control. This will allow flexibility in handling localized rainfall events to ensure that all City storage capacity is utilized prior to any CSD.

Figure 11 shows the layout of the ICT Intertie Connection relative to the existing ICT and the proposed CHTL. The ICT Intertie Connection typically comprises an 8-foot-tall-by-8-foot-wide box connecting the ICT to a drop shaft, a 21-foot ID, 140-foot deep ICT Intertie Drop Shaft and an 8-foot-tall-by-8-foot-wide mined adit connecting to the CHTL.

The 8-foot × 8-foot connecting box and mined adit have been sized to meet the hydraulic needs of the intertie. Likewise, the 21-foot ID final liner is sufficiently sized to allow the installation of a plunge drop assembly, a flow diverter wall and a dedicated access shaft with a 7-foot × 12-foot clear opening access cover at grade. The 21-foot ID Intertie Drop Shaft will be constructed within a 27-foot ID slurry diaphragm wall utilizing a 3-foot thick final liner. The 27-foot ID slurry diaphragm wall was determined to be constructible given the geologic and geometric constraints at the site. The 21-foot ID final liner is sufficiently sized to allow the installation of a plunge drop assembly, a flow diverter wall and a dedicated access shaft with a 7-foot × 12-foot clear opening access cover at grade.

The Intertie Drop Shaft location has also been identified to minimize the need for, and impact of, removal of existing pile assemblies. The slurry wall comprising shaft shoring will be located so as to sit between existing pile assemblies and allow the initial support to be installed without the need for any pile cap or pile removal. One pile cap and

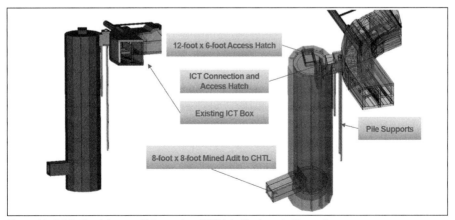

Figure 11. Islais Creek transport/storage box (ICT) intertie drop shaft overview

an assumed 12 timber piles will require removal as part of the shaft excavation. The exact location of the pile cap and piles will need to be identified during future design stages. Since the piles are not planned to be removed until after the installation of the diaphragm slurry walls, it is important that the wall excavation does not encounter any existing timber piles.

The intertie connection to the ICT, shown in Figure 11, is pile-supported. The intertie connection includes a short conveyance box section and shaft between the ICT and the Intertie Drop Shaft. An access hatch will be included in the roof of the Intertie Drop Shaft for access to the connection. The ICT Intertie Shaft will include a weir connection to regulate flows from the ICT.

The connection to the ICT is 30 feet deep and can be installed by open cut methods. The ICT connection structure will be cast-in-place on concrete piles directly adjacent to the ICT. The 8-foot by 8-foot connection will be about 10 feet long and essentially be a cast-in-place continuation of the connection structure to the Intertie Drop Shaft. This work would be completed after the installation of the Intertie Drop Shaft.

The lower connection is designed as a reinforced concrete adit, planned to be installed in a mined tunnel from the Intertie Drop Shaft with a breakthrough into the CHTL. Ground improvement is required along its 60-foot length of tunnel, approximately 150 feet below ground surface. Additional ground improvement will be required at the break out of the shaft and at the CHTL connection.

Contract Packaging and Project Schedule

The project is likely to be procured under three packages: the main tunnel (CHTL), the connections and the build-out for the CBS. The CHTL and connection elements of the project will be delivered using traditional design-bid-build procurement methods. Various alternative delivery procurement methods are under consideration for the build-out of the CBS. The project has been developed to the 35% design level with final (100%) design of the CHTL and associated shafts and connections anticipated to conclude by the end of 2020. Design of the CBS permanent pump station build-out is likely to continue beyond this date. The first construction package of the CHTL and associated Launch and the Retrieval Shafts are expected to begin in mid-2021. The Connection contract will lag the CHTL contract with connections to the CBS and CHTL Retrieval shafts coordinated with the CHTL contractor possibly in 2023. Start of the CBS build-out contract will follow the completion of the CHTL contract.

REFERENCE

Deutscher, M, Sadek, S. Wong, M., 2018, Design Challenges of the San Francisco Public Utilities Commission's Channel Tunnel, Proceedings of the 2018 North American Tunneling Conference, Society of Mining, Metallurgy & Exploration, Inc. pp. 503–515.

PART

Geotechnical Considerations

Chairs

David Watson
Mott MacDonald

Steve Lottie
FKCI

Jonathan Steflik
Black & Veatch

Jaidev "Jay" Sankar
HNTB

Anacostia River Tunnel Project Groundwater Drawdown Effects from the Deep Well Depressurization System at the CSO-019 Drop Shaft

E. Gregory McNulty ▪ Parsons Corporation
Pooyan Asadollahi ▪ Parsons Corporation

ABSTRACT

The Anacostia River Tunnel with a length of 12,300 feet (2.33 miles) and internal diameter of 23 feet will be excavated using an Earth Pressure Balance Tunnel Boring Machine (EPBM) in the hard/stiff clays and silts/sands of the Potomac Formation. This paper presents the design of a depressurization system at the CSO-019 Drop Shaft and shows that the use of laboratory derived properties will result in excessive estimates of settlement. This paper shows that more reasonable settlements can be obtained with pressuremeter data; furthermore, when the predicted settlements are compared to field measurements, it is found that the rapid vertical movement of the ground monitoring points in response to depressurization changes come from settlement or rebound in the aquifer, not the aquitard.

INTRODUCTION

This paper describes the effects of sampling disturbance on analyses to evaluate the settlement produced during pumping from the deep well depressurization system at the CSO-019 Drop Shaft and to estimate the time required for recovery once pumping has stopped. These results show that despite the large zone of influence produced, negligible settlement has occurred to bridges, roads, and other structures that are important to the District Department of Transportation (DDOT) and that the recovery time of the groundwater table may be up to several months once all pumping at CSO-019 and surrounding areas has stopped.

The deep well depressurization system at the CSO-019 Drop Shaft has operated for nearly two years and produced a zone of influence greater than the one initially estimated in part because the parameters used to estimate the zone of influence in the Patuxent (PTX) have been derived from pump tests conducted over very short time (about 2 days). Furthermore, properties derived from laboratory testing of stiff clay of the Patuxent/Arundel (P/A) aquitard have been heavily influenced by sample disturbance. Use of time drawdown data has shown that the storage coefficient for the PTX aquifer is much lower than that previously used. Finally, recovery of the water table has been delayed because of the apparent continued pumping at nearby ART wells, and possibly because of increasing groundwater withdrawal at facilities such as Fort Meade for cooling several new large data centers.

The immediate practical effect of a lower storativity also means that there is less settlement associated with the drawdown of the water table. Consequently, the straightforward application of boring, sampling, and laboratory testing programs can lead to an inadequate understanding of aquifer properties and the settlement induced by drawdown of the groundwater table. In the Sections that follow, we show how empirical relationships and review of the geologic literature can reveal problems with sampling disturbance and laboratory testing, and how modeling of the aquifer system can derive accurate in situ aquifer properties of the PTX aquifer through the matching of drawdown water level data with total heads predicted with a finite element groundwater model. Finally, we find that because of the extremely low hydraulic conductivity of the P/A aquitard and higher conductivity of the PTX aquifer, comparisons of vertical movements from surface points at CSO-019 suggest that most of the movement comes from changes in the water table within the PTX, not the P/A Formation.

COMMON INVESTIGATION PROCEDURES AND THEIR RELIABILITY IN DESIGN AND CONSTRUCTION

In 1975, the second Nabor Carrillo Lecture, Dr. Ralph Peck addressed "the practitioners who have accustomed to following, conscientiously but undeviatingly, what has become the standard routine procedure." His purpose was to show geotechnical engineers how parameters obtained by these procedures may be wrong because of biasing from sample disturbance and to suggest procedures to arriving at more satisfactory parameters. For example, one may estimate the compressibility parameter for undisturbed soil using soil properties such as water content, liquid limit and void ratio (Peck, 1984). This correlation can also be useful for checking of quality control of the soil sample.

Figure 1 presents the effects of disturbance on shear strength from use of the standard Shelby tube in comparison with other methods after Tanaka (2000).

Consequently, on some projects, the straightforward application of geotechnical investigations is inadequate to lead to a proper design. This shortcoming must be recognized before we can obtain appropriate properties by using the example of ART CSO-019. For the CSO-019 zone of influence and settlement due to consolidation of the P/A aquiclude during dewatering, this paper shows how the parameters can be determined with greater reliability on the basis that includes no soil borings at all when all available information is used, including geologic reasoning and evidence and information on the behavior of existing structures can lead to less costly geotechnical

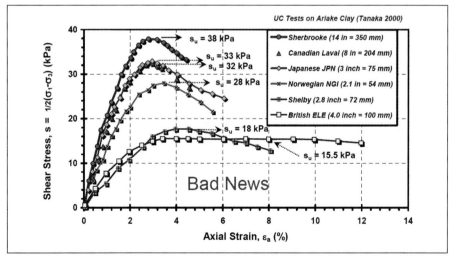

Figure 1. Sample disturbance effects on Su (and other properties) can result in a 60-percent reduction when Shelby tubes are used (Tanaka, 2000; Mayne et al., 2009).

solutions. Index tests help compare data from laboratory incremental load and constant rate of strain consolidation tests with those from field tests and the geological literature. The use of this approach can correct for large errors caused by disturbance in measured soil compressibility and hydraulic conductivity and reduce the time and cost of conducting geotechnical investigations in the future.

As demonstrated by Peck (1984), more accurate observed soil stiffness can be observed through use of empirical procedures using the liquid limit rather than using laboratory consolidation tests on soil specimens disturbed by drilling, sampling with a Shelby tube, extruded, and trimmed in the laboratory to determine Cc. For example, disturbance typically reduces the value of Cc and the preconsolidation stress and increases the compressibility of the soil specimen before it is tested. Furthermore, the plotting of consolidation data with void ratio versus log of vertical-effective stress, p', can quickly show the discrepancy between the initial void ratio in-situ and the resulting e-log-p' curve. In our case, the consolidation tests were plotted only as vertical strain versus log p'; consequently, we will rely simply on the Terzaghi and Peck (1967) relationship for the compression index, Cc:

$C_c = 0.009 (w_L - 10)$,

where, w_L represents the water content at the liquid limit, in percent.

Thus, for the soils in the P/A layer, we have measured w_L of about 68 percent yields a C_c of about 0.53 while the same soil yields a C_c laboratory of about 0.17; therefore, as Peck (1984) concluded, because the higher C_c is associated with a higher preconsolidation stress and a stiffer sample, the lower C_c for the laboratory indicates that a large amount disturbance was incurred during sampling, extrusion, and specimen preparation and that the derived values of stiffness should be expected to overpredict settlement. In addition, the hydraulic conductivities derived from the laboratory consolidation test c_v would also be non-representative

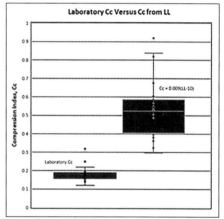

Figure 2. Compression indices measured in laboratory consolidation tests versus those derived from the Terzaghi and Peck (1967) empirical expression for liquid limit

on in-situ conductivities. The next section presents the derivation of the stiffness and conductivity by procedures that are more accurate because they avoid the biasing from sample disturbance.

DERIVATION OF APPROPRIATE PARAMETERS FROM THE CSO-019 DEWATERING EFFORT

This section will compare two different approaches for deriving the hydraulic conductivity and compressibility from a:

1. combination of laboratory and short-term pump tests; and
2. calibration of a groundwater model with long-term water levels.

The next few paragraphs therefore discuss the hydrogeology of the site, the setup of the deep well depressurization system, and how the hydraulic conductivity and compressibility properties are derived from the consolidation versus those properties derived from the matching of drawdowns at P2-17A deep well.

GEOLOGIC AND HYDROGEOLOGIC CONDITIONS

Based on data provided in the project Geotechnical Data Report (GDR) and Geotechnical Baseline Report (GBR), the CSO-019 Drop Shaft site is underlain by the following subsurface units:

- Holocene Artificial Fill
- Quaternary Alluvial Fill
- Cretaceous Potomac Group Deposits:
  - Patapsco Arundel (KP P/A) Deposits
  - Patuxent Deposits (PTX) Deposits
- Pre-Cretaceous Crystalline Bedrock

Figure 3 shows the subsurface profile at the CSO-019 Drop Shaft taken from the project Geotechnical Baseline Report.

Figure 3 shows that the P/A layer constitutes the main layer present below ground surface at the CSO-019 Drop Shaft site; it consists predominantly of very stiff to hard clay. Previous studies performed for the project and submitted with the well permit application for the CSO 019 site show that the P/A layer acts as an aquitard between the upper aquifer in the alluvial deposits and the lower aquifer in the PTX layer. The PTX layer is a confined aquifer with a total head of approximately 24.7 feet.

DEEP WELL DEWATERING SYSTEM AT CSO-019

The deep well dewatering system at CSO-019 consisted 6 depressurization wells installed to depths of around 195 feet below grade around the perimeter of the shaft. The wells extended 60 foot-long and well screens installed within the PTX Layer, with a 135-foot long solid riser pipe rising from the screen to the ground surface. An outer

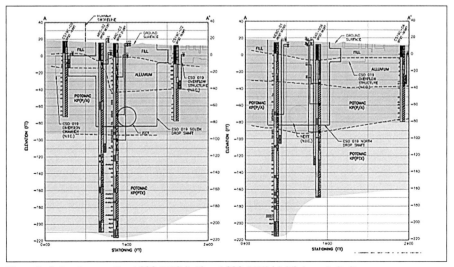

Figure 3. Subsurface profile at CSO-019S (left) and CSO-019N (right) drop shaft site

casing was socketed 10 feet into the P/A clay layer, and grouted in place to prevent cross-contamination between the Alluvium and the P/A.

The wells were designed to lower the piezometric levels in the PTX layer to around Elevation -96 feet, approximately 86.4 feet below current levels measured in the project piezometers at the site. The deep well dewatering system operated for more than one year to allow for construction of the two shafts and launching the tunnel boring machine (TBM) at the CSO 019 site.

GROUNDWATER MODELING

The groundwater modeling analysis of the CSO-019 shaft excavation has used a 2-dimensional finite element flow and transport code to design the depressurization system and to evaluate the effects of the dewatering operation (Zhu and McNulty, 2005). Originally, the zone of influence (ZOI) of the depressurization wells was estimated to be around 2000 feet from the CSO-019 Drop Shaft site. This original analysis of the ZOI used a hydraulic conductivity and compressibility derived from short-term aquifer testing of the PTX and consolidation tests on the P/A clay. However, drawdown in the observation wells showed later that the ZOI would exceed this range by an order of magnitude or more. Consequently, this paper uses the actual field water levels from nearby piezometer P2-17A and around the shafts to back-calculate the coefficient of storage and hydraulic conductivity required to produce the observed water levels versus time. The ZOI and predicted settlements of both these analyses are discuss below in the next few paragraphs. With these back-calculated properties, we predict the drawdowns that are likely incurred over the first 180 days to estimate the settlements produced in the Patuxent Arundel (P/A) aquiclude.

First, we modeled the CSO-019 site with a 2-dimensional (2D) planar model of the PTX aquifer. Using symmetry, we divided the modeling domain in half and only modeled the details on one CSO-019 shaft penetration of the PTX as shown in the zoomed in view of Figure 4. The modeling domain extended 40,000 feet in both the x and y directions. Fixed head conditions were applied at the far left and top boundaries of the model domain; no flow boundary conditions were applied at the remaining sides. Water levels for the pumping wells were approximated by piezometers located within the walls of the CSO-019 shaft.

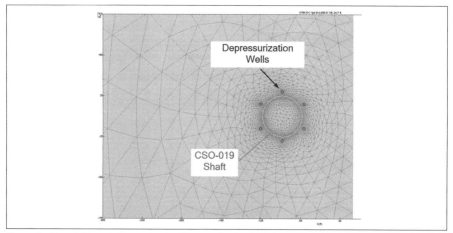

Figure 4. Near-field modeling region for 2D groundwater model

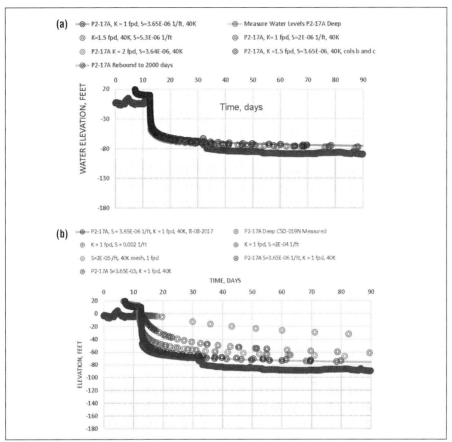

Figure 5. (a) Measured water table elevation from P2-17A versus assumed hydraulic conductivities; (b) measured water elevations at P2-17A deep versus predicted with various specific storage

MODEL CALIBRATION

The groundwater model was calibrated with the measured water levels in Well P2-17A Deep over several months. Figure 5(a) shows the change in predicted heads with time when the hydraulic conductivity of the PTX aquifer is changed for a given storativity. Figure 5(b) gives the predicted heads when the storativity is changed for a given value of hydraulic conductivity.

For the P/A aquitard, no in-situ tests were conducted. The consolidation tests on disturb soil specimens from Shelby tubes yielded hydraulic conductivities of about 1.0E-07 cm/sec. These proved to be too large. In accordance with Peck's admonition (Dunnicliff and Deere, 1984) for geotechnical engineers to think deeply before applying the standard procedures, we went to the hydrogeologic literature to review what properties other investigators had found for the P/A aquitard and PTX aquifer. Chapelle and Kean (1985, Page 30) recommend a hydraulic conductivity of 1E-09 to 1E-11 ft/sec for the P/A aquitard. Similarly, they recommend a storage coefficient of 2.326E-006. Table 1 summarizes the hydraulic parameters for the P/A and PTX formation.

Table 1. Characteristics of the P/A and PTX Formations

Formation	Description	Hydraulic Conductivity (ft/day)	Storage Coefficient 1/ft	Porosity
P/A	Hard/stiff clay or silt Plasticity Index between 18 and 70 Extremely low permeability	5.9e-006	2.326e-006	0.577
PTX	Silty/clayey sand Artesian groundwater pressure, 24.7 ft	0.3–1	3.65e-006	0.25

CONCLUSION—SETTLEMENT ESTIMATE

With these properties, we can then evaluate the drawdown induced by the depressurization wells within the PTX aquifer and the change in water pressures in the P/A aquitard that may lead to settlement. We can then compare the predicted settlement to those observed at the ground monitoring points ART-1 and ART-2. We estimate the settlement in the P/A by superimposing the change in the artesian pressure on the bottom of the P/A, calculating the new distribution of pore water pressure in the P/A, and then calculating the resulting settlement. Because of the excessive disturbance induced with the use of Shelby tubes to sample the stiff P/A layer, we have chosen to use the pressuremeter data to obtain a soil Young's modulus, E_s, more representative of undisturbed conditions. If we use a Es from the Pressuremeter of the following relationship given by the GDR for the P/A layer,

$E_s = 400 + 50*\sigma'_v$ (ksf) or 1000 ksf

The settlement induced by the depressurization wells will be quite small as shown in Figure 6. The predicted settlements for consolidation in the P/A appear to be not at all like those observed at ART-1. These movements occur too quickly to be caused by the slow consolidation response of the P/A layer.

Consequently, we surmise that these movements may be caused the depressurization changes in the PTX aquifer. Using a soil Young's modulus derived from the previously backed out storativity for the PTX aquifer, we obtain an E_s of 2.68E+07 psf. This modulus gives the settlements caused by the PTX as show in Figure 7. Therefore, the settlements being measured at ground monitoring points such as ART-1 apparently have been measuring the settlement and rebound of the PTX aquifer, not the P/A aquitard.

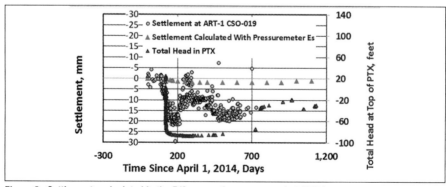

Figure 6. Settlements calculated in the P/A versus those measured at ART-1

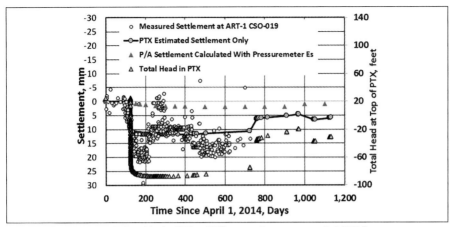

Figure 7. Settlements calculated in the PTX and P/A versus those measured at ART-1

ACKNOWLEDGMENTS

The authors thank their employer, Parsons Corporation, as well as IHP JV and DC Water for permission to publish this paper.

REFERENCES

Chapelle, F.H., and T.M. Kean, 1985, Report of Investigations No. 3, A Hydrogeology Digital Solute Transport Simulation, and Geochemistry of the Lower Cretaceous Aquifer System Near Baltimore, Maryland, U.S. Geological Survey, Prepared in cooperation with the United States Department of the Interior.

Dunnicliff, John D., and Don U. Deere, 1984, Judgment in Geotechnical Engineering, The Professional Legacy of Ralph B. Peck, BiTech Publishers Ltd., Vancouver, B.C. Canada.

Mayne, P.W., Coop, M.R., Springman, S., Huang, A-B., and Zornberg, J. (2009). State-of-the-Art Paper (SOA-1): GeoMaterial Behavior and Testing. Proc. 17th Intl. Conf. Soil Mechanics & Geotechnical Engineering, Vol. 4 (ICSMGE, Alexandria, Egypt), Millpress/IOS Press Rotterdam: 2777–2872.

Tanaka, H. 2000. Sample quality of cohesive soils: Lessons from three sites: Ariake, Bothkennar, and Drammen. Soils and Foundations 40 (4): 54–74.

Terzaghi, K. and R.B. Peck, 1967, Soil Mechanics in Engineering Practice, Wiley & Sons, Incorporated, John, New York.

Zhu, Junlin, and Greg McNulty, 2005, AquiFEM-Vision, A Two-Dimensional Finite Element, Two Phase Flow and Transport Model for Aqueous and Non-Aqueous Fluids, May.

Digging Deeper—Supplemental Geotechnical Site Investigation for Parallel Thimble Shoal Tunnel

Amanda Wachenfeld ▪ Mott MacDonald
Frank Perrone ▪ Mott MacDonald
Scott Kibby ▪ Mott MacDonald
Jose Ballesta ▪ Dragados USA

ABSTRACT

The Chesapeake Bay Bridge and Tunnel (CBBT) is a four-lane bridge—tunnel crossing consisting of a series of trestles and two approximately 1-mile-long tunnels beneath the Thimble Shoal and Chesapeake navigation channels. The CBBT Authority is expanding the existing corridor by building a parallel tunnel under the Thimble Shoal navigation channel utilizing a design build contract. As part of the D/B contract a supplemental investigation was planned and executed to support the new design. This paper explores challenges of working in a marine environment around one of the east coast's busiest commercial and Naval ports, the rationale behind the supplemental investigation and integration of the data into the design.

INTRODUCTION

The Chesapeake Bay is an estuary that is bounded by the North American mainland to the west and the Delmarva Peninsula to the east. The Chesapeake Bay Bridge Tunnel (CBBT) was originally constructed between September 1960 and April 1964. To accommodate an increase in traffic volumes, a second parallel bridge was constructed in the mid-1990s however tunnels were not expanded at the time. Today, the CBBT is a four-lane bridge crossing consisting of a series of low level trestles and two approximately 1-mile long tunnels beneath the Thimble Shoal navigation channel (southern tunnel) and the Chesapeake navigation channel (northern tunnel). Four man-made islands are located at each end of the two tunnels facilitating the transition from bridge to tunnel. The CBBT crossing is one of the longest bridge-tunnel networks in North America at 17.6 miles in length from shore to shore.

The existing Thimble Shoal Tunnel is an immersed tube tunnel, crossing beneath a navigation channel/shipping lane with a maximum water depth of approximately 55 feet. The navigation channel is 1,000 feet wide and is flanked on both sides by a 450-foot wide auxiliary channel. In addition to private boat and commercial vessel traffic, ships from the nearby Norfolk Naval base, ranging from patrol ships to aircraft carriers, and the US Coast Guard Base in Portsmouth also use the Thimble Shoal channel. Refer to Figure 1 for the general location of the Thimble Shoal Tunnel Project.

The single tunnel configuration often results in significant congestion from the merging traffic, especially in the summer months. Over the years, traffic utilizing the facility has consistently increased and the expected traffic increases only serves to exacerbate an already constrained facility, resulting in increased congestion and driver delays.

Although temporary relief of congestion was provided by the expansion of the bridges for the Parallel Crossing Project, the facility is still constrained by the original two-lane tunnels under the Thimble Shoal and Chesapeake Channels. A loss of operation for

Supplemental Geotechnical Site Investigation for Parallel Thimble Shoal Tunnel 513

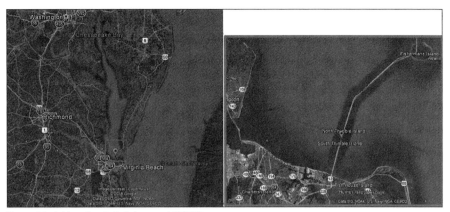

Source: Adapted from Google Earth (accessed 12/8/18)
Figure 1. Project location

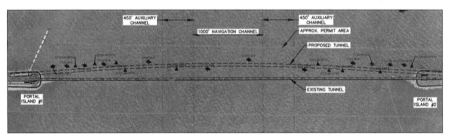

Figure 2. Proposed tunnel alignment

either of the two tunnels results in closure of the entire CBBT, and the existing configuration poses adverse impacts on traffic due to capacity constraints, safety issues, and a lack of alternative routes during tunnel closures.

The Parallel Thimble Shoal Tunnel (PTST) Project is Phase II-A of the Parallel Crossing Project, originally authorized by the Virginia General Assembly and Governor in 1990. Once completed, the PTST Project will improve regional mobility through reduced traffic congestion and corridor safety, while reducing maintenance constraints. The PTST Project will be followed by Phase II-B, the Parallel Chesapeake Tunnel, which is proposed for completion between 2030 and 2040.

PROJECT BACKGROUND

The PTST Project includes construction of a new 1.1-mile-long tunnel located west of the existing immersed tube tunnel (ITT) crossing under Thimble Shoal Channel. When completed, the new tunnel and approach structures will carry two traffic lanes southbound and the existing tunnel will carry two traffic lanes northbound. The new tunnel will be constructed using an Earth Pressure Balance (EPB) Tunnel Boring Machine (TBM). Refer to Figure 2 for an illustration of proposed tunnel alignment.

Construction has been procured via a Design-Build Contract between Chesapeake Bay Bridge and Tunnel of the Commonwealth of Virginia and the Chesapeake Tunnel Joint Venture (CTJV) comprising Dragados USA and Schiavone Construction Company, LLC. Mott MacDonald (MM) is acting as lead designer for the project.

The proposed reference design included the expansion of the existing portal islands to allow for construction of a straight alignment, using an immersed tube tunnel (ITT). During the bid phase, the CTJV team's preferred tunnel construction methodology of a bored tunnel negated the need for island expansion.

The tunnel portion of the project consists of a 6,525 foot-long, 39-foot internal diameter, (41-foot external diameter) segmentally lined bored tunnel carrying two lanes of vehicular traffic and a dedicated egress corridor. At its deepest point, the crown of the new tunnel will be roughly 55 feet below mudline. The proposed alignment curves slightly between the existing portal islands with a maximum distance of 250 feet from the existing tunnel and as little as 85 feet between the new and existing tunnels at the approaches.

GEOLOGIC HISTORY

The PTST Project lies within the Atlantic Coastal Plain physiographic providence and the historical Chesapeake Bay comet or meteorite impact crater. Bedrock in this area was compressed by the impact, and subsequent sediment deposition has buried the impact crater approximately 1,000 feet below ground surface (Jacobs GDR 2015).

The Chesapeake Bay depositional environment includes estuarine and marine conditions resulting in shallow, relatively thin layers of soft, compressible soil seaming out over short distances. In addition to the high variability of natural deposition, historic construction activities including dredging and fill placement have altered the soil layering near and along the navigation channel. Fill placement for the creation of the original portal islands has compressed the underlaying soft material for over 50 years, resulting in varying consolidation properties and soil densities over short distances directly below and adjacent to the portal islands. This variable geology presents a number of design and construction challenges.

Geology in the Chesapeake Bay area generally consists of recent deposits of organic and granular material at the bay bottom, underlain by Holocene-aged sand deposits. Channel fill deposits from the Late Pleistocene Epoch underlay the sand deposits, and contain numerous layers, including a cohesive layer, an organic layer, and a granular layer. Channel fill deposits are underlain by a Late Pleistocene granular deposit which is underlain by the Yorktown formation. The upper portion of the Yorktown Formation, deposited during the Pliocene Epoch, is a granular deposit. The lower portion of the Yorktown formation is a cohesive unit deposited during the late Miocene to early Pliocene Epoch. The Yorktown formation is underlain by a late Miocene granular deposit followed by a late Miocene cohesive deposit. The presence and thickness of the layers is highly variable over short distances which along with the changed alignment were contributing factors in performing the Supplemental Site Investigation.

The soil types identified for the project area and the general description of each soil type based on the pre-bid GDR and GBR are summarized in Table 1. The thickness of each soil type is highly variable across the site with higher volumes of the fine-grained channel fill deposits encountered at the north end of the proposed alignment.

SUPPLEMENTAL INVESTIGATION

Mott MacDonald completed a supplemental investigation to mitigate risks associated with: (1) a highly variable geology within the project area and (2) alignment changes from the conceptual design.

Supplemental Geotechnical Site Investigation for Parallel Thimble Shoal Tunnel

Table 1. Summary of soils based on GDR and GBR descriptions

Engineering Class	Major Soil Types	Fill Materials	Organics	Granular Materials	Silt	Clay with LL<50	Clay with LL>50	General Geologic Description from GBR
af	Artificial Fill Sand (SP, SP-SM)	x						• Hydraulically filled clean sands with fines content generally less than 6%
Qfg	Baymouth Fine-Grained Clay (CL, CH) Organics		x				x	• Recent estuarine fine-grained deposits • Soft clay (CL, CH) with shell fragments and slight sulfur odor due to presence of organic material
Qb	Baymouth Shoal Deposits Sand (SP)			x				• Holocene loose to medium dense fine sand with trace to some silt and occasional clay seam
Qcs	Channel Fill Deposits Silty Sand (SC-SM)			x	x			• Late Pleistocene loose to medium dense silty sands with varying amounts of clay and shells
Qcfu	Channel Fill Deposits Clay (CL, CH) Organics	x				x		• Late Pleistocene soils deposited within incised channels during high sea level periods • Qcfu is upper deposit: over consolidated, soft to medium stiff clay with varying amounts of sand, silt, shell fragments • QcfL is lower deposit: medium stiff to stiff clay, high fines content and high plasticity, normally consolidated • Organic material may be source of pockets of biogenic gas in this area
QcfL		x					x	
Organic	Organics Organics (OH) Peat (PT)		x					
Qt	Tabb Formation Silty Sand (SP-SM, SM)			x	x			• Late Pleistocene deposits composed primarily of fine to medium grained sand with some silt, medium dense to dense
Tys	Yorktown Formation Silty Sand (SM) Clayey Sand (SC)			x	x		x	• Pliocene Epoch, Tertiary Period deposits primarily composed of medium dense, silty to clayey fine to coarse sands with • Occasional pockets and lens of silty clays
Tyfu	Yorktown Formation Clay (CL, CH)						x	• Late Miocene to early Pliocene deposits composed to stiff to hard sandy shelly over consolidated clays and occasional sand seams • Tyfu has a higher plasticity and higher fines content • TyfL is either a fat or lean clay
TyfL							x	
Te	Eastover Formation Silty Sand (SM)			x	x	x		• Late Miocene deposits composed of medium dense to dense silty fine sand
Tm	St Mary's Formation Silty Clay (CL, CH)					x		• Late Miocene deposits, upper portion composed of stiff to very stiff silty clay transitioning to soft clays with sand lenses

A geotechnical baseline report and geotechnical data report were completed prior to the change in project scope and provided as part of the RFP documents. The original project scope consisted of expanding the islands by up to 300 feet on the west side of the existing portal islands and a straight tunnel alignment approximately 250 feet west of the existing tunnel. The majority of the borings presented in the GDR/GBR were completed over 100 feet west of the final alignment selected by CTJV. Limited data was available along the selected alignment, particularly below or directly adjacent to the portal islands.

The supplemental investigation was designed to target areas of known risk based on data presented in the GDR/GBR and areas with limited existing data. Particular focus was placed on refining soil parameters associated with specific project components including settlement analysis of the existing ITT, need and extent of ground improvement, tunneling methodology, and other design components. The supplemental investigation was used to refine the soil profile and soil parameters along the updated alignment in areas of concern such as on and directly adjacent to the portal islands. Information gathered from the supplementary investigation allowed Mott MacDonald to more clearly define and minimize design and construction risks.

A two-phase supplemental investigation program was designed due to the time required to obtain permits for on water work. Phase 1 addressed the portal islands and Phase 2 addressed the subsea portion of the proposed alignment. Additional supplemental investigation work included environmental sampling and both marine and land based geophysical work.

Phase 1 Land Based Investigation

As part of Phase 1, a total of 3,825 feet of soil borings were completed with 14 borings on Portal Island 1 and 11 borings on Portal Island 2. Between 20 and 80 feet of continuous sampling was completed for each boring to define the soil layering and properties within the tunnel horizon or throughout the depth of work associated with the approach structures.

Additional in-situ testing included cone penetrometer tests (CPTs), dilatometer tests (DMT), pore water dissipation tests, and seismic CPTs. Six CPTs and two DMTs were completed on each island. Boring locations, in-situ test locations and zones of continuous sampling were selected to assist in clearly defining the thickness of soil layers within target areas.

In addition to the geotechnical investigation, a targeted supplemental environmental program was undertaken consisting of 12 borings each to a depth of 100 feet, using sonic drilling methods. Samples collected were utilized to evaluate disposal options, presence or absence of contaminants and preliminary bench scale testing for TBM conditioning agents.

In order to provide long-term baseline monitoring prior to the start of construction, instrumentation was installed on both portal islands in conjunction with the supplemental site investigation. An observation well, four vibrating wire piezometers, a deep bench mark, and an inclinometer were installed on each island.

The Phase 1 investigation was completed in roughly 3 months, between November 2016 and February 2017, including the soil borings, in-situ testing, instrumentation installation, and island and marine geophysical work.

Island Geophysical Investigation

A potential construction risk of encountering obstructions from the original construction of the tunnel approach structures was identified prior to completion of the supplemental investigation. Plans provided as part of the background information indicated that sheet piling was used as temporary excavation support during the original construction but was unclear of exact location and if any piling was left in place or removed. Due to the proximity of the new approach structures to the existing, and the selection of slurry wall construction accurately identifying the presence and location of sheet piles was a critical item to minimize construction risk. As a result, identifying the presence and location of these sheet piles was a focus of the supplemental geophysical portion of the investigation.

A two-pronged approach was taken to detect sheeting, utilizing both surface and borehole geophysical methods. Mott MacDonald procured the services a geophysical subcontractor to perform the surveys on both islands. The surface method used was a magnetic survey and borehole geophysical methods consisted of Electromagnetic Induction (EM) and magnetic logging.

The surface magnetic survey was done on each island in accordance with ASTM D6429, covering an area approximately 600 feet long and 40 feet wide, west of the existing tunnel approach structures. A Geometrics Model G858-G Cesium Vapor Magnetometer was used, and results presented as contour maps overlaid on plans of the area. See Figure 3 for a heat map prepared from the results of the surface magnetic survey.

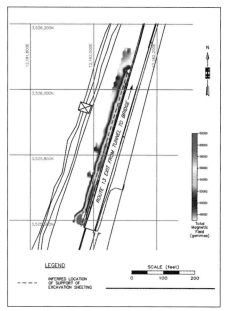

Source: Hager-Richter Geoscience, Inc., Geophysical Testing Report, December 2016

Figure 3. Portal Island 2 magnetic survey

The surface magnetic survey indicates that the steel sheeting was left in place, just west of the approach structures; although this method does not provide information on depth or length of the sheet piling. The borehole geophysical survey was required to determine those details. The downhole survey was done in 9 borehole locations (five on Portal Island 1 and four on Portal Island 2).

Three different types of downhole surveys were completed on each island: geophysical logging, electromagnetic (EM) induction logging and magnetic field logging. Surveys were performed using a Mount Sopris Matrix system for geophysical logging, a Mount Sopris 2 PLA-1000EM probe for electromagnetic induction logging, and a Mount Sopris 2DVA-1000 deviation probe.

Borehole geophysical logging measures electrical conductivity of materials which are located within a few feet of each borehole along with the earth's magnetic field and the results can be used to determine depth of the top and bottom of the sheeting.

EM logging uses an electric current from a transmitting coil to create an EM field which in turn creates eddy currents in any steel located within the EM field. The eddy currents create a secondary EM field which is sensed by a receiver. The approximate depths of the top and bottom of the steel sheeting can be determined based on where eddy currents decrease significantly.

By using a variety of geophysical methods, the locations and extents of the left-in-place sheeting were located with high confidence. EM and magnetic data were plotted versus depth to assist in determining depths to the top and bottoms of the sheet pile wall. The magnetic data was more valuable that the EM data for identifying the ends of the sheet piles. The EM results show low conductivity values above about 32 feet and

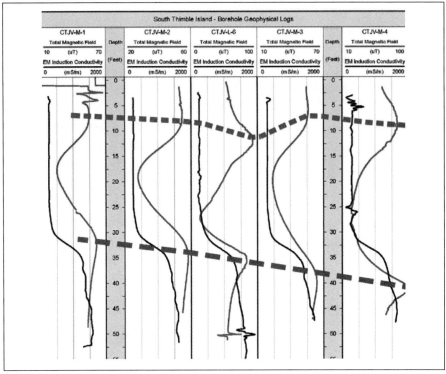

Source: Hager-Richter Geoscience, Inc., Geophysical Testing Report, December 2016
Figure 4. Downhole geophysical survey results

high values below that depth which approximately correlates to water level. The saline water is highly conductive and causes the EM induction conductivity to increase even though the sheet pile may extend beyond a depth of 32 feet.

Magnetic log data is characterized by shallow and deep magnetic anomalies and it is inferred that these anomalies indicate the top and bottom of the sheet piles. Based on the results of the magnetic logging, the top of the sheet piles is between 7 and 11 feet below ground surface on Portal Island 1 and 5 and 7 feet on Portal Island 2. The bottom of the sheet piles is between 32 and 41 feet below ground surface and 32 to 47 feet for Portal Islands 1 and 2 respectively. Figure 4 shows the results of the EM (in Black) and magnetic logging (in Red) for Island 1. The interpreted top and bottom of the sheet pile is marked by the blue and purple lines respectively.

Marine Geophysical Investigation

Prior to beginning Phase 2, a marine based geophysical survey was performed, as required by the project technical requirements, and to identify potential areas for exploration during the supplemental investigation. Marine geophysical work took place over a two-week period in late November and early December 2016.

Mott MacDonald procured the services of a subcontractor to complete bathymetric and marine high-resolution geophysical (HRG) surveys. The goal of these surveys was: develop a targeted bathymetric surface along the proposed tunnel alignment, develop a sub bottom profile, and to locate an existing inactive submarine electrical cables near the portal islands. Bathymetric survey was required to optimize tunnel design and location of the existing cables was required to minimize construction risk. The sub bottom profile was used to assist in developing the marine geotechnical investigation.

A high-resolution Compressed High Intensity Radar Pulse (CHIRP) Sub-bottom Profiler was used to develop the sub-bottom profile. Soil horizons were determined based on the strength of the reflection. The CHIRP profiler was able to penetrate up to 20 feet below the mudline, particularly in areas of soft clay, considered to be critical areas. If a hard surface, such as rock or other dense material was at mudline, the profiler would show shallower penetration depths.

A medium penetration sub-bottom system was also run along various sections of the proposed alignment and was able to map horizons up to 120 feet below mudline. Existing tunnel cover stone was also identified using this system. Both profilers were used to determine soil horizons but were unable to determine exact soil types. Therefore, the profiler results were compared with the existing marine boring and CPT logs to identify potential locations of thick layers of soft clay. See Figures 5 and 6 for profiles developed based on the CHRIP sub-bottom profiler and the medium penetration system.

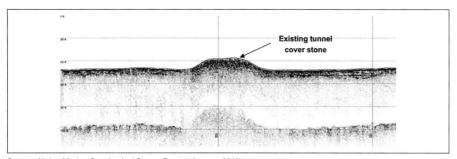

Source: Alpine Marine Geophysical Survey Report, January 2017
Figure 5. Typical profile developed by CHRIP sub-bottom profiler

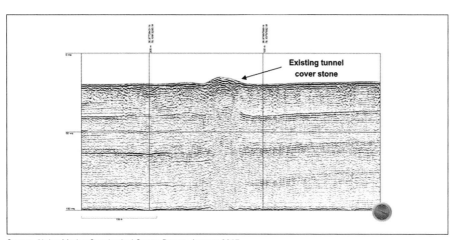

Source: Alpine Marine Geophysical Survey Report, January 2017
Figure 6. Typical profile developed by medium penetration system

The protection stone over the existing ITT could also be mapped using the sub bottom profiler. Knowing the location of the existing stone and areas of significant depths of soft clay allowed the targeted approach to the marine geotechnical investigation. Completing the sub-bottom profile before the marine investigation reduced the risk of encountering stone and allowed for targeting area of greatest concern.

Phase 2 Marine Based Investigation

Phase 2 was completed west of the existing ITT, along the proposed bored tunnel alignment utilizing a lift boat as a work platform (Figure 7).

Figure 7. Lift boat set up on boring

Twelve borings totaling 1,421 feet were completed for the Phase 2 investigation. Sixty feet of continuous sampling was completed for each boring within the tunnel horizon. In-situ testing consisted of 10 CPTs with pore water dissipation testing in select layers/zones.

Phase 2 borings and CPTs focused on gathering data in areas where the greatest amount subsurface uncertainty existed based on GDR/GBR information provided and/or area(s) with the greatest level of construction or design risk. Defining thicknesses of soil layers, such as the shallow, soft clays and organics expected near Portal Island 2. Refining soil parameters within the tunnel horizon was critical in assessing the need for buoyancy control over the shallower potions of the tunnel, the potential extent of movement of the ITT due to tunnel enabling works and berm construction, and the extent of needed ground improvement. Therefore, most of the investigation was focused along the alignment directly adjacent to existing portal island berms. Two borings were completed in the navigation channel to help correlate results from the supplemental investigation with data contained in the GDR, and to refine the lateral extents of subsurface conditions along the new tunnel alignment.

The marine investigation was focused on the Portal Island 2 "near shore" area, adjacent to the existing island and berm which was specifically the area where the extents of the soft clay and organics, originally noted in the GDR/GBR, were critical to design. This was an area of concern as the clays and organics required ground modification to facilitate tunneling and help mitigate the risk of excessive settlement of the existing ITT. Supplemental site investigation borings and CPT soundings were done at the south end of Portal Island 2 and along both sides of the proposed alignment to better determine the vertical and lateral extents and properties of the clay and organic layers. The Phase 2 investigation was completed in just under one month in April 2017. Work was completed by running two 12-hour shifts, five days a week there by minimizing the per shift lift boat cost and allowing for a quicker completion of the investigation than single shift work and still providing weekend downtime for crews.

In-Situ Testing

Various in-situ tests for both phases of the supplemental investigation were completed. CPT tests focused on obtaining a continuous profile of the subsurface to determine soil behavior and soil types, reducing the need for continuous SPT sampling, reducing overall costs and minimizing schedule. CPT quickly and cost-effectively provide a continuous picture of the behavior of the soil from ground surface to the bottom of the sounding and use of correlations allowed for determining strength and density parameters. DMTs were completed to determine in-situ lateral stress and lateral soil stiffness, and via correlations, other soil properties. These properties were used for design of new cut and cover approach structures, and related diaphragm and retaining walls. Pore pressure dissipation tests focused on the need for additional consolidation parameters, particularly in the soft, cohesive material below the islands. Dissipation tests provided data used in correlations to determine consolidation parameters and permeability of the cohesive soil. Finally, seismic CPTs gathered information for liquefaction evaluation of the portal islands.

In-situ testing required about two weeks to complete the land portion and one week to complete the marine portion and added to completing an efficient, and value-added investigation, aimed at mitigating specific risks associated with the project. However, since correlations from the in-situ tests to soil parameters were based on numerous empirical relationships, laboratory tests were also done. This approach allowed for refinement of parameters and increased confidence in the accuracy of selected values.

Laboratory Testing

As part of the supplemental investigation, 521 samples out of over 1,500 taken were sent for laboratory testing. 482 index tests were completed, including grain size analysis with and without hydrometers, moisture content, organic content, Atterberg limits, and unit weight. Index testing focused on collecting data to determine physical soil properties to confirm in-field soil classification and refine the soil profile.

Index testing allowed for refined soil interpretation and assisted in determining which soil types from the various borings had the same engineering properties. This determination helped select soil samples for strength and consolidation testing. The index tests also allowed for a re-definition of some of the soil layers. For example, organic testing was completed on the soil samples classified as "peat" during the soil interpretation. The results of the organic testing, however, indicated that the organic content in the peat was generally under 10%. These results allowed for Mott MacDonald to redefine the Organic soil layer as an organic silt rather than a peat which influenced the required extents of the jet grout and other design aspects.

A total of 39 samples were sent for strength (19 samples) and consolidation (20 samples) testing including consolidated undrained direct simple shear, consolidated undrained triaxial, and constant rate of strain consolidation tests. The direct simple shear and triaxial tests were completed to determine the shear strength of the cohesive soil. Triaxial testing also determined the lateral earth pressure coefficient of the soil. Consolidation testing was completed to determine consolidation parameters of the cohesive soils and results were used to predict settlement from berm construction. Constant rate of strain testing was used over incremental loading because it offers the advantage of continuous data collection during loading and the test results are easy to use in a conservation of work-based approach to compute maximum past pressure. Additionally, the constant rate of strain test was fully automated, taking less time than standard and automated incremental loading tests.

Because of the alignment driven, focused investigation and the data previously gathered, less than 10% of the laboratory testing was strength and consolidation testing. Results from the lab tests were used to verify correlations of soil parameters from in-situ testing and allowed Mott MacDonald to select the most relevant data for recommending parameters.

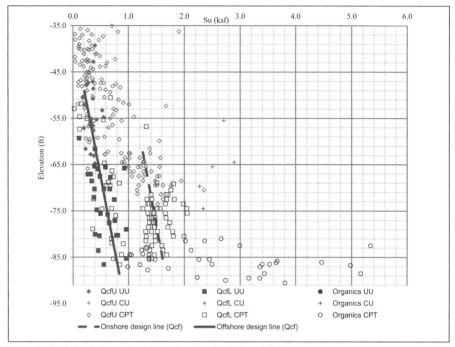

Figure 8. Undrained shear strength vs elevation for Qcfu and QcfL

Figure 8 shows a plot of the results of laboratory and CPT results for undrained shear strength vs. elevation for the Qcfu and QcfL layers and the line indicating the value used in design.

OUTCOME OF THE CTJV SUPPLEMENTAL INVESTIGATION

Benefits

Overall, the benefits of the supplemental investigation were: refinement of geotechnical parameters, reduced risk, and increased value of the existing information. By combining the findings from the supplemental investigation and GDR, the level of subsurface risk was reduced, and the data presented in the GDR augmented and refined for the proposed alignment.

Soil Profile

The combination of the borings presented in the GDR and completed during the supplemental investigation provided soil information within an approximately 400-foot wide corridor extending from the existing ITT to west of the proposed alignment and from the south end of Portal Island 1 to the north end of Portal Island 2. The extensive lateral and longitudinal reach of data allowed for the development of a detailed 3D soil profile providing additional insight into the variability of the soil layering and how soil layers seamed out over short distances. Creation of a 3D profile provided the ability to cut 2D cross sections at a desired location and allowed for a more thorough understanding of the interaction between new construction and the existing tunnel. Figures 9 and 10 show the soil profile along the proposed tunnel centerline cut from the 3D model.

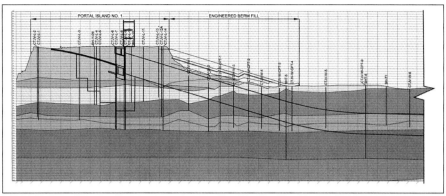

Figure 9. Soil profile—Portal Island 1 to navigation channel

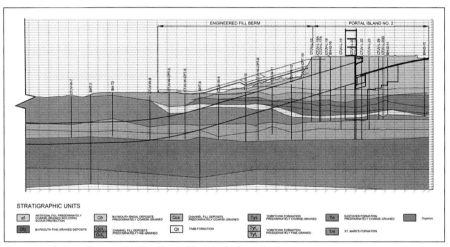

Figure 10. Soil profile—navigation channel to Portal Island 2

Soil Parameters

Because of the results of the supplementary site investigation, assumed parameters for design were revised to represent four distinct sections: Portal Island 1, Portal Island 2, Offshore Portal Island 2, and Tunnel Alignment. All data collected and correlated from the supplemental investigation and the GDR were graphed by parameter versus depth, identifying trends and developing soil parameters. Graphs showed soil parameters varied between the different project areas, likely a result of the compaction and consolidation of soils beneath or directly adjacent to the portal islands.

Focusing the marine investigation on "near shore" area of Portal Island 2 resulted in a range of data which highlighted the difference in engineering properties between the soils beneath Portal Island 2, in the channel, and offshore Portal Island 2, particularly in the soft clays. Developing the soil parameters in the off-island areas allowed for Mott MacDonald to reduce the required area of ground improvement adjacent to Portal Island 2.

CONCLUSIONS

The supplemental site investigation was completed to augment, refine and target conditions along the final tunnel alignment and allowed Mott MacDonald to focus on specific strata and areas identified as posing the greatest risk to construction. The island geophysical investigation into the location of the sheet piles showed the need to pre-excavate and remove the sheeting where it conflicted with the approach structure construction.

The outcome of the investigation provided additional detail and clarity to existing geotechnical information and helped refine soil profiles and design parameters. Compared to the GDR, the updated soil profiles also showed a reduced risk from soft, compressible soil layers and reduced the need for ground improvement for settlement mitigation. Similarly, the extent of the organic layer on Portal Island 2 was less, reducing the amount of required pre- tunneling ground improvement. Updated parameters were used for the design of slurry walls, trestle and fishing pier piles, settlement analysis, muck disposal, coastal stability, and the tunnel lining.

ACKNOWLEDGMENTS

The authors would like to acknowledge and thank the CTJV team, our Mott MacDonald colleagues and our subcontractors for their efforts during the investigation, as well as Kevin Abt and Mike Crist of CBBT.

REFERENCES

Alpine, A Gardline Company. (2017) *Marine Geophysical Survey Report: Thimble Shoal Portal Islands #1 and #2 Chesapeake Bay Bridge-Tunnel,* Chesapeake Bay, VA. Prepared for Mott MacDonald.

Hager-Richter Geoscience, Inc. (2016) *Geophysical Testing: Chesapeake Bay Bridge and Tunnel Thimble Shoals Portal Island & Mainland Substation,* Chesapeake Bay, VA. Prepared for Mott MacDonald.

Jacobs Engineering. (2015). *Geotechnical Baseline Report: Chesapeake Bay Bridge-Tunnel Parallel Thimble Shoal Tunnel.* Prepared for Chesapeake Bay Bridge and Tunnel District.

Jacobs Engineering. (2015). *Final Geotechnical Data Report: Chesapeake Bay Bridge-Tunnel Parallel Thimble Shoal Tunnel.* Prepared for Chesapeake Bay Bridge and Tunnel District.

How to Quantify the Reliability of a Geological and Geotechnical Reference Model in Underground Projects

Guido Venturini ▪ Schnabel-SWS, Global Studio TCA
Gianpino W. Bianchi ▪ EG Team
Mark Diederichs ▪ Queen's University

ABSTRACT

Today the Geotechnical Baseline Report (GBR) and the Geotechnical Database Report, (GDR), are becoming standard contractual documents in any underground project to describe the expected geological and geotechnical conditions. The GBR contains a synthesis of all the acquired geotechnical data listed in the more comprehensive and purely factual GDR, and ultimately provides a geotechnical reference model for the project. Despite this, usually few or no information about the reliability of the model are provided. The article intends to present the R-Index innovative methodology, a quantitative method to evaluate the reliability of the Geological Reference Model through a multiparametric approach and to improve the geological risk assessment. Applied examples from an alpine base tunnels (TELT) will be presented.

INTRODUCTION

Tunneling industry is daily faced worldwide with major contractual issues related to unexpected subsurface conditions and their impacts in terms of cost increases and schedule delays. The tunneling community, represented by the International Tunneling Association (ITA), has not still reached a common and homogeneous approach on this matter, also due to the laws, rules and regulations which are very often defined by each Country in which the underground project is realized. However, ITA Working Groups, made multiple suggestions in several contributions published in 2011 (WG3), 2013 (WG19), 2014 (WG4) 2016 (WG14/19) and 2017 (WG17), referring to the GBR as a common and useful practice to define the expected geotechnical conditions.

Today the literature about this subject is quite wide and articulated, vary from legal contributions to technical papers regarding specific case histories. Almost everyone agrees in the need to define the appropriate tools to clearly allocate risks and liabilities in subsurface projects between Owners and Contractors.

The tunneling community is also quite aligned in claiming the need to have a geotechnical baseline to develop the project and allocate the risks, by the Contractor to price the works during the tender phase and by both the Owner and the Contractor to determine and manage the differing site conditions (DSC).

In spite of this, only few contributions and suggested methodologies have been dedicated to the qualification and quantification of the uncertainties related to the geotechnical baselines, or, in other words, to the quantification of the reliability of the geological models that are the basis of any geotechnical baseline.

This article presents an overview of the state of the art regarding this matter in relation with the GBR, which seems to become more and more a common practice not only for the North American tunneling industry but for many other underground markets worldwide.

The ultimate goal of this contribution is to set up a certain number of concepts related to the determination of the reliability of geological modelling for underground projects, which will certainly require a more in-depth treatment in the future, with the involvement of specific Organizations to acquire the consensus and ultimately to let the quantification of the reliability (or uncertainty) becoming a common practice to decrease or efficiently manage risks in tunneling.

THE GBR APPROACH

The Suggested Guidelines to prepare Geotechnical Baseline Reports (GBR) for Construction (Essex 2007, Essex et al., 2001) states in its executive summary that "The primary purpose of the GBR is to establish a single source document where contractual statements describe the geotechnical conditions anticipated (or to be assumed) to be encountered during underground and subsurface construction. The contractual statement(s) are referred to as baselines."

While some contract models require the contractor to validate and/or interpret unwarranted geological information provided by the owner (and thereby assume some level of risk, notwithstanding the application of unit price and change order mechanisms) or to execute their own investigations, the following discussion is confined to contract scenarios in which the geological and geotechnical methods are to be fully specified by the owner at the bid stage.

"Risks associated with conditions consistent with or less adverse than the baselines are allocated to the Contractor, and those materially more adverse than the baselines are accepted by the Owner. (…) The baselines should be meaningful, reasonable and realistic, and to the maximum extent possible should be consistent with available factual information contained in the GDR."

"The greatest risks are associated with the materials encountered and their behavior during excavation and installation of support. The main purpose of the GBR is to clearly define and allocate these risks between the contracting parties."

Translating these major concepts in a graphic, we could represent them as shown in Figure 1. The line represents the state of the expected geotechnical condition during the project execution. In case of no changes, the line would remain horizontal. Any more adverse geological/geotechnical conditions would shift the line in the upper side of the graphic guiding the Owner to calculate and add some contingencies.

The GBR Guidelines recommend to clearly define a unique geotechnical profile to avoid later misunderstanding and to provide a clear set of geotechnical data to be used for the design and the correct choice of the construction methods. All the data sets acquired through the investigations, site and laboratory tests to generate the mentioned data have to be presented in the Geotechnical Database Reports.

This approach certainly allows GBR users to estimate with good detail, ideally supported by statistical parameters, the reliability of the geotechnical parametrization but does not give any information regarding the geological model that has been chosen to describe the baseline and from which the likely geotechnical parameters are based. In other words, the possible changes of the geological baseline today are not quantified

How to Quantify the Reliability of a Geological and Geotechnical Reference Model 527

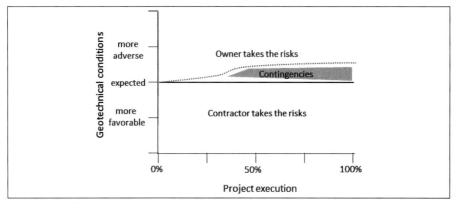

Figure 1. Graphical representation of the geotechnical baseline and the theoretical "contingency" areas

by the GBR; moreover, their qualification is not included in the report and remain an internal discussion between the Owner and the GBR authors.

The aim of this contribution is to share some consideration about possible methodologies which could be useful applied to help the GBR user to qualify and quantify the reliability of the geological model and ultimately to better quantify the contingencies which should be considered while pricing the project.

Uncertainty in Geological Modelling

Geological complexity can vary dramatically from project to project, depending from multiple factors: geological context (sedimentary, plutonic, metamorphic), tectonics (no deformation, folding, faulting, shearing), geodynamics (compression, extension, uplifting), morpho-dynamic (belts, marine, slope, plane, glacial, etc.) and several others.

Geological models are the result of the combination of many of these factors, resulting in 3D structures of different materials with different characteristics, modified during their geological evolution (4D), enabling the delineation of zones in which geotechnical parameter ranges can be specified or specific locations of discrete geotechnical concern.

Using a numeric progressive scale from 1 to 12, the simplest geological models are those constituted by unfolded and un-faulted sediments, deposited in low energy environments and located in the shallow portion of the crust (from 1 to 3). Progressively, models become more complex if affected by brittle and ductile tectonics (from 4 to 7). The combination of sedimentary and plutonic protoliths, folded and faulted generate a further increasing if the geological complexity (from 7 to 10). Finally, the chemical and physical modification of these rocks under variable conditions of temperature and pressure (metamorphism) generate a further increasing of the complexity (from 10 to 12) (Table 1). This table is certainly subjective and does not pretend to define an absolute and unique evaluation scale, also considering how geology can be complex and variable. Nevertheless, this matrix provides a useful tool to quickly evaluate the level of the complexity of the geological model considered as starting point of the geotechnical parametrization.

Complexities from 1 to 6, mainly related to sedimentary and quaternary geology, characterizes most of the urban projects worldwide, meanwhile higher complexities from 6 to 12 usually refer more to infrastructure and water projects located in the major

Table 1. Simplified numerical representation of the geological complexity

	Geological complexity (simplified from 1 to 12)					
	Sedimentary		Magmatic		Metamorphic	
	Simple	Difficult	Simple	Difficult	Simple	Difficult
Unfolded	1,2	3	2	3		
Folded	3	4	4	5,6	5	7
Folded + Faulted	4	5	6,7	7,8	8,9	10,11
Folded + Faulted + Sheared		6,7		9	9	12

Table 2. Uncertainty of the different models after a standard site investigation expenditure

	Uncertainty (after standard site investigation expenditure)					
	Sedimentary		Magmatic		Metamorphic	
	Simple	Difficult	Simple	Difficult	Simple	Difficult
Unfolded	VL	LM	L	LM		
Folded	LM	LM	LM	M	M	H
Folded + Faulted	LM	M	MH	H	H	VH
Folded + Faulted + Shared		MH		HVH	HVH	EH

VL = very low; L = low; LM = low-medium; M = medium; MH = medium high; H = high; HVH = high- very high; VH = very high; EH = extremely high

mountain belts, like Rockies, Andes, Alpes, Himalaya and several other minor but not less complex orogenic structures worldwide.

As shown in Table 2, the less complex conditions are thus characterized by lower uncertainty (VL to LM) due to the fewer number of elements that interact in the generation of the model, meanwhile the most complex models are characterized by extremely high uncertainty. The uncertainty level represented in Table 2 has been estimated considering a standard site investigation expenditure, represented by (a) bibliographic analysis, (b) comparisons with closer and similar projects (c) geological profiles extrapolated from existing large scale geological interpretative maps and (e) few boreholes and scatter geophysics.

Uncertainty and Baselines

By the definition of a Geotechnical Baseline, geological complexity and its consequent uncertainty do not have a direct impact on the baseline itself, which must be discretely specified for any geological assessment, depending from the experience and sensibility of the engineering team. Complexity and the associated geological uncertainty, rather, largely impacts on potential changes from the initial baseline conditions and, as consequence, on the contingencies and the final project execution schedule.

If, the same graphic shown in Figure 1, we consider different complexity scenarios, we obtain different drifting curves of the baseline, as represented in Figure 2. Although all data are at present qualitative and based on case histories, experience and common sense, the results are quite meaningful. These curves diverge from the baseline following a progressive increase if the geological complexity.

The drifting from the Baseline very rarely happens during the mobilization phase. Some preliminary signs of deviating conditions can appear during the ramp up phase but most of the past experiences show that the appearance of more adverse conditions (compared to the Baseline) is very often encountered between 20 and 50% of the project completion.

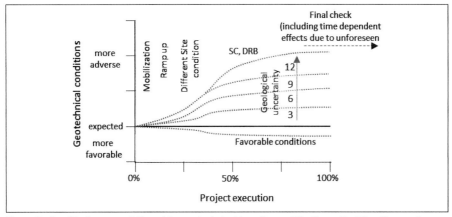

Figure 2. Graphical representation of the geotechnical baseline and the theoretical "contingency" areas: SC = Steering Committee; DR = Dispute Resolution Board; numbers 3 to 12 refer to Table 1.

For the lower complexity models, some adjustment of the Baseline can be made quite easily with minimal impact on costs and timing. In case of high complexity, the Baseline can drift from its original position almost indefinitely with dramatic impacts on the project delivery both in term of budget and schedule. Summarizing these concepts:

- the order of magnitude of potential diverge of the Baseline depends from the complexity of the geological model,
- the uncertainty in the geological forecast (or conversely, the reliability of the geological model) is directly related to the complexity of the model,
- the difference between the Baseline and its potential deviation should be defined to determine the contingencies to mitigate the geological risks,
- contingencies cannot be adequately evaluated without quantifying the uncertainty or reliability of the model.

INVESTIGATIONS AND UNCERTAINTY

This first part of this dissertation clearly states that quantify the uncertainty should be considered a major task in any project and certainly in those with complexity ranging from six to twelve.

Uncertainty in geological model leads to uncertainty in failure mode prediction and behavioral uncertainty (Vaan der Pouw Kraan et al 2014) and thereby to performance uncertainty (e.g., Langford et al. 2013, 2015). Recently, the technical and scientific community has engaged in discussions about uncertainty in geological models (Sandersen, 2008; Wellmann et al., 2011, 2014; Schweizer et al., 2017), especially with regard to *Oil & Gas* projects. In spite of that, only few contributions analyze the impact of geological model uncertainty in infrastructure projects. Carter (1992) has shown how the percentage of risk relates to unforeseen geological conditions.

Carter (1992) introduced in two major concepts: (a) the limit of practicable data acquisition, where the site acquisition expenditure curve becomes a vertical asymptote and (b) the optimum expenditure, which is defined as the intersection between the decreasing risk curve and the increasing expenditure curve. Today, in spite of hundreds of projects that have been designed and executed, an extended review of these concepts is still missing, and only general suggestions have been advanced by the ITA-AITES (1996, 2011, 2013, 2014, 2016, 2017) or by other authors (Brierley and Soule, 2009; Parnass et al., 2011; Freeman et al., 2014) at this regard.

The approach proposed by Carter (1992), which had the undeniable value to highlight the importance of correlating geological risks with investigation expenditure, has been revised hereafter, introducing the geological model complexity as one of variables that impact on the shape of these curves. Case A to D in Figure 4 show propose different curves and their consequent results by decreasing the level of complexity.

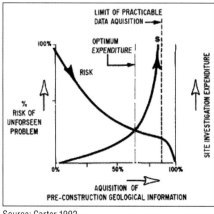

Source: Carter 1992.

Figure 3. Inverse relationship between the geological risk and site investigation expenditure

Graphics in Figure 4 show four risks curves (sloping downwards from left to right) set at complexity levels 3, 6, 9 and 12 and four site investigation expenditure curves (rising from left to right) set at the same complexity levels. Risks curves all start from 100%, in the theoretical case of fully unknown geological conditions. Their slope and the consequent risk decreasing, varies depending on the complexity. In fact, in case of simple geological models, a few general studies and investigations can considerably decrease the risk to encounter unforeseen conditions. Meanwhile for more complicated models, sophisticated and very often expensive investigations are required and can have limited success in reducing risk levels.

At the same time, the site investigation expenditure curves have been revised, also considering the complexity of the models to be analyzed. For instance, if we are facing with a shallow sedimentary and quaternary model, we can reach 100% of accuracy of the pre-construction reference model by carefully investing in borehole campaign and targeted geophysics. Such campaign will also generate an expenditure curve extremely lower that an investigation plan requested to determine high complexity models, like, for example, those that apply to long and deep transalpine tunnels.

This approach results in a revision of the uncertainty of the different models as presented in Table 3. In fact, following an optimum site investigation expenditure, all low complexity models (from 1 to 6 of Table 1) will be characterized by a very low to low uncertainty, meanwhile the most complex models, also after a detailed and expensive investigation campaign with be characterized by medium to high levels of uncertainty, which could still be also very high in some specific and localized conditions.

As a first preliminary conclusion we can suggest that necessity to qualify and quantify the uncertainty of a specific geological model, in order to understand how the Baseline could deviate from its original position, is quite negligible in case of low complexity geological environments. Such an exercise, however, becomes in our opinion mandatory for projects located within high complexity geological contexts.

RELIABILITY OF GEOLOGICAL MODELS

Reliability is defined as "how accurate or able to be trusted something is considered to be" (Cambridge Dictionary, 2018), meanwhile uncertainty is the opposite of the reliability and is define as "something that is not known or certain."

How to Quantify the Reliability of a Geological and Geotechnical Reference Model 531

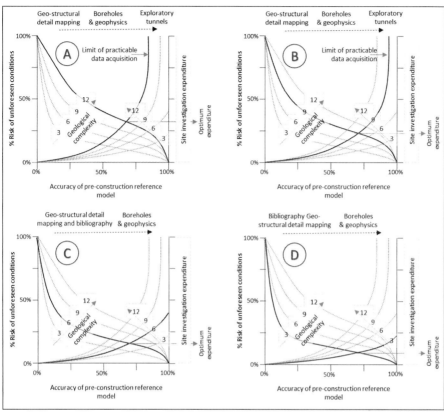

Source: modified after Carter 1992.

Figure 4. Revised inverse relationships between the geological risk and site investigation expenditure, depending from the complexity of the geological model. (A) complexity from very high to extremely high; (B) complexity from high to very high; (C) complexity from medium to high; (D) complexity from low to medium.

Table 3. Uncertainty of the different models after an optimum site investigation expenditure

| | Uncertainty (after optimum site investigation expenditure) ||||||
| | Sedimentary || Magmatic || Metamorphic ||
	Simple	Difficult	Simple	Difficult	Simple	Difficult
Unfolded	VL	VL	VL	L		
Folded	L	L	L	LM	LM	MH
Folded + Faulted	LM	M	M	MH	MH	H
Folded + Faulted + Shared		M		H	H	VH

VL = very low; L = low; LM = low-medium; M = medium; MH = medium high; H = high; VH = very high

Currently, the tunneling community seems to prefer to use the "uncertainty" instead of the "reliability" of defined geological forecast. In this paper we prefer to implement the concept of reliability, with the final goal to introduce it as a component of the GBR in its future review.

In the previous chapters we have introduced the notion of increasing complexity of geological models, depending from their geological families and deformation history.

Then we have seen how the complexity can impact in the potential divergence of the Baseline from its original state and we've discussed about how the complexity of the model impact on the geological risk percentage related to the site investigation expenditure curve.

Finally, we have stated that quantification of the reliability of geological models is a paramount process to understand and eventually quantify the possible deviation of the Baseline from its original position.

Starting from the nineties, several long and deep tunnel project through the Alps where developed to connect allow northern and southern Europe to be connected by new high-capacity and high-speed railway links (Trans European Network Transports, TEN-T). Some of these projects, like the Lotschberg, Gottard, Perthus, Frejus, Bologna-Florence tunnels have been already completed meanwhile several others like the Brenner, Lyon-Turin, Milan-Genoa, Ceneri, Trento by-pass, Koralm and Semering tunnels are all under construction at different development stages.

These projects vary from 15 to 57 kms in length and they include stretches with over 2000 meters of overburden. Initial project values range from $1B to $8B and project and construction life time varies from 15 to more than 30 years. All these projects are located within polygenetic and polymetamorphic complexes, composed by sedimentary and plutonic rocks which were lately affected by high pressure and high temperature metamorphism. The original units were subducted up to 100 kms deep and then uplifted. As consequence, most of them were affected by a wide range of ductile and brittle deformations including mylonitic and shear zones, faults, up to three generation of folding, fractures, etc.

This impressive underground development program generated the opportunity for engineering geologists, geotechnical engineers and risk engineers to face with new challenges, especially with respect to the progressive development of the geological and geotechnical baseline for each of these projects.

Extensive and systematic geological surveys were performed followed by intense campaigns of deep vertical and inclined geotechnical boreholes and in some cases deep directional boreholes, similar to those drilled for oil and gas purposes. Investigation investments in some cases largely exceed $100M. Finally, several projects were investigated by exploratory tunnels, most of them realized also with the aim to later become access galleries to the base tunnels (anticipated investments).

It can be said that, while the presence of uncertainty is understood by the cast of tunneling actors on a project, there is reluctance to adopt terminology that allows a common consideration of this uncertainty. Statistical treatment of geotechnical parameters based on measured uncertainties is more commonplace but a consideration of the model uncertainty (position of faulting, intensity of alteration, frequency of folding, etc.) leading to uncertainties in the interpolation or extrapolation of measured data is not typically expressed.

The converse terminology of reliability suffers, even within the limited exercise of geotechnical condition definition, from inaccessible language and the use of abstract multi-dimensional expressions of parameter and response space (Langford et al 2013, 2015). Very few practical developments exist to allow expression of geological model reliability.

To support the Owners to plan and develop their investigation campaign Venturini et al. (2001) lately followed by Perello et al. (2003, 2005), Bianchi et al. (2009), and Perello (2011, 2015) introduced a methodology do determine the reliability of the forecasted geological, called R-Index.

The R-Index was lately retaken by Perello (2011), which introduced the Geologic Model Rating or GMR. The index is computed on the base of two main groups of parameters which may be recognized as influencing the reliability of geological forecasts:
   a. Investigation parameters, i.e., parameters which define the quality of the investigation methods used in order to explore the rock volume to be excavated. The investigation parameters comprehend:
       – Quality of mapping process, including mapping scale, extension of the mapped area, mapping technique, outcrop percentage and depth of the tunnel from the surface along the examined area
       – Quality of the direct investigations, including number of boreholes, type of boreholes (destructive, core recovering, BHTV, sonic, etc.), distance from the tunnel, depth reached by the investigation,
       – Quality of the geophysical investigations, including number of available geophysical profiles, quality of the survey (HR vs. LR), average distance from the tunnel alignment, depth reached by the investigations
   b. System Parameters, i.e., parameters which define the geological complexity of the rock volume and therefore the system to be investigated, as stated in previous sections of this work.

Each one of elements that characterize these parameters is waited and calculated through the method of the "Interaction Matrices" or Fully Coupled Model developed by Jiao and Hudson (1995) and Hudson and Jiao (1996).

The advantage of this methodology is that the area of improvement that impact on the reliability of the model are easily highlighted and can thus be improved by additional investigations, if need. The R-Index is usually calculated for homogeneous sections along the tunnel, allowing to define which part of the project are more reliable which others need more investigation or some other particular and carefully approaches.

Figure 6 shows an extract from the geological profile of the Lyon-Turin Base Tunnel (France–Italy) presently under construction, where the R-Index was systematically applied and calculated every 500 meters. In the considered section R-Index ranges between 3.5 (low reliability) and 7.3 (medium–high reliability) in a scale from 1 to 10, depending from complexity of the geology, the number and quality of boreholes, other existing tunnels or exploratory galleries, etc.

CONCLUSIONS

GBR represents a useful contractual tool to define the boundary (baseline) of the liability between Owner and Contractor with respect to the geotechnical model. At present the suggested Guidelines to prepare Geotechnical Baseline Reports (GBR) for Construction (Essex et al., ASCE, 2007), do not indicate a methodology to define the uncertainty of the baseline and the consequent contingencies that have to be considered for the project.

At the same time the potential deviation of the baseline from its original position depends from the complexity of the geological model. In a scale from one to twelve, models with low level of complexity, usually ranging from one to six, are also

Factors contributing to DPQ (drillholes potential quality)											
Factor	Value	Rating	Factor	Value	Rating	Factor	Value	Rating	Factor	Value	Rating
Drillholes quantity in an interval of 2 km	1	3	m% of cored drillhole	0%	1	Average distance from tunnel axis (m)	2000	1	Average drillholes depth vs. tunnel depth	0,25	1
	3	5		30%	3		500	4		0,75	5
	5	8		60%	5		250	8		1,00	9
	>7	10		100%	9		0	10		1,20	10
Add the rating (10-rating) 0.5 if some of the drillholes are extrapolable with certainty to the considered stretch; rating 10 with 1 drillhole intersecting the stretch			Add the rating (10-rating) 0.5 if BHTV is available; rating=10 for 100% cored drillhole + BHTV			Add the rating (10-rating) 0.5 if some of the drillholes are extrapolable with certainty to the considered stretch;rating 10 with 1 drillhole intersecting the stretch			Rating 10 with 1 drillhole intersecting the stretch		

Factors contributing to MPQ (mapping potential quality) derivation											
Factor	Value	Rating	Factor	Value	Rating	Factor	Value	Rating	Factor	Value	Rating
Mapping scale	1:50000	1	Mapped area (km²) vs. tunnel depth (km)	2	1	Outcrop percentage	<10%	1	Field data collection method	A	2
	1:25000	3		4	4		30%	4		B	5
	1:10000	7		10	8		60%	8		C	10
	1:5000	8		>20	10		>90%	10			
This parameter is not referred to a specific section, but to the whole tunnel layout						This parameter must be evaluated over a distance of some km (0.5–3) around the considered layout, depending on tunnel depth (see also note 2 below)			See note (1) below		

Factors contributing to GPQ (geophysic potential quality) derivation											
Factor	Value	Rating	Factor	Value	Rating	Factor	Value	Rating	Factor	Value	Rating
km of sampling lines in an interval of 2 km	<0,1	1	Method resolution	A	2	Average distance from tunnel axis (m)	2000	1	Average investigated depth vs. tunnel depth	0,25	1
	0,5	4		B	4		500	4		0,75	5
	1	7		C	7		250	7		1	9
	2	9		D	10		0	10		1,2	10
Add the rating (10-Rating) 0.5 if some line is extrapolable with certainty to the considered stretch; rating 10 with 1 line intersecting the stretch			A=low resolution without validation drillholes; B=high resolution without validation drillholes; C=low resolution with validation drillholes; D=high resolution with validation drillholes			Add the rating (10-Rating) 0.5 if some line is extrapolable with certainty to the considered stretch; rating 10 with 1 line intersecting the stretch			Rating 10 with 1 line intersecting the stretch		

Source: Perello, 2011.

Figure 5. Investigation parameters considered to determine the reliability of the geological models. The variance of each factor is rated depending on the available data. The combination of these factors defines the R-Index, which represents the quantitative evaluation of the reliability of the geological model at the time of the evaluation.

characterized by low levels of uncertainties. The uncertainty and its geological associated risks also depend from the level of expenditure of the site investigation campaign.

In order to quantify the uncertainty of the model or, in other words, the reliability of the geological forecasts presented in the project geological profile, a numerical evaluation, called R-Index has been developed and applied in most of the major long and deep tunnel projects through the Alps.

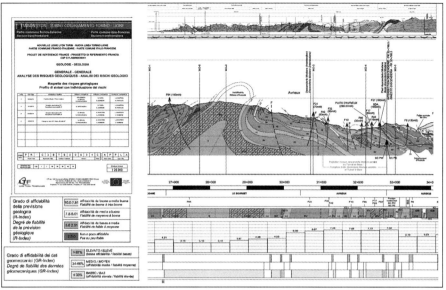

Figure 6. Geological profile of the Lyon-Turin base tunnel, 2014 (scale 1:25,000) with associated geological risks and quantification of the reliability of the geological and geomechanics forecast every 500 meters. This document represents the geological baseline for the entire project. The evaluation of the reliability level (grado di affidabilita' in Italian and degre' de fiabilite' in French) shown in the profile allows contractors to finalize the most appropriate methods to mitigate the risks. On the same time the Owners and its consultants are aware of the potential variability of the geology due to the variable complexity of the model.

The quantification of the reliability of the geological model since early stages of underground projects and its integration in the GBR as basic tool for identifying the uncertainties of the baseline model may lead to a significant improvement in the aim of cost control for underground projects.

REFERENCES

AFTES WG32. Recommendations on the characterization of geological, hydrogeological and geotechnical uncertainties and risk. Tunnels et Espace Souterrain, 232. pp. 316–355.

Bianchi, G., Perello, P., Venturini, G., Dematteis, A. 2009. Determination of reliability in geological forecasting for tunnel projects: the method of R-Index and its application on two case studies. IAEG Italia 2009. pp. 1–18.

Brierley, G., Soule, N. 2014. To GBR or not to GBR; Is that the question? Geo-Congress 2014. February 23–26, Atlanta, Georgia. pp. 2332–2244.

Carter, T.G. 1992. Prediction and uncertainties in geological engineering and rock mass characterization assessments. Quarto Ciclo di Conferenze di Meccanica e Ingegneria delle Rocce, Torino. pp. 1–23.

Ciancia, M. 2014. Geologic Considerations for Setting Geotechnical Baseline Report. Geo-Congress 2014. February 23–26, Atlanta, Georgia. pp. 2273–2289.

Dematteis, A., Soldo, L. 2015. The geological and geotechnical design model in tunnel design: estimation of its reliability through the R-Index. Georisk Assessment and Management of Risk for Engineered Systems and Geohazards 9(4). pp. 1–11.

Essex, RJ. 2017. Geotechnical Baseline Reports for Construction. ASCE. Reston, USA. 73 pages.

Essex, RJ., Diederichs, MS. and Giles, EL. 2001. Managing geotechnical and financial risk in underground construction. Int. Soc. of Rock Mech. News Journal. v6:3 pp. 38–39.

Freeman, T., Klein, S., Korbin, G., Quick, W. 2009. Geotechnical Baseline Reports—A Review. Rapid Excavation and Tunneling Conference (RETC). pp. 232–241.

Hudson, J.A. & Jiao, Y. (1996). Information audits for improving rock engineering prediction, design and performance. In Eurocl '96, Barla Ed., Balkema—Rotterdam, pp. 1405–1412.

ITA_AITES, WG3, 1996. Contractual sharing of risks in Underground Construction ITA view. pp. 1–5.

ITA_AITES, WG3, 2011. The ITA Contractual framework checklist for subsurface construction contracts. pp. 1–12.

ITA_AITES, WG19, 2013. Guidelines on Contractual Aspects of Conventional tunneling. ITA Report 2013. pp. 1–16.

ITA_AITES, WG4, 2014. Study of Access ways to underground space—Examples of spatial requirement. ITA-AITES 2014. pp. 1–73.

ITA_AITES, WG14 and WG19, 2016. Recommendations on the development process for Mined Tunnels. ITA Report 2016. pp. 1–32.

ITA_AITES, WG17, 2017, TBM excavation of long and deep tunnels under difficult rock conditions. pp. 1–64.

Jiao, Y. & Hudson, J.A. (1995). The fully-coupled model for rock engineering systems. Int. J. Rock Mech. Min. sci. & Geomech. Abstr. 32/5, pp. 491–512.

Langford, JC and Diederichs, MS. 2015. Reliable support design for excavations in brittle rock using a global Response Surface Method, Rock Mechanics and Rock Engineering. v48:2 pp. 669–689.

Langford JC and Diederichs MS. 2013. Reliability based approach to tunnel lining design using a modified Point Estimate Method, Int. Journal of Rock Mechanics and Mining Sciences, v60:6 pp. 263–276.

Lemessa, F. 2004. Reliability or Likelihood of Geological or Geotechnical Models. ITC/Delft 2004. Netherlands. pp. 1–112.

Parnass, J., Staheli, K., Hunt, S., Fowler, J., Hutchinson, M., Maday, L. 2011. An in-depth discussion on geotechnical baseline reports and legal issues. NASTT 2011. Washington D.C. pp. 1–10.

Perello P, Venturini G, Delle Piane L, Martinotti G (2003) Geostructural mapping applied to underground excavations: updated ideas after a century since the first trans-alpine tunnels. Proc Rapid Excavation and Tunnelling Conference—New Orleans 2003, pp. 581–591.

Perello, P., Venturini, G., Dematteis, A., Bianchi, G., Delle Piane, L., Damiano, A. 2005. Determination of reliability in geological forecasting for linear underground structures the Methods of the R-Index. Geoline 2005. Lyon (FR) pp. 1–8.

Perello, P. (2011). Estimate of the Reliability in Geological Forecasts for Tunnels: Toward a Structured Approach. Rock Mechanics and Rock Engineering, 44(6), pp. 671–694.

Perello, P. (2015). Are reliable geological forecasts possible for long and deep tunnels? Experience gained from some big railway project in Italy. 3rd Arabian Tunneling Conference and Exhibition, Dubai (UAE) 23–25.

Salam A., 1996. Contractual Sharing of Risks in Underground Construction: ITA Views. Elsevier Science Ltd. Great Britain 1995. pp. 1–5.

Sandersen, P. 2008. Uncertainty assessment of geological models—a qualitative approach. pp. 1–5.

Schweizer, D., Blum, P., Butscher, C. 2017. Uncertainty assessment in 3-D geological models of increasing complexity. Solid Earth 2017. pp. 1–16.

Wellmann, J.F., Horowitz, F.G., Regenauer-Lieb K. 2011. Towards a Quantification of Uncertainties in 3-D Geological Models. IAMB 2011. Salzburg. pp. 1–14.

Wellmann, J.F., Lindsay, M., Poh, J., Jessell, M. 2014. Validating 3-D structural models with geological knowledge for improved uncertainty evaluations. EGU 2014. pp. 1–8.

Van der Pouw Kraan, M. and Diederichs, MS. 2014. Behavioral uncertainty for rock tunnels: Implications for rock mass definitions or predictions in geotechnical baseline reports. Proceedings of the World Tunnel Congress 2014., Iguazu, Brazil. 10 pages.

Venturini, G, Damiano, A., Dematteis A., Delle Piane L., Fontan D., Martinotti G., and Perello P. 2001. "L'importanza dell'affidabilità del modello geologico di riferimento negli studi per il tunneling." Proceedings del 3 forum italiano di Scienze della Terra (Geoitalia) 426–427. Chieti, 2001, pp. 5–8.

Innovative Monitoring for Urban Tunneling

Paul Thurlow • Geo-Instruments
Glen Frank • Schnabel Engineering

ABSTRACT

Tunneling in soft ground for the Regional Connector Transit Corridor project in downtown Los Angeles had many challenges including typical risks of damage to buildings and utilities, community disruption, and environmental impact. A considered approach to monitoring structures and ground movement was required that had to draw on new innovations in the instrumentation industry. As well as structural monitoring, the planning and effects of compensation grouting had to be measured and quantified. Unique measures were taken to start tunneling under existing buildings immediately upon launching the earth pressure balance (EPB) tunnel boring machine (TBM) with very low ground cover. This paper provides an overview of the measures taken to safeguard all assets with particular attention to innovation in wireless instrumentation to control compensation grouting and understand structural response and visualization techniques of in-ground data that has revolutionized the way that volume loss can be viewed real time.

PROJECT OVERVIEW

The Regional Connector Transit Corridor project is a design/build, light rail underground project that will connect two existing rail lines in Los Angeles. The 3.1-km (1.9 miles) alignment with a base contract value of $ 1 billion including construction within the Little Tokyo and Bunker Hill neighborhoods, and the Financial District in the heart of downtown Los Angeles. It will provide connections between the Metro Gold Line from Pasadena and East Los Angeles; and the Metro Blue Line to Long Beach and the Expo Line to Culver City. The project includes the construction of three underground stations with depths ranging from 40 to 112 feet, approximately 1.1 miles of twin tube earth pressure balance (EPB) mining, approximately 0.9 mile of cut and cover tunnel including station excavations and a 330 feet long cavern. Running tunnels were constructed following the pressure balance tunneling method utilizing a 21.6 feet Earth Pressure Balance (EPB) Tunnel Boring Machine (TBM). The project is owned and operated by Metro with Regional Connector Constructor (RCC)—a Skanska/Traylor Brothers joint venture- as the design-builder and Mott MacDonald as the lead design engineer. Hayward Baker provided the drilling and compensation grouting package with Getec and Geo -Instruments providing the bespoke instrumentation for this work area.

BUILDING PROTECTION FOR RUNNING TUNNEL CONSTRUCTION

Elevated risks for cosmetic building damage at several locations along the alignment were identified during preliminary engineering and RCC was tasked with evaluating this risk at each of these locations in light of the proposed means and methods of construction. Most of the tunneling on the Regional Connector project will be in a very weak siltstone (the Fernando Formation) with well over two diameters of cover, which is an ideal medium for EPB tunneling. RCC was able to show that the use of relatively high tunneling support pressures would result in the anticipated ground movements less than the very stringent 0.5 inches at all of the areas of concern associated with impacts from the running tunnels constructed in the Fernando Formation. The first section of the tunneling however presented a significantly different challenge.

Reach 1A is the first 600 feet of the tunnel construction and begins in Quaternary aged Alluvium, which consists of layers (average thickness of 350 mm) of primarily clean uniform sand interbedded with silty sands, gravels, silts, and well-graded sands. The alignment is very shallow with less than a diameter of cover to the foundation of the overlying structures at the beginning of the permanent tunnel. The first 200 feet was in drier material as the alignment dips down at a 4% grade in a horizontal spiral. The water table climbs from spring line as the bottom of the tunnel starts to encounter the underlying siltstone over the next 100 feet. The final 300 feet of Reach 1A required tunneling in a mixed face with siltstone below and saturated sand and gravel above while negotiating a 600' radius curve. All of Reach 1A is directly below structures on the surface and with less than two diameters of cover, and two of these structures required preparations for compensation grouting as a mitigation to meet the settlement criteria.

COMPENSATION GROUTING

Compensation grouting involves installing horizontal ported grout pipes (Tube a Manchette or TaM consisting of PVC or metal tubing—aluminum in this case) under structures at depth in alluvial material positioned between the proposed tunnel crown and the building foundations. The TaMs are then used to deliver grout to specific locations (ports at 2 ft intervals) along the alignment first to precondition the soil, then to pressurize horizontal fractures resulting in a controlled heave at the surface. This usually requires a 3 phase process, with TaM installation and ground preconditioning being the preparatory phases that are always required, and then the compensation grouting phase which is only initiated if the overlying building or infrastructure settles more than a preset amount.

Compensation Grout Pipe Installation

A total of 13 TAM's were installed under the buildings at risk for settlement, starting below South Central Avenue and ending near 2nd Street, a total length of 500ft. The TAM's were installed using Horizontal Directional Drilling methods with the precise orientation of the holes surveyed using an Azimuth Aligner. The drill rig was positioned on a bench above the TBM launch alignment and below South Central Avenue, while decking was used to keep the road above operational. During the drilling, magnetic coils were used for guidance of tooling and to steer and control the direction. Once the HDD hole was drilled, the hole was conditioned and ready for TaM installation.

Due to easement issues these TaMs were installed in a blind hole, with a proprietary machine was designed and constructed by Hayward Baker. The locations of the TAMs were entered into the grout control software and identified in a spatial view. The ports could then be selected on screen with a digitized pen/mouse and then assigned to the grouting team with a printout /tablet view of the ports, pressures and volumes required.

Preconditioning for Compensation Grouting

Preconditioning is the process of adding grout to the formation around the TaMs such that the state of stress is modified to induce horizontal fractures when the grout is injected at pressure. When these horizontal fractures are filled with pressurized grout the result is upward movement of the ground above the grouting zone (heave). Once heave is detected at the surface above a specific grouted zone that zone is considered to be "preconditioned" and any additional grout injection will result in the buildings being raised. The use of real-time monitoring from instruments located in the structures basements are critical for the success of this phase.

The grouting instructions are prepared within the grout control software and sent by direct link to the grout containers which were located at the TBM launch work site. The grouting containers are self-contained units that houses the agitators, pump and control software. The monitoring data is downloaded to a terminal in the container so that information is available immediately. Once a particular grouting phase is complete the data is stored and the shift engineers can retrieve this information and compare it with the data from the monitoring instruments.

Preconditioning grouting injections were preformed using a "Primary, Secondary, and Tertiary" pass system in the preconditioning phase. The grouting sequence entailed the injection of Primary ports first, followed by secondary ports during preconditioning. Tertiary ports were held in reserve and were not grouted during the preconditioning phase. If necessary, they would be used for compensation grouting during the excavation by the tunnel boring machine.

INSTRUMENTATION

The structural monitoring system for the compensation grouting included a hydrostatic levelling cells system and both fixed and portable tiltmeters.

Hydrostatic Leveling Cell System

Hydrostatic levelling cells are usually installed within the basements of structures that are within compensation grouting zones or high settlement risk areas. This allows direct correlation to the grouting activities and as the cells are installed on load bearing columns or walls, providing early indication of deep ground movement. The hydrostatic levelling cells provide data at 10 minute intervals and are stable and low maintenance instruments. As the hydrostatic levelling cells installation will require the cells to be affixed to bearing walls, a detailed survey of the structure prior to installation has to be done and this allows the design to be tuned to provide the information required for the compensation grouting engineers.

Based on experiences on previous grouting sites for different European tunnel projects such as Crossrail Drive X, cell coverage varied between 25 m^2 and 40 m^2 structural area per cell. This spacing provided grouting contractors with a reliable, efficient and safe monitoring solution for the different structural layouts. Usually the spacing is designed to incorporate slope and deflection ratio parameters.

The use of hydrostatic levelling cells is dependent on a reference cells that is located outside the potential ground movement zone, and two different reference points were used at site. The system was subdivided in groups, with connections to ensure that structures are linked in the circuit.

Tiltmeters

Due to worksite restrictions the hydrostatic cell system could not be installed on all of the at risk structures, which were instrumented with a combination of mini bi-axial tiltmeters and portable tiltmeters.

The mini bi-asial tiltmeters were programmed to read every 1 hour prior to the tunneling works and during the passage of the TBM it was increased to every 15 minutes. The battery life for the tiltmeters was rated at 1.5yrs at this configuration.

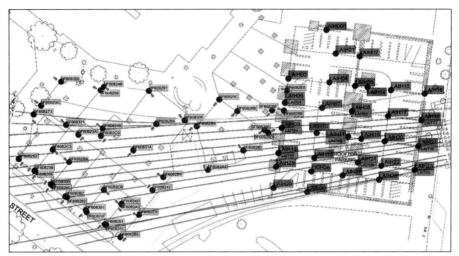

Figure 1. Layout of structural instrumentation and TAM locations

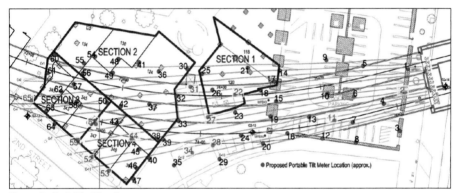

Figure 2. Portable tiltmeter sequence

The portable tiltmeters (Real-time instrumentation) were to be placed in areas directly above where grouting would take place on a day by day basis. The portable tiltmeters take measurement readings at 4-second intervals and were positioned to monitor any type of surface response during grout injection.

Installation

Due to sensitivity of the structures and the approach by Metro to the structures owners, the intended instrumentation and monitoring required a non-invasive approach. The structure owners wanted no drilling, wiring or repeated disturbance over the duration of the works. After a series of consultations with the owners it was decided that a very small wireless mesh tiltmeter network would be deployed on the structural members of the occupied buildings within the zone of the tunnel works in the high risk zone. The unoccupied structure was to be monitored with the hydrostatic water level cells with the provision that all tubing and wiring was concealed as well as possible from the public who used the facility.

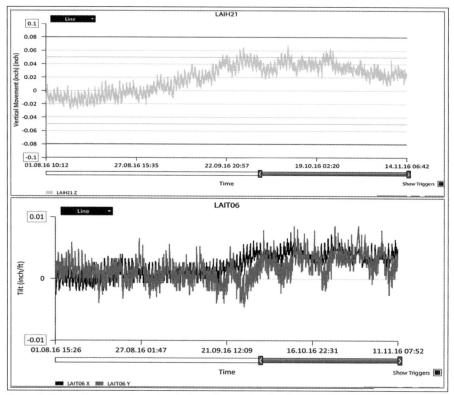

Figure 3. Effect of heave as shown by a hydrostatic water level and tiltmeter attached to a column in the unoccupied structure

The layout of the tunnel in relation to the structure allowed for two reference cells to be installed to service the structure. The structure was comprised of three water level cell circuits comprised of 40 cells including step up cells which were necessary to ensure that the systems range did not exceed 1.2 meters.

Instrumentation and Data

A total of 41,680 gallons of grout was used in pre-treating the soils from August 17th, to November 7th, 2016. The effectiveness of the grouting was measured by the reaction from the hydrostatic water cell system installed on the unoccupied building and the fixed bi-axial and portable tiltmeters in the occupied buildings.

SAA INSTALL AND DATA USE DURING DRIVE

Due to easement restrictions under the first structure, the compensation grout pipes were much closer to the tunnel (7 feet) than they were to the foundations of the building (from 12 to 15 feet). There was considerable concern that the grouting pressures required to lift the buildings would also deform the tunnel. As mitigation for this concern and other potential benefits, the team opted to install a Shape Accelerometer Array (SAA) directly above the crown of each of the two tunnels.

Shape Accelerometer Array (SAA)

The SAA instrument could be described as a "horizontal inclinometer" made up of a linear series of rods each of which has an accelerometer installed in it such that the change in orientation of each rod can be measured over time. Knowing the change in orientation and the fixed length of each rod, the location of the joint (node) between each rod can be calculated and compared to its original position. In this case the rods were each 1 meter in length resulting in a means to calculate the movement of the soil directly above the crown every few feet along the alignment in real time. Conceptually this is equivalent to having a deep settlement point installed between two feet and five feet above the crown of the tunnel every three feet along the alignment.

As with an automatically monitored deep settlement point, the data from the SAA allows for immediate (within 5 minutes or sooner) identification of lost ground as it is occurring near the TBM. This information could confirm that the excavation process is proceeding as planned and no action needs to be taken, or it could indicate that excessive lost ground is occurring and remedial measures should be initiated. Unlike a deep settlement point however, the SAA will provide immediate feedback in regards to the effectiveness of the remediation efforts, as well as providing continuous rather than intermittent data as the TBM progresses along the alignment.

Use of SAA for Building Protection

In the specific case of mitigating the potential for compensation grouting pressures deforming the tunnel, the SAA would provide information that would be used to greatly reduce the grouting pressures required to protect the buildings.

Compensation grouting as typically employed involves the very careful monitoring of the buildings during the excavation. If movement is detected then grouting is initiated at pressures high enough to lift the buildings (by lifting the soil mass they are founded on) back into place. This process continue until the building movement stabilizes such that no further grouting is required to maintain it. Thus the void created by the lost ground travels up through the ground past the grouting pipes to the surface, then the grout is used to push the soil mass and the buildings above the grout pipes back up. Very high grouting pressures may be required since the entire soil mass must be lifted and if this pressure was applied to the tunnel liner in a non-uniform manner then the tunnel would deform. Any deformation of the tunnel could lead to a reduction in the durability and or utility of the final subway structure.

With the data from the SAA the compensation grouting could be initiated as soon as lost ground occurred, potentially (depending on the nature of the lost ground event) before the created void had migrated up to the zone above the tunnel where the TaMs were located. The information from the SAA could also be used to estimate the volume of the lost ground, and thus the volume of the void that should be filled with grout from the compensation grouting. That volume of grout could be delivered to the ports directly above the area where the SAA data reported the lost ground and the void could be filled at much lower grouting pressures since the void would be very close to the grout ports when the grout is being delivered to the ports.

Preparation for SAA Data Monitoring

Once the decision to utilize the SAA in this innovative manner was made, it was critical that the entire team be aware of what was anticipated in order to minimize any misinterpretation of the SAA data. With this type of close collaboration between a rapidly advancing pressure balance tunneling system and a ground movement monitoring system it is vital to agree what is expected from the monitoring system ahead of time. Remedial actions must be implemented rapidly to minimize the negative impacts of lost ground, and there is little time to debate what the data means. Since there was very little precedent for this type of data gathering in conjunction with a pressure balance TBMs some time was spent in discussion about what was expected concerning data signatures for given potential scenarios. In this way, unexpected data resulting from sensor malfunction would be unlikely to be interpreted as ground movement, which could result in significant negative impact if acted upon.

Near crown settlement (or ground movement) monitoring for pressure balance tunneling has been common practice for several years, and recent projects have shown good correlation between near crown settlement, and surface settlements. Near crown monitoring also provides the necessary information to gauge the effectiveness of the pressure envelope around the TBM at the face, shield gap, and tail void, thus identifying the likely location of the root cause of surface settlements. Utilizing records from near crown settlement monitoring on recent projects, anticipated data signatures were developed for the SAA in the case of various potential scenarios. These scenarios included loss of pressure due to air bubble escape from the TBM plenum, lack of sufficient shield gap pressurization, and insufficient tail void grouting, among others.

In addition to modeling the data signatures that were expected from the SAA in response to different means of lost ground development, the signatures expected during the different means of mitigation were also modeled. This would include filling of the lost ground induced void with tail void grout as well as filling the void with compensation grout.

SAA Installation and Baselining

HDD for the installation of the SAAs and the installation of the instruments was accomplished from a platform created directly over the top of the TBM launch zone with the precise orientation of the holes surveyed at the drill head using an Azimuth Aligner This work was done in conjunction with the TaM installation for the compensation grouting and was completed approximately 5 months prior to the TBM launch date for the first drive. The installation of the SAA above the 1st tunnel varied in elevation from 570 mm to 1,800 mm (avg 1,100 mm) above the top of the planned tunnel excavation, and about 1 meter either side of the planned centerline.

It became apparent after a few weeks that the data from the SAA was going to be somewhat problematic. Within 3 weeks of installing the instrument the data indicated that the end node had "heaved" about 0.3 inches. After 3 months, the data indicated a "heave" of over an inch. With approximately 1 month left until the planned start of tunneling the data was re-baselined in order to closely observe this creeping behavior and try to establish some sort of work around. It was found that each of the rods that make up the instrument angled up or down slightly each time that the instrument was queried despite the fact that there was no actual movement. On average however, the rods angled up more often or with a greater magnitude than down, resulting in the upward creep over time.

Eventually the team decided that the best way to deal with the anomaly was to re-baseline the instrument again just before launch and then to use a relational approach to analyzing the data. Rather than compare the elevation of each node to zero as it the TBM passed under it in order to determine the amount of settlement at depth, it is compared to the value that it had prior to the influence zone of the TBM reaching it, which has been determined to be about 8 meters in front of the cutterhead. While this approach does not completely eliminate the potential effect of the creep behavior, especially when the TBM is progressing slowly, it proved to be a satisfactory work around.

SAA Monitoring at First TBM Launch

The first three nodes of the SAA were installed in a zone with less than eight feet of cover over the top of the TBM. The following four nodes were under the sidewalk and two of these nodes were in close proximity to the bottom two sensor of an multipoint borehole extensometer (MPBX).

There was considerable upward movement of the first several nodes, with the first node briefly showing heave of over one inch. The signature of the movement of these nodes was very similar to what was expected based on the earlier planning efforts involving data signature modeling

The high displacement is the SAA data being recorded as the TBM is advancing into the launch chamber area. The ballast at this location is providing a small amount of overburden resulting in heave above the crown of the tunnel.

In order to gain some confidence that the magnitude of the movement being reported by the SAA was correct the data was compared to the data from an MPBX that was installed near the start of the tunnel. The distance between the bottom anchor of the

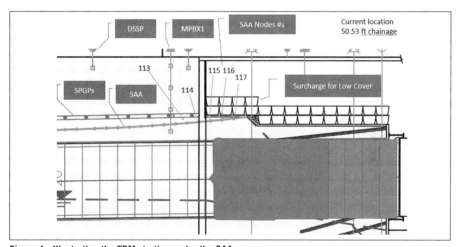

Figure 4. Illustrating the TBM starting under the SAA

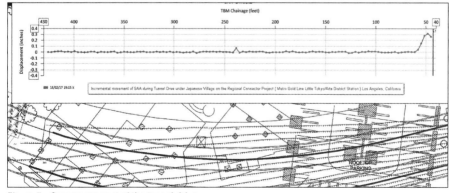

Figure 5. Commencement of the tunnel drive

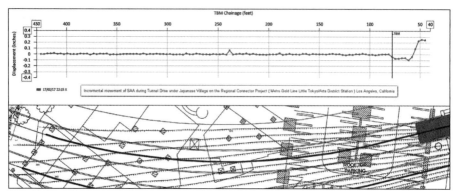

Figure 6. TBM exits the launch shaft and starts under the structure, the in ground settlement is clearly recorded

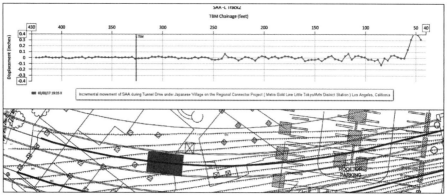

Figure 7. TBM advances away from the launch shaft and is able to apply higher support pressures, the settlement is clearly reduced

multipoint borehole extensometer (MPBX) and the SAA is approximately 300 mm laterally and 400 mm vertically. The values reported by the SAA and the values reported by the MPBX varied by less than 2 mm as the TBM passed under them. This indicates that the SAA is not only giving an indication of where and when movement is occurring as the TBM goes under each node, but that the magnitude of the movement being reported is accurate as well.

The maximum settlement reported by the SAA on the first drive was approximately 0.4 inches, which occurred over the cutterhead as the tail shield passed out of the area of very low cover. The estimated volume loss associated with this near crown settlement measurement is approximately 0.35%. Once a couple of rings had been grouted in beyond the low cover area the near crown settlement was reduced to around 0.25 inches and the corresponding volume loss was estimated to be less than 0.2%. The plenum pressure was also limited by the ground conditions, which were granular material above the water table. In these conditions it is very difficult to condition the muck such that it will flow through the system and this difficulty increases with increasing pressure. The TBM operator kept the plenum pressure at the minimum allowed until the water table reached around springline of the TBM. At this point the advance rate increased dramatically and the plenum pressure was raised above 2 bar resulting in an estimated volume loss of well under 0.1%.

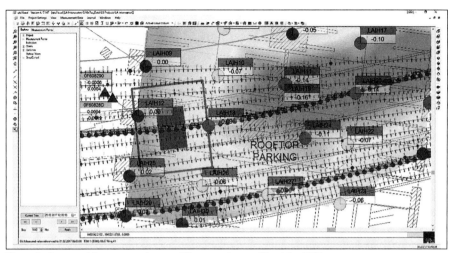

Figure 8. Structural instrumentation providing a contour profile of the TBM passage

In general, the data from the SAA shows four general patterns as the cover (confinement), plenum pressure, and tail void grouting pressure increased:

1. Marked heave prior to TBM arrival, marked settlement across the shield and marked heave at the grouting location, then minimal movement.
2. Slight heave prior to TBM arrival, some settlement across the shield, slight heave at the grouting location, then slight settlement again.
3. Very slight heave prior to TBM arrival, slight settlement across the shield arrested at the grouting location then minimal movement.
4. Very slight heave prior to TBM arrival, then minimal movement for several tunnel diameters, followed by a very slight settlement.

CONCLUSION

The use of innovative structural and in-ground instrumentation to provide tunnel engineers and designers real time ground movement information is essential for urban or other tunnel projects having high risks associated with volume loss and the associated ground movement. In addition to the data it is also important that the method of visualization is also crucial in order to minimize the time to initiate mitigation, which is critical for preventing damage to at-risk buildings and infrastructure. These innovative monitoring techniques provides a means for applying the observational approach, long used in traditional underground excavation techniques, to the pressure balance tunneling method. Integrating TBM position, in-ground (near crown) instrumentation and structural monitoring allows for greater confidence when the pressure balance method is being applied in locations where the risk associated with excessive ground movement is high. This higher level of confidence often results in better production and was likely a contributing factor to the impressive production of the TBM on the second drive of the Regional Connector Project.

REFERENCE

Hansmire, W., McLane, R., Frank, G., Bragard, C.Y., 2017. Challenges for Tunneling in Downtown Los Angeles for the Regional Connector Project. ITA AITES World Tunnel Congress, Bergen, Norway, June 2017.

Packer and Long-Term Aquifer Testing to Estimate Tunnel Inflows: A Case History for the Lower Meramec Tunnel, St. Louis, MO

Kenneth A. Johnson ▪ WSP USA, Inc.
Xiaomin You ▪ WSP USA, Inc.
Kyle D. Williams ▪ WSP USA, Inc.

ABSTRACT

Estimating groundwater inflow in tunneling projects is one of the most challenging and important elements of tunnel design and construction. For the Lower Meramec Tunnel (LMT) in St. Louis, MO groundwater inflow estimates utilized data from both packer tests and long-term aquifer testing to develop hydraulic conductivity estimates for various reaches of the tunnel alignment. Methods of Heuer (2005) and Raymer (2005) were used for initial inflow estimates and adjustments were made to reconcile the two types of hydraulic tests. The data and analysis suggest the two testing methods for hydraulic conductivity characterization display systematic differences that should be considered in the development of groundwater inflow estimates. The combination of the two types of testing can provide a broader and deeper understanding of the hydrogeologic response at critical locations along a tunnel alignment.

INTRODUCTION

The LMT Project, formally known as the Lower Meramec River System Improvements— Baumgartner to Fenton Wastewater Treatment Facility (WWTF) was initiated by the Metropolitan St. Louis Sewer District (MSD) based upon the 201 Facility Plan prepared in 1979 and updated in 1985 with a goal of removing WWTF discharges to the Meramec River. Phase I of the plan became operational in 2007 with startup of the Lower Meramec WWTF and construction of the Baumgartner Tunnel. Phase II of the Facility Plan consists of an extension of the Baumgartner Tunnel to the Fenton WWTF. The final phase, Phase III of the Facility Plan, would extend the tunnel to the Grand Glaize WWTF. Figure 1 provides an overview of the overall program.

The LMT project consists of an approximate 35,932 feet (6.8 miles) extension of the Baumgartner Tunnel that traverses through unincorporated St. Louis County, the City of Sunset Hills, and the City of Fenton. The tunnel diameter is anticipated to be approximately 12 feet with an eight-foot nominal diameter tunnel carrier pipe sloping at 0.1% towards the Baumgartner Tunnel. Six drop structures, a receiving shaft, and an optional construction shaft are also part of this project.

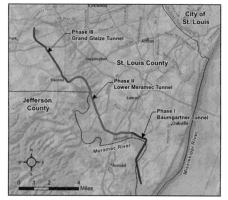

Figure 1. Program vicinity map (Litton and others, 2017)

The LMT project traverses upland and lowland areas of the Meramec River valley and the tunnel is entirely within the

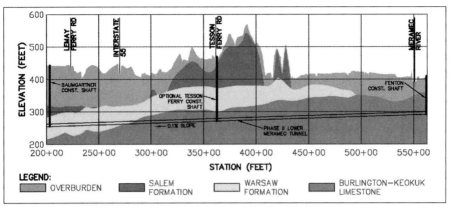

Figure 2. Geologic profile (Litton and others, 2017)

underlying bedrock units. The lowland valleys are primarily characterized by alluvial deposits ranging from silts to lean clays and from sands to gravels at greater depths. The lowland valleys are generally present at the beginning and end of the alignment near Baumgartner and Fenton, respectively. The upland area, which is higher in elevation and within the middle portion of the LMT alignment, generally consists of a veneer of loess overburden except within channels flowing to the Meramec River.

The LMT will connect to the Baumgartner Tunnel with an invert elevation of 255.1 feet and terminate at the Fenton WWTF with an invert elevation 291.1 feet relative to NAVD88. The depth of the LMT places it entirely within the Mississippian-age rock that comprises the near-surface bedrock of the project area. The near-surface bedrock consists of the Salem Formation, underlain by the predominantly limestone Warsaw Formation which is in turn underlain by the Burlington-Keokuk Limestones. Figure 2 presents a conceptual geologic profile along the alignment (see Figure 1) showing the tunnel horizon. The geologic profile was prepared based upon the interpretation of the information gathered during the LMT geotechnical investigation.

OVERVIEW OF THE INVESTIGATION

Exploration of subsurface conditions along the proposed LMT alignment began in 2012. Throughout the subsequent design phases, an exploration program was implemented in four phases between 2014 and 2018.

Borings and Core Investigation Methods

A total of 77 borings were drilled between the Baumgartner Lagoon and the Fenton WWTF along the proposed LMT alignment. Twenty-two of the deep rock borings along the tunnel alignment were inclined 15 degrees from vertical to improve the likelihood of encountering vertical or near-vertical joints. The remaining borings were drilled vertically. Deep rock borings were typically extended beneath the anticipated tunnel invert ranging from 146 feet to 312 feet below the ground surface.

Deep rock borings were typically advanced through the overburden using tri-cone roller bits with temporary steel casing, then continued into the rock using wire-line, HQ-sized tools. Coring was typically completed in 10-foot long runs. After drilling was completed, packer tests and downhole geophysical surveys using an optical or acoustical televiewer were completed in each of the deep rock borings.

Laboratory testing was performed on samples obtained from the borings. Rock tests included porosity, density, slake durability, free swell, point load, uniaxial and triaxial compression, Brazilian tensile, direct shear, chert content, Mohs hardness, Cerchar abrasivity, punch penetration, TBM Suite (composed of the Abrasion Value Steel Test, Brittleness Value S20 Test, and Sievers J-number Drillability Test) and thin-section petrographic analysis.Packer Tests

To evaluate the hydraulic conductivity of the rock mass, 571 packer tests were performed in 63 deep rock borings generally using 10-foot long test intervals. The tests began from the bottom of the borehole and ascended to approximately two tunnel diameters above the crown.

High water takes during packer testing occurred in areas close to the contact between the predominantly limestone Warsaw Formation and Burlington-Keokuk Limestone and areas near the alluvium-filled river valley near the Meramec River. Downhole televiewer (ATV) logs show open fractures in the vicinity of the expected contact between the Warsaw Formation and the Burlington-Keokuk Limestone. These fractures provide potential conduits for groundwater flow. In the alluvium-filled river valley near where the LMT crosses the Meramec River, poor to fair RQD (<50% to 75%) observed at boreholes, high water takes during packer testing, and downhole ATV logs indicate intensely fractured rock. Rod drops observed during drilling near the Meramec River suggest the possible presence of karst features in this vicinity.

Aquifer Tests

Based on the construction of the Baumgartner Tunnel (Abkemeier and Groves, 2011), high groundwater inflow is a potential concern along the Phase 2 alignment. The Baumgartner Tunnel was completed in 2007, and the majority of its alignment is located beneath the Meramec River floodplain and passed through the Warsaw Formation and Burlington-Keokuk Limestone, which are also anticipated for the LMT Phase 2 project. During construction of the Baumbartner Tunnel, a water-bearing zone was encountered within the screen shaft near the Lower Meramec WWTF. The inflow into the shaft was estimated as 200 to 250 gpm through approximately eight drill holes at the screen shaft, indicating the presence of a near-horizontal water-bearing feature. This water-bearing feature was located within the transitional zone between the Warsaw Formation and Burlington-Keokuk Limestone.

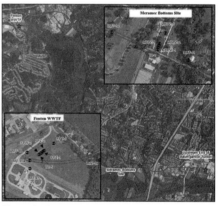

Figure 3. Aquifer test sites and LMT alignment (Shannon & Wilson, 2018)

To address risks related to groundwater inflow, two long-term constant-discharge aquifer tests were performed. Figure 3 presents an overview of the alignment with the aquifer test sites, and a detailed inset of each aquifer test site showing well and piezometer layouts. Additional details about the test procedure and results analysis will be provided in subsequent sections. One test was performed at the Meramec Bottom site where the transition zone between the Burlington-Keokuk and the Warsaw Formations is expected); the second test was performed at the Fenton shaft site in fractured rock close to the alluvium-filled river valley near the Fenton WWTF. The objectives of the aquifer tests were to:

- Better quantify the hydraulic conductivity of the rock mass in the identified areas of concern.
- Utilize the aquifer test results along with packer testing results to refine the groundwater inflow calculations to estimate the groundwater inflow anticipated into the LMT tunnel during construction

PACKER TESTING

Setup Description

The straddle packer system shown in Figure 4 was used for packer tests. The system includes a vibrating-wire piezometer (VWP), located approximately at the center between the two packers, to measure water pressure within the test interval. At each test interval, two inflatable packers were used to form a watertight seal at the top and bottom of the interval.

All packer tests were conducted below the water table. Before packer testing was started, either air or clean water was used to flush the borehole and remove the drilling fluids remaining. Each test typically included five pressure steps (i.e., test stages) completed in order of ½, ¾, 1, ¾, and ½ of the maximum water pressure. The maximum water pressure was determined based on the test interval depth and the condition of the rock in accordance with the Engineering Geology Field Manual criteria (USBR, 2001). Maximum water pressure for the first few borings was limited to 50% of the overburden pressure or 50 psi, whichever was smaller. For the remaining borings, maximum allowable water pressures

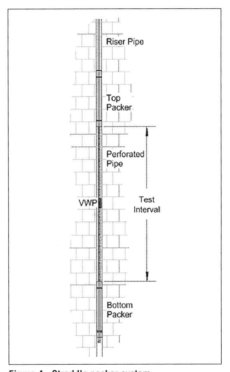

Figure 4. Straddle packer system

were calculated based on actual boring depth using a unit weight for soil of 120 to 130 pcf and a maximum allowable rock pressure based on the condition of the bedrock. The maximum allowable rock pressure was limited to 0.5 psi/ft for relatively flat-lying bedded rock, 1.0 psi/ft for relatively homogeneous but fractured rock, and 1.5 psi/ft for relatively un-fractured rock.

Test Duration

Each pressure stage water was generally held for at least 10 minutes until steady flow had been achieved, except for a few initial borings where pressure stages were held only 5 minutes. This additional pumping time for each pressure allowed a better evaluation of water takes where the rock mass had a relatively low permeability.

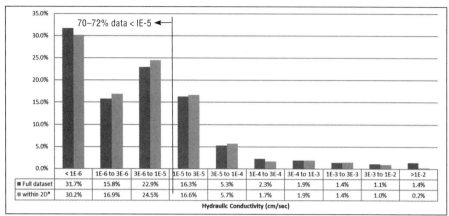

*D = tunnel excavation diameter, assumed to be 12-ft in flow estimates.
Figure 5. Histogram of hydraulic conductivity from packer tests

Two pressure readings were made for each test stage. The initial vibrating wire piezometer (VWP) transducer reading was recorded when the packers were set into position and inflated and before water was pumped or pressure applied to the test interval. The second VWP transducer reading was recorded during each test stage when the reading had stabilized, indicating a steady state water level condition had been reached under pressure. Difference between the two VWP readings was later used for hydraulic conductivity calculations.

Analysis Method and Results

Equivalent hydraulic conductivity at each test stage was estimated following the USBR Earth Manual (1974). Among 571 packer tests, about 70% of them indicate a hydraulic conductivity less than 10^{-5} cm/sec, which indicates that most of rock mass discontinuities are very tight, and 3% of the tests gave a result higher than 10^{-3} cm/sec, suggesting the existence of open or closely spaced discontinuities or voids in the test interval. Open discontinuities or voids would result in significant inflows during tunneling.

A histogram of hydraulic conductivities from packer tests is presented in Figure 5. The figure shows data from all packer tests, as well as the data within a zone of two diameters above and below the tunnel horizon. The groundwater inflow baseline conditions are based on data collected within this two-diameter zone.

An estimate of inflow quantity, based on implementation of grouting, was calculated separately using an empirical approach proposed by Heuer (1995 and 2005) and a semi-analytical approach by Raymer (2005 and 2015).

Heuer Method

An empirical approach for estimating the groundwater inflow into the rock tunnel was developed by Heuer (1995, 2005) based on his observations from tunnel projects in construction and theoretical solutions by Goodman (1965). Heuer's method uses a histogram of hydraulic conductivities, assuming that a tunnel driven through the ground will experience the same percentages of its length in different ranges of hydraulic conductivities, as did the exploratory borings.

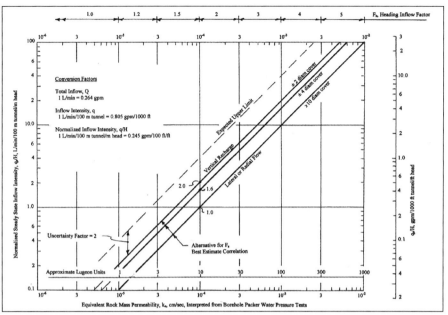

Figure 6. Relationship between steady state inflow and equivalent hydraulic conductivity (Heuer, 2005)

After plotting the frequency diagram of hydraulic conductivities, Heuer (2005) suggested a graphical correlation between the average inflow intensity and the hydraulic conductivity determined from packer tests. This graphical correlation is provided in Figure 6, which shows two sloped lines for a vertical recharge model and a line for lateral or radial flow model. The ground within LMT and above is characterized by predominantly horizontal stratified sedimentary rocks, in which very few or only short vertical fractures exist. The lateral flow along bedding features are the major source of groundwater inflow, which fits the lateral flow model as described in Heuer (2005). Therefore, the lateral flow model in Figure 6 is utilized to estimate the long term steady state inflow into the tunnel. By using Heuer's method, the estimated steady state inflow was 900 gpm for the full length of the tunnel, exclusive of the shafts.

Raymer Method

Raymer developed a lateral flow model (2005, 2015) to predict the steady state flow toward a rock tunnel based on the conservation of mass and Darcy's Law. Conservation of mass ensures the balance between inflows and outflows; while Darcy's Law controls the flow rate, which is limited by the hydraulic conductivity of the ground.

The lateral model is developed for water flow in horizontally stratified ground, in which very few or only short vertical fractures exist, and the vertical hydraulic conductivity is much smaller than in the horizontal direction. In the model, a steady water source from a leaky layer above the tunnel also contributes to the horizontal flows along the bedding. The water inflow from both tunnel sides, Q, is calculated by the equation below:

$$Q = 2Lh_o\sqrt{\frac{KbK_v}{b'}}$$

where

Q = Inflow rate to both sides of the tunnel;
L = Length of tunnel (or reach);
h_o = Initial head (depth of tunnel below water table);
K = Average horizontal conductivity;
K_v = Average vertical conductivity;
b = Thickness of the permeable zone for lateral inflow;
b' = Thickness of the leaky layer.

Using Raymer's inflow estimation approach, the LMT project was divided into three zones, i.e., stratified zone, transition zone, and river valley zone. Depending the ground conditions and Raymer's model inputs, each zone was further divided into several tunnel reaches. Appropriate parameters were determined for each reach. The hydraulic conductivities from packer tests were adjusted based on comparison to the results from aquifer tests as discussed in the next section. The hydraulic conductivities used for flow estimate were determined based on adjusted packer test results, using the methodologies as described by Raymer (2014). A log-normal distributed probability plot was used to derive the average horizontal hydraulic conductivity for the stratified zone. Vertical hydraulic conductivity was estimated by using the harmonic mean of horizontal hydraulic conductivities, assuming downward water movement perpendicular to horizontal strata and the same thickness for each layer. Water head and the groundwater table varies along the tunnel alignment. Per water table measurements in GDR, each tunnel reach was assigned its own water head depending on the elevation of the groundwater table and the tunnel elevation.

By using Raymer's method, the estimated steady state inflow was 1,120 gpm for the full 6.8-mile length of the tunnel, exclusive of the shafts.

LONG-TERM AQUIFER TESTING

The overall project risk associated with groundwater inflow during tunnel construction motivated additional testing to further refine the estimates in key locations along the alignment. Planning for conducting the aquifer testing included pretest analysis using packer testing information to evaluate the spacing and configuration of the pumping wells and the associated piezometers for each test location. Pretest analysis was performed using the forward solution option within the analysis program AQTESOLV to help evaluate appropriate spacing and layout of monitoring points for the tests.

The installation of the pumping wells and the piezometers at each of the two sites included drilling to the tunnel horizon at each location, installation of an 8-in. diameter pumping well and 3 additional 2-in. diameter piezometers at various distances from the pumping well location. Screen intervals for the wells and piezometers were determined based on the proposed tunnel horizon at each location. The layout of the wells at both locations is shown in Figure 3. An essential element of the well installation included development of the wells and piezometers to ensure good communication through the filter pack between the well and the fractured rock aquifer.

Following development, pressure transducers and dataloggers were installed in all of the wells and piezometers to be used during aquifer testing. This included the observation points installed specifically for the testing, but also other nearby piezometers installed during other phases of tunnel investigations for the project. At each of the test sites, the additional monitoring locations included a piezometer installed during earlier

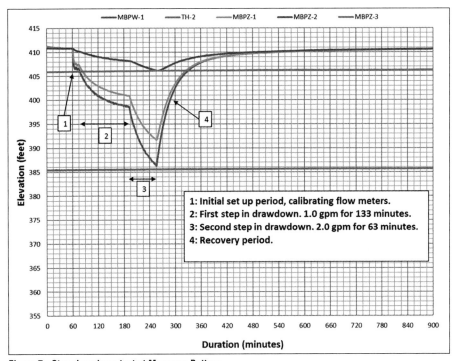

Figure 7. Step-drawdown test at Meramec Bottom

investigations that was completed in the overburden material above the rock. These piezometers provided important information about the vertical gradients at each location. Background water level measurements were then collected from all the monitoring locations for 28 days to evaluate the pre-test trends and to provide information to evaluate barometric effects on the water levels that was needed to correct test data prior to analysis. At both locations the barometric efficiency of the piezometric response was calculated and this was used to correct the hydraulic response during testing to account for changes in barometric pressure during testing.

Step-drawdown tests were performed prior to the long-term testing to evaluate the specific capacity of each pumping well and to select the pumping rate for the long-term testing. Prior to the long-term testing and following the step drawdown tests, water levels were allowed to recover to the initial water levels. The response to the step drawdown test at the Meramec Bottom location is presented in Figure 7.

The long term aquifer tests were performed at the Fenton site on February 7–9, 2017. The pumping phase of the test lasted for approximately 24 hours and the recovery phase lasted another 24 hours. The pumping rate for the Fenton test was targeted to be 0.5 gallons per minute. The uncorrected responses during the Fenton aquifer test are presented in Figure 8. The long term aquifer tests were performed at the Meramec Bottom site on February 9–13, 2017. The pumping phase of the test lasted for approximately 48 hours and the recovery phase lasted another 48 hours. The pumping rate for the Meramec Bottom test was targeted to be 3 gallons per minute. The uncorrected responses during the Meramec Bottom aquifer test are presented in Figure 9.

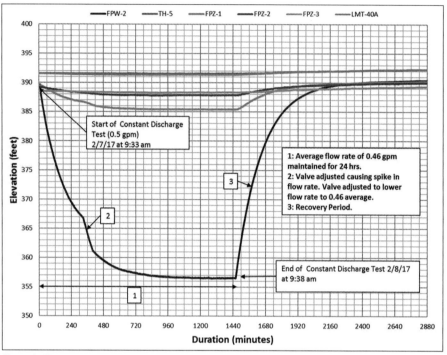

Figure 8. Response of water levels during Fenton constant discharge aquifer test

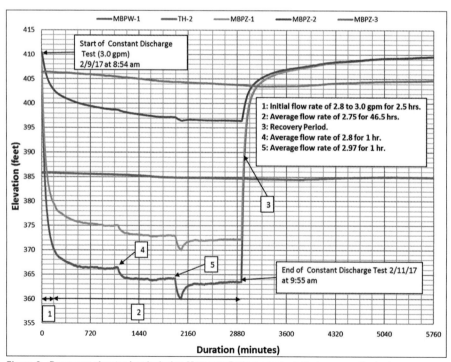

Figure 9. Response of water levels during Meramec Bottom constant discharge aquifer test

Analysis and Results

For each of the two tests, the pumping rates were adjusted slightly throughout the test to maintain the discharge rate to within about 10 percent of the target flow rate. Slight adjustments in flow rates are evident in the water level measurements as shown above. The overall average discharge rate was calculated for each of the two tests by dividing the total gallons pumped by the total duration of the pumping stage of the test. This is the pumping rate that was used to analyze the water level data.

The raw water level data was processed to account for the barometric effects that were discernable as noted above. The water level measurements were adjusted based on the average barometric efficiency calculated from the pre-test measurements. These corrected values were used in the aquifer test analysis described below.

The constant discharge aquifer tests were analyzed using the software program AQTESOLV version 4.51 Professional. This program allows the importing of the corrected water level observations, as well as information about the aquifer characteristics, pumping rate, well construction and potential boundary effects. Once all the input information is entered, type curve matching is performed to estimate the transmissivity of the aquifer being pumped. AQTESOLV incorporates a number of different solutions based on published aquifer test analysis protocols that have been developed.

The Papadopoulos and Cooper (1967), and Dougherty and Babu (1984) solutions were judged to be most representative of the testing conditions. These solutions were developed for an infinite porous medium, and while they are most often applied to porous media such as alluvial aquifers, these solutions are also applicable to fractured rock media when the fracture spacing and scale of the test approximate typical porous medium characteristics. The Papadopoulos and Cooper and Dougherty and Babu solutions are consistent with assumptions for the traditional Theis solution with the added benefit that they also account for well bore storage effects. Due to the large pumping well size and the low pumping rates, well bore storage effects were quite significant for the analysis.

During the analysis the anisotropy ratio (vertical to horizontal hydraulic conductivity) was assumed to be 1. Some test runs were made using anisotropy values of 0.1 V:H, which is typical for alluvial aquifers and sedimentary deposits, however the change in ratio had virtually no effect on the curve matched values. This is reasonable since the aquifer test solutions assume primarily horizontal flow in the aquifer, and the piezometers and pumping well are completed in the same aquifer zone, thus anisotropy would not enter into the analysis.

Fenton Site Results

Results for the combined pumping and recovery cycles at the Fenton site are presented in Table 1. Transmissivity values range from a low of 2.8 ft^2/day in the pumping well to 25 ft^2/day in piezometer LMT-40A. In general, the pumping well seems to show a lower transmissivity than the piezometers, and these values may be compromised by wellbore effects and non-laminar flow conditions due to the high levels of drawdown observed during the test. Therefore, the FPW-1 values are not considered to be as reliable as values obtained from the piezometers where the assumption of laminar flow is more valid.

The transmissivity values obtained from the piezometers are somewhat more closely clustered between 7 and 25 ft^2/day with an average value developed from the piezometers of 15.6 ft^2/day. The dimensionless storage coefficients developed from the curve

Table 1. Aquifer testing results for Fenton Site

	Papadopoulos-Cooper		Dougherty-Babu		Mean T w/ Well bore storage
	Transmissivity (ft^2/day)	Storage Coefficient	Transmissivity (ft^2/day)	Storage Coefficient	ft^2/day
FPW-1	2.8	0.1×10^{-4}	2.9	0.1×10^{-4}	
FPZ-1	7.0	2.0×10^{-4}	7.1	1.9×10^{-4}	
FPZ-2	13.3	3.1×10^{-4}	13.4	3.1×10^{-4}	
FPZ-3	17.5	2.2×10^{-4}	16.6	2.3×10^{-4}	
LMT-40A	25.0	1.8×10^{-4}	25.0	1.8×10^{-4}	
					15.6

Note: Values within the double border box are most representative because they include distance from the pumping well and well bore storage effects.

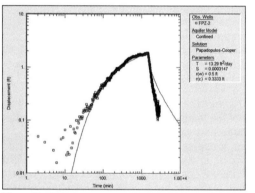
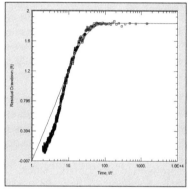

Figure 10. Type curve and residual drawdown curve for MBPZ-2

matching range between 1.8×10^{-4} to 3.3×10^{-4} which represent a reasonable range for a fractured rock aquifer. Examples of the type curve and residual drawdown curves for the Fenton site are presented in Figure 10

Review of the data also indicates that transmissivity values tend to increase with distance from the pumping well. This trend, illustrated best by the results from FPZ-3 and LMT-40A suggest that a source of recharge or "leakance" may also be contributing to the responses observed. The test site proximity to the Meramec River may provide a significant source of recharge; however, it may not be a direct source of recharge to the pumped aquifer, but rather to the overall water table conditions. The fact that the response observed in LMT-40A, located in a direction away from the Meramec River relative to the pumping well, contrary to FPZ-3 which is much nearer to the river, suggests that the water table aquifer above the zone tested may be providing the recharge that displays a stronger influence with distance from the pumping well. Additionally, indicating a source of recharge from upper zones, piezometer TH-5 which is screened in the overburden material, above bedrock, also exhibited a slight drawdown (up to 0.3 ft) during the test after correcting for barometric effects. With the Meramec River within 500 feet of the Fenton Site pumping well, it is not possible to discern from these test results whether the recharge is from the river directly or from the broader water-bearing rock and alluvium strata above the pumped zone. Nonetheless, an average value of the transmissivity of 15.6 ft^2/day is thought to be a reasonable estimate for the broader aquifer response to pumping in this area.

Table 2. Aquifer testing results for Meramec Bottom site

	Papadopoulos-Cooper		Dougherty-Babu		Mean T with Well Bore Storage ft²/day
	Transmissivity (ft²/day)	Storage Coefficient	Transmissivity (ft²/day)	Storage Coefficient	
MBPW-1	29.1	1.5×10⁻¹¹	29.1	1.7×10⁻¹¹	
MBPZ-1	25.7	2.6×10⁻¹¹	25.7	2.6×10⁻¹¹	
MBPZ-2	20.0	6.0×10⁻⁵	20.3	5.1×10⁻⁵	
MBPZ-3	18.3	1.3×10⁻³	17.0	1.3×10⁻³	
					22.9

Note: Values within the double border box are most representative because they include distance from the pumping well and well bore storage effects.

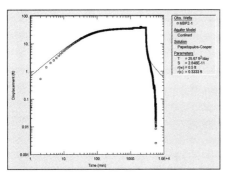
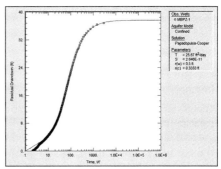

Figure 11. Type curve and residual drawdown curve for MBPZ-1

Meramec Bottom Site Results

Results for the combined pumping and recovery cycles are presented in the Table 2. Transmissivity values range from a low of 17 ft²/day in the piezometer MBPZ-3 to 25.7 ft²/day in piezometer MBPZ-1. As with the Fenton Site, the pumping well may be compromised by wellbore effects and non-laminar flow conditions due to the high levels of drawdown observed during the test. Therefore, the MBPW-2 values are not considered to be as reliable as values obtained from the piezometers where the assumption of laminar flow is more valid.

The transmissivity values obtained from the piezometers are more closely clustered between 17 and 25.7 ft²/day with an average value developed from the piezometers of 22.9 ft²/day. Storage coefficients developed from the curve matching range between 1.3×10⁻³ to 2.6×10⁻¹¹ which represent a broad range for a fractured rock aquifer. Examples of Type curve matches used to determine values summarized in Table 2 are presented in Figure 11.

Contrary to the experience from the Fenton Site, the transmissivity values do not appear to increase with distance from the pumping well. In fact, they exhibit a slight decrease; however, the results from MBPZ-3 should be considered distinct from the others as this piezometer is screened over a slightly higher elevation than the other piezometers and pumping wells. The response from this well, therefore may not represent the same strata screened by the pumping well and other MB piezometers. Regardless, the fact that no distinct aquitard separates the screen interval for these zones, and that there is clear connection and response in MBPZ-3 to pumping during the test, the transmissivity values are generally consistent with those from the other piezometers.

Monitoring results from piezometer TH-2, completed even higher within the overburden material also show a slight response during the test, however the level of response is difficult to distinguish from the barometric effects, similar to those documented during the test period. It appears there may be a distant or slight hydraulic connection between the pumped zone and TH-2, but it is not as strong or clear as that observed at the Fenton Site.

Over the course of the 48-hour pumping period of the aquifer test, the pumping well and all piezometers achieved a relatively stable degree of drawdown toward the end of the pumping stage. This would suggest that either there is a recharge boundary nearby, or that sufficient "leakage" from other portions of the formation (above or below the pumped interval) are supplying sufficient recharge to compensate for the level of pumping applied. Given that this site is approximately 0.75 miles from the Meramec River, it appears that the recharge to cause leveling off of the drawdown is most likely from water within the broader limestone above and/or below the pumped interval.

DISCUSSION

Comparison of Packer Test and Long-Term Aquifer Test Results

Test data collected during the field investigations offer an opportunity to directly compare the packer test data with the long-term aquifer test data. Figure 12 and Figure 13 present a graphic comparison between the two data sets at the Fenton and Meramec Bottom Sites, respectively. In addition to presenting the hydraulic testing data, these figures also show the lithologic profile at each site, the acoustic televiewer log for each hole, the well construction information for each hole, and the tunnel horizon. The packer test data presented are the average value for the 5 stages of testing at each ten-foot interval. The aquifer test data represent the overall average result over the screen interval.

By inspection, it is apparent that the results from packer testing and aquifer testing represent different values of hydraulic conductivity. Results from the aquifer testing are typically higher than the results from packer testing—on average about 2 to 3 times higher. In making this comparison it is important to recognize the difference between these two types of test procedures.

Aquifer testing provides a much larger representative test volume than discrete packer testing. As such, the aquifer test provides more time for conditions to be established during the test period, and reflects the type of "draining" event that is likely to occur during tunnel construction. This is distinct from a packer testing program that relies on water being pumped into a relatively smaller portion of the rock mass rather than draining water out. For these reasons, aquifer testing is thought to provide a more appropriate evaluation of the hydraulic properties of the rock mass at each of these discrete site locations. A comparison of the two methods of measuring hydraulic conductivity is presented in Table 3.

Comparison of predicted inflow estimates based on packer testing with actual inflow estimates is not often presented in the technical literature. One significant exception was by Heuer (2005). This publication presents predicted and actual inflow quantities for 5 different projects around the country. In most of these case histories the observed inflow amounts were about 1.5 times higher than the predicted values. In this article, a modification of the data analysis procedure is recommended to quantify the distribution of the hydraulic conductivity values to effect an increase in the calculated value for the predicted inflow rate to have a result about 1.5 times higher. While

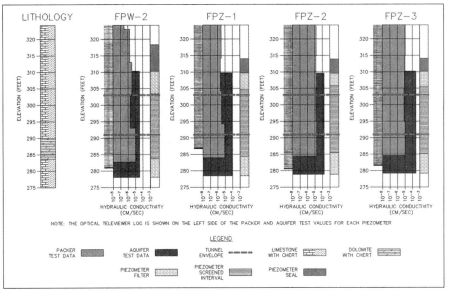

Figure 12. Comparison of packer and aquifer tests at Fenton WWTF site

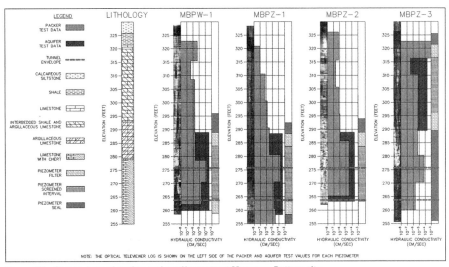

Figure 13. Comparison of packer and aquifer tests at Meramec Bottom site

Heuer suggests that this level of variability is within a reasonable uncertainty range for hydraulic conductivity values, the difference in the predicted vs observed inflow rates appears to be systematic and may be influenced by a fundamental difference between the measurement of hydraulic conductivity using packer testing and long-term aquifer testing.

As mentioned above, hydraulic conductivity values developed through aquifer testing for this project are generally about 2 to 3 times higher than average values from packer testing in the same interval and location. As a result of this systematic difference, an

Table 3. Comparison of packer testing and aquifer testing

Packer Testing	Aquifer Testing
Packer testing is frequently performed	Performed less often due to intensive effort required for a given test.
Relatively small aquifer volume tested	Tests aquifer discharge on a scale more similar to tunnel construction
Pumps fluids in	Drainage effort more similar to tunnel inflow physics
Provides reasonable discretization of data if your geology is layered	Requires special well and piezometer installation efforts
Can be performed in routine exploratory borings	Involves analysis and interpretation consistent with the conceptual hydrogeologic model
Many references available that can help understand and interpret data	

adjustment of packer test data for this particular project was considered to be warranted to result in improved predictions of tunnel inflow. In so doing, consideration should be given to adjustment of packer test values within the ± 2 tunnel diameter. Such an adjustment should be based on similar site-specific comparison of aquifer testing and packer testing results.

CONCLUSIONS AND RECOMMENDATIONS

Long-term aquifer testing provides a reliable value of hydraulic conductivity more aligned with the groundwater inflow conditions associated with tunnel construction. Tunnel construction will drain water from the surrounding rock, a condition that is simulated by the long-term aquifer test.

Packer testing evaluates the rate that water inflow can penetrate a rock mass under a range of pressures, and is commonly used to estimate a hydraulic conductivity value. While packer testing is commonly used by the industry to measure hydraulic conductivity values, interpretation of packer testing must be done carefully, especially considering the small representative zone of influence and the difference between injection during packer testing and draining during the aquifer testing. The packer test data provides the primary input for the Heuer method of estimating tunnel inflow without much interpretation being needed. The Raymer method applies the hydraulic conductivity results to a conceptual model of groundwater flow to develop the groundwater inflow estimate. Using different analytical processes is likely to yield a range of different inflow estimates, however using both can provide a broader understanding of the potential inflow rates.

It may not be practical to perform long-term aquifer testing as often or at all locations in lieu of packer testing; however, the combination of the two types of testing can provide a broader and deeper understanding of the hydrogeologic response at critical locations along a tunnel alignment. This improved understanding should help refine future groundwater inflow estimates for future projects.

Given the systematic differences between packer testing and long-term aquifer testing, as well as the documented pattern of underestimating groundwater inflows in a tunnel based solely on packer testing results, it may be appropriate to adjust the packer test conductivity values by a factor of 1.5 to 2 to provide more realistic conductivity values for estimating tunnel groundwater inflow.

This paper presents a comparison between packer testing results and aquifer testing results at two locations for one project. Additional case histories that make similar comparisons for other projects are needed to develop a better understanding of the relationship between these methods and their use to estimate groundwater inflow into tunnels.

ACKNOWLEDGMENTS

The authors would like to acknowledge the entire project team and their hard work to help conduct all the testing and analysis performed for this study. We also appreciate the support of Jerry Jung at MSD and HDR, Inc. that was essential to complete this work. Thoughtful review of the manuscript were performed by Everett Litton and Bill Hansmire. The paper was greatly enhanced by their comments.

REFERENCES

Abkemeier, T. and Groves, C., (2011) Construction Grouting of the Baumgartner Tunnel, Proceedings of the Rapid Excavation and Tunneling Conference, pp. 1500–1510.

Dougherty, D.E. and Babu, D.K., (1984) Flow to a partially penetrating well in a double porosity reservoir, Water Resources Research, vol. 20, pp. 1116–1122.

US Bureau of Reclamation, 2001, Engineering Geology Field Manual, 2nd Edition, Vol. II, p. 127.

Heuer, Ronald E.,1995, Estimating Rock Tunnel Water Inflow. Proceedings of the Rapid Excavation and Tunneling Conference, June 18–21, 1995.

Heuer, Ronald E., 2005, Estimating Rock Tunnel Water Inflow - II. Proceedings of the Rapid Excavation and Tunneling Conference.

Houlsby, A.C., (1976) Routine Interpretation of the Lugeon Water-Test, Quarterly Journal of Engineering Geology, Vol. 9, pp. 303–313.

Litton, Everett, Mark Stephani and Jerry Jung, 2018, Sewer tunnel beneath St. Louis to fulfill regional treatment plan and environmental vision, presented at North American Tunneling Conference, pp. 705–713.

Papadopoulos, I. S and Cooper, H.H, 1967, Drawdown in a well of large diameter, Water Resources Research, Vol. 3, pp. 241–244.

Raymer, J., 2005, Groundwater Inflow into Hard Rock Tunnels: A New Look at Inflow Equations, Proceedings of the Rapid Excavation and Tunneling Conference.

Raymer, J., 2014, Effect of Variability on Average Rock-Mass Permeability, Proceeding of the 48th US Rock Mechanics/Geomechanics Symposium, Minneapolis, MN.

Raymer, J., 2015, Groundwater Inflow to Tunnels, Tunneling Shortcourse at the Colorado School of Mines.

Shannon & Wilson, Inc. (2018), Geotechnical Data Report (GDR) for Lower Meramec River Systems Improvements (LMT), Baumgartner to Fenton WWTF, MSD Project No. 11746.

Underground Construction, Geology and Geotechnical Risk

Priscilla P. Nelson ▪ Department of Mining Engineering, Colorado School of Mines

ABSTRACT

Population increase means mega-cities will be growing very fast as "compact cities" for which surface space becomes a priority. This creates a particular urgency to make the underground space of the future cheaper to construct, and reliable in construction and operational performance. The cost and performance of underground projects is intimately linked to the understanding and management of geologic risk for both construction and life-cycle performance of subsurface facilities. This includes "normal" uncertainties, but also the expectation that urban growth will extend into increasingly fragile and poor geotechnical environments, and that the projects will involve larger and deeper openings.

This paper assesses the state-of-practice and future possibilities for improved management of geologic risk, including risk avoidance, new materials and methods, ground improvement, life cycle engineering for sustainability, and better subsurface characterization. Some geologic risks have plagued for centuries, e.g., ground water, shallow cover and weathered rock, subsidence and impact on structures, stresses and stress relief, progressive deterioration. New risks have arisen associated with new technologies including unexpected stress-driven ground behavior, and design for higher water inflows and pressures, increased depth, and variety of excavated shapes. In addition, a better understanding of the spatial variability of soil and rock structure is needed a priori, including application of geophysical and remote sensing techniques. Our site investigations of the future need to be increasingly confirmatory rather than exploratory.

INTRODUCTION

Sustainable urban underground development must meet current human needs while conserving spatial resources and the natural and built environments for future generations to meet their needs. This requires a systems perspective for integrated above and below ground resource use and management, and must include consideration of cost effectiveness, longevity, functionality, safety, aesthetics and quality of life, upgradeability and adaptability, and minimization of negative impacts while maximizing environmental benefits, resilience, and reliability (Bobylev 2009).

Population growth will continue, and so will the growth of our cities—but not always in predictable ways. In the past, we have lived through urban migration and growth, followed by suburbanization, and now perhaps the concept of the compact city describes how our cities will change in the future. The compact city concept is intimately wedded to increasing and intensive planned use of underground space, but engineers and planners have challenges in preparing our old and new infrastructure for the future.

The underground construction industry has consistently provided the nation with needed infrastructure, meeting schedule and scope goals. While it is generally appreciated that the nation must invest in the rehabilitation of existing infrastructure, there continues to be a lack of political and public will to do so. Our cyberphysical

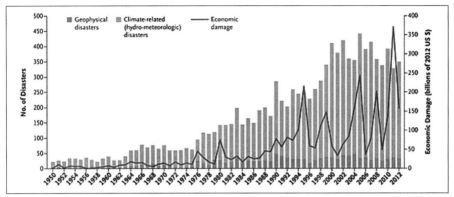

Figure 1. Recent history of disasters and impacts. http://www.accuweather.com/en/weather-blogs/climatechange/steady-increase-in-climate-rel/19974069

infrastructure systems have not been maintained, causing unexpected vulnerabilities and cascading failures (ASCE, 2017; AWWA, 2001). If our systems become increasingly unreliable, then it is likely that more of the world's leading industries will relocate headquarters to countries with more reliable infrastructure.

Significant impacts from extreme events (including climate change, earthquakes, tsunamis, floods, storms) are arguably becoming more frequent and costly (see Figure 1). Our future global cities must support the population both through disasters and for daily living, perhaps analogous to the human body's resistance and resilience to a high-grade fever and also to manage a low-grade infection (Nelson, 2016). The resilience of our urban communities depends on many factors that extend beyond the physical system complexities and interdependencies (Nelson and Düzgün, 2018). Therefore, social network research is needed to provide linked and registered metrics through crowd sourcing for event impacts, yielding change trajectories over time (Anex, et al., 2006; Bobylev, 2009, 2016). Social networks, crowd sourcing, IoT (Internet-of-Things) through location-based services potentially allow those responsible for infrastructure to access social data about impacts quickly, and software can be used to capture and analyze the public's acute reactions to extreme events (Sherrieb et al., 2010) in real time, providing an opportunity to respond and maximize the social and infrastructure performance resilience.

State-of-practice design and operation from the past has led to robust-enough systems for which we have sufficient experience to permit simplifying assumptions that enabled operation with minimal monitoring. There were sufficient reserves for acceptable service under known stress. But as we interconnect aging systems into larger networks, and observe decreasing performance levels, reductions in excess capacity and new stresses (e.g., poorly understood interdependencies, attack), we learn that our systems have lost robustness. As our system complexity has increased, many of the design simplifications are no longer acceptable, and new concepts of design and control provide an opportunity for new approaches to system management.

In many cases, the design loads used by engineers at the time of construction of our older infrastructure, may not be the loads we would use now. Our design and professional codes have always incorporated factors of safety against failure by such events, but the impacts of recent events have been more severe and complex with interdependent responses. Engineering professionals, construction contractors, and urban planners and managers must work together to identify new ways to retrofit and bolster our infrastructure against extreme event impacts. Underground engineering can be a part of effective design and solution of problems. Therefore, the underground is an important resource to enhance urban resilience, as is summarized in Table 1.

As we create and use more underground space, particularly in urban environments, we may find ourselves working in ground conditions not experienced before. This means that much of our conventional and current wisdom based on experience in a city may not be applicable. For example, we anticipate increased use of deeper underground space for many purposes. Higher ground stresses, temperatures and water pressures will likely be encountered, and soil and rock behavior may become more problematic (Fairhurst, 2017). Figure 2 contains data on the depth of shafts constructed in the U.S. over the past 150 years—clearly reflecting the trend of greater depth over time. As new needs for underground space are identified, owners and the public will request larger and more complex 3-D complex geometry for underground space applications. This may require advanced design concepts for long-term performance and stability.

Table 1. Advantages and disadvantages for underground infrastructure and extreme event impacts (modified from personal work of R.L. Sterling).

Type of Extreme Event	Advantages or Mitigations	Disadvantages or Limitations
Earthquake	Ground motions reduce rapidly below surface	Fault displacements must be accommodated
	Structures move with the ground	Instability in weak materials or poor lining backfill
Winds: hurricane, tornado	Minimal impact on fully buried structures	Damage to shallow utilities from toppling surface structures and trees
Water: Surge, flood, tsunami, sea level rise	Protection from direct impact, mass wasting and debris flows	Extensive restoration time and cost if entrances are flooded
Fire, blast, terrorism	Ground provides thermal and concussion protection, limit impact by compartmentalization	Entrances and exposed surfaces are weaknesses, confined space risk
External radiation, chemical/biological exposure	Ground provides additional protection	Appropriate ventilation system protections required

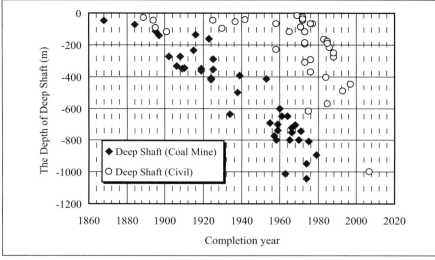

Figure 2. U.S. Data on depths of coal and civil deep shafts 1860–2010

And as our coastal cities grow, more of the new infrastructure must be placed into more challenging ground for which risks and costs may be higher. This may require new approaches to ground improvement and displacement control.

REDUCTION OF COSTS AND RISKS

Beyond the need for reliable and resilient infrastructure services, is the need to manage the budget. It is notable that infrastructure costs for construction and rehabilitation have generally and significantly increased in recent time. Innovations are needed to reduce costs and support schedule reliability, and best decisions on investments can only be made with increased use of Life Cycle Engineering (LCE) which requires data bases that by-and-large do not exist. With increased use of LCE, performance metrics can be established for integrated surface and underground infrastructure planning and design, and to support sustainable multi-hazard design and LCE trade-offs. This need may define a new profession of urban stewards—engineers who design and construct holistically integrating over x, y, z and time.

It is also imperative that the physical facilities be made more durable, and the infrastructure performance be made more reliable. Cost increases are often driven by increased risks: Risk = Probability × Consequences (or Impact). For consequence evaluation and surface vs underground placement trade-off studies, we need to know the value of underground space. However, there is no developed market for underground space value. We need an improved quantitative evaluate risks, impacts and their probability of occurrence, as well as a framework for evaluation of mitigation strategy and assignment of responsibility during construction and in operation. For example, most current flood models in urban areas fail to consider subsurface spaces in characterizing the effects of flooding, and the impacts of sea-level rise on both construction and operation of our underground systems needs to be assessed. Overall, the urban engineer must have a commitment to maintain holistic stewardship of our cities, including: (1) spatial (x, y and z) urban planning; (2) acute awareness of temporal issues (first cost, sustainability); (3) agility in integrating across physical infrastructure sectors, and across physical, natural, social, and fiscal environments and risks; and (4) the gift of communication that provides realistic expectations on cost and schedule to owners and to the public.

A majority of the risk associated with underground infrastructure construction and performance is derived from the spatial variability and uncertainty associated with geologic conditions, including soil, rock and water. Six areas of focus are discussed below:

- Risk avoidance
- New technologies and methods
- Better subsurface characterization
- Better management of water
- Risk awareness, assessment and management
- Risk communication and willingness to accept and share risk

RISK AVOIDANCE

Geologic conditions in the subsurface should be primarily managed by invoking the concept of underground zoning which provides spatial thinking and integrated planning to place above- and below-ground facilities in an optimized geologic setting. In New York City and other cities, such a consideration leads to vertical segregation of different infrastructure systems. However, much of the shallow infrastructure represents spatial chaos and project costs are strongly impacted by the need to manage the mayhem of aged near-surface systems.

The Japanese experience is a bit different (Masuda et al., 2004). The 2001 Deep Underground Utilization Law established that land ownership rights in populated areas (e.g., Tokyo, Osaka) only extend to 40 meters below ground, or 10 m below a deep foundation. The act is focused on metropolitan areas of Tokyo and Osaka Nagoya, and ensures the right of certain developers to use deep underground space regardless of surface ownership. In the case of public use of the underground space, no compensation to the land owner is required. The first projects using the law have included underground water mains in Kobe, and the Tokyo Gaikan Expressway. In 2015 Singapore adopted a similar approach by limiting ownership to a specific depth (30 m below Singapore Height Datum (SHD)) (Stones and Heng, 2016).

NEW TECHNOLOGIES AND METHODS

The underground industry has many methods that can be applied including Tunnel Boring Machines (TBMs) and shields, and the newer slurry, earth pressure balance and hybrid pressure-face equipment, but more developments are needed to decrease costs, and improve safety (e.g., avoid hyperbaric cutter replacements and other interventions). The seemingly inexorable trend is for larger and larger diameters, and this by itself drives up project costs and expands project schedule.

In many areas of research, the pipeline from fundamental research to application has been thwarted. It is imperative that industry and owners commit to partner with universities to develop new technologies and methods, including new ways to excavate and support underground openings. In addition, it is important to incentivize the application of new technologies. For example, ground improvement techniques have come a long way in the past 30 years, as is reflected by the data in Figure 3 for U.S road tunnel support over the period from 1980 to 2000. The transition from ribs-and-lagging to NATM methods is clear, and begs the issue that the long-term performance of newer method of construction need to be monitored in service so that expectations for support/lining life can be verified. It is important that advances continue, and that

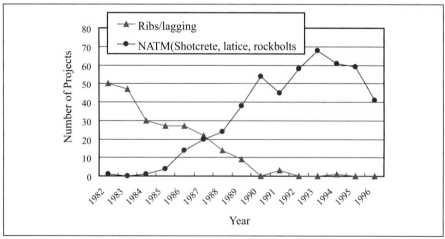

Figure 3. Data documenting the change in U.S. road tunnel support methods between 1980 and 2000 (data from FHWA)

techniques of ground improvement be proactively implemented before a project is started to change and remove identified geologic risks, rather than respond to the risks as encountered. Such postures often result in changed condition claims, litigation, increased costs and delays.

Many of our infrastructure projects are designed for low first cost and to comply with right-of-way limitations. Such systems are not necessarily designed for long-term sustainability and maintainability. Engineers must seek new materials and technologies to enhance performance and durability of our infrastructure systems, new and old. In addition, new technologies must not be just implemented—they must be assessed for short and long-term performance. Sober assessment of performance is very often forgotten in the cycle of innovation we seek for the underground industries.

Safety in the underground during construction and operation continues as a concern, and incident rates for heavy construction are considerably higher than for mining projects. Safety innovations continue to be needed, and include personal protective equipment during construction and also fire and explosion incident management, particularly when the public are involved in response.

Spatial and temporal variations in subsurface materials and conditions continue to be a risk, and a new look at integrating geophysical and remote sensing methods is warranted. Engineers should also rethink materials and methods in use. For example, development of new concrete, grout and shotcrete materials for application in the underground are needed, and engineers and contractors should revisit and dramatically improve our "old" or "conventional" technologies such as drill/blast operations.

BETTER SUBSURFACE CHARACTERIZATION

Knowledge of the underground conditions has been improving over past decades, but the combination of continuing sore points and arising new difficulties must be considered in planning. In many urban environments, previous underground works have demonstrated spatial and material property distributions to be expected, so our conventional site investigations should be confirmatory rather than exploratory.

But some geologic issues continue without full resolutions, as a low-grade infection on the industry. Examples include the following:

- Shallow cover, varying depth to rock
- Ground movements, subsidence
- Consolidation settlements
- Weathered rock and rock mass (including karst)
- Rock mass structure and variability
- Time dependency in materials behavior
- Muck abrasiveness and stickiness
- Aggregate reactions and concrete durability

Geological and Geotechnical Engineers still wrestle with scale effects as well, extrapolating from lab behavior to full scale in the field. Many rock mass rating systems have been developed. On a large number of projects, ratings applications have been uninformed and inconsistent, and there have been only limited attempts to validate their inference, or the use of a large number of empirical correlations. This observation also can be applied to the plethora of computational models available for subsurface design. We must make opportunities to validate design assumptions and performance prediction.

More urban infrastructure will necessarily be placed deeper, and the in situ stress state will likely become more important on more projects. Estimation of an in situ stress field is challenging without a clear geologic framework for interpretation, and most stress assessments are made as point measurements (interpretation of deformation measurements at a point). This can only be addressed by obtaining a better understanding of the spatial variability of rock mass structured which introduces uncertainty. The variety of excavation shapes and dimensions can be expected to vary in the future, with more gallery space rather than plane strain tunnels needed, making the predictions of displacements, strains and stress redistribution around an underground opening increasingly important. We also need to understand spatial and temporal variations that affect performance of existing facilities for sustainable design and operations.

Geologic material failure and time-depend response of geologic materials are far more likely to be observed in an underground mine than in a civil works project. Mining engineers develop a strong geologic perspective on risk that would benefit in application to civil construction projects. Such a partnership or collaboration across industries brings an enhanced potential for real spatial understanding of rock mass and water inflow and pressures variability, and for better understanding of time effects, presenting the possibility to develop sustainability performance information. The two industries also have many environmental issues in common, as do they have a mutually beneficial potential for application of automation, robotics, and big data/information systems. This is the era of information: with an expansion in sensing and measurement capabilities, how should the entire site investigation and construction process be re-thought, not to mention real-time data flows and their importance to effective management for resilience of urban infrastructure systems.

BETTER MANAGEMENT OF WATER

The presence of water in the subsurface changes the behavior of materials, and strongly influences the long-term performance of underground facilities. Full consideration of the influence of water includes knowledge and understanding of volume, flow rate, quality, pressure, and changes over time. On many tunnel projects, water is encountered but few of these parameters are assessed or evaluated for spatial variability unless a claim is anticipated. Such observations and measurements are required if we are to significantly reduce the impact of water. Research is also needed on the relationship between fracture mechanical aperture and hydraulic aperture with consideration for rock type and geologic regime, diagenesis, discontinuity fillings, normal stress and shear stress along and across fractures (Chen, 2010).

Management of water is sometimes a matter of resource conservation (e.g., impacts on a water supply), but environmental (bio-geochemical) and construction impacts are likely to be more common and profound. During construction, water management includes compressed air, grouting, and the use of pressure-faced shields. Microtunneling and trenchless methods are very flexible and work well for smaller diameter emplacements, which can be efficiently and economically reamed to larger diameters—potentially minimizing the impact of water inflows on construction. Water inflows can compromise worker safety, and in some cases may compromise the capabilities of installed support.

Some of the most active areas of new technology implementation have been related to the introduction of waterproofing into tunnel linings. The long-term performance of such installations needs to be assessed on a continuing basis. Operational impacts of seepage and inflows are incredibly important since water drives long term deterioration in the underground, and inflows can cause piping and ground loss that affects lining performance and also structures nearby. The long-term performance of waterproofing or drainage management technologies is not well documented.

RISK AWARENESS, ASSESSMENT AND MANAGEMENT

Many underground construction projects now use the three-legged stool of a Disputes Review Board (DRB) requirement for bid documents to be escrowed, and the development of a Geotechnical Baseline Report (GBR) as a part of the contract documents explicitly developed for geologic risk management. A good GBR is thoughtfully written to present a geologic analysis of expected conditions, and or "geoproblem event" frequency (temporally and spatially) to be assumed during a project. The project data collected informs designers and contractors as to behaviors and properties of geologic materials, but a statistical assessment of the probability and consequences of encountering major geotechnically-driven stoppages in underground excavations is difficult—and yet such events are the main causes of major problems on underground construction projects.

The industry as a whole should commit to building a geologically-framed data base that includes spatial information about soil and rock mass variability and impacts in a geologic context. Such a data resource can inform regarding likelihood of problems being encountered and how, for different construction means and methods, the problem conditions may be best managed. The data and information needed include:

- Type of geoproblem event
- Means and methods of excavation and equipment
- Ground and water control
- Spatial frequency: length of each encountered problem, and distance between events
- Temporal frequency: hours to handle, and time between events
- Agility and performance of the contractor in responding to each geoproblem event

Not everything encountered on a specific project needs to be considered as a "one-off," and the framework of geologic inference and analysis opens the prospect for real predictability of geotechnical event with extreme impact on a project. For this geologic effort, it is clear that the mining and civil industries can share geodata.

RISK COMMUNICATION AND WILLINGNESS TO ACCEPT AND SHARE RISK

The commitment for investment requires far more effective communication of the value of infrastructure and of underground space. The value of the nation's infrastructure may be estimated in several ways, but totals on the order of $70 to $100 trillion can be suggested for the U.S. If this number is divided by the population of the US, the per capita investment in infrastructure is on the order of $300,000, the price of a house in many areas. This $300K can be interpreted as a birthright for each person born in the U.S., a pre-investment upon which the economic engine runs, the quality of life is assured, and career potential of each individual is leveraged. Even as families reinvest in a house to retain value, so must the nation reinvest in its infrastructure. This is an example of a metric that can be meaningful to each citizen and politician.

BUILDING A FRAMEWORK MODEL OF GEOLOGIC SPATIAL VARIABILITY FOR ANTICIPATION AND MANAGEMENT OF GEOLOGIC RISK IN THE UNDERGROUND

Design in the underground is best accomplished by anticipating materials, behavior and properties needed for intelligent analysis and construction in the underground. The greatest risk for most underground project success is derived from lack of geologic knowledge, including uncertainty about groundwater, and about spatial material and property distributions. The greatest risk for long-term performance is uncertainty about as-built construction, and uncertainty concerning time-dependent behavior. What is warranted is a "Grand Campaign" to provide the knowledge base to address these risks.

Underground construction and tunnel engineers should graduate from curricula that include much more training in geology. Such training (especially field training and experience for students and professors) is mandatory for the geotechnical engineering profession to address geologic uncertainty by enhancing knowledge and application of the fundamentals of geologic knowledge and interpretation. Many geologic issues continue to be encountered and have problematic impacts like a thorn in the side, such as shallow cover and weathered rock, progressive deterioration, piping, and caving. Ground loss consequences include construction settlement, subsidence, impact on structures, consolidation with water table changes, and differential settlement associated with a varying depth to top of rock. These are perhaps the "low-grade infections" in comparison to the "high fever" of geoproblems that cause extensive stoppages. In addition, there is a growing overreliance on (and misuse of) rock mass ratings—RQD on steroids.

We should be systematically accessing any and all surface and underground exposures of geologic materials, and acquire 3D and temporal information about the spatial distribution of material characteristics in different geologic regimes of formation and stress history. This includes field work at exposures such as road cuts and natural exposures, underground excavations, and mined openings. This work also includes recording and assessment of encountered and managed risk on real projects involving surface and underground excavations. The outcome from such an effort will be development of a rational and guided geologically-informed framework for engineers, designers and contractors to characterize geologic variability and uncertainty in a form that can be applied to project management and execution, and management of risks.

SUMMARY AND CONCLUSIONS

Engineers must partner with geologists, architects, and planners in new designs for urban underground space in the future, and such space will be much more than tunnels and stations. These professions must collaborate and prepare for the creative use of urban underground space that our society will demand in terms of excavated shapes/depths, human occupancy (social acceptance of underground space, spatial referencing, emergency response, aging population). These professions must support the development and deployment of new technologies that will serve the requirements for flexibility and quality of facilities in our finite urban spatial resources.

While geologic uncertainties and impacts are the focus here, engineers should fundamentally rethink materials and methods, including development and application of advanced methods for subsurface characterization and to extend applications for ground improvement methods. The framework for understanding risk and spatial variability of geologic conditions should be improved, and should our proficiency and understanding of assessment and redistribution of in situ stress. Improvements are also needed in excavation methods including drill/blast, lasers and other innovative technologies methods.

For engineers, professional homework is required. Data to support rational and long-term sustainable design and LCE need to be acquired, including time-dependency. In addition, the true value of underground space needs to be determined, effectively by creating a market that can establish a value for, say, a cubic meter of underground space in certain soil or rock conditions.

With underground and geologic conditions managed more effectively, we will be in a position to support development of a new understanding and acceptance of urban underground design for the public.
Acknowledgments

Appreciation is given to financial support provided by many funding agencies over the past 50 years. The opinions expressed in this paper are those of the author and not the US DOT.

REFERENCES

American Society of Civil Engineers. 2017. Report Card for America's Infrastructure, http://www.infrastructurereportcard.org/.

Anex, Robert. P., Realff, Matthew. J., and Wallace, William. A. 2006. Resilient and Sustainable Infrastructure Networks (RESIN). NSF Workshop Report, http://www3.abe.iastate.edu/biobased/RESIN.htm.

AWWA. 2001. Reinvesting in Drinking Water Infrastructure, AWWA Water Industry Technical Action Fund, http://www.mcwane.com/upl/downloads/resources/americas-water-infrastructure-challenge/reinvesting-in-drinking-water-infrastructure.pdf.

Bobylev, N. 2009. Urban underground infrastructure and climate change: opportunities and threats. *Proc.* 5th *Urban Research Symposium*, Marseille, France, http://www.urs2009.net/docs/papers/Bobylev.pdf.

Bobylev, N. 2016. Underground Space as an urban indicator: Measuring use of subsurface. *Tunnelling and Underground Space Technology.* Vol 55, DOI: 10.1016/j.tust.2015.10.024 · License: CC BY-NC-ND 4.0.

Chen, R. 2010. Groundwater Inflow into Rock Tunnels. Doctoral Dissertation, University of Texas at Austin. https://repositories.lib.utexas.edu/bitstream/handle/2152/ETD-UT-2010-08-1677/CHEN-DISSERTATION.pdf.

Fairhurst, C. 2017. Some Challenges of Deep Mining. Engineering. Vol. 3, pp. 527–537.

Hoeppe, P. 2016. Trends in weather related disasters—Consequences for insurers and society. *Weather and Extreme Climates.* Vol. 11, pp. 70–79.

Masuda, Y., N. Takahashi and T. Ojima. 2004. Utilization of Deep Underground Space in Tokyo—Urban Renewal with the City's New Backbone Lifeline. *Proc. Council on all Buildings and Urban Habitat.* Seoul. http://global.ctbuh.org/resources/papers/download/1590-utilization-of-deep-underground-space-in-tokyo-urban-renewal-with-the-citys-new-backbone-lifeline.pdf.

Nelson, P.P. and R. Sterling, R. 2012. Sustainability and Resilience of Underground Urban Infrastructure: New Approaches to Metrics and Formalism, *ASCE GSP*, 10 pp.

Nelson, P.P. 2016. A Framework for the Future of Underground Urban Engineering. *Tunnelling and Underground Space Technology.* Vol. 55, pp. 32–39.

Nelson, P.P. and H.S. Düzgün. 2018. Resilience Impacts from Integrated Above- and Below-ground Urban Infrastructure. *Proc. World Tunnelling Congress*, Dubai, UAE.

Sherrieb, K., Norris, F.H., and Galea, S. 2010. Measuring Capacities for Community Resilience. *Social Indicators Research*, Vol. 99(2), pp. 227–247.

Stones, P. and T.Y. Heng. 2016. Underground Space Development Key Planning Factors. *Procedia Engineering*, 165, pp. 343–354.

Use and Misuse of Geotechnical Baselines to Predict Soft Ground TBM Tool Wear

Ulf G. Gwildis • CDM Smith
Michael S. Schultz • CDM Smith

ABSTRACT

Cutting tool changes during pressurized-face TBM drives are a significant risk and cost factor. For contractors preparing bids the uncertainty of predicting tool wear can cut both ways, leading to non-competitive bids or to budget issues during construction. Either trend constitutes a risk to cost-effective project execution and avoidance of legal dispute. This paper describes how geotechnical baselines can be used and misused for tool wear prediction. And it tackles the question of baselining the tool wear system behavior that is driven by the geotechnical conditions as much as the contractor's machine design and mode of operation. A mission impossible?

SUBJECT INTRODUCTION

TBM tool wear is a key factor to consider when planning pressurized-face TBM drives because of the significant impact potential on construction schedule and budget if changes of the cutting tools are required during tunneling. This is usually the case for long drives in abrasive ground conditions where without a thorough tool inspection and maintenance regimen tool functionality can deteriorate to cause reduced mining efficiency, to result in secondary wear damage, or to reach the point that the TBM is not able to advance anymore (Figures 1 and 2).

Ground conditions that are considered abrasive often include complex geologies such as sequences of glacial deposits with their inherent variability of depositional environments or buried rock surfaces with weathering zones of varying thickness and strength characteristics. Increasing complexity of the geologic conditions tends to result in a higher uncertainty with respect to the contractual description of the subsurface conditions, which in North America is provided by geotechnical baseline statements and values. The geotechnical conditions of the tunnel zone, however, are only one of the components of the TBM tool wear system. Other components include the tunneling

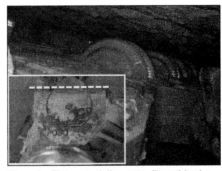

Figure 1. Completely worn ripper-type tools near cutterhead gauge position

Figure 2. Flat-spotted disc cutter (Insert) having lost rotation functionality

method (earth pressure balance or slurry TBM), cutterhead design, tool types (ripper-type tools and/or disc cutters as primary cutting tools), the way the TBM is operated, and ambient conditions.

Considering the challenges of the geotechnical baselining effort and the complexities of the TBM tool wear system, it is not surprising that tool wear tends to be a frequent focus in the case of disputes between contracting parties over financial responsibilities for schedule delays and budget overruns. This paper presents the concept of rating geotechnical baselines by their relative strength and TBM tool wear system components by their relative relevance for tool wear prediction of a specific tunneling operation. Against the background of this rating concept, project examples are discussed regarding use and misuse of geotechnical baselines. The concluding section includes conceptual ideas of how to strengthen baselining approaches related to tool wear and reduce uncertainties of project outcome.

GEOTECHNICAL BASELINE RATING

Per the ASCE Geotechnical Baseline Reports for Construction Suggested Guidelines (2007), geotechnical baselines should be (1) realistic and have a rational basis (otherwise an explanation needs to be provided why a baseline is set differently from what the available data indicates), (2) readily quantifiable/measurable, and (3) verifiable. The challenge to meet each of these requirements varies not only by specific geotechnical characteristic but also by project.

An example for the varying degree of the challenge to determine what constitutes (1) a realistic baseline would be baselining the number of boulders within the tunnel envelope of a project where the tunnel envelope is completely within a till unit and where nearby outcrops of this unit allow determining an average boulder count per volume versus a project with a deep-lying alignment perpendicular to the direction of glacial advance that crosses an unknown number of deposit types as well as an unknown number or little defined length of till boundaries with accumulations of boulders. An example for the complexities of (2) quantifying certain subsoil characteristics would be soil abrasiveness, for which laboratory test procedures such as Miller Number (ASTM G75) and SAT™ (by NTNU/SINTEF) have been developed. Because of scale issues, in the cases of coarse-grained soils, larger components need to be processed first, e.g., by crushing them to allow the test procedure to be applied, thereby translating component size into angularity. Finally, examples for the challenge to (3) verify a baseline during pressurized face mechanized tunneling would be the relatively easy task to determine if coarse-grained soils have been encountered versus the nearly impossible task to determine the size of a boulder that has been digested by the TBM.

These exemplary cases illustrate that the ability to meet the postulated requirements for a sound geotechnical baseline ranges from relatively easily achievable to almost impossible to achieve depending on the specific subsoil conditions and the extent of the geotechnical exploration effort. Consequently, these metrics can be used for defining the strength of a geotechnical baseline, e.g., by using a simple rating system where 0 is impossible, 1 is difficult, 2 is moderately challenging, and 3 is easy to achieve. A baseline may be expressed as a maximum value, a minimum value, an average value, a histogram distribution of values, or combination thereof; however, the geotechnical baseline requirements apply in each case, as do the suggested metrics of determining baseline strength.

Figure 3 illustrates applying the concept of quantifying baseline strength to select geotechnical baselines used for estimating TBM tool wear. Three past projects from the

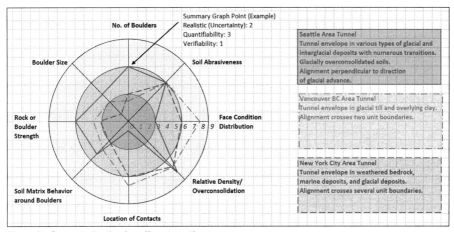

Figure 3. Concept of rating baseline strength

Seattle, Vancouver B.C., and New York City metropolitan areas are used, each of which included a TBM drive that was conducted at least in part through glacial deposits that are considered highly abrasive. While the Vancouver B.C. and New York City projects included geotechnical profiles with well-defined boundaries between glacial and non-glacial soil units, the Seattle project included deep tunnel drives perpendicular to the direction of glacial advance and retreat crossing a large number of boundaries between various glacial and non-glacial soil units whose exact distribution along the alignment between the exploratory borings was not known. This resulted in the decision to baseline the percentage of tunnel face conditions along the alignment without providing location specificity.

TBM TOOL WEAR SYSTEM COMPONENTS RATING

TBM tool wear is a complex system behavior with several system components. These interacting system components can be grouped into subsoil characteristics (grain size distribution, coarse components, relative density and shear strength, mineralogy and content of minerals with high Mohs hardness such as quartz, angularity of grains, etc.), tooling characteristics (tool type, tool materials, cutterhead design, etc.), the tunnel excavation method (TBM type using slurry method or earth pressure balance (EPB) method, the latter with variants of soil conditioning approaches), the way the TBM is operated (tool penetration rate, thrust, cutterhead revolution speed, etc.), and ambient conditions (hydrostatic head, temperature, salinity, etc.).

While the relative rating of certain system components seems practical to a degree, e.g., rating clays as generally less tool wear inducing than coarse-grained soils, any attempts of tool wear quantification is possible only for a specific wear system, for which all its components are defined. The rating of subsoil conditions alone depends to a high degree on the tool type used for excavation. While ripper-type tools when shattering boulders by impact force may experience high wear rates due to loss of carbide inserts and hardfacing followed by increased wear rates of the steel surfaces in abrasive soils (Figure 1), disc cutters may excavate boulders held in place by a high-strength soil matrix by rock chipping without significantly increased wear rates but may lose their functionality in fine-grained soils with high stickiness potential or in loose sands, which rapidly increases wear (flat-spotting) (Figure 2). Components such as soil conditioning or lack thereof can have a significant impact irrespective of the soil

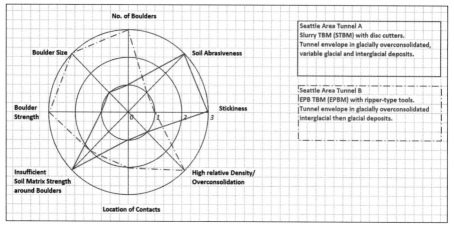

Figure 4. Concept of rating TBM tool wear system components

abrasiveness. Suboptimal driving of the TBM will increase wear and wear damage risk (secondary wear).

Consequently, for a given wear system, the individual system components can be rated regarding their impact strength, from 1 (little impact) and 2 (moderate impact) to 3 (high impact). Figure 4 illustrates this approach for two exemplary wear systems, a slurry TBM operation using disc cutters and an EPB TBM operation using ripper-type tools, both projects including the excavation of glacially overconsolidated deposits with cobbles and boulders.

USE AND MISUSE OF GBR BASELINES FOR TBM TOOL WEAR PREDICTION

During softground pressurized-face tunneling, monitoring, recording, and sharing of the TBM operational parameters is standard practice. Furthermore, some effort of tracking the geotechnical conditions encountered by sampling and by classifying the tunnel spoils in conjunction with correlation analysis if often performed. However, direct observation of the tunnel face conditions is not possible during TBM advance.

At the current state of technology, continuous measuring and tracking of TBM tool wear rates during softground pressurized-face tunneling is not conducted. Instead, cutterhead inspections, which often require pre-constructed safe havens or hyperbaric interventions, are used to collect information on the condition of the cutting tools. Depending on the anticipated wear rates when planning the project, regular inspection stops may be mandated by the construction contract, planned by the contractor in frequent or infrequent intervals, or not considered at all.

Identification of the face conditions encountered based on indirect observations and infrequent tool wear data collection during the TBM drive leaves room for intentional or unintended misinterpretation of geotechnical conditions and cause-effect relationships when it comes to tool wear. Tool wear and the need for tool changes on the other hand has a significant cost and schedule impact. Not surprisingly, discussions over responsibilities when schedule delays and budget exceedances occur often brings the subject of tool wear to the table.

The following case studies of TBM operations recently conducted in the Seattle, Vancouver B.C., and New York City metropolitan areas provide examples of conflicting interpretations and how the suggested rating approaches can provide additional insight that may be used for conflict resolution.

Case Study 1: Tunnel Face Condition Tracking

This tunnel project included two drives with a combined length of 4.1 miles utilizing two slurry TBMs equipped with disc cutters. The TBMs advanced at depths of several hundred feet through glacial and inter-glacial deposits of several glaciations. The tunnel alignments perpendicular to the direction of glacial advances resulted in a large number of different face conditions to be encountered. The baselining approach categorized the soil types based on their engineering characteristics into color-coded Tunnel Soil Groups (TSG), the teal TSG consisting of clayey soils, purple TSG of non-plastic fine-grained soils, yellow TSG of sandy soils, and red TSG of gravelly soils. The various face conditions included full faces of a single TSG as well as mixed faces of two or more TSGs. For each face condition that was expected to be encountered, a non-location-specific baseline was provided in terms of a percentage range of the overall drive length.

The soil abrasiveness of each TSG was baselined in terms of SAT™ values, Miller Numbers, and Quartz content. Further baseline statements relevant to tool functionality considerations and tool wear estimates included stickiness potential of fine-grained TSGs, if or if not boulders were to be expected within a certain TSG, and soil matrix strength in case of boulder conditions.

With respect to tool wear, in this case study the soil abrasiveness of a TSG earned a high rating and therefore the relative quantities of the less abrasive vs. the more abrasive TSGs encountered was an important factor. However, verifying the face conditions encountered was an obvious challenge and earning the verifiability aspect initially a low rating. During construction competing verification approaches were applied. One approach, conducted by the contractor, was based on a correlation of specific energy consumption—the amount of energy required to excavate a unit volume of soil—with the TSGs. Another approach was based on sampling the TBM spoils at the slurry separation plant and conducting a visual-manual geotechnical and geologic classification supplemented by geotechnical laboratory index testing in regular intervals. The second approach was conducted independently by both contracting parties.

Figure 5 illustrates a comparison of the approaches of tracking the face conditions encountered with the goal of verifying the geotechnical baseline. While the tracking based on tunnel spoils classification shows some variance between the two parties the differences are minor and could be resolved during joint data review sessions. The difference to the energy-consumption based approach is more obvious. While the latter approach has an inherent logic, the correlation may not be as clear due to factors such as the way the TBM is operated, i.e., advance rate, penetration rate, the condition of the slurry in the excavation chamber being loaded with excavated material, etc. Because the interpretation of face conditions based on the specific energy correlation approach in several instances could not be reconciled with the spoils recovered at the separation plant—a discrepancy that becomes clear during phases of homogeneous tunnel spoils output over a length of several rings—this approach was dropped. It should be noted that sampling at the separation plant recovered chunks of fine-grained and till soils with the soil structure still observable (Insert of Figure 5 shows a spoils sample of a mixed face Yellow and Teal TSG).

Conclusion: Low ratings of baseline verifiability include a high potential for misinterpretation of conditions encountered. In this case study, the initial low rating of baseline

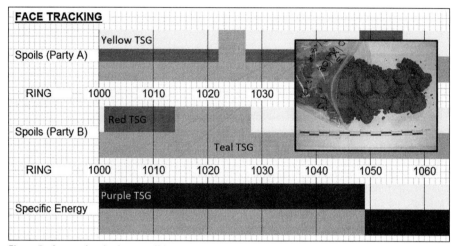

Figure 5. Approaches for face condition tracking (Insert: spoils sample of mixed face teal/yellow TSG) (Case Study 1)

verifiability in the case of face conditions was overcome by establishing a face condition tracking procedure that both contracting parties could agree on.

Case Study 2: Wear Mechanism Identification

This project included a 4-mile-long EPB TBM drive using ripper-type tools through glacially overconsolidated deposits, the first half through non-glacial deposits, the second half through glacial deposits. In addition to the differentiation between glacial and non-glacial (inter-glacial) deposits, the Geotechnical Baseline Report grouped the soils in TSGs based on their engineering characteristics and provided—as in case study 1—for each TSG baseline values of soil abrasiveness (SAT™) and statements regarding boulder numbers, sizes, strengths, and distribution.

During the TBM advance a significant difference in wear rates of the ripper-type tools was observed between the glacial and the non-glacial deposits. The contractor sampled tunnel spoils of the various TSGs and conducted soil abrasiveness tests, whose results fell within the baseline ranges, confirming them.

Tracking of the tunnel spoils including recording of coarse components and rock shards indicating the presence of boulders on a per-ring-basis revealed a rough correlation between wear rates and coarse component tracking frequency. Other observations such as the non-symmetrical wear pattern of the ripper-type tools over the face of the cutterhead, pointed to coarse component impact damage as the primary wear mechanism. Soil abrasiveness was found to be a secondary factor (Figure 6).

Conclusion: In this project establishing soil conditions tracking procedures including recording of coarse components and rock shards in the muck cars and laboratory testing of spoil samples proved the verifiability of geotechnical baselines critical for the tool wear behavior. The tracking data revealed that in this wear system the presence of coarse components (gravel, cobbles, boulders) had a high impact rating and the soil abrasiveness a relatively low impact rating.

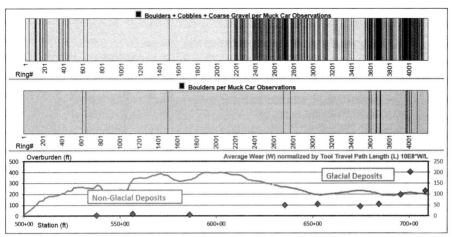

Figure 6. Correlation of tracking data per ring and wear data (Case Study 2)

Case Studies 3 to 5: Tool Wear Rate Variance

Case studies 3 to 5 were EPB TBM drives of lengths between 0.6 miles and 2.6 miles utilizing in all or most cutterhead positions ripper-type tools as primary cutting tools. In all three cases alignment length could be divided into reaches of tunnel face conditions considered highly abrasive (glacial till, weathered rock) and face conditions in alluvial soils, recent marine deposits, and fine-grained soils considered significantly less abrasive in comparison. In all three cases tool inspection stops that included tool wear measurements were conducted in irregular intervals leaving long drive lengths with no observation of wear rates. This obviously left room for varying interpretations regarding the dominant cause for tool wear and regarding the effects of tool wear on overall production rates.

The TBM of case study 3 advanced out of an alluvial valley filling into glacially overconsolidated, non-glacial and glacial deposits. Tool wear inspections during the first half of the drive length showed little wear in the sandy alluvial soils and in the glacially overconsolidated non-glacial mostly coarse-grained and partly fine-grained soils. The contractor's difficulties to conduct hyperbaric interventions resulted in the decision to allow completing the remaining drive length without the specified inspection stops. At the only additional stop after about three quarters of the drive length, just two tools were inspected and replaced, showing little wear. When the TBM daylighted, the cutterhead tools were found to be heavily worn and completely gone in the distal positions, where the tool stumps were flush with the cutterhead surface. To what degree the various geologic units, unit boundaries, and geotechnical soil groups and associated geotechnical baselines and to what degree the way the TBM was operated contributed to the high wear rate in the last drive section is speculative due to lack of wear measurements (Figure 7).

The TBM of case study 4 advanced through glacial deposits and weathered bedrock into recent marine deposits while entering glacial deposits again before reaching the reception shaft. Over the drive length two hyperbaric interventions were conducted that generated tool wear data. Tool wear consistent with tunneling in glacial deposits that include nested boulders was discovered and resulted in discussions regarding the specific weight of the various causal factors and geotechnical baseline values. The evaluation included comparing the contractor's pre-construction planning for tool

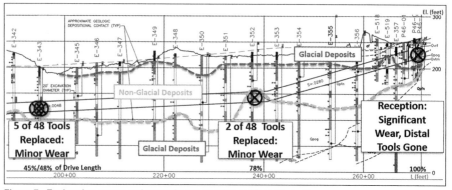

Figure 7. Tool replacements and geotechnical conditions (Case Study 3)

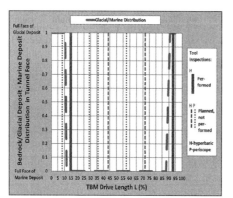

Figure 8. Reduction of tool inspection intervals (Case Study 4)

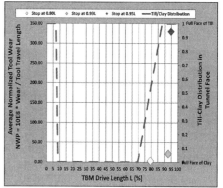

Figure 9. Tool wear rate variance and face conditions (Case Study 5)

changes, comparing estimated and actual tool consumption, and comparing geotechnical baseline values with the results of verification testing conducted during tunneling. The exceedance of a geotechnical baseline value relevant to tool wear (boulder strength) was considered and weighted in the light of less tool inspections and replacements being performed than per pre-construction planning (Figure 8).

The TBM of case study 5 utilized ripper-type tools as primary cutting tools supplemented by disc cutters in gauge positions as well as scrapers. The tunnel alignment was excavated in clay and till including two transitions (boulder contacts). Significant differences in the tool wear rate were recorded during tool inspections at 80%, 93%, and 95% of the drive length. Plotting the wear measurements over the tool travel distance indicates low wear rates at the first inspection stop and significantly increased wear rates thereafter. Plotting the average wear normalized for the tool travel distance over the drive length provides a clearer picture of this wear behavior (Figure 9). Visualizing the advance lengths through the two soil units by the graph of a till-clay distribution factor with "1" indicating a full face of till and "0" indicating a full face of clay provides the geotechnical context and identifies the drive sections through the geologic contact. The relationship between tool wear rates and geologic contact sections does not point to the contact as the main contributor to the significant wear rate increase. This happens mostly in the full face of till near the end of the drive. Reported loss of soil conditioning functionality in this section would indicate lack of the associated mitigating effects and therefore should be considered a potential factor.

Conclusion: Regular tool inspections at drives or reaches within face conditions considered highly abrasive reduce the risk of unplanned TBM stops or reduced production due to worn tools and in addition to mitigating primary wear of the cutting tool it reduces the risk of secondary wear damage to the TBM cutterhead and other components. In addition, it provides a better record of tool wear rates and thereby reduces the potential for disagreements regarding cause-effect relationships. Data collection allows better assessing ratings of geotechnical baselines and wear factors.

CONCLUSION AND IDEAS

This paper outlines the concept of rating the strength of geotechnical baselines and of rating the relevance of a geotechnical baseline for a specific tool wear system. Combining the two ratings allows better understanding the level of uncertainty of tool wear prediction for a specific tunnel project and identifying the factors with a high potential for conflict between contracting parties in case of the project falling behind schedule or exceeding budget during construction.

Because in many tunnel construction contracts the type of TBM methodology—i.e., use of a slurry TBM, an EPB TBM, or a hybrid—is not specified and the tooling is essentially always left to the contractor as the contractor's means and methods, applying this rating approach requires the designer for the owner to consider the various possible combinations of TBM types and tooling and identify the weaknesses/challenges for each option. The result, however, provides a better understanding of where the construction contract may have weaknesses and therefore provides the opportunity of developing mitigating measures at that point.

Every tunnel project has unique boundary conditions and at some TBM operations tool wear may not be an issue, in which case including mitigating measures could have unintended consequences and burden the project with additional cost that may not be considered reasonable for the reduction of risk it provides. However, the following list includes mitigating measures that may be useful to consider for tunnel contracts of long TBM drives through abrasive ground:

- Providing complete and comprehensive geotechnical baseline values and statements relevant to tool wear
- Clarifying risk-based geotechnical baseline values, e.g., where the uncertainty what would constitute a realistic baseline is high
- Requiring a detailed cutterhead maintenance plan
- Requiring data sharing of the TBM operation including tool wear measurements
- Requiring tracking procedures of the face conditions encountered
- Requiring regular joint data review meetings of tracking data collected by both contracting parties
- Requiring soil conditioning for EPB TBM drives in all face conditions
- Specifying minimum requirements for wear resistance of TBM components such as wearing plates and hardfacing
- Specifying mandatory tool inspections (which may include the ability to conduct hyperbaric interventions and/or the construction of safe havens)

- Requiring tool functionality checks (e.g., via monitoring disc cutter rotation speed)
- Requiring tool backloading functionality or tool locks (at large diameter TBMs)
- Requiring the ability to change tool types
- Requiring the contractor to have spare tools immediately available

To decide if and to what degree to include these mitigating measures in the contract documents, the designer needs to conduct an evaluation of likely tool wear rates. For generating a bid, designing the TBM, and planning the TBM operation, the contractor needs to estimate tool wear and tool replacement intervals. Both parties will use empirical knowledge of system behavior in similar geologic conditions to estimate wear system behavior. With each project the empirical knowledge base widens. With technical advances leading to increasingly instrumented TBM drives—vibration monitoring for boulder impact tracking and continuous wear rate monitoring for disc cutters are promising developments—the quality of the data base documenting these TBM drives should result in higher confidence of establishing cause-effect relationships between tool wear, TBM operation, and geotechnical conditions.

This leads to the question if at one point a functional baseline of tool wear can be considered, once the confidence of being able to determine a realistic baseline of wear has reached a high level. In conjunction with the TBM operational parameters the tool wear rates would be quantifiable. In conjunction with continuous tool wear tracking or tool inspections/replacements in regular intervals the tool wear rates would be verifiable. The authors leave it open—as a mental exercise for the reader—to contemplate the pros and cons of such an approach and if it would increase or decrease the chances of legal conflict.

REFERENCE

ASCE, 2007. Geotechnical Baseline Reports for Construction: Suggested Guidelines. Prepared by the Technical Committee on Geotechnical Reports of the Underground Technology Research Council, Randall J. Essex, Editor, ASCE, 2007.

Writing the GBR So the Contractor Can Understand It

Barry R. Doyle • Stantec

ABSTRACT

The complexity of GBR subject matter can challenge even experienced designers in conveying their understanding of subsurface conditions. This paper presents a deeply structured approach to developing a comprehensive, accurate, and clear GBR. The structure features a geological model to identify subsurface variables, a geotechnical model to define the interpretation of subsurface conditions, and a construction model to describe expected ground response to construction. A reader-oriented classic style of non-fiction writing is proposed as a better alternative to writer-oriented analytical style for achieving GBR objectives. This structure can be refined through repeated use on a range of projects, for long term continual improvement.

INTRODUCTION

The Geotechnical Baseline Report (GBR), developed by a committee of the Underground Technology Research Council (UTRC 1997), represents a significant advance in the standard of practice of tunneling. But despite our profession's 20 years of experience with the concept, writing an effective GBR remains a difficult task. It is not uncommon for the GBR to be blamed for construction problems—an ambiguity, misinterpretation, or unreasoned conclusion that led the contractor to expect something other than what the designer had in mind. The GBR concept is sound, but the complexity of its subject matter can challenge even experienced designers in conveying their understanding of subsurface conditions.

This paper presents a deeply structured approach to developing a comprehensive, accurate, and clear GBR. Order is achieved with a two-dimensional section outline. Coherence, and emphasis on consequential parts of the content, are achieved with a core structure comprised of a geological model, a geotechnical model, and a construction model. An advanced style of non-fiction writing suited to the complexity of the GBR is recommended, to achieve a reader-oriented presentation conducive to mutual understanding with the writer. Having a prescribed path with well-defined intermediate objectives relieves the writer of the burden of organizing the report while writing, allowing more time for refining. This structure provides a means to incorporate lessons learned from each GBR application, for long term continual improvement.

The author has applied this approach to four completed design-bid-build projects: (1) a 2-mile long microtunneled relief sewer, (2) a rock cavern for an underground pump station, (3) a 1-mile long 18-foot diameter rock tunnel for flood control, and (4) a small relief sewer project including a 700-foot long microtunnel and a secant pile shaft for a pump station. On these four projects there was not a single disagreement over the content of the GBR. There was not a single case where the contractor misinterpreted the ground and got into trouble. This is not to say these project were problem-free, but that their GBRs functioned as intended by the UTRC committee that developed the concept. This approach is currently being applied to a design-build project for a 30-foot diameter CSO storage/conveyance tunnel and appears to be suited to that as well.

STRUCTURE

Two-Dimensional Outline

The writer's initial task is to sort facts and ideas that will compose the GBR into sections. Of these many bits of information, a few may appear relevant to more than one section. Careful sorting is required to avoid loose organization that sets the stage for repetitive prose, ambiguities, and omissions. To promote rapid, accurate sorting of GBR content, the outline of Table 1 is used. On the left is a prescribed, linear section outline that is consistent with GBR guidelines (UTRC 2007). To eliminate sorting errors, a second dimension is added by partitioning the sections into three groups. Groups are distinguished by attributes of their content:

- The *Source Material* group accounts for all factual information used to develop sections in the following two groups. The Project Geology section is based on peer-reviewed journals and professional publications, which for engineering purposes can be considered factual. The site exploration program is summarized, and its data incorporated by reference to the Geotechnical Data Report (GDR). The section on Miscellaneous Sources of Information is for inclusion of small data sets collected outside the GDR scope of work, from third party sources; and for introducing relevant local projects, though their "lessons learned" are applied in the Construction Group.
- The *Geotechnical* group includes quantitative descriptions of existing ground conditions as interpreted from source material. For organizational purposes, ground conditions are described as they are understood to exist prior to construction. The content of this group is relevant to, but not contingent upon, construction means and methods or the shape and form of excavation.
- The *Construction* group includes topics contingent on means and methods or the shape and form of excavation. Prominent topics include expected ground response to construction, and means and methods anticipated or specified.

Inapplicable sections are omitted from the outline and the remaining sections renumbered, as when there is no information available from previous projects (section 4), standard design considerations require no further explanation (section 6), and building protection measures are adequately presented in drawings and specifications (section 8). Relevant sections only are shown in the table of contents.

Core Structure

After sorting the GBR content, the next step is to develop *coherence* by shaping the principal ideas into a story that the reader will find consistent and informative. To shape the story, the basic structure of Table 1 is complemented with a deeper structure

Table 1. Two-dimensional GBR outline

GBR Sections	Section Group	Attributes
1. Introduction 2. Project geology 3. Site exploration and testing 4. *Miscellaneous sources of information*	Source material	Facts
5. Ground characterization 6. *Design considerations*	Geotechnical	Data interpretation
7. Construction considerations 8. *Protection of existing structures*	Construction	Topics relating to means and methods

Note: Sections in italics, if inapplicable, are omitted and remaining sections renumbered.

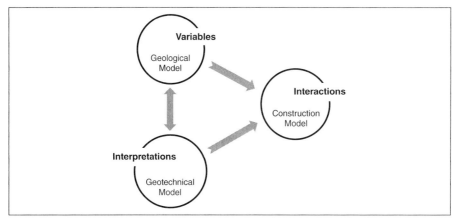

Figure 1. Core models of the GBR, their interdependency, and their emphasis

applied to the *core sections* of the GBR, those sections being Project Geology, Ground Characterization, and Construction Considerations. These are core sections because they form the basis for selecting means and methods of construction. The remaining sections tend to be narrow in scope, lend themselves to linear description, and are easier to write, so are omitted from further discussion here.

As an aid to shaping the story, each core section is conceptualized as a *model*, a complete and accurate representation of its subject matter, displaying its parts in relation to each other, and developed to a relevant level of detail. The GBR models are:

- Project Geology—*geological model*
- Ground Characterization—*geotechnical model*
- Construction Considerations—*construction model*

Core sections are interrelated, in that the geological model and the geotechnical model are mutually supportive, and both support the construction model. Their interrelatedness is obvious, but what is less obvious is the complexity that this adds to the task of writing. Model development is explained later.

To hold the reader's attention, each model should tell a story that achieves *emphasis,* meaning each story should distinguish important points from context. The model subjects of geology, geotechnical engineering, and construction are too broad to provide guidance. More specific themes are needed, so in the structure presented here the geological model emphasizes *variables*—the range of geological conditions that have a reasonable probability of occurring; the geotechnical model emphasizes *interpretations*—engineering description of existing conditions developed from geologic fact and acquired data; and the construction model emphasizes *interactions*—physical exchanges between the ground and the means and methods of construction. Model interrelations and emphases are depicted in Figure 1.

GEOLOGICAL MODEL

The Approach

The UTRC's GBR guidelines recommend that "Extended geologic descriptions and details should be limited to the Geotechnical Data Report" (UTRC 2007, 22). But this is impractical when the GDR is prepared by a geotechnical firm selected for their drilling and laboratory testing capabilities but lacking the design experience needed to develop a relevant geological model. Also, the guidelines recommendation implies that a cursory treatment of geology in the GBR is sufficient, a view consistent with general practice when the guidelines were initially published in 1997. This paper proposes a different approach—that recent developments in the application of geological models to civil works be brought to bear on the risk management function of the GBR.

Subsequent to introduction of the GBR concept, prominent engineering geologists in the UK argued that geological models should be applied to design and construction of civil works (Fookes, Baynes, and Hutchinson 2000; Knill 2003; Parry et al., 2014). A geological model drafted at the project outset can identify uncertainties to be targeted in a subsurface exploration program. A geological model provides a basis for interpolating between borings to develop a stratigraphic sequence. There should be no geological condition that comes as a surprise during construction, though its location, form, or engineering properties may not be accurately known. The papers cited are essential reading for understanding the concept. In the geological model for a GBR, emphasis is on *variables*—the range of conditions that a contractor might reasonably expect to encounter.

Traditional geotechnical investigations centered around test borings provide much detail about very little soil and rock. What these investigations reveal of the geological environment is a shadow of its true complexity, even where conditions appear to be monotonously uniform. A geological model can explain ground conditions from a broad perspective that cannot be attained from examining soil samples and rock core. Perspective is essential to perception of geological complexity and collateral risk, perception being a principal objective of the GBR.

Developing the Geological Model

The geological model is presented in two parts. The first, "Geologic Processes," is a broad history of the origin of strata; their subsequent alteration below surface by chemical, physical, and biological processes; and their alteration near-surface by weathering and erosion. A past-to-present method of development is used. The second part, "Geologic Conditions," is about the consequences of these processes—the present shape, distribution, and material properties of soil and rock. A bottom-to-top method of development is used. Topics in a geological model might include tectonics, soil-depositional environments, rock-forming environments, stratigraphy, structure, mineralogy, geomorphology, and hydrogeology. The initial draft is developed from publications on regional and local geology (maps, journal papers, government reports, and academic studies), and site walk-overs. Later it is conformed to facts acquired from geotechnical exploration, an effort that presumably has been pursued until geological uncertainties have been resolved to a manageable level of risk.

Developing the geological model is an exercise in divergent thinking, in contrast to convergent thinking employed to develop the geotechnical model (Baynes 1996; Knill 2003). Divergent thinking considers beyond the length and depth of excavation to remote yet relevant topics such as regional fracture patterns produced by interplate collision, gas generation and migration from organic-rich sedimentary rocks thousands of feet deep and miles away, and geochemical and geomicrobiological processes that affect the composition of groundwater in ways that manifest during construction. Divergent thinking eliminates "geological unknowns." The intent is not to explain the science behind such conditions but to identify their source, to give the reader context for understanding their occurrence and variability.

Geologic literature describes soil deposits and rock strata in broad terms. Such terms are appropriate to the purpose of the model and can be reiterated. These same materials will be described in engineering (quantitative) terms in the geotechnical model to follow. Unfamiliar geologic terms can be used provided they are defined in the model, or they can be replaced with other terms more familiar and appropriately descriptive; for example, *diamicton* used in a glacial context might be replaced by *till*. The decision to use technical jargon should consider both the descriptive needs of the writer, and background of the reader.

Illustrations are an effective way to communicate concepts that would be tedious to describe verbally. Examples include a geologic map, a stratigraphic column, or geological features such as type of fold or fault.

Data from geological literature are introduced in the geological model and presented in a GBR appendix. Examples from the author's work include rock stress measurements from various sources, mineralogical data from government studies, and joint mapping data from academic studies. Should a geological investigation acquire a large amount of data, as contemporary methods might (for example, see Ball, Isaacson, and Cohen 2011), such data could be reported in a separate Geological Data Report.

GEOTECHNICAL MODEL

The Approach

The geotechnical model describes stratigraphy along the alignment, material composition of soil and rock, their engineering properties, and occurrence and chemical composition of groundwater. Emphasis is on *interpretations*—characterizations developed from analysis of acquired data. Characterizations are presented in quantitative, engineering terms (descriptive terminology for soil and rock, charts, numbers) that can be applied to solving engineering problems. To conform to the report's two-dimensional outline, subject matter is confined to ground conditions as they are understood to exist prior to construction, and interpretations are independent of anticipated means and methods or the shape and form of excavation.

Characterizations are developed from boring logs, data sets generated from field and laboratory tests, and water level measurements, and are assembled into a comprehensive and accurate description of ground conditions within the influence of construction. The interpretive process is made transparent to the reader, with the expectation that an informed reader does not require every step be explained. Transparency lends credibility to the work, and enables readers to judge the uncertainty in the interpretations. Should a dispute arise, the challenger should find no cause to characterize GBR interpretations as illogical or unsubstantiated.

Developing the Geotechnical Model

The geotechnical model is presented in four parts: (1) definition of geologic units, (2) a graphic plan and stratigraphic profile, (3) supplementary text directed to the graphic, and (4) interpretations of data sets.

Geologic Units

The author's preferred method of developing subsurface profiles is to interpolate conditions between borings on the basis of geologic units of soil and rock. In some cases geologic units can be defined as lithostratigraphically described formations of relatively uniform composition, formally named in geologic literature. Soil units often require area-specific definition in terms of mode of deposition, a factor in their compositional variability and density/consistency. Differentiating soil units by depositional mode is made easier if the borehole logger is advised to record their fabric (arrangement of particle sizes), a characteristic often too subtle to register using standard descriptive procedures alone.

Geologic unit definitions are developed to serve engineering purposes. Their physical scale is made relevant to the excavation; not so big as to obscure unit variability, not so small as to be of no consequence. Anomalous features such as a lag deposit of boulders or a fault zone in rock are handled as a unique geologic unit rather than as a characteristic of the unit containing the anomaly.

Geologic units are defined prior to introducing the subsurface profile, in a tabulation of identifying characteristics such as compositional variability, lateral continuity, and dispersed features such as thin hard beds, abrasive inclusions, porous features, or cobbles.

Graphic Subsurface Profile

A graphic subsurface profile is developed by projecting stratigraphic contacts at borings onto the tunnel centerline using 2D construction, or 3D if multiple borings straddle the alignment, then interpolating between these contacts to construct a best estimate of stratigraphy. Uncertainty in the location of boundaries between geologic units is understood, and need not be reinforced by use of squiggled lines, or broken lines with question marks. The profile is developed in just enough detail to convey major geotechnical conditions relevant to making and supporting an excavation. Each geologic unit is labeled with a summary description of its predominant material character, in geotechnical engineering terms. Omitted from the summary description are localized conditions that could affect the work but that might clutter the graphic; these are described later. A graphic profile makes it unnecessary to enumerate the percentage of tunnel or shaft length expected to penetrate specific units.

Interpolating geologic units between borings can be difficult and uncertain. Latent uncertainty should not deter the effort, because achieving accuracy is not as important as conveying complexity. Cognizance of stratigraphic complexity is essential for selecting principal means and methods of construction that can advance efficiently. Where means of construction are suited to geologic complexity, the precise arrangement of strata encountered is usually of little consequence.

Numerical data obtained from borehole testing can convey complexity. Such data include SPT blowcounts, rock core recoveries and RQDs, and hydraulic conductivities from packer testing. To avoid cluttering the stratigraphic profile, these data are presented on a second profile, on stick logs. The plan view is omitted, and the vertical scale expanded to accommodate a readable font. Numerical data can be easily visualized when presented graphically, such as point plots with connecting line for SPT data, and bar charts for RQD and hydraulic conductivity data. Separating the stratigraphic and data profiles makes both presentations easier to comprehend.

Graphic subsurface profiles are *technical illustrations*, to be distinguished from *technical drawings*. The objective of the drawing is precision, whereas that of the illustration is rapid visual comprehension. An illustration is a combination of technology and art, its development requiring thought on how it will be viewed. An effective illustration is devoid of unnecessary detail. Its stylistic devices (fonts, line weights, symbols, and text) are applied with intention to achieve coherence and emphasis. Final assembly of graphic subsurface profiles is normally the task of CAD technicians, few of whom receive training in the art of illustration. Responsibility falls on the originator to see that illustrations achieve visual uniformity, contrast, balance, and symmetry.

Supplementary Verbal Description

Introduction of the graphic profile(s) is followed by supplementary text to complete the stratigraphic description. This "visual + verbal" approach is easier for a reader to comprehend than lengthy text that sequentially details conditions to be encountered down a shaft or along a tunnel. The supplementary text begins with direct reference to the graphic and a broad description of conditions portrayed, to guide the reader's eye across the illustration to familiarize the reader with its content. This broad description points out generalities, patterns, major points of interest, and may include the writer's overall impressions. Material descriptions shown on the graphic are not repeated.

This introduction is followed by description of localized conditions omitted from the graphic profile but that could affect the work. Examples of localized conditions include:

- Unconformities to be intersected by shaft or tunnel, such as the glacial-postglacial soil contact, or the soil-rock contact
- Local impediments indicated on boring logs, such as cobbles and boulders, hazardous gas, and high permeability
- Faults (See Hunt, Smith, and Moonin 2015)
- Rock discontinuities, frequency and condition of
- Boulder composition and estimated strength
- Rapid variations in geotechnical properties, such as soft stratum bounded by hard, or permeable stratum (sand) bounded by impermeable (clay)
- Soil strata sensitive to dewatering (consolidation), vibration (densification), overstress (failure), and seepage gradients (flowing)
- Man-made obstructions such as buried footings, piers, or timber piles
- Aquifers; major, minor, and perched
- Groundwater levels, average and range
- Mineralogy, where a factor in material behavior

The geotechnical model need not be exhaustive. Where the GBR is silent, the rule is "What would a reasonable contractor expect?"

Interpretations of Data Sets

Graphic and verbal descriptions are followed by characterization of those engineering properties that a contractor will need to evaluate means and methods, excavation rates, initial support systems, and groundwater control measures (Freeman et al., 2009). Quantitative characterizations are developed from objective analyses of data sets acquired from field and laboratory testing, moderated with experience and judgment to account for sample disturbance, testing errors, sample population, and geological variation. Engineering properties may be developed for each geologic unit, or for combinations of similar units. Numerical results are usually summarized in a table. The basis of any interpretation is briefly explained with accompanying text. Field-sourced graphics are placed in one appendix, followed by laboratory-sourced graphics in another appendix.

Graphical presentations that accurately portray expected range and distribution of an engineering property of a geologic unit may be sufficient. Atterberg limits plots and grain size curve compilations fall in this category. Histograms show the important data set properties of size, central tendency, range, scatter, and skew. But the histogram is unique to its data set, and unless it is shaped like a bell or smooth curve, it makes an ambiguous characterization. Irregular histograms result from data sets that are too small to represent the frequency distribution accurately, or data sets that are skewed from a central tendency.

Statistical analysis can resolve the scatter and skew of a data set into a range and distribution of the property, so supersede histograms. The normal probability functions of mean and standard deviation are used to characterize geotechnical properties that meet the requirements of a random variable, defined as independent of the properties of adjacent samples. Some random variables include strength of soil, strength of intact rock, friction angle, cohesion, unit weight, moisture content, slake durability, and abrasivity. The common assumption that random variables follow a normal distribution is not always accurate. For data sets that present a nonnormal distribution, triangular or rectangular distributions can be developed. Geostatistics is outside the scope of this paper, but references are provided for further study (Fenton 1997; Isaaks and Srivastava 1989; Raymer 2010; USACOE 1999; WES 1993).

Characterizations are developed to represent the natural variability of the geologic environment. The contractor's responsibility for selecting means and methods extends to considering the full range and distribution of ground conditions indicated. TBM cutters must be capable of cutting the hardest rock, roof support must retain the weakest rock, and TBM advance rates might reasonably be estimated from average strength of the rock.

Baselines in the Geotechnical Model

The author considers the baseline to be comprised of the full content of the GBR (excepting any data disclaimed), considered in its entirety. That portion of the baseline emphasized in the geotechnical model includes interpretive and quantitative material, that is, definitive characterizations reasonably supported by application of geologic principals, statistical analyses, and informed judgement. Ground conditions that are indicated but cannot be accurately quantified are resolved in the construction model to follow.

CONSTRUCTION MODEL

The Approach

The construction model draws attention to consequential ground conditions described in the geological and geotechnical models, and indicates construction methods considered appropriate to manage them. Emphasis is on *interactions*—the physical behavior of ground affected by excavation, dewatering, and induced vibration. If the contractor can be influenced to bring the right equipment and materials to the job, most unexpected problems can be overcome within normal measures of project success. To conform to the report's two-dimensional outline, subject matter includes topics contingent on construction taking place or on the shape and form of excavation.

The Construction Considerations section begins with an introductory statement that defines the designer's intent in subsequent references to means and methods of construction:

> *The purpose of this section is to describe how predominant types of ground are expected to behave in response to construction, and to point out localized ground conditions that could present problems. To indicate the extent and severity of ground behavior, means and methods of construction considered appropriate to control the ground within acceptable limits of risk may be stated. No detailed constructability analysis has been performed for any anticipated method of construction. In selecting means and methods, the Contractor is expected to consider the full range and distribution of ground conditions presented.*

The following writing tactics are advised to discourage manipulation of the text in an attempt to justify a DSC claim originating from poorly performing means and methods:

- Write express paragraphs on specific topics pertaining to specific locations.
- Avoid general statements that might appear inaccurate if applied to specific locations.
- Do not presume to educate the contractor on methods of construction.
- Do not attempt to lead the contractor's thinking toward a specific alternative.

Developing the Construction Model

Readers are likely to skim the Construction Considerations section looking for what interests them at the moment. For the reader to perceive the presentation as efficient, dispersion of material relevant to a subject of interest should be avoided. What constitutes an efficient presentation depends on the project, as their characteristics differ greatly. A project-specific model can be developed by applying a two-part approach: (1) general topics, and (2) ground behavior in specific excavations. This approach is demonstrated here assuming a water project consisting of shafts and a rock tunnel with no surface constraints.

General Topics

The Introduction to the construction model is followed by a general heading under which complex subjects common to multiple project elements are presented. This arrangement avoids repeating this material in following sections on specific excavations, as repetition can cause the reader to lose interest. General topics might include ground loads on initial support systems in soil, rock mass classifications in uniform rock, vibration constraints that affect blast design, groundwater inflow estimates based on interpretation of packer test results presented in the geotechnical model, and hazardous gas inflow characterizations.

Rock mass classifications, groundwater inflow estimates, and hazardous gas inflow characterizations are included in the construction model rather than the geotechnical model because their outcome is subject to shape, form, and location of excavation, in keeping with the two-dimensional outline previously described. How these estimates are developed are made transparent to the reader to the same degree as interpretations in the geotechnical model.

Specific Elements of Work

Following presentation of general topics, constructability descriptions are developed for each major element of work (each shaft, each tunnel drive). These elements of work are presented in some logical order, such as expected sequence of construction (shafts, then tunnels). Ground behavior and constructability issues are described for each. Where an element of work extends through unlike ground conditions (shaft excavation through soil, then rock), separate descriptions are developed for each condition.

A challenge in presenting the construction model is to describe constructability issues without appearing to direct the contractor's work. To meet this challenge, a form of cause-and-effect method is used, where ground conditions are the cause and construction methods are the effect. Each description is developed in the following sequence:

1. Describe the predominant soil (or rock) to be excavated in the shaft or tunnel drive (for example, stiff clay, dense sand, or the like), omitting characteristics unlikely to influence general behavior. Follow with expected behavior in response to construction, if not controlled by applied means and methods.
2. Point out known or potential variations from the predominant material, referring to the geological model or the localized conditions described in the geotechnical model. Follow with expected behavior in response to construction.
3. State one principal method of construction considered appropriate for controlling general behavior within acceptable limits. Most contractors will tolerate occasional, brief interruption of good progress to deal with localized adversity, so a few variations from the predominant behavior would not justify over-conservatism in stating a method.
4. Every principal method of construction has shortcomings. Point out complementary measures that might be needed, or are specified to be readily available, to overcome localized adverse behavior. A reasonably conservative approach is acceptable, meaning if a complementary measure is likely to be necessary, state that it is expected to be necessary.

Steps 1 and 2 on ground conditions and behavior are the "cause" part of the approach. Examples of ground behavior topics include:

- Soil behavior in excavations using Tunnelman's Ground Classification (For description of adverse soil behavior in tunnels, see Heuer 1976, and Conolly and Goodfellow 2012.)
- Rock behavior based on RMR, Q, or other definitions (as in Palmstrom and Stille 2007)
- Sticky clay
- Consolidation of soft clays, and densification of loose sands
- Slaking, swelling, or other time-dependent behavior
- Abrasivity
- Manner or rate of groundwater inflow to the excavation

Ground behaviors that are predominantly under the contractor's control are not described in quantitative terms, examples being surface settlements due to soil losses, and "technical" overbreak in rock that can be attributed to means and methods (ITA 2013).

Steps 3 and 4 on construction methods are the "effect" part of the approach. By presenting construction methods as an indicator of severity and extent of good or bad ground behavior in an excavation, the writer can develop the construction model as an expert in ground conditions, a point of view consistent with the designer's responsibilities. Construction means and methods are characterized as either "anticipated" (considered an appropriate choice) or "specified" (a risk management requirement). Any exceptions to the relevance of general topics to a major element of work are made apparent.

Separate constructability descriptions for multiple shafts or tunnel drives that share common characteristics, such as ground type or method of excavation and support, can become repetitive and tedious to read. Such elements of work can be grouped and the description consolidated. A consolidated description might conclude by pointing out issues unique to specific elements within that group (such as a major utility conflict at a shaft, a structure at risk of settlement from a tunnel drive, or a localized adverse ground condition).

Relevant construction experience from area projects introduced in section 4 (see Table 1), is incorporated directly into the construction model. This pointed application of "lessons learned" is less easily dismissed as irrelevant to the topic of discussion.

Baselines in the Construction Model

The construction model is where gaps remaining in the baseline are filled, *gaps* meaning adverse ground conditions or influential characteristics that are indicated but cannot be accurately quantified. Examples are boulder obstructions, suspected faults, length of mixed face in a tunnel heading, and calculated estimates of groundwater inflow rates that span such a wide range that a contractor pricing the high end would not be the low bidder. Such conditions are resolved by subjectively defining the condition as a narrow range or upper limit, in easily measurable terms. Conservative estimates are appropriate where an overrun might require replacing an entire system, such as piping to dewater a heading, or ducting to ventilate a tunnel. Best estimates are appropriate where an overrun would require more time or more of the same materials, and can be managed with unit pricing in the bid, or by negotiation during construction. Defining subjective baselines in a realistic and justifiable manner discourages their being dismissed as "overconservative."

TECHNICAL WRITING

Prior to introduction of the GBR, geotechnical reports for underground construction had been largely design-oriented, and simpler. The GBR concept added ground behavior and constructability to the recommended content. Technological advances have raised the significance of engineering parameters on which TBM designs, TBM performance projections, and other constructability matters are based. These changes have increased the complexity of the GBR to a degree that may be beyond what engineers have been trained to deliver by way of writing skills.

In the US, most engineering students are taught *analytical writing*, a style that emphasizes reasoning, organizing, recognizing patterns, making an argument, and giving evidence. Analytical writing style is suited to most subject matter that engineers write about, such as investigative reports to document acquisition of data, academic theses or evaluations to justify a conclusion, and procedural guides, that is to say, subject matter that is narrow in scope and lends itself to linear development. Writing in analytical style might be imagined as an exercise in "connecting the dots."

However, the GBR's core subjects of geology, geotechnical engineering, and construction are dissimilar, complex, and interrelated. Presenting these subjects is not so much an exercise in "connecting the dots" but more like painting a landscape, where the geological model defines breadth and depth of the scene, the geotechnical model provides detail, and the construction model draws the viewers eye to particular features to develop a theme. To paint this landscape the author employs a style of non-fiction writing called *classic style*. In contrast to the writer-oriented analytical style, classic style is reader-oriented. "The guiding metaphor of classic style is seeing the world. The writer can see something that the reader has not yet noticed, and he orients the reader's gaze so that she can see it for herself. The purpose of writing is presentation, and its motive is disinterested truth" (Pinker 2015, 28–29). A statement more aligned with the objectives of the GBR would be difficult to craft.

Engineers are as capable of skilled writing as any professional. That our writing tends to be judged as competent rather than accomplished can be attributed to limitations of our academic training, not to aptitude. Writing skills have such broad application that the effort to develop them is worthwhile. The author recommends the following books on classic style:

- *On writing well: An informal guide to writing nonfiction*, 5th ed. (Zinsser 1994); a study of exceptional examples of technical writing; humorous and inspiring, a good place to start for anyone with doubts about their ability to write skillfully
- *Style: Lessons in clarity and grace*, 10th ed. (Williams and Colomb 2010); a set of diagnostic principles with which to evaluate a draft, to judge dispassionately how a reader would perceive it, and to revise it to clearly express the writer's intentions; thorough, methodical, and highly instructive
- *The sense of style: The thinking person's guide to writing in the 21st century* (Pinker 2015); written by a cognitive scientist who researched brain functions of persons reading; includes an enlightening explanation of what Williams calls "our intractable subjectivity," our natural inability to perceive things from a totally objective point of view (Williams and Colomb 2010, 119)

It is helpful for a writer to keep in mind the intended readers and how they might judge the work. When writing a GBR, this author imagines three principal contractor readers: (1) the chief engineer, who selects means and methods and prepares the bid; (2) the geotechnical consultant, who designs initial ground supports and other specialty work; and (3) the foreman, who directs the day to day field work. The foreman may seem an odd choice, but among this group he will be the first to observe ground behavior in response to excavation, he is responsible for initiating complementary methods of construction when needed, and he is the first responder to unexpected adverse behavior. If GBR content were made accessible to the foreman, not simplified but written in a general-to-specific way so that technicalities could be skimmed without loss of comprehension, and if the foreman could be persuaded to stop in the field office before beginning a new excavation, to read the GBR's description of expected ground behavior and potential variations, serious problems could be avoided.

CONCLUDING REMARKS

The GBR was intended to portray ground conditions with certitude, the goal being not to eliminate differing site condition claims, but to eliminate disputes that arise from claims. The approach to presenting GBR content described here is designed to promote mutual understanding between the writer and the reader. Among the concepts presented are:

- A deep structure designed to promote accuracy, a coherent story, and emphasis on consequential parts; this structure can be refined through repeated use on a range of projects

- Inclusion of a geological model to identify the range of conditions that might reasonably be expected; this constitutes a change from UTRC's GBR guidelines, a change aligned with progressive efforts to apply geological models to design and construction of civil works
- Use of the classic style of non-fiction writing, a style designed to orient the reader's attention to seeing what the writer sees; this style is better suited to the complexity of GBR content

ACKNOWLEDGMENTS

The author would like to acknowledge the contributions of technical reviewers Jim Herbert and Julian Prada of Stantec, and Jack Raymer of Jacobs Engineering; and editorial reviewer Elizabeth Doyle; all of whom provided most helpful comments.

REFERENCES

Ball, R.P.A., J. Isaacson, and T. Cohen. 2011. 21st Century approach to geologic field reconnaissance for geotechnical or tunnel projects. In *Rapid Excavation and Tunneling Conference Proceedings*; 485–497. Englewood, Colorado: SME.

Baynes, F.J. 1996. Where is geotechnical practice heading—An engineering geologist's perspective. In *Proceedings of the 7th Australia New Zealand Conference on Geomechanics*, Adelaide, South Australia.

Conolly, R.L., and R.J.F. Goodfellow. 2012. A contractor's guide to Washington, D.C., Metropolitan Area geology. In *North American Tunneling, Proceedings*; 561–570. Englewood, Colorado: SME.

Fenton, G.A., ed. 1997. Probabilistic methods in geotechnical engineering. Workshop presented at *ASCE GeoLogan '97 Conference*, Logan, Utah.

Fookes, P.G., F.J. Baynes, and J.N. Hutchinson. 2000. Total geological history: A model approach to the anticipation, observation and understanding of site conditions. In *Proceedings of the International Conference on Geotechnical and Geological Engineering*, Melbourne, Australia; pp. 370–460.

Freeman, T., S. Klein, G. Korbin, and W. Quick. 2009. Geotechnical baseline reports—a review. In *Rapid Excavation and Tunneling Conference Proceedings*; 232–241. Englewood, Colorado: SME.

Heuer, R.E. 1976. Catastrophic ground loss in soft ground tunnels. In *Rapid Excavation and Tunneling Conference Proceedings*; 278–295. New York, New York: SME.

Hunt, S.W., E. Smith, and E. Moonin. 2015. Characterizing and baselining faults for tunneling. In *Rapid Excavation and Tunneling Conference Proceedings*; 600–615. Englewood, Colorado: SME.

Isaaks, E.H., and R.M. Srivastava. 1989. *Applied geostatistics*. New York: Oxford University Press.

ITA (International Tunnelling Association). 2013. *Guidelines on contractual aspects of conventional tunneling*. ITA Report No. 013.

Knill, J. 2003. Core values: the first Hans-Cloos lecture. *Bull. Eng. Geology Environ.* 62:1–34.

Palmstrom, A., and H. Stille. 2007. Ground behavior and rock engineering tools for underground excavations. *Tunnelling and Underground Space Technol.* 22:363–376.

Parry, S., F.J. Baynes, M.G. Culshaw, M. Eggers, J.F. Keaton, K. Lentfer, J. Novotny, and D. Paul. 2014. Engineering geological models—an introduction. IAEG Commission 25. *Bull. Eng. Geology Environ.* 73:689–706.

Pinker, S. 2015. *The sense of style: The thinking person's guide to writing in the 21st century.* New York: Penguin Books.

Raymer, J. 2010. Geotechnical variability and uncertainty in long tunnels. In *North American Tunneling, Proceedings*; 316–322. Englewood, Colorado: SME.

UTRC (Underground Technology Research Council). 1997. *Geotechnical baseline reports for underground construction.* Ed. R.J. Essex. Reston, Virginia: ASCE.

UTRC (Underground Technology Research Council). 2007. *Geotechnical baseline reports for construction: Suggested guidelines.* Ed. R.J. Essex. Reston, Virginia: ASCE.

USACOE (U.S. Army Corps of Engineers). 1999. *An overview of probabilistic analysis for geotechnical engineering problems.* ETL 1110-2-556.

WES (Waterways Experiment Station). 1993. *Statistical considerations.* RTH 104–93. U.S. Army Corps of Engineers.

Williams, J.M., and G.G. Colomb. 2010. *Style: Lessons in clarity and grace*, 10th ed. Chicago: The University of Chicago Press.

Zinsser, W. 1994. On writing well: An informal guide to writing nonfiction, 5th ed. New York: HarperPerennial.

PART  9

Ground Support and Final Lining

Chairs

Dawn Dobson
Barnard Construction

Rory Ball
Mott MacDonald

7.93 m Open TBM Shotcrete System Improvement and Innovation Jilin Project, China

Desiree Willis ▪ The Robbins Company
Ya Jun Guo ▪ The Robbins Company

ABSTRACT

In May 2018, a 7.93 m diameter open gripper (Main Beam) TBM completed the 24.3 km long Jilin Lot 3 tunnel under a maximum overburden of 272.9 m. The tunneling operation for the water transfer project, located in northeastern China, achieved a national record of 1,423.5 m in one month despite challenging conditions.

This paper will present an improved, innovative shotcrete system for TBM preliminary lining, developed through experience on previous projects. The shotcrete system, along with other structural design elements and a properly developed ground support program, allowed the TBM to bore successfully in variable hard rock and fault zones.

The paper will discuss how the shotcrete system and structural design increased safety and improved performance in a cost-effective manner. It will seek to define the variables that allowed the TBM to advance at rapid rates and will make recommendations for future types of projects that could benefit from the shotcrete system.

INTRODUCTION

The Jilin Yinsong Water Supply Project located in China's Jilin Province is a 736.3 km (457.5 mi) network, making it China's largest scale water diversion project to date. Once operational the water lines will divert the water from Fengman Reservoir at the upper reaches of Di'er Songhua River to central regions of Jilin Province experiencing chronic water shortages. These regions include the cities of Changchun and Siping, eight surrounding counties, and 26 villages and towns under their jurisdiction. The project will optimize water resource distribution, improve regional eco-systems, and ensure better food production and water safety for the people of Jilin Province.

About 134 km of the water supply network is underground. Three open-type TBMs were selected by the owner, Jilin Province Water Investment Group Co., Ltd., to bore about 62 km of tunnel in total (20–21 km boring per TBM contract with adits). The remainder of the underground work was excavated using conventional drill & blast techniques. Robbins supplied one 7.93 m diameter open-type (Main Beam) TBM for Lot 3 of the Jilin tunnel, and the other two TBMs were provided by Chinese suppliers. All three machines were designed to use continuous conveyor systems for muck removal.

The geotechnical baseline report showed that the rock consisted of tuff, granite and andesite with UCS up to 228 MPa and a maximum quartz content of about 43%. More than 80% of rock was predicted as class II & III, with maximum cover of 272.9 m (see Figure 1). Possible squeezing ground was also predicted given the relatively weak rock mass, as well as a total of 24 fault zones. Because of the geology an open-type TBM was selected to give the most flexibility in the expected conditions.

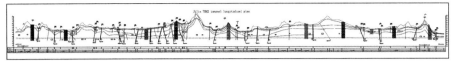

Figure 1. Jilin Lot 3 tunnel profile

Figure 2. Jilin TBM at factory acceptance

Figure 3. Rear view showing McNally pockets

TBM DESIGN

The Main Beam TBM was built in Shanghai, China, and designed for flexibility in terms of ground support. Pockets in the machine's roof shield were provided in order to use the McNally Roof Support System, designed and patented by C&M McNally. By replacing the roof shield fingers on a Main Beam TBM, the McNally system prevents rock movement in the critical area immediately behind the cutterhead support. The system has been tested and proven on projects worldwide—including the world's second deepest civil works tunnel, the 2,000 m deep Olmos Trans-Andean Tunnel in Peru—to increase advance rates while still maintaining worker safety on Main Beam machines in difficult rock conditions (see Figures 2 and 3). The cutterhead was mounted with 20-inch disc cutters, and designed with a maximum cutterhead thrust of 15,880 kN, as well as maximum torque of 9,743 kNm.

ENCOUNTERED GEOLOGY & GROUND SUPPORT

The project started in December 2013, with the first fault zone encountered after just 87 m of boring, requiring cooperation between the owner, contractor Beijing Vibroflotation Engineering Co. Ltd. (BVEC), and Robbins field service. Water inflows and collapsing ground in a section measuring 1,196 m long were resolved with a combination of McNally slats, grouting, and consolidation of the ground ahead of the machine.

Tunnel Reflection Tomography (TRT)—a method of ground prediction using seismic waves—was also used to detect changing conditions ahead of the TBM and was used largely in place of probe drilling. The TRT technique is based on the interface when a seismic wave encounters an acoustic impedance. Partial signals reflect back, while the rest passes through the medium. The change of acoustic impedance is typically observed in the interface between geologic formations or in a fractured rock mass. The reflected seismic signals are received by a highly sensitive seismic signal sensor, and the reflection coefficient is positive when the shock wave propagates from a low impedance material to a high impedance material. The opposite is true when propagating from high to low impedance--the reflection coefficient is negative. When the local seismic wave propagates from soft rock to hard surrounding rock, the deflection polarity and wave source echo are consistent. The signals can detect a number of features, such as a fracture zone inside the rock through a reversal in the echo of polarity. Analysis of the changes helps to characterize ground features including loose rock, broken rock, fault zones, and water, and the location and scale of each feature in front of the tunnel face.

In actuality the ground encountered was more difficult than thought (see Table 1). The ground support scheme was decided as follows:

Type II class ground support parameters: install rock bolt Φ 22, L = 2000.

Type IIIa class ground support parameters: install rock bolt Φ 22, L = 2500. Rock bolt position set local hang Φ 8 @ 200 * 200 wire mesh reinforcement, top injection C20 concrete at ring beam 320° range system, 10 cm thick.

Type IIIb class ground support parameters: install rock bolt Φ 22, L = 2500, 1200 * 1200, bolt spacing hang Φ 8 @ 200 * 200 wire mesh, top injection C20 concrete at ring beam 320° range system, 10 cm thick.

Type IVa class ground support parameters: ring beam adopts 16 I shape steel support, with space width of 1.8 meters, install rock bolt Φ 22, L = 2500, 900 * 1200, bolt spacing set Φ 8 @ 150 * 150 steel fabric hanging, top injection C20 concrete at ring beam 320° range system, thickness of 16 cm.

Type IVb class ground support parameters: ring beam adopts 16 I shape steel support, spacing width of 0.9 meters, install rock bolt Φ 22, L = 2500, 900 * 1200, bolt spacing set Φ 8 @ 150 * 150 wire mesh, top injection C20 concrete at ring beam 320° range system, thickness of 16 cm.

Type IVc class ground support parameters: ring beam adopts 16 I shape steel support, spacing width of 0.45 meters, install rock bolt Φ, 25 L = 3000, 900 * 900, bolt spacing set Φ 8 @ 150 * 150 wire mesh, top injection C20 concrete at ring beam 320° range system, thickness of 16 cm.

Type V class ground support parameters: ring beam adopts 16 I shape steel support, spacing width of 0.45 meters, install rock bolt Φ, 25 L = 3000, 900 * 900, bolt spacing set Φ 8 @ 150 * 150 wire mesh, top injection C20 concrete at ring beam 320° range system, thickness of 16 cm.

SHOTCRETE APPLICATION

In the 1st section of tunnel with length of 9840.187 m, there was a section of 1196 meters of tunnel in poor ground conditions. McNally ground support slats were applied in the middle of this poor section. Also in the 1st section, wire mesh was applied for 5014.87 meters and shotcrete was applied for 9155.187 meters, making shotcrete a primary means of ground support.

Shotcrete was applied at the following two zones on the TBM:

- L1 Zone
- L2 Zone

Initial support was applied at the L1 zone just behind the cutterhead support of the TBM. Te system is a manual spray system including a manual spray nozzle and piping bypass from the L2 zone. Platforms in the L1 zone give 360-degree circumferential

Table 1. Geological report of encountered rock

Chainage		Length	Rock Type	Rock Class	UCS	Quartz	Cover(m)	
26011	26402	391	Tuff	III	153.5		21	104
26402	26892	490	Tuff	IV~V	89.3		21	104
26892	27637	745	Tuff	III	89.3		21	104
27637	27987	350	Tuff, Granite	IV~V	89.3		21	104
27987	28982	995	Tuff, Granite	III	172–188		21	104
28982	29582	600	Tuff, Granite	IV~V	172–188		21	104
29582	30402	820	Tuff, Granite	III	172–188		21	104
30402	30940	538	Tuff, Granite	IV~V	172–188		21	104
30940	31415	475	Tuff, Granite	III	172–188		21	104
31415	31515	100	Tuff, Granite	IV~V	172–188		21	104
31515	32165	650	Tuff, Granite	III	172–188		21	104
32165	32365	200	Tuff, Granite	IV~V	172–188		21	104
32365	32760	395	Tuff, Granite	III	172–188		21	104
32760	32830	70	Tuff, Granite	IV~V	172–188		21	104
32830	33100	270	Tuff, Granite	III	172–188		21	104
33100	33150	50	Tuff, Granite	IV~V	172–188		104	154
33150	33515	365	Tuff, Granite	III	172–188		104	154
33515	33615	100	Tuff, Granite	IV~V	172–188		104	154
33615	34000	385	Tuff, Granite	III	172–188		104	154
34000	34350	350	Tuff, Granite	IV~V	172–188		104	154
34350	34450	100	Granite	III	186–236	37%	88	154
34450	35810	1360	Granite	II	186–236	37%	88	154
35810	35880	70	Granite	III	186–236	37%	88	154
35880	35960	80	Granite	IV~V	186–236	37%	88	154
35960	36163	203	Granite	III	186–236	37%	88	154
36523	37240	717	Granite	II	186–236	37%	69	88
37240	37540	300	Granite	III	186–236	37%	69	88
37540	37740	200	Tuff, Granite	IV~V	43–64	37%	69	88
37740	38255	515	Tuff	III	43–64		69	91
38255	38355	100	Tuff	IV~V	43–64		69	91
38355	38950	595	Tuff	III	43–64		69	91
38950	39030	80	Tuff, Andesite	IV~V	108–157	43%	69	91
39030	39530	500	Andesite	III	108–157	43%	69	91
39530	40030	500	Andesite	II	108–157	43%	69	91
40030	40490	460	Andesite	III	108–157	43%	69	91
40490	40820	330	Andesite	II	108–157	43%	69	91
40820	41400	580	Andesite	III	108–157	43%	69	91
41400	41550	150	Andesite	IV~V	108–157	43%	69	91
41550	42290	740	Andesite	III	108–157	43%	69	91
42290	42410	120	Andesite	IV~V	108–157	43%	69	91
42410	42610	200	Andesite	III	108–157	43%	69	91
42610	43110	500	Andesite	II	108–157	43%	69	91
43110	44270	1160	Andesite	III	108–157	43%	69	91
44270	44345	75	Andesite, Granite	IV~V	108–157	43%	69	91
44345	45200	855	Granite	III	108–157	43%	69	91
45200	45300	100	Granite, Tuff	IV~V	108–157	43%	69	91
45300	46700	1400	Tuff	III	90–277		69	91
46700	47010	310	Quartz Diorite	III	79–83		37	91
47010	47360	350	Quartz Diorite	IV~V	79–83		37	91
47360	47510	150	Quartz Diorite	III	79–83		37	91
47510	47519	9	Quartz Diorite	II	79–83		37	91

Ground Support and Final Lining

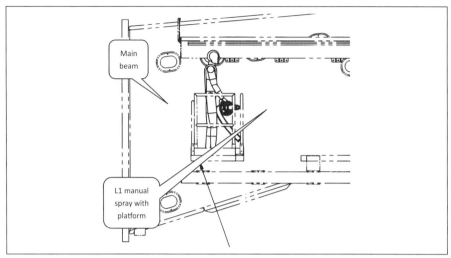

Figure 4. L1 manual spray diagram

Figure 5. L1 typical ground support on open-type TBMs

and 2 meters axial spray range. Figure 4 shows the L1 Manual Spray Diagram, used primarily to secure any loose rock and limit the convergence of the rock mass. Initial support typically consisted of rock bolts and mesh, McNally slats and ring beams. Figure 5 shows the L1 Typical Ground Support of Open-type TBMs.

Shotcrete was installed systematically at the L2 zone and formed the upper 285° of the permanent lining of the tunnel. Figure 6 shows the Jilin TBM layout.

L2 Shotcrete Zone

The L2 zone was located 50 meters behind the cutterhead. The system consisted of two boom type spray robots mounted on two independent movable rings. The boom can retract/extend 1.5 meters and the mechanical ring can move an additional 4.5 meters along the direction of the tunnel. The maximum coverage of the L2 shotcrete system was 285° of the tunnel circumference. The boom was covered by a protection roof to keep the rebound material out of the boom. Figure 7 shows the L2 shotcrete

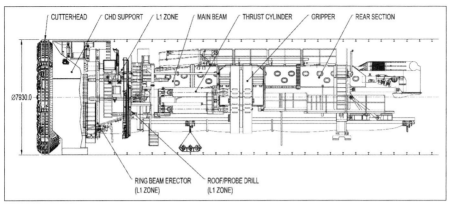

Figure 6. Jilin TBM layout

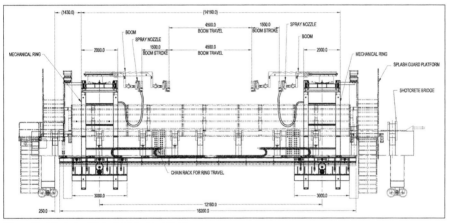

Figure 7. L2 shotcrete system arrangement

system arrangement, while Figure 8 shows the L2 shotcrete mechanical ring during shop assembly.

Stages of Shotcrete Application at the L2 Zone

The following chart in Figure 9 shows the different stages of shotcrete transport and application.

The concrete was first mixed at the TBM batching plant located in an underground assembly chamber. Because of the

Figure 8. L2 shotcrete mechanical ring during shop assembly

minus 40.2 °C extreme cold winter at the jobsite, the TBM batching plant was located inside the tunnel to ensure good conditions for concrete mixing. This required an overcut chamber and was more convenient for concrete transport to the TBM. It was then loaded into shotcrete bin. The locomotive was then driven from the assembly chamber to the TBM while the shotcrete was agitated in the shotcrete transit car. Upon arrival at

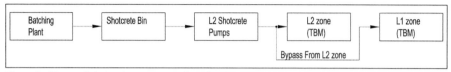

Figure 9. Stages of shotcrete transport and application

the TBM the shotcrete transit car was lifted from the locomotive and positioned above one of the shotcrete pumps on the TBM. When ready the contents of the shotcrete transit car were discharged and the shotcrete was pumped to the L2 zone where it was then applied as the final lining. In emergency cases, the L1 shotcrete could be used with the concrete material bypass from the L2 zone.

LESSONS LEARNED FROM PREVIOUS L2 SHOTCRETE & NEW SUPPLIER SELECTION

In the last 20 years, Chinese tunnel contractors have purchased more than 30 open-type TBMs. The end users are more and more familiar with the TBM and shotcrete system. They have learned based on experience and have made the following major changes:

- Higher safety requirements
- Higher reliability and performance
- More spray nozzle travel length and coverage. Normally 6 m or 8 m per robot
- More automation with less manpower
- Easy to clean and maintenance

In general, there are two primary kinds of L2 shotcrete system type used on previous projects, called bridge type and boom type (Figure 10 & Figure 11).

The following are major weakness was found for the typical telescopic boom type and bridge type on past projects:

Telescopic Boom Type With Fixed Ring Weakness

- The telescopic boom has insufficient rigidity. The deflection is more than 300 mm and even more deflection once the guiding plates are worn. This

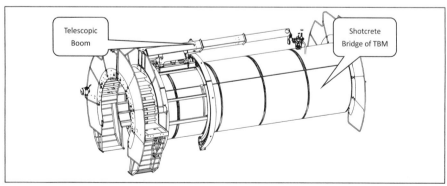

Figure 10. Telescopic boom type with fixed ring

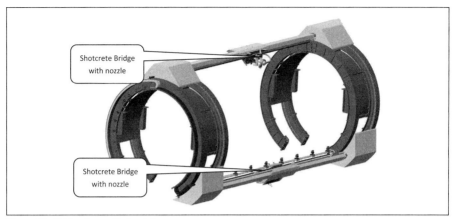

Figure 11. Bridge type with fixed ring

can be seen on the 6 m boom very often (Figure 12). Once spraying, the nozzle has a tendency to shake, which causes more rebound concrete and the decreased performance.

- There is a telescopic cylinder inside the boom. Because of the boom deflection and lower safety factor, the cylinder often leaks and can break (Figure 13).
- The motor power on the drive carriage force is insufficient when the carriage is at its max. force position (9 o'clock & 3 o'clock). Once the boom is fully extended, the carriage can become stuck, and move intermittently or be completely blocked (Figure 12). The drive motor also has a tendency to leak (Figure 14).

Figure 12. Telescopic boom deflection—Site assembly & testing

Bridge Type With Fixed Ring Weakness

- Normally two sets of robots are required by the customer. One robot is used as a spare and as a reserve capacity in case emergency conditions. If two bridge type robots are on one ring, then the customer must use both robots for spraying to achieve circumferential 270°–290° range. That means the customer has to clean both systems, and the cleaning and maintenance tasks double (Figure 15).
- The spray nozzle is often located beside the bridge. With this arrangement the rebound material always piles up on bridge, causing an additional force on the drive system and increasing clean time. The spray range is also smaller than with boom type due to the bridge structure.
- There are two carriages to support one bridge. The synchronous movements are always an issue once the encoder has failed, and the ring gear is prone to becoming damaged.

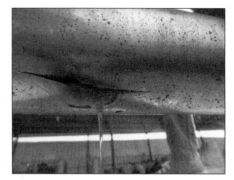

Figure 13. Telescopic cylinder broken

Figure 14. Drive motor leakage

Figure 15. Two bridges on same ring

- Another option is to use one robot per ring only, but with each additional robot a 20 meter long shotcrete deck needs to be added. This increases the back-up length, the tunnel conveyor, and all the cables and hoses for the TBM to operate, making it a very costly expense.

In previous projects, Robbins has sourced systems from European suppliers. However due to critical time constraints and customer requirements this was not possible for Jilin. Instead, Robbins China took the lead to develop an innovative system and source parts from three vendors to meet requirements and lower costs:

Imported parts from European Supplier

- Shotcrete pumps with cabinet controls
- Dosing unit
- Telescopic Boom with robot
- Partial cables & hoses

Structure and System from Local Supplier

- Movable mechanical ring with remote controls
- Carriage & drive system
- Hydraulic power unit

- Telescopic Boom with robot
- Partial cables & hoses

Accessories from Robbins China

- Piping
- Hoses
- Couplings & Elbows
- Cables

L2 SHOTCRETE SYSTEM IMPROVEMENT AND INNOVATION

Shotcrete Bin Movement to Position

The innovative shotcrete system included several key changes. The shotcrete bin is transported from the underground station via the locomotive and flat car. The empty bin on the machine needs to be moved to the storage area before the train arrival. The storage area is normally located on the opposite side of the shotcrete pump. On previous projects a winch system or electrical crane was used for lifting., but the new system makes use of a hydraulic crane (Figure 16). The new design of the hydraulic shotcrete bin lifting crane has more advantages than the original design as below:

- Fast speed with easy operation. Normally under 3 minutes time to put the bin in position.
- The crane can handle Max. 30 ton bins and the performance is more stable than a winch with wire ropes.
- The shotcrete bin car structure is simpler, and the hydraulic crane requires less headroom than using the winch to handle the shotcrete bin.

Mechanical Ring & Travel System

The project required a min. 6 meters travel length for each spray nozzle. Considering the traditional boom type and bridge type weaknesses, a new type of ring was developed as in Figure 17 and Figure 18. One set of travel track with hydraulic drive carriage is added on the bridge structure of the TBM. The circular ring is then connected

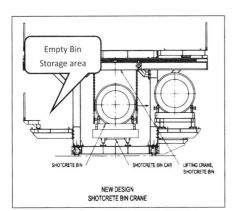
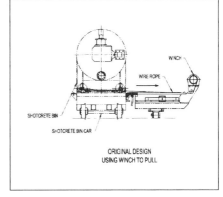

Figure 16. Shotcrete bin movement layout

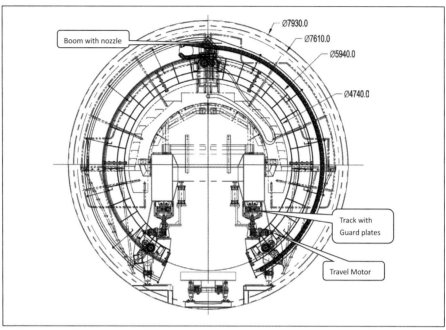

Figure 17. Mechanical ring & travel system

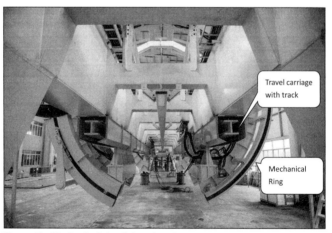

Figure 18. Mechanical ring & travel system

with the travel carriage and the travel length is set up to 4.5 meters plus a 1.5 m short boom for more flexible operation.

The track wheel assembly is installed under the bridge with side guard plates to keep dirt out of the track area. The mechanical ring travel is powered by hydraulic motors.

Considering the weakness of the circumferential drive carriage capacity, the drive motor was increased to allow for 25% more displacement and the max. continuous

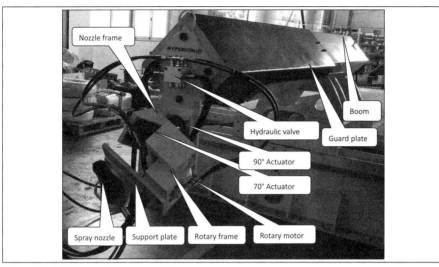

Figure 19. 1.5 m boom with spray nozzle

torque increased from 490Nm to 610 Nm. The weight of the boom was also reduced by 30%, all while achieving a high safety factor.

Shotcrete Robot

The boom and robot were redesigned as below:

- Change the longer telescopic boom to 1.5 meters (Figure 19).
- Change the drive motor and bearing to higher capacity. Increase plate thickness and spray distance adjust range. Added additional supports, etc.
- Hydraulic valve update for new drive capacity
- The spray nozzle was tested and used on the Jilin project for a short time until arrival of the imported components. The newly designed spray nozzle was used on the next section of tunnel for more than six months with very good performance (Figure 20).

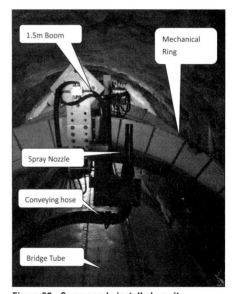

Figure 20. Spray nozzle installed on site

Shotcrete Pumps

The shotcrete pumps were located on gantry 1 and gantry 2, just behind the shotcrete bridge. The distance from the pump to spray nozzle is about 20 meters. The short conveying distance makes for easier cleaning and reduces the possibility of hose blockages. Figure 21 shows the shotcrete pump locations.

612 Ground Support and Final Lining

Figure 21. Shotcrete pump locations

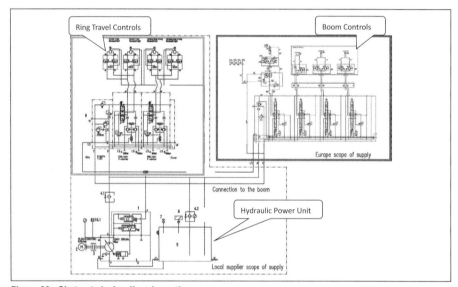

Figure 22. Shotcrete hydraulic schematic

HPU and Hydraulic System

The hydraulic system was updated as well--the power of the HPU changed from 22kw to 55kw, and new valves were added for mechanical ring travel. The schematic is as below (Figure 22).

Control System

The electrical control system governs shotcrete pump control, boom and spray nozzle control, and movable mechanical ring control. The controls come from different suppliers and Robbins merged these together with a public communication protocol. Figure 23 shows the shotcrete electrical control schematic.

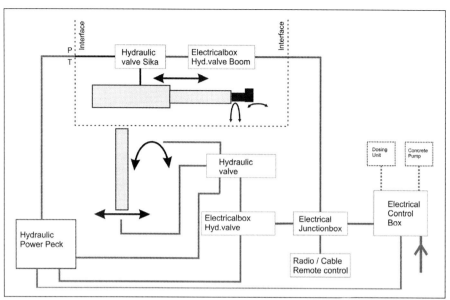

Figure 23. Shotcrete electrical control schematic

CONCLUSIONS AND RECOMMENDATIONS

In mid-May 2018 the national-record-setting 7.9 m (26 ft) Robbins Main Beam TBM at the Jilin Lot 3 Tunnel broke through. A formal ceremony followed to commemorate the stellar performance of the tunneling operation and its early completion. The project broke through nearly five months (147 days) earlier than scheduled. The project achieved the fastest monthly advance rate record—1423.5 m/4,670 ft—ever recorded for 7 to 8 m diameter TBMs in China. And the machine reached over 1000 m per month for three consecutive months. The shotcrete system played an integral part in the swift tunneling process.

The new designed L2 shotcrete system worked well on the Jilin project. Based on experiences the following recommendations should be considered for future projects:

- A hydraulic crane is a smarter system with fast speed and lower headroom suitable for shotcrete bin movement
- Considering the tunnel operation environment and shotcrete operating conditions, a drive system with a safety factor of at least 2 is recommended.
- A movable mechanical ring is a good choice for tunnel bore over 5 meters in diameter. With a 1.5–2 meter short boom. the bending moment on the robot is much smaller and the hoses can be fixed and protected easily.
- The shotcrete pump should be as near as possible to the robot as it will experience fewer stoppages and blockages in the hoses.
- Professional, experienced local supplier are needed for good communication, reduced delivery time and fast response. An experienced TBM supplier can improve the system performance and find ways to lower the cost.

Complex Tunnel Through the Abutment of the High-Risk Chimney Hollow Dam

Greg Raines ▪ Stantec
Albert Ruiz ▪ Stantec
Austin Wilkes ▪ Stantec

ABSTRACT

The Inlet/Outlet Tunnel for the proposed Chimney Hollow Dam will be 2000-ft long, six to 22-ft finished diameter with a 26-ft diameter Valve Chamber, and be excavated through siltstone and sandstone. The 330-ft high asphalt core rockfill dam and reservoir create difficult loading and design requirements for both final lining and grouting. Key design considerations include dam/tunnel integration; grout curtain continuity; large openings with high embankment surcharge loads and 300-ft of water head; and control of seepage along the tunnel below this high-risk dam. This paper reviews design concepts, requirements, and presents an overview of how project challenges were addressed.

PROJECT BACKGROUND

The Chimney Hollow Reservoir Project (the Project) is a new planned reservoir located approximately 8 miles west of Loveland, Colorado. The Northern Colorado Water Conservancy District (Northern Water) is developing the Project as a component of the Windy Gap Firming Project (WGFP). The planned reservoir will provide up to 30,000 ac-ft of firm yield to the conservancy members.

Northern Water is a public agency created to build and operate the Colorado Big Thompson (C-BT) Project. The C-BT Project collects water from the upper Colorado River basin on the Western Slope of the Rocky Mountains where water is pumped to Grand Lake. From there, the water flows by gravity through the Adams Tunnel beneath the Continental Divide and then through a series of six power plants and reservoirs. After flowing through the power generation system, water is stored in three existing reservoirs for primarily agricultural use.

Northern Water's Municipal Subdistrict (Subdistrict) is a separate and independent conservancy district formed by six Colorado Front Range communities to plan, finance, build, and operate the infrastructure built as part of the Windy Gap Project. The purpose of the Windy Gap Project is to supplement the C-BT Project for Northern Water's future needs. It is capable of diverting 48,000 ac-ft of water each year from Windy Gap Reservoir to Lake Granby. However, during wet cycles Lake Granby is often full, leaving little or no storage space for Windy Gap water. The additional storage provided by the Chimney Hollow Reservoir, contemplated since the Windy Gap Project's inception, will provide more reliable Windy Gap water deliveries.

The Project includes a main dam, spillway, saddle dam, and conveyance features connecting the Chimney Hollow Reservoir to the existing Colorado Big Thompson Project (C-BT) infrastructure. The reservoir will be formed by the main dam located at the north end of the valley, and the saddle dam at the south end built upon a topographic high point. The main dam will be an Asphalt Core Rockfill Dam, the fourth highest in

Complex Tunnel Through the Abutment of the High-Risk Chimney Hollow Dam

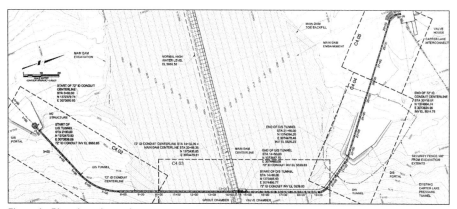

Figure 1. Plan view of tunnel

Colorado at a jurisdictional height of 332 feet and one of the first asphalt core dams constructed in the United States.

The Chimney Hollow Inlet/Outlet (I/O) Works is part of the conveyance features and consists of several components: The I/O Structure is an approximately 65-ft-high intake tower constructed just upstream of the I/O Tunnel. The 6-foot diameter I/O Conduit then travels from the I/O Structure through the Upstream Portal and into the I/O Tunnel. The I/O Tunnel consists of four different components, each with a different cross section: The Upstream Tunnel, Grout Chamber, Valve Chamber, and Downstream Tunnel. The purpose of the I/O Works is to convey water through the I/O Conduit, which connects the reservoir to the Valve House, and is capable of flow rates of 400 cfs. The total length of the I/O Works is about 2,400 ft. See Figure 1 for a plan view of the tunnel layout.

I/O Tunnel Layout

The Upstream Portal located between the I/O Structure and the I/O Tunnel has a headwall height of approximately 40 ft and is excavated through soil and rock. The Upstream Tunnel is the first section of the I/O Tunnel and is about 1,250 ft long with a depth of cover of approximately 35–136 ft. The Upstream Tunnel will be a Roman arch shape tunnel with 9 ft wide by 9 ft high minimum dimensions. The I/O Conduit will be placed within the center of the Upstream Tunnel and backfilled with concrete.

The Grout Chamber is 30 ft long and located immediately following the Upstream Tunnel and adjacent to the Valve Chamber. Since the construction of the grout curtain wall underneath the dam will be constructed axis from the ground surface before the tunnel has advanced through the grout curtain, the grout curtain must be re-grouted from within the tunnel. The Grout Chamber remains at a minimum 9 ft high but is widened to 14 ft to accommodate grouting equipment. The I/O conduit will be placed and backfilled in the Grout Chamber similarly to the Upstream Tunnel.

The 48 ft long Valve Chamber is located after the grout chamber and immediately downstream of the main dam axis with a depth of cover between approximately 87–93 ft to the existing ground surface prior to placement of the dam embankment. The purpose of the Valve Chamber is to house the valves, mechanical, and electrical equipment necessary for controlling flow within the I/O Conduit. The Valve Chamber is located beneath the dam to allow for maintenance access of the conduit between the Valve House and the Valve Chamber during reservoir operation. Additionally, it is located as close to the centerline of the dam as possible to provide the greatest length of conduit available for inspection while also taking advantage of the significant drop in pressure head downstream of the main dam centerline. Although the use of the Valve Chamber greatly increases the complexity of the tunnel design, with respect to the design of the dam it greatly improves dam safety by providing a way to shutoff the conduit in an emergency and by providing redundancy in the control of the conveyance system. The tunnel profile changes to a 30 ft excavated diameter circular shaped tunnel with a cast-in-place concrete final lining at the Valve Chamber. The I/O Conduit is supported on concrete pipe saddles on one side of the tunnel and access for personnel and equipment is provided on the other side from the Valve Chamber to the Downstream Portal.

The Downstream Tunnel is approximately 660 ft long with an approximate depth of cover of 36–114 ft. The Downstream Tunnel has a 25 ft excavated diameter circular cross section with a cast-in-place concrete final lining. At the Contractor's option, the last 417 ft of the Downstream Tunnel may be constructed as a 26-ft excavated diameter horseshoe shape. Two thrust blocks are located within the Downstream Tunnel to support the Conduit at alignment changes. One of the thrust blocks is located at the interface between the Downstream Tunnel and Portal and includes a three-dimensional change in Conduit orientation. The Downstream Portal is located at approximately STA 21+67 and includes both soil and rock excavation. The Portal will have a headwall height of approximately 50 ft.

GEOLOGY

The I/O Tunnel will be constructed under the right abutment of the main dam foundation in Fountain Formation sedimentary bedrock and Quaternary age alluvial deposits at the Upstream and Downstream Portals. The Fountain Formation was formed from alluvial fans during the Pennsylvanian era and is comprised of interbedded siltstones and sandstones with conglomerate and some shale. Well cemented and massive sandstone is sparse within the project area and occurs as interbedded lenses with lateral extents less than a few hundred feet. The sandstones also vary in grain size from fine to coarse sand and are well-sorted within the individual occurrences. These sandstone lenses are interbedded with very fine, widely-bedded siltstones. The siltstone is soft, friable, and tends to desiccate when exposed to open air. Underlying the Fountain Formation are pre-Cambrian metamorphic bedrock units that are exposed along the far western region of the project site. This bedrock is present about 1,000 ft below the tunnel alignment and will not be encountered during portal and tunnel excavations. Overlying the Fountain Formation at the top of the ridge above the portal is the Permian Ingleside Formation. This unit is more competent than the Fountain Formation and is often undercut along the crest of the hogback. This unit creates a rockfall risk to the Downstream Portal area.

Regularly spaced discontinuities were observed in the sandstone units of the Fountain Formation during surface mapping. These joint sets are not as apparent within the siltstone layers on the ground surface. Geologic structures in the project area are characteristic of massive sedimentary units, with the primary discontinuity set (bedding) being relatively flat lying and dipping towards the east, and secondary and tertiary discontinuity sets dipping vertically and trending perpendicular to each other. This structural fabric leads to small to moderate block sizes during excavation. The rock mass quality is considered to range from fair to good with core logs documenting the Fountain Formation having RQD values along the tunnel horizon ranging between

41% and 100% and is logged as ranging from moderately to highly fractured. RMR and Q values range from 55 to 68 and 2.5 to 4.9, respectively. No faulting or shearing was observed when reviewing the core samples extracted from the area. Chemical and physical weathering is present throughout the Fountain Formation, most prevalent along the top of bedrock, where meteoric water infiltrates through the overlying Quaternary soils. The alternating siltstone and sandstone beds exhibit varying degrees of cementation and weathering, and very little to no structural deformation. The weathered rock at depth usually occurs along the upper section of the siltstone subunits, along the basal contact of the relatively less weathered sandstones.

Chimney Hollow Valley is a low point in the regional topography creating a natural drainage point for groundwater within the hillslopes to the east and west. A piezometer was installed in drill hole DA-401 located near the Valve Chamber at the center of the tunnel alignment to measure groundwater levels after drilling was completed. Readings indicate water levels range seasonally from 17 to 50 feet above the tunnel. Carter Lake is located approximately 4,000 ft east of the Project site with a maximum operating water level at about El. 5,760 ft. The reservoir elevation is above the tunnel alignment, and therefore is likely a contributor to the groundwater table within the right abutment. Bedding comprising the Fountain Formation dip to the east which are expected to inhibit the flow of groundwater. However, groundwater within the right abutment appears to be a function of mass permeability and elevation head difference. The hydraulic conductivity of the Fountain Formation was highly variable with values measured during packer testing of between 1×10^{-3} cm/s and 1×10^{-7} cm/s along the tunnel alignment, with almost 90% of values less than 1×10^{-5} cm/s. The percentages of different hydraulic conductivities encountered during test of drill holes along the tunnel alignment are presented in Table 1.

Table 1. Percentages of hydraulic conductivity ranges along the tunnel alignment

Hydraulic Conductivity Range (cm/s)	% of Tests*
1.00E-07–1.00E-06	31.8%
1.00E-06–1.00E-05	30.9%
1.00E-05–1.00E-04	26.7%
1.00E-04–1.00E-03	9.7%
1.00E-03–3.00E-05	1.0%

Tunnel inflows were estimated using Heuer's methods (1995, 2005) based on packer testing performed along the tunnel alignment. Based on these calculations, the total inflow was estimated to be no more than 200 gpm for the entire tunnel. With consideration of the widely spaced drill holes and probability of intersecting water bearing joints it is expected that several distinct, high permeability zones, up to 50 feet in length, may be encountered during the excavation of the tunnel. It is estimated that these zones may have flush flows up to 100 gpm with sustained flows of up to 25 gpm, but generally will have flush flows up to 50 gpm with sustained flows of up to 15 gpm

CONSTRUCTION CONSIDERATIONS

The anticipated excavation methods for the Chimney Hollow I/O Works Tunnel are governed by the tunnel length, dimensions, schedule, site access, and subsurface conditions. Due to the varying shapes of the tunnels and cavern, the relatively short length, high mobilization cost, and the variable dimensions of the different tunnel components, a tunnel boring machine is not ideal for this project. While there are modifications that could be made to accommodate one, it is impractical from a cost perspective and considered unlikely as a chosen method by a contractor. The two most economical and feasible excavation methods for the tunnel are conventional drill and blast techniques and mechanical excavation by a roadheader.

Initial Support

Four initial support types were designed for the Chimney Hollow I/O Tunnel. Rock reinforcement and support elements were designed for each support type based on the evaluated rock loads, which were in turn, based on rock mass characterizations. Characterization of each ground type was made utilizing discontinuity data, unconfined compressive strengths, geologic mapping, and previous experience within the Fountain Formation. The initial support types are based on Bieniawski's RMR classification system, Type 1 being for very good to good ground, Type 2 for fair to poor ground, Type 3 for poor ground and Type 4 for very poor ground.

Type 1 initial support consists of a reinforced rock arch with epoxy resin-anchored rock dowels to carry ground loads and welded wire fabric to stabilize loose material between the bolts near the rock/tunnel interface. Installing rock dowels will form a coherent rock mass arch above the crown. The capacity of the rock arch is based on the compressive strength of the rock, the rock dowel length, and rock dowel spacing that creates an overlapping zone of compression. The compression arch transfers the rock loads to the rock in the sidewalls of the tunnel thereby stabilizing the crown.

Type 2 initial support consists of a reinforced rock arch with epoxy resin-anchored rock dowels to carry ground loads with shotcrete and welded wire fabric to stabilize loose material between the bolts near the rock/tunnel interface. The shotcrete and welded wire fabric will be installed to prevent raveling of the rock mass since the rock for this support type is expected to be moderately blocky and seamy with joint surface conditions in a more advanced stage of weathering than those of Type 1 ground.

Type 3 initial support consists of closely-spaced epoxy resin-grouted rock dowels to carry ground loads with shotcrete and up to two layers of welded wire fabric to stabilize loose and degraded material. A smoothing layer of shotcrete is anticipated to stabilize the ground sufficiently to install the rock dowels and first layer of welded wire fabric and shotcrete.

Type 4 ground support is intended for rock that is so heavily jointed and weathered that it resembles a soil, with minimal RQD values and flowing groundwater. Type 4 ground support is anticipated to be used at the portal turn-under and where rock dowels are not appropriate for support. Type 4 ground support consists of steel ribs and shotcrete lagging with spot bolting as necessary. A smoothing layer of shotcrete will be applied, similar to Type 3 ground support, in order to support the ground sufficiently to install the ribs.

DAM TUNNEL INTEGRATION CHALLENGES

Requirement for Valve Chamber and Maintaining Access to Pipe (Downstream Tunnel)

One challenge with respect to the design of the tunnel was that the project was a dam project and not a tunnel project, in that the design of the dam was the final driver for designing project elements. One way that this affected the design of the tunnel was that the design constraints of the dam were not always congruent with the design constraints of the tunnel with respect to the overall project. The result of this was that the tunnel design was affected and modified based upon requirements for the dam design.

One of the most obvious ways that this can be seen is in the layout of the tunnel itself, with four different tunnel cross sections, the upstream tunnel, the valve chamber, the grout chamber, and the downstream tunnel. One of the key design aspects of the dam design was placing shutoff valves for the I/O Conduit in the tunnel, directly downstream of the dam axis. Additionally, the project owner wanted to keep as much access to the pipe as possible. This necessitated the inclusion of the valve chamber in the tunnel design as well as the downstream tunnel. Similarly, and as discussed further below, the grout chamber was also required as part of the dam design to complete the curtain grouting within the vicinity of the tunnel. If not for these dam design requirements, then the full tunnel length would have been completed using the upstream tunnel design, as is typical for water supply tunnels for dam projects.

Dam/Tunnel Construction Sequencing

The sequencing of constructing the project components will place the construction of the tunnel after the completion of the dam grout curtain, but before construction of the dam itself. This sequencing added several unique design constraints and aspects to the design of the tunnel. The first was due to the installation of the dam grout curtain wall prior to the start of tunneling. The purpose of the dam grout curtain wall is to increase the flow path distance for water migration from the upstream side of the dam to the downstream side of the dam, preventing water loss and dam instability. As the tunnel is expected to be constructed using drill and blast excavation, there were concerns that the blasting would fracture and damage the rock mass in the tunnel horizon and damage or destroy the grout curtain wall in the vicinity. To mitigate this risk, the grout curtain wall in vicinity of the tunnel, will be grouted up to 10 ft radially outward from within the tunnel, immediately upstream of the valve chamber. Since the drilling equipment necessary to construct the grout curtain wall is larger than the upstream tunnel, a larger cross section had to be designed. This grouting chamber is 25 ft long by 9 ft high and 14 ft wide. Additionally, the temporary ground support for the grouting chamber, which consists of W8×67 steel ribs and a 10-in thick layer of shotcrete, had to be designed to resist a 50 psi grouting load.

Another major impact the construction sequencing had was the construction of the dam itself following the construction of the tunnel. The tunnel alignment is located beneath the right abutment of the dam. As rock fill is placed on the right abutment, it acts as a large surcharge load on the ground directly above the tunnel. At the maximum dam section that the tunnel passes underneath, the dam embankment is approximately 265 ft high, with a unit weight of 135 pcf.

Control of Seepage Along Tunnel

The dam design team was concerned that the tunnel could potentially facilitate migration of water from the upstream face of the dam to the downstream face. There was a concern that not only would water travel along the interface between the rock mass and final lining, but that blasting the tunnel would fracture and disturb the rock in the tunnel horizon and create additional flow paths along the tunnel alignment. In order to cut off these potential flow paths, in addition to the dam grout curtain wall that will be constructed from the ground surface and the tunnel, modified contact grouting will be used in place of traditional contact grouting of the final lining for the Valve Chamber and Downstream Tunnel. Traditional skin and contact grouting will still be performed around the I/O Conduit in the Upstream Tunnel. The methodology of modified contact grouting is discussed in more detail below.

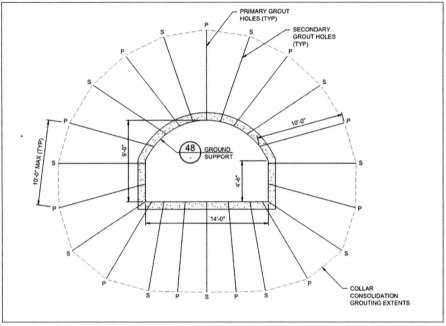

Figure 2. Extents of curtain grouting from the tunnel

GROUTING PROGRAM DESIGN

Dam Curtain Grouting and Consolidation Grouting

The curtain grouting performed from within the tunnel will be started once the tunnel has been driven and the temporary ground support installed. The curtain grouting will be performed similarly to collar grouting and grouted in radial rings spreading out from the tunnel. In order to provide adequate grout coverage, the curtain grouting will extend longitudinally approximately 10 ft past the dam curtain grouting, as seen in Figure 2. The main goal of the curtain grouting is for water control grouting of water bearing geologic features such as bedding planes, joints, fractures, and discontinuities created during blasting the tunnel excavation.

For the portion of the Conduit that is supported on concrete pedestals within the downstream tunnel, there are two alignment changes that require thrust blocks to support the hydraulic thrust loads from the pipe and to restrain the pipe from moving. Both of these thrust blocks bear on the rock mass and generate a substantial thrust force. Structural consolidation grouting will be performed at both locations in order to ensure that the rock mass has adequate bearing capacity for this trust force and to improve its competency and stiffness.

Modified Contact Grouting

A modified version of traditional contact grouting of the void between the cast in place concrete final lining and the rock mass was designed for the I/O Tunnel. Modified contact grouting involves contact grouting at higher pressures with a less viscous grout mix in order to force the grout further into the rock mass and permeate throughout fractures and bedding planes around the perimeter of the tunnel, rather than just filling the void between the cast in place lining and the rock mass. In addition, all panning and drain pipes are filled with grout to ensure a full seal of all potential flow paths along the tunnel alignment. The result of this method is a blending of contact grouting and consolidation grouting with the end result achieving the same results of both methods. The goal of modified contact grouting is to obtain a hydraulic cutoff and prevent water from migrating longitudinally along the tunnel alignment within tunnel horizon fractures.

FINAL LINING DESIGN

The Valve Chamber located just behind the dam axis on the downstream side of the dam is one of the most critical elements of the project. The valves, mechanical, and electrical equipment the chamber will house are crucial to the safe operation of the dam and the I/O Conduit. In order to protect this equipment and maintain it in good working order, the Valve Chamber was designed to be watertight. A PVC waterproofing membrane will be installed outside of the cast in place concrete final lining. This membrane is only installed around the Valve Chamber and not for the entire tunnel. The membrane will be connected to a seepage ring welded to the I/O Conduit where it penetrates the bulkhead wall at the upstream end of the Valve Chamber. This bulkhead wall separates the Valve Chamber from the Grout Chamber and the Upstream tunnel just ahead of it. A 1½-in thick expansion joint filled with compressible joint filler has been provided between the bulkhead wall and the I/O Conduit to allow for movement and differential settlement between the Valve Chamber and the Upstream Tunnel.

The Valve Chamber in particular has very high loads acting upon the final lining from two different sources. As discussed above, the dam embankment will be placed on top of the rock mass of the right abutment through which the tunnel will run, creating a very high surcharge loading. As the Valve Chamber is near the dam axis and therefore near the maximum dam height, it experiences some of the highest surcharge loading. In addition to this surcharge load, and despite the installation of the grout curtain to cutoff water migration from the upstream side of the dam to the downstream side, very high hydrostatic loading conditions were estimated to act on the lining as discussed below.

The surcharge loading condition was modeled in the finite element modeling program RocScience RS^2. Plane strain models were developed for different locations along the tunnel alignment that accurately reflected the sloping ground surface of the right abutment and then placed the dam embankment constructed at that location. The location of the tunnel alignment places the center of the tunnel alignment underneath the dam, with the sloping dam embankment reducing the surcharge loading conditions as it nears the portals at either ends. Roughly the last 300 hundred ft of the downstream end and the first 800 ft of the alignment experience no surcharge loading as they are outside of the footprint of the dam. RocScience Unwedge was used to model the joints and discontinuities within the rock mass to develop point and area loads which were then input into the RS^2 model along with the hydrostatic loading conditions.

The existing groundwater levels along the tunnel alignment vary between El. 5567 ft and 5600 ft (17 to 50 ft above the crown of the tunnel) and although mostly influenced by the adjacent Carter Lake reservoir, still changes seasonally. Once the reservoir is filled to capacity, the groundwater pressure is estimated to increase and be the highest at the upstream face of the dam, and lower as the tunnel travels downstream. The upstream tunnel is located upstream of the dam foundation curtain grouting and is therefore designed for the full potential hydrostatic force of 300 ft of water head possible from a full reservoir condition. Hydrogeologic modeling was performed to determine the groundwater pressures downstream of the grout curtain with a full reservoir

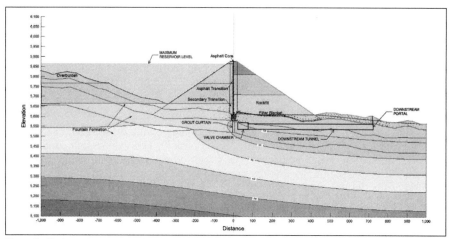

Figure 3. Hydrostatic pressure along downstream tunnel alignment for full reservoir condition

condition. Figure 3 shows the groundwater pressure head profile that was developed from the modeling for a full reservoir condition. This pressure head profile was used to design the Valve Chamber and the Downstream Tunnel. The Valve Chamber is designed for 150 ft of hydrostatic pressure head, while the Downstream Tunnel is designed for pressures ranging from 130 ft to 70 ft of head, decreasing as it travels downstream along the alignment.

The original conceptual design of the tunnel initially had the entire alignment as a roman arch or horseshoe cross section. However, once analysis and design of the final lining begin, it was quickly determined that the high embankment surcharge loads and the high hydrostatic loads acting on the lining developed extremely high stress concentrations and shear forces at the haunches of the tunnel. It was then decided that the Valve Chamber and the Downstream Tunnel would be best designed with a circular cross section to most economically handle the high loading conditions. Following construction of the circular final lining, a wedge-shaped leveling slab will be placed in the invert of the circular tunnel to provide a surface for installing the I/O Conduit on pedestals and for maintenance access.

The potential long-term differential settlement along the tunnel alignment was analyzed as well, since the center of the dam experiences the maximum surcharge loading conditions, with the load reducing as it moves to the ends of the tunnel, eventually reducing to zero where the tunnel is no longer in the footprint of the dam. Although the maximum settlement at the center is estimated to be 3.8 inches, since it acts over a distance of approximately 800 ft, the differential settlement was found to be within acceptable limits.

CONCLUSION

When planning for large dam projects with an integrated tunnel to be designed as part of the project, thought should be given to the impacts of major project components to each other. Construction sequencing and the basis of design can have a large impact on the basis of design for other project components that may not be intended. These impacts were successfully dealt with for this project, but increased foresight into basis of design decisions at the beginning of the project could have been beneficial to the project.

Design Diagrams for Fiber Reinforced Concrete Tunnel Linings

Axel G. Nitschke ▪ WSP USA

ABSTRACT

The flexural design of Fiber-Reinforced-Concrete (FRC) tunnel linings typically uses a material-specific stress strain relationship (SSR) based on beam tests. The evaluation of beam tests to gain the SSR as well as the calculation of the bearing capacity of tunnel linings using a SSR is work intensive. The latter typically requires specialized software. Getting quick results for changed input parameters is therefore a challenge.

The paper introduces design diagrams that allow, in a simple and quick manner, to either evaluate beam test results to gain the corresponding SSR or evaluate the bearing capacity of a tunnel lining for a given SSR in the Ultimate Limit State (ULS) or for the Serviceability Limit State (SLS). For given loading conditions, the diagrams can also be used to evaluate the required material properties, which can subsequently be used to assess the required fiber dosage. Therefore, the diagrams provide a powerful tool for a rough estimation and are considered very beneficial if quick results with very little effort are needed, i.e., during preliminary design stages, the bidding phase, or if the effect of changes of material properties, quality deficiencies or loading conditions need to be assessed.

The paper discusses briefly the basics of FRC design and how the diagrams were developed. In addition, the paper provides design examples to explain the use of the introduced diagrams.

INTRODUCTION

The typical governing load case of tunnel linings is a combination of bending moment (M) and thrust or normal-force (N). Just as for rebar reinforced concrete (RC) tunnel linings, the design procedure follows the classical approach for short columns [1,2], since buckling is typically not an issue for tunnel linings due to their curved shapes and continuous bedding by the ground. The stress strain relationship (SSR) for fiber-reinforced concrete (FRC) versus RC is hereby expanded on the tension side (see Figure 1). Internationally, there are multiple, but similar, beam tests that are used to develop design values for the SSR [3,4,5,6,7,8,9]. These are not discussed herein.

The biggest difference between the sectional strength of unreinforced or rebar reinforced concrete and FRC is that the concrete in unreinforced or rebar reinforced concrete has theoretically no bearing capacity in tension. In the modeling of conventionally reinforced concrete sections, all tension is supported by the rebar. Since the location of the rebar is known, the location of the resulting tensile force is also known, and this simplifies the calculation of the equilibrium compared to FRC sections. The computation of the equilibrium for FRC members is much more challenging because the location of the resulting tension force is an unknown during the computation and moves if the external load and the distribution of the strain over the cross-section changes. The design assumptions for the calculation of the sectional strength for FRC based on a SSR are presented and further discussed in detail for example in [5,6,10,11,12,13,14] and are therefore not discussed herein.

It is, however, crucial to realize, that with this approach FRC is assumed to be a macroscopically homogeneous and isotropic material [10]. The material properties of a single fiber in the model becomes irrelevant. Therefore, the fibers and the concrete are modelled using a single SSR relationship and not two, (i.e., as for steel rebar and concrete.)

After the cracking of the material under tension, the material properties in the model are based on strains rather than a discrete crack. In the model the cracked material is also viewed as homogeneous and isotropic. Since this is in the area around the crack, it is obviously not the case. This circumstance is very important to realize and understand when evaluating the sectional strength of FRC using a SSR. During the evaluation of material testing data based on beam tests (and subsequently the design of the structure), it is assumed that the crack is "smeared" over a certain length into an "equivalent strain," which is also referred to as "integral approach" [10].

Fibers influence the bearing behavior in multiple ways. However, three properties are most relevant for application in tunnels [10]: (1) they slightly increase the flexural tensile strength, which is mostly needed if improved properties under un-cracked conditions are desired, (i.e., to design for serviceability). However, for the case of ultimate bearing capacity of tunnel linings, the residual flexural tensile strength under cracked conditions (2) and the increase of the toughness (3), are the major benefits. Hence, the focus of this paper is on the performance improvements attributable to (2) and (3).

The provision of a reliable and usable post cracking tensile strength transforms the brittle failure mechanism of plain concrete into a ductile failure mode. This is a material property that provides major engineering and economic advantages, especially if utilized to facilitate system failure of a tunnel lining rather than a cross section failure at one, presumably most critical location [13].

A generic SSR and nomenclature of the variables used throughout this paper is shown in Figure 1 below. The tension side is represented by three sections, which model and control the bearing behavior in the three different phases (elastic, micro-cracking, macro-cracking). A detailed discussion about the three different phases is provided in [10]. The compression side uses a classical parabolic-constant shape.

The load bearing capacity of a cross section based on the SSR is calculated by finding the equilibrium between internal and external forces as shown in Figure 2. In addition, the reader is referred to past papers of the author and other literature regarding the basics of how the equilibrium between internal and external forces is calculated [5,6,10,11,12,13,14].

For the parametric study in this paper a generic SSR as shown in Figure 1 is used, which is identical with SSRs provided for example by RILEM in [6] and ACI in [4]. In this paper, the strains on the compression side (e_{c1}, e_{c2}) and tension side (e_{t1}, e_{t2}, e_{t3}) as well as the compressive strength (f_c) and flexural strength (f_{t1}) are kept constant with typical values shown in Table 1. Only the residual flexural strengths (f_{t2} and f_{t3}), which primarily characterize the effect of the FRC, are varied. Non-dimensional values for the compressive strength, flexural strength, and residual flexural strength in percent [%], standardized with the compressive strength, are used. Following the authors own research experience [10,11] and similar values provided by RILEM in [6], the flexural strength f_{t1} is assumed to be 10% of f_c. The residual flexural strengths f_{t2} and f_{t2} are varied within the parametric study as 0%, 1.25%, 2.5%, 5.0%, and 7.5% of the compressive strength (f_c), while the residual flexural strength f_{t2} is always greater than or

Table 1. Stress-strain-relationship (SSR) used in the parametric study

	Tension			Compression	
Stress	f_{t3}*	f_{t2}*	f_{t1}	f_c	f_c
[% of f_c]	0.0 | 1.25 | 2.5 | 5.0 | 7.5	0.0 | 1.25 | 2.5 | 5.0 | 7.5	10	100	100
Strain	ε_{t3}	ε_{t2}	ε_{t1}	ε_{c1}	ε_{c2}
[‰]	25.0	0.2	0.1	−2.0	−3.5

* $f_{t3} \leq f_{t2}$

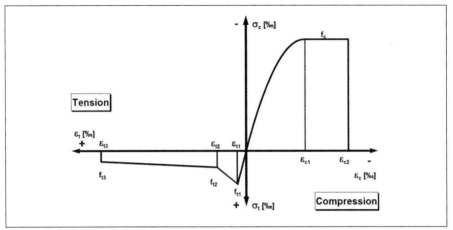

Figure 1. Generic stress-strain-relationship (SSR) for fiber reinforced concrete (FRC)

equal to the residual flexural strength f_{t3}. All assumptions of the parametric study are summarized in Table 1.

For any given combination of strain on the compression side (e_c) and strain on the tension side (e_t) a distinct bearable combination of moment (M) and axial force (N) can be calculated (see Figure 2). Within his research work [10], the author has developed software that would allow the user to compute distinct equilibriums, moment-normal-force-interaction diagrams, but also moment-curvature or moment-strain graphs for different axial forces. The moment-strain-graphs were used for the parametric study introduced in this paper. Since the input is based on a generic SSR shown in Figure 1, the results are independent of any specific fiber type or dosage.

The goal of the study was to develop simplified design tools to make the use of specialized software obsolete. To provide the widest use possible, all results are provided with non-dimensional values. As shown in both Equation 1 and Equation 2 for the non-dimensional normal-force (n) and the non-dimensional moment (m), the compressive strength (f_c), the cross-section width (b) and height (d) can be summarized in a constant factor (c). Consequently, the dimensionless results shown herein, can be easily transferred into dimensional values by multiplying the result of the non-dimensional value with the corresponding constant c_n or c_m.

Equation 1: Non-dimensional normal-force n

$$n\,[-] = \frac{N}{f_c \times b \times d} = \frac{1}{f_c \times b \times d} * N = c_n * N \text{ with } c_n = \frac{1}{f_c \times b \times d}$$

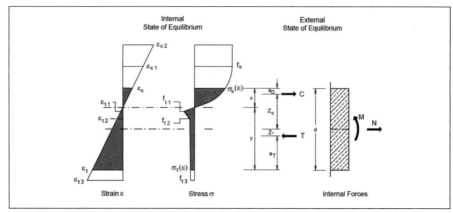

Figure 2. Calculation of equilibrium between internal and external forces

Equation 2: Non-dimensional moment m

$$m\,[-] = \frac{M}{f_c \times b \times d^2} = \frac{1}{f_c \times b \times d^2} * M = c_m * M \text{ with } c_m = \frac{1}{f_c \times b \times d^2}$$

As discussed in [10,11,12,13], the SSR on the tension side represents the three different stages a FRC lining passes through until reaching its ultimate state: (1) elastic, (2) micro-cracking, and (3) macro-cracking. For the Serviceability Limit State (SLS) the elastic and micro-cracking behavior governs the design, while for the Ultimate Limit State (ULS) the macro-cracking governs the design. Imagining different equilibriums by using Figure 2, it becomes obvious that, for the SLS, the accurate modeling of ε_{t3} and f_{t3} is irrelevant, because the actual strain (ε_t) in the SLS is far from maximum strain ($\varepsilon_t \ll \varepsilon_{t3}$). For the ULS, on the other hand, the detailed modeling of ε_{t1}, ε_{t2}, and flexural strengths (f_{t1}) is not required, because these areas provide very little bearing capacity if the tensile strain (ε_t) reaches the maximum strain ($\varepsilon_t \approx \varepsilon_{t3}$ and $\varepsilon_t \gg \varepsilon_{t2}$, ε_{t1}). Both cases can therefore, without significant loss of accuracy, be evaluated completely separately, as is shown below.

The paper starts with the discussion of the ULS in the following section, followed by the SLS. In both cases, the concept is initially developed for load cases under pure bending, without the influence of the axial compression force. In a subsequent step the concept is then expanded for different levels of axial forces.

BEARING CAPACITY FOR FRC UNDER BENDING IN THE ULTIMATE LIMIT STATE

Using the SSR introduced above, moment-strain-curves can be created by varying the compression (ε_c) and tensile strains (ε_t). Each graph shown in Figure 3 shows the result for a moment-tensile strain curve under pure bending using different levels of residual flexural strengths (f_{t2}, f_{t3}) as a percentage of the compressive strength (f_c). "Tensile strain" refers, within this context, to the strain at the side of the cross section under tension. The following figure shows moment-tensile strain graphs with a SSR with a constant residual flexural strength in the last section ($f_{t2} = f_{t3}$), varying from 0%, 1.25%, 2.5%, 5.0%, and 7.5% of the compressive strength (f_c). As can be seen in Figure 3, it is typical that the moment bearing capacity is nearly constant after reaching a certain tensile strain of approximately $\varepsilon_t > 5‰$. Since the stress is constant ($f_{t2} = f_{t3}$ = constant) and the resulting moment (m) is constant if the strain is greater than approximately 5‰, the moment bearing capacity for a strain greater than 5‰ can also be displayed as a single point for a certain residual flexural strength ($f_{t2} = f_{t3}$) in a moment-stress diagram. Figure 5 displays this effect in a moment-stress diagram for different residual flexural strengths. The arrows display the investigated residual strengths: $f_{t2} = f_{t3} = 7.5\%$, 5.0%, 2.5%, 1.25%, and 0%. Furthermore, the effect is linear for different residual flexural strengths, so that the bearing capacity can be simply displayed as a line (see solid line in Figure 5). As a result, the ultimate bearing moment for any constant residual flexural strength of a SSR ($f_{t2} = f_{t3}$) can be directly evaluated using the solid line in Figure 5.

Figure 4 shows the known moment-strain curves from Figure 3 for the constant cases $f_{t2} = f_{t3} = 2.5\%$–2.5% and 5.0%–5.0%. In addition, the effect of a trapezoidal SSR between residual flexural strenghts f_{t2} and f_{t3} is shown for $f_{t2} = 5.0\%$ with f_{t3} varying between 2.5%, 1.0% and 0% (labelled in the figure as "5.0%–2.5%," "5.0%–1.25%," and "5.0%–0.0%"). If the SSR between f_{t2} and f_{t3} is trapezoidal, the moment strain curve is typically dropping linearly after a strain of approximately 5‰ (see Figure 4).

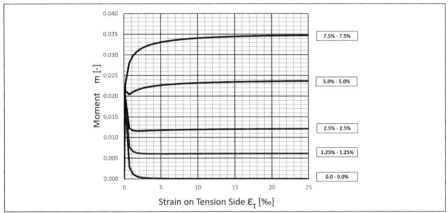

Figure 3. Moment over tensile strain for constant residual flexural strengths f_{t2} and f_{t3} [0%, 1.25%, 2.5%, 5.0%, 7.5%] for pure bending (n = 0).

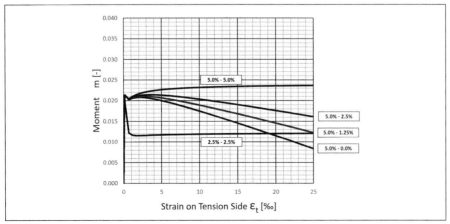

Figure 4. Moment over tensile strain for constant and trapezoidal residual flexural strengths f_{t2} [2.5%, 5.0%] and f_{t3} [0%, 1.25%, 2.5%, 5%] for pure bending (n = 0).

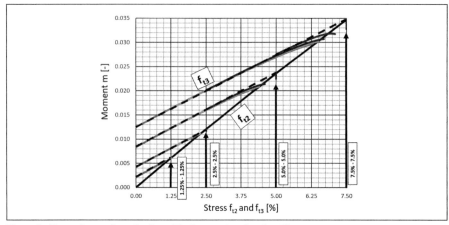

Figure 5. Moment over stress for f_{t2} and f_{t3} for pure bending (n = 0)

If the moment strain curves of Figure 4 are displayed again as a moment-stress graph by using the SSR to calculate the stress σ_t (ε) as a function of the strain on the tension side (ε_t) (see Figure 2), the curves are again linear for a strain greater of approximately 5‰. All curves with different residual flexural strengths f_{t3}, but the same residual flexural strength f_{t2}, have the same slope and are covering each other (see grey lines in Figure 5). Therefore, the trapezoidal SSRs can also be shown as lines in the moment stress diagram (see dashed lines in Figure 5).

The resulting diagram provides therefore a simple design tool for the ULS. If for example, a SSR with $f_{t2} = 5.0\%$ and $f_{t3} = 3.75\%$ is given, the bearable moment can be evaluated with the diagram to m = 0.02 (follow the $f_{t2} = 5.0\%$ arrow to the solid line and from there follow the dashed line to $f_{t3} = 3.75\%$ and read the moment m from the y-axis). The diagram can also be used in the reversed fashion: if a moment of m = 0.015 is given, it can be either born by a FRC with $f_{t2} = 0.75\%$ and $f_{t3} = 7.5\%$, or $f_{t2} = 2.25\%$ and $f_{t3} = 5.0\%$, or $f_{t2} = f_{t3} = 3.13\%$ (see Figure 5). It should be noted that the diagram can also be used to interpolate between distinct lines, as necessary.

BEARING CAPACITY FOR FRC UNDER BENDING AND THRUST IN THE ULTIMATE LIMIT STATE

In the previous section the development of a design diagram for the ULS was introduced. However, the diagram is limited to pure bending. In this section an identical approach is chosen, but it now includes the effect of different axial compression forces. Figure 6 shows moment strain graphs for different levels of axial forces (n) for different residual flexural strengths. To maintain clarity only cases with constant residual flexural strengths ($f_{t2} = f_{t3}$), labelled as 1 to 5, are shown in the figure. The lower group shows the curves for pure bending (n=0) as previously introduced in Figure 3, followed by an increasing (non-dimensional) axial force (n) chosen to 0.1, 0.2, 0.3, and 0.4. The dimensionless axial force (n) can herby also be interpreted as the bearing capacity under direct compression. For n = 1 the cross section would fail under pure compression and is not capable of bearing any moment. In tunnel designs n typically ranges below 30 to 40% of the bearing capacity under pure compression respectively in the lower third of a moment-normal-force-interaction diagram [10,13]. Compared to pure bending, the increasing influence of the axial force also leads to a change in failure mechanism. For pure bending or small axial force the maximum tensile strain

Design Diagrams for Fiber Reinforced Concrete Tunnel Linings

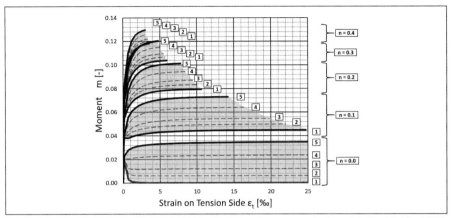

Figure 6. Moment over tensile strain for constant residual flexural stengths f_{t2} and f_{t3} [1: 0% | 2: 1.25% | 3: 2.5% | 4: 5.0% | 5: 7.5%] for different axial force

is reached before the maximum compressive strain. However, the failure mechanism switches to a compression failure with increasing thrust, reaching the maximum compression strain before the maximum tensile strain. This effect can be clearly seen in Figure 6, where the moment is shown over the tensile strain. While for pure bending the graphs reach the maximum tensile strain (ε_{t3} = 25‰), the graphs reach failure at lesser tensile strain with increased axial force. To maintain clarity in the figure, the numbers 1 to 5 refer to different residual flexural strength levels with No. 1: 0%, No. 2: 1.25%, No. 3: 2.5%, No. 4: 5.0%, and No. 5: 7.5%.

Following the same approach for the development of a design diagram introduced in the previous section for pure bending, the design diagram can then be expanded to include the axial force component. Each solid line in Figure 7 is developed for a constant SSR for the residual flexural strengths ($f_{t2} = f_{t3}$) and the dashed lines refer to trapezoidal curves between f_{t2} and f_{t3}. As can be seen in Figure 6, the influence of f_{t3} reached at 25‰ diminishes with increasing axial force n and is therefore not shown for larger axial forces (n ≥ 0.2) to maintain clarity in the figure. As emphasized in the previous section, this diagram is applicable for the ULS and tensile strains larger than approximately 5‰.

The use of the design diagram is now shown in examples for different scenarios.

Example 1: A continuous tunnel lining (b = 1 m) has a thickness of 0.2 m and a compressive strength of f_c = 35 MN/m². The lining is loaded with an axial force N = 1.4 MN and a moment of M = 0.14 MNm. What is the residual flexural strength needed in the ULS?

Solution: n = 1.4/(35*1*0.2) = 0.2
m = 0.14/(35*1*0.2²) = 0.1
Using Figure 7: m = 0.1, follow curve for n = 0.2,
result: $f_{t2} = f_{t3}$ = 6.75%
Answer: The residual flexural strength is $f_{t2} = f_{t3}$ = 35*(6.75/100) = 2.36 MN/m²

Example 2: A segmental tunnel lining (b = 1.5 m) has a thickness of 0.4 m. The compressive strength f_c is 45 MN/m² and the residual flexural strength is f_{t2} = 2.25 MN/m²

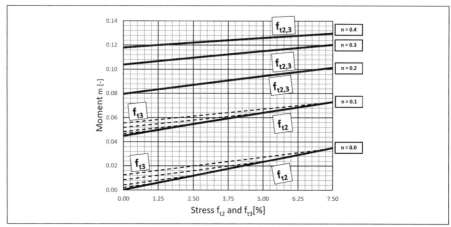

Figure 7. Moment over stress for residual flexural strengths f_{t2} and f_{t3} for different axial forces

and $f_{t3} = 1.13$ MN/m^2. What is the maximum bearable moment at ULS, if the axial force is N = 2.7 MN?

Solution: $n = 2.7/(45*1.5*0.4) = 0.1$
$f_{t2} = 2.25/45 = 5.0\%$
$f_{t3} = 1.13/45 = 2.5\%$
Using Figure 7: go to $f_{t2} = 5.0\%$ and up to intersect solid curve for n = 0.1; follow dashed line to $f_{t3} = 2.5\%$
read: m = 0.06

Answer: Maximum bearable moment is M = $0.06*45*1.5*0.4^2$ = 0.65 MNm

Example 3: A flexural beam test was conducted and a moment–strain curve was obtained. Based on the cross section geometry (b = d= 0.15 m) and the compressive strength f_c is 32.5 MN/m^2 the test result is transformed into the non-dimensional moment strain curve using the factor $c_m = 1/(32.5*0.15*0.15^2) = 9.12$ [1/MNm] from Equation 2 (for the purpose of this example assume the test result is shown in Figure 4 for "5%–1.25%").

Solution: The maximum bearable moment at e_{t3} is obtained from the curve as $m_{t3} = 0.012$.
The bearable moment at e_{t2} is obtained from the curve as $m_{t2} = 0.021$.
Using Figure 7 the solution is read as $f_{t2} = 5.0\%$ and $f_{t2} = 1.25\%$

Answer: The residual flexural strength corresponding to the test results is
$f_{t2} = 32.5 * 5.0/100 = 1.63$ MN/m^2 and $f_{t3} = 32.5 * 1.25/100 = 0.41$ MN/m^2

As the examples above show, the design diagram for the ULS introduced in Figure 7 provides a powerful tool to provide quick results easily. The design diagram can be used for different scenarios to evaluate the bearing capacity for a given SSR or evaluate residual flexural strength for given loading conditions. In addition, it can be used to quickly develop a SSR from beam test results.

In the following section a similar approach is chosen to develop a design diagram for the SLS.

Design Diagrams for Fiber Reinforced Concrete Tunnel Linings 631

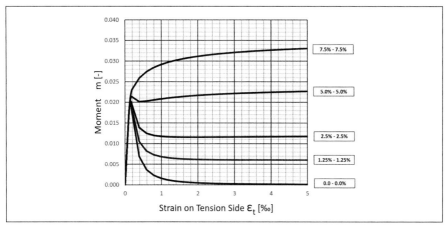

Figure 8. Moment over tensile strain for constant residual flexural strengths f_{t2} and f_{t3} [0% | 1.25% | 2.5% | 5.0% | 7.5%] for pure bending (n = 0)

BEARING CAPACITY FOR FRC UNDER PURE BENDING IN THE SERVICE LIMIT STATE

The graphs shown in Figure 8 are identical with the graphs of Figure 3, but zooms into the area between ε_t = 0 to 5‰ that is relevant for the Serviceability Limit State (SLS). As can be seen in the relevant strain area in Figure 4, showing the different trapezoidal SSRs, the results for different trapezoidal curves do not vary greatly from the constant curves in the area for $\varepsilon_t \ll$ 5‰. Therefore, and for simplification, the parametric study for the SLS uses constant SSRs only, without losing significant accuracy in the results.

The SLS design is typically governed by an allowable crack width. The allowable crack width (w) for serviceability of FRC, i.e., w = 0.2 mm, is typically reached at much lower equivalent strains compared to the equivalent strain and crack width in the ULS, which can be a couple of millimeters. In this context, it is emphasized again that in the SSR an equivalent strain on the tension side (ε_t) is used and not a crack width (w). Therefore, a relationship between crack width and equivalent strain needs to be defined. RILEM [6] provides an equation relating the crack width (w) with the equivalent tensional strain (ε_t) based on the cross section under tension (y) (refer also to Figure 2):

Equation 3: Crack-width/equivalent strain relationship

$$w = \varepsilon_t * y$$

Figure 9 shows the cross section under tension over the equivalent strain (ε_t) for the graphs shown in Figure 8, with a residual flexural strength (f_{t2} = f_{t3}) between 0% and 7.5%. The cross-section height under tension (y) is divided by the total cross-section height (d) and is shown as a dimensionless value. After the reaching the elastic limit at ε_{t1} = 0.1‰ the cross section under tension (y) grows quickly between 80 to 95% over the entire cross section height (d). It can also be observed that the increasing residual flexural strength of the FRC (shown from 0% to 7.5%) increases the cross-section area under compression (x). This respectively reduces the cross-section height under tension (y), especially if compared to f_{t2} = f_{t3} = 0%, which represents "unreinforced" concrete. This effect also reduces the tensile strain (respectively crack width), which is a known benefit of FRC.

632 Ground Support and Final Lining

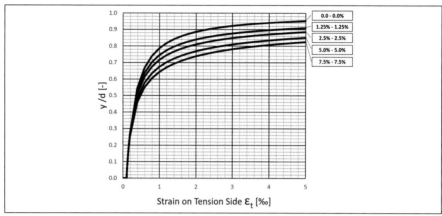

Figure 9. Cross section height under tension over equivalent tensional strain

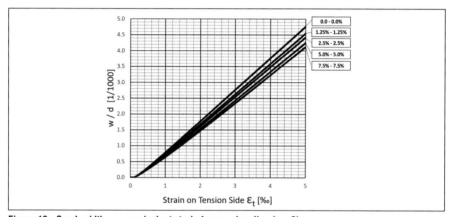

Figure 10. Crack width over equivalent strain for pure bending (n = 0)

Figure 9 can then be transformed into a crack width diagram by multiplying the height of the tension zone (y) with the related tensile strain (ε_t) using Equation 3. In Figure 10 the result of the crack width is shown. The crack width (w) is divided by the cross-section height (d) to provide a non-dimensional result over the equivalent strain (ε_t). The graph can be used in combination with Figure 9 as a SLS design diagram as shown in the following example.

Example 4: The maximum specified crack width w for the SLS is 0.4 mm. The cross-section height is d = 0.2 m, the compressive strength is f_c = 35 MN/m², and the residual flexural strength f_{t2} is 5% What is the maximum bearable moment at the SLS?

Solution: The equivalent strain ε_t for w = 0.4 mm based on Figure 10 is approximately 2.5‰ (with w/d = 0.4 mm/0.2 m = 2.0 [1/1,000]). Using this strain as an input parameter for Figure 8 the bearable SLS moment results as m = 0.022.

Answer: With a cross section width of b = 1 m and a compressive strength of f_c = 35 MN/m², the bearable moment results is M = 0.022*35*1*0.2² = 0.031 MNm or 31 kNm.

BEARING CAPACITY FOR FRC UNDER AXIAL FORCE IN THE SERVICE LIMIT STATE

The same approach developed for pure bending (n = 0) is now repeated for design diagrams including different axial forces (n = 0.1, 0.2, 0.3 and 0.4). Figure 11 shows the moment over the tensile strain over the different axial forces for n = 0 to 0.4 and different residual flexural strengths $f_{t2} = f_{t3} = 0\%$ (curve No. 1), 1.25% (curve No. 2), 2.5% (curve No. 3), 5.0% (curve No. 4), and 7.5% (curve No. 5). With an increasing axial force the entire cross section is under compression in the elastic phase, which can be seen by the negative values of the tensile strain at the beginning of curves. As discussed for the ULS, the increasing axial force also increases the bearing capacity for specific strain values.

Figure 12 shows the height of the cross section under tension for different normal forces n. For visual clarity, only the curves for $f_{t2} = f_{t3} = 0\%$ (upper boundary) and 7.5% (lower boundary) are shown. Values between the upper and lower boundary can be interpolated. The influence of the increasing axial compression is shown by a reduced cross section height under tension respectively by an increasing part of the

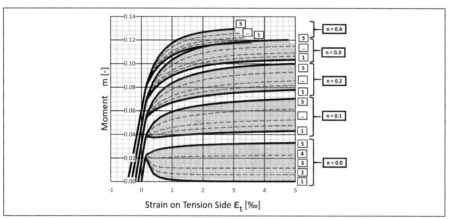

Figure 11. Moment over tensile strain for constant residual flexural strengths f_{t2} and f_{t3} [1: 0% | 2: 1.25% | 3: 2.5% | 4: 5.0% | 5: 7.5%] different axial force

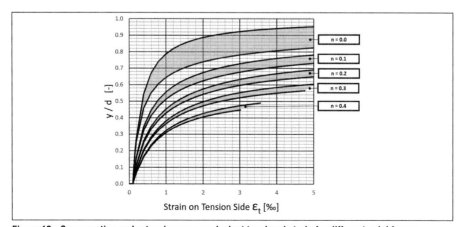

Figure 12. Cross section under tension over equivalent tensional strain for different axial forces

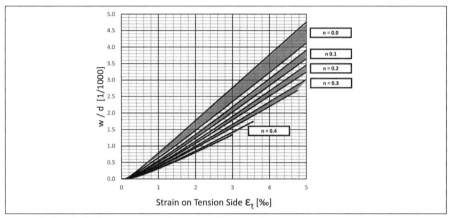

Figure 13. Crack width over equivalent strain for different axial forces

cross section under compression. For n = 0.4 and n = 0.3 the graphs are ending even before reaching the range of the diagram at ε_t = 5‰, because the cross sections fails on compression before reaching the maximum tensile strain shown.

Following the same approach introduced for pure bending the graphs in Figure 13 are displaying the relationship between the crack width w over the tensile strength for different axial forces. As above, only the upper boundary for $f_{t2} = f_{t3}$ = 0% and the lower boundary for $f_{t2} = f_{t3}$ = 7.5% are shown. Values between the upper and lower boundary can be interpolated. As expected, the increasing axial force has a positive effect on the crack width. A specific crack width is reached at a higher tensile strain.

CONCLUSION

The calculation of the bearing capacity of tunnel linings using an SSR for the ULS and the SLS is work intensive and typically requires specialized software. Parametric studies using different levels of axial compressive forces were conducted to evaluate typical load bearing behavior. Based on the typical behavior, simplified design diagrams for the SLS and ULS have been developed, which can easily be used for different types of design applications. Typical design examples using the design diagrams were discussed. The introduced design diagrams provide a powerful tool for a rough estimation and are considered very beneficial if quick results with very little effort are needed, i.e., during preliminary design stages, the bidding phase of a project, or if the effect of changes of material properties, quality issues, or loading conditions need to be assessed.

REFERENCES

[1] American Concrete Institute Committee 318: Building Code Requirements for Structural Concrete (ACI 318-14). 2015.

[2] American Concrete Institute: The Reinforced Concrete Design Handbook. Volume 1: Member Design SP-17(14). Building Code Requirements for Structural Concrete (ACI 318-14). 2015.

[3] American Concrete Institute: Report on Indirect Methods to Obtain Stress-Strain Response of Fiber-Reinforced Concrete (FRC) (ACI 544.8R-16). 2016.

[4] American Concrete Institute: Guide to Design with Fiber Reinforced Concrete (ACI 544.4R-18). 2018.

[5] American Concrete Institute: Report on Design and Construction of Fiber Reinforced Precast Concrete Tunnel Segments (ACI 544.7R-16). 2016.

[6] RILEM TC 162-TDF: Test and Design Methods of Steel Fibre Reinforced Concrete. σ-ε-design method—final recommendation. Materials and Structures, Vol. 36, October 2003, pp. 560–567.

[7] RILEM TC 162-TDF: Test and Design Methods of Steel Fibre Reinforced Concrete. Bending Test, Recommendations. Materials and Structures, Vol. 33, January-February 2000, pp. 3–5.

[8] German Society for Concrete and Construction Technology (DBV): Guide to Good Practice—Steel Fibre Concrete. 2001.

[9] Deutscher Ausschuss fuer Stahlbeton (German Committee for Reinforced Concrete) (DAfStb): DAfStb Richtlinie fuer Stahfaserbeton (DAfStb Guideline for Steel Fiber Reinforced Concrete). 2010.

[10] Nitschke, A., Tragverhalten von Stahlfaserbeton für den Tunnelbau. Dissertation. (in German. Load Bearing Behavior of Steel Fiber Reinforced Concrete for Tunneling. Doctor Thesis.) Technisch-wissenschaftliche Mitteilungen des Instituts für konstruktiven Ingenieurbau der Ruhr-Universität Bochum, TWM 98-5, 1998.

[11] Ruhr-Universität Bochum, Lehrstuhl Prof. Maidl, Nitschke, A., Ortu, M.: Bemessung von Stahlfaserbeton im Tunnelbau. Abschlußbericht. (in German: Design of Steel Fiber Reinforced Concrete for Tunneling. Final Report) Research Project funded by the Deutscher Beton-Verein E.V. (DBV-Nr. 211) and the Arbeitsgemeinschaft industrieller Forschungsvereinigungen (AiF-Nr. 11427 N). Fraunhofer IRB Verlag, 1999. ISBN 3-8167-5455-4.

[12] Nitschke, A., Bernard, E., Load-Bearing Capacity of Fiber-Reinforced Concrete Tunnel Linings. Rapid Excavation & Tunneling Conference (RETC 2017).

[13] Nitschke, A., Elasto-Plastic Design of Fiber Reinforced Concrete Tunnel Linings. North American Tunneling Conference (NAT 2018).

[14] Nitschke, A., Modelling of Load Bearing Behavior of Fiber Reinforced Concrete Tunnel Linings. Shotcrete Magazine, Volume 19, Number 2. 2017.

[15] Nitschke, A., Winterberg, R.: Performance of Macro Synthetic Fiber Reinforced Tunnel Linings, World Tunnel Congress (WTC) 2016 Proceedings, 22nd–28th April 2016 in San Francisco, CA.

Steel Fiber Segmental Linings for Mega TBMs

Tom Ireland • Aurecon, New Zealand
Shu Fan Chau • Aurecon, New Zealand
Harry Asche • Aurecon, Australia
Jack Muir • Aurecon, Australia
Ben Clarke • John Holland, Australia

ABSTRACT

The use of mega TBM's with external diameter of greater than 14 m is rapidly increasing. Up to 2017, there have been 39 tunnelling projects worldwide using mega TBMs. The 15.6 m diameter EPB machine, proposed for the $6.7B West Gate Tunnel Project in Melbourne, Australia, will the 7th largest machine.

The segmental linings for such large tunnels traditionally have steel reinforcement, however there are many benefits associated with steel fiber reinforcement. Besides manufacturing cost benefits, reduced damage during construction that has also been observed on many smaller tunnels. The segmental lining for the West Gate Tunnel will be the largest diameter steel fiber reinforced segmental lining. This paper describes the challenges with the design of a steel fiber reinforced segment for a mega TBM including massive ram loads, radial joint performance, cross passage openings and the interfaces with the proposed internal structures.

INTRODUCTION

The West Gate Tunnel Project (WGTP) in Melbourne, Australia is a city-shaping project that will deliver a vital alternative to the West Gate Bridge, provide quicker and safer journeys, and remove thousands of trucks off residential streets. The major civil works include widening the West Gate Freeway from 8 to 12 lanes, a tunnel from the West Gate Freeway to the Maribyrnong River and the Port of Melbourne, and a bridge over the Maribyrnong River.

The tunnels consist of twin 14.1 m ID TBM tunnels with length of 2436 m (Inbound) and 3675 m (Outbound) respectively. The tunnels will be constructed by two 15.6 m EPB machines, both from the Northern Portals. Internal structures include traffic deck slab, smoke duct and invert slab, all of which are precast units. Figure 2 shows the tunnel cross section. There are 21 nos. cross passages to be constructed by jacked box and mining construction methods, 9 nos. mined Egress Out and Under (EOU) from the Outbound Tunnel, and one mined Low Point Sump (LPS).

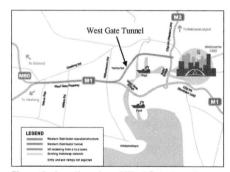

Figure 1. Location plan of West Gate tunnel project in Melbourne

The tunnel lining thickness is 500 mm and the nominal length of a ring is 2400 mm. There are 10 segments in a ring with evenly 36 degrees angle subtended by normal

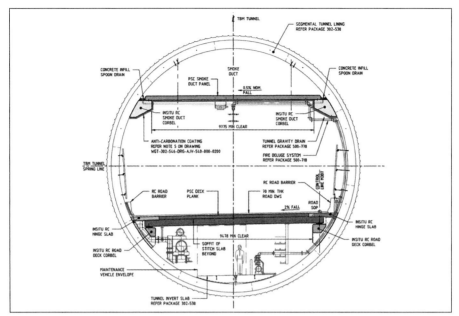

Figure 2. Tunnel cross section

rectangular segments, trapezoidal counter key segments and the key segment. The segment ring is a universal taper ring with taper of 40 mm (±20 mm) so that the key can be built in all 20 possible locations.

The segmental lining design confirmed using of steel fiber reinforced concrete (SFRC) in majority of the tunnel alignment, with short lengths of the conventional steel bar reinforced concrete at cross passages to cater for the segment opening effect, at Southern Portals where bridges are to be built above the tunnels, and at the Northern Portal with consideration of higher ovalisation due to poor ring build and low ground cover. This paper will focus on the design philosophy, methodology and challenges in designing the SFRC segmental lining for this mega TBM tunnels.

GEOLOGICAL AND HYDROGEOLOGICAL CONDITIONS

Extensive historical and project specific ground investigation reveal the geological condition along the tunnel alignments. The site geological formation varies along the alignment and include:

- General Fill
- Newer Volcanics (Basalt): Residual Soil (NVR), Weathered (NVW), Fractured (NVF), Slightly Weathered (NVS)
- Sub basaltic alluvium clay (SBA)
- Brighton Group clays (BGC) and sands (BGS)
- Newport Formation clays and sands: fine grained (NFF) and coarse grained (NFS)
- Older Volcanics (Basalt): Residual Soil (OVR), Extremely Weathered (OVX), Highly to Moderately Weathered (OVW) and Slightly Weathered (OVS)
- Werribee Formation clay (WF)

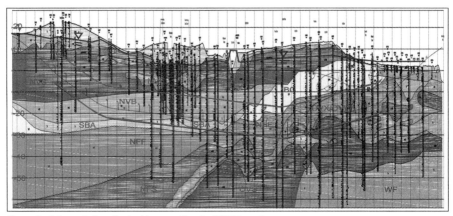

Figure 3. Interpreted geological long section for WGT tunnels

From north to south, the tunnels will mainly bore through a mixed ground (OVX/OVW/OVS), soft ground (GB/NFF), mixed ground (NVB/SBA) and rock (NVS/NVB). The groundwater table is generally 2–5 m below the ground surface. The maximum water head to the tunnel crown is about 40 m. The design groundwater levels also consider the flood levels at both portals.

DESIGN PHILOSOPHY AND CRITERIA

Segmental Lining Design Codes

Segmental linings generally have project specific design requirements and standards, which vary for different counties. Many tunnel projects refer to Bridge design codes because of the 100 years design life, and then pick and match different codes for load factors and materials factors. The WGT Project Requirements (PR) requires the tunnel design to comply with "Tunnel Reference Documents except where there is an Australian Standard equivalent to any British Standard specified in 'Specification for tunnelling, British Tunnelling Society and Institution of Civil Engineers, Third Edition, 2010', in which case the equivalent Australian Standard will prevail." This clause refers directly to the Specification for Tunnelling, a document referenced in PAS 8810: Tunnel Design—Design of Concrete Segmental Tunnel Linings—Code of Practice.

Since its publication in 2016 PAS8810 is being used increasingly for tunnel segment design. The authors would promote using PAS8810 in lieu of the bridge codes for segmental lining design, with the following considerations:

- PAS 8810 is a comprehensive and specific code for concrete segmental tunnel lining design.
- PAS 8810 allows partial factors on structural materials can be reduced if adequate controls are applied during their manufacture. Further details are given in BS EN 1992-1-1:2004+A1:2014

Steel Fiber Reinforced Concrete (SFRC)

Steel fiber reinforced concrete (SFRC) has been widely used for underground structure supports. When tensile cracks of the SFRC occur, the fibers in the concrete mix will provide bridging effect, acting as steel reinforcement, and continuously function as a load carrying element when the SFRC is loaded beyond its flexural capacity. Without

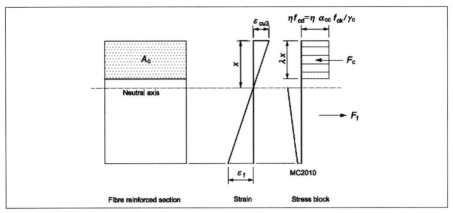

Figure 4. Stress block for SFRC

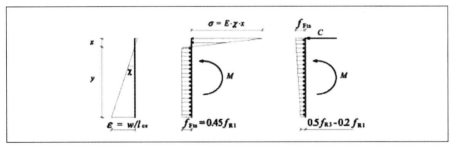

Figure 5. MC2010 residual tensile strength: (a) Stress diagrams for the determination of the residual tensile strength fFts; (b) fFtu; (c) for the linear model respectively

fibers, a plain concrete would lose its load carrying capacity when its tensile strength is exceeded and tensile cracks formed.

There are several points considered in the design to ensure the effective use of steel fibers in concrete: (1) Steel fibers should be significantly stiffer than the matrix, i.e., a higher modulus of elasticity; (2) Steel fiber content by volume must be adequate; (3) There must be a good fiber-matrix bond; (4) Steel fiber length must be sufficient; and (5) Steel fibers must have a high aspect ratio, i.e., they must be long relative to their diameter.

The design bending moments of resistance of SFRC is calculated in terms of residual tensile strengths. The stress block from PAS 8810:2016 that also references fib MC2010 has been used for the design of SFRC sections without conventional reinforcement. The residual tensile strength has been also calculated from the flexural tensile strength of concrete as per FIB MC2010.

The design compressive strength of concrete is 50MPa. The design characteristic residual flexural tensile strength of SFRC are fR1 of 3.1MPa and fR3 of 2.8MPa. The characteristic splitting strength is 5 MPa. All the design parameters for SFRC will be verified by testing. The steel fibers has been specified to be hook ended of minimum diameter 0.75 mm, have an aspect ratio (l/d) in the range 40 to 80, and a minimum tensile strength of 1050 MPa.

SFRC LINING PERFORMANCE

General Design Aspects

In general, the segmental lining associated with the TBM tunnel was designed using the following methods:

- Application of closed form solutions to calculate structural actions for different load combinations.
- Undertake two-dimensional modelling using a 'bedded beam' approach and finite element methods to determine imposed structural actions.
- Undertake FEM analyses to determine the ground loading on the lining.
- Conduct rock wedge analysis to assess the size and orientation of possible mobilised rock blocks.
- Design checks for joint capacity.
- Design checks and structural modelling for TBM thrust ram loads.
- Design checks for gaskets, bolts, taper, lifting and inserts.
- Design checks for demoulding, handling and stacking loads.
- Design checks for overall stability (floatation, heave and face stability).
- Design checks for the breakout opening considering stress re-distribution and mechanism to transfer redistributed stress on to designed structural support.

SFRC Lining Structural Capacity Check

At ultimate limit state the design resistance of members has been determined in accordance with PAS 8810 using partial factors. Derived forces and moments from the analysis have been compared with the calculated section capacity envelopes for a range of concrete grades and steel fibre dosages. Design of steel fibre reinforced concrete has been carried out as per Section 5.6 and 7.7 of the Model Code 2010, and in accordance with other relevant references, such as RILEM TC 162-TDF. For segments reinforced with conventional steel bars, the interaction chart has been based on BS EN 1992-1-1 Load combinations covered the various stages of manufacture, installation, temporary and permanent design cases. A typical M-N diagram showing the design SFRC capacity envelop and all derived axial forces and bending moment within the envelop is presented in Figure 6.

Smoke Duct Offset

Due to the offset of the smoke duct loads, additional localised moment is induced in the lining. In general cases, the lining is under high compression and the combined moments by the smoke duct offset and other load cases are within the M-N structural envelope. A conservative approach has also been used to check the lining structural capacity against the localised additional moment, by considering the low compression induced by tunnel structures only (e.g., assuming no compression by ground load and groundwater load). The results are within the designed SFRC M-N envelope.

Radial Joints

The lining is subject to bursting stresses from hoop loads on the radial joint and TBM ram forces on the circumferential joint. At ULS numerical modelling has been used to determine bursting stresses. Where SFRC is employed the resistance shall be based

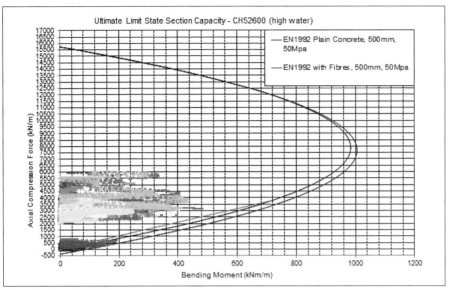

Figure 6. Typical M–N diagram

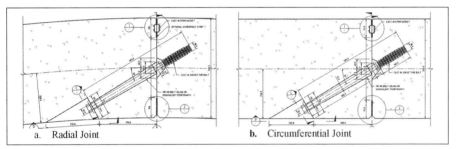

Figure 7. Radial joint and circumferential joint arrangement

on the tensile splitting strength of the concrete applied over the segment depth to resist the bursting force.

The effects of joint rotation on the load distribution at the joints shall be considered for the worst load/deflection combinations, including the effects of lipping. Such an analysis shall determine, as a minimum:

- Moments from eccentric loading.
- Stress distribution for the evaluation of bursting stresses.
- Maximum bearing stresses.

Elasticity Modulus of both short term (E_c) and long term (E_L) shall be used to calculate the centroid of the load and associated bursting stresses for the relevant ULS and SLS load combinations. A flat joint is considered in design. It may be the case where joints remain open due to the low axial force and ovalisation due to ring build and horizontal stress ratio. As a result of "birdsmouthing," bursting stresses will be developed in the lining. The stresses can be approximated using elastic theory. The method has been developed by Guyon (1972) for post-stressed concrete structures and deals with the

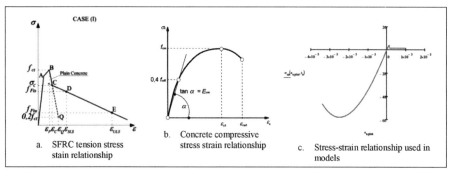

Figure 8. Non-linear FEM SLS model—material model inputs

transmission of high forces through the surface of an elastic body. Closed form analysis using the Guyon equations at SLS indicates that cracking does not occur.

FEM analysis of the radial joint capacity has also been undertaken. A non-linear analysis has been undertaken with elasto-plastic models used for the compressive (crushing) and tensile (cracking) behaviour of the concrete material. These models have been run for both the SLS case to determine crack widths and the ULS case to determine structural capacity.

The post cracking behaviour of the SFRC is defined at SLS by the FIP MC2010 (Figure 8a) and the compressive stress strain curve is derived from Eurocode 2 (Figure 8b). This combined stress strain relationship below has been used in the Elasto-plastic SLS model (Figure 8c).

The FEM model has also incorporated the guide rods and recesses. The general arrangement of the model is shown in Figure 8a and 8b. The FEM models confirmed the elastic analysis with the maximum tensile stress lower than the first crack stress in all cases. The FEM modelling has indicated that the axis joints are critical as for the squatting deformed case the load is applied to the intrados side of the joint. At the invert where the load is applied to the extrados side of the joint, the geometric shape of the lining results in lower bursting stresses. The maximum bursting stress under ULS is 3MPa which is less than the allowable stress of 3.7MPa. This is consistent with empirical checks. Output from the model for the critical design section 1d is shown in Figure 8c.

Temporary Ram Load and Circumferential Joint

The TBM will advance using thrust rams pushing against the previously installed segmental lining. The jacking load imposed by the hydraulic propulsion jacks on the bearing surface at the circumferential joint of the segmental lining ring is one of the design load cases. The TBM thrust load comprises forces due to face loads, forces due to frictional resistance of the TBM skin and weight. This force will be transferred through the ram to segmental lining. Although the TBM ram loads are temporary, the design force is huge and can be critical for the lining design (thickness and/or reinforcement).

The contact area between the shoe and the circumferential joint face shall be determined considering the following:

- Ram shoe geometry
- Joint geometry

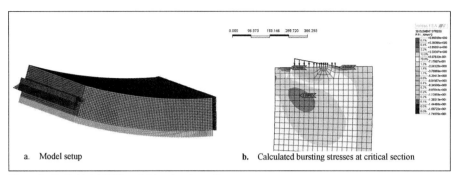

a. Model setup   b. Calculated bursting stresses at critical section

Figure 9. Radial joint FEM model input and results

- Tolerance on the position of the ring in the shield
- Build tolerance (squat or ovalisation)
- The potential for concentrated forces on the circumferential joint at the trailing edge of the key segment shall also be considered

There will be a total 20 pairs of thrust cylinders (2 pairs per segment) to be used. The total normal maximum thrust force is 194MN and the total emergency max thrust force is 212MN.

The segment has been checked for bearing, bursting, and compressive loads as a short column. Adequate resistance to crushing and splitting is demonstrated by the Guyon (1972) method, however the section is highly utilised indicating a risk of cracking due to poor build. FEM analysis of the ram loading has been undertaken, with the same analysis methodology as the radial joint. The input and outcomes are presented in Figure 9.

The SLS model confirms that the section has sufficient bursting stress capacity as predicted by the elastic model calculations (Figure 10b). The FEM model however indicates a zone of plasticity that develops between the ram shoes as a result of the elastic deformation of the concrete under ram loading. The concrete plasticity indicates cracking of the concrete under full ram load (Figure 10c). It is observed that the crack pattern is confined to a layer of concrete which is close to the loaded face. When the ram loads are removed the model indicates that these cracks close up (Figure 10d).

SEGMENT OPENING DESIGN

There are five types of segment openings to accommodate the jacked box cross passages (XP-J), mined tunnel cross passages (XP-M), Egress Out and Under (EOU) passages, smoke extraction connection (SXD) and low point sump (LPS). Steel bar reinforced concrete segment have been considered for the three rings (opening ring and adjacent rings). The temporary openings supports for the segment openings considered using bicones and shear keys, and steel props have been considered as contingency measures. Permanent collar structures will support the segment opening for long term.

The temporary opening support system and the special segmental lining was designed based on the following requirements:

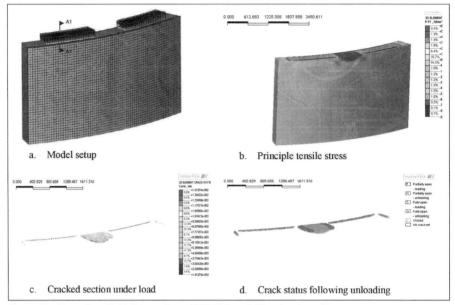

Figure 10. Circumferential joint non-linear FEM SLS model input and results

- Provide adequate support to ensure tunnel stability during opening and construction of the cross passages and other underground structures, prior to permanent opening support.
- Limit the tunnel lining deformation to acceptable levels, and joint rotation to meet watertightness requirement.
- Provide adequate structural capacity of the linings surrounding the opening that may have increased internal forces due to the opening.

3D Strand 7 models have been created for the segment opening analysis. Radial joints are not simulated in the models. The circumferential joints are simulated by point connection and shear keys/ shear cones are simulated as beams. Due to the enlargement behind the openings, the ground spring for the underground structure size behind the openings are ignored in the models. This also includes the allowance for tolerances and overbreak. The connection between rings at cross passage openings use shear keys (350 mm long, 300 mm OD 20 mm thick hollow section) and bicones (e.g., Anixter Sofrasar shear cone Sof-Shear M375).

Due to the opening, the tunnel might rotate and induce twisting moments, which have been considered in the design, by using the following equations:

$Mux = Mx + Mxy$ and $Muy = My + Mxy$

where, Mux, Muy are ultimate design moments at cross-sectional and longitudinal direction respectively; Mxy is the calculated twisting moment; Mx, My are the calculated moments at cross-sectional and longitudinal direction respectively.

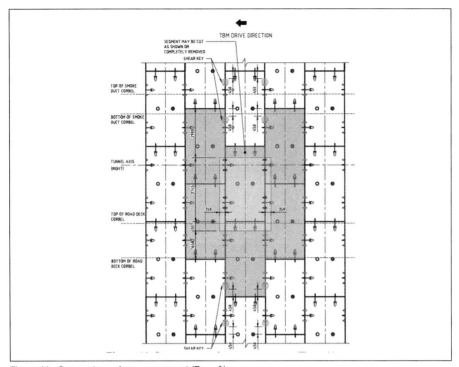

Figure 11. Segment opening arrangement (Type A)

CONCLUSIONS

Steel fiber reinforced concrete segments have been widely used in TBM tunneling projects. Normally steel ladders or reinforcement are required at radial and circumferential joints to resist the high bursting stresses. For mega TBM tunnels the segmental linings traditionally have steel reinforcement. West Gate Tunnel segmental lining design has used different empirical and numerical analysis, using PAS8810 which is a specific concrete segmental lining design code, to overcome many design challenges and confirmed the SFRC lining design for majority of the tunnel alignment. This is the first of this kind to use SFRC segmental lining with elimination of radial and circumferential joints steel ladder/reinforcement for a 15.1 m OD mega TBM tunnel. The success on the SFRC lining design in WGT project further developed steel fiber segmental tunnel lining design, which results in a cost effective and sustainable segment lining design.

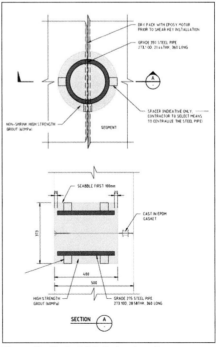

Figure 12. Typical details for shear key

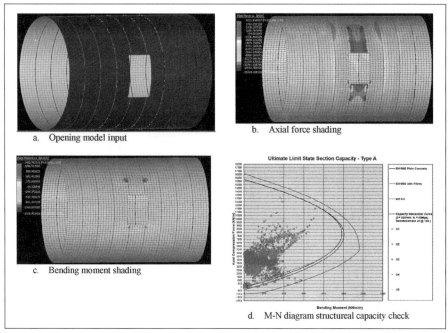

Figure 13. Segment opening models and outputs (Type A)

ACKNOWLEDGMENTS

The paper is published with the permission of the Western Distributor Authority (WDA) for which the authors would like to gratefully acknowledge. The design and construction of the WGTP tunnels is currently undertaken by CPB-John Holland Joint Venture (CPBJHJV).

REFERENCES

Asche H and Ireland T. Segmental Lining Design for Large Diameter Road Tunnels. *Rapid Excavation and Tunneling Conference*, June 2013. pp. 866–877.

Curtis DJ, 1976. Discussion of paper by Muir-Wood. *Géotechnique*, Vol 26, No 1.

Guyon, Y. 1953. Prestressed concrete, English edn. London: Contractors Record and Municipal Engineering.

Ireland TJ and Asche HR, 2011. Developments in Segmental Lining Design. *Rapid Excavation and Tunneling Conference*, June 2011. pp. 450–462.

Model Code for Concrete Structures 2010. FIB bulletin No. 65. (Sections 5.6 and 7.7 only).

Morgan HD, 1961. A Contribution to the Analysis of Stress in a Circular. Tunnel. *Geotechnique* 11(3): 37–46.

PAS 8810:2016 Tunnel Design—Design of concrete segmental tunnel linings—Code of practice. British Tunnelling Society.s

PART

Grouting and Ground Modification

Chairs

Dani Delaloye
Mott MacDonald

AG Mekkaoui
Jay Dee

Differences in Consolidation Grouting Practices Between Near Surface and Underground Applications

Adam Bedell ▪ Stantec Consulting Services Inc.
Brad Crenshaw ▪ Ground Engineering Contractors

ABSTRACT

Near surface consolidation grouting, such as dam foundation grouting, and underground consolidation grouting should not follow the same grouting practices. Contrasting methodologies, procedures, and overall general site conditions and space limitations require differences in each process. However, all too often Contract Documents contain language within grouting specifications that dictate near surface grouting processes for use in underground applications. The differences between the two procedures are outlined as well as the governing principles that drive each method. In addition to discussing varying grouting principles, grout mixes, factors affecting cost, and equipment considerations are also discussed.

INTRODUCTION

The purpose of this paper is to delineate the differences between grouting from the surface and grouting from underground for the purposes of consolidation grouting. The two methods follow different principles, while also representing end-member cases for consolidation rock grouting (consolidation of the rockmass from the surface vs underground). The need for this distinction is that, more often than not, surface consolidation grouting procedures are specified for underground applications.

The principles and methods for grouting shafts fall outside the scope of this paper. The purpose for this is because shaft grouting requires a mixture of techniques used in both dam foundation grouting and underground grouting. The authors acknowledge this distinction and wished to not confuse topics and purely focus on the more contrasting grouting styles and why one grouting style should not be specified for a grouting scope that operates under substantially different principles. This paper focuses on the end-member cases described above.

Within this paper, grouting is described as both a reactionary and proactive event. This is not to say that all grouting from underground is emergency grouting and all surface grouting is pre-planned routine work. However, underground grouting work is often performed *in reaction* to water being encountered. So it is often known that the work may need to take place, but the specifics aren't known. For example, for pre-excavation grouting ahead of a TBM, it may be known that probe hole drilling (say two or some other specific number of holes) is required through some specific zone. However, grouting will only be performed if groundwater is encountered. By contrast, for surface work, there is some minimum scope of work that is specified (e.g., at a minimum a specified number of primary and secondary holes will be drilled, water-pressure-tested and grouted) and this amount is much greater than the minimum specified for underground work. Therefore, surface work involves performing this "minimum" amount of grouting work regardless of the results for any one hole or stage, with a determination to perform additional grouting based on those results.

GROUTING PRINCIPLE DIFFERENCES

Grouting from either the surface or from underground serves the same purpose, which is to lower the overall rockmass permeability either below a dam or around an underground excavation. This in turn lowers the relative amount of groundwater infiltration either below the dam or into the underground excavated space. For underground work, grouting may be performed either pre-excavation or post excavation. For surface grouting work, grouting may be performed before the dam (or other structure) is constructed during pre-operational conditions, or grouting may be performed after the dam is in service as a remedial measure.

General Conditions

For surface grouting applications, a grout curtain is often created on the ground surface from above the rockmass to be treated. If performed during construction, the water table may be below the dam foundation and in-situ hydraulic conditions will be substantially different than during reservoir operation. Overburden pressure (both rock and soil) must be considered in the design phase of the grouting program as hydro-jacking of the rockmass is generally not a desirable outcome during the grouting process in North America. The permeability of a specific, targeted zone of the rockmass must be lowered by drilling grout holes through the overlying zone to inject grout primarily in the targeted zone.

For underground grouting applications (and from grouting within a gallery within a dam post construction), the excavation is below the water table and the intent is to lower permeability around the excavation. The excavation acts as a hydrogeologic sink, drawing groundwater into it. The permeability of the rockmass must be reduced by performing the grouting work *from within* this sink. Hydrostatic pressure, in addition to overburden pressure (both soil and rock), must be considered in the design phase of the grouting program.

Differences

Surface and underground grouting contrast each other on different, yet related fronts. Both processes involve groundwater moving from higher pressures to lower pressures, but under different conditions. Groundwater movement from a reservoir with higher head through the rockmass below a dam to an area of lesser head is not the same as groundwater movement from within an infinite sink below the water table. See Figure 1.

From a grouting perspective, grouting from the surface to treat a rockmass below the site is not the same as grouting from within an excavation, 360 degrees around you, and water *wants* to enter the excavation. See Figure 2.

Another key difference in grouting principles are recognized during the construction phase of the project (either surface or underground). Grouting from underground typically involves active fluid infiltration (either groundwater or other fluid) mitigation as the excavation progresses real-time. The excavation encounters the fluid and must immediately stop to address and mitigate the infiltration into the excavation. The critical path of the project is delayed while addressing the infiltration. Grouting from the surface, in a construction phase event, involves a planned construction activity with an associated schedule, and is typically performed off the project critical path.

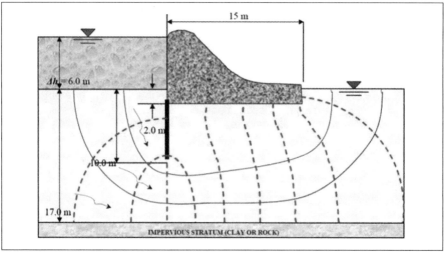

Figure 1. Flow net under a dam

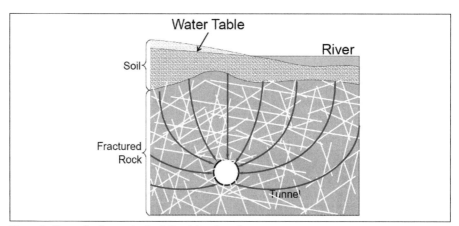

Figure 2. Generalized groundwater inflow into a tunnel

DIFFERENCES BETWEEN VARIOUS PROCESSES USED IN UNDERGROUND AND SURFACE GROUTING

Drilling

Drilling for either surface or underground grouting includes many similar characteristics, but the differences between the two associated construction types require a closer look at the mechanisms involved with each drilling process.

Surface Drilling

For surface applications, the construction site access and grout hole geometry dictates drilling equipment. Size limitations for equipment result from site access limitations, overhead clearance space, the presence of other operations, transport routes, etc. Drilling equipment is available for most any surface construction site configuration and arrangement. Another factor in drill selection is that consideration must be given to the relative number of fines that are generated by the drilling method. Surface grouting is sensitive to various drilling techniques. This is due to the multiple variables that are at play with regard to the drilling method and the relative amount of fines that are generated during the drilling process. All these variables have a response to the way that groundwater or the undesirable fluid enters the grout hole.

Surface grout drilling is initiated from above the water table and unless drilling in an area under artesian conditions, the drilled grout hole does not flush itself out. All fluids and cuttings must be evacuated from the drilled grout hole. This is accomplished through the injection of air or water. All drilling wastes may require additional management and disposal as well.

Other considerations for surface grout holes from a drilling perspective are that they are generally less expensive than underground holes. There a few reasons for this (i.e generally shallow holes), but a key reason is that this operation is typically not driving the project critical path. Surface drilling typically involves a pre-designed grouting program that may be bid separately from the main construction work, and since this grouting is typically not on the critical path of the project, nor typically is it emergency related work, lower per foot drilling rates can be received from the grouting contractors bidding the work.

Underground (from a tunnel perspective)

Underground drilling is not as sensitive to drilling techniques (as compared to surface drilling) as water is pulled into the hole helping flush the drill cuttings away from the fracture. The drilled grout hole essentially develops itself as water is pulled from the rockmass in to the drilled grout hole and excavation. While coring, rotary percussive, or various hammering techniques are suitable (depending on the head pressure at the drilling location), space, orientation, and mechanics (constructability or practicality of drilling method), of the underground excavation dictate viable drilling methods.

Drilling may be required to be modular to allow transport in a smaller diameter tunnel at long distances from the access point underground. Once details on the drilling locations are determined, then proper selection of the best drilling technique may be made. Often, these details are unknown during tunnel design, yet drilling restrictions often find their way into Contract Specifications.

Contrary to drilling surface grouting holes, grout holes drilled from underground are often expensive. The main reason for this is that most grouting performed underground affects the critical path of the project in some way. Mining may have to be stopped while all grouting (and potentially specialty drilling equipment) is moved underground. The critical path of the project is halted until the rockmass condition has been corrected, which may not have been accounted for in the schedule. This causes the drilling rate to be higher to account for excavation and associated equipment on standby while grouting is performed.

Water Testing

Water testing through the years has proven to be an invaluable tool for field testing of in-situ rockmass permeabilities. Injecting water, under the correct pressure, will suggest the likelihood of the geologic formation taking grout.

Underground (from a tunneling perspective)

A range of groundwater inflows are typically manageable during construction, though unknown and unplanned groundwater inflows do often occur. If water testing occurs from within an underground excavation, then chances are there is a sufficient groundwater inflow that excavation has stopped and grouting is the mechanism to reduce the inflow to a manageable level such that excavation may continue. The critical path of the project is now on hold, so time spent performing other activities such as grouting is expensive. Adding to the cost, additional support equipment (i.e., cranes, locis, etc.) are now utilized for grouting and not excavation.

As the process is underground, with limited resources, the typical permeability determination is reduced to a simple flow rate determined from the flow meter installed at the grout header. This information can provide a quick and easy determination of initial grout mixes and grout sequencing. Precise determination of in-situ rockmass permeability is impractical as the excavation advance and most underground resources are tied up until the groundwater inflow has been managed.

Surface Applications

Water testing from the surface can provide valuable information that allows for relatively precise determination of rockmass permeabilities and better inform the subsequent grouting program. Information provided from boring logs indicating potential discontinuities may be merged with water testing data. Space and resource limitations are not present and if necessary, multiple and various permeability tests may be run.

Typically, dam foundation grouting, pre-excavation consolidation grouting, or pre-excavation cut-off grouting is performed off (or not driving) the critical path of the project and time has already been allocated for water pressure testing and data analysis to be performed. Since other project components may move forward while the water testing and data analysis is performed, the full benefit of water testing can be realized as modifications to either the engineer-directed grouting program or the contractor-submitted and approved grouting plan.

DIFFERENCES BETWEEN GROUTING IN UNDERGROUND AND SURFACE APPLICATIONS

Grout Equipment

Grouting equipment, from a macro-perspective, is not different (to include colloidal mixers, non-pulsating pumps during injection, flow meters, and the use of agitator tanks) between surface and underground applications. This is where the similarities end. Equipment used underground must be mobile, compact, and accommodate the various dimension restrictions required by the underground excavation. Equipment must be able to fit on TBM's (including rolling stock for transport) and uneven surfaces for drill and blast operations.

Grouting from the surface allows for larger equipment and more options and flexibility concerning cement storage. While some elements may transfer to underground applications if the conditions are acceptable (shallow tunnel, shorter pumping distances, etc.), often the exact location of grouting underground will restrict the size and configuration of the grouting equipment.

Real Time Grout Monitoring

Real time grout monitoring has become an engineer-requested standard over the past 20 years. It has become such a common standard that people are requesting real time grout monitoring in underground applications, almost without thought in some cases. While real time grout monitoring is quite useful when grouting from the surface, on either a dam foundation or pre-excavation consolidation grouting perspective, there are many key differences to keep in mind.

Underground

The conditions in which grouting within an underground excavation is performed are completely different from any surface applications. Typically, real time grout monitoring is impractical for underground use. The conditions on a TBM or drill-and-blast heading are not conducive for the computers and monitors for real time monitoring. As all the grouting equipment is underground, having an area to set up a control center for constant grout monitoring is not easily possible. Communications could be established from the excavation to a surface control center, but the cost benefit is limited in most circumstances.

A more important difference is that grouting from underground is often performed in a reactive setting. There is little benefit for the added cost and schedule to transport equipment underground to set up real time monitoring. Along with this aspect is the tendency of tunneling general contractors to perform as much of the underground grouting as possible and most personnel are unfamiliar with the technology. When grouting from underground, it is typically a simple operation for which there are usually only 1 or 2 grout holes at a time available for grouting.

Surface Applications

Real time grout monitoring from the surface is quite useful as the conditions are completely different than underground. Commonly, grouting is a planned, proactive activity with a designed intent and not performed in a reactive setting. There is also typically enough site access and area to mobilize and establish the control center proximate to the treatment area.

Surface grouting is traditionally performed by specialty subcontractors whose personnel are familiar with the grouting programs. This allows the design geologists and engineers to view the data real time while grouting multiple stages with long holes across a relatively large treatment area. The real time data may be plotted and correlated with known geotechnical information.

Grout Injection Pressure

Generally speaking, the method of determining grouting pressures is similar for near surface and underground structures (for widely applied consolidation grouting principles in the United States). The intent of both processes is to overcome hydrostatic pressure to allow the grout to displace water from fractures.

However, the specifics of pressure determination usually vary in practice. For underground applications, generally a simplified approach to pressure determination is utilized (generally something like one psi per foot of depth). This is based on the simplified assumption that groundwater is present from the grouting depth to the ground surface (which, for practical purposes, is the case for many underground operations).

Often, near surface structures are more specific in pressure determination. The primary reason for this is concern that over pressurization will damage the near surface rock which is being grouted (also true for shallow excavations). This is due to several factors, some of which are that the near surface rock may be weaker (more weathered, more highly fractured), that friction loss will be less when grout reaches the weaker upper rocks, or that overburden may be damaged by the grouting. To account for specific pressure modifications, the following may be taken into account: the actual location of the surface of the groundwater table relative to the grouting depth, a reduced pressure (generally 0.5 psi) may be utilized for overburden material, and friction losses through hoses may be taken into account.

Grout Mix

This section discusses cement-based grout mixes for simplicity. There really isn't a difference between the two applications in terms of grout mix. The authors believe the typical correct course of action involves a cement-based grout without the use of stabilizing agents. Additives may be added to grouts based on the conditions for the hole being grouted and what has been learned of that specific area.

Grout mixes are thickened and thinned based on grout injection rates and injection pressure changes only. Time and volume are not considered if pressure builds and injection rate decreases. The initial grout mix should be based on known permeabilities or water testing results and desired penetrability.

Grout Injection Sequencing

Underground

As grouting from underground is typically a reactive process, choosing the proper grout sequencing generally falls under two possible options. First, one can pump on the hole connected to the largest amount (either from a velocity or volume perspective) of water first. The second option is to use the geology to deliver the grout to where it needs to go. This involves using what is known about the geology (foliation, bedding planes, fracture orientations, etc.) to deliver the grout into the rockmass. Oriented grout holes must be drilled to intersect known fracture orientations rather than use randomly oriented grout holes.

Surface

As grouting from the surface is typically a proactive process, grout injection is typically set up on a pattern progressing from primaries through either tertiaries or quaternaries. Each phase is drilled and grouted in order. Which hole is drilled and grouted first, and which is next is based on parameters specified at the start of the work.

Geologic features identified during the geotechnical investigation may be missed by the primary holes but are typically handled through the subsequent phases of drilling and grouting.

Grouting Logistics

Underground

An important consideration for grouting underground includes careful forethought and logistical planning with regards to the transport of labor, equipment, and materials. These are paramount to underground grouting operations. Potentially complicating matters is that the grouting operation must often compete with other operations for valuable ingress and egress into the excavation. This occurs when there are multiple operations working from either a single heading or shaft. Also, when looking at planning schedule, shift production must account for time to enter and leave the excavation. This could potentially could add 1–2 hours of unproductive time.

Surface

Grouting operations from the surface are not typically limited by the size or footprint of the construction site. The restrictions in logistics are through transport and general site access by labor, materials, and equipment. Unless the project is in a remote location, these restrictions are generally negligible. Unlike underground operations, travel time to different points of a grouting site typically is also negligible. A typical restriction is the presence of slopes on the site, which are often handled by appropriate surface preparation.

Grouting Results

Underground

As a lot of underground grouting is reactionary, this allows direct results to be observed. The groundwater inflow can be directly observed to have decreased in volume or perceived pressure while grouting from either a TBM probe drill deck or from 20 feet from a drill and blast heading. The relative amount of water left will be a key indicator. Verification grout holes may be drilled into the face if further confirmation of grouting is needed.

Surface

Grouting from the surface for a dam foundation or pre-excavation consolidation grouting environment prevents one from directly observing the results of grout in the ground. Even if the ground will be exposed following grouting to allow some observation, this will be after the grouting subcontractor has left the project site. Costly remobilization and subsequent delays to operations scheduled to follow grouting force the determination as to whether grouting is complete. It may be troubling for some that the results are not readily observable.

The decision to either cease or proceed with additional grouting is almost solely based on data, specifically grouting records. After the decision has been made to cease grouting, the degree of success of grouting is often unknown until after the project is complete. This underscores the importance of good record keeping, staffing, and management of all grouting data.

CONTRACTS AND PERFORMANCE

Underground

For underground applications, grouting is often performed by the tunneling General Contractor or a Specialty Subcontractor. The scope of work may be either performance based (i.e., bring the groundwater inflow to a specified level) or it also may be "as directed by the Engineer." If grouting is "as directed by the Engineer," then language in the specifications must clearly define what the Contractor may anticipate during the course of the work.

Routinely, grouting is performed as a 24-hour operation. This is because grouting is required because the excavation has exposed a ground condition that must be mitigated, but also for efficiency. The exposed ground condition almost always affects the critical path of the project. Because grouting is reactionary affecting the critical path, this manner of grouting is more expensive than surface grouting.

Surface

For surface applications, grouting is performed by a Specialty Subcontractor. The General Contractor for dam construction does not typically have the specialty equipment or staff to perform this type of work. This scope of work routinely falls under "as prescribed by" or "as directed by the Engineer." Grout designs are bid on, and the means and methods are prescribed.

Grouting for these applications is traditionally performed as a single-shift operation. Another factor is that this style of grouting does not typically interfere with driving the critical path of the project. This a proactive, scheduled operation that proceeds until completion, with other operations around the site working. A side benefit of this is that surface grouting is typically less expensive than underground grouting.

CONCLUSION

The purpose for this paper is to delineate the differences between grouting from the surface and grouting from underground. While similarities exist between these two end member grouting styles in terms of the overall lowering of rockmass permeabilities, there are vastly different variables that each process must overcome. These differences require consideration in terms of grouting scope, the language in the grouting specifications, and the understanding that surface grouting practices do not directly translate to underground grouting practices. Design engineers and geologists must keep in mind what the purpose of the specified grouting is and then build the grouting program, including drawings and specifications, around it.

Grouting a TBM Shotcrete Lining Under Challenging Conditions

M.H. Kizilbash ▪ Stantec, USA
Gary Peach ▪ Multiconsult Norge AS

ABSTRACT

Twin 8.5 m diameter, 10 km long parallel headrace tunnels under overburden of up to 1870 m with high horizontal stresses were excavated for the Neelum Jhelum Hydroelectric Project located in northeast Pakistan using two main beam gripper Tunnel Boring Machines (TBMs). Both TBMs started headrace tunnel excavation in early 2013 with completion by the first TBM in October 2016 and the second TBM in May 2017. The permanent support was a shotcrete lining applied over other support elements such as rock bolts, wire mesh, mining straps and full circular steel rings. There were separation and voids found between rock and reinforced (wire mesh) shotcrete liner. To ensure the structural integrity and also some unavoidable factors encountered during construction, the installed shotcrete lining required thorough investigations and subsequent grouting works tailoring the injection pressure and penetrability of grout mixes before water could be released into the headrace tunnels.

This paper briefly outlines the different zones on the TBM where the shotcrete was installed and some of the issues such as 1700 rockbursts encountered during construction, which led to the need for grouting works to the shotcrete lining. The paper then details the complex sequencing for implementing the grouting works due to limited access and describes the various solutions and grouting methodologies employed to finish the two tunnels. Details of quantities used for the Neelum Jhelum project are provided for use as a guideline for similar projects in the future.

PROJECT DESCRIPTION AND OVERVIEW

The Neelum Jhelum hydropower project is located in the Muzaffarabad District of Azad Jammu & Kashmir (AJK), in northeastern Pakistan within the Himalayan foothill zone known as the Sub-Himalayan Range. The terrain is rugged with ground elevations that range from 600 m to 3200 m above sea level.

The project is a run-of-river scheme, employing 28.6 km of headrace and 3.6 km of tailrace tunnels that bypass a major loop in the river system, transferring waters from the Neelum River into the Jhelum River, for a total static head gain of 420 m (Figure 1). The headrace tunnels include both single bore (31%) and twin bores (69%) sections. The tailrace tunnel consists of a single tunnel. Design capacity of the waterway system is 283 m³/s. The project, which became

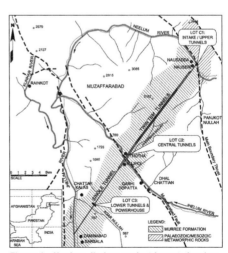

Figure 1. Neelum Jhelum project layout showing TBM Twin tunnels (in bold), major faults (dashed) and alignment geology.

operational in 2018, has an installed capacity of 969 MW, generated by four Francis turbines located in an underground powerhouse.

Tunneling commenced in 2008 using conventional drill and blast techniques. It soon became apparent that a 13.5 km section of the twin headrace tunnels (under high mountainous overburden that precluded construction of additional access Adits) would take too long to excavate. The contract was amended to incorporate two, 8.5 m diameter open gripper hard rock TBMs to excavate approximately 10.5 km of twin headrace tunnels (Figure 1). The gripper design offered flexibility for the expected conditions: possible squeezing ground given the relatively weak rock mass and overburdens up to 1870 meters, and the potential for rockbursts in the stronger beds.

GEOLOGICAL SETTINGS

The entire project was excavated in the sedimentary rocks of the Murree Formation, which is of Eocene to Miocene age, and comprises closely interbedded sandstones, siltstones and mudstones. The TBM tunnels were constructed through a zone bounded by two major Himalayan faults that trend sub-perpendicular to the tunnels: The Main Boundary Thrust, and the subsidiary Muzaffarabad reverse/thrust fault (Figure 1).

Open Gripper TBM Configurations for Support Installation

Open gripper TBMs are designed for hard, competent rock where little to no support is required.

The required thrust of the cutterhead is provided by thrust cylinders attached to grippers, which are braced against the tunnel wall. Once the thrust cylinders reach the end of their stroke, the rear cutterhead support is lowered and the grippers and cylinders are pulled in. They are then repositioned for the next boring cycle. The gripper's then re-engage the tunnel wall, the rear cutterhead support is raised and the next cycle starts.

Initial support is applied at the 'L1 zone' just behind the TBM shield and any further support is applied at the 'L2 zone'. Figure 2 shows a typical schematic of an open gripper TBM.

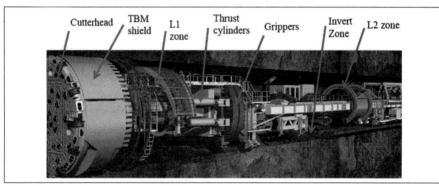

Figure 2. Schematic of open gripper TBM (indicative only)

L1 Zone

The L1 zone is where the initial support is placed (rockbolts, wire mesh, shotcrete) primarily to secure any loose rock and to limit the convergence of the rock mass. Initial tunnel support installed at the L1 zone is generally incorporated into the permanent support.

Shotcrete application is generally avoided at the L1 zone because overspray and rebound fouls the TBM's electrical and hydraulic equipment, and because the clamping pressures imposed by the grippers unavoidably damage green shotcrete. Consequently, shotcrete application is delayed where possible until the L2 zone, where dedicated robots mounted behind a shield can operate freely in a less congested environment. However, shotcrete application at the L1 zone was sometimes necessary where poor ground conditions were encountered.

Invert Installation Zone

On the Neelum Jhelum project the invert installation zone was located 20 m behind the L1 zone. Small precast concrete blocks were placed at specified spacing's to support the railway sleepers for the rails along which the TBM, back up gantries and locomotives travel. Shotcrete was then sprayed from a mobile applicator that could travel up to 12 m longitudinally and 100 degrees of the lower tunnel circumference.

L2 Zone

The L2 zone was located 65 m behind the cutterhead. Here, shotcrete was sprayed systematically over the initial support that had been installed at the L1 zone, to complete the permanent lining. The thickness of the shotcrete sprayed at the L2 zone (Figure 2) typically varied between 125 mm and 250 mm and was dependent upon the quality of the rock mass.

FACTORS INFLUENCING THE REQUIREMENT FOR GROUTING WORKS

As tunnel excavation advanced, every effort was made to complete the tunnel support and the shotcrete lining installation in one pass to avoid the need for any further grouting works. Nevertheless, a number of unavoidable factors were encountered during construction that resulted in a less-than-perfect lining, necessitating additional grouting works being required for some parts of the tunnels. The following section discusses these factors in more detail, and how they pertained to the Neelum Jhelum project.

Support Element Configuration

The tunnel support design relies on the majority of support elements being installed with full contact against the excavated tunnel profile. The presence of overbreak, which was at times difficult or impossible to control in normal operations and particularly where sub-horizontal stresses were high and the rock mass was weaker than average, precluded such a full contact from being achieved. Any overbreak over 150 mm necessitated additional support such as wire mesh, mining straps and full circular steel rings, each of which potentially prevented full contact to some degree.

Shadowing

A phenomenon related to the support element issue is shadowing, which was the largest contributor to creating voids within or behind the installed shotcrete lining on the project. Shadowing occurs when an item (such as wire mesh) obstructs the full penetration of shotcrete and creates a void or an area of poorly compacted shotcrete.

Subsequent grouting is required to fill these voids. Other factors that can result in shadowing are limited access to the tunnel periphery or poor shotcrete application.

Rebound

The application of shotcrete with differing sequences, as necessitated by variable ground conditions, sometimes precluded use of a systematic application, leading to shotcrete of reduced quality. For example, ad hoc shotcrete application in the L1 zone, if not carefully monitored, could result in rebound falling onto the sidewalls or invert. If the rebound was not removed before subsequent layers of shotcrete were sprayed in the invert or at the L2 zone, the rebound could become incorporated into the lining, leaving significant voids or honeycombing. Of course, observation of such behavior by supervising staff, and its immediate rectification, was generally implemented, but some instances occasionally slipped through, particularly early in the excavation before QA/QC procedures had been optimized.

Water Ingress

The inflow of water can be particularly problematic for a shotcrete lining as it affects the adhesion of shotcrete to the rock profile, particularly if the rock is prone to softening when wetted. Shotcrete applied under these conditions can result in the development of voids behind the lining. Fortunately, little water ingress was encountered in the project's TBM tunnels.

Specific Deleterious Geological Conditions

Squeezing ground conditions present a significant challenge to the shotcrete lining, particularly if ground movements occur over many months. Though anticipated, little squeezing ground was encountered on the project; however, localized squeezing did result in damaged shotcrete that needed repair.

A rock burst is a sudden and spontaneous release of strain energy resulting from stresses due to high overburden or tectonic forces that exceed the rock strength. This relatively unpredictable event can range in severity from very minor to major events that can inflict extensive damage on the shotcrete lining, ranging from small cracks to total destruction. On the Neelum Jhelum project there were 1700 rock bursts recorded in both TBM tunnels.

STRATEGIES TO ELIMINATE OR MINIMIZE THE REQUIREMENT FOR ADDITIONAL GROUTING WORKS

The strategies needed to eliminate or at least mitigate the issues listed above are well understood in tunneling. First, strict adherence to a Quality Assurance/Quality Control (QA/QC) system is essential to achieve the design requirements, and that the shotcrete delivered to the TBM is to the required specification. This should include an appropriate QA/QC plan for the materials being used.

Second, analysis of the application of shotcrete relative to the planned support elements to be installed should examine how the shotcrete can be applied around individual support elements whilst maintaining the required shotcrete nozzle to tunnel periphery distance and nozzle angle to ensure maximum shotcrete penetration behind the support element.

In difficult geological conditions, a permanent shotcrete lining in a tunnel excavated by TBM has a high potential for grouting works because of the factors outlined in the previous section. Therefore, before this type of final lining is selected, an initial assessment should be carried out to identify all factors that could lead to sub-standard shotcrete, which could necessitate grouting works, and how those factors can best be mitigated.

TIMING, LOCATION AND IMPLEMENTATION OF GROUTING WORKS

If ground conditions are optimal; if tunneling teams are dedicated and highly trained; and if the TBM is perfectly suited to the conditions, the no grouting works will be needed. However, in real-world conditions, particularly in harsh environments and high in situ stress conditions such as those encountered on the Neelum Jhelum project, some level of grouting works is unavoidable. Having established that grouting works are necessary, determining how delays can be minimized becomes the crucial factor. Different configurations of tunneling while grouting works could be carried out are outlined below.

1. From the TBM—TBM Still Excavating

This option would be the simplest from an operational point of view. But has two distinct disadvantages

- Depending upon progress, the shotcrete may not have reached its full strength.
- The TBM needs to have this drilling and grouting capacity setup on the gantries.

2. Behind the TBM—TBM Still Excavating

This provides greatly improved access to the shotcrete tunnel lining (as compared to from the TBM), although full access is masked by services still required by the TBM (air, electricity, water, ventilation, conveyor). However, it is logistically the most complex approach, since its implementation cannot conflict with the TBM railway supply system; although water and compressed air supply being provided to the TBM remain available. Importantly, if these works can be completed before the TBM finishes excavation, then the time penalty to the program is minimal.

3. After Completion of the TBM Excavation—TBM Operations Completed, Track in Place

This provides maximum access to the shotcrete tunnel lining. This approach requires dedicated electricity, water and compressed air supply and its own railway requirements, but since these are retained from the TBM's system, the logistics are usually not complex. However, the major disadvantage of this approach is the time penalty, since the programmed completion date needs to be put back to accommodate these works.

4. After Completion of the TBM Excavation—TBM Operations Completed, Track Removed

This type of operation is similar No. 3, with a similar time penalty, except that the operation is directed primarily at the invert.

The TBM grouting works for Neelum Jhelum project utilized a combination of (2), (3) and (4) described above.

Determining the Need for Additional Grouting Works

Two requirements need to be fulfilled to fully document shotcrete defects, and thus the requirement for grouting works.

Inspection of Completed Lining

Visual inspection may be initialized by Contract requirements (such as testing for drummy shotcrete), QA/QC procedures, or randomly inspection due to special occurrence. However, simple due diligence requires that such a visual inspection be performed regardless.

Construction Records

Provisions should be made in the TBM shift report to record shotcrete defects. This allows a grouting works register to be established that records the locations where grouting will be required. On the Neelum Jhelum project, a dedicated section for possible shotcrete grouting works was incorporated within the TBM shift reports that described the nature of the defect and its extent. An example in Table 1 has been extracted from the TBM shift reports.

Table 1. Example of grouting work item on TBM shift report

Cause	Tunnel Chainage	Location	Area	Other Details
Rockburst	9+743	11–2 o'clock	1.2 m x 1.1 m	0.32 m deep

The potential grouting work item should be documented in the format of the grouting works register, which defines the scope of work and also identifies trends and patterns to allow efficient resource and equipment deployment.

Implementing Additional Grouting Works

In practice on the Neelum Jhelum project no remediation of poor shotcrete was possible from the TBM, and nearly all the works were carried out after passage of the TBM, both while it was still excavating (termed 'Stage 1' below), after excavation had been completed but track was in place ('Stage 2'), and after the track had been removed ('Stage 3').

Stage 1

The first stage shotcrete grouting works had reduced access due to the presence of TBM services such as ventilation, conveyor, power, water and compressed air etc. plus the railway line itself. Thus, the access gantries and equipment had to be designed to fit within the existing tunnel configuration, allowing safe access to the exposed sections of shotcrete lining for repairs. The general configuration of this equipment is shown in Figure 3.

As shown in Figure 3, the shotcrete grouting works were carried out from two, 12 m long two-level access gantries equipped with a generator, air compressor, drilling equipment, a grout mixer and an injection pump. The TBM excavated tunnels had a single railway.

To allow uninterrupted train movement in the tunnel, the grouting work gantries were required to run on a separate track outside of the TBM railway track. These gantries

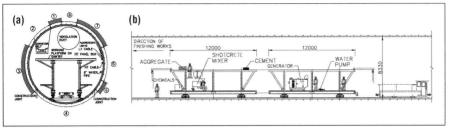

Figure 3. (a) Cross-section (looking in direction of Tunnel Drive) of grouting works equipment (available sections of tunnel periphery (during stage 1 shaded in grey) (b) Longitudinal section of grouting works equipment

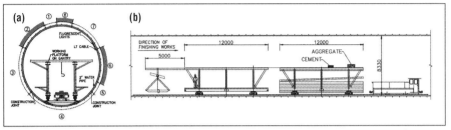

Figure 4. (a) Cross-section (looking in direction of Tunnel Drive) of grouting works equipment (available sections of tunnel periphery during Stage 2 shaded in grey) (b) Longitudinal section of grouting works equipment

were advanced by lifting the track at the rear of the gantries and transporting it forward to the front. Figure 3 (a) shows a cross-section of Stage 1 grouting works. The circumference of the tunnel was been divided into eight sections that relate to accessible sections during different stages of the grouting works operations. In Stage 1, the accessible sections were 1, 3, 5 and 7 (shown in grey in Figure 3 (a)). The remaining sections were inaccessible due to the presence of the TBM support services, such as ventilation, rails, conveyor etc. and were completed at a later stage.

Stage 2

The second stage of grouting works was carried out when the TBM had completed excavation, whilst the TBM and support services were being removed from the tunnel. This stage allowed grouting works to be carried out in previously inaccessible areas (shown in grey in Figure 4 (a)).

The general arrangement of Stage 2 grouting works is shown in Figure 4. Here the longitudinal section has the grouting work equipment removed in order to show the rail track removal process.

Stage 3

This stage was carried out after the TBM railway tracks and supporting sleepers had been removed. During this stage debris from the invert of the tunnel was removed as this area was previously inaccessible (shown in grey in Figure 5 (a)). The general arrangement of stage 3 grouting works is shown in Figure 5.

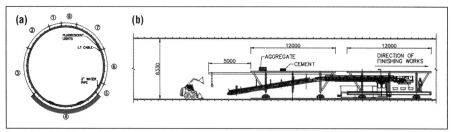

Figure 5. (a) Cross-section (looking in direction of Tunnel Drive) of grouting works equipment (available sections of tunnel periphery during Stage 3 shaded in grey) (b) Longitudinal section of grouting works equipment

GROUTING WORKS REQUIREMENTS

Contact Grouting

Intimate contact between the initial support and the final shotcrete lining is essential to maintain the integrity of the final support of the tunnel, so it was vital that there be no voids between the lining and the surrounding rock. The grouting works register was used to identify locations of potential voids although testing was carried out along the entire tunnel. The grouting works gantries allowed direct access to the full periphery of the tunnel to first carry out sounding inspections to identify suspected void locations. Soundings were performed by tapping a geological hammer at regular intervals at least once every 30 cm^2 against the shotcrete surface to listen for drumminess, above the spring line on a routine and systematic manner. Secondly, in locations suspected of significant separation were suspected, this was phase one. Phase two was performed below the spring line.

The resulting sound allowed the inspectors to determine either a potential void or delamination from a competent contact between the lining and the surrounding ground.

Areas identified for grouting were marked with spray paint and the void or delaminated area was typically contact grouted. This grouting operation was undertaken at low pressures (300 kPa) and with different grout mixes depending upon the presence of groundwater.

Consolidation Grouting

Rockbursts resulted in over break and delamination of the surrounding ground adjacent to the location of the rockburst. An average over break of 480 mm was recorded for rockbursts in general. Consequently, the consolidation grouting holes were drilled to a depth of 1000 mm to account for both the over break and delamination resulting from the rockburst. In localized zones with large overbreak holes would be drilled to 500 mm beyond the depth of overbreak. Due to the in situ stress conditions, the overbreak and consolidation grouting holes were generally located in tunnel periphery sections 1, 2, 7 and 8 described above.

GROUTING WORKS PROCEEDURES

Locating Voids Between Rock and Shotcrete and Within Shotcrete

The most challenging part of this work is to establish the location of voids in areas where shotcrete was more than 30 cm thick and reinforced, which in places includes multiple layers of wire mesh, and full ring steel supports. Drilling a series of holes at critical locations was necessary to established whether voids exist at the contact of shotcrete and rock and also within shotcrete.

After the support shotcrete lining has been installed, with passage of time cavities/voids were detected especially near close to mudstone beds and where wire mesh overlapped which reduced the effective mesh opening and restricted penetration.

When drummy sounding shotcrete was detected then the striking of the shotcrete surface was continuous radiating out from the initial drummy sound location, until competent shotcrete was located. This area was then clearly marked with spray paint. The areal extent of the marked up drummy area was then calculated and recorded on the inspection report. This was important because after grouting, the area would be inspected again, and if the area of drummy shotcrete was less than 85% of original area, then secondary and subsequent grouting would take place until, the remaining drummy shotcrete was less than 15% of the original area. Attempts were made to differentiate hollow sound.

- Shallow sound suggests the cavity is close to the shotcrete surface—These were not grouted.
- Isolated deep drummy areas less than 1 m^2. These areas were not grouted.
- Deep drummy sound suggests cavities between the shotcrete and rock interface. These areas were grouted.

Grout Mix Design

Ordinary Portland Cement (OPC) and water are the main ingredients and mixed in proportion to develop stable, non-shrink grout mixes. Grout mixes having ratios of 1:1 and 0.5:1 water: cement ratio by weight were used. Overall, the 1:1 mix was injected in dry areas. Wet areas or large hollow areas were grouted using 0.5:1 water/cement ratio along with plasticizer additive (Sika interplast Z at 1 to 2% by weight). Grout was batched in a colloidal mixer. Specific gravity, viscosity, and bleed were checked regularly to as part of the quality control process.

Closure or Refusal Pressure

Contact grouting to fill voids behind the shotcrete and rock in between shotcrete grouted at net closure injection net pressure of 300 kPa.

CONTACT GROUTING

A minimum of two short length (approximately 300 mm) holes were drilled to bedrock the lower hole(s) used as injection holes while the higher hole(s) were vents. The number of 25 mm diameter holes was increased for larger dummy areas. PVC pipes were then installed, tight fit and hammered in.

Water was circulated through the lower hole at low pressure to wet the shotcrete and rock interface to avoid caking of grout in very dry rock mass and shotcrete. The shotcrete was relatively porous so the water generally was expelled through the shotcrete and through the vent holes. This mandatory washing/wetting was not required in wet areas, which were few. Grouting starts with 100 kPa pressure and advanced up to a closure pressure of 200 kPa (occasionally injection pressure applied up to 300 kPa). Leakages through cracks and joints were controlled by lowering the pressure or doing intermittent grouting. Larger leakage is controlled by caulking when practical using oakum or quick setting mortar. After grouting, check holes were drilled and re-grouting performed as necessary based on criteria described above.

Contact Grouting Procedure

PVC pipes (25 mm diameter) were inserted into the holes in lieu of packers since plastic sleeves with disposable packers were not available. After grouting, the plastic inserts were tied back and later cut off after the grout set.

Grouting was conducted from lower to higher holes. All the grouting holes were open during grouting and closed off as grouting progressed upward.

CONSOLIDATION GROUTING SHEARED, FRACTURED AND ROCKBURST ZONES

Sections/zones of sheared and fractured rock mass and rock burst areas were adequately supported during TBM operation but could not be grouted without confinement support (lining). The boundaries of these disturbed areas were meticulously recorded from detailed logs, daily inspection reports, and Geological Record Sheet (GRS). These areas were marked (spray painted) on the shotcrete surface. Rock bursts areas were consolidated by grouting simultaneously with contact grouting operations.

Consolidation grouting was required to consolidate the fractured/sheared rock mass and new fractures and open joints caused by rock bursts. It was envisaged to create a grouted envelop that extended a distance of 10 m from the tunnel wall Drilling long grout holes through the fractured rock mass using TBM installed drilling equipment was difficult and time consuming. In addition, installing expensive pneumatic packers in the highly fractured rock mass downhole was difficult but not practical due to the risk of grout bypassing the packer and making its subsequent removal extremely difficult.

Site trials convinced all parties that instead of drilling 10 m deep holes, the consolidation grouting envelope could be conveniently developed by drilling short holes set less than 1 m into the rock and installing short mechanical packers within the shotcrete. Grout was injected to a maximum refusal pressure of 500 kPa using grout mix 1:1 (water: cement ratio by weight, Sp. Gravity = 1.4) and 0.5:1 (water: cement ration by weight Sp. Gravity 1.7). It was assumed that open cracks would be effectively filled to create the consolidated envelope extending 10 m around the tunnel, even accounting for friction and gravity losses. Mechanical packers placed near the wall (within 1 m in the holes) safely endured the 500 kPa injection pressure.

Consolidation Grouting Procedure

The arrangements for consolidation required a minimum of four, 38 mm diameter holes drilled a minimum length 1000 mm above the crown. PVC pipes were then inserted as per contact grouting procedure. Additional holes of the same length were drilled where required (see Table 1). Consolidation grouting was carried out predominantly in the crown and above spring line of the tunnel, and to the same net injection pressure criteria of 300 to 500 kPa. The first row of holes was drilled 1 m outside of the rock burst area. The second and subsequent row of holes was drilled at 1.5 m spacing along tunnel along the length of rock burst zone. The final row was drilled 1 m beyond rock burst zone.

Initially consolidation grouting carried out at pressure 300 kPa (P^{max}) or until a maximum of grout absorption of 1000 kg was achieved. However, if pressure remained significantly below 300 kPa and grout absorption of 500 kg was reached rapidly using 1:1 grout mix the intermittent grouting was carried out by pausing for 5 to 15 minutes and thickening the grout to 0.5:1 mix until minimum 300 kPa (net) pressure was achieved. The grout holes were considered grouted when grout mix of 1000 kg was injected. All

Table 2. Contact grouting for both tunnels

Grouting Work Stage	Left Tunnel (Length = 10,498 m)		Right Tunnel (Length = 9,893 m)	
	Shotcrete Surface Requiring Grouting (% of total)	Average Grout Injected (L/m^2 of treated area)	Shotcrete Surface Requiring Grouting (% of total)	Average Grout Injected (L/m^2 of treated area)
1	2.11	23	2.77	22
2	1.78	21	1.80	19
3	1.3	18.9	1.35	18.4
Total contact grouting	1.78%		2.08%	

Table 3. Consolidation grouting for both tunnels (only carried out during Stage 1)

Grouting Work Stage 1	Left Tunnel (Length = 10,498 m)		Right Tunnel (Length = 9,893 m)	
	Shotcrete Surface Requiring Grouting (% of total)	Average grout Injected (L/m^2 of treated area)	Shotcrete Surface Requiring Grouting (% of total)	Average grout injected (L/m^2 of treated area)
Total consolidation grouting	2.31	34.4	2.95	29
	2.31%		2.95%	

interconnected holes were considered grouted by communication at closure pressure. Leakages were controlled by caulking and intermitted grouting.

GROUTING WORKS RESULTS

Table 2 indicates that contact grouting undertaken was similar for both left and right tunnels. The data shows that from Stage 1 to Stage 3, the area that was affected and the amount of grout consumed decreased with each stage in both tunnels indicating grouting effectiveness. This indicates that successive two stages are the most effective.

Table 3 summarizes the consolidation grouting was broadly the same for both tunnels

CONCLUSIONS

The management of the grouting works associated with a shotcrete lining installed by a TBM may not be as glamorous as the excavation and lining installation operations, but it nevertheless forms a critical part of the tunnel completion works. The selection of a shotcrete for permanent tunnel lining offers programmed benefits over placing a cast in-situ concrete lining, but anticipated time savings can easily be lost if excessive shotcrete defects require extensive repair, particularly if the grouting works are poorly planned. In difficult conditions, a permanent shotcrete lining in a tunnel excavated by TBM has a high potential for requiring grouting works. Therefore, an initial assessment should be carried out to identify all suitable mitigation measures that will provide a high quality shotcrete lining, thereby minimizing the need for remedial grouting.

The quantities given in Table 2 and Table 3 are specific to the Neelum Jhelum project but nevertheless can be used as a general guide when planning similar projects with a final shotcrete lining for tunnels excavated by an open gripper TBM.

Grouting and Groundwater in the Greater Arncliffe Area, WestConnex New M5 Tunnels, Sydney, Australia

Jack Raymer ▪ Jacobs Engineering Group, USA
David Oliveira ▪ Jacobs Australia Pty Ltd
Harry Asche ▪ Aurecon Australasia Pty Ltd
David Crouthamel ▪ McMillen Jacobs Associates

ABSTRACT

The WestConnex New M5 includes twin four-lane tunnels as drained structures in moderately cemented sandstone. About 1.3 km crosses a faulted zone with large, open fractures beneath a 40 m deep paleochannel. Pre-excavation grouting, both from surface and underground, has been used to control groundwater inflows and depressurization of the overlying soft sediments. Each type of grouting has certain advantages and limitations; their successful combination aims to meet the project criteria while assisting the contractor to meet schedule and reduce costs. This paper focuses on the results of the work; it compares grout takes from both the surface and underground programs to what has been observed inside the tunnels during construction.

INTRODUCTION

Background

The WestConnex New M5 project includes twin highways tunnels being excavated by roadheader mainly through the Hawkesbury Sandstone Formation. The tunnels are drained structures with an average inflow rate of 1 L/s/km based on previous experience with similar tunnels in the Sydney area and as specified in the project Scope of Works and Technical Criteria (SWTC). Targeted pre-excavation grouting was required along the alignment as means to target this criterion as presented in Pelz et al (2017) and further discussed in this paper.

The study area lies roughly between Ch. 7730 and 8700, in the communities of Arncliffe and Tempe (Figure 1). The tunnels are up to 75 meters below the water table in fractured sandstone. Pre-excavation grouting, both from the surface and underground, is being used to reduce inflows. At the time of writing this paper, the temporary inflows were reported to be approximately 5 L/s/km which exceeds the expected design base case inflow without any grouting of approximately 4 L/s/km, according to the project hydrogeological model developed by others. The purpose of this paper is to summarize and evaluate the grouting program.

Grouting Design Approach

The purpose of the pre-excavation surface grouting program was to fill large fractures with ordinary Portland cement in order to build a foundation for additional pre-excavation underground grouting which was to fill finer fractures and bleed channels with ultrafine cement. The need for this two-phased approach was based on the conditions

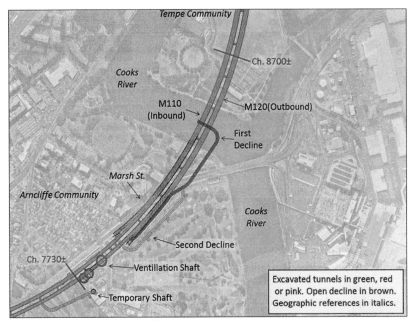

Figure 1. Site features and excavation progress

in the Arncliffe area, which contained numerous large, open, horizontal fractures at tunnel depth, some with apertures greater than 100 mm.

The holes were grouted in stages, from the bottom up, using a single packer. The stages were picked individually for each hole by using an acoustic televiewer to identify the large, open fractures and intervals of solid rock adjacent to them. The packers were typically set in the first interval of solid rock above a large fracture. This reduced the number of stages and reduced the potential for grout to bypass the packer, thus speeding up the process and reducing problems.

The primary hole spacing was set at 20 m because careful analysis of the conditions at Arncliffe indicated that grout could be pushed that far through the large, horizontal fractures. This was borne out in the early primary holes at Arncliffe, where grout migrated into newly drilled holes 45 m away from the point of injection, and groundwater flowed out of holes even farther away.

Another purpose of the 20 m primary spacing was to use the primary holes as exploration holes, to target that grouting was performed everywhere it needed to be. The result of the use of the primary holes was that the ArnWest region (see below for regions) was found to not really need grouting, whereas the area under and to the north of Marsh Street, outside of the original focus zone, i.e., the Arncliffe grouting area (Pelz et al., 2017), was found to need much grouting. Part of the area under and north of Marsh Street is in the Arncliffe Region and part is the RowClub region, depending on where the holes were collared. Once it was discovered that grouting needed to be expanded toward the Cooks River, it was decided to expand the program into the area under the river and across the river in Tempe (named in this paper as the Netball region).

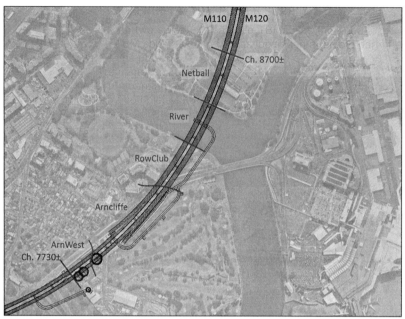

Figure 2. Five regions

Ground Conditions

The site is underlain by a thin layer of Fill, followed by an Alluvial Deposit and the Hawkesbury Sandstone Formation and intersects the NNE to NE trending Woolloomooloo Fault Zone. Ground conditions have been reported in detail in numerous project documents with a summary presented in Yim et al (2017). The following key points are of relevance to this paper.

- The Woolloomooloo fault zone runs through the area from north to south. It is a left-lateral strike-slip system. The fault zone crosses the alignment in the Arncliffe area, and again in the Tempe area. This major feature provides significant hydraulic connectivity in the rock.
- The rock contains both horizontal and vertical fractures. The horizontal fractures appear to be the biggest water producers; some have apertures in excess of 100 mm as identified during detailed design investigations.
- The rock has dual porosity. The fracture network has high permeability but low porosity whereas the sandstone matrix has high porosity and low permeability. The grouting programs were intended to fill the fractures but allow groundwater to seep through the sandstone matrix.

Regions

The study area was divided into five regions for this analysis (Figure 2). These regions were selected based mainly on site conditions and surface grouting, but are also applied to tunnel mapping and underground grouting. The five regions are listed in Table 1. The M110 Alignment is for the eastbound traffic lanes and the M120 Alignment is for the westbound traffic lanes.

Table 1. Five regions

Region	M110 Alignment		M120 Alignment		Combined Length (m)	Percent of Total
	West End	East End	West End	East End		
ArnWest	7730	7840	7730	7810	190	9.4%
Arncliffe	7840	8155	7810	8220	725	35.9%
RowClub	8155	8353	8220	8353	331	16.4%
River	8353	8523	8353	8523	340	16.8%
Netball	8523	8700	8523	8780	434	21.5%
Overall	7730	8700	7730	8780	2,020	100.0%

TUNNEL INFLOWS

Inflow Criterion

The inflow criterion for the entire alignment is an average 1 L/s/km which was set based on previous experience in the Hawkesbury Sandstone Formation. Generally, these inflows mainly consist of diffuse seepage both from fine fractures and the sandstone matrix. These fine fractures and the sandstone matrix are not generally considered practical to grout on a large-scale basis in this rock mass, particularly considering that 1 L/s/km is achieved.

Over most of the alignment, the tunnel is about 35 meters below the water table, but in the study area, the tunnel is about 70 meters below the water table. All other things being equal, doubling the head would double the inflow rate. Therefore, the experience of 1 L/s/km inflow through diffuse fine fractures and the sandstone matrix would really have translated to 2 L/s/km in the study area. On the other hand, the typical expectation in Sydney is that, at depth, the high horizontal stresses would typically promote increased joint closure, thus, reducing hydraulic conductivity which would in turn compensate for the increased water head and maintain the typical inflows at approximately 1 L/s/km. Investigations carried out during detailed design, including a pumping test (Yim et al, 2017) indicated that this was not the case in Arncliffe. The contractional duplex faulting system in Arncliffe caused significant bedding shear ramp structures with dilation which significantly increased hydraulic connectivity.

In addition, over most of the alignment, the water table is recharged by precipitation and small streams fed by precipitation. As groundwater flows into the tunnel over time, drawdown will develop around the tunnel, lowering the water table and causing the inflows to decline over time. The inflow rate will stabilize at water balance when the inflows equal the groundwater recharge along the alignment. This is not the case in the study area because the water table is fed by the Cooks River estuary, which is an infinite source of water unaffected by tunnel inflows. Because the water table cannot decline over time, the inflows cannot decline.

The factors discussed above make the groundwater hydrology of the study area fundamentally different from the rest of the project, which suggested that it was not reasonable to apply a uniform inflow criterion to the entire alignment which was considered consistent with the allowance for only achieving the 1 L/s/km as an average over the entire tunnel length.

Pumping Records During Construction

Groundwater inflow was calculated from the pumping records at the Arncliffe decline. This included inflows from all regions except ArnWest, which was excavated from

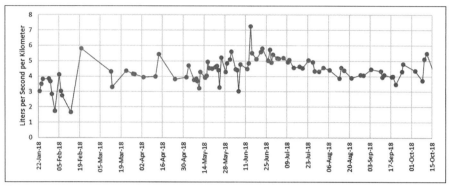

Figure 3. Inflow rate from pumping records over time

the shafts rather than the decline. It also included inflows to the First Decline and Second Decline.

The inflow rate over time is shown in Figure 3 in terms of liters per second per kilometer excavated. As of 15 October, the total inflow was around 11.5 L/s, produced from 2.54 km of tunnel, giving an overall inflow rate of 4.53 L/s/km. It is notable that the inflow per kilometer has been rather steady. The inflows as of 22 January (the first day of records) were coming from the declines.

Table 2. Inflow classes

Class	Criterion	Typ. LPM	L/s/km
A	Dry or No Inflow	0	0.00
B	<0.25 LPM	0.1	0.46
C	0.25 to 0.8 LPM	0.4	1.82
D	0.8 to 2.5 LPM	1	4.55
E	2.5 to 8 LPM	4	18.21
F	>8 LPM	10	45.54

Assessment of Inflows Based on Mapping Records

Face mapping was carried out for every tunnel advance. As of 8 August 2018, there were 393 mapping reports for the study area; The mapped intervals were typically about 3.66 m in length (average excavation advance). Each map report was assigned an inflow class for this analysis based on the narrative descriptions of groundwater in the report and the visually estimated seepage (Table 2). It should be noted that this average mapping interval equates to the average tunnel advance per excavation cycle which provides an indication that the ground conditions were generally worse than anticipated as the anticipated advance rate per cycle was in excess of 5 m. This is also consistent with the heavier rock bolt support installed.

These inflow classes are mainly useful for comparing one part of the tunnel to another. They should be used with caution when considering what the actual inflow might be. Caution is warranted because not all narratives were equally rigorous and because people tend to overestimate small inflows. Furthermore, a substantial number of reports just had "<1 L/min," which falls into Class D, and which is probably an overestimation.

Statistical Distribution on Mapped Inflows

Each alignment was divided into 1 meter increments of chainage. Each increment was assigned an inflow class from Table 2 based on the map descriptions. If more than one class could be applied to a given increment, then the class representing the highest inflow was used, which is also likely to contribute to an overestimation of inflows. More than one report could occur if parts of the heading were mapped at different times due to the excavation sequence. There were 398 increments from alignment M110 and

Table 3. Groundwater inflows from tunnel mapping

GW Class	Increments	Percent	Criterion	Typ. LPM	Local L/s/km	Weighting
A	132	16.3%	Dry or No Inflow	0	0.00	0.00
B	131	16.2%	<0.25 LPM	0.1	0.46	0.07
C	119	14.7%	0.25 to 0.8 LPM	0.4	1.82	0.27
D	286	35.4%	0.8 to 2.5 LPM	1	4.55	1.61
E	135	16.7%	2.5 to 8 LPM	4	18.21	3.04
F	6	0.7%	> 8 LPM	10	45.54	0.34
Totals	809	100.0%	Weighted Average L/s/km			5.33

411 increments from alignment M120. Table 3 shows the percentage of each inflow class in the study area. Figure 4 is a histogram of the inflow classes.

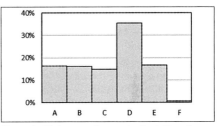

Figure 4. Histogram of groundwater inflow classes

Although the weighted average inflow from Table 3 is somewhat consistent with the pumping records, it should be noted that it does not contain the inflows from the temporary tunnels/declines whereas the pumping records do. The temporary tunnels will be backfilled and plugged at the end of construction, thus, significantly reducing their contribution to the average tunnel inflow. In addition, the weighted average inflow from mapping should not be used to determine if the tunnel inflows meet some criterion as the inflows in Table 3 are probably overestimated due to the practice of using the highest inflow per increment and because Class D is probably over-represented as described above. Actual inflows need to be measured with weirs to demonstrate compliance with the design criteria and the inflows estimates based on mapping is intended only to identify the zones of larger inflows as further discussed below.

Geographic Distribution of Inflows

Figure 5 shows where the different inflow classes occur along the alignment based on tunnel mapping. Areas with Class E and F inflows warrant special consideration and are listed specifically in Table 4. Areas with Class C and D inflows represent the general condition for the tunnel in the study area.

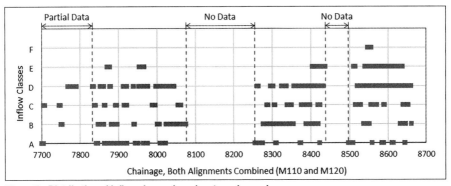

Figure 5. Distribution of inflow classes based on tunnel mapping

Areas of High Inflows

Table 4 lists the specific areas with Class E and F inflows based on tunnel mapping; their locations are shown on Figure 6. In Table 4, the chainage ranges are generalized but the length is based on the actual intervals of Class E and F inflows within those broader ranges, which makes the length less than or equal to the chainage ranges. The liters per minute (LPM) column is estimated by summing the typical Class E and F inflows from the indicated chainage intervals and dividing by 3.66 m, which is the median length of a map report/excavation advance. The L/s/km column was estimated by dividing the LPM column by the length of the interval and converting units. The value of 19.6 L/s/km represents the relative contribution of these intervals to the total tunnel inflow.

The five reaches listed in Table 4 are mainly responsible for the total inflow. Table 5 shows that if the high-inflow areas from Table 4 are subtracted out from the total inflow in Table 3, then the inflows in the rest of the study area should be significantly lower. If Class D inflows are overrepresented, as is suspected, then the rest of the area might be around 2 L/s/km, or possibly less. The analysis in Table 5 is based on incomplete mapping data and is expected to change as more reports are completed and become available.

Table 4. Areas with Class E and F inflows

From	To	Region	Length (m)	LPM	L/s/km
7870	7877	Arncliffe	7	7.7	18.2
7956	7968	Arncliffe	9	9.8	18.2
8405	8438	River	21	23.0	18.2
8512	8514	Netball	2	2.2	18.2
8541	8638	Netball	77	94.0	20.3
		Overall	116	136.6	19.6

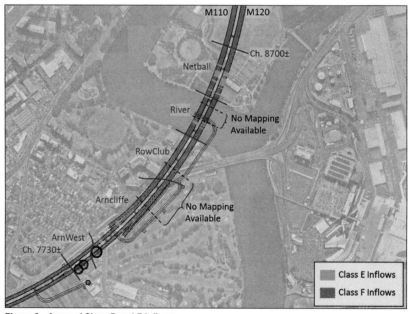

Figure 6. Areas of Class D and E inflows

Table 5. Calculation of inflows for rest of tunnel

Area of Tunnel	Inflow Rate Based on Mapping		Length		Inflow
Total Area (Table 3):	5.33 L/s/km	×	0.809 km	=	4.31 L/s
High-Inflow Areas (Table 4):	19.6 L/s/km	×	−0.116 km	=	−2.28 L/s
Rest of Area (Difference):	2.94 L/s/km	×	0.693 km	=	2.03 L/s

ASSESSMENT OF THE SURFACE GROUTING WORKS

Summary of Works

Figure 7 shows the locations and takes of the surface grouting holes. Table 6 lists the number of grouted holes in each region and whether they were primary, secondary, or tertiary holes. The few quaternary holes that were drilled are grouped with the tertiary holes. Quaternary holes everywhere and tertiary holes in the River area were generally drilled where there had been technical problems with the other holes rather than as part of the general pattern. Holes that were drilled but not properly grouted due to technical problems are not listed in Table 6.

Table 7 lists the tonnes of cement used based on the same groupings as Table 6. Table 8 lists the injected tonnes per hole calculated by dividing Table 7 by Table 6. In Table 8, the tertiary and quaternary holes in the ArnWest and River areas were combined with the secondary holes because there were too few tertiary holes to estimate a statistically meaningful tonnage per hole.

Inspection of Table 6, Table 7 and Table 8 shows that the Arncliffe region dominated the program: it had the most holes, took the most grout, and took the most grout per hole. As discussed in Pelz et al (2017), it should be noted that the surface grouting program was designed specifically for the Arncliffe region and later extended to the other

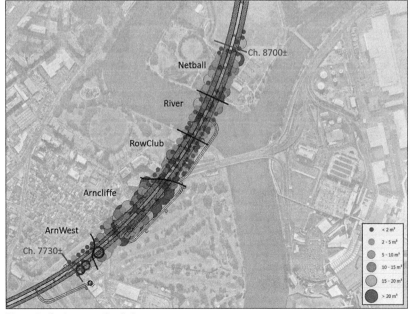

Figure 7. Surface grouting holes

Table 6. Quantity of surface grouting holes

Area	Primary	Secondary	Tertiary	Total
ArnWest	15	3	3	21
Arncliffe	85	66	104	255
RowClub	46	31	31	108
River	42	28	4	74
Netball	59	37	20	116
Overall	247	169	158	574

Table 7. Tonnage of cement in surface grouting holes

Region	Primary	Secondary	Tertiary	Total
ArnWest	14.1	1.9	2.2	18.2
Arncliffe	789.8	391.8	304.3	1,485.8
RowClub	109.0	58.5	62.1	229.6
River	106.1	40.8	3.3	150.1
Netball	277.1	59.3	31.9	368.3
Overall	1,296.0	552.2	403.8	2,252.0

Table 8. Average tonnage per hole

Region	Primary	Secondary	Tertiary	Total
ArnWest	0.94	0.68		0.86
Arcliffe	9.29	5.94	2.93	5.83
RowClub	2.37	1.89	2.00	2.13
River	2.53	1.38		2.03
Netball	4.70	1.60	1.60	3.17
Overall	5.25	3.35	2.49	3.92

areas as information from the grouting boreholes became available which indicated that more grouting was needed. However, most of the take in the RowClub region was in the part that was adjacent to the Arncliffe region and the takes dropped off considerably toward the Cooks River only increasing towards Tempe (Netball zone).

Surface Grouting Performance

Performance for the surface grouting program was gauged by comparing takes from the primary holes to the secondary and then tertiary holes in a given local area. If takes declined significantly, then it was deemed that the grouting program was working. If the takes did not decline as much, then it indicated that adjustments should be made. Furthermore, if holes took less than 1 tonne of cement, it was deemed that further surface grouting work in that location would not be economical, with final tightening being left to the underground program. Additional indication of the surface grouting performance based on field observation and comparison between packer testing before and after grouting was also discussed by Yim et al (2017).

Figure 8 shows the takes per hole based on Table 8. The expected decline is apparent in the Arncliffe, River, and ArnWest regions, but is not as clear in the RowClub and Netball regions considering that tertiary holes were drilled. The unclear decline suggests that the surface grouting program was not working quite as intended in the RowClub and Netball regions. The most likely explanation is that the primary holes may have been too far apart for the geological conditions in those areas likely

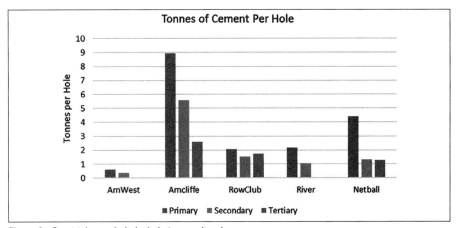

Figure 8. Grout take per hole by hole type and region

imposed by site access constraints which induced a large number of holes to have steep inclinations.

Surface Grouting with Inclined Holes

The original program in Arncliffe was designed based on vertical or slightly inclined (from vertical) holes. Vertical holes have the best probability of encountering subhorizontal fractures. As the holes become inclined, the probability of encountering subhorizontal fractures decreases within the zone of interest, especially at shallower depth. The converse is not true, however, i.e., the probability of encountering vertical fractures does not increase as holes become more inclined, unless the hole is intentionally oriented normal to the fracture plane.

The areas under Marsh Street and the River were not accessible except by drilling highly inclined holes from a limited set of locations. This significantly reduced the probability of encountering fractures, especially given the 20 m primary hole spacing. In the Marsh Street area, some holes were drilled at 45 degrees specifically to encounter shallower fractures. In the River region, beneath Cooks River, there was a large area in the center that could not be reached at all (Figure 9).

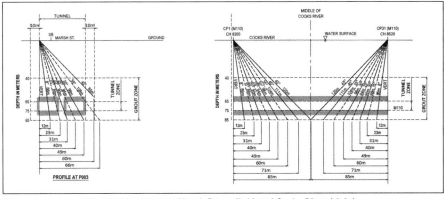

Figure 9. Pattern of inclined holes under Marsh Street (left) and Cooks River (right)

Table 9. Percentage of inclined holes (degrees from vertical)

Region	Inclined	>15°	>20°	>30°
ArnWest	14.3%			
Arncliffe	36.1%	9.8%	7.8%	2.7%
RowClub	85.2%	46.3%	34.3%	13.9%
River	97.3%	71.6%	58.1%	37.8%
Netball	95.7%	70.7%	56.9%	25.0%

Table 9 indicates the percentage of inclined holes in a given area. Holes inclined less than 15 degrees are not much different than vertical holes but the problems multiply as the inclination becomes greater.

Incomplete Stages

Many stages were not grouted to proper completion. Incomplete grouting causes bleed channels to be left open. Bleed channels are small passageways that form in the upper surface of a grout body as the excess water segregates from the cement. These passageways are pathways for water to seep into the tunnel. Because they are small, they are easily missed in subsequent holes. Incomplete grouting also means that more tertiary and quaternary holes may be needed.

Incomplete grouting was a major problem in the Arncliffe region, especially for the primary holes, which had large taking stages. Large-taking stages are more likely to be incomplete because they last several hours, which gives more opportunity for problems to occur.

Stages were terminated before they were complete for three main reasons:

- Site hours: Due to environmental conditions, the Contract required all work to stop by 6:00 pm on weekdays and 1:00 pm on Saturday. Stages had to be stopped by about 5:00 pm whether they were finished or not, to allow time to clean up and leave the site.
- Interference with other construction activities: the Arncliffe site was to be the staging area for underground operations and it required many improvements before underground construction could begin. Surface grouting was just one of many critical activities occurring simultaneously and large-taking stages were commonly stopped prematurely to allow other work to take place.
- Technical problems: technical problems occurred randomly throughout the project, but were much more common in the first few weeks when the primary holes in the Arncliffe area were being grouted. The situation improved as better equipment was brought in and crews gained experience.

UNDERGROUND GROUTING WORKS

Underground grouting was performed in "fans,," where a fan consists of several holes drilled ahead and outward from the perimeter of the tunnel face. The holes were designed to be 30 meters long and each fan was supposed to begin 10 meters before the end of the previous fan, to create a 10 meter grouted plug between the face and the ungrouted zone. Based on the tunnel profiles, the primary holes were approximately 3 to 5 meters apart at the face and 5 to 9 meters apart at the far end. Secondary holes were shown between the primary holes. The intent was to use ultrafine cement in the underground program, unless large inflows from the drill holes indicated that ordinary Portland cement should be used first, followed by ultrafine cement

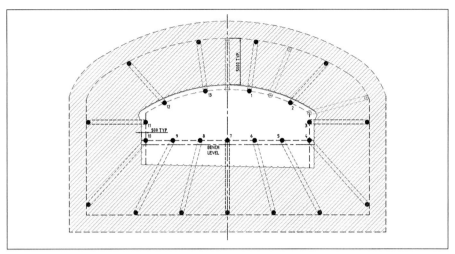

Figure 10. Typical hole pattern for underground grouting

Table 10. Underground grouting

Region	Fans	Take (m^3)	Take/Fan	Reported
ArnWest	5	10.3	2.1	50%
Arncliffe	27	343.6	12.7	94%
RowClub	13	75.0	5.8	63%
River	20	179.5	9.0	100%
Netball	21	287.9	13.7	100%
Overall	86	896.4	43.2	

in the secondary holes. Figure 10 is the typical layout from the design drawings. The locations and takes of the underground fans are shown in Figure 11 with similar colour coding to Figure 7.

The data available consisted of the locations and the total volumetric take from each fan, as posted in the project GIS system. There was no information on primary, secondary and tertiary holes, and no information on mixes or whether the grout consisted of ordinary Portland cement or ultrafine cement (both were specified in the design.)

Table 10 lists the fan takes in cubic meters of grout according to the five regions. "Reported" is the percentage of the region for which underground data had been reported when this report was written.

DISCUSSION

Comparison of Takes

The surface and underground grouting programs have to be normalized for units and coverage in order to compare them. Surface grouting was tracked by tonne of cement; underground grouting was tracked by cubic meters of grout. The surface grouting was complete and fully reported; the underground program was not.

- Surface grouting takes were normalized to cubic meters by multiplying the tonnes by 1.32 cubic meters of grout per tonne of cement. This factor is

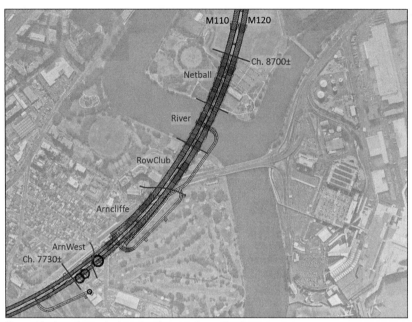

Figure 11. Location and takes of underground grouting fans

Table 11. Comparison of surface and underground takes

Region	Surface Grouting		Underground Grouting		Norm.
	Tonnes	Norm. m^3	Take (m^3)	Norm. m^3	SG/UG
ArnWest	18.2	23.9	10.3	20.6	1.16
Arncliffe	1,485.8	1,955.0	343.6	365.5	5.35
RowClub	229.6	302.1	75.0	119.1	2.54
River	150.1	197.5	179.5	179.5	1.10
Netball	368.3	484.6	287.9	287.9	1.68
Overall	2,252.0	2,963.1	896.4	972.7	3.05

  based on the most commonly used mix of 1 part cement to 1 part water by mass.
- Underground takes were normalized by dividing the takes from each region by percent reported in Table 11. This will increase the underground volumes to what would be expected if the work were fully complete and reported.

Table 12 is a comparison of the normalized takes to the size of each region. The region size is rendered as the percentage of the total length of the study area, from Table 1. The takes are rendered as the percentage that each region contributed to the overall takes, based on Table 11.

For the project as a whole, the surface program placed three times the amount of grout as the underground program. Had the surface grouting not been performed, this grout would have to have been placed from underground in order to control the inflows. In rough terms, this means that surface grouting program probably reduced the cost and time of the underground program by 75 percent.

The savings were greatest in the Arncliffe Region, where the large, open horizontal fractures took an enormous amount of cement. Table 11 provides similar conclusion based on a normalized SG/UG ratio in excess of 5 particularly considering that Table 12 also shows that the Arncliffe region took the most grout both from surface and underground grouting

Table 12. Comparison of takes to region length

Region	Length	Norm SG	Norm UG
ArnWest	9.4%	0.8%	2.1%
Arncliffe	35.9%	66.0%	37.6%
RowClub	16.4%	10.2%	12.2%
River	16.8%	6.7%	18.5%
Netball	21.5%	16.4%	29.6%

works. Grouting these from underground would have been challenging because the horizontal fractures would have been difficult to find with subhorizontal grout holes. If one of these large, open fractures had been missed, thousands of liters per minute could have poured into the tunnel.

The savings were much less in the ArnWest and River regions. In hindsight, the ArnWest Region did not really need the surface grouting program because the fractures were likely few and fairly tight. In the River Region, there was the area in the middle that could not be adequately reached by the surface program. Closer examination of the data indicates that 81 percent of the underground take in the River Region (146 m^3) was from this unreachable area.

Netball Region

The Netball Region was problematic because it accounts for two-thirds of the highest inflows for the entire study area (Table 4.) The Netball Region also had a low ratio of surface take to underground take (Table 11) and did not show the proper decline from secondary to tertiary holes (Figure 6). This suggests that the surface grouting program in the Netball region might not have performed as well as it should have.

Close examination of the mapping reports indicates that most of the inflows in the Netball region (like everywhere) come from the horizontal fractures associated with subhorizontal shears or bedding planes. These horizontal features are difficult to hit with the subhorizontal underground holes, and also with highly inclined surface grouting holes within the targeted zone. Table 9 shows that nearly all of the holes in the Netball region were inclined and about half were highly inclined which was driven by site access constraints from where the drilling works could have been collared/initiated.

CONCLUSIONS

The original intended inflow criterion of 1 L/s/km is probably not attainable for the study area with 2 L/s/km being a more appropriate expectation given the depth of the tunnel and the effects of the contractional duplex faulting in Arncliffe which considerably increased the hydraulic connectivity in the area.

The 2 L/s/km revised criterion was met where the surface grouting was performed under optimum conditions (e.g., vertical wholes). Where the surface grouting had problems, the 2 L/s/km criterion was typically not met. This indicates that successful surface grouting was essential to controlling groundwater inflows and that underground grouting alone would not have controlled the inflows.

The surface grouting program was most successful in the main Arncliffe region, where the rock conditions were the worst; two thirds of the grout in the Arncliffe region was placed from the surface. The surface grouting program was less successful in the Netball region and the River region, where the number of inclined holes was significantly higher due to access constraints, the proportion of grout placed from the surface was much lower and where the proportion of excessive inflows was much higher.

For the project as a whole, the surface program placed three times the amount of grout as the underground program. Had the surface grouting not been performed, this grout would have to have been placed from underground in order to control the inflows. In rough terms, this means that surface grouting program probably reduced the cost and time of the underground program by 75 percent.

ACKNOWLEDGMENT

The authors thank Iris Yim and John Wiley for their contribution to this paper.

REFERENCES

Pelz, U; Casado, J. Asche H.; Raymer, J; Crouthamel, D. and Fidler, S. (2017). Geologically Targeted Pre-Excavation Grouting Along the WestConnex M5 Tunnel, Sydney, Australia. Rapid Excavation and Tunneling Conference (RETC 2017), San Diego California.

Yim, I; Casado J.; Asche, H; and Raymer, J. (2017). Effect of Surface Grouting in Arncliffe Site of WestConnex Stage 2 Project. 16th Australasian Tunnelling Conference, 30 October–1 November 2017.

Quality Control of Secant Piles and Jet Grouting for the Ohio Canal Interceptor Tunnel

Stanley L. Worst ▪ Schnabel Foundation Company
Matthew J. Niermann ▪ Schnabel Foundation Company

ABSTRACT

Two circular secant pile shafts were installed along the tunnel alignment for the OCIT project in Akron, Ohio. Verticality and concrete volumes were measured on all piles to verify compliance with the design, which consisted of a finite element model with multiple openings using unreinforced concrete secant piles. Double fluid jet grouting was used at the TBM launch portal to create a soil-cement starting block. Automated grouting procedures and continuous data collection were used with confirmatory borings to ensure quality. This paper will discuss the design, construction and quality control of the secant piles and jet grouting for this project.

INTRODUCTION

Project Background

Many Midwest and Northeast cities have Combined Sewer Overflow (CSO) systems which discharge pollutants into waterways during wet weather events. One way to remedy this issue is to provide a temporary overflow tunnel. A temporary overflow tunnel connects to the combined sewer system and collects the overflow during wet weather events, storing the wastewater until the event has subsided. After the storm event, the stored wastewater can be pumped to the treatment facilities and processed as normal.

The Ohio Canal Interceptor Tunnel (OCIT) project was performed as a result of the Environmental Protection Agency's Consent Decree with the City of Akron, Ohio. This Consent Decree required modification of the existing Combined Sewer Overflow system. The resulting project consisted of a 27-foot diameter tunnel with multiple points of entry to capture the sewer overflow and prevent it from entering waterways. The tunnel is over 6,000 feet long and is as deep as 150 feet below the ground surface; see Figure 1.

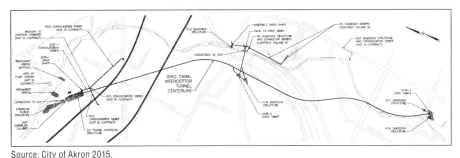

Source: City of Akron 2015.
Figure 1. Project layout

As part of the project, two circular secant pile access shafts were installed. One shaft, 46 foot inside diameter, was at OCIT-2, roughly the half-way point. The other shaft, 51 foot inside diameter, was at OCIT-3, one end of the tunnel. These were temporary Support of Excavation (SOE) structures to facilitate excavation and casting of the permanent drop structures. At the OCIT-1 location the tunnel started very close to the ground surface, daylighting in a relatively shallow excavation. Immediately behind a sheet pile wall, a 40 ft × 50 ft × 42 ft block of ground modification was installed. The purpose of the ground modification was to stabilize the surrounding soil, especially with low cover to the ground surface above, so that when the Tunnel Boring Machine (TBM) started the boring operation there was minimal disturbance to the surrounding ground.

Geotechnical Information

Akron, Ohio is a typical midwestern city which consists of a variety of glacially deposited soils and soft to hard sedimentary rock. The soils, rock and water table vary between the three different locations of the secant pile shafts and jet grout block.

At the OCIT-2 secant pile shaft location, the upper 40 feet of soil was fill material which consisted of sandy silty clay along with silt and poorly graded sand. Blow counts varied between 10 and 27 blows per foot. Underlying the fill material was a layer of outwash and lacustrine medium dense silt with sand, which sat on top of shale bedrock. The shale contained laminations of sandstone and siltstone. The RQD of the shale was 83–99% and the unconfined compressive strength averaged approximately 11,000 psi. Groundwater was encountered above the top of rock. However, water infiltrating into the secant piles during drilling was sealed off by temporarily socketing the drill casing into rock. Shale is a fine-grained, sedimentary rock which is drillable with conventional tooling.

At the OCIT-3 secant pile shaft location, the upper 20 feet of soil was fill material which consisted of sandy clay and silty gravel with blow counts varying between 0 and 69 blows per foot. Underlying the fill material was a layer of very loose organic silt, medium dense silt and poorly graded sand all below the water table. An approximate six feet layer of sandstone was the initial rock encountered, with shale underneath. The sandstone had an 85% RQD and an average unconfined compressive strength of approximately 6,000 psi. The shale had an RQD of 47–53% and an average unconfined compressive strength of approximately 9,000 psi.

At the jet grout block location, OCIT-1, rock was not within the jet grout zone, which was the upper 42' of soil. The soils consisted of mainly granular material, ranging from silty fine sand to poorly graded sand. The blow counts were anywhere between zero and 21 blows per foot. Fill material consisting of brick fragments and abandoned clay pipe, concrete, wood, metal and rock were also encountered in the upper portions.

SECANT PILES

Design Basis and Concepts

The basic design concept for the circular secant pile access shafts is to analyze a vertical cylinder with horizontal hoop stresses. Overlapping secant piles laid out in a circular pattern form a cylinder of concrete which has an effective thickness. This effective thickness theoretically varies with depth. Figure 2 shows the concept of effective thickness, which is the dimension measured across the secant piles where two adjacent secant piles overlap. Vertical drilling is subject to minor deviation in the horizontal direction. Based on previous experiences of measuring this deviation,

Quality Control of Secant Piles and Jet Grouting 685

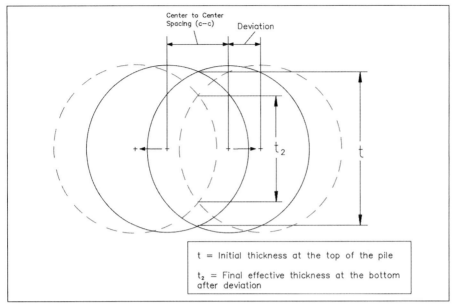

Figure 2. Secant pile deviation and effective thickness

Schnabel Foundation Company chose a verticality tolerance of 0.5%, or 1 in 200, for the design of this project. Due to this verticality tolerance, there exists the possibility of a thinner effective thickness with depth. This theoretically thinner effective thickness is accounted for in the design by analyzing the pressure on the cylinder immediately above the top of rock, where the pressure is the greatest.

Lateral forces on the cylinder are present in the form of Rankine earth pressures; linearly increasing with depth. Due to secant piles being very stiff and rigid, and that the cylinder is acting in compression, very little movement is allowed of the overall secant pile shaft. This lack of movement prevents active pressure from forming, resulting in a higher at-rest pressure. Therefore, K_o is used in the earth pressure calculations for this type of system. Hydrostatic water pressure is also added to the design and is a linearly increasing pressure with depth.

The critical case for Rankine earth pressure is where the pressure is the greatest, which is at the top of rock. For the temporary condition, rock is assumed to have a high friction angle and unconfined compressive strength, exerting a much lower lateral force on the cylinder at this elevation. The highest earth pressure at the top of rock also corresponds with the thinnest effective thickness of the overlapping secant piles. To account for this, the starting overlap of the secant piles at the ground surface is increased from what is required at depth. Using the verticality tolerance of 0.5%, which Schnabel has confirmed it can achieve through experience, the amount that the overlap is increased can be calculated.

When viewed in plan-view, the lateral force that is exerted on the cylinder can be represented by a pressure, "P." As shown in Figure 3, that pressure is distributed over the diameter of the cylinder. It is resisted by the axial thrust in the hoop at that elevation. The thrust is represented as "T" on each side of the hoop. Therefore, equation 1 represents force equilibrium.

$2T = P \times d$ (EQ 1)

The resulting thrust can be analyzed to determine the compressive stresses in the concrete. For this project, unreinforced concrete secant piles were used to resolve these stresses. The secant piles were 1,180 mm diameter above the top of bedrock and 1,060 mm diameter in bedrock.

However, this simple form of analysis is only appropriate if the loading is symmetric. Unbalanced loading as a result of large crane surcharges or localized stresses from large openings in the cylinder present different design challenges. These situations can be analyzed using the Finite Element Method. Figure 4 shows an example of how openings account for different levels of localized stresses which may or may not control the design. For the OCIT-3 shaft, there were four different break-out openings within the secant pile shaft that were analyzed with finite element software. These openings were approximately 36 inches, 84 inches and 72 inches in diameter along with one rectangular opening that was thirteen foot × eight foot.

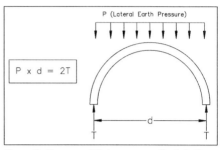

Figure 3. Force equilibrium on a vertical shaft

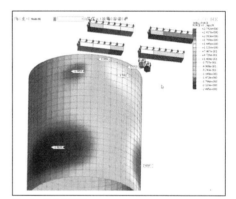

Source: Gamal 2016.
Figure 4. Finite element analysis of unbalanced loading on circular secant pile shaft

Quality Control and Construction

The verticality measurement of each individual secant pile is of utmost importance as the entire design concept is based on a certain effective thickness. Without having any additional means of lateral support, the overlap of each secant pile must be met, and therefore it must be known prior to concreting the hole. There are a few different devices which can be used to measure the verticality. For this project, Schnabel Foundation Company used the Jean Lutz Prad system. This system consists of a wireless device which is mounted to the drill tooling. It is lowered to the bottom of the hole and raised back to the top of the hole with a controlled rotation and lift rate. This produces a deviation profile every few feet which can be seen by the operator in the cab of the drill rig. It also produces a printer-friendly version for submittal and documentation purposes. An example of the in-cab display is shown in Figure 5. Once each pile's deviation is known, the effective thickness at the top of rock can be calculated to check that it is equal to or greater than what was designed. This data can also be plotted to show the deviation and resulting effective thickness at various depths.

To achieve the tight 0.5% verticality tolerance, Schnabel Foundation Company uses large, high-torque rotary drill rigs with flush-joint segmental casing. The casing is double-wall and very rigid, which makes it extremely difficult to deviate once it is in the ground. Advancing the casing along with the auger helps to maintain alignment. Also, a guidewall is used at the ground surface to act as a fixed point for the casing.

The guidewall is constructed with lightly reinforced concrete and is poured to tight tolerances, which keeps the casing from wandering off during the initial stages of drilling. The guidewall is typically a few feet thick and is poured a few feet on either side of the template. The template for the guidewall is generally constructed with either steel forms or Styrofoam.

Figure 5. In-cab display of verticality measurement

The other factor that can affect the effective thickness of the secant pile shaft is the actual diameter of concrete in each pile. This can be indirectly measured by tracking the volume of concrete placed in the pile. If a volume lower than the theoretical volume is measured, there is likely a local cave-in of soil which prevented concrete from filling up the full pile diameter. This can be avoided by extending the temporary casing full depth and using proper tremie concrete practices. Confirming the concrete mix design is fluid enough will also ensure that the concrete inside the temporary casing will fill the annular space left during casing extraction. Concrete volume measurement can also be an indicator of bulges which will need to be chipped away at later date during the shaft excavation. If volumes much higher than the theoretical volume of the pile are recorded, there exists the possibility of weak soils or voids in rock which allowed for extra concrete to be placed. On this project, there was a layer of low blow-count organic silts below the water table which yielded takes of over 40% of the theoretical volume of concrete per pile on some piles. Extra concrete does not affect the design and can be removed during shaft excavation if it encroaches into the permanent structure.

JET GROUTING

Design Basis

The overlying concept of jet grouting is to inject fluid grout into the ground at high velocity so that it erodes and mixes with the in-situ soil, producing a column of jet grouted material which is stronger and less permeable than the original soil. The resulting soil-cement strength, hydraulic conductivity, and column diameter are functions of the rate and concentration of cement injected, in addition to the time spent jetting a certain area.

The concentration of cement is controlled by the mix design, which for neat cement grout is the water/cement ratio. This ratio is the amount of water divided by the amount of cement per unit volume, as measured by weight. A 1:1 w/c ratio has equal weights water and cement. Typical jet grout w/c ratios are in the range of 0.8 to 1.8.

The amount of cement injected into the ground is controlled by two other factors; the flow rate and the lift rate. The flow rate is usually measured in gallons per minute and is a function of the pump capacity. The lift rate is usually measured in inches per minute and can be controlled at the drill rig. A zone of soil is treated for a longer time with a slower lift rate. This allows more cement to be injected as well as a potentially further radius of grout penetration into the soil.

The rotation rate during jetting is another variable which can be adjusted. To create cylindrical soil-cement columns, the rotation rate is generally calibrated to the lift rate

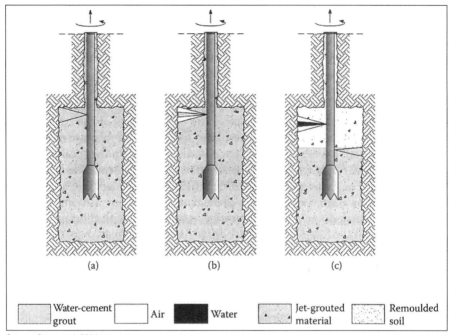

Source: Croce et al. 2014
Figure 6. (a) Single fluid; (b) double fluid; (c) triple fluid

to ensure the full circumference is treated. This can be achieved with a single nozzle and full rotation before lifting, or with two nozzles located 180 degrees apart and a half rotation before lifting.

Adjusting the w/c ratio, lift rate and flow rate will determine the amount of cement which is injected into the ground. Adjusting the lift rate, flow rate and jetting pressure will help determine the diameter of the soil-cement column. The selection of these parameters is based on the existing soil conditions. The fines content, the relative density of the soil and particle size are all factors which have an impact on the column diameter which can be achieved.s

Single, double or triple fluid jetting can be performed. Single fluid jetting is when just cement grout is injected into the ground. Double fluid jetting adds a shroud of compressed air around the grout injection point (see Figure 6). Triple fluid jetting utilizes double fluid injection of high-pressure water and compressed air, followed by injection of cement grout at lower pressure. Double fluid jetting was performed on this project.

Quality Control and Construction

On almost every project, prior to jet grout production work, a test program is performed to verify the many different parameters to be used as well as to confirm the resulting soil-cement column diameter and strength. To confirm the column diameter, it is best to excavate around a test column and visually examine and measure the geometry first-hand. However, this is usually difficult if not impossible to do when dealing with limited space on a construction site and deep columns. One alternate method of checking the column diameter is to drill and retrieve samples from the centroid of a three-column layout. In this scenario, three individual test columns are installed in a triangular layout in which all three overlap slightly in the middle. See Figure 7.

A borehole can be drilled in the centroid of this three-column layout, retrieving continuous split spoon samples. The blow counts from driving the split spoon can be compared with soil borings taken prior to jetting to confirm that the ground has been strengthened. Also, the retrieved samples, although disturbed, can be sprayed with a phenylthalene indicator, which turns a bright color in the presence of cement. This provides two ways of confirming that the column diameter has been achieved. Coring is also sometimes used to retrieve samples for lab testing. However, coring of relatively low strength material has proven to be unsuccessful in retrieving intact samples, especially in granular material containing gravel or cobbles. The rotary-wash action of the coring tends to erode low strength soil-cement; and, if any larger pieces of sand or gravel are present, they can get caught on the core barrel teeth, causing the sample to crumble. For shallow columns where exhumation is possible, visual examination is the best means of confirmation.

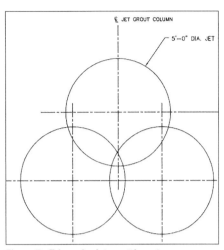

Figure 7. Triangular jet grout layout

Wet-grab samples are commonly used as a means of testing the soil-cement material for strength and hydraulic conductivity. These samples can be taken at depth in recently installed columns that are still fluid, or from spoils at the ground surface. To retrieve wet grab samples at specific depths, a device is needed to be lowered into the jet grout column. This can prove difficult if the jet grout column does not extend all the way up to the ground surface. For example, jet grouting may only be required for a 10' thick bottom plug which is 50' below the ground surface. The drill hole is typically around six inches in diameter, which makes it very difficult to insert a sampling device from the ground surface. Also, if a working slab (concrete) is present, it may be impractical to insert a larger sampling device through the drill hole made in the working slab. In these instances, wet grab samples retrieved from the ground surface have proven to be sufficient for predicting strength and hydraulic conductivity. Cylinders are made from the spoils collected at the ground surface while jetting at specific depths. Since the in-situ soil is being turbulently mixed with the cement grout being injected, it is assumed that the return spoils are of similar consistency to the jet grouted material left in the ground. In general, comparisons have shown that wet grab samples achieve lower unconfined compressive strengths than cores taken post-production of in-situ columns. This is most likely a result of the enhanced in-situ curing environment as well as fluid head pressures that may facilitate pressure filtration and subsequent lowering of the in-situ water/cement ratio of the soil-cement.

The design of jet grout column strength is based on the 28-day strength of the soil-cement, similar to concrete design. During a test program, it is impractical to wait 28 days before proceeding with production. Therefore, strength vs. time graphs are used to predict 28-day strength from 3 or 7-day strength test results. As can be seen from Figure 8, for this project the 3-day strength was up to 66% of the 28-day strength. These early strength test results enabled production jet grouting to begin

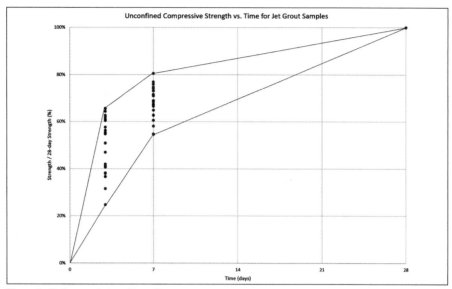

Figure 8. Soil-cement strength vs. time

with confidence that the 28-day strengths would be achieved. This was the case both in the pre-production test program as well as during production. Cylinders were always tested at 28 days to confirm these assumptions. There is also a wide range of strengths seen in jet grouting due to the inherent variability of the in-situ soils.

There are numerous input parameters that are measured for jet grouting. The following parameters are recorded with sensors and measuring devices that are mounted on the drill rig and automatically recorded for each column: lift rate, rotation speed, grout pressure, grout flow, air pressure, and air flow. These can be seen in graphical format in Figure 9. This gives an easy visual representation of the quality control for each jet grout column. Although the measurement of these parameters do not guarantee that the resulting column will be of sufficient diameter and strength, it verifies that the same parameters are being used which were shown to be effective in the pre-production test program.

CONCLUSION

Both secant piles and jet grouting are underground works which cannot be immediately inspected first-hand. They require indirect methods of sampling, testing and other quality control measurements to gain a level of confidence that what was installed was intended and meets the design requirements. For secant piles, they can eventually be visually observed during the excavation of the shaft to confirm geometry and that there are no voids or soil inclusions which might compromise the design. The jet grouting may or may not be exposed in the final condition for visual inspection. Although it was not performed on this project, soil borings can be taken post-production to confirm the elevation of ground treatment.

Due to the relatively low strengths that jet grouting can produce as well as the depths at which jet grouting is performed, some forms of quality control and testing are not always appropriate. These include coring, in-situ water-pressure testing and at-depth

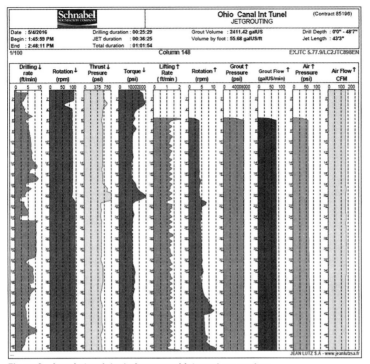

Figure 9. Jean Lutz printout of measured jet grout parameters

wet grab sampling. These may be used on a job-by-job basis depending on the ground conditions, strength requirements and depths grouted.

Secant piles can be a very effective form of earth retention and water cutoff. They can be socketed into rock to seal off groundwater and can form self-supporting circular access shafts without any internal bracing or external tiebacks. Secant piles were successfully used on the OCIT project for these purposes.

Jet grouting can be used as a form of ground modification as well as to seal off water on the sides or bottoms of excavations. Although jet grouting is not always able to be directly inspected, there are ways of providing quality control which show conformance with the design intent. Jet grouting was used successfully on the OCIT project as a means of ground modification.

REFERENCES

City of Akron. 2015. *Ohio Canal Interceptor Tunnel Contract Drawings*. Akron, Ohio. Prepared by DLZ Ohio, Inc.

Croce, P., Flora, A., and Modoni, G. 2014. *Jet Grouting*. Boca Raton, FL: CRC Press.

Gamal, Mohamed. 2016. *Design of OCIT-2 and OCIT-3 Drop Shafts*. Denver, Colorado. Prepared by Brierley Associates.

Siebert, J.A. 2015. *Ohio Canal Interceptor Tunnel Geotechnical Baseline Report*. Akron, Ohio. Prepared by DLZ Ohio, Inc.

PART  11

Hard Rock TBMs

Chairs

Jesse Salai
J.F. Shea

Nick Chen
Jacobs Engineering

Rand Park Stormwater Diversion Tunnel—Planning and Designing to Address Stormwater Flooding for Downtown Keokuk, Iowa

Mahmood Khwaja ▪ CDM Smith
Gregory Sanders ▪ CDM Smith
Michael S. Schultz ▪ CDM Smith
David R. Schechinger ▪ Veenstra & Kimm
Mark Bousselot ▪ City of Keokuk

ABSTRACT

The Rand Park Stormwater Diversion Tunnel is a proposed 910 m long, 3 m internal diameter rock tunnel that is being planned and designed to satisfy the consent decree commitments and to manage stormwater flooding for the downtown City of Keokuk, Iowa. At an approximate average depth of 24.5 m, the tunnel will be constructed through shale and will convey stormwater under Rand Park to an outfall discharging into the Mississippi River and will help address combined sewer overflow and street flooding concerns. This paper presents a synopsis of the project background, the regional geology and the subsurface conditions, challenging site conditions, evaluation of tunnel alignment/profile and the selection process for the optimal tunnel construction approach.

PROJECT BACKGROUND

The project is located in the Keokuk, Iowa. Keokuk is southernmost city in Iowa, where the Des Moines River meets with the Mississippi River—See Figure 1. The city is named after the Sauk chief Keokuk, who is thought to be buried under Rand Park. Through an agreement with EPA Region 7, the City of Keokuk (Keokuk), Iowa, is improving its combined sewer system and reducing discharges of hundreds of millions of gallons of raw sewage to the Mississippi River and its tributaries. Under the consent order, Keokuk will implement a long-term control plan (LTCP), no later than December 31, 2030.

The selected method for separation of the city's storm water and sanitary sewers is to construct a new storm water system that will convey stormwater from inlets within the city to a new stormwater diversion tunnel drop shaft located west of Rand Park. From that location, a new diversion tunnel will be constructed to discharge the stormwater into the Mississippi River. The east end of the tunnel will be located at the base of steep unstable slope along the west bank of the Mississippi river. The west end of the tunnel and the drop shaft will be located at the intersection of 14th Street and Fulton Street and a central construction/drop shaft will be located within Rand Park.

GEOLOGICAL SETTING

The project site lies within the dissected till region of the southern Iowa drift plain near the axis of the north south trending Mississippi River Arch structural feature. The Mississippi River Arch was formed during the middle to late Pennsylvanian age by tectonic activity along the Nemaha structural belt.

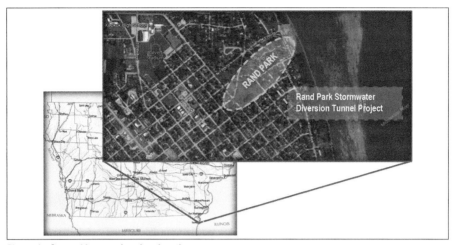

Figure 1. State of Iowa and project location

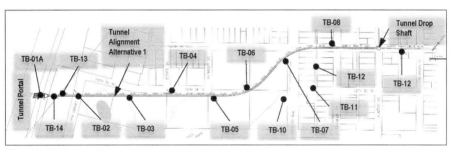

Figure 2. Boring locations

The surficial geology in the dissected till region consists of alluvial and loess soils overlying glacial till. The glacial till consist of till deposited during the Kansas and Nebraskan stage of glaciation. During the Wisconsian age, after the end of the Kansan glacial stage the surface was exposed to stream erosion and loess deposition. The result is dendritic drainage pattern is like tree branches in the of rolling hills and valleys. During this erosion period loess was deposited over the glacial till, providing additional relief particularly on leeward hillslopes and along the borders of stream valleys. The loessial soils are normally classified as low-plastic silty clays (CL). Underlying the glacial deposits is bedrock composed of the Mississippian age St. Louis, Spergen, Warsaw and Keokuk Formations of the Augusta Group

GEOTECHNICAL INVESTIGATION

As a part of the preliminary design a comprehensive geotechnical exploration and testing program was developed, consisting of 14 borings, in-situ testing and laboratory testing. The boring locations, shown in Figure 2, are strategically positioned along the initially proposed alignment, as well as around an alternate alignment. Borings locations were surveyed, and field marked, prior to drilling. All borings were advanced to the top of rock using solid stem augers or rotary wash methods. Soil samples were obtained with a split barrel sampler driven by a falling weight in general accordance with ASTM D1586 (Standard Penetration Test). After competent rock was encountered the drilling was continued with NQ2 rock coring methods. The field program was supervised on a full-time basis by a tunnel engineer or a geologist. The rock cores were described in logs and interpreted to depict the sub-surface profile.

Upon completion of the drilling a geophysical survey of the borehole profile was conducted using an acoustical tele-viewer and a mechanical caliper to map the rock structure and measure the borings deviation from plan. Televiewer images can be used to detect bedrock discontinuities in boreholes to determine their frequency, depths and orientations. All discontinuities subjected to orientation analysis were classified as fractures/joint, shale seams/inter-beds and bedding contacts and a discussion of the rock structure is provided in the following sections of this paper. The verticality of all of the boreholes were within the industry standard 2° deviation.

Geophysical testing at each boring was performed and a report was prepared, summarizing the joint spacing and orientation of the rock at the perimeter of each borehole. This information was be used during the preliminary design in developing the model to confirm the ability of the different tunnel support types relative to the expected rock conditions.

Packer testing was performed in select borings. The length of rock being tested typically consisted of 3 m intervals above, at and below the anticipated vertical tunnel alignment. The pressure was held for 10 minutes and the volume of water pumped into the test zone to maintain the pressure was monitored. Results of each test were used to calculate the permeability of the rock. This data will be used to provide anticipated groundwater inflows for the tunnel and is discussed in later section of this paper. Upon completion of the field exploration program all borings were backfilled with a cement grout or a fully grouted vibrating wire piezometer was installed.

The sub-surface profile, and the in-situ and laboratory testing were used as the basis of design for the Rand Park Stormwater Diversion Tunnel.

Bedrock Geology

The geologic unit nomenclature used for this project is based on geological descriptions developed by the Iowa Geological and Water Survey, 2010. One of the more authoritative description on the geologic formations for the Iowa region is provided by Van Tuyl [4]. Figure 3 provides a synopsis of what is provided in the report. The geologic units are assigned to a particular grouping based on our interpretation of the depositional environment, stratigraphic relationships, and engineering properties. From bottom upward in the geological depositional process the local geology anticipated to be encountered consists of Keokuk Formation, Warsaw Formation, St. Louis Formation and Glacial Deposits and Recent deposits. Although the Spergen Formation is known to be present, it could not be specifically identified within the rock cores recovered during the field investigation.

The St. Louis Formation rest on top of either the Spergan Formation or the Warsaw Formation, and ranges in thickness anywhere from 1.2 m to 27.5 m. Composition of the St. Louis Formation is primarily limestone with varying secondary lithology, which usually consists of sandstone, but can also be dolomite or chert layers.

The limestone in an unaltered condition is thin-bedded, gray, fine grained and dense. Locally it grades laterally in part into brownish massive dolomite or into sandstone. In some of its exposures it is brecciated near the bottom of the formation.

Literature review indicates that the Spergen Formation consist of varying composition ranging from crinoidal limestone, dolomitic limestone, argillaceous dolomite, and

SYSTEM	GROUP	FORMATION	SECTION	THICK-NESS	LITHOLOGIC CHARACTER
				Feet	
MISSISSIPPIAN	MERAMEC	STE. GENEVIEVE (PELLA)		55	Fine-grained bluish sandstone at base succeeded by bluish shale. Fine-grained gray limestone above.
		ST. LOUIS		40-60	Fine-grained gray limestone chiefly. Lower division locally brownish and dolomitic. In places brecciated.
		SPERGEN (BELFAST)		0-35	Crinoidal limestone grading into arenaceous dolomite.
		WARSAW		20-65	Interbedded shales and gray fossiliferous limestone. Geodes in lower part.
	OSAGE	KEOKUK		63	Gray to bluish gray crinoidal limestone interbedded with shale in upper part. Cherty in lower part.
		BURLINGTON		71	Gray, crinoidal, cherty limestone interbedded with brownish fine-grained cherty magnesian limestone.

Figure 3. Generalized columnar section (Van Tuyl [4])

sandstone. The Spergen formation in southeast Iowa grades laterally into the Warsaw or the lower St. Louis Formation due to a regional unconformity. Therefore, the formation has a limited areal extent in Iowa, thinning to the northwest. As a result of the unconformity, the Spergen Formation may rest upon either the Upper or the Lower Warsaw or be entirely absent with the St Louis Formation resting on the Warsaw Formation. The thickness of the formation thus ranges from 0 m to 10.5 m.

The Warsaw Formation is divided into a lower geode bearing argillaceous dolomite and shale unit and an upper shale unit. The upper portion of the Warsaw formation is usually represented by an argillaceous or slightly calcareous shale and a 1 m to 2.1 m dolomitic limestone layer.

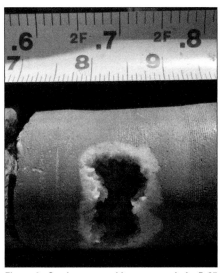

Figure 4. Geode recovered in core sample for B-05

The lower Warsaw formation is represented by fossil bearing dolomitized shale zone and a geodiferous zone that is typically barren of recognizable fossils. The geodes range in diameter from 0.2 centimeters to over 75 centimeters. In some locations the geodes are so numerous they form a continuous layer ranging in thickness from 0.15 m to 0.31 m. Figure 4 shows a geode recovered in a rock core sample from boring B-05. nodules that may be geodes that are completely infilled.

Distinction between the lower unit of the Warsaw formation and the underlying upper unit of the Keokuk formation can be difficult to discern and are frequently listed on geological profiles as the Keokuk-Warsaw interval. The Warsaw formation is fossiliferous in southeastern Iowa and contains brachiopods bryozoan, and crinoid fossils. The Warsaw formation in southeastern Iowa varies in thickness from 6.1 m to 12.2 m. A prominent unconformity occurs at the top of the formation and portions the formation may be absent in some areas.

The Keokuk Formation is divided into a lower cherty interval known as the Montrose member and an unnamed upper unit. The upper unit consist of interbedded fossiliferous limestones and shales. The Montrose member is composed of interbedded fossiliferous limestones, dolomites and nodular bedded cherts. The Keokuk formation in southeastern Iowa varies in thickness from 45 to 90 feet.

Rock Structure

The general bedding plane strike mapped by the geophysical investigation is north 7° east at approximately 9° dip. A perpendicular set of near vertical fractures were also noted. Joint surfaces varied from planar, irregular to undulating to planar smooth. No infilling was noted, and the joints were mainly tight to partly open.

Fracture data from the tunnel was used to calculate values for Rock Quality Designation (RQD), Q, and Rock Mass Rating (RMR). RQD values ranged from 40 percent to 100 percent and averaged 90 percent. Q values ranged from 0.11 to 26.4 and averaged 8.5. RMR values ranged from 35 to 70 and averaged 60.3.

Groundwater Inflow

Permeability was measured in packer testing in select borings. In the shale calculated Lugeon values ranged from 0 to 0.09 Lugeons and averaged 0.02 Lugeons. Low-volume steady-state inflow are anticipated based on preliminary assessment of the in-situ packer test results along the tunnel. Higher inflows may be anticipated during shaft construction in locations where the rock is intensely fractured. This condition will be considered in establishing inflow baseline values and was considered during the preliminary design when evaluating construction methods and the tunnel lining.

PRELIMINARY DESIGN

The project is at the preliminary design phase; two alignment alternatives and three construction approaches are currently being investigated. The hydraulic analysis dictates a minimum of 10-foot internal diameter (ID) tunnel. Two drop shafts are planned: The first shaft is positioned at the intersection of Fulton Street and North 14th Street and will serve as the transition drop shaft from the shallow new stormwater conveyance system to the proposed nearly 910 m long stormwater diversion tunnel; the second drop shaft is positioned within the Rand Park area to pick up flows from existing stormwater system.

The downstream end of the portal is located on a steep slope, captured in the photograph shown in Figure 5.

In either of the two alignment alternatives, (Alternative 1 is shown in Figures 6 with Alternative 2 marked in red dashed line), the tunnel at the outfall end daylights by mining through a reasonably steep slope (see Figure 5) before transitioning to a

Figure 5. Tunnel portal proximity

diversion/outfall structure that is navigated under the railroad tracks. The slope is considered to be unstable, however, there is commitment to stabilizing the slope as part of the project. Based on the available space at proposed launch shaft locations, power requirements, compressive strength of the rock, tunnel alignment, grade and tunnel length, three possible alternatives are proposed for excavation of the tunnel. The subsurface profile along the tunnel alignment is shown in Figure 7; subsurface conditions appear to remain relatively consistent for the two alternative alignments.

Construction Approach 1 consists of excavating the tunnel using a roadheader. The rock strength of both the limestone and shale was determined to control the size of the roadheader for the excavation. The field exploration program showed the limestone to have very good to excellent RQD values and we expect the rock UCS values to be at the upper limits of what a heavy duty roadheader can excavate. Whereas the RQD of the shale in the tunnel horizon indicates the rock is fair to good. The UCS values of this formation are expected to be significantly less than the limestone and are suitable for a light to medium duty roadheaders. To simply the tunnel construction, a vertical tunnel profile was selected that maintained the tunnel horizon (excavated) within the Warsaw Formation (shale).

Construction Approach 2 is to use Drill and Blast (D&B) excavation to construct the tunnel size. The preliminary design has determined that while feasible, the use of only drill and blast methods would result in a longer construction schedule and a higher excavation cost than currently available for the project if the completion date of Winter 2020 is to be achieved. However, the use of multiple heading could significantly reduce the anticipated drill and blast excavation schedule.

Construction Approach 3 is to use a combination of jacked pipe and roadheader. The jacked pipe would be used for construction of the tangent portion of the alignment east of the park shaft location, while the roadheader would be used for the construction of the curved segments west of the park shaft location.

Tunnel Support and Lining

The Geotechnical Evaluation focus on the anticipated ground conditions that will be encountered during tunneling operations and evaluation of appropriate design loads on the initial and final lining. Assessment of the anticipated ground conditions and associated ground behavior is important for evaluating the stability of the tunnel and for determining appropriate construction methods and equipment that will facilitate advancement of the tunnel in such a way to limit loss of ground and potential damage to existing utilities, buildings, streets, and other improvements. Ground loads, along with installation loads and hydraulic operations loads control the design of the tunnel lining section itself.

CDM Smith ran a soil-structure interaction evaluation of the tunnel system with the goal of identifying and evaluating required rock support type and preliminary tunnel liner thicknesses. To analyze the proposed tunnel excavation and lining system, Rocscience software, RS2, was used for numerical analysis. RS2 can use a variety of constitutive failure criteria, such as Mohr-Coulomb, generalized Hoek-Brown, etc. The subsurface profile shown in Figure 8 was used to develop the numerical model; project specific laboratory testing provided the basis for developing preliminary engineering parameter to be used in the numerical analysis. A summary of the preliminary average geotechnical engineering design parameters is presented in Table 1 for limestone and shale units.

Figure 6. Plan alignment Alternative 1 Rand Park Stormwater Diversion Tunnel

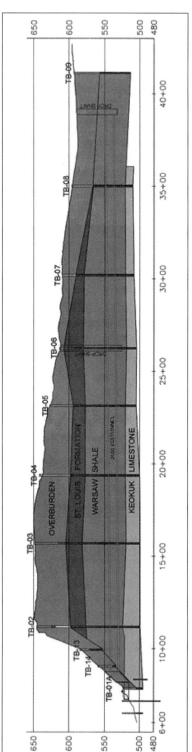

Figure 7. General subsurface profile along Rand Park Stormwater Diversion Tunnel alignment

Several iterations of the numerical analysis were run to preliminarily evaluate the initial and final support requirements for the tunnel. Once an initial selection was made the support system was cross checked using the empirical rules developed by Barton, Lien and Lunde (1974) which provide support recommendations according to the anticipated rock conditions.

The preliminary design analysis assumed the tunnel would be excavated in a horseshoe shape to an anticipated excavation width and height of 4.25 m. The analysis determined that, based on the assumed excavation width the tunnel will require a final lining consisting of a minimum 203 mm layer of concrete. However, it was determined that some initial tunnel support would also be required before the final lining is applied. The purpose of an initial support is to provide a safe working environment by until the final lining is installed, and construction is complete. Because of the variability of the rock quality and possibility of various excavation methods the initial support system would be required to consist of a combination of a number of different support types to address all the possible conditions. Each support type may use various material such as rock bolts, shotcrete and welded wire mesh for support.

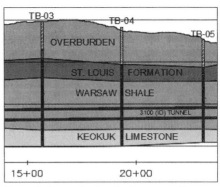

Figure 8. Generalized geologic cross section

Table 1. Preliminary geotechnical engineering design parameters

Parameter	Rock Unit	
	Limestone	Shale
Density, g/cm^3	2.29	2.41
UCS, MPa	89.3	48.4
Tensile, MPa	8.26	3.56
PLT, Is$_{50}$, MPa	3.4	1.39
CAI	0.55	2.04

To determine the required type of initial support during construction each support type will be associated with the RMR and Q rating of the rock that could be verified in the field during the construction phase. This rating has taken into account the orientation of the tunnel heading relative to the rock structure and adjusted accordingly. The RMR system also accounts for the span opening of the excavation and the excavation methods. If drill and blast methods are used during construction, it is expected that there will be more disturbances to the exposed rock from blast damage and the overall values of RMR observed in the field will likely indicate that a steel sets as initial support may be required. This will be factored into the estimated cost of the project prior to bid. Additionally, the contractor may select a larger cross-sectional excavation to accommodate means and methods and an adjustment to the RMR value will be required post bid so that additional support requirements do not result in additional cost to the Owner.

CONCLUSION

The anticipated geologic feature such as the elevation of the Keokuk limestone and the geode beds within the Warsaw formation present challenges for the appropriate construction approach and were the key factor raising the tunnel horizon to be fully within the shale unit. The requirement to meet the consent decree date for the completion of the overall City of Keokuk master plan and to keep the tunnel construction cost within the budget will be key drivers for the construction sequence and the selection of the excavation methods.

Final design for the project is planned to be completed by 2nd quarter 2019. The project is anticipated to be bid immediately afterwards with award anticipated by close of 3rd quarter 2019. Construction of the tunnel is anticipated to be completed in late winter of 2020.

ACKNOWLEDGMENT

CDM Smith (sub-consultant) and Veenstra & Kimm (prime), would like to thank the City of Keokuk for the opportunity to support their project and we look forward progressing and supporting this project through final engineering and construction.

REFERENCES

1. Doore, Keith (2016) Geology of Lee County Iowa and Its Aquifers, University of Northern Iowa, May 2016.

2. Prior, Jean (1868) A Regional Guide to Iowa Landforms, Iowa Geological Survey, October 1868.

3. Sendlein, Lyle V.A. (1968) "The Bedrock Configuration of Iowa," Proceedings of the Iowa Academy of Science: Vol 75: No.1, Article 30.

4. Van Tuyl, Francis M., The Stratigraphy of the Mississippian Formation of Iowa.

The Three Rivers Protection and Overflow Reduction Tunnel (3RPORT)—Decision-Making During Construction

Emidio Tamburri ▪ Lane Salini Impregilo
Ludovica Pizzarotti ▪ Lane Salini Impregilo
Paolo Perazzelli ▪ Pini Swiss Engineers
Roberto Schuerch ▪ Pini Swiss Engineers
Giuseppe Moranda ▪ Pini Swiss Engineers

ABSTRACT

The 3RPORT tunnel is part of the Long-Term Control Plan (LTCP) of the City of Fort Wayne (Indiana, USA) having the main goal to reduce the discharge of untreated CSOs (Combined Sewer Overflows) and to improve the water quality in Fort Wayne's CSO-impacted streams. The tunnel will have a length of 7,480 m, an internal diameter of 4.87 m and will be excavated by means of a slurry TBM. The tunnel will cross carbonate rocks at a maximum depth of 60 m and will run mainly underneath the bed of the Maumee and St. Mary's rivers. The hydraulic conductivity at the TBM tunnel elevation is expected to be extremely high, making the management of the water inflow during advance and during standstill the most important challenge of the project. As the hydrogeological conditions at the TBM tunnel sections are characterized by high heterogeneity, a deviation between the required operational TBM parameters and the design parameters has to be expected. The present paper describes a decision-making process elaborated in strict collaborations between the Contractor and the Designer for the assessment of the ground conditions and the definition of the tunnelling operations during advance.

PROJECT OVERVIEW

The 3RPORT tunnel is part of the Long-Term Control Plan (LTCP) of the City of Fort Wayne (Indiana, USA) having the main goal to reduce the discharge of untreated CSOs (Combined Sewer Overflows) and to improve the water quality in Fort Wayne's CSO-impacted streams. The tunnel passes underneath the City Fort Wayne and runs almost parallel to the St. Mary's and Maumee Rivers (Figure 1a).

The upstream tunnel boundary (retrieval shaft) is located in the vicinity of Foster Park while its downstream boundary (working shaft and pump shaft) corresponds to the City of Fort Wayne Water Pollution Control Plant (WPCP) (Figure 1a).

The tunnel will be approximatively 7,480 m long and will have a finished internal diameter of 4.87 m. The alignment runs 44% underneath rivers, 32% under green field conditions or beneath abandoned warehouses, 12% beneath existing buildings, 12% beneath roads, bridges or railway.

The tunnel will be excavated by means of a slurry TBM (Figure 1b, Figure 2) designed in order to cope with the baseline maximum water pressure of 6.5 bar (according to the Geological Baseline Report GBR, Black & Veatch 2016a). The TBM can be operated in the so-called "pressurized" or "semi-pressurized" modes. When the TBM is operated in "pressurized" mode, the support pressure compensates the undisturbed hydrostatic pressure. When the TBM is operated in "semi-pressurized" mode, the

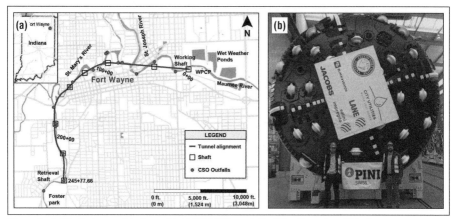

Figure 1. (a) Project area; (b) slurry shield TBM ("MamaJo") at the Herrenknecht factory

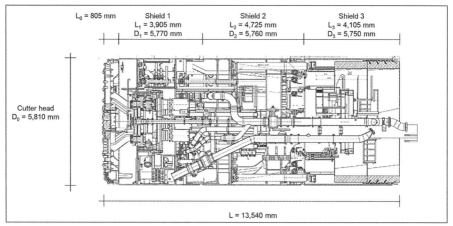

Figure 2. Geometry of the slurry shield TBM (Herrenknecht, 2017)

support pressure is lower than the undisturbed hydrostatic pressure (i.e., partial compensation of the groundwater pressure).

GEOTECHNICAL CONDITIONS EXPECTED ALONG THE TBM BORED TUNNEL

A total of 35 boreholes have been performed along, or in the close vicinity, of the TBM tunnel alignment (according to the Geotechnical Data Report GDR, Black & Veatch 2016b).

Based upon the information provided by the G=BR (Black & Veatch 2016a), the GDR (Black & Veatch 2016b) and, as confirmed by the literature on the geological setup of the project area, the TBM tunnel is expected to run through dolomite belonging to the Wabash Formation over more than 90% of the alignment and, over the remaining portion of the alignment, through dolomites belonging to the Detroit River Formation (Figure 3a).

The overburden varies approximatively from 44 m to 60 m and will consist of rock formations up to 27 m–40 m above the tunnel crown (mainly limestones of Traverse

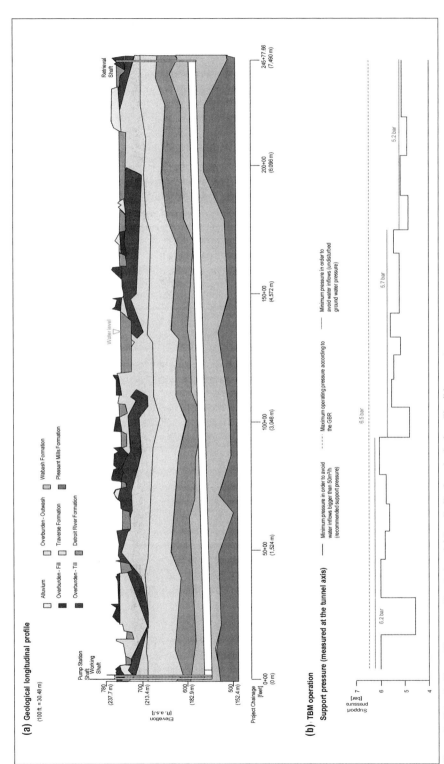

Figure 3. Extract of tunnelling operational plan: (a) longitudinal geological profile; (b) support pressure

Formation and dolomite of Detroit Formation) and soil deposits—consisting of fill, alluvium, glacial till and outwashes—above (Figure 3a).

The geology at the tunnel face elevation will be mainly characterized by medium to dark gray, dense and vuggy dolomite with possible argillaceous and with greenish-gray shale partings; less frequently it expected to be massive or highly weathered. Figure 4 shows typical rock samples of the Wabash formation expected at the TBM tunnel face elevation. Table 1 shows the mechanical parameters, boreability and abrasiveness for the intact rocks belonging to the Wabash Formation.

The rock mass crossed by the TBM is expected to be characterized by the presence of near to horizontal bedding planes and two sets of near to vertical joints. The latter may extend up to the shallows soil deposits.

No major shear and fault zones (i.e., of metric thickness) are expected along the alignment. However, minor shear and fault zones (i.e., of decimetre thickness) may be encountered during TBM advance.

A total of 46 piezometers have been executed in order to investigate the hydraulic head of the soil and bedrock aquifers in the area of the project. Based upon the executed tests and as reported in the GBR (Black & Veatch 2016a) and GDR (Black & Veatch

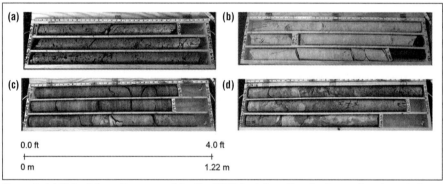

Figure 4. (a) Wabash formation, vuggy dolomite; (b) Wabash formation, fresh, massive, dense to fine grained dolomite; (c) Wabash formation, dolomite interbedded with shale; (d) Wabash formation, with sandstone with high quartz content (Black & Veatch 2016b).

Table 1. Wabash formation, properties of the intact rock (Black & Veatch 2016a)

	Properties	Symbol and unit	10th perc.	Average	90th perc.	Number of Tests
Mechanical Parameters	Unit Weight	g_i [kN/m^3]	22.9	24.3	25.8	47
	Compressive Strength	UCS [MPa]	39.3	68.3	111.7	46
	Point Load Index	PLI [MPa]	1.4	3.5	6.2	230
	Brazilian Tensile Strength	BTS [MPa]	3.1	5.2	6.9	33
	Young's Modulus	E_i [GPa]	32.3	40.4	47.2	8
	Poisson's Ratio	n_i [-]	0.1	0.2	0.4	8
Boreability Abrasiveness	Cerchar Abrasivity Index	CAI [-]	1.0	2.2	3.5	18
	Punch Penetration Index	PPI [kN/mm]	11.4	14.9	21.0	19
	Drilling Rate Index	DRI [-]	60.0	64.3	67.4	6
	Cutter Life Index	CLI [-]	64.6	77.3	93.6	6

2016b), the hydraulic head in the soil aquifers is similar to the one in bedrock aquifers. The maximum undisturbed water pressure to be expected at the tunnel elevation varies from 6.2 bar (at the end of the starter tunnel) to 5.2 bar (near to the retrieval shaft).

Taking into account that the TBM bored tunnel section runs below, or close, to the rivers and due to the presence of the vertical joints sets, it is plausible to assume that the bedrock aquifers are hydraulically connected to the soil aquifers at the TBM tunnel sections, thus making the water recharge potential almost infinite (Black & Veatch 2016a).

According to the GDR (Black & Veatch 2016b), more than 450 packer tests, 9 slug tests and 6 pumping tests have been performed in order to assess the hydraulic conductivity of the rock mass. Based upon the information provided by the GBR (Black & Veatch 2016a) and GDR (Black & Veatch 2016b) the hydraulic conductivity of the rock mass along the TBM tunnel sections ranges from 2×10^{-3} cm/s to 3×10^{-2} cm/s. In addition to that, the baseline documents highlight that regions of high hydraulic conductivity may appear spread throughout the rock mass crossed by the TBM.

MAIN GEOTECHNICAL HAZARDS OF THE TBM TUNNELLING AND MITIGATION MEASURES

According to the GBR (Black & Veatch 2016a), the large water inflows represent the main hazard of TBM tunnelling. Large water inflows are expected, at least locally, during interventions under atmospheric conditions and during TBM advance at a support pressure lower than the hydrostatic one.

Large water inflows during interventions under atmospheric conditions make the maintenance activities inside the tunnel and at the excavation chamber difficult and, in the worst case, may cause the flooding of the tunnel. Moreover, large water inflows may cause the wash-out of the annulus grouting and so, lead to a loss of the sealing of the tunnel walls and to a loss of embedment of the segmental lining. In case of asymmetric loads acting on the lining (rock blocks, asymmetric thrust force of the TBM), the loss of embedment may cause the overstress of the segmental lining.

In order to cope with the excessive high water inflows during advance, the support pressure has to be increased. During interventions under atmospheric conditions the reduction of the water inflows can be managed only by executing pre-excavation grouting.

TUNNELLING OPERATION PLAN (TOP)

A Tunnelling Operation Plan (so-called TOP) defines how the TBM has to be operated and which additional measures are required during advance.

Figure 3 shows an extract of TOP elaborated for the 3RPORT tunnel on the basis of the GBR (Black & Veatch 2016a) and the GDR (Black & Veatch 2016b). For the expected conditions, the required minimum support pressure in order to limit the water inflows during advance varies between 3 and 6 bar.

DECISION MAKING DURING COSTRUCTION

As the hydrogeological conditions at the TBM tunnel sections are characterized by high heterogeneity, a deviation between the required TBM operational parameters and the parameters assumed in the design phase (as shown in the TOP) has to be expected.

In order to cope with this uncertainty, a Decision-Making Tree has been elaborated (see Figure 5). This tool allows to assess the ground conditions during advance and so, select the most appropriated TBM parameters (and identify the most favorable locations for interventions), based upon the on-time monitoring of the TBM data.

TBM Parameters Considered for the Assessment of the Ground Conditions

Figure 5 lists the TBM operational parameters considered for the interpretation of the ground conditions ahead of the tunnel face. The parameters are divided in "raw" (i.e., control parameters of the TBM) and "derived" (i.e., data deriving from the relation between one or more raw data). The derived TBM parameters are defined as follows.

Boring Force

The boring force is the component of the total thrust used for the excavation. It is obtained by subtracting from the total thrust the force required in order to balance the slurry pressure, the pulling force of the backup and the force needed for overcoming the frictional force due to the self-weight of the TBM and the frictional force acting on the shield.

Normal Disc Force

The normal disc force indicates the contact force between disc and rock. It is obtained dividing the boring force by the number of discs. In order to avoid overload and damage of the cutters it has to be kept lower than the bearing capacity of the cutters (i.e., 250 KN for a common 17" disc).

Rolling Force

The rolling force is the cutter force component acting tangentially to the rock surface (Gertsch et al. 2007). It is calculated based upon the cutterhead torque, the weighted average cutter distance from the center of rotation, the number of discs and on the assumed overall torque system efficiency.

Field Penetration Index (FPI)

Since both the disc normal force and the penetration rate vary alongside a tunnel, their ratio represents a measure of the resistance of the ground to mechanical excavation. This ratio is the so-called field penetration index FPI and is a measure of the boreability (Tarkoy and Marconi 1991, U.S. Army Corps of Engineers 1997). The FPI typically varies between 10 and 80 kN/mm/rev (FPI = 20 kN/mm/rev indicates very high boreability).

Cutting Coefficient (Cc)

The so-called cutting coefficient Cc is the ratio between the rolling force and the normal disc force (Gertsch et al. 2007, U.S. Army Corps of Engineers 1997, Okubo et al. 2003). The cutting coefficient typically varies between 0.10–0.25 (Cc = 0.25 is typical of weak rocks).

Mass Flow Rate in the Feed and Slurry Lines

Multiplying the volumetric flow rate ($[m^3/h]$) by the material density ($[ton/h]$) one can obtain the mass flow rate ($[ton/h]$) of the bentonite in the feed line and of the mixture in the slurry. In order to avoid overload of the slurry pumps the mixture flow in the slurry line has to be kept lower than a limit value. For the selected TBM the latter is set equal

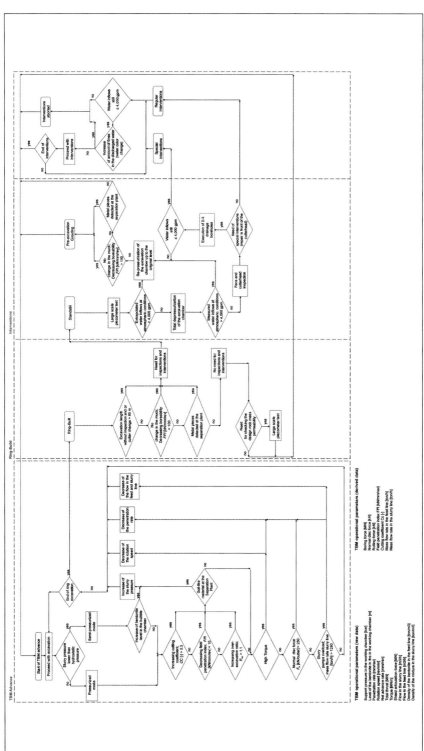

Figure 5. Decision tree for the assessment of the ground conditions and the definition of the TBM operation

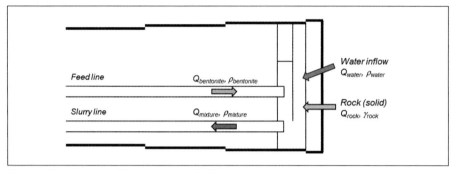

Figure 6. Mass balance of the system ground-TBM

to 1280 ton/h (i.e., value assumed by the TBM supplier for the design of the power of the slurry pumps).

Over-Excavation Ratio (R_{oc})

It denotes the ratio between the amount of the mucked-out material m_{rock} during tunneling operation and theoretical amount of the excavation material $m_{rock,th}$. The amount of the mucked-out material during tunneling operation results from the equation of the mass balance of the system ground-TBM (see Figure 6, i.e., $m_{rock} = Q_{mixture}r_{mixture} - Q_{bentonite}r_{bentonite} - Q_{water}r_{water}$). The theoretical amount of the excavation material is obtained multiplying the rock density r_{rock} by the excavation area A and the net advance rate v (i.e., $m_{rock,th} = r_{rock}Av$). If instabilities occur in front of the tunnel face, the mucked-out material will be higher than the theoretical one.

Assessment of the Ground Conditions

Interpretation of the TBM Data

Decreasing FPI indicates that the ground becomes softer. FPI values lower than 10 kN/mm/rev are not expected and may indicate the presence of a weak zone. Interaction diagrams total thrust—penetration rate—support pressure have been derived for a check in real time of the condition FPI < 10 kN/mm/rev (Figure 7a).

Increasing Cc indicates that the ground becomes softer (e.g., the sudden increase of the cutting coefficient indicates potentially unstable ground condition ahead of the TBM). Cc values higher than 0.3 are not expected and may indicate the presence of a weak zone. Interaction diagrams total thrust—torque—support pressure have been derived for a check in real time of the condition $Cc > 0.3$ (Figure 7b).

The check of the over-excavation during TBM excavation is very important in order to timely detect instabilities at the tunnel face and consequently adjust the support pressure, avoiding the occurrence of major instabilities that may block the cutterhead. R_{oc} values bigger than 1.1 are not expected and may indicate the presence of a weak zone. Interaction diagrams mass flow rate in the slurry lines—net advance rate—mass flow rate in the feed line have been derived for a check in real time of the condition $R_{oc} > 1.1$ (Figure 7c).

Analyses of the Muck

The check of the muck is crucial in order to interpret the TBM parameters correctly and in order to timely identify a change of the geological conditions. For the expected

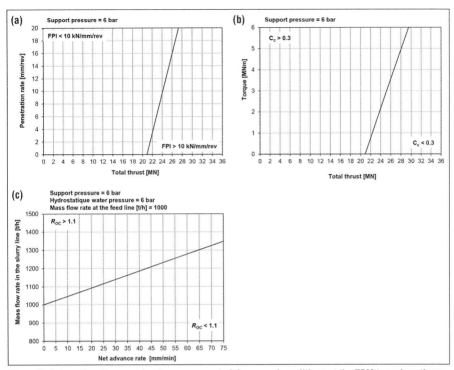

Figure 7. Interaction diagrams for the assessment of the ground conditions at the TBM tunnel sections: (a) total thrust–penetration rate for FPI = 10 kNmm/rev; (b) total thrust–torque for C_c = 0.3; (c) net advance rate–mass flow rate in the slurry line for Roc = 1.1

scenario (i.e., "regular" geological condition), the muck at the separation plant is expected to be characterized by a low content of chips and a high content of small blocks (chips will be crushed by the crusher installed in the working chamber). The change of muck to a soil-like material (see example in Figure 8) would indicate a change of geological conditions to a fault/sheared zone (unexpected according to Black & Veatch 2016a and Black & Veatch 2016b).

Large Scale Piezometer Test

The main hazards of TBM tunnelling are related to extremely high water inflows. As big uncertainties exist on the rock mass permeability along the alignment of the TBM tunnel, the estimation of the rock mass permeability is crucial in order to correctly plan the tunneling operations. As successfully developed and applied during the construction of the Lake Mead Intake No. 3 tunnel (Anagnostou et al. 2018), the assessment of the rock mass permeability (and so of the water inflows) can be safely executed by means of scale constant-head permeameter test. The test uses the TBM as a large-scale constant-head permeameter, where the slurry pressure is lowered in steps. After each step the increase of water inflow is measured by observing the change in water outflow in the slurry line, while keeping the slurry level in the bubble chamber constant. The final value for water inflow will be recorded after reaching stationary seepage flow conditions.

After several steps the relationship between water inflow quantity and slurry pressure can be determined and used for extrapolating the quantity of water-inflow under

Figure 8. Pictures of the muck on the example of the Lake Mead Intake No. 3 tunnel project

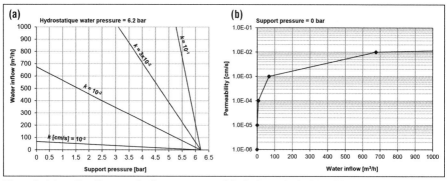

Figure 9. Interaction diagrams for the assessment of the hydrogeological conditions at the TBM tunnel sections: (a) water inflows as a function of the support pressure for different rock mass permeability; (b) rock mass permeability as function of the water inflows under atmospheric conditions

atmospheric conditions and for assessing the ground permeability. For these purposes, interaction diagrams have been derived (Figure 9).

During the water tests, observations of the force acting on the cutter head, the torque (by rotating the cutter head without advancing the TBM) and the color of the drained water have to be made, in order to detect the possible onset of instabilities and thus interrupt the test by immediately increasing the slurry pressure to its initial value. An additional indicator of local face instabilities can be the appearance of small rock blocks at the separation plant. The tests may have to be interrupted in case of unmanageable water inflows or unstable face.

Definition of the TBM Operation

Support Pressure

Depending on the encountered hydrogeological conditions, the support pressure may be increased or decreased (compared to the design support pressures—i.e., the one provided by the TOP) during tunneling operations.

Support pressures higher than the design support pressures will be required in case of weak zones (to avoid instabilities) and or of high permeability zones (in order to avoid unmanageable water inflow). The occurrence of these conditions will be indicated by high C_C, low FPI or high R_{oc} in combination with high content of soil-like material at the separation plant and by an increasing level of the bentonite in the working chamber, respectively. According to the available data, the maximum support pressure will never exceed 6.5 bar. In case of hydrogeological conditions more favorable than the one assumed in the TOP (e.g., rock mass permeability lower than the expected one), the support pressure will be decreased.

Interventions

In order to minimize the probability of needing auxiliary measures for interventions, the following approach has been elaborated:

1. Inspections of the cutterhead and interventions have to be planned with a relatively high frequency (e.g., every 80 m of excavation);
2. Large scale piezometer tests will be performed before any planned inspection or intervention activity;
3. When the water inflow under atmospheric conditions extrapolated from the water test is bigger than the assumed limit value, the inspection of the cutterhead and interventions will be postponed;
4. When regular interventions are required, the interventions will be executed without auxiliary measures and with a systematic monitoring of the water inflows and of the face conditions;
5. When special interventions are required (i.e., involving activities in front of the cutterhead), drainage boreholes may be executed in order to release the pressure acting ahead the tunnel face and so, reduce the risk of local instability. Moreover, monitoring of the water inflows and of the face conditions has to be carried-out;
6. Interventions will be required (and so not long postponed) in case metal pieces would be detected at the separation plan or the boring efficiency would significantly decrease, thus indicating an unexpected high wear of the cutting tools. The latter will be denoted by an increasing FPI without change of the geological conditions. FPI values bigger than 100 kN/mm/rev are not expected and indicate high wears of the cutters and in the worst case damage of the cutterhead. For a check in real time of the condition FPI > 100 kN/mm/rev, interaction diagrams total thrust—penetration rate—support pressure have been derived;
7. When the interventions cannot be not long delayed and the TBM is located in a high permeability zone, pre-excavation grouting in advance remains the only option for reducing the water inflows and performing interventions.

CONCLUSIONS

The main challenges of the 3RPORT tunnel are related to the high permeability and the high water recharge potential of the rock mass which may cause, at least locally, extremely high water inflows. The fundamental design aspects in the pre-construction phase are therefore the interpretation of the executed hydrogeological tests and the successive estimation of the water inflows. The latter allow to define the optimal TBM operation parameters and to preliminary identify locations over which maintenance can be safely carried-out without costly and time consuming auxiliary measures. The results of the study have been graphically summarized in the so-called TOP (Tunneling Operational Plan). An extract of the TOP is showed in the present paper.

Due to the expected heterogeneity of the hydrogeological conditions at the TBM tunnel sections, a deviation between the required operational parameters and the TOP has to be expected. Therefore, a decision-making process has been elaborated in strict collaborations between the Contractor and the Designers in order to assess the geological during construction, to select the most appropriated TBM parameters and, finally, to identify the most favorable locations for interventions.

REFERENCES

Anagnostou, G., Schuerch, R., and Perazzelli, P. 2018. Lake Mead Intake No 3 Tunnel—Design considerations and construction experience.

Lake Mead Intake No 3 Tunnel—Design considerations and construction experience. *Geomechanics and Tunnelling*, vol. 11: no. 1, pp. 15–23, Berlin: Ernst & Sohn, 2018.

Black & Veatch Corporation for the Division of Utilities of the City of Fort Wayne 2016a. 3RPORT Project, Tunnel and Shafts Package, Geotechnical Base Line Report (GBR)—Conformed to Contract Document.

Black & Veatch Corporation for the Division of Utilities of the City of Fort Wayne 2016b. 3RPORT Project, Tunnel and Shafts Package, , Geotechnical Data Report (GDR)— Conformed to Contract Document.

Gertsch, R., Gertsch, L., and Rostami, J. 2007. Disc cutting tests in Colorado Red Granite: Implications for TBM performance prediction. *Int. J. Rock Mech. Min. Sci.* 44 (2007), No. 2, pp. 238–246.

Herrenknecht Inc. 2017. Drawing number 5346-000-001-00 Shield machine—S-1132 TBM Manual.

Okubo, S., Fukui, K. and Chen, W. 2003. Expert System for Applicability of Tunnel Boring Machines in Japan. Rock Mechanics and Rock Engineering, 36(4), 305–322.

Tarkoy, P.J., and Marconi, M., 1991. Difficult rock comminution and associated geological conditions. In Proc. Tunnelling 1991, pp. 195–207. London: Elsevier Applied Science, 1991.

U.S. Army Corps of Engineers 1997. Tunnels and Shafts in Rock—Engineering and Design. Engineer Manual 1110-2-2901.

TBMs: Meeting the Challenge in Pakistan's Lower Himalayas

Gary Peach ▪ TBM, Multiconsult AS

ABSTRACT

TBMs had not been used in Pakistan for 60 years prior to the construction of the Neelum Jhelum Hydro Electric project. TBMs excavated and lined twin 8.5 m diameter, 11.2 km long parallel headrace tunnels with overburdens up to 1,870 m with high horizontal stresses. The tunnel construction experienced many challenges including soft ground faults, squeezing ground and rockbursts. The TBMs were introduced to this project to improve tunnel excavation rates and were the only realistic methodology for the portion of the tunnel alignment where Adit and shaft access locations were difficult or uneconomical due to high overburden and rugged, inhospitable terrain.

PROJECT DESCRIPTION

The project is located in the Muzaffarabad District of Azad Jammu & Kashmir (AJK), in north-eastern Pakistan within the Himalayan foothill zone known as the Sub-Himalayan Range. The terrain is rugged with ground elevations that range from 600 to 3200 meters above sea level. The project is a run-of-river scheme, employing 28.6 kilometers of headrace and 3.6 km of tailrace tunnels that bypass a major loop in the river system, for a total static head gain of 420 m. The headrace tunnels twin bores (69%) contain the TBM excavation, Figure 1.

GEOLOGICAL SETTINGS

The entire project was excavated in the sedimentary rocks of the Murree Formation, which is of Eocene to Miocene age. The TBM tunnels are being driven through a zone bounded by two major Himalayan faults that trend sub-perpendicular to the tunnels: the Main Boundary Thrust, and the subsidiary Muzaffarabad reverse/thrust fault. The Lithologies are detailed below:

Siltstones & Silty Sandstones: Uniaxial Compressive Strengths (UCS) is 50–70 MPa.

Mudstones: With UCSs in the 30 to 40 MPa range,

Sandstones: With UCS in the range 130 to 230 MPa.

In-Situ Stresses: Over-coring tests in sandstone beds in the TBM tunnels found a tectonically altered zone of high stresses (k up to 2.9) with the major principal stress oriented sub-horizontally and sub-perpendicular to the tunnel azimuth.

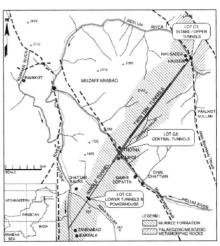

Figure 1. Project layout showing TBM twin tunnels (in bold), major faults (dashed) and simplified alignment geology

GEOLOGICAL STRUCTURES AND CONDITIONS

In addition to the expected rock types detailed above, there are also certain ground conditions which are expected to be encountered and for which the TBM will be designed to manage. Also the TBM should have further facilities to detect in advance these conditions, in order to take appropriate measures to successfully negotiate the identified conditions. These structures and conditions included:

- Unstable rock zones.
- Squeezing and swelling ground
- Soft ground
- High water inflows.
- Extensive fault zones
- Rock bursts.
- High Over burden depths (1870 m)

TBM SELECTION

20km of the headrace tunnel system was excavated by TBM in two parallel tunnels. Two Open (Gripper) TBMs were used to excavate this tunnel system (Figure 2).

A significant consideration for the TBM selection was the possibility of encountering squeezing ground, with deformations of up to 500 mm on the tunnel diameter. This would exclude many types of TBM designs due to the possibility of becoming trapped within the tunnel. The Open (Gripper) TBM is best suited to deal with this potential condition, due to the short length of the front shield and its ability to collapse inwards various sections of the front shield, depending upon ground conditions, and still maintain the ability to excavate forward.

The second ground condition which was indicated to be present in the higher overburdens and more brittle rock was rock bursting. Again, the Open Gripper TBM configuration allows for equipment to be installed to detect and mitigate potential rock bursts.

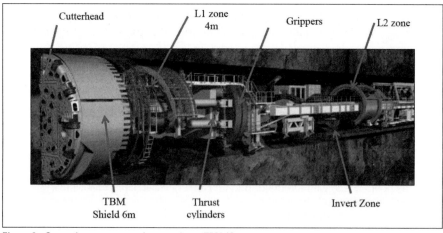

Figure 2. General arrangement of open gripper TBM (Courtesy of Herrenknecht)

The two TBMs were referred to by their model manufacturing model number namely TBM 696 and 697. TBM 697 was launched first and for the majority of the Tunnel excavation was the lead TBM and TBM 696 was the trailing TBM.

ROCK SUPPORT DESIGN

The initial design for the rock support consisted of four categories of support designated as Q2, Q3, Q4 and Q5. These support designs were to be installed according to the observed geology or Q class, which was revealed at the rea of the TBM shield, as the TBM advanced. The most favorable rock class being Q2 and the least favorable being Q5. The support requirements in terms of quantity of shotcrete and rockbolts, wire mesh mining straps and full steel rings increased with the increase of Q class. The main component being shotcrete started at 125 mm thickness and increase with class until in Q5 class where the thickness was 350 mm.The balance of support installation and rock class is an area which still relies heavily on Human intervention, skill and experience to balance support installation with safety and production.

CHALLENGE I—DELIVERY OF THE TWO TBMS

With the two TBMs procured and manufactured the first major challenge was the delivery of all the TBM and associated parts to the construction site. The two TBMs were manufactured in Germany and China and delivered to the port of Karachi in Pakistan. The manufacturer number was used for individual identification the two number being 696 and 697, The TBMs then were loaded onto road transport and travelled 1,777 km to the construction site located in north-west Pakistan (Figure 3). The route used major road systems in the south and towards the end of the journey the route climbed into the lower Himalayas following cliff edge roads and passing through towns and villages, shown in Figure 4.

The enabling works, road strengthening and improvements and temporary bridge construction commenced over 1 year prior to the TBM delivery. There was also a key date in 2012 that the TBM should be delivered by the end of May, prior to the commencement of the summer monsoon.

Figure 3. 1,777 km in land transportation route for TBM delivery to Neelum Jhelum project

Cliff edge roads

Villages and towns

Temporary bridges

Figure 4. Transportation route for TBM delivery to Neelum Jhelum project

CHALLENGE II—TBM POWER SUPPLY

The consequence of having two such very capable TBMs was the need for a dedicated power supply. The TBM construction site was located in a remote part of Azad Kashmir and did not have the power supply meet the TBM requirements, and regularly experience up to 16 hours a day of load shedding. Therefore a complete 19.6 MW power station had to be constructed on a hillside near to the TBM access adit. The power station consisted of four number duty 4 MW generators and one 3.6 MW standby generator, powered by heavy fuel oil (HFO). The full power station is shown in Figure 5.

Figure 5. TBM power station

CHALLENGE III—FAULT ZONE

Commencement of Tunnel Excavation

Both TBMs commenced the planned 11.2 km twin headrace tunnel excavation in early 2013.Both TBM launches were in stages to allow the continuous conveyors to be installed after excavation of 100 m. The first TBM to be fully installed and operational was TBM 697 and from early 2013 steady excavated and increased monthly production. The second TBM number 696 followed suit and progressed approximately 500 m behind TBM 697. The alignment of the twin tunnels encountered an existing access tunnel some 1700 m from the TBM launch location. This tunnel known as Adit 2 had been completed prior to the planned arrival of both TBMs.

Fault Zone

However some 90 m before this adit the lead TBM 697 encountered an extensive fault zone of sheared mudstone over 80 m in length. The first indication of this poor ground came when the thrust pressures dropped rapidly and large quantities of soft material came through the cutterhead and onto the TBM conveyor system. A cavity rapidly developed in front of the TBM and the TBM was stopped to access the situation. The TBM was then started and advance was attempted however, the ground being so soft flowed into the cutterhead and the cutterhead tripped electrically and stopped. A crew of tunnel workers was sent into the cutterhead to manually remove the buildup material this operation taking 8 hours. In an attempt to reduce the flow of soft material into the TBM, all six of the buckets had metal components welded into them to reduce the opening and restrict the free flow of material, Figure 6a. A further three attempts were made to advance the TBM in the poor ground conditions with limited success. It was then decided to stop any further attempts and to install a 15 m pipe roof canopy, Figure 6b over the TBM shield and in front of the TBM allowing stabilization by way of ground treatment with grout and chemicals. Once the canopy was completed a top heading Figure 6c was constructed to access the collapsed area and install further support in advance of the TBM.

After 9 weeks the TBM was started again and slowly advanced through the faulted ground installing full circular steel rings and 350 mm of shotcrete and breaking through into the adit in early 2014. The trailing TBM 696 having the benefit of the knowledge of the fault zone installed a systematic 15 m pipe canopy every 5 m was able to progress the fault zone relitively smoothly but at a reduced advance rate.

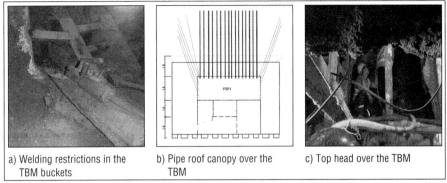

| a) Welding restrictions in the TBM buckets | b) Pipe roof canopy over the TBM | c) Top head over the TBM |

Figure 6. TBM 697 top heading and pipe roof canopy at fault zone

CHALLENGE IV—ROCKBURSTS

Good Progress

Both TBMs broke through and traversed Adit 2 in January 2014 and entered into a period of good progress reaching 460 m per month in a installing the full ranges of rock support for the ground encountered. This period of good progress lasted up until the beginning of November 2014. During this period the TBM excavation was performed with tunnel over burdens in the range of 1150 to 1350 m and the only negative experience was the occurrence of a few rockbursts which resulted in damaged to some of the TBM equipment.

Rockbursts

Rockbursts had been expected and mentioned in the geological baseline report and the expectation was that this would occur at the higher overburdens. By November 2014, with 4.7 km and 4.3 km of the tunnels excavated in the left and right tunnels respectively, regular rockbursts warranted systematic recording. Rockburst events were categorized by magnitude, from "noise only" to "major rockburst." The system aimed to correlate timing and distribution of rockbursts and facilitate selection of mitigation measures at the TBM. The total number of rockburts encountered for the two TBM during tunnel exavation was 1695 and Figure 7a shows the breakdown of rockburts by category.

The descreption of the rockbursts categories is as follows:

- Category 1: Noise only—a slight popping sound is heard no damage to the support or ejection of rock
- Category 2: Noise and weak rockburst—a popping sound is heard and there may be some light damage to the support and surrounding rock
- Category 3: Noise and medium rockburst—loud popping sounds are heard and there may be splitting, spalling or shallow slabbing to the support and surrounding rock
- Category 4: Noise and major rockburst—loud sound similar to an explosion, violent ejection of rock into the tunnel and severe damage to the installed support and TBM.

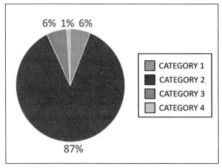 

1700 Rockbursts by category   Category 3 Rockburst at the L1 section
Figure 7. Example of rockbursts at the front of the TBM

Figure 7a shows that the majority of rockbursts are classified as Category 2, but even this category of rockburst was responsible for delays whilst repairs were under taken. Figure 7b shows the typical aftermath of a Category 3 rockbursr at the front of the TBM.

ROCKBURTS COUNTER MEASURES

Longitudinal Relief Holes

Drilling of longitudinal stress relief holes ahead of the tunnel face will fracture the rock mass, thereby releasing stress and reducing rockburst potential. Holes are drilled with the probe drill and should be closely spaced enough so that the rock between cracks or fractures to relieve the stress. The holes can be concentrated in highly stressed parts of the rock mass.

Radial Relief Holes

Radial stress relief holes reduce the likelihood of rockbursts by shifting the tangential stress peaks away from the excavated perimeter. The holes must be large enough and closely spaced enough so the rock between the holes cracks and breaks. This creates a stress-relieved zone around the excavation perimeter. Fewer holes are required in fractured rock.

Horizontal Side Wall Probe

A significant contributor to the 31 May 2015 severe rockburst was a local change in strike of the rock strata from perpendicular to the tunnel alignment to parallel. This hid the rockburst-prone sandstone beds behind siltstone beds. In order to detect future hidden sandstone beds, side probe holes were drilled at 5 meter intervals on both sides of the excavated tunnel at tunnel axis height. This activity began on all TBM tunnels in siltstone after the severe rockburst of 31 May 2015.

Installation of Shotcrete at the L1 Zone

Reinforcement of the rock mass begins with installation of rock bolts and wire mesh, used routinely on the TBM. Steel fiber-reinforced shotcrete can contribute significantly to the energy absorbing capability where rock conditions require less support, it is preferable to apply most of the shotcrete at the L2 zone to allow quicker installation of initial rock support and faster resumption of excavation. However, 94% of rockbursts were detected at the front 10 m of the TBM.

Full Ring Steel Supports

The original purpose of full ring steel supports was to support the tunnel at faults, large overbreak areas, and soft and squeezing ground. These supports are time consuming to install and can be installed at spacing's ranging from 0.9 to 1.6 m. The spacing directly influenced the daily advance rate. Initially these supports had been installed in large overbreak areas adjacent to sandstone beds. As the tunnel advanced, rockbursts commenced and Category 4 events caused major equipment damage and serve damage to rockbolts, mesh, mining straps. The full ring steel supports, however, remained mostly intact even when dislodged. These elements remained rigid but certainly prevented more extensive damage to rock supports and equipment and most importantly provided a degree of protection to TBM personnel.

CHALLENGE V—SEVERE ROCK BURTS OF 31ST MAY 2015

The severe rockburst referred to as the 5/31 event occurred on TBM 696 (trailing TBM) at approximately 11.35 pm on 31 May 2015. The magnitude of the event was equivalent to a magnitude 2.4 earthquake on the Richter scale and consequent damages to the TBM, ancillary equipment and rock support were without precedent on the project. Figure 8a and 8b shows the same location at the L2 zone of the TBM during normal operation and then after the 5/31 event.

The physical damage and losses were sudden and unforeseen and extensive. The rockburst occurred when the trailing TBM 696 was in mid-stroke. Visible damage was observed along the tunnel for 63 m, with the most severe damage to the TBM, excavation profile and permanent rock support in a 22 meter section some 28 to 50 meters behind the shield. The maximum impact of the 5/31 occurred at a tunnel location which was excavated some 10 days earlier. The time lag between excavation and occurrence of the rockburst was highly unusual since rockbursts normally occurred in the region of the TBM cutter head often whilst excavating. The TBM was completely blocked by ejected material in two locations Figure 9a, and at these locations the whole TBM had been buckled and twisted 800 mm anticlockwise by the rockburst. Invert heave was evident throughout the 22 meter long most affected zone, with many of the steel ring beams sheared Figure 9b, displaced into the tunnel and lifted above the invert, along with the track and sleepers.

In some areas the ring beams were also heaved out of position, causing massive secondary damage to the adjacent shotcrete (Figure 14b).This severe rockbursts.

Figure 8. (a) L2 zone during normal operations (b) L2 zone after the 5/31 rockburst

 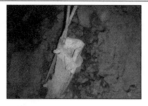 

| a) Complete blockage of the TBM | b) Shear failure of full steel ring beam | c) Repaired rockburst location |

Figure 9. Severe rockburst effects

Causing millions of dollars of equipment and severe damage to a 60 m section of tunnel as well as significant damage to the tunnel lining in the neighbouring TBM 697 tunnel.

Once the area was deemed safe to enter and the recovery plan had been developed, recovery work commenced on 16th June 2015. The most urgent activity was to start the debris removal from the top of the TBM to allow the replacement of tunnel rock support and also uncover the full extent of the rockburst zone. The exposed ground was then heavily supported with full ring steel supports, rockbolts and shotcrete. The whole recovery programme took 7.5 months. Figure 9c shows the same location as Figure 9a after the removal of ejected rock and the installation of heavy rock support, to a height of 8 m above the original excavated tunnel, prior to the 31/5 event.

CHALLENGE VI—HIGHEST OVERBURDEN AND TUNNEL COMPLETION

The 31/5 event had occurred at an overburden of approximately 1325 m and the maximum overburden of 1870 m was still some 2 km ahead. The recommencement of the lead TBM 697 some 8 days after the 31/5 event was undertaken with systematic counter measures. The tunnel rock support had been reevaluated and a special rockburst support lining designed and implemented on a permanent basis. The general design is shown in Figure 10a.

The design incorporated continuous full circular steel rings, rockbolts, heavy duty wire mesh, systematic shotcrete installed at L1 with a final lining thickness of 350 mm being installed at L2. Both forward longitudinal and radial stress relief holes were drilled in and around the sandstone beds encountered during excavation. All these precautionary measures and heavy tunnel support directly impacted upon monthly progress in particular to TBM 696 which experienced the severe rockburst event. The event and the subsequent detailed investigation resulted in new TBM operational procedures aimed at predicting and investigating future similar geological situations. The most important of these measures was the adoption of probing through the side wall of the tunnel to detect parallel, hidden sandstone beds

The use of the microseismic monitoring data and other site information enable the TBM Project team to optimize all aspects of precautionary measures and excavation operations and as such quickly start to increase monthly production up to 364 m/month please refer to Figure 10b.

The daily trend analysis and operations recording started to reveal that the microseismic activity and rockbursts occurrence did not continue to increase as the tunnels

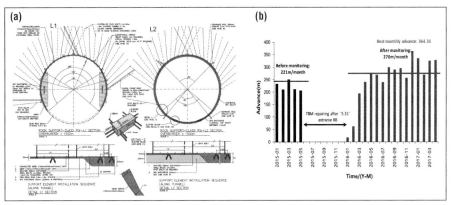

Figure 10. (a) TBM support post 31/5 event (b) TBM 696 progress after 31/5 event

headed for the highest overburdens in contracts the overall microseismic activity started to decrease. Further investigations were undertaken to record the actual in-situ stresses and within the actual rocks which firstly experienced category 4 rockburst and then secondly at varying overburdens to ascertain if the ground stresses were related to overburden. The details of these findings indicated an area of elevated horizontal stresses which was not related to overburden.

In the last 1km of each TBM tunnel the occurrence of rockbursts reduced to virtually zero and progress increased accordingly. The lead TBM 697 broke through and connected with the dam site in October 2016 and the trailing TBM followed suit and broke through in May 2017.

CHALLENGE VII—TBM OVER CUTTERS

The excavation diameter of a TBM is the cutter head width plus the protrusion of the gauge cutters. As the cutterhead turns and excavates the tunnel, the gauge cutters wear faster, leading to a reduction in the tunnel diameter. With a requirement for full circular steel rings and thicker shotcrete more space was required for these elements Overcutting was achieved by extending the cutters located on the cutterhead periphery using shims to increase the effective width of the cut and by replacing the gauge cutters with larger diameter cutter wheels (from 17 to 18 inches). Both methods were employed to increase the tunnel diameter by 100 mm. However, overcutters are normally deployed as a short term measure to address a local problem. On this project they would need to be installed for a much longer period, close to 5 km or half of the proposed TBM alignment. The perceived problem with using the overcutters for a long period would be firstly alignment control, both horizontal and vertical and rapid and excess wear of the TBM shield components. Figure 11 shows the horizontal alignment for the lead TBM before and after overcutter installation. The graph shows the difficulty of maintaining the horizontal alignment within the expected tolerances after the overcuttes were installed. The main points of deviation are numbered and the explanation is shown on the right-hand side of the graph. Control of the vertical alignment Figure 12 experienced no such control issues, the difficulty being the horizontal alignment control. The second concern of excessive wear of the TBM shield was also monitored and additional wear was detected on the grill bars, but this did not prove to be a hindrance to the TBM operation.

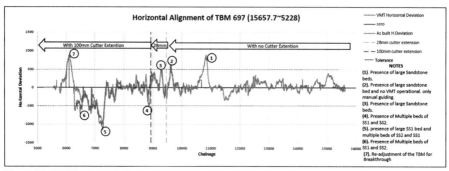

Figure 11. TBM horizontal alignment before and after overcutter installation

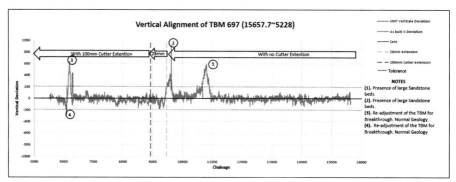

Figure 12. TBM vertical alignment before and after overcutter installation

CONCLUSIONS

This paper outlines seven significant challenges the TBM tunnels faced on the Neelum Jhelum Project, from delivering the TBMs to the site up to the unusual circumstances under which the 5/31 rockburst. These tasks were made much more difficult by the remote location of the project site, making both changes and or modifications to the TBMs or delivery of new products or equipment all the more taxing.

The successful recommencement of tunneling and modification of operating procedures and subsequent tunnel completion was only made possible by the full support of the Client and close collaborative working between the Employer, Contractor and Engineer.

Waterway Protection Tunnel: Louisville's Innovative Solution to Four Storage Basins

Jonathan Steflik ▪ Black & Veatch
Jacob Mathis ▪ Louisville and Jefferson County Metropolitan Sewer District

ABSTRACT

Louisville and Jefferson County Metropolitan Sewer District (MSD) will complete an $850 M, 20-year Integrated Overflow Abatement Plan (IOAP) by the end of 2020 to reduce combined sewer overflows (CSOs) into Southfork Beargrass Creek and the Ohio River via an off-line CSO conveyance and storage system located along the Ohio River in downtown Louisville, Kentucky. Originally scoped as four separate CSO basin projects, the Waterway Protection Tunnel (WPT) developed as challenges were encountered throughout design and construction. Three basin projects were included in the original tunnel design. During construction, a fourth basin was included in the project, requiring development of detailed design documents, subsurface investigation, natural gas exploration program, stakeholder and regulatory engagement, and contract negotiations on an accelerated three-month design schedule. The project expands the construction of an approximately a 20-foot ID, 4-mile (previously 2.5-mile) 55-million gallon (previously 37-million gallon) deep tunnel and an integrated deep pump station.

Key project elements include six drop structures (previously four) to convey consolidation flows to the tunnel, a downstream pump station shaft, and various tunnel adits connecting the drop shafts to the primary tunnel. The tunnel is located within a densely developed downtown area immediately adjacent to the Ohio River; hence, selection of the alignment, grade and construction methods required careful consideration to minimize disruption to the community and environment. Considering the accelerated design schedule and numerous technical and contractual challenges, successful project completion requires close coordination between MSD, the design team, and the contractor, along with extensive collaboration with numerous project stakeholders and regulatory agencies.

PROJECT BACKGROUND

In August 2005, the Louisville and Jefferson County Metropolitan Sewer District (MSD) entered a Consent Decree with the United States Environmental Protection Agency (EPA) and the Kentucky Environmental and Public Protection Cabinet. The Consent Decree was developed in response to an enforcement action related to the Federal Clean Water Act (CWA). Certain objectives set forth in the Consent Decree are to eliminate unauthorized discharges from MSD separate sewer systems, combined sewer overflow (CSO) systems, and water quality treatment centers.

On April 15, 2009, an Integrated Overflow Abatement Plan (IOAP) was approved by EPA and MSD and was entered into federal court. Gray solutions are identified in the IOAP to prevent and control sewer overflows. The gray solutions identified include storage, treatment, conveyance/transport, and sewer separation projects. A total of thirteen storage basins were recommended as gray solutions. Preliminary construction

Figure 1. Original waterway protection tunnel alignment

costs associated with the construction of these basins were developed as part of the IOAP (MSD Integrated Overflow Abatement Plan, 2009).

The IOAP originally defined the Waterway Protection Tunnel (WPT), also known as the Ohio River Tunnel (ORT), project as three separate CSO storage basins: Rowan CSO Basin; Story and Main CSO Basin; and Lexington and Payne CSO Basin. The IOAP required that each of these basin projects be operational by December 31, 2020. Each of the basin projects encountered issues and challenges during the design phases, eventually leading to the decision to combining the three basins into the Waterway Protection Tunnel. The horizontal alignment of the original WPT is shown on Figure 1.

The original WPT project included the construction of an approximately a 20-foot ID, 2.5-mile, 37-million gallon deep tunnel and an integrated deep pump station. Key project elements included a pump station shaft, a working shaft, four drop structures to convey consolidation flows to the tunnel, a downstream pump station shaft, and various tunnel adits connecting the drop shafts to the primary tunnel. The tunnel depth varied from approximately 220 feet to 190 feet below ground surface. (Black & Veatch, 2017).

Due to the compressed design schedule, the design team recognized the necessity of early and consistent communication with key stakeholders. Approvals were obtained by closely coordinating and quickly responding to feedback, allowing MSD and Black & Veatch to manage scope changes and accelerated design schedule throughout the design process. Detailed design of the project was completed in approximately eight months to accommodate construction completion before the IOAP deadline of December 31, 2020. Allocating essential global resources in multiple office locations allowed the design team to consolidate efforts required to complete the design deliverables.

A rigorous quality control program was essential to delivering bid-ready contract documents on time. The heightened sense of urgency due to the schedule likely contributed to timely and constructive feedback from technical experts, both internal to the design team and at regulatory agencies. Previous tunneling experience and familiarity with the regional and local geology contributed to the design team's ability to confidently make design recommendations.

The WPT project was awarded as three separate construction packages. The construction contract for the Lexington and Payne CSO Interceptor Package was awarded in November 2017 to Garney Companies for $23.1M. The Ohio River Tunnel—Tunnel and Shafts Package was awarded in October 2017 to J.F. Shea—Traylor Brothers Joint Venture (S-T JV) for $106.7M. The Rowan Pump Station and Downtown CSO Interceptor Package was awarded in June 2018 to Pace Contracting for $25.9M. The combined project construction cost of $155.7M is approximately $44.3M less than the total combined opinions of probable construction cost for the project.

I-64 & GRINSTEAD CSO BASIN

The I-64 & Grinstead CSO Basin project and the CSO 125 Stormwater Separation project were developed as part of the IOAP with a deadline to be operational by December 31, 2020. The I-64 & Grinstead CSO Basin project consisted of new consolidation sewers to capture and route combined sewer overflows from CSO 166, CSO 125, and CSO 127 to an underground storage basin and integrated pump station near the intersection of Grinstead Drive and Interstate 64. The CSO 125 Stormwater Separation project consisted of a new stormwater sewer and stream restoration to reduce the I-64 & Grinstead CSO Basin storage volume by 1.5 million gallons. The I-64 & Grinstead CSO Storage Basin was awarded by MSD in July 2017 to Thieneman Construction for $23.2M. The probable construction cost for the CSO 125 Stormwater Separation project was approximately $5.1M.

MSD and Black & Veatch evaluated alternatives for eliminating the I-64 & Grinstead CSO Basin, associated pump station, and the CSO 125 Stormwater Separation project by connecting the consolidation sewers to an extension of the WPT. Connection alternatives and respective opinions of probable construction costs were compared to the currently planned projects. A life-cycle cost analysis was also performed for each of the alternatives and compared to the collective I-64 & Grinstead CSO Basin and CSO 125 Separation projects. The life cycle cost analysis revealed the costs of the I-64 & Grinstead CSO Basin and CSO 125 Separation projects are essentially the same as the life cycle cost to extend the WTP. The evaluation also revealed that the WPT Extension would provide approximately 11 MG of additional storage, while eliminating the need for a pump station at the I-64 & Grinstead CSO Basin site. With costs being equal, MSD proceeded with additional subsurface investigations along the proposed WPT Extension alignment to determine constructability requirements.

SUBSURFACE INVESTIGATION

Over 40 test borings were drilled as part of the subsurface investigation for the original WPT project. Additional subsurface investigation efforts were necessary to confirm rock conditions and environmental considerations along the WPT Extension alignment. On April 12, 2018, MSD requested that Black & Veatch perform an additional subsurface investigation to verify subsurface soil and rock conditions directly influencing the construction of the WPT Extension. Six deep bedrock borings (B-GE-1 to B-GE-6) were drilled along the WPT Extension alignment. The borings were generally located along the WPT Extension alignment, along Beargrass Creek, shown on Figure 2.

Figure 2. Geotechnical boring locations along the WPT extension

Several key design and construction baseline criteria were evaluated during the subsurface investigation, which allowed MSD and Black & Veatch to mitigate risks and coordinate the construction schedule and cost with the contractor during the detailed design phase. These key baseline criteria include bedrock properties, natural gas, and environmental considerations.

Bedrock

The bedrock along the WPT Extension alignment consists primarily of shale, limestone, dolomite, and interlayered combinations of these three sedimentary rock types. The bedrock formations identified in the subsurface investigation in descending order include the Devonian New Albany Shale, Beechwood and Silver Creek Members of the Sellersburg Limestone Formation, and Jeffersonville Limestone Formation; the Louisville Limestone, Waldron Shale, Laurel Dolomite, Osgood, and Brassfield Formations from the Silurian Period; and the Saluda and Bardstown Members of the Drakes Formation from the Upper Ordovician Period. (McDowell, R.C. 1983; 1986).

Rock hydraulic conductivity testing with water pressure packer isolated intervals was conducted for bedrock in all borings, except for borings B-GE-5 and B-GE-6. Results indicated moderately low hydraulic conductivity in most tests and throughout most of the borehole lengths, including the proposed tunneling zones. (Black & Veatch 2018).

Natural Gas

None of the exploratory borings advanced as part of the original WPT subsurface investigation encountered natural gas in measurable concentrations. Over 40 borings were drilled to depths between 40 and 260 feet along the original 13,160-foot-long WPT alignment. Although no evidence of gas was detected, the WPT was classified as "potentially gassy" within the WPT—Tunnel and Shafts contract documents. In part, this is due to the presence of gas documented at other underground construction projects in the Louisville area.

Of the six deep bedrock borings, natural gas was encountered in four, B-GE-3 through B-GE-6, during the geotechnical investigation. Natural gas was encountered at varying depths, with fluctuating gas concentrations and pressures, and varying time intervals of venting. In addition, geophysical testing was conducted on three of the borings where natural gas was encountered.

The borings were generally located from the midpoint of the WPT Extension alignment to the proposed new Retrieval Shaft. Mitigation measures will be implemented before and during construction of the WPT Extension. Natural gas mitigation measures include electrical and monitoring upgrades to the TBM, pre-excavation venting, and the venting of natural gas along the alignment before and during tunnel construction.

To evaluate the occurrence of natural gas trapped within the bedrock, borings B-GE-3 through B-GE-6 were converted to gas monitoring and testing wells by installing a packer assembly in the borehole, with gas piping and valves leading up to the surface. A gas meter and pressure gauge was to be used to record pressure buildup within the borehole and volume of gas dissipated during the gas testing procedure. However, by the time gas testing commenced, gas emissions from the borings had nearly ceased. Pressure buildup within the borings after the gas testing wells were shut in was negligible.

To facilitate the venting of natural gas prior to construction activities along the WPT Extension alignment and at the Drop Shaft DS06 site, a series of 21 gas vent borings were drilled along Beargrass Creek Trail. Each of the gas vent borings were completed with a packer assembly, gas vent piping, and a valve at land surface; the valve made it possible to shut in the gas vent borings if needed.

Environmental Considerations

Along the original WPT alignment, environmental soil sampling and testing were performed on selected soil samples from three environmental test borings. Soil samples were analyzed for RCRA metals (arsenic, barium, cadmium, chromium, mercury, lead, selenium, and silver) via EPA Method 6010B (7471A for mercury), for PCBs via EPA Method 8082, for PAHs via EPA Method 8270C-SIM, for VOCs via EPA Method 8260B, and for SVOCs via EPA Method 8270B. Environmental groundwater sampling was completed on several piezometers along the project alignment.

Additional environmental soil samples were collected as part of the 2018 WPT subsurface investigation. Four soil samples each from two borings were analyzed. In addition to the analyses performed on the soil samples taken during the original subsurface investigation, the soil samples collected during the WPT Extension subsurface investigation were analyzed for SVOCs via EPA Method 8270B. Environmental groundwater sampling was also completed.

Impacts to soil have been previously documented at a site south of Drop Shaft DS04 along the WPT Extension alignment. Extensive environmental studies have been conducted at a former metals recycling facility at this parcel. The building structures and recycling equipment have since been removed. The WPT Extension alignment will traverse beneath this property. From the 1920s until approximately 2002, metals recycling operations were conducted on the parcel. Documented releases of contaminants have occurred at the site (Smith, 2001).

Because of metals recycling operations, polychlorinated biphenyls (PCBs), metals (such as lead, mercury, chromium, and barium), semi-volatile organic compounds (SVOCs), BTEX (benzene, toluene, ethylbenzene, and xylene), Polycyclic Aromatic Hydrocarbons (PAHs), and total petroleum hydrocarbons (TPH) have been detected in soil samples collected at the site.

Exposure to the environmentally impacted soils at the former metals recycling facility is not expected during the construction of the WPT Extension. However, as the tunnel is advanced though the bedrock units beneath the property, there is a potential for contaminated groundwater, resulting from the downward migration of contaminants, to enter the tunnel through fractures and discontinuities in the bedrock. The need to identify and mitigate this risk was identified early in the design process. As such, pre-excavation grouting is required along this portion of the WPT Extension alignment.

Select piezometers and borings situated along the WPT Extension alignment underwent groundwater sampling to characterize and baseline groundwater quality conditions along the project alignment. Groundwater samples were collected from one deep (bedrock) piezometer, one geotechnical boring and from several gas vent borings situated along Beargrass Creek Trail. The purpose of the sampling and analysis program was to establish groundwater quality baselines for construction. The data was used to determine if tunnel construction, and the long-term operation of the CSO tunnel system, may have any impact on groundwater quality over time. Data resulting from the collection and analysis of groundwater samples could be utilized in the selection of remedial measures necessary for the handling and treatment of environmentally impacted groundwater that may enter the tunnel during construction.

WATERWAY PROTECTION TUNNEL EXTENSION DETAILED DESIGN

Upon approval from MSD, Black & Veatch began the detailed design phase to extend the 20-foot diameter WPT to the I-64 & Grinstead CSO Basin site to capture flow from three existing CSO structures and eliminate the I-64 & Grinstead CSO Basin and pump station near the intersection of Lexington Road and Grinstead Drive. The WPT Extension consists of the following major components:

- Approximately 7,240 feet of additional 20-foot diameter deep storage tunnel.
- Approximately 120 feet of 10-foot diameter adit for Drop Shaft DS04 installed by drill and blast methods.
- Retrieval Shaft/Drop Shaft DS06—a new 29-foot diameter retrieval shaft, which will be converted into a drop shaft located at the I-64 & Grinstead CSO Basin site.
- Junction box and gate structure upstream of the Retrieval Shaft/Drop Shaft DS06 at the I-64 & Grinstead CSO Basin site.

The horizontal alignment of the Waterway Protection Tunnel, including the WPT Extension is shown on Figure 3.

The WPT Extension will provide approximately 11 MG of surplus storage above the required minimum storage volume of the combined four CSO Basins. The elimination of the I-64 & Grinstead CSO Basin also provides the added benefit of eliminating a remote basin, pump station, and the associated operation and maintenance of the facility.

Drop Shaft DS06 Site Coordination

The I-64 & Grinstead CSO Basin had been excavated to the top of bedrock at the time MSD decided to eliminate the basin project and extend the WPT project. The completed excavation work presented a cost-saving opportunity at the Retrieval Shaft/Drop Shaft DS06 site. The design team located the Retrieval Shaft for the WPT Extension in the footprint of the existing excavation, eliminating the need to construct secant piles through the overburden. Due and the potential for flooding at the site, MSD authorized a contactor to secure the site prior to beginning construction activities related to the WPT Extension. Sheet piling was installed between the south bank of Middle Fork Beargrass Creek and the excavated area to prevent further flooding at the site.

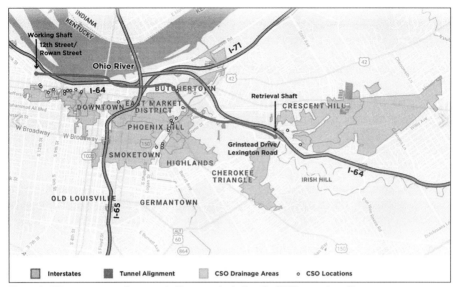

Figure 3. Waterway protection tunnel alignment, including the WPT extension alignment

MSD will bid a separate contract to perform the final site grading and landscaping at the Drop Shaft DS06 site, as well as to construct the consolidation sewers from the upstream CSO structures to the gate structure at Drop Shaft DS06. Construction scopes and schedules were carefully coordinated between design teams, ensuring smooth transitions between contractors, seamless scope requirements in contract documents, and a feasible construction schedule to meet the IOAP deadline.

Design Execution

Detailed design of the Waterway Protection Tunnel Extension was completed in approximately five months. Black & Veatch allocated global resources in multiple office locations to allow more work to be accomplished on the project each day than would typically performed in an eight-hour work day and 40-hour work week.

Design execution required frequent communication both on a consistent and as-needed basis with the internal design team, the MSD design and operations team, key stakeholders, permit reviewers, and regulators. A rigorous quality control program that was established at the beginning of the detailed design phase was followed and updated as design progressed.

In addition to the quality control program, MSD and Black & Veatch coordinated design development efforts with the contractor to control scope and construction costs before requesting a change order to the project.

CONCLUSION

The Waterway Protection Tunnel (WPT) Extension project combines the IOAP storage volume requirements from four previously identified CSO storage basins: The Rowan CSO Basin; the Story and Main CSO Basin; the Lexington and Payne CSO Basin; and the I-64 & Grinstead CSO Basin. Detailed design of the project was completed in approximately five months to accommodate construction completion before the IOAP deadline of December 31, 2020.

A detailed subsurface investigation was necessary to baseline key criteria along the WPT Extension alignment. The field work and lab testing considered the rock quality, groundwater, natural gas, and environmental considerations. Mitigation measures were implemented early as risks became apparent. MSD and Black & Veatch coordinated design criteria and risk mitigation efforts with the contractor to manage cost and schedule impacts.

The compressed design schedule required the design team to communicate early and consistently with key stakeholders and teaming partners, including regulatory authorities, contractors, and multiple consulting engineering firms. A concentrated team effort drove the detailed design phase of the WPT Extension to successful completion in approximately five months.

The Waterway Protection Tunnel, including the extension to the I-64 & Grinstead CSO Basin site, will allow MSD to meet the IOAP deadline of December 31, 2020, while increasing the storage capacity of the CSO storage and conveyance system.

ACKNOWLEDGMENT

The authors thank Greg Powell for his contribution to this paper.

REFERENCES

Black & Veatch (2017), Ohio River Tunnel Basis of Design Report—Revised Draft.

Black & Veatch (2018), ORT and ORT Extension Groundwater Quality Summary—Draft.

Smith Management Group. (2001), Site Characterization Summary, River Metals Recycling LLC. Smith Management Company, Author.

Louisville and Jefferson County Metropolitan Sewer District (2009), Integrated Overflow Abatement Plan, 2012 Modification.

McDowell, R.C. (1983), Stratigraphy of the Silurian Outcrop Belt on the East Side of the Cincinnati Arch in Kentucky, U.S. Geological Survey Professional Paper 1151-F, p. 24.

McDowell, R.C. (1986), The Geology of Kentucky—A Text to Accompany the Geologic Map of Kentucky, U.S. Geological Survey, Professional Paper 1151-H, online Version 1.0.

PART

Large Span Tunnels and Caverns

Chairs

Geoff Fairclough
Schiavone Construction

Alston Noronha
Black & Veatch

Construction Considerations for Garage Cote Vertu in Montreal

Jean Habimana • Hatch, Montreal, Quebec

ABSTRACT

The Montreal Transportation Agency (Société de Transport de Montreal—STM) has undertaken the construction of the Cote-Vertu underground storage and maintenance garage that will facilitate daily operations and offer parking spaces for the new Azur trains. The project includes three parallel tunnels that are 25 ft wide and a length of 600 for one tunnel and 1200 ft for the other two, a pillar with of 16.5 ft; a 2,000 ft long connecting tunnel to the existing tunnel. The junction that connects the existing tunnel to the new tunnel has large caverns with very shallow rock cover that had to be excavated in well-defined sequence to ensure its stability. The contract required mechanical excavation with the use of a roadheader to excavate the garage facilities to ensure the stability of the pillars and for the junction to guaranty the structural integrity of existing tunnel. The project is located near known major faults zones that were crossed during excavation of existing facilities.

The paper addresses construction considerations and provides updates on ongoing construction activities that started in May 2017.

PROJECT DESCRIPTION

The Metro of Montreal which is operated by STM was built in early 60s and its first line has been in operation since 1966. It uses rubber-tired rolling stock and was inspired by the Paris Métro. It has expanded since the 1960s from 26 stations originally on three separate lines to currently 68 stations on four lines with a total length of approximately 43 miles to link the Montreal Island to its northern and southern suburbs. The latest extension program was finished in 2007 for the Laval metro Extension that added three more stations and provided a direct link to the northern suburb of Laval via the Orange Line. It is Canada's busiest rapid transit system, and North America's third busiest by daily ridership delivering an average of approximately 1,300,000 daily unlinked passenger trips per weekday. Figure 1 shows the metro layout with its four lines.

With the increasing ridership on system and expected expansions of the offering, the construction of the Garage Cote Vertu Project that is currently underway addresses four objectives for the STM, namely:

- Provide underground track space for parking new trains that are being acquired by STM
- Allow the STM to increase frequency of service by adding more trains on the Orange Line during peak period
- Provide additional trains that will be able to handle expected ridership growth along the Orange Line, which is the backbone of the metro system, in coming years
- Ensure that the offer of service will expand to match the expected increase in demand from the future Blue Line Extension that will extend the current offering to the East of Montreal by adding five stations. This $3.9 billion Extension

Construction Considerations for Garage Cote Vertu in Montreal

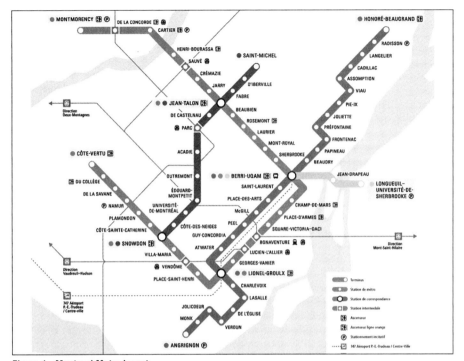

Figure 1. Montreal Metro layout

was recently approved by authorities and STM is gearing up to start preparing tender documents for its construction.

The Garage Cote Vertu Project will in fact house additional ten new third generation Azur Metro trains that are being acquired by STM. The underground storage and maintenance facility is located 1.5 km to the west of the Orange line Côte-Vertu station. Planning and feasibility stage were carried out in 2014 to 2015. The localization of the facility was selected to be able to accommodate the needs of the future extension of the Orange Line towards the north. Land availability led to the selection of an underground layout with limited surface footprint. Figure 2 shows the location and layout of the facility and its connection via a tunnel to the existing Orange line tunnel. Detailed design and preparation of bid documents were conducted during 2016 and the project was tendered under Design-Bid-Build delivery method in winter of 2016 with notice to proceed in Spring 2017. At the time of writing this paper the excavation work is almost completed and concrete works for permanent structures are well underway. It is expected to be in service by end of 2021.

Planning and design considerations were addressed in a separate paper (Osellame and Habimana 2018). This paper will focus on construction considerations.

Key components of the project include as illustrated in Figure 2:

- An area serving as the junction between the connecting tunnel and the existing Orange line tail track, which is referred to as Junction hereafter;

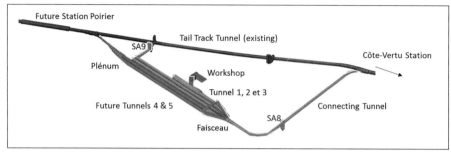

Figure 2. Projected infrastructure of the Garage Cote Vertu Project

- An 28 to 33 ft wide double-track 0.5-mile-long horseshoe shape tunnel connecting (called hereafter Connecting Tunnel) the facility to the existing Orange line tail track. Once completed this portion will house four trains;
- An open-cut area linking the three tunnels to the connecting tunnel that is called hereafter Faisceau. It was designed to accommodate two other tunnels (tunnels 4 and 5) that are planned for the future extension. It is currently used as main work area;
- Three parallel 28 ft wide double-track slightly horse shoe shape tunnels, of which tunnel 1 is 500 ft in length and will serve as the maintenance area and used for small repairs while tunnels 2 and 3 that are 1000 ft length will serve for the storage of the metro trains for 4 cars each;
- A 70 ft by 85 ft open-cut workshop and maintenance facility that will house trains for non-major repair and maintenance work. This workshop is located at the tail end to Tunnel 1;
- Two adit structures: SA9, which is 52ft by 16 ft and will house ventilation fans and emergency exit building and SA8, which is 26 ft by 20 ft and will house an emergency exit. Both structures were positioned to comply with NFPA 130 requirements for fire life safety.

GEOLOGICAL CONDITIONS

As outline in Osellame and Habimana (2018), the geology of Montreal Island consists of a variety of Pleistocene and Recent deposits that overlie early Paleozoic sedimentary rocks and Precambrian rocks, both of which are cut by Mesozoic intrusions. The principal events affecting the near surface rock assemblage were faulting, gentle folding, and minor metamorphism in the Mesozoic era and multiple glaciations and isostatic movements during the Pleistocene epoch (Grice 1972).

The geology along the alignment consists mainly of shaley limestone of the Tretrauville Formation. A series of East-west normal faults that are considered to be related to the Monteregian intrusive event during the Cretaceous and/or to the Taconic orogeny were predicted and encountered during excavation.

The limestone is moderately 4 to 15 inch thick greyish regular beds composed entirely of very finely textured micrite and interbedded with 0.5 to 4 inch thin shale beds. Several intrusive that consists of dykes or sills with thicknesses varying from a few inches to several feet are distributed within the rock mass at all depths. Rock mass discontinuities consist of a series of vertical to sub-vertical joints and horizontal joints related to the bedding planes. The main challenge was to predict the location of faults and the quality of the intrusive for stability of the excavation.

The overburden is composed of compact to dense glacial deposits consisting of a well graded mixture of gravel, sand, silt and a trace of clay. Cobbles and boulders are anticipated, and recent excavations have encountered boulders up to 1 m in size. The top of the rock is between 25 to 30 ft deep. The ground water table varies seasonally from a 5 to 8 ft deep.

The average mechanical properties of the limestone are: unconfined compression (UCS) of 91 MPa, Brazilian Tensile strength (BTS) of 7.6 MPa, Young's modulus 28 GPa and Poisson's coefficient 0.13. Cerchar abrasivity index (CAI) and quartz of 1.1 and 6 percent respectively.

The shale component of the rock mass exhibited lower properties in the range of 44 MPa for the UCS, 4.7 MPa for the BTS, 17 GPa for Young's modulus, 0.12 for Poisson' coefficient, and CAI 0.5.

The rock mass is dissected by a conjugate set of vertical joints, planar joints or discontinuities corresponding to the bedding planes and discontinuous surface joints. Spacing of the vertical joints varies from 3 to 10 ft. Joint surface characteristics are generally described as closed with no alteration and undulating to planar.

Estimated ground water inflows for the tunnels part of the project of 110 gallons/min with flush flows rate of the order of 254 to 317 gallons/min for the connecting tunnel and for the three garage tunnels.

SELECTION OF EXCAVATION METHODS

Contract Requirements

For the underground excavations of the three tunnels of the garage, mechanical excavation by roadheader was mandated by contract documents. This was due to the need to ensure the integrity of a tight pillar width required to remain within the available site right of way and reduce the footprint of the garage as explained before. In addition, the presence of vertical joints and low in situ stresses added to the concern of the stability of the pillars if drill and blast was to be used. As outlined in literature (Restner and Reumueller, 2004), the above-mentioned rock mass properties showed the limestone strength is on the upper range of the currently available on market roadhearders so specification included tight requirements to make sure prospective bidders select the heavier machine. In fact, it is known that the size and the power rating play a significant role in cutting the rock economically. During the preparation of the Contract documents and the Geotechnical Baseline Report (GBR), the performance of the roadheader was based on the experience gained on the Laval Metro Extension project in 2003–2005, in New York City Second Avenue Subway as well as in Ottawa Light Rail Phase 1 Project. Of particular interest and usefulness is the data published by Restner U and Reumueller B. (2004) on the performance of the ATM105 roadheader used on the Laval Extension Project. The analysis which was undertaken compared the ratio of UCS to BTS as a measure for the evaluation of rock toughness evaluated as low to average while the fracture energy was used as an additional factor for the evaluation of rock cutting behavior. It must be noted that the test results presented above do not account for the positive effect of the softer shaley beds on the excavation prediction rate, which are present for the Garage Cote Vertu. This was taken into accont in predicting the roadheader performance based as they are favorable to roadheader performance.

Table 1. Class of excavations and types of initial support outlined in bid documents

Class of Excavation	II	III	IV	V
RQD	Good	Fair	Poor	Very Poor
RMR	61 to 80	41 to 60	21 to 40	<21
Garage Facility Tunnels 1,2 and 3	52%	44%	3%	1%
Connecting Tunnel	51%	31%	13%	5%
Support System	Pattern grouted bolts, 10 ft long, 6 ft spacing and 3 inch SFRS	Pattern grouted bolts, 10 ft long, 5 ft spacing and 3 inch SFRS	Pattern grouted bolts, 10 ft long, 4 ft spacing and 4 inch SFRS	W 200 x 52 and W250 x 45 steel sets at 3 ft spacing, 8 inch SFRS. Pre-excavation support with spiles as required
Unsupported Length	Garage facility: 16.4 ft mechanical	Garage facility: 16.4 ft mechanical	Garage facility: 8.2 ft mechanical	All tunnels: 3 ft mechanical
Advance	Connecting tunnel: 8.2 ft if drill and blast	Connecting tunnel: 6 ft if drill and blast	Connecting tunnel: 5 ft if drill and blast	

Figure 3. Plan of projected infrastructure of the Garage Cote Vertu Project

For the connecting tunnel, the choice of construction method between drill and blast or mechanical excavation method was left to the contractor. Table 1 outlines the design of initial support for different parts of the project. Ground support systems and sequence of excavation were imposed for junction and ventilation tunnel whereas for tunnel 1, 2, 3 and the connecting tunnel an observation method was preferred. This was due to complexity of the junction and the ventilation tunnel geometry as explained further in the paper.

Contractor Selection of Excavation Methods

The successful Contractor elected to use roadheader for the tunnels 1, 2 and 3, the ventilation tunnel, and the connecting tunnel as well. Roadheader was also used for a big part of the open cut of the Faisceau.

Drill and blast was primarily used for open cut excavations of the SA8, SA9 and the Workshop. Two Sandvik roadheaders model MT720 were mobilized. Their weight is 135 tonnes with installed power of 300 kilowatts and a maximum cutting height of 6.5 m (21.3 ft). Its cutter head type is R-400-TC87 with 22 mm (0.87 inch) insert picks. Figure 4 shows one the roadheader excavating the Faisceau.

Figure 4. Roadheader model MT720 excavating the open cut of the Faisceau

There are three main factors that affect the performance of the excavation by the roadheader:

- The machine parameters, such as its installed power, the type of cutter head and its mounted rock cutting tools
- The rock mass properties, which depend not only if intact rock properties but also of the discontinuities such as bedding and joint
- The workmanship, i.e smooth operation and permanent maintenance will contribute to a successful performance over the duration of the project

There are two parameters that are typically used to assess the excavation performance of the roadheaders:

1. Roahdheader cutting performance that is directly related to the machine installed power and the rock compressive strength, toughness (deformability), and the jointing. It is typically provided using cutting chart that are provided by the machine manufacturers based on past experience. One of the way of expressing the cutting performance is the Net Cutting Rate (NCR), which corresponds to the volume of cut rock to the effective cut time counted in net cutting hours. The effective net cutting time is defined by the time the cutter head is in contact with the rock and actually cutting. This excludes the time spent on other activities such as profiling and loading or idling time.

2. Roadheader cutting tool wear, which is measured by cutting bits consumption. It is directly related to the Cerchar abrasivity Index (CAI), which is in turn related to parameters such as quartz content and the degree of mineral interlocking. There are several ways to measure the cutting tool wear in literature and the most used is the Specific Picks Consumption (SPC), which corresponds to the number of picks replaced per cubic meter of solid rock excavated.

As illustrated on Figure 5 the NCR is higher for rock mass of lower RMR and diminishes exponentially as RMR increases. Conversely SPC is lower for rock mass of lower RMR and increases exponentially as RMR increases.

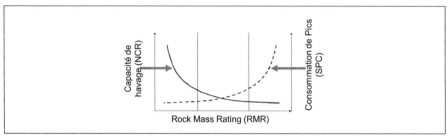

Figure 5. Comparison of NCR and SPC with RMR

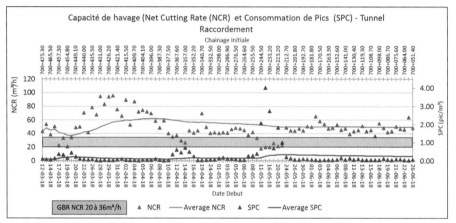

Figure 6. Measured roadheader performance for the connecting tunnel

Achieved Performance

Based on previous experience, the GBR specified NCR varying between 30 and 36 m³/h for a rock mass with a UCS of 90 and 79 MPa respectively, which corresponds to an advance rate of approximately 16.5 ft per day for most of the tunnels. For the rock mass with a higher UCS and possible igneous intrusions, predictions of NCR were 20 m³/h. Figure 6 shows NCS and SPC values that were achieved for the Connecting Tunnel, which shows the following:

1. The predicted NCR value in GBR was cautious and represent the lower range of measured values for the value along the 0.5 mile length of the tunnel

2. The contractor was able to achieve good NCR with low SPC in most of the cases which translated to an average advance rate of 17–20 feet per day versus 16.5 feet that were predicted

3. An altered dyke was encountered between stations 700+440.00 and 700+396.00 and the contractor was able to achieve very good advance rates of approximately 20–25 ft per day with a NCR ranging between 70 to 95 m³/h. The UCS for the soil-like altered dyke is approximately 10–20 MPa

4. SPC spiked when the excavation encountered a hard intrusion at mileage 700+244.5. The UCS for this intrusion is approximately 200–250 MPa and it was 3 to 5 feet thick. Figure 7 shows a photo of the intrusion in the tunnel face. Advance rate dropped to less than 10 ft per day.

Table 2. Summary of achieved roadheader performance for the different tunnels of project

	Connecting Tunnel	T1	T2	T3	Ventilation Tunnel
Average SPC (pic/m³)	0.25	0.12	0.31	0.42	0.34
Average NCR (m³/hr)	49.3	56.5	40.1	36.2	34.9
Average Excavation Rate	42%	36%	34%	33%	33%

Table 2 summarized achieved performance for all the tunnels. There is a good correlation between the rock mass quality and the achieved performance. The achieved performance generally reflects encountered geology for each of the tunnel. Notably an altered dyke was encountered in T1 over approximately zzz ft length, which translated to lower SPC and higher NCR.

Drill and blast method was used for all surface open-cut excavations and the main challenge being the need to limit vibration through the highly populated and urban areas and to comply with the City of Montreal maximum peak particle velocity (PPV) vibration criteria of 25 mm/sec. The techniques employed to limit vibrations consist of closely spaced large diameter line drilling holes using a ratio of diameter to spacing of 0.56 combined with cushion blasting to achieve the final walls. Mass excavation uses 5 m benches, a 0.6 kg/m³ powder factor, electronic delays with a limit of one hole per delay. To date, the vibration criteria has been met.

Figure 7. High strength intrusion encountered in the Connecting Tunnel

Predicted Versus Encountered Ground Conditions

Figure 8 illustrates a comparison of predicted versus encountered ground conditions. It shows that there were closely matched except that the expected extremely poor rock of Class V was never encountered along the alignment of the Connecting Tunnel. This matches most of excavation except for T1 where a fault zone that was expected to be outside the tunnel footprint ended up in the tunnel for approximately 70 ft. That zone affected also the excavation of the Workshop where longer bolts had to be used to go beyond the altered layer. The contractor used passive bolts in this area because they could not be able to torque.

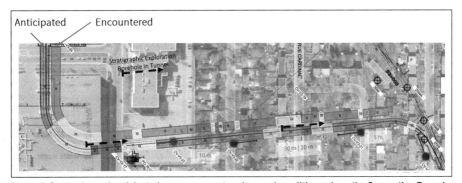

Figure 8. Comparison of anticipated versus encountered ground conditions along the Connecting Tunnel

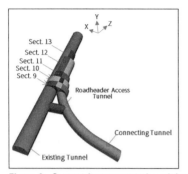 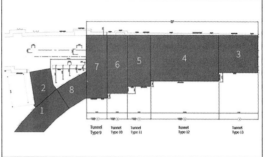

Figure 9. Geometric representation of the junction and mandated sequence of excavation

CHALLENGE OF EXCAVATING THE JUNCTION AREA

The excavation of the Junction represented a geometric challenge that had caverns of varying spans to connect to existing tail track tunnel. To maintain the structural integrity of existing tunnel, it was mandated to use roadheader to excavate the Junction and based on a series of 2D and 3D numerical modeling an excavation sequence was also imposed to the contractor. Figures 9 and 10 illustrate the sequence that was mandated and implemented successfully. The sequence used an roadheader access tunnel and a sequence that first used a pillar in the middle for the large span cavers that could reach as much as 61.7 ft for the largest cavern with a rock cover of 4 ft only. The geometry of the excavation sequence accounted for the roadheader working capabilities and limitations. Additional investigation was conducted to confirm the quality of rock above the caverns before excavation. Initial support consisted of a 3 inch steel fiber reinforced shotcrete applied in 2 layers and a series of 13 ft long grouted rock bolts at 3 ft spacing. Spiles and steel sets were specified in case the ground conditions could become poor than anticipated but they were not used. An instrumentation program was specified and implemented to follow the stability of the excavation and an action plan was pre-established in case trigger levels were reached but they were not implemented as the caverns were stable.

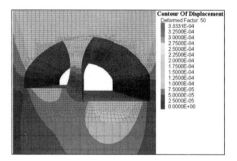

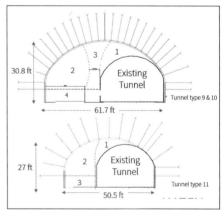

Figure 10. 3D Numerical modeling of excavation sequence and initial support for the junction

EXCAVATION OF THE VENTILATION TUNNEL

Unlike the Tunnels 1, 2, 3 and the Connecting Tunnel that have a span ranging between 28 and 29 ft with a hight ranging from 20 to 23 feet, the Ventilation Tunnels that are at the end of the Tunnels 1 and 2 have a span of 36 ft and a height ranging

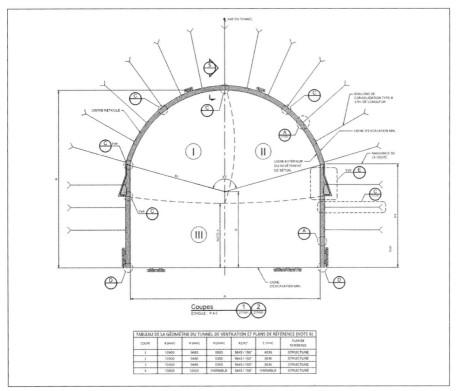

Figure 11. Excavation sequence and initial supports elements for the ventilation tunnel

Figure 12. Photo showing excavation sequence for the ventilation tunnel

from 32 to 36 ft. This means that while the roadheader excavated Tunnels 1, 2, 3 and the Connecting Tunnel in full face, sequential excavation had to be used for the excavation of the Ventilation Tunnels because of geometrical limitations of the roadheader. Figure 11 shows the sequence that were indicated on contract documents and implemented by the contractor for the Ventilation Tunnel. Initial support consisted of a 10 inch steel fiber reinforced shotcrete applied in 3 layers, 13 ft long grouted rock bolts at 4 ft spacing, and lattice girders. Figure 12 shows a photo taken during the excavation of the left top heading with installation of initial support elements.

CONCLUSIONS

The STM has undertaken the construction of the Cote-Vertu underground storage and maintenance garage that will facilitate daily operations and offer parking spaces for the new Azur trains. The project is expected to be in service by end of 2021. Selection of the project site was based on the need to accommodate future extension of the Orange Line and land availability that led to an underground facility with limited surface footprint.

The excavation of the garage facilities is almost complete at the writing of this paper. The contract documents mandated the use of roadheader for some facilities to ensure the integrity of pillar between the three parallel tunnels and the structural stability of existing structure at the Junction with existing tail track tunnel. The contract left however open the choice of excavation method for the 0.5 mile Connecting Tunnel and the large open cut excavation called Faisseau. The successful contractor elected to mobilize two Sandvik roadheaders model MT720 and limit drill and blast to few area of open cut excavations on the project.

The paper outlined achieved performances of the roadheaders and their correlation to observed geological conditions. In general, the encountered geological conditions were close to anticipated conditions except for one tunnel where a layer of altered dyke that was expected to be outside the project alignment was encountered within the excavation area. Careful planning and design was used to account for the capability and limitations of the existing roadheaders. This led to realistic sequences that were implemented successfully for both running tunnels and at critical sections such as the Junction to existing tunnel and the Ventilation Tunnel.

REFERENCES

Grice, R.H. (1972). Engineering Geology of Montreal, International Geological Congress, Canada, B-18, 15pages.

Osellame G. and J. Habimana (2018). Subsurface Investigations, Design and Construction Considerations for the Montreal Transportation Agency Côte-Vertu Underground Storage and Maintenance Garage, North American Tunneling Conference, Washington DC., 739–745.

Restner, U., Reumueller, B, « Métro Montreal—Successful operation of a state-of-the-art roadheader—ATM 105-ICUTROC—competing with drill and blast operation in urban tunnelling » EUROCK 2004 and 53rd Geomechanics Colloquim Schubert et al (ed.) 2004.

First Large Diameter Hard Rock CSO Chamber in St. Louis

Patricia Pride ▪ Metropolitan St. Louis Sewer District
Kevin Nelson ▪ Black & Veatch
Clay Haynes ▪ Black & Veatch
John Deeken ▪ Black & Veatch

ABSTRACT

The Maline Creek Tunnel (MCT) will be the first large diameter chamber to store combined sewer overflows in St. Louis. The MCT is a key feature of Project Clear, Metropolitan St. Louis Sewer District's (MSD) Long Term Control Plan to address sanitary and combined sewer overflows to local streams and rivers. Project Clear's estimated cost is greater than $4 Billion dollars making it the largest public works project to date for the state of Missouri. The MCT includes the construction of a 40-ft diameter, 12.5 MGD submersible pump station, a 28-ft diameter × 2,700-ft long cavern, a 580-ft long × 6-ft lined connecting tunnel, three deaeration chambers, three intake structures, a shallow connector sewer constructed by microtunneling, and 1,000 feet of 12-inch to 30-inch diameter near surface sewers. Bids were opened on March 10, 2016. The successful bidder was SAK/Goodwin JV with a bid of $82.3 M dollars. The Engineer's Estimate was $87.7 M dollars. The project was awarded in May 2016. The contractor mobilized on site in June 2016. The project is allocated 1567 calendar days to complete the construction.

BACKGROUND

St. Louis MSD serves approximately 1.3 million customers in St. Louis City and St. Louis County. Some of the oldest parts of MSD's sewer system were constructed in the 1850s and are combined sewers that convey both sanitary sewage and storm water in one conduit to treatment plants. During moderate to heavy storm events, storm water and wastewater discharges into local waterways from overflow points in the sewer system. These sewer overflow points act as relief valves when too much storm water enters the sewer system. Without these overflow points, many basements would experience backups and some streets would flood.

In 2007, the State of Missouri and the United States Environmental Protection Agency (USEPA) filed a lawsuit against MSD to reduce the sewer overflows. In August 2011, the Department of Justice filed a consent decree requiring MSD to spend approximately $4.7 billion dollars over the next 23 years to reduce the sewer overflows, increase water treatment capacity and provide other sewer system improvements. The Consent Decree between MSD, the USEPA and the Missouri Coalition of the Environment went into effect on April 27, 2012.

The MCT is located on the downstream end of the Maline Creek Watershed as indicated on Figure 1. The Consent Decree stipulated that the MCT would reduce the overflows at CSO BP 051 to four events or less, and 6 million gallons of untreated overflow volume in a typical year and that the MCT would reduce the overflows at CSO BP 052 to four events or less and 20 million gallons of untreated overflow volume in a typical year. The critical Consent Decree milestone for both CSO BP 051 and CSO BP 052 is 12/31/2020; Source: Consent Decree.

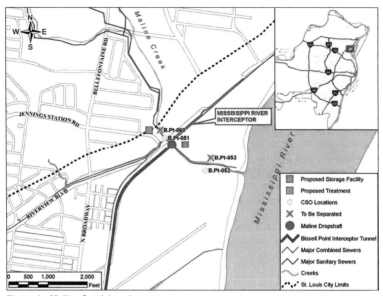

Figure 1. Maline Creek location map

MAJOR PROJECT FEATURES

The following project features were designed to meet the aforementioned performance criteria for CSO BP 051 and CSO BP 052:

- 2,700 ft long, 28 ft diameter concrete-lined cavern
- 177 ft deep, 40 ft diameter concrete-lined shaft serving as a 12.5 million gallon per day (MG) pump station
- 580 ft long, 6 ft diameter connecting tunnel
- three de-areation chambers
- three intake structures
- 213 ft of 24-inch diameter shallow connector sewer constructed by microtunneling
- 1,000 ft of 12-inch to 30-inch diameter near surface sewers

PREQUALIFICATION AND BID RESULTS

There was a very good response to the request for prequalification. Twelve domestic tunneling contractors and two European tunneling contractors were prequalified for the project. Six domestic tunneling contractors and one European tunneling contractor submitted bids. SAK/Goodwin Joint Venture submitted the successful bid at $82,828,282. The Engineer's estimate was $87,700,000.

The contractor was granted 1507 calendar days to substantially complete (SC) the project and an additional 60 calendar days beyond SC for final completion.

PUMP STATION (WORKING) SHAFT CONSTRUCTION

A Notice to Proceed was issued on May 13, 2016 and construction began on the working shaft and site in July 2016. Secant pile (SP) wall construction was required for the soil strata in the upper 40 ft plus or minus 2 ft. Nicholson was the subcontractor selected to construct the SP wall. Nicholson employed a Bauer BG-39 drill rig to install seventy eight 880 mm diameter unreinforced secant piles in a circular configuration. Some typical challenges with SP wall construction were encountered including protrusions, voids and windows.

Figure 2. Bauer BG-39 in center of photo

After careful observation of the protrusions after exposure, it appeared that the protrusions emanated predominantly from the secondary piles. One plausible explanation was that some ground was lost during drilling of the secondary piles and concrete flowed into the voids during pumping. The protrusions were removed from the interior of the SP wall with a hydraulic hoe ram mounted to a small backhoe during soil excavation.

Some voids developed in the secant piles above the protrusions. These voids were successfully repaired with shotcrete and welded wire fabric, upon agreement with the SP designer and Nicholson engineer.

One secant pile was improperly located at the shaft collar resulting in a full depth window. This pile was redrilled to remediate this problem. Nicholson elected to redrill two other piles as a precautionary measure even though these piles were within verticality tolerances according to instrumentation. The resulting cold joints continued to leak groundwater for the duration of the project. After these standard repair measures were implemented, the SP wall proved to be a very robust and nearly watertight support system. Only minor seepage was observed on the secant pile wall where repairs were not attempted.

SAK/Goodwin designed blast plans that only excavate one-half of the shaft footprint in each blast. While monitored vibrations were less than the allowable 0.5 in/sec tolerance, the overpressures recorded were at the allowable limit, which prevented full-width shaft blasting. Production blasting removed the rock in 10 to 14 foot lifts, with the rock supported by an 8'× 8' grid of 12' rock bolts. Welded wire mesh and shotcrete were added for stability and safety. A heavy blasting lid and mats were placed over the shaft for each blast to reduce overpressure and protect from fly rock.

Stub Tunnel and Main Cavern Construction

Horizontal tunneling using drill and blast techniques began in March, 2017. The 32 foot excavated diameter was driven in two stages, the 22 foot top heading and the 10 foot lower bench. Approximately 1,400 lbs of explosives were used in a 3-foot by 3-foot grid pattern with 18 inch line drilled trim holes. Intact limestone rock was excavated by loading every other trim hole with half the main powder charge (or 7 lbs/hole). Where thinly bedded limestone or shale was encountered in the excavation limits, a charging pattern of up to 3 to 5 hole produced reasonable results without excessive overbreak. Typical charges were 7 to 13 lbs per delay, using 14 foot long production holes.

Burn holes were generally placed within the hardest rock layer of the current exposed face. Poor production resulted from blast energy absorbed by shale seams, thinly bedded limestone, or soft dolomite seams. Great care was taken by the contractor to ensure the burn holes were drilled exactly parallel into the hardest rock within the face.

Stub Tunnel Ceiling Crack

During a shotcrete spraying operation in the 58' long stub tunnel, a longitudinal crack was observed to suddenly appear with a loud audible concussion. The crack extended from the main shaft opening for approximately 50 feet into the stub tunnel. The event registered on a nearby seismograph at 0.2 in/sec.

Figure 3. Drilling production blast holes in the south half of the shaft footprint

Work was stopped in the tunnel for 4 days while investigations and options were considered. Cooperatively working with the owner, contractor, engineer, shoring design, project engineer, and construction managers, it was decided to proceed forward with a repair consisting of 12' rock bolts securing 15' of mine straps running transverse to the tunnel along the ceiling. A second layer of 4×4 welded wire fabric was also installed for worker protection. Survey points were secured to the new reinforcement to monitor movements, and production work was re-started within 2 days. Shotcrete was not re-applied to provide means for observation of the crack. Later surveys revealed less than ¼" of movement of the repaired rock ceiling. Minor spalling was observed over the next 12 months.

Figure 4. Production blast pattern on top heading

Further investigation of the crack revealed that the movement occurred soon after removal of the bench heading in the stub tunnel. The application of shotcrete within 24 hours of excavation likely provided initial rigid support, preventing any stress relief or relaxation of the rock strata. Removal of the massive bench strata provided the deformation needed to crack the thin shotcrete, and fully load the rock bolts. The competent rock observed in the side walls lead to the project team taking an observational approach to shotcrete for the next 100 feet of excavation.

Close inspection of the main cavern ceiling during bench removal did not indicate any measurable movement in other portions of the cavern ceiling, and along changing headings. The engineer proposed installing rock bolts and wire mesh, then delaying shotcrete until needed for structural support or worker safety, allowing the CM and contractor to provide observations and work as needed. To this date, the cavern is completely excavated, 750 feet of cavern lining concrete placed, and there has been no need for subsequent shotcrete.

DROP AND VENT SHAFT EXCAVATIONS

Kiewit Foundations Group (KFG) was the subcontractor selected to drill the drop and vent shafts on the project. KFG selected Ziegenfuss Drilling to drill pilot holes for the drop and vent shafts using a percussive down hole drill. Pilot drilling began in June 2016 and was completed in August 2016.

Table 1 shows the dimensional parameters for the shaft drilling work.

KFG used a Liebherr LB36 drill with soil augers to drill the upper section of the shafts that was comprised of soil materials. A polymer-based drilling additive was added to the water to create a viscous drilling fluid that supported the shaft walls until drilling progressed to bedrock and a steel casing was installed. A steel casing was installed after soil excavation was completed. A concrete mud slab was constructed in the shaft base by tremie methods. The soil and the steel casing was backfilled with grout by tremie methods.

After the steel casing was installed and grouted, KFG attached Berminghammer drill pipes and air swivel and a Center Rock, Inc. (CRI) cluster rock cutterhead to the LB36 to drill the bedrock. Cuttings were removed by compressed air supplied by seven 1450 cfm Sullair air compressors. CRI supplied four different cutterheads including 6 ft and 7 ft full diameter cutterheads and 9 ft and 11 ft diameter overreamers. The 11 ft diameter excavations were to be drilled in three passes. The 6 ft diameter cutterhead was to be used on the first pass; the 9 ft diameter cutterhead was to be used on the second pass and the 11 ft diameter cutterhead was to be used on the third pass.

The 6 ft and 7 ft diameter cutterheads plugged with cuttings on several occasions requiring removal from the shaft excavation to unplug the cutterheads. Overall, drilling progress in the rock was favorable with an average excavation

Figure 5. 12' casing set in drilled hole

Table 1. Drilled shaft data

Shaft Name	Casing Dia. in Soil (ft)	Excavated Dia. in Rock (ft)	Approx. Depth to Rock (ft)	Total Depth (ft)
COR drop shaft	12	11	43	136
COR chamber vent shaft	8	7	43	136
COR cavern vent shaft	7	6	42	120
CHS drop shaft	12	11	32	155
CHS chamber vent shaft	8	7	32	155
CHS cavern vent shaft	7	6	33	140
CSO 52 drop shaft	8	7	45	138
CSO 52 vent shaft	7	6	45	138
NI shaft	11.5	N/A4	45	45

Notes:
COR = Chain of Rocks
CHS = Church Street
NI = North Interceptor
N/A = Not Applicable

rate inclusive of all down time, plugs, maintenance, etc. of about 2 to 3 ft per hour. With approval from CRI, additional passages were added to the cutterheads which reduced plugging and clogging. The discrete Pennsylvanian Shale and Mississippian Shale layers found within the Limestone units absorbed much of the drill's energy, and also contributed to the plugging.

Figure 6. Center Rock Inc. cutterhead

Spoil handling proved difficult due to the compact site layouts demanded by the project constraints. A small primary tank prevented larger cuttings from settling and to be re-circulated into the shaft, which were re-ground and absorbed energy that otherwise would have cut intact rock. Frequent cleanouts of the tank were needed and constant circulation of fluid was required to prevent freeze up.

DIVERSION STRUCTURE EXCAVATIONS

Three flow diversion structures will be constructed to convey combined sewer overflows to the MCT. The largest flow diversion structure (Chains of Rocks) is also the deepest diversion structure. The deepest section the Chain of Rocks structure is constructed in the upper section of the bedrock approximately 43 to 45 ft below grade. Nicholson was selected to install secant piles for the lower section of the Chain of Rocks diversion structure.

Figure 7. Installing a beam in a secondary pile

The majority of the secant pile wall was non-circular requiring the installation of W24 × 104 beams in the secondary piles and walers and struts inside the SP wall. In addition, toe pins were installed in the base of the secondary piles. There were some problems with cross communication of concrete into adjacent piles that were corrected promptly. Grout protrusions were observed during subsequent excavation, generally on the down-gradient side. As the site was located near (150 feet) from a cutoff flood wall, there was a buildup of groundwater near the site.

The other excavations are shallower and will be supported by sheet piling. Sheets were initially stuck and driven by a large vibratory hammer. Large seepage volumes forced the use of diesel impact hammers to better seat the sheets into rock, with good success. Obstructions were encountered during the diversion and intake structure work, which necessitated two jet grouting operations to seal off the excavations and to seal off below the removed sewer section. Where the jet grout rig was unable to reach due to rock or concrete obstructions, a soil drill rig utilizing a core bit was employed to penetrate the obstructions and pump a slurry of bentonite. The combination of techniques proved successful.

SUMMARY

The MCT is the first large diameter hard rock CSO cavern to be constructed in the St. Louis, MO area. SAK/Goodwin JV was given a Notice to Proceed with construction on May 13, 2016. The working shaft, drop and vent shafts and the Chain of Rock diversion structure are completed now The cavern lining work is 31% completed. The project is on schedule and no major problems have been encountered to date. The MCT is a key part of Project Clear's program to reduce water quality impacts to local streams and rivers in the St. Louis, MO area.

ACKNOWLEDGMENTS

Sincere thanks are due to the many team members including the Metropolitan St. Louis Sewer District (Owner), Jacobs Engineering Group Inc. (Design Engineer), SAK/Goodwin JV (General Contractor), Ziegenfuss Drilling (Pilot Hole Drilling Subcontractor), Kiewit Foundations Group (Vent and Drop Shaft Drilling Subcontractor), Nicholson (Secant Pile Wall Subcontractor), and Black & Veatch with Kwame Building Group Inc., (Construction Management Team).

REFERENCES

United States District Court for the Eastern District of Missouri Eastern Division, April 27, 2012, No. 4:07-CV-1120 (CEJ) Consent Decree United States of America and the State of Missouri (Plaintiffs) and Missouri Coalition for the Environment Foundation (Plaintiff/Intervenor) v. The Metropolitan St. Louis Sewer District (Defendant).

Rock Load Estimation for Shallow Rock Caverns

Charles A. Stone ▪ HNTB
Changsoo Moon (Kevin) ▪ HNTB

ABSTRACT

An empirical method for estimating geotechnical rock loads in shallow mined caverns is presented. The extreme variation of rock loadings on mined cavern linings with highly localized geotechnical conditions requires these underground structures to be designed in probabilistic terms with respect to ground classes, thereby minimizing costs for rock support. The method is based on previously measured rock loading data from final lining instrumentation and empirical geotechnical data from previously constructed mined subway station caverns in Washington, D.C. and New York, correlated with geological data and cavern geometry using the Q index value and scaled crown span equation.

INTRODUCTION

This publication is intended to provide the background for development of a new method for a preliminary estimation of rock loads that would likely be imparted on shallow mined cavern final linings permanently supporting underground cavern excavations using industry-accepted methodologies. This analysis has been carried out based on the following published methods of rock load assessment: Karl Terzaghi (1946); Deere (1970); Rose (1982); Cording (1972); Bieniawski (1973); Barton (1974); and Carter (1992). The new method is presented at the end of this paper.

Rock load estimates must be derived from expected geotechnical rock conditions estimated from spares borings and from related historical cavern excavation rock load data. Three methods available for rock engineers to assess ground loads are empirical, finite and discrete numerical modelling, and wedge analysis. For this paper, empirical methods and historical cavern lining load data are employed. The resulting methodology provides a method of estimating rock loads on shallow caverns.

Methodology

Definition of Rock Load

Rock load is the amount of weight from the overlying and lateral rock mass exerted upon the lining system for a given cavern excavation. Rock load is required as an input parameter in the design of the shotcrete initial linings (Desai, et al. 2007) and for cavern final linings. Specifically, rock load summarizes the impacts of the ground conditions, cavern geometry, and represents a convenient measure of ground behavior. Rock load is first estimated by rules of thumb and then verified by empirical, numerical modeling (2D and 3D, Continuum and Discontinuum), closed form solutions, and rock wedge analysis methods. The magnitude of the rock load carried by the support is determined by the support stiffness and the rock arching capacity to support the hoop stresses generated by the excavation. In shallow cavern excavations that require stiff support, the rock load has been shown based upon measured cavern load data to range from 13% to 61% of the overburden depth from the cavern crown to the ground surface after a comprehensive rock bolting program. (Cording, et al., Vol 1 1983) In shallow excavations that require stiff support and lack a sufficient crown pillar stability to develop arching action in the rock mass, the rock load can be expected to approach 80% to 100% of the vertical ground load. The goal of the design process is to verify the chosen ground support system, which is based on empirical rules and subsequently verify them with numerical and structural calculations along the cavern alignment.

Good Engineering Practices in Cavern Design

Good engineering practice requires that certain key parameters be determined in advance of the design of a mined cavern in rock. Key parameters include, but are not limited to;

- Reasonable estimate of the competency of the rock mass surrounding the cavern
- Stable cavern configuration geometry
- Compatible ground support strategy for the prevailing geotechnical conditions
- A minimum rock cover thickness over the cavern
- A minimum pillar adjacent to the cavern

After determining the prevailing geotechnical conditions at the planned cavern locations and establishing representative geotechnical design parameters, the key parameters must be evaluated, and determined to be compatible with the contemplated cavern design. Specifically, there must be a sufficient rock cover of stable rock over the cavern to allow for safe rock mining methods to be carried out. This single parameter will lower the risk and cost of mining the cavern and prevent adverse impacts to overlying and adjacent existing structures if the crown pillar does not perform in a manner to control settlement above the cavern. Good engineering practice further dictates that design of a cavern excavation should ideally be based upon multiple methods to establish rock loads. Cavern design features represented by large, shallow, non-circular openings, jointed rock masses, random shear zones, and variable rock covers required a robust design procedure to define the broad range of ground support conditions that might be encountered. (Desai, et al. 2005)

Four currently available methods of rock mechanics to establish rock loads for station caverns are typically considered; numerical modeling (2D and 3D, continuum and discontinuum), closed form solutions, rock wedge analysis, and empirical design methods. None of these methods is considered an inferior or superior analytical tool, but the methods are applied in appropriate circumstances. (Desai, et al. 2007) The method developed in this report is based on empirical design methods exclusively.

Numerical Methods (2D and 3D, Continuum and Discontinuum)

Numerical analysis is one of the most useful tools to evaluate ground-structure interaction precisely. Complex geometry and multi-stage construction sequences can be practically modelled considering various types of material models and support types. The available numerical methods include Finite Element Method (FEM) and Finite Difference Method (FDM), which are continuum-based models. The Discrete Element Method (DEM) is based on discontinuum analysis, which is able to approximate the effects of discontinuities within the rock mass directly. Numerical modeling via a modeling program such as UDEC and 3-DEC have been performed during final engineering to verify ground support systems around penetrations upon further clarification and/or confirmation of geotechnical conditions.

Closed Form Solutions

The state of stress due to tunnel excavation can be calculated from analytic closed-form solutions. The interaction between support and surrounding ground is described by the ground reaction curve, which relates internal support pressure to tunnel convergence. (Hoek, et al. 1995)

Rock Wedge Analysis

One of the major concerns for tunnels and caverns in jointed rock masses is the kinematic failure of discrete blocks in the excavation. This type of instability can lead to progressive rock failure which can destabilize the crown pillar. Therefore, the local stability of the rock mass between the rock support necessary for the global stability can be checked using a program such as UNWEDGE by Rocscience. Based on engineering geology input data (dip and dip direction) and excavation geometry and direction, the maximum geometrically formable blocks requiring rock support can be calculated. Probabilistic methods are also available. (Stone, et al. 1996)

EMPIRICAL DESIGN METHODS

Several empirical design methods have been developed for use in design of mined caverns as well as determining applicable rock support measures. The methods of Terzaghi, Deere, Rose, Cording, and Barton are directly applicable to the estimation of rock loads for underground structures. These empirical methods form the basis for the currently proposed rock load estimation method.

Empirical design methods are based on descriptions of the anticipated geotechnical conditions of the rock mass and cavern geometry. They are used to compare underground conditions between one project and another. Several of these, including the methods of Terzaghi, Bieniawski, and Barton have been used over many years to compare thousands of tunneling reaches in widely varying geotechnical environments. The portability of these methods makes them well-suited for estimating rock loads during the conceptual design of a cavern excavation project.

Empirical design methods were developed to assess the stability of underground cavern excavations via statistical analysis of the observed behavior of underground structures subjected to rock loads. The engineering rock mass classification systems are the best known empirical approach for estimating geotechnical loads in rock, which will ultimately be expressed as final lining loads. Although numerical techniques have advanced significantly in rock mechanics engineering, empirical design approaches are still preferable for a wide spectrum of design applications. These systems provide a systematic approach to correlate observed underground geotechnical conditions with observable behaviors on past projects in rock such as arching action, standup time, deformation, support requirements, probability of failure, and rock loads. These classifications are intended to be used as guides, and in conjunction with analytical studies, field observations and measurements, rather than solely as ultimate design solutions. However, the available data bases are huge, and the systems have been developed to a point where their track records are very reliable.

Several of the most relevant rock mass rating systems for designing mined excavations in rock are summarized as follows;

- Equivalent Rock Load, Karl Terzaghi (1946)
- Deere, et al. (1970)
- Rose (1982)

- University of Illinois at Urbana-Champaign, Cording, et al. (1972)
- Geomechanics System, Penn State University, Bieniawski (1973)
- Rock Tunneling Quality Index, Norwegian Geotechnical Institute, Barton, et al. (1974)
- Scaled Crown Span, Golder Associates, Carter (1992)

Equivalent Rock Load, Karl Terzaghi (1946)

The Terzaghi Rock Load method was developed by Karl Terzaghi in 1946, for use in designing steel sets for rock tunnel supports in widely varying rock conditions. (Proctor and White 1946). This is a method of predicting the degree to which an arching action is developed over the crown of the tunnel due to prevailing geotechnical conditions. The equivalent rock load is the predicted weight of the broken ground resulting from the excavation of the tunnel. This "was a landmark paper in tunneling literature, and for many years it provided the basis for the rational design of tunnels, particularly those constructed in North America. There are still many valuable lessons to be learned from this work, and it is recommended reading for anyone seriously interested in the practical aspects of tunnel design and construction." (Hoek 2000)

One of the important lessons from this method was the concept of correlating guidelines for estimating the rock load with observable differing geotechnical conditions. More importantly, this method recognized and incorporated a range of ground conditions from essentially stable to immediate collapse which are represented by a range of rock loads from 0 in hard intact rock to 250 feet in swelling conditions. Although the steel set method has fallen out of favor by designers due to its high cost, it is still used in North America as a last resort for addressing the toughest ground conditions.

Deere et al. (1970)

The Terzaghi concept for designing steel sets based on rock loads determined by research and experimentation was used extensively for designing steel sets. This system has experienced two major modifications after 1946. The first update was published by Don Deere, Ralph Peck, Harvey Parker, of University of Illinois, with J. Monsees and B. Schmidt from the construction industry in 1970. This study was a comprehensive evaluation of the design procedures for tunnel lining systems. This evaluation lead to the provision of new guidelines for the selection of support systems for 20-ft to 40-ft diameter tunnels in rock based on Rock Quality Designation (RQD) rather than qualitative rock descriptions. It should be noted that irrespective of the additions and refinements to this rock load estimation method, the range of rock loads remained 0 to 250 feet.

Rose (1982)

The Terzaghi method was further modified by Rose in 1982. The paper published by Rose was primarily concerned with reducing the cost of tunnel supports in tunneling practice in the United States, which continued to use steel sets after European practice had shifted toward shotcrete linings. Rose cited several references concluding that the rock loads for three classes could be lowered to eliminate some of the conservatism that was originally acknowledged by Terzaghi. The rock load method Rose published provided a 50% reduction in the estimated rock load for three rock classes. Rose's recommendation is to use the lower rock loads for rock conditions that are not affected by groundwater. This method continues to be employed in tunneling projects for tunnel design. A case in point is the design of the final lining for the PR-53 highway tunnels from Maunabo to Yabucoa, Puerto Rico during 2005. Exploration was carried out by horizontal core logging along the tunnel alignment. Rock loads based on the Rose method were used as input for STADD modeling of the cast-in-place concrete final lining.

University of Illinois at Urbana-Champaign, Cording, et al. (1972)

During the construction of the Subway Station Caverns for the Washington, D.C. metro system in the 1970s, Ed Cording and others at the University of Illinois carried out an extensive geotechnical study which included observations and measurements of geotechnical data as related to geologic conditions at the sites of nine 59 to 80 foot wide station caverns. (Cording and Mahar 1974) (Mahar, et al. 1972) (Cording, et al., Volume 1 1983) (Cording, et al., Volume 3 1983) The results of these cavern studies indicated that foliation shears and shear zones were the geological features causing the most significant geotechnical rock loads in the underground construction projects. Cording's team developed a procedure for estimating rock loads imposed on the structural support by considering the critical wedges surrounding the opening that ultimately require support. These studies produced a method of estimating rock loads based on rock blocks formed by foliation dip angle, bedding, and high angle joint planes. (Cording, et al., Volume 3, Page 11 1983). This system can be used whether the primary geotechnical feature is foliation or bedding with high angle joint planes or faults. It is important to note that the range of rock loads conveyed by this method ranges from zero to several 10s of feet and are consistent with measured rock loads in both Washington, D.C. and in New York City.

Geomechanics System, Penn State University, Bieniawski (1973)

The Geomechanics classification system, or Rock Mass Rating (RMR) was developed by Bieniawski in 1972 and was updated in 1976 to clarify the significance of some of the input parameters. The system is based on 351 case histories in various applications in hard rock mining. The RMR classification is an empirical method of rating relative rock mass qualities for mining and construction activities. Initially, it was intended to represent a structural region of a discontinuous rock mass by providing a single index value. Based on this value, uniform appropriate limitations on the excavation sequence may be correlated with other mining operations. It has been widely used and modified in over 1000 mining and tunneling case histories worldwide. The RMR value can be used to develop an estimate of the degree of ground support required, expected rock failure modes, estimates of prudent ground stand-up times, and appropriate limitations on excavation sequences.

Bieniawski has stated that considering the main design approaches to the design of underground structures, rock mass classifications form an integral part of the most predominant design approach, the empirical design methods. (Bieniawski 1989) One of the objectives of rock mass classification is to relate the experience of rock conditions at one site to the conditions and experience encountered at others. It has furthermore been shown in the last two decades that both the RMR system and the Q system have been used on thousands of rated tunnel intervals to describe the rock conditions in terms of a single index value, based on the prevailing geotechnical conditions, and relate those conditions to mining performance and rock behavior. The numerous rated tunnel intervals in various rock conditions that were reported in geotechnical reports allow the RMR and Q systems to be used interchangeably on various projects.

Rock Tunneling Quality Index, Norwegian Geotechnical Institute, Barton et al. (1974)

The Norwegian Geotechnical Institute's Rock Tunneling Quality index (Q) system developed by Nick Barton is a classification system used worldwide in design of rock support for civil and mining projects. It was first used in hydropower projects in Norway and in a water transfer project in Peru in 1974. (Barton and Grimstad 2014). The system was developed based on hundreds of case studies and provides a simple means of communication for geologists, rock engineers, mining engineers and lawyers. The Q system is used in thousands of tunnel reaches around the world and in all principal mining countries.

The Q-system is an empirical method of predicting probable ground behavior considering discontinuous geotechnical and stress-strength relationships. As required for the design of support systems, the Q-system has been developed with a view to determining the mechanism and mode of failure in the rock mass based roughly on the block size, inter-block shear strength, and the active stress regime, with the aim of evaluating stability as one of the first steps in designing an underground excavation. The rock tunneling quality index, Q value and rock load (roof pressure) are calculated by the following equations; (Barton, et al., p. 209 1974)

Rock Tunneling Quality Index: $Q = RQD/J_n \times J_r/J_a \times J_w/SRF$

Roof Pressure [kg/cm^2]: $P_{roof} = (2/J_r) * Q^{-1/3}$

where:
  RQD = rock quality designation
  Jn = joint set number
  Jr = joint roughness number
  Ja = joint alteration number
  Jw = joint water reduction factor
  SRF = stress reduction factor.

The Q system provides an evaluation in terms of both rock quality and cavern width. The Q index has been employed to provide a first indication of initial ground support, as well as final support for tunnels and caverns to put these designs on the same page as other tunneling projects worldwide. The Q system was updated in 1993, 2003, and 2007. Since the development of this system in 1974, the number of tunneling case studies is now in the thousands.

Scaled Crown Span, Golder Associates, Carter (1992)

The Scaled Crown Span empirical method has been developed by T.G. Carter over a period of two decades. Relevant publications on this method are provided in (Carter 1992; Carter and Miller 1995; Carter 2000, Carter 2014). This method uses case studies of both stable and unstable crown pillars of mines in Canada for various known crown pillar conditions and normalizes these cases with respect to a parameter called the scaled span of the crown pillars. Although general in nature, this method provides a realistic assessment of the stability of a crown pillar. Carter develop a method of stability analysis in which crown pillar instability occurs when the scaled crown pillar span (C_s) is greater than the critical span (S_c). These parameters are defined as follows; (Carter 1992)

Scaled crown pillar span (m): $C_s = S \times [S.G./(T(1 + S/L)(1 - 0.4 \cos \theta))]^{0.5}$

Critical span (m): $S_c = 3.3 \times Q^{0.43}$

Minimum crown pillar thickness (m): $T_{min} = 5.11 \times Q^{-0.19} \times [\sinh^{0.0016}(Q)]$

Crown pillar probability of failure (%): $P_f = 100/[1+440 \times \exp(-1.7{*}C_s/Q^{0.44})]$

Critical Span Line (50% Failure) (m): $S_c = 3.58\, Q^{0.44}$

where:

    S = actual crown pillar span (m)
    L = actual crown pillar strike length (m)
    T = actual crown pillar thickness (m)
 S.G.= rock specific gravity
    θ = foliation dip (degrees)
    Q = rock tunneling quality index.

The critical span represents the cavern span in the given geotechnical and geometrical conditions at which the span over the cavern has a 50% probability of failure if left unsupported. The critical span dimension can be considered the minimum acceptable rock cover over the crown of the excavation at which the excavation is considered stable. The minimum crown pillar thickness equation gives the minimum required rock cover in terms of the rock quality as expressed by the Q index as explained above. (Carter, p. 81 1992). In 2008 the scaled crown span method was updated to include a generalized model for probability of failure of a crown pillar based on empirical methods. (Carter 2014)

This approach was developed by seriously considering the back calculated thickness to span ratio for failed crown pillars and comparing those to crown pillars which remained stable given the prevailing geotechnical conditions and excavation geometry. This data, from the case records of over 200 crown pillars with 30 failed cases, is used to predict the stability of crown pillars based on rock quality and excavation geometry. Since the introduction of the original Scaled Span chart in 1989, further updates have included up to 500 case studies including 70 analyzed failures. The resulting probability chart allows very rapid assessment of possible risk for any known crown pillar geometry and rock mass quality.

CROWN PILLAR STABILITY ANALYSIS

A crown pillar is the zone of rock directly above the limits of a mining, tunnel, or cavern excavation. The stability of a crown pillar can be analyzed by various empirical and numerical methods. The crown pillar thickness is the rock cover, excluding the soil zone over the cavern, that forms the crown pillar to the outside edge of the cavern excavation. This information can generally be obtained only from borehole data. The unsupported rock mass in a crown pillar can be either stable, subject to failure over time when unsupported, or subject to immediate collapse upon excavation. This crown pillar condition is determined by four factors; rock mass quality, excavation width, excavation method, and proximity to the surface.

The rock mass quality can be expressed by the numerical indices RMR and/or Q. With respect to both of these systems, as well as other rock mechanics systems, the stability of the excavation decreases as the rock quality decreases. If less rock cover is provided than the critical span, ultimate failure of the crown span is likely, and the resulting rock loads will necessarily need to be accommodated by initial lining support and final lining structure. Likewise, an equivalent thickness of rock should be provided in each rib on the two sides of the cavern. If one of the ribs has been removed for some reason, such as excavation for a nearby building foundation, instability of the crown pillar becomes a definite construction risk.

The proximity of the excavation to the ground surface adversely affects the cavern stability. This is because as the cavern approaches the surface, the ability to employ arching action diminishes, and the support is required to develop beam action. Effects of this phenomena are considered by the scaled crown span equation. There are two basic cavern configurations available, which are suited to construction under shallow rock cover. An arched cavern configuration is typically constructed where there is sufficient crown pillar thickness to allow for the development of arching action in the rock mass. Alternatively, a binocular cavern configuration should be used where the cavern excavation would otherwise preclude development of a stable crown pillar.

The method of excavation is also a factor affecting stability of the crown pillar during excavation. Blasting will tend to loosen up rock blocks around and cause blast damage to the periphery of the excavation to a greater extent as compared to excavation by roadheaders. Drill and blast excavations will tend to be less stable and require more support for a given cavern width, rock quality and proximity to the surface. This effect is expressed in some systems used in rock mechanics today. However, it is not currently expressed in a practical system for evaluating the crown pillar stability.

The basis for the design acceptability of new cavern excavations for civil engineering structures is achieving a very low degree of risk. In particular, subway station caverns typically feature public access over the structure and buildings directly over near-surface underground excavations. Tolerance to risk in such cases is limited and the acceptable degree of risk must be essentially zero. To the extent possible, and where economically feasible, the crown pillar geometry and rock quality should provide for a cavern crown pillar probability of failure for an unsupported cavern excavation less than 5%. To the extent that any probability exists, the crown pillar requires rock reinforcement and structural support.

HISTORICAL DATA FOR WASHINGTON D.C. SUBWAY STATION CAVERNS

Ed Cording and others at the University of Illinois carried out a geotechnical study during construction of nine shallow mined station caverns in Washington D.C. in the 1970s which included observations and measurements of geotechnical data as related to site geologic conditions. (Cording, et al., Vol 1 1983) This study included field observations, displacement measurements, and cavern lining load estimations that increase over a wide range with respect to varying geotechnical conditions.

These caverns were driven with rock reinforcement and steel set-shotcrete structural linings as initial support and experienced an equivalent rock load on the structural lining of 12 to 49 feet after a heavy rock bolting program. Since the initial support served as the final structural lining in these cases, these caverns are very well suited to provide an indication of mined cavern lining loads. Rock loads were lowest, and correlated with elastic theory, where there was an absence of shear zones. Rock loads and rock displacements were highest where the geotechnical conditions included shear zones.

With these caverns serving as a basis for providing actual measured rock loadings on shallow mined cavern linings, and the use of the Q and scaled crown span classification systems for correlating the geotechnical conditions of the crown pillar with the lining loads for shallow caverns, an excellent rock load prediction methodology for shallow caverns can be developed.

The descriptions of geotechnical conditions for Washington D.C. subway station caverns have been condensed from the information provided in the reference study

Table 1. Estimated parameters for Washington D.C. subway caverns

Station Cavern	Actual Cavern Span, ft	Estimated Scaled Crown Span, ft	Assumed Q Parameter Values						Actual Rock Load, ft
			RQD, %	Jn	Jr	Ja	Jw	SRF	Q	
Medical Center	62	30	75	9	5	3	.5	2.5	2.78	12
Rosslyn	80	36	85	12	1	4	.66	5	0.23	16
Bethesda	62	44	75	12	1	5	.5	5	0.1	20
Cleveland Park	58	55	75	12	3	2	1	7.5	1.25	23
Van Ness	58	31	75	15	2	3	.66	5	0.44	23
Zoological Park	58	24	75	12	3	3	.66	5	0.83	30
Dupont Circle	76	45	75	15	1.5	5	.66	7.5	0.13	30
Tenley Circle	60	22	75	15	2.5	2	.66	7.5	0.55	28
Tenley Circle*	60	22	75	45	2.5	2	.66	7.5	0.18	41
Friendship Heights	67	29	75	15	1.5	6	.66	10	0.08	42
Friendship Heights*	67	29	75	45	1.5	6	.66	10	0.03	49

*Cavern intersection

report. (Cording, et al., Vol 1 1983), (Cording, et al., Vol 1, p. 235 1983), (Cording, et al., Vol 1, p. 27 1983) Equivalent rock loads on the station caverns linings back calculated from strain gage measurements have been estimated by Cording et al. to be as provided in Table 1 below. (Cording, et al., Vol 3, p. 126 1983) (Cording, et al., Vol 1, p. 235 1983) It was necessary to assume Q value parameters based on the study report, which are also shown in the table.

ROCK LOAD ESTIMATION METHOD

Based on the geotechnical conditions for the subway station caverns in Washington, D.C., as discussed above the rock loads are plotted with respect to the estimated Q and C_s parameters for the subway station caverns in Figure 1. The slopes of the rock load lines in the figure are adapted from the scaled crown span probabilities of failure. (Carter 2014) The resulting chart of rock loads has the rock tunneling quality index, Q, plotted on the x-axis and the estimated scaled crown span plotted on the y-axis. The locations for the various station caverns are plotted with visual symbols of decreasing intensity as shown on the figure. What is important to note is that the caverns with the largest measured rock loads are concentrated at the lower end of the rock quality scale (Q).

The measured data from the Washington Caverns plots loosely with the rock quality index as explained above. The trend of data is for low rock loads at the higher quality end of the scale to progress to heavy rock loads at the lower end of the quality scale. Note that the station with the highest rock quality, Q=3, for Medical Center Station has the lowest measured rock load at 12 feet. On the other end, the station with the lowest rock quality, Q=.03, for Friendship Heights station has the highest measured rock loading at 49 feet. The progression of measured data supports an increasing trend of rock loads with decreasing rock quality. Also note that the dispersion of rock data is very wide. The trend in the data is not smooth due to the widely varying nature of geotechnical conditions and variable excavation sequences employed.

As a predictive method, therefore, this method is not adequate to predict the actual values of measured rock loads. However, this method can be used to provide a good estimate of design loads for future caverns if the estimated loads are set at the high end of the variance in data. To do this, the rock load for the 100% probability of failure

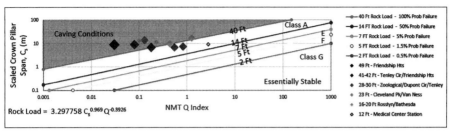

Figure 1. Estimated rock loads based on scaled crown span and Q value

can be set at 40 feet of rock, which corresponds with the data for Friendship Heights and Tenley Circle Station at 41 to 42 feet. The line corresponding to a 0.5% probability of failure is then set at 2 feet corresponding to periphery control rock loads. This allows an equation of the following form to be established.

Estimated Rock Load (feet): $P(C_s,Q) = 3.297758 \times C_s^{0.969} \times Q^{-0.3926}$

where:

$P(C_s,Q)$ = estimated Rock Load (feet)
C_s = scaled crown span (meters)
Q = rock tunneling quality index

This equation was established by assuming an equation of the form $P = A\, C_s^B\, Q^C$. Solution of the following three equations; $P(.8,.001) = 40$ feet, $P(100,150) = 40$ feet, $P(9.8,1000) = 2$ feet, leads to the establishment of the constants as follows; $A = 3.297758099$, $B = 0.969083869$, $C = -0.39259$. The precision of these constants is to best fit the risk rating divisions used in the scaled crown span method.

CONCLUSION

The methods for estimating rock loads on shallow mined cavern excavations has been developed over a period of eight decades, with many different approaches to the problem. Geotechnical conditions remain hidden underground and difficult to determine with sparse geotechnical borings. Nevertheless, principles can be developed to provide estimates of rock loads on shallow cavern excavations.

REFERENCES

Barton, Nick, 2002, Some New Q-Values Correlations to Assist in Site Characterization and Tunnel Design, International Journal of Rock Mechanics, 39, Pergamon Press, pp. 185–216, February 5, 2002.

Barton, Nick and Eystein Grimstad, 2014, Forty Years with the Q System in Norway and Abroad.

Barton, Nick, R. Lien, and J. Lunde, 1974, Engineering Classification of Rock Masses for the Design of Tunnel Support, Rock Mechanics 6, Springer Verlag, p. 13, August 31, 1974.

Bieniawski, Z.T., 1989, Engineering Rock Mass Classifications, John Wiley and Sons.

Bock, Carl, G., 1974, Rosslyn Station, Virginia: Geology, Excavation, and Support of a Large, Near Surface, Hard Rock Chamber, RETC, pp. 1373–1391.

Brady, B.H.G. and Brown, E.T., 1985, Rock Mechanics for Underground Mining, London, Allen and Unwin.

Carter, T.G., 1992, A New Approach to Surface Crown Pillar Design, Proc. 16th Canadian Rock Mechanics Symposium, Sudbury, pp. 75–83.

Carter, T.G., 2000, An Update on the Scaled Crown Span Concept for Dimensioning Surface Crown Pillars for New or Abandoned Mine Workings, Proc. 4th North American Rock Mechanics Conference, Seattle, pp. 465–47.

Carter, T.G., 2014, Guidelines for use of the Scaled Span Method for Surface Crown Pillar Stability Assessment.

Carter, T.G. and R.I. Miller, 1995, Crown-Pillar Risk Assessment-Planning Aid for Cost-Effective Mine Closure Remediation, Trans. Inst. Min. Metl, Vol 104, pp. A41–A57.

Cording, E.J. and J.W. Mahar, 1974, The Effect of Natural Geologic Discontinuities on Behavior of Rock in Tunnels, RETC.

Cording, E.J., M Van Sint Jan, C. Rodriquez, J.W. Mahar, G. Fernandez, and J. Ghaboussi, 1983, Ground and Lining Behavior of Station Chambers in Rock for the Washington, D.C. Metro, Volume 1, Observations During Construction, University of Illinois at Urbana-Champaign, U.S. Dept. of Transportation, Federal Highway Administration, October 1983.

Cording, E.J., M Van Sint Jan, C. Rodriquez, J.W. Mahar, G. Fernandez, and J. Ghaboussi, 1983, Ground and Lining Behavior of Station Chambers in Rock for the Washington, D.C. Metro, Volume 3, Analysis of Rock Displacements and Lining Performance, University of Illinois at Urbana-Champaign, U.S. Dept. of Transportation, Federal Highway Administration, November 1983.

Deere, D.U., R.B. Peck, H.W. Parker, J.E. Monsees, and B. Schmidt, 1970, Design of Tunnel Support Systems, High Res Rec no 339, pp. 26–33.

Desai, Drupad, Madan Naik, Karel Rossler, and Charles Stone, 2005, New York Subway Caverns and Crossovers, Rapid Excavation and Tunneling Conference Proceedings, Seattle, Washington, pp. 1303–1314.

Desai, Drupad, Hannes Lagger, and Charles Stone, 2007, New York Subway Stations and Crossover Caverns—Update on Initial Support Design, 2007, Rapid Excavation and Tunneling Conference Proceedings, pp. 32–43.

Grimstad, E., 2007, The Norwegian Method of Tunnelling—A Challenge for Support Design, XIV European Conference on Soil Mechanics and Geotechnical Engineering, Madrid.

Grimstad, Eystein, Rajinder Bhasin, Anette Wold Hagen, Amir Kaynia, and Kalpana Kankes, 2003, Q-System Advance for Sprayed Lining, Tunnels and Tunneling International, pp. 44–48, January, 2003.

Hoek, E., Kaiser, P.K. and Bawden, W.F., 1995, Support of underground excavations in hard rock, Balkema, Rotterdam.

Hoek, Evert, 2000, Big Tunnels in Bad Rock, ASCE Civil Engineering Conference and Exposition, Seattle, October 18–21, 2000.

Mahar, J.W., F.L. Gau, and E.J. Cording, 1972, Observations During the Construction of Rock Tunnels for the Washington, D.C. Subway, RETC.

Muir Wood, A.M., 1975, The Circular Tunnel in Elastic Ground, Geotechnique, 1, pp. 115–127.

Proctor, R.V., T.L. White, with introduction by Karl Terzaghi, 1946, Rock Tunneling with Steel Supports with an Introduction to Tunnel Geology, Commercial Shearing and Stamping Company, Youngstown, Ohio.

Rose, Don, 1982, Revising Terzaghi's Tunnel Rock Load Coefficients, Proceedings of the 23rd U.S. Symposium on Rock Mechanics, AIME, New York, 1982, pp. 953–960.

Rossler, Karel, and Charles Stone, 2007, New York Second Avenue Subway—Initial Support Design of Shallow Rock Caverns, ITA-AITES World Tunneling Congress, Prague, Czech Republic.

Stone, Charles, Joel Kuszmaul, Amarin Boontun, and Dae Young, 1996, Comparison of a Theoretical and Numerical Approach to Probabilistic Keyblock Analysis, North American Rock Mechanics Symposium 96, Montreal, Canada, June 1996.

Use of Spray Applied Waterproofing on the Downtown Bellevue Tunnel Project

Mun Wei Leong ▪ McMillen Jacobs Associates
Ied DePooter ▪ McMillen Jacobs Associates
Jacob Taylor ▪ McMillen Jacobs Associates

ABSTRACT

The use of spray applied waterproofing for tunnels in the Unites States is relatively new. Historically, PVC sheet membrane has been utilized with varying degrees of success for waterproofing tunnels. Sound Transit deliberately chose to offer an alternative to PVC sheet membrane and included a specification for spray applied waterproofing in its most recent project, the Downtown Bellevue Tunnel, which is a large-diameter SEM tunnel for light rail to the eastside of the Seattle region. This paper discusses the decision to allow the contractor the option of utilizing either spray applied waterproofing or sheet membrane, the testing of the product, required specification revisions, field testing, application, and performance of the waterproofing to date. Lessons learned are offered to the reader throughout the paper.

INTRODUCTION

The Downtown Bellevue Tunnel (DBT) is part of Sound Transit's East Link Extension Program, a 14-mile extension of the Sound Transit (ST) light rail transit system from downtown Seattle, across a floating bridge over Lake Washington, to the cities of Mercer Island, Bellevue, and Redmond. The DBT project consists of a 250-foot-long cut-and-cover portal structure at the southern end, a 1,983-foot-long SEM tunnel, and a midtunnel access shaft with a short adit connecting it to the tunnel. At the north end of the tunnel is a 200-foot-long cut-and-cover structure and a station that is being constructed under a separate contract package. See Figure 1 for a plan and profile view of the tunnel profile.

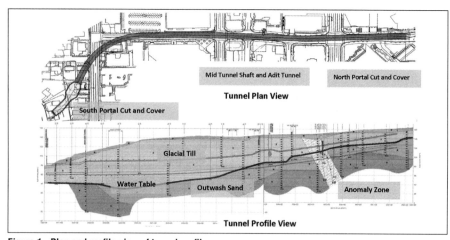

Figure 1. Plan and profile view of tunnel profile

The DBT was constructed using the Sequential Excavation Method (SEM), also known as the New Austrian Tunneling Method (NATM). The DBT is an ovoid shaped tunnel with a typical excavated cross-section of 37.7 feet wide by 30.5 feet high (Figure 2). The tunnel cross section is enlarged to 42.3 feet wide by 37.7 feet high, near the tunnel's midpoint, to provide space for an emergency ventilation fan room above the tracks.

A 12.8-foot-wide by 13.2-foot-high adit and the 20.7-foot-diameter by 50-foot-deep shaft provides maintenance access to the fan room from the surface.

The initial lining for the SEM tunnel was 10 inches of macro-synthetic polymer reinforced shotcrete. In addition, a minimum 1.5-inch-thick shotcrete smoothing layer was placed prior to installation of the waterproofing, as shown in Figure 3. Following installation of waterproofing, a 1-foot-thick final lining was then placed.

Figure 2. Downtown Bellevue Tunnel with temporary center wall and invert backfill

Figure 3. Crews patching waterproofing and preparing to install reinforcement for the mid-tunnel shaft

WATERPROOFING SPECIFICATION AND PRODUCT

The contract documents allowed for the use of either elastomeric sheet membrane waterproofing or spray applied membrane waterproofing (Figure 4), both of which require a 10-year warranty. The contractor, Guy F. Atkinson, decided to use spray applied waterproofing for the DBT, specifically MAPELASTIC TU SYSTEM (MAPELASTIC), and a MAPEI product. F.D. Thomas was chosen as the waterproofing subcontractor. Quality control was also performed by the installer. A representative from MAPEI was also present to supervise the installation of the product. McMillen Jacobs Associates was the Construction Management Team representing the Owner, Sound Transit for the project.

Various factors led to the contractor's choice of using spray applied waterproofing for the DBT. It was difficult to find a manufacturer and installer that would provide the contract-specified 10-year warranty. Cost and ease of installation was another factor driving the decision. The spray applied waterproofing has the benefit of not requiring the installation of propriety anchors for reinforcing support penetrations and has a less stringent requirement for surface waviness. The amount of water anticipated and encountered in the tunnel was also factored into the decision.

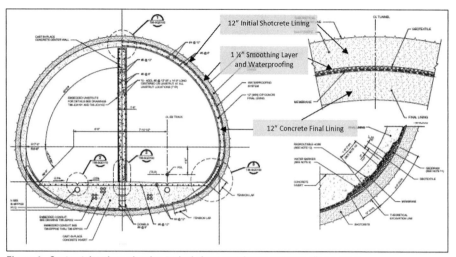

Figure 4. Contract drawings showing typical sheet membrane waterproofing system

The contract drawings required the waviness for the waterproofing to be less than W = L/60. With spray applied waterproofing, the waviness requirement was reduced to W = L/12 after a large-scale demonstration that this was adequate. The application of the spray applied waterproofing appeared to be much faster than installation of sheet membrane. Additionally, spray applied waterproofing bonds with both the shotcrete initial lining and the final lining, potentially allowing for a composite system, although this was not a factor on the design of DBT.

The waterproofing system depicted in the contract consisted of a 1.5-inch-thick smoothing layer placed on the initial lining to cover the original specified steel fiber reinforced shotcrete. Geotextile would then be then placed on the smoothing layer with the sheet membrane on top of this. WA anchors would be installed to support the reinforcement. After the sheet membrane was installed, at each construction joint (approximately 30 feet to 40 feet apart), a water barrier would be welded to the membrane and a regroutable hose would be installed at the joint. When the membrane was completed, reinforcing and 12-inch-thick cast-in-place final lining would be placed.

There were no contract drawings depicting the spray applied waterproofing system; therefore, the majority of the drawings for waterproofing do not provide useful details for spray applied membrane. The contract specification included the option to use spray applied waterproofing and required the waterproofing system to:

1. Resist water pressures up to 15 bar in combination with support from the initial shotcrete and final concrete lining.
2. Have a nominal thickness of 0.2 inch (5 mm).
3. Bond to clean, particle free, cementitious materials on both sides of the membrane with a bond strength of 145 psi (1 MPa).
4. Achieve a Shore A Hardness of 75.
5. Have a tensile strength of 580 psi (4 MPa).
6. Have an elasticity of 80%–120%.
7. Bridge a crack of 0.12 inch (3 mm) before failure.
8. Have the capability of being applied to damp surfaces.

SPECIFICATION AND DRAWING CHANGES

With the choice of the MAPELASTIC TU System, several changes were required to the original spray applied membrane specifications which were based on MasterSeal 345 by BASF. In order to modify the specification to more closely match the proposed waterproofing product the following requirements were changed:

1. The required Shore A Hardness of the cured product was revised down from 75 to 50. The specification was further revised to allow the placement of the final lining once a Shore A Hardness of 30 was achieved. Additional field and laboratory testing was performed by the contractor and successfully demonstrated that the spray applied product continues to cure after the material was sandwiched between the initial shotcrete and final concrete lining.
2. Tensile strength was revised to 290 psi (2 MPa).
3. The elasticity requirement was determined to be incorrect. A requirement for 80% elongation was added.
4. The surface waviness requirement was relaxed from L/60 to L/12.
5. The PVC water barrier requirement at the tunnel/portal transition and at every tunnel construction joint was eliminated. It was determined that the spray applied waterproofing bonded with the concrete but did not bond well to the PVC water barrier. Use of a poorly bonded water barrier would not have increased the reliability of the system.
6. The use of hand spraying was allowed instead of requiring placement with robotic equipment.
7. Two colors of membrane were required to allow visual confirmation that a second coat had been consistently applied, but provisions were included to use one thicker coat in the invert if deemed beneficial by the contractor.

TESTING REQUIREMENTS

Field trials were required to determine if the product was suitable for use (Figure 5). The testing included qualification and certification of the applicators and QC personnel from the applicator. Several test panels were shot on site and transported to the applicator's facility in Oregon. The applicators practiced applying the spray applied product on the test panels. Three of the panels were then cured and transported back on site, where bond and thickness tests were performed.

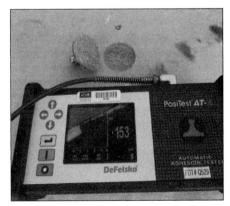

Figure 5. Bond testing on test panels performed on site

Figure 6. Spray applied waterproofing in the adit tunnel

Figure 7. Crews applying spray applied waterproofing at the mid-tunnel shaft

The bond tests were performed per ASTM D7234.* The bond was required to achieve a strength of 145 psi (1 MPa). The thickness of the membrane was measured using a micrometer and depth gauge during testing and installation. In addition to the panels, Contractor was required to conduct mockups for overhead shotcrete application, as part of allowing the use of shotcrete as final lining. As part of these mockups, waterproofing was spray applied prior to the shotcrete application. This allowed for visual inspection of the bond between the waterproofing and both the smoothing layer and the final lining shotcrete.

Finally, the midtunnel access shaft and adit were constructed prior to having the SEM tunneling completed and ready for waterproofing. The team took the opportunity to use the spray applied waterproofing product in the shaft and adit tunnel (Figure 6 and Figure 7) and applied those lessons learned to the revised specifications and application requirements of the product in the tunnel. The shaft and the adit were treated as a full-sized mockup of the upcoming tunnel work.

APPLICATION OF SMOOTHING LAYER

The tunnel hole through occurred on July 18, 2018. Immediately after tunnel hole through, contractor worked on cleaning up the tunnel invert and crown, removing tights and installing the shotcrete smoothing layer (Figure 8). The installation of the smoothing layer started in mid-August 2018 and continued through October 2018 in the crown. Contractor placed the smoothing layer shotcrete by hand. It consisted of a sand cement mixture with up to 4% accelerator used in the crown. In areas where there were leaks, panning was also placed to divert the water to the invert. A couple of large

Figure 8. Crews placing smoothing layer and panning at each girder location

---

*Standard Test Method for Pull-Off Adhesion Strength of Coatings on Concrete Using Portable Pull-Off Adhesion Testers.

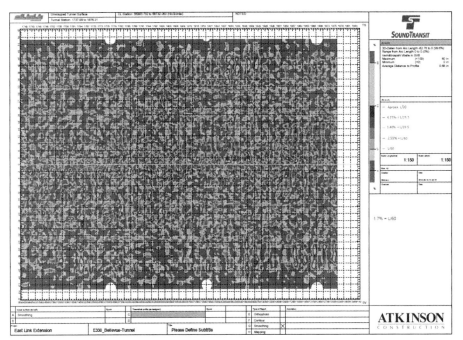

Figure 9. Typical scan of the smoothing layer

leaks contractor used drainage pipes to also divert leakage. Admixtures were later added to the smoothing shotcrete to help seal small leaks.

During placement of the initial lining and smoothing layer, nozzleman should avoid placement resulting in sharp ridges, such as when placing a thin layer over the edge of panning material. The application of the waterproofing does not easily fill in the "shadow" of this ridge.

As required for the installation of the initial lining shotcrete, exclusion zone management was utilized to prevent workers from entering the area where shotcrete was just placed. A minimum of 100 psi (689 kPa) shotcrete needed to be achieved prior to lifting the exclusion zone. Testing was done on a test panel outside the exclusion zone with a needle penetrometer.

In the invert, no accelerator was typically used. The lack of accelerator required additional time prior to allowing workers and vehicles to enter the zone to avoid risk of damage to the smoothing shotcrete.

Once the smoothing layer was placed, a surface scan survey (Figure 9) was done to verify the smoothing layer did not encroach on the final lining thickness and the waviness requirement was achieved. In some areas, where too much smoothing shotcrete was placed, the contractor had to regrind the surface and reapply smoothing layer or patch those areas with hydraulic cement or nonshrink grout.

LESSONS LEARNED

Installation of spray applied waterproofing has its limitations, and is not yet suitable for use in a very wet tunnel.

The use of the spray applied waterproofing would have been easier had it been integrated into the design, including integrating the anticipated use of waterproofing admixtures in the design of the shotcrete mix. It became apparent, partially through the smoothing shotcrete application and in the initial phases of the spray applied waterproofing, that many of the leaks could have been mitigated with better placement of the shotcrete or by the use of admixtures.

Most of the leaks were observed to be at the location of the lattice girders. This is likely because of congestion around the girder and connection plates and the necessary acceleration of the shotcrete leading to consolidation issues. A girder design that reduces the size of the connection plates might reduce the frequency of those incidents and result in better consolidation of the shotcrete.

Lattice girders could have been eliminated, or they could have been placed near the middle of the excavated round length. This would have allowed better access for the shotcrete robot, helping consolidation around the lattice girder and connection plates. The additional shotcrete in front of the girder would have better protected the brittle lattice girder from accidental contact from the excavation equipment when the subsequent round was excavated.

The Fine Print Discoveries

During initial installation, the criteria for an acceptable surface to apply the waterproofing were not understood by the contractor, the waterproofing installer, and waterproofing manufacturer. The following are lessons learned through this process.

- This spray applied waterproofing relies on drying to cure since it is a water-based product. This precluded it from being placed on a damp surface as it does not cure if there is any water present. The membrane that was used is suitable for dry surfaces.
- The typical thickness of the waterproofing is 3 to 4 millimeters. When the membrane is applied as a single coat in thicknesses much greater than the 3 to 4 millimeters required, cracking was observed during curing. However, these cracks have been easily repaired.
- Placement temperature has not been an issue; however, the colder the climate and the more humid the environment result in additional time to cure the waterproofing material.
- The protection layer should be installed as soon as possible to protect the waterproofing from damage.
- If the surface of the shotcrete is not sufficiently cleaned, the waterproofing does not bond well and will peel off the substrate.
- Smoothness of the substrate does not necessarily affect the bond. It has more to do with the cleanliness of the surface, and when the surface is smooth, cleanliness becomes more imperative.
- If the surface of the shotcrete is too rough, consumption and thickness develop. Significant time was spent preparing the smoothing surface for the application of the waterproofing.

Wet Surfaces Require Remediation

Contractor also used chemical grout product to attempt to stop the leaks prior to placing waterproofing. Various products were used to mitigate the leaks and cracks in the shotcrete lining (see Figure 10).

Use of Spray Applied Waterproofing on the Downtown Bellevue Tunnel Project

Figure 10. Crews patching leaks and smoothing surface

Figure 11. Drainage mat in the invert and drainage pipe to drain water

The design for spray applied membrane should require the use of panning or drainage mats under the smoothing layer to deal with the leaks. The use of panning was not anticipated in the design or in the contractor's initial plans. On the southern three-quarters of the tunnel, a 1-inch slotted pipe was placed under the invert panning to provide additional drainage space.

To mitigate larger leaks, drain pipes were used to pipe the water until waterproofing could be applied (Figure 11).

Installing the drainage mats after the initial lining was placed reduced the thickness of the smoothing layer placed on top

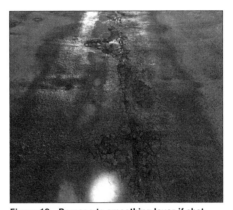

Figure 12. Damage to smoothing layer if shotcrete did not achieve required strength

of the drainage mat location. As a result, water being transmitted on the drainage mats would leak out of the shrinkage cracks or thin areas, and the running surface in the tunnel invert was thinner and cracked easily under vehicle loading.

We recommend that future applications consider including a drain system in the invert that is outside the waterproofing and allows water pressure to be relieved into the invert drain. In this way issues such as debonded bubbles causing a failure of the membrane system can be mitigated.

No traffic should be allowed on the unprotected panning material. This crushes the dimples, thereby reducing the thickness and leading to reduced flow capacity and also reducing the ability of the smoothing shotcrete applied over it to be fully supported on cementitious dimples to a firm invert (Figure 12). The flexing of the panning when the smoothing shotcrete cover on it was fully cured (but was not in sufficient contact with the initial lining behind it) allowed movement, leading to cracking and failure of sections of the shotcrete. This required repairs to the panning and smoothing shotcrete.

Consider using accelerator on the smoothing surface in the invert if time is a constraint. In some instances, the smoothing layer was damaged because of inadequate cure when the contractor drove its equipment on the smoothing layer too soon after application.

Excavation of the tunnel should have included allowances for items such as panning, removal of tights, and drainage in the planning of the works. The thickness of the protection layer above the panning should be considered as thin areas tend to be damaged during construction traffic loading.

Application of Spray Applied Waterproofing

Waterproofing was placed by the applicator under the supervision of the waterproofing supervisor from the manufacturer. Equipment used was a specialized airless sprayer to apply the waterproofing. An oil-less air compressor to blow down the surfaces and air powered pumps to move the waterproofing from 55-gallon drums to the airless sprayer. Contractor provided the surface prep required for the waterproofing, including panning, smoothing shotcrete, tight removal and repair, and surface smoothing of gunned shotcrete.

The original understanding from discussions with the supplier and the contractor was that it was acceptable to place the waterproofing product on damp surfaces (i.e., surfaces that appeared to be damp but where there was no visible moisture present on the fingers after touching). Surfaces that had moisture bleeding through and were wet on your fingers needed to be sealed. However, during placement it was discovered that areas that had any visible dampness needed to be sealed.

It was discovered that the control of water and leaks was critical to the installation of the waterproofing product. At the north portal, the contractor had constructed a dam with a pump to divert stormwater so as to not impact the installation of the waterproofing in the invert. Unfortunately, on one of the rain events in October, the pump at the dam failed and water streamed onto the invert just after a waterproofing placement (Figure 13). The product had not had enough time to cure (approximately 2 to 3 days), and the waterproofing was damaged. This resulted in the removal and repair of damaged areas.

At the area north of the midtunnel shaft, a leaking storm drain pipe was very close (approximately 5 feet) to the tunnel crown. The perforated 12-inch-diameter pipe allowed the pea-gravel-filled trench to hold the water right above the tunnel. The leaks from the pipe caused issues during tunneling and the installation of waterproofing in this area. Leaks from the crown and water diverted into the drainage panning to the side walls and invert also impacted waterproofing in the invert (Figure 14). Damaged

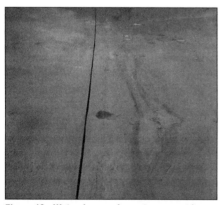

Figure 13. Water damage from storm event in October 2018

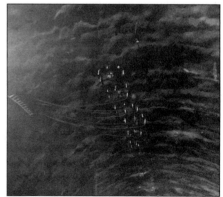

Figure 14. Leaks in the crown being diverted into drain pipes to avoid impacting the invert waterproofing

waterproofing and smoothing shotcrete were removed, panning and shotcrete/concrete were installed, and the leaks were sealed using grout or epoxy resin This work had to take place prior to replacing the damaged waterproofing.

Particular attention needs to be paid to waterproofing at the locations of expansion joints (Figure 15), such as where the SEM tunnel meets the cut-and-cover tunnel sections. For this project, the waterproofing for the cut-and-cover is applied to the outside of the cut-and-cover and wrapped into the tunnel to join to the spray applied waterproofing. However, the sheet membrane pooled water and failed prior to installation of the support provided by the final concrete lining. The solution was to cut the membrane to allow the water behind the membrane to be removed and patch the membrane prior to placing the final lining. Additionally ports will be installed to remove water from behind the membrane.

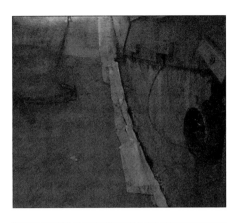

Figure 15. Expansion joint between the North Portal and the tunnel

PRODUCTION

Waterproofing installation was originally anticipated to begin in mid-September 2018, but only started in mid-October 2018. The original plan was to complete cleaning the initial lining and prepare the surface for installation of the smoothing layer 1.5 months after hole through.

At this time, waterproofing of the tunnel invert continues, and the crown has yet to be started.

RECOMMENDATIONS

We are still performing and monitoring the work. Based on our experience, in order for the membrane to be properly installed and function as intended, water ingress has to be managed and attention must be paid to preparation of the substrate. The product cures only if there is no water present when the material is placed. Plan on having the contractor apply panning at each lattice girder and panning the invert prior to placing the smoothing layer. There should be an expectation of grouting to seal larger leaks that are typical in the shafts or tunnels. The use of reinjectable hoses at construction joints has not yet been tested, but we do not anticipate reaping the full benefits of this without the use of waterstops in the construction joints to inject grout against. All parties that are planning on using this system should understand the significant surface preparation work that is involved.

PART

New and Innovative Technologies

Chairs

Bianca Messina
Skanska

Adam Hingorany
Traylor Bros., Inc.

Artificial Intelligence Technique for Geomechanical Forecasting

M. Allende Valdés ▪ SKAVA Consulting
J.P. Merello ▪ SKAVA Consulting
P. Cofré ▪ SKAVA Consulting

ABSTRACT

Rock conditions are evaluated after every blast, performing a rock mass classification. However, no person can see beyond the face. There are three methods most commonly used for this purpose: core drilling, measurement while drilling and probe holes.

Artificial Intelligence (AI) is the simulation of human intelligence processes by machines. In this project machine learning techniques, a type of AI, were applied to geotechnical information from probe hole drilling and face mappings, in order to find patterns and inferred functions based on data used for training.

Information from both sources had to be standardized, labelled, organized and stored in a way to be easily accessed by the machine learning method. The model only learns from that training data, it is understood that non-experienced situations cannot be predicted. Because of this, it was assumed that every project would need its own training process.

A testing tunnel in a Peruvian project was considered, 1692 registers were used to train a model, and 423 registers were used as test set, all from the same tunnel. Once the model was trained, it was used as a forecasting tool during the performance of new Probe Holes. Results show that the model has an accuracy of +85% forecasting rock mass classification beyond the face in a way that is suitable for advising a decision maker.

INTRODUCTION

For the construction of every underground facility in rock, the site's geology is the most important condition to consider for every stage of the project, from feasibility studies to detail engineering and construction. Prospecting and studies are performed following the level of certainty required for the project subjected to technical requisites and budget. For tunnels, in the early stages of engineering, core drilling is by far the most used method of prospecting. Depending on the overburden of the tunnel, a large total of meters of core drillings will be needed to reach a single point of the axis of the tunnel. However, there are other forms of prospecting available and used worldwide.

The geology to consider for the design of a project is the result of the interpretation of all prospections, review of references, studies and all available information, generally resulting in a Geotechnical Baseline Report. However, this information gives a general overview and does not cover the drastic variations that geology can have within a matter of meters during the construction of underground facilities resulting in changes in support conditions and in most severe cases, tunnel collapses.

This is the reason why during the construction of tunnels the forecasting of rock mass is always wanted in order to perceive in advance the changes in the geology that could result in construction delays and in worst case scenarios, damage to equipment or loss of life. In other words, a trustworthy forecast of the rock mass to be excavated will result in the reduction of geological risk.

This paper describes some of the current methods for rock mass forecasting and how the application of Artificial Intelligence (AI) improves rock mass forecasting.

ROCK MASS FORECASTING METHODS

Currently, rock mass forecasting is performed by several methods. Each method has advantages and disadvantages.

Core Drillings

Core drilling can be precise, but expensive and slow (Chapman 2000). Only specific conditions will justify the investment, because advancing face has to be halted to perform the task. If it is done in parallel, then the advance rate of the face will probably be higher than that of the core drilling, returning useless information. This occurred during the construction of the tunnel detailed in this paper. At the end, a special niche had to be constructed to perform the core drilling without stopping the excavation face advance. Due to the complex logistics and numerous interferences between core drilling and advance teams, the drill & blast cycle had important delays while coexisting with each other. The advance rate of the core drilling was only a couple of meters a week faster than the advance face, so the useful information provided was good only to forecast one or two rounds. Also, a joint system subparallel to the core drill was not detected, producing small deviations in the forecasting information.

At the end, core drilling was a very expensive tool (a special niche had to be built) that contributed with limited forecasting information. Nevertheless, this was high quality information.

Measurement While Drilling

Measurement while drilling (MWD) is a very powerful tool. The interpretation of digital drilling parameters can give useful information about what lies behind the face, but the investment in computerized equipment, and training of personnel imposes a huge constraint (Rivera 2012). Even when the MWD is considered for tunnel construction, the interpretation of the parameters must be done by a geologist with special training in such tools. In most of the cases the geologist is in charge of the drilling equipment while the MWD is in use. This can generate difficulties when the geologists in charge of the definition of the rock mass support belong to an independent third party, because the contractor must allow an external member to use their equipment.

In the case study presented in this paper, the drilling equipment (Jumbo) did not offer the option to include automated MWD tools, so this option was not considered.

Probe Holes

Probe holes, on the other hand, are cheap, fast and easy to perform (Bilgin 2016). The problem is that the data interpretation is very subjective, generally leaving the responsibility in the hands of the operator. The analysis will vary depending on which shift, equipment and operator is drilling. Probe holes are destructive perforations (i.e., no core recovery), so all information is gathered during the drilling process. Also, in old drilling equipment like the one used in the project listed in this paper, maintenance operations can modify the jumbo's performance.

All the information resulting from probe holes will be summarized and interpreted by the operations manager or geologist in charge. This will create a large scattering of information if no unified criteria are applied to the project. Also, as mentioned before, the handling by the drilling operator plays a very important role, because the perforation with old equipment like that used in the study case depends on 2 variables: drilling pressure and rotation speed. So, the operator has to modify these 2 variables to drill most effectively.

Others

Nowadays, there are other methods available for the forecast of rock mass. Just to name one, Tunnel Seismic Prediction (TSP) uses seismic induced waves and geophysics to interpret properties of the upcoming rock mass and significant changes (i.e., faults, shear zones, etc.).

DEFINITION OF THE DATA

In order to work on new forecasting options, a set of data was needed. A project located in Perú was selected for this purpose. The project consists of a river diversion tunnel of 7.7 km in length and with a cross section of 23 m^2. The construction method selected for the project was Drill & Blast.

During the engineering stages of the project, the rock mass support was to be qualified using Barton's Q Index. In total, 6 support types were defined, assigning support type 1 to the best rock mass competency and support type 6 to the weakest rock mass. The following table shows the support classes limits.

Table 1. Rock mass support types limits (Q index)

Support Type	Lowest Q Value	Highest Q Value
Type 1	10	
Type 2	5	10
Type 3	0.4	5
Type 4	0.1	0.4
Type 5	0.03	0.1
Type 6		0.03

The support consisted in combinations of sprayed shotcrete (with and without fiber), passive rock bolts, steel mesh and lattice girders.

The geology in the project area was largely dominated by igneous rocks, mainly granodiorite and riolite. There was an altered zone near the project, which produced a large number of geologic faults with very poor geotechnical conditions. The variation of geotechnical conditions were abrupt in many cases, changing from support type 3 to support type 6 in a matter of meters. These variations had a high impact in the coordination of construction directly impacting the advance rates and stability of the tunnel between blasting and support installation. Also, during rainy seasons, a significant amount of infiltration waters entered the excavation section, producing a decrease in the geotechnical rock mass quality.

Rock Mass Face Mapping and Probe Holes

Due to contractual requirements and the site conditions previously described, a geological face mapping had to be performed after every blast during construction by an independent third party (this was SKAVA's role in this project). In such face mapping, a geologist had to evaluate the geological conditions of the rock mass and calculate the Barton's Q Index. Also, in the face mapping a full register of joint systems, shear zones, infiltrations and pictures were included, giving a very complete set of data on the excavated rock mass.

Also due to contractual requirements and site conditions, the contractor had to perform Probe Holes (destructive) 30 meters long with a minimum overlap of 5 meters. The contractor used 3-meter rods for perforation, so each probe hole had a total of 10 perforation rods. The independent third party geologists were responsible for the parameter measurements and interpretation.

While performing the probe holes, the following data were recorded:

- Initial chainage of the probe hole.
- Date and time of the probe hole.
- Orientation of the probe hole.
- Equipment used for the probe hole (jumbo)
- Boring pressure for each rod.
- Advance speed for each rod, in meters per minute
- Advance speed for each rod, in seconds per meter
- Detritus colour
- Water inflow

After the total completion of the probe hole, the third party geologist had to write a report and inform the Owner and the contractor of the project about the findings obtained from the probe hole. Such report was purely factual and not interpretative.

FIRST APPROACH OF FORECAST

During the construction of the tunnel an important number of overbreaks resulted after blasting, the drastic changes in the geology being one of the most important root causes. In this scenario, the forecast performed through probe holes was studied in detail by the third party geologist in order to give this database a more useful purpose.

The first action was to gather, standardize, label, organize, and store all the existing information obtained from all historic probe holes of the project. A total of 560 sets of information were analysed, each dataset comprised all the information corresponding to one perforation rod.

Since the focus of this research was to improve the information behind the excavation face, a first examination of the data showed that for this purpose the boring pressure and the advance speed of the rod were the most useful information.

After selecting what information to consider, an historical review of the existing information was performed. As a way to simplify the forecast expected, the first correlations were made looking for support types to be installed, thus all information selected was compared with the actual support type installed. In other words, a linear comparison was performed, aligning information from the probe holes and the rock mass support type effectively installed for the same chainages.

In this line, the following Figures 1, 2 and 3 present, graphically, the history of the database from the probe holes of the project (the aforementioned 560 data sets) and the effectively installed rock mass support type.

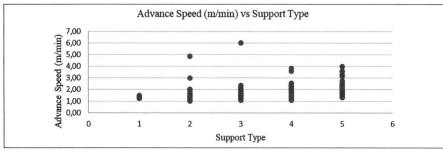

Figure 1. Advance speed (m/min) vs support type

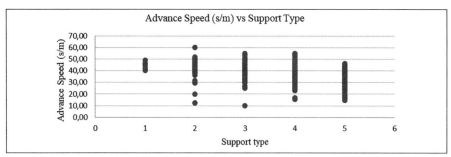

Figure 2. Advance speed (s/m) vs support type

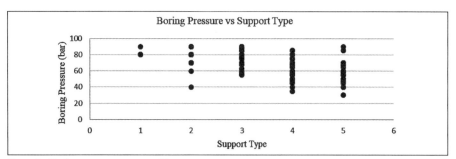

Figure 3. Boring pressure vs support type

Through visual and numerical processing of the information plotted, it was decided that the boring pressure was the most representative information obtained from probe holes to relate to support type. Figure 4 shows how the simple average of the boring pressures relate to the support type for the 560 data sets considered. Also, a simple linear regression was included in this graph to show that the average boring pressure and the support type relation has a R2 value of 0,692. As expected, the boring pressure increased with the highest rock mass quality. An important fact is that for all the data sets considered, no support type 6 had been installed, thus this support type was not statistically considered in this first approach. As shown in Figure 4, the average boring pressures are within a range of 20 bars, from 60 bars for support type 5, to 80 bars for support type 1.

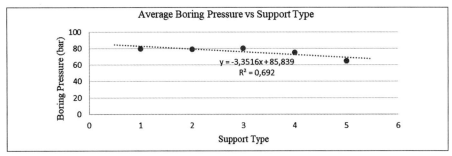

Figure 4. Boring pressure vs support type

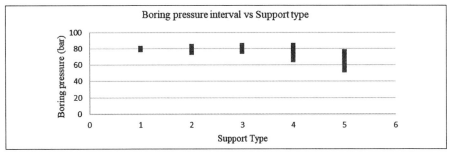

Figure 5. Boring pressure interval vs support type, "Abacus"

The next step was to generate an simple tool for rock mass forcasting. So, an "abacus" that had a pressure interval for each support type was generated. Each interval consisted of the average boring pressure plus and minus a single standard deviation (Figure 5). The abacus was applied as follows: The geologist supervising the probe hole recorded the average boring pressure for each rod and then through the abacus determined the support types that are contained in such boring pressure. Since the boring pressure intervals have common values, simple boring pressure data could forecast 2 or 3 support types. And so a single boring pressure measurement was used for a single forecast, but such forecast included 2 or 3 support types.

After some time forecasting with the "Abacus" a review of the forecast results was analyzed. A total of 213 forecasts were made using the "Abacus." Table 2 shows the results obtained with its usage. From all the forecasts, a 74,2% proved to be correct, the forecasted support types with the real mapped support type were a perfect match.

Table 2. First results using the "Abacus"

Forecasts	Unit	(%)
Correct	158	74,2%
Incorrect	55	25,8%
Total	213	100,0%

The incorrect forecasts were analyzed to realize whether the errors were conservative or non-conservative. A conservative error was defined as when the forecasted support type having a lower Q value than the real Q value mapped. Table 3 shows the analysis of the errors made.

Table 3. Error analysis

Incorrect Forecasts	Unit	(%)
Conservative	5	9,1%
Non-conservative	50	90,9%
Total	55	100,0%

MACHINE LEARNING APPROACH

After the first approach for forecasting the support type using single average values, the idea of including Artificial Intelligence (AI) was implemented.

Method Used and Data Considered

Supervised Learning is the machine learning task of inferring a function from labelled training data (Mohri et al., 2012). The goal of this project was to apply Supervised Learning to geotechnical information, in order to find patterns and infer functions based on training data. The chosen technique was Support Vector Machine (Kotsiantis 2006), and the data from Geotechnical Mapping and Probe Holes drilling was used for training.

Information from both sources had to be standardized, labelled, organized and stored in such a way that it would be easily accessed by the supervised learning method. The model only learns from that training data, and it is understood that non-experienced situations cannot be predicted. Because of this, it was assumed that every project would need its own training process.

As mentioned before, the testing tunnel was located mainly in a Granodiorite region from Superior Cretaceous-Tertiary, with a Rhyolite unit from Cretaceous-Tertiary, and some presence of Dacite. Geomechanical stability was structurally dominated, with a periodical appearance of two fault systems. Water and stresses were not dominant factors.

First Models

After the selection of the AI method to use, a model needed to be developed. This time the model should forecast the Q index value, not the support type. Once the model was complete and ready to forecast, in order to get an idea on how fast the model learned, 4 different models were trained with different sample sizes. Figure 6 shows,

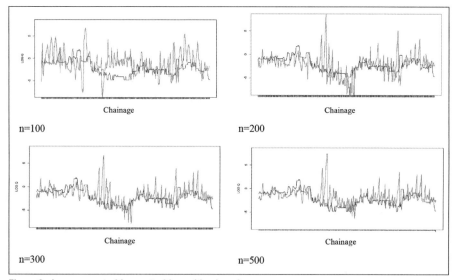

Figure 6. Improvement of forecast with machine learning

in red, the Q value forecasted and in black the real Q value mapped for four different training models with sample size 100, 200, 300 and 500.

As can be seen in Figure 6, more data available (i.e., sample size increases), the more accurate the trained model becomes. For the same models analyzed at this time, a support type forecast was made considering the forecasted Q value that would assign a support type. This was compared with the real support type mapped in the tunnel. The result was found correct within the exact support type or within 2 consecutive support types from the real ones mapped. Table 4 shows the results from this initial 4 models.

Table 4. First machine learning results

Sample Size	Exact Support Type	2 Support Types Range
100	35%	45%
200	50%	69%
300	51%	71%
500	54%	75%

Trained Model

The final model considered 1692 registers that were used to train it, and 423 registers were used as test set, all from the same tunnel. Once the model was trained, it was used as a forecasting tool during the performance of new Probe Holes. Results show that the model has an accuracy of +85% forecasting rock mass classification behind the face in a way that is suitable for advising a decision maker. Figure 7 shows the results obtained with the final model.

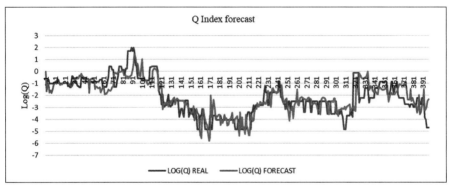

Figure 7. Final forecast with trained model'

APPLICATION TO CONSTRUCTION

Since the results from the first trained model were really accurate, the next step was to create a direct way of using trained models during construction of drill & blast tunnels.

Following the current developments for geotechnical works in the tunnels, SKAVA developed a mobile application that can be used in a tablet PC by the geologists on site.

As was described previously, in order to train a model two datasets are needed, one from the probe holes and the other from the face mappings. So the application is meant to gather this information provided by the Geologist and connected via internet, can update a web portal so real time information from face mapping, support and tunnel photos can be accessed worldwide.

The application has the option of training as many predictive models as the user wants, always keeping in mind that the more information to train the model, the better forecast will be obtained. Also, the application has the option of training different models for different tunnels in the same project considering all the information of the project and the data loading of different geologists at the same time.

Once a model is trained in the application, it can be used directly in the excavation face to forecast the rock support to be installed. The forecast requires only the information from the probe holes and takes less than 3 minutes to forecast.

With this tool, short time planning becomes much easier and early detection of weak rock mass conditions (i.e., shear zones, faults, etc.) provides risk control and avoids possible tunnel collapses.

CONCLUSIONS

Forecasting the rock to be excavated has been always a crucial tool during tunnel construction. Current methods offer some predictions, but all of them have certain limitations. Data used from an existing project showed that an historical analysis was a good tool to forecast the rock mass quality, but the inclusion of Artificial Intelligence offers more accurate results.

REFERENCES

Bilgin, N. and Ates, U, 2016. Probe Drilling Ahead of Two TBMs in Difficult Ground Conditions in Turkey. *Rock Mechanics Rock Engineering* Volume 49 (Issue 7): pages 2763–2772.

Chapman, R.E., (ed) 2000. Petroleum Geology, Developments in Petroleum Science. Amsterdam: Elsevier.

Kotsiantis, S.B., 2006, Supervised Machine Learning: A Review of Classification Techniques. *Artificial Intelligence Review* Volume 26 (Issue 3): pages 159–190.

Mohri, M., and Rostamizadeh A. & Talwalkar, A., 2012. Foundations of Machine Learning. Cambridge, MA, USA: MIT Press.

Rivera, D., 2012. Aplicación de la tecnología Measurement While Drilling en Túneles. Santiago, Chile: Universidad de Chile Repository.

Evaluating Impact of Water Content on EPB Machine Performance Based on Laboratory Experiments

Wei Hu ▪ Colorado School of Mines
Jamal Rostami ▪ Colorado School of Mines

ABSTRACT

Water content of the soil and muck has a notable impact on performance of Earth Pressure Balance (EPB) shields. Ignoring or misjudging the water content and soil condition in the field have caused severe consequences such as severe tool wear, severe secondary wear on machine components, clogging of the cutterhead opening, high power consumption, and increased machine maintenance requirement, all of which contribute to delays and lower machine utilization. In this paper, the Soil Abrasion Index Testing Machine currently available at Colorado School of Mines (CSM) was used to test a variety of soils under different water contents. The new laboratory testing show that water content dominates some fundamental aspects of muck behavior, including soil abrasiveness, compaction, and clogging potential, and consequently, EPB machine performances. Therefore, adjusting water content of the muck can work as a quick and effective soil conditioning approach in some projects and site conditions.

INTRODUCTION

Earth Pressure Balanced (EPB) TBM tunneling is a dominant method in soft ground tunneling(Alavi Gharahbagh 2013). The basic function of the EPB machine is to utilize the excavated mucks as the supporting medium for maintaining face pressure to assure its stability while transferring part of the excavated material through excavation chamber and screw conveyor to the tunnel, which is under atmospheric pressure. To optimize this system, the muck should have certain permeability, compressibility, viscosity, and strength. Meanwhile, to minimize downtime due to tool wear and machine clogging, the muck would also need to be workable and less abrasive. However, it is very rare for soil to have the desired properties in its natural state, meaning that it must be conditioned to modify its behavior to a more workable mix based on the grain size distribution, mineralogy, clay content, water content, and pressure. Some of the common approaches of soil conditioning include adding water, slurry, polymers, and foam to the excavated muck. Among them, foam has been most extensively used, so possibility of water being conditioner, or generally speaking the role of water is sometime ignored. On the other hand, there are also numerous reported cases where the addition of water to suppress dust at the cutting face turned out to deteriorate machine performance such as increasing tool wear and machine plugging. For a given project, it is worthwhile to investigate the role of water content on some of the fundamental aspects in EPB machine tunneling, including soil abrasion and soil clogging.

This paper presents the impact of water content on the soil abrasion investigation results of testing on several different soil types in the last few years, using the Soil Abrasion Index (SAI) testing method/machine currently available at Colorado School of Mines (CSM). Using the same machine, the impact of water content on clay clogging potential was also studied, in conjunction with other two commonly used clay clogging evaluation approaches, namely the Consistency Index approach and the Empirical Stickiness Ratio method.

SOIL ABRASION INDEX TESTING MACHINE

Unlike rock abrasion evaluation practice, soil abrasion studies are relatively new. The attempt to assess soil/weak rock abrasiveness did not emerge until the past decade or so due to the emerging soft ground TBM tunneling projects around the world. To date, several soil abrasion evaluation systems have been developed (Jakobsen et al., 2013; Peila et al. 2012; Barzegari et al., 2013; Rostami et al., 2012), but no standard test has been adopted. Among them, SAI is one of the most versatile devices that is capable of mimicking real working conditions, including mineralogy, large spectrum of particle dimensions ranging from clay up to gravel, water content, ambient/pore pressure, overburden pressure, tool hardness, among others. As shown in Figure 1a, the device consists of a cylindrical testing chamber 350 mm in diameter and 450 mm in height, enabling testing of soil with various particle size distribution up to gravle size particles. A propeller with three pitched blades covered by metal covers, shown in Figure 1b and 1c, is mounted at the lower end of the drive shaft and spins inside the chamber containing the soil under desired conditoins (moisture content, conditioning, pressure, etc.). The propeller is powered by a five hp drive unit. When testing, the chamber is filled with soil to roughly 300 mm of the chamber height, rendering 150 mm of soil depth both above and below the propeller. The soil can be prepared with different water content and conditioning, and tested under selected surcharge loading and ambient pressure, representing the column of soil above the tunnel, and pore pressure. The torque is directly measured by two force sensors attached to the chamber. Compressed air valves are connected to the top lid of the chamber and can simulate various pressures, mimicking tunneling conditions with pore pressure or surcharge

Figure 1. Configuration of the preliminary design of soil rheometer device: (a) overview; (b) pitched propeller with 10 deg pitch angle (c) selected propeller cover with selected hardness

loading up to 10 bar. Details of the device can be found in literatures (Rostami et al., 2012; Hedayatzadeh et al., 2017).

For a standard SAI test, a propeller with three 10 deg pitched blades covered by 17 HRC steel covers spins at 60 rpm inside the chamber. The cumulative overall weight loss of the three metal covers for 60 min of testing is recorded and compared with the soil abrasion index classifications shown in Table 1.

Table 1. Soil abrasion index classifications (Alavi Gharahbagh 2013)

Classification	Weight loss after 60 min of testing, g
Non to very low abrasiveness	<2
Low abrasiveness	2~5
Medium abrasiveness	5~10
High abrasiveness	10~15
Extremely high abrasiveness	>15

The original design of the machine allowed for operation at 60 rpm, designed for expediting tool wear measurement in the laboratory. However, to allow for examining diferrent speeds and shear rates, the unit has recently been modified and is currently capable of rotating at speeds from zero to 1,000 rpm. The original design was not feasible for study of muck flow inside the testing chamber, and current system allows the device to work as a testing component of a large-scale soil rheometer. This is accomplished by measuring torque at various rotational speed of the propeller, thus allowing for establishing a relationship between torque as representative of shear strength, and rpm that is the indicator of shear rate, for different soil settings and machine operational parameters. The corresponding soil rheology parameters (i.e., yield stress and viscosity) can be back calculated in CFD models.

IMPACT OF WATER CONTENT ON SOIL ABRASION

Testing on CSM Sand

To investigate the impact of water content on soil abrasion, two types of sandy soils were tested under the standard soil abrasion testing conditions at CSM. The first sand, namely the CSM sand, is a poorly graded sand (Figure 2) that is produced in a quarry by crushing process and commercially available in Golden, CO. Some basic soil mechanics properties of the sand were tested to offer a benchmark for comparison to other soil types. As shown in Figure 3, the sand contains 40% of quartz and 29% of plagioclase. The air-dried water content of the sand was found to be 0.2%, meaning the sand in the lab can be considered as completely dry. The specific gravity of the sand, defined as the ratio of the density of the soil solid particles to that of the water, is estimated to be 2.69.

Forty kg of dry sand was used to fill up the testing chamber for the standard test in dry condition. For the wet condition, a concrete mixer was used to mix dry sand and water to reach designated water content before testing. The abrasion testing results for the CSM sand, as shown in Figure 4, demonstrate the significance of water content on soil abrasion. When the sand is dry, the weight loss of the three 17 HRC steel covers was 11.67 gr, classified as high abrasiveness soil. When moisturized to 10% water content, the wear increased to the point that within 10 min of testing more than 11 gr weight loss was registered, meaning anticipated wear at 60 minute of nearly 69.08 gr. Compared with the soil abrasion index classification in Table 1, this level of abrasion is deemed as extremely high.

Testing on River Sand

SAI tests were also carried out on a well sorted, finer, and more rounded river sand. The testing was first conducted at low water content, as shown in Figure 5 (a), with

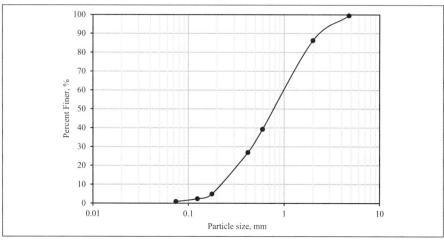

Figure 2. Particle size distribution for the CSM sand

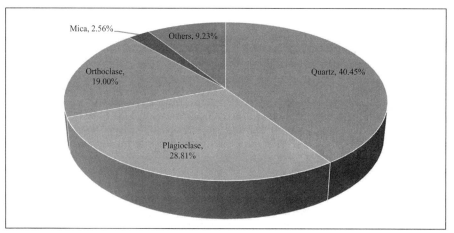

Figure 3. Mineralogy for the CSM sand via QemScan

water content measured at 15.6%. Subsequently, the sand and the water were mixed to saturated condition, as shown in Figure 5b. Note that the actual amount of water added was 0.5 kg more than calculated to ensure full saturation, and therefore, a layer of free standing water was observed on top of the soil in the testing chamber. At last, the saturated sand was air-dried to 0.5% of water content and tested, shown in Figure 5c.

The testing results of the fine river sand, as shown in Figure 6, further demonstrates the impact of changing water content on the soil abrasiveness of this sand, which is even more significant than previously found on CSM sand. That is, the weight loss at 60 min was 0.66 gr for the saturated condition, and slightly increased to 1.56 gr for the air-dried condition, and exponentially rose to 30.53 gr for the moist condition. Consenquenty, the weight losses of the saturated and air-dried conditions are classified as non to very low abrasiveness, while that of the moist condition goes to the extremely high abrasiveness zone.

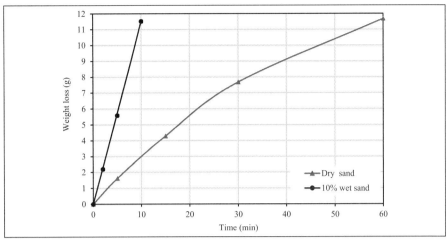

Figure 4. Measured weight losses of 17 HRC steel covers on the CSM sand under two water contents

Figure 5. Standard soil abrasion index testing on the river sand at three water contents: (a) moist, ω=15.6%; (b) saturated, ω=26%; and (c) dry, ω=0.5%

Testing on Other Soils

The impact of water content on soil abrasion has been tested and confirmed in a variety of soils, as shown in Table 2. For a variety of soils, the relative magnitude of soil abrasiveness at different water contents follow the relationship shown in Figure 6, with the lowest abrasion at the saturated condition and the most severe abrasion at

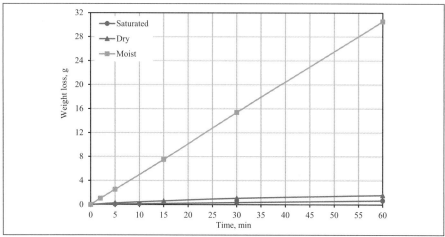

Figure 6. Measured weight losses of 17 HRC steel covers on the river sand under three water contents

Table 2. Measured weight losses of 17 HRC steel covers on various soils (updated from (Alavi Gharahbagh 2013))

Soil Type	Water content	Weight loss, g	Soil Type	Water content	Weight loss, g
CSM sand	Dry	11.67	ASTM 20/30 sand	Saturated (ω=22.5%)	1.19
River sand	Dry	1.56	CSM sand	Moist (ω=10%)	69.08
PSU silica sand	Dry	22.55	River sand	Moist (ω=15.6%)	30.53
PSU limestone sand	Dry	1.27	PSU silica sand	Moist (ω=10%)	131.94
ASTM graded sand	Dry	0.48	PSU silica sand	Moist (ω=15%)	20.02
ASTM 20/30 sand	Dry	0.48	PSU limestone sand	Moist (ω=15%)	11.94
PSU silty sand	Dry	1.37	PSU silty sand	Moist (ω=45%)	0.22
River sand	Saturated (ω=26%)	0.66	PSU silty sand	Moist (ω=60%)	0.29
PSU silica sand	Saturated (ω=22.5%)	3.52	Washington D.C clay	Dry	13.08
PSU limestone sand	Saturated (ω=22.5%)	9.12	Seattle SB clay	Dry	4.93
ASTM graded sand	Saturated (ω=22.5%)	1.63	Seattle NB clay	Dry	6.44

Note: all under standard testing conditions, i.e., 10 deg pitch propeller, 17 HRC covers, 60 min of testing time

a certain moist condition. In other words, the soil abrasiveness peaks near the optimum water content measured by Proctor compaction test (Alavi Gharahbagh et al., 2013). Depending on other influencing factors on soil abrasion such as soil mineralogy, particle size, and shape, the contrast of soil abrasiveness due to changing water content vary.

Practical Implications

The above-presented soil abrasion testing results are reminder of the critical impact of water content on soil abrasion and machine tool wear. To fully understand how water content can affect tool wear in a proposed tunnel alignment, and hence, to accurately predict the most severe tool wear scenario, tailor-made laboratory soil abrasion testing for each soil formation can be included in geotechnical investigation program to allow the contractors assess the related risks. Based on the soil abrasion testing results, one can slightly adjust the opeational water content at the face and the cutting chamber to avoid the most severe water content condition while maintaining other desired muck properties. This includes the optimization of the proper conditioning system for reduced abrasion and improved muck flowability.

IMPACT OF WATER CONTENT ON SOIL CLOGGING

While soil abrasion can cause trouble for tunneling in abrasive sand, soil clogging is one of the biggest concerns for mining fine-grained sticky soils. Water content has also proven to impose remarkable impact on clay clogging potential, as will be describe in the following.

Clay Characterization

The clay used in this study, namely the blue/gray clay from Denver metropolitan area, has high plasticity, as shown in Table 3. Particle size analysis by both sieving and hydrometer methods shows that 80% of the clay is dominated by fine particles smaller than 0.075 mm, while less than 10% of the soil sample is randomly scattered gravels, as shown in Figure 7. The mineralogy of the clay is also analyzed via XRD, as shown in Figure 8. Note that XRD is optimally suited for mineral identification and not best option for quantifying the mineral content of the sample but offers good approximation for percentage of constituent minerals. The accuracy of estimated mineral content by XRD depends on many factors including grain size, preferred grain orientation, solid solution, order-disorder, etc., thus the estimates provided herein should be considered semi-quantitative.

Table 3. Basics properties of the Denver clay

Specific Gravity, Gs	Air-Dried Water Content, ω, %	PL, %	LL, %	PI	USCS
2.72	3.9	23	50	27	CH, clay of high plasticity/fat clay

Evaluating Clay Clogging Potential by Consistency Index Approach

The clay clogging potential was first evaluated via the Consistency Index approach proposed by (Hollmann and Thewes 2013), as shown in Figure 9. The key concepts and boundaries in this approach include natural water content (ω_n), plastic limit (ω_P), liquid limit (ω_L), plasticity index (I_P) and the consistency index (I_C). I_P and I_C are defined by the following equations:

$$I_P = \omega_L - \omega_P \quad \text{(EQ 1)}$$

$$I_P = (\omega_L - \omega_n)/I_P \quad \text{(EQ 2)}$$

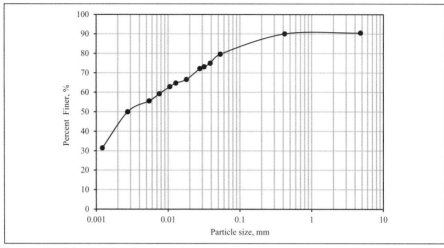

Figure 7. Particle size distribution for the Denver clay

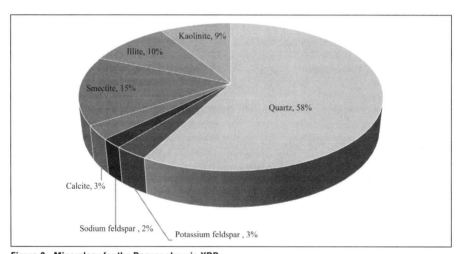

Figure 8. Mineralogy for the Denver clay via XRD

All of these concepts are associated with water content, demonstrating the dominating impact of water content on clay clogging. As the water content of the clay increased from the plastic limit of 23% to the liquid limit of 50%, as shown in Figure 9, the clogging potential initially increased, and then reached the peak between 30% and 35% of water content, followed by a reduction afterward. From pure anti-clogging perspective, the best water content is above 35%. The muck at this stage, however, is too liquid and does not live up to other muck requirements such as face pressure control. A common practice to balance anti-clogging and other muck requirements is to control the water content to the range which falls within "strong clogging" category but add anti-clay agent or foam to control the other properties of the mix.

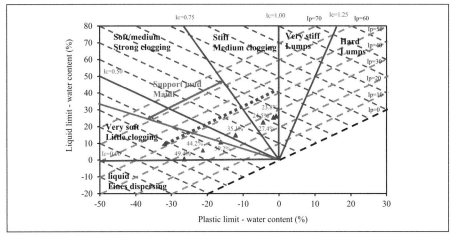

Figure 9. Clay clogging potential evaluation based on Consistency Index approach for EPB support mud (based on Hollmann and Thewes 2013)

Evaluating Clay Clogging Potential by Empirical Stickiness Ratio Approach

The clogging potential of the Denver clay was also assessed via the Empirical Stickiness Ratio (λ) approach proposed by (Zumsteg and Puzrin 2012). It is based on mixing test on the dough mixer and defined by the following equation:

$$\lambda = G_{MT}/G_{TOT} \qquad \text{(EQ 3)}$$

In this equation, G_{MT} is the soil sticking to the mixing tool and G_{TOT} the total weight of soil in the mixer. The bigger the ratio, the higher the clogging potential is. It was found that clay clogging evaluation by this Empirical Stickiness Ratio method had good agreement with that by the Consistency Index approach, while advancing the definition to a quantitative level (Zumsteg, Plötze, and Puzrin 2013). In current study, a 4.5-quart KitchenAid mixer, as shown in Figure 10a, was used to mix the pulverized clay powders with calculated amount of water (to achieve desired water content) until a homogeneous state is reached. The mixing time to ensure this satisfactory homogeneity was observed to be within 2 min for preliminary testing. To ensure parallel testing conditions, the standard mixing time thereafter was fixed at 2 min. After mixing, the flat beater was taken out for weight measurement for each water content, as shown in Figure 10b. The testing results, shown in Figure 10c and Figure 10d, correspond to λ versus water content "ω" and λ versus I_C relationships, respectively. These charts echo the clay clogging potential evaluation results by the Consistency Index method. It can be seen that two water content thresholds for the Denver clay exist and they are 20% and 40%. The ratio, , is negligible when the water content is lower than 20%, followed by a drastic increase as the water content rises from 20% to 30%. λ remains at a stable high values as the water content further increases from 30% to 40%. Subsequently, the ratio plunges as the water content keeps increasing towards the liquid limit. That being said, it is recommended that EPB operation shall avoid getting into the problematic region with high empirical stickiness ratio values.

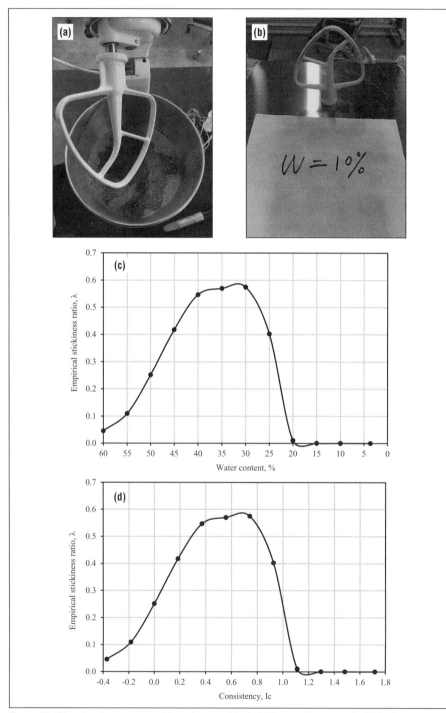

Figure 10. Clay clogging assessment based on empirical stickiness ratio approach: (a) testing device; (b) weight measurement; (c) stickiness ratio versus water content; (d) stickiness ratio versus consistency

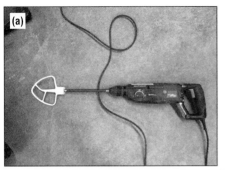 

Figure 11. Preparation of homogeneous clay with designated water content for SAITM testing: (a) mixing tool; (b) satisfactory homogeneity after mixing

Clay Clogging Potential Evaluation Using SAI Testing Method

Note that the Empirical Stickiness Ratio approach apparently equates stickiness to clogging. To further extend, both the Empirical Stickiness Ratio and the Consistency Index methods only consider the water content in the evaluation of clay clogging. The feasibility of this practice is uncertain because in addition to water content, there are other factors that influence clogging of the machine such as in situ stress conditions and machine operational parameters. However, the appropriateness of equating stickiness to clogging is beyond the scope of this paper and will be investigated in details separately. Herein, the impact of water content on the clay clogging potential will also be demonstrated by SAI testing results.

One of the greatest challenges for conducting SAI testing for clay is to prepare roughly 40 kg of homogeneous clay-water mixtures. After different ways of mixing trials, a setup with a flat beater mounted to the drill was found to offer satisfactory mixing effect, as shown in Figure 11. The well-mixed clay-water mixture with designated water content was then placed into the SAI testing chamber for testing. The propeller with 10 deg pitched blades was used to test the clay at 60 rpm. A total of six SAI tests wear conducted for six designated water contents. At each water content, the testing lasted until a steady torque was seen.

The maximum torque was obtained for each water content and plotted as shown in Figure 12. The curve is bi-linear with $\omega=27.4\%$ being the peak point. Starting from liquid limit state at the right end of the curve, the maximum torque increases very gently from 80 N·m to 205 N·m for $\omega=27.4\%$, and then drastically rises to 1,111 N·m for $\omega=23.8\%$, which is the plastic limit state. In EPB machine tunneling, the cutterhead torque can be considered as an indicator for clay clogging. This index system is now proven to be suitable for our SAI testing, in which higher torque means higher clay clogging potential. This idea was further confirmed by the observation of the soil after testing, as shown in Figure 13. As seen in these two pictures, the soils in both scenarios endured compaction due to the high contact stress imposed by the propeller with 10 deg pitched blades. The difference is, however, that the plastic yet stiff clay at $\omega=23.8\%$ showed significant compaction and would cause severe clogging issues, while the compaction for $\omega=49.4\%$ is low resulting in significantly reduced clogging potential.

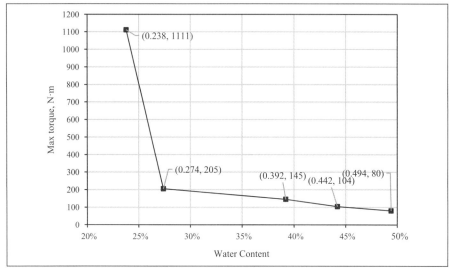

Figure 12. Measured maximum torque versus water content curve for the Denver clay via SAITM testing

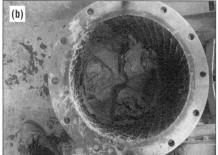

Figure 13. Compaction of the Denver clay at the end of SAITM testing: (a) significant compaction at $\omega=23.8\%$; (b) slight compaction at $\omega=49.4\%$

Practical Implications

The implications of the SAI testing are interesting. First of all, it shows the capability of evaluating the impact of water content on wear in coarse-grained and abrasive soils, and clogging in fine-grained soil and clay. The testing allows for adjusting the water content to see if the use of water alone can mitigate some of the issues, while also facilitating the use of various soil conditioning systems as a counter measure in both coarse-grained and clayey soil. It is important to note that the most severe clogging region revealed by SAI testing does not directly overlap the areas identified by Consistency Index and Empirical Stickiness Ratio approaches. While this observation requires additional investigation and comparison with field observations, it still shows the possibility of encountering clogging even outside the ranges predicted by the above mentioned methods. This difference implies that more influencing factors shall be considered for evaluating clay clogging although water content is one of the dominant factors. One advantage of SAI testing is to allow for including the soil conditioning and in situ stresses (pore pressure and surcharge loading) in the testing.

CONCLUSIONS

Water content plays a significant role in EPB machine tunneling. This paper discussed the latest laboratory testing studies to demonstrate the crucial impact of water content on two fundamental EPB machine performance aspects, including soil abrasion and soil clogging. The major findings are as follows.

- The soil abrasion testing were conducted in more than ten types of soils in dry, wet and saturated conditions. The results show that wet soils are most abrasive at water content near Proctor Compaction test peak, followed by dry soils and saturated soils. Depending on the water content of the in situ soils, a practical soil conditioning approach to reduce tool wear can be adding water to exceed the peak point at the curve but not to saturation point where it can cause free flow.

- The soil clogging evaluation were carried out by three approaches, including the Consistency Index, the Empirical Stickiness Ratio, and the SAI methods. The results confirm the existence of a critical water content region for problematic clogging potential. Due to other expected muck behaviors required by EPB TBM tunneling, the machine may have to be operated within this problematic clogging region and anti-clogging agents are needed to reduce clogging risk simultaneously.

- The testing in clay also revealed the agreement between the Consistency Index and the Empirical Stickiness Ratio methods, while there is mismatch with the SAI method relative to the water content at measured peak torque. This is due to the inclusion of compaction effect on clogging that can be observed and measured in SAI testing.

More studies need to be conducted to define, offer quantitative measures, identify influencing factors, and examine the general trends in dealing with clay clogging.

ACKNOWLEDGMENTS

This study is funded by Earthern Mechanics Institute at Colorado School of Mines. Several CSM staffs gave their extensive support for the laboratory testing in this study, particularly Brent Duncan and Omid Frough. Appreciation also goes to several individuals for their support on some aspects of this study. This includes Bruce Yoshioka, Ray Johnson, Marcelo Simoes, Richard Wendlandt, as well as several undergraduates. The use of previous soil abrasion testing database would not be possible without the courtesy of Dr. Ehsan Alavi Gharahbagh.

REFERENCES

Alavi Gharahbagh, Ehsan. 2013. "Development of a Soil Abrasion Test and Analysis of Impact of Soil Properties on Tool Wear for Soft-Ground Mechanized Tunneling." *Pennsylvania State University.*

Alavi Gharahbagh, Ehsan, Tong Qiu, and Jamal Rostami. 2013. Evaluation of Granular Soil Abrasivity for Wear on Cutting Tools in Excavation and Tunneling Equipment. In *Journal of Geotechnical and Geoenvironmental Engineering*, 139:1718–1726.

Barzegari, G., a. Uromeihy, and J. Zhao. 2013. A Newly Developed Soil Abrasion Testing Method for Tunnelling Using Shield Machines. In *Quarterly Journal of Engineering Geology and Hydrogeology*, 46:63–74.

Hedayatzadeh M , Rostami J , Peila D , Forough O, Salazar C. 2017. Development of a Soil Abrasion Test and Analysis of Impact of Soil Properties on Tool Wear for Soft-Ground Mechanized Tunneling Using EPB Machines. In *Proceedings of the World Tunnel Congress 2017*, 1–6.

Hollmann, F.S., and M. Thewes. 2013. Assessment Method for Clay Clogging and Disintegration of Fines in Mechanised Tunnelling. In *Tunnelling and Underground Space Technology*, 37:96–106.

Jakobsen, Pål Drevland, Amund Bruland, and Filip Dahl. 2013. Review and Assessment of the NTNU/SINTEF Soil Abrasion Test (SAT™) for Determination of Abrasiveness of Soil and Soft Ground. In *Tunnelling and Underground Space Technology*, 37:107–114.

Peila, Daniele, Andrea Picchio, Alessio Chieregato, Monica Barbero, Enrico Dal Negro, and Alessandro Boscaro. 2012. Test Procedure for Assessing the Influence of Soil Conditioning for EPB Tunneling on the Tool Wear. In *World Tunnel Congress*.

Rostami, Jamal, Ehsan Alavi Gharahbagh, Angelica M. Palomino, and Mohsen Mosleh. 2012. Development of Soil Abrasivity Testing for Soft Ground Tunneling Using Shield Machines. In *Tunnelling and Underground Space Technology*, 28:245–256.

Zumsteg, R., M. Plötze, and A.M. Puzrin. 2013. Reduction of the Clogging Potential of Clays: New Chemical Applications and Novel Quantification Approaches. In *Géotechnique*, 63:276–286.

Zumsteg, R., and A.M. Puzrin. 2012. Stickiness and Adhesion of Conditioned Clay Pastes. In *Tunnelling and Underground Space Technology*. 31:86–96.

The Expanding Capabilities of Microtunneling Demonstrated in Washington DC

Todd Brown • Bradshaw Construction Corporation

ABSTRACT

DC Water is improving water quality in Washington DC by replacing aging pipelines. The Oregon Avenue NW Sewer Rehabilitation project was constructed by Bradshaw Construction to replace and improve sewers in Rock Creek Park and Bingham Drive. Of the project's 4,300 feet of new 24" sewer, 2,700 feet was installed by microtunneling through the variable rock conditions at up to 90' deep. Work areas were particularly challenging as the project was located between a quiet, residential neighborhood and Rock Creek Park. Access had to be maintained for residents, emergency services and visitors to the National Park. To minimize public impact, Bradshaw reduced the seven designed tunnels to three including a single 1,860' drive and a 430' S-curve with a 625 foot radius. Both drives represented extremes for rock microtunneling the in United States for both length and the curved alignment, respectively. The collaboration and flexibility of the Oregon Avenue team allowed the project to show how far we can push the push the capabilities of microtunneling.

INTRODUCTION

From the first site visits during the pre-bid phase of the project, the first thing that the Contractor noticed about the Oregon Avenue NW Sewer Rehabilitation project was the constrained work areas. As the projects alignment traverses through a National Park and the residential neighborhoods of Barnaby Woods and Chevy Chase, the restrictions were not the typical ones, such as providing access for businesses or heavy vehicular and pedestrian traffic. They were more about maintaining the integrity of the quiet, forested landscape while still being able to build a major sewer infrastructure project.

The project's alignment was an "L" and was split into two sections. The first section was the north-south leg which was within Oregon Avenue, owned by the District of Columbia Department of Transportation (DDOT), consisted of 2,000 feet of the new sewer with 1,866 feet of which being installed by microtunneling. The west side of Oregon Avenue is confined by overhead power lines which limited crane usage. Additionally, with the expectation of a full closure at the northern terminus of the project, required one lane of traffic to be maintained throughout operations for local residents and emergency services. These two constraints requires the Contractor to conduct all excavations on the east side of the road, which was not exactly wide open as it had restrictions of its own. The east side of Oregon Avenue is bordered by Rock Creek Park, owned by the National Park Service (NPS) and in this section of the project, excavation was not permitted to cross into the park property and mature trees rooted in the park overhang the street. Thus the work zone consisted of a single lane of Oregon Avenue and the apron between the pavement and the park boundary, which is typically less than 20 feet in total.

The section leg of the "L" was within Bingham Drive, a local access road completely within NPS property that connected Oregon Avenue to Beach Drive, the main road that traverses Rock Creek Park. Of the 2,300' of new 24 inch sewer to be installed

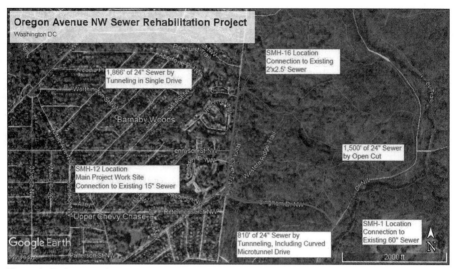

Figure 1. Project overview map

within the section, 810 feet would be installed by trenchless methods, including microtunneling and hand-mining. While Bingham Drive was completely closed to the public during construction, it was actually the most constrained portion of the project. There was only one access point to the work, from Bingham Drive's intersection with Oregon Avenue. While the new sewer would cross Beach Drive to its connection point to the existing 60" sewer interceptor on its east side, no other construction traffic was allowed on Beach. All hauling and excavation had to occur along the 18 foot wide Bingham Drive within the curb lines. The road had two existing parking lots and one opening in the curb for an unpaved park access road which were used for passes around shaft excavations. Otherwise, all construction traffic for both tunneling and open cut operations had to take turns using the same narrow lanes as NPS required operation to stay within the curb lines to preserve the flora and fauna of Rock Creek Park.

Once Bradshaw Construction was awarded the project and given notice to proceed in September 2016, all efforts went to finding the best options to overcome these confinements so that the project could be effectively built while minimizing disruption to the public. To that end, Bradshaw Construction, with the support of the DC Water project team of Construction Manager AECOM and Design Engineer Johnson, Mirmiran and Thompson (JMT), performed a number a firsts to successfully complete the project.

NARROW SHAFT CONSTRUCTION

Typical support of excavation systems for tunneling shafts are either circular, constructed steel ribs and/or liner plates, or rectangular, using steel sheeting or soldier piles and lagging. As the west side of Oregon Avenue was not available for use due to overhead power and local traffic, circular shafts were ruled out as an options in most locations. To achieve the necessary length for the microtunneling operations, a circular shaft would have encroached on the power line set backs and the traffic lane. A rectangular shaft initially appeared to be the best options, though steel sheeting was also ruled out due to the underlying rock in the geologic profile and, again, with the proximity to the overhead power lines. Soldier pile and lagging seemed to be the best option for the main tunnel Launch Shaft at SMH-12. However, as the Contractor

Figure 2. SMH-12 shaft collar near traffic

Figure 3. SMH-12 shaft excavation in rock

began the process of verifying the location of existing underground utilities, it became increasingly clear that the surrounding storm drains, water lines and gas lines would provide very little access for pile installation, so a new option would have to be found.

In past projects, Bradshaw had utilized elliptical shafts with shotcrete support where a long but narrow shaft was needed with penetration in the shaft walls for existing utilities. The SMH-12 shaft on Oregon Avenue fit this concept as it needed to be long enough to pipe jack 20 foot long steel casing without getting any wider than it needed to be. Additonally, SMH-12 was the location of the project connection to the existing 15" sewer which traversed Bingham Drive that this project was replacing. This elliptical shape would give the footprint needed for the work and would allow the existing sewer to be supported in place within the shaft excavation, but the shotcrete support part of the plan gave the Contractor some pause. The excavation for this shaft was scheduled to occur in the February of 2017 to achieve the project's schedule. Given the amount of time available between the project award and that start date, there was not enough time to perform the typically site specific testing protocols for shotcrete. Shooting shotcrete in the winter was also a concern regarding quality control of application in freezing temperatures.

Bradshaw discussed these concerns with shaft and tunnel support manufacturer, the Jennmar Corporation. The decision was made to pursue a design with the elliptical dimensions necessary for the work to be performed within, but using steel ribs and liner plates for structural support. These materials could be quickly produced and installed without the quality control burdens of a shotcrete operation. The end result was a 33 foot long by 22 foot wide elliptical shaft, supported by W10×68 rolled steel ribs and 8 gage, two-flange liner plates. This 42 vertical foot deep shaft would house the microtunneling equipment needed for 20 foot long casings. It would also allow the shaft walls to be penetrated by the existing 15" sewer, which was then flumed around the inside perimeter of the shaft so its flow could be maintained for the projects duration without the cost of impact of bypass pumping. The shaft supports were manufactured to different radii to fit the shaft geometry, which is not typical for liner plate shaft and thus assembling the plates specifically to the design drawings was an important point of emphasis during construction. The Contractor's crews found that managing this myriad of different plate dimensions was a short learning curve and the assembly during excavation was effective.

In fact, this shaft design concept was so effective that it was used at two other location on Oregon Avenue where the width of access was exceedingly tight. The drops shafts for manholes SMH-13 and SMH-14 were even more constrained than that of SMH-12, as the SMH-12 site had a green space provided by NPS to house the surface support equipment for the mircotunneling operations, these shafts had only one lane of traffic

Figure 4. SMH-13 shaft collar near traffic   Figure 5. SMH-13 shaft with manhole base

and a narrow strip on ground between the edge of pavement and the NPS boundary. They were also deep, up to 48 vertical feet, with rock in the profile and no blasting allowed on the project. The excavations needed to be narrow enough to fit the work area and long enough for a mini excavator to work within them. Again, the elliptical liner plate shaft made the most sense and given previous success at SMH-12, it was an easy decision. For these shafts, the design was for 18 foot long by 12 foot wide, 5 gage liner plate shafts. The soil conditions and dimensions did not require additional support from steel ribs, which helped gain as much space as possible inside the support of excavation to perform the necessary work. With no learning curve, these two shaft were installed with even better efficiency than previously, proving to the Contractor that elliptical liner plate shaft were a great option to have for shaft construction.

1,866' MICROTUNNEL DRIVE ON OREGON AVENUE

The tunneling on Oregon Avenue was originally designed as three tunnels with four tunnel access shaft and five corresponding manholes. The longest design tunnel was 1,345 linear feet at depths up to 90 vertical below the road surface between SMH-13 and SMH-14. This tunnel stretch was shortened as much as possible by design, but could not be shortened any more as no additional manhole locations would have been shallow enough to be used for DC Water maintenance crews over the lifetime of the sewer. As noted in the section above, the sites on either end of this long stretch did not contain enough working room for a short distance microtunneling operation, much less a 1,345 foot long drive. The best location for microtunneling was at SMH-12 at the intersection of Oregon Avenue and Bingham Drive. It would have been possible to mine from the northern extremity of the project as well, but without another intermediate shaft that would have only reduce the total drive length by less than 300 feet, so the decision was made to pursue a single, 1,866 foot long drive for the entirety of Oregon Avenue, launched and operated from SMH-12.

While this alignment was generally a straight one, there were slight azimuth adjustments at the designed manhole locations avoid existing utilities. Bradshaw's solution was to straighten the drive, favoring the east side of Oregon Avenue from SMH-12 to SMH-16, adjacent to the NPS boundary. This was such a narrow area that an existing storm drain was supported in place within the liner plate recovery shaft. The intermediate drop shafts at SMH-13 and SMH-14 would be installed after tunneling was complete for access to the new 24" sewer only. SMH-12 was the most spacious site on the project and would serve as the working locations for all microtunneling operations, but it still had restrictions. Nothing could be permanently placed ahead of the north end of the shaft at Bingham Drive as that was the only access point for the NPS section of the work and could not be obstructed. And even though this section of the road was

Figure 6. Oregon Ave. Tunnel alignment looking north

Figure 7. Herrenknecht AVN-1200 MTBM

closed to through traffic, the southbound lane of Oregon Avenue had to be kept clear to provide through access to emergency vehicles.

Bradshaw had completed multiple microtunneling drives longer than 1,000 feet, however pushing that single drive distance to 1,866 feet would exceed the longest one by over 600 feet. Tunnel utility design thus was placed under much more scrutiny in this case. While every tunnel setup can be slight differences, shorter drives have typical arrangements. With the distance scheduled for the Oregon Avenue drive, every individual component of the system was examined to ensure capacity enough to reach the recovery shaft, nearly two dozen hoses and cables in total. Critical design components were for the slurry circuit, tunnel guidance, jacking capacity and simply tunnel access.

In the tunnel, the slurry piping and pumps were upgraded to maintain slurry flows through the eventual 4,000 feet of the circuit. The feed slurry line was upsized to reduced pressure losses to the MTBM and booster tunnel pumps were added to the return line to maintain flow velocity and keep the spoils from settling out of the slurry, causing wear and blockages, prior to reaching the surface. The contractor also made the assumption that the main tunnel slurry pump would not survive the full length of the drive with the abrasivity of the rock mined and the sheer volume that would have to be process. At launch a spare tunnel pump would be pre-installed in the tunnel to replace the main pump if and when it was required. Above ground, access and footprint was the biggest constraint. The typical separation plants used by Bradshaw to this point were side dumping plants, however in this case, only a single traffic lane was available and was not wide enough to accommodate them. An end dumping plant would be required and a Herrenknecht HS-150 Separation Plant was utilized because this plant was designed as to only dump spoils from the narrow end. The spoils were then collected into a purposed built muck bin to contain the pile from spreading into the road and park lands, from which they were periodically loaded out into the triaxles for disposal offsite.

The remaining surface systems for lubrication, power distribution, etc. were wedged into the remaining space available around the separation plant and control container with every available inch used. For long tunnel drive, the space needed for the tunnel utilities alone required an enormous laydown footprint, not including the area needed to storage of the tunnel casing, and the project's proximity to Bradshaw's main office approximately 40 miles away in Eldersburg, Maryland was likely the main reason the tunnel was not delayed due to the lack of storage space.

At drive lengths of greater than 800', secondary guidance systems to the tunnel laser begin to enter the conversation in the tunnel design. At 1,866, it was a requirement

Figure 8. Tunnel utilties

Figure 9. Casing pipe change

and the inherent extrapolation errors with a laser only system would be too great to be solely relied upon. The contractor considered several options the SLS-LT Navigation System manufactured and supported by VMT USA. As there would be a clear line of sight from the shaft to the MTBM, the system would not have to be installed until the after the drive passed the 600' mark. Thereafter, VMT would perform survey updates at approximately 250 foot intervals throughout the drive to ensure the MTBM reaches the recovery shaft as intended. The elevation relief between the launch and recovery shafts was the most time consuming part of the project with a 50 foot hill in between the two shafts which blocked line-of-sight and required frequent survey resets to connect the two locations. A north seeking gyro navigation was also considered, but was ultimately no chosen and the required survey update would be required nearly twice as often for alignment. Additionally, the gyro would require a secondary water level system for grade control, which Bradshaw's experience had proven to need time consuming maintenance to ensure accuracy. The SLS-LT system, while having a more complicated initial setup, was most efficient during tunnel operations.

In theory, no manned access is required during a microtunnel drive and that may be true in certain, typical tunnels. The contractor knew that this tunnel was not typical and manned entry would be nearly constant due to the length and soil conditions. Without special consideration, trips to the MTBM to perform simple maintenance tasks such as checking the gear would take the crew hours as the tunnel lengthened just to traverse the distance between the shaft and the machine. This tunnel would require in tunnel survey and disc cutter changes in addition to basic maintenance tasks. To ease access, Bradshaw designed utility organization racks to keep all hoses off of the casing invert. This provided both head room for crews in the tunnel, reduced tripping hazards in the small opening and aided in housekeeping during mining as every line had a place. Wear parts were placed in storage trialing tubes so they would be available at the MTBM when needed and would not have to be laboriously dragged to the machine each time. The access considerations proved extremely useful throughout the entirety of mining, but one event in particular show the worth of the efforts made. 1,500 Feet into the drive, the hydraulic pump in the MTBM's power pack failed and stopped mining. Using umbilical hydraulics to power the machine at that distance was not practical, or in this case possible due to the availability of 4,000 feet of hose needed. A rescue shaft to repair it was also not feasible due to the location of the MTBM deep in the hillside, under active traffic. The pump would have to be replace from within the tunnel and the replacement parts would need to be taken through the tunnel. An organized tunnel with a clean invert allow the crew to repair the pump in a matter of days without stripping any utilities to access it.

The thrust needs for jacking 1,866' feet of casing obviously had to be considered as well. The Contractor took significant and detailed efforts in design the tunnel drive's utility setup. Specific considerations made for jacking capacity and lubrication, which based experience should have been conservative. The original plan utilized an 850 ton jacking frame, a telescoping jacking section to propel the MTBM independent of the pipe string friction forces, and two intermediate jacking stations (IJS) with 560 tons of thrust a piece, giving the systems a combine thrust capacity of over 2,000 tons. An enclosed, dedicated lubrication system was purchased and used which was capable of high pressure, high volume delivery of bentonite lubrication to two independent sections of the tunnel: one to the MTBM and the first 100 feet of tunnel to keep the overcut void filled, one for the rest of the tunnel to provided secondary lubrication between the pipe string and the bentonite already pumped.

Despite these efforts, the biggest challenge in completing Bradshaw's longest microtunneling drive to date was managing those jacking forces. The amount of friction generated exceeded the Contractors best and most conservative estimates. At completion of the drive, an additional two IJS were added to the pipe string (four total) and were all used. This included one that was rebuilt underground during the mining operations to increase its thrust capacity when the section of casing ahead required more for than the base 560 tons. In total the system provided a total of 3,000 tons of capacity, which were need to overcome the 1,400 tons of resistance encountered, though that total pressure was not evenly spread throughout the tunnel, but localized.

The first 1,200 linear feet of tunnel progress largely according to preconstruction estimates. However, pressure spikes started to show themselves in the last 600 feet of the drive's length, the vast majority of which occurred in the section of the pipe string closest to the launch shaft. It is the Contractor's belief that this section of "older" tunnel created more thrust resistance because the decomposed rock through which it was mined would continue to deteriorate slowly as more and more pipe was thrust through the overcut. This generated fine rock particles which over time contaminated the lubrication present and increased the frictional forces. This should be more prevalent in decomposed rock than either hard rock or soft ground tunnels. Evidence of this contaminate was found in the lubrication from the outside of the tunnel casing once the drive was complete. In samples taken through the grout ports near the launch eye, where lubrication had been continuous pumps during the entire four month drive, the fluid was thick and dark with rock fines, but still will the bentonite's viscosity. In hard rock, the overcut will remain open and there would be minimal additional fines created due to the intact structure of the rock. In soft ground, though the overcut may partially collapse around the pipe and the surrounding soils will ravel into the overcut void, those soils can be displaced by pressurized lubrication and even the casing itself. Decomposed rock is the worst of both world because the rock is weak enough to create fines, but strong enough that they cannot be displaced, thus the rock fines create the thrust forces encountered.

Figure 10. SMH-12 mining site at the intersection of Oregon Avenue (right) and Bingham Drive (left)

Mining was done on a single day shift, 12 hours per day, 6 days per week as that was the maximum hours allowed by Washington DC in this residential neighborhood. The restricted works hours were a critical factor in the efficiency needed on the project

Figure 11. MTBM Launch at SMH-12   Figure 12. MTBM hole-out at SMH-16

and the forethought discussed above. Keeping the pipe string moving is an enormous factor in maintaining low jacking forces. Nightly shutdowns and meant the pipe was static more than it was moving. Any downtime for guidance or maintenance would further reduce the time the string was moving. While the Contractor would have vastly preferred a multiple shift operation, the local ordiances and location of a project can be a major factor in the tunnel design that has nothing to do with the shafts, access or ground conditions.

The Herrenknecht AVN-1200 MTBM was launched in early May of 2017. Even though the final carrier pipe was a 24 inch PVC sewer line, which would easily fit in a smaller casing, the Contractor utilized a 60 inch steel casing manufactured by Permalok, a Northwest Pipe Company, due to the tunnel's geology of Tonalite in various stages of degradation from competent to highly fractured and weathered. The 60 inch diameter MTBM is the smallest available where the cutter wheel can be accessed to change the tooling during a tunnel drive. A smaller machine would have had to be rescued once the disc cutters were worn out. Despite the relatively weak rock present, any rescue shaft was not a viable option given depth of the tunnel between manholes and SMH-13 and SMH-14 and the site constraints that existing on that alignment. In total, 12 disc cutters where replaced during the drive. Though that number indicates minimal cutter wear for the distance mined, it proves that the tunnel still could not have been completed with a smaller MTBM. Despite the challenges, Bradshaw successfully completed the 1,866' long tunnel in 15 weeks. Countless lessons were learned regarding utility setup, jacking forces, cutter wear, among other things, which would be used to improve tunnel installations on this project and future jobs.

432' S-CURVE ON BINGHAM DRIVE

With the Oregon Avenue section of tunneling complete, the project's focus turned into Rock Creek Park and the mining on Bingham Drive. The next tunnel installed was from SMH-9 to SMH-8 and was installed concurrently with open cut sewer installation being performed further east on Bingham Drive. The launch shaft at SMH-9 was a 20 foot diameter liner plate shaft that filled the entirety of Bingham Drive's width, but it was located at the stationing of one the roads parking lots, so construction traffic could pass it with minimal restriction. The recovery shaft at SMH-8 was in a similar condition, located at Bingham's intersection with a park access road, again allowing traffic to pass. The microtunneling support equipment remained in the same location as for the initial tunneling to reduce the Contractor's footprint in the park. The steel casing used for this 330 foot long drive was in 10 foot lengths to be used in the smaller launch shaft and was complete successfully from September to October of 2017.

The remaining 432 feet of microtunneling would be much more complicated. The length of Bingham Drive between open excavations at SMH-9 and SMH-12 was a short but meandering section of road with no locations at which a passenger vehicle, much less an excavator or wheel loader, would be able to pass safely around a tunnel access shaft. And since Beach Drive at the far eastern end of Bingham could not be used for construction traffic, no other operations could occur on Bingham Drive once a shaft was opened. It was designed with three short, straight tunnels so that the pipe and manholes would all reside in final locations within the Bingham Drive curb line which was a NPS requirement. Two of these manholes, SMH-10 and SMH-11, would have block Bingham Drive. At the beginning of the project, investigations were made by the Project Team to determine if an alternate straight alignment could be found, to no avail. The only solution that could be determined was to utilize a curved tunnel and eliminate those two shafts.

The soil conditions through the section of the project were similar to all of the other tunneling and therefore the MTBM itself would need to remain the same. However most other components of the tunnel drive required adjustment. Permalok steel casing could not be used as the mechanical joints would not permit the anticipated deflection in this curve. 48 Inch RCP jacking pipe, manufactured by Vianini, would have to be used instead as a tunnel support. For tunnel guidance, Bradshaw again enlisted VMT USA to provide their SLS-LT Navigation System which is specifically designed for such alternate guidance, particularly in a curved pipe jacking application. In the case of the curved tunnel this system was required for any guidance at all, not just as an improvement, as there would never be a line-of-sight from the shaft to the MTBM. The last additional supplier brought onto the team was JackControl AG, who manufactures joint packers that eliminate point loading of concrete pipe in jacking operations. JackControl AG was also integral to the design of the curved tunnel's alignment needed to stay within the Bingham Drive boundaries.

The final design was for a 432 foot long, S-curved tunnel with two 625 foot radii in opposite directions, first turning to the right, then the left. Eighty-five percent of the tunnel was within the curves with only short straightaway for the MTBM launch and

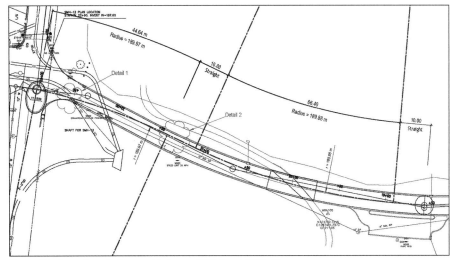

Figure 13. Curve tunnel alignment between SMH-9 and SMH-12

Figure 14. MTBM Launch at SMH-9

Figure 15. JackControl hose in RCP joint

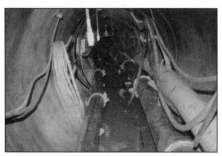

Figure 16. RCP traversing the curver

Figure 17. MTBM hole-out at SMH-12

another short section in the transitions zone between the two curves. The MTBM was launched shortly after the New Year in January 2018 and completed in early February. Tunnel excavation and jacking force exceeded the Contractor's expectations for the tunnel operation. The most challenging portion of the work proved to be the guidance. Though it was anticipated to be difficult and was completed successfully, the realities of the tight S-curve and performing the corresponding surveys within the 48 inch open space were a cumbersome experience. At the same time, it was an experience Bradshaw is ready to repeat. Not only did the change allow for better access to the other works on Bingham Drive and help the project's schedule continue to progress, it also allowed the Contractor to provide the Owner a cost savings for the removal of SMH-10 and SMH-11 manholes.

ACKNOWLEDGMENT

The author thanks Ryan Purdue for his contribution to this paper.

CONCLUSIONS

In total the $16.8 million Oregon Avenue NW Sewer Rehabilitation project was indeed a challenging one. It was at the same time a very rewarding one. The collaborative team allowed for innovative solutions to complex access problems. Those solutions provided for several firsts for Bradshaw Construction, those being the longest pipe jacking microtunnel drive at 1,860 linear feet and their first curve microtunnel drive. These firsts also demonstrated the expanding capabilities of microtunneling as a whole. New technologies and experience of contractors are allowing the boundaries to be pushed farther and farther on projects like this one are successfully completed, proving that long and complicated alignments are not only possible but can be efficient and a cost benefit of municipal owners throughout the industry.

High-Speed 3D Tunnel Inspection

Heiner Kontrus ▪ Dibit Messtechnik GmbH
Michael Mett ▪ Dibit Messtechnik GmbH

ABSTRACT

In tunnel condition assessment, the trend is towards comprehensive 3D measurement combined with high quality image texture to document and quantify damage to the tunnel surface and to create a cohesive data base of the structural conditions throughout its lifespan.

The faster such measurements can be made on site, the less it impacts shut down times to the regular traffic. The costs for the operating companies can be significantly reduced as well as the personnel efforts and risks of accidents. Furthermore, tunnel safety is increased due to faster test cycles which result in a more comprehensive overall monitoring.

Dibit has developed a photogrammetric high-speed 3D measuring system, which is unique worldwide.

The 3D surveying system can record the tunnel with high quality in terms of geometry and resolution at a speed of up to 60 mph. The system uses state-of-the-art high-speed industrial cameras and a specially developed flash technology to illuminate the (predominantly dark) tunnel structures. From the overlapping photos in the direction of travel, the 3D geometry of the tunnel is reconstructed using photogrammetric and lidar technologies. The photorealistic texturing of the 3D model enables the identification and analysis of even the smallest material damage (e.g., cracks ≥ 0.3 mm wide). The 3D tunnel data and measurements of tunnel features (cracks, damaged areas, installations, etc.) can be managed in a structured way in the Dibit-TIS (Tunnel Information System), which is the "proto"-BIM (building information modeling) for tunnel applications. The Paper introduces the technique of the novel measuring system and illustrates the quality and advantages of the measurement's results.

TECHNIQUES OF MODERN TUNNEL INSPECTION

According to DIN 1076 (1999), engineering structures are to undergo a main test every sixth year. This is mainly conducted manually by surveying critical cracks and other features, reducing the usability and reliability of the tunnel structure. With the help of scanning systems additional 3D data can be raised to support tunnel testing. Georeferenced, colorfast 3D data delivers objective information regarding tunnel deterioration and deformation. Tunnel inspection can be executed more often because of the lower effort compared to tunnel testing. Modern tunnel scanning systems improve and simplify such inspections extremely.

The main task of tunnel scanning systems is the continuous survey of tunnel surfaces and the subsequent 3D reconstruction of tunnel structures. In addition, modern scanning systems provide information about the surface in the form of a photo texture. The 3D data, or textured 3D data, are used for a variety of documentation and inspection aspects. Essentially, modern laser scanning in tunnels pursues the approaches

(1) laser scanning, (2) photogrammetric systems and (3) hybrid systems, which are described below.

Laser Scanning

One feature of laser scanners is that they cannot detect the tunnel surface color-fast with RGB values. Laser data are available as intensity images in gray scale. Detection of crack profiles of relevant crack widths (0.3–1.0 mm) is possible due to intensity and contrast differences. Established laser scanning systems can achieve geometrical accuracies of up to 5 × 5 mm, for measurement of clearance profiles, etc.

Photogrammetric Systems

In photogrammetric systems, 3D geometry and photo surface are created purely from the photos of digital cameras. Photogrammetric systems can be used variably for 3D reconstruction (e.g., tunnel construction; Paar et al. 2006; Bauer et al. 2015) and can also be operated at high speeds. However, a bright illumination and a flash technique is mandatory for its use in tunnel inspection, when measured at speed. Therefore, such devices are currently not established in the market. In general, only few photogrammetric systems are used in the field of tunnel construction, which are stationary operated from a fix point without movement like 3GSM shapematrix (with commercially available cameras) and the Dibit Handheld 3D-complete system (Dibit 2018a).

Hybrid Systems

Hybrid systems combine geometry data from laser measurements and photo textures from digital cameras. They can be either operated as "stop & go" systems, which scan one tunnel section from a constant position and then proceed to a further position. Or they can be operated as "kinematic" systems, which scan the tunnel whilst continuous movement. Well established systems are for example the SPACETEC TS2 and the Dibit LSC 4100-SRMF2 (Figure 1). Both are configurable for road and track use, achieve geometric accuracies of around 10 × 10 mm and photo resolutions of up to 1 × 1 mm. Both systems reach speeds of up to 2,5 mph (4 km/h) at walking speeds.

Figure 1. The hybrid, kinematic measuring system Dibit LSC 4100-SRMF2 in use. At the front is the laser unit, which continuously measures the geometry of the tunnel. Behind it, there are digital cameras, which make path-related images of the tunnel surface.

Delineation From Competition Systems

Dibit´s highspeed-system deviates significantly from the above described state of the art. Currently about 1,8–3,1 mph (3–5 km/h) survey velocity can be reached by competing systems in the tunnel, but with lower resolution, accuracy and image quality. Dibit´s highspeed-system reaches up to 60 mph (approx. 100 km/h), 1 mm image resolution and so it is able to detect even the finest cracks ≥0.3 mm in width.

DIBIT`S NEW HIGH-SPEED SYSTEM FOR 3D TUNNEL INSPECTION

Within the scope of a development project, Dibit designed a photogrammetric high-speed system for full surface measurements and subsequent 3D modeling of tunnel structures. The up to date system will revolutionize tunnel tests with unparalleled speed, resolution and precision.

As a unique innovation, the system includes a specially developed high-performance LED flash technology for illuminating the tunnel surface. The images are captured simultaneously by high-speed industrial cameras of the newest standard. In combination with the extremely bright and fast-acting LED flashes, it is therefore possible to achieve shake-free photos also in dark tunnels at high speeds of up to 60 mph.

The 3D measuring system is controlled by means of an embedded controller unit in combination with sophisticated, non-contact position measurement system and IMU units including odometer and tilt sensors. As a result, this system can be operated at a frequency of up to 300 photos (or flashes) per second and the positions of the individual captured image can be precisely determined. The image position will be later utilized for high accuracy photogrammetric 3D reconstruction. Overall, an absolute accuracy of about 5 mm is achieved and a geometric resolution of 1 mm. This allows the possibility to detect cracks up to 0.3 mm in width. The measuring system can be used variably in road, rail and water tunnels thanks to various platforms (Figure 2).

Figure 2. High-speed 3D measurement system FSC 6100-SRmF10 mounted on a trailer in an unlighted tunnel

The volume of data resulting from the high amounts of images captured is enormous and met the development work with major challenges. Highspeed data interfaces were created as well as sophisticated backup schemes and computer facilities. Since the system has to work autonomously, a separate power source had to be developed for the 3D system, computer and sensor technology.

The measuring unit is modular. If required, it can be combined with laser scanners and/or thermal or multi-spectral sensors. As a result, in addition to the (1) geometric 3D information and (2) colorfast photo-surfaces, data are also collected on (3) thermal and (4) material-specific properties of the tunnel surface.

In terms of hardware, software control and subsequent image analysis or surveying possibilities, the high-speed measuring system is a completely new application. (Dibit 2018b). The patent for this newly developed technology is currently pending.

ADVANTAGES OF THE NEW APPROACH

The innovative scanning system creates different advantages for the future tunnel inspection.

Technical Advantages

The main advantage of the system is the unprecedented survey speed of up to 60 mph (approx. 100 km/h). Tunnel inspections which used to take days or weeks can be done in a few minutes or hours in the future. This means that scanning operations have to be reassessed in the course of tunnel tests concerning logistics, efficiency and impairment (see below).

Furthermore, the highspeed device provides a matchless resolution in terms of surface quality of the 3D model. For example, crack widths of up to 0.3 mm can be detected and subsequently analyzed. Such fine sensitivities enable to trace crack status in an early stage before damages become hazardous or have to be rehabilitated expensively.

Another bonus of the system is the expandability with additional sensors, such as laser scanner and thermal, respectively multispectral survey units. With these, information about thermal and material characteristics can be raised in addition to 3D and textural properties of the tunnel surface.

At last, the raised data can be processed by software from Dibit in a way, that it is usable with BIM (Building Information Modeling) software, which makes it suitable for future planning and management processes.

Advantages in the Inspection Process

Traditionally, tunnels have to be closed for surveys and inspections. The time of shut downs occurs usually at night, with less traffic than during day operations. On larger projects, crews still have to work several days and weeks at night, which takes up resources and leads to additional cost due to work performed outside of regular work hours.

The new system enables minimum tunnel closure times due to enhanced surveying speeds. The tunnel is either completely closed for a few minutes during surveys or possibly the system can be operated during traffic and "co-flow" with just minimal

High-Speed 3D Tunnel Inspection

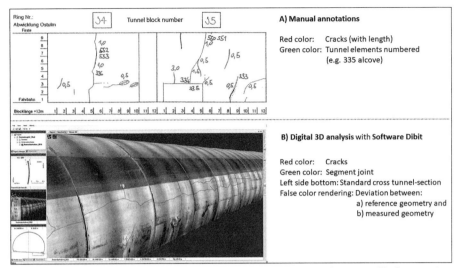

Figure 3. Upper part shows results of the traditional, manual tunnel inspection report. The lower part shows 3D analyses in software "Dibit." The data was acquired with the new highspeed system. The texturized 3D model can be exported to BIM software. The annotated tunnel characteristics contain information about 3D position, length, area, volume, etc.

impact. Vehicles in front and in the back of the measuring truck just have to be kept at a distance to enable clear line of sight. Therefore, large-scale, energy-consuming traffic diversions or traffic jams are eliminated (see below).

Due to the relatively much less effort of the new inspection method, structural health monitoring can be intensified. More frequent measurements increase the overall building safety of tunnel systems. The tunnel operator receives precise and comprehensible 3D models and thus data for maintenance, revision, construction, and so on.

The system is economical and efficient. Time, personnel and therefore cost consuming tunnel closure can be reduced or even avoided. Also, the system only needs a small amount of personnel (2-man crew). Time of measurement and its preparation are short as well.

Another aspect that has a positive effect on the efficiency of the system is the fact that most of the inspection time is shifted from the tunnel to a virtual 3D environment in an office. Cracks, tunnel installations and others can be analyzed independently from the rough and partially dangerous tunnel environment.

The digital analysis of the tunnel surface allows more details to be captured than previously possible and to increase the overall quality of the tunnel inspection. In contrast to this, it was hardly possible in the past for manual tunnel testing to detect damage in the correct position and amount (Figure 3). The new technology provides objective and comprehensible results for tunnel surveys (e.g., orthophotos with marketed cracks and relevant annotations) and interfaces in the form of data formats into other software (e.g., CAD, GIS, BIM) for the further evaluation of tunnel characteristics (Mett et al. in press).

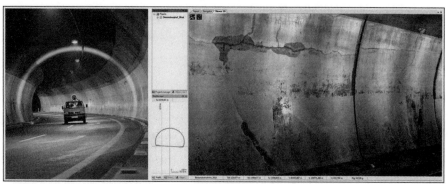

Figure 4. Left side shows the highspeed system during its first 3D measurement in the "Sonnenburghoftunnel." The right side shows the processed 3D model in which cracks and displacements are clearly visible

Environmental, Social and Safety Advantages

The new measurement technique has a strong environmental, social and safety relevance. Tunnel closure times and route diversion generate traffic jams, delays and long detours. This wastes fuel and time and releases emissions. Overall, the situation leads to burdens and displeasure among motorists, service providers (especially truck transport) and local residents. The novel 3D measuring system enables such short shut down times that the above-mentioned problems do not arise or occur in the slightest. If the measuring unit is "co-flowed" in traffic, it may be possible to dispense entirely with tunnel closures.

The collapse of the motorway bridge in Genoa (14th of August 2018) has shown that security checks and continuous health monitoring are crucial for transport infrastructure. Dibit's measurement system allows for tunnel assessments in shorter time periods than previously possible (e.g., annual inspections, rather than every 6 years as required by federal maintenance guidelines). As a result, the safety of tunnel structures can be significantly increased because damage mechanisms are already recognizable at an early stage.

EXAMPLE OF A MEASUREMENT

The first measurement for a commercial project was executed in September 2018 in the southern city limits of the city Innsbruck, Austria. The surveyed tunnel was the "Sonneburghoftunnel" with its 180 m length along one strategical section of the north-south axis of the "Brenner" highway. It is operated by the Austrian infrastructure company "ASFiNAG." For comparative reasons, a baseline measurement was performed with a conventional stationary laser scanning equipment and tunnel fix points. The actual measurement with the highspeed system was executed with a speed of 31 mph (50 km/h), which took around 13 seconds (Figure 4).

With the measurement images, a 3D model with a grid of ca. 2 × 2 cm was calculated. The 3D geometry was in perfect agreement with the previous baseline measurement. With the help of the fix points, the 3D model was georeferenced to global coordinates. Afterwards the image texture was applied to the 3D model with a resolution of 1 mm per pixel.

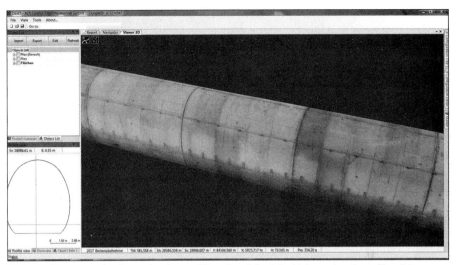

Figure 5. Prototypes of software development based on data of the highspeed system. The picture shows the automatic classification of tunnel blocks based on the occurrence of cracks (green = good, red = rehabilitation is required).

The results of the first measurement campaign are very successful in regard to geometrical accuracy (5 mm) and image resolution (1 mm). Within the processed 3D models, cracks ≥0,3 mm are to be clearly seen. With the help of the results, effective planning tools and approaches were created for the following refurbishment of the tunnel.

SUMMARY AND FUTURE PROSPECTS

Dibit has developed an unrivaled highspeed measurement system for 3D tunnel inspections. In regard to measurement velocity, measurement accuracy and image resolution, the system defines new technical standards. With the help of Dibit's software, tunnel characteristics, such as cracks, surface damages, tunnel installations, etc. can be analyzed in a virtual 3D environment. The results are valuable information for objective tunnel analysis by proving engineers.

Due to the high measurement velocity, tunnel closure times are reduced as well as traffic disturbances and related environmental impacts. Furthermore, the high velocity results in cost savings enables more frequent tunnel surveillance for the sake of tunnel safety.

The combination of the newly developed measuring system and the powerful DIBIT software will make it possible in the future to use applications that were previously thought to be impossible in tunnel inspection. This includes thermal and multispectral investigations of the tunnel surface, as well as BIM suitable data formats for future tunnel operation. Dibit is also actively working on software routines for fully-automatically crack and feature detection (Figure 5). Semi-automatic software routines are already implemented.

REFERENCES

Bauer, A., Gutjahr, K., Paar, G., Kontrus, H., and Glatzl, R. (2015). Tunnel Surface 3D Reconstruction from Unoriented Image Sequences. Austrian Association for Pattern Recognition (OAGM) Workshop. 28.05–29.05.2015. Salzburg, Austria.

Dibit (2018a). https://www.dibit.at/fileadmin/user_upload/Dibit_Struktur/07_Downloads/2017_Produktblaetter/Produktblaetter_DE/TSC_Neubau_Rohausbruch_de.pdf.

Dibit (2018b). https://www.dibit.at/fileadmin/user_upload/Dibit_Struktur/07_Downloads/2017_Produktblaetter/Produktblaetter_EN/TSC_allgemein_engl.pdf.

DIN 1076 (1999). Engineering structures in connection with roads—inspection and test. DIN Deutsches Institut für Normung e. V, Berlin.

Mett, M., Kontrus, H., and Holzer, S., in press. Dibit TIS—Das „Proto"—BIM für den Tunnelbau. Proceedings of the 20th international geodetical week Obergurgl. 10.02.–16.02.2019 Obergurgl, Austria. Edited by K. Hanke and T. Weinold. Arbeitsbereich für Vermessung und GEOinformation. Universität Innsbruck.

Paar, G., Caballo-Perucha, M., Kontrus, H., and Sidla, O. (2006). "Optical crack following on tunnel surfaces," Proc. SPIE 6382, Two- and Three-Dimensional Methods for Inspection and Metrology IV, 638207 (20 October 2006); doi: 10.1117/12.685987.

Implementation of Automation and Digitization in Tunnel Waterproofing and Grouting Practices

Stefan Lemke • Renesco Inc.
Andreas Heizmann • Renesco Inc.
Tim Kearney • Renesco Inc.

ABSTRACT

Groundwater intrusion into tunnels is one of the most formidable challenges that face the service life of all underground structures. This dilemma becomes more profound when one considers a tunnel's service life requirements of more than 100 years. While several legacy factors contribute to the overall success of these structures like design, quality, materials, engineering, construction, and environmental considerations. The tunnel industry in North America has been slow to adapt to new key materials, automation, digitization and a multidisciplinary approach in the performance of this type of tunnel work

In Europe, automation and digitization are entering the tunneling industry for Tunnel Waterproofing and Grouting. Systems such as automatic sheet-waterproofing application via hot-melt technology, spray-on waterproof coatings robotically applied and automated monitoring and associated digital reporting systems control the grouting placement procedures thus insuring that grout quantity and quality meet project specific standards. Real time monitoring incorporates the data acquisition material workflow, processing of information and the immediate visualization of gathered information to enable in timely decision making and enabling the contractor to make immediate project specific adjustments to the construction process.

INTRODUCTION

Steps towards mechanization and automation in the construction industry are predominantly aimed at in-creasing work speed, cutting costs, and to some extent, addressing the decreasing supply of skilled labor. Such measures also seek to improve safety and quality or provide a more comfortable working environment.

Along with the measures described above, there is also a higher demand of a control, monitoring and reporting systems to ensure/secure the quality, in which the digitization is of crucial importance to data processing, storage and transmission, because it allows information of all kinds in all formats to be carried with the same efficiency and also intermingled.

SHEET WATERPROOFING

Tunneling processes at all are privileged for industrialization, mechanization and automation, which is also valid for the integrated waterproofing works. This contribution was the initiation of developing a full automatized equipment for sheet waterproofing application at the 56km long Swiss AlpTransit (NEAT, Gotthard) project using a hook & loop fastener system instead of the traditional disc/washer system, in which the one side is spot-wise fixed to the substrate via nail and the other side thermo-welded to the geomembrane keeping the waterproofing temporary in place until the concrete works of the inner lining is finished.

In consequence also the system components were undergo a developing process. The relevant geomembrane was laminated with a protection geotextile, formed to a single geo-composite, as a contra part (loop) for the hook, wherein all components separated has to fulfil the specific project requirements for 100 years durability acting as a loose-laid sheet waterproofing system, as described in Lemke 2016.

Figure 1. Full automatized equipment, hot-melt technology

At the same time, as an alternative to the hook & loop fastening system the hot-melt technology was established in the Oensberg and Islisberg tunnel (Figure 1) in Switzerland. Hereby the hook disc/roundel, which was still spot-wise mechanically fixed to the substrate was replaced to a simple linear hot-melt application (chemical bonding) between laminated geotextile and substrate.

In consequence the substrate condition and the environmental aspects became for the fixation/application an important system component. Substrate roughness vs. hot-melt consumption, application speed vs. curing behavior, surface contamination, temperatures, water-spots, etc. were evaluated in view to the bonding characteristics, the effectiveness of the application, the safety and finally the overall costs.

Minimizing these risks, an automatic air-brush system for cleaning purposes and a thermo-cell for substrate drying were adopted to the full automatized application equipment.

SPRAY APPLIED WATERPROOFING

In the last decade spray-applied membranes have been undergo a revival (Girnau 1969) and have been pushed hard to the tunnel industry accompanied with the introduction of robotic applied application.

One of the arguments regarding robot spray was the quality component to guarantee a uniform, constant 3 mm thick—in situ formed—sealing layer, in addition to aspects of safety, cost controls on the used material and increase of application speed.

The material itself takes over the sealing effect and at the same time acts as a glue component. Similar to the hot-melt technology, the system must be adjusted to various underground requirements, environmental conditions and applications, which embed this spray-applied technology in a tight system frame, dedicated to change orders, substrate preparation and pre-injections/grouting.

GROUTING

Grouting works have become a key essential part of temporary and permanent works in the fields of tunneling and complex demanding construction projects, mainly in the forefront of construction to sufficiently prepare weak ground to meet the prerequisites for construction. Grouting measures are hereby often carried out as a specific measure with the emphasis to increase the bearing capacity (consolidation), stabilize or to seal the ground. Real time monitoring incorporates a high work flow of data acquisition, processing and visualization to enable in time decision making and steering of the construction.

The value of real time data access and the ability to steer the operation is evident and vital for the achievement of grouting targets and quality control (Bruce 2012, Gularte 2007, Gularte 2012).

A web based platform was established in Feuerbach tunnel/Stuttgart21/Germany, which allows for a defined control of the main operational and technical aspects of the grouting works. The established platform includes all aspects of data acquisition, modelling and visualization of the grouting works.

Grouting Control

State of the art grouting control and reporting systems ensure and secure the quality of the grouting process in order to meet specific grouting criteria and required projects specific standards. The grout control system provides data record, processing and a comprehensible steering of the process even in case of variable ground conditions. Figure 2 illustrates the flow of data of the within and throughout the communication and documentation system.

Human interactions in terms of different stakeholders as well as machine interactions from grouting control units (GCU), drilling control units, and other types of sensors, are being shown.

Grouting Control Systems

The Grouting Control Systems gathers and records data like grouting pressure, the flow rate and the volume of injected material versus time.

The recording of these parameters is mainly facilitated by weight cells, pressure and flow sensors. Pressure sensors are equipped with a membrane, which is deformed by acting pressure.

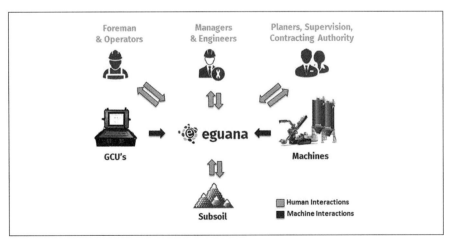

Figure 2. Data and information flow

Flow meter devices are generally used for measuring the flow rate and respectively the injected volume over time. The allocation of the flow meter device needs to be specifically addressed to the type of grouting material and shall be installed on the feeding side of the pumps. Electro-magnetic sensors are being used for conduction fluids, especially for cement based suspension types.

Due to the lack of electrical conductivity most chemical grouts are being measured using proximity switches which simply count the number of pump strokes. Flow meter devices based on ultrasonic or Coriolis measuring principles haven't been used in a wider range due to high acquisition costs, but definitely result in a better resolution. Coriolis sensors provide in addition information on the temperature and density of the fluid.

Flow meter and pressure cells do remain as the main source of information collection equipment, despite recent efforts to measure in line the rheology. The rheology of the material is generally not that highly affected to such a high degree, that obvious changes are to be expected from base mix.

Data Processing

The grouting data is being stored in the local storage of the grouting control unit (GCU). On common GCUs the grouting data can be exported to flash devices and be further analyzed and evaluated through specific software installed on the computer of the site manager.

The developed system allows a direct and secure upload of grouting data to a server in real-time. Therefore, the local storage system of the GCU is being constantly monitored by an additionally developed application, which ensures consistency and integrity of grouting data.

To ensure usability in harshest environments the developed system doesn't rely on a specific communication technology. The uplink could be performed through traditional 2G/3G/4G, Wifi, LoRa, and other communication technologies. The system simultaneously stores the data on several data servers to ensure data secured even in break down situations and data storage over years as required and demanded by many clients.

Grout Control Steering

The set-up of the main grouting operations is a rearward focused process. Readjustments of the grouting process are in many cases operator driven.

The intention of p/v criteria was to fulfil the criterion for a specific operation step by step. Readjustment within the grouting process or criteria is conducted within the subsequent step.

The RTGC method offers herby a comprehensible methodology, due to the structured input parameters for real time monitoring and guidance.

Kobayschi (2008) introduced the dimensionality factor "DIM" as a function of the volume and the flow rate. The dimensionality is an easy interpretable tool for machine based learning by a by modern grout control system.

The dimensionality indicates if the system constant flow conditions, which applies for free surface flow or otherwise the conditions allow for completion of the designed process in time. The system mainly demands for a readjustment of the grouting time, which is triggered by unsteady pressure conditions.

The grouting data management system identifies minor pressure fluctuation due to precise monitoring equipment and recalibrates the flow rate to avoid potentially hydro fracturing, which would lead basically to uneconomical conditions.

Goal oriented prediction and iterative adjustments are seen as the key development for future grouting projects.

Process Visualization

The visualization of grouting process is a core subject of the grouting work (Wilson 2002; Dreese 2003; Hang 2001). In general, the documentation of the work is the main indicator, if a work from the perspective of a contractor was delivered according to the clients and designers specification.

Additional testing will mainly confirm if the consolidation or sealing target is reached locally. The evaluation of the grouting documentation indicates if similar conditions based on the grout curve characteristics are to be expected.

Data Handling and Management

The collected and uploaded grouting data is being stored in an object-relational database system. It is being processed and analyzed automatically in terms of integrity and process specific criteria. If there are any deviations, faults or errors, managers are being informed through automated notification processes.

The grouting data management system is being accessed through web browsers not requiring specific software licenses or local software installation. It allows remote access for several users on various devices, at any location and 24 hours per day. Further it enables the access to grouting data of several construction sites, which enables managers and parties involved to keep track of grouting processes of various projects.

Visualization of Grouting Processes

The developed system offers besides grouting process analysis and evaluation, process visualization, grouting process management, bill of quantities and dashboards with several widgets, which allow a fast overview of the most important data of several construction sites.

Being able to visualize the complete construction site and navigating down to a specific grouting process, the focus of the visualization is on the transportation of information and fast navigation.

Figure 3 illustrates the grout take of 8 boreholes by color and size of the circles within a developed view. Each point is linked to the specific grouting process with interactive p-v and GIN-diagrams.

Throughout all levels of visualization, the status of grouting work is being illustrated through specific color codes.

Thereby the status of all subsections are being analyzed and forwarded to the upper level of visualization. This enables fast and simple identification of problem areas, and gives an abstract overview about grouting progresses.

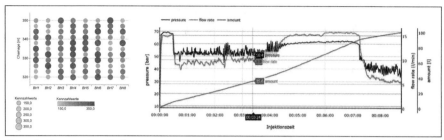

Figure 3. Process visualization

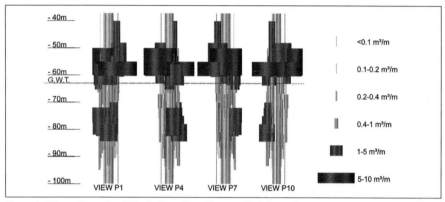

Figure 4. Three-dimensional illustration of grouting specific information

Visualization of Grouting Works

The identification of areas where specific grouting criteria haven't been met or the grouting process has been aborted demands for a specific analysis.

The system highlights these areas by color-coding within 3-dimensional diagrams (Figure 4) indicating specific information as grout take or grouting pressure. The system withholds an interactive character and allows for a zooming from global view to a more detailed view of the work.

Operational Aspects

Through the approach of a single-source system, all grouting data is being stored in a single high performant data base. This allows a comprehensive approach in terms of areas of application and feature development.

Thus the developed system does not only focus on technical aspects, but also on operational aspects. It frees up to the engineers time to concentrate on the evaluation as the documentation, organization and data handling is automated. It also forms a discussion board for all parties involved as the platform follows a self-explaining easy to use and evaluate approach.

The comprehensive data collection through the grouting data management system allows the user to evaluate the grouting data and drilling data in a detail and speed which has not been possible in the past.

Grout Process Management

The collected data is being processed and analyzed in terms of grouting and sub- or side processes. Since most of the sub- or side processes can't be digitally recorded, a semi-automated process association (Figure 5) has been developed to generate accurate data of sub- or side processes and their distributions.

Due to this approach, managers only need to edit data in case of special events, like downtimes. Thereby the grouting process is documented in an accurate and adequate way.

The gathered process data is invaluable in terms of operational aspects of grouting works. Detailed analysis of grouting processes, sub- and side processes as well as their distributions and trends can be performed. Figure 6 illustrates grouting process distributions as well as trend analysis. Detailed information about the grouting processes as well as sub- and side processes can give valuable information for planning of similar future projects, and can be seen as a feedback loop.

Bill of Quantity

Further the collected and processed data is being used to automatically generate the bill of quantity. Therefore, not only the grouted volumes and grouting process times, but also sub- or side-process times, like arrangement of equipment or downtimes, the number of packers etc. is being calculated. Specific bill of quantities and protocols can be generated, which allows for a detailed, transparent and comprehensible grouting process documentation.

This feature is going to be further enhanced through the implementation of specific interfaces to planning and controlling tools. So the gathered and processed grouting data can then be forwarded to other BIM-Software.

Integration of drilling data for more comprehensive analysis of grouting processes in terms of their success is also possible.

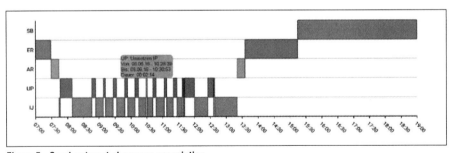

Figure 5. Semi-automated process association

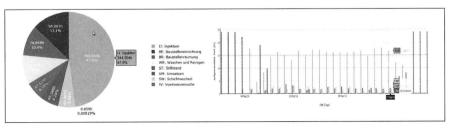

Figure 6. Process distributions and trend analysis

CONCLUSION

The developed system has been designed in terms of user interfaces (UI) and user experience (UX) to meet the specific needs of geotechnical engineers and construction managers at best. Thereby reducing the efforts spent on evaluation and analysis up to 72%, compared to common grouting software.

Due to the development of algorithms for the calculation of grouting process distributions and resource consumptions, bill of quantities can be generated automatically. This leads to an improvement of accuracy of 2.1% in average.

Through the development of specific user roles and corresponding access permissions, all stakeholders (contractor, planers, supervision and the contracting authority) are able to access data through a web browser, remotely and in real-time. This leads to an ease of communication and reduced reaction or response time.

Due to automated control the quality of the grouting process itself can be ensured, further the production efficiency is being increased, and the risks which come along with manual grouting operations can be reduced. The data uplink allows engineers and site managers to access data remotely in real-time, which decreases reaction and response times and eases communication between different stakeholders.

REFERENCES

Dreese TL, Wilson DB, Heenan DM, Cockburn J, 2003. State of the Art in Computer Monitoring and Analysis of Grouting. Grouting Treat ASCE Geotech Spec Publ 120:1440–1453.

Girnau, G. and Haack, A. 1969. Tunnelabdichtung, 6 Forschung + Praxis, Studiengesellschaft für unterirdische Verkehrsanlagen (STUVA), Alba Buchverlag, Düsseldorf, Germany: 35–39.

Gularte FB, Taylor GE, Shao LS, 2007. Observational monitoring methods for improved grouting and soil improvement. ASCE Geo Institute, GeoDenver, Denver, CO18–21.

Gularte FB, Ringen DA, Shao LS, 2012. Advances in monitoring and control systems for high mobility grouting. In: Grouting and deep mixing 2012, ASCE, pp 1238–1247.

Hang W, Zhao CH, 2001. Development and application of grouting monitoring system of Xiaolangdi grouting project. Water Res Hydropower Eng 32(11).

Lemke, S., Eckl, M. and Londschien, M. 2016. 120-Year Design Lifetime of Plastics, WTC 2016, San Francisco, California, USA.

Lemke, S. and Moran, P. 2015. A controversial discussion regarding the use of spray-applied waterproofing for tunnel applications, RETC, New Orleans, USA.

Kobayashi S, Stille H, Gustafson G, Stille B, 2008. Real time grouting control method: development and application using Åspo HRL data, SKB. Technical Report R-08-133. Swedish.

Wilson DB, Dreese TL, 2002. Advances in computer monitoring and analysis for grouting of dams. US Society on Dams Annual Conference Proceedings, San Diego, CA. USSD, Denver, USA.

Sandwich Belt High Angle Conveyors Exclusively at Paris Metro Expansion—2019

Joseph A. Dos Santos ▪ Dos Santos International LLC

ABSTRACT

Developed between 1979 and 1981, Dos Santos Sandwich Belt high angle conveyors began commercial operation in 1983. Since then some 150 systems have gone into successful operation throughout the world with most still in productive operation many years later. They have been successful with many, varied bulk materials from light, friable wood chips and grains to heavy, coarse copper and iron ores. Throughput rates have varied widely, from 272 kgs/hr to 6000 t/h. Conveying angles have varied from 30 degrees to 90 degrees (vertical). Indeed 35% of the installations elevate the bulk continuously vertically. These characteristics make the system ideal for haulage from underground, particularly at tunneling projects where the excavated muck must be elevated within the tight spaces of an access vault or shaft from the underground tunnel to the surface. Though there have been successes over the years in the tunneling industry the use of Sandwich-Belt high angle conveyors did not grow dramatically to a position of dominance.

Muck disposal from the largest TBMs (Tunnel Boring Machines) is now predominantly by conventional trailing conveyors. The large continuous volumes must be elevated to the surface then conveyed and discharged at a holding area, to be removed to permanent fill locations. For high volumetric rates, the elevating systems to the surface have been pocket belts and sandwich belt high angle and vertical conveyors. The simplicity of the pocket belts, hanging vertically between the lower and upper terminals has been compelling. They have the disadvantage however, of great difficulty in fully discharging the muck from the pockets. The sandwich belt systems on the other hand suffer the perceived complexity of equipment; idler rolls that must be accessed and occasionally replaced throughout the beltline. The resounding advantage however, is that sandwich belt high angle conveyors use conventional smooth surfaced belts that discharge the material completely as they can be scraped clean. The very sticky muck to be excavated at the Paris Metro Extension project led the client to specify the exclusive use of sandwich belt high angle conveyors for their vertical continuous haulage requirements. There is a recurrence of other tunneling projects specifying the exclusive use of Sandwich-Belt high angle conveyors for the same reason, very wet and sticky materials. The present writing retraces the Sandwich-Belt high angle conveyor milestones in vertical haulage, then describes how Dos Santos International provided two (2) DSI GPS sandwich belt high angle conveyors for the Paris Metro expansion and how they solved the daunting challenges of handling the sticky muck.

INTRODUCTION

What Are Dos Santos Sandwich Belt High Angle Conveyors?

This article deals predominantly with the Dos Santos Sandwich Belt High Angle Conveyors, a technology developed by this writer nearly 40 years ago. The first commercial installation went into operation in Wyoming, USA in 1984. This is to clarify that when referring to Dos Santos Sandwich-Belt high angle conveyor systems in this writing I am referring to the work of J.A. Dos Santos since 1979 while at the various companies of employment:

- Dravo Corporation, Pittsburgh, PA USA (1975–1982): During the period 1979–1981, under a US Bureau of Mines study; the writer developed the sandwich belt high angle conveyor technology, rationalized in the conventional conveyor technology. This also produced the landmark publication "Evolution of Sandwich Belt High Angle Conveyors,"* a writing that is complete in defining the theory and design rules and in the conceptualization of the designs that went on to commercialization.
- Continental Conveyor and Equipment Company, Winfield, AL USA: The HAC Systems from 1982–1997.
- Since the founding of Dos Santos International: The DSI Snake Sandwich and GPS (Gently Pressed Sandwich) High Angle Conveyors, as well as the Adder Snake, from 1997 until the present.

Sandwich Belt High Angle Conveyors

Development of the sandwich belt high angle conveyor concept has come a long way since its first introduction in the early 1950s. Over the approximate thirty year period until 1979, significant advances were few and only came in spurts. Such advances did not build on previous developments. Rather, they were independent developments which soon reached their technical limitations.

The latest significant development of this technology, beginning in 1979, is the first to take a broad view of the industries to benefit from high angle conveying and of all previous developments. As a result these latest developments know few technical limitations, address a broad range of applications, and offer a forum for continued logical development or evolution.

Sandwich Belt Principle

Dos Santos Sandwich Belt high angle conveyors represent logical evolution and optimization of the sandwich belt concept. The sandwich belt approach employs two ordinary rubber belts which sandwich the conveyed material. Additional distributed force on the belt provides hugging pressure to the conveyed material in order to develop sufficient friction at the material-to-belt and material-to-material interface to prevent sliding back at the design conveying angle.

Figure 1 is a realistic model of the belt sandwich. An ample belt edge distance assures a sealed material package during operation even if belt misalignment occurs. This model also illustrates the interaction of forces within the sandwich. The applied or induced hugging load is distributed across and along the carrying belt sandwich. Of that hugging pressure, only the middle pressure hugs the material load while the outer pressure merely bears against the material free edges of the belt. Both belt surfaces apply their frictional traction on the material. From this model one can calculate the required material hugging pressure that will ensure the material does not slide back due to the tangential gravity loads. This is expressed by Equation 1:

---

*Dos Santos and Frizzell

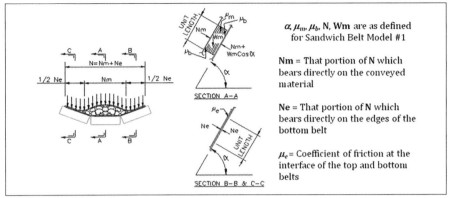

Figure 1. Sandwich belt model

$$Nm \geq \frac{Wm}{2} \left(\frac{\sin \alpha}{\mu} - \cos \alpha \right) \quad \text{(EQ 1)}$$

where: $\mu = \mu_m$ or $\mu = \mu_b$, whichever is smaller

Dos Santos Sandwich Belt High Angle Conveyors

When investigated anew in the late 1970s, it was clear that the sandwich belt concept offered the greatest potential for a cost effective, operationally appropriate high angle conveying system to address the broad needs of the mining and bulk materials handling industries.

Following the extensive study of past sandwich belt conveyors, the governing theory and constraints, and development of the governing design criteria, a broader scope effort was undertaken in 1982 to develop the first sandwich belt high angle conveyor to meet these needs. The resulting Dos Santos Sandwich Belt high angle conveyors are truly evolutionary in judiciously selecting and advancing desirable features while avoiding the pitfalls of the past. They conform entirely to the governing theory, to the constraint equations and to the development criteria.

These Sandwich-Belts fulfill all established operational requirements. The profiles can conform to a wide variety of applications.

Advantages of Sandwich Belts

Dos Santos Sandwich Belt high angle conveyors offer many advantages over other systems including:

- *Simplicity of Approach:* The use of all conventional conveyor hardware, for very high availability and low maintenance costs
- *Virtually Unlimited in Capacity:* The use of all conventional conveyor components permits high conveying speeds. Available belts and hardware up to 3000 mm wide make possible capacities greater than 10000 t/h.
- *High Lifts and High Conveying Angles:* High lifts to 300 m are possible with standard fabric reinforced belts. Much higher single run lifts are possible with steel cord or aramid fiber belts. High angles up to 90 degrees are possible.

- *Flexibility in Planning and in Operation:* Dos Santos Sandwich-Belts lend themselves to multi-flight conveying systems with self-contained units or to single run systems using externally anchored high angle conveyors. A system may be shortened or lengthened or the angle may be altered for the requirements of a new location.
- *Belts are Easily Cleaned and Quickly Repaired:* Smooth surfaced rubber belts allow continuous cleaning by belt scrapers or plows. Smooth surfaced belts present no obstruction to quick repair by hot or cold vulcanizing.
- *Spillage Free Operation:* During operation the material is contained within the belt sandwich from loading to discharge. Well centered loading and ample belt edge distance result in no spillage along the conveyor length.

Select Sandwich-Belt Installations

Dos Santos Sandwich-Belt high angle conveyors are well established in the industry with the first commercial unit beginning operation in 1984. Since then more than 100 units have gone into operation throughout the world. Table 1 lists only select installations that make the distinct case for vertical continuous haulage from construction and tunneling projects.

The select installations of Table 1 are not all in the construction industry but each makes its case in different ways.

Indeed, DS 005 was the first commercial Dos Santos Sandwich-Belt high angle conveyor for construction and it was the first vertical installation. The system was part of the Los Angeles metro expansion of the late 1980s. It was located downtown at the corner of 7th and Flower Streets. The excavation in this area was an open cut with timber lagging on steel beams to support the busy city street above while excavation proceeded below. Material movement was by load-and-carry with front end loaders to a grizzly covered hopper. The hopper loaded the tail of the high angle conveyor through a vibratory feeder. The high angle conveyor elevated the excavated earth continuously from under the street to a truck loading bin above. Though it was designed to load directly into the bin, for reasons of truck access and traffic flow, a connecting conveyor was added in order locate the bin further away from the intersection of the two

Table 1. Select sandwich-belt high angle conveyor installations

DS #	Location	Material/ Rate (t/h)	Ang (°)	Elev (m)	Lgth (m)	Width (mm)	Speed (m/s)	Top/Bott (kW)	Year
005	Construction/ Los Angeles, CA, USA	Excav. Earth/ 272	90	32.3	42.5	914	1.6	22.4/ 22.4	1988
009	Coal Prep/ West VA, USA	Clean Coal/ 762	90	76.2	90.2	1372	2.79	112/ 112	1991
024	UG Gypsum Mine/ New York, USA	Gypsum Rock/ 363	90	36.6	48.5	1067	1.52	37.3/ 37.3	1992
030	UG Coal Mine/ Illinois, USA	Raw Coal/ 1361	90	104	115	1524	4.57	298/ 298	1993
037	Tunneling/ Illinois, USA	TBM Muck/ 1266	90	70	8308	1372	3.56	186/ 186	1993
098	Refinery/ Muzkiz, Spain	Pet Coke/ 475	90	21.2	32.3	1400	3.5	45/ 45	2012
108	Tunneling/ Paris, France	TBM Muck/ 800	90	24.6	33.5	1400	3.0	75/ 75	2018
109	Tunneling/ Paris, France	TBM Muck/ 800	90	26.3	35.2	1400	3.0	75/ 75	2018

busy streets. The surge capacities of the hopper below and the bin above allowed independent discontinuous excavation and truck loading without interrupting the continuous elevating of the high angle conveyor. The system was designed to begin operation during the early excavation, requiring only 25 meters of lift. Then it was extended down as the depth increased, in 1.219 meter increments until reaching the design maximum depth and the corresponding design lift of 32.3 meters.

Valuable lessons learned during this early project included:

- Though the project was of short duration the duty was harsh
- Oversized material consisting of large rocks and large clumps easily passed through the hopper's grizzly which consisted of parallel 51 mm wide bars spaced at 203 mm
- Such material was too large for the 914 mm belt width and corresponding equipment
- Though not detected by the layman, continuity of hugging lapse along the vertically straight elevating section prompted a review and revision of the continuity of hugging criteria

Figure 2. DS 005 represents two firsts in one Sandwich Conveyor; first vertical and first in a construction project

Though the system completed its task successfully, we decided that future Sandwich-Belt high angle conveyors for such projects would use wider belts (not less than 1200 mm belt width), thicker damage resistant wear covers and rubber disc center rolls at idlers to soften the indents of the large lumps as they travel along the transition curves.

Though elevating coal to a silo, DS 009 is cited for its significant lift of 76 meters during the early development. Additionally this system featured a modularized intermediate structure with independent channel stringer tables that were anchored to the wall of the silo. The belt sandwich passed through these with all belt tension anchored at the lower and upper end terminals.

DS 024 elevated crushed rock from an underground gypsum mine. It replaced the then existing underground transfer conveyor and elevating bucket elevator which had always been high maintenance and unreliable. It was the first vertical Sandwich-Belt system to elevate primary crushed rock from underground. The crushed rock was transferred by a reciprocating feeder to the tail of the high angle conveyor. From there it was elevated to the surface without any additional transfer.

As with DS 009, DS 030 set a milestone, the highest vertical lift at 104 meters. This single 1524 mm belt width Sandwich system replaced the then existing twin pocket belts, each of 1524 mm belt width which together could not accomplish the design rate.

DS 037 was the first Sandwich-Belt installation to elevate tunnel muck from a TBM (tunnel boring machine). This was part of the Chicago TARP (Tunnel and Reservoir Plan) project. The entire excavating and muck haulage system consisted of the TBM, a trailing conveyor below ground, the vertical Sandwich-Belt conveyor system to lift the material to the surface and finally a transfer and stacking system consisting of a grasshopper conveyor and radial stacker. As with subsequent systems the major equipment (drives and take-up systems) was located on the surface, at the head end so it could be easily accessed and serviced. Only the intermediate structure and tail pulleys were located underground. Additionally the intermediate structure served as the support and guidance for the trailing conveyor's return belt strands that were elevated to its belt storage unit located on the surface, 70 meters above.

DS 098 was the predecessor to the Paris units utilizing the same belt width and similar equipment though the material and operating conditions were not quite as harsh. Like the others it demonstrated successful vertical conveying.

SANDWICH-BELT HIGH ANGLE CONVEYORS THE CHOSEN SOLUTION AT PARIS METRO EXPANSION

From the first Sandwich-Belt application at a TBM (DS 037) in 1993, many years passed before the next. In 1997, the writer founded Dos Santos International and continued to offer the Sandwich-Belt high angle conveyors. Continental Conveyor of Winfield, Alabama continued to offer their Sandwich-Belt high angle conveyors as well. Though they were of my invention, from July of 1997 they were no longer engineered under my direction. In 2005, Continental was finally successful in supplying three vertical Sandwich-Belt high angle conveyor units for the construction at the London Heathrow airport, and in 2006 (23 years after DS 037) Continental supplied two Sandwich-Belt units to elevate muck from a TBM project in London. There may have been some others since that were not as well publicized, but throughout this entire time pocket belt systems dominated the duties of elevating tunnel muck from the TBMs to the surface.

The Paris Metro expansion project finally acknowledged the distinct advantage of Sandwich-Belt high angle conveyors and specified them exclusively. That advantage is the ability to handle and fully discharge very wet and sticky material. This is because the smooth surfaced rubber belts can be continuously scraped clean. This advantage had long been demonstrated in the many Dos Santos Sandwich-Belt installations in other industries, particularly in municipal waste where many units elevated municipal sludge, sludge and sawdust mix, and some industrial sludges. Nearly all of these installations conveyed the sticky material vertically. By contrast, the tunneling industry had continued to struggle with handling tunnel muck at their pocket belt elevating systems. The pocket belt required at its discharge end a long overlap distance with the surface receiving conveyor and a series of eccentric rolls that beat the back of the pocket belt in order to dislodge caked material from the pockets and onto the outgoing conveyor below.

Figures 3 and 4 provide a depiction and technical summary for each of the two Dos Santos units supplied thus far at the Paris Metro expansion. These units share a common basic design but there are minor differences. For the common design rate of 800 t/h, the belt width is 1400 mm and the belt speed is 3 m/s. This belt width is also compatible with the specified material size which it has handled very nicely.

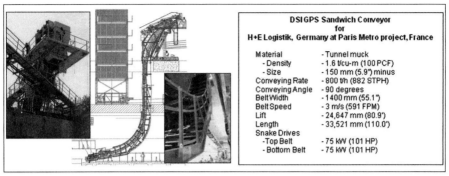

Figure 3. DS 108 depiction and technical summary

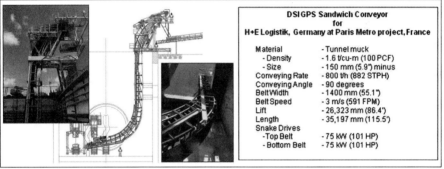

Figure 4. DS 109 depiction and technical summary

The Belt Line and Material Flow Path

By following the carrying path from loading at the bottom to discharge at the top it can be seen that the bulk material loads onto the troughed bottom belt before it enters the sandwich. From the loading point, the bottom belt with the bulk material travels into the sandwich that is formed when it is joined by the top belt. At this point and beyond, the bottom belt, now in suspension, urges itself and material up against the top belt which is supported against closely spaced inverted troughing idlers. The bottom belt with material is urged upward by a radial load which is due to the belt tension and the curved profile according to the equation **P** radial = **T**ension/**R**adius of the curve. This radial load must overcome the lineal weight of the bottom belt and conveyed material and must additionally provide the needed hugging pressure to develop the internal friction that will resist the gravitational slide back forces. In this way, the bulk travels from the sandwich entrance through the lower transition curve to the start of the vertical profile. Along the straight vertical profile the needed hugging pressure is provided by the GPS sections. Each GPS section consists of two equalized idler like assemblies of four equalized rolls so that all eight rolls are equalized. The equalized idler-like assemblies at the right are positioned between the troughing idlers at the left and they urge the outer (right side) belt with material against the inner (left side) belt with a calculated pressure that is provided by a compression spring. Continuity of hugging without lapse is achieved by close spacing of the fully equalized idler rolls which have no preferred orientation and follow perfectly the arbitrary topography of the material surface at the outer (right side) belt. As with the radial load, the hugging pressure must be sufficient to develop the needed internal friction that will resist the gravitational slide back forces. Beyond the straight profile, the belt sandwich with material travels through a short transition curve, a point of curvature reversal, then through another transition curve to discharge. Through the transition curves, as with the lower curve, the outer belt radially urges itself with the bulk up against the inner curve which is supported by closely spaced troughing idlers. At the discharge point the bottom belt is deflected over its head/drive pulley and the material is released into the discharge chute. The top belt travels a little further to its head/drive pulley. Both belts return independently through their take-up paths then to their respective tail pulleys.

Equipment Summary

In anticipation of the larger material, the belts are armored with 10 mm thick carrying covers of Grade 1 rubber, the best resistant grade against wear, impact and cutting. Additionally, along the transition curves all CEMA D6 idlers have rubber disc center rolls and steel wing rolls. The rubber disc center rolls soften the ride of the larger lumps as they are urged radially by the outer belt against the inner belt. All pulleys are rubber lagged; plain at non-drive pulleys and diamond grooved for traction at the head drive pulleys. The non-drive pulleys are also crowned. The combination of lagging and crowning promotes good belt alignment. The lagging is also softer and more forgiving on those occasions when material enters the pinch between the belt and the pulley face. Both belts are driven equally by shaft mounted drives at the head/discharge pulleys. Driving both belts shares the drive tension equally and facilitates better belt alignment than driving only one belt while the other merely follows. Variable frequency control at both drives facilitates the equal load sharing, provides soft starts, and allows variation of the belt speed as may be deemed appropriate in response to the actual material and flow characteristics. Tension control is by near constant pressure hydraulics. A tensioning cylinder at each belt pulls on the take-up pulley carriage and operates within a narrow band of hydraulic pressure, pumping at the lower operating limit and stopping at the higher operating limit.

Arrangement and Structure

The common design is for the current requirements and for the long term. Because tunneling and construction projects tend to be of short duration, lasting several years at most and typically less than two years, it is difficult to justify the cost of custom, dedicated equipment. It was thus important that we provide a common design that would address the current requirements and the requirements of future projects. Accordingly, we designed the system to easily decrease or increase the vertical lift, to a possible low of 19.6 meters and a possible high of 43 meters. This can be done by subtracting or adding vertical structure (and belt length) in lengths that are multiples of 1676 mm, the length coverage of each GPS (Gently Pressed Sandwich) hugging pressure module. Thus the system and equipment must be designed for the maximum lift. To facilitate the extension and contraction, the system is modular with most equipment on the surface for easy access and service. The intermediate structure is of simple parallel channel sections and hangs down freely, suspended from the head end. At the bottom, the loading station and approach structure is supported at the loading end (left side) on grade and at the vertical (right side) by the hanging intermediate structure. Strategic bolted splice points along the vertical facilitate the adding and subtracting of vertical structure as required for each subsequent installation. With this arrangement and structure the lower loading section and vertical intermediate sections are kept simple with most vital equipment located at the head end.

The Differences

Though having the same common design the units differ slightly in adaptation to their respective requirements. DS 108 required less vertical lift by 1676 mm, the length of one GPS section.

DS 109 had a different requirement related to the TBM's initial development. During the initial excavation, a temporary trailing conveyor was used. It was aligned along the tunnel's center so that the discharge from this conveyor was located over the return of the top belt. For this a loading table with impact idlers and loading skirts was designed for the return of the top belt. Thus the muck could be conveyed back to the loading area of the bottom belt. Once onto the bottom belt, the muck reverses direction and travels into the belt sandwich then on to the discharge at the top. Figure 5 shows loading of the top return belt during the initial tunnel development. It is especially convenient that with VFD control of the drives during this early development the Dos Santos Sandwich-Belt could be run at a reduced speed mitigating the effects of the sudden reversal of the material flow. This feature was only used at DS 109 but it is available at either unit when needed in a future installation.

Early Operation, Adjustments

The Paris Metro Expansion is well vindicated in their exclusive specification of the Sandwich-Belt high angle conveyors for the muck elevating duties. It is not to be construed that operation went smoothly immediately. This is not at all the case. We knew the muck would be sticky but no one anticipated how wet and sticky. Indeed it is baffling that such material is the foundation for Paris-the iconic City of Lights.

Project specifications did not provide sufficient guidelines to cope with the material found. I must conclude that no one was prepared, including those who supplied conventional conveyors for the project. The muck could best be described as goop or very wet goop. Figure 6 illustrates the muck. When stacked at the surface containments the muck becomes submerged under the water that was conveyed with it. At the third frame of Figure 6, muck thrown at the underside of the top belt drive base stayed there.

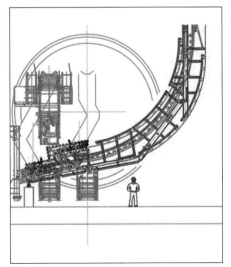

Figure 5. Sketch and photo of muck loading onto the return top belt during initial tunnel development

Figure 6. From left: TBM muck; muck stacked in containment at surface; muck stuck to underside of drive base; chute covers removed to allow water spraying

Figure 7. From left: profiled deflector in place; then removed; new hanging chain curtain; chain curtain in operation

Flow through the chutework dragged or plugged. Ironically, more water was sprayed on the material to keep it moving. Chute covers were removed to allow water spray with hoses. The original supply included a profiled deflector plate as is common to guide the discharge. This too became an obstacle and was replaced with a hanging chain curtain. The flailing motion of the latter made it self-cleaning. Figure 7 shows the deflector plate and its removal then the chain curtain that replaced it.

Figure 8. Muck in a bath from the TBM to the vertical sandwich-belt conveyor

Figure 8 shows the muck flowing in a water bath as it approaches the vertical Sandwich-Belt conveyor.

Though elevating the muck was never the problem, scraping the belts clean proved challenging. Field adjustments, including by the manufacturer's representative did not sufficiently improve their performance. Ultimately these were replaced with better scrapers that did clean the belts. The successful scrapers have individually sprung blades with positive attack angle. They have sufficient range of motion to follow the belt's dirty surface. This is especially important at the top belt as its surface is deflected upward at the middle by the material load.

The changes and adjustments mentioned greatly improved the Sandwich-Belt operation and it continues operating without interruption. Production and material flow problems continued at the TBM, especially at the inclined section of the discharge conveyor where globs of wet clay tended to slide and stagnate. With our customer, we observed that the design capacity of the vertical Dos Santos Sandwich units was not being challenged. We thus took advantage of the VFD speed control and reduced the operating speed from 3 m/s to 2.4 m/s (from 50 Hz to 40 Hz). This proved to handle the production very well while reducing the wear and tear associated with belt speed.

SUMMARY AND CONCLUSIONS

Despite the long tradition of pocket belts for vertical muck haulage from TBMs, the distinct advantage of Sandwich-Belt high angle conveyors was recognized and specified exclusively at The Paris Metro Expansion. This writing documented the success in the face of a most adverse material. There have been other projects that have followed suit and have specified Sandwich-Belt high angle conveyors for the elevating duties when the material is especially sticky. With the remaining work at the Paris Metro Expansion and the others that have followed suit, the opportunity now presents itself for a quantum leap in Sandwich-Belt high angle conveyors at construction and tunneling projects, especially for the high volumes produced by the largest TBMs.

Future Direction, Vertical Continuous Haulage

The wide use of longwall systems in the 1980s required upgrading all of the underground conveyors to larger belt widths that could keep up with the longwall production. In deep coal mines, this resulted in choking the flow at the main haulage shafts where skip hoist systems could not meet the increased production requirements. This created opportunities for vertical high angle conveyors in the 1990s. Studies of that time developed single and multi-flight systems as alternates to the traditional skip hoist systems. These proved to easily handle the large throughput rates continuously through a mere conveyor-to-conveyor transfer chute, precluding the large terminal storage and feeding systems that are required for the skip hoist batch haulage systems. The economics are overwhelmingly in favor of the continuous haulage systems with Sandwich-Belt high angle conveyors along the vertical shafts from underground.

Figure 9 shows two variations of continuous vertical haulage from underground; Scheme-A and Scheme-B. The original basis for these schemes is an underground nickel/copper mine with a net vertical lift to surface of 1381 m.

Scheme-A consists of eight (8) main sandwich conveyors along the main vertical shaft and seven (7) small connecting sandwich conveyors at excavated pockets. A main shaft of 7 meters finished diameter is able to accommodate the haulage system as well as all mine equipment access. The continuous vertical haulage system occupies half of the main shaft while the travel path of the 1.8 m × 4.9 m mine equipment cage occupies the other half.

Scheme-B consists of only (8) eight main sandwich conveyors, four (4) along the main vertical shaft and four (4) along independent vertical shafts that are solely dedicated to each sandwich conveyor. The four (4) independent shafts are of 3.7 m finished diameter as this accommodates the haulage system as well as a 1.1 m × 1.7 m service cage. Scheme-B requires the additional local excavation between the main shaft and the ends of each independent shaft in order to accommodate the transfers between the alternating Sandwich-Belt conveyors.

Such ambitious multi-flight systems as described above are yet to be realized but single flight vertical sandwich belt high angle conveyors, the basis for these systems,

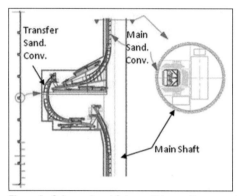 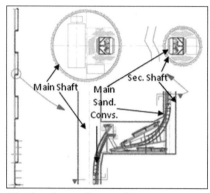

Figure 9. Scheme-A and Scheme-B for continuous vertical haulage from underground

were commercially utilized at vertical shafts from underground coal mining, gypsum mining and tunneling projects. Of these, the most impressive is DS 030 of Table 1.

REFERENCES

1. Dos Santos, J.A. and Frizzell, E.M., Evolution of Sandwich Belt High-Angle Conveyors. CIM Bulletin. Vol.576, Issue 855, July 1983, pp. 51–66.

2. Dos Santos, J.A. "High Angle Conveyor-HAC Provides Shortest Route to Train Loading Silos," Coal Prep 87, April 27–29, 1987, Lexington, Kentucky, U.S.A.

3. Dos Santos, J.A., Sandwich Belt HAC®s Broad Horizons—1992. Bulk Solids Handling, Vol. 12 (No. 3), September 1992 (8 pages).

4. Dos Santos, J.A., Continuous Vertical Haulage. Published in Bulk Solids Handling, Vol. 13 (No. 3), September 1993 (6 pages).

5. Dos Santos, J.A., The Cost/Value of High Angle Conveying. Published in Bulk Solids Handling, Vol. 18 (No. 2), April/June 1998, pp.253–260.

6. Mevissen, E.A., Siminerio, A.C. and Dos Santos, J.A., High Angle Conveyor Study by Dravo Corporation for Bureau of Mines, U.S. Department of the Interior under BuMines Contract No. J0295002, 1981, Vol. 1, 291 pages, Vol .II, 276 pages.

7. Scott, D.W., Dos Santos, J.A., Straight Up—The Best Way To Go. Presented at The AMC (American Mining Congress) Coal Convention '92, Cincinatti, OH, USA, May 3–5, 1992, published in proceedings (12 pages).

SCMAGLEV—Fast and Innovative Mode of Transportation in the Northeast Corridor—Tunneling Challenges

Vojtech Gall ▪ Gall Zeidler Consultants
Nikolaos Syrtariotis ▪ Gall Zeidler Consultants
Timothy O'Brien ▪ Gall Zeidler Consultants
Cosema (Connie) Crawford ▪ Louis Berger
David Henley ▪ Baltimore Washington Rapid Rail

ABSTRACT

The Northeast Corridor Superconducting Maglev Project (SCMAGLEV) entails construction of a high-speed train system between Washington, D.C. and New York City, with the first leg between Washington and Baltimore, MD. The system operates using an electromagnetic levitation system developed and deployed in Japan that achieves an operating speed of 500 km/h. SCMAGLEV is an innovative project that will shorten travel times between Washington D.C. and Baltimore to 15 minutes, and connect to New York City in under an hour. In 2019, the design is being advanced to an engineering level sufficient to support an Environmental Impact Statement. A preliminary geotechnical investigation campaign was also underway. This paper provides an update on the project and discusses construction of the predominantly soft ground EPBM driven tunnels.

INTRODUCTION

The Superconducting Maglev (SCMAGLEV) Project is a proposed high-speed train system between Washington, D.C. and the City of Baltimore, approximately 60 km (37 mi) in length (Figure 1). The Washington, D.C. to Baltimore segment is the first leg of a route that eventually would be between Washington, D.C and New York City. Additionally, the envisioned route would include stations at Wilmington and Philadelphia, PA, as well as additional major airports along the route. With the Northeast Corridor (NEC) home to 17% of the US population and travel between the major cities of the NEC predicted to increase 115% by 2040, the SCMAGLEV is the technology that can help reshape transportation and commerce in the NEC and US.

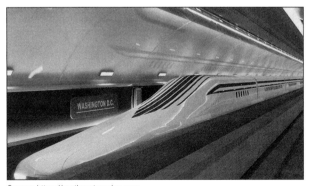

Source: https://northeastmaglev.com
Figure 1. Conceptual rendering of the SCMAGLEV

The SCMAGLEV system operates using a combination of electromagnetic levitation (support), propulsion and lateral guidance rather than flanged wheels, axles and bearings as in conventional high-speed rail systems. The train system will cross several transportation corridors including interstate highways (I-95, I-195, MD295 Baltimore Washington Parkway, I-595, I-695, I-895), several state, city and local routes, and railroad lines, as well as BWI Airport, with all crossings grade separated. The project developer is the Northeast Maglev/Baltimore Washington Rapid Rail (TNEM/BWRR) with Louis Berger as the prime consultant and Gall Zeidler Consultants as the tunneling sub-consultant.

An environmental impact statement was initiated in the fall of 2016 in accordance with the National Environmental Policy Act (NEPA) with the Draft Environmental Impact Statement (DEIS) expected to be published in Fall-2019 and the Final Environmental Impact Statement (FEIS) and Record of Decision (ROD) by mid-2020. Construction is envisioned to commence in 2021 with an estimated total cost over $10 billion.

TECHNOLOGY

The Baltimore-Washington Maglev Project provides new infrastructure, stations and facilities for a SCMAGLEV train system. The project will build on the safety practices and culture of system developer Central Japan Railway Company (JRC), which has operated the Tokaido Shinkansen bullet train between Tokyo and Osaka without a single fatality since 1964. JRC applied a similar safety approach to the development of the SCMAGLEV system. SCMAGLEV was certified by the Japanese government and has been in commercial operation since 2014. Safety systems for the Baltimore-Washington Maglev project will be developed through a collaborative process with the Federal Rail Administration (FRA) Office of Safety and local emergency response forces.

The primary elements of the project include superconducting magnetic levitation rolling stock and systems, using a proprietary technology developed by JRC, and two guideways, borne by tunnel and viaduct structures. The system deploys technologies that are new to the U.S. or of previously limited application, including most notably an electromagnetic propulsion system. This technology is capable of accelerating trains to a top cruising speed of 500 km/h (311 mph) in two minutes and allows for a driverless train operation. Additionally, the energy consumption for the train is far less than that of other ultra-high-speed travel options, such as a Boeing 777, or the operational Maglev in Shanghai China (Figure 2). The train utilizes superconducting magnets for acceleration and lateral guidance of the train. It is estimated that the total trip duration

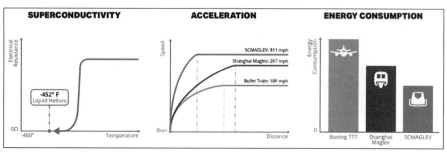

Source: https://bwrapidrail.com

Figure 2. Depiction of key aspects of the SCMAGLEV technology that utilizes superconductivity, allowing for greater acceleration and less energy consumption of competing technologies

from Washington D.C. to Baltimore, MD will be 15 minutes and Washington, D.C. to New York will be one hour, at a speed of 311 mph (511 km/h).

ALIGNMENT ALTERNATIVES

The project is located in Washington, D.C. and Maryland, traversing a distance of approximately 60 km (37 mi) with three stations in Washington D.C., at BWI Airport and in Baltimore, M.D. The Washington, DC and BWI Airport station options are underground. There are above and below ground options for the Baltimore station. The SCMAGLEV system requires an independent and secured grade-separated right-of-way. Further, assuring the safety and comfort of passengers requires use of predominantly straight geometry with limited horizontal and vertical curvature consistent with the physical dynamics of ultra-high-speed travel. To accommodate the range of topographical and surface features, existing dense urban areas, utility mains, and existing structures, the proposed construction is expected to consist of below-ground (tunnel) for at approximately 75–80% of the route, and elevated structures (viaduct) for the remainder. The train system incorporates two main guideways, three stations, one rolling stock depot, electrical substations, tunnel ventilation plants and emergency egress.

The environmental review process narrowed two alignment alternatives which generally follow Baltimore Washington Parkway (MD 295) (Figure 3a), with a preferred alternative to be named in the DEIS in 2019. Preferred alternatives for station locations in Washington, D.C. and Baltimore are still being assessed and will be named in the DEIS as well. At BWI Airport, the existing hourly parking garage will be demolished for station construction, with a replacement parking garage to be constructed prior to demolition of the existing one. The primary stations (Washington, D.C., Philadelphia, New York City) will have a platform length of approximately 400 m (1312 ft.), enabling accommodation of 16-car trains, while local stations (BWI, Baltimore, Wilmington, etc.) will have 300 m (980 ft.) long platforms to accommodate 12-car trains.

GROUND CONDITIONS

A preliminary geotechnical exploration program was conducted along the two alignment alternatives in the spring-summer of 2018 and included 22 boreholes and

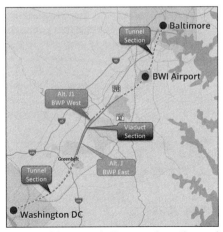
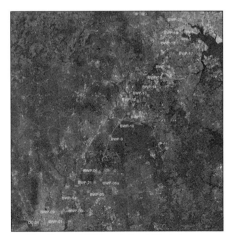

Figure 3. (a) Proposed alignment alternatives under consideration and (b) boreholes drilled along alignment alternatives for the preliminary boring program

Figure 4. Borehole drilling for preliminary ground investigation program

geotechnical testing (Figure 3b, 4). The program provided a preliminary look at the anticipated ground conditions for construction and tunneling along the alignments, and identified target areas of interest for the next phase of geotechnical investigation.

The proposed alignments are located within the Coastal Plain Physiographic Province, consisting of relatively soft strata. These strata lie on top of crystalline bedrock and thicken to the southeast on the order of approximately 150 m (500 ft.) per 8 km (5 mi). The strata consist of sedimentary deposits of the Cretaceous-age Potomac Group, which includes clays, sands, and gravels, and Holocene-Pleistocene terrace gravels and loose granular soils. All sedimentary formations sit unconformably atop each other, with the oldest sediments (Patuxent Fm.) sitting unconformably atop bedrock. Bedrock ranges in age from Jurassic to Cretaceous and is exposed or encountered at shallow depths at the Fall Line, which is the boundary between the Coastal Plain and Piedmont physiographic provinces.

Groundwater conditions are expected to vary widely across the alignments, from dry conditions to groundwater levels ranging from relative shallow depths of less than 10 ft. (3 m), to depths in excess of 40 ft. (12 m). Fluctuations in groundwater levels across the alignment will occur seasonally due to variations in rainfall, evaporation, construction activity, surface runoff and proximity to adjacent streams and the Chesapeake Bay shoreline. Localized perched groundwater and isolated water-saturated sediment lenses can also be expected. Connectivity of the aquifers to rivers and creeks has been identified in various locations.

Tunneling is expected to occur primarily through the Patapsco, Arundel, and Patuxent formations of the Potomac Group soft ground for most of the alignment. Bedrock was encountered in boreholes at portions of the alignment closest to the Fall Line, which includes the Washington, D.C. and Baltimore stations, as well as the central portion of the alignment alternatives approximate to the viaduct segment. Locally higher sections of the bedrock cannot be excluded, and would be a target of the next phase of geotechnical investigations. No unanticipated or abnormal geological features were encountered during the preliminary ground investigation program.

TUNNELING CHALLENGES

The proposed alignment alternatives include approximately 28 to 30 miles (45–40 km) of generally deep tunnel sections. Considering the length of the tunnel sections and the required uniform geometry, it is anticipated that mechanized tunneling will be implemented for the majority of the alignment that will need to address the following challenges:

- Tunneling in soft ground, consisting of sands, silts, clays and gravels.
- High groundwater level
- Tunneling across urban areas and therefore under major infrastructure

Considering the soil types and groundwater conditions expected along the deep tunnel sections, which require an active face support, the use of a closed face Tunnel Boring Machine (TBM) will be required. Based on the available preliminary information on the geological and hydrogeological conditions and the critical impact of groundwater to the tunneling activities, implementation of Earth Pressure Balance Machines (EPBM) is considered, at this stage, most appropriate for the anticipated subsurface conditions. Alternatively, Slurry and/or Mix Shield TBMs could be considered, as the alignment could encounter sections of mixed geology with hard rock potentially shallower at the two ends of the alignment, pending the next phase of ground investigations. The information acquired from the additional ground investigation program will be used to evaluate and select the TBM type and refine specifications.

TBM tunnels in soft ground are generally supported by pre-cast segments, which are erected at the tail end of the TBM producing a continuous lining over the tunnel length with a circular, uniform geometry. Segmental linings will be equipped with gaskets in the joints between the segments to inhibit groundwater inflow into the tunnel.

To minimize the construction footprint of the project, minimize surface disturbance and construction impact, while taking into consideration the spatial requirements for the train operation, a single bore TBM tunnel with an outside diameter of approximately 15 m (50 ft.) is being considered as optimal compared to twin bore tunnel configuration (Figure 5a). Although tunneling with a large bore TBM is a challenge in itself, the technology and capabilities of present day TBMs allows for unimpeded tunneling and enhanced risk management. Within the past 10 to 15 years, large-bore TBM tunnels are an increasingly common option being utilized for major transportation projects, with recent successes on the Port of Miami Tunnel in Miami, Florida (Bauer et al. 2013), Barcelona Metro—Line 9 in Barcelona, Spain, and the Shanghai River

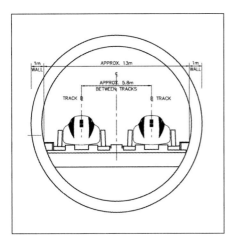 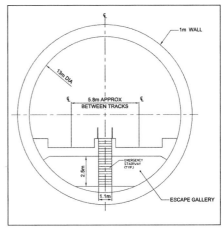

Figure 5. (a) Typical TBM tunnel cross section and (b) cross section showing emergency escape gallery

Crossing in Shanghai, China. Additionally, the alignment had been determined such that TBM tunneling will be performed under at least 1 tunnel diameter of ground cover to minimize surface impact.

Subdivision of the TBM tunnel alignment into sections with a length of 5 to 6 km (3 to 4 mi) is currently considered for enabling concurrent boring along various sections and providing flexibility for contacting and packaging of the project. This requires construction of additional launch sites, which are typically cut-and-cover structures. In areas where space restrictions do not allow for construction of launch boxes, launch shafts of adequate size are considered as an alternative. Ventilation shafts are planned to be used as launch shafts where possible, to minimize cost and streamline construction. As the launch sites will be also used for stockpiling of the spoils, implementation of additional launch sites along the alignment will allow more efficient storage and transport of the spoils to the areas designated for disposition.

Short sections of cut-and-cover tunneling will be used for the stations and the transitions between the viaduct and TBM tunnel sections including portals and TBM launch locations. Implementation of cut-and-cover tunneling requires installation of support of excavation, such as slurry walls, bored pile walls, soldier pile and lagging or shotcrete. The method of SOE chosen will also be dictated by the local ground conditions, with slurry walls a likely choice where dewatering of the sediments is to be avoided to prevent settlement of any adjacent existing structures. Depending on the limits of disturbance, generally tie-back support or internal strutting is expected for deeper excavations. A waterproofing system will be installed to prevent groundwater inflow into the tunnel in the final permanent stage.

Due to the dense urban environment in Washington D.C. and Baltimore and the relatively deep alignment, construction of the stations with minimal surface impact and disruption to the city activities will be challenging and will require a well-thought design. Similarly, construction of the station under the BWI airport without disrupting airport operations will pose a significant undertaking.

FIRE AND LIFE SAFETY

Design, construction and operations for the SCMAGLEV will be planned with a safety focus: safety of the traveling public, the construction and operations workforce, and the adjoining communities that are impacted by the construction and operations of the system. Each area will be addressed in the planning and design of the infrastructure, core systems, facilities, and operating and maintenance practices for the SCMAGLEV system.

Fire and life safety have been given full consideration and attention in the design since the inception phase of the project. Fire and life safety considerations factor into all aspects of the system design, including linear infrastructure (viaducts and tunnels), passenger stations and operations and maintenance facilities. The fire and life safety include elements and layout of egress and access paths in the tunnel system; definition of design fires for vehicles, cables, etc.

The design standards and guidelines addressing fire and life safety requirements for the structures of the project are:

- NFPA-130
- NFPA-101

- NFPA-502
- Americans with Disabilities Act
- Accessibility Guidelines (ADAAG)
- Maryland Building Performance Standards
- Maryland State Fire Prevention Code
- Washington D.C. Building code and Construction Code

To meet the requirements of the standards for Fire and Life Safety and NFPA-130 in particular, a safe emergency egress for passengers to a point of safety will be provided in the underground sections. This would be achieved by utilizing an escape walkway/gallery inside the tunnel envelope located below the guideway (Figure 5b, 6a). The considered tunnel cross section provides sufficient space below the guideways to be used as an emergency evacuation chamber. The escape gallery would have an independent ventilation system in the event of a fire or other emergency and would have surface access via ventilation plants and shafts envisioned along the underground section of the alignment at an approximate distance of 5–6 km (3–4 mi) (Figure 6b).

Due to the unique characteristics of the SCMAGLEV system, the standards and guidelines listed above will be supplemented by Japanese codes and practices that have contributed to that country's exemplary safety record. Safety systems and practices researched and developed by JRC specifically for the SCMAGLEV system will be proposed for incorporation into the proposed project to ensure that the highest standards for safety are deployed.

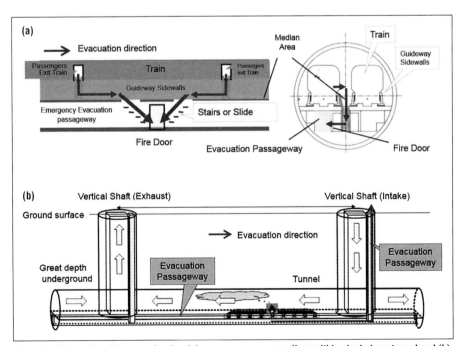

Figure 6. Schematics diagrams showing (a) emergency egress gallery within single bore tunnel and (b) connection of egress gallery to ventilation plants for surface access

OUTLOOK

With the successful completion of the preliminary ground investigation program, the Draft Environmental Impact Study is underway with Record of Decision anticipated in 2020. A positive ROD would mean forward movement on the next phase of ground investigation and preliminary engineering, with construction envisioned to commence in 2021 at an estimated total cost over $10 billion. Long-term maintenance costs of the system are minimal because there is no mechanical contact and wear between the train and the guideway.

The proposed SCMAGLEV is a technically challenging but innovative project that will shorten commuting time between Washington D.C. and Baltimore, and later to New York City. The project will enhance mobility along the northeast corridor and could spur development and economic growth in the region.

REFERENCES

Baltimore-Washington Rapid Rail. 2018. https://bwrapidrail.com.

Bauer, A., Gall, V., and Bourdon, P. 2013. Comparison of Predicted Versus Observed Structural Displacements of Existing Structures at the Port of Miami. In *Proceedings of the Rapid Excavation and Tunneling Conference 2013,* Edited by M.A. DiPonio and C. Dixon. Englewood, CO: SME.

The Northeast Maglev. 2018. https://northeastmaglev.com.

Smart Office: A Data-Driven Management Tool for Mechanized Tunneling Construction

Kamran Jahan Bakhsh ▪ Lane Construction Corporation
Jim Kabat ▪ Lane Construction Corporation
Roberto Bono ▪ Salini Impregilo

ABSTRACT

Mechanized tunneling construction with TBM is among the most innovative areas in the construction industry, however, tunnel construction process remains severely under-digitized. While vast data is recorded by construction equipment (e.g., TBM, conveyor system) and human interaction (e.g., site engineers, operators), leveraging the power of data to increase productivity and improve the construction process is overlooked. In this paper, the concept of a unified analytics center to gather, integrate, and analyze data from disparate databases on a construction site and then contextualize this information into a visual, meaningful representation provides a robust tool to enhance instant, and far-off decision-making for all levels of site and office personnel is presented. Dugway Storage Tunnel project is considered as a case study. Throughout the construction period of the tunnel, vast quantities of data from TBM sensors is streamed directly from TBM's PLC. Data is then pushed into Power BI, which is an analytics service provided by Microsoft. This system connects the project personnel (e.g., Project manager, Construction manager, Project engineer, and site Engineers) to a broad range of data via easy-to-use dashboards, interactive reports, and meaningful interactive visualizations that bring data to life.

INTRODUCTION

Like other industries, today construction firms are capturing more data than ever before through information sensing devices (e.g., job site sensors, smartphones, and heavy equipment tracking devices). Though there is an enormous potential of leveraging the power of captured data to increase productivity, much of the data is siloed without being utilized. In the mechanized tunneling construction with TBM, utilization of data to improve the construction process is limited to small portion consumed by machine operator. Indeed, much of that data is stored and filed away once a project is completed. With todays' ease of data acquisition, in addition to TBM data tunneling construction firms can capture more data than ever before through other information sensing devices in the job-site. Transforming these heterogeneous, seemingly unrelated data into coherent, visually immersive, and interactive insights will enhance every aspect of the project execution.

Data-driven modeling system (e.g., BIM, CIM) become more popular in the tunnel engineering construction. The Crossrail tunnel in London (Heikkila, R. & Makkonen, T. 2014), State Route 99 tunnel in Seattle (Lensing, R. 2016, Trimble, 2011), Hallandsas tunnel in Sweden (Smith, C. 2014), Hangzhou Zizhi tunnel in China (Wang, J. et al., 2015), and Mikusa tunnel in Japan (Sugiura, S. 2015) are all examples of projects that executed by utilizing data-driven methods (Figure 1). Despite the several advantages of BIM models in the design field, due to the disruptive nature of this system, there are several challenges of implementation in the construction firm (Davidson, A. 2009). For

Figure 1. Example of BIM model. The Crossrail tunnel in London (top left), State Route 99 tunnel in Seattle (bottom left), Hallandsas tunnel in Sweden (top right), and Mikusa tunnel in Japan (bottom right).

instance, BIM doesn't help to facilitate communication in the construction field, and it should be implemented with the full capacity to be effective.

Sporadic attempts have been made to utilize power of data in the job site and few companies (e.g., Babenderede Engineering and Tunnelware) have developed data management and visualization software compatible for tunneling construction market. This commercial software can be used as a tool to make better decisions, increase productivity, control the quality of materials, and improve safety in job sites. For instance, a modular software has developed by Babenderede Engineering to support tunneling construction process. They simply equip contractor's TBM with an external data acquisition system to pull relevant data from the machine PLC, and then by processing raw data they can generate helpful information. They also developed software exclusively for tracking segmental lining through the lifecycle of the project (Cicinelli, L. et al., 2017). Tunnelware which is still at the developing stage tries to consolidate data from several sources (e.g., TBM, site personnel reports, and job site sensors). According to the Tunnelware experts, the software will provide a robust tool for the constructors to visualize and analyze the tunnel excacavation processes on a 5D format. There are also adding other features such as cutter tool management, and virtual meeting rooms to their software package.

Although available commercial services have several benefits, there are drawbacks that make contractors hesitate to utilize these services in their projects. For example, the software visualization is old-fashioned and dissociated. Indeed, it is very crucial to prepare an easy-to-use interactive visualization platform that can facilitate construction process for the project personnel; otherwise, it will be abandoned throughout the construction period. In addition, since every tunnel project is unique, existing software in the market are not flexible enough to meet the uniqueness of the project.

Smart Office: A Data-Driven Management Tool

In this paper the concept of unified analytics center compatible for tunnel construction is presented. First, Dugway Storage Tunnel project which is considered as a case study for testing our model is introduced. Later The structure of the conceptual model, data collecting, and required hardware and software is explained in the Afterward, application of the proposed model throughout the construction period of the tunnel is partially examined. The advantages of implementing such a data management system as a tool for project managers, engineers, safety superintendents, etc. are also addressed.

DUGWAY STORAGE TUNNEL

Project Description

The Dugway Storage Tunnel (DST) is the second of seven tunnel projects that will reduce 4 billion gallons of pollution discharged into the Lake Erie due to the seasonal overflows. The tunnel alignment is approximately 2.8 mile in length with seven curves of variable radius, excavated using a single shield hard rock TBM with 27 ft. excavation diameter. The finished internal tunnel lining diameter is 24ft. using concrete segments of 12 in. thickness. Depths of the tunnel invert is ranged from 180 to 230 ft. below ground surface. The project includes a total of six deep shafts along the path of the tunnel with an internal lined diameter between 16 to 50 ft. and four Adit connections between these shafts and tunnel of variable length between 50 and 1000 ft. (Figure 2). The 46ft. diameter shaft known as DST-1 is the TBM launch shaft where all the main conveyor systems will be installed. Part of the shafts was constructed through soft ground and part also encountered chagrin shale bedrock. The project includes the construction of an additional structure as Diversion Structures, Gate Structures, Control Vaults, Ventilation Vaults, Drop Manholes and modifications to existing regulatory structures. The project area is mainly older residential, i.e., pre-1950s interspersed with commercial properties and urban parks.

The Tunnel Boring Machine used to excavate the 27ft. diameter tunnel through the Chagrin Shale was a hard rock single shield Herrenknecht machine type S-684 (Figure 3). The machine was reconditioned on site by the contractor after excavating the first phase of the tunnel i.e., Euclid Creek Tunnel. The TBM was partially assembled for the launch with only three of the six gantries in the starter tunnel which was previously excavated by employing drill and blast method. At this stage, materials were hauled out using locomotives and muck boxes. After the first 300ft. of tunnel has been excavated, the TBM was assembled in its final setup with six gantries. The tunnel conveyor system was comprised 5 sections transporting materials from the tunnel, to the vertical conveyor to overland belt and finally to the stacker. The total rings installed at

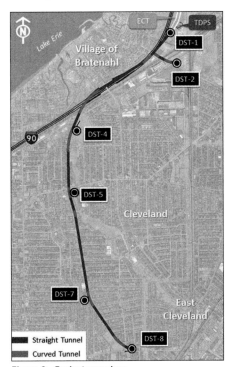

Figure 2. Project overview

the completion of the tunnel were 2911 for a total length of 14,840ft. The production average was 17 rings per day. The average of the excavation parameters were as follows: (1) Penetration 10.3 mm/rotation, (2) Cutter-head rotation 6 rpm (3) Advance of the TBM 66.75 mm/min (4) Thrust force = 10,000 kN (5) Torque = 2,500 kNm. There was no presence of water in the material, but the quantity of methane trapped between the layers was sometimes relevant.

Figure 3. Dugway storage tunnel single shield Herrenknecht TBM

UNIFIED ANALYTICS CENTER

Conceptual Model

Unlike traditional ways of managing construction process, with an extensive amount of data managing this data and using it as a management tool to make better decisions during the construction period is of utmost importance. The purpose of the conceptual model of our unified analytics center is to provide a robust tool to enhance decision-making for all levels of site and office personnel. The conceptual model is comprised of (1) a physical smart office equipped with digital display screens (2) sensors, equipment, and digital tools to gather data from project site and (3) a platform to integrate, analyze, and contextualize this information into a visual, meaningful representation. Figure 4 shows the Data Flow Model (DFM) of the proposed conceptual model. Data captured from TBM, conveyors system, equipment operating, inventory, material used, site personnel reports, labor hours, project documents, and even online weather reports are structured and pushed into a cloud-based collaboration and sharing system. Microsoft Power BI which is a suite of analytics tools is utilized to create interactive visualization and dashboards and share insight with individuals across the project and company. Depends on the end-user, model can provide customized dashboard that can get updated either in real-time, near real-time or over a longer period. For instance, real-time monitoring of critical parameters of the TBM (e.g., Thrust force, torque) can be visualized in form of interactive dashboards to assist optimizing excavation process. As another example, historical data that have collected over the years on the project can be presented to the headquarter to give them an overall insight on project progress and assists them to gain a competitive advantage when estimating and bidding on the new project with similar characteristics.

Application of the Model in Construction of the Dugway Storage Tunnel

Throughout the excavation phase of the tunnel, data from TBM PLC is pulled out and pushed into the Power BI platform. This system provided a means to connect the project personnel (e.g., project manager, construction manager, project engineer, tunnel shift engineers, and TBM operator) to a broad range of data via easy-to-use dashboards, interactive reports, and meaningful interactive visualizations. Prepared customized-dashboards eased up the decision-making process for everyone that involved directly and indirectly in the project. Figure 5, 6, and 7 show examples of crafted dashboards. For instance, as shown in Figure 5, by integrating a real-time interactive map of the TBM location and data recorded from geotechnical instrumentations we can give a better understanding of the behavior of the ground subjected

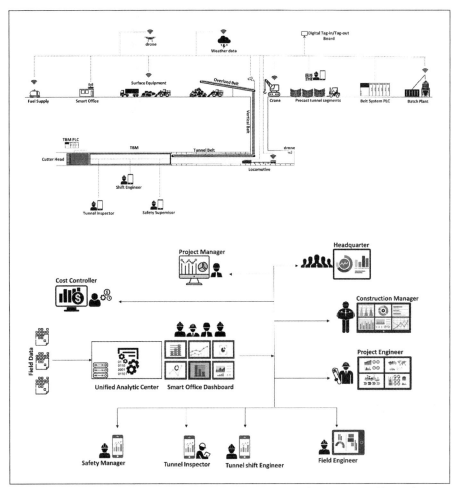

Figure 4. Data flow model of the conceptual model. Data flow from various sensors and human interaction to the unified analytics center (top), and flow of processed data from unified analytics center to end-user personnel (bottom).

Figure 5. Real-time location of the TBM on satellite map

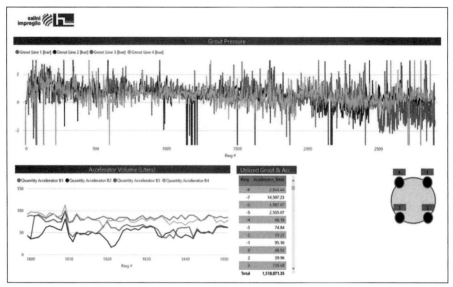

Figure 6. Real-time dashboard of accelerator admixture parameters

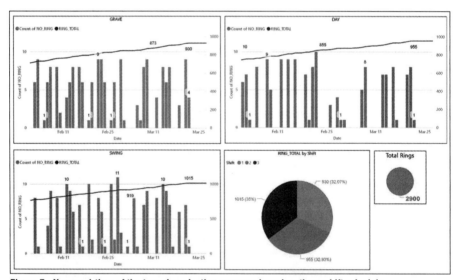

Figure 7. Near real-time of the tunnel production progress based on three-shift schedule

to excavation. This interactive dashboard can assist the TBM operator in adjusting steering parameters. Figure 6 outlines the volume of the injected accelerator admixture from each port for each ring separately. With this data available in the hand of the engineer or TBM operator, in case of any problem related to accelerator injection, they can easily examine all data to solve the problem rather than just relying on the last ring data. Collected data can be formatted in a customized easy-to-use dashboard to monitor production per shift (Figure 7). Having access to historical data of shift production can help us to increase efficiency by switching between working shift patterns (e.g., rotating three 8-hour shift schedule and two 10-hour shift schedule).

Previous examples are just a few out of many helpful dashboards that can be generated to get the right information to the right person at the right time. In order to have a better picture of the advantages of employing such a data-driven system, it is useful to address the model's application as follows:

As a Tool for Engineers

With a Wi-Fi system in the tunnel, job-site engineers have a permanent access to the real-time dashboard of all critical parameters of excavation, grout utilization, and navigation. As shown in Figure 8 grout utilization parameters (e.g., grout and accelerator admixture pressure and volume) are integrated into a customized easy-to-use dashboard which can be used as a robust tool by engineers to monitor grout utilization per each installed ring. As another example, we integrated daily production data with push and ring build time to get a better real-time picture of the TBM utilization throughout the excavation phase (Figure 9). We also formulated several other dashboards to facilitate monitoring TBM cylinders pressure and extensions, gas infiltration location and values, ring installation effect on navigation, and the variability of tendencies during the advance and many other parameters and behaviors.

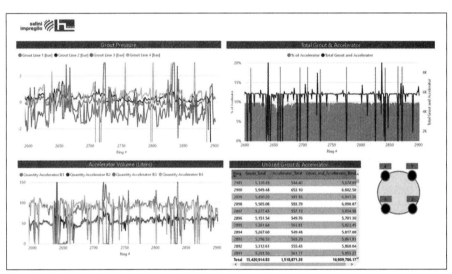

Figure 8. Real-time dashboard of the grout utilization parameter

As a Tool for Project Managers

The physical smart office where people can join to discuss ongoing tunnel construction process with all critical data in the shape of interactive dashboards, provided a unique tool in the hand of project managers to hover over all aspects of the project from a different viewpoint. For example, a dashboard is generated to illustrate the production progress on shift, daily, weekly, and monthly bases. As shown in Figure 10, monitoring the project progress through this dashboard is far more convenient than exploring archived hard-copies of Personnel report.

As a Tool for Reporting

Unified analytics center allowed us to generate digital custom reports for construction process based on different time scales (e.g., shift, daily, weekly, monthly, full project)

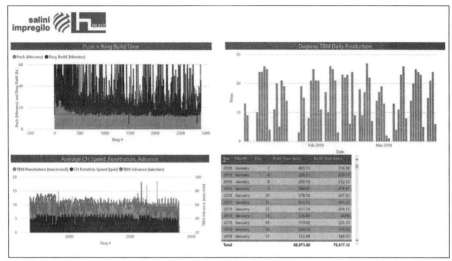

Figure 9. TBM utilization dashboard

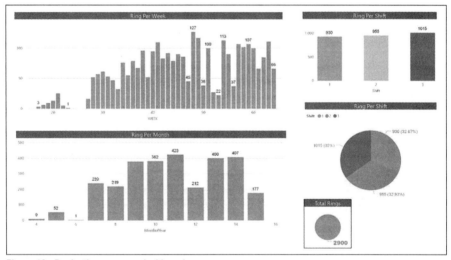

Figure 10. Production progress dashboard

automatically. This platform also let us share automatic reports on predefined timeframes with co-workers and any other recipients outside of the company just by including the list of email addresses in the system. For instance, a dashboard comprised of production progress, grout utilization, TBM utilization, and the average of TBM excavation parameters can be an informative means for program managers in headquarter and estimators when estimating and bidding on the new project with similar characteristics (Figure 11).

CONCLUSIONS

In this paper, we presented the fundamental principles of a unified analytics center as a robust tool for the tunneling industry. Moreover, we examined the application of

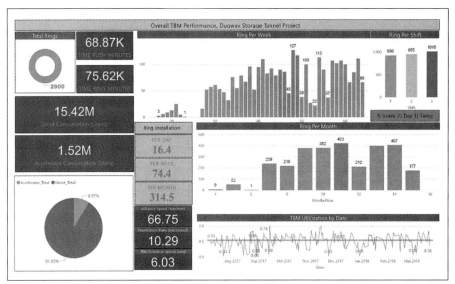

Figure 11. Overall TBM performance dashboard

a data modeling system in the construction of the Dugway Storage Tunnel. Although, data management in mechanize tunneling construction is not a new concept, there are several features that make our proposed model different. Interpreting data into meaningful easy-to-use, real-time, visual dashboards that give all individuals an insight into the project is the foremost advantage of the model. In addition, Microsoft Power BI as a backbone of the model is an analytics service entrenched in the Microsoft stack. Indeed, unlike other inflexible pre-defined data management and visualization platforms, Power BI compatibility increases adoption of our data modeling for contractors. Even though, we implemented the system just by using TBM data, due to the flexible characteristic of the model we could take advantage of the system in our decision-making process. The proposed system also allows to automatically be generating consistent reports on shift, daily, weekly, monthly scales. As a future work, we have planned to examine the system as a whole by inducing data from other sources (e.g., job site sensors, smartphones, and heavy equipment tracking devices). The all-inclusive system will allow improving decision-making and accordingly increasing efficiency and reducing construction costs.

REFERENCES

Cicinelli, V., Stahl, F., & Gronbach, T. 2017. Copenhagen Cityringen Project: Big Data to Manage Quality Control in Megaprojects. *RETC San Diego*, June 4–7, 2017 pp. 190–202.

Davidson, A. 2009. A Study of the Deployment and Impact of Building Information Modelling Software in the Construction Industry, *England University of Leeds, 2009.* Available: http://www.engineering.leeds.ac.uk/eengineering/documents/AndrewDavidson.pdf.

Heikkilä, R., Kaaranka, A., & Makkonen, T. 2014. Information Modelling Based Tunnel Design and Construction Process. *The 31st International Symposium on Automation and Robotics in Construction and Mining (ISARC Proceedings 2014).* Sydney, Australia, July 2014.

Lensing, R. 2016. BIM and construction process data in mechanized tunnel construction; Milestone control for tunnel construction sites using automatically created process data in comparison with 4D BIM. *Master of Science thesis* (Geographical Information Science & Systems). MSc(GIS) Heidelberg, 20.12.2016.

Smith, C. 2014. BIM at Sweden's Hallandsås Tunnel: Planning pioneer [Online]. New Civil Engineer. Available: http://www.newcivilengineer.com/features/geotechnical/bim-at swedenshallandss-tunnel-planning-pioneer/8663146.article.

Sugiura, S. 2015. First Application of CIM to Tunnel Construction in Japan. *Proceedings of International Conference on Civil and Building Engineering Informatics (ICCBEI 2015)*, 82. Tokyo, Japan, 2015.

TRIMBLE. 2011. Seattle's Massive Tunnel Makes travel safer—with Tekla BIMsight [Online]. Available: https://www.teklabimsight.com/references/seattles-massive-tunnel-makes-travel-safer [Accessed 24.07.2016 2016].

Wang, J., Hao, X., & Gao, X. 2015. The Application of BIM Technology in the Construction of Hangzhou Zizhi Tunnel. 3rd *International Conference on Mechatronics, Robotics and Automation (ICMRA 2015)*. ISBN 978-94-62520-76-9.

Tunnel Survey Control in Small Segmentally Lined Tunnels

Pete DeKrom ▪ Michels Corporation
Edouard Whitman ▪ Michels Corporation

ABSTRACT

The smaller the diameter of a tunnel, the greater the challenges, particularly from a survey standpoint. Usually tunnels in the 3 meter (10 ft.) inner diameter range extend for no more than a few thousand ft. to about a mile and typically have design tunnel alignments (DTA) with a minimal amount of curves. The Blacklick Creek Sanitary Interceptor Sewer (BCSIS) Tunnel Project in Columbus, Ohio, offered numerous survey challenges as the tunnel extended 6.88 km (4.3 mile) and had 13 curves, some with radii of less than 335 meters (1100 ft.). To ensure the stringent accuracy requirements of the project were met, Blacklick Constructors, LLC implemented high accuracy survey techniques such as precision total stations, redundant observations, Control checks at drop shafts, as well as implementing DMT Gyromat 5000 gyro-theodolite high-order gyro-azimuth observations.

BACKGROUND

The Blacklick Creek Sanitary Interceptor Sewer Tunnel (BCSIS) was awarded to the Blacklick Constructors, LLC, a joint venture of Michels Corporation and Jay-Dee Contractors in early 2016. The project location was located on the eastern outskirts of Columbus, Ohio, and is intended to support the sanitary infrastructure needs of future Metro-Area growth.

The tunneling operations on the (BCSIS) were completed in September 2018. The purpose of the project is to extend an existing 1.7 meter (5.5 ft.) sewer tunnel, located south of Blacklick Ridge Boulevard, with a 3.3 meter (10 ft.) internal diameter gravity sewer. The recently completed tunnel commences at Blacklick Ridge Boulevard and extends 6.88 km (4.3 miles) in a northerly direction along the Right of Way (RoW) of Reynoldsburg-New Albany Road (RNA), ending at Morse Road. Figure 1 shows the location of the project and its alignment.

The starting point of the tunnel, south of Blacklick Ridge Boulevard, is referred to as Manhole 12 (MH12). The total length of the alignment is 7,016 meters (23,020 ft.), of which 6,894 meters (22,620 ft.) was constructed using an Earth Pressure Balance (EPB) Tunnel Boring Machine (TBM). The tunnel alignment has thirteen curves with radius lengths of 1100' or less, and includes a total of 2,081 lineal meters (6,828 lineal ft.) of curves representing 30 percent of alignment.

The initial 122 lineal meter (400 ft.) was constructed by open cut, with a 36 meter (120 ft.) section installed by hand tunneling under Blacklick Ridge Boulevard. The tunnel alignment was chosen to allow the outer diameter of the EPB TBM tunnel to be constructed within 1 ft. of the extents of the RNA RoW. The tunnel alignment shortly deviated from the RNA RoW to accommodate the tunnel connections at the intermediate shaft locations. In addition to MH12 and Shaft 1 (launch shaft), three additional shafts were installed at Site 2 (Shaft 2A, 2B and 2C), and three more shafts were located at site 3 (Shafts 3A, 3B, and 3C), although the DTA only intended the TBM

Figure 1. Location and alignment of Blacklick Creek Sanitary Interceptor Sewer (BCSIS)

to intercept the "A" shafts at both locations. The remaining two shafts were located at Site 4, which included Shaft 4A (TBM receiving shaft) and Shaft 4B.

The survey challenge of the BCSIS Project was to ensure that the tunnel was bored within the prescribed tolerance to ensure the construction of the tunnel within the designated RoW. The design specifications only allowed for up to 15 cm (6 inches) of lateral deflection in either direction. As with many tunnel projects, the tunneling operation was performed concurrently with the shaft construction. This further constricted the prescribed tolerances, as even a slight lateral drift in the tunneling boring operation could have major impacts on the verticality of the installation of the shafts final liner. However, unlike most tunnel projects, the small 3 meter (10 ft.) internal diameter of the tunnel conceded very little room for tunnel Survey Control. In addition, the tight radii curves compounded already challenging survey conditions.

This paper investigates the survey challenges encountered on this project, and analyzes the survey data gathered at every step of the Control process, to determine whether the seemingly disproportionate means engaged to ensure the utmost survey accuracy and dependability were in fact justified, or whether a more traditional survey approach and methodology would have sufficed. This paper will examine the following critical components of tunnel surveying, including:

- Analysis of the existing surface Control, transfer of Control through the shafts, and design of tunnel Survey Control network;
- Implementation of a TBM guidance system;
- Use of adapted survey equipment, namely the Gyromat 5000; and
- Evaluation of survey data and analysis of the effect of refraction.

SURFACE CONTROL

As with most tunnel projects, Survey Control is typically provided (to a certain extent) by the Owner and presented to the contractor for verification and use on the construction of the tunnel and structures. Such was the case with the BCSIS Project. While the initial verification by the Contractor's surveyor of the Project Survey Control Points (PSCP) is an important step, along with determining the internal accuracy of the Survey Control obtained; this verification of accuracy, stability, and reliability of the surface Control, only minimizes complications during the construction phases. In addition to those verifications, the Project Surveyor must also ensure that the Survey Control is referenced to a geodetic datum (properly geo-referenced). If not, further transformations to a geodetic datum must be performed. This last step is particularly critical when implementing gyro-azimuth observations (as explained later is this paper).

On the Blacklick Project, the surface Control obtained from the design plans was verified using Differential Global Positioning System (DGPS) survey techniques. Historically, Differential-GPS has proven to be very reliable method for this purpose, when performed correctly.

However, DGPS should not be confused with other more common GPS survey methods of lesser accuracy. DGPS benefits from substantial reliability and accuracy enhancements over traditional GPS observations, and allows for precise measurements to be performed with the traditionally more approximate technology that is the GPS. In a nutshell, DGPS establishes and applies the necessary corrections to lesser accurate GPS signals. DGPS requires that a base GPS receiver be placed in a known location, while common satellite data is collected concurrently between multiple GPS receivers in the vicinity. The data collected is then post-processed to ensure that errors in the GPS signal caused by the ionospheric and atmospheric conditions are minimized, as the errors are similar between the common GPS data collected. On the BCSIS project, GPS data was recorded between existing and newly established Control points over multiple hours, at different times of the day, and over several days, to maximize the accuracy of the data by using multiple satellite configurations. The data was then post-processed and adjusted in a least squares adjustment package and used for the surface Survey Control.

Figure 2 shows a sub-set of the DGPS surface network, between MH12 and Shaft 1, performed by Blacklick Constructors prior to starting the tunneling operation. This resulted in a relative accuracy of the surface Control between the Launch Shaft (Shaft 1) and the Receiving Shaft (Shaft 4) of less than 5 mm (0.02 ft.). In minimizing the overall effects of GPS data on the Survey Control accuracy of the project, this level of precision provided for greater survey tolerances during both the shaft Control transfer activities, and the tunnel Survey Control throughout the entire tunneling operation.

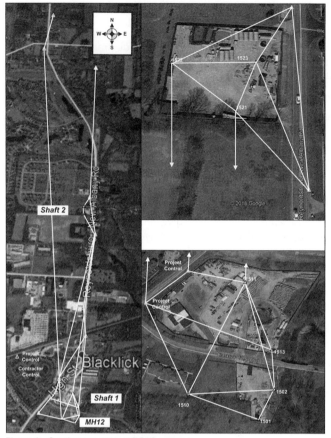

Figure 2. Southern portion of DGPS network

CONTROL TRANSFER AT SHAFTS

The BCSIS Project took advantage of its 18 m long by 12 m deep shaft (60 × 40 ft.), followed by a 122 meter (400 ft.) tail tunnel, for the launch of the EPBM. This allowed the surface Survey Control to be transferred to the tail tunnel using conventional methods (i.e., using a precise Leica MS60 total station, observing direction sets, slope distances, etc.). The tail tunnel provided a long backsight which was used for the initial orientation of the EPBM. Surface Survey Control was dropped in from both ends of the tail tunnel, with points at MH12 and Shaft 1 intervisible from each location, which greatly enhanced initial orientation of the Control. Figure 3 shows a profile of the methodology implemented to obtaining the initial Control for the BCSIS EPBM launch.

It is not uncommon for surface Survey Controls to have to be transferred down a shaft that is significantly deeper than it is wide. Such a geometry greatly complicates the survey task, as it likely precludes the direct observation of surface survey points from the shaft bottom, and conversely, the observation of shaft bottom survey points from the surface (the intervisibility mentioned above). Furthermore, the benefit of a long backsight through a tail tunnel is not available, and other techniques for transferring Survey Control are required.

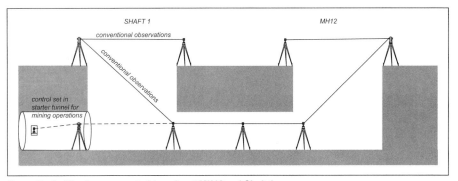

Figure 3. Profile of survey control transfer at MH12 and Shaft 1

This was such the case on the Shaft 2A and Shaft 3A Survey Control drops, where the depth of those shafts was in excess of 24 meters (80 ft.), with shafts internal diameters of 3 meter (10 ft.) or less. A methodology for transferring Survey Control from the surface to the tunnel was originally developed in the early 1990s on the Superconducting Super Collider (SSC) (Robinson, et al, 1995); and has been successfully implemented on numerous tunnel projects with shaft depths in excess of 275 meters (900 ft.). This methodology also integrates both the horizontal and the vertical Control transfer to the tunnel, in a simultaneous procedure.

This methodology requires the use of specially designed targets and spherical prisms, known as Taylor Hobson spheres and Baechler prisms, temporarily installed over the shaft in specialty brackets that provide clear vertical visibility from the shaft bottom to the brackets on the surface. The Taylor Hobson spheres and Baechler prisms are interchangeable, and carry a centering error of less than 0.1 mm in any orientation.

First, the coordinates of the temporary point(s) are determined with conventional methods from at least two locations on the surface. The Taylor-Hobson spheres are then oriented downwards. Temporary tripods are then placed directly under the Taylor Hobson spheres at the bottom of the shaft using a high precision vertical optical plummet; effectively transferring the horizontal position of the Taylor-Hobson spheres underground, and used for the tunnel Survey Control. Elevations are transferred by directly observing the vertical distance from the MS60 in the bottom of the shaft to the Baechler Prism, substituting the Taylor-Hobson spheres for this observation. The elevations are then transferred to nearby tunnel Survey Control. Figures 4 and 5 give a visual representation of the various steps of this procedure.

Figure 4. Baechler prisms and Taylor Hobson spheres

Figure 5. Shaft transfer procedure (vertical plummet on a translation stage)

On the BCSIS project however, the vertical visibility limitations induced by the 3 meter (10 ft.) internal diameter of Shafts 2A and 3A only allowed for a single survey point to be transferred down at each location. Despite this limitation, the information gathered was sufficient as it only intended to confirm the tunnel Survey Control position, relative to the surface Survey Control; and enabled both horizontal and vertical position adjustments to be performed at each location. Figure 6 shows the methodology in greater detail.

TUNNEL SURVEY CONTROL

While the space available for Survey Control in tunnels is quite often limited, this impediment is often further exacerbated in smaller diameter tunnels. For the BCSIS tunnel, the designated location for the Survey Control and the window for the guidance system was established on springline on the left side of the tunnel, as shown in Figure 7. At the early stages of the project, the type of TBM guidance system, as well as the design of the underground survey network used to establish the tunnel Survey Control were carefully evaluated to: minimize the propagation of survey error by adjusting the observation intervals, and optimize the survey methodology to provide the least disruption possible to the tunneling operations while maintaining the necessary accuracy.

While the high percentage of curves along the alignment poses its own sets of survey challenges, the overall length and congestion of the TBM 120 meter (400 ft.) ancillary support equipment, precluded any line-of-sight observations along the trailing gear. Consequently, BCL decided that a gyro/hydrostatic based Measuring While Drilling (MWD) TBM guidance system was a better suited system given the project's apparent survey limitations. In addition, specially designed survey brackets were conceived and fabricated to allow the tunnel Survey Control to be established and maintained without impeding normal tunneling operations.

Considering that seven of the thirteen curves along the DTA were left hand curves with very tight radii, the survey line-of-sights between Control brackets decreased to an interval of 23 meters (75 ft.) or less. This resulted in more setups, and consequently, the propagation of more error. Furthermore, albeit on opposite springline, the proximity of the survey brackets to the tunnel ventilation line due to the small inner diameter of the tunnel, the likelihood of Lateral Refraction (LR) influencing the survey observations became a concern. The initial tunnel Survey Control design utilized survey brackets located along left springline, with additional Survey Control points installed on the right side of the tunnel consisting of an expansion anchors fitted with special adapter compatible with Leica Circular Prisms (LCP). Since LCPs can be rotated in

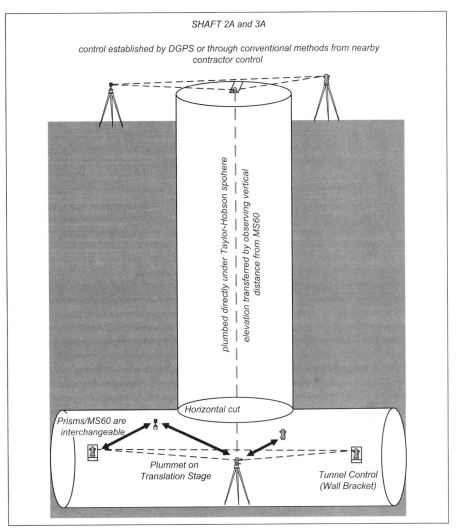

Figure 6. Surface control check at Shaft 2A and 3A

any direction without affecting their geometrical center, this allowed for redundant observations, further enhancing the accuracy, and allowing for outlier (error) of survey observations detection in the tunnel.

Lateral Refraction is a well-known, and possibly one of the largest concerns for seasoned tunnel surveyors working underground. LR affects the line-of-sight of the tunnel traverse, and is caused by the turbulent airflow induced by forced ventilation systems. On the BCSIS project, as the tunneling operation progressed, the airflow was measured and adjusted regularly, to ensure compliance with OSHA Standards. But the tunnel ventilation and consequently LR, also negatively affected survey observations in the vicinity of the 0.9 meter (36 inch) ventilation duct, leading to additional error in the tunnel Survey Control.

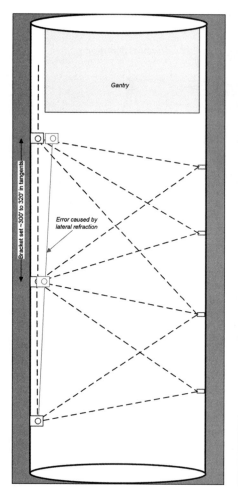
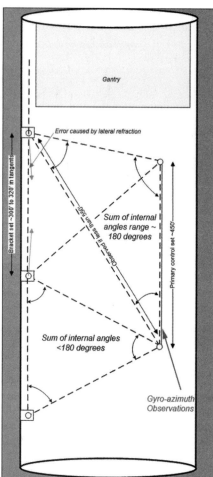

Figure 7. Initial survey control with brackets and expansion anchors

Figure 8. Centerline traverse

Preliminary indications that LR may had been affecting the survey observations was first demonstrated when gyro-azimuth observations were performed, and a discrepancy between the reduced gyro-theodolite azimuth and the tunnel Survey Control was identified, which indicated a possible rotation of the Survey Control of 15 arcseconds. To further improve upon the geometry of the Survey Control network, additional observations incorporating a centerline traverse on a regular basis with gyro observations using the DMT Gyromat 5000 gyro-theodolite were performed.

The centerline traverse was performed in approximately 330 meter (1000 ft.) intervals, during inactive tunneling periods. Additional Survey Control was set every 120~150 meter (400 to 500 ft.) near the tunnel invert. Observations were performed to the nearby brackets that were within 150 meter (500 ft.) of each centerline Survey Control point from multiple setups for redundancy. Every third or fourth setup, gyro-observations were incorporated into the centerline traverse. Figure 8 shows the typical network configuration of the centerline traverse incorporating the wall brackets.

When the centerline traverse was completed, one of the adopted procedures, performed by the BCSIS survey team, was to calculate the internal angles of enclosed triangles in the tunnel network. This method confirmed whether or not the tunnel survey observations were subject to a bias effect. This analysis concluded that internal triangles of observations directly between consecutive wall brackets, had an internal angle consistently less than 180 degrees by 1 to 3 arcseconds. Internal triangles that included wall brackets, but this time, observed from the centerline traverse points, were both greater and lower than 180 degrees. This led to the elimination of observations along the left springline, as they appeared affected, possibly by LR, which caused the tunnel Control network to appear rotated (or oriented) towards the right.

Upon identifying the probable deleterious effects of LR on the Survey Control, the possible consequences this LR could have on the reliability of the Survey Control, and by extension, to the installation of the final structures at Shafts 2A and 3A, the Survey Control design was altered: the use of the brackets were subsequently limited to the weekly advance of the Survey Control, and supplemented by a centerline traverse complemented by gyro-theodolite observations every 450~600 meters (1,500 to 2,000 ft.), for increased reliability and accuracy.

The survey observations were reduced and adjusted in a least squares adjustment package (a Microsearch™ Three-Dimensional Adjustment package known as GEOLAB) on a daily basis, and the horizontal coordinates of the survey brackets were updated on a regular basis. Furthermore, as mentioned above, the observations along the springline were not used in those adjustment so as to not introduce additional unnecessary errors in the final adjusted survey results.

Throughout the duration of the tunneling operations, 11 gyro-azimuth observations using a DMT Gyromat 5000 were performed on the centerline traverse. These gyro-azimuth observations are a critical step of the Survey Control as they minimized the propagation of random errors and attenuate the influence of LR in the tunnel. The Gyromat 5000 was pre and post calibrated before and after each underground survey, on two intervisible surface Control points established using the DGPS surface Control network. Gyro-azimuth observations in the tunnel were performed in both directions for over 80 percent of the observations. This further allowed an appraisal of the effect of LR. In order to properly reduce gyro-observations, Observations must be performed on a geodetic datum (geo-referenced datum). Consequently, a surface Control, referenced to a proper geodetic datum, as mentioned in *Surface Control*, must be the basis of the project survey. When gyro-azimuth observations are properly reduced to the 'grid' azimuth or state plane azimuth, the average of the two azimuths should then minimize or cancel the effect of refraction. This is further depicted in Figure 9.

Gyro-azimuth observations were performed during weekends and when no other maintenance activities were being performed in the tunnel, to provide the best possible observing conditions, and the highest accuracy. When properly observed, and after calculation reduction, it appears that the Gyromat 5000 accuracy is within the 3 to 4 arcsecond range.

GYRO BASED AND HYDROSTATIC LEVEL TBM GUIDANCE SYSTEM

There are many different types of TBM guidance systems available for tunneling applications. However, due to small internal diameter of the BCSIS tunnel, and the length and congestion of the TBM ancillary support equipment, BCL was concerned by the probable sub-optimal performance that a more traditional theodolite or a laser based system could exhibit, if limited and at-times non-existent line-of-sight conditions were

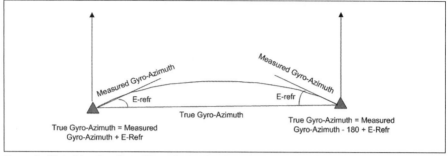

Figure 9. Minimizing the effect of lateral refraction through gyro-observations

encountered during tunneling operations. Consequently, a VMT gyro based guidance system seemed better adapted to the challenges ahead was selected for the guidance of the EPBM. The VMT Measuring While Drilling (MWD-II) Guidance System has been successfully used on directional drilling projects as well as previous tunnel projects, but these drives were typically smaller in length and bore fewer curves. BCL had concerns on how well this TBM guidance system would perform in a longer tunnel with numerous curves. Therefore, early in the project where long straight sections of the alignment would provide an acceptable line of sight, Blacklick Constructors elected to supplement the MWD-II guidance system with a proven Theodolite based guidance system, which ran concurrently with the MWD-II Gyro based system (this lower order Gyro is not to be confused with the high precision DMT Gyromat 5000).

Figure 10. Gyromat 5000 used the BCSIS Project

Initial results on the BCSIS indicated that the MWD-II system was in fact better suited for this project. It allowed uninterrupted tunneling operations with only minor (if any) disruptions during the operation. The only limitation of the MWD-II system is that it appeared to require regular adjustments to the DRIFT constant throughout the length of the operation to function satisfactorily. This DRIFT necessitated repeated unpredictable adjustments when certain conditions in the tunnel changed. These conditions, according to VMT, included changes in geological conditions, and changes in the direction of the alignment (i.e., going through curves as opposed to straight tangents). Although slightly cumbersome, this impediment did not significantly affect the operations on the Blacklick Project, as the tunneling operation was performed in two, ten hour shifts. This gave the surveyors a four hour window to update the position of the guidance system on a daily basis; and since tunneling progressed approximately 30~45 meters (100 to 150 ft.) daily, surveying and updating the MWD-II system had to be performed daily, regardless of this limitation. The Guidance System calculated position, with respect to the tunnel Survey Control, was verified daily, and typically resulted in adjustments of a few millimeters.

However, as mentioned above, whenever changing geological conditions were anticipated or encountered, or during changes in the azimuth of the DTA, significant variances between the MWD-II guidance system's calculated position and the actual position at which the tunnel Survey Control found the MWD-II system were observed; some differing by as much as 30 mm (1 ft.). Typically, such a large deviation prompted in a revision of the DRIFT parameter. Early on in the tunneling operation, the surveyors had observed that the proposed DRIFT value, which should be based on predemined factors, did not correspond to a realistic estimate of the actual drift of the guidance system. The MWD-II update procedure was subsequently modified to adjust the DRIFT by small increments when a large discrepancy between the calculated vs. actual positions were encountered, as to not overcompensate this seemingly unpredictable behavior.

ANALYSIS OF SURVEY DATA ON THE BCSIS PROJECT

From a survey perspective, one of the advantages of the survey data collected on the BCSIS Project, is that it allows an analysis of the entire tunnel Survey Control and its effect on the accuracy and the predictability of the breakthroughs at the Shafts 2A, 3A, and ultimately 4A. The survey data was reduced and adjusted on a regular basis in a three dimensional least squares adjustment, to analyze residuals of the observations, and eliminate outliers from having influence on the Survey Control.

The Microsearch™ least squares adjustment package incorporates all forms of survey observations and allows different weighting to be applied to various observations, while also incorporating Gyro-Azimuth Observations into the adjustment, allowing proper reductions (i.e., Laplace Corrections) to be applied for accuracy. As the tunneling operation progressed, and holethroughs occurred at Shafts 2A and then later at Shaft 3A, the surface Survey Control was transferred to the tunnel as discussed earlier in the *Control Transfer at Shafts* section of this paper, and the information collected was incorporated into the least squares adjustment.

Shortly after the tunnel hole-through into Shaft 4, in September 2018, the survey observations were combined, and different scenarios were perform using the actual data collected throughout the entire project to analyze the effectiveness of the methodology at key locations along the DTA, and establish whether some of the measures implemented on the project were actually necessities, or simply redundancies. The survey data collected was evaluated, withholding supplementary observations such as Survey Control transfer at Shaft 2A and 3A, and gyro-azimuth observations. The discrepancy between the tunnel Survey Control horizontal coordinates and the surface Controls transferred to the bottom of Shaft 4A was then compared. The results are presented on the graph below.

The results indicate that performing the observation of the wall brackets, with no gyro-azimuth observations, and without any survey ties to the surface, resulted in a (significant) difference of 609 mm (1.86 ft.) in horizontal coordinates, between the tunnel Survey Control and the surface Control transferred at the bottom of Shaft 4A (after 4.3 miles of tunneling).

A second scenario was developed using all Survey Control discussed in this paper, which included observing wall brackets (excluding direct observations between wall brackets), and centerline traverse with gyro-azimuth observations. This analysis resulted in a discrepancy of 67 mm (0.205 ft.).

Finally, a scenario in which the actual tunnel Control and its associated least squares adjustment were applied, including the use of all centerline survey observations, gyro-azimuth observations, and Control transfer surveys at Shafts 2A and 3A, resulted in in difference of 23 mm (0.069 ft.) over 4.3 miles of tunneling.

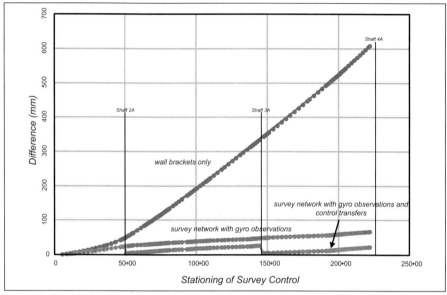

Figure 11. Graphical representation of the three scenarios of survey control

Figure 12. Hole-through on the BCSIS project

In each of the above scenarios, it appears that the discrepancy between the actual and the calculated tunnel coordinates appear to always favor the "right" side of the tunnel. This may further corroborate the possible influence of LR, and its effects on tunnel survey observations. When least squares adjustments were performed on a regular basis, the adjusted horizontal Survey Control was always adjusted "left"' of the tunnel alignment. This was also further observed after holethrough in Shaft 4A, when the final as-built of the last ring installed indicated a slight tendency to the "right" of the DTA.

CONCLUSIONS

The prescribed tolerances and construction specifications for the BCSIS required stringent survey accuracy. It should be noted that, in addition to the standard set of survey challenges they inherently present, smaller diameter tunnels are typically more susceptible to LR errors; and as such, the tunnel Survey Control design should be elaborated with particular attention and care, to mitigate or minimize this influence, as early as possible in the tunneling process.

Furthermore, the use of the Gyromat 5000, as well as the use of a pertinent tunnel survey network geometry proved very effective at minimizing LR, as well as abating or minimizing propagation of random error.

It is important, not just for tunnel surveyors, but for tunnel industry professionals in general, to understand the limitations and benefits of carefully elaborated tunnel survey networks, methods, and designs; the advantages of having access to proper survey equipment such as the Gyromat 5000, as well as acquiring sufficient knowledge on the proper establishment of surface Control, and the transfer of those Controls, to ensure the contractually prescribed tolerances are met, without any remedial work. Since the majority of tunnel projects are concluded with the installation of seals and other structures bearing strict tolerances in the receiving shafts, such strict tolerances are greatly affected by the adequacy of the tunnel survey design, its performance and ultimately, its prosecution.

REFERENCE

Robinson, G.L., Greening, W.J.T, DeKrom, P.W., Chrzanowski, A., Silver, E., Allen, G., Falk, M. 1995. Surface and Underground Geodetic Control for Superconducting Super Collider. *Journal of Surveying Engineering, American Society of Civil Engineering Division,* Vol. 121, No. 1.

PART  14

Pressure Face TBM Case Histories

Chairs
Darren VonPlaten
Traylor Bros., Inc.

Frank Huber
Metro Vancouver

Effectiveness of Risk Mitigation Strategies for Large Diameter One-Pass Tunnels in Mixed Ground from Ohio Canal Interceptor Tunnel Construction

Mike Wytrzyszczewski ▪ City of Akron Bureau of Engineering
Christopher Caruso ▪ McMillen Jacobs Associates
David Rendini ▪ Parsons
Geary Visca ▪ DLZ Ohio, Inc.
Dan Dobbels ▪ McMillen Jacobs Associates

ABSTRACT

The City of Akron, OH is implementing a $1.4 Billion CSO Long Term Control Plan, which includes the construction of a TBM-excavated 27-foot ID 6,212 LF tunnel through rock and soil, the Ohio Canal Interceptor Tunnel (OCIT). In this paper, major risk mitigation measures considered for tunnel excavation through soft ground, mixed face and hard rock will be reviewed from planning and preliminary design through completion of tunneling. Items to be discussed, include specification of reaches for open and closed mode tunneling, probing, management of potentially gassy ground conditions, and other measures. The effectiveness of these measures will be assessed retrospectively, and lessons learned will be presented.

PROJECT OVERVIEW

Akron's existing sewer system consists of combined sewers that convey both sanitary and storm flow in mid to large diameter interceptors. During dry weather flows, these larger interceptor sewers convey raw sewage through regulator structures called "racks" then into a smaller diameter sewer below to eventually be treated at Akron's sewage treatment plant or Water Reclamation Facility (WRF). When large rain events occur, the smaller sewers under these "racks" become overwhelmed with combined sewage and must be relieved through an overflow pipe to nearby bodies of water like the Cuyahoga River, Little Cuyahoga River and Ohio and Erie Canal.

As required by a 2009 Federal Consent Decree, The City of Akron developed a Long-Term Control Plan (LTCP) Update Report which in turn laid out the "roadmap" for Akron's CSO overall $1.2 billion-dollar program called "Akron Waterways Renewed" (AWR). The $185 million OCIT project is an integral component of the AWR citywide program. The intent of the OCIT project is to design and construct a conveyance and storage system to mitigate CSOs from several racks in the downtown area. The OCIT project is a part of the City of Akron's Combined Sewer Overflow (CSO) Long Term Control Plan (LTCP) Update, also known as "Akron Waterways Renewed!" (AWR)

In accordance with the LTCP requirements, the OCIT Project routes the combined sewer flows that are tributary to nine different Racks into proposed diversion structures and junction chambers, where the flow can be diverted to a new 27-foot diameter tunnel (OCIT).

The alignment for the OCIT and locations of major diversions and drop shafts are shown in Figure 1. The OCIT is approximately 6,212 feet in length with a finished inside diameter of 27 feet. The downstream end is at the connection point to the

Effectiveness of Risk Mitigation Strategies for Large Diameter One-Pass Tunnels 871

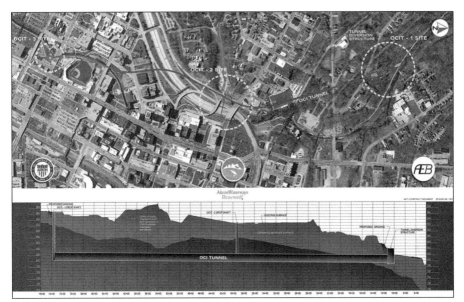

Figure 1. OCIT Alignment and basic geotechnical profile. Black denotes rock, and brown denotes soil. TBM launched from OCIT-1 site (right-most yellow circle) and retrieved from OCIT-3 drop shaft (left-most yellow circle).

Tunnel Diversion Structure (TDS) to the north of downtown Akron and adjacent to the Little Cuyahoga River. The upstream termination is at the OCIT-3 Drop Shaft, which is located on West Exchange St in downtown Akron across from Canal Park, a minor-league baseball facility and adjacent to the historic Ohio and Erie Canal.

Diversion structures are constructed near the existing "racks" to divert dry and wet weather flows to the OCIT. Dry weather flow is conveyed through the tunnel to the existing Little Cuyahoga Interceptor (LCI), which takes the wastewater to the Akron WRF. Wet weather flow up to a specific typical year (TY) storm event flow rate is stored in the OCIT. When wet weather flow exceeds the total available tunnel storage volume, flow will be discharged to the Little Cuyahoga River after passing through the Tunnel Diversion Structure (TDS) located at the downstream end of the OCIT. The OCIT will provide approximately 26.7 MG of storage. The dry weather flow conditions were evaluated for minimum velocities since they were the most critical flow conditions. Results from this evaluation showed that during average daily flow conditions the OCIT was achieving a minimum of 2 fps flow velocity for the average dry weather flow and in most cases was either exceeding or close to 3 fps during dry weather peak flow conditions. These goals were met for both conditions without the use of a cunette or steepening the slope of the tunnel.

The project was contracted under a design-bid-build methodology. The OCIT project final design began April 8, 2013 and 100% design was completed April 22, 2015. The project was advertised 5/14, 2015 and bid July 23, 2015 with a notice to proceed to the contractor of November 1, 2015. As part of a consent decree the achievement of full operation (AFO) date was established as December 31, 2018.

RISK MITIGATION STRATEGY

The overall strategy for identifying, eliminating, and mitigating risks on the OCIT project centered on taking advantage of Akron's optimizing the horizontal alignment of the tunnel to minimize unfavorable ground conditions to the extent possible and to minimize risk of surface disruption due to tunneling while still providing a gravity-drained system with minimal maintenance. Specification of tunnel boring machine (TBM) technologies, such as multi-mode design and compressed-air intervention capabilities, was also central to mitigating tunneling risks in a cost-effective manner.

Owner's Perspective on Risk Mitigation

The OCIT project is the largest single construction project undertaken by the City to date. As such, the project cost was a primary factor in the risk mitigation process. Additionally, completion of the OCIT project by the Consent Decree dates was a critical concern for the City because EPA fines would be assessed for each day of delay beyond the agreed AFO date. This would also have a negative impact with the regulatory agencies and public perception.

Safety of the public and the workers on the project were also top priorities for the City throughout the project and factored into every decision from planning through construction. The City also considered minimizing public impact and disruptions to businesses a high priority in the planning and execution of the OCIT project. To achieve this, the City undertook extensive public outreach and education efforts during the design and construction phases of the project. The City is also responsible for undertaking a variety of other public improvement projects concurrent with OCIT; minimizing disruption to those projects was also a concern.

Regarding minimizing risks arising from the OCIT construction contract, the City's strategy centered on transparency and continuous engagement of the Contractor. The City sought to minimize construction claims by conducting a comprehensive geotechnical investigation program during the planning and preliminary design phases and developing a balanced Geotechnical Baseline Report for inclusion in the OCIT Contract Documents. The project risk register was also shared with the Contractor to facilitate greater cooperation between the project team members in mitigating project risk.

Other specific concerns of the City included avoiding contaminated material in landfills along the alignment, maximizing usage of publicly owned land work sites and structures, keeping the Ohio and Erie Canal Towpath open and obtaining permits from various stakeholders impacted by the project's construction, including the National Park Service, USACE (United States Army Corp of Engineers), and Ohio EPA. These stakeholders were engaged from scope through construction.

Primary Risks Associated With OCIT Construction

The major risks that were identified by City and the Designer along with potential losses and primary mitigation measures are presented in Table 1. "Mitigation 1" represents a risk-avoidance or risk elimination strategy, where possible, and "Mitigation 2" represents a risk-management or risk-control strategy.

In the following sections, the mitigations presented in Table 1 are reviewed and discussed through the lens of their effectiveness on the OCIT project. Additionally, Contractor performance and specific issues encountered during construction are considered in a retrospective assessment of each risk mitigation measure's effectiveness on the project.

Table 1. OCIT construction risks, potential impacts, and mitigation measures

Risk	Potential Impacts	Mitigation 1	Mitigation 2
TBMs are difficult to optimize for soft ground, mixed face and rock conditions,	Low advance rates in some ground conditions lead to delays	Set horizontal alignment to minimize soft ground and mixed face reaches	Specify TBM capable of open and closed (pressure balance) mode operation
Settlement under railroads in soft-ground reach cannot be managed as anticipated	Settlement mitigation measures incur additional costs and delays and impact stakeholders	Set horizontal alignment to avoid railroad crossings in soft ground to the extent possible	Require TBM to be operated in closed mode Specify real-time settlement monitoring
Contaminated groundwater is encountered underneath landfill	Contaminated groundwater handling and treatment incurs additional costs and delays and is a safety hazard	Set alignments to clear the landfill	Require probing in open mode to determine if contaminated water is present. Specify a pre-excavation grouting program to minimize contact with contaminated water, if present.
High groundwater inflows encountered in rock	TBM stoppages to control groundwater lead to schedule delays	Require probing and pre-excavation grouting to control high groundwater inflows	Require TBM to be operated in closed mode
Gas is encountered during tunneling in rock	TBM stoppages to ventilate incur schedule delays	Specify Class 1 Div 1 TBM equipment to permit open mode excavation in gassy conditions	Require TBM to be operated` in closed mode when gas is encountered and specify Class 1 Div 1 for essential services only

HORIZONTAL AND VERTICAL ALIGNMENT OPTIMIZATION

Development of the final horizontal and vertical alignments took place during the planning and preliminary design phases of the OCIT project with the goals of achieving the hydraulic requirements of the project and addressing the City's concerns relative to cost, schedule, public impact, and risk mitigation as previously discussed. In setting the horizontal and vertical alignments, key factors included:

- Elevation of top of rock/rock cover
- Elevation of upstream and downstream connection points to permit a gravity-drained system
- Minimization of soft ground and mixed face tunneling reaches
- Availability of large City-owned parcels near the downstream connection point
- Use of existing Right-of-way (SR-59 "Innerbelt")
- Locations of existing sewers

It was the City's strong desire that the vertical alignment of the tunnel system be established so that it could drain by gravity into the existing interceptor system during dry weather and into the Little Cuyahoga River during wet weather, this avoiding long term costs associated with operating and maintaining a pump station at the downstream end of the tunnel.

A geotechnical investigation program was planned and executed to help evaluate tunnel alignment options. A key focus of the program was to evaluate top of rock elevation and rock mass quality. Based on this investigation program it was determined that it was hydraulically feasible for a gravity drained system to be built mostly in rock. However, constraints at the downstream end of the tunnel necessitated that a small portion of the alignment would pass through soil. These constraints were driven by mining site requirements, property requirements for the TDS and overflow outlet structure, and tie-in to the existing interceptor for conveyance to the Akron WRF. Due to the relative flexibility in the horizontal alignment, the soil and mixed face conditions were able to be minimized and the alignment only needed to pass beneath one sensitive feature at the ground surface, the Cuyahoga Valley Scenic Railway tracks, which is part of the Cuyahoga Valley National Park and is owned and operated by the National Park Service.

MULTI-MODE TBM—OPEN AND CLOSED MODE EXCAVATION

Several options were considered for facilitating TBM excavation through the short soft-ground and mixed face reaches at the start of the drive including specification of a multi-mode TBM and specification of ground improvement throughout the soft soil to permit excavation entirely in atmospheric conditions. The multi-mode machine that could operate in pressure balanced (i.e., closed) mode in soil and mixed face condition and non-pressure balanced (i.e., open) mode in rock conditions was ultimately selected rather than ground improvement in soil and mixed face. This became apparent during the design phase because ground improvement work would have required extensive surface impacts beneath 2 public streets and some sections of the alignment passed through wooded areas that would be difficult to access.

Design Considerations

Thorough characterization of ground conditions was required to allow development of appropriate TBM requirements, including limits of closed and open mode operation specified by the Designer and selection of specific TBM means and methods by the contractor. The geotechnical investigation program confirmed that rock quality was generally good in the zones deeper than approximately 1 tunnel diameter below top of rock, but rock quality was variable, even poor in some instances, within 1 tunnel diameter of top of rock. A layer of completely to highly weathered shale, up to 20 feet thick in some locations, drove the specification of closed mode operation in reaches where rock cover was less than one tunnel diameter (~30 feet) to minimize risk of ground loss or excessive surface settlement.

The Contract Documents divided the entire 6,212 LF tunnel drive into three separate reaches and 2 low rock cover zones to clearly define where the Contractor was to operate the TBM in open and closed mode. Table 2 describes the reaches and provides descriptors, TBM mode, and linear footage per the OCIT Contract Documents.

Construction Considerations

The Contractor elected to excavate the tunnel using a 30-foot diameter Crossover TBM manufactured by The Robbins Company (TRC) for the entirety of the drive. This TBM met the project specifications for multi-mode excavation, a manlock for compressed air excavation chamber entry, probe drilling and pre-excavation grouting capabilities up to 200 feet in front of the machine from both sides.

Table 2. OCIT contract tunnel reaches and low rock cover zones

Reach	Description	TBM Mode	LF
1	Soil	Closed	213
2	Mixed Face	Closed	600
3	Low Rock Cover	Closed	800
—	—	Open	2,350
—	Low Rock Cover	Closed	550
—	—	Open	1,698
Total Tunnel Length			6,212

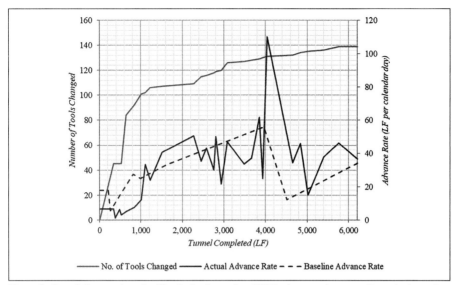

Figure 2. Cumulative number of cutting tools changed and TBM advance rate versus linear footage of tunnel completed

Prior to TBM launch, the Contractor did not propose to deviate from the contract requirements for open mode and closed mode reaches. The Contractor also did not anticipate performing any cutterhead maintenance until attaining Reach 3. However, after launch the Contractor found they could not effectively condition the ground through the mixed face ground zone (Reach 2) and had considerably lower advance rates than anticipated in their baseline schedule. This difficulty in conditioning the ground compounded with the TBM crew's learning curve, exacerbated the low production. *Through Reaches 1 and 2, the first 813 LF of the drive, the Contractor averaged an advance rate of 5.30 LF per calendar day, which is substantially lower than their baseline advance rate of 19.8 LF per calendar day.* In addition to low advance rates, the Contractor also stopped for unplanned cutterhead maintenance 5 times before reaching the first planned maintenance stop at the start of Reach 3, approximately 847 LF into the drive. To this point, the Contractor had changed 92 cutting tools, which is approximately 66 percent of all the tool changes that would take place on the entire 6,212 LF OCIT tunnel drive. Figure 2 shows the actual and baseline advance rates as a function of total completed linear footage of tunnel compared to the cumulative number of cutting tools changed.

Because of their low advance rates and unplanned cutterhead maintenance, the Contractor elected to switch to open mode operation at the start of Reach 3 within the low cover zone instead of 800 LF into Reach 3 at the end of the low cover zone. Rock cover was expected to be less than 5 feet at the location where the Contractor switched to open mode, and the condition of the rock cover was expected to be highly weathered.

The City and the Construction Management Team (CMT) collaborated with the Contractor to weigh the risks of switching to open mode in low rock cover versus remaining in closed mode and continuing with low advance rates and accelerated tool wear. Since the alignment passed through a wooded area where surface impacts were considered low, the City and the CMT decided to permit the Contractor to switch to open mode at their own risk instead of issuing a stop work order and enforcing the closed mode requirement. This decision was made between the City and the CMT and considered the Contractor's performance to date in closed mode, the schedule delays that had already been accrued to date because of poor performance in closed mode, wear and tear on critical TBM components including articulation and main bearing seals from closed mode operation, and the absence of sensitive features at the surface where the Contractor requested the switch to open mode.

As expected, the Contractor was able to sustain considerably higher advance rates after switching to open mode, >30 LF per calendar day compared to 5–7 LF per calendar day. A zone of sand was encountered near the crown of the excavation shortly after switching to open. The Contractor over-excavated this sand zone while operating in open made and promptly switched back to closed mode for the 50 feet following. The surface area above the over-excavation was fenced off, but exploratory surface probing and grouting were not undertaken due to inaccessibility of the area and safety concerns for the drilling subcontractor. This over-excavation generated a surface depression about a month later, which the Contractor immediately backfilled with flowable fill material. Note that since the surface depression was in an inaccessible wooded area, it resulted in no impacts to the public or any surface structures.

The Contractor, the City, and the CMT understood there was a risk for over-excavation and surface impacts from running the TBM in open mode with low cover through weathered rock, but the risk of potentially damaging a critical component of the TBM by continuing in closed mode was deemed to be far more costly to the project than an isolated event of lost ground in undeveloped land. Through the remainder of the drive, this same reasoning heavily influenced later situations where enforcement of the contract requirements for open vs closed mode operation was debated between the City and the CMT. By the end of the drive, the Contractor had operated in open mode for 86% of the total linear footage compared to 65% required in the Contract (see Table 3).

Table 3. Contract vs actual TBM operation modes

	TBM Mode	LF	% of Total
Contract	Open	4,048	65%
	Closed	2,163	35%
Actual	Open	5,348	86%
	Closed	863	14%

PROBING AND PRE-EXCAVATION GROUTING

The geotechnical investigation program showed that the Cuyahoga Shale formation, which constituted the majority of the tunneling ground for the OCIT, was unfavorable jointed in certain areas and may create high groundwater inflows during tunneling. While the probability of hitting significant water inflows (> 30 GPM) was considered low, the impacts to the project of such an event were considered high due to the sensitivity of the project schedule and the size of the TBM. In addition, the alignment passed underneath a landfill. Even though the vertical alignment was such that there was about one tunnel diameter of cover underneath the landfill, the probing requirement provided a "belt and suspenders" approach to ensuring that the Contractor could identify potential leachate migration from the landfill early and give the project team time to react and mitigate as needed.

Design Considerations

To proactively avoid a high groundwater inflow event and the associated potential delays, a probing program was required in the Contract throughout open-mode operation. The Contractor was expected to meet the following requirements in his probing scheme:

- Probe the entirety of the open-mode reaches
- Maximum single probe length was 200 linear feet
- Minimum overlap between subsequent probes was 30 feet
- TBM must be capable of probing from both sides (not concurrently)
- Probes shall be advanced through openings in the cutterhead with a look-out angle of 3–5 degrees.

Since the tunnel drive was short and the impacts of a high groundwater inflow were considered high, the continuous probing program was considered an appropriate proactive risk mitigation to include in the Contract. Specific provisions of the probing requirements, like the 30-foot minimum overlap, were chosen to minimize risk of leaving un-probed windows in the ground while attempting to accommodate the Contractor's anticipated production and maintenance cycles. The 200 LF max probe and 30-foot overlap requirements permitted approximately 4 days of uninterrupted production, assuming 40LF per day advance rate and maintenance stops once per week.

Construction Considerations

Problems with TBM procurement created a schedule slip in the TBM's launch of over 7 months, prompting the Contractor to press for relaxation of the probing requirement to help recover schedule. Prior to TBM launch, the Contractor anticipated that minimal water would be encountered during tunnel excavation because of their experience sinking shafts on the project through the Cuyahoga Shale and encountering minimal water. The Contractor also proposed that the TBM could switch to closed mode operation if high groundwater inflows were encountered instead of maintaining a continuous probe.

Once the Contractor began tunneling and found that closed mode operation was both slow and taxing on the TBM components, his preference shifted to avoiding closed mode operation wherever possible. Additionally, the City and the CMT agreed that dry conditions at the shafts did not necessarily mean dry conditions throughout the tunnel drive, thus the probing was still considered the best mitigation for leachates beneath the landfill, the City and CMT agreed not to relax the probing requirements and to see if it really slowed the Contractor down.

As the Contractor worked through the learning curve in open mode and established his production and maintenance cycles, it became clear that maintaining the continuous probe was not affecting the critical path of tunnel excavation as other TBM and conveyor maintenance activities, both planned and unplanned, created sufficient windows of downtime for the TBM crew to advance the probes without impacting production. Even though minimal water was encountered throughout the open-mode operation and the inflow limits to trigger pre-excavation grouting were never hit, the continuous probing requirement was still considered a worthwhile risk mitigation measure on this project.

On a longer drive or for a smaller bore diameter, the continuous probe requirement might consume excessive schedule and cost. In these cases, other measures of mitigating water inflow risks might be preferred, such as specifying higher pumping capabilities on the TBM or specifying probes for specific high-risk sections of the drive only.

MANAGEMENT OF POTENTIALLY GASSY GROUND

Shale formations are known for creating gassy tunneling conditions. A study of prior construction experiences in the Cuyahoga Shale suggested that gassy conditions might be possible for the OCIT tunnel. Management of the gas risk was done through several means during both design and construction.

Design Considerations

Gas monitoring within the test borings was performed as part of the geotechnical investigation program, and methane gas was encountered in one boring near the OCIT-3 shaft at the upstream end of the tunnel. This observation combined with the fact that gas had historically been encountered in the Akron the design team and the City to agree that all tunnels on the project should be classified as "Potentially Gassy" as defined in OSHA 3115-06R 2003.

The strategy for creating Contract requirements related to gas sought to ensure the Contractor's safety and provide means for dealing with isolated, small gas events while recognizing that results of the investigation program did not suggest that gas would be an ongoing problem for the Contractor. It was determined that requiring the Contractor to provide a fully Class 1 Div 1 compliant TBM was impractical and would provide little benefit beyond a Class 1 Div 2 compliant TBM on this project. Thus, the Contract Documents required the TBM to be capable of operating ventilation and essential (life safety) systems in gassy conditions while all other non-essential systems would deactivate in a gas event.

Additionally, the Contract Documents required continuous monitoring for explosive gases in the excavation chamber during open mode operation. If levels exceeded 20% of the lower explosive limit (LEL), the Contractor had to stop and ventilate the chamber until the levels returned below 20% LEL. A separate allowance was built into the contract to compensate the Contractor for downtime due to gas.

Construction Considerations

The Contractor generally did not encounter gas while excavating in open mode, but he did report gas on several occasions during unplanned entries into the excavation chamber in Reach 2 under atmospheric conditions. In the first instance, the Contractor felt the gas levels could not be brought down to safe levels by ventilating the chamber alone and that compressed air was required to create safe conditions in the chamber for entry. Following this incident, gas was encountered during two other similar entries under atmospheric conditions, but the Contractor reported the chamber could be effectively ventilated and compressed air was not needed for explosive gas mitigation. The first incident resulted in submittal of a differing site condition claim by the Contractor, while the following two incidents were compensated to the Contractor through the contract allowance for downtime due to gas.

CONCLUSIONS

Schedule

Schedule was identified at the start of the project as a primary risk, and significant delays were accumulated through the completion of tunneling. A summary of facts related to the slip in the Contractor's tunneling schedule are presented in Table 4. In addition, the OCIT baseline schedule is compared to the actual schedule in the line graphic in Figure 3.

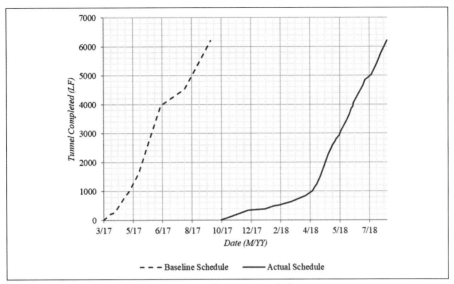

Figure 3. Baseline vs actual tunnel completion of excavation schedule

Once the Contractor was able to operate the TBM in full rock face open mode he was able to achieve his baseline production rates. Despite a slip in the tunnel excavation completion date of approximately 11 months, the Contractor has been able to rearrange the sequence of work for the project and work with the City and the CMT to reduce total delay to Substantial Completion. While the AFO date will be missed, the City is working with the EPA to permit the delay to AFO after consideration of the City's performance in meeting the Consent Decree date.

Table 4. Baseline vs. actual tunneling schedule statistics

Tunneling Schedule Item	Baseline (months)	Actual (months)
Duration	6.7	10.4
Start date slip	—	7.4
End date slip	—	11.0

Collaboration

The success of the risk mitigation strategy for the OCIT project hinged on transparency and flexibility between the City, the CMT, and the Contractor. As the Contractor began tunneling and new limitations came to light, specifically with the Contractor's ability to operate the TBM in closed mode and the significant delays incurred during TBM procurement, the CMT and the City were challenged to re-evaluate some contract requirements and adjust to the Contractor in some cases. The successful evaluation and development of the GBR was also a key to success for this project. Each of the parties clearly understood the constraints on the others and had the common goal of delivering a quality product safely and with minimal disruption to the public. In conclusion, barring the schedule delay in AFO and substantial completion that are currently being negotiated between the City and the EPA, the tunneling work is viewed by all project parties and the public as a success.

EPB Clogging Through Mixed Transitional Ground, Lessons Learned at the Ohio Canal Interceptor Tunnel, Akron, OH

Elisa Comis ▪ McMillen Jacobs Associates
Peter Raleigh ▪ McMillen Jacobs Associates
Wayne Gyorgak ▪ McMillen Jacobs Associates

ABSTRACT

In Earth Pressure Balance (EPB) tunneling, the correct approach to soil conditioning can make a considerable difference for the success of a tunneling project—both in highly permeable grounds as well as in sticky clays. To maintain the required face, support a delicate balance has to be obtained between the quality of the excavated muck within the excavation chamber and the cutterhead clogging. Soil conditioning laboratory tests are often completed during the planning phase prior to TBM excavation in an effort to develop a baseline for soil conditioning, reduce risks, costs and improve EPB performance. This paper will present the observations made during the excavation of the Ohio Canal Interceptor Tunnel in Akron, Ohio, USA particularly with regards to very slow advanced rates, rapid tool wear and clogging of the cutterhead. It also discusses the limits of desktop versus empirical clogging evaluations with respect to actual clogging experienced during tunneling and observations made after test carried out with the Hobart Mixer Methodology.

INTRODUCTION

In EPB tunneling, the correct choice and use of soil conditioners can make a considerable difference for the success of a tunneling project—both in highly permeable grounds as well as in sticky clays.

To date, there is no standard procedure to evaluate clogging in either the field or laboratory, however stickiness and clogging potential has been well documented and investigated by Thewes et al. While this research is not new, contractors continue to struggle in the field despite having adopted several countermeasures which include: the use of water to reduce stickiness, reduction in advancement rate, use of compressed air to provide face support, design of the TBM cutterhead to avoid creating large lumps of soil, and of course through the use of conditioners and additives.

The Ohio Canal Interceptor Tunnel (OCIT) Project involved the construction of a 6,200 ft long conveyance and storage tunnel with a finished inside diameter of 27 ft (8.23 m) to control combined sewer overflows for several regulators in the downtown Akron area. The OCIT Project was awarded to Kenny/Obayashi, a Joint Venture with a Notice to Proceed dated November 4, 2015.

The section of the GBR related to the OCIT identified three major reaches that were defined as distinctly different ground conditions: Reach 1 which primarily consisted of soft ground sandy soils, Reach 2 was a transitionary zone with soft ground overlying bedrock, characterized as "mixed ground conditions," and Reach 3 which was comprised of bedrock with two sections of low rock cover. The tunnel was excavated using a 30 ft (9.26 m) bore dual mode type "Crossover" (XRE) Rock/EPB Tunnel Boring

Machine (TBM), manufactured and supplied by The Robbins Company. The TBM design features adopted for the expected operating conditions are fully discussed in the paper, Comis, E. et al., 2017. As a combined rock/EPB TBM the machine could operate in Closed Mode (EPB) or Open Mode.

The working definition for Closed Mode is operation of the TBM with active face support above the ambient pressure and Open Mode is operation with some or no active face support below ambient pressure. The choice between the two modes of operation is dependent upon the stability of the excavation face, hence in granular materials below the ground water table Closed Mode was specified (Reach 1 and 2) and in the rock (Reach 3) open mode was permitted except in the Low Bedrock Cover Zones.

The TBM launched in October 2017 and completed the drive in August 2018. Reach 1 and 2, were a total of 810 ft (247 m) and presented a challenge for the contractor as these reaches were plagued by low penetration rate, high cutter consumption, difficulties in maintaining face pressure and low TBM utilization. Reach 3, was a total of 5,400 ft (1,647 m), and in contrast, distinguished by high TBM utilization and consistent advance rates.

This paper will discuss the soil conditioning tests that were carried out prior the beginning of the mining and the problems observed during Reach 1 and 2, mainly related to the "mixed ground conditions" or what is better described as a soil to rock transition zone.

SOIL CONDITOINING ASSESSMENT

As excavated materials enter the excavation chamber through the cutterhead they need to be rapidly conditioned from relatively stable in-bank material to an excavated material which is consistent with the EPB method including the following characteristics:

- Reasonable fluidity to pass through the muck handling system beginning at the cutterhead openings
- A level of homogeneity whereby water may not easily pipe or pass through open pores
- Sufficient density so that support pressure may be imparted to the excavation face

These characteristics of excavated soil are critical to maintaining active face support pressure and the metered discharge of excavated soil at a rate equal to the advancement of the TBM.

In the case of OCIT, a soil conditioning assessment was conducted by BASF prior to the commencement of tunneling to identify the baseline conditioners necessary to:

- Enhance "plastic flow" condition of the excavated materials
- Reduce permeability and promote homogeneity of the excavated materials
- Prevent adhesion of excavated materials to the cutterhead structure and openings

Soil conditioning is determined by the parameters defined in EFNARC (2005), as follows:

- Foam Expansion Ratio (FER): is the ratio between the volume of foam at working pressure and the volume of the solution. The range of FER is typically between 5 and 30
- Foam Injection Ratio (FIR): is the ratio between the injected volume of foam at working pressure and the in-bank volume of material to be excavated. The range of FIR is typically between 10% and 80%
- Concentration of agent (C_f): is the volume of agent used as a percentage of water in the preparation of the conditioning mixture. The range of C_f is typically between 0.5% and 5%

The anticipated ground conditions for the OCIT were the following:

- Reach 1 consisted primarily of Unit 3 Silty Sand (brown, silty fine sand with trace amounts of clay) interbedded with Unit 4 Silt (brown silt or clayey silt) underlain by Glacial Till (brown to gray clay with varying amounts of gravel, cobbles, and boulders)
- Reach 2 ranged from a full face of Units 3, 4, and weathered rock to a full face of bedrock
- Reach 3 consisted of Shale and Siltstone, with minor amounts of Sandstone.

Soil conditioning testing varied according to the soil type. For the sandy soils, testing for mobility was made using a tradition slump test method. For the silty soils, testing included adhesion (using a variation of the Langmaack adhesion test (NAT 2000, Langmaack)), slump, static flow, dynamic flow and rheometry. There were no tests carried out for the mixed ground conditions or for the shale.

Based on these tests the recommendation for the excavation was the following:

- Soft Ground Closed Mode (Reach 1): A polymer-based foaming agent. If more than an expected portion of clay was found in the excavated soil, an anti-clay polymer could be added.
- Mixed Face Closed Mode (Reach 2): The same polymer-based foaming agent was considered suitable for the mixed face ground with the change of ground conditions closely monitored and conditioners to be adjusted as follows;
  – If the ratio of fine particles in the ground increased, water could be injected into the chamber to increase fluidity and to adjust viscosity of the excavated materials.
  – If ratio of fine particles in the ground decreased, foam, polymer or bentonite could be injected into the chamber to keep necessary fluidity and viscosity of the excavated materials. Liquid type polymers were also considered suitable for this purpose.
- Rock (Reach 3) Open Mode: The final setup for Reach 2 ground conditions were considered sufficient for the bedrock zone. Additionally, the excavation chamber could be filled by a mix of bentonite and sand before entering the second Low Bedrock Cover Zone to obtain necessary earth pressure balance for Closed Mode operation. The mix could be determined with the actual soil conditions when the TBM entered the low cover zone.

The suggested soil conditioning parameters were FER 15, FIR varying from 30 to 80% depending on moisture content, and C_f of 3.0.

IMPACT OF MIXED TRANSITIONAL GROUND IMPACT ON TBM PERFORMANCES

The OCIT TBM weekly production for the whole drive is shown in Figure 1. An average weekly advance rate of 34.6 ft/week was recorded for reach 1 and 2, and 242 ft/week for reach 3. Reach 1 and 2 were plagued by low penetration rate, high cutter consumption, and difficulties in maintaining face pressure.

Lack of production (high thrust with low penetration rate) prompted several cutterhead inspections, which showed material packed in the rear of the housings as well as the buckets. Buckets were filled with clay and sand, easily removable with a pressure washer. Most of the outer radial buckets were hard packed and needed the use of pneumatic tools to remove hard packed material. Material was also found compacted in the disk cutter housings preventing rotation of the disk cutters and causing their failure due to flat-spotting. The cutterhead continued to clog developing high temperature in the plenum, baking the clay and sand material and steaming muck which was observed being conveyed outside the tunnel (Figure 2).

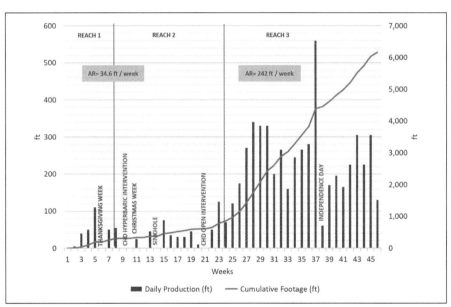

Figure 1. OCIT weekly production

Figure 2. Clogging inside the TBM cutterhead and steam in the tunnel

Figure 3. Foam quality vs generator mediums

The initial foam produced was of low-quality when checked at the foam generators on the TBM with a generally liquid consistency and no real "stand-up" time. The conditioner supplier ran multiple tests to detect if there was some defect with the product or the with mixing ratios being recommended and the results of these tests did not show a problem in the laboratory. Next, the foam generation medium in the generators was evaluated and modified from the initial setup of loosely packed larger metal cylinders to a tightly packed plastic top hat and then to a tightly packed steel wool. The result was a foam of good consistency and quality (Figure 3). It was also decided to add water directly in the excavation chamber. Despite getting both the desired quality and quantity of foam, the cutterhead still experienced clogging.

Focus then shifted to the foam parameters and the adjustment of the various ratios. The initial soil conditioning evaluation suggested an FER of 15 with C_f between 3% and 5%. However, even with a FIR that has been as high as 200–300% the clogging was not mitigated. The FER was then reduced to 2 and the FIR set between 60% and 80%. This finally allowed the completion of Reach 2 however at an unsatisfactorily

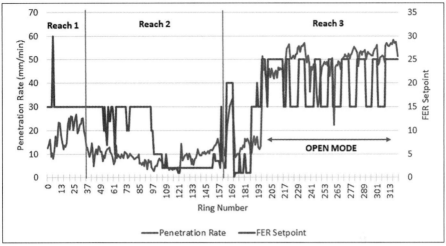

Figure 4. TBM penetration rate vs. FER setting

slow pace, the penetration rate increased from 5 mm/min to 15 min/min suggesting that this approach to treatment was improving the characteristics of the excavated materials. However, the increase in penetration rate coincided with an increase in rock ratio at the excavation face (Figure 4) making that finding somewhat dubious.

Lastly even if Reach 3 was characterized by higher penetration rate and high TBM utilization, and considered overall a successful drive, clogging of the cutterhead buckets could be still observed at the breakthrough (Figure 5).

HORBART MIXER METHODOLOGY TESTS

Following the breakthrough, and in light of the TBM drive experience it was decided to carry out additional laboratory tests to characterize the mixed ground condition

Figure 5. CHD at the breakthrough

encountered in Reach 2 using the Hobart Mixer Methodology by Zumsteg and Puzrin (2012). The method empirically estimates the clogging potential of soils by using a HOBART mixer and a B-flat beater. Initially, the mixer was equipped with a 20-liter capacity bowl. It was later changed to a smaller mixer with 5-liter capacity (Zumsteg, Puzrin and Anagnostou, 2016).

The proposed methodology consists of initially measuring the total weight of the soil, mixing it for three minutes with the speed 1 (around 100 rpm), and, afterwards, measuring the weight of the soil that remained stuck in the beater, obtaining the parameter λ.

Table 1. Sample mixture proportions

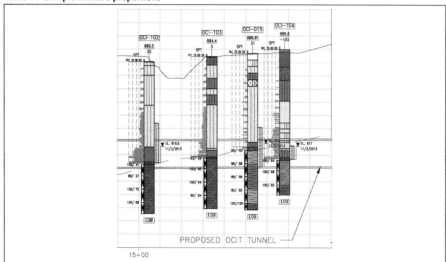

	Sample 1	Sample 2	Sample 3	Sample 4
Outwash & Lacustrine (Silty Sand)	40%	30%	0	0
Glacial Till Deposit	40%	20%	20%	0
Shale	20%	50%	80%	100%

This empirical stickiness parameter (λ) is the ratio of the weight of the soil stuck in the beater over the weight of the entire sample (Equation 1). Depending on the λ value obtained, the tested sample can be classified as low ($\lambda < 0.2$), medium (λ between 0.2 and 0.4) or high ($\lambda > 0.4$) clogging potential. Samples from the main geological formations were collected, pulverized and mixed in the proportion showed in Table 1. Atterberg limits (liquid and plastic limits) were measured and determined for each mixed sample.

With the addition of water the test showed that in general the increase in quantity of shale increased the stickiness and the clogging potential.

Samples were then conditioned with a polymer-based foaming agent with c.f. 3 and FER 15 and FER 5. Samples 2, 3, and 4 plots are shown in Figure 7.

In both simulated and observed mixed conditions the use of the foam generally helped to reduce the stickiness. Interesting to note based on the graphics below is how the FER influenced the empirical stickiness. When mixed ground conditions are simulated the lower FER reduced the stickiness to the optimum range for EPBM operation. In contrast to the mixed ground condition results, in the 100% shale condition an *increase* in FER resulted in a decrease in stickiness.

The test results were in accordance with the observations made during TBM excavation i.e., a reduction in FER was required to increase the TBM penetration rate in mixed ground.

In contrast, when excavating a full-face of shale, the FER was increased to 15 to reduce stickiness which resulted a consequent reduction of torque and thrust required to advance the TBM.

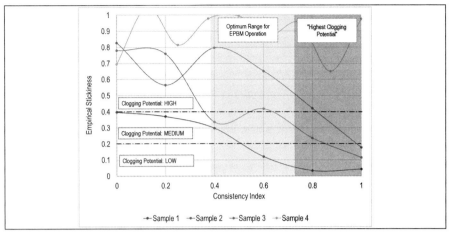

Figure 6. Empirical stickiness vs consistency index—Plot 1

Further, an increase of the water content can be beneficial to reduce stickiness, optimally it should be introduced in the tool gap i.e., the space created between the cutterhead structure and the intact soil, where the greatest amount of mixing occurs during excavation, as it was not effective being solely introduced into the excavation chamber during the advance.

CONCLUSION

The OCIT project involved excavation in dry sandy soil and transitional mixed ground. In these conditions the TBM experienced low penetration rates caused by the clogging of the cutterhead. The foam agent selection and injection/expansion parameters suggested by the preliminary soil conditioning assessment proved to be ineffective and the contractor was forced to perform experimental trials during production excavation in a bid to improve TBM advance rates. To date, there is no standard procedure to evaluate clogging under operational conditions, however stickiness and clogging potential has been well documented and investigated by Thewes et al. While this research is not new, contractors continue to struggle in the field. The Hobart mixing technology has potential to become an effective way to quickly evaluate the clogging potential and to help in selecting the optimum conditioning agent and for setting foam injection and expansion parameters, especially in transitional mixed ground conditions where these parameters may have to be adjusted on an advance by advance basis. In mixed transitional ground conditions, additional injection ports on the cutterhead would be beneficial to inject a combination of both water and foam conditioner. Further, the authors suggest including a more extensive stickiness and clogging potential analysis using Hobart mixing technology in preparing the preliminary soil conditioning assessments for mixed transitional ground, including the conditioning of clayey rock conditions such as the shale encountered at OCIT.

ACKNOWLEDGMENTS

The Authors would like to acknowledge BASF for supporting material and manpower to execute the testing made in their laboratory in Cleveland. Special thanks to Joel Joaquin and Jeffrey Champa for the time they dedicated to this research.

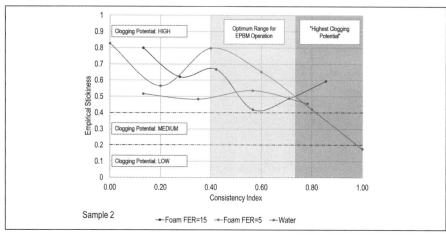
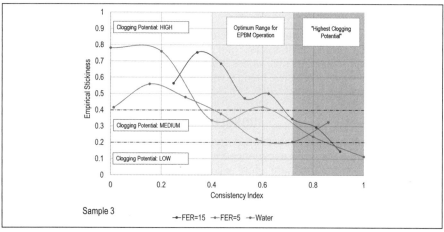
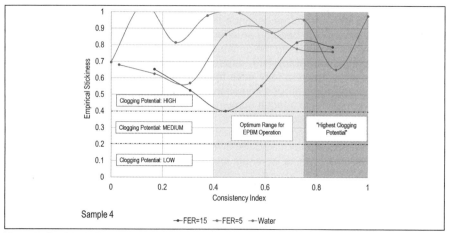

Figure 7. Empirical stickiness vs consistency index—Plot 2, 3, and 4

REFERENCES

Comis, E. & Chastka, D. 2017. *Design and Implementation of a Large-Diameter, Dual-Mode "Crossover" TBM for the Akron Ohio Canal Interceptor Tunnel; Proc. RETC 2017*: 488–497.

Comulada, M., Maidl, U., Silva, M.A.P., Aguiar, G. & Ferreira, A. 2016. *Experiences gained in heterogeneous ground conditions at the twin tube EPB shield tunnels in São Paulo Metro Line 5; Proc. ITA 2016 WTC, San Francisco*: 1–11.

Della Valle, N. 2001. *Boring through a rock-soil interface in Singapore; Proc. RETC 2001*: 633–645.

EFNARC, 2005. *Specification and Guidelines for the Use of Specialist Products for Mechanized Tunneling (TBM) in Soft Ground and Hard Rock.*

Langmaack, L. & Lee, K.F. 2016. Difficult ground conditions? Use the right chemicals! Chances–limits–requirements. *Tunnelling and Underground Space Technology 57*: 112–121.

Oliveira, D.G.G. & Diederichs, M. 2016. *TBM interaction with soil-rock transitional ground; Proc. TAC 2016, Annual Conference, Ottawa*: 1–8.

Thewes, M. & Burger, W. 2004. Clogging risks for TBM drives in clay. *Tunnels & Tunnelling International, June 2004*: 28–31.

Zhao, J., Gong, Q.M., & Eisensten, Z. 2007. Tunnelling through a frequently changing and mixed ground: a case history in Singapore. *Tunnelling and Underground Space Technology 22*: 388–400.

Zumsteg R., Puzrin A.M. 2012. Stickiness and adhesion of conditioned clay pastes. *Tunnelling and Underground Space Technology 31*: 86–96.

Planning, Design and Construction of the Regional Connector Bored Tunnels—An EPB Tunneling Case History

Richard McLane ▪ Traylor Bros., Inc.
William Hansmire ▪ WSP
Ron Drake ▪ EPC Consultants, Inc.
Derek Penrice ▪ Mott-MacDonald
Darren von Platen ▪ Traylor Bros., Inc.

ABSTRACT

Los Angeles County Metropolitan Transportation Authority's (Metro) Regional Connector Transit Corridor (RCTC) is a $1.75 billion (US) design-build, underground light rail project, connecting two existing rail lines (Metro Contract No. C0980). The project alignment runs through the heart of Downtown Los Angeles, with construction ongoing in the city's Little Tokyo, Bunker Hill, and Financial Districts. It will connect the existing Metro Gold Line to the Blue and Expo Lines, providing one-seat rides in the North-South direction from Azusa to Long Beach and East-West from East Los Angeles to Santa Monica, allowing passengers to bypass Union Station for transfers to the Red and Purple Lines.

This paper presents a case history of planning, design and construction efforts to achieve the successful completion of the Bored Tunnels. Further, this paper outlines the pro-active partnership between the owner, Metro, and the Design Builder, Regional Connector Constructors (RCC, a joint Venture of Skanska USA and Traylor Bros., Inc.) with design services provided by Mott-MacDonald (MM) and Hatch, to address project schedule and technical challenges, including various differing site conditions and critical settlement control done in low-cover EPB tunneling.

INTRODUCTION

RCTC will be the first tunnel project for Los Angeles County Metropolitan Transportation Authority (LA Metro) to connect between existing light rail operating systems at either ends of the project. The 1.9 mile underground light rail in downtown Los Angeles shown in Figure 1 connects to the Gold Line on the east and the Blue and Expo Lines on the west, and includes three cut and cover stations: Little Tokyo/Arts District, Historic Broadway, and Grand Ave Arts/Bunker Hill. In addition to EPB tunneling, the alignment includes a track crossover cavern mined by sequential excavation methods (SEM), and 2,100 ft of cut and cover guideway construction. The $1.75 billion project utilizes the design-build delivery method commencing with a notice to proceed in July 2014 and with completion expected during the winter of 2021/2022.

Types of Construction

The several types of construction required along the project alignment are shown in more detail in Figure 2. The tunneling portion of the project consists of a total of 5,795 feet of twin 21-ft-diameter tunnels, initially four, but then reduced to three cross passages, and a 300-foot-long cavern excavated and supported using the SEM method. The tunnel alignment is generally below the ground water table and the

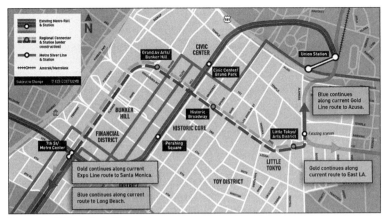

Figure 1. RCTC project alignment

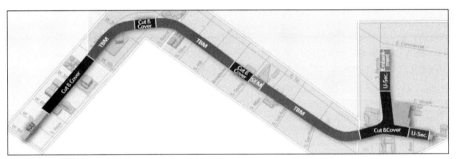

Figure 2. RCTC alignment elements

ground conditions range from alluvium soils with the potential for boulders to siltstone of the Fernando formation, all with the potential for methane and hydrogen sulfide gas. Pressure balance tunneling utilizing either an EPB or SPB TBM with double-gasketed, precast concrete segmental lining system were required. The tunnel depth (cover) generally ranges from 25 ft in alluvium soils under the buildings and structures in the Little Tokyo district to approximately 120 ft in the Fernando formation.

Some of the specific tunneling challenges found along the alignment and presented by this paper include:

- 4th Street Overpass foundation piles
- Tiebacks known to be within the tunnel alignment
- Low cover for launch
- 583-ft radius curves
- Steep grades up to 4.5%
- Existing Metro Red Line tunnels undercrossing with less than 7 ft clearance
- Highly sensitive receptors to ground borne noise and vibration (Disney Concert Hall and the Colburn School of Music)

Original Plan Challenges

The alignment of the Regional Connector under streets, buildings, utilities and facilities in urban downtown Los Angeles required a plan to minimize impacts to the city, businesses, and the public. Settlement control and protection of structures was a critical project requirement. A large number of high-rise buildings had been constructed with deep basements with excavations laterally supported with tie-backs extending into the project's subsurface alignment. Consequently, the guideway design involved a 1350 ft reach of cut and cover construction along Flower Street to the tie-in point at 7th Street/Metro Center Station to avoid the risk of tunneling through the tie-backs.

The original project plan proposed tunneling concurrently with cut and cover station construction such that the stations would be excavated prior to walking the TBM through the stations. Due to complexities and delays associated with utility relocations for the cut and cover stations, the original plan was changed considerably to re-sequence work and implement different techniques to minimize impacts to the project's overall schedule. The changed conditions and encountered differing site conditions required a collaborative effort between Metro's project staff and the contractor's team to address the schedule delays and technical challenges. To mitigate impacts, Metro's initiatives included acquisition of TBM performance monitoring system that permitted continuous reporting of tunneling status, providing as a change, installation of state-of-the-art horizontal inclinometer geotechnical instrumentation, and apportioning risk to adopt schedule resequencing and delay recovery measures.

PROJECT GEOLOGY

The project Geotechnical Baseline Report (GBR) identified four primary geological units: (1) Artificial Fill, (2) Alluvium, (3) Colluvium, and (4) Fernando Formation. The Alluvium is further subdivided into two sub-units, referred to as Qal1 and Qal2. The regional groundwater table existed 20–80 feet below the surface, with the possibility of local perched groundwater in the alluvial deposits above the highly impermeable Fernando Formation.

The Qal1 Alluvium is a variable formation with roughly equivalent occurrence of silty sand, well-graded sand, and lean clay as shown in Figure 3. Qal2 is also a variable formation, beneath the geologically younger Qal1. As shown in Figure 4, it is predominantly comprised of well graded sand, with more frequent occurrence of poorly-graded sand, and fewer deposits of silty sand than Qal1. Qal2 is coarser, with most samples having less than 10% passing the #200 sieve and typically denser than the Qal1 formation.

The Fernando Formation is described by the GBR as "a poorly bedded to massive clayey siltstone to silty claystone that is poorly cemented and extremely weak to very weak." (Hansmire et al. 2013). The GBR makes a distinction between weathered and fresh siltstone where fresh siltstone is characterized by much higher blow-counts based on sampler penetration tests. The Fernando formation has been known to contain sporadic calcium-carbonate cemented materials, or concretions. Concretions found during the previous Metro Red-Line tunnel excavation ranged in thickness from inches, up to nominally 2 ft blocks. While the Fernando Formation is often referred to as "bedrock" in local geotechnical documents—as it does exhibit substantial strength in its fresh condition and is typically stable on a vertical excavation face—it will rapidly degrade to a soft a highly plastic clayey-silt when disturbed in the presence of water. The Fernando formation is composed almost entirely of fines passing the #200 sieve. Figure 5 illustrates the character of the Fernando formation in terms of plasticity and the potential for EPB clogging.

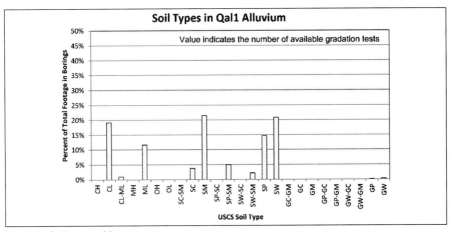

Figure 3. Qal1 composition

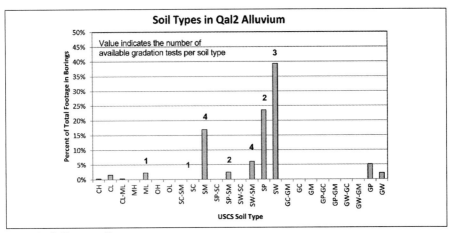

Figure 4. Qal2 composition

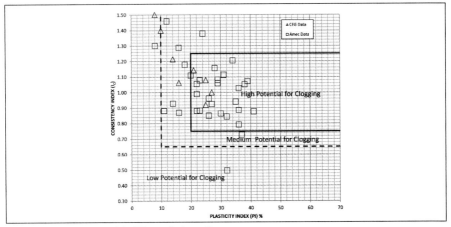

Figure 5. Clogging potential of Fernando formation

For the purpose of the GBR, the tunnel alignment is divided by the limits of the project's cut and cover excavations into three reaches, numbered from East to West. For most of the tunnel alignment, a full face of Fernando Formation was anticipated. The exceptions to this are at the beginning of the drive in Reach 1, where a full face of Qal2 Alluvium was expected, and a short stretch of Reach 3, where Qal1 was expected in the crown. The tunnel alignment was entirely beneath the water table except at the beginning of Reach1 where the face of excavation was fully drained, and Reach 3, where the water table rested slightly below the crown of the planned tunnel alignment. Reach 2 was extremely consistent, entirely below the water-table, with a full face of fresh Fernando Formation for the entire length of the Reach.

TUNNEL DESIGN REQUIREMENTS

Metro Rail Design Criteria (MRDC) recognizes that tunnel design is not explicitly addressed in 'standard' design codes, which are typically developed specifically for use with buildings or bridges. Correspondingly MRDC provides guidance on the codes, standards and procedures to be followed in the design of precast concrete tunnel linings (PCTL) for temporary, static, and seismic design conditions. The MRDC also includes requirements to demonstrate that the PCTL would remain durable and serviceable both during and after a design fire event. Compliance with MRDC must be demonstrated to obtain Metro design approvals.

Temporary Loadings

PCTL's are subject to several temporary loading conditions—handling loads, TBM thrust loads, annular backfill grouting pressures, and forces arising from imperfect liner erection, which can represent controlling conditions for the PCTL design. Handling loads from segment lifting and stacking occur at a time when the PCTL concrete is immature. TBM thrust loads are highly concentrated and can be applied eccentric to the PCTL, resulting in large concentrated stresses within the lining. The magnitude of these forces and application thereof are not defined in the MRDC, as they are means and methods dependent. It was the responsibility of RCC's engineer to prepare the design to accommodate these temporary conditions. PCTL temporary loading criteria were agreed with RCC—such as 2,000 psi minimum strength for demolding and lifting of segments, whereas TBM thrust loads and the potential eccentricity of their application were derived from the physical characteristics of the TBM and the PCTL at 500 kips typical thrust per shoe and 600 kips exceptional thrust per shoe, which could each be applied at a maximum eccentricity of 1.25 inches. These temporary loading conditions—intensity, application, load factors etc., were documented in a project-specific Basis of Design document appended to the RCC Design Work Plan submittal for Metro review and approval.

Static Design

For long term in-service loading conditions, MRDC requires the design to follow Load and Resistance Factor Design (LRFD), utilizing the Federal Highway Administration (FHWA) FHWA-NHI-09-010 Technical Manual for Design and Construction of Road Tunnels—Civil Elements, and the Caltrans Bridge Design Specifications, which is the California State amended version of the American Association of State Highway and Transportation Officials Bridge Design Specifications. The static conditions typically do not control the design.

Seismic Design

MRDC prescribes a two-level approach to seismic design, comprising an Operating Design Earthquake (ODE) with a 150-year return period, which could reasonably be expected to occur over the design life of the project and Maximum Design Earthquake (MDE), with a 2,500-year return period, which has a small probability of occurrence over the life of the structure. MRDC requires that the PCTL perform in an elastic manner for the ODE event but allows inelastic behavior in the MDE event provided there is no collapse, life safety is maintained, and the tunnel will be repairable.

For seismic analysis, MRDC permits the use of empirical closed form solutions and numerical modeling, with recent modifications to MRDC prescribing pseudo-dynamic time history analysis as a minimum requirement for the latter. MRDC further states the pseudo-dynamic analysis should be performed by performing a wave scattering analysis to determine the time-histories of seismic displacements of the tunnels from a two-dimensional site-response analysis, followed by a soil-structure interaction analysis whereby the time-histories of the displacements are applied to the structure in a static analysis to determine the maximum structural response. However, several concerns with this approach were identified relative to the PCTL design:

- The soil-structure interaction analysis, comprising a beam-and-spring model would not be able to accurately capture effects caused by skewed and convex radial joints.
- Results from the beam-and-spring models would be superimposed on initial static stresses. As such, non-linear response of Mohr-Coulomb material that may be predicted during earthquake motion would not be captured.
- Scattering analysis or beam-and-spring model would be unlikely to include effect of any ground plasticity in the limited pillar between the twin tunnels.
- Scattering and beam-spring analysis cannot be applied to locations where significant 3D effects exist, such as at cross passages

These concerns were put forward to Metro and their Technical Advisory Panel (TAP), and a revised analysis approach was agreed:

- Perform a full dynamic time-history analysis of a one-dimensional (1D) soil column subject to ground motion using SHAKE software, to obtain the maximum free-field displacement response of the 1D soil column when subject to outcropping motion applied at the base of the column.
- Perform pseudo-static soil-structure interaction analysis. The envelope of maximum free-field shear strain responses are applied as a monotonically increasing static displacement at the vertical boundaries of three-dimensional (3D) soil-structure interaction models to obtain the structural response of the tunnels under these imposed shear strains.
- Design linings in accordance with MRDC, based on the obtained response.

To adequately envelope the stresses in the PCTL the seismic analysis should consider the built-out condition of the tunnel lining, including the emergency walkway and invert slab, which can locally limit the seismic ovaling of the PCTL.

Fire Performance Design

MRDC requires that the tunnel lining consider the effects of fire on the lining, and that the lining be able to withstand the heat of a Metro specified fire intensity and duration without loss of structural integrity, with protection measures to be provided by concrete cover to reinforcement and a suitable concrete mix design. Given the relatively thin PCTL cross-section (10.5 inches), section loss arising from spalling in a fire event could have potentially significant implications for the performance of the tunnel. However, as specific methods of achieving the fire performance of the PCTL are not prescribed in the MRDC, demonstrating compliance with this requirement proved to be the most contentious aspect of the PCTL design.

General and Technical Performance Requirements

In addition to MRDC, the General and Technical Performance Requirements of the design-build Contract Documents provided supplemental, mandatory criteria to be incorporated into the PCTL design:

- Minimum internal diameter of tunnel of 18 foot-1 inches
- Minimum PCTL thickness of 10½ inches
- Minimum concrete strength of 6,500 psi
- PCTL to have radial convex to convex joints to allow flexing during a seismic event
- PCTL to be have double gasketed system to assure no leakage under static and seismic conditions

FINAL DESIGN

Geometric Design

The PCTL geometry was established during the project bid phase. As noted above, several requirements were prescribed by the contract documents. To facilitate more installation options with the key above tunnel springline, a left/right ring configuration was adopted, with each ring consisting of six segments: four 67.5-degree parallelograms, one 67.5-degree trapezoid and a 22.5-degree key. The ring has a nominal length of 5-feet, with a 3-inch taper provided (±1.5 inches) to accommodate alignment curvature as low as 583 feet. The radial segment joints have a convex shape with a radius of 12 feet.

Metro requires that segments be double-gasketed—with continuous EPDM gaskets near both the intrados and extrados faces of segment—to provide grout confinement in the event remedial sealing is required. To limit the travel path of remedial grout, or groundwater which breaches the first gasket row, Metro also mandated the use of cross-gaskets. To accommodate all possible segment positions and ensure proper gasket mating, 16 cross-gaskets were required on the circumferential face of each ring.

The key segment includes an additional M24 bolt across the circumferential Joint to the previous ring. This was requested by RCC based upon previous industry experience, as a redundant safety measure to ensure the key stays in place during shield jack retraction. While only one bolt pocket was included on the circumferential joint, 16 bolt receivers were required to accommodate segment clocking. Ring-to-ring connection was otherwise provided by friction dowels.

Steel Reinforcement Design

The preferred PCTL configuration—ring geometry, gasket, and 'furniture' type was established early in the design process, based upon RCC preferences and previous experience. This enabled the PCTL geometric drawings to be finalized as an

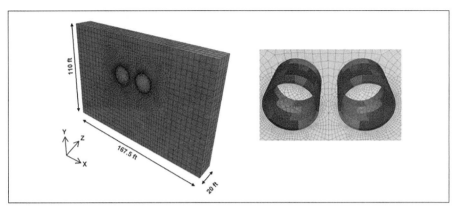

Figure 6. PCTL finite difference model

Table 1. RCTC representative geological locations chosen for analysis

Analysis Location	Rationale
Reach 1, Station 19+00	Shallowest section fully embedded in the Fernando Formation and representative of cross passage at Sta. 24+00.
Reach 2, Station 36+80	Deepest section in the Fernando Formation and representative of cross passage at Sta. 41+10.
Reach 3, Station 62+50	Section representative of cross passages at this station and at Sta. 69+00.
Reach 3, Station 73+00	Mixed face condition in Fernando Formation/Alluvium.
Reach 3, Station 76+80	Shallowest section in alluvium.

advance partial design unit, in turn accommodating the timely ordering of segment molds. Thereafter the final design was primarily focused upon establishing the PCTL reinforcement requirements.

The determination of the PCTL primary reinforcement was performed using a combination of two- and three-dimensional (2D and 3D) ground-structure interaction modeling involving the Fast Lagrangian Analysis of Continuum (FLAC) software. Ground behavior was evaluated using an elastic-plastic stress-strain relationship in conjunction with Mohr-Coulomb failure criterion. The PCTL was modeled as linear elastic, with properties based upon the gross concrete section, and the radial joints between segments were modeled as pins. The tunnel lining-ground interface was modeled as linear elastic with Mohr-Coulomb shear strength.

To envelope the long-term service conditions in terms of tunnel depth and geologic conditions, analysis was performed at five locations encompassing the range of conditions representative of project conditions for design analysis as described in Table 1.

The MRDC mandated use of LRFD required the development of 'average' geotechnical parameters, to be used in the PCTL design, rather than a more traditional approach of enveloping the design with maximum/minimum soil parameters. However, at each location the static analysis considered both a "most representative" and a "worst credible" scenario of ground parameters, with the worst credible condition considered as an 'extreme' event, with corresponding load and resistance factors. As such, the worst credible ground conditions were not applied in conjunction with the MDE seismic event, as an overly onerous design condition would have ensued. The design also considered significant variations in groundwater elevation, considering a design

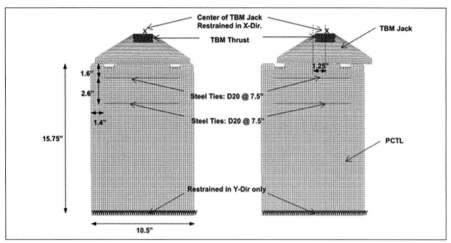

Figure 7. PCTL circumferential joint evaluation

ground water level, a historical high groundwater elevation, as required by MRDC, to address the possibility of future major changes in groundwater elevation, and a hypothetical low groundwater level due to potential adjacent dewatering.

Beyond the FLAC modeling, additional forces in the PCTL arising from the skewed joint configuration were estimated using structural analysis software STAAD.Pro and combined with the force outputs from the FLAC modeling. The numerical modeling was supplemented by Empirical Methods as required by MRDC. Hand calculations were also prepared to demonstrate the adequacy of bolted and doweled connections, gaskets, tie bars and the shear strength of the segments.

Due to the significant variation in the topography and geology over the extent of the TBM drive, two separate lining designs were initially developed, taking advantage of the greater confinement provided at depth within the Fernando Formation to economize the PCTL reinforcement requirements. Subsequent to the development of the two PCTL designs, a third PCTL design was introduced within the footprint of Japanese Village Plaza (JVP) area near the TBM launch location at Metro's direction to increase the MRDC minimum surcharge of 400 psf to 1,000 psf over the JVP footprint to accommodate potential future development at this site.

Analysis of the circumferential and radial joints was performed to determine additional local reinforcement requirements to ensure that the PCTL can sustain the applied TBM thrust loads and radial compressive stresses respectively. Two-dimensional (2D) finite difference analyses using FLAC 7.0 were used to assess the structural capacity of the joints. The typical finite difference model mesh for the circumferential joint is shown in Figure 7. Within the model, the PCTL is restrained at its bottom boundary in the vertical direction only. This allows a crack to develop and open anywhere within the PCTL. The TBM jack is restrained in the horizontal direction at one point at the centerline of the jack, allowing the jack to rotate freely about its centerline. Compression-only frictional interface elements were provided at the jack/PCTL interface with an assumed friction angle of 45°. Steel ties in the PCTL were explicitly modelled using cable elements. The FLAC strain-softening constitutive model was implemented to allow the properties of the concrete to vary more realistically as functions of deviatoric plastic strains.

Table 2. Results of numerical modeling for circumferential joint assessment under normal (typical) and eccentric (exceptional)

Eccentricity	TBM Thrust Condition	TBM Thrust (kip/shoe)	Conc. Stress (psi)	Rebar Stress (ksi)	Crack Width (in)
0.0"	Typical	510	2,787	4.1	0.0
0.0"	Exceptional	600	4,335	42.9	0.028
1.25"	Typical	510	6,380	58.1	0.042
1.25"	Exceptional	600	5,504	48.5	0.052

Table 3. Reinforcement demand by PCTL type

PCTL Type	Primary Reinforcement	Secondary Reinforcement	Circumferential Joint	Radial Joint
Type 1	14 D20 wires per face (A_s 0.199 inch^2)	D11 wires at 6" maximum spacing each face (A_s 0.110 inch^2)	2 rows D20 wires at 7.5" maximum spacing each face (A_s 0.199 inch^2)	2 rows D20 wires at 4" maximum spacing each face (A_s 0.199 inch^2)
Type 2	14 D14 wires per face (A_s 0.140 inch^2)			
Type 3	14 D28 wires per face (A_s 0.280 inch^2)			

With concentrically applied thrust load and typical TBM thrust the PCTL is predicted to remain un-cracked. A crack of limited width is predicted to develop with the exceptional TBM thrust load, but the steel ties will provide adequate control of the crack. A similar cracked condition was predicted for the typical TBM thrust load with 1.25-inch eccentricity. Damage to the PCTL was predicted under the exceptional TBM thrust load. However, as this load represents an extreme case with a low probability of occurrence, which is also temporary in nature, the potential risk of local concrete crushing was deemed acceptable, with the steel ties provided at the circumferential joint serving to maintain the stability of the segment. However, under this condition, some post-installation repair of the PCTL would be required. In the field, cracking occurred mainly due to solid rubber corners for the gaskets and minimally due to thrust loads from the TBM.

Similar modeling was performed for the radial joints. Instead of the TBM thrust load, tunnel lining load from the ground in the form of the axial compressive stresses from the primary numerical modeling for the PCTL were introduced into the joint model. The maximum unfactored axial compressive force was predicted to be 566 kips/ring and 867 kips/ring for the non-seismic and seismic load cases, respectively. The PCTL was predicted to remain uncracked under non-seismic load conditions. Under seismic load cases, limited cracking of the segment was predicted to occur, with an estimated maximum crack width of 0.027 inch. No concrete crushing was predicted. Reinforcement stresses were well within yield strength and the ties helped in controlling the cracking and maintaining the structural integrity of the segments.

Reinforcement demand for the 3 PCTL types is summarized in Table 3. All reinforcement within the PCTL was Grade 80. The primary, secondary and radial joint reinforcement demand was controlled by either long term static or seismic (MDE) conditions. The circumferential joint demand was dictated by the TBM thrust load.

The reinforcement arrangement for the Type 2 PCTL is shown in Figure 8. A similar bar arrangement was maintained for the other PCTL types, though in each case the reinforcement layout was modified by the fabricator to provide a more uniform primary bar spacing.

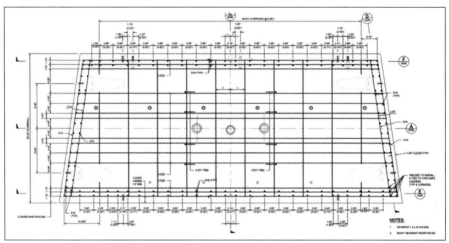

Figure 8. Typical PCTL reinforcement layout

A number of supplemental analyses were performed at critical locations on the alignment where the bored tunnels passed in close proximity to the Metro Red Line Crossing, the 2nd Street Tunnel and the Fourth Avenue Bridge Structure, but in each case the focus of these analysis was on ensuring that the tunneling did not result in adverse impacts to this critical infrastructure and to confirm the stability of the narrow pillar between the TBM bores, which varied along the alignment from approximately 10 feet to a little as four feet, as opposed to developing yet another PCTL design. In each case 3D-modeling using FLAC, with the existing structures explicitly modeled, demonstrated negligible effects of the tunnel construction. There was no need for another specially designed segment.

Fire Design

As stated previously, the most contentious aspect of the PCTL design was the fire design. The contract scope of work established a medium fire growth rate as a basis of design, which differed significantly from the super-fast (arson) fire growth specified in the MRDC Fire/Life Safety Criteria. The requirement to design for the medium fire growth rate did not include key criteria including the fire intensity and duration. As a consequence, Mott-MacDonald (MM) was left to assume these criteria using the peak train fire heat release rate and time of tenability respectively. MM also assumed the station ventilation fans would be operating, which is reasonable in the case of a tunnel fire, which would limit the peak PCTL temperature. MM ultimately performed several analyses to demonstrate the satisfactory performance of the PCTL during and after a 'design' fire event.

To improve PCTL fire performance, a Metro directed change required the inclusion of polypropylene microfiber in the PCTL concrete mix. The microfiber does not prevent spalling from prolonged exposure of the PCTL to a fire event but can help mitigate explosive spalling that may occur during the first 20–30 minutes of a fire when the PCTL is exposed to rapid temperature rise. During this time period the microfiber melts, creating a network of microcracks that permits the movement of steam through the concrete, relieving the stresses created by the steam generation and helping mitigate against explosive spalling. Accurate determination of the minimum fiber dosage can be established by large scale fire testing of the PCTL. As the testing is time consuming and segment production was fast approaching, the dosage rate had to be established from precedent. Comparable LRT projects commonly specify a fiber dosage of 1.7 lb/cy. As research suggests that dosages of more than 1.7 lb/cy can have an adverse effect on concrete workability and strength, necessitating revisions to mix design, a dosage of 1.7 lb/cy was adopted, with fibers specified to be between 32–54 microns in diameter and between ½-inch and ¾-inch in length for maximum performance.

BUILDING AND UTILITIES PROTECTION

RCTC required an extensive instrumentation program to minimize settlement risks and to monitor excavation performance using a variety of geotechnical instruments and other settlement monitoring controls. MRDC defined settlement limitations, which were incorporated into an extensive program of instrumentation that included Multi Point Borehole Extensometers (MPBX), Building Monitoring Points (BMP), Deep Surface Settlement Points (DSSP), Ground Surface Settlement Points (GSSP), Inclinometers, liquid levels, Deep Bench Marks, observation wells, Piezometers, strain gauges, tiltmeters, and utility monitoring points (UMP). The instruments were placed strategically along the project alignment to maximize the ability to monitor ground and structure movement. Sensors on these instruments combined with survey checks produced data that was documented in a web-based GIS and instrumentation system (ICSGIS) that provided real time data collection and the ability for the project team to monitor performance of settlement control during excavation on the entire project.

Action and maximum levels for angular distortion and total settlement/heave for buildings, utilities and other structures were defined in the zone of influence for the TBM tunnels and are summarized in Table 4. The action and maximum levels for settlement/heave directly above each tunnel, as measured by MPBX's are summarized in Table 5.

A TBM action plan was developed to address the EPB tunneling methods and parameters (discussed later) and methodology to respond to action levels during tunnel operations.

Specific alignment sections required focused geotechnical instrumentation and monitoring. The most critical zones included the tunnel launch and first 500 feet from the Little Tokyo Station, crossing under existing Metro Red Line tunnels, and avoiding piles supporting the 4th Street Bridge.

Table 4. Action & maximum levels for total settlement or heave

	Action Level	Maximum Level
Angular distortion*	1/1000	1/600
Total Settlement/ Heave (in.)	0.25	0.5

*Average settlement slope or slope between building walls or columns which ever less.

Table 5. Action & maximum levels for settlement or heave of MPBX

	Action Level	Maximum Level
3 ft. above tunnel (in.)	0.5	0.75

Compensation and Permeation Grouting

The vertical and horizontal tunnel alignment for the first 500 feet of tunnel drive commencing at the Little Tokyo station posed significant challenges for the tunnel launch and advance through alluvium soils in a downward sloping curve with little cover under buildings in Japanese Village Plaza. See Figure 9. There was minimal cover (starting at less than 15 feet) from the tunnel crown to building footings.

Metro prescribed permeation and compensation grouting measures to mitigate any settlement due to tunneling under the buildings in this initial tunnel reach. As shown

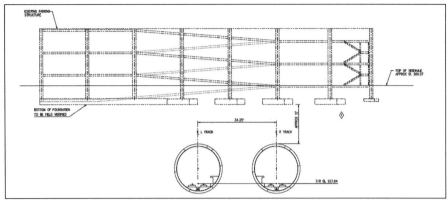

Figure 9. TBM tunnels to be constructed below the JVP parking garage

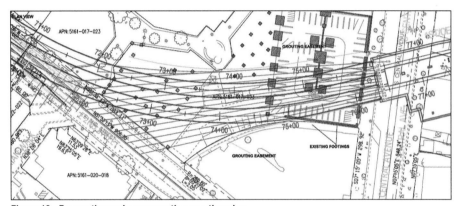

Figure 10. Permeation and compensation grouting plan

in Figure 10, a total of 13 compensation grout pipes averaging 300 feet in length were installed in a directionally drilled, fanned array below the buildings approximately 8 feet above the tunnel crown. Due to access, feasibility and easement limitations, permeation grouting was only performed under one of the Little Tokyo buildings. See Figure 11.

Horizontal Inclinometer

Due to the risk of ground loss or heave in the alluvium soils under sensitive buildings in the Little Tokyo area, more controls were identified by the project team to help monitor ground movement (settlement/heave). A horizontal inclinometer was installed approximately 3 to 5 feet above each tunnel crown for the first 400 feet of tunnel below the compensation grout pipes. The tool used was part of the Shape Array Accelerometer (SAA) line of instruments, manufactured by Measurand Inc., which consisted of multiple longitudinal sensors spaced 1 meter apart resulting in a Longitudinal Settlement Sensor (LSS). The two SAA instruments were installed inside Sleeve Port Grout Pipes (SPGP), drilled with the same directional drill rig as used for the compensation grouting. The pipes were installed above the tunnel centerlines, and the instruments were installed within a 1-inch plastic pipes shoved inside the SPGPs.

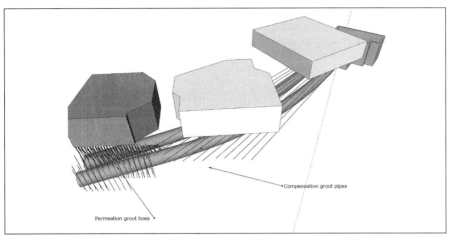

Figure 11. Rendering of grouting fixtures

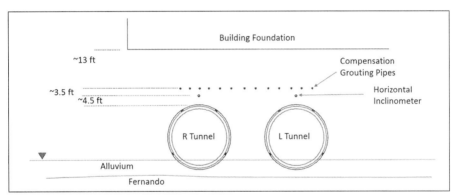

Figure 12. Location of SAAs

The SAA was used to lower the risk of ground loss and building settlement damage through advanced TBM performance observation. It provided near real time measurements every meter above the tunnels similar to the data collected from the bottom anchor of MPBX instruments. The data was used to help monitor and adjust EPB pressure, bentonite injection around the shield, and to monitor and adjust pressure of backfill grout. The SAA data also provided information to set the bounds and triggers for compensation grouting to fill potential voids caused by ground loss.

A TBM action plan was devised by RCC to set the pressure limits for EPB, steering gap bentonite and backfill grout to

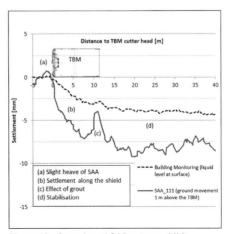

Figure 13. Snapshot of SAA output exhibiting real-time reactivity to TBM's operational parameters

address the difficult low-cover tunneling through alluvium soils under buildings. The plan was adjusted as needed to respond to ground movement data provided by the collection of instrumentation data.

All the instrumentation measurements and TBM performance data was integrated into sophisticated dash boards which the project team used to manage the EPB tunneling operation. The use of the horizontal inclinometer (SAA) provided unique measurements of ground movement directly above the tunnels and facilitated a very successful tunnel launch and drive through the critical reach in the alluvium soils under the JVP buildings.

GENERAL APPROACH TO BORED TUNNEL CONSTRUCTION

TBM Selection

The technical requirements from the contract required a pressurized face TBM, either Earth Pressure Balance (EPB) or Slurry Pressure Balance (SPB). The GBR showed that 90% of the tunnel would be driven through the finely graded, and highly plastic Fernando Formation. While a SPB Machine may have offered some advantages in the Coarse Qal2 formations encountered at the very beginning of Reach 1, it would certainly struggle with the fines over the majority of the drive. RCC elected to choose the technology which would be best suited to deal with the Fernando Formation, and settled on an EPB Machine.

In fact, the TBM used on the Regional Connector was one of the original TBMs used on the Metro Gold Line Eastside Extension in 2005, (HK serial number S-298). Its specific design elements made it well suited to the geologic conditions and potentially gassy tunnel classification; however, some modifications were required to the TBM and trailing gear for the TBM to navigate the very tight, 583 ft radius curves. These modifications included the addition of a copy-cutter to the cutting wheel, a re-designed articulation joint with increased stroke at the middle and forward shields, and modifications to the screw conveyors.

Site Setup and Logistics

The contract provided a construction staging area for tunneling, referred to as the Mangrove Property, adjacent Metro's Integrated Project Management Office, and enclosed by the current Metro Gold Line alignment as it rounds the corner of Alameda and 1st St. The staging area afforded the tunnel contractor some operational advantages:

1. A relatively large space for staging tunnel operations, compared to the very small work area provided for excavation of the future Little Tokyo/Arts District Station.
2. Access for Muck and Segment Hauling because traffic was removed from busier intersections, with easier access to highways.
3. An operational site more isolated from the businesses and residences of Little Tokyo.
4. A long open-cut excavation, which would facilitate a long-mode machine launch

In order to take full advantage of the site, the contract required the project team to design and construct an additional excavation, called the "TBM Launch Pit," beneath the intersection of Alameda and 1st St. and the existing Gold Line Tracks, connecting the Mangrove property to the location of the Little Tokyo Station Excavation. Site Set-up also involved a critical Metro service shut-down to re-route the existing Gold Line tracks onto a shoofly over the excavation. This work, and all the associated utility interface would need to be competed in an expedited manner to facilitate a timely tunnel launch.

One of the key features of the operational set-up was the choice of a muck-conveyor over more traditional rail mounted muck boxes. With the relatively long and shallow excavation at Mangrove, an economical incline conveyor could be used to continuously and efficiently extract muck. Further, the tunnel profile has grades of up to 4.5%, thus making it nearly impossible to excavate with boxes. Despite the relatively short mined distance, the efficiencies gained from simplified train traffic and reduced crane demand proved an overall benefit to the project, especially considering that a single tunnel muck conveyor system could service both tunnel drives.

The constraints of a small urban worksite produced the need for an efficient use of underground space. The two component backfill grout was batched and pumped down the entire length of the tunnel from a central grout batch plant. To save space on the surface, the plant was located at the bottom of the Little Tokyo station excavation, along with the ground conditioning plant, and bulk material storage for both plants. The take-up unit for the continuous tunnel conveyor was installed in the bottom of the TBM Launch Pit. The use of excavations for installation of plants and equipment left the most surface space possible for muck and segment storage, as well as staging for vent-line and other utilities.

Schedule

The combination of the potential for rapid excavation rates over relatively short tunnel drive distances, and the difficulties associated with locating and supporting utilities and traffic above complex cut and cover excavations in an urban setting led to several scheduling challenges. While both Metro and RCC took many proactive measures to mitigate these schedule risks, the project team was, and is still, constantly challenged

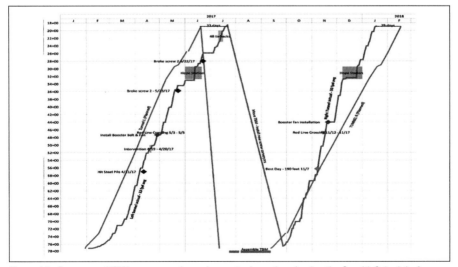

Figure 14. Summary of TBM progress, planned vs. actual, produced using the GraphicSchedule App

to adapt the project schedule to project conditions as they evolve to maintain the planned delivery dates.

Many of these schedule challenges took root with the addition of advance utility relocation scope to RCC's design-build contract. Early in the life of the C0980 contract, the project team realized that difficulties experienced with advance utility relocation contracts by another Metro contractor would lead to overall project delays. An agreement was forged to transfer the scope to the RCC's design-build team to expedite solutions for the utility relocation problems.

Resequencing of Bored Tunnel Construction

Originally, the first drive of bored tunnels was to be constructed after excavation of the Broadway Station was completed, with the second drive after completion of the crossover cavern excavation. In order to expedite the overall project schedule while clearing utilities at Broadway, Metro issued a change to drive both bored tunnels through the unexcavated Broadway Station and crossover cavern and, when the time came, to remove the tunnel liner as the station and cavern excavations progressed; eliminating a station crossing and adding over 1,200 feet of TBM-driven distance to the project scope.

Progress and Production Rates

The first tunnel drive was completed in approximately 25 weeks, with an average production rate of approximately 52 feet per working day (excluding startup). The majority of tunneling for this drive was performed on a 2 × 10hr shift schedule, with 5 production days per week. Noted challenges to production were the difficult launch in the dense sands of the Qal2 alluvium, an unexpected steel pile strike and related mechanical failures, and 48 unexpected steel tie-backs encountered in Reach 3 (described in further detail below). The first drive was completed 32 days behind schedule projections, mainly due to the Differing Site Conditions experienced.

Tracking tunneling production and accommodating the associated delays, the project team undertook intense planning and preparations to expedite the retrieval and re-launch of the TBM for the second drive. Planning efforts paid off, and—despite the added step of screw conveyor refurbishment due to the steel pile and tieback strikes—the machine was re-launched for the second drive approximately 10 weeks after completion of the first drive, with roughly half of the 32-day delay recovered during the re-launch activities.

With all cut and cover excavations complete, tie-backs cleared, and crews trained, the project team sought further schedule recovery by implementing an additional production shift; for 3 × 8hr shifts, with 5 productions days per week. Experience with the difficulties of the Qal2 during the first drive also led to some ground conditioning improvements, which resulted in speedier start-up to production. By the end of the 6th

Figure 15. Mangrove staging area on the brink of "muck-bound"; the combined result of weather closures at land-fills and high TBM production rates

Planning, Design and Construction of the Regional Connector Bored Tunnels

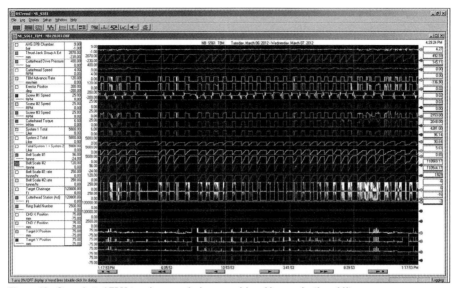

Figure 16. Snapshot of TBM trend screen during record-breaking production shifts

week of the second drive, all 32 days of delay accumulated during the first drive had been recovered.

The 7th week of the drive marked the best production week, with 750 feet of advance over 15 × 8hr shifts. Tuesday, 7-Nov-2017, of the same week, was the project's best day, with 192 feet driven over 3 × 8hr shifts. Mid-way through the best week of the second drive, the TBM passed the location of the steel pile strike from the first drive.

The second drive was completed in 17 weeks, 29 days ahead of the schedule projection, with a total of 61 days gained since the completion of the first drive. The average production rate realized during the second drive was approximately 90 feet per working day (excluding startup and walking through the station)—more than a 50% improvement from the second drive, even after compensating for the added shift.

UNIQUE PROJECT CHALLENGES

The congested underground setting of Los Angeles presented many unique challenges to the project team. Nearly every underground structure is installed in close proximity to deep foundations of nearby buildings or, beneath live buried utilities. Limiting disruption to the flow of city traffic and neighboring residences is also a primary concern of Metro and the City. While the choice of bored tunnels naturally circumvents many of these obstacles, there were still several unique challenges facing the tunnel team.

TBM Launch and Reach 1 Tunneling

One of the most challenging elements of the tunnel construction presented itself immediately at the TBM Launch. The first 500 feet of the tunnel drive would be excavated immediately below a multi-story parking structure at the historic Japanese Village Plaza (JVP) shopping center. The limits of the cut & cover excavation for the future Little Tokyo Station are immediately adjacent to the structure, and the tunnel profile left slightly more than half a TBM cut diameter of clearance between the crown of the machine and the building footings.

In order to prove the TBM's capabilities to limit settlements prior to tunneling beneath the JVP, the design-build contract included a requirement for a 60-ft-long "TBM Demonstration Zone." To satisfy this requirement, a shotcrete soil-nail wall was constructed within the station footprint, 60 ft away from the pile-line nearest the building, allowing the end of the station to remain unexcavated. The TBM would be launched and driven through the soil-nail wall, and up to the pile line, to test and prove all systems, thereby fulfilling the contract requirements, and mitigating potential building settlements caused by TBM troubleshooting during launch.

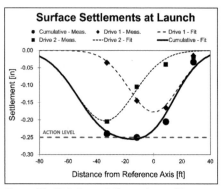

Figure 17. Plot of cumulative surface settlements as measured by MPBX's at the beginning of Reach 1

As discussed above, the contract also required a compensation grouting program in the Qal1 alluvial soils beneath the JVP. The preliminary design drawings proposed an access shaft in 2nd St. for the installation and use of grouting pipes. During the design phase difficulties with traffic control, obstruction of buried utilities, including the historic Zanja Madre, and easement constraints were assessed in detail. To minimize impact outside the primary jobsite, and limit utility and easement interferences, the project team elected to use the station excavation for installation and injection of grout pipes, in lieu of a separate grouting shaft.

In order to create access and a work area for the compensation grouting activities, the aforementioned 60 ft demonstration zone was excavated to a depth of approximately 15 ft and capped with a temporary concrete slab. This in turn created an extremely technical setting for launch, where the TBM would begin advance in the demonstration zone with only 3 to 5 ft of cover, then experience a sudden increase in overburden to approximately 20 ft. The extremely low cover of 3 ft was found to be insufficient to confine face pressure during excavation through the demonstration zone, and eventually out from the confines of the station. Prior to launch and after compensation grouting pipes had been installed and pre-treated, bulk bags of excavated station spoils were stacked above the tunnel alignment to contribute an additional approximately 8 ft of overburden to confine face support pressure and backfill grouting pressure.

With very little overburden, the range of allowable face pressures for advance through the demonstration zone was extremely narrow. The machine launched with interlocks which would only allow nominal face pressures between 0.80 and 1.00 bar for the advance through this zone. As the TBM progressed through the future station wall, and beneath JVP, the operating range for pressure stepped up to a range of 1.00 to 1.30 bar. As the alignment deepened, and the overburden gradually increased to about 1 × cut diameter, the range for allowable face support pressures incrementally increased, eventually allowing for a more forgiving operating band of 1.5 ± 0.5 bar. See Figure 18.

The careful control of EPB pressures, real-time settlement feed-back that was provided by SAA's and pro-active grouting program proved quite successful in mitigating settlements while tunneling beneath the JVP. Surface settlements measured near the TBM launch location did not exceed the action limit of 0.25 in. As a result, no compensation grouting was required after either of the two tunnel drives.

Planning, Design and Construction of the Regional Connector Bored Tunnels 909

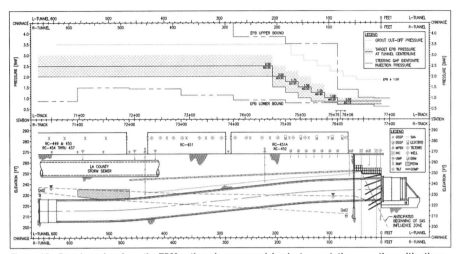

Figure 18. Drawing taken from the TBM action plan summarizing instrumentation, grouting mitigations and TBM parameters at launch

The combination of a relatively light soil-nail wall, low overburden of granular material, and a settlement sensitive structure gave need for a means of generating complete and immediate face pressure at launch. Typically, this is done by simply using bentonite slurry. However, slurry produces little or no resistance as it flows through the screw and is more difficult to handle with tunnel conveyor systems than with muck-boxes. The project team developed an improved priming material which included bentonite slurry and aggregate to mimic the behavior of properly conditioned muck. The tunnel eye-seal, the excavation chamber and the first screw were pre-filled with the material prior to launch, and the TBM was able to apply complete face support from the beginning of the first advance through the soil nail wall.

Despite a 24-hour, 7-day production schedule below the sensitive structures, the first drive took off to a very slow start, with the first 100 ft of advance taking nearly 3-weeks. In addition to the tight operating bands, restricted advance pressures, typical learning curve for ring-building, and the cautions approach to the overhead structure, the Qal2 formation initially proved to be quite a challenge for tunneling. The coarse sand was found to be dense and drained, requiring significant volumes of conditioner, and tooling load. A surfactant + cellulose polymer type ground conditioning was used on the first drive. For the second drive, RCC prepared more ground conditioning options, including modifications to the machine to facilitate bentonite injection if it was found to be useful. After some trials during initial advances, surfactant without polymer additives seemed the most effective, and allowed the team to advance over 200 ft in the first 2-weeks; a marked improvement over the first launch.

Conditioning the Fernando Formation

The proper conditioning of clay and clay-like soils is a common challenge faced on EPB tunneling projects. To mitigate clogging potential, and high cutterhead torque, excavated spoils with a relatively liquid consistency are desirable to keep the muck moving through the TBM. In order to effectively transmit support pressure to the face of excavation and maintain control over material flow through the screw-conveyor, a paste-like consistency is ideal. Yet, efficiency in disposal, hauling and handling operations can be gained by keeping excavated spoils as dry as possible.

RCC faced all these common challenges in conditioning the Fernando Formation. Lessons learned at University Link Light Rail project in Seattle, Washington were applied and further refined during construction of the RCTC. While the Fernando Formation was siltier than the fat clays encountered in Seattle, it responded well to a similar ground conditioning approach. The material demanded moisture to achieve and maintain a workable consistency in the excavation chamber and screw. The addition of Anti-Clay treated surfactants reduced the clogging potential of the spoils and reduced the water content required for conditioning. The careful management of foam expansion prevented the accumulation of trapped air and helped to maintain a full excavation chamber.

Station Crossings

As mentioned previously, the TBM tunneled through Broadway Station prior to the station being excavated. Once at the Bunker Hill Station, the TBM was to be "walked" across the station. RCC designed and built a specialized cradle to facilitate the movement of the TBM through the station. This cradle was a low-profile cradle utilizing jacks to lift the TBM shield structure, and jacks to move the cradle forward.

To facilitate this motion, through the constructability review process, the RCC Team revised the typical station design from square hammerheads to tapered. The tapering of the hammerheads to the main box structured allowed for the navigation envelope of the TBM to pass through the station without modification or disassembly, greatly reducing the time for the station crossing.

The combination of the station hammerhead tapering and the 'walking cradle' allowed the project team to move the TBM through the station in a matter of days in lieu of weeks.

TBM Extraction

Key to recovering the schedule delays resulting from the pile-strike, screw breaks, tie-backs, and other difficulties experienced during the first drive, would be careful planning for two successful crossings of the Bunker Hill Station, a successful TBM reception and recovery at the Flower Street cut and cover excavation, and a timely assembly and re-launch from the Little Tokyo Station.

The TBM would break-in to the cut and cover excavation beneath Flower St. immediately south of the 4th St. overpass; a location that presented some challenges. The extraction window was adjacent to the 4th Street Overpass to the North-East, pedestrian overpasses to the South-West, and flanked by high-rise buildings. Not to mention, that Flower Street is one of the busiest thoroughfares in the Financial District.

Figure 19. Photo taken during TBM retrieval; view from 4th St. overpass

Surface work related to the machine extraction and TBM shield transport would require a full closure of Flower Street between 3rd and 5th Streets, and a closure of the ramp from 4th Street. To limit traffic impacts, intersection closures were only permitted during weekends, with further restrictions that limited hauling of the 22-ft-diameter TBM shield sections to 6-hour windows on Saturday and Sunday nights. Each reception activity would be allowed 9 sequential weekend closures; two for preparation, three for extraction and transport, two for demobilization, and two for delay contingency. Work was completed in a total of 6 weekends for both receptions.

The shield weight, proximity of overpasses, nearby soldier pile and lagged excavation, and need to re-open traffic lanes between extraction weekends eliminated the option of a heavy mobile

Figure 20. Hoisting the cutterhead during TBM retrieval; taken from the cut & cover excavation below Flower St.

crane for hoisting operations. Working with the City and Metro, RCC elected to use a gantry mounted strand jack system to extract the shield components. The project team incorporated a planned window for TBM extraction through the traffic decking, and integrated with the SOE design, structural support for the gantry system and a 350-ton mobile assist crane, provided by Bigge Crane & Rigging Company. Because gantry mobilization and demobilization would take longer than a single weekend, the system was designed to straddle the extraction window and allow public traffic to pass underneath during the weekdays.

Obstacles Along the Tunnel Alignment

The choice of bored-tunnels for urban underground construction projects is often, in itself, a mitigation for existing obstacles; either buried, or on the surface. Still, the alignment of the bored tunnels was highly constrained by several challenging underground obstacles.

Undercrossing of the Existing Redline Tunnels

Immediately west of the Broadway Station were the Metro Red Line tunnels that had been constructed in the 1990s with two-pass cast-in-place concrete final linings the project setting did not afford space for the bored tunnels of the RCTC above the existing tunnels. With the existing Red Line tunnels lying roughly 70' below grade at their inverts, crossing beneath the tunnels would result in the Historic Broadway (2nd & Broadway) and Grand Ave Arts/Bunker Hill (2nd & Hope) stations being two of the deeper underground stations within Metro's system. The minimum clear cover from the RCTC tunnels to the inverts of the Red Line tunnels was set by design at 7 ft to limit the risk of settlement of the existing tunnels. Further complicating the alignment design, the very next station, only 1200 ft down the line from the Red Line Tunnels would be accessed from near the top of Bunker Hill. The design constraint presented by the red-line tunnels was overcome by balancing station depth with a steep tunnel grade of 4.6%.

4th Street Overpass Foundation Piles and Sewer

Immediately before the termination of the bored tunnels at the cut & cover excavation below Flower St., the TBM would be required to pass between a series of side-by-side overpasses of 4th Street. This resulted in an extremely narrow pillar between the two tunnels. To further complicate the matter, a 37-ft-deep sewer manhole encroached within the plan alignment.

The contractor was given the option to re-align and replace the sewer or, to avoid, protect and if necessary, repair the sewer after tunneling below it. The latter of the options was selected, and the alignment of the second drive tunnel was lowered by 2.3 ft and shifted closer toward to the first drive tunnel, until the pillar width between the tunnels was 4.5 ft. Despite the re-alignment, the sewer structure remained within approximately 8.5 in. of the TBM navigation envelope.

A supplementary mitigation and monitoring plan was developed and implemented for tunneling beneath the sewer. Some of the key features of the plan were: (1) Implementation of narrower and more conservative TBM ground support and grouting parameters; (2) The addition of a UMP to monitor movement of the base of the sewer manhole; (3) Additional pre-condition, post-condition and leak-rate surveys; and (4) Isolation and internal pressure equalization of the sewer in question, by bulk-heading and flooding the affected section with water in advance of the TBM crossing.

Tie-Back Removal Shaft

Near the beginning of Reach 3, for the second tunnel drive, a number of tie-backs, left behind from the foundation construction of the adjacent high-rise building, were known to be unavoidably within the Tunnel alignment. The contract presented a need to mitigate the risks the tie-backs posed to construction of the bored tunnels and proposed a roughly 30 ft × 100 ft open cut excavation to pre-clear the tie-backs in advance of tunnel construction. The open cut excavation would have presented a significant disruption to Downtown traffic and further complications with the temporary support of underground utilities.

The project team developed an alternative approach to the removal of the tie-backs. With the assistance of BIM modeling, and the application of SEM excavation methods, RCC was able to replace the large open-cut excavation with a 24-ft-diameter shaft and 11-ft-diameter horseshoe-section adit, to remove all the tie-back obstructions. The

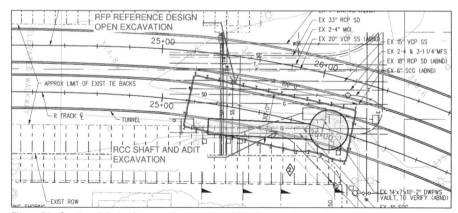

Figure 21. Overlay of tie-back removal excavations; bid-phase vs. construction

combined effort of all parties in the redesign and construction of the Tie-Back removal excavation resulted in a number of project benefits, not the least of which were: (1) Reduced impact to Downtown traffic; (2) Reduction in the risk of settlements through the reduction in volume of excavation; (3) Reduced interface with buried utilities; and (4) Cost and Schedule Savings from the reduced excavation and backfill volumes. (Bragard et al. 2017).

Encountering Buried Steel

At the beginning of April of 2017, while tunneling through what would eventually become the Crossover Cavern, the TBM struck an obstacle. This obstacle was eventually discovered to be the tip of a steel pile, but only after several pieces of steel were retrieved from the discharge of the screw conveyor. Tunneling had already advanced 200 feet beyond the pile, when the project team staged an intervention to inspect the cutterhead of the TBM. During the intervention remnants from approximately 7 ft of a 21 in steel pile were retrieved from the cutterhead and excavation chamber, and several disk cutters were replaced.

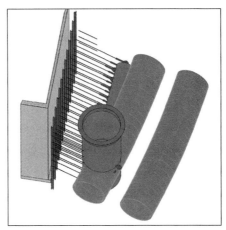

Figure 22. BIM model of tie-back removal works

Figure 23. W21×132 pile struck by TBM; photo taken after pile was recovered from SEM cavern excavation

After the intervention, tunneling continued, though with slightly elevated resistance noted in the screw conveyors. Eventually, several hundred feet after the steel pile was encountered, a final piece of steel was found at the screw discharge, screw pressures stabilized, and TBM advance proceeded as normal. Shortly after the final piece of steel was retrieved from the screw, late in May 2017, the auger stem of the Screw Conveyor #2 sheared, marking the beginning of a sequence of events that culminated in the replacement of all three of the augers in the TBM's screw conveyor system.

Three separate temporary repairs were required to complete the First Tunnel drive:

1. On 23-May-2017, the auger of Screw #2 was found to have sheared near the drive. A temporary repair was made by welding bolting flanges on either side of the break.
2. On 22-June-2017, the stem of the auger sheared again in the same location, at the root of one of the repair flange welds. The design of the repair was improved to give better access for a more robust weld, and some modifications to ensure the bolts would fail before the welds.
3. On 17-July-2017, immediately before the end of the first tunnel drive, the temporary repair failed for a third time, but was quickly repaired by replacing the sheared repair bolts.

The third and final field repair occurred after a separate encounter with abandoned tieback excavation supports along the alignment—where the TBM advanced thru 48 anchor rod and bearing plate style tie-backs, during the middle of July. This was not planned, as documentation of past construction in that area depicted a more frequent pattern of shorter tie-backs. The alignment which was designed to avoid the tie-backs as documented, did not avoid the longer, heavier anchor-rod style tie-back which were actually encountered.

While the temporary repairs were clearly more susceptible to failure than a replacement, or full refurbishment of the screw conveyors would have been, they were quickly implemented, and facilitated the timely completion of the first tunnel drive. During reception and re-launch activities, the entire screw conveyor system was removed from the TBM, so the compromised augers could be replaced with newly fabricated augers. Since this work was performed concurrently with scheduled preparations for the second tunnel drive, it had minimal impact on the overall project schedule.

Figure 24. Steel tie-back found wedged between auger flights; photo taken during screw conveyor refurbishment between tunnel drives

CONCLUSION

The Regional Connector Project's location, in the heart of Downtown Los Angeles, created many unique challenges the project team would face through all its phases; procurement, design and construction alike. Design professionals surmounted many obstacles to create a tunnel alignment to safely clear underground structures including the operating Metro Red Line. Incorporating stringent vibration control elements into the project was vital for Metro to meet its commitments to adjacent performing arts and recording studios. The extensive building and utility protection system was also essential in managing settlement to within predicted values, which was important in maintaining trust with stakeholders.

Similarly, during the mining operations, the team overcame multiple challenges, beginning with launching the TBM with extremely shallow cover in alluvium beneath public facilities. The ability to maintain the operation, with minimal down-time for repairs after striking two abandoned steel piles and multiple undocumented tie-backs, is a credit to the experienced team.

The success of the tunnel construction can largely be attributed to the partnership between LA Metro, with support from their consulting partners, and the Joint Venture of Skanska USA and Traylor Bros. Inc., with support from Mott MacDonald and Hatch. Not only were technical challenges met with innovative solutions, the tunneling schedule was maintained as commercial issues were addressed in a proactive manner. With the tunnel construction completed, the Regional Connector Transit Corridor is poised to be completed as scheduled, improving the mobility of transit passengers within Los Angeles County.

ACKNOWLEDGMENTS

The design/build contractor is Regional Connector Constructors (RCC), a joint venture of Skanska USA Civil and Traylor Bros., Inc. Chemical and compensation grouting was by Hayward Baker, Inc. Geotechnical instrumentation was by Geocomp. Final Design is by Mott MacDonald (MM). Preliminary Engineering was undertaken by CPJV, a joint venture of AECOM and WSP | Parsons Brinckerhoff. During construction, CPJV has provided engineering support services and Arcadis has provided construction support services to Metro. EPC Consultants, Inc. is part of the Arcadis team. The authors thank Gary Baker for his contriubtion to this paper.

REFERENCES

Bragard, C., Hadley, B., Antonelli, M. 2019. Unexpected underground obstructions: Challenges and lessons learned from Angeli, the Regional Connector TBM. In *RETC: 2019 Proceedings*. Englewood, CO: SME.

Broadbaek, C., Penrice, D., Kusdogan, E., Bragard, C. 2017. Removal of Interfering Tiebacks Using SEM in Advance of TBM Mining on the Regional Connector Project. In *RETC: 2017 Proceedings*. pp. 45. Englewood, CO: SME.

Hansmire, W.H., Choi, J., and Vu, T. 2013. *Geotechnical Baseline Report Rev. 1a*. Report. Los Angeles: Regional Connector Transit Corridor Project CN E0119.

Hansmire, W.H., McLane, R., Frank, G., Bragard, C. 2017. "Challenges for Tunneling in Downtown Los Angeles for the Regional Connector Project," Proceedings, World Tunnel Congress, 2017.

McLane, R. 2014. Automatic Soil Conditioning Through Clay. In *North American Tunneling: 2014 Proceedings*. pp 195. Englewood, CO: SME.

Ulm, F., Acker, P., and Lévy, M (1999). "The Chunnel Fire II: Analysis of Concrete Damage," Journal of Engineering Mechanics, Vol. 125, No. 3, March 1999, pp. 283–289.

Vitek, J.L. (2008). "Fire Resistance of Concrete Tunnel Linings," Tailor Made Concrete Structures: New Solutions for Our Society, Editor: Walraven, J.C. and Stoelhoerst, D., CRC Press, pp. 207–210.

Preparation for Tunneling, Don River & Central Waterfront Coxwell Sanitary Bypass Tunnel Project in Toronto, ON

William Hodder ▪ Jay Dee Contractors, Inc.
Ehsan Alavi ▪ Jay Dee Contractors, Inc.
Daniel Cressman ▪ Black & Veatch

ABSTRACT

The Don River & Central Waterfront Coxwell Sanitary Bypass Tunnel Project in Toronto will extend approximately 10.5 km from Ashbridges Bay Treatment Plant west under Lake Shore Boulevard, north under Bayview Avenue and east through the Don River park area to the Coxwell Ravine Park. The tunnel will be mined by using a shielded tunnel boring machine capable of operating in EPB mode and lined by using a precast concrete segmental tunnel linear with an internal diameter of 6,300 mm and external diameter of 6,900 mm. In addition to the TBM excavated tunnel, the scope of this project includes two sections of tunnel using hand tunnel construction methods, five shafts corresponding to the locations of 5 sewer structures, eleven drop shafts, vent shafts, deaeration and adit tunnels to be designed and constructed, two diversion structures and two consolidation sewers to be constructed using mainly open cut and shored excavation methods, 8507 Connection from the existing wet weather flow outfall to one of the shafts and North Toronto Sanitary Trunk Sewer from the existing sewer to one of the shafts. This paper summarizes the preparation work that has been done prior to the launch of the TBM.

INTRODUCTION

The Don River and Central Waterfront Coxwell Bypass Tunnel is part of a 2 Billion Dollar development plan in City of Toronto. The approval process started as far back as 1999, with the initial approval of the Ashbridge's Bay Treatment Plant Environmental Assessment. The Toronto City Council then approved the Wet Weather Flow Master Plan in 2003 and the Don River and Central Waterfront Class Environmental Assessment in 2012, both precursors to the Project. The overall project is supposed to be completed over 25 years and will increase wastewater storage capacity and protect against accidental discharges into the Don River, Taylor-Massey Creek and Inner Harbor. The Coxwell Bypass Tunnel Contract includes the installation of five intake shafts connected by 10.4km of 6.3 m internal diameter tunnel and eleven drop shafts and vent shafts, connected to the tunnel by a series of adits. The tunnel will provide a bypass form Coxwell Ravine Park to the North Toronto Treatment plant, and utimately the Ashbridges Bay Treatment Plant. The Coxwell Bypass Tunnel contract is scheduled to be complete in 2023. This contract was awarded to North Tunnel Constructors LLC, a joint venture of Jay Dee Canada, Michels Canada and C&M McNally Tunnel Constructors in 2018 for 448.9 Million dollars.

OVERALL WORK PLAN

The Coxwell Bypass Tunnel along with a new Integrated Pumping Station at Ashbridges Bay Treatment Plant are the first projects to be constructed in the City of Toronto's $ 2.0 billion Don River and Central Waterfront (DR&CW) program. The DR&CW program has been designed to provide redundancy to the City of Toronto's

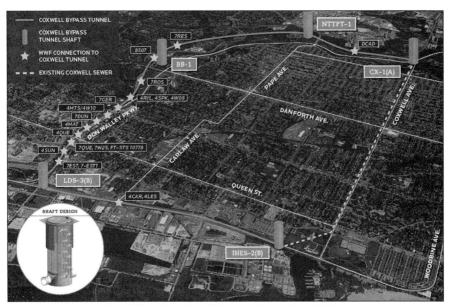

Figure 1. Don River & Central Waterfront Coxwell Sanitary Bypass Tunnel Project

Table 1. DR&CW tunnel summary

DR&CW Tunnel Description	Length (meter)	Diameter (meter)	Geology
Coxwell Bypass (CBT)	10.6	6.3	Rock
Taylor-Massey (TMT)	6.0	4.4	Soft Ground
Inner Harbour West (IHW Tunnel)	5.6	6.3	Rock

existing Coxwell sanitary trunk sewer and reduce combined sewer overflows into Lake Ontario, the Don River and Taylor Massey Creek.

Completion of the Coxwell Bypass Tunnel and Integrated Pumping Station will allow the Coxwell Bypass Tunnel to be put in operation with the Integrated Pumping Station allowing both sanitary and wet weather flow to be pumped out of the Coxwell Bypass Tunnel through Ashbridges Bay Treatment Plant. The Coxwell Bypass Tunnel has been designed to accommodate connection of Wet Weather Flow overflow sewers to the tunnel drop structures and connection of future phases of wet weather flow tunnel at the storage shaft locations. Future phases of tunnel include the Taylor Massey Tunnel and Inner Harbour West Tunnel, Table 1 provides a summary of the three phases of tunnel construction on the Don River and Central Waterfront Project.

DESIGN CONSIDERATIONS

The DR&CW tunnel system has been sized to provide 1-CSO level of wet weather flow control. 1-CSO level of control is a Ministry of the Environment and Climate Change (MOECC) standard that refers to the allowance of, on average, one overflow per year per wet weather flow outfall. Hydraulic design of the tunnel system has required a minimum of 26% of the wet weather flow storage volume be provided in the tunnels and shafts to mitigate the risk of geysering and ensure safe operation of the system. The hydraulic design requirement has been satisfied with placement of five wet weather

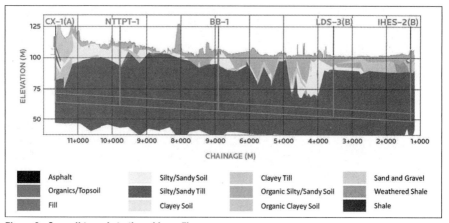

Figure 2. Coxwell tunnel stratigraphic profile

flow storage shafts along the Coxwell Bypass tunnel alignment. The shafts have been strategically placed at locations sufficient land was available for construction and at approximately equal distance along the tunnel alignment to facilitate construction.

Tunnel Geology

The vertical profile of the Coxwell Bypass Tunnel is provided in Figure 2. The CBT is expected to be excavated entirely through shale bedrock of the Georgian Bay Formation, at approximately 50 meters deep. The majority of the tunnel alignment runs parallel to a buried bedrock valley located adjacent to Bayview Avenue. The tunnel crosses this valley in two locations, in the southern portion of the alignment, adjacent to the Don River, and at the northern portion of the alignment approaching the CX-1 shaft. In these two locations, the bedrock cover is expected to be less than 6 meters.

The Georgian Bay Formation, within the tunnel alignment, typically consists weak to medium strong, fresh, grey, fine grained shale with widely spaced jointing and sub-horizontal bedding planes. The Georgian Bay Formation is characterized by its high horizontal stresses and has been observed to exhibit time-dependent deformation (TDD). During design, geotechnical investigation measurement of the Major Horizontal Stress (P) provided a range of 1.6 to 5.6 megapascals (MPa) and a Minor Horizontal Principle Stress (Q) range of 0.1 to 3.8 MPa. The term "Time-Dependent Deformation" has been used to describe the concurrent behavior of stress relief and the swelling mechanism which causes shale of the Georgian Bay Formation to squeeze into the excavation of not supported with an appropriate swell suppression pressure.

Tunnel Lining

For the CBT, the use of a one-pass, precast concrete tunnel lining (PCTL) was evaluated against the use of a two-pass lining system. The typical approach in the Greater Toronto Area has been to excavate the shale of the Georgian Bay Formation with an open-faced main-beam type rock TBM and install temporary support directly behind the main shield. The final lining, cast-in-place concrete, is typically installed a minimum of 100 days after excavation. This two-pass system allows for the full elastic stress relief and the majority of the TDD of the rock mass to take place prior to the

Table 2. Tunnel lining considerations

Tunnel Lining	Advantages	Risks
Two-Pass	• Cost-Effective • Typical Approach in Georgian Bay Formation	• Ground water seepage (risk increased in reduced rock cover) • Overbreak in tunnel crown • Slaking in tunnel invert • Buried valleys • Gas (methane) • Degradation of exposed shale
One-Pass (PCTL)	• Schedule Efficient • Cleaner Tunnel	• Support pressure required from lining is substantial • Time-dependent deformation and elastic stress relief places stress on lining • Squatting of PCTL rings • Grouting TBM in place • Washout of annular grout

installation of the final lining. As an alternative, the design team looked at the use of a single shield rock TBM with a PCTL to mitigate certain risks associated with the two-pass system. As previously discussed in the section titled Tunnel Geology, the CBT alignment crosses a buried bedrock valley with reduced rock cover in two locations, and in addition to these two locations, the tunnel alignment runs parallel to the buried valley on Bayview Avenue. The risk of encountering weathered bedrock, significant ground water inflows or surficial soil deposits through these buried valleys was thought to be a significant risk. In an attempt to mitigate this risk as well as others, the design team analyzed the use of a PCTL, as shown in Table 2.

Selection of a PCTL was considered to mitigate some risks and take certain risks associated with a two-pass lining system off the table. However, use of a one-pass system introduces risks of its own. The biggest risk identified was TDD. To evaluate this risk, a detailed numerical analysis to estimate the loads that would be exerted on the PCTL was conducted. The results of the analysis indicated that the displacements that will occur after lining installation will range between 0 percent and 10 percent of the total displacements, corresponding to up to approximately 2 millimeters (mm). Through selection of an appropriate annular grout and segment design, the displacements, after lining installation, can be accommodated in the PCTL design. The one-pass PCTL was selected as the preferred tunnel lining method because the inherent risks associated with the use of PCTL can be mitigated through incorporation of the appropriate design measures and the uncertainty associated with ground condition risks, apparent in the one-pass system, are avoided.

SITE WORK

The project is Broken into five primary shaft sites and 11 drop/vent shaft sites. The Ashbridges Bay Treatment Plant Site (IHES-2(B)) is the primary site where the Tunnel Boring Machine will be launched and all excavated material will be hauled from. The other primary sites, in the order that the TBM will tunnel to them, are the Lakeshore and Don Valley Parkway (LDS-3(B)) site, the Bayview and Bloor (BB-1) site, the North Toronto Treatment Plant (NTTPT-1) site and the Coxwell Ravine Park (CX-1) site. There are an additional 11 drop shaft/Vent shaft sites along the alignment. 20 m Diameter (Final Structure) shafts will be excavated at the IHES-2(B), LDS-3(B), BB-1 and NTTPT-1 sites and a similar 22 m diameter (Final Structure) shaft at CX-1. Each shaft consists of a soil zone, that will utilize secant piles as the initial support of excavation, over to shale, which will be temporarily supported by rock bolts and welded wire mesh.

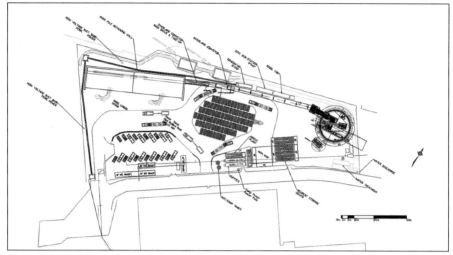

Figure 3. Coxwell bypass tunnel IHES shaft layout

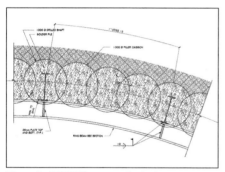

Figure 4. CBT IHES secant pile design

Figure 5. CBT IHES shaft excavation

IHES-2(B) Site

The IHES-2(B site was prepared previously by a separate contract. This site will serve as the launch site for the TBM. All the excavated material along the 10.4 km of the tunnel will be transferred to this site and will be hauled off site. This site will serve as the main site supporting the tunneling operation from Shaft IHES-2(B) to Shaft BB-1 site. The first year of the project is devoted to excavation of the launch shaft with diameter of 20 m and depth of 57.6 m, excavation of a 115 m starter and 71 m tail tunnel to be utilized for assembly of the TBM and preparing this site for the acceptance, assembly, launch, and support of the Tunnel Boring Machine. The site design and operation have been divided into several stages in order to slowly progress the site based on the activities. Several utilities were laid out for office trailers that were delivered to the site early on. Temporary power was also established to power up the trailers and support some of the early on activities. The water treatment system was set, a haul road was designed and paved to provide access to remove shaft spoils, access the future TBM muck bin, and separate haul Trucks and Segment Trucks. The main purpose of the provided haul road was to reduce the amount of tracking on and off site.

IHES-2(B) Support of Excavation and Shaft Excavation

The shaft at the IHES site was designed as 57.6 m deep (surface grade to bottom of excavation). The top 14 to 17.5 m depth is soil and shale for the remaining depth. Initial support of excavation in the soil consisted of 79 secant piles, at 1.18 m diameter, pocketed 3 m into competent shale, with the purpose of cutting off ground water in the transition zone between soil and rock. Rock support, below the secant piles, consists of rock bolts and welded wire mesh and a collar of shotcrete ties the bottom of the secant piles to the rock. The depth of the shaft is currently at 30 m with a planned finish excavation finish date in late March 2019.

LDS-3(B) Site

The Lakeshore and Don Valley Parkway site is located between a live railway running parallel to Lakeshore Blvd and a prominent movie production studio, north of the site and east of the shaft location. There is a 500 mm gas main running across the site that needs to be protected throughout the work.

LDS-3(B) Preconstruction Work and Utility Installation

As a result of the proximity of the movie production studio, constraints such as noise and vibration limitations and working hours were put in place for work on the site. The original design had an Eastern entrance on Booth Avenue with a narrow 250 m long road to access the site. A secondary access, coming through a neighboring business' parking lot and onto the site from the West, was established to allow the removal of the road and reduction of the eastern end of the site to the bare minimum. This was done to reduce the potential noise and vibration impacts to the adjacent movie production studio. Water and Sewer connections were both located offsite, across the existing railway. To mitigate this, an agreement was made with the neighboring business to use their existing, private water line. Establishing a sewer connection required drilling under the railway. Detectible Volatile Organic Compound (VOC) levels in the existing soil paired with high water recharge, created a condition where a starter pit was not possible, so the line had to be directional drilled from the surface.

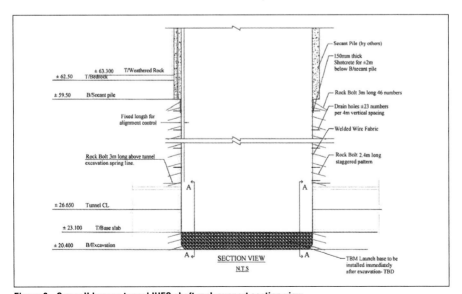

Figure 6. Coxwell bypass tunnel IHES shaft rock support section view

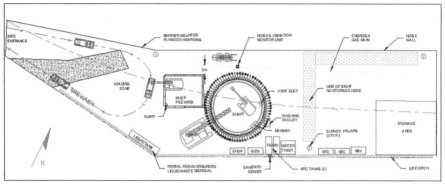

Figure 7. Coxwell bypass tunnel LDS site plan

LDS-3(B) Support of Excavation and Shaft Excavation

The presence of High VOC levels in the existing soil and high-water recharge required special attention during the installation of the Support of Excavation at the LDS site. A level of contamination was expected at the site, but the materials saturation, paired with VOCs created layers of material that could only be disposed of as liquid waste. The material presented a strong odor, so testing was performed to make sure the material was not hazardous and Environmental and Health and Safety Consultants helped establish a plan to safely monitor the soil emissions released while drilling and stabilize the liquid waste with Cement Kiln Dust (CKD) in order to haul the material.

The LDS secant piles were drilled to a depth of 14.2 m. They embed 3 m into competent rock. All personnel working near the drilling operation were outfitted with air monitors and half face respirators. Procedures were established to vacate the area anytime VOC concentrations rose to a level deemed unsafe to work in bare faced. Half face respirators were provided and would be donned for evacuation purposes only, no production work was to be performed using a respirator. Secant pile installation for this site started in October of 2018 and finished in December of 2018. Shaft Excavation is scheduled to start in Jan 2019. Boreholes and Support of Excavation (SOE) drilling suggest that the VOC concentrations and high saturation in the soil will be isolated to the top 7.1 m. Due to the conditions of the soil, the top 7.1 m of material will be excavated from the surface. Only after this material is removed will an excavator will be lowered into the shaft and excavation will continue from within. Total depth of the LDS shaft is 52.8 m.

BB-1 Site

The Bayview and Bloor (BB-1) site is located on an existing track field that was used by a neighboring school. Only a portion of the track was intended for the site, but an agreement was made with the school to allow the full site to be utilized. This will allow the site to act as the secondary tunnel support site in between shafts BB-1 and CX(1). After the TBM passed through the shaft, the grout plant and segment storage facility will be moved up so that the tunnel from

Figure 8. Coxwell bypass tunnel LDS shaft SOE installation

IHES-2(B) to BB-1 will become available for adit excavation work however muck will be transported via conveyor system back to shaft IHES-2(B).

BB-1 Preconstruction Work and Utility Installation

Site grading and installation of hoarding has been performed at the BB-1 site. One of the major issues on this site is that there is no power or water source available close to this site. Required water for SOE installation and shaft excavation must be trucked in to this site.

BB-1 Support of Excavation and Shaft Excavation

The BB-1 support of excavation started in January of 2019. This shaft will be 53.1 m deep. The secant piles at this shaft will be drilled to a depth of 23.5 m to socket 3 m into competent rock. Excavation of this shaft is scheduled to start in late March of 2019.

NTTPT-1 Site

The North Toronto Treatment Plant (NTTPT-1) site has been graded and hoarding is being installed around the site. The support of excavation is planned to start in March of 2019. The depth of the shaft will be 54.0 m and the overburden is expected to be 19–26 m deep.

CX-1 Site

The Coxwell Ravine Park (CX-1) site preparations will start in January of 2019. This site is located adjacent to an existing dog off leash area and access is limited by a 4.4 m tall underpass. The shaft will be 55.9 m deep with 33–37 m of overburden. The site contains an abandoned gas main that will require removal, as well as overhead hydro lines within 4 meters of the shaft. This will be the final shaft for the TBM and TBM removal needs to be planned around these constraints.

SHAFT FINAL LININGS

Each of the five intake sites will have a permanent lining poured into them, and shaft IHES-2(B), LDS-3(B), NTTPT-1 and CX-1 will contain a demising wall and baffle structure. An estimated 34,000 m^3 of concrete will be poured into the structures. The LDS-3(B), NTTPT-1 and CX-1 structures will be poured prior to the TBM getting to the shafts. THE LDS-3(B) structure will start in July of 2019 and finish in Jan 2020. Block outs will be left at the locations of the tunnel eyes, so the TBM can pass through the shafts unhindered.

DROP SHAFTS

There are eleven drop/vent shaft sites: 4CAR/4LES, 4SUN/4QUE, 7QUE, 4MAT, 7DUN, 4MTS, 7GER, 4SPK, 7ROS, 7RES, and DCAD. Each site contains a drop shaft and vent shaft that drop to a deaeration chamber at the elevation of the tunnel and that chamber connects to the tunnel via Adit. The adits will be excavated from within the tunnel. After the TBM passed through BB-1, the TBM supply will be moved up the BB-1 site so adit excavation may commence between the IHES-2(B) and BB-1 sites. The Drop and Vent shafts will be drilled and lined prior to excavating the Adits. Drilling of the shafts is planned to start in May of 2019.

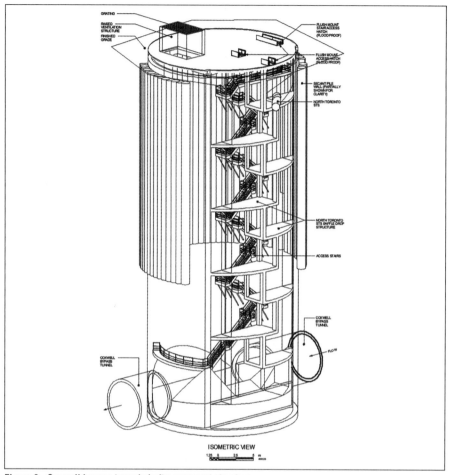

Figure 9. Coxwell bypass tunnel shaft permanent structure

PREPARATION FOR TUNNELING

Tail Tunnel/Starter Tunnel

Construction of the starter/tail tunnel is planned between March and August of 2019. The final assembly of the TBM and backup gantries will be completed in the starter tunnel, allowing the TBM to launch and not have to stop to reconfigure the tunnel belt, ventilation, utility lines, and segment supply and exclude the need to launch with umbilicals. The starter and tail tunnels will be excavated by using a road header and supported by rock bolts and shotcrete. A shorter tail tunnel will also be excavated as part of the requirement of the contract to tie-in to the future structure. The available space in tail tunnel will be utilized for storage purposes during the tunneling operation and also to allow early on traffic of Rubber Tired Vehicles (RTVs) during the launch process. The supply of segments to the tunnel heading will be facilitated by utilizing Metalliance Rubber Tired Vehicles (RTVs). Initial cycle studies suggested that switches for the RTV's could be avoided by giving the TBM and RTV's a 2 ring (or four stacks of three segments) capacity. This required long RTVs that will require the space provided by the tail tunnel to maneuver during the launch.

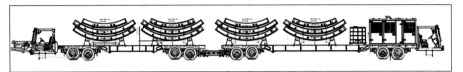

Figure 10. Coxwell bypass tunnel RTV

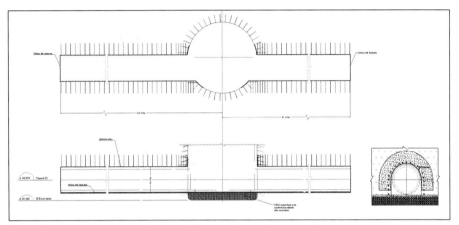

Figure 11. Coxwell bypass tunnel starter/tail tunnel plan profile and section

Grout Plants

A Sagami Servo Grout and Bentonite plant are set up at the IHES-2(B) site to batch and provide grout to the TBM. The plants were purchased from a past tunnel job and are being refurbished on-site. Initial inspection for refurbishment requirements were performed in January of 2019 and final commissioning will be performed in the spring of 2019.

Electrical Substation

Electrical conduits and grounding grids have been designed and laid for a future electrical supply that will be provided by the City of Toronto. At this time, all site power is being run off of a temporary supply from the Ashbridges Bay Treatment Plant and also by utilizing generator power.

Water Treatment

A large water treatment plant, capable of treating and discharging all construction water and shaft water that will run down the tunnel from all five shafts, is being constructed at the IHES site. The treatment plant is being designed to discharge to storm or sanitary sewer.

Conveyors (Vertical Belt, Take Up Units, Shuttle/Tripper)

The conveyor system is being designed and scheduled to be installed prior to the arrival of the TBM. A continuous conveyor will transfer muck from the tunnel to a vertical belt in the shaft. An overland conveyor will transfer the material to the muck bin where muck will be separated into multiple piles by a shuttle conveyor. The conveyor design is in the final stages and conveyor footings are in the early stages of design.

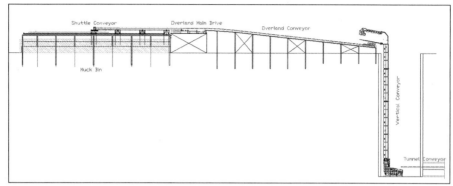

Figure 12. Coxwell bypass tunnel conveyor concept drawing

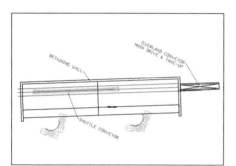

Figure 13. CBT IHES muck bin plan   Figure 14. CBT IHES muck bin section

Muck Bin

The preliminary muck bin designs were based on maximizing storage capacity to mitigate any trucking delays and allow space for trucks to que and load on site. The final capacity of the muck bin is approximately 4500 cubic meters. This volume will allow storage of approximately 4 shifts worth of material.

Segments

The tunnel segmental liner is a universal ring, comprised of six segments including a key and counter key. The segments will be transported in the heading by vacuum systems, both for hoisting and erecting. The design of the rings is to be finalized in early 2019 and production of segments is set to begin in early summer of 2019. Segments will be delivered to site and stored in full ring stacks. Two storage areas were laid out for segments, both a heated storage area adjacent to the haft, for segments that will be utilized short term, and an overflow area for additional segments.

Tunnel Boring Machine Preparations

The Don River and Central Waterfront Coxwell Bypass Tunnel TBM will be a Lovsuns Hard Rock TBM designed to convert into an Earth Pressure Boring Machine if needed. The bulk of the TBM design was completed in late 2018 and manufacturing of TBM components began as early as June of 2018. The TBM is designed in Toronto and manufactured at Lovsuns facility in China. The TBM will be fully assembled and factory tested in April of 2019 and is expected to arrive in Toronto in August of the same year. Once the TBM is in its final location at the headwall, a steal jacking frame will be erected behind the TBM for launch.

Remaining Work

Most of the initial design for the TBM supply and launch has been started or finalized, but installation prior to the arrival of the TBM still needs to be implemented. The TBM will have two phases of supply. The first phase is the tunneling from IHES-2(B) to BB-1 shaft. For the duration of this run, the tunnel will be supplied fully from the IHES shaft. The drive from BB-1 to CX-1 shaft will be supplied from the BB-1 site. Initial designs for moving up the TBM supply have been created but not finalized. The BB-1 site does not have a local water or electrical supply so the logistics of moving up the utilities is ongoing. By moving up the operation the operation, the adit excavations between the IHES-2(B) and BB-1 sites become available, allowing us to gain schedule. Designing the drop and vent shaft structures, so that they can be undercut and tied into from below, is ongoing as well as sequencing the final structures of the LDS-3(B) and NTTP-1 shafts. The primary challenge with the LDS-3(B) and NTTP-1 shaft structures is leaving a support system for the TBM to pass through the shafts in a curve at the bottom of the shaft. Drilling and lining the drop and vent shaft structures prior to excavating the adits, and pouring the final structures in shafts LDS-3(B), NTTPT-1 and CX-1 will allow a gain in schedule for an expected finish is October of 2023.

Tunneling in Mixed Face Conditions: An Enduring Challenge for EPB TBM Excavation

Jim Clark ▪ The Robbins Company
Paul Verrall ▪ Afcons

ABSTRACT

EPB TBM tunneling in mixed face conditions—partially in both rock and soil—is inherently problematic for even the most experienced crews. Over-excavation, excessive damage to cutter tools and regular cutterhead interventions are major challenges when negotiating mixed face geology.

This paper will draw from real field experiences, including successful bores in abrasive rock and soil at India's Chennai and Bangalore metro projects, to determine the optimal operational parameters for TBMs in such conditions. It will also address reduction of air losses to facilitate cutterhead interventions under hyperbaric conditions when installation of safe-haven grout blocks is not an option due to surface structures.

INTRODUCTION: CHENNAI METRO

Contract UAA-01 is part of the Chennai metro line 1, which runs from the Washermanpet area of the city to the airport. It consists of 5500 meters of twin tunnels, five stations and 17 cross passages. AFCONS was awarded the contract and subsequently chose The Robbins Company to supply a TBM to bore the Chennai Central Station to Mayday Park station section of the project. This section consists of two tunnels: an up line and a down line of approximately 1032 meters each. The works schedule was for the machine to be launched from Mayday Park station, disassembled and retrieved at central station, then relaunched at Mayday park to complete the second drive. The machine was actually used to complete a third drive but that is not in the scope of this paper as the geology did not contain mixed face conditions.

Geological/Hydrological

The entire area of the project is situated within a flood plain, with relatively minor deviations in surface levels. The geology along the alignment of the Chennai Central Station to Mayday Park station section is made up of dense silty sands interbedded with layers of clay and silt and granite with weathering grades varying from highly fractured and weathered (Grade V) to fresh rock (Grade I). The water table varies seasonally but remains above the tunnel alignment all year round. The overburden along the alignment ranged between approximately 11 meters and 16 meters (see Figure 1).

TBM Selection

Based on the geological information and discussions with AFCONS, The Robbins company supplied a mixed face EPB machine fitted with active articulation. The cutterhead was designed with the option of changing out soft ground tools to hard rock tools for excavation in weathered granite, sand, silt, and clay, and to cope with boulders up to 300 mm in diameter. The TBM was equipped with small grippers to allow for cutterhead stabilization in harder ground and to provide the reaction forces required to

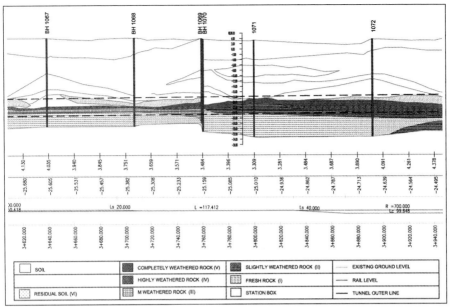

Figure 1. Geological section showing mixed face in the tunnel cross section

Table 1. Key machine specifications

Machine Type	Mixed Face EPB
Design Parameters	
Shield Diameter Bore diameter (Soft ground tools) Bore diameter (Hard rock tools)	6.6 m 6.63 m 6.65 m
Disc cutters	17," back-loading
Cutterhead power	6 x 210 kW = 1,260 kW
Cutterhead speed Cutterhead working torque	0–4.5 RPM 7,038 kNm
Maximum starting torque	9,149 kNm
Maximum main thrust Articulation type Trust cylinder stroke length	45,000 kN Active 2.2 m
Screw conveyor diameter Screw conveyor speed	0.9 m 0–22 rpm
Segment Backfill	Bi-component grout

pull the cutterhead back from the face in difficult conditions. The key technical specifications of the machine can be seen in Table 1.

TBM Launch

The machine was launched into a full face of silty sands/silty clays in January 2012. It completed the initial drive with a short start-up procedure utilizing umbilical cables without any problems. After completion of the initial drive the machine was stopped, the remaining gantries installed, the temporary rings and reaction frame removed, and the shaft bottom rail system installed. Boring operations then recommenced. Upon completing 160 m of boring (ring No. 109) granite was encountered in the invert of the bore. This was much sooner than expected but the rock was highly weathered and

Figure 2. Damaged cutter worn flat   Figure 3. Damaged cutter ring

estimated to be weathering grade IV or V, so it didn't pose a significant challenge to the boring operations. The level of the weathered granite gradually increased in the face until 188 m of boring had been completed (Ring No. 125). At this point the whole of the lower 30% of the face was made up of weathered granites, and furthermore, fresh granites were now present in the invert of the alignment. It was not possible to access/assess the strength of the rock in the invert as interventions were being carried out under hyperbaric conditions and site experience dictated that at least 40% of the chamber needed to remain filled with material to hold hyperbaric pressure.

Increase in Cutter Consumption

After encountering the fresh rock in the invert, operations were adversely affected by excessive cutter damage. Production time was lost carrying out cutterhead interventions to facilitate cutter changes. A total of five cutterhead interventions were undertaken between ring No. 134 and ring No. 259 (187 m of boring). During these interventions 144 cutters needed replacing. These interventions were all carried out under hyperbaric conditions. A substantial percentage of the damaged cutters were deformed and jammed into the housings; therefore, major difficulties were faced during the removal process. The total time taken to complete these interventions was over 90 days. The impact on the project schedule was significant so AFCONS requested that Robbins investigate the reason for the damage and offer a solution.

Analysis of Cutter Failure Mode and Boring Parameters

Analysis of the mode of cutter failure revealed that almost 90% of the cutter consumption was due to abnormal wear. A significant percentage of cutters that had been replaced had either chipping or radial cracks in the cutter discs. Many of the cracked discs had been displaced completely either by cracking or worn flat, leaving the cutter hubs exposed to the rock and resulting in the hubs being worn through to the bearings and shaft (see Figure 2 and Figure 3).

To further understand the reason for this damage the historical information on boring parameters was downloaded from the machine's data logger. Comprehensive information was available for each ring that had been bored since the machine was launched. Table 2 shows the typical boring parameters that had been used while boring in the silty sands and clays up until around ring No. 109. No major problems were faced during this time but the information is useful as a baseline for analysis.

Table 3 shows the typical boring parameters that had been used from ring 109 through to ring 125 while boring through a mixed face of soils and weathered granites.

Table 2. Typical boring parameters in silty sands/clays

TBM Operational Parameters in Full Face of Silty Sands/Clays	
Cutterhead Speed	1.0–1.4 RPM
Main Thrust pressure	19,000 kN
CHD torque	3,500 kNm
Earth pressure (centre of face)	2.0 Bar
Advance rate	25 mm/min

Table 3. Typical boring parameters in mixed face of soils and weathered granite

TBM Operational Parameters in Mixed Face of Silty Sands/Clays and Weathered Granite	
Cutterhead Speed	1.7–2.0 RPM
Main Thrust pressure	21,000 kN
CHD torque	3,800 kNm
Earth pressure (centre of face)	2.0 Bar
Advance rate	12 mm/min

Table 4. Typical boring parameters in a mixed face of soils and weathered to fresh granite

TBM Operational Parameters in Mixed Face of Silty Sands/Clays, Weathered & Fresh Granite	
Cutterhead Speed	1.7–2.6 RPM
Main Thrust pressure	24,000 kN
CHD torque	3,600 kNm
Earth pressure (centre of face)	2.0 Bar
Advance rate	5–9 mm/min

It was observed that the penetration rate was reduced by over 50% despite the cutterhead speed being increased by 35%, thrust increased by 22% and a minor increase in cutterhead torque; however, this was within acceptable ranges when boring through mixed face conditions with highly weathered, low-strength rock. Cutter consumption was within original estimations at this point, hence, there was no concern regarding these operating parameters.

Table 4 shows the typical boring parameters that were used from ring 125 through to ring 138. The geology through this stretch was made up of a mixed face including soils in the upper 60% of the face, weathered granite below this and fresh granite in the invert. These parameters needed to be carefully analysed as an intervention was carried out at this point and 33 cutters needed replacing. Also, similar boring parameters had been used from ring No. 138 to ring No. 259 resulting in a further 111 cutters needing replacement.

The initial observations and analysis of the chipping and radial cracks pointed to the failure mode of the cutters being caused by impact damage. Impact damage is caused when disc cutters are rotating through soft material or voids in the geology before coming into contact with hard rock. In general terms, the higher the cutterhead speed and the smaller the percentage of rock in the face, the more likelihood there is of impact damage to the cutters. However reasonably hard rock is usually present when impact damage occurs, so to confirm this theory Robbins deputed a geologist to site. The mixing chamber of the machine was emptied as low as possible to enable face mapping and rock identification was carried out. The granite in the invert was found to be fresh with weathering grade I and an estimated UCS value of 180 mPa (see Figure 4).

The information from the face mapping along with data showing that the cutterhead had been rotating at speeds of up to 2.6 rpm confirmed beyond any reasonable doubt

932 Pressure Face TBM Case Histories

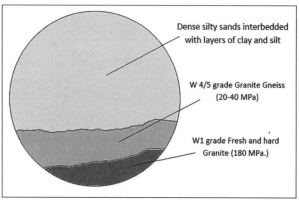

Figure 4. Face map showing mixed face conditions

that impact damage was the cause of many of the cutter failures. However, it was not the cause of all cutter failures.

The reason that 33 cutters needed replacing in a single intervention was most certainly caused by the failure of a single cutter initially, which then resulted in a cascade type wipe-out of adjacent cutters. When failure of a single cutter was not immediately identified it caused the adjacent cutters to take more load, ultimately causing failing due to overload. The overload situation is especially critical in the centre cutter positions, as their cutter tracks overlap. When failure of a centre cutter occurs, it is almost a certainty that a cascading failure will continue until the machine is stopped.

Cascade type wipe-outs are not uncommon when boring in rock or mixed face conditions containing rock. With open type hard rock machines, they are generally caused by cutter bearings failing and subsequently the cutters becoming blocked. In the case of broken discs, the damaged cutter parts are discharged almost immediately so boring is halted as soon as these parts are spotted and this prevents a wipe-out from occurring.

Wipe-outs due to broken discs are more difficult to prevent with EPB machines, as the displaced cutter discs can remain in the mixing chamber for extended periods of time before being discharged through the screw conveyor. The operator has to rely on being able to interpret changes in the operating parameters, especially cutterhead torque and TBM rate of penetration, to identify cutter damage and to cease boring before a wipe-out occurs. Interpretation of TBM parameters isn't always straightforward even when boring in a full face of relatively homogenous strata. In mixed face conditions it is far more difficult as the operating parameters can vary significantly during the length of a single stroke.

Solution

Although it was counterintuitive the solution to improving overall production was to reduce the cutterhead speed, restrict penetration per revolution and carry out interventions more frequently. The aim was to prevent or at the very least minimise impact damage to the cutters and to also reduce the risk of wipe-out failures. In the long term this would result in less time spent carrying out lengthy interventions to replace large amounts of cutters. Based on Robbins previous global experience and discussions with AFCONS It was decided that while boring through the mixed face conditions a baseline rotational speed of 15.0 meters per minute would be used for the outer

Table 5. Final operating parameters to prevent impact damage in mixed face conditions

TBM Operational Parameters, Mixed Face Rock/Soils to Prevent Impact Damage to Cutters	
Cutterhead Speed	0.8 RPM (Maximum)
Main Thrust pressure	19000 kN (Maximum)
CHD torque	3,600 kNm (Maximum)
Earth pressure (centre of face)	2.0 Bar
Advance rate	5–7 mm/min (Maximum)

profile of the cutterhead. This equates to a cutterhead speed of around 0.75 rpm. The advance rate of the machine would also be restricted. To enable machine parameters to be refined to suit rock conditions, face mapping was carried out during every intervention. The agreed upon operating parameters are shown in Table 5.

Over the following 150 m of boring, the geology along the alignment gradually changed from a mixed face containing silty sands, clays and rock to a full face of granites. During this stretch the parameters were revised accordingly from the values detailed in Table 5.

Results

In the course of boring 187 m in mixed face conditions a total of 144 cutters needed to be changed due to impact damage and wipe outs. This equates to the consumption of a cutter for every 1.29 m of bored tunnel. From the introduction of refining boring parameters through to the breakthrough of the machine, a total of almost 650 m of boring, 253 cutters were replaced which equates to 2.57 m of boring for each cutter. This was an improvement of almost 100%. The first 388 m of boring had taken 11 months. The initial drive of 150 m, setup of the shaft and installation of the remaining TBM gantries was completed by mid-May but the following 240 m of boring was not completed until mid-December. A month was lost due to issues unrelated to the geology; this means that it took six months to complete 250 m of tunnel, an average of 40 m boring per month. From the change in operational parameters and increased interventions the machine completed the remaining 640 m in 6.5 months, an average of almost 100 m per month, which equates to an increase in production rates of 250%.

The total time taken to complete the first drive was 17 months. The lessons learned regarding the most suitable operating parameters were utilized on the second drive, the face mapping from the first drive aided better planning of interventions for the second drive, and hence less interventions were required. The second drive was completed in 12 months despite being driven in almost identical geology to that of the first drive.

BANGALORE METRO

The Chickpet to Majestic Station section of the North-south corridor of Bangalore Metro Phase I consists of 750 m of twin bored tunnels. The Two EPB TBMs that had been utilized to bore the sections leading up to Chickpet had suffered major delays mainly due to mixed face conditions on the alignment of the twin 432 m drives between City Market station and Chickpet station. The average time taken to complete these two drives was 17 months per drive, averaging only 25 m per month of boring progress. The slow progress on this section of the tunneling operations was delaying completion of the whole project so the project owner, Bangalore Metro Rail Corporation (BMRC) and its contractor JV, approached The Robbins Company with a proposal to overhaul the machines and accept a contract to bore the remaining two drives. Although Robbins had not supplied the machines, they agreed to an operational contract. The agreement covered all tunnel and site works apart from muck transportation

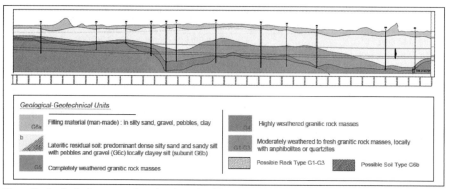

Figure 5. Geological cross section showing mixed face conditions

from site and segments casting/delivery. This was an industry first, wherein a TBM manufacturer had utilised their in-house expertise and knowledge to take on this level of responsibility for a project.

Geological/Hydrological

The geology between Chickpet and Majestic station is made up of fill material, residual soils, dense silty clay and varying grades of granite ranging from completely weathered through to fresh un-weathered rock. The piezometric level ranges from between 5 m to 10 m above the crown of the tunnel. The overburden along the alignment ranges from approximately 10–14 meters. Approximately 60% of the drives were expected to be driven through difficult mixed face conditions. In addition to the difficult geology the vast majority of the alignment ran beneath a densely populated area crammed with surface structures, many of which were poorly constructed. The geological long section can be seen in Figure 5.

Machine Launch

The first machine was launched in March 2015 into strata consisting predominantly of residual soils and dense silty sand. While boring through the residual soils over the initial 160 m the cutterhead rotational speed had ranged between 1.75 to 2.5 rpm and advance rates of up to 16 mm per evolution/40 mm per minute were achieved.

Mixed Face Conditions

From chainage 160 m weathered granite was encountered in the invert of the bore. The weathered granite acted as a conduit for water—the mixture of excavated clayey silt and water caused blockages of the cutterhead openings and a build-up of material on the plenum bulkhead. This resulted in high cutterhead torque, greatly reduced advance rates, and a substantial increase in muck temperature in the mixing chamber due to heat caused by friction. Interventions under hyperbaric conditions were undertaken to clear the choked cutterhead and material from the plenum bulkhead. Typically, an intervention needed to be carried out every 15 rings (22.5 m of boring) and each intervention would take over 24 hours. Up to 12 hours of this time was consumed waiting for the temperature in the chamber to drop from up to 65° Centigrade to below 35° Centigrade before personnel could enter. Cautious trials with various TBM operating parameters and foam injection/expansion ratios were carried out without any significant improvement. The problem was of course rooted in the ground conditioning regime. A good ground conditioning regime can be equally as important as the machine design and logistical aspects on any EPB project [Roby et al., 2014]. The ground conditioning agent that was being used up until this point was a basic foaming agent. After consulting various industry sources trials with an anionic water-soluble polymer (Mapedrill M1) were carried out. The product has an encapsulating effect on active clays, as well as acting as a lubricant. The polymer was mixed into the bentonite holding tank at a dosage of 5 kg per 1000 litres of water and pumped directly into the cutterhead and mixing chamber at a rate of 2–3 litres per cubic metre of excavated material. Cutterhead torque was reduced by approximately 40% within 20 minutes of injection. The overall effect of introducing the polymer into the ground conditioning regime was the reduction of cutterhead blockages to the point where cleaning was only carried out during standard interventions for cutter changes. It also reduced the friction and subsequent temperature in the chamber down to a maximum of 55° Centigrade.

There were no problems regarding cutter consumption while boring through the residual soils and weathered granites. At chainage 215 m the machine encountered fresh granite in the invert of the bore. Similar operating parameters to those detailed previously on the Chennai project were implemented successfully. The results were that cutter damage due to radial cracking was not excessive and consumption was close to pre-project estimates.

Charted and Uncharted Wells

At approximately the same chainage as the machine encountered fresh granites, it also encountered both charted and uncharted wells. Numerous wells had been identified in close proximity and directly along the alignment of the tunnels. The concentration of wells increased noticeably in the 140 m stretch of the alignment beginning at ring No. 215. There are 31 charted wells (Indicated by the blue and pink spots) in the vicinity of the twin drives through this stretch (see Figure 6)

All charted wells had been filled and capped to prevent loss of face pressure and/or ground conditioning foams rising to the surface. The uncharted wells were only identified by the presence of anomalies such as bottles and clay pots in the excavated material discharged through the screw conveyor (see Figures 7–8). By the time these items were discovered the machine had already passed through the wells.

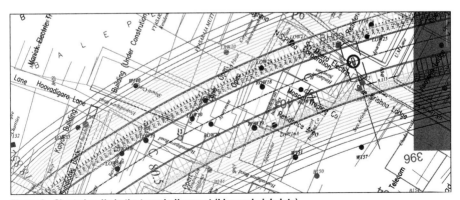

Figure 6. Charted wells in the tunnel alignment (blue and pink dots)

Figure 7. Clay pot found in excavated material　　Figure 8. Bottles found in excavated material

Table 6. Final design mix

Bangalore Metro Design Mix for Cutterhead Intervention Ground consolidation	
Cement	130 kg
Bentonite	50
Retarder	2 litres
Sodium silicate accelerator	55 litres
Water	820 litres
Set Time	12–14 hours

Failed Interventions Due to Air Losses

Neither the charted nor uncharted wells produced problems with interventions while boring through the soils; however, problems with face pressure during interventions commenced along the alignment where there was rock in the face as well as wells in the vicinity. Interventions had to be aborted due to excessive air loss, although in most cases the air loss could not be located on the surface. Initially the procedure adopted was to continue boring for 3–4 rings in an attempt to reach a location were air losses were less severe. Although this proved to be effective regarding reduction of air losses and facilitation of an intervention, the continuation of boring after observation of boring parameters dictated that an intervention was necessary invariably resulted in additional cutter damage. Due to the close-packed nature of the surface structures it was not possible to install safe haven grout blocks to prevent air losses during interventions, so a procedure of pumping a weak grout mix solution through the cutterhead and mixing chamber was adopted. The grout was injected up to a predefined pressure limit of 2.5 bar before pumping was stopped. Face pressure was then monitored for 30 minutes. A reduction in pressure in excess of 0.2 bar dictated that further grout injection was required. This procedure was repeated until a steady pressure was achieved.

Various design mixes were laboratory tested and trialled during interventions before the most efficient mix was found. See Table 6 for details of final mix design.

Initially, pumping weak grout solution after an intervention failed but this involved either boring forward to fill the chamber with material or filling approximately 60% of the chamber with weak mix grout. Boring forward even if only for 600–700 mm with damaged cutters ran the risk of further damage, hence filling the chamber with weak mix grout was the option chosen. In this scenario approximately 35 m³ of grout was required to consolidate the ground, fill the annular gap around the shields and fill the chamber. This methodology caused delays related to longer setting times for the grout plus additional time spent cleaning grout from the cutter housings and cutters. To reduce these impacts the procedure was revised to drop the level of material in the chamber by approximately 0.75 m to allow the flow of industrial compressed air into the chamber. This involved removing approximately 3–4 m^3 or 10% of the material in the chamber. The industrial air pressure was then slowly increased until it reached the design face pressure of between 1.3–1.6 bar (dependent on the overburden and water table at each specific location). If the pressure held steady for one hour without excessive air losses the chamber was emptied to below the rotary union and the intervention commenced. If air losses were excessive, weak mix grout was injected as described above. The volume of grout required for this procedure was reduced to approximately 15 m^3, a reduction of 57% compared to injecting week grout mix after a failed intervention. This methodology also drastically reduced the time spent cleaning grout from the cutter housings and cutters as they remained encased in soils, which prevented ingress of grout. The procedure was used successfully throughout areas of the alignment where excessive air losses occurred.

Surface Settlement

Another problem related to the wells along the sections of alignment consisting of mixed face of rock and soils was an increase in surface settlement. The increase in settlement occurred despite strict control and monitoring of TBM parameters and excavation volume. Face pressure had been maintained at the design value 1.8 bar, the amount of excavated material was checked using the TBM belt scales, load cell on the shaft gantry crane (when lifting muck cars), and visual inspection of each muck car during excavation. The density of the excavated material was confirmed daily and consumption of ground conditioning agents were input into the final calculations. Injection volume of the annular backfill grouting was above theoretical volume at approximately 115%. Backfill grout was also monitored closely in respect to machine advance in real time. These measures and checks confirmed that over-excavation had not occurred and that sufficient backfill grout had been injected into the annular gap.

After discussion it was agreed that the machine should continue boring, but face pressure would be increased to 2.0 bar. Additional secondary grouting would also be carried out immediately behind the tail shield of the machine. The increase in face pressure had to be aborted almost immediately due to conditioning foams being pushed through to the surface. The machine was stopped, and a weak grout mix was injected through the cutterhead and around the TBM shields to seal the leakage. The same procedure was used as described in the intervention consolidation operations. The total weak mix grout solution injected during this incident was over 60 m^3 and yet there was no sign of grout on the surface. Face mapping during cutterhead interventions confirmed that there were no existing voids in front of the machine and therefore the only viable explanation was that over decades of drawing water from the wells, the water that was recharging the wells had washed fines out of the matrix of the surrounding geology and that the vibration of the machine combined with the high-water table was causing fines to settle back into the matrix resulting in surface settlement.

As can be seen in Figure 9 the initial settlement was within acceptable limits of less than 5 mm for almost 10 days after the machine had passed. The rate of settlement then increased and continued to increase for up to 6 weeks.

Considering the time dependent nature of the settlement and the conclusion that it was being caused by vibration and infill of fines into the matrix of the strata, the preferred solution to enable continuation of boring was to revise the annular backfill grouting procedure. The parameters for primary and secondary grouting operations were changed to a system based on reaching a predefined maximum pressure rather than maximum design volume. The primary grouting cut-off pressure was defined as 2.0 bar and the secondary grouting pressure was increased to approximately 3.0 bar

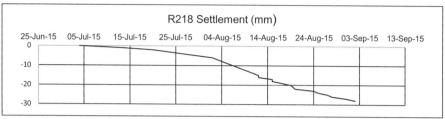

Figure 9. Measured surface settlement

depending on site conditions. An additional phase of secondary grouting approximately 30 meters behind the TBM was also introduced. This resulted in a substantial increase in grout consumption but was successful in controlling surface settlement. During the two drives the total backfill grout consumption including primary and secondary operations was over 190% of the original design volume.

Surface Vibration

As the percentage of rock in the face increased it was possible to gradually increase the cutterhead speed without risking damage to the cutters. By the time the first machine had completed almost 70% of its drive it was boring in a face consisting predominantly of rock. The cutterhead speed was operating at around 2.5 rpm, torque values were well within normal range, settlement issues and intervention problems had been resolved, and cutter consumption was not excessive; hence, the remainder of the drive was expected to be completed without any major issues. Unfortunately, this was not the case—there was one more challenge to be faced regarding machine parameters.

Up until this point there had been no complaints regarding vibration. This is probably because the alignment up until this location ran beneath small shops and other commercial buildings that had either no residential accommodation or accommodation only on the upper floors. The machine was now beneath a building that had residential accommodation on the ground floor and numerous complaints were made by panicked residents during the night shift. Due to these complaints and the possibility of further complications with local residents the boring operations had to be halted. The following day vibration monitors were utilized to determine the magnitude of the vibrations. Various cutterhead speeds between 1.0 rpm and 2.5 rpm were used during the monitoring period. Peak particle velocities were recorded ranging from at 0.3 mm/s up to 0.7 mm/s, rising steadily with a more or less constant correlation to increased cutterhead speed.

Several representatives of the residents were present during the monitoring and a comprehensive explanation of the results were conveyed to these representatives. The explanation was lengthy but conveyed in layman's terms why there was absolutely no risk of damage to the buildings. It is the unpredictability and unusual nature of a vibration rather than the level itself that is likely to result in complaints, the effect of intrusion tends to be psychological rather than physiological and is more of a problem at night when occupants of buildings expect no unusual disturbances from external sources [Schexnayder et al., 1999]. The residents remained unconvinced of the lack of risk and became irate when further discussion was attempted, so a compromise was reached wherein cutterhead speeds were restricted to 1.8 rpm during daylight hours and 1.2 rpm during the night. These speeds restricted the PPV's to less than 0.5 mm/s which was not in any way a noticeable difference to human perception, compared to PPV's of 0.7 mm/s produced at a cutterhead rpm of 2.5; however, the reduction did have a noticeable effect on production rates, reducing meters bored per day by around 40% for the following 60 m of boring.

Despite the difficult geology, charted/uncharted wells, settlement and vibration issues both drives were successfully completed by September 2016.

CONCLUSION

Mixed geology presents a multitude of challenges for mechanized tunnelling. Some but not all of these challenges are discussed in this paper. Technical advances in TBM technology have produced a generation of machines that are capable of traversing mixed geological conditions that would previously have required separate machines with different capabilities. On many projects TBMs would have been ruled out completely in favour of slower but more adaptable, traditional tunnelling methods. Although the new generation of TBMs is capable of managing a wide variety of geological conditions, the technology is only a part of the equation in successfully completing a project. Baseline parameters can be pre-calculated to suit different ground conditions, but these parameters should only be used a guideline. Experienced operational personnel need to be available on site to among other things: fine tune operating parameters, decide on the frequency of interventions, and revise ground conditioning regimes. Only then will the technology be utilized to its full potential.

REFERENCES

Roby, J Willis, D, 2014 Achieving fast EPB advance in mixed ground: A study of contributing factors, North American Tunneling Proceedings.

Schexnayder, C, Ernzen, J, 1999 Mitigation of night time construction noise, vibrations and other nuisances. NCHRP Synthesis of Highway Practice Issue Number: 218.

PART

Pressure Face TBM Technology

Chairs

Mina Shinouda
Jay Dee

Bryce Scofield
Traylor Bros., Inc.

Curved Microtunneling Alignments in the Design Toolbox

David Mast ▪ AECOM Technologies, Inc.
Patrick Dodds ▪ AECOM Technologies, Inc.
Paul Nicholas ▪ AECOM Technologies, Inc.
Rob Dill ▪ AECOM Technologies, Inc.
Alison Schreiber ▪ Northeast Ohio Regional Sewer District
Frederick (Rick) Vincent ▪ Northeast Ohio Regional Sewer District

ABSTRACT

The Doan Valley Relief and Consolidation Sewer (DVRCS) is a combined sewer overflow control project in Cleveland, Ohio, with a $13.5M awarded construction value. The project requires installation of 3,161 linear feet (LF) of 72-inch inside diameter (ID) trenchless sewer, approximately 1,523 LF of 48" ID open-cut sewers, approximately 50 LF of 48" ID trenchless sewer, and related structures. The Northeast Ohio Regional Sewer District (NEORSD, Owner) and AECOM (Designer) identified $4.5M in value engineering concepts and included significant construction flexibility in the bid documents to allow for innovative construction approaches by contractors. In one project reach, subsurface conditions and community impact considerations reduced the degree of flexibility that could be offered and resulted in a curved microtunneling construction method requirement in the bid documents, a first for NEORSD. This paper highlights the successful design optimization, construction method flexibilities, and reasons for the curved microtunneling requirement.

INTRODUCTION

The increasing success of installing new large diameter pipes on curved alignments using microtunneling methods has added a new tool for owners and designers to consider. In the Doan Valley Relief and Consolidation Sewers (DVRCS) project, AECOM and the Northeast Ohio Regional Sewer District (NEORSD) needed to install a shallow 72-inch ID consolidation sewer to convey combined sewage. Hydraulic constraints and existing utilities in the highly urbanized project area restricted the available tunnel horizons and horizontal alignments. Subsurface conditions in the available tunnel horizon included potentially flowing granular glaciolacustrine materials below the groundwater table. By incorporating curved alignments built using microtunneling methods, AECOM and NEORSD tried to address construction risks, reduce community impacts, and reduce overall bid costs. This is the first NEORSD construction bid which required microtunnel installation on a curved alignment.

PROJECT BACKGROUND

The DVRCS project is a combined sewer overflow (CSO) control project in the Easterly Wastewater Treatment Plant service area of NEORSD. Figure 1 illustrates the NEORSD service area and the approximate DVRCS project location. The project is located on the east side of Cleveland, Ohio. NEORSD entered into a Consent Decree with the United States Department of Justice, United States Environmental Protection Agency, and Ohio Environmental Protection Agency in 2011 to implement within 25 years a long-term control plan for CSOs currently impacting Lake Erie and its tributary waterways. DVRCS is part of Control Measure #8, identified in Appendix 1 of NEORSD's Consent Decree.

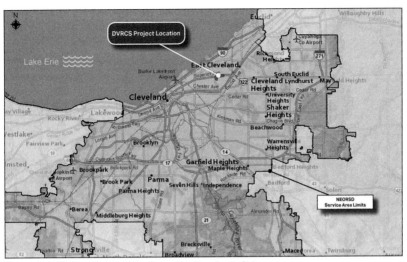

Figure 1. DVRCS project location in greater Cleveland, Ohio

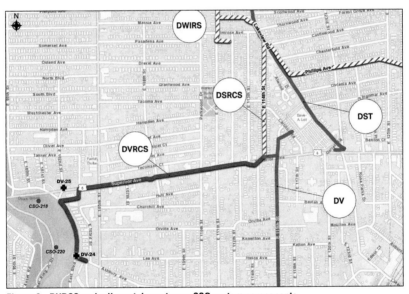

Figure 2. DVRCS and adjacent downstream CSO system components

The existing combined sewer system in the DVRCS project area, the Doan Valley Interceptor (DVI) system, experiences flow volumes in excess of the system capacity and, consequently, overflows frequently at regulator structures, discharging into local streams and rivers at the CSO permit points. The system also surcharges. Figure 2 illustrates the project area, including permitted CSOs and regulator structures in the DVI system which are addressed by DVRCS. The regulator structures being addressed include DV-24 (tributary to CSO-220) and DV-25 (tributary to CSO-218). Collection system hydraulic model simulations performed during planning tasks estimated that CSO-220 and CSO-218 activate 22 and 47 times, respectively, during the simulated Typical Year under baseline conditions. Simulations suggested the total CSO volume

Figure 3. Available sewer corridor on Superior Avenue (looking west)

from these two overflows could be approximately 130 million gallons (MG). As part of the Consent Decree, CSOs 218 and 220 are non-priority outfalls which are permitted to overflow up to three (3) times during the Typical Year after the proposed DVRCS is operational. Per final design model results, CSO-218 will activate three times in the Typical Year with a total overflow volume of 0.30 MG; while CSO-220 will activate once in the Typical Year with a total overflow volume of 0.18 MG.

The DVRCS is one of several legs of the Easterly CSO collection system. As such, it must convey flows from the local sewer systems to previously-built CSO collection system components. On Figure 2, the completed CSO collection system projects are shown with a diagonal hatch pattern (Dugway South Relief and Consolidation Sewers, DSRCS and Dugway West Interceptor and Relief Sewers, DWIRS); the components currently under construction are shown as solid gray lines (Doan Valley Tunnel, DVT and Dugway Storage Tunnel, DST); and the proposed DVRCS is labeled and shown as a solid red line. The DVRCS must convey flows from the area near CSO-218 and CSO-220 to the DSRCS, which then flows to the north.

DESIGN RISKS AND CHALLENGES

The DVRCS presented a number of challenges for hydraulic design and alignment optimization which ultimately dictated feasible construction methods. Significant project layout constraints are listed below:

- The downstream invert of the new DVRCS system was vertically constrained by the invert elevation of the existing DSRCS receiving sewer. The upstream invert of the new DVRCS system was vertically constrained by the invert elevations of the regulators DV-24 and DV-25 and the associated interceptors being relieved. This left an approximately 30-foot vertical window to build the entire DVRCS system with sufficient grade to achieve both storage and flowrate requirements.
- Regulator DV-25 is located on an existing 84-inch ID brick sewer which extends east to west underneath the center turn lane of Superior Avenue (see Figure 3). Combined sewage currently flows from east to west with relief at CSO-218 during high flow periods, so the existing brick sewer is deeper on

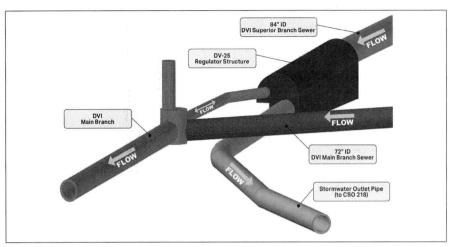

Figure 4. Existing DV-25 regulator with connecting structures

the west end of the project. Since the DVRCS will flow west to east and then turn north to DSRCS, the new DVRCS sewer had to cross underneath the existing 84-inch brick sewer at the east end of the project and extend only on the south side of Superior Avenue.

- The new sewer must convey both dry weather sanitary flows and wet weather flows without overwhelming the downstream receiving system. This meant the DVRCS Superior Avenue pipe must have a grade sufficient to produce acceptable flow rates at low flow periods, must accept and convey enough flow during the design storm to meet CSO control requirements and prevent surcharging of local sewers, but also restrict flow to less than 100 million gallons per day (MGD) to the downstream DSRCS system to optimize performance of large downstream CSO storage tunnels. The impact of this requirement was that the new regulators incorporated orifice plates for flow control, and the DVRCS Superior Avenue branch had to be built with a relatively shallow slope of 0.2% and a 72-inch I.D.
- The existing DV-25 and DV-24 regulator structures are 50-foot long brick structures with internal side spill weirs. The piping network in and out of the structures is very complicated in 3-dimensions, as shown in Figure 4 for DV-25. NEORSD and the design team identified that working near these brick structures would be a significant risk. As a result, the design incorporated new flow control structures into the project, both of which had to be built in the middle of the existing roadway upstream of the existing regulators. Figure 5 illustrates the location of new flow control structure DV-25A.

ALIGNMENT CONDITIONS—SUPERIOR AVENUE

The Superior Avenue portion of the DVRCS project is an urbanized corridor with commercial and residential structures built up to the right-of-way line. Figure 3, presented previously, is a photograph of Superior Avenue, looking west. Superior Avenue is a five-lane state route with sidewalks on both sides. Utilities extend underneath the sidewalks on both sides. The existing 84-inch sewer discussed earlier extends beneath the turn lane, as shown, and is the northern boundary of the new DVRCS alignment window. Overhead power lines on the south curb line are present for the entire proposed DVRCS Superior Avenue alignment, effectively becoming the southern boundary of the new DVRCS alignment window. As a result, NEORSD was faced with the potential for either 3,000 feet of deep open cut sewer trenches centered on the two eastbound lanes of Superior Avenue, or, at a minimum, three (3) or more tunnel construction shafts in that part of the road.

ALIGNMENT CONDITIONS—EAST 105TH STREET AND EAST BOULEVARD

The DVRCS system must relieve flows in sewers which overflow to CSOs 218 and 220. The optimal point to intercept the existing sewers is near their overflow regulators, which are at the west end of the project area. The sewer regulator leading to CSO-218 was already located in Superior Avenue and the new DVRCS was also in Superior Avenue, so picking up these flows was fairly simple (See Figure 5).

The sewers overflowing at CSO-220 were a bit more complicated. The flows in this area are conveyed by two (2) different sewers, located at different vertical positions, and these sewers were the ones which were most likely to surcharge. Early planning studies suggested the flows could be intercepted at the intersection of East 105th Street and Ashbury Boulevard, and then conveyed north to Superior Avenue (see Figure 6). However, this alignment was constrained by underground and overhead utilities, was in the City of Cleveland's plan for major redevelopment, and ground conditions were challenging.

AECOM and NEORSD identified a potential alternate route along the west side of East Boulevard in an existing City of Cleveland Park (see Figure 7). This alignment opened up the possibility of a shallow, open cut alignment, if sewer flows could be captured at a higher elevation. However, the area where the new DVRCS pipe would be required was a complicated configuration of crossing sewers and overflow pipes. Figure 8 is a 3-D rendering of the existing DV-24 regulator located in Ashbury Boulevard, the stormwater outlet pipe (SWO) which crosses through the intersection in Figure 7, and two existing interceptors that are also present in the intersection. AECOM designers found a solution wherein flow in the DVI Ashbury Branch Sewer was relieved upstream of DV-24 and the new DVRCS pipe could be installed shallow, passing over the DVI

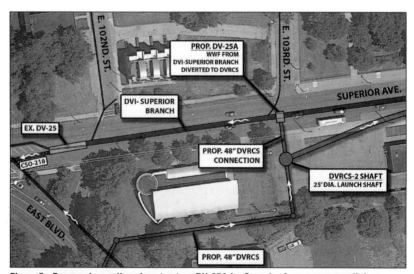

Figure 5. Proposed new diversion structure DV-25A for Superior Avenue sewer relief

Figure 6. East 105th Street, looking north. Intersection with Ashbury Boulevard is at the nearest stoplight.

Figure 7. East Boulevard, looking north. Ashbury Boulevard intersects from the right-hand side of the photo. City of Cleveland Park borders East Boulevard on the west side.

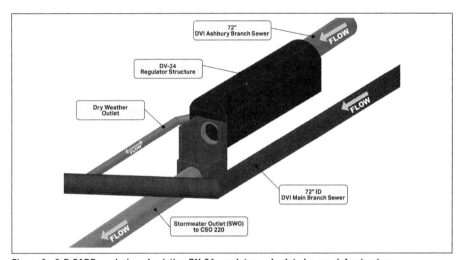

Figure 8. 3-D CADD rendering of existing DV-24 regulator and related sewer infrastructure

Main Branch sewer and the SWO pipes before turning to following East Boulevard north towards Superior Avenue.

RESULTS OF GROUND INVESTIGATIONS

As the vertical and horizontal constraints were being identified, AECOM and their subconsultant NTH Consultants completed a multi-phase geotechnical investigation along several potential sewer alignments, including Superior Avenue, East 105th Street, and East Boulevard. The East 105th Street and East Boulevard sewer reaches were necessary to relieve the existing sewers contributing flows to CSO 220. Early investigations suggested shale bedrock would be relatively shallow at the east end of the project, and the top of bedrock would drop in elevation to the west. AECOM and NEORSD initially hoped to keep the entire DVRCS project mostly in glacial tills and silty clay lacustrine soils, both of which are relatively good tunneling ground. However, AECOM found on both the East 105th Street alignment and the west end of the Superior Avenue alignment granular glaciolacustrine soils with a groundwater table above the granular layers. The ground conditions present beneath East 105th

Street and the presence of relatively expensive existing buildings at the intersection of East 105th Street and Superior Avenue (where a shaft would be needed to connect to the Superior Avenue reach) contributed to the decision not to build a new DVRCS consolidation sewer on that alignment.

Figure 9 illustrates the baseline ground conditions anticipated along the east end of the Superior Avenue segment. The Reach 1 tunnel will connect the DVRCS to the completed DSRCS construction shaft. This tunnel is anticipated to be mined through Chagrin Shale, a horizontally-bedded argillaceous shale predominant in the Cleveland area and through which many tunnels have been mined. Mining with a shielded open-face TBM and a two-pass lining system is very common in this type of ground.

Figure 9 and Figure 10 illustrate the baselined ground conditions anticipated in Reach 2 West and Reach 2 East, respectively. Reach 2 is the next upstream reach of the project extending from shaft DVRCS-2 at East 113th Street, west to shaft S-1 near East 108th Street. Reach 2 ground conditions are anticipated to be mixed ground of very stiff to hard cohesive glacial till, Weathered Shale and Sound (Shale) Bedrock. The tunneling face conditions are anticipated to range from slow raveling to firm, therefore open face tunneling is feasible. For baseline purposes, in naturally occurring soils, the contractor was told to anticipate encountering 10 boulders within the bottom portion of the glacial till deposit, with a size of 18 inches in maximum dimension. The baseline maximum UCS of boulders in Reach 2 was established as 18,000 psi, which is representative of competent sandstone or siltstone rock. An established Specific Allowance was used to compensate the contractor for time spent dealing with removal of boulders that obstruct and completely halt microtunneling or TBM-bored tunnel progress.

Figure 11 is an illustration of the baseline ground conditions anticipated in Reach 3. For this reach, NEORSD and AECOM determined that a requirement to use MTBM construction methods should be incorporated into the contract documents. The requirement for microtunneling was driven by the presence of coarse-grained glaciolacustrine sub-unit soils below the groundwater table, potentially within the tunnel face and above the tunnel crown. The baseline report identified this layer as having potential to behave as flowing ground. AECOM and NEORSD had initially hoped to lower the DVRCS sewer profile to stay in the glacial till soils and/or weathered shale materials. However, the downstream connection point and required dry weather flow velocities would not allow Reach 3 to be lowered any further.

SELECTION OF CURVED MICROTUNNELING

NEORSD has utilized microtunneling construction methods on projects since 2005 (project SEA 2A). In 2014, a NEORSD contractor proposed using a MTBM for a curved alignment on the DWIRS project. This was the first curved tunnel completed for NEORSD using microtunneling construction methods. Since then, two (2) more projects have incorporated contractual substitutions by contractors involving curved microtunneling construction, including a double-curve single drive which set a new record for North America. However, until this project, NEORSD had not specified curved microtunneling in the contract bid documents.

Curved microtunneling was thoroughly evaluated by NEORSD and AECOM to assess the apparent advantages and risks. This included evaluating relevant and available data for the longest single MTBM drives to minimize the number of intermediate jacking manholes along Superior Avenue. Ultimately, NEORSD and AECOM decided to specify the use of curved MTBM construction methods for the western approximately

Curved Microtunneling Alignments in the Design Toolbox

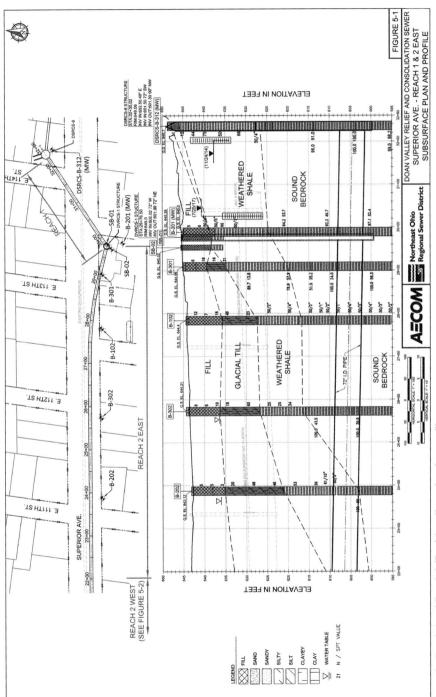

Figure 9. Reach 1 and Reach 2 (East) baseline subsurface profile

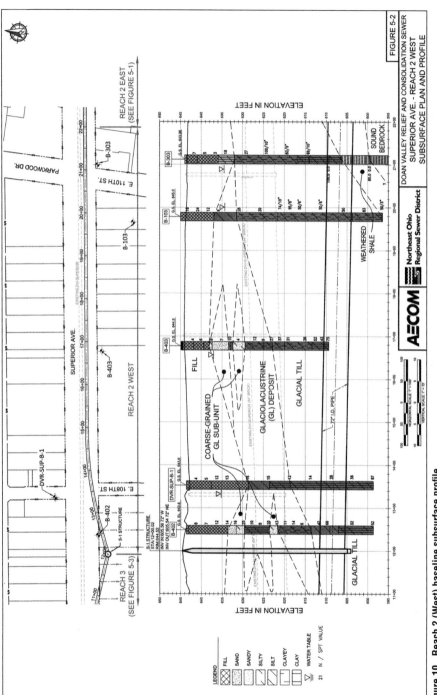

Figure 10. Reach 2 (West) baseline subsurface profile

Curved Microtunneling Alignments in the Design Toolbox 951

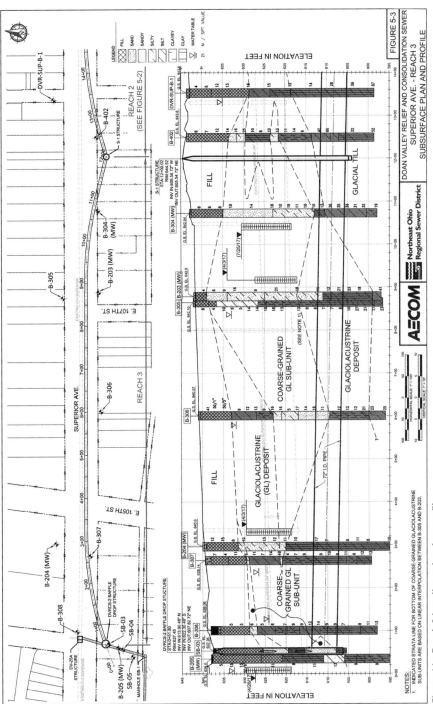

Figure 11. Reach 3 baseline subsurface profile

1,130 feet of the Superior Avenue segment. The advantages of the curved MTBM construction method on this project are listed below.

- Curved alignments would allow the main portion of the tunnel to be in the existing road right of way (ROW) but launch and retrieval shafts could be placed outside the road ROW, avoiding the overhead utility lines and eliminating the need for a shaft in the eastbound travel lanes (likely resulting in a full closure of the lane for several weeks or months).
- Pressurized face tunneling methods were expected to reduce the risk of ground movement in potential running ground conditions along the reach. This was particularly important to protect existing utilities in the road ROW, including electric duct banks, water lines, and the existing 84-inch ID brick combined sewer which this project was designed to relieve.
- The direct jack installation of final sewer pipe may improve the project schedule.
- Contractor's work and laydown areas could be located on property to the south of the ROW along with the launch and receiving shafts, providing construction site security and reducing impacts to the surrounding community.
- The MTBM receiving shaft will be finished as a 17-foot ID baffle drop structure for flow control. Placing this structure out of the ROW will allow NEORSD Sewer System Operations and Maintenance safer access to this structure.
- By leaving the MTBM launching shaft size to the contractors and requiring only a 60-inch ID precast manhole riser structure, the cost of shaft excavations would likely be reduced.

Ultimately, the contract documents required that an approximately 1,130-foot-long microtunneling pipe be installed in a single run between Shaft S-1 and Shaft DVRCS-2. The presence of a public library, multi-story apartment buildings, and a local park led NEORSD and AECOM to require the contractor to launch from the downstream shaft (S-1) and mine to the west. This reduced the impacts of the MTBM construction operation to the areas around the public library. The contractor was required to mine uphill at approximately a 0.2% grade. The proposed alignment began with an approximately 100-foot straight segment out of the shaft, followed by a 200-foot long curved segment with an internal radius of 800 feet. The alignment extended another 570 feet on a straight line before entering a 300-ft long curve (800-foot radius) and finishing at Shaft DVRCS-2. The alignment enters shaft DVRCS-2 on a curve. Subsurface property easements were obtained on properties which the tunnel crossed. Surface easements were obtained for properties with both a tunnel and shaft.

CONSTRUCTION METHOD FLEXIBILITIES IN BID DOCUMENTS

NEORSD has been a leader in the use of trenchless construction methods (tunneling) for several decades. As such, a large number of contractors have gained local tunneling experience. In order to manage the risks associated with the anticipated subsurface conditions, while at the same time structuring the contract documents to encourage competitive bid prices, several key decisions were made related to the construction requirements on this project. The key decisions are discussed in the paragraphs below.

Allowed Open Cut Sewer or Trenchless Construction Methods Along East Boulevard

As discussed earlier, ground conditions and existing buildings along East 105th Street suggested this route could be very difficult to construct. AECOM and NEORSD revised the hydraulic design of the system to allow for a shallow open cut sewer to be built along East Boulevard instead of requiring a trenchless method. This decision reduced the construction cost and allowed for local open cut contractors to participate in the project.

Contractor-Designed Temporary Earth Retention Systems

In general, NEORSD has seen increased competition when the contract documents allow the contractors to propose their own methods for shaft construction and earth/rock support. Exceptions to this practice are when the shaft is anticipated to be left open for an extended period of time (2+ years) or another contract may need to use or connect to the construction shaft, among others. On this project, the Contract Documents did not show any temporary earth support designs, but the specifications provided NEORSD expectations for several common types of shafts, including steel liner plate and wales, soldier pile and lagging, drilled secant pile shafts, and driven steel sheeting.

Multiple Trenchless Methods Permitted for Reaches 1 and 2

Reach 3 was the only long reach of the project for which the Contract Documents required that a specific trenchless construction method, microtunneling, be used. The Contract Documents allowed the Bidders to utilize one-pass or two-pass tunneling methods in Reaches 1 and 2, but they had to indicate their methods in the Bid Documents. NEORSD anticipated that microtunneling Reach 1 and 2 might be competitive. At bid time, NEORSD did receive a bid which assumed microtunneling for the entire Superior Avenue tunnel.

BIDDING

The DVRCS project was advertised for construction bids on March 13, 2018. NEORSD received three (3) bids. The total bid prices, which included several Owner-specified allowances, ranged from $13,524,335 to $21,159,000. The second lowest bid was at $13,868,280. It is worth noting that NEORSD had several similar tunneling projects advertised at approximately the same time, which may have influenced the number of bidders. The construction contract was awarded to Triad Engineering and Contracting, Inc. (Triad), who also submitted the lowest bid. Triad's bid was based on curved microtunneling in Reach 3 by their subcontractor, Super Excavators, and utilization of a shielded open-face TBM with ribs and lagging as a primary liner for Reaches 1 and 2. The second lowest bidder proposed to use an MTBM for Reaches 1, 2, and 3.

CONCLUSIONS

NEORSD and AECOM were able to optimize this project and identify over $4.5 million in cost savings during design by picking up flows on Superior Avenue upstream of the existing regulator, by moving and raising the East Boulevard sewer to an elevation where open cut construction was feasible, and by choosing alignments and vertical profiles that took advantage of good tunneling conditions. Specifying curved alignments built using microtunneling construction methods allowed construction shafts and permanent structures to be placed outside the ROW and away from overhead power lines, while also addressing risks from potential flowing ground conditions below the groundwater table. The construction flexibility permitted in the Bid Documents allowed contractors to bid their preferred construction methods for the tunnel and shafts and resulted in two of the three bids coming in below the published Engineer's Estimate but within approximately 3% of each other.

Forrestfield Airport Link—Project Challenges and TBM Solution

Karin Bäppler ▪ Herrenknecht AG
Michael Strässer ▪ Herrenknecht AG

ABSTRACT

A large number of successfully completed tunnelling projects in sensitive environments and airport areas show the highest technical and quality standards of mechanized tunnelling technology mastering project challenges and individual tasks in the interest of customers, clients and the environment. The Forrestfield Airport Link, currently under construction in Perth Australia, is one of these projects tackling challenges in a sensitive airport area being sensitive to any kind of settlement and subsurface conditions. The paper highlights the project challenges that require a well-adapted TBM concept and the experiences with the use of variable density TBM technology from the manufacture's point of view.

INTRODUCTION

The Forrestfield-Airport Link (FAL) project is a significant investment in the future development of Perth in Western Australia that will boost public transport for the surrounding areas. The project is a new passenger rail service that will link the eastern and foothill suburbs with the existing suburban rail network through a 7.6-kilometer long twin bore tunnel crossing beneath the physical barrier created by the airport. The tunnels cross beneath the Swan River, both runways of Perth airport and a series of rail lines with little impact to the land above, minimizing the environmental impact and delivering a sustainable transport solution for the community. The Forrestfield Line will add three new stations to the suburban rail network, Airport West, Consolidated Airport and Forrestfield. As much of the alignment crosses beneath sensitive areas such as the Perth airport precinct, including the airport runways, taxiways and buildings, there is a great demand to avoid disturbing airport operations and thus for safe and proven technology. The subsurface conditions along the tunnel alignment comprise a significant variation in ground conditions with weak rock underlying interbedded sands/silts and soft to firm plastic clays. This publication will describe in detail the solution of applied tailored machine technology for these specific project conditions.

SUBSURFACE CONDITIONS AND TUNNEL DESIGN

The total length of the bored tunnels of the Forrestfield Line is around 7,145 m. The tunnels have an internal diameter of 6.17 m and are lined with segments with a thickness of 300 mm and a 5+1 ring configuration. The centerline separation between the twin bore tunnels varies between 10 m to 20 m but is typically around 13 m to 15 m.

The expected ground conditions along the alignment during bored tunnelling comprise various geological units. The tunnel alignment between the Forrestfield Launch Box to the Bayswater Retrieval Box is split into three tunnel drives for the twin bore tunnels:

- Forrestfield Launch Box to Airport Central Station (1,908 m per drive)
- Airport Central Station to Redcliffe Station (2,515 m per drive)
- Redcliffe Station to Bayswater Retrieval Box (2,718 m per drive)

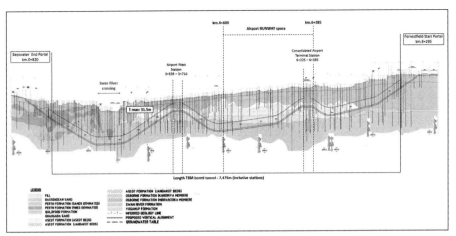

Figure 1. Geological longitudinal section FAL project

The subsurface conditions at tunnel face will generally comprise of weak rocks of the Osborne Formation such as sandstone, siltstone and mudstone underlying interbedded sands/silts and soft to firm plastic clays. The interpretation of the geological profile indicates that a minority of the TBM excavation will be in what can be considered as "uniform" tunnel face conditions with face excavation in a single geological unit where the material characteristics are reasonably consistent. Therefore, one of the key construction considerations was that the design and operation of the TBMs must accommodate variable excavation conditions for face stability, muck handling and discharge. To assess possible abrasiveness of the FAL geological units, a number of laboratory tests have been carried out. The results indicate a low abrasivity potential based on the abrasivity classification systems using Cerchar Abrasivity Index (CAI) and Laboratoire Central des Ponts et Chausee (LCPC). But, as it was noted that the quartz content of the various units, measured either by microscopic examination or by X-Ray diffraction, is generally in the range 40% to 60% and reaches over 90% in tests carried out on Guildford Formation samples, the potential for excessive wear on the cutter tools was considered for the TBM design.

The tunnel is along its entire length beneath the groundwater table. Groundwater levels vary seasonally but generally fluctuate within a few meters of the ground surface for much of the alignment.

The design of the tunnel lining is based on the predicted geological conditions and water pressure along the alignment and consists of five precast concrete segments plus a key segment having an inner diameter of 6.17 m. The segments have a thickness of 300 mm with double sealing gaskets, one at the inside and one at the outside, to mitigate the risk of water ingress. The annular gap between the outside of the segmental lining and the excavated surface of the ground is backfilled as the TBM advances to provide the bedding of the lining and to prevent subsidence. The backfill material is composed of a two-component grout and is injected through grout lines incorporated in the tailskin at the rear of the shield structure. The component A is pumped via a pipeline from the batching plant at the surface to the TBM; the B-component is transported into the tunnel in containers by the Multi Service Vehicles (MSVs).

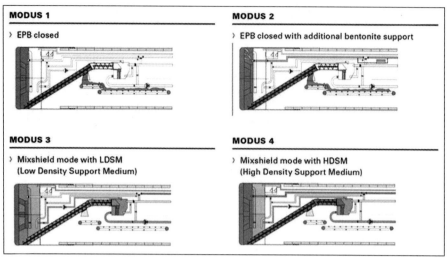

Figure 2. Variable density TBM, different tunnelling modes for optimum adaptability in difficult soft ground

In total two TBMs were planned and are now in use for the twin tunnel drives between the portals at Forrestfield and Bayswater. Tunnelling started at the launch box in the Forrestfield dive structure and will be retrieved from a retrieval box in the Bayswater dive structure. Intermediate working sites for TBM assembly, disassembly or spoil handling are not permitted within the Perth airport precinct.

TBM LAYOUT AND PROJECT-SPECIFIC DESIGN FEATURES

The Scope of Work and Technical Criteria (SWTC) required the TBMs procured for the FAL Project to be dual mode and capable of operating in both Slurry and Earth Pressure Balance (EPB) modes. The choice of operating mode is heavily influenced by the particle size composition and distribution of the excavation material. In the tender documents, a dual mode machine was the preferred variant. The sensitive section underneath the airport should be equipped with fluid support of the face, for the other sections also EPB mode would be possible. The two machines were designed and manufactured based on the SWTC requirements. The two identical shield machines with a diameter of 7.05 m are Variable Density TBMs designed with an operational pressure of four bar. Salini Impregilo S.p.A. and NRW Pty Ltd JV was contracted to construct the twin tube rail tunnels for the Forrestfield Airport Link project. The Variable Density TBM technology was considered for the excavation of the twin bore tunnels based on the predicted variable geological units of varying strength. This technology was developed based on the demand of today's specific project conditions often characterized by prevailing frequent changes of soils, rocks and mixed face conditions of soils and rocks at the tunnel face beneath the groundwater table. With this technology, it is possible to change between operation modes from earth pressure supported face to a slurry supported face with full control of face pressure. The transition between the operating modes can be achieved without the need for chamber interventions and without any need of mechanical modification in the excavation chamber or in the gantry area in the tunnel behind. The Variable Density TBM technology combines the individual advantages of both systems in one machine. It can also be operated using a high density material in the excavation chamber that would be too dense for classical slurry operation but that would be too fluid for a classical EPB operation.

The basic version of a Variable Density TBM is always equipped with a slurry circuit and its corresponding operating modes always function. Depending on the specific project conditions, an additional dry mucking system can be selected subject to a comparison of the investment against additional performance and cost for muck handling.

If the Variable Density TBM is fully equipped for both operating modes it would require two muck transportation systems in the tunnel. In a classical Mixshield mode (with air bubble for active face pressure control), a closed slurry circuit is required and if operated in a full EPB mode (dry system) muck cars or continuous conveyors are required to transport the excavated material. The Variable Density TBM can also be operated with classic earth pressure support in the excavation chamber and closed slurry circuit to convey the spoil from the tunnel. For all operation modes of the Variable Density TBM the muck is extracted from the excavation chamber by a screw conveyor. The further processing of the muck depends on the operation mode in place and the choice of logistics for material conveying. This can be changed from hydraulic transportation through pipes to belt-conveyor transport or muck car haulage. [1]

In the case of the two Variable Density TBMs in use for the FAL project, both TBMs are operated along the entire tunnel alignment in closed pressurized mode with a filled excavation chamber and a controlled support of the tunnel face pressure. This to guarantee ground stability during excavation. In both EPB and slurry mode, the muck is extracted from the pressurized excavation chamber by a screw conveyor. From there the muck is further processed with liquid mucking or hydraulic material transport via a closed, pressurized slurry circuit in the slurry mode or HDSM operation to a slurry treatment plant on the surface. For the hydraulic muck transport, the muck is transferred at the end of the screw conveyor into a slurryfier box to liquefy the excavated material. A roller crusher is installed in the slurryfier box to process the material to a size suitable for hydraulic mucking through the slurry circuit.

Based on the scope of works and technical criteria, the TBMs need to be operated in slurry mode when beneath the Airport Land but should be designed and manufactured to operate in EPB mode after Redcliffe station.

The predicted variable geological conditions of abrasive nature demanded an adapted cutting wheel design, dressed with 17-inch double disc cutters, cutting knives and buckets. The cutting wheel is designed with wear protection and wear detection systems. The wear protection is composed of two rows of grillbars, hardox plates in the face and gauge area and protection wedges. Wear detection systems comprise five sensors (pins) and one duct on the cutting wheel. The machines are also designed with 16 inclined drill port around the circumference of the shield body with an inclination angle of the drill pipes of 12° and having a nominal diameter of 100 mm. Four horizontal drill pipes are to be installed in the front shield for covering the tunnel face with nominal diameter of 100 mm.

SITE EXPERIENCE

The two 7.05 m-nominal diameter Variable Density TBMs were designed and manufactured by Herrenknecht in Germany and assembled at the Herrenknecht premises in Guangzhou in China.

The machines were launched at Forrestfield launch portal extending westwards to where the TBM drives will end at the Bayswater dive portal. Since there was not enough space in the area of the assembly and start shaft, the TBM was pre-assembled in the area of the site facilities. From there it was transported to the launch shaft

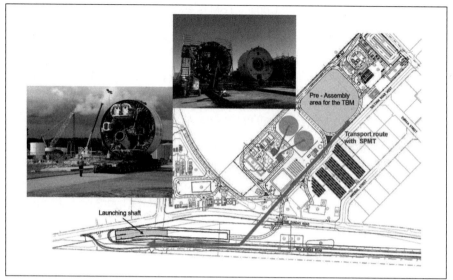

Figure 3. TBM transport to the launch shaft with a Self-Propelled Modular Transporter (SPMT)

over a section of about 500 meters with a Self-Propelled Modular Transporter (SPMT) and then the parts were lifted into the shaft with a mobile crane.

The first TBM (Grace) had its factory acceptance test on 10th February 2017 and arrived on site on 1st May in Perth with TBM assembly starting mid May 2017. The second TBM (Sandy) had its factory acceptance test on 10th April 2017. The TBMs started in the ramp area at Forrestfield station. The geological conditions along the first section comprised Guildford Formation, a sand-clay formation, that can be supported with a fairly simple medium (water). The JV achieved relatively high performances within the first 5 weeks of operation of up to 160 meters per week working 24 hours a day and 7 days a week without any incidents. After nine weeks of operation, there was a planned intervention. TBM operation then continued towards Airport Central station operating 24 hours a day for 5.5 days a week, as the station box was not ready.

After the Christmas break, during the excavation of the first tunnelling section of about 1.9km between the Forrestfield launch portal towards Airport Central Station, TBM Sandy experienced a small slurry/grout frack-out (with 75 mm settlement) when the machine reached the transition zone between the Guildford and Ascot formation (called the "scarp") that had been previously disturbed (by TBM Grace). The focus on the tunnel drives in general is on maintaining face stability and/or control of deformations within acceptable limits. Along the first section, the ground surface settlements were within the design tolerance of 10–15 mm.

The first tunnelling section indicates a change from full face Guildford Formation to mixed face geology comprising Guildford Formation (soils of generally fine to medium sand), Ascot Formation (generally coarser soil and gravel size particles), Gnangara Sand (fine uniform sand) trending towards full face Ascot formation.

Based on grading curves and borehole information the soils from Ascot Formation are predicted to be highly variable with respect to gravel content. These subsurface conditions indicate a sensitive tunnelling ground that requires the face pressure and slurry quality to be adapted in order to avoid settlement. The high settlement recorded in the

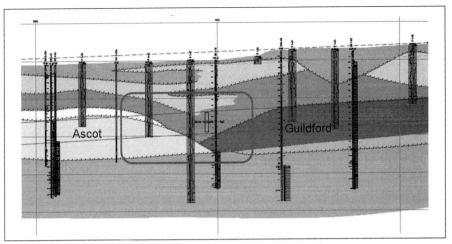

Figure 4. Transition zone between Guildford and Ascot formation ("Scarp")

scarp area most likely occurred due to the slurry penetrating into fine washed coarse grained soils, in combination with a variation in water pressure which resulted in insufficient support of the tunnel face and thus settlement. These variable ground conditions, respectively mixed face conditions with coarse grained soils content, demand an adequate slurry mix with an adjusted density of the slurry to be able to fill the pores and be able to support the tunnel face.

Following the above events Grace was halted for about 8.5 weeks, while Sandy was stopped for two weeks, before TBM operations resumed in mid-April 2018 (calendar week 16) with an adapted slurry mix. Both tunnelling machines arrived at the Airport Central Station in May 2018 where two concrete blocks had been built for TBM arrival in the station box.

The TBMs were then pushed over slideways to the other end of the station. Before the TBM restarted to continue operating to Redcliffe Station, various refurbishment and repair works and minor modifications to the TBM were carried out in the station box. These modifications included the installation of additional manual valves in the communicating pipeline in the working chamber, calibration bars at the inlet of the pump to limit grain sizes for density measurement and extension of flushing nozzles in the working chamber.

The lessons learned from the first tunnelling section, in respect of adapted slurry mix and monitoring of sensitive settlement relevant data, were implemented in the preparation of the follow-on tunnelling drives and in particular for the section beneath the sensitive airport precinct.

TBM Grace continued operation for the second section on 18 June 2018, and Sandy resumed on 16 July 2018. The subsurface conditions along this sensitive tunnel 2.5km section beneath the airport runway towards Airport West Station are mainly characterized by Ascot Formation for about the first half with a short mixed face section (Ascot and Osborne Formation) before then entering full face in the sand dominated Osborne Formation, with a fines content of about 20–30%. With particular focus on this section, the Joint Venture is strongly focusing on an adapted slurry concept with defined slurry properties and slurry mixes for the respective formations.

CONCLUSION

The Forrestfield Airport Link Project is one of the latest challenging tunnelling projects where mechanized tunnelling technology of the highest technical and quality standards is being successfully applied to a sensitive tunnelling area. A major part of the tunnel alignment crosses beneath the Perth airport precinct with sections beneath the runways and taxiways demanding safe and proven technology. The applied variable density TBM is not only advantageous in respect of safe and secure tunnelling processes, with less impact to the surroundings, but also show an advance strategy in respect of variable tunnelling sections where a flexible adaptation of operation principles is needed. This technology offers great opportunities for future challenging tunnelling projects in sensitive environments.

REFERENCE

[1] Bäppler, K., Burger, W. Battistoni, F., Tunnelier à densité variable—associant deux technologies de tunnelier pour sol meuble, AFTES Congres Lyon 2017.

Interpretation of EPB TBM Graphical Data

Keivan Rafie ▪ Santec
Steve Skelhorn ▪ McNally

ABSTRACT

Tunnel construction using a tunnel boring machine (TBM) involves a highly complex operation. Such processes generate large amounts of data that can be used for monitoring, reporting and analysis. Major TBM manufacturers have developed software systems to support tunnel contractors and their site teams in both data management and analysis. These programs are mostly web-based and have many advantages.

Data acquisition cannot prevent breakdowns from occurring but can facilitate forensic investigations to quickly determine the root cause of a breakdown and provide basis for implementing corrective actions. This paper analyzes these data acquisition tools and presents case studies, primarily involving earth pressure balance (EPB) TBMs, to illustrate how the formation of critical interpretations can be made from user-defined charts and diagrams to diagnose issues and optimize TBM operational parameters.

INTRODUCTION

The storage and visualization of measured values acquired by sensors and recorders is a crucial element of TBM tunneling. All of the work being performed by the machine is documented in terms of the recorded data to allow the complete or partial tracing of the tunnel construction in real-time or after completion.

This information could help engineers and operators examine a vast and complex set of data related to TBM operation that cannot be ascertained in the field by the TBM engineers or work crews, particularly when visualized in a graphic format. The examples of measured data and sample graphs presented in this paper are mainly taken from Earth Pressure Balanced (EPB) TBMs, but the logic behind the interpretation of these examples can also be applied to data from hard rock or slurry TBMs.

TBM DATA ACQUISITION AND VISUALIZATION SYSTEM

The purpose of a TBM data and acquisition system can be summarized as the "acquisition, processing, storage, display and evaluation of all data connected to the tunneling machine." A TBM data acquisition system continuously records and visualizes all measured data in a pre-determined cycle. Logging, however, occurs only at specific times. The average time period between logs can be individually selected for each measuring point but is set to 10 s for most parameters.

The operating phases of the tunneling machine are generally classified according to three periods: advance, ring building and standstill. These three phases form a unit called a ring. The data for each ring are usually stored in separate consecutively numbered files. An immediate correlation to the respective construction phase can be made based on the ring number, file date and file time. The measured data acquisition program automatically opens after each system restart and loads all required program components into its memory. It then acquires, stores, and visualizes the currently available measured data. (See Figure 1)

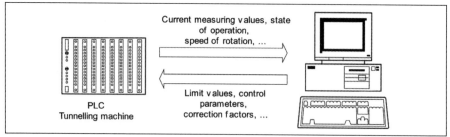

Figure 1. TBM PLC and data acquisition system

INTERPRETATION OF TBM OPERATIONAL GRAPHS IN CASE STUDIES

Some of the most common graphs representing the general status of TBM operations are ram extension, rate of advance (ROA), thrust force, cutter head torque, EPB/slurry pressures, weight/volume of excavated material, and grout volume. Of course, illustrating too many parameters on one chart makes interpretation more difficult, so there must always be a compromise between amount of information given and the clarity of the graphics.

EPB Pressure Graphs

Case Study 1

In successful EPB operation, face pressures should be maintained at all times and monitored with the data acquisition system. Pressure of material in the chamber could be assessed by information available from EPB cells. TBM operator closely monitors excavated material and adjusts the type and amount of water, bentonite, polymers, and foam to ensure that the material is properly forming a plug to resist piezometric and ground pressures.

Below graphs from pressure cell data are among the most used graphs in EPB tunneling and demonstrate the difference between correct (A) and incorrect (B) operation. Excavations similar to graph (A) result in safe and steady progress while performances similar to graph (B) are usually linked with significant loss of ground and surface settlement.

Case Study 2

The EPB pressures for the top, middle and bottom sensors used in this case study are presented in the Figure 3, which shows that the bottom sensors record higher pressures due to the higher density and greater compaction of the excavated material. The top sensors record the least pressure and fluctuation because they have less direct contact with the soil and mud in the chamber but middle and bottom sensors have more fluctuations as shown in Figure 3.

The proper estimation of material contact and density in an excavation chamber is important. It is common practice for the operator to perform and complete an excavation with full level of material in chamber when using EPB TBMs. However, TBMs must sometimes be operated in semi-open, in which only a portion of the face is balanced by excavated material. These operating conditions are generally determined by engineers based on the ground conditions and stoppage time. Compared to the semi-open mode, full material contact in a chamber requires more thrust and torque from the TBM and increases the equipment wear and the cost of replacing excavation tools

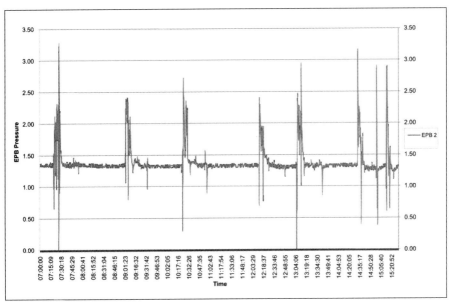

Figure 2A. Maintaining face pressures in proper TBM operation

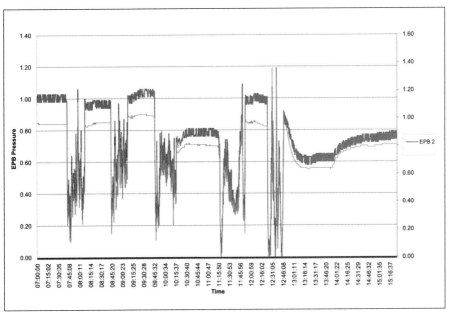

Figure 2B. Pressure drop during excavation

on the cutter head. Working in a semi-open mode could alleviate these issues but is not advisable if there is a high risk of ground collapse and overexcavation. Other scenarios, such as preparation for cutter head maintenance or leaving the TBM unused for long periods of time, could also influence decisions regarding the level of material that should be present in a chamber during tunneling operations.

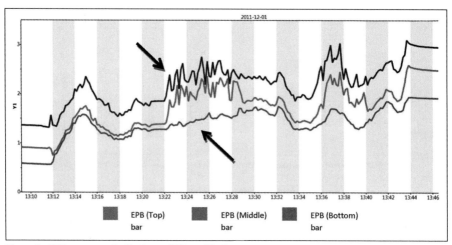

Figure 3. EPB pressures for the top, middle and bottom sensors

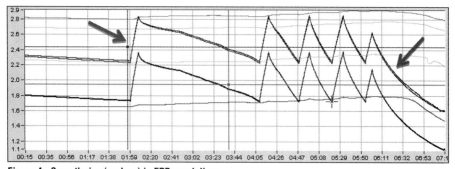

Figure 4. Smooth rise (or drop) in EPB graph lines

Case Study 3

The smooth rise (or drop) in EPB graph lines (Figure 4) indicates the passage of gaseous or liquid material into (or out of) the excavation chamber which occurs mostly during the TBM ring build phase. Soft rising curves may be the result of ground water filling the chamber or the injection of ground conditioning material (foam, water or compressed air). A smooth drop in cell pressure suggests the leakage of air or water through porous ground, a tail shield, purge line or screw conveyor.

Thrust, Cutterhead Torque, RPM and Rate of Advance

Higher advance rates in TBMs are generally achieved in two ways.

1. A higher cutterhead rotation speed, which increases the distance that cutters or rippers travel and thus their work per unit time (mm per min). In this case, the cutterhead torque will increase. An increase in cutterhead torque can also result from other factors such as poor ground conditioning or high material density in the excavation chamber.

2. Higher forward forces in TBM cylinders to make cutters and rippers excavate more intensively, thereby increasing the cut depth per cutterhead rotation. In this case, both the TBM thrust force and its torque will increase. The TBM thrust can also be increased due to shield friction with the ground or the

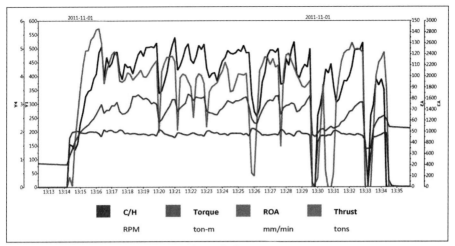

Figure 5. Higher thrust forces and increased rate of advance

TBM's pulling force due to its weight (a factor discussed later in conjunction with contact force).

Case Study 4

Scenario (B) is illustrated in the graph below. The cutterhead rotation speed is set at approximately 2 rpm, so the occasional increase in torque is due to higher thrust forces exerted by the propulsion cylinders at that moment and increases the rate of advance.

It should be noted that higher efficiency is usually achieved in soft ground with a lower RPM and higher thrust forces for deeper excavations, whereas cutters break into hard rock by rolling on it. Therefore, better advance rates occur with higher RPMs.

Cutter Head Contact and TBM Thrust Forces

The graphical representation of the relationship between TBM thrust force and contact force of cutter head is mainly used to identify any opposing forces to the TBM other than the excavation face. In general, the TBM thrust is used to maintain EPB pressure, push the material in the chamber, and pull the gantries and frictional forces of the shield.

The thrust left over from propulsion energy is consumed by the cutterhead in the form of the contact force required to cut through the ground. Because the parameters other than contact force are relatively constant during normal TBM operation, TBM contact and thrust forces are typically synchronized in their fluctuations. Therefore, any mismatch in the graphical patterns between these two forces suggests a status change in other parameters and usually indicates an obstacle during operation.

Case Study 5

The theoretical graphs shown in Figure 7 show a sudden drop in contact force despite a constant increase in the thrust force (Graph A). These data could indicate collapsed ground around the TBM shield or an entrapped gantry back in the tunnel. Variations

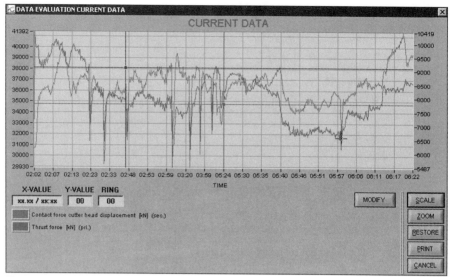

Figure 6. TBM contact and thrust forces synchronized in their fluctuations

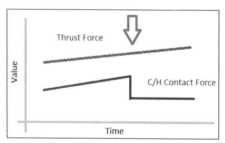

Graph A. Sudden drop in contact force

Graph B. Gradual drop in contact force

between the contact and thrust force that are more gradual could result from a change in tunnel slope or the accumulation of heavy, dense material in the chamber (Graph B).

Identifying Overexcavation

Most EPB machines today are equipped with weight sensors and laser scanners to estimate the weight and volume of excavated ground. The theoretical weight and volume that a TBM data acquisition system is expected to show is usually calculated manually based on the TBM dimensions, ground properties and advancing distances. These figures are compared with the quantities shown on TBM graphs to check for overexcavation. This information is also useful in the analysis of excessive volume loss and settlement.

Case Study 6

TBM advance with overexcavation can generally be recognized on TBM data graphs by a higher- than-normal grade in the excavation weight or volume line. For example, the following theoretical graph illustrates three sets of data from different advances. Line A, which is the typical advance at a constant rate of excavation, is usually the

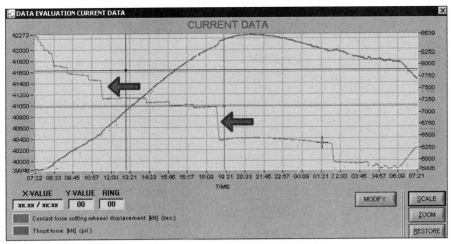

Figure 7. Sudden drops in contact force while TBM thrust force is increasing

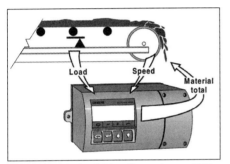

Figure 8. Typical weight scaling system on TBM conveyor

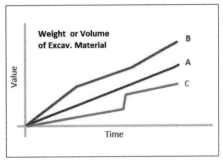

Figure 9. 3 scenarios for excavation weight/ volumes

preferred scenario and ensures the consistency of other parameters, such as ground conditioning and screw conveyor speed during the shove.

Compared to Line A, Line B has a steeper increase at the beginning and end of excavation period and ends at a higher value. This condition can be interpreted as general overexcavation. Line C represents a normal extraction scenario, except for a very sudden increase over short period of time that could indicate overexcavation with a sudden rush of material through the screw conveyor and out of the chamber. However, other aspects, such as those stated below, must be considered to achieve a realistic understanding of the data.

1. The higher slopes in the graph lines based on material weight and volume can be a result of higher advance rates. Thus, the final excavation values must be checked. The screw conveyor rotation speed in those time periods can provide insight as to whether the high extraction rates were intentional or due to ground conditions.

2. The theoretical material weight depends on the advance distance of the TBM and the density of the ground, but other factors, such as added water or ground conditioning agents, should also be considered. In regard to ground conditioning material, only the liquid portion will affect the material weight, so the foam expansion ratio (FER) should not be considered in calculations. The

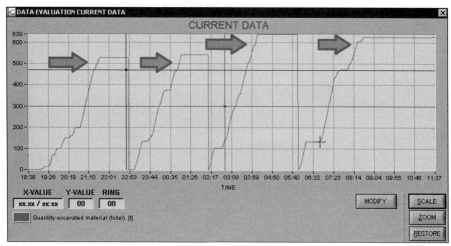

Figure 10. Weight scale data showing higher quantity of excavated material for last 2 advances

FER of the ground conditioning material added to the chamber has no effect on the weight calculations but should be considered with regard to the material bulking factor when scanned by a laser on a conveyor belt for volumetric data.

Figure 11. TBM operator checking TBM parameters to assess theoretical weight/volume

3. Comparing the weight and volume data/graphs of different advances only makes sense if the level of material in the chamber after each advance remains full or relatively constant. For example, if an operator has started an advance with a half-full chamber and decides to fill the chamber to its maximum level, less material will be extracted and shown in the data, even though the same amount of material has been excavated from the ground. On the other hand, when a TBM chamber must be emptied during an advance (e.g., the last advance before cutterhead maintenance), the TBM data will show a higher amount of extracted material than average. To eliminate this problem, engineers also look at the rolling average of values for several consecutive rings, which eliminates the effect of chamber space and gives a more realistic picture of the scenario to identify possible overexcavation.

4. Added water should be considered in theoretical calculations. Occasionally, depending on ground conditions, most of the injected water is absorbed by the ground, and sometimes added water only replaces the water in saturated soils.

5. The calibration of weight scales and laser scanners must be part of a contractor's regular maintenance program. Some weight sensors are very sensitive to misalignment and curves in the TBM conveyor belt, while laser scanners could have inaccurate readings depending on their position and air/light interference, such as dust. Utilizing two belt scales and observing their averages can also aid in identifying errors and obtaining more realistic results.

Grouting System

Two-component grouting (A+B) systems through the tail shield have been one of the most problematic areas in TBM tunneling. Proper grouting is important to prevent ground movement and surface settlement due to volume loss at the tail void. Grouting also stabilizes segmental lining in the ground and improves a tunnel's watertightness.

Information available in TBM data acquisition systems can show early signs of system malfunctions and indicate which components require attention or which control settings need to be modified.

TBM data loggers typically record flow, pressure and volume parameters for each grout line. To check the quality of grouting behind segments and ensure that the correct dosage of accelerator (B) is mixed with part (A), TBM data for injected volumes should be checked against the theoretical volume of voids behind the segments. Gauge cutter wear should be considered in theoretical calculations, particularly for larger TBMs. Understanding the bore and cut diameters in hard and soft ground types can also lead to more accurate calculations.

If the grouting volumes are lower than their theoretical values, other data must be checked to identify and solve the problem. The simultaneous spike in grout pressure and halt of grout flow in the Figure 12 is commonly an indication of blockage in the line. If grout volumes cannot be achieved when all lines are in operation, then the pre-sets and cutoff levels should be checked. Generally, grout pressures must overcome hydrostatic pressures by 1–2 bar behind segments.

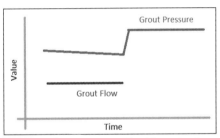

Figure 12. Grout pressure without flow, showing the blockage in lines

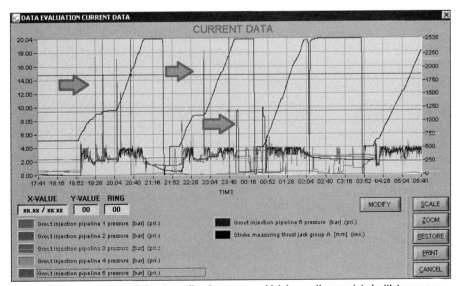

Figure 13. Grout lines 5 and 6 showing spikes in pressure which is usually associated with temporary blockage

Case Study 7

If grouting volumes exceed their theoretical values, assessments must be performed to identify any channeling of grout to the surrounding environment or leakage through the tail shield. In some instances, high-pressure grout finds its way to the excavation chamber, mixes with the excavated material and exits through the screw.

Propulsion Cylinders and Ring Build

Information and graphs derived from TBM data on propulsion cylinders can be used to analyze several aspects of their operation, including ring build and steering. These data can also explain damage

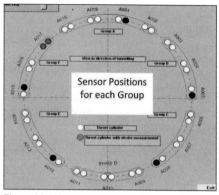

Figure 14. 19 rams collected in 6 groups (A-F); the location of a representative ram from each group is shown in black

to segments that occurs after installation. TBM data acquisition systems generally display the pressures and extension of ram groups using the sensors on a representative ram from that group. Figure 14 shows information collected on 19 rams in 6 groups (A-F). The location of each group's representative ram is shown in black.

Case Study 8

Figure 15 shows the pressures applied to group of segments during ring erection. The lines representing each group show a sudden jump from zero, indicating that the rams have been extracted to hold each segment after its erection. The lines also indicate common slow pressure loss due to micro-movement of the TBM and ring compression in the tunnel. However, an excessive loss of pressure in any group could loosen the adjacent segment and cause vertical (step) and horizontal (gap) misalignments.

On the other hand, excessive pressures on cylinders can cause damage, particularly around the circumferential joints of the segments in front of the cylinders. Ring build reports from the guidance system (as shown below) must be studied in conjunction with ram pressure graphs to confirm the location of segments in relation to the propulsion cylinders and explain the damage incurred.

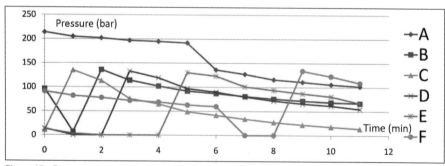

Figure 15. Ram pressures applied to segments during ring erection

CONCLUSION

The graphical representation and measured values of TBM data can assist contractors by providing information that helps TBM crews increase the reliability and productivity of TBM operations. Such an advantage would ultimately lead to fewer breakdowns and lower tunneling expenses. Data acquisition and visualization alone does not benefit the contractor unless the data is accurately interpreted. The utilization and correct interpretation of data acquisition systems' outputs could greatly enhance the control of the excavation and operation of various tunnel boring machine systems. As TBMs grow in size and complexity, advances in data monitoring and presentation to optimize TBM parameters will likely continue as well. The correct interpretation of these data is essential for the effective utilization of these tools and to ensure efficient and productive tunneling operations.

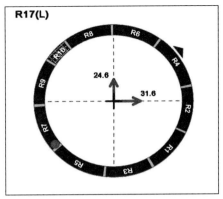

Figure 16. Sample ring build report from the guidance system

The key to success in EPB tunneling is proper engineering combined with experienced operators. Data acquisition cannot prevent breakdowns from happening but allows the rapid identification of the root cause of a breakdown and the timely implementation of corrective actions.

Mechanized Tunneling at High Pressure—More Than Just a Stronger Bulkhead

Werner Burger • Herrenknecht AG

ABSTRACT

Mechanized tunneling at face pressures far above 3.5bar has become common practice in recent years. Closed mode TBM layout and operation at very high face pressure affects a large number of TBM design aspects from seal systems to chamber interventions. The paper will cover consequences, technical solutions and project examples of TBM design for high pressure applications.

INTRODUCTION

In mechanized closed mode TBM tunneling two tendencies could be observed throughout the last two decades. There was a strong market demand towards larger diameters and parallel to that high pressure applications due to deeper tunnel alignments or deep river and strait crossings became more common practice (Figure 1). There is obviously a relation between diameter increase and required overburden or depth of alignment (Burger 2015) but also technical developments of seal systems and the introduction of professional diving technologies (Burger 2011) into the field of tunneling opened the door for the planning of deep tunnels with higher face pressures.

Whereas in the past TBM structure and especially seal system design have been the key areas of concern for TBM applications beyond 3.5 bar an additional focus had to be put on TBM design features for high pressure chamber access employing professional divers using mixed gas and saturation diving technologies for face pressures above 4.0–5.0 bar.

At the same time such professional diving technologies entered into the TBM tunneling world significant efforts have been made in developing technologies reducing the need for or frequency of pressurized chamber interventions. Such technologies comprise tool condition monitoring systems, improved cutter tool seal systems or designs for atmospheric cutter change and chamber installation maintenance.

SHIELD STRUCTURE

Common practice in TBM design today is to use Finite Element (FE) software for the structural calculation of TBM shields and bulkheads. Resulting stresses and deflections can be calculated and predicted with a high accuracy, however the resulting values are only as good as the input data or load assumptions they are based on. The load conditions for shield structures under operation can be quite complex and typically a number of different load cases or load combinations must be considered.

In general, there are the outer ground and water loads acting on the structure. At the same time the inner loads created by shield thrust or articulation cylinders, cutter head torque or thrust, ring erection, dead weights etc. have to be taken into account (DAUB 2005). For the majority of the tunnels there is a number of different outer load and bedding conditions together with their anticipated operational loads along the

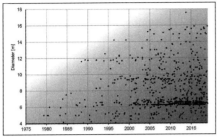 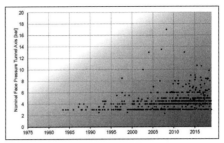

Figure 1. Development of diameter and layout pressure of Herrenknecht TBMs

tunnel alignment. All normal scenarios of operation including chamber access as well as accidental situations like for example shield steering over-reaction, excess shield thrust due to unforeseen friction resistance or local overpressure caused by backfill or pre excavation grouting must be taken into account.

To confirm the structural shield design for the anticipated tunneling conditions not only the safe stress limits but also the allowable maximum deflection can become the decisive factor. The limitation of structural deflection for functional reasons is typically one of the important criteria for tailskin designs or any structures adjacent to seal systems. Staying within still safely acceptable variations of ring build clearance or seal system gaps can become, (especially for larger diameters) the leading argument.

TAILSKIN SEAL SYSTEM

Wire brush tailskin seal systems have become the state of the art solution for pressure face TBMs over the last few decades. Previously also employed single or double lip rubber tail seal designs mostly disappeared from the market. Depending on the anticipated operational pressure two to five rows of wire brush seals are installed, creating a 1–4 chamber system. The chambers between the individual rows of wire brush seals are filled and continuously supplied by a special tailskin sealant mastic that creates in combination with the individual wire brush barriers an efficient seal system against the outer water and backfill grout pressure.

The pumping and distribution network of the sealant mastic is of major importance for the function and durability of the entire system. Automatic interlocks must stop the TBM advance in case of distribution network failures. System variants replacing the last (rear) row of wire brush seals against a multi-layer spring steel plate design are also common in the industry with the rear row providing then a more robust barrier against the backfill grout or mechanical impacts. Of major importance especially in combination with initially fluid two-component grout mixtures are segment design features restricting grout penetration under the brushes into the sealant chambers along the longitudinal joints.

Four row, three chamber systems have successfully been used for long duration high pressure operations above 10 bar at the Istanbul Strait Crossing or the Lake Mead Intake #3 projects.

Depending on the allowable gap variation or radial working range of the individual wire brush seals, a single chamber length can become 400 to 600 mm. The requirement for a multiple chamber arrangement of 4 or even 5 chambers (5 or 6 rows of wire brush) will result in a very long tailskin design with associated disadvantages for

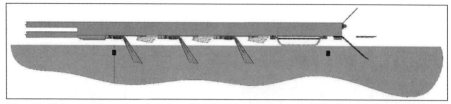

Figure 2. Three chamber tail seal design with inflatable emergency seal (blue) and double shell tailskin structure (Istanbul Strait Crossing TBM)

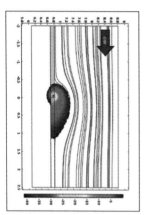  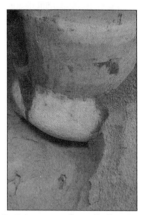

Figure 3. Tailskin freeze channel development: theoretical simulation—test rig preparation—test result → freeze barrier at the rear end of the tailskin mock-up

curved drives. Also, even for long 2 m segments the last built ring will still be sitting inside the tailskin at the end of the next stroke.

A frequent requirement for high pressure applications is the provision of a standby emergency system within the tailseal arrangement in order to stop leakage and allow replacement and repair of the front rows. Typical solutions for this is the installation of an inflatable seal in one of the rear chambers of the system. However, as a result of this additional feature the length of the seal system will be further increased. In addition, the most frequent reasons for tailseal leakages are either mechanical damage from reduced clearance of the rear seal rows or contamination with hardened grout in the rear sealant chambers, both of these present some risk for the proper function of the emergency seal once activated.

In order to overcome this conflict and create a reliable technical solution a system integrating freeze channels in the rear end of the tailskin has been developed and tested by Herrenknecht. The system does allow to create a freeze barrier around the annular gap just behind the tailskin without the need to increase the tailskin length and is independent from any potential grout contamination in the mastic chambers.

MAIN BEARING SEAL SYSTEM

The main bearing or cutter head drive seal system is the most important seal system on any TBM. The system separates the main bearing cavity from the pressurized excavation chamber. Even a small amount of leakage caused by a malfunction of the system that finds it way into the bearing cavity will result in a main bearing contamination and in the worst case to a main bearing failure with severe consequences for the TBM or entire project. Even so main bearing failures have become a very rare incident in TBM tunneling, it has to be noted that the majority of such failures have not been caused by main bearing overload, but by contamination, as a consequence of seal failures.

It is obvious that the requirements and workload for the main bearing seal system is increasing with higher face pressure. Also, geometrical aspects resulting from larger TBM diameters like seal/raceway diameter and gap variation due to deflection present additional challenges to the system.

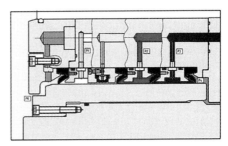

Figure 4: Example of a 4 lip main drive seal arrangement (yellow: P0, face pressure excavation chamber; white: P1, flushing grease pressure chamber 1; red: P2, constant fill lube oil pressure chamber 2; blue: P3, air pressure leakage detection chamber 3; grey: P4, bearing cavity pressure)

The most common solution in the industry today is the installation of multiple lip seals running on hardened seal races combined with grease or oil lube systems including associated monitoring systems for pressure, flow etc. However seal materials, details and geometries as well fixation systems are specific and different for different TBM manufacturers.

For high, and very high pressure Herrenknecht TBMs are equipped with multi row rubber compound lip seals supported by a pressurized cascade lube system in order to control and limit the work load or pressure drop for each individual seal (Figure 4). The different pressures in the cascade system are controlled by automatic systems following the face pressure.

The controlled pressurization of the individual chambers assures that none of the individual seals has to operate with a pressure difference exceeding its safe working range. Figure 5 shows the pressure development in the individual chambers during a system test of approximately 60 min duration increasing the face pressure from atmospheric to 14bar. In addition to the dynamic lip seal system a static seal can be an integrated part of the seal system that additionally isolates, when activated, the

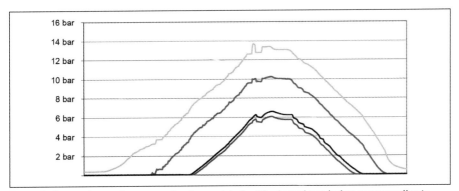

Figure 5. Face and cascade pressure development during 60 min seal test (colors corresponding to chamber colors in Figure 4)

chambers 2, 3 and the bearing cavity from the face pressure. The static seal can be activated during extended TBM standstill periods with no cutter head rotation.

It is mandatory that even for normal, non-pressure cascade systems as used for face pressures up to 4 bar, all relevant parameters of the seal system and its supporting grease supply and lube oil system are permanently monitored and recorded in in the data recording system of the TBM. Alarms and control interlocks have to be in place stopping the TBM advance and cutterhead rotation as soon as preset values cannot be maintained or sensor failures are detected. Also, a defined start and stop sequence for the cutterhead rotation has to be part of the PLC programming. Options for manual overrides should be restricted. Especially for TBM refurbishments, the full knowledge of such PLC program details is of major importance.

CHAMBER ACCESS

With higher face pressures, the influence of installations and design features for chamber access becomes a decisive element for the TBM and gantry design. As soon as the anticipated face pressure exceeds 3,6 bar and in any case above 5 bar, preparation for mixed gas breathing, rescue—or even crew shuttle transfer have to be taken into account (Burger 2011).

Extended decompression times of more than 60–90 min as a consequence of higher face pressures may be acceptable by special permit as long as sufficient man lock space or an additional pressurized chamber for stretching during the airbrakes of the oxygen decompression procedure is available.

Depending on the TBM diameter man lock designs with headroom space exceeding the required minimum of 1,6 m for Europe or 5 ft for US can become difficult to install in the classical way attached directly to the shield bulkhead. In such cases, a three chamber system with an additional third pressurized chamber that is integrated part of the shield structure but not a decompression chamber can be a solution. A three chamber system provides benefits not only for extended decompression time but also for work organization during chamber interventions since it provides an additional, larger size room for chamber work preparation and support independent from the decompression chamber as well as a first safe area of escape in case of fast withdrawal from the excavation chamber.

Especially when oxygen decompression is foreseen, housekeeping and cleanliness of the decompression chamber becomes an even more important requirement, therefore the presence of such a third support—or staging chamber provides significant practical benefits in such cases.

As soon as shuttle handling for rescue or crew transfer becomes part of the chamber intervention procedure, it has a major impact not only on the shield but also on the entire gantry design. Crew shuttle sizes can reach easily dimensions of 1.6 m × 1.6 m × 6.0 m and weights of 4 to or more asking for significant available space for safe passage and fully mechanized means of handling. In addition, its safe transfer has to be guaranteed at all times and therefore all involved functions have to be considered to be part of the essential services (ITA 2015).

For large diameter TBMs with an inner tunnel diameter above 7 m the shuttle transfer through the gantry normally can be realized without major compromises. For inner diameters between 4 m and 7 m gantry preparation activities to clear the shuttle passage that can, depending on the TBM type, take several days are most likely required.

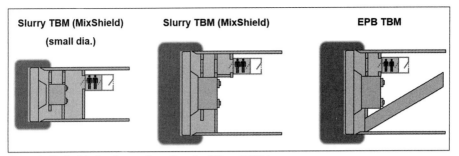

Figure 6. Typical 3 chamber configurations for different TBM types

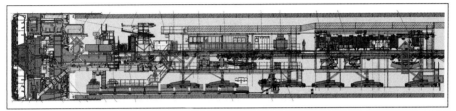

Figure 7. Crew shuttle transfer at the 13.7 m Mixshield for Istanbul Strait Crossing

For inner diameters below 4 m an engineered solution for safe crew shuttle transfer through he gantry has to be considered critical to impossible.

Once the shuttle passage through the gantry has been solved, the next hurdle is presented by the ring build or erector area. Here the influence of the machine type, if Slurry TBM or EPB becomes significant. Whereas the piping for the slurry circuit on a Mixshield can be arranged in a way to provide as much as possible free space in that area, the screw conveyor structure of an EPB presents a much more significant obstacle for the shuttle passage. On the other hand, this restriction is eased to some extend due to the fact that the application of the EPB technology still has its limitations when it comes to higher face pressures.

For a special designed Mixshield of an OD of 6.8 m or more a permanently available shuttle docking position and passage way through the ring build area to the shield bulkhead can be achieved. To realize a comparable configuration for an EPB it does need a shield diameter exceeding 11.0 m. In order to overcome the passage restriction through the ring build area access tube structures for temporary installation have been designed and provided in special cases. However, it has to be clear that the shape and length of such access tubes has to fulfill minimum requirements for safe access and egress as well as rescue that have to be demonstrated for each individual case. In addition, it has to be accepted that the installation and commissioning of such access structures prior to use will need major preparation time and it is therefore obviously difficult to be considered as a solution for frequent chamber interventions.

NON EXPOSURE MAINTENANCE AND CONDITION MONITORING SYSTEMS

At higher face pressure the technical effort, cost and time for chamber interventions is increasing. Therefore there is a strong tendency within the mechanized tunneling industry for developments to reduce the frequent need for or even eliminate pressurized chamber interventions.

Project layouts or tunnel alignments that take advantage of natural safe havens for interventions is and has been the target and will be for sure in the future. However, the majority of the "high pressure" projects just cannot offer such options due to ground conditions or alignment restrictions.

Increasing the durability of cutterhead structures and cutting tools in combination with a better understanding of the TBM ground interaction, soil conditioning or chamber muck flow are R&D activities that have been ongoing for decades and that will continue to provide positive results in the future.

The availability of reliable cutting tool condition monitoring systems for function, wear or load must be considered as a big step towards reducing the need for interventions in closed mode. Access need for tool inspection is reduced to access need for tool replacement once worn out or damaged. Other means for remote inspection like chamber camera systems are additional helpful features for this approach.

The availability of ground prediction systems can also help to reduce the risk of major damage to cutting tools or cutterhead structure resulting from unforeseen ground conditions or man made obstacles.

Structural cutterhead damage as a consequence of man-made obstacles and/or undetected tool failure have been reasons for major long term chamber interventions in the past.

The introduction of accessible cutting wheel technology with the possibility for tool change under atmospheric conditions was a major step in reducing or even eliminating the need for pressurized chamber access for cutting tool change. At the Istanbul Strait crossing project all tool changes at high face pressure were done under atmospheric conditions from inside the cutterhead (Burger 2016).

However, it is a fact that this technology is still limited to larger machine diameters and requires compromises for chamber muck flow and soil mixing due to geometrical requirements for the cutterhead structure caused by sufficient space for interior man access.

The development for next generations of the accessible cutterhead technology is still ongoing and the present limitations related to TBM size and type may be reduced in the future. However, safe working and rescue conditions for the confined workspace conditions inside the cutterhead structure cannot be compromised.

The use of robotic technology inside the cutterhead chamber already had its first experimental application at the Bouygues/Dragages TMCLK Project in Hong Kong.

Table 1. Present state of the art for the applicability of Herrenknecht accessible cutterhead technology

TBM Type	Mixshield	Variable Density TBM	EPB
TBM diameter	>11.5 m	>11.5 m	>14.0 m
Rock condition	++	++	--
Soft ground condition	++	++	++
Mixed face condition	+	+	+
Clogging	+	+	--
Chamber muck flow	++	+	--
Chamber mixing	n.a.	±	--
Applicability	yes	yes	no

This development is ongoing and may open up additional options for application in the excavation chamber itself or in combination with the accessible cutterhead technology.

But, even in light of all such successful developments targeting the reduction or elimination of chamber interventions, it is mandatory that especially high pressure, closed mode machines are still prepared for pressurized chamber interventions to account for unforeseen events or conditions. How far such preparation has to be permanently on board has to be decided based on a project risk analysis, but in any case features and adaptations that cannot be installed or modified once the TBM is underground should be provided and installed from the beginning. As a minimum, engineered plans and solutions for remaining adaptation for high pressure intervention should be available.

CONCLUSION

High pressure mechanized tunneling in excess of 8 bar is safely possible today given the right team and equipment. The affected elements of the entire tunneling process are far more than just the seal systems reaching from TBM design aspects to intervention procedures and technologies to aspects of digitalization. For all involved parties, it has to be clear that the higher the pressure the smaller is the bandwidth of tolerance for mistakes and the more important is a close collaboration of all involved from an early design stage to completion of excavation.

REFERENCES

W. Burger, 2015. Super diameters—Design aspects for very large TBMs. RETC 2015, pp. 1223–1231.

W. Burger, 2011. Interventions and chamber access in pressurized face TBMs. RETC 2011, pp. 1036–1047.

DAUB (German Committee for Underground Construction), 2005. Recommendations for Static Analysis of Shield Tunneling Machines. www.daub-ita.de.

ITA (International Tunnelling and Underground Space Association), 2015. Guidelines for good working practice in high pressure compressed air. ITA Report No 10-V2 (2019 revision under preparation). www.ita-aites.org.

W. Burger, 2016. Istanbul Strait Road Crossing Tunnel. WTC Conference 2016.

Proposal of Some Cuttability Indexes for Evaluating the Performance of Mechanical Excavators Using Conical Picks

Okan Su • Bulent Ecevit University
Xiang Wang • Chongqing University of Technology

ABSTRACT

In this study, rock cuttability indexes are evaluated based on peak cutting force, specific energy and rock strength. For this purpose, a number of samples were collected from different parts of China and the mechanical properties of the samples were initially determined. Then, a series of cutting tests in unrelieved cutting mode were conducted at the linear rock cutting rig. The tests were performed at the cutting depths ranging from 3 mm to 18 mm. As a result of the tests, the effect of a significant cuttability index, which is the ratio of cutting force to cutting depth (FC/d), on the uniaxial compressive strength was investigated. Based on the acquired data from linear cutting test rig, the correlations between the ratio of FC/σ_c and cutting depth and also between the ratio of (SE/σ_C) and cutting depth were examined and their validity to be as new cuttability indexes were checked. High correlation coefficients among the cutting variables verified that the new cuttability index values can be successfully used for designing the cutterheads or the drums of mechanical excavators as well as evaluating their cutting efficiency.

INTRODUCTION

Mechanized equipment has gained popularity within the past 50–60 years since the tunneling has developed rapidly. Beforehand, drilling and blasting method was applied by using efficient rock drills. However, technological developments led to excavate the rock by different type of machines. Although there were attempts to improve mining machines for various geological conditions, their cutting conditions were very limited at the beginning of 19th century. The first mechanized machines were not even able to excavate the full face and they were essentially used for cutting soft rock. The average advance rate was very low and also downtime of the machine was very high. Automation and mechanization systems of the machines provided practical applications of muck remove, continuous production and supporting, and safe working conditions. However, these machines involve high capital costs and the cutterheads play an important role so as to ensure the maximum cutting performance.

Rock cuttability, which means the ease or difficulty of excavations, is determined to make a performance assessment of a machine. It is also defined as the resistance to cutting by mechanical tools such as pick cutters and roller cutters (Copur et al. 2003). In general, the cutting performance of mechanical excavators are predicted by cuttability index values as well as other governing factors summarized below.

- Machine utilization time (%)
- Cutter consumption (cutter/m^3)
- Average advance/progress rate (m/day)
- Instantaneous cutting rate (m^3/h)

- Specific energy (MJ/m^3)
- Tool forces acting on the pick (Normal, Cutting, and Side forces) (N)

There are a number of researchers who proposed different types cuttability/excavabilty indexes (Scoble & Muftuoglu 1984; Bilgin et al.1988; Tsiambaos & Saroglu 2010). A detailed review of the cuttability indexes can be found in Dey & Ghose (2011). Some researchers also proposed different indexes based on RMR (Bieniawski 1974) and Q rock mass classification systems (Barton 1974). However, these indexes require a number of inputs such as RQD, stress conditions, quartz content, and so on. In this context, it is clear that geological parameters such as rock strength, hardness, abrasivity, rock quality designation (RQD), jointing, and bedding can influence the cuttability of rock, in other words the performance of the machines. Although there are a number of cuttability indexes, no particular method is universally accepted.

LABORATORY STUDIES

Conical picks, which are also known as point attack picks, have been widely applied in underground and surface excavations. The complex interactions at the rock and pick interface are of importance for the engineers since the cutting behavior of the pick is affected by rock hardness, toughness, abrasivity, and so on. These properties would not be predictable due to heterogeneous nature of rock and may result in incorrect estimation of machine performance.

Among all performance prediction methods, linear cutting tests avoid basic assumptions and do not take into account geometrical complexities as in the empirical models. They offer the most straightforward way about the mechanisms of rock cutting and they provide significant contribution for understanding the cutting characteristics of picks.

The cuttability behavior of a conical pick was monitored in the linear cutting test in order to propound new cuttability indexes. For this purpose, a number of sandstone blocks were collected from the fields around the Chongqing, China. The mechanical properties of the samples were determined according to ISRM (1981) standards and the results are summarized in Table 1. Following the mechanical tests, an experimental study by using conical pick was carried out. The cuttability behavior of the pick was evaluated by the gathered data from the linear rock cutting rig shown in Figure 1.

The samples with dimensions of 150 mm × 150 mm × 200 mm can be accommodated in the linear cutting test rig. The maximum load capacity of the rig is about 50 kN in the normal direction and 100 kN in the cutting direction. The cutting depth is fixed at the beginning of the test. The sample is placed in steel box and clamped by bolts. A 3D load cell, mounted between the pick holder and the stiff base, records the normal and cutting forces. Moreover, a displacement transducer monitors the cutting distance and the recorded data is transferred to the computer.

Table 1. Mechanical properties of rocks

Rock Name	ρ	σ_c	σ_t	E
Sandstone 1	2.22	17.91±4.18	1.64±0.10	3.00
Sandstone 2	2.43	79.20±5.43	4.97±0.48	15.94
Sandstone 3	2.36	52.99±2.08	3.67±0.25	5.07
Sandstone 4	2.35	59.80±6.14	3.93±0.44	5.50
Sandstone 5	2.59	85.98±9.33	3.69±0.42	6.31

where ρ is the density of the rock (g/cm^3), σ_C is uniaxial compressive strength (MPa), is Brazilian tensile strength (MPa), E is the elasticity modulus (GPa).

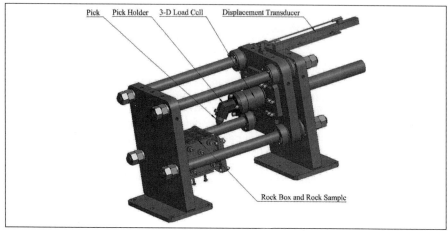

Figure 1. Linear rock cutting test rig

Table 2. Cutting test results in unrelieved cutting mode on five sandstone samples

Rock	d	FC_p	FC_m	FN_p	FN_m	FC_p/d	SE
Sandst. 1	3	0.89	0.55	0.82	0.59	0.30	6.47
	6	2.01	1.26	1.76	1.22	0.33	6.07
	9	3.40	2.03	2.94	2.10	0.38	3.88
	12	4.77	2.74	4.06	2.53	0.40	3.4
	15	6.77	3.42	4.64	3.19	0.45	2.61
Average						0.37	4.49
Sandst. 2	3	2.71	1.41	2.30	1.59	0.90	29.24
	6	5.66	2.66	4.43	2.89	0.94	11.67
	9	8.86	4.11	6.75	4.54	0.98	11.21
	12	13.55	6.60	9.93	6.54	1.13	8.7
	15	17.99	8.25	11.94	8.05	1.20	6.7
Average						1.03	13.50
Sandst. 3	3	1.52	0.93	1.54	1.10	0.51	14.6
	6	3.26	1.61	2.59	1.73	0.54	7.13
	9	5.46	2.99	4.78	3.38	0.61	6.02
	12	8.82	4.60	7.38	4.69	0.73	5.28
	15	11.66	6.22	9.61	6.26	0.78	5.16
Average						0.63	7.64
Sandst. 4	3	1.84	0.93	1.57	1.07	0.61	14.05
	6	4.67	2.37	3.83	2.60	0.78	8.55
	9	8.85	3.71	5.93	3.82	0.98	5.86
	12	12.94	5.71	8.87	5.80	1.08	5.82
	15	18.18	8.18	13.79	7.75	1.21	4.84
Average						0.93	7.82
Sandst. 5	3	2.68	1.35	2.57	1.69	0.89	10.89
	6	4.90	2.22	4.05	2.69	0.82	7.72
	9	8.40	3.28	5.93	3.52	0.93	4.67
	12	13.85	5.58	7.79	5.39	1.15	3.94
Average						0.95	6.81

where d is the cutting depth (mm) FC_p is the peak cutting force (kN), FN_p is the peak normal force (kN), FC_m is the mean cutting force (kN), FN_m is the mean normal force (kN), SE is the specific energy (kWh/m^3).

The cutting tests were performed with a conical pick having a cone angle of 80°. In the course of experimental tests, an attack angle of 55°, a rake angle of −5°, and a clearance angle 15° were applied. Besides, a wide range of cutting depth from 3 mm to 18 mm was selected in unrelieved cutting mode. The measured forces are then evaluated for the performance assessment of mechanical excavator. Additionally, the specific energy was calculated and the results are presented in Table 2.

PROPOSAL OF NEW CUTTABILITY INDEXES

It is necessary to take into account of the fact that cuttability of rock is not evaluated by itself. A series of laboratory tests must be conducted and the entire test results should be assessed together. In this sense, an increasing number of cuttability-rock properties relations have been formulated so far. Bilgin et al. (2006, 2014) carried out linear cutting tests on 22 different rock specimens varying from 10 MPa to 170 MPa in order to estimate the cutter forces and specific energy from the rock properties. They emphasize that uniaxial compressive and tensile strength of rocks are the most important rock properties affecting the cutting behavior of the conical picks. Based on the relieved and unrelieved cutting tests, they proposed Equations 1–2 for unrelieved cutting conditions.

$$FC/d = 0.826\ \sigma_c + 21.76 \quad \text{(EQ 1)}$$

$$FC/d = 12.625\ \sigma_t + 8.78 \quad \text{(EQ 2)}$$

Furthermore, Liu et al. (2009) determined the cutting force as a function of compressive strength, tip diameter and cutting depth for coals. Williams and Hagan (2014) performed the cutting tests at 2.5 mm and 5 mm depth of cut and they found reasonable linear correlations between the strength and cuttability of rocks for 44 rock samples obtained from coal measure rocks. They also pointed out that their results are also consistent with the results of Roxborough (1987). According to these correlations, if rock strength is known, cutting force acting on a pick can be predicting based on the variation in cutting depth.

In this study, the applicability of the FC_p/d ratio as a cuttability index was also investigated on five sandstone rocks and the average values are summarized as follows (Table 3).

Data from this table can be compared with the data in Bilgin (2006) which shows a linear relationship between the ratio of FC_p/d and the uniaxial compressive strength. As given in Figure 2, it is apparent that both of the trend lines have approximately the same slope. Thereby, the results of Bilgin (2006) is validated with the results of this study and confirmed that the ratio of FC_p/d can be successfully employed as a cuttability index for the performance assessment of excavation machines once a conical pick is used.

On the other, we also propounded alternative and new cuttability indexes based on the test results obtained from experiments. As shown in Figure 3, the ratio of peak cutting force to uniaxial compressive strength (FC_p/σ_c) and the ratio of specific energy to uniaxial compressive strength (SE/σ_c) can be taken into account for the performance assessment of mechanical excavator.

Table 3. The average values of the FCp/d ratio for various cutting depths

d (mm)	FC_p/d (kgf/mm)
3	37.1
6	103.2
9	63.4
12	93.3
15	94.9

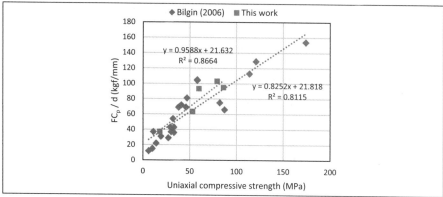

Figure 2. The relationship between the ratio of *FC/d* and uniaxial compressive strength

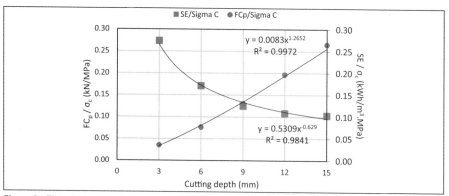

Figure 3. The relationships between the ratio of $FC_p/\sigma_C - SE/\sigma_C$ and cutting depth

As can be seen, the variation of two different cuttability indexes would be crucial in a rock cutting operation. If the compressive strength of rock is determined, the peak cutting forces and specific energy can be predicted by Equation 3 and Equation 4, respectively.

$FC_p/\sigma_c = 0.0083 \, d^{1.2652}$ (EQ 3)

$SE/\sigma_c = 0.5309 \, d^{-0.629}$ (EQ 4)

According to Equations 3–4, it is observed that the relationship between the ratio of FC_p/σ_C and cutting depth is exponential with a determination coefficient (R^2) of 0.99. Moreover, it was found that the relationship between SE/σ_C and cutting depth is also exponential on a downward trend. The ratio of SE/σ_C decrease with increasing cutting depth. The reason is that lower amount of energy is required to remove the rock from rock surface in deeper cutting conditions. However, it is a fact that the dataset should be expanded to increase the confidence of test results.

CONCLUSIONS

While a variety of index test have been proposed so far, this paper gives an account of new definitions of cuttability that can provide a comprehensive measure of machine performance in conjunction with uniaxial compressive strength. In accordance with the unrelieved cutting tests performed on five sandstones by using a conical pick, the cuttability characteristics by means of cuttability indexes were discussed.

Experimental results revealed that the ratio of FC_p/d is reliable for the performance assessment of excavation machines. In addition, it was found that there are significant correlations between the ratios of $FC/\sigma_C - SE/\sigma_C$ and cutting depth. Thus, it can be attained that these cuttability indexes might contribute to the performance assessment of a machine and its selection. However, it should be noted that the cutting forces are found to be relatively more significant than that of specific energy.

REFERENCES

Barton, N. 1974. *Engineering classification of rock masses for the design of tunnel support.* Oslo: Norwegian Geotechnical Institute.

Bieniawski, Z.T. 1974. Geomechanics classification of rock masses and its application to tunelling. *Proceedings of the 3rd Congress of ISRM*, 1: 27–32. Denver.

Bilgin, N., Seyrek, T. and Shahriar, K. 1988. Golden horn clean up contributes valuable data, tunnels and tunnelling. *Tunnels and Tunneling*, 41–44.

Bilgin, N., Demircin, M.A., Copur, H., Balci, C., Tuncdemir, H., and Akcin, N. 2006. Dominant rock properties affecting the performance of conical picks and the comparison of some experimental and theoretical results. *Int. J. Rock Mech. Min. Sci.* 43(1):139–156.

Bilgin, N., Copur, H. and Balci, C. 2014. *Mechanical Excavation in Mining and Civil Industries.* 1st edn. CRC Press, New York.

Copur, H., Bilgin, N., Tuncdemir, H., and Balci, C. 2003. A set of indexes based on indentation tests for assessment of rock cutting performance and rock properties. *The South African Institute of Mining and Metallurgy*, 589–600.

Dey, K., and Ghose, A. 2011. Review of cuttability indexes and a new rockmass classification approach for selection of surface miners. *Rock Mech. Rock Eng.* 44:601–611.

ISRM 1981. *Rock Characterization, Testing and Monitoring. International Society of Rock Mechanics Suggested Methods*, Pergamon Press, Oxford, 211 p.

Liu, S.Y., Du, C.L. and Cui, X.X. 2009. Research on the cutting force of a pick. *Mining Science and Tech.* 19:514–517.

Roxborough, F.F. 1987. The role of some basic rock properties in assessing cuttability. *Proceedings on Seminar on Tunnels: Wholly Engineered Structures*, Sydney, Australia.

Scoble, M.J., and Muftuoglu, Y.V. 1984. Derivation of a diggability index for surface mine equipment selection. *Min. Sci. Technol.* 1:305–322.

Tsiambaos, G., and Saroglu, H. 2010. Excavatability assessment of rock masses using the Geological Strength Index (GSI). *Bull. Eng. Geol. Environ.* 69(1):13–27.

Williams J.L. and Hagan P. 2014. An assessment of the correlation between the strength and cuttability of rock. Coal Operators' Conference, The University of Wollongong, 186–192.

Tuen Mun—Chek Lap Kok Link in Hong Kong—Innovative Technologies and Methodologies for an Outstanding Project

Antoine Schwob • Dragages Hong Kong Ltd.
Bruno Combe • Bouygues Travaux Publics

ABSTRACT

On Tuen Mun Chek Lap Kok Link subsea tunnels project, the difficult ground conditions as well as the tunnel depth reaching almost 60 m made the tunneling works particularly challenging. Numerous innovative technologies and methodologies were thoroughly developed and implemented in order to overcome these challenges and to make the TBM drives a success. Two 14 m diameter TBMs were used along the 4.5 km long drive below the sea. One of them had started its drive with the world's largest diameter (17.63 m) along the first 630 m before being reconfigured into a 14 m diameter TBM. For hyperbaric maintenance interventions, saturation diving system and full automatic robotic arms were developed to change a total of about 2000 disc-cutters on both TBM cutterheads. Another major challenge was the construction of 41 subsea cross passages for which the pipe jacking methodology was adapted for the project. Apart from TBM drives, innovative design solutions were developed for the construction of Shafts, their crossing with TBMs and Cut and Cover sections including a 500 m-long and 43 m-deep Caterpillar-shaped cofferdam with 15 cells.

Figure 1. Tuen Mun

INTRODUCTION

In August 2013, Dragages Hong Kong Ltd and Bouygues Travaux Publics were awarded the largest construction contract ever awarded from Hong Kong Government: Design and Built of the Tuen Mun–Chek Lap Kok Link, Northern Connection Subsea Tunnel Section.

This New Road will provide alternative access route to the Hong Kong International Airport, located on Chek Lap Kok Island, which, until now, is accessible only via the Tsing Ma suspension bridge. TM-CLKL is also part of a cross-border project connecting the Northwest New territories (few kilometers away from the mega-city of Shenzhen, entrance gate to Mainland China) to Zuhai and Macau via the new Hong Kong–Zuhai–Macau Bridge. The artificial island of Hong Kong Boundary Crossing Facilities (HKBCF) on which the tunnel lands on its south end will be the heart of this new cross-border route.

Figure 2. TM-CLKL project layout

The 4.5 km long subsea section of the tunnel is made of two tubes built with two Slurry Tunnel Boring Machines (14 m excavation diameter). Each tunnel has a 2-lane carriageway. On both sides, the North (630 m) and South (670 m) approach tunnels were built in freshly reclaimed land (Figure 2). While the construction of the HKBCF artificial island was part of a separate contract, the completion of the North reclamation in a very short time-frame was the first challenge faced by Bouygues Construction's teams.

NORTH APPROACH TUNNELS—MAIN CHALLENGES

Northern Landfall

After less than a year of intensive day and night work, a new piece of land emerged from the sea constituting phase 1 of the north reclamation. With an area of about 16.5 hectares for a total length of 1.2 km and width of 140 m (Figures 3 and 4), the north reclamation allows the tunnels to deepen and pass below the existing seabed with a gradient of 5%.

Two large shafts were built within the North reclamation:

- The 3-cells caterpillar-shaped Launching shaft is located at the northern end. Relatively shallow (20 m) but large (80 m long, 40 m wide), its main function was to give space to assemble the two TBMs used to build the North Approach tunnels. A short section of Cut and Cover Tunnels, forming the North Entrance of the Tunnels, is also located there.

- The circular-shaped Ventilation Shaft (56 m in diameter), is located at the other end of the reclamation. Much deeper (45 m), it forms the transition between the North Approach

Figure 3. North reclamation

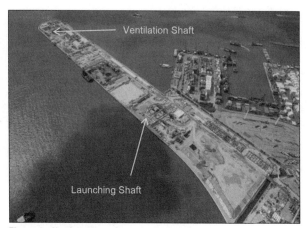

Figure 4. North reclamation

tunnels and the subsea tunnels. During construction stage, it was used for receiving and reconfiguring the TBMs for them to start the subsea drive. In permanent stage, it is filled with many large concrete ventilation ducts connecting the tunnels to the North Ventilation Building constructed right above.

The World's Largest Tunnel Boring Machine

At the early stage of the Project, the Reference Design for the North Approach Tunnels was a Cut and Cover structure. The construction of such a structure (630 m-long, 45 m-deep and 40 m-wide) in a freshly-reclaimed land was one of the major challenges faced initially. The construction programme for this Cut and Cover section was very tight as the land above the North Approach tunnels had to be available partly for handover to third parties but also for site installations required for a timely start of the subsea TBM drives.

The teams of Bouygues Construction therefore proposed to replace this Cut and Cover by bored tunnels with slurry TBMs. This could reduce the risks during construction, minimize earth movements and secure the programme. It was approved by the Employer's Supervising Officer as it satisfied all the functionalities of the structure defined in the contract while reducing the risk profile of the project.

Starting from the surface at the northern end of the reclamation and in order to reach sufficient cover underneath the seabed at the southern end, the tunnels have to follow a steep gradient of 5%. With such a gradient, the regulation in Hong Kong imposes a 3-lane carriageway for the ramp up. To accommodate this third lane and the required ventilation duct area, a tunnel with an internal diameter of 15.6 meters is needed (Figure 6) leading to the design and construction of the largest TBM ever used in the World with a diameter of 17.63 meters (Figure 5). The transition from 2 to 3 lanes takes place at the Ventilation Shaft. The other tunnel (ramp down) remains identical to the subsea section with an excavation diameter of 14 meters and an internal diameter of 12.4 meters.

The high variability of the ground encountered by the TBMs, from sand and clay to the hardest granite, was taken into account in the TBMs specifications. On site, the engineers decided to prioritize power and robustness for the cutter head and imposed to the TBM Manufacturer, Herrenknecht, a large number of disc cutters to cross the

Figure 5. World's largest TBM (S-880)

Figure 6. North approach tunnel (ramp up)

rock without any difficulty. This allowed the completion of the North Approach tunnels and associated works a few weeks ahead of the initial programme, 30 months after Project award.

Enhancement to the North Reclamation

Driving TBMs in a freshly reclaimed land was a major challenge that required some enhancement to the North reclaimed land.

In addition to the geotechnical data available from earlier investigation campaigns, a thorough survey of 157 Cone Penetration Tests (CPTs) was conducted along the tunnel alignment to map the geological conditions under the reclaimed land. The Marine deposits and the clayey Alluvium layers are fine-grained soils. These layers would be disturbed by the backfilling works of the North reclamation, then by the tunneling activity, and would consolidate under the new loads. The design of the reclamation had therefore to take into account the pre-existing state, its own requirements, but also prepare the ground for the subsequent boring of the tunnels.

In order to bring strength parameters of the compressible soils—namely, the Marine Deposits and the clayey Alluvium—to the minima given by the design of the tunnels, a classical scheme of consolidation under surcharge was chosen.

Due to the surcharge imposed on the compressible soils, in which drains of high permeability were placed, the water was squeezed out of the soil matrix and the grains were packed more tightly against each other, leading to an increase in the shear strength of the material. The height of surcharge (up to 12.5 m) was defined in order to reach the required shear strength for tunneling. From the schedule of works that allowed for a consolidation period of 4 to 6 months, the pattern and spacing of the band drains were defined using the radial consolidation theory (Barron, 1948), to reach the final efficiency of 90% consolidation under the weight of the surcharge at the end of the consolidation period.

Plastic band drains were installed by a derrick lighter using a hydraulic hammer. After the installation of the band drains, the reclamation was backfilled up to the levels defined for the consolidation, and monitoring of the consolidation begun (Figure 7).

The monitoring of the consolidation was of paramount importance: what was at stake is to guarantee that the target degree of consolidation of 90% under surcharge weight was achieved, which in turn is a *sine qua non* condition to the success of the tunneling works.

Figure 7. (a) Installation of plastic band drains from barge; (b) reclamation and surcharge in progress

In order to check the efficiency of the consolidation on the strength parameters of the ground, a post-consolidation CPT campaign was conducted.

Where the success conditions were not met, a local ground improvement by jet grouting was performed. The reason for this lack of performance was investigated, and was often correlated to difficulties during the installation of the band drains.

Figure 8. Flooding of ventilation shaft

After the end of the consolidation and surcharge period, the twin TBM tunnels under the newly constructed Northern landfall were excavated.

Crossing the North Ventilation Shaft

After a 630 m long journey below the North reclamation, TBM S-880 (17.63 m diameter) reached the North Ventilation Shaft. To secure the break-out of this mega machine, the shaft was flooded to balance the hydrostatic pressure inside the shaft with the outside ground pressure (Figure 8). After pumping the water out, TBM S-880 was reconfigured into S-881 (Figure 9). Its diameter had to be reduced to 14 m to bore the 2-lane carriageway subsea tunnel.

This modification was thoroughly prepared in order to re-use a maximum of noble pieces and to limit the changes to the cutterhead and the shield elements only. This change-diameter operation required the use of the 580 t capacity

Figure 9. Change diameter operation and crossing steel bell

gantry crane that was used to assemble the TBMs in the launching shaft 8 months earlier.

The second TBM (S-882) crossed the North Ventilation Shaft with no interruption to its production cycle. For this purpose, a crossing Steel bell weighing 1500 tons, filled with concrete, was installed at the bottom of the Shaft (Figure 9). This technique eliminated the need for any ground treatment outside the shaft as the confinement pressure of the TBM could be maintained throughout the crossing. In addition, this arrangement avoided any disruption to the adjacent on-going change diameter operation while crossing with S-882. Furthermore, it saved significant time as S-882 was not slowed down by the Break-out/Break-in operations that are always heavy and time consuming for a TBM of this size. It allowed an early start for the main section of the project: the subsea tunnels and associated Internal structures and cross passages.

SUBSEA TUNNELS

Adverse Geology

An extensive geotechnical campaign was carried out immediately after the project award in order to get an accurate knowledge of the geology, starting point for the design of the structures and equipment. The subsoil of this region of Hong Kong is extremely disturbed and shows a high variability (Figure 10). Along the Northern part of the alignment, the Granite gives way progressively to CDG (Completely Decomposed Granite). Through this section, the TBMs had to encounter mixed face conditions which was a challenge for cutterhead maintenance and hyperbaric interventions. Along the southern half of the alignment, the ground conditions change to alluvium with interbedded layers of sand and clay. While driving within alluvium, the wear of the disc cutters was much reduced. However, ensuring the face stability when hyperbaric intervention was required became an issue. The seabed is composed of a thick layer of Marine Deposits, very soft and of low density.

Subsea Tunnels Drive and TBM Maintenance

In this adverse geology, the two Tunnel Boring Machines specifically designed for the project were driven in parallel. With an outside diameter of 14 m (Figures 11 and 12), the face was maintained by a Slurry Pressure system ensuring a confinement pressure up to 5.8bar at the deepest sections along the alignment.

In order to be able to complete the subsea section on time, barely more than a year was allocated to the excavation of the twin 4.5 km long subsea tunnels. The TBMs had therefore to progress at a steady pace allowing the progress of various work

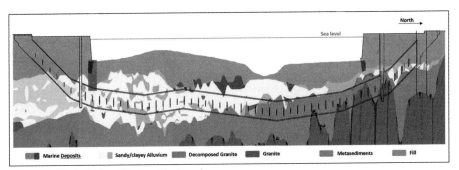

Figure 10. Geological profile along the tunnels

Figure 11. Subsea tunnels driven from northern landfall

fronts immediately at their back: installation of Tunnel internal structures, Cross passages construction, etc. The production rates required for TBMs were very demanding and required a well mastered logistics. Along the last section of the alignment, while the two TBMs were progressing in alluvial soils, each TBM could erect 7 to 10 rings per day. A total of approx. 35 m of tunnel were erected daily.

In order to avoid breakdowns and maintain these rates of TBM excavation, ensuring an efficient maintenance of the TBMs cutter tools was crucial. The geology along the first part of the alignment was particularly demanding for the machines: the disc cutters worn out quickly in the abrasive granite and CDG and had to be regularly replaced. It was therefore required to have daily intervention under hyperbaric conditions, in the excavation chambers of the TBMs allowing access to the disc cutters and tools. Working in a hyperbaric environment reaching a pressure up to 5.8bar was certainly the most difficult challenge faced by the teams of Bouygues Construction.

Figure 12. TBM excavation

Saturation Technique

In order to avoid exposing workers to the risks associated to daily compression/decompression cycles and to increase the effective time of interventions, the saturation technique was developed and implemented for the first time at such a scale on an underground project. It consists in lengthening the cycles by maintaining in a pressurized environment a team composed of 4 workers, specialized in hyperbaric works. The duration of the cycles is 28 days. For that purpose, a living habitat was built on surface. It included, among other things, the main hyperbaric chambers in which the workers were living, compression/decompression chambers and medical caissons equipped with the necessary equipment to provide care in case of injury (Figure 13).

To ensure the proper functioning of the installation, many parameters had to be constantly monitored: the pressure of the various chambers but also the temperature, humidity and the mixture of gas including helium that the workers were breathing.

Figure 13. Living habitat for saturation diving

From a fully equipped control room, a surface team was constantly checking the proper functioning of the installation. A doctor certified for hyperbaric works was also full-time available to intervene when needed.

Every day, the teams were transferred via a pressurized shuttle (Figure 14) from the living habitat to the TBMs. This shuttle was then connected directly to the hyperbaric chambers of the TBMs. Replacement of the disc cutters could then start.

Figure 14. Shuttle transfer from living habitat

TBM Monitoring System

In order to make full benefit of the time allowed for each intervention, it was essential to have in advance a good knowledge of the condition of the cutter head. Hyperbaric workers could then prioritize the most critical areas. For this reason, the cutterheads of the TBMs were equipped with innovative technologies developed by Bouygues Travaux Publics:

- The Mobydic system performs a real-time mapping of the cutterhead. Each disc cutter is identified and any abnormal behavior caused by excessive wear is tracked and immediately reported to the system. Mobydic is also able to assess the type of ground encountered based on the pressure applied to the disc cutters. It allows getting an accurate view of the environment in which the TBM operates.
- The Snake system is a remote-controlled, poly-articulated arm equipped with a camera and a high pressure water jet to clean the cutterhead and carry out its inspection.

These systems are proven to be crucial to ensure the optimal operation of Tunnel Boring Machines while minimizing human interventions. They now equip most of the Company's TBMs.

Telemach Robot

The TM-CLKL project made a step further in the automation of maintenance systems with Telemach robot, developed by Bouygues Travaux Publics. Telemach is an articulated robot located in a dedicated air locked chamber in the TBM Shield (Figure 15). It is able to access the excavation chamber under hyperbaric conditions, deposit a worn or damaged disc cutter and replace it by a new one. Telemach is similar to robots used in the car industry but with greater difficulty. On a car assembly line, the car frames always stop precisely at the same place, facing the robots. Here, the disc cutter to be replaced can be anywhere. It was therefore necessary to develop a system able to locate the disc cutter and adapt the movements of the robot accordingly in a compressed air environment in the presence of spoil. In January 2016, for the first time, Telemach performed a complete cycle of disc cutter replacement in a fully automated way (Figure 16). This system proved its effectiveness on TM-CLKL Project and has, since then, been used on several projects by Bouygues Construction.

Figure 15. Telemach chamber

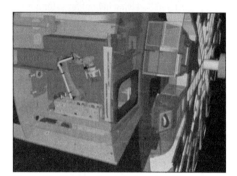

Figure 16. Telemach robot changing disc cutters

CROSS PASSAGES WITH PIPE JACKING TECHNOLOGY

Context

To provide means of egress for pedestrians in case of emergency, cross-passages linking the two tunnels are provided every 100 m. The typical length of a cross-passage is approximately 13 m. Value engineering was conducted at the early stages of the project to standardize the diameters and levels, given the high number of 57 cross-passages to be built.

The initial design was based on ground freezing (using brine) and conventional excavation. However, it quickly became obvious that this option was bringing significant threats to the project, because of risks in construction and potential delays.

As a result, a decision was made, during an early optioneering & value-engineering stage, to opt for a mechanized solution, with the clear intention to industrialize as far as possible the realization of the cross-passages.

Design and Methods of Construction

An innovative solution was developed using, for the first time in the World, small diameter (3.665 m) slurry TBM for construction of Cross Passages (Figure 17). The well-known pipe jacking technology was used but this technique had to be fully revisited in order to adapt to this specific context.

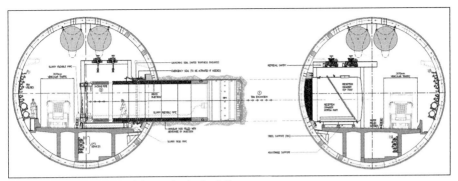

Figure 17. General concept of the mechanized solution

Figure 18. Launching side—pipe jacking equipment

Figure 19. Receiving side—steel bell

The general concept of the solution is as follows:

- Construction of a concrete structure acting both temporarily and permanently
- On one side, installation of the pipe-jacking equipment (Figure 18)
- On the other side, installation of the receiving steel bell (Figure 19)
- Excavation using the pipe-jacking technique, with precast concrete pipe as lining.

A key element of the mechanized solution is the use of a multi-purpose concrete tympanum, cast in the main tunnels, and serving both temporary and permanent purposes. This tympanum is designed to

- stiffen the main tunnel lining against the squat around the opening
- withstand the loads brought by the pipe-jacking machine
- act as a permanent structure on the long term, with embedded waterproofing and second-phase concrete for the door.

On top of the time constraint, the cross passages works were concurrent to all the other main tunnel activities such as the main TBMs logistics but also all the internal structures works on-going at the back of CP works areas in each tunnel. This meant that half of the tunnel section had to be kept free for traffic at all times and all the equipment were designed accordingly.

Achievement

The greatest achievement was to make 39 cross passages in less than 13.5 months, which represents more than 3 CPs per month in variable geology under high water pressure.

This groundbreaking mechanized solution allowed controlling the production schedule while minimizing exposure to geological risks. From design to production, the guiding thread was to make the solution as industrial as possible, and to use permanent elements for temporary stages to minimize the number of operations.

CATERPILLAR COFFERDAM FOR SOUTH CUT AND COVER TUNNELS

Southern Landfall

Upon their arrival below the HKBCF reclamation tip, the two TBMs have crossed the 2 circular South Ventilation Shafts (55 m deep—future connection to the adjacent South Ventilation Building—Figure 20). Immediately southwards, the South Approach tunnels are made of a 250 m long section of TBM tunnels with the same diameter as the subsea section, followed by a 420 m long Cut and cover structure (Figure 21).

Figure 20. South ventilation shafts

Context

On the South reclamation (HKBCF Island), the geology is particularly adverse, with very thick (more than 30 m) and very deep (up to 50 m below ground level) layers of marine deposits and alluvium clays. In these conditions, building a large cofferdam for construction of South Cut and Cover tunnel is a major challenge.

A conventional straight D-wall cofferdam would have required up to 9 layers of strutting with large steel members as well as an extensive ground treatment below the final excavation level to maintain toe stability of the D-walls (soil layers could not provide sufficient passive resistance)

A very positive experience was gained from the design and construction of the 3-cell caterpillar shaped North Launching Shaft. Therefore, the project's technical team decided to push the concept to a world's first 15-cells caterpillar-shaped cofferdam of 500 m long.

Figure 21. Southern landfall

Design of the Caterpillar Cofferdam

The proposed structure consists of 15 numbers of truncated circular cells ranging from approx. 25 to 37 m long and approx. 44 to 57 m (diameter of cell) wide each. Each cell is formed by perimeter D-walls in an arc shape, which resists lateral pressures by hoop force. The hoop forces induced on the perimeter arc D-walls are then transferred at the junction of the cells to heavy duty D-walls called "Y-Panels," which are transversally supported by reinforced concrete struts and cross walls.

Figure 22. Y-panel cage

The Y-Panels transmitting vast loadings (up to 44 m long to 56 m deep of lateral pressure), a special type of D-wall panel had to be invented. It had to be cast from a single trench of 3.6 m wide by 6.5 m long with varying geometry. In-situ reinforcement installation method had to be adopted for the heavy reinforcement cage (Figure 22).

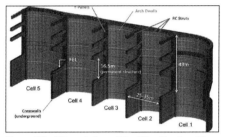

Figure 23. 3D modeling of the 15-cell caterpillar

The design modelling required 3 different 3D Models (Plaxis 3D, SAP2000, Strand 7) and 4 consultants to validate properly the design down to even the D-wall joints behavior (Figure 23).

Overall, the caterpillar scheme went beyond a simple cofferdam proposal and became a fully integrated solution involving innovations on the design modelling, the construction methods, and the permanent works design

Main Advantages of the Caterpillar Cofferdam

The principle of the caterpillar-shaped structure is that it uses the arch effect instead of vertical bending for retaining the earth and water pressures acting on the excavated trench. This makes the D-walls much more efficient as the main forces are transferred in compression inside the D-wall arches.

A main advantage is the absence of steel struts and associated king posts and waler beams. It offers a drastic optimization to the construction programme both during:

- Excavation stage: whilst the total amount of excavation volume is increased, the absence of steel struts allows for ease of plant movement, ease of vertical lifting leading to a much faster excavation overall.

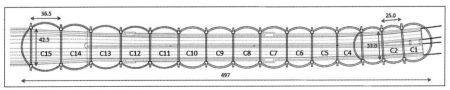

Figure 24. Layout of the caterpillar

Figure 25. Excavation of first 3 cells Figure 26. Caterpillar cofferdam

- Permanent structure construction stage: The permanent structure can be cast as if the works were conducted in open air, instead of being confined by multiple strut layers.

The caterpillar cofferdam also removes the necessity for ground treatment below the excavation level as the arch effect still exists below the excavation level and toe stability is therefore not an issue.

CONCLUSION

In order to meet the numerous technical challenges for the construction of the Tuen Mun–Chek Lap Kok Link–Northern Connection Subsea tunnels in Hong Kong, a large number of innovative design and technologies were put in place over the past five years.

Launching the World's largest Tunnel Boring Machine in a freshly reclaimed land was the starting point of a 5.5 km long adventure.

In adverse ground conditions, at more than 55 m below sea level, state of the art technologies were used for facilitating the daily hyperbaric interventions on the TBM cutterheads including the use of a robotic arm able to change disc cutters without human interventions.

Another major milestone of the project was the construction of the cross passages for which small-size TBMs were launched more than 40 times from one tunnel to the other.

The Tunnels will then end up with a cut and cover section built within a gigantic caterpillar-shaped cofferdam (500 m long and 43 m deep).

These innovative technical solutions, developed through a thorough preparation of the design and construction methods, laid the foundations of recent developments in underground works, pushing further the limits of feasibility of the techniques used.

Figure 27.

PART

SEM/NATM

Chairs

Steve Price
Walsh Group

Ken Dombroski
McMillen Jacobs Associates

Christian Heinz
J.F. Shea

Tom Peyton
WSP

Culvert Construction Under I-89 in Vermont Using the Sequential Excavation Method

Jon Pearson ▪ Stantec
Julian Prada ▪ Stantec
Anil Dean ▪ Stantec
Eric Eisold ▪ Bradshaw Construction

ABSTRACT

Following the deterioration of two large culverts under I-89 in Vermont, the Vermont Agency of Transportation (VTrans) issued a request for design-build proposals for replacing and enlarging the culverts. The winning team of JA McDonald/Bradshaw/Stantec selected the sequential excavation method of tunneling for the 21 foot wide × 13 foot high excavation which was used as temporary support for the culvert installation with permanent works consisting of both cast-in-place and precast structures. This paper will detail tunneling excavation support for the culvert replacement including an overview of the geological and geotechnical features of the project, a summary of modeling that was performed, and an overview of the methodologies used during construction.

PROJECT BACKGROUND

Interstate-89 (I-89) is a vital thoroughfare in Vermont and provides easy access from Montreal to Boston, and links the state's largest city, Burlington, with the state capital Montpelier. The project included replacement of culverts at two locations along I-89. This paper focuses on replacement of the two culverts in the vicinity of the Georgia, VT crossing. After 50 years of service, these two 6-foot diameter drainage culverts located below a section of I-89 embankment had deteriorated and visible perforations in the pipes along with sinkholes in the embankment slopes suggested there may be voids in the embankment threatening the integrity of the roadway. Given this concern, the Vermont Agency of Transportation (VTrans) issued a request for design-build proposals for replacing and enlarging the culverts. Limiting or eliminating impact to the public was of primary importance to VTrans. As such, the culvert installation called for the use of tunneling methods as opposed to open-cutting the roadway. The Design-Build project delivery system was chosen by VTrans with the winning design build team consisting of JA McDonald as the prime contractor, Bradshaw as the tunneling subcontractor, and Stantec as the design engineer.

Figure 1. Sinkholes above existing culvert

GEOLOGICAL SETTING AND GEOTECHNICAL INVESTIGATION

The culvert site is located just east of Lake Champlain in Northern Vermont. Based on the Surficial Geologic Map of Vermont dated 1970, the local surficial geology is

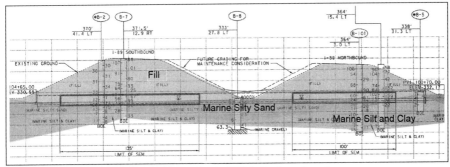

Figure 2. Expected geological profile (approximate SEM limits shown in red)

mapped as a pebbly marine sand. Fine grained lake bottom sediments from Glacial Lake Vermont and a marine clay deposit from the Champlain Sea are also known to exist in the area. Ten existing borings around the culverts were performed and provided to the prospective design-build teams in the request for proposal. The borings were performed to depths of 18 ft to 63 ft below ground surface. An additional geotechnical boring was performed by Stantec after project award to confirm ground conditions. Based on the borings available, the primary strata expected to be encountered during culvert construction were as follows:

- Existing embankment fill consisting of medium dense to very dense sandy gravel and silty sand with gravel
- Marine silty sands consisting of very loose to dense silty sand
- Marine silt and clays consisting of very soft to medium stiff silty clays and clayey silts.

The geological profile is presented in Figure 2 with approximate limits of the SEM installation shown in red. The SEM installation transitioned to an open cut installation on each side of each embankment that was shored with a headwall. While obstructions such as boulders and man made debris were not encountered during the investigation, the construction method for the culvert installation would still need to account for the possibility of obstructions in the fill. The potential for very loose silty sands and very soft silty clays and clayey silts were also an important consideration when designing the SEM tunnels.

During the investigation, groundwater levels in wells installed were observed to be slightly higher than the existing water levels in the stream channel. Given the groundwater levels observed during the investigation and the stream channel generally flowing at the same elevation as the existing ground, groundwater was expected to be encountered during the culvert installation within the tunnel top heading.

ALTERNATIVE TECHNICAL CONCEPT

The base technical concept in the request for proposal called for the installation of an approximately 15 ft by 7 ft concrete box below the interstate using box jacking methods. Additional requirements were to maintain both lanes of traffic in each direction of the interstate at all times during construction, limit settlement to less than ½ inch, and to provide a 100-year design life for the culvert. The Design-Build team identified potential risks with the box jacking method, which included the potential for the box to lock in place due to the large jacking forces required and difficulties maintaining line and grade with the technique. An additional challenge with the box jacking configuration is that it would be difficult to remove large obstructions such as boulders, tree trunks, or man-made debris that may have been present in the fill while maintaining adequate heading stability.

Given the potential risks with a jacked box installation, the Design-Build team proposed to use the Sequential Excavation Method (SEM) for temporary support of the tunnel. The SEM method would provide better control of line and grade and would not require any jacking system. Additionally, obstructions are more easily handled as the SEM method is better suited for changing ground conditions or obstructions given the availability of different support and excavation methods under the technique. The final structure would consist of a pre-cast concrete arch and cast in place bottom slab placed within the temporary SEM support and the area between the pre-cast concrete arch and temporary SEM support would be backfilled with flowable fill. The final structure would be approximately 15 feet wide and up to 8.5 feet tall as shown on Figure 3.

SEM TEMPORARY SUPPORT BASE DESIGN

The SEM tunnels were designed to be approximately 100 feet long below the I-89 southbound lanes and approximately 135 feet long below the I-89 northbound lanes. Cover between the SEM crown and the ground surface ranged from 26 feet to 33 feet below the center of the embankment to only a few feet at the portals. The SEM cross section was designed to be approximately 21 ft wide by 13 ft high to provide enough room for the installation of the final structure. Our team's alternative technical concept design called for a heading and bench excavation whereby the heading would be excavated two to three rounds ahead of the bench at all times. The planned round length was three feet on average with a maximum round length of four feet if ground conditions permitted. Planned support measures were as follows and are shown in Figure 3:

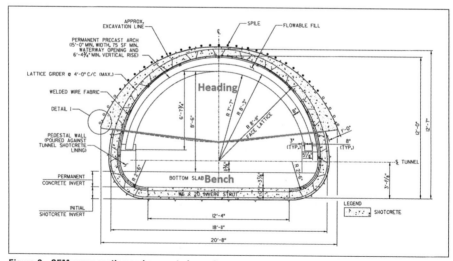

Figure 3. SEM cross section and support elements

- Pre-support measures proposed for the design prior to excavation included grouted hollow bar spiles above the tunnel crown and fiberglass face dowels to be installed as needed. During extended periods of downtime, the face of the excavation was to be supported with a thin layer of shotcrete and/or additional soil temporarily placed at the face to stabilize the heading.
- Initial Support Measures in Heading: Following each heading excavation round, lattice girders and welded wire fabric would be installed on the heading crown and sidewalls. Shotcrete would be installed over the welded wire fabric and lattice girders to achieve a final shotcrete thickness of nine inches in the heading crown. A temporary two to four inch layer of shotcrete without any additional reinforcing may also be installed at the heading face and invert in locations determined in the field based on the ground conditions encountered.
- Initial Support Measures in Bench: Following bench excavation, lattice girders and welded wire fabric would be installed in the bench side walls and a steel I-beam and welded wire fabric in the bench invert. A 12 inch thick shotcrete layer would then be installed sprayed over the welded wire fabric, lattice girders, and steel I-beam.

The tunnel portals were designed to be excavated and supported using sheet piling with the sheet pile face eventually cut to the shape of the tunnel when breaking in and breaking out of the tunnel.

SEM Finite Element Model

In order to verify the tunnel shape, excavation sequence, and support element selection, modeling was performed using two-dimensional (2D) finite element analysis (FEA) with the RS2 software (Version 9.0, formerly called Phase2) from Rocscience. Various cross sections were modeled to assess the affects of SEM tunneling on ground deformation of the embankment and roadway and loading on the SEM support under varying amounts of cover. Cross sections were also selected where worst-case ground conditions were expected based on the geotechnical investigation. Overall, two cross sections were most helpful at each embankment: one at the center of each embankment under maximum cover and beneath the roadway and one at one of the portals at each embankment where minimum cover would be encountered. The model set up for the cross section in the middle of I-89 Southbound lanes can be found in Figure 4 for reference.

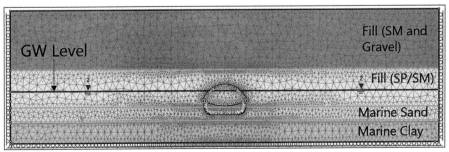

Figure 4. Finite element model set up for cross section beneath I-89 southbound lanes

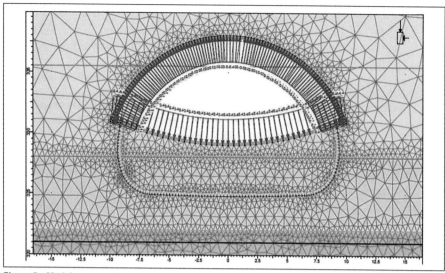

Figure 5. Model cross section showing induced stresses to model ground relaxation between excavation and support installation

The modeling process accounted for ground relaxation, or ground stress release, during the sequential excavation process in accordance with the convergence-confinement method (Panet, 2001). The ground relaxation due to the excavation of each drift was modeled by adding an induced stress in the model normal to the boundary of the excavation perimeter prior to placement of the liner. The ground is then allowed to relax and deform in this stage until equilibrium is reached. The induced stress is removed in the next stage and the remaining stresses are supported by the liner. An induced stress of 40% of the natural ground stress was added to the models to account for ground relaxation after excavation and prior to adding the SEM temporary support elements. The induced stress magnitude was chosen as it resulted in a ground loss of approximately one percent of the excavated volume which is in agreement with the volume loss that would be expected for good quality construction (Thomas, 2009). A close up of the modeled cross section with induced stress shown can be found in Figure 5.

Overall, a seven stage analysis was performed at each cross section. The seven stages are summarized as follows:

Stage 1: Existing conditions and in-situ stresses.

Stage 2: Groundwater table lowered to 4 feet below tunnel invert.

Stage 3: Heading excavated and ground relaxed using estimated confinement loss parameter in heading.

Stage 4: Shotcrete, welded wire mesh, and lattice girder liner added to heading crown as part of a single layer in the model, shotcrete liner added to heading invert.

Stage 5: Ground excavated and relaxed in bench using estimated confinement loss parameter.

Stage 6: Shotcrete, welded wire mesh, and lattice girder liner added to bench sidewalls as part of a single layer in the model, shotcrete and I-beam added to bench invert as part of a single layer in the model.

Stage 7: Groundwater table restored to pre-existing condition.

Cross sections were initially modeled using the most likely ground conditions expected. In order to confirm that the design would be sufficient should ground conditions be worse than expected, a parameteric analysis was performed on the model input parameters that were found to produce the greatest impact on ground deformations and loading. The parameters that were varied included induced stress, horizontal stress coefficient, and the soil elastic modulus.

Based on the results of the model, the appropriate shotcrete thicknesses were found to be nine inches in the heading and 12 inches in the bench assuming use of 5,000 psi compressive strength shotcrete. The reasoning for the higher shotcrete thickness in the invert had to do with the concentrated bending stresses at the haunches on either side of the SEM cross section. The modeling also showed that the planned cross section dimensions and shape and planned heading and bench excavation sequence were sufficient to maintain adequate stability during construction. Settlements beneath the I-89 roadway surface were estimated in the model to be approximately 1-inch for the base case expected ground conditions.

Spiling Design for Pre-support

Spiling was included in the SEM design to provide ground support during excavation and prior to the installation of shotcrete and welded wire fabric. Based on the results of the geotechnical investigations, varying ground conditions were a concern within the embankment fill in which the spiling would be installed due to the granular nature of the material, expected ground loads and the potential for ravelling. The spiling was designed to address various cross sections considered in the finite element modeling. Two types of spiling were considered during the design: rebar spiles and grouted hollow bar spiles. The grouted hollow bar spiles were able to resist higher ground loads within the variable embankment fill material and the added benefit of having the ability to grout the spiles was also seen as a positive as it would mitigate against the possibility of material loss between the spiles prior to shotcrete installation. A photo of the hollow bar spile installation above the tunnel heading can be found in Figure 6.

Figure 6. Installation of groutable hollow bar spiles in tunnel heading

CONSTRUCTION

SEM Construction Overview

Construction of the tunnel pits and portals began in the fall of 2017 with tunneling commencing beneath the I-89 Southbound lanes in May 2017. Sump pumps were placed in the corners of the each tunnel pit excavation including between the two embankments to aid in groundwater management in advance of the tunneling operations. In order to lower the groundwater table in advance of the top heading excavation, dewatering was performed at the portals using vacuum lances advanced approximately 20 to 25 feet within the excavation and just outside of the excavation though the portal walls. The lances outside of the excavation zone would be left to continuously drain the surrounding ground during tunneling and additional vacuum lances were installed within the advancing bench excavation approximately every third round. This technique was effective at controlling groundwater levels and enhancing face stability.

Spiling was installed through the sheet pile portals prior to cutting the sheet piles for the mass excavation. Groutable hollow bar spiles were initially installed using a drill mounted to a track crawler. After installation at the portals, ground conditions along the tunnel were observed to be better than expected based on the boring logs, and the team was able to change the presupport spiling design to standard rebar spiles, which were shown to provide adequate presupport for the remainder of each drive. This illustrates one of the advantages of the SEM system, whereby support elements can be adjusted as required during construction to provide the desired level of support matched to the actual ground conditions encountered. In this case, as ground conditions were better than expected, the capacity of the presupport elements could be reduced, which made for a more efficient construction process without compromising stability or safety of the work. Had the ground been observed to be worse than expected, additional presupport using the designed groutable bar spiles would have been provided instead. The solid rebar spiles were were installed by a portable jackleg operated drill.

With the spiles installed and the groundwater lowered to the extent possible from the excavated zone, the sheet piling was cut to the shape of the tunnel. The initial heading excavation was advanced carefully to observe how the ground reacted. The bulk of the excavation was performed using a small excavator and front loader to remove the tunnel muck. Once the bulk of the material had been excavated for each round, hand mining was performed at the tunnel perimeter to prevent over-excavation. Following each round of excavation, the requisite initial support previously described including welded wire fabric, lattice girders, next round of spiles and shotcrete were applied prior to the next round of excavation. Photos of the tunnel portal beneath I-89 Southbound lanes and the application of shotcrete to the tunnel heading are included in Figure 7.

Large man-made objects and boulders were not encountered during tunneling, but organic material such as roots, fibers and large wood debris including a tree trunk

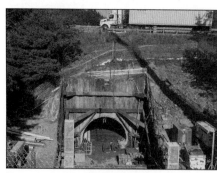 

Figure 7. Tunnel portal beneath I-89 southbound lanes (left) and application of shotcrete to tunnel heading (right)

was encountered during excavation of the tunnels. However, the tree trunks and wood debris did not hinder overall progress of the work significantly as they could be cut with a chain saw at the excavtion profile since the sequential excation method was being applied which was an added benefit with respect to the box jacking method. The tunnels were advanced at an average rate of four feet per day and were completed in 4 months, approximately 1 month ahead of schedule. Following the completion of the temporary SEM support, the bottom slab and pedestal walls were poured against the shotcrete. Once, the pedestal walls and bottom slab had set, a precast concrete arch was placed on rollers and rolled into the tunnel. The precast arch was then jacked up, rollers were removed, and the arch was placed on top of the pedestal walls. The annulus was subsequently backfilled with flowable fill. The bottom of the culvert was ultimately lined with gravel and rock to simulate a stream bed and provide a passage way for wildlife to cross undereneath the embankments. The project was commissioned in Spring of 2018.

Modifications to SEM Design During Construction

One of the advantages of the SEM method is the ability to adapt the construction methodology to the actual ground conditions encountered. In the event ground conditions are worse than expected, additional support can be added, and excavation rounds can be shortened. In the event ground conditions are better than expected, modifications can be made to the excavation process to reduce the amount of support and/or increase the round length. During the early stages of the excavation for the project, it was observed that overall ground conditions were more favorable than the geotechnical investigation had indicated. While the marine silt and sand and marine clay layers were at a higher elevation than expected, they were generally denser and stiffer than anticipated which resulted in better overall stand up time for unsupported sections of the tunnel.

The observation of ground conditions that were more favorable than that assumed during the design allowed for two key changes to the excavation process. First, the spiling was changed from a 2-inch grouted hollow bar spile to a smaller #8 solid rebar spile. This resulted in cost savings due to the material cost difference between the larger hollow bar spile and the smaller rebar spile. The change also allowed for quicker installation of the spiling, as the hollow bars proved difficult to install in the dense soil conditions and solid rebar spiles could be installed using a portable jackleg drill as opposed to the track mounted drill. The grouting process of the hollow bars also added time to the installation process, which was eliminated with the use of solid bars.

Figure 8. Ground conditions observed to have good standup time

An additional change to the planned excavation process was the increase in round lengths. The initial SEM design called for an average round length of three feet with up to four feet advances in favorable ground conditions. Given the presence of material that exhibited good stand up time, the round lengths were increased to four feet for a majority of the tunneling. These changes contributed to faster excavation and the SEM portion of the work was completed ahead of schedule. Both of the changes

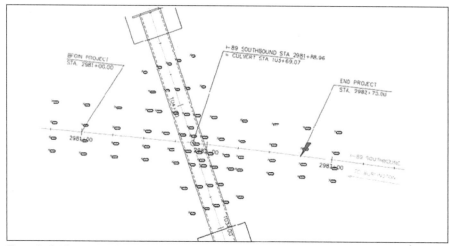

Figure 9. Settlement monitoring plan for southbound I-89 embankment

described here demonstrate how the SEM design was adapted during construction to fit the actual ground conditions encountered.

Monitoring and Instrumentation

A comprehensive monitoring and instrumentation program was critical to the success of the project and was used to verify ground movements were within acceptable limits. Over 150 monitoring points were placed during construction on the shoring system, embankment, and roadway surface to monitor vertical and horizontal movement of the ground and structures. Monitoring points were placed up to 120 feet on either side of the tunnel centerline to evaluate the potential settlement trough. The settlement monitoring plan for the southbound I-89 embankment can be found in Figure 9. Monitoring occurred daily during tunneling and was performed periodically until spring of 2018 after the SEM support was completed and the precast arches were grouted in place to verify no further ground movement had occurred. In addition to the surface monitoring points, an inclinometer was placed at one portal at each embankment to monitor any horizontal movement of the slope.

Overall, settlements were very close to those predicted by the finite element modeling. Settlements at the road surface were generally one inch or less while those in the embankment were greater in some areas due to the low ground cover as well as ground disturbances associated with the installation of the sheet piles for the portals. While these settlements were greater than the original proposal requirement of 1/2 inch, it was agreed based on the design results and in advance of construction that settlements up to a predicted one inch of settlement would be acceptable if confirmed by extended monitoring results provided repaving was undertaken at project completion since the contract required repaving at the end of construction. Horizontal movements observed were up to 0.75 inches toward the portals indicating minor ground disturbance from portal and tunnel construction and were within expected limits.

CONCLUSIONS

This paper highlights the application of the SEM design and construction methodology which was used to provide temporary support and successfully complete a culvert installation beneath a vital thoroughfare in northern Vermont. The Design-Build team of JA McDonald, Bradshaw and Stantec successfully proposed the use of SEM as an alternate to jacked box installation of the culvert. Modifications to the SEM presupport were made during construction based on the actual ground conditions observed, demonstrating the flexibility of the SEM approach and ability to modify the excavation process and support measures as needed to be compatible with actual ground conditions encountered. Construction of SEM temporary support was completed without impact to the public and was completed ahead of schedule, allowing for successful installation of the final culvert structure and completion of the Georgia Culvert portion of the project.

REFERENCES

Panet, Marc. Recommendations on the Convergence-Confinement Method. Tech. N,p,: AFTES, 2001.

Stantec Consulting Ltd. 2017. VTrans Culvert No. 83-1N and 83-1S SEM RS2 Analyses. Prepared for Vermont Agency of Transportation. March 15, 2017.

Stantec Consulting Ltd. 2016. Geotechnical Engineering Report. Georgia IM CULV(24) I-89 Culverts 83 1N & 83-1S. Prepared for Vermont Agency of Transportation. July 1, 2016. Revised September 2, 2016.

Terracon Consultants, Inc. 2014. Geotechnical Data Report. Georgia IM CULV(24) I-89 Culverts 83-1N & 83-1S. Prepared for Vermont Agency of Transportation. February 25, 2014.

Thomas, Alun. 2009. Sprayed Concrete Lined Tunnels. New York, NY: Taylor & Francis.

Delivering Value Through the Innovative Contractor Engagement (ICE) Model at London Underground Bank Station Capacity Upgrade Project

Enrique Fernandez ▪ gGravity Engineering
Alejandro Sanz ▪ gGravity Engineering
Juan Ares ▪ Dragados S.A.
Andy Swift ▪ London Underground
Bethan J. Haig ▪ Dr. Sauer & Partners Ltd.

INTRODUCTION

Bank is one of the oldest railway complexes within the London Underground scheme. The station has been developed in a piecemeal manner from 1884 onwards as additional lines have been built, reaching its present form in 1991 when the DLR extension opened. As areas of the station are close to saturation point and the demand trend is expected to continue, there was a requirement to upgrade the station (BSCU), in order to increase capacity.

London Underground decided to upgrade the station through an Innovative Contractor Engagement (ICE) procurement model, focusing on adding value and long-term social benefit. Sprayed Concrete Lining (SCL) and traditional timber headings were the tunnelling methods selected for this Design & Build contract. This paper describes the ICE procurement model, the challenges encountered at design stage and the solutions applied during construction in order to minimise the impact on buildings and the operational railway.

BSCU BACKGROUND AND PROJECT REQUIREMENTS

Bank-Monument station is the fourth busiest station in the London Underground (LU) network, located in the heart of the City of London financial district. During the 3 hour morning peak period over 100,000 passengers travel through the station when boarding, alighting or interchanging.

The two adjoining stations Bank and Monument, commonly referred to as 'Bank' or 'Bank station' form a large underground structure that is one of the most complex subterranean railway stations in the world. It is a strategic network interchange serving six separate underground lines at different levels and is important to the UK's economy as a major gateway to the City for employees and visitors.

Since 2013, Bank station has experienced a significant increase in passengers with a growth of 25% in entry, 29% in exit and 41% growth in interchange demand. As the station has become busier, the congestion problem has increased and a station upgrade became inevitable.

To address these issues the Bank Station Capacity Upgrade (BSCU) project's design is driven by four key technical requirements, namely:

- Increase passenger capacity within the station;
- Reduce passenger journey times through the station;

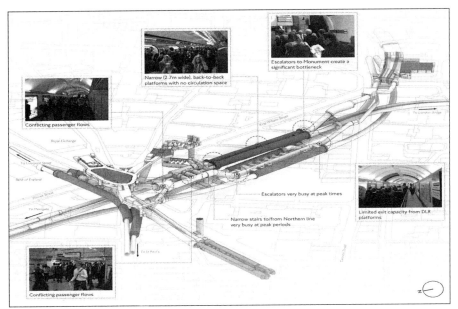

Figure 1. 3D view of the existing Bank Station

- Provide step-free access to Northern line (NL) and Docklands Light Railway (DLR) levels; and
- Provide compliant emergency fire and evacuation protection measures for NL and DLR passengers.

THE INNOVATIVE CONTRACTOR ENGAGEMENT PROCUREMENT MODEL

There is a widely held acceptance that the UK Construction industry needs to innovate and integrate in order to deliver significant value improvements on infrastructure projects. A key lesson learnt by London Underground from the award of previous design and build contracts (Bond Street and Victoria, both in 2010) was that supply chain innovation had to be identified at the early stages of the procurement process in order to maximize the potential benefits. This lesson was one of the factors that led LU to implement the Innovative Contractor Engagement (ICE) model.

The ICE model was designed to incentivise early innovation, with the intent that imaginative and innovative supply chain ideas could be incorporated into the design in a timely manner. The model consists of two strands to promote the contractors involvement and the innovation:

- Commercial protection of competitive advantage from bidder's innovation, in the sense that only the proponent can use the unique and original proposal. Ideas and proposals are secret until the award date. From that date, the winner team can purchase ideas from other bidders in such a way that even if a team is not the winner, it can take some advantage from sharing those innovations.
- Competing performance delivery against outputs (not pricing of a common outline design)

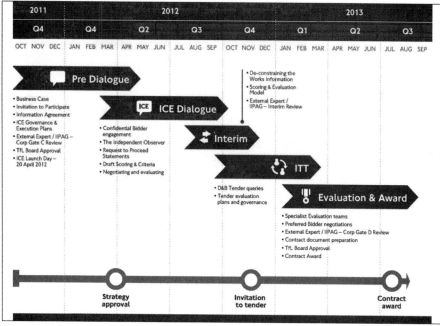

Figure 2. BSCU ICE implementation phases

The BCSU project provided an opportunity to codify and implement ICE on a large, complex capital investment project. By Nov 2011, BSCU was already underway with the client led RIBA D design stage. However a change in the passenger modelling parameters on which the engineering development was based resulted in a need for interchange capacity greater than that achieved under the RIBA C/C+ design stage.

The resulting concept design (known as the Base Case) met the additional demand requirements but was less than optimal in the following areas:

- It pushed the required completion date and exceeded the established budget in excess of £150 m;
- The operational proposal was suboptimal as it contained passenger hotspots throughout the life of the business case;
- The design was based on construction from a constrained 10 King William Street worksite. Construction from a larger worksite was likely to result in delivery benefits; and
- Development of the project would require an order from the Secretary of State for Transport under the Transport and Works Act (TWA). The project was no longer confident that the Base Case could be promoted through the TWA Order process as the optimum scheme for development

A review of the project towards the end of the RIBA D Stage concluded that further client led design development was not the most efficient way to address these shortcomings. Instead, it suggested that involving the market was more likely to result in a solution that delivered the strategic objectives at a price closest to the project budget.

ICE would effectively reverse the typical sequence of works, by procuring a D&B contractor first and then seeking the TWAO on the D&B contractor design (which would already include the contractor innovation). While this had the impact of deferring the original TWA submission date by up to two years, it was recognised that early market involvement could produce time and cost savings and encourage innovative thinking including early contractor definition of the project to help design, plan and deliver the works.

Four pre-qualified bidders were selected for the ICE and they provided four different schemes with significantly different approaches. In 2013, Dragados was awarded the BSCU D&B contract with a score of 72.5, 16.2 points above the second bidder. The tender winning provides a more "Effective Product," increasing the benefits within the business case, and provides a more "Efficient Method," delivering it faster and cheaper compared to the original LU Base Case. This value is made up of:

- An increase of 45.1% on the benefit/cost ratio (from 2.4:1 in the LU Base case to 3.5:1)
- A 19.2% increase in Journey Time Social Benefit over the 60 year project life
- A 9.8% reduction in the estimated final cost
- A 5 week (22.7%) reduction in closure duration of the Northern line, to 17 weeks.
- An increase of 15.6% in induced revenue throughout the life of the project
- A more effective Step-Free Access solution direct from street to platform on both the Northern & DLR lines
- A more efficient fire and evacuation strategy throughout the whole station

To win this bid, Dragados partnered with key players in the supply chain, including designers and subcontractors that were maintained and formed part of the delivery team to ensure the innovative ideas were implemented in the delivery stage.

THE DRAGADOS SOLUTION FOR BSCU

To achieve the project's requirements while minimising the impact on the operational station, the principle features of the project works are:

- A new ticket hall at ground level with access from Cannon street and connecting to the Northern line and DLR platforms through new triple escalators and two new lifts.
- Construction of new Northern line southbound platform tunnel, 82 feet to the west of the existing southbound platform, with a new running tunnel. The new tunnel cross section is 15.2 feet diameter, circular shape, being 32.5 feet diameter at the new platform tunnel.
- A new passenger concourse at Northern line level, formed from conversion of the existing southbound platform into circulation space, and linked with new cross passages to the new southbound platform.
- Four new triple escalator links from street level to Northern line, Northern line to DLR and from Northern line level to Central line.
- A new link tunnel, 400 feet in length, between the Northern line and the new Central line escalators. This new moving walkway located in a new large tunnel, dramatically improves passenger flows and journey times.

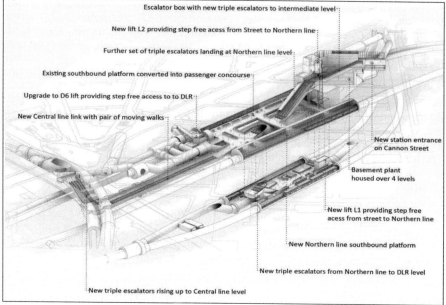

Figure 3. 3D view of the new Bank Station

In total, the new project requires some 4,265 feet of new galleries with cross sections from 107 to 968 sq.ft., including all the E&M works, security, sanitation as well as the architecture for the new structures.

Most of the design, planning and preliminary works were developed from 2013 to 2016, while the construction works were scheduled from 2016 to 2022.

BSCU TUNNELLING WORKS

Tunnelling Works

The BSCU strategy was based on the delivery of an efficient design, providing a substantial increase to the capacity of the station, while minimising the impact to the nearby structures and the daily operations of the station.

The majority of tunnelling works are carried out from the Arthur Street site at the southern end of the project extents. For this purpose Arthur Street is closed and a temporary SCL access shaft constructed to current Northern line level, through an historic station structure at mid-depth. This access, allowed for an early commencement of the underground works, originally planned to occur after the demolition of six buildings where the future ticket hall sits.

Most of the tunnels are built using the Sprayed Concrete Lining method (SCL)—also known as Sequential Excavation Method (SEM)-. The excavation of the cross passages and adits behind the existing tunnels is completely carried out from the new tunnels but without breaking through, minimising the interface with the station operations. The final breakthrough is made during the engineering hours (night shift) while the station is not in operation.

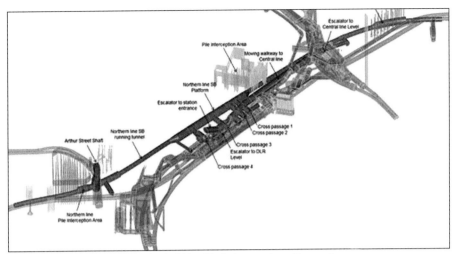

Figure 4. BSCU project layout—the dark blue illustrates new tunnelling works

Ground Conditions

The BSCU site is located within 'The London Basin' geology which is typical for central London. It comprises of a chalk bedrock overlain successively by the Thanet Sand Formation, Lambeth Group, London Clay Formation, River Terrace Deposits and Made Ground.

The majority of the tunnels are located at the Northern line level at approximately 100 feet below ground level. The deepest structures are at DLR level at approximately 130 feet below ground level whilst the shallowest points will be new tunnels at Central line level at approximately 65 feet below ground level.

All SCL structures will be constructed in the London Clay, which is around 150 feet thick, and very suitable for the tunnelling works due to its mechanical properties and low permeability. The encountered London Clay predominantly consists of stiff to very stiff, closely to very closely fissured, dark bluish to brownish grey clay, containing variable amounts of fine sand and silt. The groundwater table is assumed to sit 26 feet below the ground surface.

Selection of Tunnelling Method

The selection of the tunnelling method for the new tunnels took into account the four project requirements, the groundwater conditions, the site logistics, the existing London Underground infrastructure and the presence of buildings of the City of London in the zone of influence. Taking into consideration these constraints, together with the geometry of the new structures, the tunnelling method selected to construct the majority of the BSCU works was SCL, a mechanical and sequential excavation and temporary support based in sprayed concrete. Squareworks is used in specific areas where mechanised tunnelling is not feasible due to restricted space. The selected designer is Dr. Sauer and Partners, which formed part of the ICE process as part of the Dragados' team, and are well known for their expertise in delivering cutting-edge SCL designs in urban environments such as London.

While SCL can be applied using an observational method (R. Peck), which means that monitoring (in-situ-measurements) of deformation in the ground and stress in the

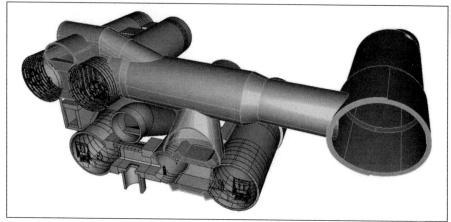

Figure 5. 3D view shown the new Northern line southbound platform tunnel and its connections to the existing Northern Line and DLR

initial lining (shotcrete) is essential to the actual support means, in a location such as Bank the design is fully developed in advance and monitoring is used only to verify performance is in line with expectations. The basic rules are:

- The excavated cross section should always be an ovoid shape.
- Installation of immediate and continuous smooth support around its perimeter (and, if required, smooth support at the face) is a significant factor in minimizing initial movement in the surrounding ground.
- It is also essential to structurally close the supporting ring (shotcrete) as quickly as possible within one tunnel diameter of the advancing excavation face.

Should any stability issues be encountered the design includes a so-called "toolbox" to address them, whilst allowing installation of the structure as designed. The toolbox includes but is not limited to: dewatering of the excavation area; spiling with rebar, pipes or metal sheets; stabilisation of the face with earth wedges; stabilisation of the face with flashcrete and pocket excavation to minimise the time the ground must support itself.

During construction, the monitoring system allows for the evaluation of the ground parameters and other assumptions made during the design stage to confirm their validity. Should a variance be identified back analysis can be completed to allow adjustment of the excavation sequence and support for further excavations.

After completion of the excavation and primary lining and once the monitoring shows that tunnel movements have ceased without further deformations, the waterproofing and secondary lining can be installed.

Composite Shell Lining Design

The design assumes a combined double shell lining system with both linings considered part of the permanent load bearing structure. The combined shell tunnel system comprises the following layers, as shown in Figure 6:

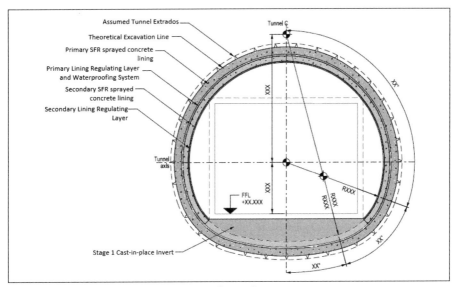

Figure 6. SCL Temporary and permanent support in tunnel

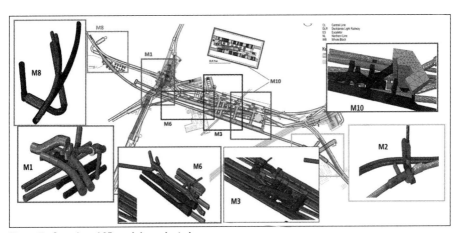

Figure 7. Samples of 3D models evaluated

- Primary Lining: a steel fibre reinforced concrete lining installed as part of the excavation and support process. This includes the initial lining, a thin layer protecting the surface of the excavated face from deterioration and initial loosening.
- Regulating layer: A layer of unreinforced shotcrete, of equal or smaller aggregate size, applied to the surface of the primary lining to cover steel fibres and to smoothen sharp edges protruding from the initial lining.
- A spray applied waterproofing membrane
- Secondary lining: a steel fibre reinforced sprayed or cast-in-place concrete secondary lining

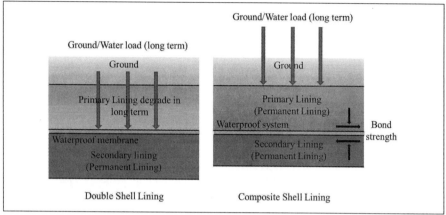

Figure 8. Comparison between double and composite shell lining

- Secondary lining regulating layer: when the secondary lining is sprayed, an inner unreinforced concrete regulating layer is then placed for operational and safety reasons to avoid fibre exposure to the public.

The design considers two different situations, the short term (5 years) and the long term; after the construction of the final lining for the structure life time of 120 years.

The primary support is designed to provide stability with a limited safety factor and similar durability criteria to the final lining. In the combined shell both layers are part of the permanent support and the load sharing is defined by a 3D finite elements analysis.

The non-lineal behaviour of the concrete linings, both primary and secondary, allows an elasto-plastic analysis, which is very adequate for fibre reinforced concrete under stress and consequently fissure width criteria are considered as the way to evaluate its tensile behaviour (max tensile stress of 10‰). As a result, steel reinforcement can be minimized and concentrated only in the connections between galleries.

By applying this combined shell lining design approach, substantial savings in lining thickness have been achieved. The reduced lining thickness also includes the following advantages:

- Lower risk of sprayed concrete fallouts due to less volume of material needed
- Reduced exposure to risk for the workforce overall through shorter construction time.
- Reduction of tunnel cross-section, hence minimising impact on existing infrastructure due to ground movement.
- Reduced excavation volumes and lining thickness

The Radial Joint Design

In recent years in UK tunnelling there has been a lot of focus on the fallout of sprayed concrete and the risks associated with this. During the construction of the lower connection of a traditional joint it was common to use hand excavation methods for the final trimming or to drill bars into the top heading concrete. This required entry into the face area. While there is no exposed ground overhead there are still hazards

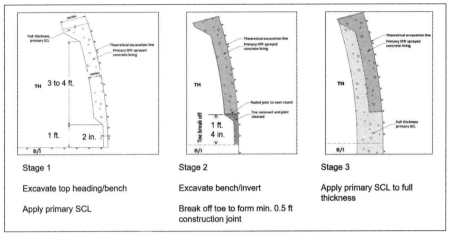

Figure 9. Radial joint construction process

associated with this. Managed exclusion zones combined with a 2–3" sealing layer on the excavation surface have been typically applied for protection for this short-term entry into the face area. Although a proven method of risk mitigation, the project viewed sealing an excavation face as insufficient protection for the workforce during tunnelling and as creating a lack of clarity on exclusion zones. On BSCU, the construction team has sought to eliminate this risk through avoiding the need for personnel to prepare the joints and therefore enter the face. Dr Sauer & Partners, Dragados and London Underground, endeavoured to deliver an innovative design solution that allowed for a complete change in behaviour on the project, ensuring that exclusion zones are enforced at all times.

The solution consists of a development of a radial joint, which is employed in all sprayed concrete lined tunnels with a subdivision of the face. The joint is designed in the way that the bottom of the joint is trimmed with an excavator; this eliminates the need for hand excavation and the requirement for post-drilled bars.

As radial joints are known to be sensitive elements of the design, the team specified that during preconstruction trials the joint should be sprayed and tested. The tests carried out showed that there was minimal difference between the strength of the cores for full lining and for the joint. Testing has been continued during construction, which has shown that not only is the joint detail reproducible but also that the anticipated behaviour is being observed. All of the cores tested to date exceed the required compressive strength for the concrete.

This inspired a change to the detail for installation of monitoring of SCL tunnels, which resulted in complete elimination of entry into the sealed face. It is envisaged that this will change the norm related to working in proximity to sealed excavation faces for future tunnelling projects. This initiative has proven that not only can solutions be safer but that safety does not have to come at the expense of production, and vice versa.

Pile Interceptions

Some of the buildings in the Bank Zone of Influence are founded on piled foundations; others are either on strip footings or raft foundations. The alignment of the new Northern line southbound running tunnel intercepts with existing piled foundations of

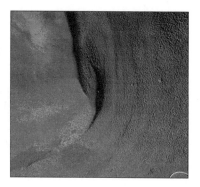

Figure 10. Radial joint stages 2 and 3

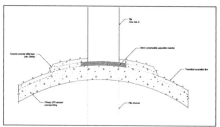

Figure 11. Intercepted pile before cutting and pile separation detail

four buildings. The way the project addressed this interception varied depending on pile loads, noise & vibration requirements both at construction and operational stages.

During concept and detailed design phases, a 3D model was created with the as-built information provided by the building owners and the checks undertaken by the survey team to assess the accuracy of those as-builts. This way, the most likely location of the piles was identifed and the tunnel alignment and excavation and support sequences designed to fit those best estimate positons, accommodating up to 21" tolerances in plan in any direction.

In three of buildings whose piles are intercepted, assessment found that removal of the lower section of the pile and separation from the new tunnels would not adversely affect the building above. In these circumstances, the pile was exposed, then cut and a 2" compressible material installed between the pile toe and the lining extrados to ensure separation and therefore no future load is transferred to the tunnel.

On the fourth building, while the damage assessment concluded that removal of the piles would not cause damage to the building above in its current form, the capacity of the building's foundations for future development would be compromised. As a result a pile transfer structure was provided to allow the load of four piles to be carried into the ground adjacent to the tunnel and maintain the current level of capacity. In addition, the design accommodated the potential for partial removal of 3 further piles depending on tolerances.

The transfer structures are housed within a sprayed concrete primary lining and for fully intercepted piles are comprised of a cast-in-place bar reinforced concrete (RC) arch, which is connected through an RC invert. The transfer structures are designed to support the current design load of the piles and an additional 25% load. Within

 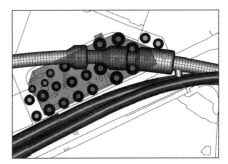

Figure 12. Pile transfer structures

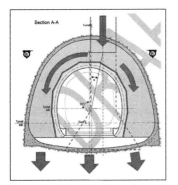 

Figure 13. Pile fully intercepted section and plan view

the transfer structure is a layer of compressible material that will allow for changes in loading and movement of the pile without impact on the running tunnel. Inside the compressible layer will be the running tunnel structure, made up of a waterproofing layer and a combined and steel fibre reinforced secondary lining.

For partially intercepted piles within tolerance, a contingency detail was produced that was made up of two reinforced concrete columns that transfer the pile load into the existing bell footing. The use of the contingency detail was dependent on whether the shaft of the pile encroached into the minimum lining thickness for the tunnel lining. As for fully intercepted piles, a separation layer will be installed outside of the tunnel lining minimum lining thickness, to ensure no load transfer between the structures.

Where required, the bell of the pile could be removed without a transfer structure if the shaft of the pile remained intact; in these cases a separation material was still installed.

ASSET PROTECTION, INSTRUMENTATION, MONITORING AND INSPECTION

Bank station is located in the heart of the UK's financial district, and an extensive arrangement of new tunnels and shafts will be constructed, with many in close proximity to existing LU tunnel assets.

67 buildings are in the zone of influence of the excavation works, 38 of which are of heritage interest and 6 are Grade 1 listed (maximum status in the UK, for exceptional interest), notably St Mary Abchurch and The Mansion House.

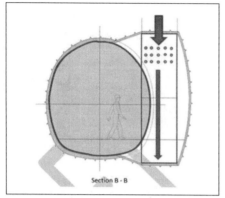 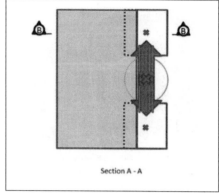

Figure 14. Partially intercepted pile transfer structure

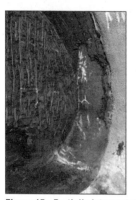   

Figure 15. Partially intercepted pile separation

There are 287 civil engineering assets in the existing LU station within the zone of influence of the works. The station was constructed incrementally, with the first line opening in 1884, until 1991 when the station was largely changed by the construction of the Docklands Light Railway (DLR) new terminus. Accordingly, there is a variety of construction materials and forms in use in the station, with each responding to and tolerating the effects of ground movement differently.

A staged and risk based approach was applied to the damage assessment process, that ensured buildings and LU assets requiring particular care and attention received it without expending unnecessary effort on those that were more tolerant or affected in a benign way.

In conjunction with a light scheme of mitigation measures, a scheme of monitoring and inspection was proposed to address the risks, refer to Figure 16. From an initial list of 67 buildings and 287 civil assets at Bank, only 5 buildings and 13 civil assets in directly affected locations required strengthening to withstand the changes in loading and deformation to them despite the substantial tunnelling planned in the area.

The monitoring system is primarily in place to validate that ground movements within the zone of influence are in accordance with design assumptions and that the infrastructure remains within acceptable limits. Inspections are focused on serviceability

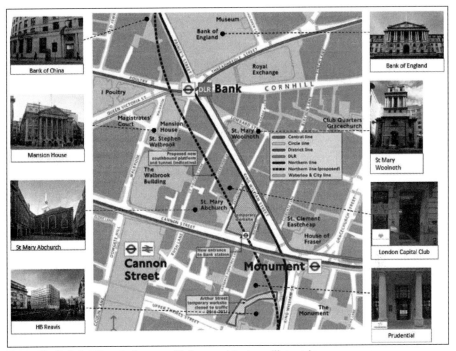

Figure 16. Main buildings in the zone of influence of the tunnelling works

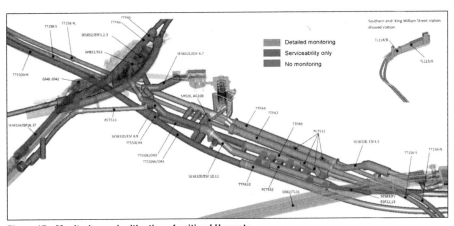

Figure 17. Monitoring and mitigation of exiting LU assets

concerns, namely steps/trip hazards from cracked, lose or deformed finishes and water ingress.

The predictions obtained from the damage assessments and the 3D model are inputted into the monitoring plan and specific action plans, that define the instruments, reading frequencies, trigger levels and associated actions when any of these is breached.

 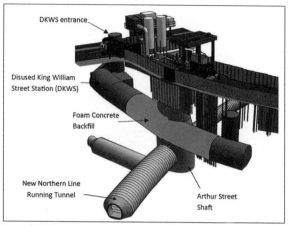

Figure 18. Arthur street shaft aerial view and 3D image

Figure 19. Excavation using ITC-210 (L) and 120N (R)

TUNNEL CONSTRUCTION

The Arthur street shaft is located in a narrow street, 33' to 46' wide, which has been closed to the traffic for a length of 300', keeping pedestrian and vehicle parking access to the nearby buildings. It is a small site with one-way construction traffic accessed via Upper Thames Street. A 38 Tonne Gantry crane has been installed to supply materials in and out the shaft, while two 110 Tonne silos provide SCL dry mix.

The available space on surface for specific facilities and stand-by equipment is extremely limited. This reduced space this been mitigated by using the disused King William Street station –DKWS—a previous terminus to the Northern line, which is intercepted by the temporary access shaft on its way down, during the construction stage. Pedestrian access to the tunnels is via DKWS's original spiral stair shaft from 1890, which remains in the overlying building's basement.

Once the shaft was completely excavated, the first activity was to build the logistics chamber, which is a 200' long tail tunnel (back shunt) for manoeuvring and temporary equipment storage. This tail tunnel will be part of the permanent NL southbound tunnel in the future. The next step consisted of the excavation of a 100' long muck storage chamber to allow regulation of the mucking activity through the very constrained site. These two spaces gave enough resilience to the team to proceed with the excavation of the tunnels and galleries included in the project, minimising the impact of having such a restricted site on surface.

Figure 20. Muckaway method

Figure 21. Spray works

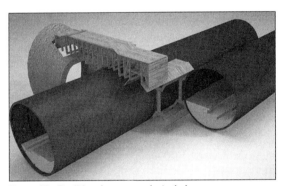 

Figure 22. Traditional squareworks techniques

Excavation of the running tunnels has been carried out mostly using a Terex Schaeff ITC-120N F8, fitted with a loading system, easing the mucking activity on these narrow tunnels. A Terex Schaeff TE-210 was used for the larger tunnels, such as the Central line link that will house the moving walkway, the main passenger interchange and the Northern line platform tunnels. 5.5T, 3.3T & 1.6T excavators were used to support excavation of the smaller tunnels.

All excavation material has been removed via the shaft, using a 14.4 cu.yd. Conquip's Bulk X bottom discharging skip tipping system.

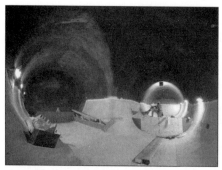 

Figure 23. Spray applied waterproofing membrane

Tunnel	Diameter	Shift Pattern	Productivity (ft/day)	Excavation Sequence	Note
NL Running Tunnel & Pilot Tunnel	16 ft	2 × 12h	6.5	Full Face	
		3 × 8h	8.8		
NL Platform tunnel	32 ft	2 × 12h	3.9	Pilot Tunnel + Top Heading/ Invert (enlargement)	Rates are for Enlargement from pilot tunnel
		3 × 8h	4.9	Top Heading/ Invert	
CL Moving Walkway Tunnel	25.2 ft	3 × 8h	5.9	Top Heading/ Invert	

Figure 24. Productivity achieved during excavation and support

Due to the limited space, a Meyco Oruga shotcrete robot with Suprema pump was selected for spraying the running tunnels, while a Meyco Potenza robot was used for the larger tunnels Four Dieci L4700 mixer trucks were used to deliver the concrete from pit bottom to the tunnel face.

Whilst efforts have been made to maximise mechanical excavation and robotic installation of concrete support, some specific areas have required hand mining excavation techniques due to the proximity to existing tunnels and reduced dimensions.

In order to minimise the risk of future water ingress a double bonded spray applied waterproofing system has been used between the primary and the secondary lining, which prevents the passage of water along the lining interfaces. Two Piccola spray pumps have been used for regulating layer application purposes, while the spray applied membrane uses a hand-held pneumatic/electric airless pump.

At the preparation of the paper, December 2018, the project is 80% complete in terms of excavation and lining. The overall completion is expected for 2022. The Figure 24 summarises the main productivity rates achieved to date.

CONCLUSIONS

Early contractor engagement during the ICE procurement process at BSCU led to a product that increased the benefits within the business case, delivering it faster and more cheaply when compared to the original LU Base Case.

In partnership with London Underground and Dr. Sauer & Partners, Dragados as D&B Contractor is delivering a world class tunnel design and construction with significant programme efficiencies while minimising risks to the health and safety of the workforce, impact on surrounding buildings and the operational railway.

SCL remains the most appropriate method to excavate short galleries, ramps and shafts with variable geometry in London Clay. The innovative design, supported by controlled execution, supervision and monitoring systems during construction stage, has allowed the delivery of these new structures in the vicinity of the existing assets and under sensitive stakeholders, therefore creating one of the biggest subterranean railway infrastructure complexes in the world.

ACKNOWLEDGMENTS

The authors truly appreciate the BSCU project for their review and input into this technical paper. The authors also are grateful to the members of BSCU Tunnelling team (London Underground, Dragados and Dr. Sauer & Partners) for their joint efforts to deliver a world class tunnelling project.

REFERENCES

London Underground (September 2014). Innovative Contractor Engagement. Project report on the implementation of Innovative Contractor Engagement for the appointment of a Design and Build Contractor for the Bank Station capacity upgrade scheme.

A. Nasekhian, A. Onisie-Moldovan, J. Ares, D. Kelly. SCL vs Squareworks—Timberless Tunnelling in Future LU Station Upgrade Projects? Proceedings of the World Tunnelling Congress 2019.

A. Nasekhian, P. Spyridis, 2017. Finite Element modelling for the London Underground Bank Station Capacity Upgrade SCL design and deep tube tunnels assessment. Proc. of IV Int. Conf. on Computational Methods in Tunnelling and Subsurface Eng. (EU-RO:TUN).

C.A. Anthony, S. Kumpfmüller, B. Lyons, J.Ares. Improving safety through design at London Underground Bank Station Capacity Upgrade project, London, UK. Proceedings of the World Tunnelling Congress 2017.

P. Spyridis. Recent experience in the design and construction of underground metro works by use of sprayed concrete. Greek National Concrete Conference 2017.

Dragados, Dr. Sauer & Partners and London Underground. SCL Radial Joint Design—Bank Station Capacity Upgrade. Proceedings of the International Tunnelling Awards 2017, category Safety Initiative of the year.

B.J. Haig, P. Dryden. Keeping London Moving: a risk management approach to assessing existing underground assets at Bank. Proceedings of the World Tunnelling Congress 2017.

SEM—Single Shell Lining Application for the Brenner Base Tunnel

Thomas Marcher • SKAVA Consulting ZT GmbH

ABSTRACT

The Brenner Base Tunnel is a railway tunnel between Austria and Italy through the Alps with a length of 64 km. The construction lot Tulfes-Pfons was awarded to the Strabag/Salini-Impregilo consortium in 2014. The construction lot includes 38 km of tunnel excavation work and consists of several structures such as the 9 km long Tulfes emergency tunnel. Such a service (non-public) tunnel does not necessarily require a tunnel lining system with two shells, but under certain boundary conditions can be achieved by a single shell lining approach. The required conditions and limitations for the single lining approach are reflected. A proposal for structural verification is provided. For verification approach a novel constitutive model for the shotcrete design is used.

OVERVIEW

The Brenner Base Tunnel (BBT) is a flat railway tunnel between Austria and Italy. It runs from Innsbruck to Fortezza. Including the Innsbruck railway bypass the entire tunnel system through the Alps is 64 km long. It is the longest underground rail link in the world.

This tunnel consists of a system with two single-track main tunnel tubes 70 meters apart that are connected by crosscuts every 333 metres (see Figure 1). A service and drainage gallery lies about 10–12 metres deeper and between the main tunnel tubes. It is constructed ahead of the main tunnels and will be used as an exploratory tunnel for them. Four connection tunnels in the north and south link to the existing lines and

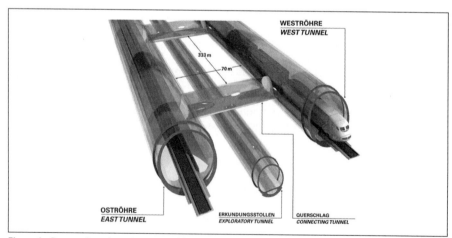

Figure 1. Brenner Base tunnel system with two main tunnels and the exploratory tunnel (Eckbauer et al., 2014)

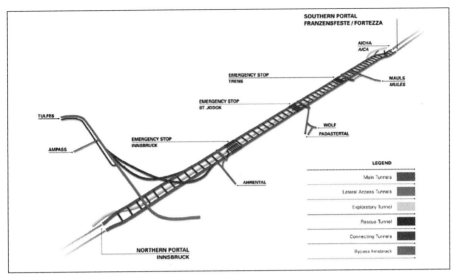

Figure 2. Brenner Base Tunnel System including the connection to the Inn Valley Tunnel (Eckbauer et al., 2014)

also belong to the tunnel system, with a total length of approx. 230 km. Three emergency stops, each about 20 km apart, are planned in Ahrental, St. Jodok and Trens (see Figure 2).

The emergency stops serve to rescue passengers from trains with technical difficulties. In addition, all emergency stops are accessible through driveable approach tunnels. A detailed project description is provided e.g., in (Eckbauer et al., 2014).

The two-track Inn Valley Tunnel has been already completed in 1994 and will be integrated into the overall tunnel system. This will reduce the travel time from Germany greatly. The 8 km long section of the Inn Valley Tunnel between the Tulfes portal and the link with main Brenner tubes is a part of the BBT system and will, for reasons of safety, be retrofitted with a rescue tunnel.

Excavation works of the construction lot "Tulfes-Pfons" will last until Spring of 2019. The construction lot includes 38 km of tunnel excavation work. It consists of several structures such as the Tulfes emergency tunnel, the Connection Ahrental access tunnel—Innsbruck emergency stop, the Innsbruck emergency stop with central tunnel and ventilation structures, the Main tunnel tubes, various Connecting tunnels and Ahrental-Pfons exploratory tunnel (see Figure 3).

The paper focuses on the experiences with the Tulfes emergency tunnel. The emergency tunnel is being driven parallel to the existing Innsbruck railway bypass; it will be 9 km long and the excavation cross-section is 35 m^2 (see Figure 3). The drill-and-blast excavation work (SEM) on this tunnel starts from three points at the same time: from Tulfes westwards (already completed), from the Ampass access tunnel eastwards and again westwards. The excavation works for this emergency tunnel were completed in Summer of 2017.

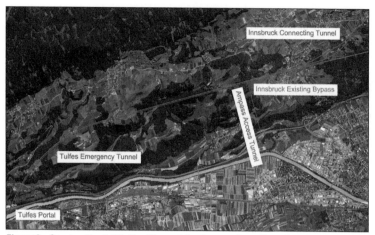

Figure 3. Construction lot Tulfes-Pfons of the Brenner Base Tunnel project

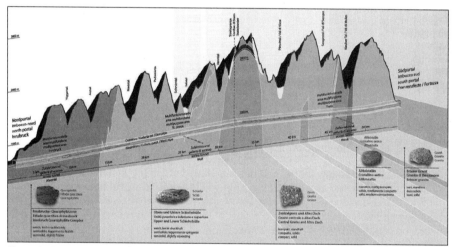

Figure 4. Longitudinal profile of geology with main lithological units (Eckbauer et al., 2014)

GEOLOGICAL CONDITIONS

This tunnel system within the Central Eastern Alps is crossing the collision zone of the European and the Adriatic (African) plate. The main lithological units are the Innsbrucker Quarzphyllites, the Bündnershists, the central gneisses and the Brixener granites (see Figure 4).

From the hydrogeological point of view the water ingress is very limited in the homogenous sections of the phyllites, schist, gneisses and granites. The amount of water is expected to increase only with advance through some fault zones. A lowering of the water table at the surface is prohibited in these fault zones because of environmental aspects.

The overburden varies between 1,000 and 1,500 m over most of the tunnel. The maximum of 1700 m will be reached in the central gneisses.

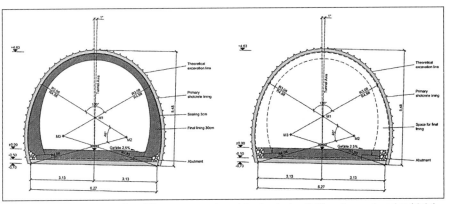

Figure 5. Regular cross section of the emergency tunnel tender design (left), value engineering (right)

OPTIMISATIONS

Such an emergency tunnel does not necessarily require a tunnel lining system with two shells (double shell lining system), but under certain boundary conditions can be achieved by a single shell lining approach. This is especially valid for non-public tunnels where lower levels of water tightness are acceptable. Single shell lining systems offer the most efficient lining design as they take both the temporary and long-term loads. Additionally, the construction is very fast compared to a double shell (or even composite lining systems) where multiple stages of construction are required.

The tunnel cross-sectional area remains unchanged compared to the tender design (see Figure 5). The permanent shotcrete lining shall have a minimum thickness of d = 25 cm with two layers of reinforcement. The effectively applied shotcrete thickness shall be determined based on back-calculations taking into account all available on-site information. Additionally the following requirements have been defined:

- increased shotcrete quality
- design for service life of 200 years (same as for main tunnels)
- increased concrete cover of 55 mm
- maximum crack widths ≤0,3 mm
- alkali-free accelerator to be used

In addition the applied monitoring concept has been adopted:

- five measuring bolts per cross section with installation interval 5–10 m in weak rock/fault zones and 10–30 m in competent rock formations
- timely implementation of the first convergence measurements
- measurement of the shotcrete thickness by means of tunnel scan
- continuous tests for stiffness and strength development of shotcrete

For the permanent shotcrete lining the verification analyses have to be carried out for the ultimate limit state (ULS) and the serviceability limit state (SLS) as well as for the durability of the shotcrete. This verification assessment is done in two steps:

1. Design of the shotcrete lining on the basis of hydrogeological, geological and geotechnical forecasts.

2. Verification on the basis of the measured (monitored) deformations, the actual shotcrete thicknesses and the determined shotcrete properties (modulus of elasticity, strengths).

The proof of suitability for the permanent use is provided by the proof that the analysed crack widths do not exceed the maximum crack width as defined for the project.

TUNNEL BOUNDARY CONDITIONS

The reference calculation (see Figure 6) considers a service tunnel with a width and height of W/H = 14,4/10,9 m and an overburden of 650 m above the crown. The rock mass has been modelled as Mohr-Coulomb material with a rock mass stiffness of E_{rm} = 12,000 MPa, Poisson ratio ν = 0,2 and a rock mass strength of c_{rm} = 2,0 MPa and ϕ_{rm} = 34,3°.

The shotcrete itself has been modelled using two different approaches: (1) using a Mohr Coulomb model based on the concept of "hypothetic stiffness" which is in fact based on experience and has been introduced e.g., [John et al., 2003], which considers a reduced stiffness of young shotcrete of 5 GPa and 15 GPa for hardened shotcrete, and (2) the shotcrete model mentioned above which has been implemented in the model code PLAXIS as a user defined shotcrete model [Schädlich et al., 2014 and Saurer et al., 2014].

CONSTITUTIVE MODEL OF SHOTCRETE

The theoretical background to the novel constitutive model has already been presented in [Schädlich et al., 2014] and validated against a practical engineering task, which is an executed high-speed railway tunnel project in Germany, built according to the SEM design philosophy [Saurer et al., 2014]. Fundamental equations are provided in [Schädlich et al., 2014]. The model has been implemented in the finite element

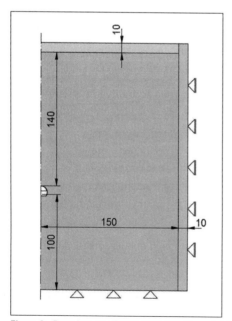
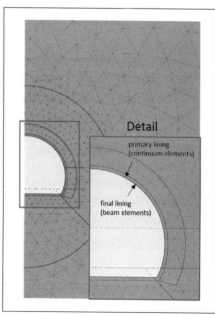

Figure 6. Boundary conditions (left) and FE mesh (right)

software PLAXIS 2D 2012 [Brinkgreve et al., 2012]. Note that compression is defined as a negative number.

MODEL PARAMETERS

The model parameters for the shotcrete model are presented in Table 1.

Table 1. Parameters of the shotcrete model and values considered in the example calculation

Name	Unit	Value	Remarks
E_{28}	[GPa]	30	Young's modulus after 28 days
ν	[—]	0,2	Poisson's ratio
$f_{c,28}$	[MPa]	33	Uniaxial compressive strength after 28 days
$f_{t,28}$	[MPa]	3,3	Uniaxial tensile strength after 28 days
ψ	[°]	0	Angle of dilatancy
E_1/E_{28}	[—]	0,6	Ratio of Young's modulus after 1 day and 28 days
$f_{c,1}/f_{c,28}$	[—]	−2*	Ratio of f_c after 1day and 28days
f_{c0n}	[—]	0,15	Normalized initial yield stress (compr.)
f_{cfn}	[—]	1	Normalized failure strength (compr.)
f_{cun}	[—]	1	Normalized residual strength
ε_{cp}^p	[—]	−0,03, −0,002, −0,001	Plastic peak strain in uniaxial compression at shotcrete ages of 1 hour, 8 hours and 24 hours
$G_{c,28}$	[kN/m]	6,1	Fracture energy in compression after 28 days
f_{tun}	[—]	1	Normalized residual tensile strength
$G_{t,28}$	[kN/m]	0,03	Fracture energy in tension after 28 days
ϕ^{cr}	[—]	2,0	Ratio of creep vs. elastic strains
t_{50}^{cr}	[days]	3,0	Time at 50% of creep
ε_∞^{shr}	[—]	0,0005	Final shrinkage strain
t_{50}^{shr}	[days]	70	Time at 50% of shrinkage

SHOTCRETE LINING CALCULATIONS

Initially, calculations using a linear elastic model, a Mohr-Coulomb Model and the shotcrete model without applying the time dependency, which means that using the common approach with reduced stiffness have been performed to benchmark the result. The results are marked as 'calculations with hypostatic stiffness'.

The following calculation phases have been considered in the model (see Table 2).

The shotcrete strength has been considered according to the J2 curve described above. In order to consider the effect of time in the model, two additional analyses have been performed marked with 'calculations with time-dependent stiffness'.

Initially, a benchmark model using linear elastic material, Mohr-Coulomb material and the UDM shotcrete model with the hypothetic stiffness approach (i.e., with E = 5/15 GPa) has been calculated to compare the resulting effects of actions in the elastic mode. (i.e., ignoring implicit time dependency).

Table 2. Calculation phases

Phase	Name/Description
1	In-situ stress state
2a	Stress release of the top heading
2b	Excavation of the top heading, installation of the shotcrete in the top heading
3a	Stress release of the bench
3b	Excavation of the bench, installation of the shotcrete in the bench

As can be observed in Figure 7 the effects of actions remain within the elastic limits and results from all three models are comparable. Hence the model allows comparing the results in terms of time effects shown hereafter as a next step.

As a next step, the time dependency of the shotcrete strength and stiffness have been included in the model, while still considering E = 15 GPa as maximum stiffness instead of consideration of implicit creep and shrinkage. Here, the boundary conditions as shown in the first chapters have been considered.

As a last step, the effects of shrinkage and creep have been modelled using the above expressed equations, considering realistic parameters for both, material model parameters and construction time. For this model, the parameters as stated in Table 1 have been considered.

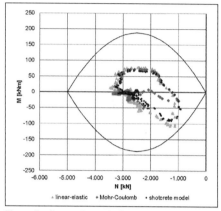

Figure 7. Comparison of effects of actions using the hypothetic stiffness approach

However, note that for fully time dependent calculations, boundary conditions in terms of construction time and the time dependent behaviour of shotcrete are important factors that have a major influence on the calculation results.

INTERPRETATION OF RESULTS

Shotcrete exhibits a significant time dependent behaviour, in particular during the initial hours of curing. This is important, because once applied as primary lining, the shotcrete is immediately loaded due to the excavation process.

In practical engineering crude simplifications are usually adopted with respect to modelling the mechanical behaviour of shotcrete in numerical analysis, The present paper presents the application of a novel constitutive shotcrete model using realistic boundary conditions for a shotcrete tunnel lining excavation. The model has first been verified by applying a classical application of a hypothetic E-modulus for shotcrete, which has been manually changed during calculation (see Figure 7). Only the strength has been calculated implicitly based on elastoplastic strain hardening/softening plasticity.

Thereafter the shotcrete stiffness has been determined automatically as a function of time. An automatic increase of shotcrete stiffness with progressing excavation has been applied. The resulting stresses and strains show a realistic range of values (see Figure 8 and 9).

The benefit of such calculation is that there is no longer need for manual adaptation of strength and stiffness with time. Of course, such an advanced constitutive model requires both a higher number of parameters and more detailed knowledge about construction time.

CONCLUSION

The present paper illustrates the required boundary conditions and limitations for the single shotcrete lining approach. A proposal for geomechanical and structural verification is provided. For the verification approach both, the use of a novel constitutive

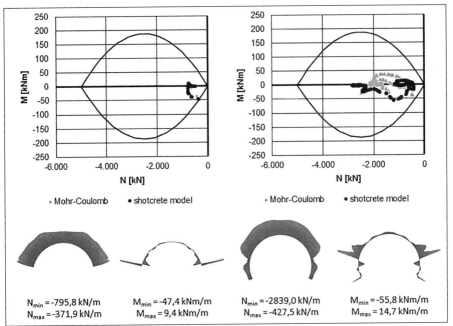

Figure 8. Effects of actions using the UDM for shotcrete using the hypothetic stiffness approach (lower part) and comparison with MC results as an interaction plot (upper part) for both excavation of crown and crown + bench

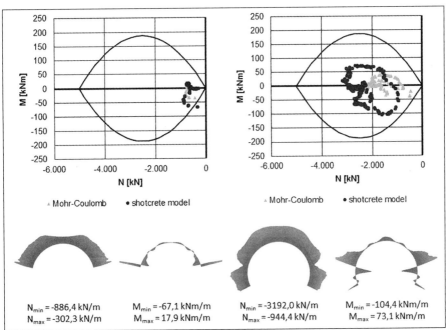

Figure 9. Effects of actions using the UDM for shotcrete using full stiffness approach with creep and shrinkage (lower part) and comparison with MC results as an interaction plot (upper part) for both excavation of crown and crown + bench

model with elastoplastic strain hardening/softening plasticity and time dependent strength and stiffness, creep and shrinkage are considered.

The paper illustrates a successful implementation of such a "SEM—Single Shell Lining Concept" for the Emergency Tunnel of the Brenner Base Tunnel Project.

REFERENCES

Bergmeister, K. 2015. Life cycle design and innovative construction technology. Swiss Tunnel Congress.

Brinkgreve R.B.J., Engin E., and Swolfs W.M.. 2012. Finite element code for soil and rock analyses. User's Manual. Plaxis bv, The Netherlands.

Eckbauer, W., Insam, R. and Zierl, D. 2014. Planning optimisation for the Brenner Base Tunnel considering both maintenance and sustainability. Geomechanics and Tunnelling, No. 5.

Eurocode 2: EN 1992-1-1. 2004. Design of concrete structures—Part 1-1: General rules and rules for buildings [Authority: The European Union Per Regulation 305/2011, Directive 98/34/EC, Directive 2004/18/EC].

Insam, R. and Rehbock-Sander, M. 2016. The Brenner Base Tunnel. RETC Conference Proceedings, San Francisco.

John M., Mattle B., Zoidl T.. 2003. Berücksichtigung des Materialverhaltens des jungen Spritzbetons bei Standsicherheitsuntersuchungen für Verkehrstunnel, in: DGGT (Hsg.): Tunnelbau, S 149–188.

Saurer E., Marcher T., Schädlich B., and Schweiger H.F.. 2014. Validation of a novel constitutive model for shotcrete using data from an executed tunnel, Geomechanics and Tunnelling 7, No. 4, 353–361.

Schädlich B., and Schweiger H.F.. 2014. A new constitutive model for shotcrete, Numerical methods in Geotechnical Engineering, Hicks, Brinkgreve, Rohe (eds.) 103–108.

SEM Cavern Construction in Downtown LA

Carlos Herranz ▪ Mott MacDonald
Christophe Bragard ▪ Regional Connector Constructors (Skanska-Traylor Brothers Joint Venture)
Ivan Hee ▪ Regional Connector Constructors (Skanska-Traylor Brothers Joint Venture)
Dominic Cerulli ▪ Arcadis, CM

ABSTRACT

The Regional Connector Transit Corridor is a 1.9-mile-long light-rail (LRT) line that will link existing LRT and subway lines to improve mobility within Downtown Los Angeles and throughout the greater City area. A challenging and risky adverse element of this $1-billion design-build project consists of a 287 ft long, 58 ft wide and 36 ft high track crossover structure which is being constructed at relatively shallow depth in soft rock beneath existing building foundations and major utilities using Sequential Excavation Methods (SEM). Through a combination of monitoring data, test results and field observations this paper correlates construction performance with design assumptions and analysis and demonstrates how collaboration between designer, contractor and owner is key to successfully adapt to changes during construction.

INTRODUCTION

The Los Angeles County Metropolitan Transportation Authority's (Metro) highly prioritized Regional Connector Transit Corridor (RCTC) project will bring light rail transit through downtown Los Angeles and link multiple existing lines to improve public mobility within the City and County. A US$927-million design-build contract for the project was awarded in 2014 to Regional Connector Constructors (RCC), a joint venture of Skanska and Traylor Brothers. Mott MacDonald is RCC's principal designer. The project totals 1.9 miles of new light rail infrastructure with 1 mile of twin bored tunnels, 0.7 miles of cut-and-cover tunnels, three cut-and-cover stations, 1000 ft of at-grade alignment, and a crossover cavern. The expected construction completion is in 2020. A location map showing the alignment, stations, and crossover cavern is shown in Figure 1.

A crossover structure is needed for operational flexibility to permit trains to switch from one track to the other. A limited right-of-way at the crossover location, walled by existing basements and structures, including underneath the sidewalks, made a cut-and-cover option infeasible. Thus, the project adopted one of its biggest design and construction challenges: a mined crossover cavern measuring 58 ft wide by 36 ft tall and 287 m in length, the overall design of which is described in depth in Herranz et al (2017). Once completed, the cavern will have the largest tunneled cross-section in Los Angeles.

The cavern comprises a twin sidewall and center drift configuration, with each of the side drifts enveloping and removing a previously driven bored tunnel. The sidewall drifts are excavated in a top heading and invert sequence, and the center drift in a top heading-bench-invert sequence. Cavern construction started in June of 2018 and at the time of writing the side drifts have been completed and the center drift top heading has reached half its total length. Center drift bench and invert excavation started

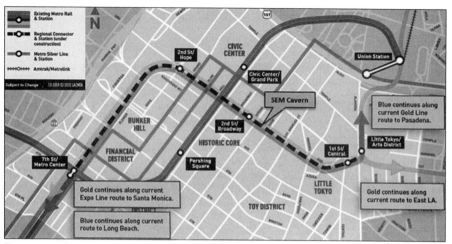

Figure 1. RCTC location map showing SEM cavern location

in November 2018. Demolition of the temporary sidewalls is expected to begin before the end of 2018.

DESIGN SUMMARY

The geologic setting in the SEM cavern area, according to the geotechnical investigation, was expected to consist of 6 to 10 ft of artificial fill (A_f), underlain by about 10 to 16 ft of coarse-grained alluvial deposits (Q_{al2}). The coarse-grained alluvium is underlain by the Fernando Formation bedrock, an extremely weak to weak massive clayey siltstone, that comprises an upper layer of moderately to highly weathered material, with a slightly weathered to fresh layer underneath. The entire cavern will be excavated in the more competent material. Face stability was checked during the design stage, showing no issues with the drift sizes planned. The Fernando Formation has relatively low hydraulic conductivity (average value of 4×10^{-7} in/sec), so groundwater infiltration during excavation was expected to be limited to seeps that originate primarily from rock discontinuities.

The cavern is constructed below portions of three adjacent buildings and directly below a large LA County storm drain and operating roadway. A critical cross-section, shown in Figure 2, has the cavern between the Higgins Building, a 10-story Los Angeles Historical-Cultural Monument that was built in 1910, and the 10-story headquarters of the Los Angeles Police Department (LAPD) that was built in 2009.

The initial design of the cavern called for a two-drift sequence with a single sidewall. Two major accommodations that led to this configuration were that the left track bored tunnel would be completed prior to the SEM cavern excavation and the hammerhead at 2nd/Broadway station, both of which essentially precluded the possibility of a three-drift configuration. Revisions to the project global schedule to account for delays at the Second and Broadway Station caused by the advance utility contract required both bored tunnels would need to be excavated prior to SEM construction. A three-drift scheme with two temporary sidewalls was selected for the cavern excavation, with excavation advance lengths of 3'-4" in top heading and bench in the sidedrifts and 3'-4" and 6'-8" in the center drift for the top heading and bench/invert, respectively. The advance length was selected to accommodate removal of the bored

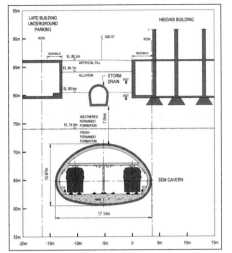 

Figure 2. Site cross-section (left) and site 3D view (right)

tunnel precast tunnel lining, which had a 5'-0" ring length, thereby accommodating the removal of 2 lining rings with every three rounds of excavation. The first 60-feet of cavern excavation are protected by a 53-unit pipe canopy system. The pipe canopy was installed from the 2nd/Broadway station excavation prior to the commencement of the SEM cavern excavation.

The initial lining comprises a minimum of 12 inches of 5,000 psi fiber-reinforced shotcrete, reinforced lattice girders, supplemental local bar reinforcement in corners, and mesh in specified locations in the temporary sidewalls.

The detailed design of the SEM cavern required extensive numerical modeling efforts due to the scale of the excavation and its interaction with existing site infrastructure. The design of the drift and heading sizes, sequencing, and temporary ground support required both two-dimensional (2D) and three-dimensional (3D) numerical models to evaluate the response of the ground, lining and overlying structures. Results from these models, mainly in terms of expected ground, buildings, utilities and initial lining displacements, have been used as a baseline to compare with actual monitoring values.

SEM TEAM AND CONSTRUCTION PROCESSES (RESS MEETINGS)

Because of its nature, SEM construction requires close collaboration between Designer, Contractor, Construction Manager (CM) and infrastructure Owner, as well as other Stakeholders (e.g., utility owners, City Board of Engineers, residents and business owners). Even though representing different entities, representatives of MM, RCC, Metro and CM are referred to as the SEM Team in this paper, as they share the common goal of safely building this critical part of the project, with each party fulfilling different responsibilities (Figure 3).

RCC and MM collaborated during the design stage through workshops and design reviews with the aims of minimizing constructability issues and providing a safe and cost-effective design that is tailored to RCC's preferred means and methods.

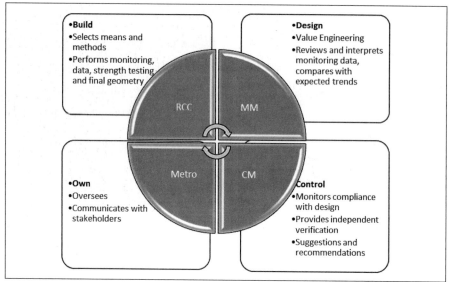

Figure 3. SEM team, construction stage

Collaboration during construction needs to be based on mutual trust and adequate processes. In preparation for the cavern, construction of a tieback removal shaft and adit and three cross-passages using SEM served as a test to set up the processes and engage the SEM Team.

Daily SEM meetings are held to discuss constructions progress, monitoring data (ground deformation and convergence data), gas monitoring, groundwater flow, geologic observations, shotcrete test results and construction contingencies applied, as well as to plan for the following construction activities. The outcomes of the meetings are documented on RESS (Required Excavation and Support Sheets). This continuous communication and engagement between all parties involved serves to face and react effectively to any issue, discuss well in advance potential design modifications and streamline decision making and documentation.

ACTUAL GROUND CONDITIONS

Actual ground conditions encountered have confirmed the design assumptions: the SEM cavern has been excavated thus far in fresh Fernando Formation, a weak massive clayey siltstone. There has been no groundwater infiltration at the time of writing this paper. The Fernando Formation shows good stand-up time and there have been no face stability issues (Figure 4).

During the excavation of the bored tunnels, two embedded steel piles, approximately aligned with the basement perimeter wall of the LAPD underground parking structure were encountered by the Tunnel Boring Machine. Available as-built information did not suggest a support of excavation system reaching to the depth at which the pile was encountered.

While drilling the pipe canopy tubes designed to protect the initial 60-feet of the SEM cavern excavation, three of the pipes could not be completed because of obstructions in the ground, one of which was aligned vertically with the location where the steel profile had been hit by the TBM. This raised the concern in the SEM Team of the

Figure 4. Exposed Fernando Formation in left drift break out (left picture) and steel profile encountered (right picture)

possibility of a continuous issue throughout the length of the LAPD underground parking structure, potentially affecting the whole length of the SEM cavern.

Correspondingly, RCC and MM developed a procedure to excavate and support around piles encountered during the excavation, with the aim of isolating the pile from the initial lining.

Three piles were encountered during the excavation of the left side drift (Figure 4, right), which was a great relief for the team, as the potential schedule impact to cut and remove the conflicting steel sections would have been significant.

GROUND SUPPORT AND DESIGN MODIFICATIONS

Verifying design assumptions and predictions through observed actual ground conditions and monitoring and test data is a key aspect of SEM construction. Based on actual behavior of the ground/structure system, feasibility of design modifications can be assessed to overcome unexpected situations, improve advance rates or address safety concerns.

Shotcrete Strength

The initial lining was designed for minimum 5,000 psi 28-day compressive strength and a residual flexural strength of 290 psi. Panel tests were shot prior to construction as part of the mix design confirmation and approval. During construction, panels where taken on site and compression strength tests were performed at 1, 3, 7, and 28 days. Figure 5 shows distribution of results obtained, as well as minimum strength required at each age. For tests with lower strength than the required 28-day strength, additional tests were performed.

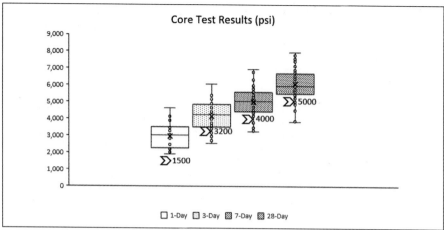

Figure 5. Core test results

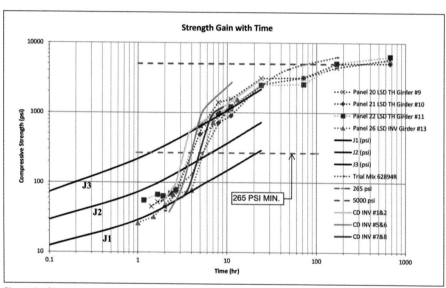

Figure 6. Shotcrete early strength test results

With respect to early age shotcrete strength (less than one day), RCC performed penetration tests to confirm the shotcrete was gaining strength appropriately and to verify the 72.5 psi minimum strength requirement to enable re-entry into a personnel exclusion zone, representing good safety practice. Due to construction sequence, the design included specific shotcrete strengths to be achieved before excavation advance. To confirm early shotcrete strengths RCC performed stud driving method tests, as needle penetration tests do not have enough capacity for the ranges required. Figure 6 shows a sample of comparison between required (265 psi for center drift invert) and obtained results along with J1, J2 and J3 curves (Austrian Concrete Association, 2004), reaching the required strength at approximately 3–5 hours after placement.

Excavation Profile and Shotcrete Thickness

After every round of excavation RCC surveyed the excavated surface profile and resurveyed the shotcrete surface after placement so that minimum design shotcrete thickness can be verified and the minimum required clearances for the installation of the final lining thickness can be confirmed.

Probe Drill Holes

Investigating the ground ahead of the excavation face through probe holes is typically performed to limit the risk of water and gas inflow and worse than expected ground condition. The design required probe drill holes to be performed as the excavation progressed, with the probe maintained a minimum of 20-feet ahead of the excavation face. Three probe holes were performed from the station (one for each drift). The homogeneity of the Fernando formation observed during the excavation of the left drift and lack of water or gas inflow led to the SEM Team agreeing that no further probing was necessary for the right or center drifts.

Flashcrete

The design required that all exposed ground be flashcreted immediately after excavation, which is a common practice in SEM construction in soft ground. Due to the observed condition of the Fernando formation, this requirement was waived for the side drifts invert and center drift bench and invert. For the side drifts the full invert shotcrete thickness was placed in one single layer, and no flashcrete was required on the invert face due to the relatively rapid infilling of the invert with temporary backfill to accommodate excavator reach.

Side Drift Round Length

Per project schedule and design, demolition of the existing precast tunnel liner (PCTL) had to occur concurrently with the SEM side drift excavation.

Due to the size of the cavern and the inherent risk of ground movement and damage RCC and MM collaborated extensively on the optimum excavation advance length. However, the ability to safely remove the PCTL was also a significant factor in the selection of advance length. The nominal width of the PCTL ring was 5-feet. An advance length of 3'-4" was selected, which permitted the radial joint of the PCTL to align with the face of the SEM excavation every three SEM rounds, or at 10'-0" intervals.

During the design stage it was decided to limit the SEM excavation round length as a mean to control risk. Similar size excavations in other projects in urban environments in soft rock have comparable round lengths.

RCC started the excavation of the side drift top headings with a 3-step sequence (A, B, C) in which PCTL segments were removed in steps A and B, with an over-excavation around the PCTL to reach the end of the segment.

The Step B excavation resulted in an effective unsupported ground length greater than the as-designed excavation sequence. The excavation of Step B also posed construction difficulties related to the local over-excavation, due to the limited space to install the lattice girders and wire mesh in the center sidewall, which resulted in longer time than desired to complete the excavation and support sequence. These difficulties were also resulting in construction quality issues related to placement of shotcrete and appropriate encapsulation of the lattice girders and reinforcement. As a result, RCC

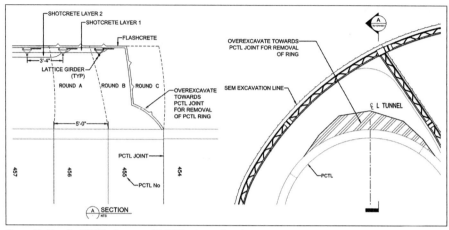

Figure 7. Ten-foot SEM advance with PCTL removal cycle

systematically contact grouted the crown of top headings of the left, right and center drifts, at the end of each week with, on occasion, significant volume takes.

To improve productivity RCC proposed that the round length be extended from 3'-4" to 5' so that the excavation and support could be completed faster, and the excavation face geometry could be kept on a more regular shape, eliminating the need for local over-excavation around the PCTL and eliminating the overhang on Round B. In turn, the proposal meant a greater extent of ground would be exposed on each round. However, with a faster cycle time, the exposed ground would be supported with shotcrete sooner than with the 3–4" advance length. RCC would also continue to systematically contact grout the crown.

MM assessed the proposed modifications considering the actual ground conditions, construction activities duration, and potential impacts on initial lining, settlement and face stability. MM agreed with RCC's proposed 5' round length for the side drifts excavation provided that the following performance requirements were met:

- The ground conditions remain similar to those observed to date without instability, wedge failure or significant over-break occurrence.
- Trends of ground settlement observed are similar or flatter than those observed to date, and continue to develop, as the tunnel progresses, without sharp or sudden changes in magnitude ahead of the tunnel face or transverse to the tunnel.
- There shall be a cavern convergence monitoring array installed in the round immediately prior to the change to the 5-ft round length.
- RCC will continue to systematically contact grout the top headings on a weekly basis and will closely monitor and evaluate contact grout volumes.
- Timely delivery and application of shotcrete is maintained.
- RCC shall monitor and maintain close field control over actual versus estimated construction activity durations including delays and discuss deviations in terms of production and the time ground is exposed and unsupported in the SEM daily meetings.

Center Drift Excavation and Temporary Walls Demolition

The center drift excavation and support design maintained the excavation of the center drift bench and invert 60' behind the center drift top heading, with a minimum required invert strength of 265 psi before proceeding with the demolition of the sidewalls. Side wall demolition would follow the invert placement once it reached 265 psi. This design approach was driven by RCC's proposed equipment selection for the excavation of the center drift bench and invert, as the CAT 328 excavator would not fit between the sidewalls.

RCC revisited this excavation and demolition sequence during construction and proposed using smaller excavation equipment, an ITC 312 that was being used for the center drift top heading excavation, that would fit between the temporary sidewalls. This allowed to stagger the sidewall demolition from the invert placement, allowing it to gain strength beyond minimum required per design.

Figure 8. Center drift bench and invert excavation

Another modification proposed for the center drift was the use of cast-in-place concrete instead of shotcreting the invert. The concrete mix slump was controlled so that the curved geometry of the invert could be obtained to avoid the final lining invert thickness being reduced by a flat temporary invert. Also, monitoring points were placed at the center drift invert to inform the SEM Team of displacements once the load transfer from the sidedrifts is completed with the temporary sidewalls demolition (Figure 9).

Instead of proceeding with the demolition of the sidewalls at the same time, RCC proposed removing the left side drift wall only, which would allow RCC to build an access to continue with the excavation of the center drift and perform activities on the right drift such as smoothing layer placement. This introduced a temporary asymmetry that had not been considered in the design of the initial lining. It was agreed to proceed with the proposed change in construction sequence after a successful demonstration of a 10-ft 'test zone' of demolition and verification of lining response before proceeding further.

Figure 9. Center drift invert; side drift connection (left), cast in place invert and monitoring point (right)

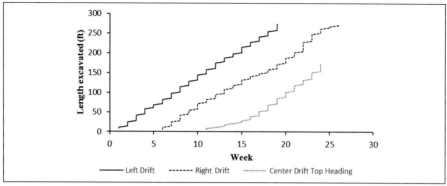

Figure 10. SEM excavation progress

CONSTRUCTION PROGRESS

The SEM cavern construction started at the beginning of June 2018, with the excavation of the left side drift, preceded by the pipe canopy installation and probe holing from 2nd/Broadway Station. Figure 10 shows progress so far. Excavation of the center drift bench and invert has recently started, as well as demolition of the temporary sidewalls.

RCC is using a CAT 328 as main excavator for the side drifts and top heading, a John Deer 80 excavator for the side drifts invert, and an ITC 312 for center top heading and invert excavations. Initial lining placement is being performed with a MEYCO Potenza shotcrete robot.

As discussed previously, the side drifts excavation length was modified at week 17 from 3'-4" to 5' to ease the PCTL demolition. Even though it was a longer excavation span, RCC was able to support the ground faster to minimize settlements and had the best production week.

Grout takes performed between lattice girders varied from o to 185 gallons, with an average of 20 gallons and a general trend of lower takes as construction progressed.

MONITORING

A rigorous instrumentation and monitoring program has been implemented for protection of buildings and utilities near the cavern, for ground response evaluation, and to verify the performance of the initial shotcrete lining. Instrumentation systems include arrays of convergence monitoring points, ground surface settlement points, multi-point borehole extensometers, inclinometers, piezometers, tiltmeters and monitoring points on buildings, and utility monitoring points on the overlying storm drain.

Based on the numerical modeling results, action and maximum instrumentation reading levels were established during the design stage at each construction stage. Actual displacements measured in the field during construction have been continuously compared with the contract limits and those obtained in the numerical models.

To facilitate these comparisons RCC has maintained a web portal-based instrumentation data management system, in which all monitoring data is stored, and provided the SEM Team with access to all information collected.

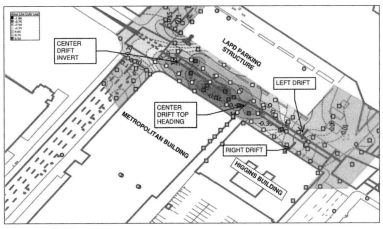

Figure 11. Settlement contour maps

A key monitoring feature that RCC proposed for the SEM cavern was the use of Automated Total Stations (AMTS) for surface monitoring, which allows for virtually real-time readings of all ground surface settlement points and building monitoring points. For the rest of the RCTC project, surface monitoring is being performed with traditional surveying, but it was decided that due to the size of the cavern, its construction methodology and sensitivity of surrounding structures, a more intensive control was necessary.

As part of the preparation for construction, the SEM Team discussed how the monitoring information should be made available to ease interpretation of the raw data and agreed on specific plots to be provided daily.

Ground Monitoring

In addition to daily readings of all instruments, settlement contour maps as shown in Figure 11 are provided weekly to the SEM Team showing the location of the excavation faces, such that ground displacements can be easily compared to excavation progress.

RCC provides monitoring data post-processing to ease interpretation and comparison with design values. As an example, Figure 12 shows right drift vertical displacements as measured in Ground Surface Settlement Points (GSSPs). Each data series corresponds to a monitoring point and the vertical dashed lines show the position of each excavation face (left drift, right drift, center drift top heading and center drift invert, from left to right in the figure) with respect to the GSSP location. Even though there is some variability depending on the location of the monitoring point in the alignment, all series follow a similar "S" trend. Similar plots were provided daily for the MPBXs, which showed higher response of the ground (up to 0.1 inches) at depth to the excavation of the center drift than what was observed at the surface. This confirms the need to monitor the ground at depth to get early warnings in case excessive ground deformations occur before they reach the surface.

Surface settlements have also been plotted in cross-section so that measured and predicted values at each stage can be compared, as shown in Figure 13. So far measured displacements are slightly higher than expected from the numerical models in some sections at the corresponding construction stage, but with values considered

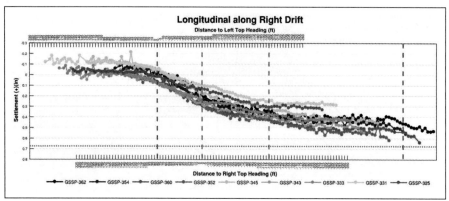

Figure 12. Settlement in GSSPs along the right drift

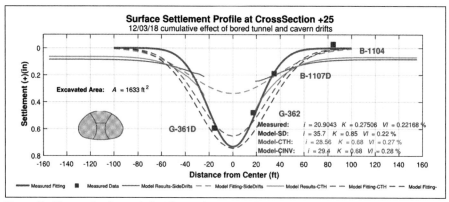

Figure 13. Measured and expected settlement in cross section

within the sensitivity of the numerical models and survey tolerance. Actual volume losses range so far between 0.2 and 0.3%, as compared with 0.3% estimated as part of the cavern design at the end of construction.

Building Monitoring

Building displacements observed so far have been within contract limits—0.5" maximum displacement and 1/600 maximum angular distortion, as well as within predicted displacements per models, even though displacements are ultimately expected to reach values close to the maximum settlement limit once the temporary sidewalls are demolished. Movements are being monitored through Building Monitoring Points (BMPs), Tiltmeters and Crack Gauges.

Cavern Monitoring

Monitoring arrays in the cavern are surveyed at the end of each shift. Figure 14 shows results for the left side drift at cavern station 12+00 as measured from 2nd/Broadway Station, with dates on the horizontal axis and convergence measured on the vertical axis. It can be observed how initially, with the left side drift excavation, convergence of horizontal chords increased (negative values indicate closure). This trend changed when the excavation of the right drift started and was further modified as the ground

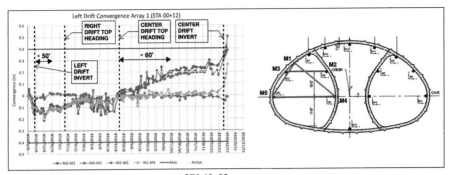

Figure 14. Left side drift convergences at STA 12+00

behind the temporary sidewalls was removed with the center drift top heading excavation, with the distance between monitoring points M2-M3 and M4-M5 increasing. Finally, with the center drift invert, an additional spike in horizontal chord extension occurs. This seems obvious, and was the behavior observed in the numerical models performed during the design stage, but the expected soil/structure response requires documented confirmation. Another important point of check, besides absolute values, is how they develop. The 3D numerical model results were used during the design stage to define the minimum distance between drifts, with a 60-foot stagger. As can be observed, the actual response of the structure confirms that this was an adequate value, as convergences tend to stabilize on each stage beyond that distance.

CONCLUSIONS

The construction of a large SEM cavern in Downtown Los Angeles poses a number of significant design and construction challenges. Design-build contracts promote close collaboration of contractors, designers, construction managers and owners, which is a key advantage in this case. Daily coordination between all parties involved allows for streamlining key decisions during the construction of this high-risk structure.

SEM is a flexible construction approach which can be adapted to suit the actual conditions encountered which in turn provides opportunities for increased productivity. However, decisions requiring changes to the design must be based upon sound judgement and must be confirmed by instrumentation data related to lining performance and ground and building response. Modifications to the design can be borne out of necessity due to unexpected conditions or can be driven by actual behaviors.

As an example of design modifications due to unexpected conditions, three steel piles were encountered during the excavation of the left side drift, which led to a local modification of the excavation sequence and support with the aim of isolating the remaining pile sections from the cavern initial lining.

Other modifications to the design were agreed during construction based on actual ground conditions and soil/structure behavior observed with the aim of improving the construction sequence, including the following:

- Probehole requirement waived based on observed homogeneity of the ground
- Flashcreting of all exposed surfaces requirement waived locally on inverts based on observed conditions of the Fernando formation

- Increased round length on side drifts from 3'-4" to 5' to ease bored tunnel liner demolition
- Cast-in-place concrete invert on the center drift instead of shotcrete
- Staggered temporary walls demolition

Meanwhile as a precaution, systematic grouting through and behind the liner at every round was implemented following observed instances of local shotcrete voids.

Measured ground, utilities and building displacements, as well as cavern convergences are consistent with predictions obtained from numerical models, with volume losses varying currently between 0.2 and 0.3%. Monitoring data post-processing is key to the interpretation of measured values and comparison with expected soil and structure response.

ACKNOWLEDGMENTS

The authors would like to thank Los Angeles County Metropolitan Transportation Authority for providing permission to publish this paper.

REFERENCES

Herranz, C., Z. Horvath, Lianides, J., and D. Penrice. 2017. "Cavern Configuration." *Tunnels and Tunneling International*. December 2017. 29–33.

Österreichischer Betonverein [Austrian Concrete Association]: Guidelines for shotcrete. January 2004.

SEM Tunnel Herrschaftsbuck: A Geological Challenge in Germany's Southwest

Roland Arnold ▪ BeMo Tunnelling GmbH

ABSTRACT

The Herrschaftsbucktunnel is located near the city of Rheinfelden (Germany) close to the swiss border. This highway tunneling project with two tunnels and a length of approximately 480 meters is excavated by applying the new Austrian tunneling method (NATM). Due to the geological situation and a cross section of 160 m² in the southern tunnel, it was required to make use of a complex exploratory drilling program.

The paper describes the challenges of constructing this NATM Tunnel in difficult geological conditions, low overburden and big cross sections.

INTRODUCTION

General

The Herrschaftsbucktunnel is part of the new construction of the highway A98 Waidhof—Rheinfelden Karsau (Germany). The Tunnel is located in the triangle of Germany, Switzerland and France and crosses the forest of the Nollingerberg-Mountain near the well-known swiss city of Basel. The project consists of two separated, almost parallel tubes. The North tube is designed as a two-lane highway tunnel with an excavated cross section of about 100 m². The South tube is designed as a three-lane highway tunnel with an impressive cross section of more than 160 m². Due to the difficult geology both tunnels were excavated with a deep invert over the whole tunnel length of approximately 480 m. The chosen excavation method was the new Austrian tunneling method (NATM). The Tunnel was basically built in three sections top heading, bench and invert. The top heading was built first, bench and invert followed in variable distances, depending on the geological conditions.

Geology

The tunnel is primarily crossing the geological formation of Gipskeuper and Lettenkeuper (middle trias formation formed of clay and gypsum). The base consists

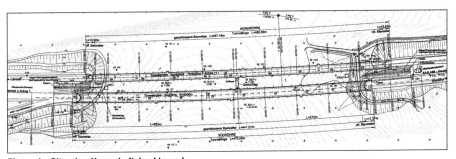

Figure 1. Site plan Herrschaftsbucktunnel

of shell limestone (trias limestone formation). Both Exploration cuts (West and East) were built more than 10 years ago and were therefore exposed to atmospheric conditions for a long time.

In the tender the prognosed geology was classified into two different ground zones. The location of the socalled leaching front in reference to the tunnel cross section is the main difference between these two ground zones:

In ground zone I the tunnel cross section is mainly located below the leaching front.

In ground zone II the cross section is mainly located above the leaching front.

The geological forecast shows ground zone two from the portal zones to a depth of approximately 110 m. The remaining, central portion of the tunnel is shown as ground zone one.

Hydrogeological Conditions

The hydrogeological conditions are corresponding with the inhomogeneous ground structure. There are four points of ground water monitoring in the area of the Herrschaftsbucktunnel. While two of the groundwater monitoring points clearly show groundwater levels above the top of the tunnel, the water levels in the other two monitoring points are below the tunnel invert.

Any water flowing into the tunnel during excavation will be guided to the temporary construction drainage pipe (partial drain pipe, nominal diameter 300 mm) below the invert. After passing a water treatment plant (sedimentation and ph adjustment) at the portal the water gets discharged into a nearby creek.

The final stage of the project does not foresee any groundwater drainage. After installation of the inner lining groundwater

Figure 2. Visualization east-portal

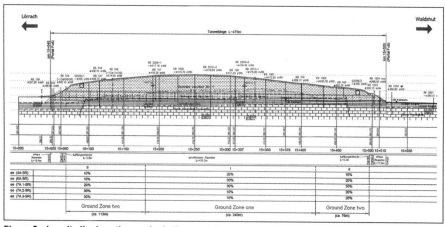

Figure 3. Longitudinal section geological prognosis

SEM Tunnel Herrschaftsbuck: A Geological Challenge in Germany's Southwest

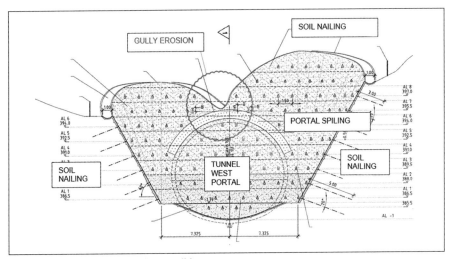

Figure 4. West portal in-situ surface conditions

levels around the tunnel will reach their initial elevation. The 40 to 60 cm thick inner lining in combination with a PVC waterproofing membrane has to resist groundwater pressures of up to 20 m (2 bar) water column.

DIFFICULTIES AT PRECUT WEST

The portal precuts at both portals were excavated more than ten years ago. The slopes at these portal areas were for many years exposed to the atmospheric conditions and showed very deep gullies caused by surface run-off, especially at the west portal. A drone flight was executed over this area to be able to quantify the effect of the atmospheric influence. After evaluation of the aerial field survey the whole extent could be determined. Some of these surface channels were almost reaching the top of the tunnel crown, which made a conventional start situation impossible.

The BeMo design department elaborated two constructable alternatives:

A simple method, although connected to enormous overbreak, would be to build an enlarged pilot cut with a subsequent cut & cover Tunnel section. The second—more elegant—solution was to build a tunnel shaped concrete cover on a tunnelshaped earth-form. This turned out to be the technically and economically most reasonable method, which was the also presented to the client. This method was previously successfully used in different variations on tunnel projects in Austria and Germany and it is called "Carinthian cut & cover method," after the first application which was done at tunnel Wolfsberg in Carinthia, Austria. After construction of the concrete cover the tunnel excavation works can be done in a safe way below the cover like shown in Figure 5.

Unfortunately, the construction works of the pilot cut west began in the late autumn and for this reason the works in this soft, cohesive ground had to be realized in winter until the following spring.

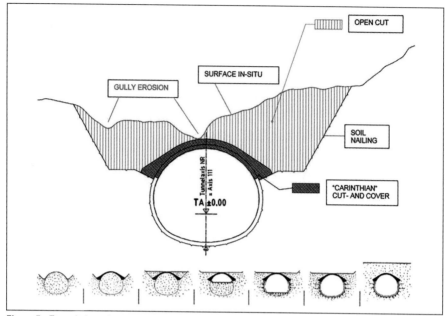

Figure 5. Tunnel shaped concrete cover—"Carinthian cut & cover method"

SITE INVESTIGATION

The geological descriptions of both ground zones considered it to be possible that cavernous carst areas were encountered. Due to this an extensive exploration campaign was planned by the client.

These underground exploratory drillings served the investigation and early detection of possible cavernous phenomena. These drillings are performed in addition to the regular drillings necessary for the excavation works (blastholes, anchor- and spile-drillings). Already the regular drillings are able to give a first indication of cavernous phenomena. The more data can be gathered out of the combination of drilling characteristics (drilling resistance, water circulation loss, pressure, encountered voids, geological interferences) the clearer becomes the picture of the ground laying ahead of the tunnel face.

When the drilling is in progress the project team and the client's geologist are assessing the available information. Further data is documented by the MWD-system (Measurement While Drilling) which is installed on the drilling jumbo. To avoid an uncontrolled approach of caverns the personnel of the construction site is urged to stop excavation works in case of any suspicion. Then the following steps will be determined by the expert team. The expert team is composed of representatives of the contractor and the client.

The tender documents predetermined a series of exploratory drilling arrangements which must be put in action in case of encountering carst phenomena. These exploratory drilling arrangements are mainly designed for the safety of the completed tunnel construction. As shown in Figure 6 the drillings are only located around the tunnel. But to be able to guarantee safe excavation works it is necessary to explore the ground under each construction step (top heading, bench, invert). This means that the exploration must happen for each working level. The surroundings can be explored

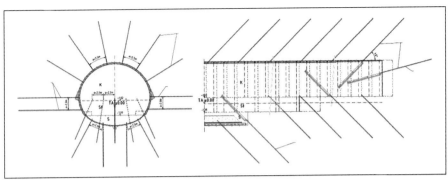

Figure 6. Carst exploration program

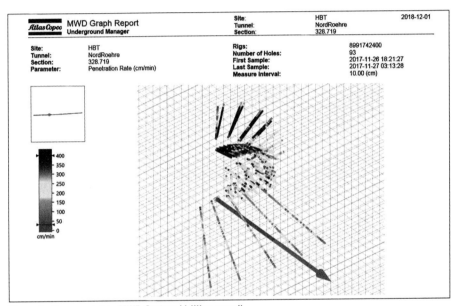

Figure 7. Documentation MWD-System (drilling speed)

sufficiently with the regular drillings around the tunnel (anchors and spiles), the invert requires additional exploration. All the realized drillings (regular and exploratory) were documented by the MWD-System and evaluated by the project team that is built by the owner and the contractor. In Germany and Austria, the responsibility for the ground is in the sphere of the owner. Due this reason the owner must be involved into the evaluating process.

To get the best result of exploration, the ground was explored by 3-meter-long spile drilling at the roof, 4 to 6-meter-long radial anchor drillings at each excavation step and 5 exploratory inclined drillings with a length of 10 m at each third step. Figure 7 shows an example for the documentation of the MWD-System. One of the most significant parameters shown in this example is the drilling speed in centimeters per minute.

Figure 8 shows the corresponding geological documentation at Station 337.48 in the north tube.

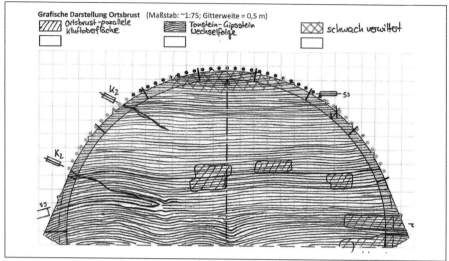

Figure 8. Geological documentation TM 337.48

The comparison of the documented parameter of the MWD-drillings and the corresponding geologic documentation shows the correlation between the higher drilling speed (hash in Figure 8) and weak zone at the top of the tunnel. To get secured Information about existing karst caverns it is necessary to observe all parameters in combination. The drilling speed, resistance and pressure are the main parameters helping to forecast karstified and cavernous phenomena in this area. The project specifications include different possibilities for grouting and backfilling options, depending on the specific phenomena encountered. The subsequent course of action will be decided by owner and contractor together.

In addition to the possibility to forecast caverns the MWD provides good information about the ground conditions ahead of the face. This enables the team to choose the right amount of spiles and provides support for the decision of the length of the next advance.

EXCAVATION PROCESS NORTH TUBE

The excavation of the north tube began on August 7, 2017. Due to some problems with the blasting permit the first excavation meters had to be done mechanically by using a tunnel excavator. Unfortunately, the ground in this phase was already really hard and did not correlate with the forecast. Also, the level of the leaching front was like shown in Figure 9 not as low as predicted and the lettenkeuper reached higher than expected. After a few days the first mechanical problems and damages of the excavation equipment caused by the very hard dolomitic rocks were registered. The blast permit was issued in the middle of September 2017.

The mixed-face conditions presented a big challenge for the excavation crews. The condition of the leached lettenkeuper was much worse than expected. Due to the fact that the face-stability was not given during the excavation works the further excavation progress had to be done with in multiple divided headings (pocket excavation). In addition to the pocket excavation it was necessary to install face bolts and shotcrete support on the face.

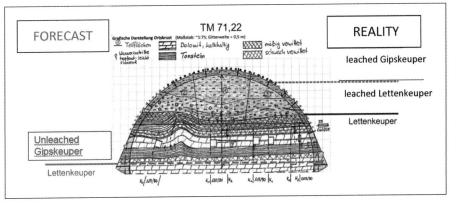

Figure 9. Forecast–reality: comparison at TM 71.22

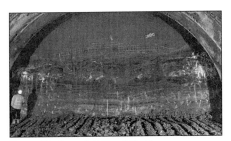

Figure 10. Face at station 188.72

Figure 11. Typical face conditions north tube

The lower part of the top heading was situated in the stable rock of the Lettenkeuper and later on in the unleached Gipskeuper, so there were no settlement problems and it was therefore not necessary to install a temporary invert in the top heading. However, next to the pocket excavation in the upper part of the face it was necessary to blast the lower part of the tunnel face. With increasing tunnel length, the leaching front moved higher and the number of pockets got less.

Ground zone two which was expected only in the portal near areas and for about 40 percent of the tunnel was in reality encountered on more than 95 percent of both tubes. The picture shown in Figure 11 is representative for big sections of the North tube.

Ground zone one was predicted with homogeneous conditions over the whole face. Such conditions were encountered only on about 20 meters of the whole North tube. Ground Zone one was expected for more than 240 meters. When the leaching front was close to the top of the tunnel it was possible to manage the excavation without pocket excavation. As shown in Figure 10 and 11 the leaching front accompanied the excavation works wavelike in the tunnel top heading. The excavation works were supported by spile umbrellas at each lattice girder (at least 35 spiles at each location).

Starting from tunnel station 350 meter the leaching front started to decline as predicted. Up to this station it was possible to execute the excavation works without settlement problems. The leached Keuper began to dominate the excavation face so it was also necessary to increase the amount of pockets.

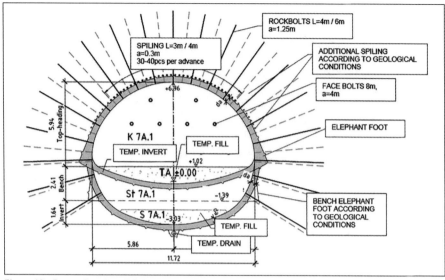

Figure 12. Excavation class 7 A.1

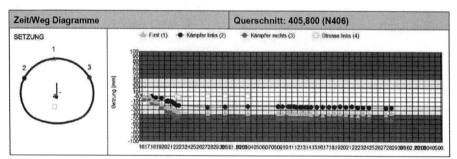

Figure 13. Settlement curve north tube TM 405.80 after change from excavation class 6A.2. to 7A.1

Another negative fact was that the elephant feet of the top heading were now situated in the soft and cohesive ground of the leached Gipskeuper.

The settlement curve in Figure 13 shows the effect of the change from excavation class 6A.2 (Figure 14) to 7A.1 (Figure 12). The construction of the temporary invert in the top heading which is part of Excavation class 7 A.1 stabilized the tunnel in the encountered ground and the settlement were reduced significantly.

EXCAVATION PROCESS SOUTH-TUBE

Due to the close proximity to the North tube there were no significant geological differences to be expected when excavating the South tube. One essential difference was the much bigger cross section of the South tube (160 m^2 vs. 110 m^2).

Due to this cross-section design with a top heading of over 70 m^2 some face stability issues were encountered in an effort to reduce settlements the project team decided to install a temporary invert in the top heading over about 120 m of the tunnel 480 m long tunnel. Pocket excavation was applied extensively. The number of subareas was fixed daily by the project team (owner and client) based on the knowledge of the past

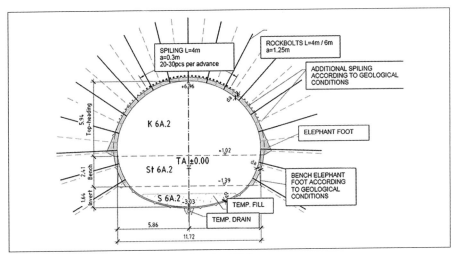

Figure 14. Excavation south tube—excavation class 6A.2

excavation steps and the information gathered from the exploratory drilling program. Depending on this evaluation there were up to 12 (!) pockets. In addition to pocket excavation there were also up to 12 facebolts and at least 10 cm of shotcrete support on the tunnel face.

In contrast to the North tube the face of the South tube could not be classified completely as an unbleached gypsum-keuper. The leaching front was always in the tunnel-face, resulting in were mixed-face conditions over the whole tunnel Due to these conditions a 3-layered spile umbrella was required to guarantee safe conditions for excavation.

Figure 15. Pocket excavation with 10 pockets

Close to the portal area both tunnel excavations passed a critical area with low overburden. The overburden of the west end of the southern tube was only about 6 to 8 m. When the tunnel excavation passed this area considerable settlements on the surface was monitored. The settlements were in the range of about 20 cm. The project team decided to observe the settlement at the surface with a complex automatic monitoring system which was already installed to monitor the portal pre-cut. It only had to be extended by 5 monitoring points. Measurements were repeated every 15 minutes and in case of exceeding the critical value (of 300 mm in total or 15 mm between two measurement interval) the project team got informed via text message. Based on this information the excavation could be continued safely.

It is assumed that the relative high settlement values are caused by small surface-near cavities collapsing when the tunnel works passed this area. The exploratory drillings did not detect cavities, so the reason for this high settlement could not be fully explained. This course of action could only be realized because there was no risk of damage to the surroundings. The South tube of the Herrschaftsbucktunnel was successfully broken through on March 1, 2018.

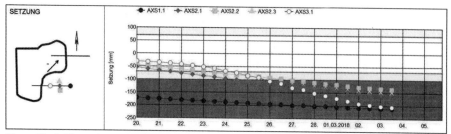

Figure 16. Settlement curve surface

SUMMARY

The Herrschaftbucktunnel was due to geological and geotechnical conditions a big challenge. This challenge could be mastered with a combination of extensive exploratory drillings and high expertise of the Project team. This combination allowed to realize the excavation works in an efficient way with an economic result. At this point we can say that the team work between the owner and the contractor was excellent and necessary for the success of the project.

Figure 17. Break through south tube on March 1, 2018

Shallow SEM Tunneling with Limited Clearance to Existing Structures: Design, Construction and Observations

Hong Yang ▪ Mott MacDonald
Derek Penrice ▪ Mott MacDonald
Walter Klary ▪ Gall Zeidler Consultants
Vojtech Gall ▪ Gall Zeidler Consultants

ABSTRACT

The excavation and initial lining has just been successfully completed for the Downtown Bellevue Tunnel using Sequential Excavation Methods (SEM). The soft ground tunnel, approximately 2,000 feet long and 38 feet wide with very shallow cover, faced several unique challenges and design refinements that were managed and implemented during final design and construction. This paper provides an overview of the tunnel construction, then focuses on three design refinements; tunneling under an existing utility trench with 4 feet of vertical clearance, tunneling within 4 feet of an existing building basement, and eliminating pipe canopy and replacing 12-foot thick overburden soil with controlled low strength material (CLSM) at the tunnel's north portal. Details of the analysis, design, and construction of these design refinements, together with the comparisons between the predicted and observed settlement results are presented.

INTRODUCTION

East Link is a 14-mile extension of the existing Sound Transit light rail transit system from downtown Seattle, across Lake Washington, to the cities of Mercer Island, Bellevue and Redmond. The Downtown Bellevue Tunnel (DBT) is the only tunnel on the extension and, as its name suggests, extends through the downtown of Bellevue with relatively low cover between the at-grade East Main and Bellevue Downtown stations (Figure 1). The DBT consists of 250-foot-long south cut-and-cover portal structure, the 1,983-foot-long SEM tunnel, the mid-tunnel access shaft and connecting adit, and the 200-foot-long north cut-and-cover structure. The DBT was originally planned

Figure 1. Plan view of project alignment

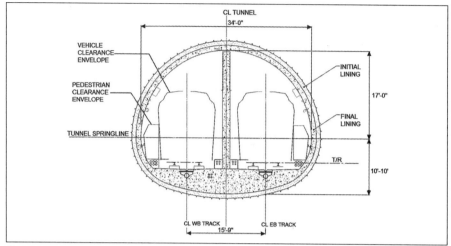

Figure 2. Tunnel geometry for typical tunnel section

as a cut-and-cover tunnel, but this was changed to reduce surface disruption and community impacts (Penrice etc, 2017).

The excavated cross-section for the typical SEM tunnel is a large 38-foot-wide by 30.5-foot-high ovoid, sized for twin rail tracks (Figure 2). The cross section is enlarged near the tunnel's mid-point to provide additional space for an emergency ventilation fan room over the tracks. An adit and shaft, also constructed using SEM provide maintenance access to the fan plant from the ground surface.

SITE GEOLOGY

The geologic profile along the tunnel indicates glacial deposits consisting of glacially overconsolidated stratigraphic sequence that includes Vashon till, Vashon advance outwash deposits, and pre-Vashon glacio-lacustrine deposits (Figure 3). North of the enlarged tunnel section the profile indicates an "anomaly zone." During the design an extensive ground investigation program was executed but no conclusive geological model could be established for this zone. During excavation of the tunnel no change of the ground behavior was observed in this zone, however offsets in the stratigraphy were encountered. The design groundwater table generally follows the top of the advance outwash, and was expected to be encountered in the tunnel face in the northern half of the tunnel. During excavation of the tunnel the ground showed more favorable conditions than anticipated. In particular, the groundwater table was much lower than expected and the planned dewatering measures which included dewatering with surface wells and vacuum dewatering from within the tunnel were not required. However perched ground water with a water inflow rate of approximately 0.2 gallon per minute was periodically encountered from within sand layers in the till.

CONSTRUCTION OVERVIEW

The excavation of the DBT started on February 3, 2017 and was completed on July 17, 2018, several months ahead of schedule. Prior to the start of the tunnel excavation a pipe arch canopy with a length of 70 feet was installed at the south portal. At the north portal, in lieu of a pipe arch canopy, the area was excavated from the surface to approximate springline elevation and the in-situ ground was replaced with CLSM

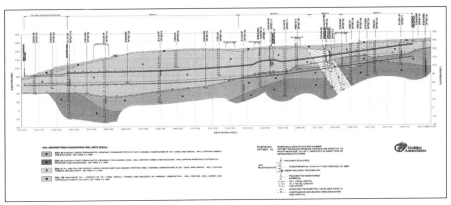

Figure 3. Geologic profile with tunnel alignment

and concrete. This significantly reduced the interference with the adjacent contractor for the cut-and-cover tunnel at the north portal. This design change is described in greater detail later.

The DBT was designed as a single side-drift excavation with a 6-heading sequence and with five Ground Support Classes, and with round lengths from 3 feet to 4 feet and with systematic spiling. To help mitigate delays caused by delayed right of way acquisition, the Contractor proposed changing the excavation from the 6-heading sequence to a 3-heading sequence from the south portal to the start of the enlarged tunnel section (approximately 50% of the tunnel length). The 3-heading sequence included a top heading, bench, and invert. This change was implemented for construction after concept validation by performing detailed analyses, and agreement on criteria that would result in a return to the as-designed 6-heading sequence (see Brodbaek etc. 2018).

From the enlarged tunnel section to the north portal, a single side-drift excavation with the 6-heading sequence was implemented, which consisted of a temporary center-wall, and each drift consisted of a top heading, bench and invert. Figure 4 shows the transition for the 3-heading to the 6-heading sequence.

For the excavation using the 6-heading sequence, two Liebherr 950 tunnel excavators were utilized. The majority of the tunnel was excavated with a bucket. Only in very short stretches of the tunnel, e.g., where lean concrete was present, a roadheader attachment or a hoe ram was used. Mucking and limited invert excavation was performed using a John Deere 135D. Figure 5 shows some of the typical activities during tunneling using the 6-heading sequence.

It was anticipated that the tunneling would encounter tiebacks and soil nails, left in place from adjacent building basement excavations. In areas with tiebacks the utilization of a bucket made it possible to facilitate a more precise excavation around the tiebacks. This was especially important in the vicinity of the Skyline

Figure 4. Transition from 3-heading to 6-heading sequence

Tower, where it was imperative that the small pillar between the tunnel and the existing building not be disturbed. The tiebacks and soil nails were included in the project Geotechnical Baseline Report as obstructions. However, their removal was relatively straightforward, with most of the elements being cut out with a torch such that the Contractor did not maintain a count of their number. For the removal of the centerwall a concrete shear was utilized, and the last foot was ground down to the required clearance with the Liebherr 950 and a roadheader attachment.

Figure 5. Excavation in left drift and preparation for setting girder in right drift

Due to favorable ground conditions encountered, only a small number of the designed spiles were installed. However, pre- and during-construction investigations of utilities and basements of buildings showed that pre-treatment of the ground was required in the proximity of the Skyline Tower and at the intersection of 110th Ave and NE 4th Street, as also described herein.

In order to have shotcrete available as required over the 24-hour work day, a batch plant was setup on site. Wet-mix shotcrete with designed 5,000-psi 28-day compressive strength was applied using a Normet shotcrete robot. As a contingency a second setup with a Reed pump and a shotcrete manipulator mounted on a tractor was used. Since the Reed pump is not an integrated system like the Normet and requires a separate accelerator pump, the dosing of the accelerator had to be carefully monitored.

Several changes to the design were made to adapt to the favorable ground conditions encountered. In addition to the 3-heading excavation sequence and the deepened CLSM at the north portal, modifications to the design included the following:

- Use of macrosynthetic fiber in lieu of steel fiber in the initial lining shotcrete
- Elimination of prescriptive spiling
- Elimination of 2 probe holes per cycle
- Elimination of bench probe holes in last 300 feet of tunnel approximately
- Elimination of continuous core holes over tunnel extent
- Increased round length (typical) from 4'-0" to 4'-3"
- Use of 4'-0" advance length in anomaly zone
- Elimination of invert excavation face shotcrete, provided invert was backfilled after shotcrete placement
- Elimination of bench sidewall flashcrete, allowed placement of 10" at once
- Elimination of convergence monitoring points in tunnel invert
- Reduced frequency of shotcrete testing based upon favorable results of previous testing (time dependent)
- Elimination of 28-day shotcrete strength testing should 7-day strength test results exceed 28-day strength requirement
- Elimination of the surface-based dewatering system

During daily SEM meetings in early mornings, the ground conditions, monitoring data and other relevant data of the previous 24 hours were reviewed and discussed. In conjunction with designer's representative, construction management and Contractor, ground support classes for the next 24 hours were agreed on. The change from a robust prescriptive design to a more adaptable design required experienced personnel from all parties attending the daily SEM meetings. Effective communication and collaboration between the designer, construction management, Contractor and owner's representatives was essential for the successful implementation of this more adaptable design.

A number of challenges and refinements were managed and implemented during the final design and the construction, in particular in the northern section of the project. Three such refinements, from south to north of the alignment, are tunneling under an existing utility trench with 4 feet of vertical separation at NE 4th Street; tunneling within 4 feet of the existing Skyline Tower basement; and eliminating pipe canopy and replacing 12-foot thick overburden soil with CLSM at the tunnel's north portal, discussed hereafter in greater details. The locations of these three refinements are shown in Figure 1.

TUNNELING UNDER NE 4TH STREET UTILITY TRENCH

At the intersection of NE 4th Street and 110th Ave (Figure 6), the Contractor's pre-construction investigation indicated that an existing storm drain utility was in poorer condition than anticipated during the design. Inspection of video survey and still photographs revealed holes in the invert and offsets at the pipe joints of a 12-inch diameter concrete drainage pipe which had likely resulted in leakage. The duration and extent of the leakage was unknown. It was postulated that the leakage may travel along the bedding of the pipes and accumulated at low points, resulting in saturation and degradation of the surrounding soils. The soil cover underlying the utility trench above the tunnel excavation was limited, with a minimum estimated thickness of approximately 4-foot of glacial till. The potential degradation of the soil between the tunnel and utility trench and the risk of the presence of sand lenses in the till between the tunnel and trench, coupled with the small soil cover, created significant risk to the tunneling. It was further discovered during the Contractor's exploratory work comprising field investigations and potholing that the utility trench was up to 6-foot wide, significantly wider than the 2-foot trench originally anticipated. Furthermore, the trench had been backfilled with pea gravel, which would have exhibited running behavior if encountered during tunneling. Correspondingly ground stability was a principal risk.

The original design of the SEM tunnel at the 4th Street crossing was based on assumptions that the utilities were in relatively good condition, that the groundwater table was located much deeper (close to tunnel springline), and that groundwater would be removed by surface dewatering measures applied in advance of tunnel excavation. In addition, the existing utility trench was assumed to be 2-foot wide. A trench of this width was considered in the design and was shown to not impact or be impacted by the tunneling.

Because of the new information provided during the construction, prior to tunneling reaching the intersection of 4th Street and 110th Ave, a number of mitigation measures were proposed and implemented, including local excavation of the trench and removal of the pea gravel by vacuum extraction; permeation grouting of the pipe bedding using 150-psi grout; then backfilling the trench with CLSM. As part of the design service during construction, 2-dimensional numerical modelling was performed to determine the minimum thickness of till that would support the utility trench without CLSM. Where

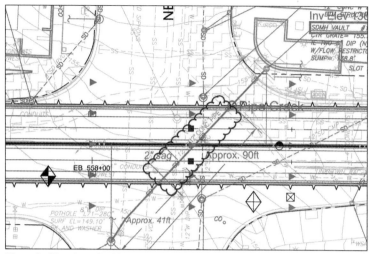

Figure 6. Plan view of utility trench crossing tunnel alignment

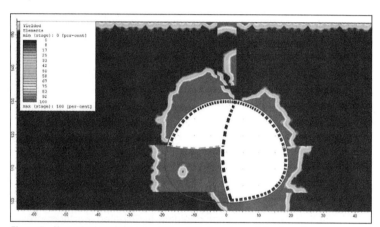

Figure 7. Numerical modeling result for utility trench overlying tunnel

the till thickness was less than this minimum, the trench would be filled with CLSM to the ground surface. The locations where sufficient till cap existed served as the limits of the CLSM replacement. The modeling steps and methods involved the use of the ground relaxation method for SEM (FHWA 2010) and was performed using the Rocscience RS2 software. An example of the model result is shown in Figure 7. Parametric studies were performed by varying the till cap thickness, soil strength and stiffness parameters, Ko values, and ground relaxation factor to envelope the potential conditions which could be encountered at the crossing location.

The modeling results showed that the various conditions modelled resulted in ground volume loss ranging from 0.16 to 0.4%, and estimated ground surface settlement ranging from 0.3 to 0.7 inches. The effectiveness of the CLSM mitigation was also demonstrated by the modelling results. The tunnel shotcrete lining capacity showed the as-designed lining would have adequate capacity under the various conditions modelled. Based on the modeling results, it was concluded that tunneling under the utility trench was feasible with the implementation of the proposed pre-tunneling mitigation measures.

A number of additional mitigation measures were implemented during tunneling at the crossing, including: reducing round length and lattice girder spacing from 4 feet to 3 feet; excavating the top headings utilizing pocket excavation with a minimum of two pockets of similar sizes; installing 9-foot long spiles with a maximum of 18-inch spacing under the utility pipe, with the spiles extending to 45 degrees to either side of the tunnel centerline; and maintaining a face wedge (Figure 8) in each of the top headings. These agreed measures further reduced the risks associated with the existing utility trench during tunnelling. During construction, water inflow of maximum 0.5 gallon per minute was observed inside the tunnel when crossing the trench, and this water inflow continued for several months. Water encountered drained off inside the tunnel, and no surface dewatering was performed based on the Contractor's engineer's understanding of the groundwater regime in the project area. No significant instabilities or overbreak were observed during excavation under the utility trench.

Instrumentation and monitoring results verified that the ground and tunnel were stable during the tunneling. An example of monitoring results is shown in Figure 9, which shows a maximum ground surface settlement of 0.30 inches and also shows good correlation with the lower end of the range of the predicted settlement.

Figure 8. Face wedge formed during tunnel excavation under the utility trench

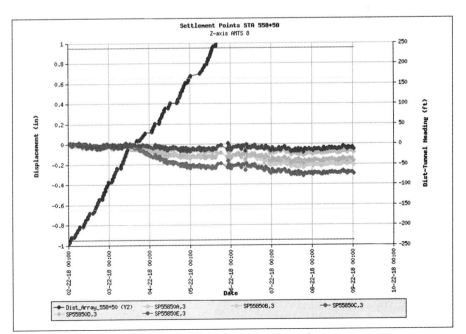

Figure 9. Ground surface settlements measured at utility trench crossing

TUNNELING NEXT TO EXISTING BASEMENTS

Several existing building basements are located in close proximity to the tunnel. In particular at the Skyline Tower basement, the minimum separation between tunnel extrados and outside of basement excavation support was limited to approximately 4 feet (Figure 10). The existing permanent basement wall construction comprised a composite system of concrete encased steel soldier piles at 7-foot spacing with 7-inch thick cast-in-place concrete facing that was lightly reinforced.

The interaction between the tunnel, ground, and the existing basement was studied in depth during the final design. The key objectives were to confirm the stability of the relatively narrow soil pillar between the tunnel and the basement, determine the movement of the ground and basement structure in response to tunnel excavation, and assess the change of ground stresses and soil pressure on the basement walls as a result of SEM tunnel construction sequence, among others.

Two-dimensional finite element analyses were performed using the Rocscience RS2 software. The effects of ground relaxation ratio were used to simulate three-dimensional effect ahead of the heading. Soil moduli, Ko and soil cohesion were parametrically studied. The basement wall(s) and floor slabs were added to the 2D numerical models. The basement walls were modeled as vertical beam elements. The basement floor slabs were modeled as spring supports for the basement wall. To replicate the existing conditions of stress in the ground prior to tunneling, modeling and analysis steps simulated the support of excavation (SOE) installation, basement excavation sequence, and the application of building loads to the foundation soil. An example of the modeling results in terms of ground displacement is graphically shown in Figure 11.

Parametric studies evaluated the sensitivity to the modeling techniques and input parameters, such as the mathematical representation of the floor slab (as beam elements vs spring supports), depth of the first lift of basement excavation, the level of prestressing force in the temporary tiebacks, the interface between the basement wall and soil (slip, no-slip, or with friction angle), Ko value, and the stiffness relaxation of

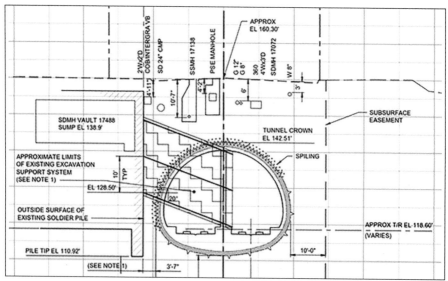

Figure 10. Cross section at Skyline basement showing small clearance from tunnel

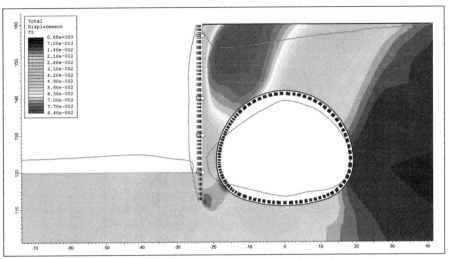

Figure 11. Numerical modeling result for post-tunneling condition at Skyline basement

the soil pillar between the basement and the tunnel. For these parametric studies, the change in lateral soil pressures on the basement wall after SEM tunneling were found to be sensitive to the assumed stiffness relaxation of the soil pillar, Ko value, and the soil-wall interface friction angle. The changes in soil pressures on basement wall after SEM tunneling did not appear to be sensitive to the variation in the tieback prestressing level or the depth of the first lift of basement excavation.

The modeling results showed that lateral soil pressures on the basement wall would increase significantly after tunneling. The structural capacity of the adjacent basement was evaluated for the change in lateral soil pressures and was found to be inadequate. Given the unusual basement design, and limited capacity to absorb any changes in stress arising from the tunneling, structural strengthening of the basement wall, comprising an additional 6-inch of bar reinforced shotcrete was performed in advance of tunneling. During this work a water filled void behind the basement wall was encountered. Due to the close proximity to the tunnel and the risk of water causing further deterioration, the thin pillar and void was grouted extensively. Additional monitoring points were installed in the garage to measure the performance of the building structure during tunneling, including tiltmeters, strain gages and structural monitoring points.

Tunneling adjacent to the Skyline Tower basement was successfully completed based on the design that was supported by the extensive analyses, and 4-foot round length excavation was adopted. The as-designed excavation sequence was modified to include pocket excavation during bench excavation for the left tunnel drift next to the basement due to the close proximity to the basement wall. During this excavation, one particular challenge was the removal of the existing tiebacks within the tunnel excavation limit. Figure 12 shows a tieback

Figure 12. A tieback encountered during excavation next to Skyline basement

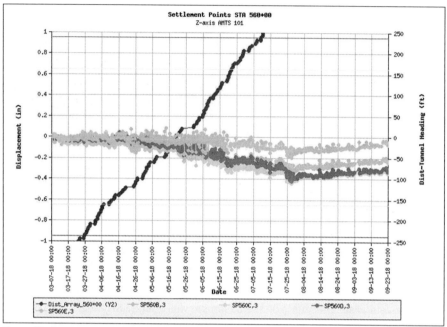

Figure 13. Ground surface settlements measured at Skyline

encountered during tunnel excavation. Another challenge was to remove boulders encountered. Tieback cutoff and boulder removal had to be performed with precision to minimize the disturbance to the ground.

The stability and safety of the ground, tunnel and basement were monitored and confirmed by the results of the extensive instrumentation and monitoring. The largest ground surface settlement observed was 0.45 inch (Figure 13). The strengthened basement wall was observed to move only 0.07 inch laterally. The observed maximum ground surface settlement value compared well with the predicted settlement that was estimated ranging from 0.5 to 1.0 inch.

TUNNELING AFTER GROUND REPLACEMENT AT NORTH PORTAL

At the northern section of the SEM tunnel, the soil cover over the tunnel crown was approximately 12 feet, which was shallow relative to the tunnel excavation size. Originally it was perceived that the material overlying the tunnel crown was principally Vashon Till, overlain by a limited depth of fill and roadway pavement. A pipe canopy, comprising a double row of 6-inch diameter pipes was to be installed over the tunnel over a length of approximately 70 feet to help support the ground in the area of limited cover. These soil conditions, with the inclusion of the canopy, were demonstrated to be sufficient to maintain a stable excavation. However, greater scrutiny of existing borings during the final design phase and removal of borings more remote from the tunnel alignment suggested that the fill was far deeper than originally anticipated, presenting significant risk to the tunneling.

After a series of mitigation options were studied during the final design, the as-bid design required that the fill material be removed over the width of the tunnel and over a length of 125 feet from the north SEM tunnel section and that the pipe canopy be

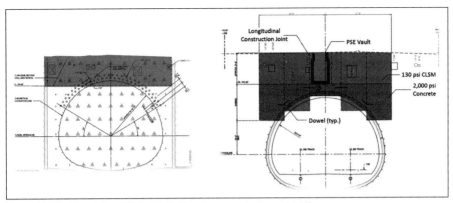

Figure 14. North portal as-bid design (left) and proposed scheme (right)

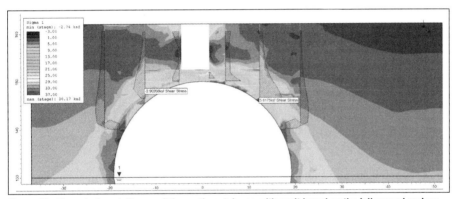

Figure 15. Numerical modeling result for north portal area with vault based on the full ground replacement scheme

installed (Figure 14). The installation of the pipe canopy at the north SEM limit would require the Contractor to coordinate possession of the north cut-and-cover excavation site with an adjacent contract, introducing risk of delays and conflicts for both contracts. As a mitigation, the Contractor proposed deepening the existing CLSM to create a 'structural' arch over the crown of the tunnel (Figure 14), and thereby avoided the need to install the pipe canopy.

A further complication in the north portal area was that a large electrical utility vault was found to be deeper than previously identified and could conflict with the pipe canopy installation. As part of the evaluation of the deepened CLSM, a separate structural concrete arch capable of spanning the tunnel excavation was developed for the vault.

To maintain surface traffic, the Contractor further proposed to install the structural concrete and CLSM in a series of 5 longitudinal trenches, as indicated in Figure 14 and the concept was evaluated. The details of the extended CLSM replacement, construction joints, material properties, and required analyses and modeling inputs and parameters were agreed by all parties in advance. The strengths of the CLSM and concrete arches were 130 psi and 2,000 psi, respectively. The design was confirmed by a series of 2D numerical models using the Rocscience RS2 software, as shown in Figure 15 for result, for example. For the structural concrete arch, dowels were required at each longitudinal joint to accommodate shear transfer. For the CLSM, only roughening and cleaning of the vertical cold joints was necessary.

Since the numerical modeling demonstrated the feasibility of the deepened CLSM solution, and as the solution eliminated a contract interface and reduced costs on two separate construction contracts, this change was implemented for construction with the agreement of all parties.

Figure 16. Concrete of 2000 psi from ground replacement exposed during tunneling

One challenge during tunneling under the CLSM replacement area was to remove a limited amount of the 2000-psi structural concrete (Figure 16) that was installed to protect the electrical vault; the concrete that protruded into the tunnel excavation line had to be carefully trimmed off to form the tunnel profile, while the disturbance to the underlying soil had to be minimized. Tunnel excavation and initial lining construction for the tunnel under the CLSM replacement area was completed safely in July 2018, with hole-through occurring in mid-July.

During and after the tunneling, the ground movements observed were small; the largest ground surface settlement observed was 0.33 inch (Figure 17), which was comparable with the predicted settlement of 0.5 inch.

SUMMARY AND CONCLUSIONS

Three cases involving unique challenges and design refinements facing the design and construction of the DBT have been presented and discussed. The mitigation measures implemented prior to or during tunnelling and the monitoring results have also

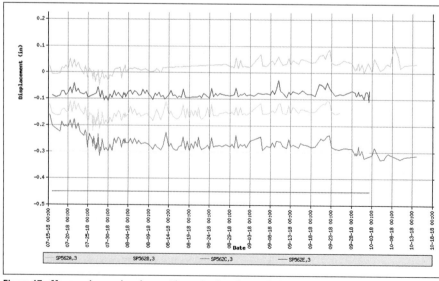

Figure 17. Measured ground surface settlements at north portal after tunneling

been discussed. These cases show that numerical modeling has proven to be efficient as part of the analyses, designs and design refinements, even during the construction. Instrumentation and monitoring results confirmed the validity of the designs and analyses and ensured the safe proceeding of construction. The cases presented also demonstrate the flexibility of the SEM tunnelling method to adapt to the changes in field conditions, provided the issues are identified sufficiently in advance. The cooperation of the project parties, including the design engineer, the construction management team, the Contractor, and Sound Transit, was also the key in achieving a successful outcome in each case.

ACKNOWLEDGMENTS

The authors would like to thank Sound Transit for the permission to publish this paper. In addition, the contributions of the design team, i.e., the H-J-H team (HNTB Corporation, Jacobs Engineering, Mott MacDonald, and Golder Associates), the Construction Management team (HDR and McMillen-Jacobs Associates), and the Contractor (Atkinson Construction, and its consulting engineer—Gall Zeidler Consultants) are gratefully acknowledged.

REFERENCES

Brodbaek, C., Penrice, D., Coibion, J., and Frederick, C. 2018. Downtown Bellevue Tunnel—Analysis and Design of SEM Optimization. North American Tunneling Conference.

FHWA, 2010. Technical Manual for Design and Construction of Road Tunnels—Civil Elements. Federal Highway Administration.

Penrice, D., Yang, H., Frederick, C. and Coibion, J. 2017. Downtown Bellevue Tunnel—Concept Optimization Through Team Collaboration. Rapid Excavation and Tunneling Conference, San Diego 2017.

Soft Ground Tunnelling Techniques for Mine Access Development

Ben Ablett ▪ Golder, Doncaster
Joe Anderson ▪ Golder, Saskatoon
Jack Nolan ▪ Golder, Saskatoon

ABSTRACT

The use of wet-mix structural sprayed concrete support has become increasingly prevalent in the tunneling industry over the last 50 years, being used to support both soft ground and hard rock conditions in temporary and permanent cases. In contrast, the mining industry has typically concentrated on using shotcrete, bolts and mesh as ground support in hard rock applications. However, applying techniques successfully implemented in tunneling was valuable when establishing a new hard rock development in South America where initial development through approximately 70 m of Saprolite and weak rock was required prior to driving 2 km of traditional hard rock drifts. Saprolite is a heavily weathered rock material, that behaves similarly to clay, and is common in humid and tropical climates. It inherits some of the relic structure of the decomposed rock, along which ground water flow can be highly variable. The material properties typically improve with depth prior to intersection with the fresh rock mass. The Sequential Excavation Method (SEM) was identified during the feasibility stages of the project as the most efficient and safe method of construction for the initial 70 m development from surface, prior to transition into hard rock development. The final lining system was designed to accommodate long term ground loading as well as significant out of balance loads due to the weakness of the Saprolite material, low overburden, high ground water table and seismic activity. Prior to commencing tunnelling works, a cut and cover approach was used to establish a steel lined portal and provide an engineered face in which to start SEM tunnelling activities. This paper presents the design and construction considerations of the project, including discussions of the challenges met in implementing such a scheme in a remote location, with difficult ground conditions and with a hard rock mining team.

INTRODUCTION

Wet-mix structural sprayed concrete is commonly used in the civil tunnelling industry for both temporary and permanent support in a variety of ground conditions and applications; from use as support for the entire development to complex elements of larger schemes, such as station cross-passages and escalator barrels.

There has been limited cross-over of the use of this approach into the mining industry, which has typically concentrated on using thin shotcrete, bolts and mesh as ground support in hard rock applications in both new developments and repair to existing infrastructure.

Therefore, when establishing a new development in South America, where initial development through approximately 70 m of Saprolite and weak rock was required prior to driving 2 km hard rock drifts, there was an opportunity to bring civil tunnelling methodology and application to the mining industry in order to access the deposit.

Once operational, the mine will exploit a high-grade gold deposit between depths of 100 and 400 meters below surface using conventional underground methods.

This paper will focus on how two portal accesses were established through the Saprolite and weak rock, rather than the drill and blast development.

For this project, Golder formed part of the team from the second feasibility study onwards, providing expertise during feasibility before leading the detailed design and construction management of the development prior to drill and blast mining.

Success in this project was achieved by providing specialist design and construction teams to be involved in the project from its early stages and maintaining this team from feasibility through to the end of the construction phase. This allows for a fluid approach and adaptive solutions to be developed to match the conditions encountered, in this case:

- Optimising the design to utilize the available material
- Providing expert oversight and training to local operatives
- Providing specialist teams to undertake construction

PROJECT BACKGROUND

Following discovery of the deposit, which is situated in a remote location of the Andean mineral belt, a significant drilling program was undertaken to map the extent of the ore-body. The bulk of the deposit was found to be microscopic and associated with quartz, carbonates and sulphides, with some coarse-grained veins in higher grade areas of the deposit.

Whilst exploration drilling work was being undertaken, concurrent feasibility work established that there was a strong economic case for the mine.

Subsequently, a detailed design was prepared to establish access to the ore-body via a decline (a tunnel driven downward to access the ore body). For the purposes of this paper, the terms are considered interchangeable.

Initial Construction

The initial development involved the establishment of a decline through a polymictic conglomerate formation, intermixed with clay, using drill and blast excavation methods. Mining progressed a few hundred meters through these strata, but the project was hampered by this difficult geology. This geology was reasonably competent when first exposed but was prone to swelling and sloughing when exposed to water resulting in sidewall and crown failure. The lack of a tunnel lining and water inflows along the length of the decline compounded this problem, as any water which entered the excavation followed the decline to the face resulting in increasingly difficult working conditions.

The increased risk of ground failures and associated health and safety mitigation measures led to a slowing advance rate to manage the ground conditions. The decline was halted part way through construction and left to flood.

Second Feasibility Study

Following the abandonment of the initial decline, the mine plan was revisited, and a second feasibility study, including a trade-off exercise, was undertaken on the deposit, including site abandonment, rehabilitation of the initial decline or accessing the deposit from an entirely new development.

The initial decline was in poor condition and in need of extensive remedial work, including lowering the ground water table to gain access and installation of a sprayed concrete hydrostatic lining to seal off the water inflows. This would allow the decline to be continued through the conglomerate to the deposit.

Construction of a second decline on the close to the initial development was also investigated, requiring a similar sprayed concrete hydrostatic lining to that proposed in the rehabilitated decline. This was discounted as it was estimated the excavation would advance at a rate approximately 40% slower than the rehabilitation option.

The third option proposed was the construction of a new means of access from elsewhere on the mine site. This comprised of an initial bulk excavation, followed by a relatively short section of decline through formations of Saprolite (a soft, tropical residual soil), Sap-Rock and weak broken rock, before transitioning into traditional drill and blast mining construction (i.e., thin shotcrete, bolts and mesh.) The decline needed to be capable of resisting: wedge failures, active soil pressures, hydrostatic pressures and forces induced during seismic events.

From previous experience at this site, and at others, the client was advised that utilizing a civil tunnelling construction methodology would provide safe and structurally competent mine access. This would also provide improved working conditions at the face. These ground conditions are common in near surface civil developments, where the necessary techniques to successfully and safely traverse poor, wet ground conditions have been developed.

Once in competent hard rock, the development of the decline can be advanced via drill and blast methods.

Following assessment of the feasibility study and the economic proposal, the third option, a new decline was chosen to be progressed to detailed design.

DETAILED DESIGN

The detailed design phase began with further investigation work undertaken at the chosen decline location to define strata changes along the decline length, determine ground water levels and improve the definition of geotechnical parameters required.

Characterization of the strata indicated the declines would pass through three zones before encountering Hard Rock; Saprolite, Sap-Rock and Weathered Rock.

Ground Conditions

Saprolite is a heavily weathered rock material, that behaves similarly to clay, and is common in humid and tropical climates. It inherits some of the relic structure of the decomposed rock, along which ground water flow can be highly variable.

Ground material properties typically improve with depth, becoming more compacted and improving in shear resistance, although there is no set boundary for when the Saprolite transitions into stiffer Sap-Rock. Similar loading regimes through the

Table 1. Soil design parameters

Zone	c' (kPa)	φ (°)	UCS (MPa)	Permeability (m/s)
Saprolite	2.5+ 1.7*Depth	28–30 (Decreasing with Depth)	5.5*Depth	5E-8
Sap-Rock	2.5+ 1.7*Depth	30	5.5*Depth	3E-7

Table 2. Rock design parameters

Zone	UCS (MPa)	GSI	m_i	E (MPa)	Permeability (m/s)
Weathered Rock	2.4	75	29	588	5E-7
Fresh Rock	40	30	29	5222	1E-7

Saprolite and Sap-Rock were expected and as such were considered as soils for design purposes (Table 1).

The weathered rock mass is broken granite with low uniaxial compressive strength (UCS) and rock mass rating (RMR) values (Table 2). Maximum and minimum horizontal stress are also likely to be larger than vertical stresses due to the reverse thrust faulting regime present. The Fresh Rock conditions are significantly improved, with UCS values more than 40MPa. Once these conditions are noted on site, the profile of the decline transitions from the civil tunnelling envelope to that of hard rock mining.

Bulk Excavation and Portal Design

To establish a suitable face to begin tunnelling, a bulk excavation of approximately 14 m depth was required. This allowed for the weakest material to be excavated and the establishment of approximately one and a half tunnel diameters of cover above the first excavation advance.

Design of the temporary excavation support was undertaken as a 2D slope analysis (Figure 1), with the support capable of resisting active pressure from the Saprolite. Horizontal earthquake loading, and hydrostatic pressures were not applied to the temporary slopes and were to be observed during excavation to ensure competency. An overall factor of safety of 1.3 was considered acceptable due to the temporary nature of the slope. Pressure relief holes were installed through the sprayed concrete, steel mesh and soil nail support to limit hydrostatic pressure build up.

A 7.3 m internal diameter circular portal formed from prefabricated corrugated steel liner plates, was designed to provide access from the entrance to the headwall and sprayed concrete liner (SCL), traversing the bulk excavation. The benefit of the steel portal is to enable the excavated embankment to be backfilled, thereby ensuring the supported bulk excavation slope face is temporary, whilst providing protection at the entrance to the portal. The steel portals are designed for temporary loading conditions from heavy equipment during fill compaction and permanent loading from active ground pressures and seismic events.

Structural Lining Design

The structural design of the decline combined both analytical methods and finite element modelling techniques. Empirical methods were used to predict stand up time and allowable face advance rate.

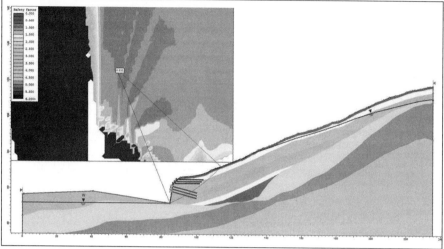

Figure 1. Slope stability analysis

As outlined in section 2, the decline passed through four distinct strata which formed the basis of the design cases; these were Saprolite, Sap-Rock, Weathered Rock and Fresh Rock.

The design approach utilized Mohr-Coulomb parameters, with equivalent values determined for the rock strata from the Hoek-Brown failure criteria. Limited ground investigation and laboratory test data at the chosen declines led to selecting conservative, lower bound, geotechnical parameters in the design.

To ensure economy of design, account for ground arching effects and relaxation, ground loads were established the Bierbaumer theory with Mohr-Coulomb parameters, rather than designing for the full overburden pressure.

A wedge or block failure load was also analyzed for the Weathered Rock and Fresh Rock, considering the joints spacing in determining the block size (Figure 2). The magnitude of the side and roof wedges were determined using stereographic projection and a 1% area contour as the joint data was highly scattered.

In addition, the lining was designed to resist the full hydrostatic pressure, however, it was not intended to be completely water tight. Some water seepage through the liner and damp patches would be expected, although this would not impact long term mine operation, as such, no additional membrane was required. Water proofing was advised at the headwall interface between the SCL and steel pipe as the ground is highly permeable at this location; hydrophilic strips were detailed along the joints which expand on contact with water and plug the cold joints.

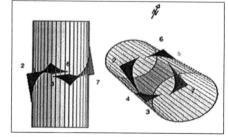

Figure 2. 3-D wedge analysis based stereographic projection of wedges

With the decline located in a seismic region, loads acting on the lining during

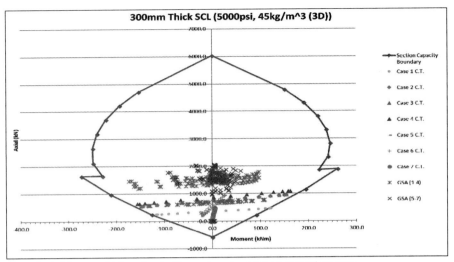

Figure 3. Axial-moment interaction for fibre and mesh reinforced sprayed concrete

a seismic event were considered. The specified national standard for seismic actions references the Mononobe-Okabe method which was used to determine the applied loads.

The sprayed concrete lining was analyzed based on an elastic response to the design loadings. The analytical method given by Carranza-Torres (2013) for the analysis of a circular tunnel, was corroborated with a beam and spring model of the lining for the various load cases. The sprayed concrete section capacity, reinforced with steel reinforcing bars and steel fibers, was determined using provisions set out in ACI 318-08 and the methodology presented by Rilem, TC (2003) for the flexural contribution of the steel fibers. This approach takes account of the concrete in compression and the steel reinforcement and mesh in tension to evaluate the ultimate capacity of the section for combined bending and axial force. The results were presented as a capacity envelope on axial moment interaction diagram (Figure 3), so that all the results from the analyses could be plotted and verified efficiently.

It was determined that in the Weathered and Fresh Rock, the capacity of the lining with steel fibers alone was sufficient to resist the applied forces. Therefore, a Type 2 liner was developed without steel reinforcing bars, as required as per the Type 1 liner through the Saprolite and Sap Rock (Figure 4). This would significantly speed up construction rates for this portion of the decline.

To produce the required flexural strength and durability in the SCL section (Table 3), a dosage of 45kg/m^3 steel fibers with an aspect ratio of 40 was specified for the concrete mix, to achieve the required tensile strength of 1.78MPa. To ensure the early age strength gain within the J2 curve and set times are achieved, the use of accelerator and silica fume was specified.

CONSTRUCTION

The initial bulk excavation stage of the project was constructed via benching down in approx. 1.5 m high benches and providing temporary support of the exposed slopes via a combination of sprayed concrete, steel mesh and soil nails. This allowed a near

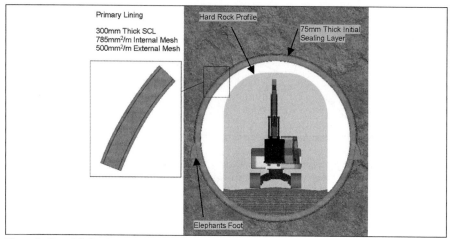

Figure 4. Type 1 lining design

Table 3. Sprayed concrete liner summary

Liner	Strength (MPa)	Thickness (mm)	Steel Fibre (kg/m^3)	Steel Reinforcement (mm^2/m)
Type 1	35	300	45	500 (outer face) 785 (inner face)
Type 2	35	300	45	—

vertical face to be established from which to commence tunnel excavation and minimized the extent and time for the cut.

The Sequential Excavation Method

The design of the circular profiled, hydrostatic tunnel liner was based on the use of the Sequential Excavation Method (SEM), whereby the excavation is carried out in sectional advances called rounds (Figure 5). This method allows flexibility in the construction and management of the ground conditions by splitting the profile into a top and bottom portion. The top is called a heading and the bottom is called an invert.

The top is typically further sub-divided into pockets as required, depending on the ground conditions. This allows for establishment of an initial layer of sprayed concrete to support the excavation against the short-term ground loading (desiccation, unravelling and block failure), before a permanent lining system to resist ground, hydrostatic and seismic loads—as such, sprayed concrete is required to gain strength immediately to resist the short-term ground loads, as well as gain the long-term strength required to resist overburden and hydrostatic loads.

The sequence for excavating and lining the tunnel was initially developed to ensure the ring was closed as soon as possible to minimise the risk of swelling and uplift of the invert, this was seen as one of the reasons that the previous attempted decline was not successful. Due to the ground conditions and large tunnel diameter, a full-face advance was not possible, so a top heading and bottom bench approach was adopted with the invert being completed only a short distance behind the heading to provide full support.

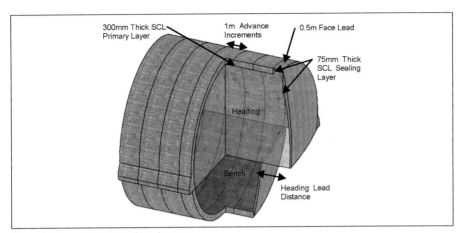

Figure 5. Sequential excavation advance

Civil tunnelling methods are based on having larger diameters than required for the operating profile, to both improve operations during excavation and ensure a near circular profile is utilized for optimal structural design. Whilst the adoption of a larger profile, when compared to traditional hard rock mining methods, which are based on a limited excavation to minimize the mining of none ore material, is compensated for by the higher capacity of the larger equipment and the greater loading and hauling efficiency possible. Local in line enlargements can be built into the circular lining during the excavation cycle as needed to allow two trucks to pass. The construction of an enlargement is simpler than a re-muck bay, as the lining design would be complicated at a junction.

Pre-Construction Mix Trials

Sprayed concrete is required to gain both short and long-term strength to ensure stability of the excavation. As such, the quality of the sprayed concrete that ends up lining the decline requires a rigorous quality control process, with checks throughout the material procurement, concrete mixing and application process. The specifications are, therefore, written to reflect these differences when compared to cast in place structural concrete.

Upon arriving at site, initial testing on the supplied concrete showed strength results to be highly variable; resulting in a wide distribution curve and lower than required strengths, based on a 5th percentile analysis. As such, more testing was instructed and uncertainties over the material supply and storage were queried.

Limited laboratory results on the mix design had led to a poorly understood mix and the remoteness of the site had led to delayed delivery of materials and equipment, leading to replacements being made to elements of the mix. There was also a lack of stringent quality control at some suppliers, leading to numerous issues with the sprayed concrete quality, including; oversized and poorly graded aggregates, variable aggregate moisture content due to poor storage, a volumetric batch plant being used in place of a weigh batch facility, high early age strength cement not being used and an untested (poor) chemical reaction between the aggregate, cement and additives. Combination of these issues contributed to the poor concrete strengths. Overdosing of accelerator by the appointed contractor to gain early age strength was also severely limiting the long-term strength build up.

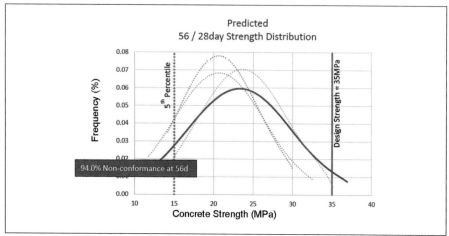

Figure 6. Initial concrete testing distribution

Initial testing on the concrete provided was conducted and compared to the design specification to indicate the scale of the problem. To save some time, a predictive analysis of the concrete results at 7-days was undertaken to determine whether the specification was likely to be met at 28 days. The analysis looked at the sample strength distribution for 1 day, 3 days, 7 days and predicted 28 days for the sprayed panel cores and determined the non-conformance rates (%). The distribution curves were affected by the low number of samples which resulted in a lower predicted achievable strength, based on the established method of utilising the 5th percentile as the design strength of a concrete mix (Eurocode 2; Design of Concrete Structures. Part 1-1,) than specified, based on an acceptable non-conformance rate.

As can be seen in Figure 6, the predicted concrete conformance at 28 days to the specified strength was 6%, showing the mix design would have to be refined and the quality control and assurance procedures from material selection to application be improved prior to construction.

To ensure that sufficient strength was achieved during construction, the issue of the poor-quality sprayed concrete was tackled in four identified areas:
1. Material Procurement
2. Weight Batch Plant establishment and training
3. Sprayed concrete operator training and accelerator dosage monitoring
4. Liner design optimisation based on predicted concrete strengths

To allow the above to take place, the construction sequence was altered to allow work to continue by completing the entire portal structure before advancing the tunnel. This gained a window of approximately 4 weeks to allow for the improvements to be made. In this period, the design team performed a sensitivity analysis on the sprayed concrete liner and developed several lining options related to the fibre flexural strength and the area of steel mesh used in construction.

The initial design called for a lighter gauge mesh to be placed to the outside face of the liner and a heavier gauge to the inside face, however, when a heavier gauge mesh was used in both faces, a lower concrete strength was considered acceptable.

Improved tensile properties of the fibre reinforced concrete than those considered in the design stage also helped to achieve this, although after increasing the tensile strength above 2MPa, the benefit became nominal. Increasing the liner thickness has limited benefit, as this creates a stiffer lining and consequently attracts more load.

Following completion, an optimised design, utilising 21MPa concrete, 785 mm^2/m of mesh in each face and a 2MPa tensile strength was detailed. During the four-week window, the weigh batch plant was established, all aggregate material arriving on site was regraded, an experienced SEM tunnel crew were hired to lead the development, whilst training the operatives of the mining contractor and the concrete sprayer dosage was calibrated. The concrete sprayer used to establish the structural lining through the soft ground was a mining machine, developed to apply immediate support to hard rock strata. The requirement for immediate support, which is then followed up with bolts and mesh to support wedges, means the concrete flow rate and accelerator dosage can be controlled independently, so any spalling rock can be managed. However, this relationship is critical in civil tunnelling applications to ensure that the accelerator dosage does not exceed 10% and there is no impact to the long-term concrete strength. Civil tunnelling machines tie these two functions together, by setting the accelerator dosage at the concrete pump.

To ensure accurate dosing of the machine used on this project, a table was produced so the operators could identify the required accelerator dosage based on the concrete flow rate. A prolonged testing period resulted in a final round of 7-day testing prior to the commencement of the tunnelling works that indicated 21MPa was likely to be achieved after 28 days, achieving 95% conformance (Figure 7).

These 7-day results still left uncertainty in the mix, as limited information was available to produce an accurate prediction. Although the mine operator, based on the data provided agreed to continue at their own risk, it was critical to understand that getting the 28-day strength in production and establishing the long-term integrity of the tunnel structure was still required. A testing regime using sprayed panels was specified during the construction of the lining and analysed throughout the construction phase. This test data, along with detailed information on chainage and qualitative details on each advance was tracked and recorded which formed the basis of the operations and maintenance manual.

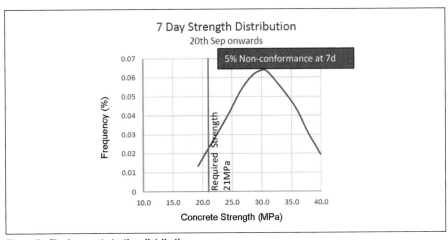

Figure 7. Final concrete testing distribution

Instrumentation and monitoring of the lining was another means to control the affects of the poor material quality by validating the predicted structural behaviour. The monitoring scheme was designed to measure the movement and ovalisation of the tunnel lining and trigger alerts if they exceeded safe limits. The limits were based on predicted values for serviceability state, ultimate state and at full structural capacity.

Management and Monitoring of the Excavation

As previously discussed, the tunnel was to be constructed through formations of Saprolite, Sap-Rock, and weak broken rock, before transitioning into hard rock (Figure 8).

During the bulk excavation phase, the Saprolite conditions were found to be relatively stable when benching down, however, several relic structures were encountered. Exposing these relic joints, which were in-filled with organic material, and acted like a fissure through the Saprolite, led to several small wedge failures as the excavation deepened. These were managed through mapping of the face to predicted areas of instability and then reducing the opened bench length whilst draining any encountered water away from the face.

Prior to commencing tunnelling works, the faces of the tunnels were drilled to 16 m behind the face to lower the ground water level and reduce the potential for wedge failures during the initial tunnelling works. The top of the excavated slope was covered to increase surface run off and reduce the amount of water entering the system.

The initial tunnelling works conducted through Saprolite were found to be dry and have an improved stand-up time across relic structures when compared to the bulk excavation works. These conditions were similar to working in other clay like materials whereby stand up time is limited but can be effectively managed by opening up a pocket of the heading and applying a sealing layer to the strata shortly after it is exposed. This limited the potential for drying out, allowing the material to retain some additional shear strength as a result of cohesion from water suspended within the soil matrix.

As the excavation progressed into the Sap-Rock strata, which was below the lowered ground water level, the amount of relic structures acting as a conduit for water increased considerably, leading to wedge failures in the crown and shoulders of the tunnel and instability in the face of the excavation. These issues were managed through

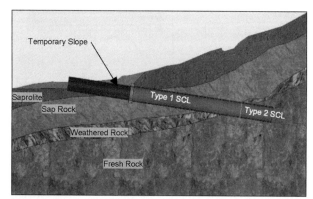

Figure 8. Strata along the tunnel

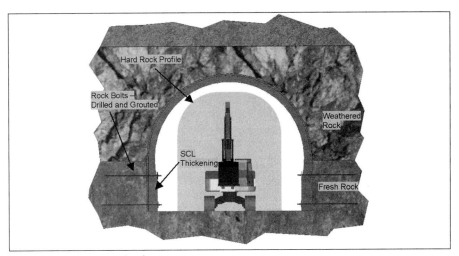

Figure 9. Transition to hard rock

a combination of increased spiling to limit the height of any wedges forming in the crown, excavation of each heading in pockets and controlling any water encountered.

Sprayed concrete does not perform well in wet conditions, with any water on the sprayed surface creating a high water-cement ratio at the critical concrete-ground interface, thereby reducing bond strength. This leads to local fall ins during spraying. Due to the water running along relic faults within the Sap-Rock, rather than percolating through strata, in this application, short sections of pipe were placed through the liner to manage the water, before being capped following 28-day strength gain to seal off the inflow.

The weathered rock, although susceptible to large wedge failures, was generally more competent than the Sap-Rock, but carried more water and its broken nature did tend to lead to spalling at the face. The hard rock material was very competent, with some pockets found in the transition zone, from weathered rock to hard rock. This required some small-scale drill and blasts to remove the hard rock and allow the excavation to continue. A transition zone, utilizing a horse-shoe shaped profile was designed to carry any loads experienced in the crown to the hard rock material, whilst reducing the profile to limit the amount of blasting required, see Figure 9.

Invert construction was particularly challenging due to the amounts of water ponding at the invert excavation face and weakening the ground supporting the heading. To combat this, drain walls and pumps were placed along the length of the tunnel to limit the build up of water at the face, the distance between heading and invert increased and the advance length of each invert was increased from the recommended design advance to allow for a temporary sump to be placed ahead of the invert.

The total heading lead was limited to 10 m however, as this was a trade-off between keeping the invert as stable as possible and closing the lining as early as possible. During construction, five monitoring stations were installed in the waist, shoulders and crown of the tunnel to monitor movement every 5 m. This was an increase from every 10 m considered at the design stage due to the variability of the concrete and only having 7-day strength information available. It was therefore critical to conduct both survey and visual monitoring of the tunnel lining to ensure that a competent liner was being constructed, as well as undertaking an increased number of panel tests until the 28-day strength could be accurately established.

As the trigger limits had been established based on a 35MPa concrete strength, it was anticipated that larger deformations would occur in areas where concrete strengths were below the design strength, therefore an active approach to monitoring the completed liner was required. Surveys took place daily, to check movement against trigger levels, until a level trend was established. Visual inspections of the tunnel were undertaken at the start of each shift to determine integrity of the structure and any breaches of the trigger levels from the previous shift were investigated, to determine if the information was reliable.

Following completion of the tunnels, all monitored stations showed a steady trend and only minor circumferential cracking was noted where different rounds connected. This is a known area of weakness in sprayed concrete tunnels and cracking at this joint is acceptable providing that the structure continues to provide adequate radial support pressure.

Sprayed concrete test panels were taken for each heading for the duration of the tunnelling works, as analysis undertaken following the initial 28-day strength still showed a high-variability in the cores. Comparison to the UCS tests performed on cast concrete cylinders, and inspection of the cored samples indicated that this was a problem with the spraying technique of the mining contractor, including; shadowing and poor compaction.

Inspections did not show any significant signs of structural distress; however, hydrostatic pressure and active pressures are likely to build up over time. As such, when completing the operations and maintenance manual, the 28-day strengths were noted for each heading, thereby outlining all non-conforming rounds so that effective long-term monitoring of the tunnel lining could be undertaken as there is an increased risk of movement and deterioration of the liner. This is more in line with a traditional mining approach on infrastructure which has a limited design life and can be more easily maintained compared to civil tunnelling works.

CONCLUSIONS AND RECOMMENDATIONS

Following successful completion of two 7.3 m internal diameter tunnels through a combination of residual soils and weathered rock, it has been shown that the use of civil tunnelling techniques in soft ground could be of substantial benefit in the development of mining infrastructure.

Due to the remote location of many mine sites, it is critical that the solution remains adaptable to account for changing ground conditions, available materials and personnel who are not familiar with this technique.

Success in this project was achieved by providing specialist design and construction teams to be involved in the project from its early stages and maintaining this team from feasibility through to the end of the construction phase. This allows for a fluid approach and adaptive solutions to be developed to match the conditions encountered, in this case:

- Optimising the design to utilize the available material
- Providing expert oversight and training to local operatives
- Providing specialist teams to undertake construction

With key personnel on site, these changes can then be made with the approval of all stakeholders; mine operator, engineer of record and contractor to ensure that a solution is developed that meets both the long-term design requirements and the immediate needs of the project.

ACKNOWLEDGMENT

The authors thank Ed Gillard and Gabino Preciado for their contribution to this paper.

REFERENCES

ACI 318-08. 2008. *Building Code Requirements for Structural Concrete and Commentary*, American Concrete Institute, USA.

Carranza-Torres, C., Rysdahl, B. and Kasim, M. 2013. Elastic analysis of circular lined tunnel considering relaxation of ground stresses prior to installation of the support, *International Journal of Rock Mechanics and Mining Sciences*, 61: 57–85.

Eurocode 2: Design of Concrete Structures. Part 1-1: General Rules and Rules for Buildings. Brussels, 2004.

Mononobe N, Matsuo H 1929, On the determination of earth pressure during earthquakes. In Proc. of the World Engineering Conf., Vol. 9, str. 176.

Okabe S., 1926 General theory of earth pressure. Journal of the Japanese Society of Civil Engineers, Tokyo, Japan 12 (1).

Rilem, TC. 2003. Test and design methods for steel fibre reinforced concrete, *Materials and Structures*, 36: 560–567.

Schezy, K. 1973. Determination of Vertical Rock Pressures, The Art of Tunnelling, 214–217. Akadémiai Kiadó; 2nd revised & enlarged edition.

Steep Inclined SEM Excavation—Successful Execution Applying Drill and Blast

Richard Griesebner ▪ BeMo Tunnelling GmbH

INTRODUCTION

Since the 1950s electricity from hydroelectric power has been generated at the upper Inn-valley. The emerging power plant "GKI" is a milestone to achieve the objectives of European and regional energy strategies. "GKI" stands for the German acronyms "**G**emeinschafts **K**raftwerk **I**nn." A corporation of three different power suppliers in Europe made it possible to build a new river power plant.

Extending over two countries the project reaches from the community of Prutz in the Tyrolean Oberinntal-valley to the Swiss community of Valsot.

Water of the river Inn passes into an intake structure in Switzerland and runs through a penstock tunnel with an inner diameter of 5.76 m over 24 km to the power house in Austria.

The project includes also the reservoir and a massive concrete intake structure. As part of the project the penstock tunnel is being constructed with two double shield TBM's each commencing in the middle and driving to opposite directions, one in North direction and one in South direction. Also a vertical surge shaft with an excavation diameter of 15.30 m and a depth of 101 m was essential for the project. Excavation of the surge shaft commenced at the high point applying Drill & Blast.

Prior to the surge shaft a mucking shaft 101 m deep with a diameter of 1.84 m was needed. It was located in the centre of the surge shaft and cut with a raise boring unit. The device chamber 15.3 m high, 15.9 m long and 10.5 m wide was driven with SEM. Access tunnels and the inclined penstock tunnel were also driven with SEM. With a cut and cover method using slurry walls no tunnelling was needed for the casting pit of the power house and the discharge channel.

The penstock is 145 m in total and the flow volume will be 75 m³ per second. In the year 2021 production of 400 Giga Watt hours a year will start with two Francis turbines.

One of the challenging construction components was the excavation of an inclined penstock tunnel. To build this 385 m long penstock tunnel the following objects had to be established in advance: the access road to the portal (1), the access tunnel to the device chamber (2) and the device chamber (3) itself—see Figure 2.

Figure 1. Map GKI

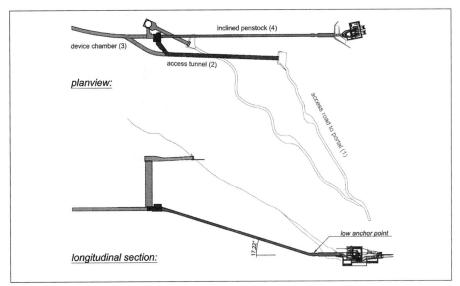

Figure 2. Overview penstock tunnel GKI

Excavation of the penstock tunnel commenced with the 56 m long horizontal tunnel at the lower anchor point. The regular excavation cross section of this tunnel was split into a crown excavation with an area of 41 m^2 and an invert section with 9.5 m^2. The first meters have been a good opportunity to get to know the existing ground in situ. It was also a good exercise to optimize the equipment and to consolidate the team spirit within the crews.

EXCAVATION METHOD—THE DECISION PROCESS

Before starting the 385 m long tunnel several "ways" of construction were investigated. First a comparison TBM vs SEM was done. It quickly showed the disadvantages of the TBM solution because of the following three reasons:

1. The inclination change at the low anchor point,
2. The set up time for a TBM was too long and
3. The costs of a TBM were too high considering the short tunnel length.

The decision to go with SEM was made and new tasks followed this decision.

One of the major questions to be considered was if the excavation could be done in a declined way beginning at the high point using a self-made gantry. But set up time for this doable solution would not have fit the time schedule of the GKI project. The time necessary for adapting and optimizing an existing gantry was not available. The other question was how effective it could be to excavate this steep inclination from the low anchor point upwards.

To begin the excavation at the low anchor point was linked to benefits like using commonly available equipment and machinery and having a simple system of gravitational dewatering. Initial investigations came to the result that it was possible to use commonly available equipment.

Vehicles on tracks and wheels were considered for the equipment concept at this stage of the project. Logistics was important and had to be considered for each step. The wet shotcrete method was chosen and supply of shotcrete by using concrete truck mixers on wheels or tracks was considered doable. Finally, almost all circumstances were looked at and the decision was made for SEM commencing at the low point.

REQUIREMENTS AND WORKMANSHIP

The design included support measures like grouted self-drilling spiles and steel spiles, shotcrete and wire mesh, lattice girders, face support with reinforced shotcrete and pocket excavation. Radial installation of rock bolts was an essential completion of the tunnel support. Geotechnical deformation experienced already at the beginning of the horizontal tunnel made it necessary to install a domed shotcrete invert. The design allowed 9.0 m maximum distance between the face of the crown and the closure of the domed invert. After driving through non-cohesive gravel and weathered rock at the beginning of the tunnel the first explosives were needed at tunnel station 22 m.

The Drill & Blast excavation continued with drilling boreholes followed by loading explosives and blast one advance length, in this case 1.0 m to 1.3 m length. In average 66 no. boreholes were needed. A cone cut was drilled in the centre of the profile. The spacing of the profile holes at the circumference was 0.5 m. For the sericite phyllite and limestone phyllite along the inclined direction 2.25 kg emulsion explosives (Emulex 2) were used in average per cubic meter of rock.

The next step after mucking and scaling was the application of a 30 mm flashcrete layer on all excavated areas (face and profile). This was followed by installation of the first layer of wire mesh. Spacing of the wires in the mesh was 100 mm with a diameter of 5.0 mm. Lattice girders (type 70/20/30) were put up and the first layer of shotcrete with a thickness of 100 mm got sprayed. At the crown area where spiles were required the application of shotcrete was temporarily left behind. After driving spiles through the girder the second layer of wire mesh was applied, followed by spraying the second layer of shotcrete (including the crown which was initially left behind). The final step for each excavation round was the installation of radial rock bolts at the previous advance. The type of rock bolt used was a drill hollow bar system R32-280 grouted through the centre with a drill diameter of 51 mm. Five respectively four bolts were installed in each advance.

When reaching the low anchor point of the penstock tunnel at station 56.30 m the excavation team faced two major changes: the cross section to be excavated was reduced to 23 m^2 and the start of the 31% inclination. The geological situation required mandatory installation of lattice girders at each advance. The lattice girders were placed rectangular to the inclined axis of the tunnel. Due to the inclination of 17.22° with a vertical tunnel face the advance length at the top of the crown was significantly longer than in the invert. Due to full face excavation this additional advance length measured 1.07 m, see figure 3. Initial layer of shotcrete combined with 4 m long steel spiles ahead of the girder gave sufficient support for the next excavation. The regular spacing of the spiles was 200 mm, locally the geology at hand made it necessary to reduce the spacing to 150 mm. The encountered dominating sericite phyllite and limestone phyllite were partly fragile and kept the advance length at a maximum of 1.30 m. The regular excavation process needed advance cutting in front of the lattice girder of 0.75 m to allow the required overlap of the first layer of wire mesh, see Figure 3.

The overburden along the inclined penstock tunnel was between 80 m and 140 m. Every 10 m five survey points have been installed along the distance in the tunnel.

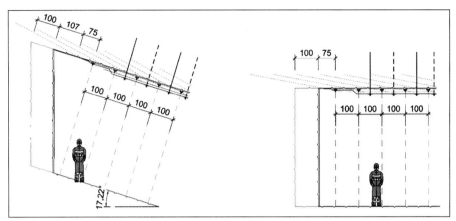

Figure 3. Comparison inclined and horizontal advance length of 1.0 m

Daily survey showed geotechnical deformations in the range of maximum 2.0 mm and a quick stabilisation.

Because of the steep inclination dewatering had to be treated very sensitive and deviated immediately. One reason for a careful dewatering treatment was that the walking and driving surface became dangerously slippery when it got in touch with water. Although some near miss incidents due to the slick invert occurred at the beginning major injuries caused by the wet surface have been avoided by installing simple things like for example a handrail made of steel cables for pedestrians. Generally foot traffic was kept to a minimum.

Also the use of equipment was very tricky. Combined with water the roadway became weak and the profile in the invert was demolished especially by the chains of the track loader.

PLANT AND EQUIPMENT

Figure 4 makes it clearly visible that space for any equipment was very limited.

The first sequence of the Drill & Blast excavation is drilling. A suitable drill rig witch fits into the limited cross section was found with the French drill rig Robodrill© Pantofore 504. It was equipped with one boom and a man basket mounted on a track carrier, see figure 5, figure 6, and figure 7. The following range dimensions were available: length of the boom was 3820 mm, length of rod 2435 mm, drilling length 2292 mm. A common roto percussion hammer with 36 kW input power was sufficient to manage all drill works. The regular 3 m long sections of the grouted hollow bar system R32-280 had to be extended once to achieve the locally required length of 6.0 m. The machine came with a diesel engine to drive and an electrical engine to work at the face.

The vehicle with its man basket allowed safe drilling of the boreholes at the face, radial drilling of bolts and spilling in tunnel direction.

Mucking with track loaders with self-loading buckets was only done briefly. The many drives necessary for mucking each advance combined with the poor traction at the steep inclination the crawler chains destroyed the invert of the tunnel, see Figure 9. The slight water egress in the range of approx. 2.0 l/s along the tunnel also had an

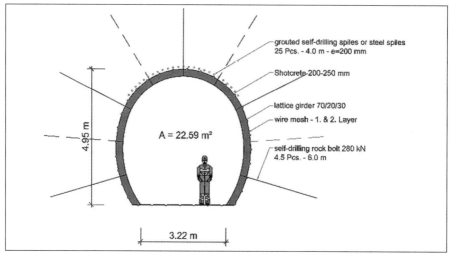

Figure 4. Regular cross section penstock tunnel

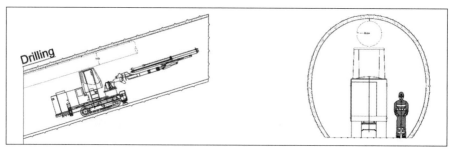

Figure 5. Drilling

Figure 6. Drill rig

Figure 7. Drill rig with man basket

adverse effect on the traction of the steel chains. This meant that a different solution for the mucking process had to be developed.

The new idea was to use equipment on wheels instead of tracks. The obvious challenge was to find a machine capable of dealing with a 31% inclination in a wet rock tunnel. Common mucking equipment available on the market cover usually an inclination of up to 25%. Talking to several long-time suppliers we ended up with a Paus 10000

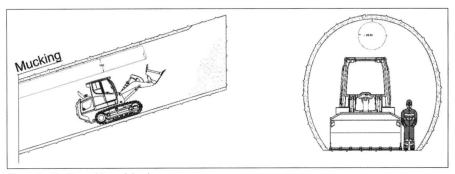

Figure 8. Mucking with track loader

Figure 9. Invert

Figure 10. Track loader

dump truck. Equipped with a hydrostatic drive a fully loaded truck can handle downwards moves under these circumstances. All operators had to go through intense and periodically repeated inductions on how to safely handle the vehicles including respecting the speed limit of 5 km/h. Daily service inspections by responsible mechanical staff and intense checks of the vehicles were obligatory. The traction issue with the pneumatic tyres was resolved by mounting a standard rock chain to the wheels.

Figure 11. Excavator ITC 312

Changing the mucking system required to find a new way of loading the trucks. One of the joint venture partners had an excavator ITC 312 with a steel conveyor belt available, see Figure 11. The size of the excavator was ideal and fit into the cross section of the inclined tunnel. Fortunately the excavated material was fractured enough for the conveyor belt. A hydraulic hammer was assembled to the boom using a quick connector to deal with rocks not suitable for the conveyor belt. This equipment was well known by the team and has been used on former projects. The excavator ITC 312 is generally equipped with two engines, one for travelling on tracks and one electrical generator for excavation.

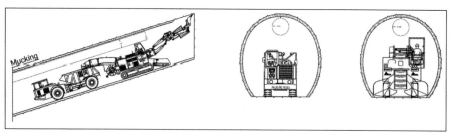

Figure 12. Mucking

Figure 13. Excavator commencing work

Figure 14. Excavator in tunnel

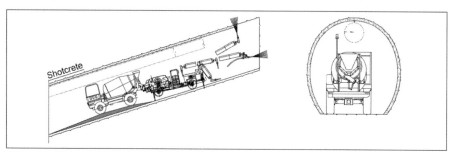

Figure 15. Shotcreting

For the sequence of applying shotcrete a four-wheel-drive sprayed concrete manipulator Sika PM407 PD was used, see Figure 15. Provided with additional hydraulic support legs at the rear, the plant could level itself. To fill the hopper a ramp was necessary. This ramp was built either with excavated material when stable or with wood wedges. After positioning the truck filling the hopper took place in a usual way.

Shotcrete was delivered from an external batching plant. Transferring the shotcrete into four wheel drive Dieci F7000 truck mixers was done on site. A real safety advantage of this vehicle is the possibility to turn the steering unit including the driver seat by 180°. This allows the driver to always operate the machine forwards and saves the spotter in a confined space. The standard capacity of 5 m^3 volume per load could not be fully utilised due to the steep inclination. The 31% made it impossible to empty the mixing drum completely. Fortunately, there was no need for a bigger amount of shotcrete per operation.

CONCLUSION

The steep inclination was a real challenge.

Looking at allocation of different advance lengths along the inclined part of the penstock tunnel 40% of the total length was 1.3 m regular advance length and 60% was 1.0 m regular advance length. All works were performed in a 24/7 operation with 3 shifts per day. In average a daily progress of 3.64 m was achieved with a team of 5 people working in the tunnel. Excavation of the inclined penstock tunnel was completed in 92 days, which was about 50% of the estimated time, which emphasises the success of the chosen method and the good performance of the team on site.

When building tunnels with steep inclination applying SEM excavation there is a limit for using standard tunnel equipment. The safety risk for man and machinery rises with every degree of inclination.

Considering the summary of technical expertise building this particular inclined penstock tunnel with SEM was finally the right decision.

PART

Shafts and Mining

Chairs

Phil Backers
Jay Dee

Phil Rice
WSP

Construction of Combined Sewer Overflow 021 Diversion Facilities in the District of Columbia

Moustafa Awad ▪ Aldea Services
Paul Leduc ▪ Aldea Services
Louise Headland ▪ Aldea Services
Ryan Mains ▪ James G. Davis Construction
Ryan Payne ▪ PCO/JCK Underground

ABSTRACT

The CSO 021 project is located within the John F. Kennedy Center for the Performing Arts (KCPA) property at its southern end, just north of the Theodore Roosevelt Bridge. The project involved the construction of diversion facilities (Diversion Chamber, Drop Shaft and Ventilation Control Facility) in soil and rock ground conditions approximately 50 feet and 90 feet deep, respectively. Construction activities included secant pile walls support of excavation (SOE), soldier piles and lagging (SPL) SOE, rock curtain grouting, rock blasting, micropiles, and rock bolting. The purpose of this paper is to provide an overview of the construction activities and challenges associated with the diversion facility.

BACKGROUND

The Combined Sewer Overflow (CSO) 021 Diversion Facilities Project is part of the District of Columbia Water and Sewer Authority (DC Water) Long Term Control Plan (LTCP). The LTCP provides a strategy to control combined sewer overflow (CSO) discharges to the Potomac River, Anacostia River, and Rock Creek within the District of Columbia (DC). Approximately one-third of Washington DC's sewer system is a combined sewer system, which carries both stormwater run-off and sewage. During dry weather conditions, the combined water is conveyed to the Blue Plains Wastewater Treatment Plant (Blue Plains WWTP). Once at the plant, the waste water is treated and discharged to the Potomac River. When the capacity of a combined sewer is exceeded during a storm event, the excess flow is discharged to the nearest outfall into the Potomac River, Anacostia River, or Rock Creek. Part of the LTCP solution to controlling the amount of discharge into the rivers and streams is the construction of diversion facilities and a series of tunnels. The diversion facilities would redirect the wastewater to be stored and conveyed through tunnels to the Blue Plains WWTP. A conceptual illustration of this obtained from the DC Clean Rivers Project Potomac River Tunnel Environmental Assessment report is presented in Figure 1.

Part of the LTCP is the construction of the CSO 021 project, which provides a diversion structure to divert flow from the existing 108-inch combined sewer (Structure 035 at CSO 021) and convey it to the future Potomac River Tunnel. Figure 2 obtained from the DC Clean Rivers Project Potomac River Tunnel Environmental Assessment report provides the control plan for CSO 21.

The CSO 021 project is located within the John F. Kennedy Center for the Performing Arts (KCPA) property at its southern end, just north of the Theodore Roosevelt Bridge

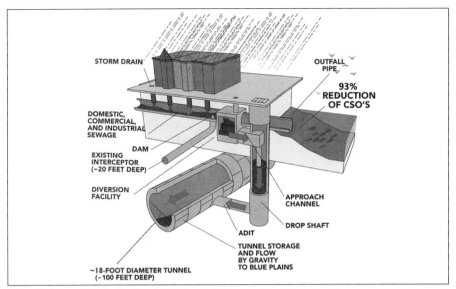

Figure 1. Conceptual illustration of diversion facility connecting to tunnel system

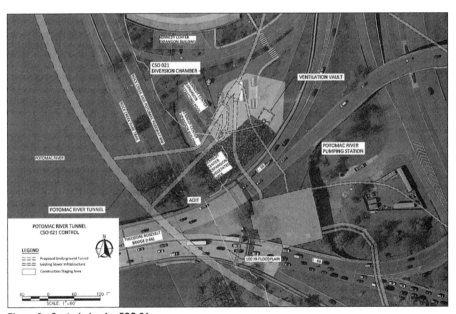

Figure 2. Control plan for CSO 21

and west of the US Interstate 66 (I-66) bridge ramp. Figure 3 provides the location of the CSO 021 project.

The primary structures constructed for the project were a Diversion Chamber, Drop Shaft, and Ventilation Control Facility. Configuration of these structures are presented in Figure 4. The Drop Shaft is approximately 23 feet in diameter and 91 feet deep. The Diversion Chamber is a trapezoidal shape with a length of approximately

Figure 3. Project location

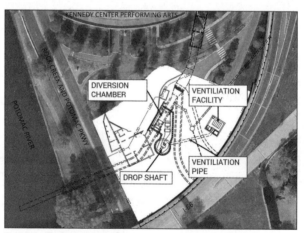

Figure 4. Project structures

46 feet, a width ranging from 21 feet to 44 feet, and approximately 51 feet in depth. The Ventilation Control Facility is rectangular approximately 31 feet wide by 30 feet long and 25 feet deep. Additionally, the Drop Shaft includes a 10-foot diameter Cast Fiberglass Reinforced Polymer Mortar (CFRPM) pipe and a 4-foot internal diameter (ID) Ventilation Pipe. The Ventilation Pipe which is approximately 140 linear feet and runs from the Drop Shaft to the Ventilation Facility. Other minor piping was constructed as well as two 36-inch diameter manholes in the Diversion Chamber and one in the Ventilation Control Facility.

During the construction of the CSO 021 project, the KCPA building was expanding and the construction of both projects occurred simultaneously. Figure 5 shows the proximity of both the CSO 021 Diversion Facilities Project and the Kennedy Center Performing Arts Expansion (KCPA Expansion).

REGIONAL GEOLOGY

The project site is located within the 7.5 Minute Series Washington Quadrangle (Washington West, DC-MD-VA) (USGS, 2011) on the eastern bank of the Potomac River. Original ground surface elevations at the site generally sloped from east to west from approximately elevation El. 27 feet close to the I-66 bridge ramp to elevation El. 20 feet close to the KCPA Expansion. A surface construction grade elevation of El. 16 feet was proposed for construction of the Drop Shaft and Diversion Chamber.

Figure 5. CSO 021 diversion facilities project and Kennedy Center Performing Arts expansion

The project site is located at the interface of the Piedmont Physiographic Province and the Atlantic Coastal Plain Physiographic Province. The boundary between the two provinces is known as the Fall Line and starting at the Fall line and thickening eastward, a wedge of Coastal Plain sedimentary deposits overlies the older Piedmont residual soils and crystalline bedrock. Geology within the vicinity of the site is shown on the Geologic Map of the Washington West Quadrangle (DC, MD, VA GQ-1748) (USGS, 1995) and comprises artificial fill, Quaternary-age alluvium overlying Pre-Cretaceous-age crystalline bedrock. The Rock Creek Shear Zone is located near to the site, possibly extending beneath the site. The shear zone is a north trending belt of metamorphosed rocks that has an exposed length of 25 km and a width of 3 km (Flemming and Drake, 1998). This zone of ductile deformation forms a system of oblique-slip and thrust faults and much of the deformation is considered to have occurred in Cambrian- and Ordovician-age crystalline rock. Figure 6 shows the geology of the Washington DC and Capital Beltway area (Means, 2010).

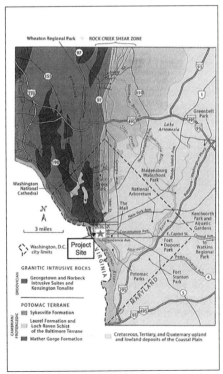

Figure 6. Regional geology of Washington DC and Capital Beltway area

Table 1. Baseline properties of soil and rock

Parameter/Property	Fill	Alluvium	Decomposed Rock	Bedrock
Total unit weight (pcf)	100–120	100–120	140	170–175
Effective friction angle (deg.)	28	25	36	45
At-rest earth pressure coefficient	0.5	0.5	N/A	N/A
Active earth pressure coefficient	0.33	0.4	N/A	N/A
Passive earth pressure coefficient	3	2	N/A	N/A
Undrained shear strength	N/A	200–500 (Inorganic) 0–250 (Organic)	N/A	N/A
Unconfined compressive strength (psi)	N/A	N/A	15,500 (maximum)	

N/A = Not Applicable

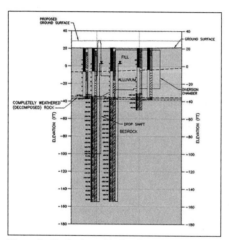 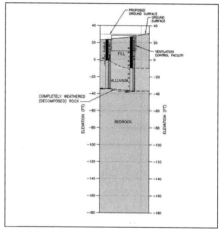

Figure 7. Geologic profile for diversion chamber (left) and ventilation facility (right)

SUBSURFACE GROUND CONDITION

The subsurface materials at the site comprised of Fill, Alluvium, Decomposed Rock and Bedrock, as shown on Figure 7. The Fill was highly variable, ranging in thickness from approximately 20 to 40 feet below the ground surface and consisted of fragments of asphalt, wood, concrete, metal, plastic, slag, brick, in a soil matrix with variable amounts of organic matter and cobbles. The Alluvium ranged in elevation from approximately El. 0 feet to El. −10 feet and consisted of very soft to medium stiff, high to low plasticity clay and silt with varying amounts of organics and lenses of sand and gravel and ranged in thickness from approximately 5 to 30 feet. Completely weathered or decomposed rock was encountered at the interface between the soil and underlying bedrock ranged in thickness from less than 0.5 feet to approximately 3 feet and generally consisted of silty sand and gravel size fragments of bedrock. Bedrock at the site was recovered as a good to excellent quality, medium strong to strong, slightly weathered to fresh, moderately fractured mica schist and was encountered at elevations ranging from approximately El. −32 feet to El. −38 feet. Baseline properties for the soil and rock are presented in Table 1.

Discontinuities within the rock mass were characterized from the rock core recovered during rock coring. The majority of discontinuities were observed to range from 30 to 60 degrees from the horizontal. Occasional joints were observed to range from 5 to 15 degrees and 75 to 85 degrees from the horizontal. The discontinuities were described as slightly rough to rough, and occasionally slickensided, with clay, silt and calcite infilling. The discontinuities were observed to be more closely spaced in the upper 30 feet.

Groundwater flow at the site was expected to occur from the interface between the soil and bedrock and bedrock discontinuities. Groundwater levels at the site were evaluated using two monitoring wells installed in two of the borings as 120-foot long open rock wells. Groundwater levels measured in the wells ranged in elevation from approximately El. +2 feet to El. +3 feet. For the project site a baseline piezometric level of between elevation El. 0 feet and El +3 feet was proposed. Equivalent hydraulic conductivity of the rock mass was evaluated based on the results of water pressure (packer) tests and revealed an equivalent hydraulic conductivity ranging from about 1×10^{-6} to 7×10^{-5} cm/s.

SUPPORT OF EXCAVATION

Two different support of excavation (SOE) systems were used for the Diversion Chamber and Drop Shaft and the Ventilation Facility. For the Diversion Chamber and Drop Shaft, a secant pile wall SOE system was used in soil and rock and consisted of a combination of rock bolts, welded wire fabric, ring steel and shotcrete. Solider Pile and Timber Lagging (SPL) was the preferred SOE system for the Ventilation Facility.

Secant Pile System

Secant piles were the preferred method of SOE for the Diversion Chamber and the Drop Shaft due to the excavation being below the water table. Because of the proximity to the Potomac River, the water table was located at an elevation of approximately El. 0 feet. With the construction grade at an elevation of approximately El. 16 feet, the Diversion Chamber extended to an elevation of El. −35 feet and the Drop Shaft extended to elevation of approximately El.−75 feet. The secant pile wall provided a dry work environment for excavation purposes.

The secant pile wall was formed by constructing alternate "male" and "female" piles; the "female" piles were unreinforced while the "male" piles were reinforced with W18×17 steel beams. The "male" piles and "female" piles had concrete compressive strengths of 5000 psi and 3000 psi, respectively. An exception to this was the portion of the secant pile wall designed to support the KCPA Expansion with both "female" and "male" piles composed of 5,000 psi concrete compressive strength. The piles were 30 inches in diameter with a centerline-to-centerline spacing of 22 inches. Cased bored piles were used to form the secant piles. A total of 172 secant piles were installed and were embedded a minimum of five feet into bedrock.

Secant piles were drilled into place using a combination of two different drill rigs. First, an auger bore drill rig was used for the overb9urden portion to remove the soil down to the top of rock. As the auger removed the soil, the rig would periodically stop to push the casing further into the overburden. Casing lengths varied between five to ten feet in length, so new segments of casing were added as the drill rig advanced. Once the operator and the representative engineer on site acknowledged the transition from weathered rock to bedrock, a down-the-hole air hammer, shown in Figure 8, was lowered into the borehole.

The air hammer increased the depth of the borehole by an additional five feet, the minimum embedment depth into bedrock as required by the specifications. This embedment was necessary to securely socket the secant piles in place. Water was flushed down the hole to prevent excess dust from permeating the construction site and the surrounding area. Prior to pouring the concrete, the borehole was blown out by the air hammer to remove excess water. The engineer on site ensured that no more than two inches of water was present at the bottom of the hole before concrete was poured.

After all the secant piles were drilled and poured, the excavation of the Diversion Chamber and the Drop Shaft commenced. The overburden was removed to approximately El. 7 feet, two feet below the initial bracing level. Lateral bracing was then installed at El. 9 feet to support the earth retaining secant piles. Below the first bracing level, an existing 108-inch CSO pipe had to be uncovered and supported before further excavation. Once a system was in place to support the existing utility (CSO 021 Overflow Sewer), excavation continued down to two feet below the second bracing level at EL. −8 feet. This process was repeated for all five bracing levels until the top of rock was reached. The secant pile wall was braced as shown on Figure 9. The five levels of bracing were located at elevations of approximately El. 9 feet, El. −8 feet, El. −16 feet, El. −24 feet, and El. −32 feet. The bracing system consisted of W24×176 steel beam walers and HP 14 × 102 struts as shown on Figure 10. For the Drop Shaft a 25-inch steel box beam was used as the lateral bracing support, as seen below on Figure 11.

Rock Blasting and Rock Support

Rock drilling and blasting along with rock supports was the method utilized to excavate and support the rock in the Drop Shaft. Rock blasting was performed from top of rock to the bottom of the shaft, which was approximately 40 feet. Figure 12 depicts the blasting arrangement within the Drop Shaft. Rock support used to stabilize the excavation consisted of rock dowels, wire mesh, and shotcrete. Ten foot #8 hot rolled

Figure 8. Down-the-hole air hammer (left) and auger borer drill rig (right)

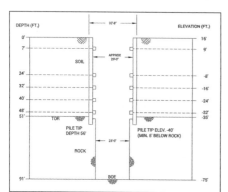
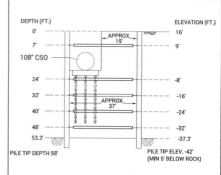

Figure 9. SOE cross sections for the drop shaft (left) and the diversion chamber (right)

Figure 10. Diversion chamber SOE

Figure 11. Bracing for drop shaft

thread bars were spaced at four feet centers horizontally and vertically within the excavation. Welded wire fabric (WWF) of 4×4—W2.9 × W2.9 and approximately four inches of shotcrete cover was used as needed to ensure stabilization of the rock.

To prevent any structural damage to nearby utilities and structures it was imperative to keep the blasting controlled and to maintain a peak particle velocity lower than the maximum threshold of 2.0 inches/second. An exception to this was the Theodore Roosevelt Bridge and I-66 retaining wall which had a more stringent maximum threshold and depended on the frequency of the blast, for frequencies less than 10 hertz a maximum threshold of 1.5 inches/second and for frequencies greater than 10 hertz a maximum threshold of 2.0 inches/second had to be maintained. Controlled blasting was performed through regulating the amount of dynamite and the use of blast mats and a timber lagging cap. Since the site is surrounded by government buildings, blast mats were key to suppressing noise and preventing flying rock fragments. Table 2 provides the design and as-built excavation depths from blasting. The amount of dynamite was modified after each shot to keep the Peak Particle Velocities (PPVs) within the contract requirements. Table 5 presents the PPVs result obtained after each blast shot.

Figure 12. Overhead view of blast charges inserted into rock

Table 2. Design and as-built excavation depths due to blasting

Blast #	Design			Actual		
	Start Elevation (ft)	End Elevation (ft)	Excavation Depth (ft)	Start Elevation (ft)*	End Elevation (ft)*	Excavation Depth (ft)*
1	−40	−45	−5	−38	−46	−8
2	−45	−51	−6	−46	−53	−7
3	−51	−58	−7	−53	−63	−10
4	−58	−66.5	−8.5	−63	−69	−6
5	−66.5	−75	−8.5	−69	−76	−7

*Elevations are approximated

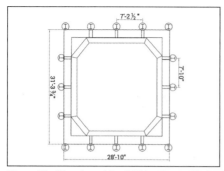 

Figure 13. Plan view of the SOE for the ventilation facility (left) and completed bracing for ventilation facility (right)

Ventilation Facility SOE

Prior to the start of construction, the area was surveyed and cleared of existing utilities and structures and the proposed footprint cleared of vegetation and graded. Once the site was ready, the piles were drilled in the layout shown below on the left in Figure 13. The cap waler and handrail system were installed at the top section of the piles. Excavation began by removing five feet of soil at a time and installing the timber lagging until two feet below the planned location of the lower intermediate brace at elevation El.12 feet. The lower waler was installed as shown below in Figure 13 and the excavation continued down to the bottom of the slab and the bottom of the sump pit at elevations of El. 6 feet and El. 0 feet, respectively.

Figure 14. Drill rig equipment on site at drop shaft location

The SOE for the ventilation facility consisted of a soldier beam and timber lagging shoring method. This method for SOE was chosen because most of the excavation was above the water table. The soldier pile elements were made up of a total of sixteen W18×71 steel beams. Pile lengths ranged between 27.5 feet and 35 feet and the timber lagging consisted of 3-inch thick (nominal) untreated mixed hardwoods. Bracing was achieved through the use of two sets of walers, a W24×104 cap waler and a W18×97 intermediate bracing waler, at an elevation of El.12 feet. Furthermore, since the mat foundation of the Ventilation Facility was planned to be located in soil rather than bedrock, a total of twenty micropiles were drilled to transfer the load to the more stable underlying bedrock.

ROCK GROUTING

With the CSO 021 project located along the banks of the Potomac River, groundwater was an issue that needed to be addressed prior to excavating. Groundwater control during construction of the Drop Shaft was achieved by grouting around the perimeter of the structure footprint from the existing ground surface to a minimum depth of 30 feet below the planned base of excavation (shaft invert). Figure 14 presents the grouting operations on site. The total drill length per grout hole was approximately 121 feet.

The rock grouting plan initially called for a total of 27 grout holes drilled around the outside perimeter of the secant pile shaft. However, due to the closure criteria defined in the specifications (less than 0.5 sacks (47 lb) per foot of hole is injected and grout refusal occurs), an additional 13 secondary and tertiary grout holes were added to the primary grout holes. The total number of grout holes drilled was 40, seven of which were drilled inside of the shaft. The closure criterion was used to determine when an area of rock had been sufficiently treated and in the case that closure criteria was not met, additional grout holes were drilled on both sides of the tested hole. Figure 15 illustrates the general layout of each grout hole surrounding the perimeter of the future drop shaft.

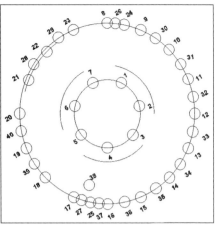

Figure 15. Grout hole locations for drop shaft

Rock grouting operations included drilling, pressure testing, and grouting. The drilling consisted of advancing 4-inch casing through the overburden and then using a water hammer to drill a 4-inch diameter hole down to the desired elevation of approximately El. −108 feet. Each hole was then water tested no longer than 72 hours after the drilling was completed. Rock grouting was performed from the bottom of the hole up to the top of rock, with the exception of seven grout holes that were grouted to the shaft invert. Grouting was performed using the split space sequence technique whereupon a packer was lowered into the hole to isolate the particular stage for injecting the grout, and the grout was injected and ramped up to a predetermined "target pressure" where it was held at this pressure until refusal criteria was met. The grout mix consisted of ultra-fine cement with a 1:1 water to cement ratio and one percent of a superplasticizer by weight. The pre-excavation grouting allowed for verification of the top of rock, that was useful for the design of the SOE later in the project as it provided a better understanding of the immediate rock conditions.

The shaft grouting specifications mandated that the maximum allowable total inflow into the partially and fully excavated shaft was 5 gallons per minute (gpm). Also, the maximum allowable inflow at any arbitrary location within the excavation was limited to 1 gpm. Groundwater inflow was monitored throughout the duration of the excavation works. The grouting operation was deemed successful as it sufficiently halted the advance of groundwater into the excavated Diversion Chamber and Drop Shaft during the latter stages of the project. A summary of the grout take volumes at each grout hole is provided on Figure 16.

PROTECTION OF EXISTING STRUCTURES AND UTILITIES

Numerous existing structures and utilities were located on the project site as illustrated on Figure 17. To prevent damage to the structures and utilities, a geotechnical instrumentation monitoring program was executed to evaluate the deformations of the SOE systems and surrounding ground. Additionally, during blasting operations, noise and Peak Particle Velocities (PPVs) were closely observed.

In total, there were two main structures and ten utilities within the limits of disturbance of the project site. The KCPA Expansion and Theodore Roosevelt Bridge westbound

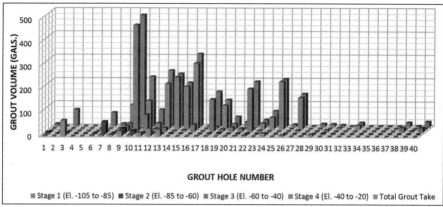

Figure 16. Summary of rock grout take volumes

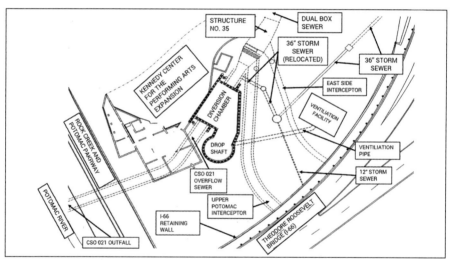

Figure 17. Structures and utilities within project area

approach ramp and retaining wall were the two main structures located adjacent to the CSO 021 diversion facility structures and out of the ten utilities five were small diameter shallow storm sewers and five were larger diameter sewers or diversion structures.

Structures

The KCPA Expansion was being constructed concurrently alongside the CSO 021 diversion facilities structures. The three-level KCPA Expansion is located adjacent to the Drop Shaft and Diversion Chamber. A portion of the KCPA Expansion foundation was coordinated to be supported by the temporary SOE secant pile wall of the Diversion Chamber which resulted in the KCPA Expansion sharing a wall on top of the secant pile wall constructed for the Diversion Chamber SOE system. The Theodore Roosevelt Bridge westbound approach ramp and I-66 retaining wall is supported by two rows of concrete filled pipe piles driven to bedrock. Figure 18 demonstrates the close proximity of the approach ramp and retaining wall to the site.

Utilities

Located to the north of the Diversion Chamber is a dual box sewer and Structure No. 35. The dual box sewer consists of two large rectangular sewers stacked on top of one another. Structure No. 35 is an automated diversion structure that splits flow into three separate lines: East Side Interceptor, Upper Potomac Interceptor, and Overflow Sewer. The 66-inch East Side Interceptor sewer (66-inch sewer) and the 108-inch Upper Potomac Interceptor sewer (108-inch sewer) extend below the Theodore Roosevelt Bridge to the Potomac Pump Station (see Figure 17). The 108-inch Overflow Sewer (CSO 021 Overflow Sewer) extends south of the site beneath Rock Creek and Potomac Parkway and discharges into the Potomac River at CSO outfall No. 021. When a storm event occurs, the excess flow is guided to the CSO 021 Overflow Sewer via an inflatable dam located within Structure No. 35. The East Side Interceptor, Upper Potomac Interceptor, Overflow Sewer, and Structure No. 35 are made of reinforced concrete supported on H-piles driven into bedrock.

Figure 18. KCPA expansion shared wall

The five additional smaller utilities within the limits of disturbance of the project included two 36-inch and one 12-inch storm sewer pipe and two 36-inch sewer pipes that were relocated as part of the project.

Geotechnical Instrumentation Plan and Results

A variety of instruments were installed to monitor the SOE Systems, KCPA Expansion, I-66 retaining wall, existing utilities, and ground behavior during construction. Movement of the structures, utilities, and ground behavior was analyzed by a variety of instruments. Table 3 provides the total number of instruments used to monitor the movement on site. Real-time and manual data was collected throughout the duration of the project. Real-time data was possible through the use of Automated Motorized total stations and Automated Vibration Monitors.

Table 3. Total instruments used to monitor the movements on site

Instrumentation	Quantity
Borehole Extensometers	5
Piezometer (P)	2
Inclinometers (I)	3
Observation Wells (W)	2
Utility Monitoring Points (UMP)	10
Tiltmeters (T)	5
Seismic Monitoring (VM)	4
Structural Monitoring Points (SMP)	62
Crack Gauges (CG)	11

Support of Excavation Systems (Diversion Chamber and Drop Shaft, and Ventilation Facility)

The Diversion Chamber and Drop Shaft SOE system was monitored through the use of structural monitoring points (SMPs) and Inclinometers. A total of 38 SMPs were mounted on to the secant piles to measure wall movements. Two inclinometers were installed near the SOE system, one near the Diversion Chamber and the other near the Drop Shaft to measure the lateral ground movement as a function of depth. The maximum secant pile horizontal wall movement was approximately 0.3 inches, while the maximum lateral ground movement obtained from the inclinometer near the Drop Shaft was roughly 0.6 inches. A surface settlement of approximately 0.5 inches was obtained through a borehole extensometer near the Drop Shaft.

tThe Ventilation Facility SOE was monitored during construction with four SMPs and one inclinometer. The maximum horizontal wall movement was less than 0.1 inches and the maximum lateral ground movement was approximately 0.1 inches. The observed surface settlement around the SOE was less than 0.1 inches.

Utilities

Ten Utility Monitoring Points were installed on the selected nearby utilities around the SOE system. During the duration of the project the maximum settlement of 0.45 inches occurred at CSO 021, adjacent to the north end of the Diversion Chamber.

Structures (I-66 Retaining Wall and Kennedy Center Expansion Shared Wall)

The I-66 retaining wall and KCPA Expansion shared wall were outfitted with multiple types of sensors to record all movement throughout the duration of the project. The I-66 wall included six SMPs, one tiltmeter, and 11 crack gauges, eight of which were manual and three were automated. The horizontal movement of the wall was less than 0.5 inches and the existing cracks that were monitored indicated movements less than one mm. Fourteen SMPs were installed on the KCPA Expansion shared wall to monitor the movement during construction. The maximum horizontal and vertical movement of the wall was approximately 0.65 and 0.44 inches, respectively.

Blasting Operations

Four seismic triaxial geophones were installed to monitor vibrations during construction activities, with particular focus on the vibrations due to blasting operations. To record real-time signals from the geophones during construction, automated vibration monitors (VM) were used. Geophone one and four were located near the Theodore Roosevelt bridge abutment, geophone three was located between the Drop Shaft and the KCPA Expansion and geophone two was place north of the Diversion Chamber near the CSO 021 Overflow Sewer. Table 4 provides the approximate ground surface distance from the center of the Drop Shaft to the geophones. The results obtained from the vibrations due to each blast shot are presented in Table 5.

Table 4. Approximate distances between geophones and the center of the drop shaft

Geophone	Distance From Center of Drop Shaft (ft)
1	137
2	132
3	29
4	84

Instruments were closely monitored to ensure that the Action and Maximum Level thresholds set forth by the contract were not exceeded. Mitigation measures were in place in the case that the readings from the instruments came close to or exceeded the threshold levels. Such measures included reducing the amount of dynamite, adjusting

Table 5. Results from vibration monitors due to each blast

Blast Shot #	Vibration Monitor 1	Vibration Monitor 2	Vibration Monitor 3	Vibration Monitor 4	PPV_{MAX} (in./sec.)
			PPV (inch/second)		
1	N/A	0.103	0.900	0.280	0.900
2	0.385	0.124	0.855	0.310	0.855
3	0.780	0.284	2.000	0.510	2.000
4	0.390	0.171	0.875	0.300	0.875
5	0.190	0.119	0.420	0.190	0.420

the boreholes spacings, depths, and diameters. Post construction visual surveys were performed on the KCPA Expansion and the I-66 retaining wall after each blast shot. The visual surveys concluded the structures sustained no damage.

None of the blasts conducted, exceeded the Maximum Vibration Levels with the exception of blast shot three, in which VM-3 mounted to the CSO-021 outfall structure downstream of the diversion structure indicated a PPVs of two inches/second. As shown in Table 4, this was the closest in proximity to the blast and a post-construction Closed Circuit Television Camera's (CCTVs) performed indicated no damage occurred downstream of the diversion structure to the CSO-021 outfall pipe.

CONSTRUCTION CHALLENGES

During the construction of the CSO-021 diversion facilities, numerous constructability issues arose that required coordination efforts. Some of the most prevalent issues dealt with by the project team during construction included a small project footprint, water intrusion within the secant piles during construction, and coordination of the blasting operations.

The total area of disturbance for the project was approximately 41,000 square feet, or about one acre. When taking into account the footprint of the Diversion Structure, the Ventilation Control Facility, installed site utilities, site instrumentation, right of ways and easements, as well as the construction entrance, the total footprint in which to conduct work and maneuver construction equipment was much less than the total limits of disturbance. As such, site access coordination between subcontractors was of critical importance. The general contractor, had to limit the number of activities on-site at any given time while also ensuring the project was meeting scheduled milestones and limiting the number of mobilizations for each subcontractor. Construction of the CSO-021 diversion facilities was taking place concurrently with the KCPA Expansion on the same project site creating an additional layer of coordination. For example, decisions were made early on during the planning stages of the CSO-021 project to redesign certain aspects of the Diversion Structure, to avoid conflict with the KCPA River Pavilion. Implementation of a tremie system for pouring concrete within such a small footprint was not feasible. Instead, upon reaching the appropriate elevation for the secant pile, steel was set in the borehole utilizing a crane (if the secant pile was a male pile), concrete was poured utilizing a free fall method, and the steel outer casing was removed.

Boring immediately adjacent to the Potomac River meant that groundwater intrusion was commonplace while drilling secant piles. In some instances, the groundwater would intrude through the steel casings and normalize to the water table elevation. The specifications for the free fall concrete placement method required that the borehole would contain no more than two inches of water to prevent mixing of concrete and groundwater. Utilizing pumps to try to remove water from the secant pile borehole was not only time-consuming, but it was costly as well. As such, some basic changes in drilling methodology were implemented to prevent future water intrusion, which had profound impacts on the overall success of the drilling process. These changes included adding rubber gaskets and sealant to the temporary steel casing joints and bolts that connected the joints. Additionally, while drilling, ensuring that the auger did not advance past the steel casing to maintain an earth plug was always in place to prevent water intrusion. This drilling methodology may have resulted in longer drill times, but it eliminated any need for pumping water after drilling was completed.

Two major coordination efforts took place during blasting operations. First, a significant public relations campaign was carried out between DC Clean Rivers and the general contractor with external entities. Blasting permits were obtained from DC Fire and Emergency Medical Services (FEMS), who made several trips to the site prior to blasting, accompanied by the DC bomb squad. During blasting, the Metropolitan Police Department Special Operations Division (MPDSOD) maintained traffic control on I-66 Roosevelt Bridge and Rock Creek and Potomac Parkways. For all five blasts, traffic was impacted for no longer than three minutes. Coordination also occurred between the general contractor, DC Clean Rivers, and District Department of Transportation (DDOT). Reports were provided to DDOT prior to blasting that provided a baseline of normal movements for I-66 Roosevelt Bridge from SMPs mounted directly to the I-66 wall.

The second major coordination effort occurred in regards to the secant pile system in the Drop Shaft. According to the design, a rock shelf was to be maintained between the secant piles and blasted diameter of rock. However, after blasts one and two, there was nothing to support the rock shelf, and the remaining rock still attached to the secants was excavated for safety purposes. This resulted in concerns over the potential of undermining the toe of the secant piles from the subsequent blasts. In response, an inspection was performed to determine potential locations of undermining and a concrete block with rock bolts was constructed to prevent the potential undermining of the secant piles. Additionally, the blasting design was revised and the number of explosives in the next blast was reduced and diameter of the blast holes was tightened. As such, the blasting for the drop shaft was completed ahead of schedule and without undermining of the secant piles or any significant quality assurance issues.

CONCLUSION

The CSO 021 Diversion Facilities project overcame unique challenges to maintain schedule and budget constraints. Immense coordination took place due to the project's small footprint, a shared support of excavation secant pile wall with the KCPA Expansion, and blasting operations in Washington DC. Adjacent to the Potomac River, controlling water was significant to the success of the project. To safeguard the numerous existing structures and utilities within the limits of disturbance of the project, a geotechnical instrumentation program was implemented.

Coordination efforts between the general contractor, subcontractors, and government agencies was required to accomplish the construction of the SOE systems for the Diversion Chamber and Drop Shaft, and Ventilation Control Facility. With a small project footprint and a constricted schedule, both rock grouting and secant pile construction operations occurred simultaneously which required day-to-day communication. Blasting operations involved significant public outreach and coordination with DC Clean Rivers, DDOT, MPDSOD, and DC FEMS.

With the project site located approximately 200 feet from the Potomac River water control was of critical importance. Early in the project, secant pile operations encounter some water intrusion which was mitigated through adjustment to construction methodologies. The pre-excavation rock grouting program provided additional assurance that a dry excavation would be achieved during construction of the Drop Shaft. The program was successful and water was not an issue inside the shaft.

The geotechnical instrumentation program provided continuous observations of the existing structures and utilities. In instances where the contract movement thresholds were exceeded, mitigations measures were implemented immediately. Throughout the duration of the project, visual surveys were performed on the existing structures and no damage was sustained. CCTV scans were performed on utilities of major concerns and the inspections concluded no damage occurred due to the construction activities on site. Through careful planning and continuous instrumentation monitoring, surrounding structures and utilities were successfully protected.

ACKNOWLEDGMENTS

The authors would like to thank DC Water for providing permission to publish this paper. The projects success was due to the contributions of many people from various companies, a few of which, James G Davis Construction, Berkel & Company Contractors, Moretrench, Corman Construction, Geo-Instruments, Bradshaw Construction Corp, JCK Underground, and Aldea Services. The authors are particularly thankful for a few of the team members and would like to acknowledge the efforts of Robert Goodfellow, Joe Olmstead, Tim Fedder, Todd Brown, Jeff Paddon, and Amanda Morgan, whose work contributed to this paper and the success of the CSO 21 Diversion Facilities Project.

REFERENCES

DC Water and Sewer Authority. (2018). DC Clean Rivers Project Potomac River Tunnel. (p. 69). National Park Service U.S Department of Interior.

Fleming, A.H., & Drake, A.A. (1998). Structure, age, and tectonic setting of a multiply reactivated shear zone in the piedmont in Washington D.C., and Vicinity. Volume 37.

Means, J. (2010). Roadside Geology of Maryand, Delaware and Washington DC. (p. 178). Mountain Press Publishing Co.

U.S. Geological Survey. (1994). Geological Quadrangle Map, Washington West Quadrangle, D.C., MD, VA GQ-1748. In A. Fleming, A. Drake, & L. McCartan.

U.S. Geological Survey. (2011). *USGS UIS Topo 7.5-minute map for Washington West, DC-MD-VA 2011.* USGS National Geospatial Operations Center (NGTOC).

Design and Construction Considerations for Large Shafts in Hudson River Tunnel Project: Hoboken Shaft

Arman Farajollahi ▪ AECOM
Paul Roy ▪ AECOM
Young Jin Park ▪ AECOM

ABSTRACT

As part of the Gateway Program, the Hudson River Tunnel project is a new tunnel alignment running parallel to the existing North East Corridor (NEC) tunnel, and consisting of various surface structures and tunnels from east of Frank R. Lautenberg station in Secaucus, New Jersey, to existing rail complex at Penn Station New York (PSNY). The proposed profile and alignment of the corridor passes as a bored tunnel under the Palisades area in New Jersey and beneath Hudson River through the Hoboken shaft. The Hoboken shaft will be used to receive the Tunnel Boring Machines (TBM) from the Palisades and launch the Hudson River TBMs. The final shaft configuration will is intended to provide a ventilation shaft when completed. The ventilation fans and equipment will be located directly above the tunnel alignment within the Hoboken shaft. At the surface, the ventilation structure will be extended to include the tunnel ventilation operations. The Hoboken shaft is the division between the excavation in rock and in soil. The internal diameter excavation of the shaft is approximately 110 feet with a depth of more than 100 feet. The shaft bottom is located in a rock and very soft soil mixed face condition. The rock is declining from west to east resulting in founding the shaft base at different rock levels and creating an unbalanced pressure loading. It is envisioned that shaft construction will use diaphragm slurry walls as support of excavation (SOE); however, the Contractor would have the option to consider other methods for SOE, including secant walls and grouting or ground freezing construction techniques. Design and construction of such a large shaft in difficult ground conditions are challenging. This paper reviews these challenges and design methodology for the use of diaphragm slurry walls, connections between the shaft wall and the tunnels, and founding the walls in the rock; as well as, construction challenges and issues.

INTRODUCTION

Existing Northeast Corridor (NEC) tunnels between New Jersey and New York beneath the Hudson River were built more than 100 years ago and new tunnels are required to provide more reliable connections between these two states and AMTRAK's Northeast Corridor alignment. Several alignment alternatives were considered during preliminary design stage. The selected alternative alignment (approximate) is shown in Figure 1.

The extensive lengths of new tunnel alignments beneath the Hudson River required the project to have two large diameter ventilation shafts in New Jersey (Hoboken Shaft) and New York (12th Ave Shaft in Manhattan) as shown in Figure 2. The alignment shown in this figure is approximate.

Also Figure 3 shows the approximate location of Hoboken shaft and the tunnels under Hudson River.

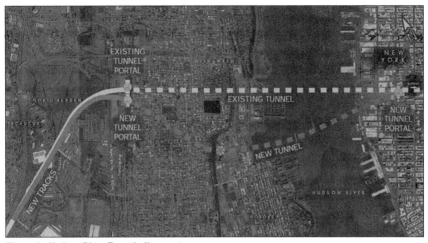

Figure 1. Hudson River Tunnel alignment

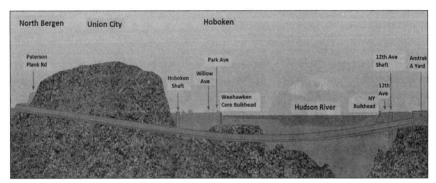

Figure 2. Profile alignment of the Hudson River Tunnel project

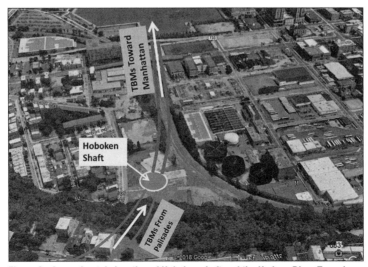

Figure 3. Approximately location of Hoboken shaft and the Hudson River Tunnels

GEOLOGY OF THE HOBOKEN SHAFT AREA

The Hoboken shaft site geology is founded on land made during the third quarter of the nineteenth century. Prior to that time, this area consisted predominantly of tidal salt marsh, within Hoboken Creek, a tidal waterway, bordering the northern limit of the site. By 1819, earthen sea walls and open ditches were constructed to drain the area and reclaim the land for development. The geology of the area consists of the following:

Artificial Fill: A large percentage of soils in the area have been altered by excavation or filling for residential, commercial, or industrial purposes. Earth and manmade materials that have been emplaced include gravel, sand, silt, clay, trash, cinders, ash, timbers and construction debris. The groundwater is very shallow and a few feet below the ground surface.

Estuarine Deposits: A thick sequence of post-glacial estuarine deposits overlies the glacial lake deposits, or the tidal marsh deposits if present. The estuarine deposits consist of gray clayey silt and silty clay, with traces of fine sand and shells.

Glacial Till: The glacial till is the surficial deposit directly overlying bedrock in most of Hoboken shaft area. It forms nearly continuous blanket overlying bedrock. It is an unstratified, compact deposit with pebbles, cobbles, and boulders in a reddish-brown matrix of poorly sorted sand, silt, and clay.

Stockton Sandstone: Consists of Arkosic sandstone, interlayered with conglomerates and siltstones.

SHAFT DESIGN

Geotechnical Considerations

The shaft would be excavated in soil and rock and also the bedrock slopes at the shaft sites. Since the soil and rock have different geotechnical properties, the lateral earth pressure as well as subgrade reaction modulus of the ground would dramatically changes at the SOE structure and the ground interaction. The differential lateral loads would cause significant asymmetrical stress distribution in the SOE structure.

The main part of the shaft would be excavated in very soft and compressible soils. Because of high ground water elevation any water withdraw due to construction could cause ground settlements at the exist structures nearby.

Design Considerations

Factors to be considered during design of the shaft as follows:

- Circumferential stress in the circular shaft. The design-build contractor's designer shall develop proper number of panels and thicknesses, somehow providing sufficient effective thickness (shown in Figure 4) which would provide for hoop stress induced in the SOE shaft slurry wall. The design should consider the effective thickness of the wall in the analysis and design of the shaft.
- The circumferential stress challenge requires resolution of the soft-eye holes penetrated by the TBM's requiring large TBM openings within the SOE walls (Virollet et al., 2006). Figure 5 shows the stress distribution due to a circular opening is a cylindrical shaft. As it is shown the opening could cause increase in the compression and tension stresses around itself. Our preliminary design considers the addition internal ring beams in the shaft at

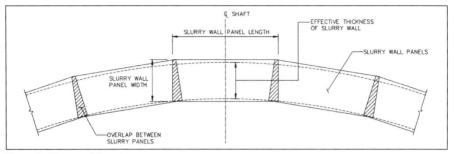

Figure 4. Slurry wall layout

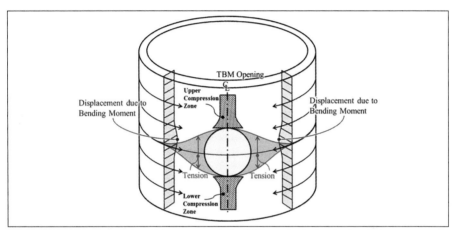

Figure 5. Changes in circumferential stresses around a TBM opening in the circular shaft

the compression zones. However, extending the wall into the bedrock and creating full circular rock-socket may eliminate requirements for an internal ring beam below the TBM openings at the lower compression zone. The tension zones at both sides of TBM soft-eye could also be satisfied by either increasing slurry wall reinforcement at the tension zones or considering collar structures at the openings. It should be noted that the detail to create soft-eye penetrations around the openings will be the responsibility of the design-build contractor's designer.

- The slurry wall panels may be placed either Partial Depth (Shown in Figure 6) or Full Depth (Shown in Figure 7) in the bed rock. Because of the bedrock slope at the shaft area, the lateral earth pressure would be different around the shaft as shown in Figures 6 and 7. Since only Full Depth shaft option could provide a full circular rock socket at the bottom of the shaft, the designer may be required to consider post tension walls (or other construction methods) to supplement and increase bending resistance which would cause more complicated construction procedures for Partial Depth shaft option (Xanthakos 1994). The post tension panels may not have rectangular shapes and they could be designed with T-shape or L-shape geometries to provide cantilever racking systems.

- Uplift pressure due to high hydrostatic pressure beneath the working slab shown in Figures 6 and 7 could cause instability at the bottom of the shaft.

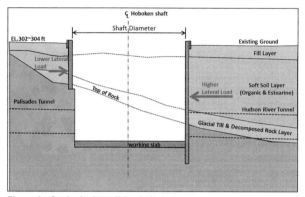

Figure 6. Geological condition in Hoboken shaft for partial depth option

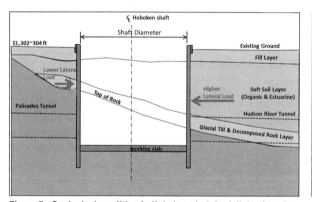

Figure 7. Geological condition in Hoboken shaft for full depth option

The contractor's design may be required to consider tension piles or anchors at the bottom of the shaft. In this case the connections between piles and the working slab would be required to be waterproofed and perform as permanent structures and anchors to resist hydrostatic uplift pressure for the Hoboken shaft structure.

- Above all of these challenges, the design-build contractor's designer needs to consider construction tolerances during construction of the slurry wall panels and apply these verticality tolerances to the final design. These tolerances could occur during construction due to very deep slurry wall panels as well as heavy reinforcement cages. Figure 8 shows the reducing the wall thickness which could happen due to tolerances at two adjusted panels. This could

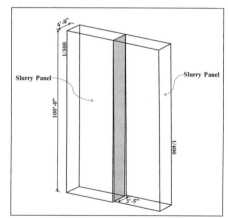

Figure 8. Reducing wall thickness due to verticality tolerance of panels

cause a reduction in effective thickness of slurry wall (See Figure 4) at the deeper part of the shaft. The vertical tolerances shown in Figure 8 are only examples.

CONTRACTIBILITY CONSIDERATIONS

Factors to consider during the SOE construction:

- Because of the tolerances the groundwater control would be an issue due to missed alignment between the panels. Consequently grouting behind the panels may be required.
- Seepage flow may occur at the invert of the shaft from the bed rock. The contractor may require to perform rock mass grouting at the invert of the shaft to avoid groundwater withdraw at the surrounding area.
- The contractor requires a large lay down area for staging, cage assembly, slurry desanding, spoil storage, cranes, and slurry plant. Other SOE construction method solutions would require similar conditions.
- The slurry panels would be encountered several types of soil stratums including fill layers, very soft soil layers and granular glacial deposits. That would make it challenging to provide appropriate type of slurry to stabilize the slurry trenches. Also during excavations in the fill layers, soil tends to lose its strength by vibrations caused by excavating machines.
- To avoid any water leak from the beneath of the SOE wall, the contractor may be required to perform rock mass grouting underneath the some parts of the SOE wall as an additional sealing to increase the impermeability (Seitz et al., 2017).
- The contractor may require considering special foundations for heavy vehicles and machineries such as cranes used for mucking. The induced lateral loads due to these construction activities should be considered in the SOE design. (Shown in Figure 9).
- During concrete pouring the concrete should replace the bentonite slurry. If it does not, cavities filled with bentonite occur at the joints between the panels (Shown in Figure 10). These bentonite cavities are difficult to detect. However, contractor would encounter the bentonite cavities during excavation of the shaft. Because of high hydrostatic pressure behind the slurry wall in lower part of the Hoboken Shaft, the cavities may become unstable and cause groundwater leakage into the shaft. Therefore, contractor may be required to apply proper types of grouting as ground treatments to fill out the cavities.
- The design-build contractor's designer is required to provide instrumentation and monitoring programs during construction of the SOE shaft. The programs should include Surface and Subsurface, as well as Construction Operation (or Performance) monitoring (Schwamb 2014). Surface program would be included monitoring of the existing structures such as rail tracks and buildings, as well as, ground settlement and slope monitoring. Subsurface program mainly provide monitoring for groundwater control and lateral and horizontal displacements due to construction of the shaft. Construction Operation (or Performance) monitoring is mainly to provide quality control of slurry wall structures such as soil collapsing during concreting (Abdelhadi et al., 1996).
- While the design-build delivery method has not been officially selected by the sponsor of the project, we have assumed design-build delivery method

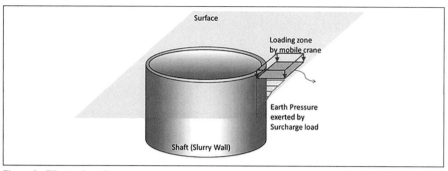

Figure 9. Effects of surcharge

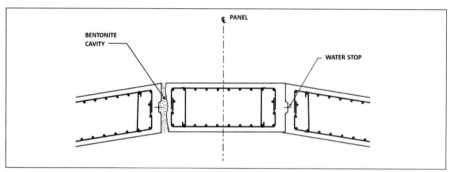

Figure 10. Bentonite cavity at the joint between two panels

in the paper as it provides, in our opinion, the most efficient delivery solution for this aspect of the project.

CONCLUSIONS

The paper discusses the overall Preliminary Engineering required to provide a buildable construction with the information available at this stage of the project. Final design and construction methods will be the responsibility of the design-build contractor. The following points should be considered during the final design of the shaft:

- The slurry wall panel joints are formed by overcutting of a secondary panel into two adjacent primary panels. Therefore, there would not be reinforcement continuity between the panels. Consequently no moment would be transferred at the joints between the panels and only transferred shear and axial forces will exist. In the final design the panel joints should be modeled with releases to ensure that moment transfer does not occur at all (Theophilou 2001).
- The design-build contractor's designer needs to consider the contract packaging, scheduling and construction sequences as well as means and methods of the contractor for design of some structural elements such as TBM soft-eyes, ground improvement for receiving and launching the TBMs, ground improvement for watertightness of the shaft, working slab, and internal ring beams.
- The design-build contractor's designer may be required to consider some temporary treatments such as reinforced shotcrete skin walls for as treatments for the slurry panel tolerances during construction to obtain the

required effective thickness as mentioned before (See Figures 4 and 8) as well as curing possible bentonite cavities (Lubach 2010).
- The design-build contractor's designer may be required to consider effects of temperature fluctuations during construction.

ACKNOWLEDGMENTS

The authors would like to thank the Gateway Trans-Hudson JV Partners WSP and STV for support in the preparation of this paper. The views and opinions expressed in this article, and assumptions made within this article, are those of the authors and do not necessarily reflect the official policy or position of the project sponsors including, Amtrak, Port Authority, Gateway Corporation, NJ Transit or any public or government agency related to this project.

REFERENCES

Abdelhadi, N., Abdelhadi, M. and Gotoh, K. (1996) "Problems Encountered in Diaphragm Wall Excavation" Reports of the Faculty of Engineering, Nagasaki University, Vol. 26, No. 46.

Bruce, D.A., Chan, P.H.C. and Tamaro, G.J. (1991), "Design, construction and performance of a deep circular diaphragm wall" ASTM International Symposium, Atlantic City, NJ.

Lubach, A. (2010), "Bentonite cavities in diaphragm walls—Case studies, process decomposition, scenario analysis and laboratory experiments" M.Sc. Thesis, Delft University of Technology.

Schwamb T. (2014), "Performance Monitoring and Numerical Modelling of a Deep Circular Excavation" Ph.D. Thesis, University of Cambridge.

Seitz, J. and Pengyu, B. (2017), " Challenging construction of two huge shafts by diaphragm walls-New dimensions for Africa" Proceedings of the 19th International Conference on Soil Mechanics and Geotechnical Engineering, Seoul 2017.

Theophilou, A.I. (2001), "Structural design of an underground cylindrical shell" M.Eng. Thesis, Massachusetts Institute of Technology.

Virollet, B., Gilbert, C. and Deschamps, R. (2006), "Recent Advances in Large Diameter Diaphragm Wall Shafts" Deep Foundations Institute, 31st Annual Conference on Deep Foundations.

Xanthakos, Petros P. (1994), "Slurry Walls as Structural Systems" Second Edition, 1994.

Design Challenges of Deep Underground Shafts

Daniel Garcia ▪ WSP
Ravi Jain ▪ WSP

ABSTRACT

Deep shafts serve varying purposes in urban development: from egress, tunnel boring machine (TBM) launch/recovery to combined sewer overflow drop shafts and wastewater storage. The construction of a deep shaft may similarly involve a broad range of support of excavation (SOE) inclusive of ground freezing, slurry walls and secant piles. The geotechnical and structural parameters and analyses involved often require unconventional considerations and creative solutions. The installation of shaft structures is often a critical path activity necessary to commence other elements of the project. The analyses of these shafts must account for constructability without compromising efficiency. The design of shaft structures should consider unforeseen circumstances, uncertainty in ground conditions and provisions for subsequently constructed elements. The analysis of these structures is undertaken using creative closed-form solutions, soil-structure interaction and innovative finite element solutions.

INTRODUCTION

Deep shafts pose unique challenges which must be evaluated during design. Consideration should be given to the benefits and limitations of each shaft type and support of excavation based on required water tightness, ground stability and client-specific mandatory requirements prior to selection. Contrary to a shallow shaft, a deep shaft is usually always circular or oval-shaped more efficiently resist significant hydrostatic and earth pressures by hoop compression. In many cases, deep shafts may reach several hundreds of feet below grade. A deep shaft exhibits variable structural behavior relative to depth. Regions close to the interface with the base slab are affected by lining-to-slab connection details, tunnel openings or other specific loading such as TBM thrust frame support loads. Regions close to the ground surface are controlled by unbalanced construction surcharge loads and seasonal temperature variations. Bending strains are caused by temperature gradients across the lining. Mid-height regions are typically compression-controlled zones. The structural behavior of the lining must be considered for the installation of subsequently constructed internal structures to prevent significant increase in out-of-plane forces.

Design considerations should also account for deep shafts installed in mixed-ground conditions, which may require the use of different ground support systems and/or a soil-rock interface measures. Additionally, large diameter deep shafts installed in soil typically require large lining thicknesses, often impractical, for permanent ground stability or flotation resistance. Specific installation conditions must be considered to allow the SOE system to be utilized as a permanent structural element. Construction considerations are important due to unforeseeable conditions, quality control issues or general constructability. This includes special considerations for SOE repair work due to damage or improper installation, waterproofing, base slab construction, and utilization of instrumentation and monitoring systems.

SHAFT LOADING AND DESIGN CONSIDERATIONS

The selection of an adequate SOE system is often associated with the investigation of previous case histories and experience, as well as site-specific subsurface conditions. A reasonable assessment and selection may be achieved using relevant successful reference projects that have compatible characteristics including: subsurface conditions, shaft geometry (depth-to-diameter ratio), and construction site constraints such as equipment operation and mobilization, community impacts and disturbance limitations. In most cases, project-specific guidelines, the contractors experience, preferred method of construction and codes also influence the selection of shaft type. Therefore, it is imperative to understand the principal characteristics and limitations of each shaft type. Table 1 summarizes general considerations for shaft design.

Table 1. Description of different shaft types

Type	Description
Slurry Wall (Temporary and Permanent Walls)	A slurry wall shaft involves the placement of a bentonite or polymer slurry to retain earth as it is excavated using a hydromill or clamshell. Once excavated, concrete is placed from the bottom up, with the tremie pipe(s) submerged in concrete as it is placed, displacing the slurry. Considerations involve: a. Minimum panel overlap. The overlap should allow for maximum allowable vertical deviation while maintaining contact with the adjacent panel. b. Verticality requirements, which add eccentric bending moments c. Horizontal hoop force has sufficient bearing at panel ends—minimum panel-to-panel contact. d. Glass fiber reinforced polymer (GFRP) may be used for any penetrations and TBM openings. e. Base slab and intermediate floor/roof slab connection requirements
Secant Pile Wall (Typically used only for Temporary SOE)	A secant pile shaft wall involves installation of primary and secondary circular piles with either steel core beam or conventional reinforcement. The depth limitations of a secant pile shaft are less compared to slurry walls. Considerations include: a. Pile alignment for deep shafts. b. Weaker primary piles than secondary due to overlap required. c. Difficulty keying base and intermediate slabs. The base slab must either rely on friction or on welded connections, or a temporary jet grout plug/tremie slab must be placed with tie-downs. d. GFRP may be used for any penetrations and TBM openings.
Ground Freezing (Temporary SOE)	Ground freezing involves a circulation of brine or nitrogen via a freeze plant and pipe system. Nitrogen may freeze ground quicker, but is generally more expensive. Pipes are installed vertically to rock to provide groundwater cutoff. Considerations include: a. Alignment of freeze pipes for deep shafts b. Energy costs for maintaining frozen ground for the duration of shaft construction and operation c. Sacrificial concrete for the face cast against frozen ground. 3" is the minimum recommended. d. Mass concrete placement, such as the base slab, may generate heat enough to melt some frozen ground and diminish concrete quality. Water may also pool on top of the slab as it cures. e. Skin friction is minimal and primarily from any dense sands—must be tested.
Slip-Formed (Permanent Shaft Structure)	A slip-formed shaft is typically done with fiber-reinforced concrete. The shaft is constructed away from the SOE as a free-standing concrete cylinder. A concrete infill is poured to fill the gap between the SOE and permanent shaft structure, putting the lining directly into compression. The forms slide up to expose the concrete just as it begins to harden. The lining is constructed top to bottom with no construction joints. Concrete mixes for the lining and for infill concrete need to be carefully developed.
Shotcrete (Temporary SOE)	Excavation support in rock will likely consist of shotcrete and rock anchors. In soft soil with ground improvements, it might consist of lattice girders and wire mesh. Consideration include: a. Method and details for ground improvement. b. If friction needs to be relied upon for buoyancy, the shotcrete smoothness criterion must be considered when determining the friction coefficient. c. Lattice girders and wire mesh requirements for temporary loading. d. Testing for shotcrete properties if in frozen ground.

Shaft Structural Behavior and Design at Varying Depths

The design of the shaft permanent lining considers temporary loads (e.g., construction surcharge loads and TBM thrust loads) and long-term permanent. Generally, long-term loads in the lining include earth and water pressures that are uniform around the lining perimeter that increase with depth. For shaft excavation in rock, the rock pressure is typically a constant value but applied as unbalanced loads that vary around the perimeter of the lining based on potential rock wedge failure. During construction, the shaft lining is designed to resist construction surcharge loads imposed by equipment near the shaft. The depth of this lateral surcharge depends on the type of equipment, proximity to the shaft and any project-mandatory requirements. During the structural analysis, this load should be applied to half of the shaft or on opposing sides to simulate a "squeeze" effect.

The design of the permanent lining requires an understanding of the shaft's variable structural behavior relative to depth. The lining may be divided into three regions based on its response to exterior loading. The bottom-most region is mainly influenced by the connection to the base slab and by the lining discontinuities for the excavation of a tunnel or adit. As shown Figure 1, significant vertical bending develops at the bottom of the lining due to the restraint of the base slab connection. This connection may be either fully rigid or pinned. For deep shafts with large diameters, a simple dowel connection is recommended to prevent large redistributed moments from the stiffer base slab. Tunnel openings also influence the behavior of the shaft. In cases where multiple openings are proposed, the two-way action of the lining is eliminated, increasing bending moments and decreasing the horizontal hoop forces. This condition may cause the lining to behave as a flexure-controlled member in the vertical direction. For the shaft shown in Figure 1, the bottom-most region extends for approximately 20% of the depth.

The lining's structural behavior at the upper-most region is affected by unbalanced construction surcharge loads and temperature gradient loads due to seasonal variations during temporary condition. During construction, the shaft cover is not considered as a restraining member for the shaft lining and, instead, the shaft is analyzed considering a free end at the top. Horizontal bending deformation is expected to occur due to unbalanced construction surcharge loads, unbalanced surface live loads, irregular ground surface, and temperature loads. Additionally, at shallower depths, the hoop force decreases, which may cause the lining to behave as a flexure-controlled member in the horizontal direction. Bending strains due to the temperature gradients develop less vertical bending stresses than horizontal because of the free-end at the top which allows for movement.

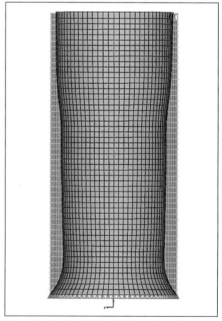

Figure 1. Profile of typical deformed shaft

The mid-region over the lining height may be classified as the region of the lining less impacted by lining discontinuities and/or boundary conditions. In this region, the lining behaves as a compression-controlled element with minimal out-of-plane deformations. For compression-controlled regions, minimum reinforcement should suffice. The reinforcement requirements for shaft structures vary by code interpretation. For example, the minimum reinforcement may be per Chapter 14, Walls, of ACI with a slight modification. Vertical walls of buildings have vertical compression and horizontal tension due to potential wind loads. Since the lining would have horizontal compression, the requirements must be flipped. Similarly, it may be interpreted that the shaft lining is a compression member requiring minimum 1% reinforcement. This is a common overconservative consideration as this excerpt of the code also requires tie reinforcement, which is typically not provided for shaft linings outside of beams. Further, the code allows minimum reinforcement requirements to be waived if the reinforcement provided is one-third greater than required by analysis. Reinforcement optimization must be performed during final design.

Loading Due to Temperature Variations

Shafts that are open at ground surface for long periods of time, such as TBM retrieval/launch shafts, are subject to seasonal temperature changes at the ground surface. These temperature variations cause deformations that induce out-of-plane bending stresses. The temperature loading due to seasonal variations is applied as a gradient; a function of change in temperature in the inner and outer faces of the shaft divided by the thickness of the lining as shown in Figure 2. The temperature drop across the concrete lining is calculated by considering material parameters such as coefficients of thermal conductivity, thickness and average seasonal temperature variations. The following considerations may be used to derive the temperature gradient variation along the lining height:

a. Consider seasonal average maximum and minimum ground temperatures specific to the region where the shaft is constructed.

b. Outside the shaft, it may be assumed that the seasonal temperature decreases/increases linearly as a function of depth. Existing scientific research shows that at depth of approximately 30-ft below the surface, the temperature outside the shaft becomes constant based on average ground

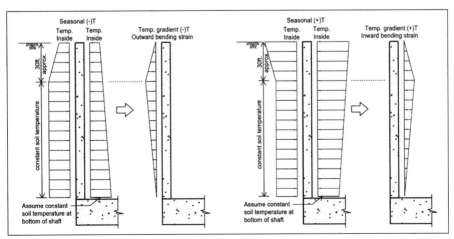

Figure 2. Derivation of temperature gradient along shaft lining

temperature and any fluctuation of the earth temperature may be considered negligible.

c. Inside the shaft, it may be assumed that the temperature decreases/increases linearly as a function of depth along the full height of the shaft. The temperature at the bottom of the shaft may be assumed to be equal to the average normal ground temperature to maximize the temperature gradient effect at regions closer to the surface during the temporary condition.

d. Thickness of the SOE wall may considered as part of the overall thickness used to calculate the temperature gradient as this is a temporary condition. The resultant temperature gradient is applied to the permanent lining only using a finite element software.

Internal Structures

Deep shafts may have a variety of applications such as combined sewer overflow, stairs for egress, utility/ventilation chase, among others. They often require the installation of permanent, wall-mounted internal structures (ventilation chambers, drop pipes, floor slabs etc.). These installations may contribute to the increase of localized out-of-plane stresses due to forces transferred from the internal structures' connection into the lining. To preserve the simplicity of the lining structural design, the following general considerations may be made:

- Detail connections that reduce unfavorable out-of-plane forces transferred between the lining and the internal structures, and
- Evaluate the locations (shaft elevations) at which the internal structures will be mounted to the lining.

Wall-mounted internal structures may be connected to the lining using reinforcement with mechanical couplers able to resist tension and shear forces due to load eccentricity. This connection not only prevents transferring additional bending forces from internal structure but, concurrently, prevents external loading on the shaft such as unbalanced surcharge loads, seismic deformations or lining deformations due to temperature gradients, from influencing the internal structures design. If the shaft lining is analyzed using a finite element model and including mounted internal structures, their connection should be modeled as pinned connections, constraining only translational degrees of freedom (DOFs) to prevent any out-of-plane bending moment to be transferred between the lining and the internal structures. In finite element software, these connections may be modeled as "weld constraints." These constraints generate multiple body constraints that are applied locally to a set of joints within a given tolerance. As shown in Figure 3, the connection of the internal structure is modeled at a distance less than one inch away from the CIP lining shell elements while the constraint tolerance is set to one inch. Only translation DOFs are constrained, preventing any other out-of-plane forces to be transferred between the CIP lining wall and the internal structures.

While the location of the internal structures depends on its function within the shaft, as part of a holistic approach at preliminary shaft planning and early stage design, its construction may benefit from properly detailing its structural behavior at the internal structure connection points. For instance, shaft elevations near the surface or close to the base slab, are typically designed with suitable flexural reinforcement to resist out-of-plane bending moments induced by external loads. These are regions with positive reinforcement that can accommodate such connections without altering the overall design at those elevations. Conversely, at mid-elevations of the shaft, or

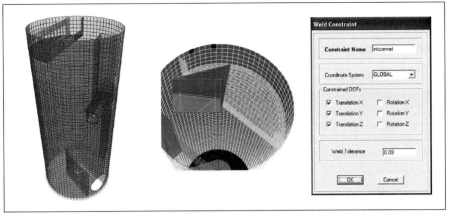

Figure 3. Example of shaft with internal structures for combined sewer overflow (left); modelling the connection of the internal structures (middle & right)

compression-controlled shaft elevations and locations where flexural reinforcement might not be required by design, out-of-plane forces should be limited. These considerations are of greater importance for deep shafts where properly defined regions along the shaft height may ultimately reduce the cost and time of advanced design and commencing construction.

TBM Thrust Support Loads

A major consideration in TBM launch shafts is the analysis of TBM thrust frame loads. Conventionally, TBM thrust frames use a base slab for support as shown in Figure 4(a). This requires a slab suitable for high concentrated uplift forces and with enough tensile capacity to resist pull-out forces from the anchorage connections. This condition may be more significant in the design of a base slab for a deep shaft as it acts in the same direction as hydrostatic pressures. An alternative installation method, shown in Figures 4(b) and 4(c), consists of using the lining to resist the TBM thrust load with a set number of support points depending on the lining strength and project specific TBM thrust frame capacity requirements. Shear forces are prevented from transferring into the base slab by introducing a concrete block at the bottom of the support confined with the lining. The analysis of the lining capacity under this loading condition includes checking for flexural strength, punching shear and bearing capacity. In most cases, project-specific provisions allow the SOE system to be accounted for to resist the TBM thrust load in conjunction with the permanent lining as this is a

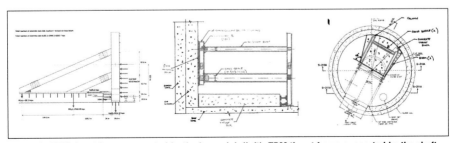

Figure 4. TBM thrust frame supported by the base slab (left); TBM thrust frame supported by the shaft lining—elevated view (middle) and plan view (right)

temporary loading condition. Additionally, the TBM thrust load should be combined with actual external loads during construction to obtain an accurate net load acting in the lining to prevent unconservative results. A finite element model of the shaft is the preferred method of analysis to include additional resistance from the surrounding ground stiffness.

The SOE System as a Permanent Structural Member

Deep, large-diameter shafts, especially those shafts excavated in deep soils, often require a large permanent lining thickness to ensure long-term strength and serviceability requirements are met. Project-specific codes allow the use of the SOE system as a permanent structural member if specific provisions are followed. This allows for decreasing the permanent lining thickness significantly by sharing the external permanent load between the SOE and the CIP inner lining proportionally based on their stiffness. Alternatively, the SOE may be designed and installed as a permanent component to resist the earth pressure while the permanent cast-in-place lining resists the water pressure only.

An important consideration in the design of any shaft is checking it against buoyant forces. For large-diameter and depth shafts, the buoyancy capacity may rely on the SOE wall. If a slurry wall SOE system is utilized, block-outs may be used in the slurry wall panels to form a key between the invert slab and the SOE wall. For other shaft types, concrete cast against concrete, provided the surface is roughened to a ¼-inch amplitude in accordance with ACI, will generally obtain a satisfactory bond. Where the surface would not properly adhere, a positive doweled connection may be provided as well. The relative deflection of each concrete medium be it initial support or permanent lining, if utilized long term, should be checked relative to stiffness to determine if they will deflect away from each other as the support of excavation may extend deeper to provide groundwater cutoff. A support of excavation system installed prior to excavation such as slurry wall or secant piles may also utilize skin friction with the surrounding ground, which is a function of the friction angle of each stratum. In the case of constructing a shaft with frozen ground, testing is required to consider and demonstrate skin friction of the thawed ground, which would otherwise be zero. Ground freezing in this instance would be a vertical application around the shaft perimeter. Soils with large void ratios and inherently higher water contents are most susceptible to impact due to ground freezing. A dense stratum, like sand, is less impacted due to a low void ratio. A staged finite element analyses is needed to simulate this analysis with test results. When frozen ground is used for support of excavation, shotcrete may be applied for short-tern structural stability and to prevent water ingress. Testing is needed to verify the elastic modulus when shotcrete application is to be considered. When other options are not possible, resistance against buoyant forces may also be achieved using tiedown ground anchors; either temporary for construction or permanent (Parkes and Castelli 2017).

Shaft Design Under Mixed Ground Conditions

Any support of excavation system installed from the surface in mixed ground involving rock does not need to necessary extend the depth of the shaft so long as rock provides adequate groundwater cutoff and is sound. The SOE can be keyed into rock and set back from the face of the rock excavation. A perimeter ring beam (or toe anchors) may be installed to provide lateral stability as needed based on bedrock quality. The permanent lining design at soil to rock interface may involve a stepped transition or a smooth transition as shown in Figures 5(a) and 5(b). In a stepped transition, an

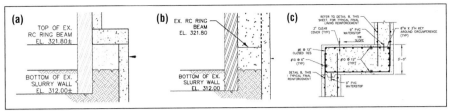

Figure 5. (a) Stepped transition section; (b) smooth transition section; (c) permanent ring beam for a steeped transition

Table 2. Types of lining soil-to-rock interface design

Type	Description
Smooth transition	a. Simpler shaft construction
	b. Increased lining durability (increased outer reinforcement cover)
	c. Eliminates "ring beam" (step) for lining transition at soil-rock interface; fewer joints at interface will reduce the potential for water infiltration from the rock-soil interface
	d. Shortens span of shaft cover
	e. Allows contractor to use one set of forms for the entire height of the shaft
	f. May not be a practical solution for large variations in shaft diameter
	g. Requires additional concrete which may increase material costs
Stepped transition	a. Analysis of shaft along soil and rock may be done independently.
	b. Lining reinforcement provided may be independent for each section (soil and rock).

additional permanent ring beam can be designed as shown in Figure 5(c). This flat ring beam must resist the reactions from the dowels connecting the lining in the soil and rock. Waterstops should be provided at each construction joint. Table 2 lists some characteristics and limitations for each type of interface design.

SHAFT CONSTRUCTION CONSIDERATIONS

Shafts designs often need to be re-evaluated during construction due to unforeseen circumstances, quality control issues or general constructability. Evaluations may consider provisions of the American Concrete Institute (ACI) code that deals with existing structures based on the as-built conditions and concrete strengths of shaft elements that have already been placed.

A shaft lining or SOE system may require repair. However, the structure, typically under high compression forces, must be assessed prior to performing repairs. Part of the assessment involves using data from available instrumentation and monitoring data. A vibrating wire piezometer allows for determination of the groundwater table elevation while a stress cell will provide the soil and pore pressure readings. Based on these two values, the lateral pressure and equivalent at-rest earth pressure coefficient may be determined. Using this information, the circumferential hoop force may be calculated. If part of the SOE requires concrete removal, the extent of that removal will be governed by the ability of this force to span around the opening. The control of groundwater inflow and ground loss must be considered when developing repair measures.

For large diameter shafts, reinforcement traversing the circumference (horizontal hoop bars) may not be exactly plumb. The reinforcement may jog in and out of horizontal construction joints. For reinforced concrete to maintain continuity, bond strength and adhesion, cementitious paste must be placed within the area where reinforcement protrudes from the joint up to a point where the clear distance between the reinforcement and construction joint is 1.5 times the maximum aggregate size as illustrated in Figure 6. For example, for ¼" aggregate, the required distance is 1⅛". This must be done prior to the subsequent pour. Spacers or standees may be utilized, but due to the variability of placement height in large diameter shafts and concealment of the top behind formwork, often this issue is unavoidable.

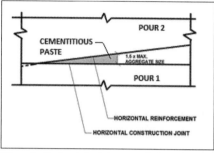

Figure 6. Horizontal reinforcement at construction joint

Base Slab Installation

Base slabs cast in multiple lifts due to thickness or temperature control requirements must act as a composite flexural member. A positive connection should be provided between lifts to transfer horizontal shear. While dowels may be used, a more efficient concept would be to utilize standees, or reinforcement cage chairs. Standees should be considered similar to ties providing shear reinforcement. Surfaces need to be clean, free of laitance and roughed to ¼" amplitude in accordance with guidelines for surface preparation provided in American Concrete Institute (ACI) 318 or 350. A shear key may be used to provide support for the base slab connecting it to the shaft. It is recommended to have the top of the base slab be a minimum 1-ft above the top of the shear key to account for required concrete consolidation within the keyway. The depth of the keyway depends on the required bearing area. It is recommended to utilize keyways, when possible, as opposed to a more rigid connection. Depending on the diameter of the shaft, the moment transferred to the SOE may be high and inefficient towards reinforcement requirements on a temporary structure. In the temporary condition, where the SOE provides support for the base slab, the shear design should

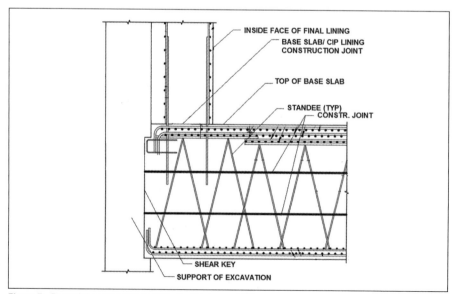

Figure 7. Base slab connection to SOE, final lining and lifts

be based on the height, or keyway height, applicable, for maximum direct shear at the face of the support. In the permanent condition, the shear plane is shifted to the inner face of the final lining and, therefore, the full thickness may be considered for shear capacity. Shear-friction reinforcement may be provided as required.

Waterproofing

Most projects mandate a maximum allowable ingress or dampness criterion. Examples of typical waterproofing accessories are presented in Table 3. A waterstop should be placed at every shaft permanent construction joint. It is generally preferred to not form a shear key as, often, it is mound-formed by hand and allows for water to circumvent the waterstop if done incorrectly. At the interface between the shaft lining and base slab, detailing becomes critical. Ideally, the reinforcing mat is depressed or bent around the outer perimeter to allow for the waterstop installation with required vertical clearance from the mat. However, since base slabs are often governed by the hydrostatic uplift force, it is not always efficient to do so. An option here is a retrofit waterstop as shown in Figure 8(a). At shaft penetrations, a hydrophilic waterstop, as shown in Figure 8(b), should be added to a PVC waterstop for added protection. At TBM openings, hydrophilic waterstops are recommended in addition to a grout hose to seal the interface and provide a remedial measure in case there is still leakage. Support of excavations, such as slurry walls or secant piles, maintain some degree of water tightness through the structural interlock formed by overbites.

Table 3. Shaft waterstop types

Type	Description and Purpose
Standard Waterstop	Made of PVC and available in a variety of sub-categories: ribbed, ribbed with center bulb or flexible 'U'. Generally used at construction joints
Retrofit Waterstop	Made of PVC with flat bottom anchored into concrete and a vertical ribbed piece embedded in subsequently cast concrete.
Hydrophilic Waterstop	Made of chloroprene rubber, the waterstop expands up to 8 times when exposed to water. Placing it within a slight groove will help against displacement when concrete is cast on top or against.
PVC Membrane	Membrane placed against a smooth surface and compartmentalized via water barriers. For example, it would be placed against SOE prior to casting the permanent lining. Water barriers would outline sheets of membrane, welded together, with a prescribed square footage to not diminish the post-grouting capability in case of leakage.
Re-injectable Grout Hose/ Contact Grout Tube	PVC Hose system embedded in concrete to allow grouting seal at construction joints.

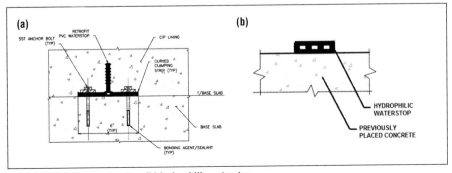

Figure 8. (a) Retrofit waterstop; (b) hydrophilic waterstop

Table 4. Instrumentation types and considerations

Type	Purpose
Inclinometer	May be installed either within the SOE or just outside of it (unexcavated side). It is used to monitor and measure movement of wall elements and the surrounding ground.
Multiple Point Extensometer	Installed adjacent to SOE to measure vertical movement (settlement) of the surrounding ground.
Earth Pressure Cells	Installed adjacent to SOE to measure earth pressure or pressure due to fresh concrete placement for slurry walls or secant piles.
Concrete Stress Cells	Installed within the SOE and used to measure stress in concrete.
Vibrating Wire Strain Gauge	Installed on steel reinforcement to measure strain and to used estimate concrete stresses.
Vibrating Wire Piezometer	Installed adjacent to SOE to measure hydrostatic pressure.

Instrumentation and Monitoring

Any deep shaft construction should include robust instrumentation and monitoring. Prior to construction, baseline readings must be taken and instruments calibrated. Instrument readings are particularly important for mitigating risk to any adjacent structures, but also critical for construction activities that may arise: a slurry wall panel or secant pile may not be constructed with the proper overlap, an incorrect concrete strength may be used due to poor quality control, or an unforeseen penetration may be required within the shaft. While undertaking any construction phase mitigation/design modification it is critical to understand earth and water pressures, differences between current readings and baseline readings and their relation to design load considerations. For example, while a slurry wall panel or secant pile may have deviated more than allowable, for the temporary system, earth pressures of groundwater readings may be lower than anticipated required less bearing. These conditions should not be relied upon, but are tools to utilize. Table 4 presents typical instruments that may be utilized for shaft construction.

CONCLUSION

The design process of a deep underground shaft involves a multitude of considerations. The constructability, while not entirely predictable, may be mitigated through careful consideration of anticipated issues and utilizing available data to efficiently adapt the design to suit means and methods. The general recommendations presented are adequate for shaft sizing and additional considerations for shaft type. However, limitations on lift lengths due to temperate control or concrete truck availability, attachments for shaft access, TBM launch thrust block and frame layout among other construction issues should be considered in any final design.

REFERENCES

1. American Concrete Institute (ACI) 318—Building Code Requirements for Structural Concrete.

2. American Concrete Institute (ACI) 350—Code Requirements for Environmental Engineering Concrete Structures and Commentary.

3. Castelli, Raymond and James Parkes. "The Use of Permanent Tiedown Anchors for Underground Structures." American Society of Civil Engineers. Geotechnical Frontiers, 2017. 536–545.

Merry Christmas!—Emergency Repair of the PCI-12A Interceptor Collapse in Macomb County Michigan

Nicholas Kacynski ▪ FK Engineering Associates
Fritz Klingler ▪ FK Engineering Associates
Zachary Carr ▪ FK Engineering Associates
Evans Bantios ▪ Office of the Macomb County Public Works Commissioner
Louis Urban ▪ Anderson, Eckstein and Westrick, Inc.

ABSTRACT

On the morning of Christmas Eve, 2016, a family awoke to creaking and cracking noises throughout their home as a massive sinkhole began to daylight under their basement, as well as under the adjacent 5-lane thoroughfare and a major utility corridor serving over a million people. The cause laid 65-feet beneath the ground surface as an 11-foot diameter sewer interceptor that services most of Macomb County had collapsed. Following Christmas Eve and Christmas Day evacuation and efforts to protect the public; measures were immediately enacted which lead to emergency construction of a 150 cfs (95 MGD) pump station and bypass, a 30' × 320' × 70' deep recovery shaft, grouting to stabilize the collapsing sewer, and relining of over 3700 feet of failing interceptor. This paper will provide a case history of this monumental repair, including discussion of the many challenges, such as emergency evaluations using conventional and remote geophysical exploration; design and bidding the emergency repair, addressing numerous unknowns during the repair, and completion of this $75 million repair within about 10 months from the failure.

SYSTEM OVERVIEW AND HISTORY

The Macomb Interceptor Drain (MID) is located within Macomb County in the State of Michigan. The MID flows into the Oakland-Macomb Interceptor Drain (OMID) forming the Oakland-Macomb Interceptor System (OMIS). The MID was constructed by the Detroit Water and Sewerage Department (DWSD) in 1970. It is referred to as the PCI-12A sewer, corresponding to the original construction contract number and is also referred to as the Romeo Arm Interceptor. Since the original construction, ownership of the interceptor was transferred to the Macomb Interceptor Drain Drainage District. The Macomb County Public Works Office (MCPWO) oversees maintenance, repairs and capital improvements on the system. Figure 1 depicts the communities served by the interceptor upstream of collapse.

The 11-foot diameter interceptor was originally constructed between 1971 and 1973 using a two-pass tunnel system.

Figure 1. Aerial image of PCI-12A interceptor repair

The primary liner consisted of steel rib and wood lagging, with the secondary liner consisting of 16 to 20-inch-thick unreinforced concrete that was cast in place. In general, the interceptor was constructed at the interface of two soil layers, with silty sand encompassing the lower portion and a stiff clay mantle above extending to the ground surface.

The collapse observed on December 24, 2016 was the third collapse of the 11-foot diameter interceptor within the Romeo Arm reach. The first collapse occurred in 1978 as a contractor attempted to construct a drop manhole. Difficulties were experienced with a high-water table and a large amount of sediment was allowed to enter the interceptor, removing all confining pressure around the sewer lining, leading to collapse. In 2004, a second collapse occurred approximately 1,000-feet downstream of the 2016 collapse. An independent investigation revealed that the 2004 collapse mechanism was similar to that of 1978, except that the loss of ground was through an infiltrating crack or joint.

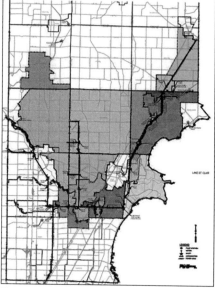

Figure 2. Macomb County communities serviced by PCI-12A

PHASE ONE: EMERGENCY RESPONSE

Upon receipt of the distressed resident's 911 phone call on Christmas Eve morning, City of Fraser officials were notified, who in turn contacted their municipal engineer, Anderson Eckstein and Westrick (AEW), to investigate.

By 11AM that morning, it became clear that the cause of the sinkhole was the PCI-12A sewer, and by noon, the MCPWO assembled the core of an emergency response team, with key personnel from MCPWO, AEW and FK Engineering, Inc. (FKE). The focus of the initial plan was to immediately formulate an emergency response plan. Time was critical, and the consequences great—as sewage level readings throughout the collection system were rising upstream due to the collapsed portion of the interceptor blocking flow. The rising sewage level upstream meant that hundreds of thousands of basements could be flooded with sewage during Christmas morning. On top of that, electric, gas, and water service to tens of thousands of homes were cut off, as the area's most critical utility corridor was engulfed by the growing sinkhole. To make matters worse, the County's 911 service relied on a fiber optic cable that extended through the sinkhole area. By late afternoon on Christmas eve, hundreds of emergency responders, engineers, construction workers, utility workers, and public officials were struggling to gain control of the increasing disaster.

Sewer monitoring points were set up at key locations along the 20-mile length of the system, and contractors and heavy equipment were mobilized. By 10PM on Christmas Eve, manhole cones were removed at strategic upstream locations and large pumps were on-site and being prepared to be placed into the interceptor, in an attempt to pump the increasing backup into surface drains and rivers. By 11:30PM, as the team fought to prevent widespread basement flooding (among other critical issues), the interceptor upstream was surcharged about 37 feet, and just inches from flooding thousands of basements in the lower lying areas of the sewer service area. At 11:50PM on Christmas Eve, as crews were still struggling to get emergency pumping in place, the blocked portion of the sewer below the sinkhole washed free, and the backwater dropped about 25 feet in a matter of minutes. The threat of sewage flooding hundreds of thousands of basements was narrowly averted, for the time being. Merry Christmas!

Figure 3. Sinkhole, December 28th, 2016

Emergency Pumping to Surface Waters

In the hours and days that followed, emergency pumping was established at multiple locations, which initially provided the ability to discharge about 15 cfs (10 MGD) of flow into nearby rivers and streams (later increased to about 25 cfs (16 MGD)). Emergency pumps were installed as far as several miles upstream of the collapsed portion of the interceptor. The locations were chosen based on the depth of the interceptor below the ground surface at each location and the proximity of manholes to county drains, where the sewage could be discharged. At each location, a pump was installed, along with a generator. Catch nets and disinfection systems were put in place in the county drains downstream of each discharge location to minimize solid refuse discharge and to sanitize the flow following any event.

Short Term Bypass

Following the initial emergency response and installation of emergency pumping capacity to pump into surface waters if necessary, the majority of the teams' focus lay in stabilizing the interceptor to prevent further damage and developing and implementing a short-term bypass plan (to be followed by long term bypass and repair). It was quickly determined that about 40 cfs (26 MGD) of sewage was, at that time, flowing through the collapsed portion of interceptor (enough to carry most of the dry weather flow), it was not clear how long this situation would last due to the threat of soil migration and further collapse. Additionally, it was very clear that the sewer would be unable to handle any wet weather flow, and certainly could not carry peak wet weather flow of 150 cfs (95 MGD).

The threat of even a small rain event loomed, as almost any event would result in raw sewage being pumped into surface waters, together with widespread basement flooding.

Within 5 days, short-term bypass pumping was developed, capable of bypassing about 15 cfs (10 MGD) of flow around the collapsed portion of sewer. Within a few more days, the short-term bypass was enhanced to be able to carry about 50 cfs (32 MGD) of flow. As part of this plan, pumps were installed in the two manholes immediately upstream of the collapse, and 32-inch High Density Polyethylene (HDPE) bypass piping was fused around the clock to connect the bypass pumps with discharge locations up to ½ mile downstream.

Long Term Bypass

As a short-term bypass system was established, and flow around the collapsed portion was restored work began on the development and construction of a long-term bypass system. An existing flow control structure, designated CS-3, was located approximately 4,000 feet upstream of the sinkhole, and was designated to operate as a long-term bypass pumping station. The goal of the long-term bypass system was for the pumping station to be able to bypass the 150 cfs (95 MGD) capacity of the system (the flow associated with a 25-year/24-hour rain event) and satisfy the MDEQ permit requirement. To accomplish this feat, seven 25 cfs (16 MGD) pumps were installed, and dual 48-inch diameter HDPE bypass pumping lines extending approximately 6000-feet along 15-Mile Rd were constructed. To power these large pumps, generators were initially used. Over time, the pumps were transitioned off of the generators and onto the electric grid, with generators on standby. A bulkhead was installed downstream of the pumps and all flow through the collapsed reach of the interceptor was blocked. A screen gate was used immediately upstream of the pumps to catch solid debris in the sewage flow and prevent clogging of the pumps.

Figure 4. Completed long-term bypass pumping station at CS-3

Figure 5. Dewatering well installation

Dewatering

Within hours of mobilizing to the site, an emergency headquarters was established in the community center immediately adjacent to the collapse area, and all available records were brought to the site. It was immediately apparent that the general ground condition through which the interceptor was built, was fine sand and silt, with groundwater level about 35 feet above the sewer invert. On that basis, it was clear that the sinkhole would continue to grow, as groundwater washed sand and silt into the sewer system. As such, the groundwater level needed to be lowered as soon as possible. As the initial stages of the dewatering plan were implemented in about a week, close attention was given to how the site would develop, to ensure that all wells would remain operational and out of the way of future excavation and repair efforts.

Drilling rigs worked around the clock to drill three-foot diameter, 90 feet deep gravity wells into the sand layer beneath the interceptor. One-foot diameter well casings were installed into the bores and the annulus was filled with 2NS filter pack. Pumps, up to 50 horsepower in size, were installed and activated. All in all, 29 active dewatering wells were installed, with an average combined discharge of over 2,500 gpm. Over the course of 10 weeks, the water level was drawn down from an average elevation of 585, corresponding to 27-feet below the ground surface, to an average elevation of 548, about 64-feet below the ground surface, at which point it was below the interceptor invert. Groundwater removed was routed to two 12" header pipes, located on each side of 15 Mile Rd. The flow from these pipes was then discharged into separate holding (frac) tanks, where any fines within the removed water could be monitored. Constant monitoring of the individual wells was performed, to ensure that pumping excessive fines were quickly identified and addressed. Several such wells were sleeved, and one well had to be abandoned throughout operation. Initially, all 29 active dewatering wells were powered by generators, transitioning to line power over time. The dewatering system served a secondary purpose as well; after the ground stabilized the interceptor would need to be removed and replaced, requiring a drawn-down water table for construction.

Compaction Grouting

In the first few days following the collapse of the sewer, the sinkhole on the surface grew considerably and formed into a nearly 300-foot long depression over the interceptor alignment. As installation of the dewatering wells was still in progress and the subsequent dewatering would take a substantial amount of time, a more immediate action was needed to stymie the expansion of the sinkhole, and to prevent a zipper-collapse of the sewer upstream and downstream of the collapse. It was feared that as time passed, additional ground loss would occur upstream, removing the confining pressure around the unreinforced concrete interceptor and compromising the intact interceptor. The engineering team implemented a low mobility grouting (compaction grouting) program along the interceptor alignment.

The planned compaction grouting was performed beginning at a depth of approximately 75-feet below the prevailing grade on the site, beneath the invert of the interceptor, and extended vertically to 35-feet below the ground surface. Primary and secondary compaction columns were used, with tertiary columns installed based on the results of the primary and secondary columns. Strict pressure cutoff levels were set to minimize the stress induced by compaction grouting and to prevent overstressing the walls of the interceptor.

Primary grouting locations were placed at 10-feet on center along the tunnel alignment and offset from the interceptor centerline 16-feet (9-feet from springline). Secondary grouting locations were placed between each primary location and moved closer to the interceptor, such that the offset from centerline was 14-feet. To monitor the impact of compaction grouting on the soils surrounding the interceptor, additional inclinometers were installed to observe soil movement and vibrating wire piezometers fixed to the casing being used to detect increases in pore water pressure. Tell tales and ground monitoring points were used to detect any unwanted movement of the intact sewer and overlying ground. After instrumentation detected movement during the on-going compaction grouting activities, additional measures were implemented to further limit

Figure 6. Compaction grouting operation and soil exploration

the cutoff volume of grout and the pressure that would be considered refusal (the program was continually monitored, and adjustments were discussed and implemented in real time). Furthermore, an alert system was installed with targets placed on points of interest, to immediately detect any excessive movement. The compaction grouting program was performed at the terminal ends of the collapsed zone and was effective in slowing/eliminating the zippering expansion of the sinkhole.

PHASE TWO: CONSTRUCTION OF RECOVERY SHAFT

As long-term bypass pumping was established and the ground stabilized, attention was fixed on reaching and repairing the collapsed portion of the interceptor. From the start of the sinkhole, the repairs were largely performed by contractors on a Time and Materials (T&M) emergency response basis, as decisions were being made and activities implemented without design drawings and specifications. However, once the immediate emergency had been abated, the engineering team began to develop a comprehensive plan set that could be released for bidding before spring. The primary component of this second phase was the construction of a recovery shaft and the replacement of the collapsed sewer with glass fiber reinforced polymer mortar (GFRPM) pipe. Secondary components consisted of cementitious grouting of the interceptor liner upstream of the collapse (taking advantage of the access now available), and removal of debris downstream of the collapse. Based on a topographic survey of the surface depression of the sinkhole, an estimated 4,400 cubic yards of material was lost into the sewer system. The contractor was tasked with the momentous job of removing this soil and debris by manual and conventional methods to restore the downstream interceptor.

Awarding the Contract

After nearly 4 months and approximately 20 million dollars expended on a T&M basis, a "Recovery Shaft" contract was awarded to Dan's Excavating (DEI) on March 3rd, 2017, for just over $35 million. Once awarded, it was a race to the finish, as an aggressive deadline of reopening 15 Mile Rd. by the end of 2017 had been set. Work immediately began on demolishing two condemned houses within the footprint of the sinkhole, followed by a 10-foot precut to the elevation where the recovery shaft would be constructed.

The recovery shaft was designed as a 3-foot diameter tangent pile shaft with 4 levels of internal steel bracing and W21× 182 beams placed in every other drilled pier. The

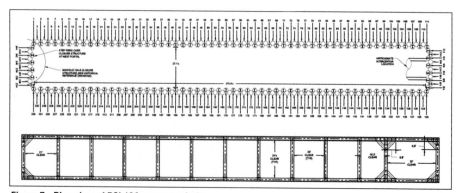

Figure 7. Plan view of PCI 12A recovery shaft

tangent pile design offered a low vibration installation method and the ongoing dewatering process meant that groundwater flow into the shaft would not be a concern.

Installation of Piers

To install the drilled piers, DEI subcontracted specialty foundation contractor Malcolm Drilling. Malcolm Drilling began their drilling operations on Friday, April 24th. The installation of the drilled piers was an intricate process, as it required many parts to be synchronized to provide an efficient process. At any given time, there was one drilling rig boring holes and advancing casing, an oscillator removing casing, a mobile crane setting pier reinforcement, and a pump truck delivering concrete to each location.

An elaborate orchestration was conducted daily as a drilling rig moved between five different drilled piers. Simultaneously a crane was setting a reinforcing beam into the drilled pier, and assisting the pump truck in placing the concrete, while the casing was being extracted by an oscillator. The operation moved seamlessly as a section of casing was removed from one completed pier and set directly on another that was being advanced. Six reinforced drilled piers had inclinometers anchored to the piles, allowing for movement to be monitored upon excavation of the recovery shaft.

Figure 8. Drilling of piers

While Malcolm advanced ahead of schedule, the operation was not without hurdles. Due to the nature of the sinkhole, there were locations where a substantial amount of concrete was lost out of the pier as the casing was extracted. It was initially theorized, and confirmed during excavation, that large voids remained present within the soil strata surrounding the sinkhole. Up to an additional 17 cubic yards of concrete was lost in some locations, nearly doubling the planned quantity of 23 cubic yards per pier.

On April 10, the bulkhead was installed at CS-3, signaling the true beginning of the long-term bypass pumping plan, as no flow was coursing through the constricted interceptor. With the flow stopped, the interceptor was set to be inspected by manned entry upstream of the collapse. Initially, steel ribs were to be installed at the east end of the temporary earth retention system (TERS), proceeding upstream, to provide additional support during the installation of the drill piers as they were stepped over the interceptor. However, after the manned entry into the interceptor on April 25, the condition of the remaining interceptor was found to be much worse than originally thought. Due to numerous fractures within the secondary liner of the interceptor, it was decided to extend the length of the TERS by 20-feet to encompass this compromised portion of the interceptor. The timing of this decision posed a significant challenge. As of April 30th, Malcolm had installed and completed over half of the planned piers, and while the expansion of the shaft only required the installation of an additional 12 piers, it also required the procurement of an additional 6 steel beams to be set in the piers. As a cost cutting and schedule enhancing measure, the owner had pre-ordered and procured the 122 originally planned steel beams, prior to awarding the contract. The designed W21× 182 beams were not readily available, and to wait for production would have postponed the project. Instead, the owner opted to use larger reinforcement that was readily available. The scramble at the end lead to steel beams arriving on site and being placed into their designated piers within that same hour. To aide in the progress of the job, field representatives were in constant contact with the testing laboratory to test concrete cylinders to verify compressive strength minutes before drilling began on adjacent piers.

Over the course of 32 days, from April 24th to May 25th, Malcolm drilling had installed 256 piers, completing the walls of the TERS that was yet to be constructed. The task was completed well under the projected duration even with the expansion of the shaft. It was a significant milestone for the project, as work to excavate down towards the interceptor would follow shortly.

Excavation of Recovery Shaft

Before Malcolm demobilized from the site, DEI was placing crane mats around the drilled piers to provide a surface for heavy duty, off-road dump trucks to transport removed material. On June 6th, an excavator was positioned in the east end of the TERS footprint, and routinely worked from east to west quickly ladling out the stiff clay soil and loading dump trucks.

As both the contractor and engineer desired to advance the project quickly, DEI was allowed to excavate past the prescribed depth of 6 feet, two feet beneath the first level of walers, without the installation of struts. Struts were cut to length and set aside for rapid installation if needed. The 6 inclinometers placed into the drilled piers were monitored on a daily basis and survey points were established on piles prior to excavation. As the excavation progressed, the drilled piers cantilevered nearly 16-feet. After observing appreciable movement in an inclinometer, which was confirmed by survey points, DEI was directed to install the first level of struts prior to proceeding. Excavation continued, and bracing was threaded between struts to be installed at lower levels.

Figure 9. Excavation of recovery shaft

The same process continued until approximately 8-feet above the level of the interceptor, where the excavator could no longer reach the excavation bottom. At this point, mini-excavators and muck-buckets were used to haul out soil, significantly slowing the excavation process. After the installation of the third level of bracing, the top of the interceptor was reached. As excavation proceeded and demolition of the collapsed interceptor progressed, the full collapse portion was exposed. The north side of the secondary concrete liner had curled in on itself at springline, creating a massive fracture that was present for approximately 200-feet upstream from the west end of the shaft.

Lining of the Interceptor

Due to the severely compromised state observed during the manned entry and in consideration of the available access, the engineers determined to line the entire 3,700-foot stretch of sewer from the collapsed area to CS-3. This was a substantial addition to the scope of the already executed DEI contract. MCPWO had two options to proceed: (1) They could elect to release a separate bid and manage the two concurrent contracts. This would have had several downfalls, as there would be two prime contractors on one site, each working to complete their portion of the work under separate schedules; or (2) the owner could create a three-party contract between themselves, DEI, and the selected contractor for the lining. DEI would remain the prime contractor and a revised schedule would be created that worked for all three parties. MCPWO chose the three-party agreement and released a set of plans for bid in May. Similar to the reinforcing steel used to construct the shaft, MCPWO opted again to pre-purchase the GFRPM pipe directly to save on the markup cost and provide for schedule enhancement of approximately 6 to 8 weeks.

Figure 10. Collapsed PCI 12A interceptor

Cementitious Grouting Program

As the decision to line the shaft was made, this increased pressure on the contractor to prepare the compromised interceptor for lining. Inland Water's and Pollution Control (IWPC) was working to perform cementitious grouting through the interceptor liner under contract with DEI. This effort was necessary to address potential loosened soil conditions or voids behind the host pipe that was to be lined, in order to limit future distortion of the lined tunnel in such areas.

IWPC worked from the upstream end of the lining reach and progressed towards the shaft, managing high hydrogen sulfide levels within the sewer and overcoming difficult grout injection operations. On numerous occasions, as the contractor drilled through the liner, sand was ejected like a slurry from behind the interceptor wall. Originally, the pressure to inject the grout was set at 0.5 psi per foot of depth beneath the ground surface, corresponding to approximately 35 psi at the level of the interceptor. However, it was observed on numerous occasions that the volume of grout pumped through a packer was less than the volume of sand ejected during its installation. Knowing that a small void existed which was not able to be filled, engineers determined it was necessary to first counteract the existing water pressure by pumping water through the grout port at a relatively high pressure of 70 psi. After the water began to be injected to the surrounding soils cement grout was pumped immediately, without allowing time for the external water pressure to flow back into the interceptor.

Progress was slow and began to threaten the lining timeline. The contractor began to work around the clock, as grout takes near the shaft were expected to be greater than those encountered further upstream. Fortunately, the anticipated large volume of grout takes near the shaft never materialized, and the schedule was not impacted. Through the months long grouting operation over 100 cubic yards of cement grout was placed through the 3,700-foot stretch to stabilize the host pipe.

COMPLETION OF REPAIR EFFORTS

After installing the fourth and final set of bracing, DEI worked to remove the soil and collapsed interceptor from the eastern end of the shaft, as this would be where the lining of the interceptor would be conducted. Jay Dee Contractors (JDC) had been awarded the lining contract, at nearly three million dollars, and a proposed schedule of 33 days (which was met, despite an unanticipated shaft flooding event).

Interceptor Cleaning Downstream of Collapse

Just 2 days before JDC's construction period was slated to start, DEI completed demolition and removal of the interceptor from their work area and poured a base slab to work from, allowing JDC to progress as planned. Simultaneous with JDC's lining, DEI worked to remove the remainder of the soil and debris from the shaft and allow cleaning of the downstream eight-foot diameter interceptor through the west portal of the TERS.

The downstream 8-foot diameter interceptor was found to be nearly full of sand from the collapsing of the interceptor. The material needed to be hand-mined out over 100-feet to allow for an additional cementitious grouting program of the eight-foot interceptor. Removing material by hand was slow, and soon became the critical path. In response, the cleaning contractor, Doestch Environmental, fabricated a specialized mini-excavator that would fit within the eight-foot pipe to remove soil. Cleaning operations continued slowly but steadily as the remaining base slab was poured within the shaft by DEI. On August 22nd, a near catastrophe occurred as sewage from the downstream interceptor backed up (essentially blowing out the collapsed sand material that was acting as a plug) and flooded the recovery shaft. Numerous pieces of equipment, including JDC's proprietary GFRPM pipe carrier, were damaged due to the flooding.

Figure 11. Completed recovery shaft

Ironically, the flooding of the interceptor was a blessing in disguise. With the flooding, a large volume of soil from the 8-foot interceptor was eroded by the sewage, into the recovery shaft. The volume present in the 8-foot interceptor after the flooding was sufficient to allow for flow from the 25-year 24-hour storm and did not require further substantial cleaning downstream (for now). A temporary structural bulkhead was installed approximately 120-feet downstream from the west portal of the shaft and Doestch continued cleaning efforts from the bulkhead to the west portal of the shaft, exposing apparently fresh fractures within the reinforced concrete liner. Due the new discovery, FK Engineering recommended applying a fiber reinforced cementitious liner after the completion of the cement grouting program.

Restoration of Flow

The next weeks marked significant milestones on the project. After Labor Day weekend, JDC completed the lining upstream of the Recovery Shaft, and DEI was installing the GFRPM pipe within the shaft, working alternately from both sides. On September 9th, DEI and JDC worked together to place the final piece of GFRPM pipe within the recovery shaft, marking the most significant milestone on the project yet. The following day, DEI's subcontractor, GEO-cell, began filling the shaft with cellular grout. While the cellular grout was placed around the anchored interceptor on the outside, spray lining of the downstream interceptor was occurring from within. By September 18th, cellular grout covered the GFRPM pipe and spray lining downstream was complete. From a structural standpoint, the repair was complete. Reinforcement of the bulkhead at CS-3 was removed, as was the downstream temporary structural bulkhead.

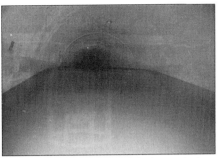

Figure 12. First flow of sewage in the repaired PCI-12A interceptor

On September 22nd, there was a collective sigh of relief as the bulkhead at CS-3 was lifted and flow restored to the interceptor for the first time in 8-months. With the restoration of flow the perpetual threat of an environmental and economic disaster was averted.

REFLECTIONS

This project was monumentally complex, but in the end, was a great engineering success, in terms of completing the repair without significant discharge of sewage to the environment, without basement flooding, and without injury to the public or workers. Were this to have been a planned repair, the design and permitting process alone may have taken years to complete, before construction could even be started. The extraordinary efforts by the contracting group, owner and engineers in an effort to serve and protect the community from disaster were commendable. Each company, public servant, contractor down to the individual man and woman gave more than ever expected to ensure success. Utilizing local contractors and engineers, this endeavor was a community repairing a community. The residents of Fraser and Clinton Township suffered the most but were appreciative of the efforts to minimize impacts and restore their community by the MCPWO, Engineers and the Contractors.

MCPWO has been very proactive in the inspection and maintenance of other portions of Macomb County's underground infrastructure. An inclusive multichannel analysis of surface waves (MASW) exploration has been conducted, highlighting potential areas of concern. Coupled with the MASW program, investigatory borings were performed along sewer alignments to analyze soil conditions, and CCTV footage was taken to identify areas of distress. MCPWO is currently working to develop a design for the rehabilitation of several thousand feet of interceptor downstream of the collapse. And though with the completion of this project the interceptor is no longer in peril, it is certainly not out of mind for MCPWO.

The Role of Mechanized Shaft Sinking in International Tunnelling Projects

Andrea Fluck • Herrenknecht AG
Peter Schmäh • Herrenknecht AG

ABSTRACT

Tunnelling projects require shafts, either as start and reception shafts for the tunnelling process or for inspection, ventilation and rescue purpose. The current trend towards installations in growing depths is driven by deep sewers to avoid pumping stations and the need to build new installations below existing infrastructure.

Inner-city shaft structures demand safe working principles for surrounding buildings and environment, especially regarding potential settlements. In addition, deep shafts need special attention for the safety of the operating personnel. The VSM method procures safety for the surrounding environment and for all personnel. As the water level in the shaft equals the groundwater level outside the shaft, there is no water flow which can cause ground movement. All installations including the lining erection are completed from the surface. No personnel have to enter the shaft until it has reached the final depth and is fully secured. The lining consists of either precast segments or cast in place concrete. As the lining installation is completed on the surface, a high quality installation can be reached. In most cases, a secondary lining is not required.

The first VSM equipment was put into operation more than 15 years ago. Today, nearly 80 shafts of up to 85 m depth have been sunk worldwide using VSM. This paper will describe the role of shaft sinking in tunnelling projects along with all necessary safety and planning aspects when VSM is being discussed.

INTRODUCTION

Almost all tunnelling projects require shafts, either as start and reception shafts for the tunnelling process or for inspection, ventilation and rescue purposes. Also, a current trend towards infrastructure installations in growing depths can be observed. It is driven, among other things, by deep sewer construction projects that aim to avoid pumping stations as well as the need to build new installations below existing infrastructure. Our paper discusses the benefits of the mechanized shaft sinking technology and presents a selection of worldwide references from a variety of tunnelling projects.

The Vertical Shaft Sinking Machine (VSM) was originally developed by Herrenknecht for the mechanized construction of deep launch and reception shafts for microtunnelling. After starting design and testing in early 2004, the first Herrenknecht VSM equipment went into operation in Kuwait and Saudi Arabia in 2006. The machine concept, fully remote-controlled from the surface, as well as its implementation on site proved to be an efficient solution right from the start for the safe and fast realization of shafts especially in difficult, inner-city environments without lowering the groundwater table. To date, approx. 75 shafts have been successfully installed worldwide with the Herrenknecht VSM technology, reaching depths of up to 85 m. They serve today, for

Figure 1. Overview of VSM applications, from left to right: Ventilation/emergency shaft, microtunnelling shaft, sewage collector shaft, U-Park® shaft

example, as ventilation shafts for metro systems, maintenance or collector shafts for sewage, or as temporary microtunnelling shafts (Figure 1).

BENEFITS OF VSM TECHNOLOGY

The VSM technology masters the main challenges associated with shaft sinking: inner-city shaft structures demand safe working conditions for surrounding buildings and the environment, especially regarding potential ground settlement. There is an increased requirement to avoid the lowering of the groundwater during the construction of shafts in order to avoid the associated settlement, which can affect a wide area. Deep shaft construction companies often encounter difficult geological conditions such as high groundwater pressure combined with layers of hard and soft material. In addition, deep shafts need special attention for the safety of the operating personnel.

Cost and Time Benefits

Simultaneous excavation and ring building facilitate high advance rates and shortened overall project duration. At the same time, continuous performance ensures overall high planning reliability for all stakeholders.

The lining of a VSM shaft consists of either precast segments or cast-in-situ concrete. As the lining installation is completed on the surface, high quality installation can be accomplished, leading to greater accuracy of the shaft structure. In most cases, a secondary, time-consuming lining is not required, resulting in a reduced wall thickness of the shaft and, thus, less soil excavation.

Furthermore, each VSM type is very flexible as its excavation diameter can be adjusted within a specific range. A VSM10000, for example, can cover an inner shaft diameter range from 5.5 m to 10.0 m and is, therefore, a one-time investment for multiple use.

Construction and Occupational Safety Benefits

As the water level in the shaft is maintained close to the groundwater level outside the shaft, water flow is prevented which otherwise could cause ground movement and lead to a high risk of settlement. The Herrenknecht VSM can be applied below groundwater with a hydrostatic pressure of up to 10 bar and in heterogeneous soil and hard rock of up to 140MPa compressive strength.

All installations, including the lining erection, are remotely controlled from the surface. No personnel have to enter the shaft until it has reached the final depth and is fully secured. In general, mechanized shaft sinking requires less personnel and machinery on site, which leads to minimized risk exposure.

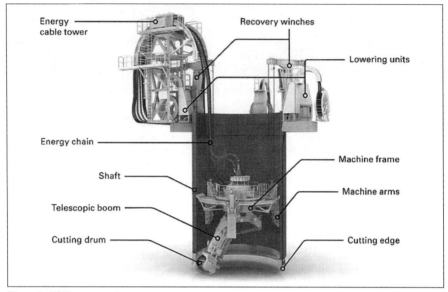

Figure 2. VSM components

Environmental Benefits

Measures for groundwater lowering are not necessary, as the VSM machine concept is designed for operation under groundwater. As the VSM technology applies a high degree of accuracy of shaft construction, the shaft lining thickness can be reduced to a minimum, which reduces the amount of excavated soil.

VSM MACHINE COMPONENTS

The VSM consists of two main components (Figure 2): the excavation unit and the lowering unit. The excavation unit systematically cuts and excavates the soil and consists of a cutting drum attached to a telescopic boom that allows excavation of a determined overcut. The lowering unit on the surface stabilizes the entire shaft construction against uncontrolled sinking by holding the total shaft weight with steel strands and hydraulic jacks. When one excavation cycle is completed, the complete lining can be lowered uniformly and precisely.

A slurry discharge system removes the excavated soil and a submerged slurry pump is located directly on the cutting drum casing. It transports the water and soil mixture through a slurry line to a separation plant on the surface. The whole operation takes place from the surface and is controlled by the operator from the control container on the surface. All machine functions are remote-controlled without the necessity to view the shaft bottom or the machine. Power supply for the submerged VSM is secured by the energy chain. After reaching its final depth, the VSM is lifted out of the shaft by the recovery winches and the jobsite crane.

VSM JOBSITE PREPARATION

Jobsite Layout

Depending on the space conditions on site, the VSM components can be positioned flexibly to suit local circumstances (Figure 3). As most sites are located in heavily

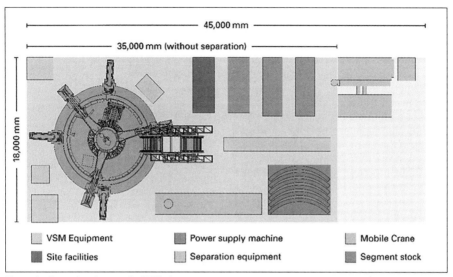

Figure 3. Exemplary layout of a VSM site

built-up urban areas, the access for logistics, e.g., trucks, ring segment stock or soil disposal, is limited. Special concepts to relocate components such as the separation plant already exist for this purpose, and can be discussed if required.

Ring Foundation

After preparation of the required site surface, a concrete ring foundation has to be installed in a pre-excavated pit. This foundation bears the loads of the VSM and serves as a support for the lowering units, which guide and hold the shaft at all times. The size of the ring foundation depends on the ground conditions and the size of the shaft. Connection bolts for various VSM components such as lowering units, recovery winches and energy winch tower are also integrated into the ring foundation.

Cutting Edge

After pre-excavation and ring foundation, the installation of the cutting edge and the first segment rings are the next steps. The cutting edge is designed to cut the shaft profile in soft and loose soil conditions. Its design depends on the shaft diameter and wall thickness. It can be integrated as the first concrete segment ring or welded onto the shaft lining as a separate steel ring.

Start Section and Machine Attachment

The first five meters of the shaft lining constitute the so-called start section (Figure 4). The start section has a stronger steel reinforcement to be able to take the loads and reaction forces of the machine during excavation and shaft sinking. Furthermore, the shaft lining is equipped with cast-in steel plates, onto which the brackets for the machine arms of the VSM are welded. As the VSM employs a sequential partial face excavation technique there is no torque transmitted into the shaft structure.

INSTALLATION OF VSM EQUIPMENT

The excavation unit arrives on site in three parts: the telescopic boom with the cutting drum, the machine main body, and the adapter parts to the required shaft internal diameter. The lowering unit consists of the strand jacks and the coiled steel strands on a drum. The number of strands depends on the total predicted weight of the shaft including the machine weight and the estimated buoyancy and friction forces. The strand jacks are bolted to the ring foundation by anchor bolts. Coming from the strand drum, the strands are fed through the strand jack, lowered through the outer annulus of the shaft wall and connected to the cutting edge. When all strand jacks are installed and connected to the cutting edge the strands can be tensioned and carry the loads.

Figure 4. Completely installed start section with attachment brackets

Now, the preassembled excavation unit can be lifted into the start section. The VSM is secured by hydraulically activated locking bolts. When the VSM is in place, the recovery winches are installed and connected to the three arms of the machine. The recovery winches are used to recover the VSM for required maintenance or for final machine recovery.

Next, the energy chain tower is installed and the excavation unit is connected to the hydraulic and electrical supply as well as to the feed and discharge lines. The energy chain tower with its winch has bolted connections for easy assembly.

As a final step, all the electrical and hydraulic connections are done, and the equipment is now ready to operate. Before starting the excavation, a calibration of the VSM in reference to the projected alignment of the shaft is required to ensure the accurate action of the cutting boom.

VSM OPERATION—SHAFT SINKING PROCEDURE

Excavation is completely remote-controlled from the operator cabin at the surface. Stored data, together with the position of the cutting boom, is shown on a graphic display, giving the operator full control of the excavation and sinking process. The excavation unit can be operated in three different overcut options, which requires the installation height measured from the level of the cutting edge to be adjustable. In its highest position, the excavation unit is not able to create an overcut under the cutting edge, which is important in soft or unstable soil to maintain surrounding stability. When an overcut is required, e.g., in stable or cohesive soil, the excavation unit works in its lowest position. In this case, the annulus should be stabilized by a bentonite-water suspension. In addition, each segment can be equipped with bentonite nozzles for lubrication, which can also later be used to grout the

Figure 5. VSM excavation unit working below groundwater

annulus. The standard installation is to connect the segments in ring number 3 and 5 with the bentonite mixing unit right from the beginning and to lubricate from these two segment rings at the shaft bottom during the sinking. The stabilization of the annulus together with the controlled sinking of the shaft by the strand jacks minimizes the risks of settlement.

During the excavation and sinking process the shaft is kept full of water to balance the level of the groundwater table in the surrounding geology (Figure 5). The cutting drum cuts and crushes the material to a granular size that can be handled by the pumps (pump capacity: 200–400 m^3/h). A slurry circuit transports the excavated material from the shaft to a separation plant on the surface.

The telescopic boom allows varying diameters: flexible use is possible up to 12 m and up to 14 m with a reinforced frame structure. Diameters of up to 18 m are realized through an enlarged concept. Excavation is possible even in water depths up to 85 m. The cutting arm moves radially from the center to the outside of the shaft with an additional telescopic extension of 1 m. With a rotation of ±190° the cutting boom covers the whole cross section of the shaft. The cutting speed and the movement of the boom can be varied to achieve the best excavation rate.

Shaft Lining

In most cases, the shaft lining consists of precast concrete segments installed at the surface. This so-called ring building is comparable to segmental lining in tunnelling. The ring is built at the surface by crane. The number of segments depends on the shaft diameter. Ring building work includes the proper connection of the rings by anchors and bolts, which can be handled from outside the shaft. The excavation process of the VSM is not affected by the ring building process. This increases the shaft sinking performance significantly.

Alternatively, in-situ concrete casting of the shaft walls is another solution, especially for larger shaft diameters where segment handling becomes more difficult. In this case, the progress of shaft construction works is slowed down by the necessary time to build the formwork and the setting time of the concrete structure. The benefit of in-situ casing is the "continuous" structure without joints and the possibility to integrate entire entry and exit structures, e.g., for microtunnelling activities in the shaft walls.

COMPLETION OF THE SHAFT

After reaching the final depth, the bottom plug has to be installed. Usually, the VSM is used to excavate the required overcut. When this final excavation is done, the VSM can be disconnected, recovered by the recovery winches and lifted up to the surface and out of the shaft by a crane. The bottom plug is cast with underwater concrete. In a next step, the shaft annulus is grouted through the lubrication lines to stabilize and anchor the shaft to the surrounding ground. Finally, the shaft water can be pumped out and the shaft is completed (Figure 6). Personnel access is now possible.

Figure 6. VSM reaching final depth

Figure 7. Confined space conditions in Girona   Figure 8. VSM site in Naples

DEEP SHAFT APPLICATIONS AND REFERENCES

Ventilation and Emergency Shafts

Girona and Barcelona, Spain

For the high-speed rail link from Barcelona to the French border, a total of four shafts were built in Girona by a Herrenknecht VSM as ground stabilization shafts before tunnelling works (4 shafts, ID 5,250 mm, depth 20 m). The same VSM built one additional shaft in Barcelona for ventilation and as an emergency exit (ID 9,200 mm, depth 47 m).

The major challenge for the construction contractors were the extremely confined working conditions. For example, one of the shafts in Girona was located between two rows of houses with a spacing of only 12 m (Figure 7). Here, the VSM's ability to work under limited space conditions in inner cities proved to be a major benefit.

Due to the lack of ground stability in the center of the city of Girona, only small, lightweight cranes could be used and this led, in turn, to a complete reorganization of assembly logistics: The main components of the VSM10000 were delivered just in time and assembled directly in the shaft start section. The average daily performance was 3.0 m.

Naples, Italy

A total of 13 ventilation and emergency shafts (ID 4500/5500, depth up to 45 m) for the subway line were sunk in Naples, Italy, by a Herrenknecht VSM. The site was located in a densely built-up area in the inner city with high traffic (Figure 8). The required jobsite footprint was approx. 300 m^2 in the narrow streets of Naples. Noise exposure for the residents had to be kept at a low level. Because the excavation of the shafts and their lining with precast concrete segments could be realized simultaneously, the production of the shafts could be finished quickly with performance rates of up to 5 m per day. Due to its modular setup, the VSM was rapidly disassembled and transported to the next site after completing one shaft.

Upcoming: Grand Paris Express

In August 2018, a Herrenknecht VSM will be installed on a shaft sinking site in France for the first time. In the context of Grand Paris Express, currently the largest infrastructure project in Europe (200 kilometers of automatic metro lines, circulating around Paris), a VSM12000 will sink emergency and ventilation shafts for the Line 15 South tunnels excavated by Herrenknecht tunnel boring machines. Four shafts are to be

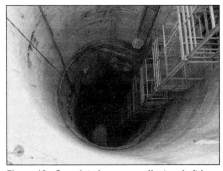

Figure 9. VSM site in Honolulu

Figure 10. Completed sewage collector shaft in St. Petersburg

constructed with inner diameters of 8,300 mm, 10,300 mm and 11,900 mm and depths of up to 53 m.

Microtunnelling Shafts

Hawaii, USA

Two large shafts with a 10 m inner diameter were sunk in Honolulu, Hawaii. These 36 m deep shafts were to be used as launch shafts for a pipe jacking project (Figure 9). Cast-in-situ was the preferred lining method in order to handle the necessary thrust forces in the shaft wall when launching the pipe jacking machine. Moreover, fiberglass reinforcement simplified the launch process for the TBM. In Hawaii, the VSM successfully handled a challenging geology comprised of hard basalt as well as coral that would have been problematic for conventional methods. Best daily performance with the Herrenknecht VSM was 2.3 m.

Sewage Collector Shafts

St. Petersburg, Russia

The deepest VSM shaft under groundwater to date was sunk in St. Petersburg, Russia, where a total of four shafts were realized to depths ranging from 65 to 83 m. In St. Petersburg, the Herrenknecht VSM technology proved to be especially efficient in the face of tight time schedules. Together with the customer, Herrenknecht assembled the machine in nine days following site preparation, and successfully finished the first shaft of 83 m depth and an internal diameter of 7.7 m in 50 working days (Figure 10).

Launch Shaft and Sewage Collector Shafts

Upcoming: DTSS Phase 2 Singapore

In the autumn of 2018, the first VSM project in Asia will see the use of VSM technology for the Deep Tunnel Sewer System (DTSS) Phase 2 in Singapore with a total of approx. 100km of main sewer tunnels and link sewers. Seven shafts with inner diameters of 10 and 12 m will be sunk down to depths of up to 60 m. A project-specific feature will be the combination of segmental lining in the upper section for fast construction progress and in-situ concrete casting in the lower section for the connection of the tunnel. The shafts will be used first as launch shafts for the tunnelling operation and later as collector shafts.

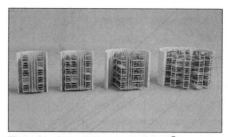

Figure 11. Shaft sizes used for U-Park®

Outlook: U-Park® Shafts

Especially in large cities, new parking concepts have to be developed because space above ground is extremely built-up and expensive. Therefore, new parking solutions are being designed that make use of underground space. One of them, called U-Park®, was conceptualized as a combination of VSM technology for creating shafts and automatic parking systems, which are accommodated in these shafts (Figure 11). The number of parking lots per system depends on the diameter and depth of the shaft.

SUMMARY

The current and worldwide trend to construct more and more infrastructure underground, e.g., metro, road and railway as well as a large variety of utility lines, promotes a growing demand for the construction of shafts.

With increasing depth and groundwater levels, conventional shaft construction methods reach their technical and economical limits. Herrenknecht has developed the solution: the Vertical Shaft Sinking Machine (VSM). Its efficiency and benefits in terms of budget, construction time and occupational safety have led to a total of approx. 75 VSM projects with a total depth of 3,8km where shafts have been successfully sunk e.g., in inner-city environments with tight space contraints and a requirement to avoid all settlement. Since its first design in 2004 and its first deployment in 2006, Herrenknecht has continuously developed the VSM machine design to a proven technology with a growing range of applications.

South Hartford Conveyance and Storage Tunnel Project—Drop and Vent Shaft Construction

Andrew Perham ▪ The Metropolitan District
Jim Sullivan ▪ AECOM
Scott Jacobs ▪ Case Foundation
Mark Careyva ▪ Case Foundation
David Belknap ▪ Black & Veatch
Clay Haynes ▪ Black & Veatch

ABSTRACT

The South Hartford Conveyance and Storage Tunnel (SHCST) is a major component of the Hartford Metropolitan District's Clean Water Project (CWP). The tunnel will capture and store Combined Sewer Overflows (CSO) from the southern portion of Hartford, CT and Sanitary Sewer Overflows (SSO) from West Hartford and Newington, CT. The project includes 21,800 ft of 18 ft final diameter tunnel, several miles of consolidation sewers, eight hydraulic drop structures and a 50 mgd tunnel dewatering station. This paper describes the challenges and methods used to construct the drop and vent shafts.

PROJECT DESCRIPTION

The purpose of the SHCST project is to eliminate West Hartford and Newington SSO, eliminate Franklin Area CSO discharging to Wethersfield Cove and to reduce CSO discharges to the South Branch Park River.

The SHCST will collect and store CSO and SSO during wet weather events and ultimately convey the overflows to the Hartford Water Pollution Control Facility (HWPCF) for treatment and discharge to the Connecticut River. The eight hydraulic drop shafts which are the subject of this paper are relatively small diameter shafts varying from 2.5 ft to 6.0 ft finished diameter that were constructed by a Bauer BG-39 drill rig in the soil section and a Wirth pile top drill rig in the rock section with exception of the DS-RS shaft which used a down hole cluster drill in the rock section of the shaft. These hydraulic drop shafts will convey flows from near surface combined sewers and sanitary sewers to the 18 ft finished diameter tunnel constructed below in the bedrock. The overall layout of the tunnel system is provided in Figure 1.

GEOLOGIC SETTING

The project site lies in the Central Lowland physiographic province that extends in a north-south direction in the middle of the state of Connecticut. The central lowland area consists mainly of the sedimentary rock and the associated igneous basalts of Triassic and Jurassic age. The Hartford Basin of Connecticut and southern Massachusetts is a half graben structure, 90 miles long, and filled with approximately 13,000 ft of sedimentary rocks, and basaltic lavas and intrusions (Hubert, et al., 1978). The drop and vent shafts are being constructed entirely within the Portland Arkose which consists of mainly siltstone with some shale and sandstone.

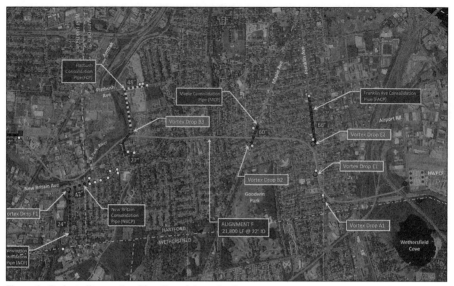

Figure 1. Project alignment and location of drop structures

The region has undergone periods of glaciation that has greatly influenced the soil overburden. Glaciers laid down a heterogeneous layer of ground-up rock (glacial till). This till layer is present over much of lower lying bedrock surfaces. The sediments of the Glacial Lake Hitchcock filled in the deeply-incised Connecticut River Valley. The lake deposits are present in varying forms from Rocky Hill, Connecticut to northern Vermont. Glaciers shaped the topography and left the area with much of the topographic relief present today. More recent alluvial deposits are common along the Connecticut River and Park Rivers and their tributaries.

In the project area, the following soils are present overlying the bedrock, in general order of sequence from the ground surface downwards: artificial fill, alluvium, glaciolacustrine deposits, glaciofluvial deposits and glacial till.

GEOTECHNICAL CHALLENGES

The geotechnical challenges associated with construction of the shafts were as follows:

- The glaciolacustrine deposits are extremely weak and unconsolidated sediments deposited in a glacial lake bottom. These deposits are up to 30 ft thick in the area of four of the eight shafts. These formations required the following characteristics of the initial support system in the shafts.
    - These extremely weak deposits have no ability to stand after excavation. Thus, the initial support system must be installed before removal of water or drilling fluids from the excavation of the soil sections of the shafts.
    - These unconsolidated deposits will compress and settle if dewatered or disturbed. Thus, the initial support system has to be nearly water-tight to prevent soil consolidation if the shaft were dewatered before installation of the carrier pipe.

- The artificial fill deposits have a tendency to contain manmade obstructions such as concrete and wood. Soil with these obstructions can be difficult to drill.
- The alluvium deposits tend to be coarser grained soils that readily convey groundwater. Although the alluvial deposits could be dewatered with wells, this approach seemed incompatible with pre-excavation support methods.
- The glacial till deposits have cobbles and boulders which makes excavation difficult for drilling machinery.
- The glaciofluvial deposits were only present at the western end of the project. The glaciofluvial deposits have a tendency to contain cobbles, boulders and wood debris which makes excavation difficult for drilling machinery.
- The bedrock is layered and dipping to the east. This condition has a tendency to cause drills to deviate from vertical while drilling.
- The depth to bedrock varied from less than 20 feet to nearly 115 feet over the alignment of the tunnel. This required multiple installation methods for the initial support casing.

DESIGN OF INITIAL SUPPORT IN THE SOIL SECTION OF THE SHAFTS

Due to the relatively unstable nature of the soil, the design engineer specified that a steel casing was to be used as initial support in the soil section of the shafts. A final liner or carrier pipe was designed inside of the steel casing initial support. From the geotechnical information obtained during design, the design engineer specified that the steel casing be socketed at least five feet into the bedrock to reduce the potential for soil intrusion into the shaft if drilling fluid were lost rapidly during excavation. It was determined from geotechnical investigations of the bedrock that rock support would not be needed in these shafts below the rock socket. The design engineer required the steel casing to be designed for a full hydrostatic load of groundwater from the ground surface to the bottom of the casing. The contractor submitted for this requirement to be waived because they intended to install the carrier pipe in the "wet." However, the design engineer felt that it was prudent to retain the original design criteria in the event of rapid drilling fluid loss. To promote the complete encapsulation and backfill of the carrier pipe, the design engineer specified a minimum annular space thickness.

Due to the challenging soil conditions and lack of current and relevant experience in the local marketplace, the designer required minimum relevant project experience for the contractor to construct the shafts.

PERTINENT DETAILS OF EACH OF THE SHAFTS

A summary of the shaft parameters is included in Table 1.

BASIC OVERVIEW OF THE DROP SHAFT SYSTEM DESIGN

There are many publications in the tunnel industry that describe the different drop shaft systems for tunnels. It is not the purpose of this paper to summarize the advantages and disadvantages of the many different drop shaft systems.

During preliminary design, the design engineer selected a vortex drop system (with exception of DS-RS as noted in Table 1) based on many different design system requirements. A general rule of thumb for the vortex drop systems is the drop shaft shaft diameter is 0.5D and the vent shaft is 0.25D where D is the diameter of the influent sewer. As noted in Table 1, the Owner wanted 5 ft diameter vent shafts to provide

Table 1. Shaft parameters

Shaft	DS-RS	DS-1	DS-2	DS-3	DS-4	DS-5	DS-6	DS-7
Total Depth (ft)	216	192	200	232	199	197	197	175
Soil Depth (ft)	75	113	99	21	38	18	20	79
Rock Depth ft	141	79	101	211	161	179	177	96
Excavated Diameter (ft) in Soil	6	7	7	8	9	7	8	8
Excavated Diameter (ft) in Rock	4	6	6	7	8	6	7	6
Final Shaft Liner Diameter (ft)	2.5	4	3.5	4.5	6	3.5	4.5	4
Air Vent Diameter (ft)	N/A	5	5	5	5	5	5	5

Notes:
1. Five ft diameter vent shafts were provided in the deaeration chambers except for DS-RS which vented to the retrieval shaft. Vent shaft diameter was driven by Owner access requirements to the deaeration chambers.
2. Some of the excavated diameters of the shaft exceeded minimum requirements due to the size of the available rock cutter heads and to account for verticality tolerances.
3. The typical plan and cross section of the drop shaft and vent shaft is depicted in Figure 2.

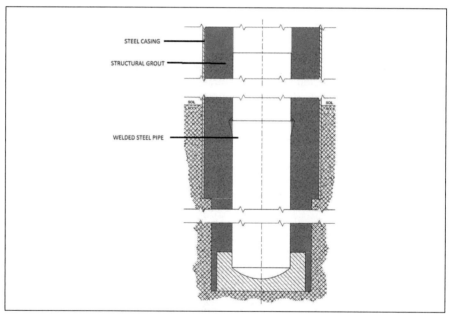

Figure 2. Cross section of the drop shafts and vent shafts

maintenance access to the deaeration chamber. Thus, the vent shafts were much larger than needed based on air venting requirements only. Figure 3 shows the various components of the typical vortex drop system.

SELECTION OF VERTICALITY TOLERANCES

During design development, the design engineer strove to develop reasonable verticality requirements for the drop shafts and vent shafts that struck a reasonable balance between constructability and hydraulic performance. Using reasonable care, a drilling contractor can start the drop shaft and vent shaft excavation within an inch of theoretical center line at the ground surface. The contractor's challenge is maintaining the position of the drop shaft and vent shaft relative to each other as the drilling progresses downward to the deaeration chamber. If the drop shaft and vent shaft deviate substantially at the deaeration chamber interface, this can result in construction problems and rework.

From a hydraulic performance perspective, the drop shaft should be as close to vertical as reasonably possible. However, engineering judgment would suggest that the vortex drop shaft is relatively insensitive to variations in verticality. Verticality requirements for the vent shafts were driven primarily by the access man cage configuration.

Thus, the verticality tolerance for the project was primarily driven by the deaeration chamber construction considerations. A verticality tolerance of ±6 inches from top of shaft to bottom of shaft and all points between was established for the project. During construction, the design engineer did consider slight variations from this requirement on a case by case basis.

OVERCOMING CONSTRUCTION CHALLENGES

The soil excavation phase consisted of drilling the soil strata column and five feet into the bedrock surface. The soil excavation went relatively well.

Figure 3. Typical drop shaft system

Shafts DS-3, DS-4, DS-5 and DS-6 were constructed in a relatively shallow soil column varying from 18 to 38 ft in soil thickness. A Bauer BG-39 drill rig was used to excavate the soil column and five feet into the bedrock in Shafts DS-3 through DS-6. Additional rock drilling below the casing was required with the BG-39 drill rig at DS-3, DS-5 and DS-6 to reach the required depths for insertion of the Wirth Drill String and for proper performance of the reverse circulation drilling system. No significant problems were encountered while drilling these soil columns and rock sockets. Drilling the ground, installation of the steel casing and backfill grouting of the annular space between the steel

Figure 4. Bauer BG-39 drilling the soil section of DS-6

Figure 5. Wirth cutter head

casing and the ground was completed in approximately one week each for Shafts DS-3 through DS-6.

Shafts DS-RS, DS-1, DS-2 and DS-7 were constructed in a deeper soil column varying from 75 to 113 feet in soil thickness. The Bauer BG-39 drill rig was used to excavate the deeper soil column and five ft rock socket. These excavations took longer to excavate averaging about two calendar weeks to complete. The most significant challenge occurred at DS-1 where drilling was inadvertently terminated before socketing five feet into solid bedrock. This was rectified before drilling the full rock column by the insertion of a steel sleeve that was grouted into the bedrock and the lower few feet of the soil casing.

Figure 6. Wirth drill rig at DS-4

The rock sections were drilled with a Wirth reverse circulation drilling machine in the following order: DS-7, DS-4, DS-2, DS6, DS-3, DS-5 and DS-1. DS-7 was drilled with no significant problems within the specified verticality tolerance of plus or minus 6 inches from top to bottom of the shaft. However, when drilling DS-4, Case discovered it was drifting out of tolerance at approximately 100 ft below grade requiring the drill to be removed and the shaft backfilled with concrete. This deviation was not expected as the Wirth drilling method has a proven track record of drilling straight holes through the pendulum effect created by the use of the 75+ ton bottom hole assembly and a series of tight stabilizers on the drill string. After redrilling the shaft through the concrete, DS-4 was eventually drilled to depth within specified tolerances. The DS-2 shaft was drilled in accordance with the original work plan and within the verticality tolerances.

Due to the difficulties associated with achieving the required verticality tolerances at DS-4, Case decided to change the work plan for the remaining shafts that had longer lengths of rock to drill (DS-3, DS-5 and DS-6). Case decided to subcontract drilling of pilot holes with a down hole hammer drill. While still achieving the desired results, the layering of the rock and inherent dipping also caused greater than expected deviations with the pilot hole drilling. A custom built Wirth Bit Plate with a stinger attachment was used to follow the pilot holes and guide the cutter head during drilling DS-3, DS-5 and DS-6. This provided assurance that the DS-3, DS-5 and DS-6 bores would be constructed within verticality tolerances.

Due to its relatively small size (4 ft diameter), DS-RS was drilled with a cluster drill with air-powered down hole hammers that was attached to the BG-39 drill and utilized reverse circulation methods. A pilot hole and stinger attachment were also used for DS-RS to maintain the specified verticality tolerances.

In addition to creating difficulties for maintaining verticality, the siltstone bedrock also proved to be challenging to process the rock cuttings. A custom closed loop processing system was designed for the project to process the cuttings from the Wirth with a series of shakers and de-sanders. However, the siltstone rock was broken down into silt and clay sized particles during drilling and could not be removed with conventional shakers and de-sanders. This required the use of a centrifuge with polymer injection to remove these silt and clay sized particles from the drill fluid to provide a clean shaft for placement of the final liner.

CONCLUSION

Case completed all of the drop and vent shaft excavations successfully within vertical tolerances well ahead of completion of the TBM excavation. The BG-39 drill rig performed well drilling the soft ground section, rock sockets and the rock section of DS-RS using a cluster drill attachment. The Wirth drill rigs performed well drilling the rock sections of seven of the shafts although there was a tendency for them to drift out of verticality tolerances on the shafts with the longer rock sections.

It is recommended that detailed physical modeling of drop structures be performed to more accurately define hydraulic performance of the structures at various inclinations from vertical. Computational fluid dynamics (CFD) is an excellent tool for modeling hydraulic performance of drop shafts but CFD modeling experts disagree on how verticality of the drop shaft affects hydraulic performance. Physical modeling should help answer these questions more definitively.

The addition of pilot holes and stinger attachments on relatively deep sections in rock should be considered where verticality is important. Alternatively, the minimum excavated shaft diameter in rock could be increased to account for verticality problems.

REFERENCES

Hubert, J; Reed, F.; Dowdall, W. and Gilchrist, J. 1978. Guide to the Mesozoic Redbeds of Central Connecticut, Guidebook No. 4, Connecticut Geology and Natural History Survey, Hartford, CT.

Nasri, Verya; Bent, William; Hogan, William 2015. South Hartford CSO Tunnel and Pump Station, Rapid Excavation and Tunneling Conference 2015, New Orleans, LA, June 7 to 10, 2015, pp. 140–151.

Structural Design of Large Diameter Shafts for the Coxwell Bypass Tunnel

Gorki Filinov • R.V. Anderson Associates Limited
Tyler Lahti • R.V. Anderson Associates Limited

ABSTRACT

To reduce wet weather flow into Lake Ontario and the Don River watershed, the City of Toronto is implementing the Don River and Central Waterfront Wet Weather Flow Project. The project involves roughly 22km of tunnel, 27 connections and 12 shafts and has been split into five phases that are expected to be completed over the next 25 years. Black & Veatch in association with R.V. Anderson Associates Limited were contracted to perform the detailed design of the first phase of the project, called the Coxwell Bypass Tunnel, which includes 10.5km of 6.3 m inner diameter rock tunnel and five large diameter storage shafts within the Georgian Bay Shale rock formation.

The five storage shafts range in diameter from 20–22 m and are up to 58 m deep extending through both soil and rock. The shafts feature a composite roof slab design consisting of precast concrete panels and a cast-in-place concrete topping to aid in constructability. Unique unbalanced soil loading conditions were evaluated at one site to accommodate a planned future development. Large hydrostatic water pressures at the base of the shaft led to the use of an arched base slab design to help resist the uplift forces applied over the long spanning shaft diameters. This paper will provide a comprehensive overview of the unique design conditions and challenges encountered during the detailed design of the shafts, a brief background on the analysis techniques utilized, and a summary of the resultant final shaft design.

INTRODUCTION

Authors' Note

The goal of this paper is to provide a general overview of the procedures followed during the detailed structural design of large diameter shafts for the Coxwell Bypass Tunnel. Design approaches, formulae, analyses and calculations presented herein are provided for information—they do not encompass all carried out components of the design and are not intended to be used as guidance.

Project Background

Over the next 25 years, the City of Toronto will be implementing a multi phase 1.5 billion dollar project aimed at controlling wet weather flow runoff into local tributary water systems and Lake Ontario. The first phase of the project is currently under construction and comprises of a 10.3km long, 6.3 m diameter rock tunnel called the Coxwell Bypass (refer to Figure 1). In addition to improving water quality, the tunnel will provide redundancy and future capacity to the city's existing Coxwell Sanitary Trunk Sewer, which serves roughly 30% of the city's population and currently has no redundancy to convey flows to the city's Ashbridges Bay Treatment Plant. The Coxwell Bypass tunnel will be accompanied by five large storage shafts, ranging from 20–22 m in diameter and reaching depths of up to 58 m below grade. Table 1 provides a detailed

Structural Design of Large Diameter Shafts for the Coxwell Bypass Tunnel

Figure 1. Tunnel alignment and shaft locations for the first phase of the project

Table 1. Summary of below grade conditions

Shaft	Diameter, m	T/Grade, m	T/Weathered Rock, m	T/Sound Rock, m	Tunnel Inv., m	Depth, m
CX-1(A)	22	91.9	59.3	57.2	39.05	52.9
NTTPT-1	20	87.7	61.0–66.7	60.0–65.0	36.22	51.7
BB-1	20	82.0	70.5	62.3	31.91	50.1
LDS-3(B)	20	79.0	65.9	65.2	26.88	52.1
IHES-2(B)	20	78.0	63.3	59.0–63.3	23.49	54.5

breakdown of each shaft's size parameters. The tunnel and shaft system will provide temporary storage for large influx volumes of stormwater and combined sewer overflows (CSO) caused by heavy rain and snow melt events. These stored volumes will then eventually be treated at a new high rate treatment facility planned for Ashbridges Bay Wastewater Treatment Plant and subsequently discharged into the lake.

Special Design Considerations

Shaft Site Location

The five shafts are spaced at approximately equal lengths along the tunnel alignment (refer to Figure 1). Due to the significant distance between shaft sites, unique design conditions exist including, but not limited to variances in: hydraulic grade line of the system, soil stratigraphy, rock elevation, ground water elevation, surrounding future developments, regional flood levels, and above grade conditions. Although similar methodologies and procedures were utilized for the structural design of all shafts, each one had to be thoroughly analyzed to account for the many discrepancies between sites and their effect on the applied loads.

Operational & Maintenance Requirements

Operational and maintenance requirements had a significant influence on the structural design of the shafts. Ventilation was required to accommodate the large exhausting of air that would occur during the filling of the storage system. These vents had to be placed above the regional flood plain to isolate the system from flood waters, which at some shaft sites were several meters above the surrounding grade and top of shaft elevation. The ventilation openings were designed to serve a dual purpose and provided access for maintenance. Access stairs were requested by the city's operation staff to aid in their ability to access and maintain the shafts and tunnel over their service life. These large scale openings complicated the roof slab design, thereby necessitating unique solutions such as the use of customized precast roof panels elements.

Crack Control

Due to the harsh exposure conditions expected in sanitary sewage, CSO flows, and stormwater, the reinforced concrete design was checked in accordance with the more onerous crack control requirements from the ACI 350 design code. This design approach was taken to maximize the long term durability of the concrete. Crack control limits were relaxed for concrete components that were not expected to be in direct contact with stormwater, such as the roof slabs.

Subsurface Conditions

All shafts will extend through both soil and rock profiles, with the majority of each being embedded within the underlying bedrock. The soil subsurface varied in stratigraphy, but was often comprised of fills, sand, silt and occasional clay materials. The top of the Georgian Bay Shale, a rock formation that extends across much of the Greater Toronto Area, ranged anywhere from 11.5–32.6 m below grade. The sound shale bed rock was typically located another 0.5–2.0 m below the weathered surface. Geotechnical investigation reports indicated that ground water elevations were found to be a few metres below grade. Table 1 provides a summary of the below grade conditions at each shaft site.

CODES, LOADS, LOAD COMBINATIONS & DESIGN PARAMETERS

Numerous codes were consulted throughout the structural design of the shafts. Primarily, Canadian codes and standards were adhered to, however codes from the American Concrete Institute (ACI) were used to supplement the Canadian codes for special applications. The main codes and references that were utilized during design are as follows:

- Ontario Building Code (OBC) 2012
- CSA A23.1—Concrete Materials and Methods of Concrete Construction
- CSA A23.3—Design of Concrete Structures
- CSA A23.4—Precast Concrete—Materials and Construction
- CSA S6—Canadian Highway Bridge Design Code
- PCA—Circular Concrete Tanks Without Prestressing
- ACI 350—Code Requirements for Environmental Engineering of Concrete Structures
- ACI 224R—Control of Cracking in Concrete Structures

Load factors and load combinations for strength design were developed from the approach taken in the OBC 2012. Design coefficients and material unit weights were applied as per the requirements of applicable CSA codes and the geotechnical investigation report recommendations for each shaft site, respectively.

DETAILED DESIGN

Roof Slabs

Due to the depth and large diameter of the shafts, one of the key goals for the design was to utilize a precast concrete system to aid in the constructability of the shaft covers. This approach eliminated the construction time and effort associated with scaffolding the entire depth of the shaft, or erecting elaborate falsework to pour a conventional cast-in-place roof slab. This goal was a big challenge to the design, as several shafts were subjected to very high surcharge loads. To limit the mass of the precast concrete elements, the shaft roof was designed as a composite section consisting of custom precast concrete panels with a 200 mm thick cast-in-place bonded structural concrete topping. The use of a cast-in-place topping will aid in the long term durability of the roof slabs.

The roof slab spans varied among the five shafts anywhere from 15 m, where intermediate baffle walls were present, to 20 m where no internal walls were present. For some shafts, large live loads had to be considered during design. In particular, NTTPT-1 was located within a regional flood plain and was subjected to a hydrostatic live load of 33.4kPa. Another shaft, LDS-3(B), was located within a large development site and would be buried below future roads subjecting it to vehicular traffic loads and soil overburden loads. These types of live loads were considered in combination with concrete deadloads and phased construction loads under both serviceability and ultimate limit states.

The design of the composite roof slabs became an optimization problem in which the end goal was to reduce the concrete thickness of the precast panels so that they could be placed with a crane, while ensuring the following criteria were met:

- Enhanced crack control requirements due to the harsh exposure conditions
- Sufficient section bending and shear capacity
- Immediate deflection and long term creep requirements

In order to accomplish this, a high density of steel reinforcement was used. The precast panels were designed individually under immediate construction live loads and as composite sections, in combination with the cast-in-place concrete topping, under the more severe long term loads. The interface between the precast panels and cast-in-place concrete topping had to resist the resulting shear flow loads. CSA A23.3 was referred to for shear interface design. It provided guidance with regards to roughening the interface and using shear reinforcement ties to provide a sufficient bond between the precast panels and cast-in-place concrete topping.

Ventilation, stair access, and large vehicle access openings had to be provided within the roof slabs, adding a degree of complexity to their design. Where possible, these openings were strategically positioned in areas of low bending moments to reduce reinforcement requirements. Larger width precast panels were used to accommodate the openings and ensure sufficient section capacity in the highly stressed regions around them. For the LDS-3(B) shaft in particular, where applied loads and a full 20 m span called for 900 thick precast panels, wider custom panels could not be used due to excessive weight and lifting concerns during construction. As a solution, two side by side standard width precast panels were designed to be entirely removable. The

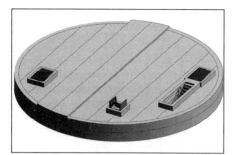 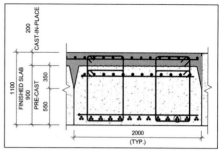

Figure 2. Precast panel arrangement and cross section through composite slab for LDS-3(B)

precast panel arrangement and cross section details for LDS-3(B) can be seen in Figure 2.

Lifting hooks for precast panels were designed using standard steel rebar embedded into the concrete. After the application of a 1.25 impact factor to account for dynamic loads during lifting and a 1.5 live load safety factor, the factored tensile load of the largest precast panels amounted to 2000kN, or roughly 200 tonnes. The lifting hooks were designed accordingly for concrete breakout, concrete pullout, development length, tensile strength and shear strength based on CSA code formulas and requirements.

Shaft Walls

The critical load cases considered during the design of the shaft walls were internal and external water pressures in combination with unbalanced soil loading conditions that varied in severity based on the shaft's surroundings. All loads were considered under ultimate limit states for strength design and serviceability limits states for crack control requirements. Time dependant rock swelling loads provided by the geotechnical investigation reports were also evaluated, however, these did not to govern the design.

The wall thicknesses were preliminarily sized using formulas from the PCA method for design of circular tanks, which are aimed at limiting the concrete stress under direct tension to prevent excessive cracking. Upon determining a suitable wall thickness, circumferential/ring steel design was performed and found to be governed by the internal water pressure load case within the overburden region. This load case neglects the confinement provided by the external soil mass as an allowance for any planned future excavations adjacent to the shaft sites. Below the top of rock elevation, the effects of the confinement provided by the rock mass were considered in the design to counteract the internal water pressure loading, allowing for a reduction in reinforcing steel. Figure 3 depicts the circumferential tensile axial force results under the internal water pressure load case. The CSA crack control parameter formula referenced in CSA A23.3 was used in combination with the more onerous environmental exposure crack limits provided by ACI 350 to determine the allowable steel stress under direct tension service loads. Based on this allowable stress, circumferential steel was sized and spaced appropriately to ensure there was enough capacity within the walls to resist the calculated tensile ring forces.

Concrete compressive capacity was determined by CSA A23.3 and checked against the maximum factored compressive forces calculated near the base of the shaft walls. This force was due to the external water pressure load case applied along the full height of each shaft wall. Figure 3 shows a visual of the resulting circumferential

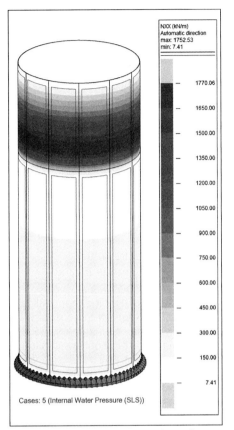 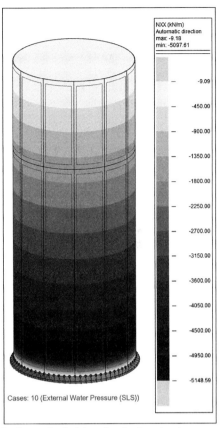

Figure 3. Circumferential axial tension and compression forces due to internal and external water pressure load cases for LDS-3(B)

compressive axial forces under this load case. The predetermined wall thickness was found to be sufficient against the maximum compressive ring forces calculated.

Using Autodesk Robot Structural Analysis, shell Finite Element Models (FEM) of the shaft walls were generated. In order to simulate the stiffness of the rock support below the respective top of rock elevation for each shaft, compression only springs were applied to the wall panels. Lower bound spring stiffness coefficients of 250 kPa/mm were provided by the geotechnical engineer and inputted into the model's parameter settings. To account for likely variations in rock stiffnesses, an upper bound (very stiff) rock coefficient was also considered during design in order to gauge the adverse effects on the outputted model results. All relevant load combinations were applied to each model under both ultimate and service limit states. Within the shaft wall load combinations, an unbalanced soil load was typically considered to account for possible future construction or connections being made directly adjacent to the shaft. For shaft LDS-3(B), which would be located immediately adjacent to a future large scale development, this unbalanced load consisted of a soil excavation that extended down to the top of rock elevation and could be applied along any extent of the shaft's perimeter. This unusual loading requirement was necessary due to specific negotiations between the city and the developer. Model results were checked against hand calculations to confirm their suitability for continued use. Circumferential bending moments,

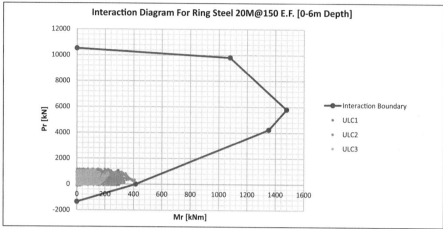

Figure 4. FEM circumferential bending moment & axial force results under unbalanced soil loading conditions for NTTPT-1 (1,968 nodal results plotted between 0–6 m depth of the shaft wall)

shear force and axial force results at each finite element node were exported into a spreadsheet and sorted for each applied load case. The bending moments and axial tension/compression forces at each point along the height of the shafts' walls were plotted against the moment-axial interaction diagram for the reinforced section at that location (refer to sample graph shown in Figure 4). This analysis confirmed whether the previously detailed circumferential reinforcement was sufficient. If any plotted points exceeded the boundaries set by the moment-axial interaction diagram of the designed wall section, additional steel reinforcement was provided. The same procedure was followed for the longitudinal moments and forces along the shaft walls.

Base Slabs

At a depth of 52 m for the critical shaft, LDS-3(B), the base slab was subjected to an approximate factored hydrostatic uplift pressure of 760 kPa. A design uplift pressure of this magnitude over a large 20 m diameter span length produced significant bending and shear forces that were difficult to reinforce for using a conventional slab design. In an attempt to reduce the applied bending forces, a unique shallow arch base slab design was utilized. A curved base slab has a smaller cross section than a conventional flat base slab, which allows for greater deflection to occur. This increased deflection allows a compression arch to develop within the slab and transfers a percentage of the pure bending stresses to compressive stresses within the section. The newly introduced compressive axial force component counter acts the applied bending within the section, thereby reducing the reinforcement requirements under the applied loads and creating a more efficient design.

Both axisymmetric and shell FEMs were created using RS2 & Robot Structural Analysis software. Model results were verified with hand calculations based on flat circular plate moment and shear formulas. The approximate hand calculated results were within a 10% margin of error of the model results, confirming the suitability of the model. Using the principles of solid mechanics, the calculated model stresses were converted to bending moments, axial forces and shear forces along the length of the slab. Sample model results for radial stresses, along with calculated bending moments and axial forces can be found in Figure 5 and Table 2, respectively. Upon

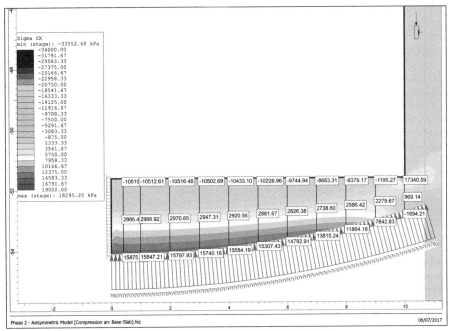

Figure 5. Axisymmetric FEM radial stress results for typical shaft base slab

Table 2. Approximate calculated factored bending moments and axial forces

Distance Along Radius, m	0	1	2	3	4	5	6	7	8	9	10
Approx. Mf, kNm	13,742	13,729	13,120	12,356	11,280	9,891	8,220	6,244	3,965	1,392	−1,944
Approx. Nf, kN	6,706	6,668	6,460	6,224	5,875	5,476	5,061	4,683	4,429	4,433	8,661

determination of the bending moments and axial forces, moment-axial interaction was employed and the appropriate steel reinforcement was provided across the slab.

The final design consisted of bundled bar reinforcement in both directions along the top face of the slab. With such high reinforcement requirements, an orthogonal reinforcement pattern was used, as radial and circumferential bar placement could have led to rebar congestion and placing difficulties during construction. In addition to the high bending forces, the slab was subjected to large shear forces under the applied loads. Headed shear reinforcement studs were provided circumferentially to provide sufficient capacity to the concrete section.

Wall-Base Slab Joint

Due to the intended operation of the system, the shafts will usually be empty and only fill up after a rain event. This would mean that under the majority of its life, the base of the shaft would be subjected to an unbalanced uplift pressure due to the external groundwater.

Detailed design calculations and FEM results demonstrated that under a full depth external groundwater pressure load, a fixed wall-base slab joint resulted in large bending moments near the base of the shaft walls. These forces were too large to design for using conventional reinforcement, or reasonable wall thicknesses. To mitigate the high transfer of stress at this joint, the decision was made to design and detail the joint as a pinned connection and allow for rotation. Under a pinned joint condition, the moments along the base of the shaft walls were greatly reduced and reinforcement could be provided accordingly.

Movement at this joint would also be influenced by the stiffness of the surrounding rock mass. To mimic the behaviour of the joint and its effect on both the shafts' wall and base slab designs, an axisymmetric FEM was created using Rocscience's RS2 analysis software. The Geotechnical investigation reports provided a range of rock mass properties and parameters, which were used within the model to simulate the rock-structure interaction.

At the wall-base slab interface, a roughened exposed aggregate condition was specified to maximize shear resistance between the two concrete components. A heavy duty waterstop capable of withstanding the expected deformation across the joint was specified to maintain a watertight seal.

Baffle Drop Structures

Multiple shafts had to accommodate flows from adjacent sewer systems. Baffle drop structures (as illustrated in Figure 6) were used to convey water to the base of four of the five shafts. To deal with the large turbulent volumes of water expected within the baffle structures, several precautions were taken for the design of these elements. High performance concrete of C-XL exposure class with an increased compressive strength, low water cement ratio, high quality curing and a type E finish for resistance to wear were specified per CSA A23.1. The baffle steps were designed for a factored hydrostatic load of 4 m, which equates to the vertical distance between each baffle step and is a realistic load under peak flow conditions.

It is expected that these baffle slabs will be subjected to erosion over the long term life of the structure. To combat this, additional measures were put in place to counteract these effects. An increased concrete clear cover was specified for the top layer of reinforcing steel within each baffle step, where erosion due to falling water will be most significant.

The dividing baffle walls were designed for an unbalanced 2 m water level differential along their full height to account for hydraulic losses due to obstructions, or other unusual circumstances during operation.

FINAL DESIGN SUMMARY

The final design of each of the five shafts varied due to site specific conditions. In general, the roof slab designs consisted of a composite section utilizing both a precast panel and a cast-in-place topping system. Heavy reinforcement was provided within these panels as a means to minimize the precast panel weights, while still being able to resist the large surcharge loading required at specific sites. Bundled bar reinforcement was used to limit congestion and transverse shear reinforcement ties were utilized to provide the required capacity of the composite roof slab system under both construction and long term loading. The shaft walls varied between 700 mm and 800 mm in thickness and were reinforced with longitudinal and circumferential steel bars along their full height. Below the top of rock elevation, steel reinforcement within the shaft walls could be reduced due to the confinement provided by the rock mass. To handle the applied loads within the overburden soil, circumferential steel reinforcement density was increased to counteract the high tensile stresses caused by the internal water pressure. The specified wall reinforcement assisted in resisting

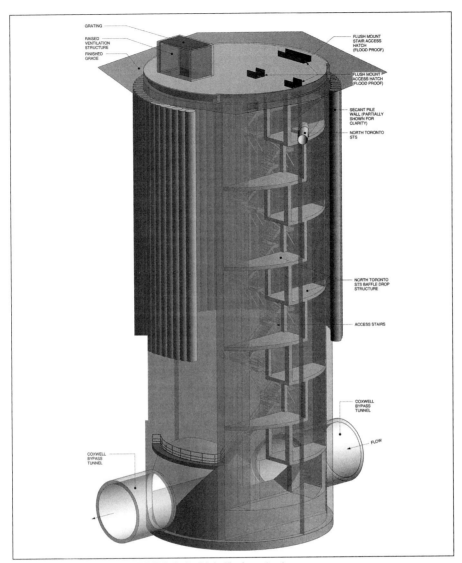

Figure 6. Isometric view of NTTPT-1 shaft with baffle drop structure

the global and local wall stresses caused by the various load combinations considered during analysis, such as unbalanced loading conditions. Shaft base slabs were designed as shallow arches to help reduce the large bending moments by allowing for a compression arch to develop within the concrete. The base slabs were 2.5 m thick at their midpoints and tapered down to a thickness of 0.9 m at the exterior wall face. Bundled bar reinforcement was used to limit congestion along the top face of the base slabs and to resist the high groundwater uplift pressures. Analysis indicated that a fixed wall-base slab joint connection produced bending moments in the base of the walls that were too large to provide a reasonable design for under the applied loads. As a result, the joint was designed as a pinned connection, which reduced the forces transmitted from the base slab to the wall and allowed for a more economical design.

PART  18

Tunneling for Sustainability

Chairs

Keith Ward
Seattle Public Utilities

Samer Sadek
Jacobs Engineering

Assessment and Remediation of Gas Utility Tunnels in Chicago

Chad Gailey • Black & Veatch Corporation
Brian Selph • Black & Veatch Corporation
Aswathy Sivaram • Black & Veatch Corporation
Cary Hirner • Black & Veatch Corporation

ABSTRACT

The Peoples Gas Light & Coke Company (PGL) owns 17 critical utility tunnels located in the City of Chicago, constructed during the late 19th century and early to mid-20th century, that house medium and high-pressure gas pipelines. The project includes assessment, inspection and remediation of these tunnels and associated pipelines. Initial assessments have been conducted to determine the priority ranking of the tunnels for inspection. This paper discusses the inspection procedures to determine if remediation is required and summarizes the findings and recommendations from the first two tunnels inspected to date.

INTRODUCTION

The Peoples Gas Light & Coke Company (PGL) owns 17 critical utility tunnels located in the City of Chicago, constructed between the late 19th century and early to mid-20th century, that house medium and high-pressure gas pipelines. A location map of the 17 tunnels is shown on Figure 1. The Tunnel Remediation Program (TRP) includes assessment, inspection and potential remediation of these tunnels and associated pipelines. The TRP has been divided into three phases for planning purposes.

- Phase 1—Initial Assessments
- Phase 2—Detailed Inspections
- Phase 3—Tunnel Remediation/Replacements

Phase 1 assessments activities included data collection, site reconnaissance, and initial inspection of all 17 tunnels. The reconnaissance work included observing surface conditions, opening the tunnel manholes/hatches/doors at each shaft, observing shaft conditions, sounding the shaft to confirm invert depth, and taking water level readings. The initial site inspection included performing a cathodic protection potential survey and environmental sampling of water in the tunnel shafts.

Phase 2 used the findings of the Phase 1 tunnel risk rankings developed using Criterium DecisionPlus (CDP) software to prioritize the tunnels for detailed inspections. The prioritized list generated the Phase 2 detailed inspections schedule. This manuscript will discuss the requirements and procedures for the ongoing Phase 2 detailed inspections and will summarize the findings and recommendations for the first two tunnel inspections. Phase 3 is just being planned and hence, beyond the scope of this paper.

The key aspects of each of the three TRP phases are listed in Table 1.

Assessment and Remediation of Gas Utility Tunnels in Chicago

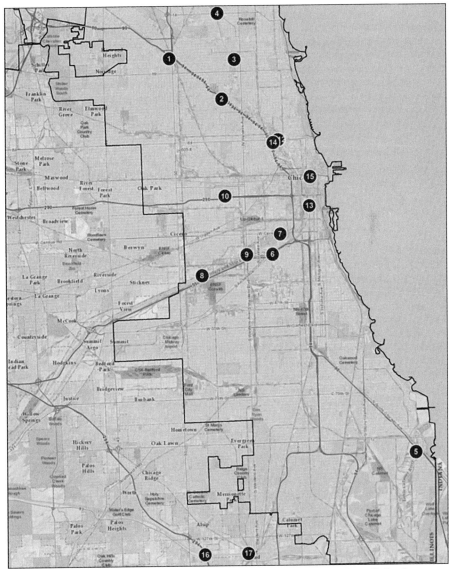

Figure 1. Location map showing 17 PGL tunnels assessed as part of the Tunnel Remediation Program

Table 1. TRP planning and delivery process

Initial Assessment	Detailed Inspections	Tunnel Rehabilitation
• Plan, initiate QA/QC/safety protocols, and scope of work • Review historical data, drawings and perform site surveys • Perform initial testing to understand and characterize tunnel water infiltration • Assess and prioritize tunnel detailed inspection schedule	• Develop detailed inspection program • Develop schedule and budget estimates, and safety plans for each tunnel site • Gather site-specific data and establish an accurate reporting framework • Develop remediation recommendations based on data gathered	• Obtain permits and develop safety and compliance plans • Develop master schedule and budget • Complete tunnel and pipeline remediation design • Assist with contract procurement and bidding; and construction phase services

PROJECT REQUIREMENTS

The project scope includes detailed inspection and documentation of the tunnel and pipeline. The inspection includes documenting the condition of tunnel, tunnel shafts, lining if present, pipeline and pipeline support, and access components to establish tunnel and pipeline condition assessment ratings. The tunnel and pipeline conditions and ratings are captured in a Tunnel and Pipeline Inspection Report.

Key milestones for each tunnel inspection include:

- Tunnel Inspection Plan finalized
- Tunnel Inspection Memorandum submitted
- Tunnel Inspection Report submitted with remediation recommendations and opinion of probable construction cost

PHASE 1 DATA—SITE LOCATION, CHARACTERISTICS, AND CONSTRUCTION LIMITATIONS

Based on the information collected during Phase 1, a detailed site location and characterization of the tunnel groundwater infiltration is estimated, cathodic protection system health is assessed, and water corrosivity is estimated. The laboratory analysis performs two tasks; ensures groundwater will meet the requirements of disposal permit and provides inputs to assess cathodic protection system remaining life, health, as well as possible corrosion rates.

Phase 1 data collection included the following items:

- Environmental water samples—Collected for laboratory analysis to characterize the water for current and future discharge needs per the Metropolitan Water Reclamation District of Greater Chicago.
- Estimation of Infiltration—Performed a modified drawdown test.
- Corrosion Potential Tests—This included testing the tunnel water for conductivity, chlorides, sulfates, sulfides, and oxidation reduction potential (ORP). The down shaft pipe-to-water potential surveys included "on" Direct Current (DC) potential measurements performed with the cathodic protection current applied and measurement of Alternating Current (AC) Root Mean Square (rms) potentials with the cathodic protection current applied. The results of this test were used to characterize the presence (or absence) of cathodic protection on the pipeline.

PHASE 2—DETAILED INSPECTION REQUIREMENTS

Permitting and Access Agreement Requirements

Permitting and access agreements vary with each tunnel location. In general, four permits and four types of access agreements are obtained.

- Metropolitan Water Reclamation District of Greater Chicago
- Chicago Park District
- City of Chicago
- Private parties

Traffic Control Plan

Depending on location, the inspection requires traffic control for lane closures and sidewalk re-routing. The traffic control setup adheres to IDOT Standard 701601-09; with traffic cones, an arrow board, and appropriate signage.

CONDITION ASSESSMENT RATING SYSTEM

Based on the conditions encountered, tunnels, pipelines and appurtenances will be classified based on reliability, support type and condition, defects, seepage, recommendation for repairs, maintenance, re-inspection intervals and other items agreed upon with PGL. Attachment A Table 1 shows the tunnel and shaft condition assessment rating system, and Attachment A Table 2 shows the pipeline condition assessment rating system.

SITE PROCEDURES

Pre-Site Inspection Activities

The following actions are required prior to any individual being allowed to enter any tunnel or shaft not normally accessible:

- Pre-site inspection activities
  - Participation in tunnel inspection planning, including formulation of the details of the safety plan.
  - Access contractor will supply emergency escape oxygen equipment, and training in the use thereof, for all tunnel entry personnel.
- Site inspection support
  - Rescue personnel and equipment positioned at the two access shafts for the duration of the inspection.
  - Inspection team will enter the shaft and tunnel, once the rapid assessment is performed and identified the tunnel and shaft as safe.
  - Inspection team will be equipped with four points harness, but will not be required to be attached to life lines. Life lines will be available at each shaft with retrieval equipment.

Emergency Safety Coordination and Communication Meeting

Inspection Safety Team Lead coordinate with City of Chicago Fire Department Battalion Chief of the Special Project Platoon, rescue staff and inspection staff following a site-specific health & safety plan.

Rapid Assessment Procedures

Prior to the detailed inspection, a rapid assessment is required to verify the shafts and tunnel are safe for detailed inspection. A rapid assessment team consisting of a Geologist and a Structural Engineer access both shafts and tunnel. Rapid assessment requires the following items to be evaluated:

- Air quality
- Tunnel and shaft integrity
- Ladder and platform integrity
- Pipeline integrity

Tunnel Access Requirements

The tunnel access requires several detailed procedures defined in the inspection plan and safety plan, including a pre-site inspection safety meeting, permits, safety equipment, and lines of communication ventilation, lighting, and tunnel access.

Entry Procedures

Confined space entry according to the OSHA guidelines of CFR1926.1200 are utilized for the rapid assessment entry and the inspection phase of the tunnel, and include underground construction CFR 1926.800 tag-in and tag-out procedures.

Confined space entry procedures proceed as follows:

- One week prior to mobilization the Chicago Fire Department Special Operations Unit is notified.
- On the day of rapid assessment, confirm with PGL's Gas Control the pipeline is at the reduced pressure of 22 psi and lock out procedure is used (only the entrant can remove the lock out).
- Daily onsite safety meeting with PGL and access contractor providing onsite entry rescue as needed for the rapid assessment and inspections.
- Complete job hazard analysis with entire team (access contractor, PGL, Black & Veatch, and onsite rescue team).
- Test of all communication lines between both attendants and the rapid assessment team.
- Setup fall protection, ventilation, test lights, calibrate air monitors, and test air horn.
- Execute confined space entry permit per the site-specific safety plan.
- Proceed with donning personal protection equipment, check fit of harness, and fall protection.
- Enter tunnel and perform rapid assessment.

Additional Entry Controls

Shaft are guarded by fencing to ensure employees and equipment have safe means to access. Adjacent areas are secured to prevent loose materials from endangering shaft access areas.

Fall Protection

Fall protection procedures and requirement are defined in the site-specific safety plan and meet or exceed the requirement of CFR Subpart M—Fall Protection CFR 29 1926.500 through 1926.503.

Emergency Response

The first line of defense is the onsite access contractor competent person and trained onsite five-person rescue team able to perform entry rescue. The second line of defense for tunnel rescue is the Chicago Fire Department.

Gas Line Operation Procedure and Notification

Tunnel inspections are performed with the gas pipeline at a distribution pressure of 22 psi. This activity is coordinated through PGL's Gas Control Group before setting up of all safety requirements.

Pre-Pipeline Assessment

The piping integrity assessment procedures cover evaluation of threats including but not limited to external corrosion, internal corrosion, stress corrosion cracking, external damage, manufacturing defects, and construction defects. Before tunnel dewatering, a water sample for corrosion panel testing is collected.

Dewatering Requirements

The tunnel is pumped dry for the duration of the inspection period meeting the requirements of MWRD, unless inclement weather requires dewatering to stop.

Testing Requirements

The MWRD permit application and permit require water samples to be collected for BOD and TSS. The tunnel and pipeline assessment require water samples be analyzed for corrosion panel and microbiological and non-microbiological testing parameters.

Tunnel and Pipeline Hazardous Material Assessment

A tunnel hazardous material assessment is performed after the rapid assessment is completed and prior to any additional inspections for asbestos.

INSPECTION PROCEDURES

The inspection team consists of a geologist, corrosion/pipeline inspector, safety personnel and a structural engineer to perform a methodical and documented visual inspection of the tunnel, pipeline and appurtenance conditions. Each discipline follows industry standards to assess the tunnel and pipeline.

General Inspection Procedures

- Stationing is put in place in 25-foot intervals by the access contractor as reference points for all inspections.
- Temporary portable lighting is installed by the contractor to assist with tunnel and pipeline inspection of not less than four times the OSHA requirement.
- Each type of deficiency is photographed noting the date, station location, type of deficiency (i.e., geological, structural, pipeline, or pipeline support).
- Pipeline testing assessment is performed to determine if guided wave testing (GWT) is feasible or ultrasonic testing (UT) is appropriate.

Shaft and Tunnel Inspection and Mapping

- Geological mapping is performed on any unlined shaft and tunnel indicating the geological formation/member and type of discontinuity. Each geological feature is noted by the station location, strike and dip measured with a Brunton compass.
- A two-dimensional geological map is created showing all geological features on a scaled drawing with a north arrow showing orientation.

- The geological formation feature is identified indicating the overall orientation.
- Sounding of tunnel walls is performed along the length of the tunnel to identify voids, and delamination and areas of weak material are noted following ASTM D4580/D4580M.
- All tunnel support systems are identified, measured, located by stationing, counted, mapped and visually assessed for structural integrity. Rock bolts, dowels, mine straps, ribs, wood lagging, and wire mesh are compared to applicable ASTM standards for material integrity, size, and grade.

Ladder Platform and Pipeline Support Inspection

During the ladder, platform and pipeline support inspection, a reference table is used to rate each item listed as satisfactory, non-satisfactory, not applicable, or not inspected.

Pipeline Inspection

The pipeline inspection is performed by a qualified pipeline inspector and covers the following items:

- Pipeline characteristics
- Ambient conditions
- Cathodic protection
- General coating assessment
- Mechanical defects assessment
- Individual coating defect assessment
- Guided wave and or ultrasonic testing
- Coating removal and application

Emergency Repair Procedures

If emergency repairs and demolition are identified during the inspection, they will be performed by PGL and access contractor. A recommendation for repair or demolition will be provided by the inspection team to PGL in writing.

LiDAR Surveying of Tunnel

A 3-D LiDAR survey is performed to document current as-built configuration and dimensions of the access shafts, tunnel, pipeline, pipeline supports, ladders, and platforms.

SUMMARY OF INSPECTION FINDINGS AND NEXT STEPS

Rapid assessment is the first level of inspection to determine if the tunnel is in acceptable condition to perform the detailed inspection. The two tunnels inspected to date have been acceptable for detailed inspection except for the compromised structural integrity of ladders and platforms that were deteriorated due to anode bed high voltage output. The ladders and platforms were supplemented by installing temporary scaffolding to perform the detailed inspection.

The detailed inspection started within 24 hours after dewatering collecting microbiological and non-microbiological samples to assist in the pipeline and coating deterioration assessment. The major issue has been iron bacteria with growth exceeding eight inches in depth on exposed steel. The next level of assessment is the pipeline coating mapping of defects, thickness and adhesion. In all cases to date, the mapping has been complex with tubercle groupings covering 60 to 70 percent of the pipeline. The coatings have multiple patches and locations of failed adhesion, rendering guide wave testing impossible to perform and increasing the pipeline inspection time with hundred percent inspection using visual and multi phased array ultrasonic (UT) testing. The results of the UT testing have shown the pipeline to be in good condition with no loss of sections to the pipeline.

The tunnel inspections found localized concrete spalling, honeycombing, and/or shrinkage cracking throughout the shafts and tunnel. Another defect found was the deterioration of beam and anchor stubs in the shaft walls that were replaced in the past, leading to concrete shaft cracking. The recommended fixes include the use of epoxy injection and poured epoxy adhesive concrete grout.

The structural inspection found some components of the ladders, platforms and pipeline supports to be compromised. Those will be repaired or replaced. Recommendations include fiber reinforced plastic for replacement ladders and platforms, stainless steel for replacement pipeline supports, and bonding all the components to the cathodic protection system. Components of the existing cathodic protection system will also be repaired or replaced.

Design of the remediation recommendations for the first tunnel is ongoing and will commence in early 2019 for the second tunnel. Tunnel remediation construction is scheduled for 2019 for both tunnels.

ACKNOWLEDGEMENT

The authors thank Scott Zanoni for his contribution to this paper.

Table 1. Tunnel and shaft rating system

Condition Rating	Reliability	Tunnel Support Type	Typical Defects	Seepage	Repairs	Maintenance	Re-Inspection Interval
Excellent	Tunnel is functioning as intended. Confidence is high that the tunnel will remain operationally reliable for 20+ years.	Unlined	Little to no rockfall debris accumulation on invert. No rock or excavation deformation observed; Bedding is non-vertical/horizontal and massive (No slabbing); No rock fissures (Water flowing through rock mass); Rock mass strength very good (Joint spacing wide to very wide); No hanging blocks (Cross jointing producing a wedge), Joints have tight aperture; No faults or shear zones.	Minimal to no observed seepage into tunnel	Not warranted at this time	Not warranted at this time	Detailed inspection recommended in 10 to 20 years.
		Timber and/or Steel Supports	Timber/steel support elements are in good condition with no visible defects. Little to no rockfall debris accumulation on invert between sets. Timbers are sound, no longitudinal cracking, or separation from rock face; Steel sets are tight, no corrosion or delamination.				
		Concrete Lined	Minor hairline liner cracks with no observed offsets. Cracks are scattered (not systematic); no displacement, jacking, scouring or seepage.				
Good	Tunnel is functioning as intended. Confidence is high that the tunnel will remain operationally reliable for 10 to 20 years.	Unlined	Potential for localized small volume (<1 yd^3) rockfalls. Minor accumulation (<1 yd^3) of rockfall debris locally on invert. Medium to massive bedding between 20 and 60 degrees; Rock strength good (joints moderately wide to wide); No hanging blocks (Cross jointing showing infinite wedges); Joints have tight to narrow aperture; shear zones with no presence of brecciated material.	Minimal observed seepage into tunnel	Not warranted at this time	Some minor maintenance items may be required during next annual outage	Detailed inspection recommended within 10 years.
		Timber and/or Steel Supports	Timber sets/lagging exhibit isolated minor cracking and decay. Steel sets/lagging exhibit isolated minor surface corrosion.				
		Concrete Lined	Scattered (not systematic) circumferential liner cracks with no offset and minimal aperture (<1/16 inch). Minor concrete spalling or scour present locally.				

Assessment and Remediation of Gas Utility Tunnels in Chicago

Table 1. Tunnel and shaft rating system (continued)

Condition Rating	Reliability	Tunnel Support Type	Typical Defects	Seepage	Repairs	Maintenance	Re-Inspection Interval
Fair	Tunnel is functioning as intended. Confidence is moderate that the tunnel will remain operationally reliable for 10+ years.	Unlined	Potential for localized rockfalls up to 2 yd³. Accumulation of rockfall debris locally on invert (up to 2 yd³). Medium to thin bedding < 20 degrees or > 80 degrees; rock strength fair (joint occurrence moderately wide to close) rock wedges non-finite and non-tapered blocks; Joints have moderately wide aperture; shear zones with in place brecciated material.	Seepage into the tunnel is minor, and may be accompanied by minor amount of fine material.	Repairs not required, but will likely be necessary within 10 years.	Maintenance items may be required during next annual outage.	Detailed inspection recommended in 5 years.
		Timber and/or Steel Supports	Timber sets/lagging exhibit numerous minor defects and isolated moderate decay or cracking. Isolated timber members are split. Steel sets/lagging exhibit surface corrosion but with no significant section loss.				
		Concrete Lined	Systematic circumferential or longitudinal liner cracks with minimal offset (<1/16 inch) and minimal aperture (<1/8 inch). Localized zones of poor quality concrete, spalling, or scour may exist. Structural supports behind/within the lining may be locally exposed, and exhibit minor deterioration.				
Poor	Tunnel is functioning as intended; however, may have localized partially collapsed.	Unlined	Potential for localized rockfalls which could potentially obstruct tunnel flow. Accumulation of rockfall debris on invert. Bedding thin horizontal to vertical (producing slabs); rock strength poor (joints close to very close), rock wedges are finite tapered blocks; Joints have wide to very wide aperture; shear zones and faults present with accumulation of debris greater than 2 yd³, fissures with water infiltration.	Moderate to substantial seepage into the tunnel. Water inflow may include substantial fines.	Repairs not required immediately, but will likely be necessary within 5 years.	Maintenance items will likely be required during the next scheduled outage.	Detailed re-inspection within 1–5 years to assess the rate of deterioration. Frequency of detailed inspection should be re-assessed following each inspection. Further study, such as coring of the liner and tunnel stability analyses, may be warranted.
		Timber and/or Steel Supports	Timber sets/lagging exhibit numerous moderate defects including decay, cracking, splits, or localized failures. Steel sets/lagging exhibit severe corrosion and deterioration.				
		Concrete Lined	Circumferential or longitudinal liner cracks with offsets >1/16 inch, and aperture >1/8 inch. Localized zones of poor quality concrete, spalling, or scour. Structural supports behind/within lining are exposed and may be substantially deteriorated.				
Critical	Tunnel does not appear to be functional and/or critical defects are present which could jeopardize the continued operation of the pipeline or structural integrity of the tunnel. Immediate action is required and should include implementing an operations contingency plan and repairs, as appropriate. The tunnel should not be put back into service prior to repairs. If the tunnel must be placed back into service, a contingency plan for operations should be developed.						

Table 2. Pipeline rating system

Condition Rating	Reliability	Hoop Stress at Maximum Allowable Operating Pressure (MAOP)	Typical Defects Upon Direct Examination	Cathodic Protection	Repairs	Maintenance	Re-Inspection Interval
Excellent	Pipeline is functioning as intended. Confidence is high that this pipeline segment will remain operationally reliable for 20+ years without significant repairs.	Operating Above 50% of SMYS.	Minor defects in the coating without significant metal loss, mechanical damage, or coating disbondment around the defects. No blisters in the coating. No evidence that microbes are impacting corrosion.	Interrupted coupon tests indicate cathodic protection meets 49 CFR 192 appendix D criteria.	Not warranted at this time.	Not warranted at this time. Removal of all asbestos coatings may be considered to reduce cost of future inspections.	Detailed inspection, including Direct Assessment or In-Line Inspection (ILI), recommended within 20 years with at least one hydrostatic test to re-validate pipeline integrity within 10 years. Additional Direct Assessments or ILI within 10 years eliminates the need for validating hydrostatic test.
		Operating Above 30% of SMYS, Not Exceeding 50% of SMYS.	Minor defects in the coating with minimal metal loss corresponding to a predicted failure pressure not less than 230% of the current MAOP but no coating disbondment around the defects. No blisters in the coating. No evidence that microbes are impacting corrosion.				Detailed inspection, including Direct Assessment or ILI, recommended within 20 years with at least one hydrostatic test to re-validate pipeline integrity within 15 years. Additional Direct Assessments or ILI within 15 years eliminates the need for validating hydrostatic test.
		Operations Do Not Exceed 30% of SMYS.	Minor defects in the coating with minimal metal loss corresponding to a predicted failure pressure not less than 330% of the current MAOP but no coating disbondment around the defects. No blisters in the coating. No evidence that microbes are impacting corrosion.				Detailed inspection, including direct Assessment or ILI, recommended within 20 years.

Table 2. Pipeline rating system (continued)

Condition Rating	Reliability	Hoop Stress at Maximum Allowable Operating Pressure (MAOP)	Typical Defects Upon Direct Examination	Cathodic Protection	Repairs	Maintenance	Re-Inspection Interval
Good	Pipeline is functioning as intended. Confidence is high that this pipeline segment will remain operationally reliable for 15+ years without significant repair.	Operating Above 50% of SMYS.	A few moderate sized coating defects with little or no disbondment around the defects may be present. Metal loss or mechanical damage creating a predicted failure pressure not less than 150% of the current MAOP or no metal loss with isolated coating blisters across the pipeline may be present. No evidence that microbes are impacting corrosion.	Interrupted coupon tests indicate cathodic protection meets 49 CFR 192 appendix D criteria or is achieving some level of cathodic polarization but does not meet criteria.	Not warranted at this time	Some minor coating repairs may be required if asbestos material is not removed. Complete recoating and removal of all asbestos materials encountered is recommended when permitted by budget. Adjustment of nearby cathodic protection rectifiers may be required to meet regulation.	Detailed inspection, including Direct Assessment or ILI, recommended within 15 years with at least one hydrostatic test to re-validate pipeline integrity within 10 years. Additional direct assessments or ILI within 10 years eliminates the need for validating hydrostatic test.
		Operating Above 30% of SMYS, Not Exceeding 50% of SMYS.	A few moderate sized coating defects with little or no disbondment around the defects and isolated coating blisters containing minimal corrosion may be present. Metal loss or mechanical damage creating a predicted failure pressure not less than 200% of the current MAOP may be present at coating defects or on the interior of the pipeline. No evidence that microbes are impacting corrosion.				Detailed inspection, including Direct Assessment or ILI, recommended within 15 years.
		Operations Do Not Exceed 30% of SMYS.	A few moderate sized coating defects with little or no disbondment around the defects and isolated coating blisters containing minimal corrosion may be present. Metal loss or mechanical damage creating a predicted failure pressure not less than 270% may be present at coating defects or on the interior of the pipeline. No evidence that microbes are impacting corrosion.				Detailed inspection, including Direct Assessment or ILI, recommended within 15 years.

Table 2. Pipeline rating system (continued)

Condition Rating	Reliability	Hoop Stress at Maximum Allowable Operating Pressure (MAOP)	Typical Defects Upon Direct Examination	Cathodic Protection	Repairs	Maintenance	Re-Inspection Interval
Fair	Pipeline is functioning as intended. Confidence is moderate that the pipeline will remain operationally reliable for 10+ years with only minor repairs.	Operating Above 50% of SMYS.	Wide spread blistering of the tape coat containing minimal corrosion and large areas of coating damage may be present. Metal loss or mechanical damage creating a predicted failure pressure not less than 140% of the current MAOP may be present at coating defects or on the interior of the pipeline. Microbes known to influence corrosion may be present in blisters or coating defects but are not associated with locations of severe corrosion.	Interrupted coupon tests indicate cathodic protection is achieving some level of cathodic polarization but 49 CFR 192 appendix D criteria is not met.	Repairs not required, but may be necessary within 10 years.	Complete recoating and removal of all asbestos materials encountered is recommended within 10 years. Adjustment of nearby cathodic protection rectifiers may be required to meet regulation.	Detailed inspection, including Direct Assessment or ILI, recommended in 10 years.
		Operating Above 30% of SMYS, Not Exceeding 50% of SMYS.	Wide spread blistering of the tape coat containing minimal corrosion and large areas of coating damage may be present. Metal loss or mechanical damage creating a predicted failure pressure not less than 170% of the current MAOP may be present at coating defects or on the interior of the pipeline. Microbes known to influence corrosion may be present in blisters or coating defects but are not associated with locations of severe corrosion.				Detailed inspection, including Direct Assessment or ILI, recommended in 10 years.
		Operations Do Not Exceed 30% of SMYS.	Wide spread blistering of the tape coat containing minimal corrosion and large areas of coating damage may be present. Metal loss or mechanical damage creating a predicted failure pressure not less than 220% of the current MAOP may be present at coating defects or on the interior of the pipeline. Microbes known to influence corrosion may be present in blisters or coating defects but are not associated with locations of severe corrosion.				Detailed inspection, including Direct Assessment or ILI, recommended in 10 years.

Table 2. Pipeline rating system (continued)

Condition Rating	Reliability	Hoop Stress at Maximum Allowable Operating Pressure (MAOP)	Typical Defects Upon Direct Examination	Cathodic Protection	Repairs	Maintenance	Re-Inspection Interval
Poor	Pipeline is functioning; however, may have severe damage requiring regular monitoring and repair sooner than 5 years.	Operating Above 50% of SMYS. Derating to a lower operating pressure may need to be considered until scheduled repairs are possible.	Blisters and other coating defects that contain aggressive microbiologically influenced corrosion may be present. Large areas of the pipe may be completely bare. Metal loss or mechanical damage creating a predicted failure pressure not less than 120% of the current MAOP or metal loss may cover large areas and be difficult to repair without completely replacing sections of pipe.	Interrupted coupon tests indicate cathodic protection does not meet 49 CFR 192 appendix D criteria inside the tunnel and is not achieving any level of cathodic polarization on this segment or stray current interference is confirmed to be unmitigated inside the tunnel.	Repairs to pipe are not required immediately, but will likely be necessary within 5 years. Repairs to cathodic protection system inside tunnel or additional cathodic protection capacity in this area may be required to meet regulation.	Complete recoating of the pipe may be required within 5 years.	Follow-up Direct Assessment inspection within 1 year to assess the rate of deterioration and need for more immediate repairs. Further study, such as effect of reduced operating pressure or use of reinforcing sleeves, may be warranted to determine alternate frequencies of detailed repairs.
		Operating Above 30% of SMYS, Not Exceeding 50% of SMYS. Derating to a lower operating pressure may need to be considered until scheduled repairs are possible.	Blisters and other coating defects that contain aggressive microbiologically influenced corrosion may be present. Large areas of the pipe may be completely bare. Metal loss or mechanical damage creating a predicted failure pressure not less than 130% of the current MAOP or metal loss may cover large areas and be difficult to repair without completely replacing sections of pipe.				Follow-up Direct Assessment inspection within 1–3 years to assess the rate of deterioration and need for more immediate repairs. Further study, such as effect of reduced operating pressure or use of reinforcing sleeves, may be warranted to determine alternate frequencies of detailed inspection.
		Operations Do Not Exceed 30% of SMYS.	Blisters and other coating defects that contain aggressive microbiologically influenced corrosion may be present. Large areas of the pipe may be completely bare. Metal loss or mechanical damage creating a predicted failure pressure not less than 160% of the current MAOP or metal loss may cover large areas and be difficult to repair without completely replacing sections of pipe.				Follow-up Direct Assessment inspection within 1–5 years to assess the rate of deterioration and need for more immediate repairs.
Critical	Pipeline was found to be operating under abnormal operating conditions, defects exist that require a response time in accordance with B31.8S Figure 7.2.1-1 of much less than 5 years, or the tunnel condition is found to be critical. Immediate repair is required for all defects creating a predicted failure pressure under 110% of MAOP. Repair within 90 days is required for all defects creating a predicted failure pressure above 110% of MAOP and below or equal to 120% of MAOP. For predicted failure pressures equal to 120% of MAOP or greater than 120% of MAOP but less than the applicable predicted failure pressure under classification "POOR," an operations contingency plan shall be in place and repairs shall be scheduled in accordance with B31.8S Figure 7.2.1-1 prior to resuming normal operation of the pipeline at the current MAOP. Appropriate contingency plan for operations, reductions in MAOP, and strict controls on the operating pressure should be developed and then verified to meet regulatory compliance if the pipeline must resume normal operation before repairs are completed. All defects other than corrosion anomalies shall be brought to the attention of the PGL representative upon discovery for evaluation in accordance with the appropriate Pipeline Integrity Program. This PGL evaluation may be in addition to any items contained within the Pipeline and Pipeline Support check sheet and shall be done at the discretion of the PGL representative. Defects other than corrosion anomalies that may fall within classification "CRITICAL" may include dents having indications of metal loss, cracking or other stress risers within the affected zone of the dent as well as any indication of stress corrosion cracks.						

Design and Construction of Lockbourne Intermodal Subtrunk Sewer

Steven Thompson ▪ AECOM
Irwan Halim ▪ AECOM
Michael Nuhfer ▪ Aldea Services
Jeremy Cawley ▪ City of Columbus

ABSTRACT

The City of Columbus is currently constructing the 3.27-mile long Lockbourne Intermodal Subtrunk (LIS) project in order to provide sanitary sewer service to the Northern Pickaway Joint Economic Development District and other surrounding areas. The LIS is an extension of an existing 168-inch sewer (BWARI) that passes north of the Village of Lockbourne and terminates at the Southerly Wastewater Plant on State Route 23 (SR 23). The project begins near the intersection of SR 23 and Rowe Road, and ends at the existing Norfolk Southern Intermodal Facility, just south of the Rickenbacker International Airport. The finished sewer will include 1.94 miles of 78-inch and 1.33 miles of 60-inch diameter pipe. It is being constructed of centrifugally cast fiberglass reinforced mortar pipe, manufactured by Hobas Pipe USA, installed by microtunneling and open cut methods through a complex glacial deposits. Critical shallow tunnel crossings below the Big Walnut Creek and multiple railroad tracks are also parts of this project challenges. Geotechnical and hydrogeological studies were performed to investigate the ground conditions in order to select the best tunneling and shaft construction methods that minimize impacts to nearby residential wells. This paper describes the project design and construction considerations. Some construction challenges and their mitigation measures are also presented.

PROJECT OVERVIEW

The purpose of the LIS is to initially provide sanitary sewer service to the Norfolk Southern Intermodal Facility near the Rickenbacker International Airport, and ultimately to provide service to developing areas in the vicinity. The LIS is located about 11.5 miles south of the center of Columbus, Ohio, and spans the Franklin-Pickaway counties border. The downstream end of the LIS connects to the existing bulkhead at the Big Walnut Augmentation/Rickenbacker Interceptor (BWARI). The upstream end is located near the Intermodal Facility and will connect to an existing pump stations on Ashville Pike, which is just south of the southwest corner of the Rickenbacker Airport (see Figure 1).

The LIS consists of approximately 3.27 miles of sanitary sewer. The sewer includes a 60-inch diameter pipe from the upstream end for a distance of 1.33 miles downstream to Shaft/Manhole-13 (S/M-13). At S/M-13, the invert drops by 20 feet, the pipe turns 90 degrees to the north, increases in diameter to 78 inches, and proceeds for a distance of 1.94 miles to the downstream end of the LIS, where it crosses under the Big Walnut Creek immediately west of the Village of Lockbourne and connects to the existing BWARI. The 60-inch pipe is being installed by the open cut method, except for a portion that will be installed by a Guided Boring Machine (GBM), under a 300-foot wide rail corridor. The 78-inch pipe is being installed by the microtunneling method. A total of nine shafts are being constructed within the microtunnel portion of the project, with

Design and Construction of Lockbourne Intermodal Subtrunk Sewer 1187

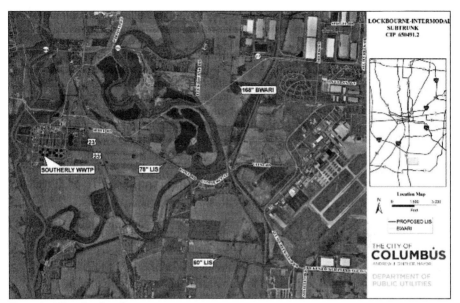

Figure 1. LIS project location

permanent manholes to be installed in six of the shafts for future flow connections and tunnel access points. The Contractor elected to use not as many shafts as designed, and instead perform longer microtunnel drives (i.e., up to 1,900 feet in a single drive). A total of 17 manholes are being constructed within the open cut portion of the sewer. Figure 1 shows the project location.

The sewer is being constructed through sand and gravel deposits and clayey till of glacial origins. The downstream third of the project is being constructed in and adjacent to the floodplain of Big Walnut Creek. The remainder of the project's alignment east of the creek and south of the Village of Lockbourne is being constructed below the higher ground of the till plains, which are about 20 to 30 feet higher in elevation than the Big Walnut Creek floodplain. Figure 2 and Figure 3 show the complex subsurface profiles and conditions along the project alignment with interlayered deposits containing numerous boulders.

SUBSURFACE EXPLORATION

Geologic Setting

The LIS project lies entirely within the Columbus Lowland division of the Till Plains section of the Central Lowland physiographic province that includes most of Franklin and Pickaway Counties. Bedrock will not be encountered during the construction of the LIS, but consists of a sedimentary sequence of limestone and shale of Devonian age. The overburden materials above the bedrock are glacial and post-glacial soils. The glacial soils were laid down by the continental ice sheets and meltwater that moved across the region over the last two million years. The repeated advance and retreat of glaciers has resulted in a very complex subsurface with interlayered soils and sudden changes in soil types. The post-glacial soils, which comprise the uppermost stratigraphy along the northwest section of the project alignment, were deposited by the Big Walnut Creek and the Scioto River. Station 65+00 south of the Village

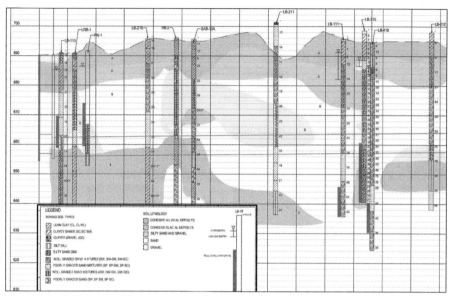

Figure 2. Partial tunnel profile

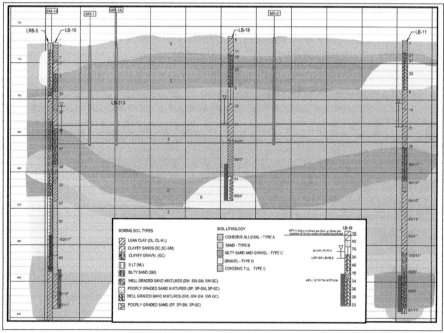

Figure 3. Partial open cut profile

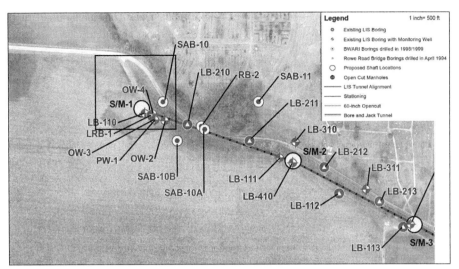

Figure 4. Partial boring location plan

of Lockbourne marks the transitions from alluvial deposits to glacial moraine that rise approximately extend to the south and east of the LIS alignment.

Geotechnical Investigations

To characterize the complex subsurface conditions, extensive geotechnical investigations were performed and evaluated for the purpose of the tunnel and shaft design consisted of:

- Five phases of test borings
- Rotosonic borings
- Past exploration programs in the project area including borings completed as part of the previous BWARI and Rowe Road bridge projects
- Nearby quarry boulder study to characterize the anticipated tunnel impact, and
- Hydrogeological investigation and study of dewatering and impact

Test borings included conventional hollow stem auger borings with split spoon sampling. The hollow stem augers used carbide tipped and/or conical teeth. For test boring Phases 1 through 3, split spoon samples were collected at 2.5-foot intervals for the first 10 feet and at 5-foot intervals or change in strata thereafter. For test boring Phases 4 and 5, borings were continuously sampled. Groundwater monitoring wells were installed in selected test borings after completion. Undisturbed thin-walled Shelby Tube samples were attempted several times. However, the tubes typically collapsed without advancement in the hard cohesive till layer. Collection of undisturbed soil samples was more successful using a 4-inch diameter CME continuous core sampler, which was rotated rather than pressed into the soil. Figure 4 illustrates the high density of borings along the project alignment and shaft areas.

Hydrogeologic Investigation

The purpose of hydrogeologic investigation was to evaluate groundwater conditions in the project area. Water levels were measured in piezometers installed in selected

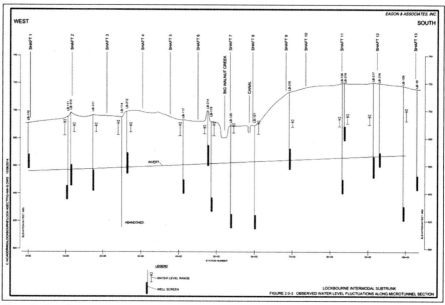

Figure 5. Baseline microtunnel groundwater profile

geotechnical borings. The baseline microtunnel groundwater profile is shown on Figure 5. In addition to slug tests and mini pump tests that performed in most of the piezometers, four sites were selected for extended pumping tests. The pumping wells at three of these sites were installed in a sonic-drilled test hole with a 6-inch well screen and casing, while at one site a 12-inch pumping well was installed using a 36-inch diameter bucket auger. 24-hour constant-rate pumping tests were performed at each of these sites. The pump tests results were used to estimate the impacts of dewatering on nearby water wells.

DESIGN CONSIDERATIONS

Microtunnel Shaft and Site

Microtunnel shafts were estimated during design to be 25-foot diameter circular shafts. Circular shafts were assumed to be more economical than rectangular shafts, and have the ability to launch and receive a microtunneling boring machine (MTBM) from any direction. The Contractor ultimately selected to use varying sized shafts for launching and receiving shafts based on alignment changes and incorporation of additional flow connection manholes. Figure 6 shows an MTBM launching shaft site that must accommodate not only the shaft but a range of other supporting equipment on the site surface.

Shaft Structures

Shafts can either be constructed as watertight or dewatered structures. Usually in the glacial deposits found in the Columbus area, dewatering does not cause substantial settlement and is the standard method used for temporary microtunnel shaft construction. Additional dewatering analysis concluded that a number of shaft sites would pose dewatering difficulties due to subsurface conditions. As a result of that analysis 6 shafts were designed to be watertight. However during construction the

Contractor elected to dewater most the shaft for expediency.

Access Manholes

Shafts allow the economical placement of an access manhole, if necessary. However, manholes are not required at every shaft unless a change in direction takes place. Additionally, the Contractor was allowed to eliminate construction shafts they found to be unnecessary. This increased the drive lengths but eliminated four previously designed shafts. On the alignment and with the eliminated shafts directional changes occur at every shaft except for one, and a manhole will then be required at each of those shafts. A single shaft was considered as a temporary construction shaft only and was backfilled to the surface without manhole.

Figure 6. MTBM launching shaft site

MTBM Launch

When launching an MTBM below the groundwater level, ground and groundwater losses must be minimized. For an MTBM of approximately 90 to 100 inches in diameter, a low strength concrete block or "eye" is cast against the launch area. A launch seal is bolted to the concrete block to prevent bentonite slurry leakage. The MTBM then cuts through the concrete block, with the launch seal providing a water tight seal to prevent the ground and groundwater losses. The Contractor elected to inject Acrylate grout in addition to the proposed seal to further prevent soil and water loss.

Construction Dewatering

Dewatering during tunneling operations will not be necessary as the MTBM is pressurized and the presence of water aids in lubrication of the pipe. Dewatering during shaft construction will be required at most shaft sites. All Contractor elected shafts did require dewatering or impermeable construction. The pumped water was discharged to the Big Walnut Creek and/or the Buhlen Ditch near the open cut portion of the pipe. Shafts located furthest away from the creek required approximately one mile of piping to reach the discharge point. Groundwater from shafts along Ashville Pike may be discharged to any suitable drainage ditch along the side of the road.

Power Availability

The MTBM was provided with a dedicated generator and power pack and does not require power from a public utility. Power drops, however, were installed at individual shaft locations to provide power for dewatering pumps and lighting. There were power poles along the entire length of the alignment and thus adequate power was readily available.

Tunnel Muck

Slurry muck from the microtunneling operations and shaft construction was stored on site temporarily to accommodate work hour restrictions, scheduling, and logistics. Adequate space was available on LIS easements and in the open cut areas for sample and stockpile storage. Muck from microtunneling operations contains bentonite used as part of the slurry system for the MTBM. A slurry separation system was used to remove the aggregate and sand from the bentonite slurry. This aggregate and sand as

Table 1. Shaft dewatering analysis overview—Lockbourne intermodal subtrunk

Shaft #	Base of Aquifer	Site Conditions	Estimated Dewatering Rates (gpm)	Dewatering Shortfalls	Remarks	Lockbourne Landfill Impact	Wells Potentially Impacted
1	Above Invert	Difficult	3400–1400	Can't get totally dry/req sumping	Consider sinking caisson?	Minimal	5**
2	30'+ Below Invert	Ideal	5500–2900		High K	Minimal	
3	5' Below Invert	Somewhat Difficult	1900–1600		Consider sinking caisson?	Minimal	
4	Above Invert	Difficult	440–340	Can't get totally dry/req sumping	Consider sinking caisson?	Minimal	0
5	Above Invert	Difficult	2000–1400	Can't get totally dry/req sumping	Consider sinking caisson?	Minimal	
6	2' Above Invert	Difficult	1100–730	Can't get totally dry/req sumping	Consider sinking caisson?	Minimal	
7	At Invert	Difficult	2600–1700	Can't get totally dry/req sumping	Consider sinking caisson?	1–2'	8
8	14'+ Below Invert	Ideal Moderate	2400–1900		Some fine zones	1–2'	
9	11' Below Invert	Ideal	2400–1600		Some perched zones	1–2'	
10	18' Below Invert	Ideal	1600–1500		Some perched zones	1–2'	
11	At Invert	Difficult	860–710	May need wells below clay layer	Consider sinking caisson?	1–2'*	4
12	14' Below Invert	Ideal	1400–970			1–2'*	
13	11' Below Invert	Good-Moderate	680–640		Some tight zones	1–2'*	

*Minimal if 7–10 are not dewatered
Dewatering does not significantly alter flow directions from landfill.
**Two of these wells were replaced by BWARI Project.

well as muck from the shaft and open cut excavations was reused within the project or taken to nearby quarries as was done on the BWARI project. Hydrocyclones further separate the slurry into water, bentonite solution (to be recirculated), and waste solution (including bentonite to be disposed). The waste solution is then sent to a settlement system or settlement tank where the solids are settled out of the solution. These solids are then disposed as non-hazardous waste at a local landfill, as they cannot be used for fill material.

Permits and Approvals

The following are the permits and approvals that were obtained as part of the design process. No additional permits are known to be required at this time.

- Ohio EPA Permit to Install (PTI)
- 401 Water Quality Certification
- Nationwide Permit 12 Approval
- Franklin County Floodplain Permit
- Federal Aviation Administration-Approval of project work in Rickenbacker Airport clear zone.

Floodplain Considerations

The proposed alignment shows that approximately 6,400 feet of the proposed sewer will be tunneled under the 100-year floodplain of Big Walnut Creek. One proposed shaft is located in the FEMA Floodway, but was eliminated by the Contractor and seven proposed shafts are located within the 100-Year Floodplain of Big Walnut Creek. Bolted and gasketed watertight manholes will be used at these locations and a 2 foot free board will be required at the shaft locations.

Right-of-Way and Maintenance of Traffic

East of the Village, construction at the shaft sites along Rowe Road occur within the roadway right-of-way and existing easements. Construction within the Rowe Road right-of-way required the closure of at least one travel lane. Barriers and work zone impact attenuators were placed around each work site in the right-of-way. It was possible to maintain local traffic on Rowe Road with signaling and/or other temporary traffic control devices. This required coordination with and approval by the Franklin County Engineer, but it did not have major impacts on the construction schedule or cost because the road is not heavily used.

Anticipated Soil Types

Five soil types were defined along the project alignment as follows:

Type A—Cohesive Alluvial

Type B—Sand

Type C—Silty Sand and Gravel

Type D—Gravel

Type E—Cohesive Till

Type A soil is typically encountered during shaft excavations only, whereas the remaining soil types are all present during microtunneling. The anticipated behaviors of the ground during construction for each soil type were described using the Tunnelman's Ground Classification system.

While the soil types describe soil textural differences, the soils are frequently present in horizontally layered systems in close proximity to each other. Thus, excavation for tunnels and shafts more often "encountered" several layers with different behavioral characteristics than a full face of one soil type. In addition, for most soil types, numerous cobbles and boulders occurred either in isolation or in horizontally layered concentrations. And although not encountered nor identified specifically in borings for the LIS project, a cemented gravel conglomerate material has been observed during investigation for the nearby BWARI project, and was also expected. The ground behavior described by the Tunnelman's Ground Classification system is what would occur in an open, unprotected tunnel face or in shaft and open cut excavations. Much of the described behavior was effectively mitigated by means of a pressurized face MTBM during tunneling. Figure 7 shows the baseline ground

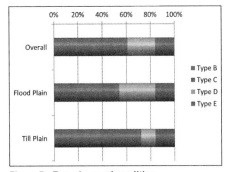

Figure 7. Tunnel ground conditions

conditions for microtunneling, consisting of different soil type fractions to be encountered along the tunnel horizon, with each of the soil type can occur at any locations along the alignment.

Other Tunnel Design Considerations

Tunnelling Method:

- Slurry MTBM with CCFRPM Jacking Pipe.
- Drive range of 530 feet to 1880 feet.
- Shaft locations and tunnel alignment outside of roadways.

Tunnel Impacts and Monitoring:

- Action level ground movements on roadways—0.3/0.6 inch.
- Action level ground movements on bridge and South Central Power (SCP) substation foundations—0.25/0.25 inch.
- Action level ground movements on utility and railroad crossings—0.25/0.25 inch.
- Action level ground movements in other areas (including shaft excavations)—0.5/1.0 inch

CONSTRUCTIBILITY CONSIDERATIONS

A key objective of the project is constructability in order to ensure that the proposed design can be built using the anticipated construction methods while remaining within the operational, schedule, cost, and other constraints of the project. Two particular aspects of constructability were identified and pursued during the project study, design development, and value engineering phases: firstly, constructability of the proposed alignment with all of the shafts included; and secondly, access to the face of the machine in cases where surface easements were not available in several critical locations such as the creek crossing.

As a part of this review process, an independent value engineering team was engaged to provide input to the project team during the design development. The prime intent was to identify any fatal flaws or areas for refinement with the design team's approach, and no such fatal flaws were identified. The review also confirmed that the work could be constructed with the proposed approach. Many useful refinements were noted and incorporated into the final design.

A critical constructability issue identified was the potential for obstruction occurrence and removal of numerous cobbles and boulders anticipated along the tunnel alignment. This was viewed as a major risk and uncertainty. In an effort to establish that risk tolerance and limit the potential for cost overruns a baseline for cobbles and boulders was developed and included in the project documents. These values are shown in Table 2. For payment purpose, obstruction is defined to completely stop tunneling or excavation progress by more than 1 hour. Obstruction removal in the tunnel is to be paid for by a per crew hour basis.

Another critical constructability issue identified was the potential for dewatering uncertainty as a result of the complex subsurface conditions. This was viewed as a major risk and uncertainty. The decision was made to specify watertight shafts for locations where dewatering is anticipated to be difficult. In an effort to establish that risk

Table 2. Cobbles and boulders baseline for tunnel excavation

| Excavated Tunnel Diameter, ft | Tunnel Reach | Distance, lft | Diameter of Clast, ft ||||||||
| | | | Obstructions ||||||||
			0.5–1	1–1.5	1.5–2	2–3	3–4	4–5	5–6	>6
7	Reach 1	1110	4196	123	16	55				
	Reach 2	940	3553	104	14					
	Reach 3	940	3553	104	14					
	Reach 4	730	2759	81	11					
	Reach 5	730	2759	81	11					
	Reach 6	855	3232	94	13					
	Reach 7	590	2230	65	9					
	Reach 8	875	3308	97	13					
	Reach 9	530	2003	59	8					
	Reach 10	950	3591	105	14					
	Reach 11	935	3534	103	14					
	Reach 12	983	3715	109	15					

tolerance and limit the potential for cost overruns an allowance for temporary water supply and well rehabilitation was included in the Contract bid form in the case that dewatering impact nearby residential wells.

In order to provide a means for conflict resolution a special provision to the Contract was included for a dispute review board. This provision allows the Contractor or the Owner to seek a three person panel review of an issue that is not able to be solved at the project administration level.

Lastly, due to the uncertainty associated with underground projects the Contract was required to include a 15% construction contingency as part of his bid. This additional funding is accessible to the Contractor through the successful implementation of a Request for Proposal and Change Order process that must be mutually agreed upon between the Contractual parties.

CONSTRUCTION UPDATE AND LESSONS LEARNED

Construction Update

Michels Tunneling was selected as the LIS contractor in January 2017. They purchased a new Akkerman SLP82-84 MTBM and constructed circular launch and retrieval shafts with liner plates and dewatering, except for S/M-1 which was constructed as a rectilinear, water-tight secant pile shaft. Construction of the LIS is approximately seventy-five percent complete as of December 2018. Construction of all nine of the microtunnel shafts has been completed, and backfill has started in five of the shafts. Seven out of the eight microtunneling runs have been completed. Nearly 8,800 feet of 78-inch Hobas pipe has been installed out of the 10,200 feet of the microtunnel section. The last microtunnel run is scheduled to be complete by February 2019. Nearly 4,800 feet of 60-inch Hobas pipe has been installed out of the 6,700 feet of the open cut section. The GBM beneath the railroad tracks is scheduled to be complete by April 2019. Connection to the existing BWARI sewer is scheduled to occur in February 2020. The LIS sewer is expected to be ready to take flow in August 2019, and final completion is scheduled for February 2020, prior to the contracted milestone completion date of March 15, 2020.

Figure 8.

The final and potentially most critical work remaining is the shallow river crossing at the Big Walnut Creek that will occur on Run 8. During design a lot of thought went into connection points, various diameters and slopes, and potential solutions to a known problem of shallow cover at the river crossing. Ultimately, a design that has less than one tunnel diameter of cover at the creek crossing was selected. The Contractor understands the problem and has performed additional bathometric surveys on the river bottom to confirm the cover. They also plan to actively manage their lubrication system while under the river to limit the potential of fracking out. The Contractor has been very active in developing a protocol that allows the greatest opportunity for success.

Lessons Learned

Project risks have been tracked throughout construction using a detailed construction risk register. Monthly updates to the risk register have been made, and periodic in-depth risk update workshops between the contractor and owner's representatives have been held. Mitigation activities for each risk were identified at the start of construction, and specific names were assigned as a responsible party for each of the mitigation. The risk register has often been the starting point in the resolution of project challenges.

A Disputes Review Board (DRB) has been retained since the start of construction. A panel of three qualified experts was selected by the owner and contractor, and has been visiting the project once a quarter to maintain an awareness and understanding of the progress and challenges of the work. The presence of the DRB has been a motivating factor in the amicable resolution of construction disputes. To date, no claims have been brought before the DRB for adjudication.

As can be seen in Figures 3 and 4, the geology for this project is complex. When the MTBM has encountered cohesive till in the invert with sands and gravels above, the MTBM has sometimes been difficult to keep on grade, and ground loss has sometimes occurred. Through a concerted effort, the contractor, MTBM manufacturer, project designers, and construction managers identified MTBM and operational changes to try to limit grade deviation and ground loss. MTBM modifications included:

1. The number of slurry cutting ports was reduced to increase pressure to better process clayey soils,
2. The orifice openings of the chamber ports were modified to better manage the flow of slurry through the slurry system, and
3. The height of some collection ports was raised to allow the crushing arm wipers to clear away cohesive materials from clogging.

In addition to modifying the MTBM, the contractor made operational changes, in an attempt to maintain face and slurry pressures at the start and end of each push cycle. Also, in order to better anticipate the volume of ground loss, inspectors performed frequent visual inspections and logging of muck volumes by counting backhoe scoops of muck removed from the slurry separation plant, and the contractor began weighing each muck truck to better correlate muck volume and weight.

Changes in the microtunnel procedures were envisioned and ultimately implemented after many open and honest meetings with the owner, machine manufacturer, designer, construction manager, and contractor. The meetings proved to be a constructive forum for the identification and vetting of ideas from all parties, who shared an aligned interest for the successful completion of the project.

The Contract Documents required the use of an airlock on two of the microtunnel drives. This was specified as a means to rescue the MTBM if it encountered an obstruction while beneath the substation, the creek, or the abandoned canal. The contractor selected an MTBM with a cutterhead designed to move through ground with many cobbles and boulders. To date, the MTBM has not been stopped by any obstruction. This success is likely due the design of the cutterhead, the power of the machine (400 HP), the use of Intermediate jacking stations (IJS's), and the operational parameters selected by the machine operator. Rescue shafts and/or the use of an airlock would come at some expense and cause a delay to the project.

Shaft dewatering was optional for each of the microtunnel shafts. The Contract Documents included estimated inflow rates for each shaft, which ranged from a few hundred to a few thousand gallons per minutes. The contractor chose to dewater all of the microtunnel shafts, except S/M-1. Due to the complex layering of the soils at the shafts, a simple uniform solution for all shaft sites was not possible. While it would have been easiest to implement a single dewatering design for all the shafts, ultimately each shaft dewatering solution was unique. Auto-switching of backup power for the dewatering pumps was shown to be critical to maintaining dry shafts.

At the start of the project, the contractor submitted a request to eliminate one shaft and move another, converting shorter three microtunnel drives into two longer drives. The owner accepted the proposal, on the conditions that it be a no cost change, the contractor waive their right to a differing site condition claim at the new shaft location, and the contractor was required to construct the new shaft in the existing easement. Prior to accepting the conditions, the contractor conducted an exploratory drilling program, and concluded that the geology would pose no problems. The period of time from the initial request and ultimate acceptance by both parties was less than one month, and provided the contractor with a preferable microtunneling plan. This was only made possible because there were no drop shafts or other permanent structures needed in any of the unconstructed or new shafts. This process demonstrated the advantages of an owner and contractor who are both willing to make design changes after contract award.

The selection of the sewer pipe material was a key topic during design, bidding, and construction. This sanitary sewer will be subject to corrosive gases and a pipe material resistant to those gasses was selected during design. A thorough evaluation resulted in the selection of centrifugally cast fiberglass reinforced mortar pipe (FRPM), manufactured by Hobas. During the bidding phases a number of contractors expressed concerns about jacking force limitations of Hobas pipe, particularly in ground where numerous cobbles and boulders were expected. With seven out of eight of the microtunnel drives complete, there have been no problems with buckling or cracking of the pipe due to the presence of cobbles and boulders. A North American record was set on the job after a successful drive length of 1,880 feet with Hobas pipe in bouldery ground was completed on Run 6, with the aid of only two IJS's Jacking forces topped out around 550 tons on Run 6. Careful monitoring and use of IJS's is critical when microtunneling with FRPM in ground with cobbles and boulders.

Third party coordination is always a challenge and this project was no different. The nearly 17,000 feet of sewer being installed in this project crosses through two counties, a village, two townships, two railroad owners, a canal remnant, a river, an intermodal facility, and adjacent to an airport. The coordination through design and construction was not simple but was aided through the use of public meetings, mailers, press releases, and conversations with impacted stakeholders.

CONCLUDING REMARKS

The LIS project serves as a critical sanitary sewer extension from BWARI to provide service to the rapidly developing Northern Pickaway Joint Economic Development District which is located just south of Rickenbacker International Airport and other surrounding areas. The tunnel alignment goes through some of the most complex glacial depositional environment with numerous cobbles and boulders and multiple aquifers and artesian groundwater conditions. Comprehensive geotechnical and hydrogeological investigation programs were performed to characterize the subsurface conditions. Shaft locations were selected not only to place the future connections, but also to accommodate reasonable drive lengths for the MTBM. Watertight shafts were specified in locations with anticipated difficult dewatering conditions. The design also provides some flexibility for the Contractor to eliminate intermediate work shafts not needed for permanent structures, and perform full dewatering of all shafts regardless of difficulties. Contractual baselines were implemented to allocate subsurface risks between the Owner and Contractor.

Several challenges were encountered during construction including difficult shaft dewatering conditions and excessive ground loss resulting in formation of sinkholes on the ground surface. These challenges were resolved by collaboration amongst the designer, owner, construction management team and the Contractor. Regular coordination with third party stakeholders including the nearby town of Lockbourne, local farmers, the Norfolk Southern and CSX railroad operators, and others were performed diligently to ensure the successful completion of the project.

ACKNOWLEDGMENTS

The authors wish to acknowledge City of Columbus, as the Owner of the project described in this paper. This article represents the opinions and conclusions of the authors and not necessarily those of City of Columbus. This article shall not be used as evidence of design intent, design parameters or other conclusions that are contrary to the expressed provisions in the contract documents for the Lockbourne Intermodal Subtrunk Extension Project.

Great Hill Tunnel Inspection and Rehabilitation

Brian Lakin ▪ McMillen Jacobs Associates
Joe Schrank ▪ McMillen Jacobs Associates
Daniel Ebin ▪ McMillen Jacobs Associates
Lawrence Marcik ▪ South Central Connecticut Regional Water Authority

ABSTRACT

The Great Hill Tunnel, owned and operated by the South Central Connecticut Regional Water Authority, is a 100 MGD, 3,600-foot-long, 6-foot horseshoe-shaped raw water transmission tunnel that cannot be taken out of service and dewatered without installation of a bypass system. During 2017, routine tunnel leakage increased dramatically, causing concerns that known defects were getting larger and that increased leakage signaled impending tunnel failure. This paper discusses the rehabilitation program, which included manned inspection and repairs such as shotcrete, cement mortar patching, and contact grouting. It also compares the manned inspection with earlier ROV inspections of the tunnel.

BACKGROUND

The Great Hill Tunnel (GHT), originally constructed in 1927, consists of two separate sections. Section 1 is the drill and blast tunnel section and extends from the intake within Lake Gaillard approximately 2,700 feet to the southwest, where it transitions to a 48-inch-diameter cast iron pipe that extends an additional 860 feet to a control valve. The pipeline portion, Section 2, was installed using standard cut-and-cover methods and links the rock tunnel portion to the raw water transmission main and eventually to the Regional Water Authority's (RWA) water treatment plant nearby (Figure 1). The GHT is an integral asset in the RWA's water collection system. The unreinforced concrete tunnel is horseshoe shaped and is the last link in a series of connected reservoirs and tunnels that conveys raw water from remote areas to the treatment plant. The tunnel could only be accessed through the Intake Gate House within Lake Gaillard prior to this project. The original construction contract drawings for the tunnel show that a downstream access shaft was installed, but the condition of the access shaft and the watertight cover were unknown as the portal was buried beneath a quarry access road. Therefore, until the tunnel was depressurized, and the shaft uncovered, the functionality of this access point was unknown.

The GHT, operationally, is the sole conveyance of raw water between the reservoir and the water treatment plant, making routine manned inspection and maintenance nearly impossible. Available outage times are measured in hours at off-peak demand times, so putting personnel into the tunnel to inspect on a regular basis is not an option for the RWA.

GEOLOGIC SETTING

The project area is located in the central lowlands of Connecticut within the Gaillard graben (down-dropped bedrock block). This graben formed over 140 million years ago during regional extensional faulting along the Foxon Fault (west side) and Eastern Border Fault (east side). As the graben dropped, sediments eroded off the bordering metamorphic highlands and were deposited as sedimentary strata within the graben.

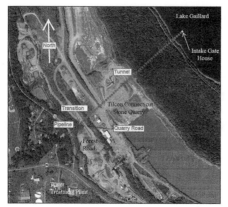 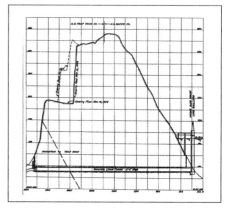

Figure 1. Plan and profile of the Great Hill Tunnel

Concurrently, thick, tabular basalt flows periodically covered the sedimentary strata within the graben and became interbedded with these strata as the graben subsided.

Differential movement along the faults that border the graben caused the interior block and overlying sediments and flows to become rotated. The bedding in the project area dips at approximately 13 degrees to the northeast. Subsequent erosion of the highlands and graben material has removed the sedimentary rock and exposed the resistant basalt flows. These exposed flows form a northwest-southeast monocline (Totoket Mountain) that dips at approximately 12 to 13 degrees to the northeast.

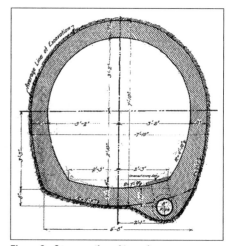

Figure 2. Cross section of tunnel

The GHT alignment is located between the inlet along the western shoreline of Lake Gaillard and the outlet portal on the west side of Totoket Mountain. The tunnel crosses beneath both Totoket Mountain and the arcuate-shaped North Branford trap rock (basalt) quarry, which is located on the west side of the mountain. The tunnel alignment crosses beneath the North Branford quarry along a 200-foot-wide restricted zone that bisects the quarry. Quarry excavation within this restricted zone is maintained at no lower than El. +211 feet to provide adequate cover (approximately 70 feet) and mitigate the impacts of blasting on the tunnel.

The tunnel was excavated through the trap rock and sandstone in a 7 foot 10 inch wide by 7 foot 10 inch high horseshoe shape (Figure 2). Based on geologic mapping, the GHT penetrates the Holyoke Basalt (trap rock) from the Intake Gate House to approximately Sta. 24+30, and the underlying Shuttle Meadow Formation (sandstone) west of Sta. 24+30.

The Holyoke Basalt, which overlies the Shuttle Meadow Formation, consists of at least two thick flows of massive, finely crystalline basalt that directly underlie the majority of Totoket Mountain, including the North Branford quarry. This lower Jurassic-age basalt is at least 350 feet thick in the project area and is generally fresh to slightly altered. The vertical columnar joints that were originally nearby now dip approximately 13 degrees to the southwest.

The Shuttle Meadow Formation is exposed along the western edge of Totoket Mountain and underlies the lowlands in the vicinity of State Route 80. This lower Jurassic-age unit consists of interbedded sandstone (arkose), siltstone, shale, and conglomerate. The thickness of each sedimentary layer is generally less than 3 feet but ranges up to about 20 feet in the project area for shale layers.

The contact between the Holyoke Basalt and Shuttle Meadow Formation is baked (hardened), welded together from the heat of the overlying basalt flow, and occurs at around Sta. 24+30 in the tunnel. No weathering is present on the upper contact of the Shuttle Meadow Formation where it underlies the Holyoke Basalt in the project area.

GROUNDWATER CONDITIONS

Based on available groundwater information, the Holyoke Basalt and Shuttle Meadow Formation are considered to be poor aquifers. The interbedded siltstone and shale of the Shuttle Meadow Formation most likely act as aquitards within this unit, restricting vertical groundwater flow. Additionally, the contact between the Shuttle Meadow Formation and the overlying Holyoke Basalt is tight and relatively impermeable. Artesian groundwater pressure may be present at the contact between the gently dipping, truncated sandstone beds in the Shuttle Meadow Formation and the Holyoke Basalt.

The contact between Holyoke Basalt flows observed in the quarry is distinct but not open. No water has been observed flowing from the contacts between the two exposed basalt flows. Additionally, no holes or cavities were observed in the basalt. The contact between the basalt and the underlying Shuttle Mountain Formation is baked and tight. No permeable zones were observed along the upper and lower contact of the Holyoke Basalt (JA, 2011).

PAST INSPECTIONS AND MAINTENANCE

The GHT is the sole raw water transmission line supplying water to the RWA's nearby water treatment plant, the largest in the RWA system. This treatment plant supplies potable water to most of the 16 towns within the RWA's distribution system and, because of pressure zone configurations and existing legacy restrictions within the distribution network, only local clearwell storage is available during water treatment plant outages. In a total shutdown of the treatment plant, RWA has only a couple of hours of supply, necessitating any inspection work to be done while the tunnel remains in operation. Performing inspections at low demand times, such as overnight during winter months, would allow for the facility to operate at low flows, thus enabling personnel and/or equipment to enter and function within the tunnel. These time restrictions and safety rules in place to protect workers make entry for maintenance purposes nearly impossible. In fact, such entry has not been done within the tunnel itself. Portions of the system upstream and downstream of the tunnel do have regular maintenance, but the tunnel itself is generally skipped.

The first documented inspection of GHT was conducted in 1973 with a commercial diver (JA, 2011). Diver entry, at the time, did not follow any Occupational Safety and Health Administration (OSHA) standards. OSHA, established in 1971, did not adopt its first commercial diving safety standard until July 22, 1977 (OSHA, 2011). For the 1973 inspection, the diver entered the watered tunnel with both a tether to the surface and air bottles within the tunnel, which were moved around during the inspection on a hand truck. As the diver neared the point at which the tunnel transitions to a pipeline, around Sta. 29+61, the diameter of the conduit narrowed. Flow, which was still ongoing, increased in velocity, causing the air bottles and hand truck to be pulled into the pipeline, dragging the diver with it. The diver was unable to climb his way out of the tunnel, so the diver's attendants obtained the aid of some of the miners at the nearby quarry and physically dragged the diver back to the surface. To aid in his rescue, the diver removed his weight belt and left it in the tunnel. He was then rushed to the submarine base in Groton, CT, for decompression treatment. The diver survived this incident; however, the owner has been reluctant to repeat this method.

Apart from some localized inspections of the intake chamber (the entry point into the tunnel), the next documented inspection of GHT did not happen until 2011, 34 years after the failed diver inspection. For the 2011 inspection, RWA opted to use an ROV equipped with high-definition cameras, halogen lighting, and sonar. This inspection became the baseline for the RWA, noting that the tunnel was generally in good condition. During this 2011 inspection, two main defects were identified by the ROV. The first, located in the crown at approximately Sta. 23+15, did not appear to pose a risk to the operation of the tunnel. Although the liner had failed in a small area of the crown, the area was still watertight, even though the void was the full thickness of the lining. The defect was estimated to be about 2.5-feet wide and 3-feet long. Just upstream of the transition point at approximately Sta. 29+61, however, a large liner defect was identified, measured, and monitored for leakage. This defect in the lining was estimated to be approximately 6-inches wide and 8-inches long with a larger cavity behind the lining approximately at least 1.6 feet deep, 6 feet wide, and 3 feet long based on sonar measurements. Although the observed exfiltration was noted, the team was unable to quantify it.

In the spring of 2017, RWA performed a second ROV inspection of the GHT (MJA, 2017). This was both a reinspection of those portions of the tunnel that were inspected in 2011 and an inspection of the 48-inch-diameter cast iron pipeline and associated 36-inch-diameter cast iron access portal riser between the tunnel transition point and the gate house located near Forest Road. Because of a combination of low reservoir levels and excessive wind conditions at the time of the inspection, water quality within the tunnel was poor, with high turbidity. This made for poor visibility in the water column for the ROV throughout the course of the inspection. In addition, there appeared to be more organic material coating the interior surface of the tunnel than in the 2011 inspection, making detection of some surface anomalies difficult.

The previously identified crown defect at approximately Sta. 23+15 was found to have little to no change in appearance, and it again appeared that no water was exfiltrating the tunnel at this location. This defect was again classified as a narrow and fully penetrating structural feature defect because it locally interrupted the concrete lining but did not appear to have a cavity behind the lining, nor was it considered to affect operational reliability at the time.

What was of concern to the inspection team was the apparent increase in size of the defect at the transition point at approximately Sta. 29+61 (MJA, 2017). There was also concrete debris in the invert that was not previously seen in 2011. While exfiltration was observed, the inspection team was unable to quantify the exfiltration at this location. In addition to optical cameras, side scanning sonar was used to map the extent of the defect. Finally, this inspection was able to continue through the 48-inch pipeline to the gatehouse.

During fall 2017, increased leakage and a secondary leakage site were noted at the surface above the tunnel alignment. The RWA grew concerned that the condition of the known defects had changed or that a new defect was now expressing leakage, either of which could lead to failure of the lining. Therefore, the RWA commenced an emergency ROV inspection.

The emergency ROV inspection found that the dimensions of the defection at the transition point at Sta. 29+61 had not significantly changed since the previous inspection in the spring (SeaView, 2017). However, cracking was observed propagating from the defect, and it was noted that part of the liner could fail. It was unknown if this cracking was new since the last inspection or was just now being seen because the visibility was considerably better than the previous inspection. The ROV continued down the pipeline in an attempt to determine the source of the newly noticed secondary leakage. With the use of saline dye, the secondary leak was found in the invert of the 48-inch pipeline at approximately Sta. 30+05. It was also in this area that the 1973 diver's weight belt was found.

MANNED INSPECTION AND REPAIRS

Based on the ROV results from the 2011 inspection, the planned 2017 inspection, and the emergency 2017 inspection, RWA began procurement for the design and installation of a temporary bypass system, to allow for the continued operation of the water treatment plant while taking the GHT out of service. Separate from this procurement, RWA also procured contracts for the Design/Resident Engineering Inspection for repairs to the tunnel and construction of the necessary repairs.

The temporary bypass system, designed and installed by others under a separate contract, was built to carry approximately 30–40 MGD from the reservoir to the raw water transmission line, downstream of the tunnel repair work area. It was installed, maintained, and operated throughout the duration of the repair work by a separate contractor, hired by the RWA specifically for this piece of work.

Contracting the repair work for GHT was somewhat different than a normal design-bid-build or design-build format because of the short timeline from the start of design through to the end of construction. The owner initiated the start of design on November 1, 2017, and had to have a completed repair and be able to return the tunnel to operation on or before May 1, 2018, a total of 6 months. In addition to the relatively short timeline to complete design, procurement, and construction, available information on the tunnel condition was limited to what was collected remotely during the previous ROV inspections discussed previously. With a final goal of providing a high-quality repair to the client and providing contract documents that were fair to both owner and contractor, the design team approached the project atypically.

Tunnel Repair Design

Design for the tunnel repair began on November 1, 2017, and was based on the information collected during the three previous ROV inspections as no other data were available. McMillen Jacobs Associates completed the design, aided in procurement, and inspected the tunnel repairs to meet RWA's operational deadline of May 1, 2018. This six-moth timeline, in conjunction with the relatively unknown condition of the tunnel liner outside of the few areas identified in the ROV inspections, created the need for a design that was fair to both RWA and the contractor, and also provided a mechanism to monitor construction costs, time, and quality.

Because of the limited amount of information available to the design team following the various ROV inspections, the owner and engineers determined that it was best to provide standard design repairs for four classes of anticipated defect types. These defect types, ranging from surficial damage to the existing concrete liner to full depth penetration of the liner creating a structural deficiency, were detailed as typical repair details. Figure 3 shows an example of one such standard design repair for a Type IV defect. It was noted that during an initial manned tunnel inspection, the engineers would work with the contractor to identify which repair type would be used at each identified defect.

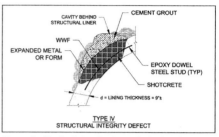

Figure 3. Typical standard repair design detail

After liner repairs were completed, the void space behind the lining was to be grouted, first with a sanded Type 1 grout and then with a lean Type 2 grout, as required. As with the defect repair design, it was understood by the designers and owner that there was great uncertainty in the amount of void space behind the lining and subsequent need for grouting. As such, the designers provided a "worst-case" grout volume estimate for the bid quantities, attempting to prevent overruns and allow proper budgeting for the owner.

Pay quantities needed to be both flexible and fair to the contractor while still providing control to the owner for budget management. Generally, unit prices were used for all work items, with typical lump sum dollars used for mobilization, demobilization, preparation of a health and safety plan, and the project maintenance bond. The rest of the pay quantities were structured differently. While all were unit prices, the remaining work items were set up for payment such that the contractor would receive payment per each work shift for fielding a crew with equipment, performing site work, and other general access-related preparation; and then would receive an additional amount per unit of work done above the basic tasks. Additionally, there were some items that the design team was unable to estimate ahead of time (like determining how much leakage water was to be expected) for use in sizing a temporary water treatment facility. Instead of asking contractors to develop a price for these items, the owner agreed to provide allowance values in the bid table.

As an example, Figure 4 shows the blank price proposal sheet that was included with the contract documents. It is important to note that the final negotiated price eliminated the call for liquidated damages that is discussed at the top of the form. The contractors involved in bidding this work were adamantly opposed to this requirement and were able to get that removed from the final contract. The unknown nature of the work did not lend itself to being able to fairly assess liquidated damages in the event that work took longer than initially estimated. The client always requires liquidated damages to be assessed in its contracts but was willing to forego this requirement in this particular contract.

Procurement

On November 29, 2017, the RWA and McMillen Jacobs met with two preselected contractors with the 90% draft design and specifications. Both contractors were well qualified to perform the repair work in question and were capable of meeting the owner's

Great Hill Tunnel and Pipeline Restoration – Price Proposal

The Contractor should complete this form in conjunction with the Pay Item Description document following this form. The Contractor will also propose a liquidated damages daily fee, to be negotiated with the Owner at the time of contract negotiation. The proposed liquidated damages fee should be based on the Owner's daily expense of $75,000/day to run the temporary bypass system.

Item	Description	Quantity	Units	Unit Cost	Extended Cost
1	Mobilization	1	LS		$ -
2	Health & Safety Plan	1	LS		$ -
3	On-Site Rescue Team, Base Rate	120	Shifts		$ -
3a	Differential for 24-hour, 6-day	24	Shifts		$ -
4	Base Shift Rate, 24-hour, 5-day	120	Shifts		$ -
4a	Differential for 24-hour, 6-day	24	Shifts		$ -
5	Tunnel Cleaning/Scaling	36	Shifts		$ -
6	Probe Holes	145	Ea		$ -
7	Grout Holes	2,200	Ea		$ -
8	Grout Type 1	3,000,000	LBS		$ -
9	Grout Type 2	600,000	LBS		$ -
10	Shotcrete	10	CY		$ -
11	Concrete Repair Mortar	200	SF		$ -
12	Rock Anchors	50	LF		$ -
13	Structural Pipe Repair	200	LF		$ -
14	Temporary Water Treatment Allowance				$ 100,000
15	Demobilization	1	LS		$ -
16	2-Year Maintenance Bond	1	LS		$ -
17	Pipe/Access Manhole Connection Allowance				$ 30,000
18	Miscellaneous Allowance				$ 30,000
				Base Cost	$ 160,000
				6-Day Cost	$ 160,000

Figure 4. Contract bid table

tight timeline. The project, including the proposed method for bid preparation and standard repair design, was explained to both contractors, and time was provided for them to ask questions and comment on the design. Those questions and comments were then used by the design team to finalize the contract document package. These final documents were revised and issued to the preselected construction firms on December 4, 2017. These contractors then had until December 8, 2017, four days, to prepare their cost proposals for consideration by RWA. By the following Monday, December 11, 2017, RWA had elected to enter into negotiations with one of the contractors. These negotiations were completed, and a preconstruction meeting held on January 9, 2018. The original contractor mobilization date was set for mid-January; however, because of complications with the bypass installation work this was delayed a few weeks.

2018 Initial Manned Inspection

The initial manned tunnel inspection of the GHT was performed on Friday, February 16, 2018, and included representatives from the contractor, McMillen Jacobs, and RWA. A rescue team provided by a subcontractor was present throughout the initial inspection and for all of the in-tunnel work. The inspection team accessed the tunnel from the intake shaft, and the rescue team provided a safety attendant who remained at the bottom of the intake shaft during the tunnel inspection. The remaining four members of the rescue crew were stationed in the intake building at the surface.

The purpose of the manned inspection was to visually evaluate the stability of the liner for rehabilitation and document major defects. There were concerns about the stability of the liner at the downstream end where the major defects through the liner had been observed by the ROV, and whether some initial repairs were required before the rehabilitation program could begin.

The major observations from the initial manned inspection included:

- Seepage under pressure was observed entering the tunnel through construction joints in the first 71 feet of the tunnel (section within Lake Gaillard), but no seepage was observed coming through the concrete tunnel liner between the joints. The inflow was approximated at 50 to 100 gpm.
- Holes in the liner (other than minor seeps) started appearing at Sta. 15+50. The holes went through the entire thickness of the concrete liner, so any void space and the bedrock behind the liner were visible. In one hole at Sta. 17+29, a half-barrel from a blast hole from tunnel construction was visible in the bedrock.
- The holes in the liner consistently showed void spaces between the extrados of the liner and the bedrock. The void spaces meant that the bedrock was self-supporting and not putting weight on the liner, and was interpreted as the best case scenario.
- Some of the void space behind the liner was extensive, including areas greater than 2 feet thick and extending upstream and downstream from the holes in the liner.
- Some honeycombed concrete was visible in the tunnel liner exposed by the liner defects, particularly where the largest holes in the liner occurred.
- The biggest holes in the liner were at the downstream end near the transition to the pipeline. Also at the downstream end, the cracks in the liner were more open and wider than they were upstream. There were also several deep concrete spalls that did not fully penetrate through the liner.
- One of the largest holes in the liner, at Sta. 23+15, was 38 inches long, 18 inches wide, and 18 inches deep with seepage into the tunnel occurring through it. Concrete debris—including one piece 33 inches long by 11 inches wide by 7 inches thick—was observed in the tunnel invert below this hole. Another hole, at Sta. 24+90, was 4 feet long, 12 inches wide, and 19 inches deep. A piece of wood was visible behind the liner and was likely a piece of formwork from original tunnel construction.
- At the start of the transition zone at Sta. 29+61, the largest hole in the liner was present from the 1:00 to the 3:00 o'clock position in the tunnel arch, with water flowing out of it at the base. The hole was 2 feet 3 inches wide, 4 feet 11 inches long, 9 inches deep, and penetrated through the liner. There was significant cracking of the liner adjacent to the hole with a couple of large pieces of concrete (approximately 2 feet long by 1 foot wide) that were clearly deteriorated and appeared ready to fall out. Extensive void space and self-supporting sandstone bedrock were visible behind the liner. Concrete debris was present on the tunnel invert.

Defect Classification and Repair

The main conclusion from the 2018 initial manned inspection was that, in general, the concrete liner was in good condition considering the age of the tunnel and the concrete technology available at the time of construction. There were some honeycombing and segregation of the concrete visible in the liner around the larger defects, which could explain why the holes occurred where they did (in the areas of poorest quality concrete); however, the majority of the tunnel had a limited number of defects. No bulging or squeezing of the liner was observed, which indicated that the liner was not taking any weight from the surrounding rock mass, and the bedrock observed behind the liner was self-supporting.

In accordance with the contract documents, defects were classified from I to IV. Type I defects were defined as surficial features that are minor, localized, and penetrate less than one-quarter of the liner thickness. Type II defects were nonpenetrating structural features that penetrated up to half of the liner thickness. Type III defects were fully penetrating structural features that were narrow and did not compromise the structural integrity of the liner. Type IV defects were defined as structural integrity defects that fully penetrated the liner.

Based on the initial inspection, the first priority after the contractor installed the required utilities in the tunnel was to scale the hole at Sta. 29+61 and install 5-foot-long epoxy resin rock bolts into the bedrock through the hole for temporary support since all of the manpower, equipment, and supplies for the tunnel rehab needed to pass through this section of the tunnel. Once the rock bolts were installed, it was recommended that the concrete liner be carefully scaled back to sound concrete to remove any deteriorated concrete and debris, and as the liner was scaled, shotcrete and additional rock bolts were to be installed as needed as the bedrock was uncovered. The additional shotcrete and rock bolts were recommended since there was concern that the concrete across the entire crown of the tunnel at this location may have been of very poor quality and deteriorated. Once scaling began, the extent of the deteriorated concrete was limited and did not extend even to the top of the crown.

Another recommendation resulting from the initial inspection was for the heavy seepage at the construction joints in the first 71 feet of the tunnel to be injected with chemical grout. At each joint, the injections were supposed to chase the seeps around the joint and then seal them off. This was attempted with limited success and was not considered critical for the operation of the tunnel (Figure 5).

The other holes, defects, and major cracks were repaired in accordance with the design. The initial manned inspection allowed a general classification of the number and types of repairs to be completed, and their prioritization for the rehab program. This included 54 defects ranging from Type I (surficial feature) to Type IV (structural integrity defect). This initial classification was further defined once work in the tunnel began and additional inspections could occur. Once construction access and utilities were established within the tunnel, an additional approximately 15 defects were identified, which were generally initially classified as Type I and Type II.

Figure 5. Before (left) and after (right) chemical grouting on the left arch

The repair process generally entailed first sawcutting around the defect. Sawcutting ensures that while chipping out the defect, the size of the repair area is contained. Additionally, the cut can be flared to provide a mechanical locking mechanism of the repair material. Poor-quality concrete was then chipped out until only sound concrete remained. Often, what appeared to be a small Type I defect on the surface of the lining turned out to be a larger Type II or Type III defect after chipping out the poor-quality concrete. Most of the poor-quality concrete consisted of coarse aggregate with washed out fine aggregate and cement. It is suspected that this resulted from poor water management during the placement of the concrete. Finally, the defect was repaired with cement mortar or fiber reinforced shotcrete with or without welded wire mesh, as required.

Probe Drilling and Grouting

Construction access for the repair and grouting work was established by cutting out the leaking section of the 48-inch-diameter pipeline, approximately 150 feet downstream of the transition point. From this access point, workers, equipment, and materials could be winched into the tunnel with a system devised and specifically built by the contractor for this job.

In addition to repair of identified defects, a major component of the work was to grout behind the lining. The contract documents initially called for probe hole arrays, consisting of five probe holes, to be drilled every 100 feet in the mined section of the tunnel. Because of space and equipment limitations, the array arrangement was changed to three holes, one each at the 10:00, 12:00, and 2:00 o'clock positions. The purpose of the probe holes was to quantify the extents of the voids behind the lining and refine the grout quantities.

After an initial cleaning of the tunnel and establishing of the work zone, the contractor proceeded to drill 129 probe holes to investigate the extents of the voids behind the lining. With a better understanding of the extents of the voids, production grout holes were drilled on 8-foot spacings along the length of the mined tunnel. While the construction documents called for Type 1 sanded grout to be used for primary, Type 2 neat cement grout was used for both primary and secondary grouting. A total of approximately 350,000 pounds of Type 2 neat cement grout were used.

CONCLUSIONS

Facility owners, responsible for the ongoing monitoring and maintenance of our country's aging infrastructure, are faced with many difficulties. Rising costs, shrinking budgets, and increasingly stringent safety standards are just a few of the many issues faced when attempting to provide for reliable facilities and uninterrupted operations. As such, it is important for all players—owners, designers, and contractors—to work cooperatively in completing projects, even in the case where a contracting method is not one that they are used to seeing. Although it is important to work together collaboratively, all parties must also tend to their own interests. They need to be aware of contract terms, both for their own protection and for the protection of the success of the project. Regardless of the speed necessary to get a project done within the known constraints, contracts still must be written, and drawings and specifications must still be developed and enforced such that the owner, designer, and contractor all know what is expected from the work. Payment provisions should be fair to the contractor and the owner. Penalties should be enforceable and should only be included when they are necessary and effective. Open lines of communication are imperative to keep change orders to a minimum. In short, working cooperatively can make a complicated yet short-timeline project a success. Rehabilitation projects typically have quite a few unknown conditions that will be encountered throughout the course of work, and as long as open communication continues throughout the project, the work can be completed successfully.

REFERENCES

Jacobs Associates (JA). 2011. Great Hill Tunnel Condition Assessment. Prepared for South Central Connecticut Regional Water Authority.

McMillen Jacobs Associates (MJA). 2017. ROV Inspection of the Great Hill Tunnel. Prepared for South Central Connecticut Regional Water Authority.

Occupational Safety and Health Administration (OSHA). 2011. Timeline of OSHA's 40 Year History. Retrieved from https://www.osha.gov/osha40/OSHATimeline.pdf.

SeaView Systems, Inc. 2017. Regional Water Authority Great Hill Tunnel Investigation Report. Prepared for South Central Connecticut Regional Water Authority.

Hecla's Mobile Mechanical Vein Miner

Clayr Alexander • Hecla Ltd.
Mark Board • Hecla Ltd.
David Berberick • Hecla Ltd.
Wes Johnson • Hecla Ltd.
Marcus Eklind • Epiroc

ABSTRACT

Hecla—Lucky Friday is an underground silver, lead and zinc mine operating at depths over 2.3 kilometers (7,500 feet) below the surface. Technical issues at this depth include high rock temperature, seismicity and logistics. Lucky Friday has partnered with Epiroc to design, build, test and deploy a mechanical miner for primary ore production. In this paper, Lucky Friday and Epiroc provide a case study discussing lessons learned from the final design and of this machine.

INTRODUCTION

The Lucky Friday mine of Hecla Mining Company (Figure 1) is the oldest active (76 years) and deepest mine in the Coeur d'Alene mining district, producing approximately 725 tonne per day (800 tons per day) from narrow, vertical silver, lead, and zinc veins. Cut and fill stoping is currently employed at depths to 2.3 km (7,500 ft) below ground surface, making it one of the deepest mines in the western hemisphere (Figure 2). The #4 Shaft (Sturgis et al., 2017), a winze collared at approximately 1.8 km (5,900 ft) below surface and extending to a depth of nearly 2.9 km (9,600 ft), was recently completed to exploit the high grade 30 Vein, whose resource has been identified to at least 3 km (9,700 ft) in depth.

As with other deep mines worldwide, several challenging engineering issues occur, including: high in situ rock stress, squeezing ground, seismicity, high in situ rock temperature that requires refrigeration, limited ventilation airflow rates due to high water gage resistance, and long travel times for worker transport to mining areas.

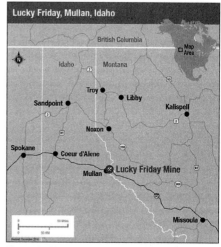

Figure 1. Location of the Lucky Friday Mine

GEOLOGY AND IN SITU STRESS

The host rock mass is an extremely thick sedimentary sequence (the Belt Series) that has been tilted vertically by mountain building. The rock mass in the current section of the mine consists of weak and heavily foliated argillites and somewhat stronger, bedded silty argillites termed siltite. The vertical veins are parallel to the foliation with width around 3.0 m (10 ft). There are two regional, vertical faults that parallel the vein—one forming the hangingwall of the orebody and one approximately 90 m (300 ft) in the footwall. Additionally, there are a number of thin and rough, north dipping faults that occur in the footwall of the vein and terminate on the major vertical faults.

Both the argillite and vein "package" rock have moderate strength, with uniaxial compressive strength averaging about 100 MPa (14,700 psi with a range from about 70 to 125 MPa), and is moderately abrasive, with CERCHAR abrasivity index of two to three. The major in situ stress is horizontal and oriented in a NW direction with magnitude about 1.5 times the vertical stress. Consequently, the Lucky Friday stress magnitude at depth is roughly similar to significantly deeper South African gold operations.

MINING METHOD EVOLUTION

Overhand Cut and Fill with Hydraulic Fill

The initial mining method employed at the Lucky Friday mine since its inception was captive cut and fill (jacklegs and slushers) mining. Prior to the 1980s, overhand cut and fill with hydraulic sand backfill was used. Horizontal production levels were driven every 60 m (200 ft) vertically from the shaft, and five or more slusher stopes established on each level. As the horizontal line of stopes progressed upward toward the previously mined level above, a sill pillar of diminishing height was created. When this pillar was reduced in height to less than around 15 m (50 ft), pillar bursting would begin and continue until the pillar eventually yielded. To minimize the size and occurrence of the seismicity, the advance line of the stopes was altered to a staggered pyramid or stair step shape in the mid 1970s (Board and Crouch, 1977). This change in geometry, though simple, tended to minimize the volume of highly stressed pillar, reduced the overall energy release rate and tended to shunt stresses to the abutments.

Underhand Mining with Paste Fill

Although the overhand stope advance geometry change reduced the magnitude of seismic events, it did not totally prevent them as pillars were still created. A study aimed at reduction in the seismic hazard was conducted in the late 1970s, with a recommendation to convert the method to a mechanized underhand approach (Board and Voegele, 1981). In the early 1980s, the mining method was changed accordingly, utilizing an engineered cemented paste fill back (Peppin et al., 2001). The method, termed the Lucky Friday Underhand Longwall (LFUL) converted the stope access from captive in vein raises to a series of footwall spiral ramps with slotting to the vein. The original objective of the method was to create a downward advancing longwall, to eliminate pillars and to continuously shunt the stresses to abutments in advance of mining. This method is still currently in use and has been successful in reducing seismic occurrence and event magnitude through minimizing the creation of pillars and the reduction in the exposure of miners to the stope face area. The engineered cemented paste fill back has been very successful in improving stope ground conditions. In the mid 1990s, a major new vein (the 30 Vein) was discovered at depth beneath the historic Gold Hunter mine about 1.6 km (1 mile) northwest of the Lucky Friday mine. Contrasted with the host rock mass at the Lucky Friday which is a brittle and strong siliceous quartzite (the Revett formation), the 30 Vein is located in the weaker and heavily foliated argillite (the Wallace formation) described previously.

Underhand Mining Using Continuous Mechanical Cutting

All mining activity was transferred from the Lucky Friday Vein to the 30 Vein and associated parallel veins in the late 1990s, with access to the 30 Vein via horizontal haulage drives from the Silver Shaft at the 4900 and 5900 Levels. The shallower depth

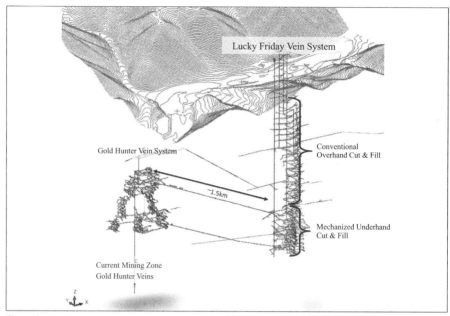

Figure 2. Isometric veiw of Lucky Friday Mine—Lucky Friday and Gold Hunter vein systems and workings

and lithologic change to a weaker, more deformable rock mass resulted in a significant reduction in seismicity. As mining continued with depth, seismicity began to recur, primarily at some distance from the vein, as the result of slip on the north dipping fault structures. At the present time, mining has progressed down dip to approximately the 6300 Level, 2.2 km (7,300 ft) below surface (Figure 2). Exploration drilling below the advancing mining front has identified a prominent widening 30 Vein resource with increasing silver and base metal grades. The current reserve has been identified to about 7600 level and has been intercepted below the 8000 level. To exploit this deep resource, a decision was made in 2009 to begin construction on a winze from the 4760 Level to the 8620 Level (2.9 km (9,590 ft) at depth). This shaft was completed in January 2017 (Sturgis et al., 2017).

To ensure viability of the mining of the 30 Vein to such great depth, a revision of the current underhand mining method was necessary to minimize seismic potential as well as exposure of workers to increased heat and diesel particulates. An internal Hecla study was conducted to first identify overall criteria for the method, and to then develop a mining concept. The criteria identified were as follows:

- Minimize seismic potential by elimination of pillars using a true underhand longwall to top slice the 30 Vein, and, through use of an incremental, continuous mechanical mining method to eliminate in stope blasting that may trigger remote fault slip movement.

- Eliminate diesel particulates and reduce heat rejection to the ventilation by implementing battery and electric equipment to the extent possible. Ultimately, this goal will minimize refrigeration requirements and minimize an increasing ventilation demand.

- Remove workers from in stope conditions to the extent possible by using state of the art tele remote and autonomous operations.

Based on these criteria, Hecla entered into discussions with numerous equipment OEM's for development of a purpose built mechanical Mobile Miner that could cut the relatively weak and moderately abrasive vertical veins. Mechanical cutting is facilitated by the relatively uniform width of the 30 Vein and the predictable verticality and straightness (no offsets) of the vein over its 760 m (2,500 ft) strike length. Epiroc was selected as a partner in this project due to its long term development of steel disc based (TBM) Mobile Miner technology (including two similar units currently in testing) and Epiroc's willingness to share in development engineering and testing costs. A set of design and operational criteria were developed for the Mobile Miner based on:

- Advance rate (ft/day/machine)
- Mining width and height
- Maneuverability (turning radius) and negotiable grade
- Operation within the constraints of the current underhand mining system with an engineered paste fill back
- Tele remote (tethered) operation and automated face cutting operations
- On board bolter with an operator for wall bolting and meshing (roof pre supported) with goal of eventual automation
- Wall collision detection and guidance assist
- High definition cameras for face viewing

Confirmation of the cutability of the 30 vein "package," which consists of sulfide ore, gangue material of siderite (iron carbonate) and argillite was confirmed via linear cutter testing on large blocks of vein package material at the Earth Mechanics Institute of the Colorado School of Mines (CSM, 2017). This testing, using standard 17" steel disc cutters showed relatively easy cutting and spalling of rock chips using 75 to 100 mm (3 to 4 in) cutter spacings and moderate applied cutter forces (Figure 3). Additionally, due to the moderate abrasivity of the vein, cutter wear is expected to be low.

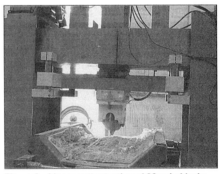

Figure 3. Linear cutter testing of 30 vein blocks at Colorado School of Mines Earth Mechanics Institute—chip formation between cutter tracks

An approximate 1½ year design program was initiated in 2015, culminating in completion of the design in 2018 with manufacturing scheduled to be completed by early 2019. Currently, the Mobile Miner is estimated to undergo a 5 month testing phase that will be conducted in a test mine at the manufacturer's site, followed by shipment, reassembly and proof of concept at the Lucky Friday mine starting in late 2020.

HECLA MOBILE MINER DESCRIPTION

The Hecla Mobile Miner, otherwise referred to as the Remote Vein Miner (RVM), is a machine for mechanical rock excavation of tunnels in hard rock. It uses standard steel disc ring rolling cutters, placed on the circumference of a "wheel" or cutterhead, to crack and fracture the rock so that the rock can be excavated and transported with a conveyor to the back of the machine. The cutterhead is suspended on a boom and the rotating axis is horizontal so the cutters are rolling with vertical kerfs. The RVM is a "partial face machine" which means the cutterhead must be repositioned to be able to excavate the whole tunnel face.

The RVM main benefits are:

- High mobility—multi face development
- High productivity compared to drill and blast
- No need for blasting and therefore no need for charging, ventilating and scaling
- Electric hydraulic powered utilizing 4160v for mining and 480v for tramming thereby reducing DPM emissions
- No secondary cracks in rock due to blasting which improves ground support application and stability
- Continuous production mining process; excavation, muck handling/removal, ground support, and backfill placement during the mining cycle

General Arrangement

The RVM consists of two major parts; the rear "power unit" and the front "miner" (Figure 4). The miner section incorporates the cutterhead and everything needed to excavate ore; the power module contains all the motors, pumps, ground support and control systems. Each unit is roughly 10 m (30 ft) in length. The two parts are connected by an articulation joint to improve maneuverability for negotiation of tight corners as well as to minimize vibration at the operator's cab. The miner has two track units and the power unit has one twin track that can be controlled separately. To obtain better steering control of the power unit, the track is mounted on a slewing bearing allowing the trailer to swing sideways when the tracks are slewed 90 degrees.

Figure 4. Side view of the Mobile Miner (RVM) showing major machine components

The RVM's cutterhead is capable of slewing 25 degrees left and right from center and its boom lift ranges from 26.5 degrees up to 6.5 degrees down. This results in a minimum drift excavation width of 3.3 m (10.6 ft) to a maximum width of 5.2 m (16.9 ft) and a minimum excavation height of 3.6 m (11.8 ft) to a maximum height of 5.0 m (16.3 ft). The typical maximum excavation without repositioning is 5.2 m (16.9 ft) in width and 4.5 m (14.7 ft) in height (Figure 5).

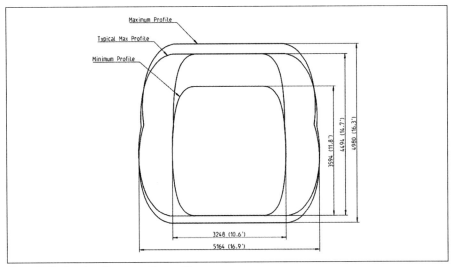

Figure 5. Various heading excavation profiles

PROJECT PHASES AND MILESTONES

The Hecla Remote Vein Miner Project is divided into six Phases that were also used as Stage Gates or decision points:

- Phase 1—Prefeasibility Study
- Phase 2—Feasibility Study
- Phase 3—Design
- Phase 4—Manufacture and Testing
- (Pending) Phase 5—Test Trial at Epiroc test mine
- (Pending) Phase 6—Field Trial at Lucky Friday mine

The project is currently in Phase 4. Manufacturing will be complete in 2nd Quarter 2019 with machine deployment and field trial at the Lucky Friday in 2020.

DEPLOYMENT STRATEGY

As stated previously, the RVM will be used to create an underhand longwall by top slicing the Gold Hunter 30 Vein. The miner will enter one end of the vein and advance horizontally on strike, mining beneath the engineered paste backfill until it exits the mineable vein at the end of its strike length.

The transition from the current cut and fill mining method to a method (with rubber tire mobile equipment) that accommodates the RVM will require development of larger drift profiles and turning radius. The RVM has been specifically designed for the Gold Hunter 30 Vein orebody at Lucky Friday. The initial work area selected is the 12 Stope as shown in Figure 6.

The implementation plan of the RVM is envisioned to be in two phases. The first phase of implementation is to establish a true longwall face across the entire 30 Vein by advancing 12 Stope (yellow in Figure 6) to the same elevation as the adjacent

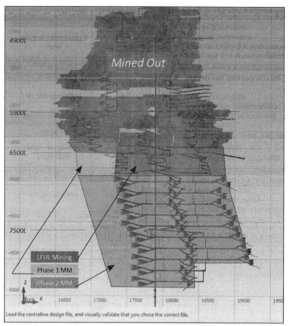

Figure 6. Phased deployment of the remote vein miner (current LFUL mining consisting of three stopes is shown in blue) initial proof of concept (Phase 1) in 12 stope (yellow) followed by underhand longwall (tan)

stopes. The second phase will be the implementation of a second RVM and with two machines top slicing the full minable strike length of the 30 Vein.

Phase 1 Proof of Concept Deployment

The Phase 1 implementation will involve development of the infrastructure necessary to assemble and maintain the RVM. An assembly shop approximately 10 m × 10 m cross section by 30 m in length (33 ft × 33 ft × 100 ft) with overhead crane, is currently in development in the vicinity of the 12 Stope and 6500 level. To keep large spans at a minimum shop depth is approximately 2,300 m (7,500 ft), the crane bay will be limited to travel on one side of the RVM power module or mining module. Large excavations have previously been mined at this elevation to support refrigeration system expansion. Lessons learned at Lucky Friday have shown that aligning the drift perpendicular to foliation in the argillite enhances stability significantly.

Transportation of components underground will be a significant logistical task as the mine does not intend to curtail production while the machine is being assembled. The Lucky Friday mine has a history of successfully managing logistical challenges. The #4 Shaft project is one of these examples, (Sturgis et al., 2017). The transport path for parts is approximately 6.4 km (4 miles) long and must first pass down the Silver Shaft to the 5900 Level. The Silver Shaft is limited to a cross section of 1.8 m × 1.8 m (6 ft × 6 ft) and a weight of 14 tonne (31,000 lbs). Considering that the machine is over 150 tonnes fully assembled, the disassembled machine must be reduced to much smaller components for transport. Large components such as the main frame and cutter head boom will require special slinging arrangements to enable transport down the shaft.

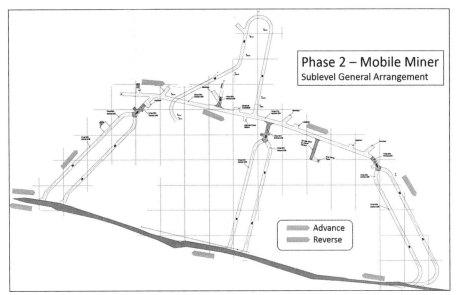

Figure 7. Phase 2 deployment—full 30 vein implementation

Following RVM assembly, commissioning and initial mechanical excavation will commence in the lagging 12 Stope, starting on the east side and mining a series of horizontal cuts west to the economic limit of the 30 Vein, a strike length of approximately150 m (500 ft). The RVM will then reverse to the sublevel to allow for backfilling of the stope before taking the next cut. This mining sequence will eventually connect to the other 30 Vein stopes (15 and 16).

Phase 2 Full Deployment Along Entire Vein Strike Length

Phase 2 will involve regular mining of the underhand longwall along the entire strike length. The following major activities will be performed (Figure 7):

- Establish the sublevel ahead of mining
- Combine the current three ramps that service the drill and blast stopes into one central ramp with centralized utilities
- Use an advancing and retreating sequence shown in Figure 7

The mining method to utilize the RVM across the full strike length of the 30 Vein 760 m (2,500 ft) requires a different development layout and production strategy. Conventional rubber tire mechanized cut and fill mining is designed to minimize development in waste. Therefore, the LFUL cut and fill layout has central access into the vein and mines to an efficient ending limit of 90 m (300 ft) per side. The RVM will be deployed at one end of the vein and mine to the extent of economic mineralization. Access to the vein will be via a series of slots that range in grade from +20% to –20% (Figure 7). One sublevel, as shown in this figure, will service ten total stope cuts. Phase 2 implementation will allow for a higher utilization of the ore face by dedicating the full strike length to the machine and allowing non mining activities to occur behind the advancing ore face. As the miner exits the vein, it will move out through the slot to the sublevel and loop back around to start the next cut.

Two machines mining will be utilized on successive, staggered horizontal cuts with a lead/lag of approximately 300 m (1,000 ft). This means that the lower RVM will be excavating beneath paste fill placed behind the upper miner. Backfilling of the vein will be done behind the advancing RVM after it has passed each slot access point and a new stope entry/exit is established, thereby removing backfill activity from the critical path of ore mining.

The sublevel spacing will be every 45 m (150 ft) vertical and require an increase of waste development over the current LFUL development of about 25% due to the enlarged drift size and sublevel configuration. Development will be mined using drill and blast methods with a strategy to complete development just in time so that the RVM is not delayed by development activity. The mining method used in Phase 2 will allow for the RVM to remain at the ore face while backfilling occurs behind the forward excavation. This fundamental method change allows for an estimated 20% more time at the face as compared to the current LFUL method.

LESSONS LEARNED

The development of mechanical excavation systems to replace the conventional drill and blast cycle in a range of applications including shafts, tunnels, development, and ore production for the underground hardrock mining industry historically has produced various levels of success and therefore many lessons learned. Some of the learnings are generic such as the importance of change management and others are specific to individual prototype trials. Below are some of the historical learnings that have been considered in order to improve the Hecla program.

The RVM has a development history that links back to the three successive Robbins Mobile Miners developed and tested from 1983 to 1993. The Mobile Miners were successful in many ways and provided learnings that were incorporated into the Epiroc portfolio of excavation machines (Hartwig & Delabbio 2010). These learnings include:

- Groove Deepening: The RVM utilizes "groove deepening" with disc cutters following the path of previous cutters which is the same as in civil tunnel boring machines (TBMs). This improves the performance and wear compared to the approach grooved deepening trialed with the original Mobile Miners.
- Prototype Trial Focus: The RVM trial plan has a clear and focused approach that will not have significant delays due to conflicts with existing operations.
- Ability to clean invert: The RVM utilizes proven material handling components to collect cut material from the invert and to prevent buildup.

The RVM program incorporates many learnings from the use of civil TBMs (Zheng et al., 2016) which includes.

- Understand the rockmass: The RVM has been developed based on significant rockmass characterization of the target material. The result of this is a simplified machine that is custom designed for the Lucky Friday operation. This design approach will improve the performance and decrease equipment and maintenance costs.

The overall RVM program has also utilized generic learnings from other technology development and implementation projects such as those described by (Willis, et. all., 2004) and (Delabbio, 2003) which includes:

- Long term approach: The program has been developed to incorporate all the stakeholders and to address the multi phase approach which will require many years to complete.
- Organizational and change management: The Hecla organization along with the RVM project team has been developed with the understanding that success is both organizational and technical to enable sustainable application of disruptive technology such as mechanical excavation.
- Labor skills development: Identify needs and provide training and/or retraining of the workforce on the necessary mechanical, electrical, operating and supervisory skill sets required for mechanical mining.
- Partnership: Hecla has experience working with partners and has developed an approach such that all stakeholders and partners have aligned motivated goals on the shared RVM success.

Hecla engaged their corporate Project Management Group to manage the overall Remote Vein Miner Project utilizing onsite support from Lucky Friday's staff. Learnings from this structure includes:

- Building an early team that includes representations from corporate, maintenance, operations, engineering, finance, and safety has promoted buy in and ownership throughout the company and mine where the RVM will be tested and deployed.
- Once manufacturing began there has been limited contractual opportunities based on the original development agreement and contract for Hecla to control manufacturing cost and schedule.
- Conferencing with Epiroc on a weekly basis has helped to identify issues and opportunities for corrective action. The owner and supplier must communicate on a frequent basis.
- A CPM schedule of design and fabrication activities was not developed early in the project. Without the schedule it was challenging to track progress and take corrective actions on a timely basis. The RVM fabrication has been delayed by the CSM cutter tests, design work slower than planned and procurement delays that were not in the original scope of work.
- Implement a comprehensive risk management program both internal to Hecla and jointly with Epiroc across the design, deployment and operation of the RVM.
- A detailed budget of scope activities spread across time was not developed early in the project. Without the budget it was challenging to track progress and take corrective actions on a timely basis.

The development agreement between Epiroc and Hecla provides stage gate reviews at predetermined project decision points where either party can backout of the development agreement or suggest changes to the agreement. Lessons learned from each of these stage gates includes:

- Prefeasibility Study: Identify key performance indicators early to set expectations for the final product. The needs and application of the technology need to be clearly identified during this stage. What the machine will do and what it won't do will become the project boundaries.

- Feasibility Study: Vetted cutting technology to ensure that the assumptions about excavation were accurate and achievable. Integration of the machine into the mine plan was done. Details of how the machine is to be commissioned and deployed were defined. Study should develop complete cost and schedule estimate for machine deployment. This is to ensure that all Stakeholders understand timing, scope and cost for machine and constructed infrastructure to test machine.
- Design: Change management and version control of design is very important. These should be checked back to the original specification to ensure design is still within operational parameters.
- Manufacture and Testing: Owner must maintain a physical presence in the manufacturing and testing facility to ensure that assembly and functional testing are proceeding according to design criteria. Owner should consider third party verification for quality control.
- Test Trial at Epiroc test mine in Sweden: Project phase has not been executed at this time prior to paper submission for publication.
- Field Trial at Lucky Friday mine in Idaho: Project phase has not been executed at this time prior to paper submission for publication.

SUMMARY

The Lucky Friday mine is one of the deepest mines in the western hemisphere. As the depth has increased, engineering issues typical of deep mines such as seismicity and heat become important aspects of the mine design. A number of revisions to the mining method have been made to increase safety and maintain productivity. Hecla Mining is currently making a step change in deep vein mining technology by developing and employing remote continuous mechanical mining to ensure the continued safety and productivity of the mine. Figures 8 and 9 provide photos of the manufacturing process as of late 2018.

Figure 8. Remote vein miner module in assembly   Figure 9. RVM cutter boom

REFERENCES

Board, M.P., and S.L. Crouch. (1977) "Mine Planning to Control Rockburst in Cut and Fill Excavations," in *Design Methods in Rock Mechanics*, pp. 249–256. New York: ASCE, 1977.

Board, M.P., and M.D. Voegele. (1981) "Examination and Demonstration of Undercut and Fill Stoping for Ground Control in Deep Vein Mining," in *Application of Rock Mechanics to Cut and Fill Mining*, pp. 300–306. London: Institute of Mining & Metallurgy.

CSM (Colorado School of Mines) Earth Mechanics Institute. (2017), "DRAFT—Results of Laboratory Linear Cutting Tests and Performance Estimates for Proposed Atlas Copco Mobile Miner/Vein Miner at Lucky Friday Mine," Golden, CO.

Delabbio, F. (2003) "Hardrock Excavation Alternatives—Present Status and Future Options?" *Conference Proceedings Expanding the Limits of Mechanical Excavation* World Rock Boring Association (WRBA).

Hartwig, S., and F. Delabbio. (2010) "New Tunnel Boring Machine Accelerates Tunnel Construction in Deep Mining Operations," *Mining Engineering*, July, pp. 28–34.

Peppin, C., T. Fudge, K. Hartman, D. Bayer and T. Devoe. (2001) "Underhand Cut-and-Fill Mining at the Lucky Friday Mine," in Underground Mining Methods: Engineering Fundamentals and International Case Histories, Ed. W. Hustrulid and R. Bullock, SME.

Sturgis, G., D. Berberick, W. Strickland and M. Swanson. (2017) "Hecla Mining Lucky Friday #4 Shaft Challenges and Possibilities," Rapid Excavation and Tunneling Conference: 2017 Proceedings.

Willis, R.P.H., Dizon, J.R., Cox, J.A., and A.D. Pooley. (2004) "A Framework for the Introduction of Mechanized Mining," *International Platinum Conference 'Platinum Adding Value*,' The South African Institute of Mining and Metallurgy.

Zheng, Y.L., Zhang, Q.B., and J. Zhao. (2016) "Challenges and Opportunities of Using Tunnel Boring Machines in Mining," *Tunnelling and Underground Space Technology* 57, pp. 287–299.

LED Construction Lighting in Tunnel Projects

Brian Astl ▪ Lind Equipment Ltd.
Michael Cook ▪ SJ Louis Construction
Zach West ▪ SJ Louis Construction
Jim Bresnen ▪ Jennmar Corporation (Civil and Tunnel Division)

ABSTRACT

LED lighting is commonly used in households, commercial settings, and increasingly often as permanent lighting in completed tunnels. However, the use of LED lighting during the tunnel construction phase is less prominent. This paper outlines the benefits of using LED construction lighting in tunnels for contractors and other stakeholders. Implementation of LED construction lighting for a specific project will be detailed, with realized quantitative benefits, challenges overcome, and lessons learned to assist readers with future implementations of LED construction lighting.

INTRODUCTION

In the spring of 2018, the SJ Louis project team constructing the Parmer Lane Interceptor tunnel in Austin, TX was considering whether the status quo of tunnel construction lighting was the right path to follow. LED lighting had become so prominent in other applications, so there was a view that many of the same benefits that LED lighting provides in consumer applications could also apply for the tunnel industry. After a search for the appropriate LED lighting solution and consultations with the City of Austin, the SJ Louis team selected a vendor and started to implement the LED tunnel construction lighting part way through the project. This paper provides details on the LED tunnel construction lighting experience in general, using data from the Parmer Lane experience to highlight salient information.

The Parmer Lane Interceptor project was the construction of a 42" diameter, 11,300' wastewater pipeline in Austin, TX. Standard tunnel construction lighting was used on the first 2,000' of tunnel, with LED tunnel construction lighting used on the remaining 9,300'. The implementation of LED construction lighting led to many quantitative and qualitative benefits, as described below.

CURRENT TUNNEL CONSTRUCTION LIGHTING STATUS QUO

Description

Tunnel construction lighting is defined as the temporary lighting used during construction of the tunnel and is separate from the permanent lighting that may be installed in the completed tunnel. Construction lighting includes above-ground site lighting, lighting around the shafts and lighting inside the tunnel.

The status quo on tunnel construction lighting is not much different than the status quo for commercial or institutional building construction. The vast majority of projects still use bulb-based light source technology such as metal halide, fluorescent and even incandescent. Even while permanent lighting fixtures that go into tunnels are moving to LED quickly, the construction lighting applications lag in adoption of the LED lighting technology.

In terms of form factors, these technologies above are typically deployed in two main ways. Fluorescent and incandescent technologies are typically deployed in a stringlight form factor, with a main electrical cable suspending a light bulb even 10' feet (Figure 1).

These stringlights are typically limited in the voltage that they need to operate properly, and thus are deployed throughout a tunnel in limited daisy-chain lengths that are fed from transformers along the length. Sometimes they are also deployed by tapping into a single-phase of a three-phase circuit, rotating which phase is used each time.

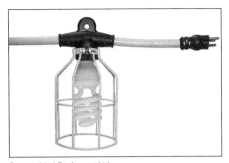

Source: Lind Equipment Ltd.
Figure 1. A typical stringlight form factor

The other typical form factor for deploying fluorescent and metal halide technologies is a floodlight type construction light (Figure 2). These lights are stand-alone light sources that are fed by a common electrical circuit. They can typically operate at higher voltages than a stringlight and are favoured for the length that can be achieved on a single circuit (as higher voltage allows electricity to travel greater distances along an electrical cable). They either tap into rotating single-phases of a three-phase circuit (as described above), or they are fed directly by a higher voltage (e.g., 480V) circuit that is boosted along the length of a tunnel by transformers.

Choosing between which form factor to use typically comes down to the diameter of the tunnel. Smaller diameter tunnels will typically use the stringlight form factor, where larger diameter tunnels will require a brighter light to reach the full

Source: Lind Equipment Ltd.
Figure 2. A typical floodlight form factor, hung from the ceiling in this tunnel

area of the tunnel, and thus will use the flood light form factor. Lighting technology typically follows this pattern as well, with smaller diameter tunnels using incandescent or fluorescent bulbs, and larger diameter tunnels using larger fluorescent tubes or metal halide bulbs, which can put out more light than incandescent or fluorescent bulbs.

Identified Challenges

There are many challenges to working with bulb-based technology that have been identified. They range from cost concerns to environmental concerns to safety and productivity concerns. Many of these concerns come from the fact that bulb-based lighting form factors and performance are a result of the bulb technology and form factor itself. Lights are built *around* the constraints of the bulb's light output and physical attributes, meaning that the lighting outcomes are constrained by those factors as well. As we will see with LED lights below, the lighting outcomes are much more varied as the lighting technology itself is easier to manipulate for the application and environment at hand. Here are many of the myriad challenges identified with traditional bulb-based lighting by the SJ Louis team and other contractors.

Bulb-based lights often don't meet OSHA minimum lighting requirements. The OSHA standard for lighting requirements in "tunnels, shafts, and general underground work areas..." is a minimum of 5 foot-candles (Occupational Health and Safety Administration). Many tunneling projects, as well as above-ground construction projects, struggle to achieve this standard while using bulb-based lighting. One study that examined several above-ground construction sites showed that they didn't meet the OSHA requirements for over 50% of the site area, even during daylight hours (Smith, 2007).

Bulb-based lights require significant installation efforts. A typical stringlight has one hanging point every 10', with a bulb at that hanging point for illumination. This means a worker has to go through the effort of hanging something every 10'. In addition, on most tunnel stringlights, there are exposed connectors for attaching the next stringlight, meaning that electricity (and thus light) needs to be shut down for the duration of the installation of the next stringlight. SJ Louis estimates this to be 15 minutes of total downtime every other day of work.

Bulb-based lights require constant maintenance. The lifetime of bulb-based technology is limited. Everyone has experienced a fluorescent or incandescent bulb 'burn out' over time, let alone being broken through incidental contact. Tunnels take a long time to build, and as such it is expected that all the bulbs will need replacing at least once during the course of construction, increasing material costs and labor.

Bulb-based lights have poor quality of light. The yellowish hue of an incandescent bulb may feel 'warm' in our households, but on a jobsite that hue means poor visibility. In addition, the heat given off by bulb-based technology requires them to have mechanical protection so that workers do not come in contact with the bulbs. These guards create shadows, further impacting the ability for workers to see properly on the job.

Bulb-based lights are cumbersome and prone to breakage. Bulb-based lighting technology has become a commodity over time, and manufacturers have continually brought quality down in order to meet customer pricing demands. As such, many contractors break bulb-based lighting during a project and throw it away afterwards. Bulb-based lighting has become a disposable product rather than an asset, creating increased waste and costs. SJ Louis has found stringlights to be cumbersome and difficult to store between projects. Additionally, when there is a problem with a stringlight, it is difficult to determine which part of the stringlight is broken, leading to expensive and time consuming repairs.

Bulb-based lights use a significant amount of energy. Particularly when compared to LED lights (see below), bulb-based lights use significantly more energy, increasing costs (especially if electricity comes from an on-site generator) and environmental impact. In addition, this larger use of electricity then requires larger gauge electrical circuits to move the electricity to the lights, increasing material costs and installation time.

LED TUNNEL CONSTRUCTION LIGHTING TECHNOLOGY

Description

Just as there are options for LED lighting in residential, commercial and industrial permanent lighting fixtures, there are options for LED tunnel construction lighting as well. And while these solutions will differ in shape, quality, and price, the common use of LED technology to generate light instead of fluorescent, incandescent or metal halide technology provides many benefits.

LED lighting technology uses electrical components to generate light. Unlike bulb-based technology, which use filaments or gases to create the light, LEDs are solid-state. They have no moving parts or filaments to break, and no gases to contain. LEDs can also be developed to provide a range of light output at different colors, including anywhere along the spectrum of 'white' light (from bluish-white to yellowish-white), all at a much higher energy efficiency than bulb-based lighting.

Source: Lind Equipment Ltd
Figure 3. A typical LED tunnel construction light, close-up on the LED array (in yellow)

Espoused Advantages

With this new technology comes several advantages for the tunnel construction industry. These are the factors that SJ Louis considered when choosing to move to LED tunnel construction lighting. This section will discuss the general advantages espoused by the LED industry, while the following section will detail the actual benefits SJ Louis received over the course of the project.

Decrease in electricity consumption and environmental impact. Depending on the bulb-based technology that is being compared, LED technology can use up to 90% less energy to emit the same amount of light. This decrease not only saves money on electricity consumption, but also lessens the environmental impact the project has. Additionally, lower electrical consumption means lower requirements for electrical cable gauge, generator capacity, transformers, etc.

Voltage flexibility. Many LED tunnel construction lights will operate across a range of voltage from 90V–277V without any need for modification of the product. Voltage drops as the distance increases along an electrical circuit. Tunnels are particularly prone to voltage drop as they are long runs of cable without the option for multiple generation points along the path. With bulb-based lighting that runs on a specific voltage level (rather than a range), voltage transformers need to be used to bring the voltage back up as the tunnel progresses. LED tunnel construction lights can continue to be operated as the voltage drops, showing no degradation in performance until the bottom of the voltage range.

Increased durability. LED tunnel construction lights can be built to be extremely durable, and with the solid-state construction mentioned above, there is little chance of damaging or breaking well-built LED lights. Bulb-based lights are much more fragile

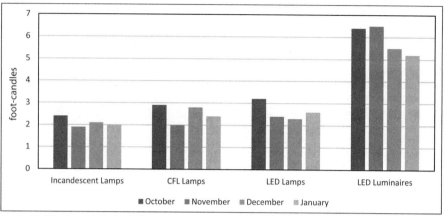

Source: (Lighting Research Center, 2013)

Figure 4. Average horizontal illuminance on the ground by technology and date (excluding daylight to the degree possible)

simply due to the nature of the technology. And with certain technologies like compact fluorescent lighting, breaking a bulb not only reduces light levels, but also exposes users to dangerous compounds like mercury.

Better light levels. As mentioned earlier, bulb-based technology often fails to reach OSHA minimum lighting levels due to the form factor of the bulbs themselves. LED tunnel construction lighting can be built to provide more light in a variety of form factors. Thus, purpose-built LED construction lights perform better than bulb-based lights and even better than LED bulbs.

A study of different technologies of construction lighting at a project in New York showed that purpose-built LED lights provided much better average horizontal illuminance (Lighting Research Center, 2013) (Figure 4).

Increased productivity. With higher light levels and better 'white' color (i.e., not yellowish), LED construction lighting has also been shown to increase worker productivity. In one study, construction workers were asked to build two walls in an 'L' shape in different levels of overall light. The experiments showed that the average time taken to complete the build was faster for teams in the higher lighting level scenario: 34.1 minutes on average versus 39.2 minutes on average for the lower lighting level scenario (Smith B.W., 2008).

Decreased accidents. The higher levels of better quality light also decrease accidents on the jobsite. A seminal 1995 study showed clear downward trends of many different types of injuries, from sprains to burns, to cuts and fractures, as light levels improved on a jobsite (Volker, 1995).

REALIZED BENEFITS OF LED CONSTRUCTION LIGHTING

As mentioned earlier, SJ Louis incorporated LED tunnel construction lighting into 9,300' of an 11,300' tunnel. Every 30', a 50W LED tunnel construction light was placed. Each light came complete with the cord, plug and connector needed to daisy-chain it to the previous light.

Table 1. Comparison of lighting technology wattage

75W Incandescent Bulbs			LED Tunnel Construction Lighting		
# of Bulbs	Watts/Bulb	Total Watts	# of LED Lights	Watts/Light	Total Watts
930	75	69,750	310	50	15,500

Source: SJ Louis, Lind Equipment

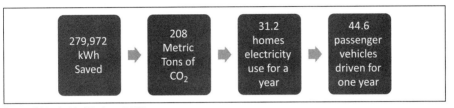

Source: SJ Louis, Lind Equipment, EPA
Figure 5. Environmental impact of adopting LED lighting in Parmer Lane Tunnel

At the time of writing, the LED lights had been in place since July of 2018, with an estimated removal date of June 2019. SJ Louis noticed and quantified several benefits that match with the espoused advantages of LED lighting. Not all potential benefits were measured. For example, productivity and accident levels were not measured for comparison purposes between different lighting technologies.

Decrease in electricity consumption. The specification from the City of Austin was to have a 75W incandescent bulb placed every 10'. SJ Louis used a 50W LED tunnel construction light placed every 30'. The result was that electricity consumption decreased by over 77% (Table 1).

Over the duration of the project (assuming a conservative removal date of June 1, 2019), total electricity saved will reach 279,972 kWh. At current Texas rates of $0.084/kWh, this will result in a saving of over $5000 in electricity costs alone.

Decrease in environmental impact. This electricity saving has a direct impact on environmental impact. Using the USEPA Greenhouse Gas Equivalencies calculator (Environmental Protection Agency, 2018), 279,972 kWh of electricity savings is shown in Figure 5.

Decrease in maintenance. Without the need to change light bulbs throughout the project, the labor and materials associated with construction lighting maintenance disappeared. This was a particularly welcome result given the complexities in the other parts of the project that needed to be attended to.

Decrease in installation time and downtime. The LED tunnel construction lighting used is a plug-and-play type system, with each light easily and quickly connecting into the previous one. There is no need to shut down the lighting in the tunnel to install, and the lights are easier to handle and install than cumbersome stringlights.

Decrease in overall initial spend. In addition to the electricity and labor savings detailed above, the overall initial spend for materials and equipment was less for the LED tunnel construction lighting than it would have been for traditional bulb-based lighting. Exact data has been kept confidential.

Increase in light levels. The City of Austin asked that the LED tunnel construction lighting be spaced such that the minimum light level at any point in the tunnel be equal

Table 2. Comparison of light levels between LED and bulb-based tunnel lighting

Light Source	Minimum fc	Average fc	Maximum fc
75W Incandescent bulb spaced 10'	0.91	5.95	42
50W LED light spaced 30'	0.93	14	300

Source: Lind Equipment

to or greater than it would be with 75W incandescent bulbs spaced 10' apart. When this was accommodated, the overall average light level in the tunnel was higher than it would have been with the incandescent lights. This makes for a safer and more productive environment, although quantitative outcomes were not measured. A lighting simulation program called Dialux was used to model the light levels created by a 75W incandescent bulb hung every 10' versus the 50W LED tunnel construction lights in the 42" diameter Parmer Lane tunnel.

Expected increase in reusability. The salvage rate of bulb-based stringlights from one project to the next is not very high. As mentioned earlier, the stringlights are often broken or damaged in the project and then discarded. The LED tunnel construction lights are expected to be used on multiple projects, which would multiply the return on investment for these lights versus traditional bulb-based lighting. If there are any repairs that need to be done, the parts are easily identified and replacement parts are readily available.

LESSONS LEARNED

SJ Louis was satisfied that the switch to LED tunnel construction lighting was a beneficial one and was both financially and operationally better than traditional bulb-based lighting. There were some specific learnings that can be translated to other projects.

LED tunnel construction lighting can be used for multiple purposes. The versatility of the LED tunnel construction lighting that SJ Louis used was not initially taken into account but proved very helpful over time. Unlike stringlights, the LED tunnel construction lighting could also be used effectively as surface lighting in more remote sites that don't have permanent power. They are bright enough that daisy-chaining a couple together, powered by an on-site generator, can avoid the need to rent a large generator-fed light tower, saving money and time.

LED savings increase linearly. The more bulb-based lights that are replaced with LED lighting, the greater the savings. The longer the LED lights are used, the greater the savings. The savings increase linearly, so any given project that is larger or lasts longer will show greater total savings than smaller and shorter projects. But at all points there is a payback for moving to LED tunnel construction lighting.

LED tunnel construction lighting could be spaced further apart. If the goal was to reach the OSHA 5fc minimum (Occupational Health and Safety Administration), then the 50W LED tunnel construction lights could have been spaced 40' rather than 30' apart. This would have increased electricity savings by another 7% and reduced initial materials spend.

Any project considering LED tunnel construction lighting should undergo a lighting simulation to determine the best lights and spacing for the specific dimensions of their tunnel.

Educating clients on benefits of LED is important. As LED tunnel construction lighting is still relatively novel, educating the client on the potential benefits is important. Through lighting simulations and prospective energy consumption calculations, the benefits can easily be seen before the lights are installed. Not only does this help to increase likelihood of adoption, but the client can also incorporate the benefits into ongoing communication with its stakeholders.

Not all LED tunnel construction lighting is equal. Like any other product, there are varying degrees of quality between different brands and types of LED tunnel construction lighting. Future projects should seek out and verify that the proposed lighting is:

- Durable enough for the environment
    - Lighting should be a reusable asset, not a disposal product
    - Ensure water resistance meets the project environment. Look for labeling such as "wet located rated" rather than "damp location rated."
- Able to meet the lighting requirements
    - Perform lighting simulation studies to ensure the pattern and level of light emitted by the LED construction lights meet OSHA and the client's requirements. Most manufacturers can perform this free of charge.
- Easily installed and interconnectable
    - Reducing labor, downtime, and storage requirements are important factors that can save a lot of money and time

REFERENCES

Environmental Protection Agency. (2018). *Greenhouse Gas Equivalencies Calculator.* Retrieved from EPA: https://www.epa.gov/energy/greenhouse-gas-equivalencies-calculator.

Lighting Research Center. (2013). Alternative Technologies for Construction Lighting. *Field Test Delta Snapshots, Issue 6.*

Occupational Health and Safety Administration. (n.d.). *1926.56 Illumination.* Retrieved from United States Department of Labor: https://www.osha.gov/laws-regs/regulations/standardnumber/1926/1926.56.

Smith, B.W. (2007). Inadequate and Unsafe Temporary Lighting in Buildings under Construction: Risks, Challenges, and Solutions. *Proceedings of Fourth International Conference on Construction in the 21st Century (CITC0IV) Accelerating Innovation in Engineering, Management and Technology.* Gold Coast, Australia.

Smith, B.W. (2008, May-August). Temporary Construction Lighting of Buildings: An Evaluation of Four Techniques, both Qualitative and Quantitative. *The International Journal of Construction Education and Research, Volume 4.*

Volker, S.R. (1995). Beleuchtung und Unfallgeschehen am Arbeitsplatz. *Zeitschrift fur die Berufsgenossenschaften.*

Pumped Storage Projects in Switzerland—Challenges and Solutions

Jürg Künzle ▪ Marti Tunnel Ltd.
Bruno Gisi ▪ Marti Tunnel Ltd.

ABSTRACT

Pumped storage projects are big and complex. The Kraftwerk Linth Limmern and the Nant de Drance plant in Switzerland illustrate that schedule coordination and logistics management are among the most critical success factors for this type of hydroelectric power scheme.

INTRODUCTION

Pumped storage plants store energy by pumping water from a lower reservoir into an upper reservoir and generating power by releasing stored water through turbines located at a lower elevation. There is a need for these types of "hydroelectric battery" facilities where the demand and supply of energy are unbalanced, including the demands arising from public transportation networks during energy rush hours or the energy supply from weather-dependent forms of power generation.

Pumped storage plants involve the design and execution of large-scale excavation and concrete works above and underground, steel lining works, electromechanical works, and various kinds of permanent and temporary ancillary works. Large dams have to be built to create reservoirs, as well as huge caverns to accommodate power generation units, and miles or kilometers of access galleries and water tunnels to connect the different project components. The technical challenges associated with these jobs are indeed fascinating—each of which would be an interesting subject in its own right. However, the *characteristic* challenges of pumped storage projects are those relating to the task of building a multidimensional and multidisciplinary big-ticket project above and underground. Accordingly, this paper concentrates on the construction of such projects and three interrelated challenges: size, complexity, and uncertainty, based on the experience from two recently completed projects in Switzerland: the Kraftwerk Linth Limmern and the Nant de Drance plant.

The Kraftwerk Linth Limmern (hereinafter "KLL") is located in the canton of Glarus, in the southern part of the Linth river valley, a one-and-a-half hour drive from Zurich. The project is based on an existing 480 MW plant. KLL has now a maximum power generation capacity of 1,480 MW and an energy storage capacity of 30,000 MWh. KLL consists of an upper reservoir (*Muttsee*) with a water storage capacity of approximately 30 million yd^3 (23 million m^3), which was created by a 3,458 ft (1,054 m) long concrete gravity dam, and a lower reservoir (*Limmernsee*) with a water storage capacity of 120 million yd^3 (92 million m^3); the difference in elevation between the two reservoirs is 1,942 ft (592 m). Moreover, KLL consists of various underground components, including a 502 ft (153 m) long machine cavern to accommodate four pump turbines with a power production capacity of 250 MW each, a 430 ft (131 m) long transformer cavern, several access galleries, two headrace tunnels, two inclined pressure shafts, two tailrace tunnels, and a surge shaft. The execution of civil and underground works by Marti Group companies began in 2009 and was completed in 2016.

The Nant de Drance plant (hereinafter "NDD") is located in the canton of Valais, close to the border between Switzerland and France, a one-and-a-half hour drive from Geneva. NDD has a maximum power generation capacity of 900 MW and an energy storage capacity of 17,000 MWh. NDD consists of an upper reservoir (*Lac du Vieux-Emosson*) with a water storage capacity of approximately 35 million yd^3 (27 million m^3), which was created by elevating an existing concrete arch-gravity dam, and a lower reservoir (*Lac d'Emosson*) with a water storage capacity of 297 million yd^3 (227 million m^3); the difference in elevation between the two reservoirs ranges from 820 to 1,295 ft (250–395 m). NDD also consists of several underground components, including a 637 ft (194 m) long machine cavern to accommodate six pump turbines with a power production capacity of 150 MW each, a 417 ft (127 m) long transformer cavern, a 18,372 ft (5,600 m) long main access gallery, a 7,875 ft (2,400 m) long system of further access galleries, two headrace tunnels, two vertical pressure shafts and two tailrace tunnels (with a total length of water tunnels and access galleries of 10.5 mi. or 17 km. The execution of the civil and underground works by a Marti-Implenia joint venture called "Groupement Marti Implenia" (hereinafter "GMI") began in 2008 and will be completed in September 2019.

CHALLENGES

Size

The sheer size of a pumped storage construction project can be very challenging. As far as pumped storage plants are concerned, the scope requires contractors to perform a great deal of demanding activities and works in more or less suitably connected areas above and underground. Contractors are required to deal with massive quantities of all kinds of materials; for example, NDD required civil and underground contractor GMI to handle 2.2 million yd^3 (1.7 million m^3) of excavated rock and 562,000 yd^3 (430,000 m^3) of concrete, with truck hauling distances in excess of 3 mi. (5 km) and differences in altitude of up to 1,476 ft (450 m); moreover, GMI operated a Marti Technics conveyor system connecting the portal area of the main access tunnel (ZTH) in Châtelard with the underground batching plant and the area around the machine cavern; the conveyor had a transportation length of 18,175 ft (5,540 m) and a capacity of 1,000 tons per hour. Pumped storage project contractors are also required to move and operate heavy machinery and sophisticated systems of equipment in different project areas in a coordinated fashion; for example, KLL required the civil and underground contractor to perform their activities and works not only above and underground but also in different locations and at different altitudes, including the installation area *Tierfehd* at 2,660 ft (811 m) above sea level, the working area between station *Kalktrittli*, the main machine cavern, and station *Ochsenstäfeli* at 6,168 ft (1,880 m) above sea level, and the vast working area at the upper reservoir at 8,117 ft (2,474 m) above sea level.

Complexity

KLL and NDD are design-bid-build projects, which means that both owners entered into separate individual contracts with multiple designers and multiple contractors for the performance of technically and logistically interdependent activities and works. The complexity of pumped storage projects relates primarily to the fact that the activities and works of the different contractors are interdependent and that, accordingly, their individual time schedules are interrelated.

Interdependencies Between Works of Multiple Contractors

Pumped storage plants are multidimensional and multidisciplinary big-ticket projects. Different categories of works are performed by multiple contractors in multiple working

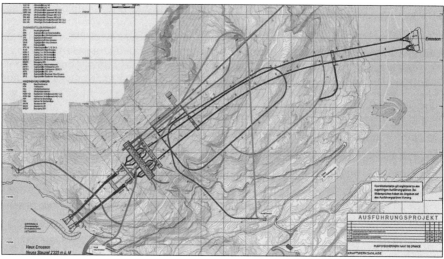

Figure 1. NDD overview of underground components

areas. In the beginning, the construction site is primarily under the civil and underground contractor's control. With some part of his works completed, however, other contractors start to perform their own onsite operations, in parallel with and typically in the immediate vicinity of ongoing concrete works performed by the civil and underground contractor. To some extent, the different contractors are required to perform their work shoulder-to-shoulder, sometimes even hand-in-hand, particularly in the interface areas between the works of the different contractors.

The most demanding interdependencies are those relating to interfaces that are located in the machine and transformer caverns, and along the water tunnels and shafts, that is, where the concrete works, steel lining works, and electromechanical works intertwine with one another.

As far as pumped storage projects are concerned, the management of interdependencies between the activities and works of multiple contractors is among the most characteristic challenges. This applies particularly to activities and works that are located on the overall critical path of a project.

Figures 2 and 3. NDD steel lining interfaces

Scheduling

Managing such interdependencies between the activities and works of multiple contractors is first and foremost a matter of coordinating their substantively interrelated time schedules, which, however, relate to separate individual contracts. The coordinator is required to ensure that there will be no conflicts between activities and works of the different contractors and, if conflicts cannot be avoided, that their impact on the project will be reduced.

Site Access and Possession

It goes without saying that access to and possession of working areas are basic prerequisites for allowing contractors to perform the works in accordance with their various contractual commitments. To be able to give multiple contractors site access and possession, the owner is required to coordinate their individual needs, and their rights and obligations, particularly to the extent that contractors are supposed to share the use of any means of access or working areas. For pumped storage projects this can be very challenging because hundreds of workers, heavy machinery, all kinds of bulky steel pipes, parts of power generators, and other goods need to be brought to different installations and working areas. They must all be transported in accordance with multiple interrelated time schedules.

Uncertainty

The third characteristic challenge of pumped storage projects is that there can be a high degree of uncertainty. The uncertainty with regard to the natural circumstances of the project relates to any weather and geologic conditions. The weather conditions may be very challenging because sites are located at high altitude; for example, the upper reservoirs of KLL (*Muttsee*) and NDD (*Lac du Vieux-Emosson*) are located at altitudes of 8,117 and 7,234 ft (2,474 and 2,205 m) above sea level. Accordingly, from November through April, huge quantities of snow made the performance of work in outside areas above 5,000 ft (1,500 m) largely impossible. This had a considerable impact on the sequence and timing of the different contractors' activities and works. The uncertainty with regard to geological conditions relates to discrepancies between the anticipated and encountered circumstances and their impact on the overall project time schedule. For example at KLL, the civil and underground contractor encountered an unforeseeable 53 ft (16 m) long fault zone called *Mörtalbruch* when excavating the two inclined pressure shafts with a gripper Tunnel Boring Machine (TBM), which delayed work on the critical path of the project by more than three months per shaft. During design, geologists retained by the owner had expected the fault zone to be between 8 in. (20 cm) and 65 ft (20 m) long. Because the solution for overcoming the actual geotechnical challenges was largely unknown when the TBM entered the fault zone, the ultimate impact on the overall time schedule was very uncertain.

Similarly, when excavating the main access tunnel (ZTH) with a 31 ft (9.4 m) gripper TBM, NDD civil and underground contractor GMI encountered a series of unforeseeable ground conditions, including a 395 ft (120 m) long fault zone called *La Veudale* with water ingress in excess of 400–800 liters per second (6,300–12,600 gallons per minute). These circumstances delayed work on the critical path of the project by one year. The owner thus instructed GMI to continue their drill and shoot excavation in the upper part of the tunnel system, which it had begun in parallel with the TBM drive. As a result, the owner was forced to instruct GMI to transport the excavated material from the drill and shoot operation with trucks from *Collecteur Ouest* to the dump area *La Gueleaz* near the lower reservoir instead of conveying it through ZTH down to the dump area in Châtelard. This change had a cost impact.

SOLUTIONS

Schedule coordination and logistics management are the most critical success factors for overcoming the three challenges described above.

Schedule Coordination

Pumped storage plants are multidisciplinary projects that involve multiple contractors performing interdependent activities and works under multiple contracts.

There are different ways of structuring pumped storage and other large and complex infrastructure projects. The owner may want to have a single point of contact for the entire project, that is, for all the design and execution of the civil works, steel lining works, electromechanical works, and ancillary works. In this case, the owner enters into a single contract with a design-build contractor, so that the design-build contractor and not the owner is responsible for coordinating the design and the execution of the works. Alternatively, the owner may elect to divide the project into discrete design and/ or construction packages and enter into separate contracts with different designers and contractors. In such instances, the owner bears the coordination responsibility, though they may want to delegate all or part of that responsibility to one or more of the designers, contractors, or other involved parties; the same applies to the design-build contractor vis-à-vis the designers and subcontractors retained by them.

At KLL and NDD, the owners entered into separate individual contracts with the designers and contractors, because both projects were delivered by traditional design-bid-build methods.

Irrespective of whether the owner elects to divide the project into discrete design and construction packages, someone must coordinate the design and execution of the different works, particularly with regard to all the temporal interdependencies. Thus the coordinator is required to bring the individual time schedules of designers and contractors together to ensure that the interfaces between the multiple contracts are integrated. For example, the commencement and completion dates and milestones specified in the civil contract must be consistent with those specified in the steel lining and electromechanical contracts.

KLL

At KLL, the owner created an overall project time schedule, which is a schedule that includes all major activities and works of all the contractors during the project. When creating the overall schedule, the owner used the civil and underground contractor's sequence and timing of activities and works as a basis for determining the sequence and timing of time critical activities and works of any subsequent contractors, particularly those of the steel lining contractor and the electromechanical contractor. This is how the owner was able to ensure that the milestones and interim or final completion dates set forth in contracts yet to be negotiated would be consistent with the civil and underground contractor's schedule. The overall schedule, which was updated on a quarterly basis and sometimes every month, was used as a guideline for daily decision-making on the part of the owner. The KLL owner devoted several full-time positions to the task of managing the overall project time schedule.

However, a continuously updated overall schedule is only as valuable as the way in which it interacts with the underlying schedules of the individual contractors. This is where another success factor comes into play: flexibility. The KLL scheme includes a 12,467 ft (3,800 m) long permanent access tunnel (ZS1) under a separate contract with another consortium; the tunnel has a diameter of 26 ft (8 m) and a declivity of 24% (!); it accommodates a cable car with load capacities of 120 tons (regular maximum) and 215 tons (exceptional), which provides access to the machine cavern and the transformer cavern. At some point, the execution of the underground works in ZS1 were several months behind schedule, which threatened to delay subsequent works in the caverns, including those to be performed by the electromechanical contractor. In order to avoid this delay, the owner initiated a change order under the main civil and underground contract, so that the Marti-led consortium would excavate the upper part of ZS1 from the top (in the cavern) down towards the delayed gripper TBM so that it could be disassembled in the upper part of ZS1 and not in the cavern. This is how the owner managed to disentangle the delayed excavation of ZS1 and the subsequent works in the excavated machine cavern. To be in a position to do so, however, the owner had to have the right to issue such a change order to a contractor, which contractor did not only have to be willing but also able to perform the demanding additional work on short notice. This capacity should be a consideration for an owner when awarding the main civil and underground contract.

Figure 4. NDD machine cavern

NDD

Schedule coordination was also important for the successful delivery of NDD. The owner awarded the civil and underground contract in 2008 to GMI and the works started in 2008. 18 months into the project, the owner decided to increase the power generation capacity from 600 to 900 MW so that large parts of the civil and underground works had to be redesigned and the contract with GMI renegotiated. With GMI the owner needed to agree on new interim and final completions dates, which lead to a first addendum to the existing civil and underground contract. At the same time, the owner and GMI agreed on a series of principles governing future contract updates and renegotiations. In a general project time schedule called "GATP," the owner started to combine the individual schedules of GMI, steel lining contractor Andritz Hydro, and electromechanical contractor Alstom, which was later purchased by General Electric. Then, in 2013, the owner invited GMI to participate in another renegotiation round to agree on a series of new milestones and to settle a series of compensation and extension-of-time claims submitted by GMI, which led to the second addendum. The NDD owner reset the time-extension counter to zero and brought the sequence and timing of the civil works into accordance with the other contractors' schedules. A similar process was adopted between 2015 and 2017, which led to five complementary agreements, by means of which the owner and GMI agreed on new milestones, the full and final settlement of various claims, and on the scope of and compensation for minor additional works. The owner followed this strategy to repeatedly adjust the civil and underground contract in accordance not only with GMI's own progress but also with that of the steel lining contractor and the electromechanical contractor. Additionally, the owner and GMI agreed that GMI shall perform the concrete works in the caverns as instructed by the owner's project manager ("auf Zuruf") rather than in accordance with a detailed sequence and timing of its activities and works, so as to ensure that GMI would perform its works in a flexible fashion.

Logistics Management

The other solution to tackle the challenges that are associated with pumped storage projects relates to the important task of moving hundreds of workers, heavy machinery, many pieces of equipment, and all kinds of materials to different working areas. The logistics management is typically very challenging because it has to be done in accordance with a great number of substantively interrelated commencement and completion dates set forth in separate contracts, and with each individual contractor's underlying sequence and timing for performing particular activities and works.

KLL

During construction, access to the working areas of KLL was provided by two temporary cable ways. The first was from the terminus to a first intermediate station, and had an inclined length of 6,306 ft (1,922 m), a difference in altitude of 3,448 ft (1,051 m), and load capacities of 25 tons (regular maximum) and 40 tons (exceptional)., The second cable way went from a second intermediate station to the upper reservoir (*Muttsee*), and it had an inclined length of 5,810 ft (1,771 m), a difference in altitude of 1,955 ft (596 m), and load capacities of 25 tons (regular maximum) and 30 tons (exceptional). The two-cable-way system, which included a 1.86 mi. (3 km) long access connection gallery, was the only means of access for all personnel, equipment, and contractor's materials until the 12,395 ft (3,778 m) long access tunnel (ZS1) was finally completed to provide permanent access.

The cable way system was designed, procured, operated, and maintained by the owner which was a key success factor in the successful logistics management at KLL. The owner operated the system 24/7 during 350 days of the year, with departures every 12 minutes between 6am and 10pm (and a little less frequent during the night). Use of cable way was defined by the owner based on the different contractor's particular needs. Their shared use of the cable way system required the owner to follow a stringent coordination process. The owner arranged weekly logistics meetings with the different contractors as well as requiring them to notify them of their transportation needs not less than one week in advance. Additionally, the owner held daily coordination meetings with the logistics managers of the different contractors to clarify the details regarding particular transports. What matters is that it belonged to the owner to set the priorities and that, therefore, decisions have been made from a neutral perspective and with regard to the overall project time schedule. The cable way system was and remained under the owner's control, during the whole project; this allowed him to ensure that decisions regarding the shared use of the cable way system would be made not only in due consideration of the entire project, but also that the different contractors would be treated equally. Had it been for one of the prime contractors to operate and maintain the system, it would have been his responsibility to coordinate his own practical

Figure. 5. KLL cable way system

needs and legitimate interests with those of other contractors and, more importantly, without seeing the big picture. Therefore, as far as design-bid-build pumped storage projects are concerned, the logistics management is a responsibility that the owner should not, and cannot, delegate to one of the involved prime contractors.

NDD

The main access of NDD was provided by means of a 18,372 ft (5,600 m) long *permanent* access tunnel (ZTH). ZTH connects the public road in the village of Châtelard with the 637 ft long, 105 ft wide, and 170 ft high machine cavern. From there, the access to the other working areas (upper dam, transformer cavern, shafts, and intakes) was provided through spiral-shaped system of access galleries (ZTE/ZTVE). From November through April, ZTH and ZTE/ZTVE were the only means of access to the different project areas for all personnel, equipment, and materials of the different contractors. There was no temporary access like at KLL. During the summer, however, a second access via a serpentine road was available to allow the civil and underground contractor to start to excavate the tunnel system from the lower reservoir (*Emosson*), that is, independently from the TBM drive at ZTH.

To facilitate the excavation and concrete works in the machine cavern, and to reduce the risk of conflict and interference arising in connection with the shared use of ZTH and ZTE/ZTVE, the logistics concept of NDD included an underground batching plant, an underground workshop operated by GMI, and the already mentioned conveyor system through ZTH. The underground batching plant was located in the upper part of ZTH, at a distance of approximately 655–1,640 ft (200–500 m) to the machine cavern and the transformer cavern; this allowed GMI to produce concrete regardless of weather conditions and to pump the concrete directly into the working areas since it was a short distance to the caverns.

LESSONS LEARNED

The challenges associated with pumped storage projects relate to their size, their inherent complexity, and the uncertainty that arises from weather and geologic conditions. The solutions to these challenges relate to schedule coordination and logistics management.

There are quite many lessons one can learn from KLL and NDD. In our opinion, the most noteworthy ones are the following:

1. When creating the overall project time schedule, the owner should use the civil and underground contractor's sequence and timing of activities and works as a basis for determining the sequence and timing of time critical activities and works of any subsequent contractors, particularly those of the steel lining contractor and the electromechanical contractor.

2. The overall project time schedule, and each contractor's individual sequence and timing of work, must be updated regularly (for example, quarterly), in consideration of the actual progress. The updating is the result of an iterative process, which means that the owner may be required to review two or more time schedules in parallel. Based on that, the owner may be required to renegotiate the commencement and completion dates set forth in substantively interrelated contracts.

3. Delay on the overall critical path of the project does not only cause contractors to incur additional costs including direct and time-related costs, but also delays the commissioning of the entire facility. Therefore, when it comes

to absorbing critical delay suffered by one contractor, the owner must have a realistic chance to change at least one other contractor's scope and/or sequence and timing of work. This requires the owner to be legally entitled to initiate such change, and to work with contractors who are able to perform accordingly, that is, to be flexible.

4. Instead of disputing the contractors' time-related claims, the owner should from time to time settle these amicably, by renegotiating existing prime contracts in connection with an agreement on new interim completion dates. This allows the owner to reset the individual time-extension counters to zero and, hence, to ensure that under each contract the actual progress corresponds to the planned progress and, more importantly, that actual progress under one contract will be in tune with actual progress under any other contract. The ability to renegotiate interrelated contracts calls for great deal of communication, interaction, and cooperation between the owner and the individual prime contractors, particularly between the owner and the civil and underground contractor. This is a matter of developing and maintaining relationships on different hierarchical levels.

5. The logistics management should be under the owner's control, to ensure that decisions regarding the shared use of any means of access would be made not only in due consideration of the entire project, but also that the different contractors would be treated equally. As far as design-bid-build pumped storage projects are concerned, the logistics management is a responsibility that the owner should not delegate to one of the involved prime contractors.

Rebuilding TBMs: Are Used TBMs as Good as New?

Doug Harding ▪ The Robbins Company

ABSTRACT

Much has been made worldwide of the difference in performance between new and rebuilt TBMs. Worldwide, a bias exists that seems to favor new machines, but is the bias warranted? The reuse of machines can, if done to exacting standards, reduce costs and time to delivery while also reducing the carbon footprint. But guaranteeing the quality of TBM rebuilds is another issue—one that seems only minimally improved by the existence of international guidelines.

This paper will discuss the process of machine rebuilds and the use of rebuilt TBMs with performance examples from projects worldwide. It will seek to establish guidelines and recommendations based on real experiences of success in the shop and in the field.

INTRODUCTION

International guidelines have been developed to standardize the process of reusing a TBM for another project. But is standardization truly possible, and what makes a TBM perform successfully in the first place?

There is no reason why a used machine shouldn't perform as well as a new machine if the operational history of the machine is known and the geology is suitable for its design. In fact, contractors may opt for a used machine for its proven performance, although this view is certainly not the case in every market (see Figure 1).

Large metro projects worldwide often employ dozens of TBMs working simultaneously, resulting in a glut of secondhand machines on the marketplace at any given time. But contractual constraints often form barriers towards using these machines on subsequent projects. Consultants employed by project owners can over-specify technical specifications, often in an attempt to lower risk, but these specifications are not always necessary.

If owners were to instead specify a minimum quality of a rebuilt machine required, this would prevent rebuilt machines from being excluded from projects. This type of specification would require a certain level of TBM knowledge—for example, things that could be specified include main bearing requirements, either that the bearing is new or that it has to be certified for a certain number of hours. A cutter load could be specified for hard rock tunneling, but details like thrust and torque would not be necessary and could be too exclusive.

Not only can rebuilt TBMs result in significant cost savings, but they can also be the answer to aggressive delivery schedules. That, combined with their proven track record, and the reduction in carbon footprint by using an existing machine, makes them an attractive solution for the tunneling industry moving forward.

Figure 1. A main beam TBM originally built in 1979 was refurbished and successfully completed the Albany Park Stormwater Diversion Tunnel in Chicago, IL, in late 2017

QUANTIFYING TIME AND COST SAVINGS

Time and cost savings for a rebuilt machine can be highly variable, depending on the extent of the rebuild and the number of projects the machine is used on. But there is general agreement that under the right conditions, the savings can be significant.

TBMs are still in operation in the industry that have lasted over five decades—in particular a 2.7 m diameter Robbins Main Beam TBM originally built in 1968 is still in operation in Canada. The machine has been used on many projects, and with contractor-led refurbishment at the start of each project the TBM can continue boring tunnels for many more years. With each subsequent tunnel the savings in terms of time and cost multiply.

As for the savings of using rebuilt machines vs. new ones for each project, this is highly variable and can range from 75% cheaper for a simple machine and a tunnel project with tried and tested ground conditions, to around 20% cheaper for a project with more complex requirements (a high-pressure EPB for example).

The advantages of rebuilt machines aren't just in the costs, however. Contractors have stated that the time savings of using a rebuilt machine can be six months or more (as long as the TBM truly fits the project specifications and is not a compromise, and major changes aren't required).

The other benefit is in owning the machine itself: Familiarity of the TBM is a big plus, and operators and maintenance crews are familiar with the equipment, all of which can greatly improve performance during the initial learning curve.

REBUILDING TBMs

The rebuilding of TBMs—both the process and the standardization of rebuilds—has become a focus for the industry as more projects with multiple machine requirements and short time frames are being proposed. The focus has been further highlighted by the ITAtech, a technology-focused committee for the International Tunneling Association (ITA-AITES) that produced guidelines on rebuilds of machinery for mechanized tunnel excavation in 2015. While the guidelines are relatively new, Robbins has a long history of delivering robust machines, many of which are rebuilt (many are also 100% new).

In general, Robbins' experience with rebuilding machines has yielded some key insights. As long as the TBM is well-maintained, there will be jobs it can bore economically. Optimal TBM refurbishment on a used machine requires a broad knowledge of the project conditions, and there are some limitations:

- Machine diameter can be decreased within the limits set by free movement of the grippers and side/roof supports
- Machine diameter can be increased subject to the structural integrity of the machine and the power/thrust capabilities
- Propel force can be increased only to the level supported by the grippers' thrust reaction force
- Cutterhead power must be adequate to sustain the propel force in the given rock, but cannot be increased beyond the capacity of the final drive ring gear and pinions
- Cutterhead speed increases must not exceed the centrifugal limits of muck handling or the maximum rotational speed of the gage cutters

Increasing the power of the TBM is one way to make the design more robust for a longer equipment life. Strong designs have been developed in recent years, including Robbins High Performance (HP) TBMs, used on a number of hard rock tunnels. The HP TBM is designed with a greater strength of core structure and final drive components. They can be used over a much wider range of diameters, whereas older machines (from the 1970s and '80s) are typically limited to a range of less than 1 m of diameter change plus or minus their original size.

HP TBMs have the capability of operating over a broad range. For example, a 4.9 m TBM can be refurbished between 4.3 m and 7.2 m diameters—a range of 2.9 m. Main bearing designs have allowed for greater flexibility, evolving from a 2-row tapered roller bearing to the 3-axis, 3-row cylindrical roller bearing used today. This configuration gives a much higher axial thrust capacity for the same bearing diameter and far greater life in terms of operating hours or revolutions.

Overall, what determines how long a TBM will last is a function of the fundamental design, such as the thrust and gripper load path through the machine and the robustness of the core structure. On older model TBMs, the ring gear and pinions can be strengthened, and larger motors can be added. With sufficient core structure strength, it is also possible to increase the thrust capacity. The limitation is the capacity of the gripper cylinder to handle the increased power and thrust. Once replacement of the gripper cylinder and carrier are required, TBM modification costs are generally considered uneconomic (Roby & Walford, 1995).

OPTIMIZATION OF TBMs FOR MULTIPLE USES

Over the years, Robbins has built a quality assurance system that ensures when it delivers a rebuilt machine, either to the original configuration or a modified one, it still adheres to a design life of 10,000 hours. This standard also includes checks to make sure that all the components are in a functional condition of 'as new' or 'new'.

In order to guarantee the same design life and same warranties on a rebuilt machine, the initial design of the TBM will need to consider that the TBM will be used on several projects. This means that the major structures will need to be strong enough to survive even the toughest conditions and that worn parts can easily be replaced. If the machine is not properly designed for multiple projects, there will be a need to do major work to get the TBM in a working condition, either in its original or modified configuration.

One can argue that project owners typically only have one project and that the condition of the TBM and the suitability of its rebuild is therefore not essential. This is something that is also reflected in many of today's tunneling projects, where the commercial consideration is often given far more attention than the technical one. We would argue, however, that an initially sturdy and robust design of the TBM will give the project more uptime, higher production rates and better flexibility if unexpected conditions are encountered, making it a good and effective insurance against many types of obstacles. Some examples of design aspects that enable longer TBM design life are given below (Khalighi, 2015).

Robust Cutterhead and Machine Structure

A machine designed with multiple projects in mind relies on a heavy steel structure that can stand up to the harsh environments often encountered underground. Designs that take into account high abrasivity of the excavated material or the possibility of high abrasivity are even more robust. Ideally, the cutterhead should be designed with regular cutter inspections and changes in mind. It must also be built to last: this can be difficult with a back-loading cutterhead design, which is full of holes not unlike Swiss cheese. In order to build up the structure, much of the strengthening occurs during the manufacturing process. Full penetration welds are recommended for the cutterhead structure to battle fatigue loading and vibration. Rigorous weld inspections and FEA stress analysis checks can then be made for vulnerabilities in the cutterhead structure (see Figure 2).

Main Bearing and Seals

Large diameter 3-axis main bearings, with the largest possible bearing to tunnel diameter ratio have larger dynamic capacity, and therefore are capable of withstanding more load impacts and giving longer bearing life. It is important to retain as high a ratio as possible (see Figure 3).

The bearing and ring gear are in a difficult-to-access spot on the TBM, so they must be designed for longevity, with a robust structure and high safety factor. Safety factor is defined as any surplus capacity over the design factor of a given element, and overbuilding such structures is of necessity when a TBM is planned to be used over multiple, long tunnel drives.

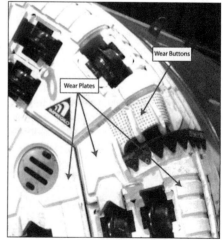

Figure 2. Example of a cutterhead designed for abrasive hard rock conditions

Robust seal design is also essential. The Robbins Company provides a proven seal design using hardened wear bands. Many other manufacturers don't use wear bands, and so as the TBM operates, it wears a groove into the seal lip contact zone. Robbins sacrificial wear bands can be switched out or replaced, making repairs easier. The abrasion-resistant wear bands, made of Stelite™, can be changed in the tunnel in the unlikely event of excessive wear, or can be relocated on the carrier to ensure that damage is not done to the TBM structure itself on long drives.

In addition to the seal design, other elements of the main bearing such as the internal fasteners must be designed to be durable and of high reliability, as these fasteners are difficult to access and are not easily replaceable. The studs connecting the cutterhead to the main bearing seal assembly must also be closely analyzed for strength, deflection, and adequate fastening/clamping force, and protection against abrasive muck must be provided for the fasteners.

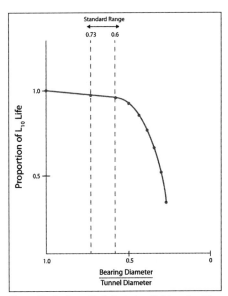

Figure 3. As the ratio falls below 0.6 bearing life is reduced

Lubrication

Dry sump lubrication is a critical way of keeping the main bearing cavity clean by filtering and recycling the oil at a constant rate. Any contamination is cleaned from the cavity, prolonging bearing life. The system also has an added benefit: The oil can be monitored and analyzed for any indications of distress in the main bearing or gears. This monitoring has the potential to allow for correction or intervening maintenance of critical structures/components before a failure occurs.

Drive System

The right drive system is also important for heavy TBM usage. Variable Frequency Drives (VFDs) and planetary gear reducers allow for infinitely adjustable torque and speed control based on the encountered ground, which optimizes the TBM advance rate and reduces damage to machine components (see Figure 4). This is in comparison to older style drives: In older model TBMs, often the drive system was single speed or 2-speed. If a machine bored into a fault zone, for example, there would

Figure 4. VFD setup on a hard rock TBM

be no way to slow down the cutterhead. Such drives would often result in undue wear to the TBM, or even damage to structural components.

Drive motors must also be designed to withstand high vibration as a result of excavating through hard rock conditions. Cantilevered motors must be able to withstand the high g-forces applied to them by violent machine vibration, which is induced by the rock cutting action.

Load Path

A uniform load path, from cutterhead to main bearing to cutterhead support, is always desirable. However, for long distance tunnelling or for multiple uses, the load path can be crucial as high stresses occur wherever the load path shifts. A cutterhead with a cone-shaped rear section can help with this problem by evenly distributing the load across the circumference of the main bearing. In general, everything must be designed in a more robust fashion, and the loads generated by the cutterhead must also translate into a heavier overall structure of the machine.

ON THE IMPORTANCE OF MAINTENANCE

Regular service, good housekeeping and efficient organization of maintenance periods on site are essential to maximize a TBM's performance, its availability and safe employment on a project. When it is planned to use a machine on multiple projects or on long tunnel drives, this is all the more important. In general, the total life cycle of a TBM should be considered and care can be taken during a tunnel drive above and beyond what is considered 'normal'. Gearboxes, for example, may be designed for long tunnels but if it is known that the tunnel length will exceed the life of the gearboxes then planned refurbishment should occur during tunneling. This procedure has been done on several tunnels including India's AMR tunnel—what will be the longest tunnel without intermediate access at 43.5 km once complete.

It is important to remember that the basic structure of a TBM is metal—as long as the structure is intact, one can then check on the bearings, conveyor, hydraulics, and other components. Particular attention should be paid to components that are hard to reach. The main bearing is one of those parts that is difficult to replace during tunneling.

When developing a maintenance plan, it is critical that TBM crews are properly trained on how to operate the machine in the entire gamut of ground conditions that may be encountered on a given tunnel project. Plans must be in place to deal with a wide range of ground conditions as well (e.g., fault zones, water inflows), with protocols as to how the machine should be operated in such conditions. Once the machine has been launched, regularly scheduled maintenance based on tunnel length and geological conditions is also essential. While there are no special guidelines for long-distance tunnels or machines being used on multiple tunnels, crews must be diligent and conduct more detailed inspections the longer a TBM is in operation.

Planned cutter inspections are a regular part of maintenance, which is recommended daily. Checking of oil levels, and all fluids, greases and hydraulics, is also of primary importance. Daily logs are recommended for monitoring of all major systems on the TBM. A daily maintenance regime typically involves routine checks without TBM downtime. Protocols for more in-depth monthly, semi-annual, and annual checks of systems should also be in place. These full checks of various systems do require downtime but are all the more critical when tunneling over a long distance or in variable conditions. These checks are also typically based on the rigors of the project schedule—in hard rock, a week is assumed to be equivalent to 100 m of advance while a month is assumed to be equivalent to 500 m as a baseline.

Maintenance while storing the TBM between projects can also maximize equipment life—such as storing components indoors, coating the equipment with anticorrosive spray, and making sure the main bearing is filled with oil. Owning and using a new TBM has added hidden benefits including familiarity of machine operation and proven performance for that particular piece of equipment.

CASE HISTORIES

So is newer really better? In many cases the record shows that they are equivalent. If the age and number of projects bored by a TBM is seen by some as an issue, a history of record-breaking projects achieved using rebuilt machines does exist. More than one third (36%) of currently standing world records have been broken using a refurbished TBM, some of them in service for decades.

In cases where it is believed that new TBMs will perform or have performed better, this is often an experience bias based on the result of a TBM employed where it wasn't suitable or where it wasn't rebuilt properly. For example if an older machine was initially built for sandstone, it will not have enough power to work well in granite 25 years later without modifications.

A custom design, for a project's specific requirements and geology, is just as important on a rebuilt machine as a new one. For example, a contractor may wish to save money by purchasing a used TBM and rebuilding it to its original specifications. A 3 m diameter Main Beam TBM, rebuilt to the same diameter and specifications, will cost less than rebuilding the same machine but increasing the size to 4 m and adding custom elements. But are the savings truly obtained if the original TBM specifications do not fit the geology? Cutterhead configurations are a particularly important example, with cutter spacing, cutting tools, cutterhead geometry, and muck openings all coming into play and greatly affecting the rate of penetration.

The DigIndy Tunnel System

A good example of custom modifications resulting in success can be seen at the DigIndy project in Indianapolis, Indiana, USA. The TBM, originally manufactured in 1980 for New York's East 63rd Street Subway, had then gone on to bore at least five other hard rock tunnels including New York City's Second Avenue Subway. The 6.2 m diameter Main Beam TBM was chosen for the Deep Rock Tunnel Connector, the first phase of DigIndy, with design updates that included a new back-loading cutterhead with 19-inch disc cutters, variable frequency drive (VFD) motors, and a rescue chamber. The TBM made a record performance for TBMs in the 6 to 7 m diameter range, including "Most Feet Mined in One Day" (124.9 m), "Most Feet Mined in One Week" (515.1 m), and "Most Feet Mined in One Month" (1,754 m). The machine is currently boring the next phases of the DigIndy network—a further 28 km in addition to the 12.5 km of the DRTC already completed (see Figure 5).

Túnel Emisor Poniente II

What about shielded machines—is the rebuild process equally applicable? A good example of this process can be seen at Mexico City's Túnel Emisor Poniente (TEP) II. The TBM was originally manufactured as a 7.23 m diameter Single Shield Hard Rock TBM for Morocco's Abda Doukkala project in 1995. It was then converted to a Double Shield machine for Cleveland, Ohio's Mill Creek Tunnels in the early 2000s and then

Figure 5. The DigIndyTBM, originally built in 1980, was fitted with a new back-loading cutterhead and VFD motors, among other upgrades

back to a Single Shield at 8.7 m diameter for a hydropower tunnel in Laos. In 2015 the machine underwent another transformation when it became a hybrid-type Crossover TBM for the TEP II project (see Figures 6–8).

The 8.7 m diameter Crossover (XRE) TBM was designed for a 5.6 km long tunnel in ground conditions including andesite and tuff with major fault zones containing water-bearing ground. Design components included a convertible cutterhead that could be changed from a Hard Rock to EPB design, a removable

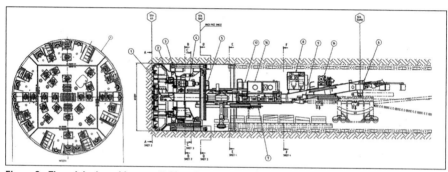

Figure 6. The original machine as a 7.23 m diameter Single Shield Hard Rock TBM in 1995

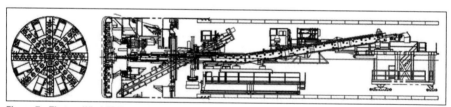

Figure 7. The modified TBM as an 8.7 m diameter Crossover XRE for the TEP II project in 2015—shown in EPB mode

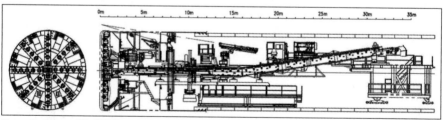

Figure 8. The modified TBM as an 8.7 m diameter Crossover XRE for the TEP II project in 2015—shown in hard rock mode

belt conveyor and screw conveyor, and multi-speed gearboxes to increase torque for tunneling through difficult ground. The machine's performance was highly successful, achieving national records for TBM tunneling after boring 57 m in one day and 702.2 m in one month despite difficult conditions.

CONCLUSIONS

Is a used TBM as good as a new one? In short, the answer is yes, with qualifications. The machine's rebuilt specifications should fit that project's geology and unique requirements. With a proper design and rebuild, a used machine has advantages: The design is proven, the cost is usually lower and there is an advantage in faster delivery times. The risks are only when the TBM is not properly built or when a machine is put into geology where it is not suitable.

Overall, there are many benefits, both obvious and hidden, to using a rebuilt machine, but the rebuild should be done within certain design restraints to remain economical. There is always the possibility to upgrade power and thrust on a machine but there are strict engineering limits. When increasing the cutterhead drive motor power, the gear reducers and final drive ring gear and pinions must have the capacity to take that increase in power. When increasing thrust, the bearing life must be checked to make sure that the bearing can take the increased forces. If the project requires exceeding gripper capacity on a hard rock TBM, then another machine must be considered. The type of TBM and whether it is shielded or not also matters. For example, if an EPB is being used, changing the diameter of an EPB such that it requires new shields may not be the best choice economically. Purchasing a larger EPB would make better sense in that application.

Overall TBM design and usage for the long haul is simply a cost effective, energy efficient, and sustainable way of thinking about tunnel boring. Used machines can and have shown their ability to excavate projects at world-class rates of advance and complete many kilometers of tunnel with success.

REFERENCES

Khalighi, B. 2015. TBM Design for Long Distance Tunnels: How to Keep Hard Rock TBMs Boring for 15 km or More. Proceedings of the ITA-AITES World Tunnel Congress 2015, Dubrovnik.

ITATech. (2015). ITAtech Guidelines on Rebuilds of Machinery for Mechanized Tunnel Excavation.

Roby, J & Walford, C. (1995). Selecting, repairing and modifying a secondhand TBM. International Journal of Rock Mechanics & Mining Sciences: Vol 32, Iss 27. 39, 41–42.

Index

A

Akron, OH
    Ohio Canal Interceptor Tunnel (OCIT), 120–131, 683–691, 870–889
Alexandria, VA
    RiverRenew, 440–452
Alexandria Renew Enterprises, 440–452
Anacostia River Tunnel (Washington, D.C.), 504–511
aquifer testing, 548–563
artificial intelligence (AI), 776–784
Asia
    Bangalore Metro Project (India), 933–939
    Bheri Babai Diversion Multipurpose Project (BBDMP) (Nepal), 242–249
    Chennai Metro Project (India), 928–933
    Jilin Yinsong Water Supply Project (China), 600–613
    NBAQ4 Project (Manila, Philippines), 178–198
    Neelum-Jhelum Hydroelectric Project (Pakistan), 277–289, 657–667, 715–724
    Tuen Mun–Chek Lap Kok Link Northern Connection Subsea Tunnel Section (Hong Kong), 986–998
Atlanta, GA
    Plane Train Tunnel West Extension Phase 1 (Atlanta, GA), 60–61
Austin, TX
    Parmer Lane Interceptor Tunnel, 1222–1229
Australia
    Forrestfield Airport Link (FAL) (Perth), 954–960
    Metropolitan Mine (New South Wales, Australia), 401
    West Gate Tunnel Project (Melbourne), 230–239, 636–646
    WestConnex New M5 Project (Sydney), 668–682
Austria
    Brenner Base Tunnel, 1028–1036
    fire safety rehabilitation, 370–379
    GKI Project, 1088–1095
    Metropolitan Mine (New South Wales), 401

B

Baltimore, MD
    Northeast Corridor Superconducting Maglev Project (SCMAGLEV), 837–844
Bangalore Metro Project (India), 933–939
Bank Station Capacity Upgrade (BSCU) (London, England), 1010–1027
Bellevue, WA
    Downtown Bellevue Tunnel (DBT), 764–773, 1061–1073
Bheri Babai Diversion Multipurpose Project (BBDMP) (Nepal), 242–249
Blacklick Creek Sanitary Interceptor Sewer Project (BCSIS) (Columbus, OH), 250–260, 855–867
Blacksnake Creek Stormwater Separation Improvement Tunnel (St. Joseph, MO), 46–54
Boring Company, 6
Boston, MA
    Metropolitan Tunnel Redundancy Program, 477–489
Brenner Base Tunnel (Austria-Italy), 1028–1036
British Columbia
    Canada Line Rapid Transit Project (Vancouver), 7–8
Brunswick, GA
    Brunswick Glynn Joint Water and Sewer North Mainland Sewer Transmission System, 57–58
Brunswick Glynn Joint Water and Sewer North Mainland Sewer Transmission System (Brunswick, GA), 57–58
building information modeling (BIM), 174–175
Bypass Tunnel (New York, NY), 78–86

C

California
    Central Bayside System Improvement Project (San Francisco), 490–502
    Hawthorne Test Tunnel (Los Angeles), 6
    Mountain Tunnel Improvements Project, 464–476

Regional Connector Project (Los Angeles), 199–218, 299–311, 538–547, 890–915, 1037–1050
Silicon Valley Clean Water Gravity Pipeline Project (Redwood City), 58–60
Canada
   Canada Line Rapid Transit Project (Vancouver, BC), 7–8
   Don River & Central Waterfront Coxwell Sanitary Bypass Tunnel Project (Toronto, ON), 916–927, 1160–1169
   Garage Cote Vertu Project (Montreal, QC), 734–744
   Réseau Express Métropolitain (REM) (Montreal, QC), 166–177
Canada Line Rapid Transit Project (Vancouver, BC), 7–8
cavern excavation
   Garage Cote Vertu Project (Montreal, QC), 734–744
   Maline Creek Tunnel (MCT) (St. Louis, MO), 745–751
   Regional Connector Project (Los Angeles, CA), 1037–1050
   shallow rock, 752–763
Central Bayside System Improvement Project (San Francisco, CA), 490–502
Chennai Metro Project (India), 928–933
Chesapeake Bay, VA
   Parallel Thimble Shoal Tunnel (PTST) Project, 512–524
Chicago, IL
   Des Plaines Inflow Tunnel, 342–350
   Peoples Gas Light & Coke Company (PGL) Tunnel Remediation Program (TRP), 1172–1185
Chile
   El Teniente mine, 330–341
Chimney Hollow Reservoir Project (Loveland, CO), 614–621
China
   Jilin Yinsong Water Supply Project, 600–613
Cleveland, OH
   Doan Valley Relief and Consolidation Sewer (DVRCS) Project, 942–953
   Dugway Storage Tunnel (DST), 845–854
Colorado
   Chimney Hollow Reservoir Project (Loveland), 614–621
Columbus, OH
   Blacklick Creek Sanitary Interceptor Sewer Project (BCSIS), 250–260, 855–867

Lockbourne Intermodal Subtrunk (LIS) Project, 1189–1198
combined sewer overflow (CSO) tunnels
   Anacostia River Tunnel (Washington, D.C.), 504–511
   Central Bayside System Improvement Project (San Francisco, CA), 490–502
   CSO 021 Diversion Facilities Project (Washington, D.C.), 1098–1113
   DC Clean Rivers (DCCR) Project (Washington, D.C.), 157–165
   Des Plaines Inflow Tunnel (Chicago, IL), 342–350
   Doan Valley Relief and Consolidation Sewer (DVRCS) Project (Cleveland, OH), 942–953
   Don River & Central Waterfront Coxwell Sanitary Bypass Tunnel Project (Toronto, ON), 916–927, 1160–1169
   Dugway Storage Tunnel (DST) (Cleveland, OH), 845–854
   Maline Creek Tunnel (MCT) (St. Louis, MO), 745–751
   Northeast Boundary Tunnel (NEBT) (Washington, D.C.), 408–420
   Ohio Canal Interceptor Tunnel (OCIT) (Akron, OH), 120–131, 683–691, 870–889
   RiverRenew (Alexandria, VA), 440–452
   South Hartford Conveyance and Storage Tunnel (SHCST) (Hartford, CT), 1153–1159
   Three Rivers Protection and Overflow Reduction Tunnel (3RPORT) (Fort Wayne, IN), 261–269, 357–368, 703–714
   Waterway Protection Tunnel (WPT) (Louisville, KY), 725–732
concrete
   fiber-reinforced (FRC), 623–635
   secant piles, 683–691
   shotcrete, 600–613, 657–667
   steel fiber reinforced (SFRC), 636–646
   wet-mix structural sprayed, 1074–1087
Connecticut
   Great Hill Tunnel, 1199–1209
   South Hartford Conveyance and Storage Tunnel (SHCST) (Hartford), 1153–1159
contracts and contractors. *see also* delivery methods
   Bypass Tunnel (New York, NY), 78–86
   fixed-price incentive fee (FPIF), 34–45
   Hudson Tunnel Project (Secaucus, NJ/ New York, NY), 97–108

Innovative Contractor Engagement (ICE) model, 1010–1027
understanding the Geotechnical Baseline Report (GBR), 585–598
conveyor belts, 825–836
CSO 021 Diversion Facilities Project (Washington, D.C.), 1098–1113
cut and cover excavation, 199–218
cuttability indexes, 980–985

D

Dallas, TX
   DART D2 Subway Project, 453–463
DART D2 Subway Project (Dallas, TX), 453–463
data management, 845–854, 961–971
DC Clean Rivers (DCCR) Project (Washington, D.C.), 157–165
delivery methods
   alternative, 3–14, 148–156
   design-build, 14–22, 65–77
   and fixed-price incentive fee (FPIF) contracts, 34–45
   progressive design-build (PDB), 55–64, 76
depressurization, 504–511
Des Plaines Inflow Tunnel (Chicago, IL), 342–350
design methodology
   deep shafts, 1122–1132
   42nd Street Times Square Shuttle Station (New York, NY), 132–146
   Hudson Tunnel Project (Secaucus, NJ/ New York, NY), 97–108, 1114–1121
   Northgate Link Extension Project (Seattle, WA), 109–119
   Ohio Canal Interceptor Tunnel (OCIT) (Akron, OH), 120–131
   Surface Water Supply Project (Houston, TX), 88–96
design-build delivery method
   about, 14–22, 65–77
   DC Clean Rivers (DCCR) Project (Washington, D.C.), 157–165
   NBAQ4 Project (Manila, Philippines), 178–198
   and permitting, 219–229
   progressive (PDB), 55–64, 76
   Regional Connector Project (Los Angeles, CA), 199–218
   Réseau Express Métropolitain (REM) (Montreal, QC), 175–177
   West Gate Tunnel Project (Melbourne, Australia), 230–239
District of Columbia. *see* Washington, D.C.

Doan Valley Relief and Consolidation Sewer (DVRCS) Project (Cleveland, OH), 942–953
Don River & Central Waterfront Coxwell Sanitary Bypass Tunnel Project (Toronto, ON), 916–927, 1160–1169
Dos Santos Sandwich Belt high angle conveyors, 825–836
Downtown Bellevue Tunnel (DBT) (Bellevue, WA), 764–773, 1061–1073
drill and blast excavation
   Des Plaines Inflow Tunnel (Chicago, IL), 342–350
   El Teniente mine (Chile), 330–341
   GKI Project (Switzerland-Austria), 1088–1095
   Lake Mead Intake No. 3 Project (Nevada), 314–329
   Luck Stone Inter-Quarry Tunnel (Leesburg, VA), 351–356
   Three Rivers Protection and Overflow Reduction Tunnel (3RPORT) (Fort Wayne, IN), 357–368
Dugway Storage Tunnel (DST) (Cleveland, OH), 845–854

E

earth pressure balance machines (EPBMs)
   Anacostia River Tunnel (Washington, D.C.), 504–511
   Bangalore Metro Project (India), 933–939
   Blacksnake Creek Stormwater Separation Improvement Tunnel (St. Joseph, MO), 46–54
   Chennai Metro Project (India), 928–933
   data acquisition tools, 961–971
   impact of water content on, 785–798
   and mixed face conditions, 928–939
   Northeast Corridor Superconducting Maglev Project (SCMAGLEV) (Washington, D.C./Baltimore, MD), 837–844
   Ohio Canal Interceptor Tunnel (OCIT) (Akron, OH), 880–889
   Regional Connector Project (Los Angeles, CA), 890–915
El Teniente mine (Chile), 330–341
England
   Bank Station Capacity Upgrade (BSCU) (London), 1010–1027
EPBMs. *see* earth pressure balance machines (EPBMs)
Europe
   Bank Station Capacity Upgrade (BSCU) (London, England), 1010–1027

Brenner Base Tunnel (Austria-Italy),
    1028–1036
  fire safety rehabilitation (Austria), 370–379
  GKI Project (Switzerland-Austria),
    1088–1095
  Herrschaftsbuck Tunnel (Rheinfelden,
    Germany), 1051–1060
  Kraftwerk Linth Limmern (KLL)
    (Switzerland), 1230–1238
  Nant de Drance (NDD) plant
    (Switzerland), 1230–1238
  Paris Metro Expansion Project (Paris,
    France), 830–836
  Tunnel Kaiser Strasse (Karlsruhe,
    Germany), 387–398
excavation
  cavern, 734–763, 1037–1050
  cut and cover, 199–219
  drill and blast, 314–368, 1088–1095
  and ground subsidence, 399–407
  hard rock, 261–269, 694–732, 745–751
  mixed transitional ground (MTG),
    120–131, 880–889
  mountain, 715–724
  sequential method (SEM), 206–218, 351–
    356, 387–398, 764–773, 1000–1095
  shaft, 323–329, 360–365, 490–502,
    745–751, 1098–1169
  soft ground, 109–119, 575–584,
    1074–1087
  station, 132–146, 166–177, 199–218, 462
  of underground obstructions, 290–311

F

fault zones
  Garage Cote Vertu Project (Montreal, QC),
    734–744
  Jilin Yinsong Water Supply Project
    (China), 600–613
  Neelum-Jhelum Hydroelectric Project
    (Pakistan), 277–289, 657–667,
    715–724
financing, 3–14
fires, 370–379, 387–398
Forrestfield Airport Link (FAL) (Perth,
  Australia), 954–960
Fort Wayne, IN
  Three Rivers Protection and Overflow
    Reduction Tunnel (3RPORT),
    261–269, 357–368, 703–714
42nd Street Times Square Shuttle Station
  (New York, NY), 132–146
France
  Paris Metro Expansion Project (Paris),
    830–836

G

Garage Cote Vertu Project (Montreal, QC),
  734–744
Georgia
  Brunswick Glynn Joint Water and Sewer
    North Mainland Sewer Transmission
    System (Brunswick), 57–58
  Plane Train Tunnel West Extension Phase
    1 (Atlanta), 60–61
Georgia, VT
  Interstate-89 (I-89) culvert project,
    1000–1009
Geotechnical Baseline Report (GBR),
  525–537, 585–598
geotechnical instrumentation
  Lower Meramec Tunnel (LMT) (St. Louis,
    MO), 548–563
  modeling for urban tunneling, 538–547
  Parallel Thimble Shoal Tunnel (PTST)
    Project (Chesapeake Bay, VA),
    512–524
  reliability of reference models, 525–537
  rock load estimation, 752–763
  Soil Abrasion Index Testing Machine,
    785–798
  3D surveying systems, 809–816
  and tool wear prediction, 575–584
Germany
  Herrschaftsbuck Tunnel (Rheinfelden),
    1051–1060
  Tunnel Kaiser Strasse (Karlsruhe),
    387–398
GKI Project (Switzerland-Austria), 1088–1095
Great Hill Tunnel (Connecticut), 1199–1209
ground freezing, 270–273
grouting
  automation and digitization of, 817–824
  Chimney Hollow Reservoir Project
    (Loveland, CO), 614–621
  compensation, 538–547
  consolidation, 648–656
  jet, 683–691
  Neelum-Jhelum Hydroelectric Project
    (Pakistan), 657–667
  Ohio Canal Interceptor Tunnel (OCIT)
    (Akron, OH), 683–691
  WestConnex New M5 Project (Sydney,
    Australia), 668–682

H

hard rock excavation
  Maline Creek Tunnel (MCT) (St. Louis,
    MO), 745–751
  Neelum-Jhelum Hydroelectric Project
    (Pakistan), 715–724

Rand Park Stormwater Diversion Tunnel
  (Keokuk, IA), 694–702
Three Rivers Protection and Overflow
  Reduction Tunnel (3RPORT) (Fort
  Wayne, IN), 261–269, 703–714
Waterway Protection Tunnel (WPT)
  (Louisville, KY), 725–732
Hartford, CT
  South Hartford Conveyance and Storage
    Tunnel (SHCST), 1153–1159
Hawthorne Test Tunnel (Los Angeles, CA), 6
Hecla–Lucky Friday Mine, 1210–1221
Herrschaftsbuck Tunnel (Rheinfelden,
  Germany), 1051–1060
highway tunnels
  fire safety rehabilitation, 370–379
  Hawthorne Test Tunnel (Los Angeles,
    CA), 6
  Herrschaftsbuck Tunnel (Rheinfelden,
    Germany), 1051–1060
  Jarvis River Crossing Project (Neebing,
    OH), 290–298
  Ohio River Bridges East End Crossing
    (ORBEEC) (Louisville, KY/Utica, IL),
    8–9
  Parallel Thimble Shoal Tunnel (PTST)
    Project (Chesapeake Bay, VA),
    512–524
  Tuen Mun–Chek Lap Kok Link Northern
    Connection Subsea Tunnel Section
    (Hong Kong), 986–998
  West Gate Tunnel Project (Melbourne,
    Australia), 230–239, 636–646
  WestConnex New M5 Project (Sydney,
    Australia), 668–682
Hoboken, NJ
  Hudson Tunnel Project: Hoboken Shaft,
    1114–1121
Houston, TX
  Surface Water Supply Project, 88–96
Hudson Tunnel Project (Secaucus, NJ/New
  York, NY), 97–108
Hudson Tunnel Project: Hoboken Shaft
  (Hoboken, NJ), 1114–1121
hydroelectric projects
  Bheri Babai Diversion Multipurpose
    Project (BBDMP) (Nepal), 242–249
  GKI Project (Switzerland-Austria),
    1088–1095
  Neelum-Jhelum Hydroelectric Project
    (Pakistan), 277–289, 657–667,
    715–724
  Switzerland, 1230–1238
hyperbaric conditions, 380–386, 972–979

I
Idaho
  Lucky Friday Mine, 1210–1221
Illinois
  Des Plaines Inflow Tunnel (Chicago),
    342–350
  Peoples Gas Light & Coke Company
    (PGL) Tunnel Remediation Program
    (TRP) (Chicago), 1172–1185
India
  Bangalore Metro Project, 933–939
  Chennai Metro Project, 928–933
Indiana
  Ohio River Bridges East End Crossing
    (ORBEEC) (Utica), 8–9
  Three Rivers Protection and Overflow
    Reduction Tunnel (3RPORT) (Fort
    Wayne), 261–269, 357–368, 703–714
innovation. *see also* technology
  artificial intelligence (AI), 776–784
  automation and digitization, 817–824
  surveying, 855–867
  3D surveying systems, 809–816
Innovative Contractor Engagement (ICE)
  model, 1010–1027
inspection, 809–816, 1199–1209
instrumentation. *see* geotechnical
  instrumentation
Interstate-89 (I-89) culvert project (Georgia,
  VT), 1000–1009
Iowa
  Rand Park Stormwater Diversion Tunnel
    (Keokuk), 694–702
Italy
  Brenner Base Tunnel, 1028–1036

J
Jarvis River Crossing Project (Neebing, OH),
  290–298
Jilin Yinsong Water Supply Project (China),
  600–613

K
Kentucky
  Ohio River Bridges East End Crossing
    (ORBEEC) (Louisville), 8–9
  Waterway Protection Tunnel (WPT)
    (Louisville), 725–732
Keokuk, IA
  Rand Park Stormwater Diversion Tunnel,
    694–702
Kraftwerk Linth Limmern (KLL) (Switzerland),
  1230–1238

L

Lake Mead Intake No. 3 Project (Nevada), 314–329
LED lighting, 1222–1229
Leesburg, VA
    Luck Stone Inter-Quarry Tunnel, 351–356
Lockbourne Intermodal Subtrunk (LIS) Project (Columbus, OH), 1189–1198
London, England
    Bank Station Capacity Upgrade (BSCU), 1010–1027
Los Angeles, CA
    Hawthorne Test Tunnel, 23–33
    Regional Connector Project, 199–218, 299–311, 538–547, 890–915, 1037–1050
Louisville, KY
    Ohio River Bridges East End Crossing (ORBEEC), 23–33
    Waterway Protection Tunnel (WPT), 725–732
Loveland, CO
    Chimney Hollow Reservoir Project, 614–621
Lower Meramec Tunnel (LMT) (St. Louis, MO), 548–563
Luck Stone Inter-Quarry Tunnel (Leesburg, VA), 351–356
Lucky Friday Mine (Idaho), 1210–1221

M

Macomb County, MI
    Oakland-Macomb Interceptor System (OMIS), 1133–1143
Maline Creek Tunnel (MCT) (St. Louis, MO), 745–751
Manila, Philippines
    NBAQ4 Project, 178–198
Maryland
    Northeast Corridor Superconducting Maglev Project (SCMAGLEV) (Baltimore), 837–844
    Purple Line Project, 5
Massachusetts
    Metropolitan Tunnel Redundancy Program (Boston), 477–489
Melbourne, Victoria
    West Gate Tunnel Project, 230–239, 636–646
Metropolitan Mine (New South Wales, Australia), 401
Metropolitan Tunnel Redundancy Program (Boston, MA), 477–489
Michigan
    Oakland-Macomb Interceptor System (OMIS) (Macomb County), 1133–1143

microtunneling, 799–808, 942–953
mines
    El Teniente mine (Chile), 330–341
    Luck Stone Inter-Quarry Tunnel (Leesburg, VA), 351–356
    Lucky Friday Mine (Idaho), 1210–1221
    Metropolitan Mine (New South Wales, Australia), 401
Missouri
    Blacksnake Creek Stormwater Separation Improvement Tunnel (St. Joseph), 46–54
    Lower Meramec Tunnel (LMT) (St. Louis), 548–563
    Maline Creek Tunnel (MCT) (St. Louis), 745–751
mixed transitional ground (MTG) excavation, 120–131, 880–889
Montreal, QC
    Garage Cote Vertu Project, 734–744
    Réseau Express Métropolitain (REM), 166–177
Mountain Tunnel Improvements Project (California), 464–476

N

Nant de Drance (NDD) plant (Switzerland), 1230–1238
NBAQ4 Project (Manila, Philippines), 178–198
Neebing, OH
    Jarvis River Crossing Project, 290–298
Neelum-Jhelum Hydroelectric Project (Pakistan), 277–289, 657–667, 715–724
Nepal
    Bheri Babai Diversion Multipurpose Project (BBDMP), 242–249
Nevada
    Lake Mead Intake No. 3 Project, 314–329
new Austrian tunneling method. *see* sequential excavation method (SEM)
New Jersey
    Hudson Tunnel Project (Secaucus), 97–108
    Hudson Tunnel Project: Hoboken Shaft (Hoboken), 1114–1121
New South Wales, Australia
    Metropolitan Mine, 401
    WestConnex New M5 Project (Sydney), 668–682
New York
    Bypass Tunnel (New York City), 78–86
    42nd Street Times Square Shuttle Station (New York City), 132–146

Index

Hudson Tunnel Project (New York City), 97–108
progressive design-build (PDB) projects in, 62
New York, NY
  Bypass Tunnel, 78–86
  42nd Street Times Square Shuttle Station, 132–146
  Hudson Tunnel Project, 97–108
North America
  Anacostia River Tunnel (Washington, D.C.), 504–511
  Blacklick Creek Sanitary Interceptor Sewer Project (BCSIS) (Columbus, OH), 250–260, 855–867
  Blacksnake Creek Stormwater Separation Improvement Tunnel (St. Joseph, MO), 46–54
  Brunswick Glynn Joint Water and Sewer North Mainland Sewer Transmission System (Brunswick, GA), 57–58
  Bypass Tunnel (New York, NY), 78–86
  Canada Line Rapid Transit Project (Vancouver, BC), 8–9
  Central Bayside System Improvement Project (San Francisco, CA), 490–502
  Chimney Hollow Reservoir Project (Loveland, CO), 614–621
  CSO 021 Diversion Facilities Project (Washington, D.C.), 1098–1113
  DART D2 Subway Project (Dallas, TX), 453–463
  DC Clean Rivers (DCCR) Project (Washington, D.C.), 157–165
  Des Plaines Inflow Tunnel (Chicago, IL), 342–350
  Doan Valley Relief and Consolidation Sewer (DVRCS) Project (Cleveland, OH), 942–953
  Don River & Central Waterfront Coxwell Sanitary Bypass Tunnel Project (Toronto, ON), 916–927, 1160–1169
  Downtown Bellevue Tunnel (DBT) (Bellevue, WA), 764–773, 1061–1073
  Dugway Storage Tunnel (DST) (Cleveland, OH), 845–854
  42nd Street Times Square Shuttle Station (New York, NY), 132–146
  Garage Cote Vertu Project (Montreal, QC), 734–744
  Great Hill Tunnel (Connecticut), 1199–1209
  Hawthorne Test Tunnel (Los Angeles, CA), 6
  Hudson Tunnel Project (Secaucus, NJ/ New York, NY), 97–108

Hudson Tunnel Project: Hoboken Shaft (Hoboken, NJ), 1114–1121
Interstate-89 (I-89) culvert project (Georgia, VT), 1000–1009
Jarvis River Crossing Project (Neebing, OH), 290–298
Lake Mead Intake No. 3 Project (Nevada), 314–329
Lockbourne Intermodal Subtrunk (LIS) Project (Columbus, OH), 1189–1198
Lower Meramec Tunnel (LMT) (St. Louis, MO), 548–563
Luck Stone Inter-Quarry Tunnel (Leesburg, VA), 351–356
Lucky Friday Mine (Idaho), 1210–1221
Maline Creek Tunnel (MCT) (St. Louis, MO), 745–751
Metropolitan Tunnel Redundancy Program (Boston, MA), 477–489
Mountain Tunnel Improvements Project (California), 464–476
Northeast Boundary Tunnel (NEBT) (Washington, D.C.), 408–420
Northeast Corridor Superconducting Maglev Project (SCMAGLEV) (Washington, D.C./Baltimore, MD), 837–844
Northgate Link Extension Project (Seattle, WA), 109–119, 270–273, 430–438
Oakland-Macomb Interceptor System (OMIS) (Macomb County, Michigan), 1133–1143
Ohio Canal Interceptor Tunnel (OCIT) (Akron, OH), 120–131, 683–691, 870–889
Ohio River Bridges East End Crossing (ORBEEC) (Louisville, KY/Utica, IL), 8–9
Oregon Avenue NW Sewer Rehabilitation Project (Washington, D.C.), 799–808
Parallel Thimble Shoal Tunnel (PTST) Project (Chesapeake Bay, VA), 512–524
Parmer Lane Interceptor Tunnel (Austin, TX), 1222–1229
Peoples Gas Light & Coke Company (PGL) Tunnel Remediation Program (TRP) (Chicago, IL), 1172–1185
Plane Train Tunnel West Extension Phase 1 (Atlanta, GA), 60–61
Purple Line Project (Maryland), 5
Rand Park Stormwater Diversion Tunnel (Keokuk, IA), 694–702
Regional Connector Project (Los Angeles, CA), 199–218, 299–311, 538–547, 890–915, 1037–1050

Réseau Express Métropolitain (REM) (Montreal, QC), 166–177
RiverRenew (Alexandria, VA), 440–452
Silicon Valley Clean Water Gravity Pipeline Project (Redwood City, CA), 58–60
South Hartford Conveyance and Storage Tunnel (SHCST) (Hartford, CT), 1153–1159
Surface Water Supply Project (Houston, TX), 88–96
Three Rivers Protection and Overflow Reduction Tunnel (3RPORT) (Fort Wayne, IN), 261–269, 357–368, 703–714
University Link Extension (U-Link) (Seattle, WA), 430–438
Waterway Protection Tunnel (WPT) (Louisville, KY), 725–732
Northeast Boundary Tunnel (NEBT) (Washington, D.C.), 408–420
Northeast Corridor Superconducting Maglev Project (SCMAGLEV) (Washington, D.C./Baltimore, MD), 837–844
Northgate Link Extension Project (Seattle, WA), 109–119, 270–273, 430–438

O

Oakland-Macomb Interceptor System (OMIS) (Macomb County, Michigan), 1133–1143
Ohio
　Blacklick Creek Sanitary Interceptor Sewer Project (BCSIS) (Columbus), 250–260, 855–867
　Doan Valley Relief and Consolidation Sewer (DVRCS) Project (Cleveland), 942–953
　Dugway Storage Tunnel (DST) (Cleveland), 845–854
　Jarvis River Crossing Project (Neebing), 290–298
　Lockbourne Intermodal Subtrunk (LIS) Project (Columbus), 1189–1198
　Ohio Canal Interceptor Tunnel (OCIT) (Akron), 120–131, 683–691, 870–889
Ohio Canal Interceptor Tunnel (OCIT) (Akron, OH), 120–131, 683–691, 870–889
Ohio River Bridges East End Crossing (ORBEEC) (Louisville, KY/Utica, IL), 8–9
Ontario
　Don River & Central Waterfront Coxwell Sanitary Bypass Tunnel Project (Toronto), 916–927, 1160–1169

Oregon Avenue NW Sewer Rehabilitation Project (Washington, D.C.), 799–808

P

packer testing, 548–563
Pakistan
　Neelum-Jhelum Hydroelectric Project, 277–289, 657–667, 715–724
Parallel Thimble Shoal Tunnel (PTST) Project (Chesapeake Bay, VA), 512–524
Paris Metro Expansion Project (Paris, France), 830–836
Parmer Lane Interceptor Tunnel (Austin, TX), 1222–1229
Peoples Gas Light & Coke Company (PGL) Tunnel Remediation Program (TRP) (Chicago, IL), 1172–1185
permitting, 219–229
Perth, Western Australia
　Forrestfield Airport Link (FAL), 954–960
Philippines
　NBAQ4 Project (Manila), 178–198
Plane Train Tunnel West Extension Phase 1 (Atlanta, GA), 60–61
procurement
　design-build, 14–22, 65–77
　Hudson Tunnel Project (Secaucus, NJ/New York, NY), 97–108
　Innovative Contractor Engagement (ICE) model, 1010–1027
　P3 projects, 3–14
　progressive design-build (PDB), 55–64, 76
　of TBMs, 23–33
progressive design-build (PDB) delivery method, 55–64, 76
pumped storage projects, 1230–1238
Purple Line Project (Maryland), 5

Q

quarries. see mines
Quebec
　Garage Cote Vertu Project (Montreal), 734–744
　Réseau Express Métropolitain (REM) (Montreal), 166–177

R

railroad tunnels
　Brenner Base Tunnel (Austria-Italy), 1028–1036
　Forrestfield Airport Link (FAL) (Perth, Australia), 954–960
　Hudson Tunnel Project (Secaucus, NJ/New York, NY), 97–108

Index 1257

Hudson Tunnel Project: Hoboken Shaft (Hoboken, NJ), 1114–1121
Northeast Corridor Superconducting Maglev Project (SCMAGLEV) (Washington, D.C./Baltimore, MD), 837–844
Rand Park Stormwater Diversion Tunnel (Keokuk, IA), 694–702
Redwood City, CA
   Silicon Valley Clean Water Gravity Pipeline Project, 58–60
Regional Connector Project (Los Angeles, CA), 199–218, 299–311, 538–547, 890–915, 1037–1050
remediation, 1172–1185
Remote Vein Miner (RVM), 1210–1221
Réseau Express Métropolitain (REM) (Montreal, QC), 166–177
Rheinfelden, Germany
   Herrschaftsbuck Tunnel, 1051–1060
risk management
   and alternative delivery methods, 148–156
   fixed-price incentive fee (FPIF) contracts, 34–45
   geologic and geotechnical risk, 564–574
   Ohio Canal Interceptor Tunnel (OCIT) (Akron, OH), 870–879
   P3 projects, 9–11
   and underground obstructions, 310
RiverRenew (Alexandria, VA), 440–452

S

safety
   fires, 370–379
   ground subsidence, 399–407
   hyperbaric conditions, 380–386, 972–979
   Northeast Corridor Superconducting Maglev Project (SCMAGLEV) (Washington, D.C./Baltimore, MD), 842–843
   refuge chambers, 387–398
   ventilation systems, 175–177, 421–429
   vibration, 430–438
San Francisco, CA
   Central Bayside System Improvement Project, 490–502
SAR Interferometry (InSAR), 399–407
Seattle, WA
   Northgate Link Extension Project, 109–119, 270–273, 430–438
   University Link Extension (U-Link), 430–438
secant piles, 683–691
Secaucus, NJ
   Hudson Tunnel Project, 97–108

sequential excavation method (SEM)
   Bank Station Capacity Upgrade (BSCU) (London, England), 1010–1027
   Brenner Base Tunnel (Austria-Italy), 1028–1036
   Downtown Bellevue Tunnel (DBT) (Bellevue, WA), 764–773, 1061–1073
   GKI Project (Switzerland-Austria), 1088–1095
   Herrschaftsbuck Tunnel (Rheinfelden, Germany), 1051–1060
   Interstate-89 (I-89) culvert project (Georgia, VT), 1000–1009
   Luck Stone Inter-Quarry Tunnel (Leesburg, VA), 351–356
   Regional Connector Project (Los Angeles, CA), 206–218, 1037–1050
   South American mine project, 1074–1087
   tunnel refuge chambers, 387–398
sewer tunnels. see wastewater tunnels
shaft excavation
   Central Bayside System Improvement Project (San Francisco, CA), 490–502
   CSO 021 Diversion Facilities Project (Washington, D.C.), 1098–1113
   design challenges, 1122–1132
   Don River & Central Waterfront Coxwell Sanitary Bypass Tunnel Project (Toronto, ON), 1160–1169
   Hudson Tunnel Project: Hoboken Shaft (Hoboken, NJ), 1114–1121
   Lake Mead Intake No. 3 Project (Nevada), 323–329
   Maline Creek Tunnel (MCT) (St. Louis, MO), 745–751
   Oakland-Macomb Interceptor System (OMIS) (Macomb County, Michigan), 1133–1143
   South Hartford Conveyance and Storage Tunnel (SHCST) (Hartford, CT), 1153–1159
   Three Rivers Protection and Overflow Reduction Tunnel (3RPORT) (Fort Wayne, IN), 360–365
   Vertical Shaft Sinking Machine (VSM), 1144–1152
shape accelerometer array (SAA), 542–547
shotcrete, 600–613, 657–667
Silicon Valley Clean Water Gravity Pipeline Project (Redwood City, CA), 58–60
soft ground excavation
   and liner load estimation, 109–119
   predicting tool wear, 575–584
   techniques for mine access development, 1074–1087

software
    building information modeling (BIM), 174–175
    data management, 845–854, 961–971
Soil Abrasion Index Testing Machine, 785–798
South America
    El Teniente mine (Chile), 330–341
    soft ground tunneling techniques in, 1074–1087
South Hartford Conveyance and Storage Tunnel (SHCST) (Hartford, CT), 1153–1159
St. Joseph, MO
    Blacksnake Creek Stormwater Separation Improvement Tunnel, 46–54
St. Louis, MO
    Lower Meramec Tunnel (LMT), 548–563
    Maline Creek Tunnel (MCT), 745–751
station excavation
    DART D2 Subway Project (Dallas, TX), 462
    42nd Street Times Square Shuttle Station (New York, NY), 132–146
    Regional Connector Project (Los Angeles, CA), 199–218
    Réseau Express Métropolitain (REM) (Montreal, QC), 166–177
stormwater tunnels. see also combined sewer overflow (CSO) tunnels
    Blacksnake Creek Stormwater Separation Improvement Tunnel (St. Joseph, MO), 46–54
    Interstate-89 (I-89) culvert project (Georgia, VT), 1000–1009
    Rand Park Stormwater Diversion Tunnel (Keokuk, IA), 694–702
subsidence, 399–407
Surface Water Supply Project (Houston, TX), 88–96
surveying, 855–867
sustainability
    Great Hill Tunnel (Connecticut), 1199–1209
    Kraftwerk Linth Limmern (KLL) (Switzerland), 1230–1238
    Lockbourne Intermodal Subtrunk (LIS) Project (Columbus, OH), 1189–1198
    Lucky Friday Mine (Idaho), 1210–1221
    Nant de Drance (NDD) plant (Switzerland), 1230–1238
    Parmer Lane Interceptor Tunnel (Austin, TX), 1222–1229
    Peoples Gas Light & Coke Company (PGL) Tunnel Remediation Program (TRP) (Chicago, IL), 1172–1185
    rebuilt tunnel boring machines (TBMs), 1239–1247
Switzerland
    GKI Project, 1088–1095
    Kraftwerk Linth Limmern, 1230–1238
    Nant de Drance plant, 1230–1238
Sydney, New South Wales
    WestConnex New M5 Project, 668–682

T
TBMs. see tunnel boring machines (TBMs)
technology
    artificial intelligence (AI), 776–784
    automation and digitization, 817–824
    building information modeling (BIM), 174–175
    data management, 845–854
    Dos Santos Sandwich Belt high angle conveyors, 825–836
    geotechnical instrumentation, 512–563, 575–584, 752–763
    Remote Vein Miner (RVM), 1210–1221
    Soil Abrasion Index Testing Machine, 785–798
    surveying, 855–867
    3D surveying systems, 809–816
    tunnel boring machines (TBMs), 942–998
    Vertical Shaft Sinking Machine (VSM), 1144–1152
Texas
    DART D2 Subway Project (Dallas), 453–463
    Parmer Lane Interceptor Tunnel (Austin), 1222–1229
    Surface Water Supply Project (Houston), 88–96
Three Rivers Protection and Overflow Reduction Tunnel (3RPORT) (Fort Wayne, IN), 261–269, 357–368, 703–714
Toronto, ON
    Don River & Central Waterfront Coxwell Sanitary Bypass Tunnel Project, 916–927, 1160–1169
transit tunnels
    Bangalore Metro Project (India), 933–939
    Bank Station Capacity Upgrade (BSCU) (London, England), 1010–1027
    Canada Line Rapid Transit Project (Vancouver, BC), 7–8
    Chennai Metro Project (India), 928–933
    DART D2 Subway Project (Dallas, TX), 453–463
    Downtown Bellevue Tunnel (DBT) (Bellevue, WA), 764–773, 1061–1073

Northgate Link Extension Project (Seattle, WA), 109–119, 270–273, 430–438
Paris Metro Expansion Project (Paris, France), 830–836
Plane Train Tunnel West Extension Phase 1 (Atlanta, GA), 60–61
Purple Line Project (Maryland), 5
Regional Connector Project (Los Angeles, CA), 199–218, 299–311, 538–547, 890–915, 1037–1050
Réseau Express Métropolitain (REM) (Montreal, QC), 166–177
Tunnel Kaiser Strasse (Karlsruhe, Germany), 387–398
University Link Extension (U-Link) (Seattle, WA), 430–438
Tuen Mun–Chek Lap Kok Link Northern Connection Subsea Tunnel Section (Hong Kong), 986–998
tunnel boring machines (TBMs). see also earth pressure balance machines (EPBMs)
  Bheri Babai Diversion Multipurpose Project (BBDMP) (Nepal), 242–249
  Blacklick Creek Sanitary Interceptor Sewer Project (BCSIS) (Columbus, OH), 250–260
  Chennai Metro Project (India), 928–933
  cuttability indexes, 980–985
  Don River & Central Waterfront Coxwell Sanitary Bypass Tunnel Project (Toronto, ON), 916–927
  and Dos Santos Sandwich Belt high angle conveyors, 825–836
  Forrestfield Airport Link (FAL) (Perth), 954–960
  and high pressure tunneling, 972–979
  Jilin Yinsong Water Supply Project (China), 600–613
  mega, 636–646
  in mixed transitional ground (MTG), 120–131
  NBAQ4 Project (Manila, Philippines), 178–198
  Neelum-Jhelum Hydroelectric Project (Pakistan), 657–667, 715–724
  Northgate Link Extension Project (Seattle, WA), 430–438
  Ohio Canal Interceptor Tunnel (OCIT) (Akron, OH), 120–131, 870–879
  pressure balance tunneling, 109–119
  procurement, 23–33
  rebuilt, 1239–1247
  refuge chambers, 387–398
  Regional Connector Project (Los Angeles, CA), 199–218, 299–311, 890–915

Three Rivers Protection and Overflow Reduction Tunnel (3RPORT) (Fort Wayne, IN), 261–269, 357–368, 703–714
tool wear prediction, 575–584
Tuen Mun–Chek Lap Kok Link Northern Connection Subsea Tunnel Section (Hong Kong), 986–998
Tunnel Kaiser Strasse (Karlsruhe, Germany), 387–398
University Link Extension (U-Link) (Seattle, WA), 430–438
and vibration, 430–438
West Gate Tunnel Project (Melbourne, Australia), 230–239
Tunnel Kaiser Strasse (Karlsruhe, Germany), 387–398
tunnel linings. see also grouting
  Brenner Base Tunnel (Austria-Italy), 1028–1036
  Chimney Hollow Reservoir Project (Loveland, CO), 614–621
  fiber-reinforced concrete (FRC), 623–635
  Jilin Yinsong Water Supply Project (China), 600–613
  liner load estimation, 109–119
  NBAQ4 Project (Manila, Philippines), 187–189
  Neelum-Jhelum Hydroelectric Project (Pakistan), 657–667
  Réseau Express Métropolitain (REM) (Montreal, QC), 172–174
  shotcrete, 600–613, 657–667
  steel fiber reinforced concrete (SFRC), 636–646
  waterproofing, 764–773, 817–824
  wet-mix structural sprayed concrete, 1074–1087
tunnel rehabilitation, 370–379, 1199–1209
tunnel repair, 1133–1143, 1199–1209
tunneling
  micro, 799–808, 942–953
  mixed face conditions, 928–939
  mountain, 242–249, 277–289, 614–621, 657–667
  underwater, 97–108, 219–229, 512–524, 986–998
  urban, 157–177, 199–218, 399–420, 453–463, 538–547, 564–574

U

underwater tunneling
  Hudson Tunnel Project (Secaucus, NJ/ New York, NY), 97–108

Parallel Thimble Shoal Tunnel (PTST) Project (Chesapeake Bay, VA), 512–524
and permitting, 219–229
Tuen Mun–Chek Lap Kok Link Northern Connection Subsea Tunnel Section (Hong Kong), 986–998
United States
  Anacostia River Tunnel (Washington, D.C.), 504–511
  Blacklick Creek Sanitary Interceptor Sewer Project (BCSIS) (Columbus, OH), 250–260, 855–867
  Blacksnake Creek Stormwater Separation Improvement Tunnel (St. Joseph, MO), 46–54
  Brunswick Glynn Joint Water and Sewer North Mainland Sewer Transmission System (Brunswick, GA), 57–58
  Bypass Tunnel (New York, NY), 78–86
  Central Bayside System Improvement Project (San Francisco, CA), 490–502
  Chimney Hollow Reservoir Project (Loveland, CO), 614–621
  CSO 021 Diversion Facilities Project (Washington, D.C.), 1098–1113
  DART D2 Subway Project (Dallas, TX), 453–463
  DC Clean Rivers (DCCR) Project (Washington, D.C.), 157–165
  Des Plaines Inflow Tunnel (Chicago, IL), 342–350
  Doan Valley Relief and Consolidation Sewer (DVRCS) Project (Cleveland, OH), 942–953
  Downtown Bellevue Tunnel (DBT) (Bellevue, WA), 764–773, 1061–1073
  Dugway Storage Tunnel (DST) (Cleveland, OH), 845–854
  42nd Street Times Square Shuttle Station (New York, NY), 132–146
  Great Hill Tunnel (Connecticut), 1199–1209
  Hawthorne Test Tunnel (Los Angeles, CA), 8–9
  Hudson Tunnel Project (Secaucus, NJ/New York, NY), 97–108
  Hudson Tunnel Project: Hoboken Shaft (Hoboken, NJ), 1114–1121
  Interstate-89 (I-89) culvert project (Georgia, VT), 1000–1009
  Jarvis River Crossing Project (Neebing, OH), 290–298
  Lake Mead Intake No. 3 Project (Nevada), 314–329
  Lockbourne Intermodal Subtrunk (LIS) Project (Columbus, OH), 1189–1198
  Lower Meramec Tunnel (LMT) (St. Louis, MO), 548–563
  Luck Stone Inter-Quarry Tunnel (Leesburg, VA), 351–356
  Lucky Friday Mine (Idaho), 1210–1221
  Maline Creek Tunnel (MCT) (St. Louis, MO), 745–751
  Metropolitan Tunnel Redundancy Program (Boston, MA), 477–489
  Mountain Tunnel Improvements Project (California), 464–476
  Northeast Boundary Tunnel (NEBT) (Washington, D.C.), 408–420
  Northeast Corridor Superconducting Maglev Project (SCMAGLEV) (Washington, D.C./Baltimore, MD), 837–844
  Northgate Link Extension Project (Seattle, WA), 109–119, 270–273, 430–438
  Oakland-Macomb Interceptor System (OMIS) (Macomb County, Michigan), 1133–1143
  Ohio Canal Interceptor Tunnel (OCIT) (Akron, OH), 120–131, 683–691, 870–889
  Ohio River Bridges East End Crossing (ORBEEC) (Louisville, KY/Utica, IL), 8–9
  Oregon Avenue NW Sewer Rehabilitation Project (Washington, D.C.), 799–808
  Parallel Thimble Shoal Tunnel (PTST) Project (Chesapeake Bay, VA), 512–524
  Parmer Lane Interceptor Tunnel (Austin, TX), 1222–1229
  Peoples Gas Light & Coke Company (PGL) Tunnel Remediation Program (TRP) (Chicago, IL), 1172–1185
  Plane Train Tunnel West Extension Phase 1 (Atlanta, GA), 60–61
  Purple Line Project (Maryland), 8–9
  Rand Park Stormwater Diversion Tunnel (Keokuk, IA), 694–702
  Regional Connector Project (Los Angeles, CA), 199–218, 299–311, 538–547, 890–915, 1037–1050
  RiverRenew (Alexandria, VA), 440–452
  Silicon Valley Clean Water Gravity Pipeline Project (Redwood City, CA), 58–60
  South Hartford Conveyance and Storage Tunnel (SHCST) (Hartford, CT), 1153–1159

Surface Water Supply Project (Houston, TX), 88–96
Three Rivers Protection and Overflow Reduction Tunnel (3RPORT) (Fort Wayne, IN), 261–269, 357–368, 703–714
University Link Extension (U-Link) (Seattle, WA), 430–438
Waterway Protection Tunnel (WPT) (Louisville, KY), 725–732
University Link Extension (U-Link) (Seattle, WA), 430–438
urban tunneling
  DART D2 Subway Project (Dallas, TX), 453–463
  DC Clean Rivers (DCCR) Project (Washington, D.C.), 157–165
  and geotechnical risk, 564–574
  and ground subsidence, 399–407
  modeling for, 538–547
  Northeast Boundary Tunnel (NEBT) (Washington, D.C.), 408–420
  Regional Connector Project (Los Angeles, CA), 199–218
  Réseau Express Métropolitain (REM) (Montreal, QC), 166–177
Utica, IL
  Ohio River Bridges East End Crossing (ORBEEC), 23–33
utility tunnels
  Peoples Gas Light & Coke Company (PGL) Tunnel Remediation Program (TRP) (Chicago, IL), 1172–1185

V

Vancouver, BC
  Canada Line Rapid Transit Project, 23–33
ventilation systems
  optimization of, 421–429
  Réseau Express Métropolitain (REM) (Montreal, QC), 175–177
Vermont
  Interstate-89 (I-89) culvert project (Georgia), 1000–1009
Vertical Shaft Sinking Machine (VSM), 1144–1152
vibration, 430–438
Victoria, Australia
  West Gate Tunnel Project, 230–239
Virginia
  Luck Stone Inter-Quarry Tunnel (Leesburg), 351–356
  Parallel Thimble Shoal Tunnel (PTST) Project (Chesapeake Bay), 512–524
  RiverRenew (Alexandria), 440–452

W

Washington (state)
  Downtown Bellevue Tunnel (DBT) (Bellevue), 764–773, 1061–1073
  Northgate Link Extension Project (Seattle), 109–119, 270–273, 430–438
  University Link Extension (U-Link) (Seattle), 430–438
Washington, D.C.
  Anacostia River Tunnel, 504–511
  CSO 021 Diversion Facilities Project, 1098–1113
  DC Clean Rivers (DCCR) Project, 157–165
  Northeast Boundary Tunnel (NEBT), 408–420
  Northeast Corridor Superconducting Maglev Project (SCMAGLEV), 837–844
  Oregon Avenue NW Sewer Rehabilitation Project, 799–808
wastewater tunnels. see also combined sewer overflow (CSO) tunnels
  Blacklick Creek Sanitary Interceptor Sewer Project (BCSIS) (Columbus, OH), 250–260, 855–867
  Brunswick Glynn Joint Water and Sewer North Mainland Sewer Transmission System (Brunswick, GA), 57–58
  Lockbourne Intermodal Subtrunk (LIS) Project (Columbus, OH), 1189–1198
  Lower Meramec Tunnel (LMT) (St. Louis, MO), 548–563
  Oakland-Macomb Interceptor System (OMIS) (Macomb County, Michigan), 1133–1143
  Oregon Avenue NW Sewer Rehabilitation Project (Washington, D.C.), 799–808
  Parmer Lane Interceptor Tunnel (Austin, TX), 1222–1229
  Silicon Valley Clean Water Gravity Pipeline Project (Redwood City, CA), 58–60
water tunnels. see also stormwater tunnels
  Bheri Babai Diversion Multipurpose Project (BBDMP) (Nepal), 242–249
  Bypass Tunnel (New York, NY), 78–86
  Chimney Hollow Reservoir Project (Loveland, CO), 614–621
  Great Hill Tunnel (Connecticut), 1199–1209
  Jilin Yinsong Water Supply Project (China), 600–613
  Lake Mead Intake No. 3 Project (Nevada), 314–329

Metropolitan Tunnel Redundancy Program (Boston, MA), 477–489
Mountain Tunnel Improvements Project (California), 464–476
NBAQ4 Project (Manila, Philippines), 178–198
Neelum-Jhelum Hydroelectric Project (Pakistan), 277–289
Surface Water Supply Project (Houston, TX), 88–96

waterproofing, 764–773, 817–824
Waterway Protection Tunnel (WPT) (Louisville, KY), 725–732
West Gate Tunnel Project (Melbourne, Australia), 230–239, 636–646
WestConnex New M5 Project (Sydney, Australia), 668–682
Western Australia
    Forrestfield Airport Link (FAL) (Perth), 954–960